STUDENT'S
SOLUTIONS MANUAL

JEFFERY COLE

Anoka-Ramsey Community College

PREALGEBRA &
INTRODUCTORY ALGEBRA
FOURTH EDITION

Margaret L. Lial
American River College

Diana L. Hestwood
Minneapolis Community and Technical College

John Hornsby
University of New Orleans

Terry McGinnis

D1502296

PEARSON

Boston Columbus Indianapolis New York San Francisco Upper Saddle River
Amsterdam Cape Town Dubai London Madrid Milan Munich Paris Montreal Toronto
Delhi Mexico City São Paulo Sydney Hong Kong Seoul Singapore Taipei Tokyo

Reproduced by Pearson from electronic files supplied by the author.

ISBN-13: 978-0-321-85484-1
ISBN-10: 0-321-85484-5

1 2 3 4 5 6 OPM 16 15 14 13 12

www.pearsonhighered.com

PEARSON

Table of Contents

Table of Contents

Table of Contents

Table of Contents

Preface

This *Student's Solutions Manual* contains solutions to selected exercises in the text *Prealgebra and Introductory Algebra, Fourth Edition* by Margaret L. Lial, Diana L. Hestwood, John Hornsby, and Terry McGinnis. It contains solutions to all margin exercises, the odd-numbered exercises in each section, all Relating Concepts exercises, as well as solutions to all the exercises in the review sections, the chapter tests, and the cumulative review sections.

This manual is a text supplement and should be read along *with* the text. You should read all exercise solutions in this manual because many concept explanations are given and then used in subsequent solutions. All concepts necessary to solve a particular problem are not reviewed for every exercise. If you are having difficulty with a previously covered concept, refer back to the section where it was covered for more complete help.

A significant number of today's students are involved in various outside activities, and find it difficult, if not impossible, to attend all class sessions; this manual should help meet the needs of these students. In addition, it is my hope that this manual's solutions will enhance the understanding of all readers of the material and provide insights to solving other exercises.

I appreciate feedback concerning errors, solution correctness or style, and manual style. Any comments may be sent directly to me at the address below, at jeff.cole@anokaramsey.edu, or in care of the publisher, Pearson Addison-Wesley.

I would like to thank Marv Riedesel and Mary Johnson, formerly of Inver Hills Community College, for their careful accuracy checking and valuable suggestions; Rachel Haskell, for checking the final manuscript; and Maureen O'Connor and Lauren Morse, of Pearson Addison-Wesley, for entrusting me with this project.

Jeffery A. Cole
Anoka-Ramsey Community College
11200 Mississippi Blvd. NW
Coon Rapids, MN 55433

CHAPTER 1 INTRODUCTION TO ALGEBRA: INTEGERS

1.1   Place Value

1.1   Margin Exercises

1. The whole numbers are: 502; 3; 14; 0; 60,005

2. **(a)** The 8 in 45,628,665 is in the thousands place.

    (b) The 8 in 800,503,622 is in the hundred-millions place.

    (c) The 8 in 428,000,000,000 is in the billions place.

    (d) The 8 in 2,385,071 is in the ten-thousands place.

3. **(a)** 23,605 in words: twenty-three *thousand*, six hundred <u>five</u>.

    (b) 400,033,007 in words: four hundred *million*, thirty-three *thousand*, seven.

    (c) 193,080,102,000,000 in words: one hundred ninety-three *trillion*, eighty *billion*, one hundred two *million.*

4. **(a)** Eighteen million, two thousand, three hundred five
    The first group name is *million,* so you need to fill *three groups* of three digits.

    <u>0 1 8</u>, <u>0 0 2</u>, <u>3 0 5</u> = 18,002,305

    (b) Two hundred billion, fifty million, six hundred sixteen
    The first group name is *billion,* so you need to fill *four groups* of three digits.

    <u>2 0 0</u>, <u>0 5 0</u>, <u>0 0 0</u>, <u>6 1 6</u> = 200,050,000,616

    (c) Five trillion, forty-two billion, nine million
    The first group name is *trillion,* so you need to fill *five groups* of three digits.

    <u>0 0 5</u>, <u>0 4 2</u>, <u>0 0 9</u>, <u>0 0 0</u>, <u>0 0 0</u> = 5,042,009,000,000

    (d) Three hundred six million, seven hundred thousand, nine hundred fifty-nine
    The first group name is *million,* so you need to fill *three groups* of three digits.

    <u>3 0 6</u>, <u>7 0 0</u>, <u>9 5 9</u> = 306,700,959

1.1   Section Exercises

1. False; we can also use the digit 0.

3. True; none of the numbers are whole numbers.

5. The whole numbers are: 15; 0; 83,001

7. The whole numbers are: 7; 362,049

9. The 2 in 61,284 is in the hundreds place.

11. The 2 in 284,100 is in the hundred-thousands place.

13. The 2 in 725,837,166 is in the ten-millions place.

15. The 2 in 253,045,701,000 is in the hundred-billions place.

17. Name the place value for each zero in

    302,016,450,098,570.

    From left to right: ten-trillions, hundred-billions, millions, hundred-thousands, and ones.

19. 8421 in words: eight <u>thousand</u>, four <u>hundred</u> twenty-<u>one</u>.

21. 46,205 in words: forty-six thousand, two hundred five.

23. 3,064,801 in words: three million, sixty-four thousand, eight hundred one.

25. 840,111,003 in words: eight hundred forty million, one hundred eleven thousand, three.

27. 51,006,888,321 in words: fifty-one billion, six million, eight hundred eighty-eight thousand, three hundred twenty-one.

29. 3,000,712,000,000 in words: three trillion, seven hundred twelve million.

31. Forty-six thousand, eight hundred five
    The first group name is *thousand,* so you need to fill *two groups* of three digits.

    <u>0 4 6</u>, <u>8 0 5</u> = 46,805

33. Five million, six hundred thousand, eighty-two
    The first group name is *million,* so you need to fill *three groups* of three digits.

    <u>0 0 5</u>, <u>6 0 0</u>, <u>0 8 2</u> = 5,600,082

35. Two hundred seventy-one million, nine hundred thousand
    The first group name is *million,* so you need to fill *three groups* of three digits.

    <u>2 7 1</u>, <u>9 0 0</u>, <u>0 0 0</u> = 271,900,000

37. Twelve billion, four hundred seventeen million, six hundred twenty-five thousand, three hundred ten
    The first group name is *billion,* so you need to fill *four groups* of three digits.

    <u>0 1 2</u>, <u>4 1 7</u>, <u>6 2 5</u>, <u>3 1 0</u> = 12,417,625,310

39. Six hundred trillion, seventy-one million, four hundred
    The first group name is *trillion,* so you need to fill *five groups* of three digits.

    <u>6 0 0</u>, <u>0 0 0</u>, <u>0 7 1</u>, <u>0 0 0</u>, <u>4 0 0</u> = 600,000,071,000,400

41.   6041 in words: six thousand, forty-one

43.   Seven hundred sixty million, three hundred nine thousand
The first group is *millions,* so fill *three groups* of three digits.

$\underline{7\,6\,0},\,\underline{3\,0\,9},\,\underline{0\,0\,0} = 760{,}309{,}000$

45.   1,056,720,000 in words: one billion, fifty-six million, seven hundred twenty thousand

47.   2000 in words: the year two thousand;
One hundred forty-one million, seven hundred units: The first group is *millions,* so you need to fill *three groups* of three digits.

$\underline{1\,4\,1},\,\underline{0\,0\,0},\,\underline{7\,0\,0} = 141{,}000{,}700$

49.   6,400,000 in words: six million, four hundred thousand every day. 2,336,000,000 in words: two billion, three hundred thirty-six million in one year.

51.   Four billion, two hundred million is 4,200,000,000.

Relating Concepts (Exercises 53–56)

53.   To make the largest possible whole number, arrange the digits from largest to smallest.

$97651100 \rightarrow 97{,}651{,}100$

In words: ninety-seven million, six hundred fifty-one thousand, one hundred.

To make the smallest possible whole number, arrange the numbers from smallest to largest with one exception: because we must use all the digits, start with the smallest nonzero digit.

$10015679 \rightarrow 10{,}015{,}679$

In words: ten million, fifteen thousand, six hundred seventy-nine.

54.   Answers will vary.

55.

sixty-fours	thirty-twos	sixteens	eights	fours	twos	ones
$\underline{1}$	$\underline{1}$	$\underline{1}$	$\underline{1}$	$\underline{1}$	$\underline{1}$	$\underline{1}$

(a) $5 = 4 + 1 =$ binary $\underline{101}$

(b) $10 = 8 + 2 =$ binary $\underline{1010}$

(c) $15 = 8 + 4 + 2 + 1 =$ binary $\underline{1111}$

56.   **(a)** Answers will vary but should mention that the location or place in which a digit is written gives it a different value.

(b) $8 = 5 + 3 = \text{VIII}$
$38 = 30 + 5 + 3 = \text{XXXVIII}$
$275 = 200 + 50 + 20 + 5 = \text{CCLXXV}$
$3322 = 3000 + 300 + 20 + 2 = \text{MMMCCCXXII}$

(c) The Roman system is *not* a place value system because no matter what place it's in, $M = 1000$, $C = 100$, etc. One disadvantage is that it takes much more space to write many large numbers; another is that there is no symbol for zero.

1.2   Introduction to Integers

1.2   Margin Exercises

1.   **(a)** "Below zero" implies a negative number.

$-5\frac{1}{2}$ degrees

(b) "Lost 12 pounds" implies a negative number.

-12 pounds

(c) "Deposit" implies a positive number.

$+\$210.35$ or $\$210.35$

(d) "Overdrawn" implies a negative number.

$-\$65$

(e) "Below the surface of the sea" implies a negative number.

-100 feet

(f) "Won 50 points" implies a positive number.

$+50$ points or 50 points

2.   **(a)** -2 **(b)** 2 **(c)** 0 **(d)** -4 **(e)** 4

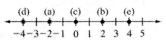

(f) $-3\frac{1}{2}$ **(g)** 1 **(h)** -1 **(i)** 3

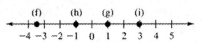

3.   **(a)** 5 is to the *right* of 4 on the number line, so 5 is *greater than* 4. Write $5 > 4$.

(b) 0 is to the *left* of 2 on the number line, so 0 is *less than* 2. Write $0 < 2$.

(c) -3 is to the *left* of -2 on the number line, so -3 is *less than* -2. Write $-3 < -2$.

(d) -1 is to the *right* of -4 on the number line, so -1 is *greater than* -4. Write $-1 > -4$.

(e) 2 is to the *right* of -2 on the number line, so 2 is *greater than* -2. Write $2 > -2$.

(f) -5 is to the *left* of 1 on the number line, so -5 is *less than* 1. Write $-5 < 1$.

4. **(a)** $|13| = 13$ because the distance from 0 to 13 on the number line is 13 spaces.

(b) $|-7| = 7$ because the distance from 0 to -7 on the number line is 7 spaces.

(c) $|0| = 0$ because the distance from 0 to 0 on the number line is 0 spaces.

(d) $|-350| = 350$ because the distance from 0 to -350 on the number line is 350 spaces.

(e) $|6000| = 6000$ because the distance from 0 to 6000 on the number line is 6000 spaces.

1.2 Section Exercises

1. "Above sea level" implies a positive number.

$$+29,035 \text{ feet or } 29,035 \text{ feet}$$

3. "Below zero" implies a negative number.

$$-128.6 \text{ degrees}$$

5. "Lost a total of 18 yards" implies a negative number.

$$-18 \text{ yards}$$

7. "Won \$100" implies a positive number.

$$+\$100 \text{ or } \$100$$

9. "Lost $6\frac{1}{2}$ pounds" implies a negative number.

$$-6\frac{1}{2} \text{ pounds}$$

11. Graph $-3, 3, 0, -5$

13. Graph $-1, 4, -2, 5$

15. **(a)** $0 < 5$ in words: zero is less than five, or, zero is less than positive five

(b) $-10 > -17$ in words: negative ten is greater than negative seventeen

17. 10 is to the *right* of 2 on the number line, so 10 is *greater than* 2. Write $10 > 2$.

19. -1 is to the *left* of 0 on the number line, so -1 is *less than* 0. Write $-1 < 0$.

21. -10 is to the *left* of 2 on the number line, so -10 is *less than* 2. Write $-10 < 2$.

23. -3 is to the *right* of -6 on the number line, so -3 is *greater than* -6. Write $-3 > -6$.

25. -10 is to the *left* of -2 on the number line, so -10 is *less than* -2. Write $-10 < -2$.

27. 0 is to the *right* of -8 on the number line, so 0 is *greater than* -8. Write $0 > -8$.

29. 10 is to the *right* of -2 on the number line, so 10 is *greater than* -2. Write $10 > -2$.

31. -4 is to the *left* of 4 on the number line, so -4 is *less than* 4. Write $-4 < 4$.

33. $|15| = 15$ because the distance from 0 to 15 on the number line is 15 spaces.

35. $|-3| = 3$ because the distance between 0 and -3 on the number line is 3 spaces.

37. $|0| = 0$ because the distance from 0 to 0 on the number line is 0 spaces.

39. $|200| = 200$ because the distance between 0 and 200 on the number line is 200 spaces.

41. $|-75| = 75$ because the distance between 0 and -75 on the number line is 75 spaces.

43. $|-8042| = 8042$ because the distance between 0 and -8042 on the number line is 8042 spaces.

Relating Concepts (Exercises 45–48)

45. Graph -1.5 as A, 0.5 as B, -1 as C, and 0 as D.

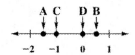

46. From Exercise 45, in order from lowest to highest: $-1.5, -1, 0, 0.5$

47. A: -1.5 is in the Below -1 range. This patient may be at risk.

B: 0.5 is in the Above 0 range. This patient is above normal.

C: -1 is in the 0 to -1 range. This patient is normal.

D: 0 is in the 0 to -1 range. This patient is normal.

48. **(a)** A patient who did not understand the importance of the negative sign would think the interpretation of -1.5 was "above normal" (range Above 0) and wouldn't get treatment.

(b) For Patient D's score of 0, the sign plays no role. Zero is neither positive nor negative.

1.3 Adding Integers

1.3 Margin Exercises

1. **(a)** $-2 + (-2) = -4$

(b) $2 + 2 = 4$

(c) $-10 + (-1) = -11$

(d) $10 + 1 = 11$

(e) $-3 + (-7) = -10$

(f) $3 + 7 = 10$

2.   **(a)** $-6 + (-6)$ Adding *like* signed integers

Step 1 $|-6| = 6; |-6| = 6;$ Add $6 + 6 = 12$

Step 2   Both numbers are negative, so the sum is negative.

$$-6 + (-6) = -12$$

(b) $9 + 7$ Adding *like* signed integers

Step 1 $|9| = 9; |7| = 7;$ Add $9 + 7 = 16$

Step 2   Both numbers are positive, so the sum is positive.

$$9 + 7 = 16$$

(c) $-5 + (-10)$ Adding *like* signed integers

Step 1 $|-5| = 5; |-10| = 10;$ Add $5 + 10 = 15$

Step 2   Both numbers are negative, so the sum is negative.

$$-5 + (-10) = -15$$

(d) $-12 + (-4)$ Adding *like* signed integers

Step 1 $|-12| = 12; |-4| = 4;$ Add $12 + 4 = 16$

Step 2   Both numbers are negative, so the sum is negative.

$$-12 + (-4) = -16$$

(e) $13 + 2$ Adding *like* signed integers

Step 1 $|13| = 13; |2| = 2;$ Add $13 + 2 = 15$

Step 2   Both numbers are positive, so the sum is positive.

$$13 + 2 = 15$$

3.   **(a)** $-3 + 7$ Adding *unlike* signed integers

Step 1 $|-3| = \underline{3}; |7| = \underline{7};$ Subtract $7 - 3 = \underline{4}$

Step 2   7 has the larger absolute value and is positive, so the sum is positive.

$$-3 + 7 = +4 \text{ or } \underline{4}$$

(b) $6 + (-12)$ Adding *unlike* signed integers

Step 1 $|6| = 6; |-12| = 12;$ Subtract $12 - 6 = 6$

Step 2 -12 has the larger absolute value and is negative, so the sum is negative.

$$6 + (-12) = -6$$

(c) $12 + (-7)$ Adding *unlike* signed integers

Step 1 $|12| = 12; |-7| = 7;$ Subtract $12 - 7 = 5$

Step 2   12 has the larger absolute value and is positive, so the sum is positive.

$$12 + (-7) = +5 \text{ or } 5$$

(d) $-10 + 2$ Adding *unlike* signed integers

Step 1 $|-10| = 10; |2| = 2;$ Subtract $10 - 2 = 8$

Step 2 -10 has the larger absolute value and is negative, so the sum is negative.

$$-10 + 2 = -8$$

(e) $5 + (-9)$ Adding *unlike* signed integers

Step 1 $|5| = 5; |-9| = 9;$ Subtract $9 - 5 = 4$

Step 2 -9 has the larger absolute value and is negative, so the sum is negative.

$$5 + (-9) = -4$$

4.   **(a)** Starting temperature in the morning is -15 degrees. A rise of 21 degrees implies a positive number. A drop of 10 degrees implies a negative number.

$$\begin{aligned} -15 &+ 21 + (-10) \\ &= 6 + (-10) \qquad \textit{Add left to right.} \\ &= -4 \end{aligned}$$

The new temperature is 4 degrees below zero or -4 degrees.

(b) The beginning balance is \$60. Deposits imply positive numbers and payments imply negative numbers.

$$\begin{aligned} 60 &+ 85 + (-20) + (-75) \\ &= 145 + (-20) + (-75) \qquad \textit{Add left to right.} \\ &= 125 + (-75) \\ &= 50 \end{aligned}$$

His account balance is \$50.

5.   **(a)** $175 + 25 = 25 + \underline{175}$
Both sums are $\underline{200}$.

(b) $7 + (-37) = \underline{-37} + \underline{7}$
Both sums are $\underline{-30}$.

(c) $-16 + 16 = \underline{16} + (\underline{-16})$
Both sums are $\underline{0}$.

(d) $-9 + (-41) = \underline{-41} + (\underline{-9})$
Both sums are $\underline{-50}$.

6.   **(a)** $\begin{aligned}[t] -12 + 12 + (-19) &= (-12 + 12) + (-19) \\ &= 0 + (-19) \\ &= -19 \end{aligned}$

(b) $\begin{aligned}[t] 31 + (-75) + 75 &= 31 + (-75 + 75) \\ &= 31 + 0 \\ &= 31 \end{aligned}$

(c) $\begin{aligned}[t] 1 + 9 + (-16) &= (1 + 9) + (-16) \\ &= 10 + (-16) \\ &= -6 \end{aligned}$

(d) $-38 + 5 + 25 = -38 + (5 + 25)$
$$= -38 + 30$$
$$= -8$$

1.3 Section Exercises

1. $-2 + 5 = +3$ or 3

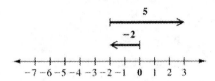

3. $-5 + (-2) = -7$

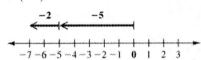

5. $3 + (-4) = -1$

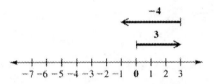

7. **(a)** $-5 + (-5)$ ▪ Adding *like* signed integers

    Step 1 Add the absolute values.

    $$|-5| = 5$$

    Add $5 + 5$ to get 10.

    Step 2 Both integers are negative, so the sum is negative.

    $$-5 + (-5) = -10$$

    (b) $5 + 5 = 10$ Adding *like* signed integers
    Both addends are positive, so the sum is positive.

9. **(a)** $7 + 5$ ▪ Adding *like* signed integers
    Both addends are positive, so the sum is positive.

    (b) $-7 + (-5) = -12$ Adding *like* signed integers

    Step 1 Add the absolute values.

    $$|-7| = 7; |-5| = 5$$

    Add $7 + 5$ to get 12.

    Step 2 Both integers are negative, so the sum is negative.

    $$-7 + (-5) = -12$$

11. **(a)** $-25 + (-25)$ ▪ Adding *like* signed integers

    Step 1 Add the absolute values.

    $$|-25| = 25$$

    Add $25 + 25$ to get 50.

Step 2 Both integers are negative, so the sum is negative.

$$-25 + (-25) = -50$$

(b) $25 + 25 = 50$ Adding *like* signed integers
Both addends are positive, so the sum is positive.

13. **(a)** $48 + 110$ ▪ Adding *like* signed integers
    Both addends are positive, so the sum is positive.

    (b) $-48 + (-110) = -158$ Adding *like* signed integers

    Step 1 Add the absolute values.

    $$|-48| = 48; |-110| = 110$$

    Add $48 + 110$ to get 158.

    Step 2 Both numbers are negative, so the sum is negative.

    $$-48 + (-110) = -158$$

15. The absolute values are the same in each pair of answers, so the only difference in the sums is the common sign.

17. **(a)** $-6 + 8$ Adding *unlike* signed integers

    Step 1 $|-6| = 6; |8| = 8$

    Subtract $8 - 6$ to get 2.

    Step 2 8 has the larger absolute value and is positive, so the sum is positive.

    $$-6 + 8 = +2 \text{ or } 2$$

    (b) $6 + (-8)$ Adding *unlike* signed integers

    Step 1 $|6| = 6; |-8| = 8$

    Subtract $8 - 6$ to get 2.

    Step 2 -8 has the larger absolute value and is negative, so the sum is negative.

    $$6 + (-8) = -2$$

19. **(a)** $-9 + 2$ Adding *unlike* signed integers

    Step 1 $|-9| = 9; |2| = 2$

    Subtract $9 - 2$ to get 7.

    Step 2 -9 has the larger absolute value and is negative, so the sum is negative.

    $$-9 + 2 = -7$$

    (b) $9 + (-2)$ Adding *unlike* signed integers

    Step 1 $|9| = 9; |-2| = 2$

    Subtract $9 - 2$ to get 7.

    Step 2 9 has the larger absolute value and is positive, so the sum is positive.

    $$9 + (-2) = +7 \text{ or } 7$$

21. **(a)** $20 + (-25)$ Adding *unlike* signed integers

Step 1 $|20| = 20; |-25| = 25$

Subtract $25 - 20$ to get 5.

Step 2 -25 has the larger absolute value and is negative, so the sum is negative.

$$20 + (-25) = -5$$

(b) $-20 + 25$ Adding *unlike* signed integers

Step 1 $|-20| = 20; |25| = 25$

Subtract $25 - 20$ to get 5.

Step 2 25 has the larger absolute value and is positive, so the sum is positive.

$$-20 + 25 = +5 \text{ or } 5$$

23. **(a)** $200 + (-50)$ Adding *unlike* signed integers

Step 1 $|200| = 200; |-50| = 50$

Subtract $200 - 50$ to get 150.

Step 2 200 has the larger absolute value and is positive, so the sum is positive.

$$200 + (-50) = +150 \text{ or } 150$$

(b) $-200 + 50$ Adding *unlike* signed integers

Step 1 $|-200| = 200; |50| = 50$

Subtract $200 - 50$ to get 150.

Step 2 -200 has the larger absolute value and is negative, so the sum is negative.

$$-200 + 50 = -150$$

25. Each pair of answers differs only in the sign of the answer. This occurs because the signs of the addends are reversed.

27. $-8 + 5$ Adding *unlike* signed integers

Step 1 $|-8| = 8; |5| = 5$

Subtract $8 - 5$ to get 3.

Step 2 -8 has the larger absolute value and is negative, so the sum is negative.

$$-8 + 5 = -3$$

29. $-1 + 8$ Adding *unlike* signed integers

Step 1 $|-1| = 1; |8| = 8$

Subtract $8 - 1$ to get 7.

Step 2 8 has the larger absolute value and is positive, so the sum is positive.

$$-1 + 8 = +7 \text{ or } 7$$

31. $-2 + (-5)$ Adding *like* signed integers

Step 1 $|-2| = 2; |-5| = 5$

Add $2 + 5$ to get 7.

Step 2 Both integers are negative, so the sum is negative.

$$-2 + (-5) = -7$$

33. $6 + (-5)$ Adding *unlike* signed integers

Step 1 $|6| = 6; |-5| = 5$

Subtract $6 - 5$ to get 1.

Step 2 6 has the larger absolute value and is positive, so the sum is positive.

$$6 + (-5) = +1 \text{ or } 1$$

35. $4 + (-12)$ Adding *unlike* signed integers

Step 1 $|4| = 4; |-12| = 12$

Subtract $12 - 4$ to get 8.

Step 2 -12 has the larger absolute value and is negative, so the sum is negative.

$$4 + (-12) = -8$$

37. $-10 + (-10)$ Adding *like* signed integers

Step 1 $|-10| = 10; |-10| = 10$

Add $10 + 10$ to get 20.

Step 2 Both integers are negative, so the sum is negative.

$$-10 + (-10) = -20$$

39. $-17 + 0 = -17$

Adding zero to any number leaves the number unchanged.

41. $1 + (-23)$ Adding *unlike* signed integers

Step 1 $|1| = 1; |-23| = 23$

Subtract $23 - 1$ to get 22.

Step 2 -23 has the larger absolute value and is negative, so the sum is negative.

$$1 + (-23) = -22$$

43. $\underline{-2 + (-12)} + (-5)$

$= \underline{-14 + (-5)}$ *Add left to right.*

$= \underline{-19}$

45. $8 + 6 + (-8)$

$= 8 + (-8) + 6$ *Commute addends.*

$= 0 + 6$ *Add left to right.*

$= 6$

47. $-7 + 6 + (-4)$
$= -1 + (-4)$ *Add left to right.*
$= -5$

49. $-3 + (-11) + 14$
$= -14 + 14$ *Add left to right.*
$= 0$

51. $10 + (-6) + (-3) + 4$
$= 4 + (-3) + 4$ *Add left to right.*
$= 1 + 4$
$= 5$

53. $-7 + 28 + (-56) + 3$
$= 21 + (-56) + 3$ *Add left to right.*
$= -35 + 3$
$= -32$

55. "Yards gained" are positive $(+13)$, and "yards lost" are negative (-17).

$13 + (-17) = -4$ yards

The team lost 4 yards.

57. The overdrawn amount is negative $(-\$62)$, and the deposit amount is positive $(+\$50)$.

$-\$62 + \$50 = \underline{-\$12}$

Nick is $12 overdrawn.

59. $88 stolen implies a loss of money or $-\$88$.

Jay received $35 back implies a gain of money or $+\$35$.

$-\$88 + \$35 = -\$53$

Jay's net loss was $53.

61. Jeff: $-20 + 75 + (-55)$
$= 55 + (-55)$ *Add left to right.*
$= 0$ points

Terry: $42 + (-15) + 20$
$= 27 + 20$ *Add left to right.*
$= 47$ points

63. $-2 + 0 + 5 + (-5)$
$= -2 + 5 + (-5)$ *Add left to right.*
$= 3 + (-5)$
$= -2$

Angela lost 2 pounds.

65. $3 + (-2) + (-2) + 3$
$= 1 + (-2) + 3$ *Add left to right.*
$= -1 + 3$
$= 2$

Brittany gained 2 pounds.

67. $\underbrace{-18 + (-5)}_{-23} = \underbrace{-5 + (-18)}_{-23}$ *Commutative property*

Both sums are -23.

69. $\underbrace{-4 + 15}_{+11} = \underbrace{15 + (-4)}_{+11}$ *Commutative property*

Both sums are $+11$ or 11.

71. $6 + (-14) + 14$

Option 1: $(6 + (-14)) + 14 = -8 + 14$
$= 6$

Option 2: $6 + (-14 + 14) = 6 + 0$
$= 6$

Option 2 is easier.

73. $14 + 6 + (-7)$

Option 1: $(14 + 6) + (-7) = 20 + (-7)$
$= 13$

Option 2: $14 + (6 + (-7)) = 14 + (-1)$
$= 13$

Option 1 might seem easier.

75. Answers will vary. Some possibilities are:
$-6 + 0 = -6; 10 + 0 = 10; 0 + 3 = 3$

77. Be sure to use the *negative* key as opposed to the *subtraction* key.

$-7081 + 2965 = -4116$

79. $-179 + (-61) + 8926 = 8686$

81. $86 + (-99,000) + 0 + 2837 = -96,077$

1.4 Subtracting Integers

1.4 Margin Exercises

1. **(a)** The opposite of 5 is $\underline{-5}$. $5 + (\underline{-5}) = 0$

(b) The opposite of 48 is -48. $48 + (-48) = 0$

(c) The opposite of 0 is 0. $0 + 0 = 0$

(d) The opposite of -1 is 1. $-1 + 1 = 0$

(e) The opposite of -24 is 24. $-24 + 24 = 0$

2. **(a)** $-6 - 5$
$= -6 + (-5)$ *Change subtraction to addition. Change 5 to -5.*
$= \underline{-11}$

(b) $3 - (-10)$
$= 3 + (\underline{+10})$ *Change subtraction to addition. Change -10 to $+10$.*
$= \underline{13}$

(c) $-8 - (-2)$

$= -8 + (+2)$ *Change subtraction to addition.*
                Change -2 to $+2$.

$= -6$

(d) $0 - 10$

$= 0 + (-10)$ *Change subtraction to addition.*
                Change 10 to -10.

$= -10$

(e) $-4 - (-12)$

$= -4 + (+12)$ *Change subtraction to addition.*
                 Change -12 to $+12$.

$= 8$

(f) $9 - 7$

$= 9 + (-7)$ *Change subtraction to addition.*
               Change 7 to -7.

$= 2$

3.   **(a)** $6 - 7 + (-3)$

$= \underbrace{6 + (\underline{-7})} + (-3)$ *Change subtraction to addition.*
                                  Change 7 to -7.

$= \underbrace{-1 \qquad + (-3)}$ *Add left to right.*

$= \underline{-4}$

(b) $-2 + (-3) - (-5)$

$= -2 + (-3) + (+5)$ *Change subt. to addition.*
                       Change -5 to $+5$.

$= -5 + 5$ *Add left to right.*

$= 0$

(c) $7 - 7 - 7$

$= 7 + (-7) + (-7)$ *Change all subt. to additions.*
                      Change 7 to -7.

$= 0 + (-7)$ *Add left to right.*

$= -7$

(d) $-3 - 9 + 4 - (-20)$

$= -3 + (-9) + 4 + (+20)$

$= -12 + 4 + (+20)$

$= -8 + (+20)$

$= 12$

1.4   Section Exercises

1.   The opposite of 6 is -6. $6 + (-6) = 0$

3.   The opposite of -13 is 13. $-13 + 13 = 0$

5.   The student forgot to change 6 to its opposite, -6.

Correct Method:

$-6 - 6$

$= -6 + (-6)$ *Change subtraction to addition. Change 6 to -6.*

$= -12$ *Add.*

7. $19 - 5$

$= 19 + (-5)$ *Change subtraction to addition. Change 5 to -5.*

$= \underline{14}$

9. $10 - 12$

$= 10 + (-12)$ *Change subtraction to addition. Change 12 to -12.*

$= -2$

11. $7 - 19$

$= 7 + (-19)$ *Change subtraction to addition. Change 19 to -19.*

$= -12$

13. $-15 - 10$

$= -15 + (-10)$ *Change subtraction to addition.*
                  Change 10 to -10.

$= -25$

15. $-9 - 14$

$= -9 + (-14)$ *Change subtraction to addition.*
                 Change 14 to -14.

$= -23$

17. $-3 - (-8)$ *Change subtraction to addition. Change -8 to $+8$.*

$= -3 + (\underline{+8})$

$= \underline{5}$

19. $6 - (-14)$

$= 6 + (+14)$ *Change subtraction to addition.*
                Change -14 to $+14$.

$= 20$

21. $1 - (-10)$

$= 1 + (+10)$ *Change subtraction to addition.*
                Change -10 to $+10$.

$= 11$

23. $-30 - 30$

$= -30 + (-30)$ *Change subtraction to addition.*
                  Change 30 to -30.

$= -60$

25. $-16 - (-16)$

$= -16 + (+16)$ *Change subtraction to addition.*
Change -16 to $+16$.

$= 0$

27. $13 - 13$

$= 13 + (-13)$ *Change subtraction to addition. Change 13 to -13.*

$= 0$

29. $0 - 6$

$= 0 + (-6)$ *Change subtraction to addition. Change 6 to -6.*

$= -6$

31. (a) $3 - (-5)$

$= 3 + (+5)$ *Change subtraction to addition.*
Change -5 to $+5$.

$= 8$

(b) $3 - 5$

$= 3 + (-5)$ *Change subtraction to addition. Change 5 to -5.*

$= -2$

(c) $-3 - (-5)$

$= -3 + (+5)$ *Change subtraction to addition.*
Change -5 to $+5$.

$= 2$

(d) $-3 - 5$

$= -3 + (-5)$ *Change subtraction to addition.*
Change 5 to -5.

$= -8$

33. (a) $4 - 7$

$= 4 + (-7)$ *Change subtraction to addition. Change 7 to -7.*

$= -3$

(b) $4 - (-7)$

$= 4 + (+7)$ *Change subtraction to addition.*
Change -7 to $+7$.

$= 11$

(c) $-4 - 7$

$= -4 + (-7)$ *Change subtraction to addition.*
Change 7 to -7.

$= -11$

(d) $-4 - (-7)$

$= -4 + (+7)$ *Change subtraction to addition.*
Change -7 to $+7$.

$= 3$

35. $-2 - 2 - 2$

$= \underbrace{-2 + (-2)} + (-2)$ *Change all subtractions to additions. Change 2 to -2 and 2 to -2.*

$= -4 \quad\quad + (-2)$ *Add left to right.*

$= \underline{-6}$

37. $9 - 6 - 3 - 5$

$= 9 + (-6) + (-3) + (-5)$

Change all subtractions to additions.
Change 6 to -6, 3 to -3, and 5 to -5.

$= 3 + (-3) + (-5)$ *Add left to right.*

$= 0 + (-5)$

$= -5$

39. $3 - (-3) - 10 - (-7)$

$= 3 + (+3) + (-10) + (+7)$

Change all subtractions to additions.
Change -3 to $+3$, 10 to -10, and -7 to $+7$.

$= 6 + (-10) + (+7)$ *Add left to right.*

$= -4 + (+7)$

$= 3$

41. $-2 + (-11) - (-3)$

$= -2 + (-11) + (+3)$ *Change subtraction to addition.*
Change -3 to $+3$.
Add left to right.

$= -13 + (+3)$

$= -10$

43. $4 - (-13) + (-5)$

$= 4 + (+13) + (-5)$ *Change subtraction to addition.*
Change -13 to $+13$.
Add left to right.

$= 17 + (-5)$

$= 12$

45. $6 + 0 - 12 + 1$

$= 6 + 0 + (-12) + 1$ *Change subtraction to addition.*
Change 12 to -12.
Add left to right.

$= 6 + (-12) + 1$

$= -6 + 1$

$= -5$

47. **(a)** The 30°F column and the 10 mph wind row intersect at 21°F. The difference between the actual temperature and the wind chill temperature is $30 - 21 = 30 + (-21) = 9$ degrees.

(b) The 15°F column and the 15 mph wind row intersect at 0°F. The difference between the actual temperature and the wind chill temperature is $15 - 0 = 15$ degrees.

(c) The 5°F column and the 25 mph wind row intersect at −17°F. The difference between the actual temperature and the wind chill temperature is $5 - (-17) = 5 + (+17) = 22$ degrees.

(d) The −10°F column and the 35 mph wind row intersect at −41°F. The difference between the actual temperature and the wind chill temperature is $-10 - (-41) = -10 + (+41) = 31$ degrees.

49. $-2 + (-11) + |-2|$

$$= -2 + (-11) + 2 \quad \begin{array}{l}\textit{|−2| = 2 because the}\\ \textit{distance from 0 to −2}\\ \textit{is 2 units.}\end{array}$$

$$= -13 + 2 \quad \textit{Add left to right.}$$

$$= -11$$

51. $0 - |-7 + 2|$

$$= 0 - |-5| \quad \begin{array}{l}\textit{Simplify the sum within the}\\ \textit{absolute value bars first.}\end{array}$$

$$= 0 - 5 \quad \begin{array}{l}\textit{|−5| = 5 because the}\\ \textit{distance from 0 to −5}\\ \textit{is 5 units.}\end{array}$$

$$= 0 + (-5) \quad \begin{array}{l}\textit{Change subtraction to}\\ \textit{addition. Change 5 to −5.}\end{array}$$

$$= -5 \quad \textit{Add.}$$

53. $-3 - (-2 + 4) + (-5)$

$$= -3 - 2 + (-5) \quad \begin{array}{l}\textit{Simplify the sum within}\\ \textit{the parentheses first.}\end{array}$$

$$= -3 + (-2) + (-5) \quad \begin{array}{l}\textit{Change subtraction}\\ \textit{to addition and}\\ \textit{change 2 to −2.}\end{array}$$

$$= -5 + (-5) \quad \textit{Add left to right.}$$

$$= -10$$

Relating Concepts (Exercises 55–56)

55. $-3 - 5 = -3 + (-5) = -8$
$5 - (-3) = 5 + (+3) = 5 + 3 = 8$
$-4 - (-3) = -4 + (+3) = -4 + 3 = -1$
$-3 - (-4) = -3 + 4 = 1$

Subtraction is *not* commutative; the absolute value of the answer is the same, but the sign changes.

56. Subtracting 0 from a number does *not* change the number. For example, $-5 - 0 = -5$. But subtracting a number from 0 *does* change the number to its opposite. For example, $0 - (-5) = 5$.

1.5 Problem Solving: Rounding and Estimating

1.5 Margin Exercises

1. **(a)** −746 (nearest ten)
Draw a line under the 4. $-7\underline{4}6$
−746 is closer to −7<u>5</u>0.

(b) 2412 (nearest thousand)
Draw a line under the leading 2. $\underline{2}412$
2412 is closer to <u>2</u>000.

(c) −89,512 (nearest hundred)
Draw a line under the 5. $-89,\underline{5}12$
−89,512 is closer to −89,<u>5</u>00.

(d) 546,325 (nearest ten-thousand)
Draw a line under the 4. $5\underline{4}6,325$
546,325 is closer to 5<u>5</u>0,000.

2. **(a)** $\underline{3}4 \approx 30$ ■ Locate the place to which the number is being rounded. Draw a line under that place. Because the next digit to the right of the underlined place is 4 or less, do not change the digit in the underlined place. Change all digits to the right of the underlined place to zeros.
Note: The symbol " $\approx$ " means "approximately equal to."

(b) $-\underline{6}1 \approx -60$ ■ Locate the place to which the number is being rounded. Draw a line under that place. Because the next digit to the right of the underlined place is 4 or less, do not change the digit in the underlined place. Change all digits to the right of the underlined place to zeros.

(c) $-6\underline{8}3 \approx -680$ ■ Locate the place to which the number is being rounded. Draw a line under that place. Because the next digit to the right of the underlined place is 4 or less, do not change the digit in the underlined place. Change all digits to the right of the underlined place to zeros.

(d) $17\underline{9}2 \approx 1790$ ■ Locate the place to which the number is being rounded. Draw a line under that place. Because the next digit to the right of the underlined place is 4 or less, do not change the digit in the underlined place. Change all digits to the right of the underlined place to zeros.

3. **(a)** $\underline{1}725 \approx 2000$ ■ Locate the place to which the number is being rounded. Draw a line under that place. Because the next digit to the right of the underlined place is 5 or more, add 1 to the digit in the underlined place. Change all digits to the right of the underlined place to zeros.

(b) $-\underline{6}511 \approx -7000$ ■ Locate the place to which the number is being rounded. Draw a line under that place. Because the next digit to the right of the underlined place is 5 or more, add 1 to the digit in the underlined place. Change all digits to the right of the underlined place to zeros.

(c) $58,829 \approx 59,000$ ■ Locate the place to which the number is being rounded. Draw a line under that place. Because the next digit to the right of the underlined place is 5 or more, add 1 to the digit in the underlined place. Change all digits to the right of the underlined place to zeros.

(d) $-83,904 \approx -84,000$ ■ Locate the place to which the number is being rounded. Draw a line under that place. Because the next digit to the right of the underlined place is 5 or more, add 1 to the digit in the underlined place. Change all digits to the right of the underlined place to zeros.

4.   **(a)** $-6036 \approx -6040$ ■ Locate the place to which the number is being rounded. Draw a line under that place. Underline the tens place. Next digit is 5 or more. Tens place changes. Add 1 to 3. Change all digits to the right of the underlined place to zeros.

(b) $34,968 \approx 35,000$ ■ Locate the place to which the number is being rounded. Draw a line under that place. Underline the hundreds place. Next digit is 5 or more. Hundreds place changes. Add 1 to 9. Write 0 and carry 1 into the thousands place. Change all digits to the right of the underlined place to zero.

(c) $-73,077 \approx -73,000$ ■ Locate the place to which the number is being rounded. Draw a line under that place. Underline the thousands place. Next digit is 4 or less. Leave 3 as 3. Change all digits to the right of the underlined place to zeros.

(d) $9852 \approx 10,000$ ■ Locate the place to which the number is being rounded. Draw a line under that place. Underline the thousands place. Next digit is 5 or more. Thousands place changes. Add 1 to 9. Change all digits to the right of the underlined place to zeros.

(e) $85,949 \approx 85,900$ ■ Locate the place to which the number is being rounded. Draw a line under that place. Underline the hundreds place. Next digit is 4 or less. Leave 9 as 9. Change all digits to the right of the underlined place to zeros.

(f) $40,387 \approx 40,000$ ■ Locate the place to which the number is being rounded. Draw a line under that place. Underline the thousands place. Next digit is 4 or less. Leave 0 as 0. Change all digits to the right of the underlined place to zeros.

5.   **(a)** $-14,679 \approx -10,000$ ■ Locate the place to which the number is being rounded. Draw a line under that place. Underline the ten-thousands

place. Next digit is 4 or less. Leave 1 as 1. Change all digits to the right of the underlined place to zeros.

(b) $724,518,715 \approx 725,000,000$ ■ Locate the place to which the number is being rounded. Draw a line under that place. Underline the millions place. Next digit is 5 or more. Add 1 to 4. Change all digits to the right of the underlined place to zeros.

(c) $-49,900,700 \approx -50,000,000$ ■ Locate the place to which the number is being rounded. Draw a line under that place. Underline the millions place. Next digit is 5 or more. Add 1 to 9. Write 0 and carry 1 to the ten millions place. Change all digits to the right of the underlined place to zeros.

(d) $306,779,000 \approx 300,000,000$ ■ Locate the place to which the number is being rounded. Draw a line under that place. Underline the hundred-millions place. Next digit is 4 or less. Leave 3 as 3. Change all digits to the right of the underlined place to zeros.

6.   **(a)** $-94 \approx -90$
Underline the first digit. Next digit is 4 or less. Leave 9 as 9. Change 4 to 0.

(b) $508 \approx 500$
Underline the first digit. Next digit is 4 or less. Leave 5 as 5. Change 8 to 0.

(c) $-2522 \approx -3000$
Underline the first digit. Next digit is 5 or more. Add 1 to 2. Change all digits to the right of the underlined place to zeros.

(d) $9700 \approx 10,000$
Underline the first digit. Next digit is 5 or more. Add 1 to 9. Write 0 and carry 1 to the ten thousands place. Change all digits to the right of the underlined place to zeros.

(e) $61,888 \approx 60,000$
Underline the first digit. Next digit is 4 or less. Leave 6 as 6. Change all digits to the right to zeros.

(f) $-963,369 \approx -1,000,000$
Underline the first digit. Next digit is 5 or more. Add 1 to 9. Write 0 and carry 1 to the millions place. Change all digits to the right to zeros.

7.   "Overdrawn" implies a negative number, $-\$3881$
"Deposit" implies a positive number, $+\$2090$

Estimate: $-\$3881 \approx -\$4000$
$ +\$2090 \approx +\$2000$

Balance $= -\$4000 + \$2000 = -\$2000$
Approximately $2000 overdrawn.

Exact:

Balance $= -\$3881 + \$2090 = -\$1791$

Pao Xiong is overdrawn by $1791.

The estimate of $2000 overdrawn is fairly close to the exact amount.

1.5 Section Exercises

1. **(a)** 3702 (nearest ten)
    Draw a line under 0. 37$\underline{0}$2
    Next digit is 4 or less, so leave 0 as 0 in the tens place.

    (b) 908,546 (nearest thousand)
    Draw a line under 8. 90$\underline{8}$,546
    Next digit is 5 or more, so change 8 to 9 in the thousands place.

3. $6\underline{2}5 \approx 630$ (nearest ten)
    Next digit is 5 or more. Tens place changes. Add 1 to 2. Change 5 to 0.

5. $-108\underline{3} \approx -1080$ (nearest ten)
    Next digit is 4 or less. Tens place remains 8. Change 3 to 0.

7. $7\underline{8}62 \approx 7900$ (nearest hundred)
    Next digit is 5 or more. Hundreds place changes. Add 1 to 8. Change 6 and 2 to 0 .

9. $-86,\underline{8}13 \approx -86,800$ (nearest hundred)
    Next digit is 4 or less. Hundreds place remains 8. Change 1 and 3 to 0.

11. $42,\underline{4}95 \approx 42,500$ (nearest hundred)
    Next digit is 5 or more. Hundreds place changes. Add 1 to 4. Change 9 and 5 to 0.

13. $-5\underline{9}96 \approx -6000$ (nearest hundred)
    Next digit is 5 or more. Hundreds place changes. Add 1 to 9 and carry 1 to thousands. Change 9 and 6 to 0.

15. $-7\underline{8},499 \approx -78,000$ (nearest thousand)
    Next digit is 4 or less. Thousands place remains 8. Change 4, 9, and 9 to 0.

17. $\underline{5}847 \approx 6000$ (nearest thousand)
    Next digit is 5 or more. Thousands place changes. Add 1 to 5. Change 8, 4, and 7 to 0.

19. $5\underline{9}5,008 \approx 600,000$ (nearest ten-thousand)
    Next digit is 5 or more. Ten-thousands place changes. Add 1 to 9 and carry 1 to hundred-thousands place. Change 5 and 8 to 0.

21. $-8,\underline{9}06,422 \approx -9,000,000$ (nearest million)
    Next digit is 5 or more. Millions place changes. Add 1 to 8. Change other digits to 0.

23. $139,\underline{6}10,000 \approx 140,000,000$ (nearest million)
    Next digit is 5 or more. Millions place changes. Add 1 to 9. Carry one to ten-millions. Change 6 and 1 to zeros.

25. $19,\underline{9}51,880,500 \approx 20,000,000,000$ (nearest hundred-million)
    Next digit is 5 or more. Hundred-millions place changes. Add 1 to 9. Write 0 and carry 1 to the ten-billions place. All digits to the right of the underlined place change to 0.

27. $\underline{8},608,200,000 \approx 9,000,000,000$ (nearest billion)
    Next digit is 5 or more. Billions place changes. Add 1 to 8. All digits to the right of the underlined place change to 0.

29. Answers will vary but should mention looking only at the second digit, rounding first digit up when second digit is 5 or more, leaving first digit unchanged when second digit is 4 or less. Examples will vary. Some possibilities are $\underline{2}7 \approx 30$, $\underline{6}41 \approx 600$.

31. $\underline{3}1,500 \approx 30,000$ miles
    Next digit is 4 or less. Leave 3 as 3. Change 1 and 5 to 0. 31,500 is closer to 30,000 than 40,000.

33. $-\underline{5}6 \approx -60$ degrees
    Next digit is 5 or more. Change 5 to 6 and change 6 to 0. -56 is closer to -60 than -50.

35. $\$\underline{9}942 \approx \$10,000$
    Next digit is 5 or more. Add 1 to 9 and carry 1 to the ten-thousands place. Change 9, 4, and 2 to 0. $9942 is closer to $10,000 than $9000.

37. $5\underline{3},500,000,000 \approx 50,000,000,000$ minutes
    Next digit is 4 or less. Leave 5 as 5. Change 3 and 5 to 0. 53,500,000,000 is closer to 50,000,000,000 than 60,000,000,000.

39. $\underline{6}98,473 \approx 700,000$ people in Alaska
    Next digit is 5 or more. Change 6 to 7. Change all other digits to 0. 698,473 is closer to 700,000 than 600,000.

    $\underline{3}6,961,664 \approx 40,000,000$ people in California
    Next digit is 5 or more. Change 3 to 4. Change all other digits to 0. 36,961,664 is closer to 40,000,000 than 30,000,000.

41. $-42 + 89$
    $-\underline{4}2$ is closer to -40 than -50.
    $\underline{8}9$ is closer to 90 than 80.

    Estimate: $-\underline{4}0 + \underline{9}0 = \underline{5}0$;
    Exact: $-42 + 89 = 47$

43. $16 + (-97)$
    $\underline{1}6$ is closer to 20 than 10.
    $-\underline{9}7$ is closer to -100 than -90.

    Estimate: $20 + (-100) = -80$;
    Exact: $16 + (-97) = -81$

45. $-273 + (-399)$
$\underline{2}73$ is closer to -300 than -200.
$\underline{3}99$ is closer to -400 than -300.

Estimate: $-300 + (-400) = -700$;
Exact: $-273 + (-399) = -672$

47. $3081 + 6826$
$\underline{3}081$ is closer to 3000 than 4000.
$\underline{6}826$ is closer to 7000 than 6000.

Estimate: $3000 + 7000 = 10,000$;
Exact: $3081 + 6826 = 9907$

49. $23 - 81$

$23 + (-81)$ — *Change subtraction to addition. Change 81 to −81.*

$\underline{2}3$ is closer to 20 than 30.
$-\underline{8}1$ is closer to -80 than -90.

Estimate: $20 + (-80) = -60$;
Exact: $23 - 81 = 23 + (-81) = -58$

51. $-39 - 39$

$-39 + (-39)$ — *Change subtraction to addition. Change 39 to −39.*

$-\underline{3}9$ is closer to -40 than -30.

Estimate: $-40 + (-40) = -80$;
Exact: $-39 - 39 = -39 + (-39) = -78$

53. $-106 + 34 - (-72)$

$-106 + 34 + (+72)$ — *Change subtraction to addition of the opposite.*

$-\underline{1}06 \approx -100; \underline{3}4 \approx 30; \underline{7}2 \approx 70$

Estimate: $-100 + 30 + (+70) = 0$;
Exact: $-106 + 34 - (-72)$
$\quad = -106 + 34 + (+72) = 0$

55. Already raised: $\$52,882 \approx \$50,000$
Amount needed: $\$78,650 \approx \$80,000$
Amount that still needs to be collected:

Estimate: $80,000 - 50,000 = \$30,000$;
Exact: $78,650 - 52,882 = \$25,768$

57. Estimate Dorene's expenses.
Rent: $\$\underline{8}45 \approx \$800$
Food: $\$\underline{3}25 \approx \$300$
Childcare: $\$\underline{3}65 \approx \$400$
Transportation: $\$\underline{1}82 \approx \$200$
Other: $\$\underline{2}40 \approx \$200$

Estimate: Dorene's total expenses:
$\$800 + \$300 + \$400 + \$200 + \$200 = \$1900$.
Estimate Dorene's monthly take home pay.
$\$\underline{2}120 \approx \$2000$
Subtract Dorene's expenses from her take home pay to estimate her monthly savings.
$\$2000 - \$1900 = \$100$.

Exact:
Total Expenses $= \$845 + \$325 + \$365$
$\qquad\qquad\qquad + \$182 + \$240$
$\qquad\qquad = \$1957$
Monthly savings $= \$2120 - \$1957 = \$163$.

59. The final temperature equals the initial temperature plus the two increases.

$-102 \approx -100; 37 \approx 40; 52 \approx 50$

Estimate: $-100 + 40 + 50 = -10$ degrees
Exact: $-102 + 37 + 52 = -13$ degrees

61. $\underline{4}12 \approx 400$ doors
$\underline{1}47 \approx 100$ windows
Total number of doors and windows:
Estimate: $400 + 100 = 500$ doors and windows
Exact: $412 + 147 = 559$ doors and windows

1.6 Multiplying Integers

1.6 Margin Exercises

1. **(a)** $\underbrace{100}_{} \times \underbrace{6}_{} = \underbrace{600}_{}$
Factor Factor Product

Equivalent forms: $\quad 100(6) = 600$
$\qquad\qquad\qquad 100 \cdot 6 = 600$
$\qquad\qquad\qquad (100)(6) = 600$

(b) $\underbrace{7}_{} \times \underbrace{12}_{} = \underbrace{84}_{}$
Factor Factor Product

Equivalent forms: $\quad 7(12) = 84$
$\qquad\qquad\qquad 7 \cdot 12 = 84$
$\qquad\qquad\qquad (7)(12) = 84$

2. **(a)** $7(-2) = \underline{-14}$
The factors have *different* signs, so the product is *negative*.

(b) $-5 \cdot (-5) = 25$
(*same* signs, product is *positive*)

(c) $-1(14) = -14$
(*different* signs, product is *negative*)

(d) $10 \cdot 6 = 60$
(*same* signs, product is *positive*)

(e) $(-4)(-9) = 36$
(*same* signs, product is *positive*)

3. **(a)** $-5 \cdot (10 \cdot 2)$ — *Work within parentheses first.*
$= -5 \cdot (\underline{20})$ — (*same signs, product is **positive***)
$= \underline{-100}$ — (*different signs, product is **negative***)

(b) $-1 \cdot 8 \cdot (-5)$ — *No parentheses. Multiply from left to right.*
$= -8 \cdot (-5)$ — (*different signs, product is **negative***)
$= 40$ — (*same signs, product is **positive***)

(c) $-3(-2)(-4)$ *Multiply from left to right.*
$= (\underline{6})(-4)$ *(same signs,*
                              *product is **positive**)*
                          (different signs,
$= \underline{-24}$ *product is **negative**)*

(d) $-2(7)(-3)$ *Multiply from left to right.*
$= (-14)(-3)$ *(different signs,*
                    *product is **negative**)*
                  (same signs,
$= 42$ *product is **positive**)*

(e) $(-1)(-1)(-1)$ *Multiply from left to right.*
$= 1(-1)$ *(same signs,*
                    *product is **positive**)*
                  (different signs,
$= -1$ *product is **negative**)*

4.   (a) $819 \cdot 0 = 0$; multiplication property of 0

(b) $1(-90) = -90$; multiplication property of 1

(c) $25 \cdot 1 = 25$; multiplication property of 1

(d) $(0)(-75) = 0$; multiplication property of 0

5.   (a) $\underbrace{(3 \cdot 3)} \cdot 2 = 3 \cdot \underbrace{(3 \cdot 2)}$
$\quad\quad \underline{9} \;\; \cdot 2 = 3 \cdot \;\; \underline{6}$
$\quad\quad\quad \underline{18} = \underline{18}$

This illustrates the <u>associative</u> property of multiplication.

(b) $11 \cdot 8 = 8 \cdot 11$
$\quad\quad 88 = 88$
Commutative property of multiplication

(c) $2 \cdot (-15) = -15 \cdot 2$
$\quad\quad -30 = -30$
Commutative property of multiplication

(d) $-4 \cdot (2 \cdot 5) = (-4 \cdot 2) \cdot 5$
$\quad\quad -4 \cdot 10 = -8 \cdot 5$
$\quad\quad\quad -40 = -40$
Associative property of multiplication

6.   (a) $3(8 + 7) = 3 \cdot 8 + \underline{3} \cdot 7$
$\quad\quad 3(\underline{15}) = \underline{24} + \underline{21}$
$\quad\quad\quad \underline{45} = \underline{45}$ *Both results are 45.*

(b) $10(-6 + 9) = 10(-6) + 10(9)$
$\quad\quad\quad 10(3) = -60 + 90$
$\quad\quad\quad\quad 30 = 30$ *Both results are 30.*

(c) $-6(4 + 4) = -6 \cdot 4 + (-6) \cdot 4$
$\quad\quad -6(8) = -24 + (-24)$
$\quad\quad\quad -48 = -48$ *Both results are -48.*

7. 27,095 fans rounds to 30,000 fans and 81 games rounds to 80 games.

Total attendance for the season:

Estimate: $30,000(80) = 2,400,000$ fans

Exact: $27,095(81) = 2,194,695$ fans

1.6   Section Exercises

1.   (a) $9 \cdot 7 = 63$ *Factors have the **same** sign, so the product is **positive**.*

(b) $-9 \cdot (-7) = 63$ *Factors have the **same** sign, so the product is **positive**.*

(c) $-9 \cdot 7 = -63$ *Factors have **different** signs, so the product is **negative**.*

(d) $9 \cdot (-7) = -63$ *Factors have **different** signs, so the product is **negative**.*

3.   (a) $7(-8) = -56$ *Factors have **different** signs, so the product is **negative**.*

(b) $-7(8) = -56$ *Factors have **different** signs, so the product is **negative**.*

(c) $7(8) = 56$ *Factors have the **same** sign, so the product is **positive**.*

(d) $-7(-8) = 56$ *Factors have the **same** sign, so the product is **positive**.*

5. $-5 \cdot 7 = -35$
(different signs, product is negative)

7. $(-5)(9) = -45$
(different signs, product is negative)

9. $3(-6) = -18$
(different signs, product is negative)

11. $10(-5) = -50$
(different signs, product is negative)

13. $(-1)(40) = -40$
(different signs, product is negative)

15. $-56 \cdot 1 = -56$; multiplication property of 1

17. $-8(-4) = 32$ *(same signs, product is positive)*

19. $11 \cdot 7 = 77$ *(same signs, product is positive)*

21. $25 \cdot 0 = 0$; multiplication property of 0

23. $-19(-7) = 133$ *(same signs, product is positive)*

25. $-13(-1) = 13$ *(same signs, product is positive)*

27. $(0)(-25) = 0$; multiplication property of 0

29. $-4 \cdot (-6) \cdot 2$
$= 24 \cdot 2$ *Multiply from left to right.*
$= \underline{48}$

31. $(-4)(-2)(-7)$
$= 8(-7)$ *Multiply from left to right.*
$= -56$

33. $5(-8)(4)$

$\quad = -40(4)$ *Multiply from left to right.*

$\quad = -160$

35. $(-3)(\underline{5}) = -15$ (*negative* product, *different* signs)

37. $\underline{-3} \cdot 10 = -30$ (*negative* product, *different* signs)

39. $-17 = 17(\underline{-1})$ (*negative* product, *different* signs)

41. $(\underline{0})(-350) = 0$ (*multiplication property of 0*)

43. $5 \cdot (-4) \cdot \underline{5} = -100$

$\quad -20 \cdot 5 = -100$ *(negative product, different signs)*

45. $(\underline{-4})(-5)(-2) = -40$

$\quad (-4)(10) = -40$ *(negative product, different signs)*

47. Commutative property: changing the *order* of the factors does not change the product.
Associative property: changing the *grouping* of the factors does not change the product.
Examples will vary. Some possibilities are
$-4 \cdot 7 = 7 \cdot (-4) = -28$,
$(2 \cdot 5) \cdot (-8) = 2 \cdot (5 \cdot (-8)) = -80$.

49. $9(-3 + 5)$ rewritten by using the distributive property is $9 \cdot (-3) + 9 \cdot 5$.

$$9(-3 + 5) = 9 \cdot (-3) + 9 \cdot 5$$
$$9(2) = -27 + 45$$
$$18 = 18$$

51. $25 \cdot 8$ rewritten by using the commutative property is $8 \cdot 25$.

$$25 \cdot 8 = 8 \cdot 25$$
$$200 = 200$$

53. $-3 \cdot (2 \cdot 5)$ rewritten using the associative property is $(-3 \cdot 2) \cdot 5$.

$$-3 \cdot (2 \cdot 5) = (-3 \cdot 2) \cdot 5$$
$$-3 \cdot 10 = -6 \cdot 5$$
$$-30 = -30$$

55. Income: $\$324 \approx \$300$
52 weeks ≈ 50

Estimate: $\$300 \cdot 50 = \$15,000$
Exact: $\$324 \cdot 52 = \$16,848$

57. Monthly loss: $-\$9950 \approx -\$10,000$
12 months ≈ 10

Estimate: $-\$10,000 \cdot 10 = -\$100,000$
Exact: $-\$9950 \cdot 12 = -\$119,400$

59. Tuition: $\$182$ per credit $\approx \$200$ per credit
13 credits ≈ 10 credits

Estimate: $\$200 \cdot 10 = \$2000$
Exact: $\$182 \cdot 13 = \$2366$

61. Hours: $24 \approx 20$
365 days ≈ 400

Estimate: $20 \cdot 400 = 8000$ hours
Exact: $24 \cdot 365 = 8760$ hours

63. $-8 \cdot |-8 \cdot 8|$

$\quad = -8 \cdot |-64|$ *Multiply the factors within the absolute value bars (different signs, negative product).*

$\quad = -8 \cdot 64$ *-64 is 64 units from 0, so $|-64| = 64$.*

$\quad = \underline{-512}$ *Multiply. (different signs, negative product)*

65. $(-37)(-1)(85)(0) = 0$;
multiplication property of 0

67. $|6 - 7| \cdot (-355,299)$

$\quad = |6 + (-7)| \cdot (-355,299)$ *Subtract within the abs. value bars first.*

$\quad = |-1| \cdot (-355,299)$

$\quad = 1 \cdot (-355,299)$ *-1 is 1 unit from 0, so $|-1| = 1$*

$\quad = -355,299$ *Multiplication property of 1*

69. The charge for each cat's shots will be
$\$24 + \$29 = \$53$.
The total for all four cats will be four times that, plus the office visit charge.

$\quad 4 \cdot \$53 + \$35$ *Do the multiplication first.*
$\quad = \$212 + \$35$
$\quad = \$247$

71. The temperature drops 3 degrees (-3 degrees) for every 1000 feet climbed into the air. An altitude of 24,000 feet would require 24 increases of 1000 feet each.

$$-3 \cdot 24 = -72 \text{ degrees}$$

The temperature at 24,000 feet is

$$50 + (-72) = -22 \text{ degrees}.$$

73. Points possible on tests:
$6(100) = 600$ points
Bonus points possible on tests:
$6(4) = 24$ points
Points possible on quizzes:
$8(6) = 48$ points
Points possible on homework assignments:
$20(5) = 100$ points
Total points possible:
$600 + 24 + 48 + 100 = 772$ points

Relating Concepts (Exercises 75–76)

75. Examples will vary. Some possibilities are:

(a) $6 \cdot (-1) = -6; 2 \cdot (-1) = -2;$
$15 \cdot (-1) = -15$

(b) $-6 \cdot (-1) = 6; -2 \cdot (-1) = 2;$
$-15 \cdot (-1) = 15$

The result of multiplying any nonzero number times -1 is the number with the opposite sign.

76.
$$-2 \cdot (-2) = \underline{4}$$
$$-2 \cdot (-2) \cdot (-2) = \underline{-8}$$
$$-2 \cdot (-2) \cdot (-2) \cdot (-2) = \underline{16}$$
$$-2 \cdot (-2) \cdot (-2) \cdot (-2) \cdot (-2) = \underline{-32}$$

The absolute value doubles each time and the sign changes. The next three products are $-2 \cdot (-32) = 64$, $-2 \cdot 64 = -128$, and $-2 \cdot (-128) = 256$.

1.7 Dividing Integers

1.7 Margin Exercises

1. (a) $\dfrac{40}{-8} = -5$ The integers have **different signs**, so the quotient is **negative**.

(b) $\dfrac{49}{7} = 7$ (*same* signs, quotient is *positive*)

(c) $\dfrac{-32}{4} = -8$ (*different* signs, quotient is *negative*)

(d) $\dfrac{-10}{-10} = 1$ The integers have the **same sign**, so the quotient is **positive**.

(e) $-81 \div 9 = -9$ (*different* signs, quotient is *negative*)

(f) $-100 \div (-50) = 2$ (*same* signs, quotient is *positive*)

2. (a) $\dfrac{-12}{0}$ is undefined; division by 0 is <u>undefined</u>.

(b) $\dfrac{0}{39} = 0$; 0 divided by any nonzero number is 0.

(c) $\dfrac{-9}{1} = -9$; any number divided by 1 is the number.

(d) $\dfrac{21}{21} = 1$; any nonzero number divided by itself is 1.

3. (a) $60 \div (-3)(-5)$
$= (\underline{-20})(-5)$ *Work from left to right.*
$= \underline{100}$

(b) $-6(-16 \div 8) \cdot 2$
$= -6(-2) \cdot 2$ *Work inside parentheses first.*
$= 12 \cdot 2$ *Now work from left to right.*
$= 24$

(c) $-8(10) \div 4(-3) \div (-6)$
$= -80 \div 4(-3) \div (-6)$ *Work left to right.*
$= -20(-3) \div (-6)$
$= 60 \div (-6)$
$= -10$

(d) $56 \div (-8) \div (-1)$
$= -7 \div (-1)$ *Work left to right.*
$= 7$

4. A loss of money implies a negative number:
$-\$2724$ in stocks rounds to $-\$3000$.
12 months rounds to 10 months.
To find an average, use division.

Estimate: $-\$3000 \div \underline{10} = \underline{-\$300}$ each month
Exact: $-\$2724 \div 12 = -\$227$ each month

5. (a) 116 cookies are separated into packages of 12 each.

$$
\begin{array}{r}
9 \\
12\overline{)116} \\
\underline{108} \\
8
\end{array}
$$

They will have 9 packages of a dozen cookies each, and there will be 8 cookies left over for them to eat.

(b) 249 senior citizens separated into buses with a 44-person limit per bus.

$$
\begin{array}{r}
5 \\
44\overline{)249} \\
\underline{220} \\
29
\end{array}
$$

If Coreen dispatches 5 buses, then 29 senior citizens will not have transportation to the game. She will need to dispatch 6 buses.

1.7 Section Exercises

1. (a) $14 \div 2 = 7$ (*same* signs, quotient is *positive*)

(b) $-14 \div (-2) = 7$
(*same* signs, quotient is *positive*)

(c) $14 \div (-2) = -7$
(*different* signs, quotient is *negative*)

(d) $-14 \div 2 = -7$
(*different* signs, quotient is *negative*)

3. **(a)** $-42 \div 6 = -7$
(*different* signs, quotient is *negative*)

(b) $-42 \div (-6) = 7$
(*same* signs, quotient is *positive*)

(c) $42 \div (-6) = -7$
(*different* signs, quotient is *negative*)

(d) $42 \div 6 = 7$ (*same* signs, quotient is *positive*)

5. **(a)** $\dfrac{35}{35} = 1$; any nonzero number divided by itself is 1.

(b) $\dfrac{35}{1} = 35$; any number divided by 1 is the number.

(c) $\dfrac{-13}{1} = -13$; any number divided by 1 is the number.

(d) $\dfrac{-13}{-13} = 1$; any nonzero number divided by itself is 1.

7. **(a)** $\dfrac{0}{50} = 0$; zero divided by any nonzero number is 0.

(b) $\dfrac{50}{0}$ is undefined; division by zero is undefined.

(c) $\dfrac{-11}{0}$ is undefined; division by zero is undefined.

(d) $\dfrac{0}{-11} = 0$; zero divided by any nonzero number is 0.

9. $\dfrac{-8}{2} = -4$ (*different* signs, quotient is *negative*)

11. $\dfrac{21}{-7} = -3$ (*different* signs, quotient is *negative*)

13. $\dfrac{-54}{-9} = 6$ (*same* signs, quotient is *positive*)

15. $\dfrac{55}{-5} = -11$ (*different* signs, quotient is *negative*)

17. $\dfrac{-28}{0}$ is undefined. Division by zero is undefined.

19. $\dfrac{14}{-1} = -14$ (*different* signs, quotient is *negative*)

21. $\dfrac{-20}{-2} = 10$ (*same* signs, quotient is *positive*)

23. $\dfrac{-48}{-12} = 4$ (*same* signs, quotient is *positive*)

25. $\dfrac{-18}{18} = -1$ (*different* signs, quotient is *negative*)

27. $\dfrac{0}{-9} = 0$; zero divided by any nonzero number is 0.

29. $\dfrac{-573}{-3} = 191$ (*same* signs, quotient is *positive*)

31. $\dfrac{163,672}{-328} = -499$
(*different* signs, quotient is *negative*)

33. $-60 \div 10 \div (-3)$
$= -6 \div (-3)$ *Work left to right.*
$= \underline{2}$

35. $-64 \div (-8) \div (-2)$
$= 8 \div (-2)$ *Work left to right.*
$= -4$

37. $100 \div (-5)(-2)$
$= -20(-2)$ *Work left to right.*
$= 40$

39. $48 \div 3 \cdot (-12 \div 4)$
$= 48 \div 3 \cdot (-3)$ *Start inside the parentheses.*
$= 16 \cdot (-3)$ *Now work from left to right.*
$= -48$

41. $-5 \div (-5)(-10) \div (-2)$
$= 1(-10) \div (-2)$ *Work left to right.*
$= -10 \div (-2)$
$= 5$

43. $64 \cdot 0 \div (-8)(10)$
$= 0 \div (-8)(10)$ *Work left to right.*
$= 0(10)$
$= 0$

45. $2 \div 1 = 2$ but $1 \div 2 = 0.5$, so division is not commutative.

47. Similar: If the signs match, the result is positive. If the signs are different, the result is negative. Different: Multiplication is commutative; division is not. You can multiply by 0, but dividing by 0 is undefined.

49. Examples will vary.

(a) $\dfrac{-6}{-1} = 6$; $\dfrac{-2}{-1} = 2$; $\dfrac{-15}{-1} = 15$

(b) $\dfrac{6}{-1} = -6$; $\dfrac{2}{-1} = -2$; $\dfrac{15}{-1} = -15$

When dividing by -1, change the sign of the dividend to its opposite to get the quotient.

51. Depth below sea level implies a negative, $-35,836$ feet. Use division to find the size of each step.
$-35,836 \approx -40,000$; $17 \approx 20$

Estimate: $-40,000 \div \underline{20} = \underline{-2000 \text{ feet}}$
Exact: $-35,836 \div 17 = -2108 \text{ feet}$

53. Overdrawn implies a negative, $-\$238$. Transfer of money into the account, implies a positive, $\$450$.
$-238 \approx -200; \ 450 \approx 500$

Estimate: $-200 + 500 = \$300$
Exact: $-238 + 450 = \$212$

55. The number of non-foggy days equals the total number of days in a year minus the number of foggy days. $365 \approx 400; \ 106 \approx 100$

Estimate: $400 - 100 = 300$ days
Exact: $365 - 106 = 259$ days

57. Descending implies a negative, -730 feet each minute. $-730 \approx -700$
Because the plane took $37 \approx 40$ minutes to land, use multiplication to find how far the plane descended.

Estimate: $-700 \cdot 40 = -28{,}000$ feet
Exact: $-730 \cdot 37 = -27{,}010$ feet

59. Use division to find how many miles were covered in each hour. $315 \approx 300; \ 5 \approx 5$

Estimate: $300 \div 5 = 60$ miles
Exact: $315 \div 5 = 63$ miles

61. To find the average, add all the scores and divide by the number of scores.
Sum of scores: $143 + 190 + 162 + 177 = 672$
Number of scores: 4 scores given

$$\text{Average score} = \frac{672}{4} = 168$$

His average score was 168.

63. Calculate the total weight from the data on the back, then use subtraction to find the difference between that and the figure on the front.

Total weight from back:
$13 \cdot 40$ grams $= 520$ grams

Difference from front: $520 - 510 = 10$ grams

The back claims 10 more grams than the front.

65. The $\$302$ already in Stephanie's account and her $\$347$ paycheck are positives. The money she paid for day care, $\$116$, and rent, $\$548$, are negatives.
$\$302 + (-\$116) + (-\$548) + \$347 = -\$15$

Stephanie's balance is $-\$15$. She is overdrawn by $\$15$.

67. Use division to convert minutes to hours.

$$\frac{1000 \text{ minutes}}{60 \text{ minutes per hour}}$$

```
        1 6
  60 ) 1 0 0 0
        6 0
        4 0 0
        3 6 0
          4 0
```

A new subscriber will receive 1000 minutes, which is 16 hours, with 40 minutes left over.

69. Use division to find the number of rooms.

$$\frac{163 \text{ people}}{5 \text{ people per room}}$$

```
        3 2
   5 ) 1 6 3
        1 5
          1 3
          1 0
            3
```

32 rooms will be full, and that leaves 3 people. So 33 rooms are needed, with space for 2 people $(5 - 3 = 2)$ unused.

71. $|-8| \div (-4) \cdot |-5| \cdot |1|$

$= 8 \div (-4) \cdot 5 \cdot 1$ *Simplify the absolute values first.*

$= -2 \cdot 5 \cdot 1$ *No parentheses, so start from the left.*

$= -10 \cdot 1$

$= -10$

73. $-6(-8) \div (-5 + 5)$

$= -6(-8) \div 0$ *Start inside the parentheses.*

$= 48 \div 0$ *Multiply and divide from left to right.*

Undefined   *Division by zero*

75. Start by entering $1\,000\,000\,000$.
Divide by 60. ($\approx 16{,}666{,}667$ minutes)
Divide by 60. ($\approx 277{,}778$ hours)
Divide by 24. ($\approx 11{,}574$ days)
Divide by 365. (≈ 31.7 years)
31.70979198 rounds to 32 years to receive one billion dollars.

Summary Exercises
Operations with Integers

1. $2 - 8$
$= 2 + (-8)$ *Add the opposite.*
$= -6$

3. $-14 - (-7)$
$= -14 + 7$ *Add the opposite.*
$= -7$

5. $-9(-7) = 63$ (*same* signs, product is *positive*)

7. $(1)(-56) = -56$ *Multiplication property of 1*

9. $5 - (-7)$
$= 5 + 7$ *Add the opposite.*
$= 12$

11. $-18 + 5 = -13$

13. $-40 - (-40)$
$= -40 + (+40)$ *Add the opposite.*
$= 0$ *Addition of opposites is 0.*

15. $8(-6) = -48$ (*different* signs, product is *negative*)

17. $-5(10) = -50$ (*different* signs, product is *negative*)

19. $0 - 14$

  $= 0 + (-14)$ *Add the opposite.*

  $= -14$ *Addition property of 0.*

21. $-13 + 13$

  $= 0$ *Addition of opposites is 0.*

23. $20 - 50$

  $= 20 + (-50)$ *Add the opposite.*

  $= -30$

25. $(-4)(-6)(2)$

  $= (24)(2)$ *Multiply from left to right.*

  $= 48$

27. $-60 \div 10 \div (-3)$

  $= -6 \div (-3)$ *Divide from left to right.*

  $= 2$

29. $64(0) \div (-8)$

  $= 0 \div (-8)$ *Multiply.*

  $= 0$ *Divide.*

31. $-9 + 8 + (-2)$

  $= -1 + (-2)$ *Add from left to right.*

  $= -3$

33. $8 + 6 + (-8)$

  $= 14 + (-8)$ *Add from left to right.*

  $= 6$

35. $-25 \div (-1) \div (-5)$

  $= 25 \div (-5)$ *Divide from left to right.*

  $= -5$

37. $-72 \div (-9) \div (-4)$

  $= 8 \div (-4)$ *Divide from left to right.*

  $= -2$

39. $9 - 6 - 3 - 5$

  $= 9 + (-6) + (-3) + (-5)$ *Add the opposite.*

  $= 3 + (-3) + (-5)$ *Add from left to right.*

  $= 0 + (-5)$

  $= -5$

41. $-1(9732)(-1)(-1)$

  $= (-9732)(-1)(-1)$ *Multiply from left to right.*

  $= (9732)(-1)$

  $= -9732$

43. $-10 - 4 + 0 + 18$

  $= -10 + (-4) + 0 + 18$ *Add the opposite.*

  $= -14 + 0 + 18$ *Add from left to right.*

  $= -14 + 18$

  $= 4$

45. $5 - |-3| + 3$

  $= 5 - 3 + 3$ *Absolute value first*

  $= 5 + (-3) + 3$ *Add the opposite.*

  $= 2 + 3$ *Add from left to right.*

  $= 5$

47. $-3 - (-2 + 4) - 5$

  $= -3 - 2 - 5$ *Parentheses first*

  $= -3 + (-2) + (-5)$ *Add the opposites.*

  $= -5 + (-5)$ *Add from left to right.*

  $= -10$

49. **(a)** If zero is divided by a nonzero number, the quotient is 0.

  (b) If any number is multiplied by 0, the product is 0.

  (c) If a nonzero number is divided by itself, the quotient is 1.

1.8 Exponents and Order of Operations

1.8 Margin Exercises

1. **(a)** $3 \cdot 3 \cdot 3 \cdot 3 = 3^4$
is read "3 to the <u>fourth</u> power."

  (b) $6 \cdot 6 = 6^2$
is read "6 squared" or "6 to the second power."

  (c) $9 = 9^1$
is read "9 to the first power."

  (d) $(2)(2)(2)(2)(2)(2) = (2)^6 = 2^6$
is read "2 to the sixth power."

2. **(a)** $(-2)^3 = (-2)(-2)(-2)$

  $= (\underline{4})(-2)$

  $= \underline{-8}$

  (b) $(-6)^2 = (-6)(-6)$

  $= 36$

  (c) $2^4(-3)^2 = \underbrace{(2)(2)(2)(2)}\ \underbrace{(-3)(-3)}$

  $= \quad (16)\quad \cdot \quad (9)$

  $= 144$

  (d) $3^3 \cdot (-4)^2 = \underbrace{(3)(3)(3)}\ \underbrace{(-4)(-4)}$

  $= \quad (27)\quad \cdot \quad (16)$

  $= 432$

3. **(a)** $-9 + (-15) - 3$

  $= -9 + (-15) + (-3)$ *Change subt. to addition. Change 3 to -3.*

  $= \underline{-24} + (-3)$ *Add left to right.*

  $= \underline{-27}$

(b) $-4 - 2 + (-6)$

$= -4 + (-2) + (-6)$ *Change subt. to addition.*
Change 2 to −2.

$= -6 + (-6)$ *Add left to right.*

$= -12$

(c) $3(-4) \div (-6)$

$= -12 \div (-6)$ *Multiply and divide left to right.*

$= 2$

(d) $-18 \div 9(-4)$

$= (\underline{-2})(-4)$ *Multiply and divide left to right.*

$= 8$

4. **(a)** $8 + 6(14 \div 2)$

$= 8 + 6(7)$ *Parentheses first*

$= 8 + \underline{42}$ *Multiply.*

$= \underline{50}$ *Addition last*

(b) $4(1) + 8(9 - 2)$

$= 4(1) + 8(7)$ *Parentheses first*

$= 4 + 56$ *Multiply.*

$= 60$ *Addition last*

(c) $3(5 + 1) + 20 \div 4$

$= 3(6) + 20 \div 4$ *Parentheses first*

$= 18 + 20 \div 4$ *Multiply and divide left to right.*

$= 18 + 5$

$= 23$ *Addition last*

5. **(a)** $2 + 40 \div (-5 + 3)$

$= 2 + 40 \div (-2)$ *Parentheses first*

$= 2 + (\underline{-20})$ *Divide.*

$= \underline{-18}$ *Addition last*

(b) $-5(5) - (15 + 5)$

$= -5(5) - 20$ *Parentheses first*

$= -25 - 20$ *Multiply.*

$= -25 + (-20)$ *Change subtraction to addition. Change 20 to its opposite.*

$= -45$ *Addition last*

(c) $(-24 \div 2) + (15 - 3)$

$= -12 + 12$ *Parentheses first*

$= 0$ *Addition last*

(d) $-3(2 - 8) - 5(4 - 3)$

$= -3(-6) - 5(1)$ *Parentheses first*

$= 18 - 5(1)$ *Multiply left to right.*

$= 18 - 5$

$= 18 + (-5)$ *Change subtraction to addition. Change 5 to its opposite.*

$= 13$ *Addition last*

(e) $3(3) - (10 \cdot 3) \div 5$

$= 3(3) - 30 \div 5$ *Parentheses first*

$= 9 - 30 \div 5$ *Multiply and divide left to right.*

$= 9 - 6$

$= 9 + (-6)$ *Change subtraction to addition. Change 6 to its opposite.*

$= 3$ *Addition last*

(f) $6 - (2 + 7) \div (-4 + 1)$

$= 6 - 9 \div (-3)$ *Parentheses first*

$= 6 - (-3)$ *Divide.*

$= 6 + (+3)$

$= 9$

6. **(a)** $2^3 - 3^2$

$= 8 - 9$ *Apply exponents.*

$= 8 + (-9)$ *Change subt. to addition. Change 9 to −9.*

$= -1$ *Add.*

(b) $6^2 \div (-4)(-3)$

$= 36 \div (-4)(-3)$ *Apply exponent.*

$= -9(-3)$ *Work left to right.*

$= 27$

(c) $(-4)^2 - 3^2(5 - 2)$

$= (-4)^2 - 3^2(3)$ *Parentheses first*

$= 16 - 9(3)$ *Apply exponents.*

$= 16 - 27$ *Multiply.*

$= 16 + (-27)$ *Change subtraction to addition. Change 27 to −27.*

$= -11$ *Add.*

(d) $(-3)^3 + (3 - 9)^2$

$= (-3)^3 + (-6)^2$ *Parentheses first*

$= -27 + 36$ *Apply exponents.*

$= 9$ *Add.*

7. **(a)** $\dfrac{-3(2^3)}{-10 - 6 + 8}$

Numerator:

$= -3(2^3)$

$= -3(\underline{8})$ *Exponent first*

$= \underline{-24}$ *Multiply.*

Denominator:

$= -10 - 6 + 8$

$= -10 + (-6) + 8$ *Change subtraction to addition. Change 6 to −6.*

$= \underline{-16} + 8$ *Add from left to right.*

$= \underline{-8}$

Last step is division: $\dfrac{-24}{-8} = 3$

(b) $\dfrac{(-10)(-5)}{-6 \div 3(5)}$

Numerator:
$= (-10)(-5)$
$= 50$

Denominator:
$= -6 \div 3(5)$
$= -2(5)$ *Multiply and divide left to right.*
$= -10$

Last step is division: $\dfrac{50}{-10} = -5$

(c) $\dfrac{6 + 18 \div (-2)}{(1 - 10) \div 3}$

Numerator:
$= 6 + 18 \div (-2)$
$= 6 + (-9)$ *Divide before adding.*
$= -3$

Denominator:
$= (1 - 10) \div 3$
$= -9 \div 3$ *Parentheses first*
$= -3$

Last step is division: $\dfrac{-3}{-3} = 1$

(d) Numerator:
$= 6^2 - 3^2(4)$
$= 36 - 9(4)$ *Exponents first*
$= 36 - 36$ *Multiply before subtracting.*
$= 0$

Denominator:
$= 5 + (3 - 7)^2$
$= 5 + (-4)^2$ *Parentheses first*
$= 5 + 16$ *Exponent next*
$= 21$ *Addition last*

Last step is division: $\dfrac{0}{21} = 0$

1.8 Section Exercises

1. Exponential Form: 4^3
Factored Form: $4 \cdot 4 \cdot 4$
Simplified: 64
Read as: 4 cubed or 4 to the third power

3. Exponential Form: 2^7
Factored Form: $2 \cdot 2 \cdot 2 \cdot 2 \cdot 2 \cdot 2 \cdot 2$
Simplified: 128
Read as: 2 to the seventh power

5. Exponential Form: 5^4
Factored Form: $5 \cdot 5 \cdot 5 \cdot 5$
Simplified: 625
Read as: 5 to the fourth power

7. Exponential Form: 7^2
Factored Form: $7 \cdot 7$
Simplified: 49
Read as: 7 squared

9. Exponential Form: 10^1
Factored Form: 10
Simplified: 10
Read as: 10 to the first power

11. **(a)** $10^1 = 10$

(b) $10^2 = 10 \cdot 10 = 100$

(c) $10^3 = 10 \cdot 10 \cdot 10$
$= 100 \cdot 10$
$= 1000$

(d) $10^4 = 10 \cdot 10 \cdot 10 \cdot 10$
$= 100 \cdot 10 \cdot 10$
$= 1000 \cdot 10$
$= 10,000$

13. **(a)** $4^1 = 4$

(b) $4^2 = 4 \cdot 4 = 16$

(c) $4^3 = 4 \cdot 4 \cdot 4$
$= 16 \cdot 4$
$= 64$

(d) $4^4 = 4 \cdot 4 \cdot 4 \cdot 4$
$= 16 \cdot 4 \cdot 4$
$= 64 \cdot 4$
$= 256$

15. 5^{10} on the calculator is 9,765,625.

17. 2^{12} on the calculator is 4096.

19. $(-2)^2 = (-2)(-2)$
$= 4$

21. $(-5)^2 = (-5)(-5)$
$= 25$

23. $(-4)^3 = (-4)(-4)(-4)$
$= 16(-4)$
$= -64$

25. $(-3)^4 = (-3)(-3)(-3)(-3)$
$= 9(-3)(-3)$
$= -27(-3)$
$= 81$

27. $(-10)^3 = (-10)(-10)(-10)$
$= 100(-10)$
$= -1000$

29. $1^1 = 1$ because 1 times itself any number of times equals 1.

31. $3^3 \cdot 2^2$

$\quad = 27 \cdot 4 \qquad$ *Apply exponents.*

$\quad = 108 \qquad$ *Multiply.*

33. $2^3(-5)^2$

$\quad = 8(25) \qquad$ *Apply exponents.*

$\quad = 200 \qquad$ *Multiply.*

35. $6^1(-5)^3$

$\quad = 6(-125) \qquad$ *Apply exponents.*

$\quad = -750 \qquad$ *Multiply.*

37. $(-2)(-2)^4$

$\quad = -2(16) \qquad$ *Apply exponent.*

$\quad = -32 \qquad$ *Multiply.*

39. $(-2)^2 = \underline{4} \qquad\qquad (-2)^6 = \underline{64}$

$\quad (-2)^3 = \underline{-8} \qquad\qquad (-2)^7 = \underline{-128}$

$\quad (-2)^4 = \underline{16} \qquad\qquad (-2)^8 = \underline{256}$

$\quad (-2)^5 = \underline{-32} \qquad\quad (-2)^9 = \underline{-512}$

(a) When a negative number is raised to an even power, the answer is positive; when raised to an odd power, the answer is negative.

(b) 15 is odd, so the sign of $(-2)^{15}$ is negative. 24 is even, so the sign of $(-2)^{24}$ is positive.

41. $12 \div 6(-3)$

$\quad = \underline{2}(-3) \qquad$ *Divide.*

$\quad = \underline{-6} \qquad$ *Multiply.*

43. $-1 + 15 - 7 - 7$

$\quad = 14 - 7 - 7 \qquad$ *Add left to right.*

$\quad = 14 + (-7) + (-7) \qquad$ *Add the opposite.*

$\quad = 7 + (-7) \qquad$ *Add left to right.*

$\quad = 0 \qquad$ *Add.*

45. $10 - 7^2$

$\quad = 10 - 49 \qquad$ *Apply exponent.*

$\quad = 10 + (-49) \qquad$ *Add the opposite.*

$\quad = -39 \qquad$ *Add.*

47. $2 - (-5) + 3^2$

$\quad = 2 - (-5) + 9 \qquad$ *Apply exponent.*

$\quad = 2 + (+5) + 9 \qquad$ *Add the opposite.*

$\quad = 7 + 9 \qquad$ *Add from left to right.*

$\quad = 16 \qquad$ *Add.*

49. $3 + 5(6 - 2)$

$\quad = 3 + 5[6 + (-2)] \qquad$ *Add the opposite.*

$\quad = 3 + 5(4) \qquad$ *Brackets*

$\quad = 3 + 20 \qquad$ *Multiply.*

$\quad = 23 \qquad$ *Add.*

51. $-7 + 6(8 - 14)$

$\quad = -7 + 6[8 + (-14)] \qquad$ *Add the opposite.*

$\quad = -7 + 6(-6) \qquad$ *Brackets*

$\quad = -7 + (-36) \qquad$ *Multiply.*

$\quad = -43 \qquad$ *Add.*

53. $2(-3 + 5) - (9 - 12)$

$\quad = 2(-3 + 5) - [9 + (-12)] \qquad$ *Add the opposite.*

$\quad = 2(2) - (-3) \qquad$ *Brackets*

$\quad = 4 - (-3) \qquad$ *Multiply.*

$\quad = 4 + (+3) \qquad$ *Add the opposite.*

$\quad = 7 \qquad$ *Add.*

55. $-5(7 - 13) \div (-10)$

$\quad = -5[7 + (-13)] \div (-10) \qquad$ *Add the opposite.*

$\quad = -5(-6) \div (-10) \qquad$ *Brackets*

$\quad = 30 \div (-10) \qquad$ *Multiply.*

$\quad = -3 \qquad$ *Divide.*

57. $9 \div (-3)^2 + (-1)$

$\quad = 9 \div 9 + (-1) \qquad$ *Exponent first*

$\quad = 1 + (-1) \qquad$ *Divide.*

$\quad = 0 \qquad$ *Add.*

59. $2 - (-5)(-2)^3$

$\quad = 2 - (-5)(-8) \qquad$ *Exponent first*

$\quad = 2 - 40 \qquad$ *Multiply.*

$\quad = 2 + (-40) \qquad$ *Add the opposite.*

$\quad = -38 \qquad$ *Add.*

61. $-2(-7) + 3(9)$

$\quad = 14 + 3(9) \qquad$ *Multiply.*

$\quad = 14 + 27 \qquad$ *Multiply.*

$\quad = 41 \qquad$ *Add.*

63. $30 \div (-5) - 36 \div (-9)$

$\quad = -6 - 36 \div (-9) \qquad$ *Divide.*

$\quad = -6 - (-4) \qquad$ *Divide.*

$\quad = -6 + (+4) \qquad$ *Add the opposite.*

$\quad = -2 \qquad$ *Add.*

65. $2(5) - 3(4) + 5(3)$

$\quad = 10 - 3(4) + 5(3) \qquad$ *Multiply.*

$\quad = 10 - 12 + 5(3) \qquad$ *Multiply.*

$\quad = 10 - 12 + 15 \qquad$ *Multiply.*

$\quad = 10 + (-12) + 15 \qquad$ *Add the opposite.*

$\quad = -2 + 15 \qquad$ *Add.*

$\quad = 13 \qquad$ *Add.*

67. $4(3^2) + 7(3 + 9) - (-6)$

$\quad = 4(3^2) + 7(12) - (-6) \qquad$ *Parentheses first*

$\quad = 4(9) + 7(12) - (-6) \qquad$ *Exponent next*

$\quad = 36 + 7(12) - (-6) \qquad$ *Multiply.*

$\quad = 36 + 84 - (-6) \qquad$ *Multiply.*

$\quad = 36 + 84 + (+6) \qquad$ *Add the opposite.*

$\quad = 120 + 6 \qquad$ *Add.*

$\quad = 126 \qquad$ *Add.*

69. $(-4)^2 \cdot (7-9)^2 \div 2^3$

$\begin{aligned}
&= (-4)^2 \cdot [7 + (-9)]^2 \div 2^3 && \textit{Add the opposite.}\\
&= (-4)^2 \cdot (-2)^2 \div 2^3 && \textit{Brackets first}\\
&= 16 \cdot 4 \div 8 && \textit{Exponents next}\\
&= 64 \div 8 && \textit{Multiply.}\\
&= 8 && \textit{Divide.}
\end{aligned}$

71. $\dfrac{-1 + 5^2 - (-3)}{-6 - 9 + 12}$

Numerator:
$-1 + 5^2 - (-3)$

$\begin{aligned}
&= -1 + 25 - (-3) && \textit{Exponent first}\\
&= -1 + 25 + (+3) && \textit{Add the opposite.}\\
&= 24 + 3 && \textit{Add left to right.}\\
&= 27
\end{aligned}$

Denominator:
$-6 - 9 + 12$

$\begin{aligned}
&= -6 + (-9) + 12 && \textit{Add the opposite.}\\
&= -15 + 12 && \textit{Add left to right.}\\
&= -3
\end{aligned}$

Last step is division: $\dfrac{27}{-3} = -9$

73. $\dfrac{-2(4^2) - 4(6-2)}{-4(8-13) \div (-5)}$

Numerator:
$-2(4^2) - 4(6-2)$

$\begin{aligned}
&= -2(4^2) - 4(4) && \textit{Parentheses first}\\
&= -2(16) - 4(4) && \textit{Exponent}\\
&= -32 - 16 && \textit{Multiply from left to right.}\\
&= -32 + (-16) && \textit{Add the opposite.}\\
&= -48 && \textit{Add.}
\end{aligned}$

Denominator:
$-4(8-13) \div (-5)$

$\begin{aligned}
&= -4[8 + (-13)] \div (-5) && \textit{Add the opposite.}\\
&= -4(-5) \div (-5) && \textit{Brackets first}\\
&= 20 \div (-5) && \textit{Multiply.}\\
&= -4 && \textit{Divide.}
\end{aligned}$

Last step is division: $\dfrac{-48}{-4} = 12$

75. $\dfrac{2^3 \cdot (-2-5) + 4(-1)}{4 + 5(-6 \cdot 2) + (5 \cdot 11)}$

Numerator:
$2^3 \cdot (-2 - 5) + 4(-1)$

$\begin{aligned}
&= 2^3 \cdot [-2 + (-5)] + 4(-1) && \textit{Add the opposite.}\\
&= 2^3 \cdot (-7) + 4(-1) && \textit{Brackets}\\
&= 8 \cdot (-7) + 4(-1) && \textit{Exponent}\\
&= -56 + 4(-1) && \textit{Multiply.}\\
&= -56 + (-4) && \textit{Multiply.}\\
&= -60 && \textit{Add.}
\end{aligned}$

Denominator:
$4 + 5(-6 \cdot 2) + (5 \cdot 11)$

$\begin{aligned}
&= 4 + 5(-12) + (5 \cdot 11) && \textit{Parentheses}\\
&= 4 + 5(-12) + 55 && \textit{Parentheses}\\
&= 4 + (-60) + 55 && \textit{Multiply.}\\
&= -56 + 55 && \textit{Add left to right.}\\
&= -1
\end{aligned}$

Last step is division: $\dfrac{-60}{-1} = 60$

77. $5^2(9-11)(-3)(-3)^3$

$\begin{aligned}
&= 5^2[9 + (-11)](-3)(-3)^3 && \textit{Add the opposite.}\\
&= 5^2(-2)(-3)(-3)^3 && \textit{Brackets first}\\
&= 25(-2)(-3)(-27) && \textit{Exponents}\\
&= -50(-3)(-27) && \textit{Mult. left to right.}\\
&= 150(-27)\\
&= -4050
\end{aligned}$

79. $|-12| \div 4 + 2 \cdot \left| (-2)^3 \right| \div 4$

$\begin{aligned}
&= |-12| \div 4 + 2 \cdot |-8| \div 4 && \begin{array}{l}\textit{Work inside the}\\ \textit{absolute value}\\ \textit{bars signs first.}\end{array}\\
&= 12 \div 4 + 2 \cdot 8 \div 4 && \textit{Absolute values}\\
&= 3 + 2 \cdot 8 \div 4 && \textit{Divide.}\\
&= 3 + 16 \div 4 && \textit{Multiply.}\\
&= 3 + 4 && \textit{Divide.}\\
&= 7 && \textit{Add.}
\end{aligned}$

81. $\dfrac{-9 + 18 \div (-3)(-6)}{32 - 4(12) \div 3(2)}$

Numerator:
$-9 + 18 \div (-3)(-6)$

$\begin{aligned}
&= -9 + (-6)(-6) && \textit{Divide.}\\
&= -9 + 36 && \textit{Multiply.}\\
&= 27 && \textit{Add.}
\end{aligned}$

Denominator:
$32 - 4(12) \div 3(2)$

$\begin{aligned}
&= 32 - 48 \div 3(2) && \textit{Multiply.}\\
&= 32 - 16(2) && \textit{Divide.}\\
&= 32 - 32 && \textit{Multiply.}\\
&= 32 + (-32) && \textit{Add the opposite.}\\
&= 0 && \textit{Add.}
\end{aligned}$

Last step is division: $\dfrac{27}{0}$ is undefined.

Chapter 1 Review Exercises

1. The whole numbers are: 86; 0; 35,600

2. 806 in words: eight hundred six

3. 319,012 in words: three hundred nineteen thousand, twelve

4. 60,003,200 in words: sixty million, three thousand, two hundred

5. 15,749,000,000,006 in words: fifteen trillion, seven hundred forty-nine billion, six

6. Five hundred four thousand, one hundred
    The first group name is *thousand*, so you need to fill *two groups* of three digits.

    $\underline{5\,0\,4}, \underline{1\,0\,0} = 504{,}100$

7. Six hundred twenty million, eighty thousand
    The first group name is *million*, so you need to fill *three groups* of three digits.

    $\underline{6\,2\,0}, \underline{0\,8\,0}, \underline{0\,0\,0} = 620{,}080{,}000$

8. Ninety-nine billion, seven million, three hundred fifty-six
    The first group name is *billion,* so you need to fill *four groups* of three digits.

    $\underline{0\,9\,9}, \underline{0\,0\,7}, \underline{0\,0\,0}, \underline{3\,5\,6} = 99{,}007{,}000{,}356$

9. Graph $-3, 2, -5, 0$

    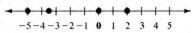

10. 0 is to the *right* of -4 on the number line, so 0 is *greater than* -4. Write $0 > -4$.

11. -3 is to the *left* of -1 on the number line, so -3 is *less than* -1. Write $-3 < -1$.

12. 2 is to the *right* of -2 on the number line, so 2 is *greater than* -2. Write $2 > -2$.

13. -2 is to the *left* of 1 on the number line, so -2 is *less than* 1. Write $-2 < 1$.

14. $|-5| = 5$ because the distance from 0 to -5 on the number line is 5 spaces.

15. $|9| = 9$ because the distance from 0 to 9 on the number line is 9 spaces.

16. $|0| = 0$ because the distance from 0 to 0 on the number line is 0 spaces.

17. $|-125| = 125$ because the distance from 0 to -125 on the number line is 125 spaces.

18. $-9 + 8$ ▪ Add *unlike* signed integers.

    $|-9| = 9$; $|8| = 8$; Subtract $9 - 8$ to get 1.

    -9 has the larger absolute value and is negative, so the sum is negative.

    $$-9 + 8 = -1$$

19. $-8 + (-5)$ ▪ Add *like* signed integers.

    $|-8| = 8$; $|-5| = 5$; Add $8 + 5$ to get 13.

    Both numbers are negative, so the sum is negative.

    $$-8 + (-5) = -13$$

20. $16 + (-19)$ ▪ Add *unlike* signed integers.

    $|16| = 16$; $|-19| = 19$; Subtract $19 - 16$ to get 3.

    -19 has the larger absolute value and is negative, so the sum is negative.

    $$16 + (-19) = -3$$

21. $-4 + 4 = 0$
    Addition of opposites is always zero.

22. $6 + (-5) = +1$ or 1

23. $-12 + (-12) = -24$

24. $0 + (-7) = -7$

25. $-16 + 19 = +3$ or 3

26. $9 + (-4) + (-8) + 3$
    $= 5 + (-8) + 3$ *Add from left to right.*
    $= -3 + 3$
    $= 0$

27. $-11 + (-7) + 5 + (-4)$
    $= -18 + 5 + (-4)$ *Add from left to right.*
    $= -13 + (-4)$
    $= -17$

28. The opposite of -5 is 5. $-5 + 5 = 0$

29. The opposite of 18 is -18. $18 + (-18) = 0$

30. $5 - 12$
    $= 5 + (-12)$ *Change subtraction to addition. Change 12 to -12.*
    $= -7$

31. $24 - 7$
    $= 24 + (-7)$ *Add the opposite.*
    $= 17$

32. $-12 - 4$
    $= -12 + (-4)$ *Add the opposite.*
    $= -16$

33. $4 - (-9)$
    $= 4 + (+9)$ *Add the opposite.*
    $= 13$

34. $-12 - (-30)$
    $= -12 + (+30)$ *Add the opposite.*
    $= 18$

35. $-8 - 14$
    $= -8 + (-14)$ *Add the opposite.*
    $= -22$

36. $-6 - (-6)$
    $= -6 + (+6)$ *Add the opposite.*
    $= 0$

37. $-10 - 10$
$= -10 + (-10)$ *Add the opposite.*
$= -20$

38. $-8 - (-7)$
$= -8 + (+7)$ *Add the opposite.*
$= -1$

39. $0 - 3$
$= 0 + (-3)$ *Add the opposite.*
$= -3$

40. $1 - (-13)$
$= 1 + (+13)$ *Add the opposite.*
$= 14$

41. $15 - 0$
$= 15 + 0$ *Add the opposite.*
$= 15$

42. $3 - 12 - 7$
$= 3 + (-12) + (-7)$ *Add the opposites.*
$= -9 + (-7)$ *Add left to right.*
$= -16$

43. $-7 - (-3) + 7$
$= -7 + (+3) + 7$ *Add the opposite.*
$= -7 + 7 + (+3)$ *Commutative property*
$= 0 + (+3)$
$= 3$

44. $4 + (-2) - 0 - 10$
$= 4 + (-2) + 0 + (-10)$ *Add the opposites.*
$= 2 + 0 + (-10)$ *Add left to right.*
$= 2 + (-10)$
$= -8$

45. $-12 - 12 + 20 - (-4)$
$= -12 + (-12) + 20 + (+4)$ *Add opposites.*
$= -24 + 20 + 4$ *Add left to rt.*
$= -4 + 4$
$= 0$

46. $2\underline{0}5 \approx 210$

Underline the tens place. The next digit is 5 or more. Add 1 to 0. Change 5 to 0.

47. $5\underline{9},499 \approx 59,000$

Underline the thousands place. The next digit is 4 or less. Leave 9 as 9. Change all digits to the right of the underlined place to zeros.

48. $8\underline{5},066,000 \approx 85,000,000$

Underline the millions place. The next digit is 4 or less. Leave 5 as 5. Change all digits to the right of the underlined place to zeros.

49. $-2\underline{9}63 \approx -3000$

Underline the hundreds place. The next digit is 5 or more. Add 1 to 9. Write 0 and carry the one to the thousands place. Change all digits to the right of the underlined place to zeros.

50. $-7,0\underline{6}3,885 \approx -7,060,000$

Underline the ten-thousands place. The next digit is 4 or less. Leave 6 as 6. Change all digits to the right of the underlined place to zeros.

51. $39\underline{9},712 \approx 400,000$

Underline the thousands place. The next digit is 5 or more. Add 1 to 9. Write 0 and carry the one to the ten-thousands place: $9 + 1 = 10$. Write 0 and carry the one to the hundred-thousands place: $3 + 1 = 4$. Change all digits to the right of the underlined place to zeros.

52. Weight loss implies a negative number:

$$-\underline{1}97 \text{ pounds} \approx -200 \text{ pounds}$$

Underline the first digit. The next digit is 5 or more. Add 1 to 1. Change all digits to the right of the underlined place to zeros.

53. Below sea level implies a negative number:

$$-\underline{1}388 \approx -1000 \text{ feet}$$

Underline the first digit. The next digit is 4 or less. Leave 1 as 1. Change all digits to the right of the underlined place to zeros.

54. $\underline{2},095,006,000$ Internet users $\approx$
$\underline{2},000,000,000$ Internet users

Underline the first digit. The next digit is 4 or less. Leave 2 as 2. Change all digits to the right of the underlined place to zeros.

55. $\underline{9},502,000,000$ people $\approx 10,000,000,000$ people

Underline the first digit. The next digit is 5 or more. Add 1 to 9. Change all digits to the right of the underlined place to zeros.

56. $-6(9) = -54$ (*different* signs, product is *negative*)

57. $(-7)(-8) = 56$ (*same* signs, product is *positive*)

58. $10(-10) = -100$
(*different* signs, product is *negative*)

59. $-45 \cdot 0 = 0$; multiplication property of 0

60. $-1(-24) = 24$ (*same* signs, product is *positive*)

61. $17 \cdot 1 = 17$; multiplication property of 1

62. $4(-12) = -48$
(*different* signs, product is *negative*)

63. $(-5)(-25) = 125$ (*same* signs, product is *positive*)

64. $-3(-4)(-3)$
$= 12 \cdot (-3)$ *Multiply from left to right.*
$= -36$

65. $-5(2)(-5)$
$= -10 \cdot (-5)$ *Multiply from left to right.*
$= 50$

66. $(-8)(-1)(-9)$
$= 8(-9)$ *Multiply from left to right.*
$= -72$

67. $\dfrac{-63}{-7} = 9$ (*same* signs, quotient is *positive*)

68. $\dfrac{70}{-10} = -7$ (*different* signs, quotient is *negative*)

69. $\dfrac{-15}{0}$ is undefined. Division by zero is undefined.

70. $-100 \div (-20) = 5$
(*same* signs, quotient is *positive*)

71. $18 \div (-1) = -18$
(*different* signs, quotient is *negative*)

72. $\dfrac{0}{12} = 0$; 0 divided by any nonzero number is 0.

73. $\dfrac{-30}{-2} = 15$ (*same* signs, quotient is *positive*)

74. $\dfrac{-35}{35} = -1$ (*different* signs, quotient is *negative*)

75. $-40 \div (-4) \div (-2)$
$= 10 \div (-2)$ *Divide from left to right.*
$= -5$

76. $-18 \div 3(-3)$
$= -6(-3)$ *Divide.*
$= 18$ *Multiply.*

77. $0 \div (-10)(5) \div 5$
$= 0(5) \div 5$ *Divide.*
$= 0 \div 5$ *Multiply.*
$= 0$ *Divide.*

78. 1250 hours are separated into 8-hour days.

$$
\begin{array}{r}
156 \\
8\,\overline{)1250} \\
\underline{8} \\
45 \\
\underline{40} \\
50 \\
\underline{48} \\
2
\end{array}
$$

It took 156 work days of 8 hours each, plus 2 extra hours.

79. $10^4 = 10 \cdot 10 \cdot 10 \cdot 10 = 10{,}000$

80. $2^5 = 2 \cdot 2 \cdot 2 \cdot 2 \cdot 2 = 32$

81. $3^3 = 3 \cdot 3 \cdot 3 = 27$

82. $(-4)^2 = (-4)(-4) = 16$

83. $(-5)^3 = (-5)(-5)(-5)$
$= (25)(-5)$
$= -125$

84. $8^1 = 8$

85. $6^2 \cdot 3^2$
$= 36 \cdot 9$ *Apply exponents.*
$= 324$

86. $5^2(-2)^3$
$= 25(-8)$ *Apply exponents.*
$= -200$

87. $-30 \div 6 - 4(5)$
$= -5 - 4(5)$ *Divide.*
$= -5 - 20$ *Multiply.*
$= -5 + (-20)$ *Add the opposite.*
$= -25$ *Add.*

88. $6 + 8(2 - 3)$
$= 6 + 8(-1)$ *Parentheses first*
$= 6 + (-8)$ *Multiply.*
$= -2$ *Add.*

89. $16 \div 4^2 + (-6 + 9)^2$
$= 16 \div 4^2 + (3)^2$ *Parentheses first*
$= 16 \div 16 + 9$ *Apply exponents.*
$= 1 + 9$ *Divide.*
$= 10$ *Add.*

90. $-3(4) - 2(5) + 3(-2)$
$= -12 - 10 + (-6)$ *Multiply left to right.*
$= -12 + (-10) + (-6)$ *Add the opposite.*
$= -22 + (-6)$ *Add left to right.*
$= -28$

91. $\dfrac{-10 + 3^2 - (-9)}{3 - 10 - 1}$

Numerator:
$-10 + 3^2 - (-9)$
$= -10 + 9 - (-9)$ *Exponent*
$= -10 + 9 + (+9)$ *Add the opposite.*
$= -1 + (+9)$ *Add from left to right.*
$= 8$

Denominator:
$3 - 10 - 1$
$= 3 + (-10) + (-1)$ *Add the opposites.*
$= -7 + (-1)$ *Add from left to right.*
$= -8$

Last step is division: $\dfrac{8}{-8} = -1$

92.
$$\frac{-1(1-3)^3 + 12 \div 4}{-5 + 24 \div 8 \cdot 2(6-6) + 5}$$

Numerator:

$-1(1-3)^3 + 12 \div 4$

$= -1(-2)^3 + 12 \div 4$	*Parentheses first*
$= -1(-8) + 12 \div 4$	*Exponent*
$= 8 + 12 \div 4$	*Multiply.*
$= 8 + 3$	*Divide.*
$= 11$	*Add.*

Denominator:

$-5 + 24 \div 8 \cdot 2(6-6) + 5$

$= -5 + 24 \div 8 \cdot 2(0) + 5$	*Parentheses first*
$= -5 + 3 \cdot 2(0) + 5$	*Divide.*
$= -5 + 6(0) + 5$	*Multiply.*
$= -5 + 0 + 5$	*Multiply.*
$= -5 + 5$	*Add.*
$= 0$	*Add.*

Last step is division: $\dfrac{11}{0}$ is undefined.

93. **[1.3]** $-3 + (5+1) = (-3+5)+1$
Associative property of addition

94. **[1.6]** $-7(2) = 2(-7)$
Commutative property of multiplication

95. **[1.3]** $0 + 19 = 19$
Addition property of 0

96. **[1.6]** $-42 \cdot 0 = 0$
Multiplication property of 0

97. **[1.6]** $2(-6+4) = 2 \cdot (-6) + 2 \cdot 4$
Distributive property

98. **[1.6]** $(-6 \cdot 3) \cdot 1 = -6 \cdot (3 \cdot 1)$
Associative property of multiplication

99. **[1.6]** 192 rounds to 200.
$11,900 rounds to $10,000.

Estimate: $\$10,000 \cdot 200 = \$2,000,000$ total value
Exact: $\$11,900 \cdot 192 = \$2,284,800$ total value

100. **[1.3]** Account balance of $185 rounds to $200.
The deposit of $428 rounds to $400.
The check for $706 rounds to $700.

Estimate: $\$200 + \$400 - \$700 = -\$100$
Exact: $\$185 + \$428 - \$706 = -\$93$

101. **[1.7]** 22 gallons rounds to 20 gallons.
880 miles rounds to 900 miles.
Divide miles by gallons to get miles per gallon.

Estimate: $900 \div 20 = 45$ miles for each gallon
Exact: $880 \div 22 = 40$ miles for each gallon

102. **[1.6]** 19 calculators rounds to 20.
12 modems rounds to 10.
$39 rounds to $40. $85 rounds to $90.

Estimate:
$(\$40 \cdot 20) + (\$90 \cdot 10) = \$800 + \$900 = \$1700$
Exact:
$(\$39 \cdot 19) + (\$85 \cdot 12) = \$741 + \$1020 = \$1761$

103. **[1.3]** Expenses imply negative numbers.

Jan.	$\$2400 + (-\$3100) = -\$700$ (loss)
Feb.	$\$1900 + (-\$2000) = -\$100$ (loss)
Mar.	$\$2500 + (-\$1800) = \$700$ (profit)
Apr.	$\$2300 + (-\$1400) = \$900$ (profit)
May	$\$1600 + (-\$1600) = \$0$ (neither)
June	$\$1900 + (-\$1200) = \$700$ (profit)

104. **[1.3]** January had the greatest loss.
April had the greatest profit.

105. **[1.7]** To find her average monthly income, add her income from each month and divide by the number of months.

$$\frac{\begin{array}{c}\$2400 + \$1900 + \$2500 \\ + \$2300 + \$1600 + \$1900\end{array}}{6}$$
$$= \frac{\$12,600}{6} = \$2100$$

106. **[1.7]** To find her average monthly expenses, add her expenses from each month and divide by the number of months.

$$\frac{\begin{array}{c}-\$3100 + (-\$2000) + (-\$1800) \\ + (-\$1400) + (-\$1600) + (-\$1200)\end{array}}{6}$$
$$= \frac{-\$11,100}{6} = -\$1850$$

Chapter 1 Test

1. 20,008,307 in words: twenty million, eight thousand, three hundred seven

2. Thirty billion, seven hundred thousand, five

The first group name is *billion*, so you need to fill *four groups* of three digits.

$\underline{0\,3\,0},\underline{0\,0\,0},\underline{7\,0\,0},\underline{0\,0\,5} = 30,000,700,005$

3. Graph 3, -2, 0, $-\frac{1}{2}$

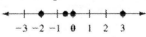

4. 0 is to the *right* of -3 on the number line, so 0 is *greater than* -3. Write $0 > -3$.

-2 is to the *left* of -1 on the number line, so -2 is *less than* -1. Write $-2 < -1$.

5. $|10| = 10$ because the distance from 0 to 10 on the number line is 10 spaces.

$|-14| = 14$ because the distance from 0 to -14 on the number line is 14 spaces.

6. $3 - 9$
   $= 3 + (-9)$ *Add the opposite.*
   $= -6$

7. $-12 + 7 = -5$

Add *unlike* signed integers.

$|-12| = 12$; $|7| = 7$; Subtract $12 - 7$ to get 5.

-12 has the larger absolute value and is negative, so the sum is negative.

8. $\dfrac{-28}{-4} = 7$ (*same* signs, quotient is *positive*)

9. $-1(40) = -40$ (*different* signs, product is *negative*)

10. $-5 - (-15)$
   $= -5 + (+15)$ *Change subt. to addition.*
   $= 10$ *Add.*

11. $(-8)(-8) = 64$ (*same* signs, product is *positive*)

12. $-25 + (-25)$
   $= -50$ *Add.*

13. $\dfrac{17}{0}$ is undefined.

14. $-30 - 30$
   $= -30 + (-30)$ *Change subtraction to addition.*
   $= -60$ *Add.*

15. $\dfrac{50}{-10} = -5$ (*different* signs, quotient is *negative*)

16. $5(-9) = -45$ (*different* signs, product is *negative*)

17. $0 - (-6)$
   $= 0 + (+6)$ *Change subtraction to addition.*
   $= 6$ *Addition property of zero*

18. $-35 \div 7(-5)$
   $= -5(-5)$ *Divide.*
   $= 25$ *Multiply.*

19. $-15 - (-8) + 7$
   $= -15 + (+8) + 7$ *Change subtraction to addition.*
   $= -7 + 7$ *Add from left to right.*
   $= 0$

20. $3 - 7(-2) - 8$
   $= 3 - (-14) - 8$ *Multiply.*
   $= 3 + (+14) + (-8)$ *Change subtraction to addition.*
   $= 17 + (-8)$ *Add from left to right.*
   $= 9$

21. $(-4)^2 \cdot 2^3$
   $= 16 \cdot 8$ *Apply exponents.*
   $= 128$

22. $\dfrac{5^2 - 3^2}{(4)(-2)}$

Numerator:
$5^2 - 3^2$
   $= 25 - 9$ *Apply exponents.*
   $= 25 + (-9)$ *Add the opposite.*
   $= 16$ *Add.*

Denominator:
$(4)(-2)$
   $= -8$ *Multiply.*

Last step is division: $\dfrac{16}{-8} = -2$

23. $-2(-4 + 10) + 5(4)$
   $= -2(6) + 5(4)$ *Parentheses*
   $= -12 + 5(4)$ *Multiply.*
   $= -12 + 20$ *Multiply.*
   $= 8$ *Add.*

24. $-3 + (-7 - 10) + 4(6 - 10)$
   $= -3 + [-7 + (-10)] + 4[6 + (-10)]$
   $= -3 + (-17) + 4(-4)$ *Brackets*
   $= -3 + (-17) + (-16)$ *Multiply.*
   $= -20 + (-16)$ *Add.*
   $= -36$ *Add.*

25. An exponent shows how many times to use a factor in repeated multiplication. Examples will vary. Some possibilities are
$(2)^1 = 2 \cdot 2 \cdot 2 \cdot 2 = 16$; $(-3)^2 = (-3)(-3) = 9$.

26. Commutative property: changing the *order* of addends does not change the sum.

One possible example: $2 + 5 = 5 + 2$

Associative property: changing the *grouping* of addends does not change the sum.

One possible example:
$(-1 + 4) + 2 = -1 + (4 + 2)$

27. $\underline{8}51 \approx 900$

Underline the hundreds place. The next digit is 5 or more. Add 1 to 8. Change all digits to the right of the underlined place to zeros.

28. $36,420,498,725 \approx 36,420,000,000$

Underline the millions place. The next digit is 4 or less. Leave 0 as 0. Change all digits to the right of the underlined place to zeros.

29. $349,812 \approx 350,000$

Underline the thousands place. The next digit is 5 or more. Add 1 to 9. Write 0 and carry 1 to the ten-thousands place. Change all digits to the right of the underlined place to zeros.

30. $184 account balance rounds to $200.
The $293 deposit rounds to $300.
The $506 tuition check rounds to $500.

Balance in her account:
Estimate: $\$200 + \$300 + (-\$500) = \$0$
Exact: $\$184 + \$293 + (-\$506) = -\$29$

31.   Loss of 1144 pounds rounds to 1000 pounds.
22 people rounds to 20 people.

Average weight loss for each person:
Estimate: $-1000 \div 20 = -50$ pounds
Exact: $-1144 \div 22 = -52$ pounds

32.   Cereal 1: 220 calories rounds to 200
Cereal 2: 110 calories rounds to 100
31 days rounds to 30

Estimate of calories saved by eating Cereal 2:
$200 - 100 = 100$ calories per day
$30 \cdot 100 = 3000$ calories in the month

Exact calories saved by eating Cereal 2:
$220 - 110 = 110$ calories per day
$31 \cdot 110 = 3410$ calories in the month

33.   The difference between the high and low temperatures on Mars is:

$-10 - (-100)$
$\quad = -10 + (+100)$ *Add the opposite.*
$\quad = 90$ degrees difference

34.   1276 books separated into cartons holding 48 books each.

$$
\begin{array}{r}
2\,6 \\
48\overline{\smash{\big)}\,1\,2\,7\,6} \\
\underline{9\,6} \\
3\,1\,6 \\
\underline{2\,8\,8} \\
2\,8
\end{array}
$$

Anthony will need 27 cartons to ship all of the books. He will have 26 full boxes and one box with only 28 books in it.

CHAPTER 2 UNDERSTANDING VARIABLES AND SOLVING EQUATIONS

2.1 Introduction to Variables

2.1 Margin Exercises

1. $c + \underline{15}$

   The expression is $c + 15$.

   The variable is $\underline{c}$. It represents the class limit.

   The constant is $\underline{15}$.

2. **(a)** Evaluate the expression $c + 3$ when c is 25.

   $$c + 3 \quad \textit{Replace } c \textit{ with 25.}$$
   $$\underbrace{25 + 3}_{28}$$

   Order $\underline{28}$ books.

   (b) Evaluate the expression $c + 3$ when c is 60.

   $$c + 3 \quad \textit{Replace } c \textit{ with 60.}$$
   $$\underbrace{60 + 3}_{63}$$

   Order $\underline{63}$ books.

3. **(a)** Evaluate the expression $4s$ when s is 3 feet.

   $$4s \qquad \textit{Replace } s \textit{ with 3 feet.}$$
   $$\underline{4 \cdot 3 \text{ feet}}$$
   $$\underline{12 \text{ feet}}$$

   The perimeter of the square table is 12 feet.

   (b) Evaluate the expression $4s$ when s is 7 miles.

   $$4s \qquad \textit{Replace } s \textit{ with 7 miles.}$$
   $$4 \cdot 7 \text{ miles}$$
   $$28 \text{ miles}$$

   The perimeter of the square park is 28 miles.

4. Evaluate the expression $100 + \dfrac{a}{2}$ when a is 40.

   $$100 + \dfrac{a}{2} \quad \textit{Replace } a \textit{ with 40.}$$
   $$100 + \dfrac{40}{2} \quad \textit{Divide.}$$
   $$100 + 20 \quad \textit{Add.}$$
   $$120$$

   The approximate systolic blood pressure is 120.

5. **(a)** Evaluate the expression $\frac{t}{g}$ when t is 532 and g is 4.

   $$\dfrac{t}{g} \qquad \textit{Replace } t \textit{ with } \underline{532} \textit{ and } g \textit{ with } \underline{4}.$$
   $$\dfrac{532}{4} \quad \textit{Divide.}$$
   $$133$$

   Your average score is $\underline{133}$.

   (b)

Value of x	Value of y	Expression $x - y$
16	10	$16 - 10$ is 6
100	5	$100 - 5$ is 95
3	7	$3 - 7$ is -4
8	0	$8 - 0$ is 8

6. **(a)** Multiplying any number a by 0 gives a product of 0.

Any number	times	zero
↓	↓	↓
a	$\cdot \quad \underline{0}$	$= \quad \underline{0}$

   (b) Changing the grouping of addends (a, b, c) does not change the sum.

   $$(a + b) + c = a + (b + c)$$

7. **(a)** x^5 can be written as $\underbrace{x \cdot x \cdot x \cdot x \cdot x}$
   x is used as a factor 5 times.

   (b) $4a^2b^2$ can be written as $4 \cdot \underline{a} \cdot \underline{a} \cdot \underline{b} \cdot \underline{b}$

   (c) $-10xy^3$ can be written as $-10 \cdot x \cdot y \cdot y \cdot y$

   (d) s^4tu^2 can be written as $s \cdot s \cdot s \cdot s \cdot t \cdot u \cdot u$

8. **(a)** y^3 means

   $$y \cdot y \cdot y \qquad \textit{Replace } y \textit{ with } -5.$$
   $$\underbrace{-5 \cdot (-5)} \cdot (-5) \quad \textit{Multiply left to right.}$$
   $$\underbrace{25 \cdot (-5)}$$
   $$-125$$

   (b) r^2s^2 means

   $$r \cdot r \cdot s \cdot s \qquad \textit{Replace } r \textit{ with 6 and } s \textit{ with 3.}$$
   $$\underline{6 \cdot 6} \cdot \underline{3 \cdot 3} \quad \textit{Multiply left to right.}$$
   $$\underline{36 \cdot 3} \cdot 3$$
   $$\underline{108 \cdot 3}$$
   $$324$$

   (c) $10xy^2$ means

   $$10 \cdot x \cdot y \cdot y \qquad \textit{Replace } x \textit{ with 4 and } y \textit{ with } -3.$$
   $$\underline{10 \cdot 4} \cdot (-3) \cdot (-3) \quad \textit{Multiply left to right.}$$
   $$\underbrace{40 \cdot (-3)} \cdot (-3)$$
   $$\underbrace{-120 \cdot (-3)}$$
   $$360$$

(d) $-3c^4$ means

$-3 \cdot c \cdot c \cdot c \cdot c$ *Replace c with 2.*
$\underbrace{-3 \cdot 2} \cdot 2 \cdot 2 \cdot 2$ *Multiply left to right.*
$\underbrace{-6 \cdot 2} \cdot 2 \cdot 2$
$\underbrace{-12 \cdot 2} \cdot 2$
$\underbrace{-24 \cdot 2}$
-48

2.1 Section Exercises

1. $c + 4$ *c* is the variable;
            4 is the <u>constant</u>.

3. $-3 + m$ *m* is the variable;
            -3 is the constant.

5. $5h$ *h* is the variable;
            5 is the coefficient.

7. $2c - 10$ *c* is the variable;
            2 is the coefficient.
            10 is the constant.

9. $x - y$ ▪ Both *x* and *y* are variables.

11. $-6g + 9$ *g* is the variable;
            -6 is the coefficient;
            9 is the constant.

13. Expression (rule) for ordering robes: $g + 10$

(a) Evaluate the expression when there are 654 graduates.

$g + 10$ *Replace g with 654.*
$\underbrace{654 + 10}$ *Follow the rule and add.*
<u>664</u> robes must be ordered.

(b) Evaluate the expression when there are 208 graduates.

$g + 10$ *Replace g with 208.*
$\underbrace{208 + 10}$ *Follow the rule and add.*
218 robes must be ordered.

(c) Evaluate the expression when there are 95 graduates.

$g + 10$ *Replace g with 95.*
$\underbrace{95 + 10}$ *Follow the rule and add.*
105 robes must be ordered.

15. Expression (rule) for finding perimeter of an equilateral triangle of side length *s*: $3s$

(a) Evaluate the expression when *s*, the side length, is 11 inches.

$3s$ *Replace s with 11.*
$\underbrace{3 \cdot 11}$ *Follow the rule and multiply.*
33 inches is the perimeter.

(b) Evaluate the expression when *s*, the side length, is 3 feet.

$3s$ *Replace s with 3.*
$\underbrace{3 \cdot 3}$ *Follow the rule and multiply.*
9 feet is the perimeter.

17. Expression (rule) for ordering brushes: $3c - 5$

(a) Evaluate the expression when *c*, the class size, is 12.

$3c - 5$ *Replace c with 12.*
$\underbrace{3 \cdot 12} - 5$ *Multiply before subtracting.*
$\underbrace{36 - 5}$
31 brushes must be ordered.

(b) Evaluate the expression when *c*, the class size, is 16.

$3c - 5$ *Replace c with 16.*
$\underbrace{3 \cdot 16} - 5$ *Multiply before subtracting.*
$\underbrace{48 - 5}$
43 brushes must be ordered.

19. Expression (rule) for average test score, where *p* is the total points and *t* is the number of tests: p/t

(a) Evaluate the expression when *p*, the total points, is 332 and *t*, the number of tests, is 4.

$\dfrac{p}{t}$ *Replace p with 332 and t with 4.*
$\dfrac{332}{4}$ *Follow the rule and divide.*
83 points is the average test score.

(b) Evaluate the expression when *p*, the total points, is 637 and *t*, the number of tests, is 7.

$\dfrac{p}{t}$ *Replace p with 637 and t with 7.*
$\dfrac{637}{7}$ *Follow the rule and divide.*
91 points is the average test score.

21.

Value of x	Expression $x + x + x + x$	Expression $4x$
12	$12 + 12 + 12 + 12$ is 48	$4 \cdot 12$ is 48
0	$0 + 0 + 0 + 0$ is 0	$4 \cdot 0$ is 0
-5	$-5 + (-5) + (-5) + (-5)$ is -20	$4 \cdot (-5)$ is -20

23.

Value of x	Value of y	Expression $-2x + y$
-4	5	$-2(-4) + 5$ is $8 + 5$, or 13
-6	-2	$-2(-6) + (-2)$ is $12 + (-2)$, or 10
0	-8	$-2(0) + (-8)$ is $0 + (-8)$, or -8

25. A variable is a letter that represents the part of a rule that varies or changes depending on the situation. An expression expresses, or tells, the rule for doing something. For example, $c + 5$ is an expression, and c is the variable.

27. Multiplying a number by 1 leaves the number unchanged. Let b represent "a number."

$$b \cdot 1 = b \quad \text{or} \quad 1 \cdot b = b$$

29. Any number divided by 0 is undefined. Let b represent "any number."

$$\frac{b}{0} \text{ is undefined} \quad \text{or} \quad b \div 0 \text{ is undefined.}$$

31. c^6 written without exponents is

$$c \cdot c \cdot c \cdot c \cdot c \cdot c$$

33. $x^4 y^3$ written without exponents is

$$x \cdot x \cdot x \cdot x \cdot y \cdot y \cdot y$$

35. $-3a^3 b$ can be written as $-3 \cdot a \cdot a \cdot a \cdot b$.
The exponent 3 applies only to the base a.

37. $9xy^2$ can be written as $9 \cdot x \cdot y \cdot y$.
The exponent 2 applies only to the base y.

39. $-2c^5 d$ can be written as $-2 \cdot c \cdot c \cdot c \cdot c \cdot c \cdot d$.
The exponent 5 applies only to the base c.

41. $a^3 b c^2$ can be written as $a \cdot a \cdot a \cdot b \cdot c \cdot c$.
The exponent 3 applies only to the base a.
The exponent 2 applies only to the base c.

43. Evaluate t^2 when t is -4.

t^2 means

$t \cdot t$ *Replace t with -4.*

$\underbrace{-4 \cdot (-4)}$ *Multiply.*

16

45. Evaluate rs^3 when r is -3 and s is 2.

rs^3 means

$r \cdot s \cdot s \cdot s$ *Replace r with -3 and s with 2.*

$\underbrace{-3 \cdot 2} \cdot 2 \cdot 2$ *Multiply left to right.*

$\underbrace{-6 \cdot 2} \cdot 2$

$\underbrace{-12 \cdot 2}$

-24

47. Evaluate $3rs$ when r is -3 and s is 2.

$3rs$ means

$3 \cdot r \cdot s$ *Replace r with -3 and s with 2.*

$\underbrace{3 \cdot (-3)} \cdot 2$ *Multiply left to right.*

$\underbrace{-9 \cdot 2}$

-18

49. Evaluate $-2s^2 t^2$ when s is 2 and t is -4.

$-2s^2 t^2$ means

$-2 \cdot s \cdot s \cdot t \cdot t$ *Replace s with 2 and t with -4.*

$\underbrace{-2 \cdot 2} \cdot 2 \cdot (-4) \cdot (-4)$ *Multiply left to right.*

$\underbrace{-4 \cdot 2} \cdot (-4) \cdot (-4)$

$\underbrace{-8 \cdot (-4)} \cdot (-4)$

$\underbrace{32 \cdot (-4)}$

-128

51. Evaluate $r^2 s^5 t^3$ when r is -3, s is 2, and t is -4, using a calculator.

$r^2 s^5 t^3$ *Replace r with -3, s with 2, and t with -4.*

$\underbrace{(-3)^2 (2)^5 (-4)^3}$ *Use the y^x key.*

$\underbrace{(9)(32)(-64)}$ *Multiply left to right.*

$\underbrace{(288)(-64)}$

$-18{,}432$

53. Evaluate $-10r^5 s^7$ when r is -3 and s is 2, using a calculator.

$-10r^5 s^7$ *Replace r with -3 and s with 2.*

$-10 \underbrace{(-3)^5} \underbrace{(2)^7}$ *Use the y^x key.*

$\underbrace{-10(-243)}(128)$ *Multiply left to right.*

$\underbrace{2430(128)}$

$311{,}040$

55. Evaluate $|xy| + |xyz|$ when x is 4, y is -2, and z is -6.

$|xy| + |xyz|$ *Replace x with 4, y with -2, and z with -6.*

   Multiply left to right within the abs. value bars.

$\underbrace{|4 \cdot (-2)|} + \left| \underbrace{4 \cdot (-2) \cdot (-6)} \right|$

$|-8| + \left| \underbrace{-8 \cdot (-6)} \right|$

$|-8| + |48|$ *Evaluate the absolute values.*

$\underbrace{8 + 48}$ *Add.*

56

57. Evaluate $\dfrac{z^2}{-3y+z}$ when z is -6 and y is -2.

$\dfrac{z^2}{-3y+z}$	*Replace z with -6 and y with -2.*
$\dfrac{(-6)^2}{-3(-2)+(-6)}$	*Follow the order of operations.*
	Numerator:
$\dfrac{36}{0}$	$(-6)^2 = -6 \cdot (-6) = 36$
	Denominator:
	$-3(-2)+(-6) = 6+(-6) = 0$
Undefined	*Division by 0 is undefined.*

Relating Concepts (Exercises 59–60)

59. **(a)** Evaluate $\dfrac{s}{5}$ when s is 15 seconds.

$\dfrac{s}{5}$	*Replace s with 15.*
$\dfrac{15}{5}$	*Divide.*
3 miles	

(b) Evaluate $\dfrac{s}{5}$ when s is 10 seconds.

$\dfrac{s}{5}$	*Replace s with 10.*
$\dfrac{10}{5}$	*Divide.*
2 miles	

(c) Evaluate $\dfrac{s}{5}$ when s is 5 seconds.

$\dfrac{s}{5}$	*Replace s with 5.*
$\dfrac{5}{5}$	*Divide.*
1 mile	

60. **(a)** Using part (c) of Exercise 59, the distance covered in $2\frac{1}{2}$ seconds is half of the distance covered in 5 seconds, or $\frac{1}{2}$ mile.

(b) Using part (a) of Exercise 59, the time to cover $1\frac{1}{2}$ miles is half the time to cover 3 miles, or $7\frac{1}{2}$ seconds. Or, using parts (b) and (c), find the number halfway between 5 seconds and 10 seconds

(c) Using parts (a) and (b) of Exercise 59, find the number halfway between 10 seconds and 15 seconds; that is $12\frac{1}{2}$ seconds.

2.2 Simplifying Expressions

2.2 Margin Exercises

1. **(a)** $3b^2 + (-3b) + 3 + b^3 + b$

The like terms are $-3b$ and b since the variable parts match; both are b.

The coefficients are $\underline{-3}$ and $\underline{1}$.

(b) $-4xy + 4x^2y + (-4xy^2) + (-4) + 4$

The like terms are the constants, -4 and 4. There are no variable parts.

(c) $5r^2 + 2r + (-2r^2) + 5 + 5r^3$

The like terms are $5r^2$ and $-2r^2$ since the variable parts match; both are r^2.

The coefficients are 5 and -2.

(d) $-10 + (-x) + (-10x) + (-x^2) + (-10y)$

The like terms are $-x$ and $-10x$ since the variable parts match; both are x.

The coefficients are -1 and -10.

2. **(a)** $10b + 4b + 10b$ *These are like terms.*

$\quad\downarrow\qquad\downarrow\qquad\downarrow$

$(10 + 4 + 10)b$ *Add the coefficients.*

$\underline{24}b$ *The variable part, b, stays the same.*

(b) $y^3 + 8y^3$ *These are like terms.*

$1y^3 + 8y^3$ *Rewrite y^3 as $1y^3$.*

$(1 + 8)y^3$ *Add the coefficients.*

$9y^3$ *The variable part, y^3, stays the same.*

(c) $-7n - n$ *These are like terms.*

$-7n - 1n$ *Rewrite n as 1n.*

$-7n + (-1n)$ *Change to addition.*

$[-7 + (-1)]n$ *Add the coefficients.*

$-8n$ *The variable part, n, stays the same.*

(d) $3c - 5c - 4c$ *These are like terms.*

$3c + (-5c) + (-4c)$ *Change to addition.*

$[3 + (-5) + (-4)]c$ *Add the coefficients.*

$-6c$ *The variable part, c, stays the same.*

(e) $-9xy + xy$ *These are like terms.*

$-9xy + 1xy$ *Rewrite xy as 1xy.*

$(-9 + 1)xy$ *Add the coefficients.*

$-8xy$ *The variable part, xy, stays the same.*

(f) $-4p^2 - 3p^2 + 8p^2$ *These are like terms.*

$-4p^2 + (-3p^2) + 8p^2$ *Change to addition.*

$[-4 + (-3) + 8]p^2$ *Add the coefficients.*

$1p^2$ or p^2 *The variable part, p^2, stays the same.*

(g) $ab - ab$ *These are like terms.*

$1ab - 1ab$ *Rewrite ab as 1ab.*

$1ab + (-1ab)$ *Change to addition.*

$[1 + (-1)]ab$ *Add the coefficients.*

$0ab$ *Zero times anything is zero.*

0

3. **(a)** $3b^2 + 4d^2 + 7b^2$

$3b^2 + 7b^2 + 4d^2$ *Rewrite using the commutative property.*

$(3+7)b^2 + 4d^2$ *Combine $3b^2 + 7b^2$.*

$\underline{10b^2} + \underline{4d^2}$ *Add the coefficients.*

(b) $4a + b - 6a + b$

$4a + b + (-6a) + b$ *Change to addition.*

$4a + (-6a) + b + b$ *Rewrite using the commutative property.*

$4a + (-6a) + 1b + 1b$ *Rewrite b as 1b.*

$[4 + (-6)]a + (1+1)b$ *Add the coefficients of like terms.*

$-2a + 2b$

(c) $-6x + 5 + 6x + 2$

$-6x + 6x + 5 + 2$ *Rewrite using the commutative property. Add the coefficients of like terms.*

$(-6+6)x + (5+2)$

$0x + 7$

$0 + 7$ *Zero times anything is zero.*

7

(d) $2y - 7 - y + 7$

$2y + (-7) + (-y) + 7$ *Change to addition.*

$2y + (-7) + (-1y) + 7$ *Rewrite $-y$ as $-1y$.*

$2y + (-1y) + (-7) + 7$ *Rewrite using the commutative prop.*

$[2 + (-1)]y + (-7 + 7)$ *Add the coefficients of like terms.*

$1y + 0$

$1y$ or y

(e) $-3x - 5 + 12 + 10x$

$-3x + (-5) + 12 + 10x$ *Change to addition.*

$-3x + 10x + (-5) + 12$ *Rewrite using the commutative prop.*

$(-3 + 10)x + (-5 + 12)$ *Add the coefficients of like terms.*

$7x + 7$

4. **(a)** $7(4c)$ means $7 \cdot (4 \cdot c)$. Using the associative property, it can be rewritten as

$\underbrace{(7 \cdot 4)} \cdot c$

$\underbrace{28 \cdot c}$

$28c$

(b) $-3(5y^3)$ can be written as

$\underbrace{(-3 \cdot 5)} \cdot y^3$

$\underbrace{-15 \cdot y^3}$

$-15y^3$

(c) $20(-2a)$ can be written as

$\underbrace{[20 \cdot (-2)]} \cdot a$

$\underbrace{-40 \cdot a}$

$-40a$

(d) $-10(-x)$ *Rewrite $-x$ as $-1x$.*

$-10(-1x)$ *can be written as*

$\underbrace{[-10 \cdot (-1)]} \cdot x$

$\underbrace{10 \cdot x}$

$10x$

5. **(a)** $7(a + 10)$ can be written as

$\underline{7 \cdot a} + \underline{7 \cdot 10}$

$\underline{7a} + \underline{70}$

(b) $3(x - 3)$ can be written as

$\underline{3 \cdot x} - \underline{3 \cdot 3}$

$\underline{3x} - \underline{9}$

(c) $4(2y + 6)$ can be written as

$4 \cdot 2y + \underline{4 \cdot 6}$

$\underline{4 \cdot 2} \cdot y + 24$

$\underbrace{8 \cdot y} + 24$

$8y + 24$

(d) $-5(3b + 2)$

$\underbrace{-5 \cdot 3b} + \underbrace{(-5) \cdot 2}$

$-15b + (-10)$ *Multiply.*

$-15b - 10$ *Change addition to subtraction.*

(e) $-8(c + 4)$

$\underbrace{-8 \cdot c} + \underbrace{-8 \cdot 4}$

$-8c + (-32)$ *Multiply.*

$-8c - 32$ *Change addition to subtraction.*

6. **(a)** $-4 + 5(y + 1)$ *Distributive property*

$-4 + 5 \cdot y + 5 \cdot 1$

$-4 + \underline{5y} + \underline{5}$ *Rewrite using the commutative property.*

$\underbrace{-4 + 5} + 5y$ *Combine constants.*

$1 + 5y$ or $\underline{5y + 1}$

(b) $2(3w + 4) - 5$ *Distributive property*

$\underline{2 \cdot 3w} + \underline{2 \cdot 4} - 5$ *Multiply.*

$6w + \underline{8 - 5}$ *Combine constants.*

$6w + 3$

(c) $5(6x - 2) + 3x$ *Distributive property*

$\underline{5 \cdot 6x} - \underline{5 \cdot 2} + 3x$ *Multiply.*

$30x - 10 + 3x$ *Change to addition.*

$30x + (-10) + 3x$ *Rewrite using the commutative property.*

$30x + 3x + (-10)$

$(30 + 3)x + (-10)$ *Add the coefficients of like terms.*

$33x + (-10)$ or $33x - 10$

(d) $21 + 7(a^2 - 3)$ *Distributive property*

$21 + 7 \cdot a^2 - \underline{7 \cdot 3}$ *Multiply.*

$21 + 7a^2 - 21$ *Change to addition.*

$21 + 7a^2 + (-21)$ *Rewrite using the commutative property.*

$\underline{21 + (-21)} + 7a^2$ *Combine constants.*

$\underbrace{0 + 7a^2}$

$7a^2$

(e)

$-y + 3(2y + 5) - 18$ *Distributive property*

$-y + \underbrace{3 \cdot 2y} + \underbrace{3 \cdot 5} - 18$ *Rewrite $-y$ as $-1y$.*

$-1y + 6y + 15 + (-18)$ *Change to addition.*

$\underbrace{(-1 + 6)} y + [15 + (-18)]$ *Add the coefficients of like terms.*

$5y + (-3)$ or $5y - 3$

2.2 Section Exercises

1. $2b^2 + 2b + 2b^3 + b^2 + 6$ ▪ $2b^2$ and b^2 are the only like terms in the expression. The variable parts match; both are b^2. The coefficients are $\underline{2}$ and $\underline{1}$.

3. $-x^2y + (-xy) + 2xy + (-2xy^2)$ ▪ $-xy$ and $2xy$ are the like terms in the expression. The variable parts match; both are xy. The coefficients are -1 and 2.

5. $7 + 7c + 3 + 7c^3 + (-4)$ ▪ 7, 3, and -4 are like terms. There are no variable parts; constants are considered like terms.

7. $6r + 6r$ *These are like terms.*

$\downarrow \quad \downarrow$ *Add the coefficients.*

$\underbrace{(6 + 6)}r$

$\underline{12r}$ *The variable part, r, stays the same.*

9. $x^2 + 5x^2$ *These are like terms. Rewrite x^2 as $1x^2$.*

$1x^2 + 5x^2$ *Add the coefficients.*

$(1 + 5)x^2$

$6x^2$ *The variable part, x^2, stays the same.*

11. $p - 5p$ *These are like terms. Rewrite p as 1p.*

$1p - 5p$ *Change to addition.*

$1p + (-5p)$ *Add the coefficients.*

$[1 + (-5)]p$

$-4p$ *The variable part, p, stays the same.*

13. $-2a^3 - a^3$ *These are like terms. Rewrite a^3 as $1a^3$.*

$-2a^3 - 1a^3$ *Change to addition.*

$-2a^3 + (-1a^3)$ *Add the coefficients.*

$[-2 + (-1)]a^3$

$-3a^3$ *The variable part, a^3, stays the same.*

15. $\underline{c - c}$

0 Any number minus itself is 0.

17. $9xy + xy - 9xy$ *These are like terms. Rewrite xy as 1xy.*

$9xy + 1xy - 9xy$ *Change to addition.*

$9xy + 1xy + (-9xy)$ *Add the coefficients.*

$[9 + 1 + (-9)]xy$

$1xy$ or xy *The variable part, xy, stays the same.*

19. $5t^4 + 7t^4 - 6t^4$ *These are like terms. Change to addition.*

$5t^4 + 7t^4 + (-6t^4)$ *Add the coefficients.*

$[5 + 7 + (-6)]t^4$

$6t^4$ *The variable part, t^4, stays the same.*

21. $y^2 + y^2 + y^2 + y^2$ *These are like terms. Write in the understood coefficients of 1.*

$1y^2 + 1y^2 + 1y^2 + 1y^2$

$(1 + 1 + 1 + 1)y^2$ *Add the coefficients.*

$4y^2$ *The variable part, y^2, stays the same.*

23. $-x - 6x - x$ *These are like terms. Rewrite $-x$ as $-1x$ and x as 1x.*

$-1x - 6x - 1x$ *Change to addition.*

$-1x + (-6x) + (-1x)$ *Add the coefficients.*

$[-1 + (-6) + (-1)]x$

$-8x$ *The variable part, x, stays the same.*

25. $8a + 4b + 4a$ *Use the commutative property to rewrite the expression so that like terms are next to each other.*

$8a + 4a + 4b$ *Add the coefficients of like terms.*

$(8 + 4)a + 4b$

$12a + 4b$ *The variable part, a, stays the same.*

27. $6 + 8 + 7rs$ *Use the commutative property to put the constants at the end.*

$7rs + 6 + 8$ *Add the coefficients of like terms.*

$7rs + 14$ *The only like terms are constants.*

29. $a + ab^2 + ab^2$ *Write in the understood coefficients of 1.*

$1a + 1ab^2 + 1ab^2$ *Add the coefficients of like terms.*

$1a + (1 + 1)ab^2$

$1a + 2ab^2,$ *The variable part, ab^2, stays the same.*

or $a + 2ab^2$

31. $6x + y - 8x + y$ *Write in the understood coefficients of 1. Change to addition. Rewrite using the commutative property. Add the coefficients of like terms.*

$6x + 1y + (-8x) + 1y$

$6x + (-8x) + 1y + 1y$

$\underbrace{[6 + (-8)]}_{-2x} x + \underbrace{(1 + 1)}_{2y} y$

$-2x + 2y$

33. $8b^2 - a^2 - b^2 + a^2$ *Write in the understood coefficient of 1.*

$8b^2 - 1a^2 - 1b^2 + 1a^2$ *Change to addition.*

$8b^2 + (-1a^2) + (-1b^2) + 1a^2$ *Rewrite using the commutative property.*

$8b^2 + (-1b^2) + (-1a^2) + 1a^2$ *Add the coefficients of like terms.*

$\underbrace{[8 + (-1)]}_{7b^2} b^2 + \underbrace{(-1 + 1)}_{0 \cdot a^2} a^2$

$7b^2 + 0$

$7b^2$

35. $-x^3 + 3x - 3x^2 + 2$ ▪ There are no like terms. The expression cannot be simplified.

37. $-9r + 6t - s - 5r + s + t - 6t + 5s - r$

Write in the understood coefficients of 1. Change to addition.

$-9r + 6t + (-1s) + (-5r) + 1s + 1t$
$+ (-6t) + 5s + (-1r)$

Rewrite using the commutative property.

$-9r + (-5r) + (-1r) + (-1s) + 1s + 5s$
$+ 6t + 1t + (-6t)$

Add the coefficients of like terms.

$\underbrace{[-9 + (-5) + (-1)]}_{-15r} r + \underbrace{(-1 + 1 + 5)}_{5s} s$
$+ \underbrace{[6 + 1 + (-6)]}_{1t} t$

$-15r + 5s + t$

39. By using the associative property, we can write $3(10a)$ as

$$(3 \cdot 10) \cdot a = 30 \cdot a = 30a.$$

So, $3(10a)$ simplifies to $30a$.

41. By using the associative property, we can write $-4(2x^2)$ as

$$(-4 \cdot 2) \cdot x^2 = -8 \cdot x^2 = -8x^2.$$

So, $-4(2x^2)$ simplifies to $-8x^2$.

43. By using the associative property, we can write $5(-4y^3)$ as

$$[5 \cdot (-4)] \cdot y^3 = -20 \cdot y^3 = -20y^3.$$

So, $5(-4y^3)$ simplifies to $-20y^3$.

45. By using the associative property, we can write $-9(-2cd)$ as

$$[-9 \cdot (-2)] \cdot c \cdot d = 18 \cdot c \cdot d = 18cd.$$

So, $-9(-2cd)$ simplifies to $18cd$.

47. By using the associative property, we can write $7(3a^2bc)$ as

$$(7 \cdot 3) \cdot a^2 \cdot b \cdot c = 21 \cdot a^2 \cdot b \cdot c = 21a^2bc.$$

So, $7(3a^2bc)$ simplifies to $21a^2bc$.

49. $-12(-w)$ *Write in the understood coefficient of -1.*

$-12(-1w)$ *Rewrite using the associative property.*

$[-12 \cdot (-1)]w$

$12 \cdot w$

$12w$

51. $6(b + 6)$ *Distributive property*

$\underline{6 \cdot b} + \underline{6 \cdot 6}$

$\underline{6b} + \underline{36}$

53. $7(x - 1)$ *Distributive property*

$7 \cdot x - 7 \cdot 1$

$7x - 7$

55. $3(7t + 1)$ *Distributive property*

$3 \cdot 7t + 3 \cdot 1$

$21t + 3$

57. $-2(5r + 3)$ *Distributive property*
$-2 \cdot 5r + (-2) \cdot 3$

$-10r + (-6)$ *Change addition to subtraction of the opposite.*

$-10r - 6$

59. $-9(k + 4)$ *Distributive property*
$-9 \cdot k + (-9) \cdot 4$

$-9k + (-36)$ *Change addition to subtraction of the opposite.*

$-9k - 36$

61. $50(m - 6)$ *Distributive property*
$50 \cdot m - 50 \cdot 6$
$50m - 300$

63. $10 + 2(4y + 3)$ *Distributive property*
$10 + \underbrace{2 \cdot 4} \cdot y + \underbrace{2 \cdot 3}$

$10 \; + \; 8y \; + \; 6$ *Rewrite using the commutative property.*

$8y + 10 + 6$ *Combine like terms.*

$8y + 16$

65. $6(a^2 - 2) + 15$ *Distributive property*
$6 \cdot a^2 - 6 \cdot 2 + 15$

$6a^2 - 12 + 15$ *Combine like terms.*

$6a^2 + 3$

67. $2 + 9(m - 4)$ *Distributive property*
$2 + 9 \cdot m - 9 \cdot 4$

$2 + 9m - 36$ *Change to addition.*

$2 + 9m + (-36)$ *Rewrite using the commutative property.*

$9m + 2 + (-36)$ *Add the coefficients of like terms.*

$9m + (-34)$ *Change addition to subtraction of the opposite.*

$9m - 34$

69. $-5(k + 5) + 5k$ *Distributive property*
$-5 \cdot k + (-5) \cdot 5 + 5k$

$-5k + (-25) + 5k$ *Rewrite using the commutative property.*

$-5k + 5k + (-25)$ *Add the coefficients of like terms.*

$\underbrace{(-5 + 5)} \, k + (-25)$

$0k + (-25)$ *Zero times any number is 0.*

$\underbrace{0 + (-25)}$ *Zero added to any number is the number*

-25

71. $4(6x - 3) + 12$ *Distributive property*
$4 \cdot 6x - 4 \cdot 3 + 12$

$24x - 12 + 12$ *Change to addition.*
                      Combine like terms.

$24x + (-12) + 12$ *Any number plus its opposite is 0.*

$24x + 0$

$24x$

73. $5 + 2(3n + 4) - n$ *Distributive property*
$5 + 2 \cdot 3n + 2 \cdot 4 - n$ *Rewrite n as 1n.*

$5 + 6n + 8 - 1n$ *Change to addition.*

$5 + 6n + 8 + (-1n)$ *Rewrite using the commutative property.*

$5 + 8 + 6n + (-1n)$ *Add the coefficients of like terms.*

$(5 + 8) + [6 + (-1)]n$

$13 + 5n$ or $5n + 13$

75. $-p + 6(2p - 1) + 5$ *Distributive property*
$-p + 6 \cdot 2p - 6 \cdot 1 + 5$

$-p + 12p - 6 + 5$ *Rewrite $-p$ as $-1p$.*
                         Change to addition.

$-1p + 12p + (-6) + 5$ *Add the coefficients of like terms.*

$(-1 + 12)p + (-6 + 5)$

$11p + (-1)$ *Change addition to subt. of the opposite.*

$11p - 1$

77. A simplified expression usually still has variables, but it is written in a simpler way. When evaluating an expression, the variables are all replaced by specific numbers and the final result is a numerical answer.

79. Like terms have matching variable parts, that is, matching letters and exponents. The coefficients do not have to match. Examples will vary. Possible examples: In $-6x + 9 + x$, the terms $-6x$ and x are like terms. In $4k + 3 - 8k^2 + 10$, the terms 3 and 10 are like terms.

81. $\underbrace{-2x + 7x} + 8$
$5x + 8$

Keep the variable part unchanged when combining like terms. As shown above, the correct answer is $5x + 8$.

83. $-4(3y) - 5 + 2(5y + 7)$ *Distributive prop.*
$-4 \cdot 3y - 5 + 2 \cdot 5y + 2 \cdot 7$

Change subtrac-
tion to adding
$-12y - 5 + 10y + 14$ *the opposite.*

Group like terms
$-12y + (-5) + 10y + 14$ *and add the*
coefficients.

$\underbrace{-12y + 10y}_{-2y} + \underbrace{(-5) + 14}_{9}$

$-2y + 9$

85. $-10 + 4(-3b + 3) + 2(6b - 1)$
Distributive property
$-10 + 4 \cdot (-3b) + 4 \cdot 3 + 2 \cdot 6b - 2 \cdot 1$
$-10 + (-12b) + 12 + 12b - 2$
Change to addition.
$-10 + (-12b) + 12 + 12b + (-2)$
Group like terms and add the coefficients.
$\underbrace{-12b + 12b}_{0b} + \underbrace{-10 + 12 + (-2)}_{0}$

0

87. $-5(-x + 2) + 8(-x) + 3(-2x - 2) + 16$
Distributive property
$-5 \cdot (-x) + (-5) \cdot 2 + 8 \cdot (-x) + 3 \cdot (-2x)$
$\qquad\qquad\qquad\qquad\qquad - 3 \cdot 2 + 16$
$5x + (-10) + (-8x) + (-6x) - 6 + 16$
Change to addition.
$5x + (-10) + (-8x) + (-6x) + (-6) + 16$
Group like terms and add the coefficients.
$\underbrace{5x + (-8x) + (-6x)}_{-9x} + \underbrace{-10 + (-6) + 16}_{0}$

$-9x$

Summary Exercises
Variables and Expressions

1. $-10 - m$ *m is the variable;*
or $-10 - 1m$ -1 *is the coefficient;*
$\qquad\qquad\qquad$ -10 *is the constant.*

3. $6 + 4x$ *x is the variable;*
$\qquad\qquad$ 4 *is the coefficient;*
$\qquad\qquad$ 6 *is the constant.*

5.   Expression (rule) for finding the total cost of a car with down payment d, monthly payment m, and number of payments t: $d + mt$

   (a) Evaluate the expression when the down payment is $3000, the monthly payment is $280, and the number of payments is 36.

$d + mt$ *Replace d with $3000, m*
with $280, and t with 36.
$\underbrace{\$3000 + \$280 \cdot 36}$ *Multiply before adding.*
$\underbrace{\$3000 + \$10,080}$

$13,080 is the total cost of the car.

   (b) Evaluate the expression when the down payment is $1750, the monthly payment is $429, and the number of payments is 48.

$d + mt$ *Replace d with $1750, m*
with $429, and t with 48.
$\underbrace{\$1750 + \$429 \cdot 48}$ *Multiply before adding.*
$\underbrace{\$1750 + \$20,592}$

$22,342 is the total cost of the car.

7. b^3cd written without exponents is

$$b \cdot b \cdot b \cdot c \cdot d$$

9. $w^4 = w \cdot w \cdot w \cdot w$ ▪ *Replace w with 5.*

$\underbrace{5 \cdot 5}_{25} \cdot 5 \cdot 5$ *Multiply left to right.*

$\underbrace{25 \cdot 5}_{125} \cdot 5$

$\underbrace{125 \cdot 5}_{625}$

11. yz^2 ▪ *Replace y with -6 and z with 0.*
If 0 is multiplied by any number, the result is 0. Thus, there is no need to make any calculations since the result is 0.

13. $x^3 = x \cdot x \cdot x$ ▪ *Replace x with -2.*

$\underbrace{-2 \cdot (-2)}_{4} \cdot (-2)$ *Multiply left to right.*

$\underbrace{4 \cdot (-2)}_{-8}$

15. $3xy^2 = 3 \cdot x \cdot y \cdot y$ ▪ *Replace x with -2 and y with -6.*

$\underbrace{3 \cdot (-2)}_{-6} \cdot (-6) \cdot (-6)$ *Multiply left to right.*

$\underbrace{-6 \cdot (-6)}_{36} \cdot (-6)$

$\underbrace{36 \cdot (-6)}_{-216}$

17. $-7wx^4y^3$ ▪ *Use a calculator. Replace w with 5, x with -2, and y with -6.*

$-7(5)(-2)^4(-6)^3 = -35(16)(-216) = 120{,}960$

19. $-3x - 5 + 12 + 10x$
$= -3x + 10x + (-5) + 12$
$= (-3 + 10)x + (-5 + 12)$
$= 7x + 7$

21. $-9xy + 9xy = (-9 + 9)xy$
$= 0xy$
$= 0$

23. $3f - 5f - 4f = 3f + (-5f) + (-4f)$
$= [3 + (-5) + (-4)]f$
$= -6f$

25. $-a - 6b - a = -a + (-6b) + (-a)$
$= -a + (-a) + (-6b)$
$= -1a + (-1a) + (-6b)$
$= [-1 + (-1)] \cdot a + (-6b)$
$= -2a + (-6b)$
or $-2a - 6b$

27. $5r^3 + 2r^2 - 2r^2 + 5r^3$
$= \underbrace{5r^3 + 5r^3}_{} + \underbrace{2r^2 + (-2r^2)}_{}$
$= \qquad 10r^3 \quad + \qquad 0$
$= 10r^3$

29. $-3(m + 3) + 3m = -3 \cdot m + (-3) \cdot 3 + 3m$
$= -3m + 3m + (-9)$
$= (-3 + 3) \cdot m + (-9)$
$= 0m + (-9)$
$= 0 + (-9)$
$= -9$

31. $2 + 12(3x - 1) = 2 + 12 \cdot 3x - 12 \cdot 1$
$= 2 + (12 \cdot 3) \cdot x - 12$
$= 2 + (-12) + 36x$
$= -10 + 36x$
or $36x - 10$

33. **(a)** Simplifying the expression correctly:

$$6(n + 2) = 6 \cdot n + 6 \cdot 2$$
$$= 6n + 12$$

The student forgot to multiply $6 \cdot 2$.

(b) Simplifying the expression correctly:

$$-5(-4a) = [-5 \cdot (-4)] \cdot a$$
$$= 20a$$

Two negative factors give a *positive* product.

(c) Simplifying the expression correctly:

$$3y + 2y - 10 = (3 + 2)y - 10$$
$$= 5y - 10$$

Keep the variable part unchanged; that is, adding y's to y's gives an answer with y's, not y^2's.

2.3 Solving Equations Using Addition

2.3 Margin Exercises

1. **(a)** $c + 15 = 80$ *Given equation*
$95 + 15 \overset{?}{=} 80$ *Replace c with 95.*
$110 \neq 80$ *110 is more than 80.*
No, 95 is not the solution.
$65 + 15 \overset{?}{=} 80$ *Replace c with 65.*
$80 = 80$ Balances
Yes, 65 is the solution.
(No need to check 80 and 70.)

(b) $28 = c - 4$ *Given equation*
$28 \overset{?}{=} 28 - 4$ *Replace c with 28.*
$28 \neq 24$
No, 28 is not the solution.
$28 \overset{?}{=} 20 - 4$ *Replace c with 20.*
$28 \neq 16$
No, 20 is not the solution.
$28 \overset{?}{=} 24 - 4$ *Replace c with 24.*
$28 \neq 20$
No, 24 is not the solution.
$28 \overset{?}{=} 32 - 4$ *Replace c with 32.*
$28 = 28$ Balances
Yes, 32 is the solution.

2. **(a)** Solve $12 = y + 5$ for y.

To get y by itself, add the opposite of 5, which is -5. To keep the balance, add -5 to *both* sides.

$$\begin{array}{rcl} 12 &=& y + 5 \\ -5 & & -5 \\ \hline 7 &=& y + 0 \\ 7 &=& y \end{array}$$ The solution is 7.

Check $12 = y + 5$ *Original equation*
$12 = \underline{7 + 5}$ *Replace y with 7.*
$12 = \underline{12}$ Balances, so solution is 7.

(b) Solve $b - 2 = -6$ for b.

Change to addition.

$$b + (-2) = -6$$

To get b by itself add the opposite of -2, which is 2, to both sides.

$$\begin{array}{rcl} b + (-2) &=& -6 \\ 2 & & 2 \\ \hline b + 0 &=& -4 \\ b &=& -4 \end{array}$$ The solution is -4.

Check $b - 2 = -6$ *Original equation*
$-4 - 2 = -6$ *Replace b with -4.*
$-4 + (-2) = -6$
$-6 = -6$ Balances

3. **(a)** $2 - 8 = k - 2$ ▪ Rewrite both sides by changing subtraction to addition. Combine like terms.

$$2 + (-8) = k + (-2)$$
$$-6 = k + (-2)$$

To get k by itself add the opposite of -2, which is 2, to both sides.

$$\begin{array}{rcl} -6 &=& k + (-2) \\ \underline{2} & & \underline{2} \\ -4 &=& \underbrace{k + 0} \\ -4 &=& k \end{array}$$ The solution is -4.

Check

$$\begin{array}{ll} 2 - 8 = k - 2 & \textit{Original equation} \\ 2 - 8 = -4 - 2 & \textit{Replace } k \textit{ with } -4. \\ \underbrace{2 + (-8)} = \underbrace{-4 + (-2)} & \\ -6 = -6 & \textit{Balances} \end{array}$$

(b) $4r + 1 - 3r = -8 + 11$ ▪ Change to addition.

$$4r + 1 + (-3r) = -8 + 11$$

Rewrite the left side by using the commutative property.

$$\begin{array}{rcll} 4r + (-3r) + 1 &=& -8 + 11 & \textit{Combine like terms.} \\ 1r + 1 &=& 3 & \textit{To get } r \textit{ by itself,} \\ \underline{-1} & & \underline{-1} & \textit{add } -1 \textit{ to both sides.} \\ 1r + 0 &=& 2 & \\ 1r &=& 2 & \\ \text{or} \quad r &=& 2 & \textit{The solution is 2.} \end{array}$$

Check

$$\begin{array}{l} 4r + 1 - 3r = -8 + 11 \\ 4 \cdot 2 + 1 - 3 \cdot 2 = -8 + 11 \quad \textit{Replace } r \textit{ with 2.} \\ 8 + 1 - 6 = 3 \\ 9 - 6 = 3 \\ 3 = 3 \qquad\qquad \textit{Balances} \end{array}$$

2.3 Section Exercises

1. $n - 50 = 8$ ▪ Replace n with 58, 42, 60, and 8.

$$\begin{array}{ll} n - 50 = 8 & \textit{Given equation} \\ \mathbf{58} - 50 \overset{?}{=} 8 & \textit{Replace } n \textit{ with 58.} \\ 58 + (-50) \overset{?}{=} 8 & \\ 8 = 8 & \end{array}$$

Yes, 58 is the solution.
(No need to check 42, 60, and 8.)

3. $-6 = y + 10$ ▪ Replace y with -4, -16, 16, and -6.

$$\begin{array}{ll} -6 = y + 10 & \textit{Given equation} \\ -6 \overset{?}{=} \mathbf{-4} + 10 & \textit{Replace } y \textit{ with } -4. \\ -6 \neq 6 & \end{array}$$

No, -4 is not the solution.

$$\begin{array}{ll} -6 \overset{?}{=} \mathbf{-16} + 10 & \textit{Replace } y \textit{ with } -16. \\ -6 = -6 & \end{array}$$

Yes, -16 is the solution.
(No need to check 16 and -6.)

5. **(a)** $m - 8 = 1$ ▪ Add 8 to both sides because $-8 + 8$ gives $m + 0$ on the left side.

(b) $-7 = w + 5$ ▪ Add -5 to both sides because $5 + (-5)$ gives $w + 0$ on the right side.

7. $$\begin{array}{rcll} p + 5 &=& 9 & \\ \underline{-5} & & \underline{-5} & \textit{Add the opposite of} \\ & & & \textit{5, } -5\textit{, to both sides.} \\ p + 0 &=& 4 & \\ p &=& 4 & \textit{The solution is 4.} \end{array}$$

Check $\quad p + 5 = 9$
$\qquad\qquad 4 + 5 = 9 \quad$ *Replace p with 4.*
$\qquad\qquad\quad 9 = 9 \quad$ *Balances*

9. $$\begin{array}{rcll} 8 &=& r - 2 & \\ 8 &=& r + (-2) & \textit{Change to addition.} \\ \underline{+2} & & \underline{+2} & \textit{Add the opposite of} \\ & & & -2\textit{, 2, to both sides.} \\ 10 &=& r + 0 & \\ 10 &=& r & \textit{The solution is 10.} \end{array}$$

Check $\quad 8 = r - 2$
$\qquad\qquad 8 = 10 - 2 \quad$ *Replace r with 10.*
$\qquad\qquad 8 = 8 \qquad\quad$ *Balances*

11. $$\begin{array}{rcll} -5 &=& n + 3 & \\ \underline{-3} & & \underline{-3} & \textit{Add the opposite of} \\ & & & \textit{3, } -3\textit{, to both sides.} \\ -8 &=& n + 0 & \\ -8 &=& n & \textit{The solution is } -8. \end{array}$$

Check $\quad -5 = n + 3$
$\qquad\qquad -5 = -8 + 3 \quad$ *Replace n with -8.*
$\qquad\qquad -5 = -5 \qquad$ *Balances*

13. $$\begin{array}{rcll} -4 + k &=& 14 & \\ \underline{4} & & \underline{4} & \textit{Add the opposite of} \\ & & & -4\textit{, 4, to both sides.} \\ 0 + k &=& 18 & \\ k &=& 18 & \textit{The solution is 18.} \end{array}$$

Check $\quad -4 + k = 14$
$\qquad\qquad -4 + 18 = 14 \quad$ *Replace k with 18.*
$\qquad\qquad\quad 14 = 14 \qquad$ *Balances*

15.
$$y - 6 = 0$$
$$y + (-6) = 0 \quad \textit{Change to addition.}$$

$$\begin{array}{r} \underline{ 6 \quad\quad 6} \\ y + 0 = 6 \end{array} \quad \begin{array}{l} \textit{Add the opposite of} \\ \textit{−6, 6, to both sides.} \end{array}$$

$$y = 6 \quad \text{The solution is 6.}$$

Check $y - 6 = 0$
$$6 - 6 = 0 \quad \textit{Replace y with 6.}$$
$$0 = 0 \quad \text{Balances}$$

17.
$$7 = r + 13$$

$$\begin{array}{r} \underline{-13 \quad\quad -13} \\ -6 = r + 0 \end{array} \quad \begin{array}{l} \textit{Add the opposite of} \\ \textit{13, −13, to both sides.} \end{array}$$

$$-6 = r \quad \text{The solution is −6.}$$

Check $7 = r + 13$
$$7 = -6 + 13 \quad \textit{Replace r with −6.}$$
$$7 = 7 \quad\quad\quad \text{Balances}$$

19.
$$x - 12 = -12$$
$$x + (-12) = -12 \quad \textit{Change to addition.}$$

$$\begin{array}{r} \underline{ 12 \quad\quad 12} \\ x + 0 = 0 \end{array} \quad \begin{array}{l} \textit{Add the opposite of} \\ \textit{−12, 12, to both sides.} \end{array}$$

$$x = 0 \quad \text{The solution is 0.}$$

Check $x - 12 = -12$
$$0 - 12 = -12 \quad \textit{Replace x with 0.}$$
$$0 + (-12) = -12 \quad \textit{Change to addition.}$$
$$-12 = -12 \quad \text{Balances}$$

21. $-5 = -2 + t$

$$\begin{array}{r} \underline{2 \quad\quad 2} \\ -3 = 0 + t \end{array} \quad \begin{array}{l} \textit{Add the opposite of} \\ \textit{−2, 2, to both sides.} \end{array}$$

$$-3 = t \quad \text{The solution is −3.}$$

Check $-5 = -2 + t$
$$-5 = -2 + (-3) \quad \textit{Replace t with −3.}$$
$$-5 = -5 \quad\quad\quad \text{Balances}$$

23. $z - 5 = 3$ ■ The given solution is −2.

Check $z - 5 = 3$
$$-2 - 5 = 3 \quad \textit{Replace z with −2.}$$
$$-2 + (-5) = 3 \quad \textit{Change to addition.}$$
$$-7 \neq 3 \quad \text{Does not balance}$$

Correct solution:
$$z - 5 = 3$$
$$z + (-5) = 3 \quad \textit{Change to addition.}$$

$$\begin{array}{r} \underline{ 5 \quad\quad 5} \\ z + 0 = 8 \end{array} \quad \begin{array}{l} \textit{Add the opposite of} \\ \textit{−5, 5, to both sides.} \end{array}$$

$$z = 8 \quad \text{The solution is 8.}$$

Check $z - 5 = 3$
$$8 - 5 = 3 \quad \textit{Replace z with 8.}$$
$$8 + (-5) = 3 \quad \textit{Change to addition.}$$
$$3 = 3 \quad \text{Balances}$$

25. $7 + x = -11$ ■ The given solution is −18.

Check $7 + x = -11$
$$7 + (-18) = -11 \quad \textit{Replace x with −18.}$$
$$-11 = -11 \quad \text{Balances}$$

−18 is the correct solution.

27. $-10 = -10 + b$ ■ The given solution is 10.

Check $-10 = -10 + b$
$$-10 = -10 + 10 \quad \textit{Replace b with 10.}$$
$$-10 \neq 0 \quad\quad\quad \text{Does not balance}$$

Correct solution:
$$-10 = -10 + b$$

$$\begin{array}{r} \underline{10 \quad\quad 10} \\ 0 = 0 + b \end{array} \quad \begin{array}{l} \textit{Add the opposite of} \\ \textit{−10, 10, to both sides.} \end{array}$$

$$0 = b \quad \text{The solution is 0.}$$

Check $-10 = -10 + b$
$$-10 = -10 + 0 \quad \textit{Replace b with 0.}$$
$$-10 = -10 \quad \text{Balances}$$

29.
$$c - 4 = -8 + 10$$
$$c - 4 = 2 \quad \textit{Simplify the right side.}$$
$$c + (-4) = 2 \quad \textit{Change to addition.}$$

$$\begin{array}{r} \underline{ 4 \quad\quad 4} \\ c + 0 = 6 \end{array} \quad \textit{Add 4 to both sides.}$$

$$c = 6 \quad \text{The solution is 6.}$$

Check $c - 4 = -8 + 10$
$$6 - 4 = -8 + 10 \quad \textit{Replace c with 6.}$$
$$2 = 2 \quad\quad\quad \text{Balances}$$

31. $-1 + 4 = y - 2$
$$3 = y - 2 \quad \textit{Simplify the left side.}$$
$$3 = y + (-2) \quad \textit{Change to addition.}$$

$$\begin{array}{r} \underline{2 \quad\quad\quad 2} \\ 5 = y + 0 \end{array} \quad \textit{Add 2 to both sides.}$$

$$5 = y \quad \text{The solution is 5.}$$

Check
$$-1 + 4 = y - 2$$
$$-1 + 4 = 5 - 2 \quad \textit{Replace y with 5.}$$
$$-1 + 4 = 5 + (-2) \quad \textit{Change to addition.}$$
$$3 = 3 \quad \text{Balances}$$

33.
$$10 + b = -14 - 6$$
$$10 + b = -14 + (-6) \quad \textit{Change to addition.}$$
$$10 + b = -20 \quad \textit{Add.}$$

$$\begin{array}{r} \underline{-10 \quad\quad -10} \\ 0 + b = -30 \end{array} \quad \textit{Add −10.}$$

$$b = -30 \quad \text{The solution is −30.}$$

Check

$$10 + b = -14 - 6$$

$10 + (-30) = -14 + (-6)$ *Replace b with −30.*

$$-20 = -20$$ Balances

35. $t - 2 = 3 - 5$

$t + (-2) = 3 + (-5)$ *Change to addition.*

$t + (-2) = -2$ *Simplify the right side.*

$\underline{\quad\quad 2 \quad\quad\quad 2\quad\quad}$ *Add 2 to both sides.*

$t + 0 = 0$

$t = 0$ The solution is 0.

Check

$$t - 2 = 3 - 5$$

$0 - 2 = 3 - 5$ *Replace t with 0.*

$0 + (-2) = 3 + (-5)$ *Change to addition.*

$$-2 = -2$$ Balances

37. $10z - 9z = -15 + 8$

$10z + (-9z) = -15 + 8$ *Change to addition.*

$1z = -7$ *Combine like terms.*

$z = -7$ *1z is the same as z.*

The solution is −7.

Check

$$10z - 9z = -15 + 8$$

$10 \cdot (-7) - 9 \cdot (-7) = -15 + 8$ *Replace z with −7.*

$$-70 - (-63) = -7$$

$-70 + 63 = -7$ *Change to add.*

$-7 = -7$ Balances

39. $-5w + 2 + 6w = -4 + 9$ *Rearrange and combine like terms.*

$\underbrace{-5w + 6w}_{1w} + 2 = \underbrace{-4 + 9}_{5}$

$1w + 2 = 5$

$\underline{\quad\quad -2 \quad\quad -2\quad}$ *Add −2 to both sides.*

$1w + 0 = 3$

$w = 3$ The solution is 3.

Check

$$-5w + 2 + 6w = -4 + 9$$

$-5 \cdot 3 + 2 + 6 \cdot 3 = -4 + 9$ *Replace w with 3.*

$$-15 + 2 + 18 = -4 + 9$$

$5 = 5$ Balances

41. $-3 - 3 = 4 - 3x + 4x$ *Change to addition.*

$\underbrace{-3 + (-3)}_{-6} = 4 + \underbrace{(-3x) + 4x}_{1x}$ *Combine like terms.*

$-6 = 4 + 1x$

$\underline{-4 \quad\quad\quad -4\quad}$ *Add −4 to both sides.*

$-10 = 0 + 1x$

$-10 = 1x$ *1x is the same as x.*

$-10 = x$ The solution is −10.

43. $-3 + 7 - 4 = -2a + 3a$

$-3 + 7 + (-4) = -2a + 3a$ *Change to addition.*

$0 = 1a$ *Combine like terms.*

$0 = a$ The solution is 0.

45. $y - 75 = -100$

$y + (-75) = -100$ *Change to addition.*

$\underline{\quad 75 \quad\quad\quad 75\quad}$ *Add 75 to both sides.*

$y + 0 = -25$

$y = -25$ The solution is −25.

47. $-x + 3 + 2x = 18$ *Rearrange and combine like terms.*

$1x + 3 = 18$

$\underline{\quad\quad -3 \quad\quad -3\quad}$ *Add −3 to both sides.*

$1x + 0 = 15$

$x = 15$ The solution is 15.

49. $82 = -31 + k$

$\underline{\quad 31 \quad\quad 31\quad}$ *Add 31 to both sides.*

$113 = 0 + k$

$113 = k$ The solution is 113.

51. $-2 + 11 = 2b - 9 - b$

$-2 + 11 = 2b + (-9) + (-1b)$ *Change to addition.*

$9 = 1b + (-9)$ *Rearrange and combine like terms.*

$\underline{\quad 9 \quad\quad\quad 9\quad}$ *Add 9 to both sides.*

$18 = 1b + 0$

$18 = b$ The solution is 18.

53. $r - 6 = 7 - 10 - 8$

$r + (-6) = 7 + (-10) + (-8)$ *Change to addition.*

$r + (-6) = -11$ *Combine like terms.*

$\underline{\quad 6 \quad\quad 6\quad}$ *Add 6 to both sides.*

$r + 0 = -5$

$r = -5$ The solution is −5.

55. $-14 = n + 91$

$\underline{-91 \quad\quad\quad -91\quad}$ *Add −91 to both sides.*

$-105 = n + 0$

$-105 = n$ The solution is −105.

57. $-9 + 9 = 5 + h$

$0 = 5 + h$ *Combine like terms.*

$\underline{\quad -5 \quad\quad -5\quad}$ *Add −5 to both sides.*

$-5 = 0 + h$

$-5 = h$ The solution is −5.

59.   No, the solution is −14, the number used to replace x in the original equation.

61. $g + 10 = 305$

$\underline{\quad -10 \quad\quad -10\quad}$ *Add the opposite of 10, −10, to both sides.*

$g + 0 = 295$

$g = 295$

There were 295 graduates this year.

63. $92 = c + 37$

$\underline{-37 \qquad -37}$ *Add −37 to both sides.*

$55 = c + 0$

$55 = c$

When the temperature is 92 degrees, a field cricket chirps 55 times (in 15 seconds).

65. $p - 65 = 45$

$p + (-65) = 45$ *Change to addition.*

$\underline{\quad 65 \qquad 65}$ *Add 65 to both sides.*

$p + 0 = 110$

$p = 110$

Ernesto's parking fees average $110 per month in winter.

67. $-17 - 1 + 26 - 38$

$\qquad = -3 - m - 8 + 2m$

$-17 + (-1) + 26 + (-38)$

$\qquad = -3 + (-1m) + (-8) + 2m$

Change all subtractions to additions.

$-17 + (-1) + 26 + (-38)$

$\qquad = -3 + (-8) + (-1m) + 2m$

Commutative property

$-30 = -11 + 1m$ *Combine like terms.*

$\underline{\quad 11 \qquad 11}$ *Add 11 to both sides.*

$-19 = 1m$

$-19 = m$ The solution is -19.

69. $-6x + 2x + 6 + 5x = |0 - 9| - |-6 + 5|$

$-6x + 2x + 5x + 6 = |0 + (-9)| - |-6 + 5|$

Change subtraction within absolute value to addition and rearrange the terms.

$1x + 6 = |-9| - |-1|$

Simplify inside absolute value bars. Collect like terms.

$1x + 6 = 9 - 1$ *Evaluate absolute values.*

$1x + 6 = 9 + (-1)$ *Change to addition.*

$1x + 6 = 8$

$\underline{\quad -6 \qquad -6}$ *Add −6 to both sides.*

$1x + 0 = 2$

$x = 2$ The solution is 2.

Relating Concepts (Exercises 71–72)

71. (a) Equations will vary. Some possibilities are:

$n - 1 = -3$

$n + (-1) = -3$ *Change to addition.*

$\underline{\quad 1 \qquad\quad 1}$ *Add 1 to both sides.*

$n + 0 = -2$

$n = -2$ The solution is -2.

$8 = x + 10$

$\underline{-10 \qquad -10}$ *Add the opposite of 10, −10, to both sides.*

$-2 = x + 0$

$-2 = x$ The solution is -2.

(b) Equations will vary. Some possibilities are:

$y + 6 = 6$

$\underline{\quad -6 \qquad -6}$ *Add the opposite of 6, −6, to both sides.*

$y + 0 = 0$

$y = 0$ The solution is 0.

$-5 = -5 + b$

$\underline{\quad 5 \qquad\quad 5}$ *Add the opposite of −5, 5, to both sides.*

$0 = 0 + b$

$0 = b$ The solution is 0.

72. (a) $x + 1 = 1\frac{1}{2}$

$\underline{\quad -1 \qquad -1}$ *Add the opposite of 1, −1, to both sides.*

$x + 0 = \frac{1}{2}$

$x = \frac{1}{2}$ The solution is $\frac{1}{2}$.

(b) $\frac{1}{4} = y - 1$

$\frac{1}{4} = y + (-1)$ *Change to addition.*

$\underline{\quad 1 \qquad\qquad 1}$ *Add the opposite of −1, 1, to both sides.*

$1\frac{1}{4} = y + 0$

$1\frac{1}{4} = y$ or $y = \frac{5}{4}$ The solution is $\frac{5}{4}$.

(c) $\$2.50 + n = \$3.35$

$\underline{-\$2.50 \qquad -\$2.50}$ *Add the opposite of $2.50, −$2.50, to both sides.*

$\$0 + n = \$0.85$

$n = \$0.85$

The solution is $0.85.

(d) Equations will vary. Some possibilities are:

$a - \$7.32 = \$9.16$ The solution is $16.48.

$5c - \$11.20 = 4c - \$2.00$ The solution is $9.20.

2.4 Solving Equations Using Division

2.4 Margin Exercises

1. (a) Solve $4s = 44$.

Use division to undo multiplication. Divide *both* sides by the coefficient of the variable, which is 4.

$$\frac{4s}{4} = \frac{44}{4}$$

$$s = \underline{11}$$

The solution is 11.

Check $4s = 44$ *Original equation*

$4 \cdot \underline{11} = 44$ *Replace s with $\underline{11}$.*

$44 = 44$ Balances

(b) $27 = -9p$

$$\frac{27}{-9} = \frac{-9p}{-9} \quad \textit{Divide both sides by } -9.$$

$$-3 = p$$

The solution is -3.

Check $27 = -9p$

$\quad\quad\quad 27 = -9 \cdot (-3) \quad \textit{Replace p with } -3.$

$\quad\quad\quad 27 = 27 \quad\quad\quad\quad \text{Balances}$

(c) $-40 = -5x$

$$\frac{-40}{-5} = \frac{-5x}{-5} \quad \textit{Divide both sides by } -5.$$

$$8 = x$$

The solution is 8.

Check $-40 = -5x$

$\quad\quad\quad -40 = -5 \cdot 8 \quad \textit{Replace x with 8.}$

$\quad\quad\quad -40 = -40 \quad \text{Balances}$

(d) $7t = -70$

$$\frac{7t}{7} = \frac{-70}{7} \quad \textit{Divide both sides by 7.}$$

$$t = -10$$

The solution is -10.

Check $7t = -70$

$\quad\quad 7 \cdot (-10) = -70 \quad \textit{Replace t with } -10.$

$\quad\quad\quad -70 = -70 \quad \text{Balances}$

2.   **(a)** $-28 = -6n + 10n$

$\quad\quad\quad -28 = 4n \quad\quad\quad\quad\quad \textit{Combine like terms.}$

$$\frac{-28}{4} = \frac{4n}{4} \quad\quad \textit{Divide both sides by 4.}$$

$\quad\quad\quad -7 = n \quad\quad\quad\quad \text{The solution is } -7.$

Check $-28 = -6n + 10n$

$\quad\quad -28 = -6 \cdot (-7) + 10 \cdot (-7)$

$\quad\quad\quad\quad\quad\quad\quad \textit{Replace n with } -7.$

$\quad\quad -28 = 42 + (-70)$

$\quad\quad -28 = -28 \quad \text{Balances}$

(b) $p - 14p = -2 + 18 - 3$

$1p + (-14p) = -2 + 18 + (-3)$

$\textit{Change to addition. Rewrite p as 1p.}$

$\quad\quad -13p = 13$

$\quad\quad\quad\quad\quad\quad \textit{Combine like terms.}$

$$\frac{-13p}{-13} = \frac{13}{-13}$$

$\quad\quad\quad\quad\quad \textit{Divide both sides by } -13.$

$\quad\quad\quad p = -1$

The solution is -1.

Check $p - 14p = -2 + 18 - 3$

$\quad\quad -1 - 14(-1) = 16 - 3$

$\quad\quad\quad\quad\quad\quad \textit{Replace p with } -1.$

$\quad\quad -1 - (-14) = 13$

$\quad\quad -1 + (+14) = 13$

$\quad\quad\quad\quad\quad 13 = 13 \quad \text{Balances}$

3.   **(a)** $-k = -12$

$\quad -1k = -12 \quad \textit{Write in the understood } -1 \textit{ as the coefficient of k.}$

$$\frac{-1k}{-1} = \frac{-12}{-1} \quad \textit{Divide both sides by } -1.$$

$\quad\quad\quad k = 12 \quad\quad \text{The solution is 12.}$

Check $-k = -12$

$\quad\quad\quad -1k = -12$

$\quad\quad -1 \cdot 12 = -12 \quad \textit{Replace k with 12.}$

$\quad\quad\quad -12 = -12 \quad \text{Balances}$

(b) $7 = -t$

$\quad\quad 7 = -1t \quad \textit{Write } -t \textit{ as } -1t.$

$$\frac{7}{-1} = \frac{-1t}{-1} \quad \textit{Divide both sides by } -1.$$

$\quad -7 = t \quad\quad \text{The solution is } -7.$

Check $7 = -t$

$\quad\quad\quad 7 = -1t$

$\quad\quad\quad 7 = -1 \cdot (-7) \quad \textit{Replace t with } -7.$

$\quad\quad\quad 7 = 7 \quad\quad\quad\quad \text{Balances}$

(c) $-m = -20$

$\quad -1m = -20 \quad \textit{Write } -m \textit{ as } -1m.$

$$\frac{-1m}{-1} = \frac{-20}{-1} \quad \textit{Divide both sides by } -1.$$

$\quad\quad m = 20 \quad\quad \text{The solution is 20.}$

Check $-m = -20$

$\quad\quad\quad -1m = -20$

$\quad\quad -1 \cdot 20 = -20 \quad \textit{Replace m with 20.}$

$\quad\quad -20 = -20 \quad \text{Balances}$

2.4 Section Exercises

1. $6z = 12$

$$\frac{6z}{6} = \frac{12}{6} \quad \textit{Divide both sides by } \underline{6}.$$

$\quad z = \underline{2} \quad \text{The solution is 2.}$

Check $6z = 12$

$\quad\quad \underline{6 \cdot 2} = 12 \quad \textit{Replace z with 2.}$

$\quad\quad\quad \underline{12} = \underline{12} \quad \text{Balances}$

3. $48 = 12r$

$$\frac{48}{12} = \frac{12r}{12} \quad \textit{Divide both sides by 12.}$$

$\quad 4 = r \quad\quad \text{The solution is 4.}$

Check $48 = 12r$

$\quad\quad\quad 48 = 12 \cdot 4 \quad \textit{Replace r with 4.}$

$\quad\quad\quad 48 = 48 \quad\quad \text{Balances}$

5. $3y = 0$

$$\frac{3y}{3} = \frac{0}{3} \quad \textit{Divide both sides by 3.}$$

$\quad y = 0 \quad \text{The solution is 0.}$

Check $3y = 0$

$\quad\quad\quad 3 \cdot 0 = 0 \quad \textit{Replace y with 0.}$

$\quad\quad\quad\quad 0 = 0 \quad \text{Balances}$

7. $-7k = 70$

$\dfrac{-7k}{-7} = \dfrac{70}{-7}$ *Divide both sides by -7.*

$k = -10$ The solution is -10.

Check $-7k = 70$

$-7 \cdot (-10) = 70$ *Replace k with -10.*

$70 = 70$ Balances

9. $-54 = -9r$

$\dfrac{-54}{-9} = \dfrac{-9r}{-9}$ *Divide both sides by -9.*

$6 = r$ The solution is 6.

Check $-54 = -9r$

$-54 = -9 \cdot 6$ *Replace r with 6.*

$-54 = -54$ Balances

11. $-25 = 5b$

$\dfrac{-25}{5} = \dfrac{5b}{5}$ *Divide both sides by 5.*

$-5 = b$ The solution is -5.

Check $-25 = 5b$

$-25 = 5 \cdot (-5)$ *Replace b with -5.*

$-25 = -25$ Balances

13. $2r = -7 + 13$

$2r = 6$ *Combine like terms.*

$\dfrac{2r}{2} = \dfrac{6}{2}$ *Divide both sides by 2.*

$r = 3$ The solution is 3.

Check $2r = -7 + 13$

$2 \cdot 3 = -7 + 13$ *Replace r with 3.*

$6 = 6$ Balances

15. $-12 = 5p - p$

$-12 = 5p + (-p)$ *Change to addition.*

$-12 = 5p + (-1p)$ *Rewrite $-p$ as $-1p$.*

$-12 = 4p$ *Combine like terms.*

$\dfrac{-12}{4} = \dfrac{4p}{4}$ *Divide both sides by 4.*

$-3 = p$ The solution is -3.

Check

$-12 = 5p - p$

$-12 = 5 \cdot (-3) - (-3)$ *Replace p with -3.*

$-12 = -15 - (-3)$

$-12 = -15 + 3$ *Change to addition.*

$-12 = -12$ Balances

17. $3 - 28 = 5a$ *Original equation*

$3 + (-28) = 5a$ *Change to addition.*

$-25 = 5a$ *Combine like terms.*

$\dfrac{-25}{5} = \dfrac{5a}{5}$ *Divide both sides by 5.*

$-5 = a$ The solution is -5.

19. $x - 9x = 80$ *Original equation*

$x + (-9x) = 80$ *Change to addition.*

$1x + (-9x) = 80$ *Rewrite x as 1x.*

$-8x = 80$ *Combine like terms.*

$\dfrac{-8x}{-8} = \dfrac{80}{-8}$ *Divide both sides by -8.*

$x = -10$ The solution is -10.

21. $13 - 13 = 2w - w$ *Original equation*

$13 + (-13) = 2w + (-w)$ *Change to addition.*

$13 + (-13) = 2w + (-1w)$ *Rewrite $-w$ as $-1w$.*

     Combine like terms.

$0 = 1w$

     1w is the same as w.

$0 = w$ The solution is 0.

23. $3t + 9t = 20 - 10 + 26$ *Original equation*

$3t + 9t = 20 + (-10) + 26$ *Change to addition.*

$12t = 36$ *Combine like terms.*

$\dfrac{12t}{12} = \dfrac{36}{12}$ *Divide both sides by 12.*

$t = 3$ The solution is 3.

25. $0 = -9t$ *Original equation*

$\dfrac{0}{-9} = \dfrac{-9t}{-9}$ *Divide both sides by -9.*

$0 = t$ The solution is 0.

27. $-14m + 8m = 6 - 60$ *Original equation*

$-14m + 8m = 6 + (-60)$ *Change to addition.*

$-6m = -54$ *Combine like terms.*

$\dfrac{-6m}{-6} = \dfrac{-54}{-6}$ *Divide both sides by -6.*

$m = 9$ The solution is 9.

29. $100 - 96 = 31y - 35y$ *Original equation*

$100 + (-96) = 31y + (-35y)$ *Change to addition.*

$4 = -4y$ *Combine like terms.*

$\dfrac{4}{-4} = \dfrac{-4y}{-4}$ *Divide both sides by -4.*

$-1 = y$ The solution is -1.

31. $3(2z) = -30$ *Original equation*

$(3 \cdot 2) \cdot z = -30$ *To multiply on the left, use the associative property.*

$6z = -30$

$\dfrac{6z}{6} = \dfrac{-30}{6}$ *Divide both sides by 6.*

$z = -5$ The solution is -5.

33. $50 = -5(5p)$ *Original equation*

$50 = (-5 \cdot 5) \cdot p$ *To multiply on the right, use the associative prop.*

$50 = -25p$

$\dfrac{50}{-25} = \dfrac{-25p}{-25}$ *Divide both sides by -25.*

$-2 = p$ The solution is -2.

35.
$$-2(-4k) = 56 \quad \text{\textit{Original equation}}$$
$$[-2 \cdot (-4)] \cdot k = 56 \quad \text{\textit{Associative property}}$$
$$8k = 56$$
$$\frac{8k}{8} = \frac{56}{8} \quad \text{\textit{Divide both}}$$
$$\text{\textit{sides by 8.}}$$
$$k = 7 \quad \text{\textit{The solution is 7.}}$$

37.
$$-90 = -10(-3b) \quad \text{\textit{Original equation}}$$
$$-90 = [-10 \cdot (-3)] \cdot b \quad \text{\textit{Associative property}}$$
$$-90 = 30b$$
$$\frac{-90}{30} = \frac{30b}{30} \quad \begin{array}{l}\text{\textit{Divide both}}\\ \text{\textit{sides by 30.}}\end{array}$$
$$-3 = b \quad \text{\textit{The solution is }}-3.$$

39.
$$-x = 32 \quad \text{\textit{Original equation}}$$
$$-1x = 32 \quad \text{\textit{Write in the understood }}-1.$$
$$\frac{-1x}{-1} = \frac{32}{-1} \quad \begin{array}{l}\text{\textit{Divide both}}\\ \text{\textit{sides by }}-1.\end{array}$$
$$x = -32 \quad \text{\textit{The solution is }}-32.$$

41.
$$-2 = -w \quad \text{\textit{Original equation}}$$
$$-2 = -1w \quad \text{\textit{Write in the understood }}-1.$$
$$\frac{-2}{-1} = \frac{-1w}{-1} \quad \begin{array}{l}\text{\textit{Divide both}}\\ \text{\textit{sides by }}-1.\end{array}$$
$$2 = w \quad \text{\textit{The solution is 2.}}$$

43.
$$-n = -50 \quad \text{\textit{Original equation}}$$
$$-1n = -50 \quad \text{\textit{Write in the understood }}-1.$$
$$\frac{-1n}{-1} = \frac{-50}{-1} \quad \begin{array}{l}\text{\textit{Divide both}}\\ \text{\textit{sides by }}-1.\end{array}$$
$$n = 50 \quad \text{\textit{The solution is 50.}}$$

45.
$$10 = -p \quad \text{\textit{Original equation}}$$
$$10 = -1p \quad \text{\textit{Write in the understood }}-1.$$
$$\frac{10}{-1} = \frac{-1p}{-1} \quad \begin{array}{l}\text{\textit{Divide both}}\\ \text{\textit{sides by }}-1.\end{array}$$
$$-10 = p \quad \text{\textit{The solution is }}-10.$$

47. Each solution is the opposite of the number in the equation. So the rule is: When you change the sign of the variable from negative to positive, then change the number in the equation to its opposite. In $-x = 5$, the opposite of 5 is -5, so $x = -5$.

49. Divide by the coefficient of x, which is 3, *not* by the opposite of 3.
$$3x = \underbrace{16 - 1}$$
$$3x = 15$$
$$\frac{3x}{3} = \frac{15}{3}$$
$$x = 5 \quad \text{\textit{The correct solution is 5.}}$$

51.
$$3s = 45$$
$$\frac{3s}{3} = \frac{45}{3} \quad \begin{array}{l}\text{\textit{Divide both}}\\ \text{\textit{sides by 3.}}\end{array}$$
$$s = 15$$

The length of one side is 15 feet.

53.
$$120 = 5s$$
$$\frac{120}{5} = \frac{5s}{5} \quad \begin{array}{l}\text{\textit{Divide both}}\\ \text{\textit{sides by 5.}}\end{array}$$
$$24 = s$$

The length of one side is 24 meters.

55.
$$89 - 116 = -4(-4y) - 9(2y) + y$$
$$89 - 116 = [-4 \cdot (-4)] \cdot y - (9 \cdot 2) \cdot y + y$$
$$\text{\textit{Associative property}}$$
$$89 + (-116) = 16y + (-18y) + 1y$$
$$\text{\textit{Change to addition.}}$$
$$-27 = -1y \quad \text{\textit{Combine like terms.}}$$
$$\frac{-27}{-1} = \frac{-1y}{-1} \quad \begin{array}{l}\text{\textit{Divide both}}\\ \text{\textit{sides by }}-1.\end{array}$$
$$27 = y$$

The solution is 27.

57.
$$-37(14x) + 28(21x) = |72 - 72| + |-166 + 96|$$
$$(-37 \cdot 14) \cdot x + (28 \cdot 21) \cdot x$$
$$= |0| + |-70|$$
$$\text{\textit{Assoc. prop. Simplify within the absolute values.}}$$
$$-518x + 588x = 0 + 70$$
$$\text{\textit{Simplify the absolute values.}}$$
$$70x = 70 \quad \text{\textit{Combine like terms.}}$$
$$\frac{70x}{70} = \frac{70}{70} \quad \begin{array}{l}\text{\textit{Divide both}}\\ \text{\textit{sides by 70.}}\end{array}$$
$$x = 1$$

The solution is 1.

2.5 Solving Equations with Several Steps

2.5 Margin Exercises

1. **(a)**
$$\begin{array}{ll}2r + 7 = 13 & \text{\textit{To get 2r by itself,}}\\ \underline{-7 -7} & \text{\textit{add }}-7\text{\textit{ to both sides.}}\\ \underbrace{2r + 0} = 6 &\\ 2r = 6 &\end{array}$$
$$\frac{2r}{2} = \frac{6}{2} \quad \begin{array}{l}\text{\textit{To solve for r,}}\\ \text{\textit{divide both sides by}}\\ \text{\textit{the coefficient, 2.}}\end{array}$$
$$r = \underline{3} \quad \text{\textit{The solution is 3.}}$$

Check
$$2r + 7 = 13$$
$$2 \cdot 3 + 7 = 13 \quad \text{\textit{Replace r with 3.}}$$
$$\underbrace{6 + 7} = 13$$
$$\underline{13} = 13 \quad \text{\textit{Balances}}$$

(b)

$$-10z - 9 = 11$$
$$-10z + (-9) = 11 \quad \textit{Change to addition.}$$
$$\underline{9 \quad\quad 9} \quad \textit{Add 9 to both sides.}$$
$$-10z + 0 = 20$$
$$-10z = 20$$
$$\frac{-10z}{-10} = \frac{20}{-10} \quad \begin{array}{l}\textit{Divide both}\\ \textit{sides by } -10.\end{array}$$
$$z = -2 \quad \textit{The solution is } -2.$$

Check

$$-10z - 9 = 11$$
$$-10 \cdot (-2) - 9 = 11 \quad \textit{Replace r with } -2.$$
$$\underbrace{20}_{} - 9 = 11$$
$$\underbrace{11}_{} = 11 \quad \textit{Balances}$$

2.   **(a)** Solve, keeping the variable on *left* side.

$$3y - 1 = 2y + 7$$
$$\underline{-2y \quad\quad -2y} \quad \textit{Add } -2y \textit{ to both sides.}$$
$$1y - 1 = 0 + 7$$
$$1y + (-1) = 7 \quad \textit{Change to addition.}$$
$$\underline{1 \quad\quad 1} \quad \textit{Add 1 to both sides.}$$
$$1y + 0 = 8$$
$$1y = 8$$
$$\text{or } y = 8 \quad \textit{The solution is 8.}$$

Solve, keeping the variable on the *right* side.

$$3y - 1 = 2y + 7$$
$$\underline{-3y \quad\quad -3y} \quad \textit{Add } -3y \textit{ to both sides.}$$
$$0 - 1 = -1y + 7$$
$$-1 = -1y + 7$$
$$\underline{-7 \quad\quad\quad -7} \quad \textit{Add } -7 \textit{ to both sides.}$$
$$-8 = -1y + 0$$
$$-8 = -1y$$
$$\frac{-8}{-1} = \frac{-1y}{-1} \quad \begin{array}{l}\textit{Divide both}\\ \textit{sides by } -1.\end{array}$$
$$8 = y \quad \textit{The solution is 8.}$$

(b) Solve, keeping the variable on *left* side.

$$3p - 2 = p - 6$$
$$3p - 2 = 1p - 6 \quad \textit{Rewrite p as 1p.}$$
$$\underline{-1p \quad\quad -1p} \quad \textit{Add } -1p \textit{ to both sides.}$$
$$2p - 2 = 0 - 6$$
$$2p + (-2) = -6 \quad \textit{Change to addition.}$$
$$\underline{2 \quad\quad 2} \quad \textit{Add 2 to both sides.}$$
$$2p + 0 = -4$$
$$2p = -4$$
$$\frac{2p}{2} = \frac{-4}{2} \quad \begin{array}{l}\textit{Divide both}\\ \textit{sides by 2.}\end{array}$$
$$p = -2 \quad \textit{The solution is } -2.$$

Solve, keeping the variable on the *right* side.

$$3p - 2 = p - 6$$
$$3p - 2 = 1p - 6 \quad \textit{Rewrite p as 1p.}$$
$$\underline{-3p \quad\quad -3p} \quad \textit{Add } -3p.$$
$$0 - 2 = -2p - 6$$
$$-2 = -2p + (-6)$$
$$\underline{6 \quad\quad\quad 6} \quad \textit{Add 6.}$$
$$4 = -2p + 0$$
$$4 = -2p$$
$$\frac{4}{-2} = \frac{-2p}{-2} \quad \begin{array}{l}\textit{Divide both}\\ \textit{sides by } -2.\end{array}$$
$$-2 = p \quad \textit{The solution is } -2.$$

3.   **(a)**

$$-12 = 4(y - 1)$$
$$-12 = \underbrace{4 \cdot y} - \underbrace{4 \cdot 1} \quad \textit{Distribute on the right.}$$
$$-12 = 4y - 4$$
$$-12 = 4y + (-4) \quad \textit{Change to addition.}$$
$$\underline{4 \quad\quad\quad 4} \quad \textit{Add 4 to both sides.}$$
$$-8 = 4y + 0$$
$$-8 = 4y$$
$$\frac{-8}{4} = \frac{4y}{4} \quad \begin{array}{l}\textit{Divide both}\\ \textit{sides by } \underline{4}.\end{array}$$
$$-2 = y \quad \textit{The solution is } -2.$$

Check $-12 = 4(y - 1)$
$$-12 = 4(-2 - 1) \quad \textit{Replace y with } -2.$$
$$-12 = 4(-3)$$
$$-12 = -12 \quad \textit{Balances}$$

(b)

$$5(m + 4) = 20$$
$$5 \cdot m + 5 \cdot 4 = 20 \quad \textit{Distribute on the left.}$$
$$5m + 20 = 20$$
$$\underline{-20 \quad\quad -20} \quad \textit{Add } -20 \textit{ to both sides.}$$
$$5m + 0 = 0$$
$$5m = 0$$
$$\frac{5m}{5} = \frac{0}{5} \quad \begin{array}{l}\textit{Divide both}\\ \textit{sides by 5.}\end{array}$$
$$m = 0 \quad \textit{The solution is 0.}$$

Check $5(m + 4) = 20$
$$5(0 + 4) = 20 \quad \textit{Replace m with 0.}$$
$$5(4) = 20$$
$$20 = 20 \quad \textit{Balances}$$

(c)

$$6(t - 2) = 18$$
$$6 \cdot t - 6 \cdot 2 = 18 \quad \textit{Distribute on the left.}$$
$$6t - 12 = 18$$
$$6t + (-12) = 18 \quad \textit{Change to addition.}$$
$$\underline{12 \quad\quad 12} \quad \textit{Add 12 to both sides.}$$
$$6t + 0 = 30$$
$$6t = 30$$
$$\frac{6t}{6} = \frac{30}{6} \quad \begin{array}{l}\textit{Divide both}\\ \textit{sides by 6.}\end{array}$$
$$t = 5 \quad \textit{The solution is 5.}$$

Check $6(t - 2) = 18$
$6(5 - 2) = 18$ *Replace t with 5.*
$6(3) = 18$
$18 = 18$ Balances

4.   **(a)** $3(b + 7) = 2b - 1$ *Distribute.*
$3 \cdot b + 3 \cdot 7 = 2b - 1$
$3b + 21 = 2b + (-1)$ *Variables left*
$\underline{-2b \qquad\quad -2b}$ *Add $-2b$.*
$1b + 21 = 0 + (-1)$
$1b + 21 = -1$
$\underline{-21 \qquad -21}$ *Add -21.*
$1b + 0 = -22$
$1b = -22$
or $b = -22$

The solution is -22.

Check $3(b + 7) = 2b - 1$
$3(-22 + 7) = 2 \cdot (-22) - 1$
$3(-15) = -44 - 1$
$-45 = -45$ Balances

(b) $6 - 2n = 14 + 4(n - 5)$ *Distribute.*
$6 - 2n = 14 + 4 \cdot n - 4 \cdot 5$
$6 - 2n = 14 + 4n - 20$ *Add the opposite.*
$6 + (-2n) = 14 + 4n + (-20)$ *Combine like terms.*
$6 + (-2n) = -6 + 4n$
$\underline{2n \qquad\qquad 2n}$ *Add $2n$.*
$6 + 0 = -6 + 6n$
$6 = -6 + 6n$
$\underline{6 \qquad\quad 6}$ *Add 6.*
$12 = 0 + 6n$
$12 = 6n$
$\dfrac{12}{6} = \dfrac{6n}{6}$ *Divide both sides by 6.*
$2 = n$

The solution is 2.

Check $6 - 2n = 14 + 4(n - 5)$
$6 - 2 \cdot 2 = 14 + 4(2 - 5)$ *Let $n = 2$.*
$6 - 4 = 14 + 4(-3)$
$2 = 14 + (-12)$
$2 = 2$ Balances

2.5   Section Exercises

1. $7p + 5 = 12$ *To get $7p$ by itself,*
$\underline{-5 \qquad -5}$ *add -5 to both sides.*
$7p + 0 = 7$
$7p = 7$
$\dfrac{7p}{7} = \dfrac{7}{7}$ *Divide both sides by 7.*
$p = 1$ The solution is 1.

Check $7p + 5 = 12$
$7(1) + 5 = 12$ *Let $p = 1$.*
$7 + 5 = 12$
$\underline{12 = 12}$ Balances

3. $2 = 8y - 6$
$2 = 8y + (-6)$ *Change to addition.*
$\underline{6 \qquad\qquad 6}$ *Add 6 to both sides.*
$8 = \underbrace{8y + 0}$
$8 = 8y$
$\dfrac{8}{8} = \dfrac{8y}{8}$ *Divide both sides by 8.*
$1 = y$ The solution is 1.

Check $2 = 8y - 6$
$2 = 8(1) - 6$ *Replace y with 1.*
$2 = 8 - 6$
$2 = 2$ Balances

5. $28 = -9a + 10$ *To get $-9a$ by itself,*
$\underline{-10 \qquad\qquad -10}$ *add -10 to both sides.*
$18 = -9a + 0$
$18 = -9a$
$\dfrac{18}{-9} = \dfrac{-9a}{-9}$ *Divide both sides by -9.*
$-2 = a$ The solution is -2.

Check $28 = -9a + 10$
$28 = -9(-2) + 10$ *Replace a with -2.*
$28 = 18 + 10$
$28 = 28$ Balances

7. $-3m + 1 = 1$ *To get $-3m$ by itself,*
$\underline{-1 \qquad -1}$ *add -1 to both sides.*
$-3m + 0 = 0$
$-3m = 0$
$\dfrac{-3m}{-3} = \dfrac{0}{-3}$ *Divide both sides by -3.*
$m = 0$ The solution is 0.

Check $-3m + 1 = 1$
$-3(0) + 1 = 1$ *Replace m with 0.*
$0 + 1 = 1$
$1 = 1$ Balances

9. $-5x - 4 = 16$ *Change to addition.*
$-5x + (-4) = 16$ *To get $-5x$ by itself,*
$\underline{4 \qquad\qquad 4}$ *add 4 to both sides.*
$-5x + 0 = 20$
$-5x = 20$
$\dfrac{-5x}{-5} = \dfrac{20}{-5}$ *Divide both sides by -5.*
$x = -4$ The solution is -4.

Check $-5x - 4 = 16$
$-5(-4) - 4 = 16$ *Replace x with -4.*
$20 - 4 = 16$
$16 = 16$ Balances

11. Solve, keeping the variable on the *left* side.

$$
\begin{aligned}
6p - 2 &= 4p + 6 \\
6p + (-2) &= 4p + 6 \quad \text{\textit{Change to addition.}} \\
\underline{-4p \qquad\quad -4p} &\qquad \text{\textit{Add }} -4p \text{ \textit{to both sides.}} \\
2p + (-2) &= 0 + 6 \\
2p + (-2) &= 6 \\
\underline{\qquad\quad 2 \qquad 2} &\qquad \text{\textit{Add 2 to both sides.}} \\
2p + 0 &= 8 \\
2p &= 8 \\
\frac{2p}{2} &= \frac{8}{2} \qquad \text{\textit{Divide both sides by 2.}} \\
p &= 4 \qquad \text{The solution is 4.}
\end{aligned}
$$

Solve, keeping the variable on the *right* side.

$$
\begin{aligned}
6p - 2 &= 4p + 6 \\
6p + (-2) &= 4p + 6 \quad \text{\textit{Change to addition.}} \\
\underline{-6p \qquad\quad -6p} &\qquad \text{\textit{Add }} -6p \text{ \textit{to both sides.}} \\
0 + (-2) &= -2p + 6 \\
-2 &= -2p + 6 \\
\underline{-6 \qquad\qquad\quad -6} &\quad \text{\textit{Add }} -6 \text{ \textit{to both sides.}} \\
-8 &= -2p + 0 \\
-8 &= -2p \\
\frac{-8}{-2} &= \frac{-2p}{-2} \qquad \text{\textit{Divide both sides by }} -2. \\
4 &= p \qquad \text{The solution is 4.}
\end{aligned}
$$

Check $6p - 2 = 4p + 6$
$$
\begin{aligned}
6(4) - 2 &= 4(4) + 6 \\
24 - 2 &= 16 + 6 \\
22 &= 22 \qquad \text{Balances}
\end{aligned}
$$

13. Solve, keeping the variable on the *left* side.

$$
\begin{aligned}
-2k - 6 &= 6k + 10 \\
-2k + (-6) &= 6k + 10 \quad \text{\textit{Change to addition.}} \\
\underline{-6k \qquad\qquad -6k} &\qquad \text{\textit{Add }} -6k \text{ \textit{to both sides.}} \\
-8k + (-6) &= 0 + 10 \\
-8k + (-6) &= 10 \\
\underline{\qquad\quad 6 \qquad 6} &\qquad \text{\textit{Add 6 to both sides.}} \\
-8k + 0 &= 16 \\
\frac{-8k}{-8} &= \frac{16}{-8} \qquad \text{\textit{Divide both sides by }} -8. \\
k &= -2 \qquad \text{The solution is } -2.
\end{aligned}
$$

Solve, keeping the variable on the *right* side.

$$
\begin{aligned}
-2k - 6 &= 6k + 10 \\
-2k + (-6) &= 6k + 10 \quad \text{\textit{Change to addition.}} \\
\underline{\;2k \qquad\qquad 2k} &\qquad \text{\textit{Add 2k to both sides.}} \\
0 + (-6) &= 8k + 10 \\
-6 &= 8k + 10 \\
\underline{-10 \qquad\qquad -10} &\quad \text{\textit{Add }} -10 \text{ \textit{to both sides.}} \\
-16 &= 8k + 0 \\
\frac{-16}{8} &= \frac{8k}{8} \qquad \text{\textit{Divide both sides by 8.}} \\
-2 &= k \qquad \text{The solution is } -2.
\end{aligned}
$$

Check
$$
\begin{aligned}
-2k - 6 &= 6k + 10 \\
-2(-2) - 6 &= 6(-2) + 10 \quad \text{\textit{Replace k with }} -2. \\
4 + (-6) &= -12 + 10 \\
-2 &= -2 \qquad \text{Balances}
\end{aligned}
$$

15. $-18 + 7a = 2a + 3 + 4$ simplifies to
$-18 + 7a = 2a + 7$.

$$
\begin{aligned}
-18 + 7a &= 2a + 7 \\
\underline{\qquad\quad -2a \qquad -2a} &\qquad \text{\textit{Add }} -2a \text{ \textit{to both sides.}} \\
-18 + 5a &= 0 + 7 \\
-18 + 5a &= 7 \\
\underline{\;18 \qquad\qquad 18} &\qquad \text{\textit{Add 18 to both sides.}} \\
0 + 5a &= 25 \\
5a &= 25 \\
\frac{5a}{5} &= \frac{25}{5} \qquad \text{\textit{Divide both sides by 5.}} \\
a &= 5
\end{aligned}
$$

The solution is 5.

Check $-18 + 7a = 2a + 3 + 4$
$$
\begin{aligned}
-18 + 7(5) &= 2(5) + 7 \\
-18 + 35 &= 10 + 7 \\
17 &= 17 \qquad \text{Balances}
\end{aligned}
$$

17. Neither side can be simplified, so solve the equation.

$$
\begin{aligned}
-3t &= 8t \\
\underline{\;3t \qquad\quad 3t} &\qquad \text{\textit{Add 3t to both sides.}} \\
0 &= 11t \\
\frac{0}{11} &= \frac{11t}{11} \qquad \text{\textit{Divide both sides by 11.}} \\
0 &= t \qquad \text{The solution is 0.}
\end{aligned}
$$

Check $-3t = 8t$
$$
\begin{aligned}
-3(0) &= 8(0) \qquad \text{\textit{Replace t with 0.}} \\
0 &= 0 \qquad \text{Balances}
\end{aligned}
$$

19. $4 + 16 - 2 = 2 - 2b$ simplifies to $18 = 2 - 2b$.

$$
\begin{array}{rcl}
18 & = & 2 - 2b \\
-2 & & -2 \qquad \text{\textit{Add} -2 \textit{to}} \\
\hline
& & \qquad\quad \text{\textit{both sides.}} \\
18 - 2 & = & 0 - 2b \\
16 & = & -2b \\
\dfrac{16}{-2} & = & \dfrac{-2b}{-2} \qquad \text{\textit{Divide both}} \\
& & \qquad\quad \text{\textit{sides by} -2.} \\
-8 & = & b
\end{array}
$$

The solution is 5.

Check $4 + 16 - 2 = 2 - 2b$
$$20 - 2 = 2 - 2(-8)$$
$$20 + (-2) = 2 + 16$$
$$18 = 18 \qquad \text{Balances}$$

21.
$$
\begin{array}{rcl}
8(w - 2) & = & 32 \\
8w - 16 & = & 32 \qquad \textit{Distribute.} \\
8w + (-16) & = & 32 \qquad \textit{Change to addition.} \\
16 & & 16 \qquad \textit{Add 16 to both sides.} \\
\hline
8w + 0 & = & 48 \\
8w & = & 48 \\
\dfrac{8w}{8} & = & \dfrac{48}{8} \qquad \textit{Divide both} \\
& & \qquad\quad \textit{sides by 8.} \\
w & = & 6 \qquad \text{The solution is 6.}
\end{array}
$$

23.
$$
\begin{array}{rcl}
-10 & = & 2(y + 4) \\
-10 & = & 2y + 8 \qquad \textit{Distribute.} \\
-8 & & -8 \qquad \textit{Add} -8 \textit{ to both sides.} \\
\hline
-18 & = & 2y + 0 \\
-18 & = & 2y \\
\dfrac{-18}{2} & = & \dfrac{2y}{2} \qquad \textit{Divide both} \\
& & \qquad\quad \textit{sides by 2.} \\
-9 & = & y \qquad \text{The solution is } -9.
\end{array}
$$

25.
$$
\begin{array}{rcl}
-4(t + 2) & = & 12 \\
-4t + (-8) & = & 12 \qquad \textit{Distribute.} \\
8 & & 8 \qquad \textit{Add 8 to both sides.} \\
\hline
-4t + 0 & = & 20 \\
-4t & = & 20 \\
\dfrac{-4t}{-4} & = & \dfrac{20}{-4} \qquad \textit{Divide both} \\
& & \qquad\quad \textit{sides by} -4. \\
t & = & -5 \qquad \text{The solution is } -5.
\end{array}
$$

27.
$$
\begin{array}{rcl}
6(x - 5) & = & -30 \\
6x - 30 & = & -30 \qquad \textit{Distribute.} \\
6x + (-30) & = & -30 \qquad \textit{Change to addition.} \\
30 & & 30 \qquad \textit{Add 30 to both sides.} \\
\hline
6x + 0 & = & 0 \\
6x & = & 0 \\
\dfrac{6x}{6} & = & \dfrac{0}{6} \qquad \textit{Divide both} \\
& & \qquad\quad \textit{sides by 6.} \\
x & = & 0 \qquad \text{The solution is 0.}
\end{array}
$$

29.
$$
\begin{array}{rcl}
-12 & = & 12(h - 2) \\
-12 & = & 12h - 24 \qquad \textit{Distribute.} \\
-12 & = & 12h + (-24) \qquad \textit{Change to addition.} \\
24 & & 24 \qquad \textit{Add 24 to both sides.} \\
\hline
12 & = & 12h + 0 \\
12 & = & 12h \\
\dfrac{12}{12} & = & \dfrac{12h}{12} \qquad \textit{Divide both} \\
& & \qquad\quad \textit{sides by 12.} \\
1 & = & h \qquad \text{The solution is 1.}
\end{array}
$$

31.
$$
\begin{array}{rcl}
0 & = & -2(y + 2) \\
0 & = & -2y - 4 \qquad \textit{Distribute.} \\
0 & = & -2y + (-4) \qquad \textit{Change to addition.} \\
4 & & 4 \qquad \textit{Add 4 to both sides.} \\
\hline
4 & = & -2y + 0 \\
4 & = & -2y \\
\dfrac{4}{-2} & = & \dfrac{-2y}{-2} \qquad \textit{Divide both} \\
& & \qquad\quad \textit{sides by} -2. \\
-2 & = & y \qquad \text{The solution is } -2.
\end{array}
$$

33.
$$
\begin{array}{rcl}
6m + 18 & = & 0 \\
-18 & & -18 \qquad \textit{Add} -18 \textit{ to both sides.} \\
\hline
6m + 0 & = & -18 \\
6m & = & -18 \\
\dfrac{6m}{6} & = & \dfrac{-18}{6} \qquad \textit{Divide both} \\
& & \qquad\quad \textit{sides by 6.} \\
m & = & -3 \qquad \text{The solution is } -3.
\end{array}
$$

35.
$$
\begin{array}{rcl}
6 & = & 9w - 12 \\
6 & = & 9w + (-12) \qquad \textit{Change to addition.} \\
12 & & 12 \qquad \textit{Add 12 to both sides.} \\
\hline
18 & = & 9w + 0 \\
18 & = & 9w \\
\dfrac{18}{9} & = & \dfrac{9w}{9} \qquad \textit{Divide both} \\
& & \qquad\quad \textit{sides by 9.} \\
2 & = & w \qquad \text{The solution is 2.}
\end{array}
$$

37.
$$
\begin{array}{rcl}
5x & = & 3x + 10 \\
-3x & & -3x \qquad \textit{Add} -3x \textit{ to both sides.} \\
\hline
2x & = & 0 + 10 \\
2x & = & 10 \\
\dfrac{2x}{2} & = & \dfrac{10}{2} \qquad \textit{Divide both} \\
& & \qquad\quad \textit{sides by 2.} \\
x & = & 5 \qquad \text{The solution is 5.}
\end{array}
$$

39.
$$
\begin{array}{rcl}
2a + 11 & = & 8a - 7 \\
2a + 11 & = & 8a + (-7) \qquad \textit{Change to addition.} \\
-2a & & -2a \qquad \textit{Add} -2a. \\
\hline
0 + 11 & = & 6a + (-7) \\
11 & = & 6a + (-7) \\
7 & & 7 \qquad \textit{Add 7 to both sides.} \\
\hline
18 & = & 6a + 0 \\
18 & = & 6a \\
\dfrac{18}{6} & = & \dfrac{6a}{6} \qquad \textit{Divide both} \\
& & \qquad\quad \textit{sides by 6.} \\
3 & = & a \qquad \text{The solution is 3.}
\end{array}
$$

41.
$$
\begin{aligned}
7 - 5b &= 28 + 2b \\
7 + (-5b) &= 28 + 2b \quad \text{Change to addition.} \\
\underline{ 5b } &\quad \underline{ 5b} \quad \text{Add } 5b \text{ to both sides.} \\
7 + 0 &= 28 + 7b \\
7 &= 28 + 7b \\
\underline{-28} &\quad \underline{-28} \quad \text{Add } -28 \text{ to both sides.} \\
-21 &= 0 + 7b \\
-21 &= 7b \\
\frac{-21}{7} &= \frac{7b}{7} \quad \text{Divide both sides by 7.} \\
-3 &= b \quad \text{The solution is } -3.
\end{aligned}
$$

43.
$$
\begin{aligned}
-20 + 2k &= k - 4k \\
-20 + 2k &= k + (-4k) \quad \text{Change to addition.} \\
-20 + 2k &= -3k \quad \text{Combine like terms.} \\
\underline{-2k} &\quad \underline{-2k} \quad \text{Add } -2k. \\
-20 + 0 &= -5k \\
-20 &= -5k \\
\frac{-20}{-5} &= \frac{-5k}{-5} \quad \text{Divide both sides by } -5. \\
4 &= k \quad \text{The solution is 4.}
\end{aligned}
$$

45.
$$
\begin{aligned}
10(c - 6) + 4 &= 2 + c - 58 \\
10c - 60 + 4 &= 2 + c - 58 \quad \text{Distribute.} \\
10c + (-60) + 4 &= 2 + c + (-58) \quad \text{Change to add.} \\
10c + (-60) + 4 &= 2 + (-58) + c \quad \text{Group terms.} \\
10c + (-56) &= -56 + c \quad \text{Combine terms.} \\
\underline{-c} &\quad \underline{-c} \quad \text{Add } -c. \\
9c + (-56) &= -56 + 0 \\
9c + (-56) &= -56 \quad \text{Add 56.} \\
\underline{56} &\quad \underline{56} \\
9c + 0 &= 0 \\
\frac{9c}{9} &= \frac{0}{9} \quad \text{Divide both sides by 9.} \\
c &= 0
\end{aligned}
$$
The solution is 0.

47.
$$
\begin{aligned}
-18 + 13y + 3 &= 3(5y - 1) - 2 \\
-18 + 13y + 3 &= 15y - 3 - 2 \quad \text{Distribute.} \\
-18 + 13y + 3 &= 15y + (-3) + (-2) \quad \text{Add the opposites.} \\
13y + (-18) + 3 &= 15y + (-3) + (-2) \quad \text{Group like terms.} \\
13y + (-15) &= 15y + (-5) \quad \text{Combine like terms.} \\
\underline{-13y} &\quad \underline{-13y} \quad \text{Add } -13y. \\
0 + (-15) &= 2y + (-5) \\
-15 &= 2y + (-5) \\
\underline{5} &\quad \underline{5} \quad \text{Add 5.} \\
-10 &= 2y + 0 \\
-10 &= 2y \\
\frac{-10}{2} &= \frac{2y}{2} \quad \text{Divide by 2.} \\
-5 &= y
\end{aligned}
$$

The solution is -5.

49.
$$
\begin{aligned}
6 - 4n + 3n &= 20 - 35 \\
6 + (-4n) + 3n &= 20 + (-35) \quad \text{Change to add.} \\
6 + (-1n) &= -15 \quad \text{Combine terms.} \\
\underline{-6} &\quad \underline{-6} \quad \text{Add } -6. \\
0 + (-1n) &= -21 \\
\frac{-1n}{-1} &= \frac{-21}{-1} \quad \text{Divide both sides by } -1. \\
n &= 21
\end{aligned}
$$

The solution is 21.

51.
$$
\begin{aligned}
6(c - 2) &= 7(c - 6) \\
6c - 12 &= 7c - 42 \quad \text{Distribute.} \\
6c + (-12) &= 7c + (-42) \quad \text{Change to add.} \\
\underline{-6c} &\quad \underline{-6c} \quad \text{Add } -6c. \\
0 + (-12) &= 1c + (-42) \\
-12 &= 1c + (-42) \\
\underline{42} &\quad \underline{42} \quad \text{Add 42.} \\
30 &= 1c + 0 \\
30 &= c
\end{aligned}
$$

The solution is 30.

53.
$$
\begin{aligned}
-5(2p + 2) - 7 &= 3(2p + 5) \\
-10p + (-10) - 7 &= 6p + 15 \\
-10p + (-10) + (-7) &= 6p + 15 \\
-10p + (-17) &= 6p + 15 \\
\underline{-6p} &\quad \underline{-6p} \quad \text{Add } -6p. \\
-16p + (-17) &= 0 + 15 \\
-16p + (-17) &= 15 \\
\underline{17} &\quad \underline{17} \quad \text{Add 17.} \\
-16p + 0 &= 32 \\
\frac{-16p}{-16} &= \frac{32}{-16} \quad \text{Divide by } -16. \\
p &= -2
\end{aligned}
$$

The solution is -2.

55.
$$
\begin{aligned}
2(3b - 2) - 5b &= 4(b - 1) + 8b \\
6b - 4 - 5b &= 4b - 4 + 8b \\
b - 4 &= 12b - 4 \\
\underline{4} &\quad \underline{4} \quad \text{Add 4.} \\
b &= 12b \\
\underline{-b} &\quad \underline{-b} \quad \text{Add } -b. \\
0 &\quad 11b \\
\frac{0}{11} &= \frac{11b}{11} \\
0 &= b
\end{aligned}
$$

The solution is 0.

57. The series of steps may vary. One possibility is:

$$
\begin{array}{rcll}
-2t - 10 & = & 3t + 5 & \textit{Change to addition.}\\
-2t + (-10) & = & 3t + 5 & \textit{Add 2t to both sides}\\
\underline{2t} & & \underline{2t} & \textit{(addition property).}\\
0 + (-10) & = & 5t + 5 & \textit{Add }-5\textit{ to both sides}\\
\underline{-5} & = & \underline{-5} & \textit{(addition property).}\\
\dfrac{-15}{5} & = & \dfrac{5t}{5} & \textit{Divide both sides by 5}\\
& & & \textit{(division property).}\\
-3 & = & t &
\end{array}
$$

The solution is -3.

59. **Check** $\quad -8 + 4a = 2a + 2$

$$
\begin{aligned}
-8 + 4(3) &= 2(3) + 2\\
-8 + 12 &= 6 + 2\\
4 &\neq 8
\end{aligned}
$$

The check does not balance, so 3 is not the correct solution. The student added $-2a$ to -8 on the left side, instead of adding $-2a$ to $4a$. The correct solution, obtained using $-8 + 2a = 2$, $2a = 10$, is $a = 5$.

Relating Concepts (Exercises 61–64)

61. **(a)** It must be negative, because the sum of two positive numbers is always positive.

   (b) The sum of x and a positive number is negative, so x must be negative.

62. **(a)** It must be positive, because the sum of two negative numbers is always negative.

   (b) The sum of d and a negative number is positive, so d must be positive.

63. **(a)** It must be positive. When the signs are the same, the product is positive, and when the signs are different, the product is negative.

   (b) The product of n and a negative number is negative, so n must be positive.

64. **(a)** It must be negative also. When the signs are different, the product is negative, and when the signs match, the product is positive.

   (b) The product of y and a negative number is positive, so y must be negative.

Chapter 2 Review Exercises

1. **(a)** In the expression $-3 + 4k$, k is the variable, 4 is the coefficient, and -3 is the constant term.

   (b) The term that has 20 as the constant term and -9 as the coefficient is $-9y + 20$.

2. **(a)** Evaluate $4c + 10$ when c is 15.

$$
\begin{aligned}
&4c + 10\\
&\underline{4 \cdot 15} + 10 \quad \textit{Replace c with 15.}\\
&\underline{60 + 10}\\
&70 \qquad\qquad \textit{Order 70 test tubes.}
\end{aligned}
$$

(b) Evaluate $4c + 10$ when c is 24.

$$
\begin{aligned}
&4c + 10\\
&\underline{4 \cdot 24} + 10 \quad \textit{Replace c with 24.}\\
&\underline{96 + 10}\\
&106 \qquad\qquad \textit{Order 106 test tubes.}
\end{aligned}
$$

3. **(a)** $x^2 y^4$ means $x \cdot x \cdot y \cdot y \cdot y \cdot y$

   (b) $5ab^3$ means $5 \cdot a \cdot b \cdot b \cdot b$

4. **(a)** n^2 means

$$
\begin{aligned}
&n \cdot n\\
&\underbrace{-3 \cdot (-3)}_{9} \quad \textit{Replace n with }-3.
\end{aligned}
$$

(b) n^3 means

$$
\begin{aligned}
&n \cdot n \cdot n\\
&\underbrace{-3 \cdot (-3)}_{} \cdot (-3) \quad \textit{Replace n with }-3.\\
&\underbrace{9 \cdot (-3)}_{}\\
&-27
\end{aligned}
$$

(c) $-4mp^2$ means

$$
\begin{aligned}
&-4 \cdot m \cdot p \cdot p\\
&\underbrace{-4 \cdot 2}_{} \cdot 4 \cdot 4 \quad \textit{Replace m with 2}\\
&\textit{and p with 4.}\\
&\underbrace{-8 \cdot 4}_{} \cdot 4\\
&\underbrace{-32 \cdot 4}_{}\\
&-128
\end{aligned}
$$

(d) $5m^4 n^2$ means

$$
\begin{aligned}
&5 \cdot m \cdot m \cdot m \cdot m \cdot n \cdot n\\
&\underline{5 \cdot 2} \cdot 2 \cdot 2 \cdot 2 \cdot (-3) \cdot (-3) \quad \textit{Replace m with 2}\\
&\textit{and n with }-3.\\
&\underline{10 \cdot 2} \cdot 2 \cdot 2 \cdot (-3) \cdot (-3)\\
&\underline{20 \cdot 2} \cdot 2 \cdot (-3) \cdot (-3)\\
&\underline{40 \cdot 2} \cdot (-3) \cdot (-3)\\
&\underline{80 \cdot (-3)} \cdot (-3)\\
&\underline{-240 \cdot (-3)}\\
&720
\end{aligned}
$$

5.
$$
\begin{aligned}
&ab + ab^2 + 2ab\\
&\underline{1ab} + ab^2 + \underline{2ab} \quad \textit{Combine like terms.}\\
&3ab + ab^2 \quad \text{or} \quad ab^2 + 3ab
\end{aligned}
$$

6.
$$
\begin{array}{ll}
-3x + 2y - x - 7 &\\
-3x + 2y - 1x - 7 & \textit{Rewrite x as 1x.}\\
-3x + 2y + (-1x) + (-7) & \textit{Change to addition.}\\
-4x + 2y - 7 & \textit{Combine like terms.}
\end{array}
$$

7. $-8(-2g^3)$ *Associative property*
$[-8 \cdot (-2)] \cdot g^3$
$16 \cdot g^3$
$16g^3$

8. $4(3r^2t)$ *Associative property*
$(4 \cdot 3) \cdot r^2t$
$12 \cdot r^2t$
$12r^2t$

9. $5(k + 2)$ *Distribute.*
$5 \cdot k + 5 \cdot 2$
$5k + 10$

10. $-2(3b + 4)$ *Distribute.*
$-2 \cdot 3b + (-2) \cdot 4$
$-6b + (-8)$ or $-6b - 8$

11. $3(2y - 4) + 12$ *Distribute.*
$\underbrace{3 \cdot 2y - 3 \cdot 4} + 12$
$6y - \quad 12 + 12$
$6y + (-12) + 12$
$6y + 0$
$6y$

12. $-4 + 6(4x + 1) - 4x$ *Distribute.*
$-4 + 24x + 6 - 4x$
$-4 + 24x + 6 + (-4x)$
$2 + 20x$ or $20x + 2$

13.   Expressions will vary. One possibility is
$6a^3 + a^2 + 3a - 6$.

14.
$$\begin{array}{rcl} 16 + n & = & 5 \\ -16 & & -16 \\ \hline 0 + n & = & -11 \\ n & = & -11 \end{array}$$
Add -16 to both sides.

The solution is -11.

Check $16 + n = 5$
$16 + (-11) = 5$ *Replace n with -11.*
$5 = 5$ *Balances*

15.
$$\begin{array}{rcl} -4 + 2 & = & 2a - 6 - a \\ -4 + 2 & = & 2a + (-6) + (-1a) \\ -2 & = & 1a + (-6) \\ 6 & & 6 \\ \hline 4 & = & 1a + 0 \\ 4 & = & a \end{array}$$

The solution is 4.

Check $-4 + 2 = 2a - 6 - a$
$-4 + 2 = 2(4) - 6 - 4$
$-2 = 8 + (-6) + (-4)$
$-2 = 2 + (-4)$
$-2 = -2$ Balances

16. $48 = -6m$
$\dfrac{48}{-6} = \dfrac{-6m}{-6}$ *Divide both sides by -6.*
$-8 = m$ The solution is -8.

17.
$$\begin{array}{rl} k - 5k &= -40 \\ 1k - 5k &= -40 \\ 1k + (-5k) &= -40 \\ -4k &= -40 \end{array}$$
Combine like terms.
$\dfrac{-4k}{-4} = \dfrac{-40}{-4}$ *Divide both sides by -4.*
$k = 10$ The solution is 10.

18.
$\underbrace{-17 + 11 + 6} = 7t$
$0 \qquad\quad = 7t$
$\dfrac{0}{7} = \dfrac{7t}{7}$ *Divide both sides by 7.*
$0 = t$ The solution is 0.

19. $-2p + 5p = 3 - 21$
$-2p + 5p = 3 + (-21)$
$3p = -18$
$\dfrac{3p}{3} = \dfrac{-18}{3}$ *Divide both sides by 3.*
$p = -6$ The solution is -6.

20. $-30 = 3(-5r)$
$-30 = -15r$
$\dfrac{-30}{-15} = \dfrac{-15r}{-15}$ *Divide both sides by -15.*
$2 = r$ The solution is 2.

21. $12 = -h$
$12 = -1h$
$\dfrac{12}{-1} = \dfrac{-1h}{-1}$ *Divide both sides by -1.*
$-12 = h$ The solution is -12.

22.
$$\begin{array}{rcl} 12w - 4 & = & 8w + 12 \\ 12w + (-4) & = & 8w + 12 \\ -8w & & -8w \\ \hline 4w + (-4) & = & 0 + 12 \\ 4w + (-4) & = & 12 \\ 4 & & 4 \\ \hline 4w + 0 & = & 16 \\ 4w & = & 16 \end{array}$$
Add $-8w$ to both sides.
Add 4 to both sides.
$\dfrac{4w}{4} = \dfrac{16}{4}$ *Divide both sides by 4.*
$w = 4$

The solution is 4.

23.
$$0 = -4(c + 2)$$
$$0 = -4 \cdot c + (-4) \cdot 2 \quad \textit{Distribute.}$$
$$0 = -4c + (-8)$$
$$\underline{\begin{array}{cc} 8 & 8 \end{array}} \quad \textit{Add 8 to both sides.}$$
$$8 = -4c + 0$$
$$8 = -4c$$
$$\frac{8}{-4} = \frac{-4c}{-4} \quad \textit{Divide both sides by -4.}$$
$$-2 = c \quad \text{The solution is } -2.$$

24.
$$34 = 2n + 4$$
$$\underline{\begin{array}{cc} -4 & -4 \end{array}} \quad \textit{Add -4 to both sides.}$$
$$30 = 2n + 0$$
$$30 = 2n$$
$$\frac{30}{2} = \frac{2n}{2} \quad \textit{Divide both sides by 2.}$$
$$15 = n$$

The number of employees is 15.

25. **[2.5]**
$$12 + 7a = 4a - 3$$
$$\underline{\begin{array}{cc} -4a & -4a \end{array}} \quad \textit{Add -4a to both sides.}$$
$$12 + 3a = 0 - 3$$
$$12 + 3a = -3$$
$$\underline{\begin{array}{cc} -12 & -12 \end{array}}$$
$$0 + 3a = -15$$
$$\frac{3a}{3} = \frac{-15}{3} \quad \textit{Divide both sides by 3.}$$
$$a = -5 \quad \text{The solution is } -5.$$

26. **[2.5]**
$$-2(p - 3) = -14$$
$$-2p + 6 = -14 \quad \textit{Distribute.}$$
$$\underline{\begin{array}{cc} -6 & -6 \end{array}} \quad \textit{Add -6 to both sides.}$$
$$-2p + 0 = -20$$
$$\frac{-2p}{-2} = \frac{-20}{-2} \quad \textit{Divide both sides by -2.}$$
$$p = 10 \quad \text{The solution is } 10.$$

27. **[2.5]**
$$10y = 6y + 20$$
$$\underline{\begin{array}{cc} -6y & -6y \end{array}} \quad \textit{Add -6y to both sides.}$$
$$4y = 0 + 20$$
$$\frac{4y}{4} = \frac{20}{4} \quad \textit{Divide both sides by 4.}$$
$$y = 5 \quad \text{The solution is } 5.$$

28. **[2.5]**
$$2m - 7m = 5 - 20$$
$$2m + (-7m) = 5 + (-20) \quad \textit{Add the opposites.}$$
$$\quad \textit{Combine like terms.}$$
$$-5m = -15$$

$$\frac{-5m}{-5} = \frac{-15}{-5} \quad \textit{Divide both sides by -5.}$$
$$m = 3$$

The solution is 3.

29. **[2.5]**
$$20 = 3x - 7$$
$$20 = 3x + (-7)$$
$$\underline{\begin{array}{cc} 7 & 7 \end{array}} \quad \textit{Add 7 to both sides.}$$
$$27 = 3x + 0$$
$$\frac{27}{3} = \frac{3x}{3} \quad \textit{Divide both sides by 3.}$$
$$9 = x \quad \text{The solution is } 9.$$

30. **[2.5]**
$$b + 6 = 3b - 8$$
$$\underline{\begin{array}{cc} -3b & -3b \end{array}} \quad \textit{Add -3b to both sides.}$$
$$-2b + 6 = 0 - 8$$
$$-2b + 6 = -8$$
$$\underline{\begin{array}{cc} -6 & -6 \end{array}} \quad \textit{Add -6 to both sides.}$$
$$-2b + 0 = -14$$
$$\frac{-2b}{-2} = \frac{-14}{-2} \quad \textit{Divide both sides by -2.}$$
$$b = 7 \quad \text{The solution is } 7.$$

31. **[2.3]**
$$z + 3 = 0$$
$$\underline{\begin{array}{cc} -3 & -3 \end{array}} \quad \textit{Add -3 to both sides.}$$
$$z + 0 = -3$$
$$z = -3 \quad \text{The solution is } -3.$$

32. **[2.5]**
$$3(2n - 1) = 3(n + 3)$$
$$6n - 3 = 3n + 9 \quad \textit{Distribute.}$$
$$\underline{\begin{array}{cc} -3n & -3n \end{array}} \quad \textit{Add -3n to both sides.}$$
$$3n - 3 = 0 + 9$$
$$3n - 3 = 9$$
$$\underline{\begin{array}{cc} 3 & 3 \end{array}} \quad \textit{Add 3 to both sides.}$$
$$3n + 0 = 12$$
$$\frac{3n}{3} = \frac{12}{3} \quad \textit{Divide both sides by 3.}$$
$$n = 4 \quad \text{The solution is } 4.$$

33. **[2.5]**
$$-4 + 46 = 7(-3t + 6)$$
$$-4 + 46 = -21t + 42 \quad \textit{Distribute.}$$
$$42 = -21t + 42$$
$$\underline{\begin{array}{cc} -42 & -42 \end{array}} \quad \textit{Add -42 to both sides.}$$
$$0 = -21t + 0$$
$$\frac{0}{-21} = \frac{-21t}{-21} \quad \textit{Divide both sides by -21.}$$
$$0 = t \quad \text{The solution is } 0.$$

34. [2.5]

$$
\begin{aligned}
6 + 10d - 19 &= 2(3d + 4) - 1 \\
6 + 10d + (-19) &= 6d + 8 - 1 \\
-13 + 10d &= 6d + 7 \\
\underline{ -6d} \quad \underline{-6d} \quad & \qquad Add\ -6d. \\
-13 + 4d &= 0 + 7 \\
-13 + 4d &= 7 \\
\underline{ 13} \quad \underline{13} \quad & \qquad Add\ 13\ to \\
& \qquad both\ sides. \\
0 + 4d &= 20 \\
\dfrac{4d}{4} &= \dfrac{20}{4} \qquad Divide \\
& \qquad by\ 4. \\
d &= 5
\end{aligned}
$$

The solution is 5.

35. [2.5]

$$
\begin{aligned}
-4(3b + 9) &= 24 + 3(2b - 8) \\
-12b - 36 &= 24 + 6b - 24 \\
-12b + (-36) &= 24 + 6b + (-24) \\
-12b + (-36) &= 6b \\
\underline{ 12b} \quad \underline{12b} \quad & \qquad Add\ 12b\ to \\
& \qquad both\ sides. \\
0 + (-36) &= 18b \\
\dfrac{-36}{18} &= \dfrac{18b}{18} \qquad Divide \\
& \qquad by\ 18. \\
-2 &= b
\end{aligned}
$$

The solution is -2.

Chapter 2 Test

1. In the expression $-7w + 6$, -7 is the coefficient, w is the variable, and 6 is the constant term.

2. Evaluate the expression $3a + 2c$ when a is 45 and c is 21.

$$
\begin{aligned}
&3a + 2c \\
&\underline{3 \cdot 45} + \underline{2 \cdot 21} \\
&\ 135\ +\ \ 42 \\
&\qquad 177
\end{aligned}
$$

Buy 177 hot dogs.

3. $x^5 y^3$ means $x \cdot x \cdot x \cdot x \cdot x \cdot y \cdot y \cdot y$

4. $4ab^4$ means $4 \cdot a \cdot b \cdot b \cdot b \cdot b$

5. $-2s^2 t$ means

$$
\begin{aligned}
&-2 \cdot s \cdot s \cdot t \\
&\underline{-2 \cdot (-5)} \cdot (-5) \cdot 4 \qquad \textit{Replace s with } -5 \\
&\qquad\qquad\qquad\qquad \textit{and t with 4.} \\
&\quad \underline{10 \cdot (-5)} \cdot 4 \\
&\qquad \underline{-50 \cdot 4} \\
&\qquad -200
\end{aligned}
$$

6.

$$
\begin{aligned}
&3w^3 - 8w^3 + w^3 \\
&3w^3 - 8w^3 + 1w^3 \\
&\underline{3w^3 + (-8w^3)} + 1w^3 \\
&\quad \underline{-5w^3 + 1w^3} \\
&\qquad -4w^3
\end{aligned}
$$

7.

$$
\begin{aligned}
&xy - xy \\
&1xy - 1xy \\
&(1 - 1)xy \\
&\quad 0xy \\
&\quad\ 0
\end{aligned}
$$

8.

$$
\begin{aligned}
&-6c - 5 + 7c + 5 \\
&-6c + (-5) + 7c + 5 \\
&\underline{-6c + 7c} + \underline{(-5) + 5} \\
&\quad 1c \quad + \quad\ 0 \\
&\quad 1c \quad or \quad c
\end{aligned}
$$

9. $3m^2 - 3m + 3mn$
There are no like terms.
The expression cannot be simplified.

10. $-10(4b^2)$

$$(-10 \cdot 4) \cdot b^2 \qquad \textit{Associative property}$$
$$\textit{of multiplication}$$
$$-40b^2$$

11. $-5(-3k)$

$$[-5 \cdot (-3)] \cdot k \qquad \textit{Associative property}$$
$$\textit{of multiplication}$$
$$15k$$

12. $7(3t + 4)$

$$7(3t) + 7(4) \qquad \textit{Distributive property}$$
$$21t + 28$$

13. $-4(a + 6)$

$$-4 \cdot a + (-4) \cdot 6 \qquad \textit{Distributive property}$$
$$-4a + (-24)$$
$$-4a - 24$$

14. $-8 + 6(x - 2) + 5$

$$-8 + 6x - 12 + 5 \qquad \textit{Distributive property}$$
$$-8 + 6x + (-12) + 5$$
$$6x + (-15) \qquad \textit{Combine like terms.}$$
$$\text{or} \quad 6x - 15$$

15. $-9b - c - 3 + 9 + 2c$

$$-9b - 1c - 3 + 9 + 2c$$
$$-9b + (-1c) + (-3) + 9 + 2c$$
$$-9b + c + 6 \qquad \begin{array}{l}\textit{Combine}\\\textit{like terms.}\end{array}$$

16.

$$
\begin{aligned}
-4 &= x - 9 \\
\underline{ 9} \quad & \qquad \underline{9} \qquad Add\ 9\ to\ both\ sides. \\
5 &= x + 0 \\
5 &= x
\end{aligned}
$$

The solution is 5.

Check $-4 = x - 9$
$-4 = 5 - 9$ *Replace x with 5.*
$-4 = -4$ Balances

17. $-7w = 77$
$$\frac{-7w}{-7} = \frac{77}{-7}$$ *Divide both sides by -7.*
$w = -11$

The solution is -11.

Check $-7w = 77$
$-7 \cdot (-11) = 77$ *Replace w with -11.*
$77 = 77$ Balances

18. $-p = 14$
$-1p = 14$
$$\frac{-1p}{-1} = \frac{14}{-1}$$ *Divide both sides by -1.*
$p = -14$

The solution is -14.

Check $-p = 14$
$-1p = 14$
$-1 \cdot (-14) = 14$ *Replace p with -14.*
$14 = 14$ Balances

19. $-15 = -3(a + 2)$
$-15 = -3a - 6$
$\underline{ 6 6}$ *Add 6 to both sides.*
$-9 = -3a$
$$\frac{-9}{-3} = \frac{-3a}{-3}$$ *Divide both sides by -3.*
$3 = a$

The solution is 3.

Check $-15 = -3(a + 2)$
$-15 = -3(3 + 2)$ *Replace a with 3.*
$-15 = -3(5)$
$-15 = -15$ Balances

20. $6n + 8 - 5n = -4 + 4$
$6n + 8 + (-5n) = 0$
$n + 8 = 0$
$\underline{ -8 -8}$ *Add -8.*
$n = -8$

The solution is -8.

21. $5 - 20 = 2m - 3m$
$5 + (-20) = 2m + (-3m)$
$-15 = -1m$
$$\frac{-15}{-1} = \frac{-1m}{-1}$$ *Divide both sides by -1.*
$15 = m$

The solution is 15.

22. $-2x + 2 = 5x + 9$
$\underline{ 2x 2x}$ *Add $2x$ to both sides.*
$2 = 7x + 9$
$\underline{ -9 -9}$ *Add -9 to both sides.*
$$\frac{-7}{7} = \frac{7x}{7}$$ *Divide both sides by 7.*
$-1 = x$

The solution is -1.

23. $3m - 5 = 7m - 13$
$\underline{ -3m -3m}$ *Add $-3m$ to both sides.*
$0 - 5 = 4m - 13$
$-5 = 4m - 13$
$\underline{ 13 13}$ *Add 13 to both sides.*
$$\frac{8}{4} = \frac{4m}{4}$$ *Divide both sides by 4.*
$2 = m$

The solution is 2.

24. $2 + 7b - 44 = -3b + 12 + 9b$
$7b - 42 = 6b + 12$
$\underline{ -6b -6b}$ *Add $-6b$ to both sides.*
$1b - 42 = 12$
$\underline{ 42 42}$ *Add 42 to both sides.*
$1b = 54$
$b = 54$

The solution is 54.

25. $3c - 24 = 6(c - 4)$
$3c - 24 = 6c - 24$ *Distribute.*
$\underline{-3c -3c}$
$-24 = 3c - 24$
$\underline{ 24 24}$ *Add 24 to both sides.*
$0 = 3c$
$$\frac{0}{3} = \frac{3c}{3}$$ *Divide both sides by 3.*
$0 = c$

The solution is 0.

26.   *Addition property of equality:* Start with a possible solution, for example, $x = -4$. Now add an abitrary number, say -5, to both sides, to give us the equation $x - 5 = -9$.

Division property of equality: Start with a possible solution, for example, $-4 = y$. Now multiply both sides by an abitrary number, say 6, to give us the equation $-24 = 6y$.

Thus, equations will vary. Two possibilities are

$$x - 5 = -9 \quad \text{and} \quad -24 = 6y.$$

Solving:

$$x - 5 = -9$$
$$\underline{5 \qquad 5} \quad \textit{Add 5 to both sides.}$$
$$x = -4$$
$$-24 = 6y$$
$$\frac{-24}{6} = \frac{6y}{6} \quad \textit{Divide both sides by 6.}$$
$$-4 = y$$

CHAPTER 3 SOLVING APPLICATION PROBLEMS

3.1 Problem Solving: Perimeter

3.1 Margin Exercises

1. **(a)** $P = 4s$ *Perimeter formula for a square*

    $P = 4 \cdot 20$ in.   *Replace s, side length, with 20 in.*

    $P = \underline{80 \text{ in.}}$

    (b)

    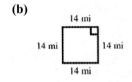

    $P = 4s$ *Perimeter formula for a square*

    $P = 4 \cdot 14$ mi   *Replace s with 14 mi.*

    $P = 56$ mi

2. **(a)** $P = 4s$ *Perimeter formula for a square*

    $28 \text{ in.} = 4s$ *Replace P, perimeter, with 28 in.*

    $\dfrac{28 \text{ in.}}{4} = \dfrac{4s}{4}$ *Divide both sides by 4.*

    $\underline{7 \text{ in.}} = s$

    The length of one side is 7 inches.

    7 in.

    7 in.   7 in.

    7 in.

    Check
    $P = 7 \text{ in.} + 7 \text{ in.} + 7 \text{ in.} + 7 \text{ in.} = 28 \text{ in.}$

    (b) $P = 4s$ *Perimeter formula for a square*

    $100 \text{ ft} = 4s$ *Replace P with 100 ft.*

    $\dfrac{100 \text{ ft}}{4} = \dfrac{4s}{4}$ *Divide both sides by 4.*

    $25 \text{ ft} = s$

    The length of one side is 25 feet.

    25 ft

    25 ft   25 ft

    25 ft

    Check
    $P = 25 \text{ ft} + 25 \text{ ft} + 25 \text{ ft} + 25 \text{ ft} = 100 \text{ ft}$

3. **(a)** $P = 2l + 2w$ *Perimeter form. for a rectangle*

    $P = \underbrace{2 \cdot 17 \text{ cm}} + \underbrace{2 \cdot 10 \text{ cm}}$ *Replace l with 17 cm and w with 10 cm.*

    $P = \underline{34 \text{ cm}} + \underline{20 \text{ cm}}$ *Multiply first.*

    $P = \underline{54 \text{ cm}}$ *Add last.*

    The perimeter is 54 cm.

    Check
    $P = \underline{17 \text{ cm}} + \underline{17 \text{ cm}} + \underline{10 \text{ cm}} + \underline{10 \text{ cm}} = \underline{54 \text{ cm}}$

    (b) $P = 2l + 2w$

    $P = \underbrace{2 \cdot 25 \text{ ft}} + \underbrace{2 \cdot 12 \text{ ft}}$ *Replace l with 25 ft and w with 12 ft.*

    $P = 50 \text{ ft} + 24 \text{ ft}$ *Multiply first.*

    $P = 74 \text{ ft}$ *Add last.*

    The perimeter is 74 ft.

    Check
    $P = 25 \text{ ft} + 25 \text{ ft} + 12 \text{ ft} + 12 \text{ ft} = 74 \text{ ft}$

4. **(a)** $P = 2l + 2w$

    $36 \text{ in.} = 2l + \underbrace{2 \cdot 8 \text{ in.}}$ *Replace P with 36 in. and w with 8 in.*

    $36 \text{ in.} = 2l + 16 \text{ in.}$ *To get 2l by itself, add*

    $\dfrac{-16 \text{ in.} \qquad\quad -16 \text{ in.}}{20 \text{ in.} = 2l + 0}$ *−16 in. to both sides.*

    $\dfrac{20 \text{ in.}}{2} = \dfrac{2l}{2}$ *Divide both sides by 2.*

    $10 \text{ in.} = l$

    The length is 10 in.

    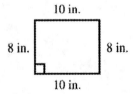

    Check
    $P = 10 \text{ in.} + 10 \text{ in.} + 8 \text{ in.} + 8 \text{ in.} = 36 \text{ in.}$

    (b) $P = 2l + 2w$

    $32 \text{ cm} = 2l + 2 \cdot 4 \text{ cm}$ *Replace P with 32 cm and w with 4 cm.*

    $32 \text{ cm} = 2l + 8 \text{ cm}$ *To get 2l by itself,*

    $\dfrac{-8 \text{ cm} \qquad\quad -8 \text{ cm}}{24 \text{ cm} = 2l + 0}$ *add −8 cm both sides.*

    $\dfrac{24 \text{ cm}}{2} = \dfrac{2l}{2}$ *Divide both sides by 2.*

    $12 \text{ cm} = l$

    The length is 12 cm.

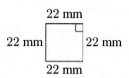

Check

$P = 12 \text{ cm} + 12 \text{ cm} + 4 \text{ cm} + 4 \text{ cm} = 32 \text{ cm}$

5. **(a)** $P = 27 \text{ m} + \underline{27 \text{ m}} + \underline{15 \text{ m}} + \underline{15 \text{ m}}$
    $P = \underline{84 \text{ m}}$

    (b) $P = 5 \text{ ft} + 4 \text{ ft} + 5 \text{ ft} + 4 \text{ ft}$
    $P = 18 \text{ ft}$

6. **(a)** $P = 31 \text{ mm} + \underline{16 \text{ mm}} + \underline{25 \text{ mm}}$
    $P = \underline{72 \text{ mm}}$

    (b) $P = 5 \text{ in.} + 5 \text{ in.} + 5 \text{ in.}$
    $P = 15 \text{ in.}$

7. "Fencing needed to go *around* a flower bed" implies perimeter.

    $P = 6 \text{ m} + 2 \text{ m} + 4 \text{ m} + 3 \text{ m} + 5 \text{ m}$
    $P = 20 \text{ m}$

    20 m of fencing are needed.

3.1 Section Exercises

1. A square has four right angles and all four sides have the same length. Drawings will vary but should have the same measurement for every side. The following drawing has sides of length 16 cm.

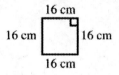

3. The perimeter formula for a square is $P = 4s$.

    $P = 4s$
    $P = 4 \cdot \underline{9 \text{ cm}}$ *Replace s with 9 cm.*
    $P = \underline{36 \text{ cm}}$

    The perimeter of the square is 36 cm.

5. The perimeter formula for a square is $P = 4s$.

    $P = 4s$
    $P = 4 \cdot 25 \text{ in.}$ *Replace s with 25 in.*
    $P = 100 \text{ in.}$

    The perimeter of the square is 100 in.

7. A square park measuring 1 mile (mi) on each side

    1 mi

    1 mi   1 mi

    1 mi

    $P = 4s$
    $P = 4 \cdot 1 \text{ mi}$ *Replace s with 1 mi.*
    $P = 4 \text{ mi}$

    The perimeter of the park is 4 miles.

9. A 22 mm square postage stamp

    22 mm

    22 mm   22 mm

    22 mm

    $P = 4s$
    $P = 4 \cdot 22 \text{ mm}$ *Replace s with 22 mm.*
    $P = 88 \text{ mm}$

    The perimeter of the stamp is 88 mm.

11. The perimeter of a square is 120 ft.

    $P = 4s$
    $120 \text{ ft} = 4s$ *Replace P with 120 ft.*
    $\dfrac{120 \text{ ft}}{4} = \dfrac{4s}{4}$ *Divide both sides by 4.*
    $\underline{30 \text{ ft}} = s$

    The length of one side of the square is 30 ft.

13. The perimeter of a square parking lot is 92 yards (yd).

    $P = 4s$
    $92 \text{ yd} = 4s$ *Replace P with 92 yd.*
    $\dfrac{92 \text{ yd}}{4} = \dfrac{4s}{4}$ *Divide both sides by 4.*
    $23 \text{ yd} = s$

    The length of one side of the parking lot is 23 yards.

15. The perimeter of a square closet is 8 feet (ft).

    $P = 4s$
    $8 \text{ ft} = 4s$ *Replace P with 8 ft.*
    $\dfrac{8 \text{ ft}}{4} = \dfrac{4s}{4}$ *Divide both sides by 4.*
    $2 \text{ ft} = s$

    The length of one side of the closet is 2 feet.

17. A rectangle is a figure with four sides that meet to form four right angles. Each set of opposite sides is parallel and has the same length. Drawings will vary, but the two longer sides should have the same measurement and the two shorter sides should have the same measurement. The rectangle shown below has length 11 m and width 6 m.

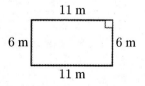

19. $P = 2l + 2w$

$P = \underbrace{2 \cdot 8 \text{ yd}} + \underbrace{2 \cdot 6 \text{ yd}}$ *Replace l with 8 yd and w with 6 yd.*

$P = \underline{16 \text{ yd}} \; + \; \underline{12 \text{ yd}}$ *Multiply first.*

$P = \underline{28 \text{ yd}}$ *Add last.*

The perimeter of the rectangle is 28 yd.

Check $P = 8 \text{ yd} + 8 \text{ yd} + 6 \text{ yd} + 6 \text{ yd}$
$P = 28 \text{ yd}$

21. $P = 2l + 2w$

$P = 2 \cdot 25 \text{ cm} + 2 \cdot 10 \text{ cm}$ *Replace l with 25 cm and w with 10 cm.*

$P = 50 \text{ cm} + 20 \text{ cm}$ *Multiply first.*

$P = 70 \text{ cm}$ *Add last.*

The perimeter of the rectangle is 70 cm.

Check $P = 25 \text{ cm} + 25 \text{ cm} + 10 \text{ cm} + 10 \text{ cm}$
$P = 70 \text{ cm}$

23.
20 ft
16 ft ⌐▭⌐ 16 ft
20 ft

$P = 2l + 2w$

$P = 2 \cdot 20 \text{ ft} + 2 \cdot 16 \text{ ft}$ *Replace l with 20 ft and w with 16 ft.*

$P = 40 \text{ ft} + 32 \text{ ft}$ *Multiply first.*

$P = 72 \text{ ft}$ *Add last.*

The perimeter of the rectangular living room is 72 ft.

25.
5 in.
8 in. ▭ 8 in.
5 in.

$P = 2l + 2w$

$P = 2 \cdot 8 \text{ in.} + 2 \cdot 5 \text{ in.}$ *Replace l with 8 in. and w with 5 in.*

$P = 16 \text{ in.} + 10 \text{ in.}$ *Multiply first.*

$P = 26 \text{ in.}$ *Add last.*

The perimeter of the rectangular piece of paper is 26 inches.

27. $P = 2l + 2w$

$30 \text{ cm} = 2l + 2 \cdot 6 \text{ cm}$ *Replace P with 30 cm and w with 6 cm.*

$30 \text{ cm} = 2l + 12 \text{ cm}$ *Multiply on the right.*

$\underline{-12 \text{ cm}} \qquad \underline{-12 \text{ cm}}$ *Add −12 cm to both sides.*

$18 \text{ cm} = 2l + 0 \text{ cm}$

$\dfrac{18 \text{ cm}}{2} = \dfrac{2l}{2}$ *Divide both sides by 2.*

$9 \text{ cm} = l$

The length is 9 cm.

Check $P = 9 \text{ cm} + 9 \text{ cm} + 6 \text{ cm} + 6 \text{ cm}$
$P = 30 \text{ cm}$

30 cm matches the original perimeter, so 9 cm is the correct length.

29. $P = 2l + 2w$

$10 \text{ mi} = 2 \cdot 4 \text{ mi} + 2w$ *Replace P with 10 mi and l with 4 mi.*

$10 \text{ mi} = 8 \text{ mi} + 2w$ *Multiply on the right.*

$\underline{-8 \text{ mi}} \qquad \underline{-8 \text{ mi}}$ *Add −8 mi to both sides.*

$2 \text{ mi} = 0 \text{ mi} + 2w$

$\dfrac{2 \text{ mi}}{2} = \dfrac{2w}{2}$ *Divide both sides by 2.*

$1 \text{ mi} = w$

The width is 1 mile.

Check $P = 4 \text{ mi} + 4 \text{ mi} + 1 \text{ mi} + 1 \text{ mi}$
$P = 10 \text{ mi}$

10 mi matches the original perimeter, so 1 mi is the correct width.

31. $P = 2l + 2w$

$16 \text{ ft} = 2 \cdot 6 \text{ ft} + 2w$ *Replace l with 6 ft and P with 16 ft.*

$16 \text{ ft} = 12 \text{ ft} + 2w$ *Multiply on the right.*

$\underline{-12 \text{ ft}} \qquad \underline{-12 \text{ ft}}$ *Add −12 ft to both sides.*

$4 \text{ ft} = 0 \text{ ft} + 2w$

$\dfrac{4 \text{ ft}}{2} = \dfrac{2w}{2}$ *Divide both sides by 2.*

$2 \text{ ft} = w$

The width of the table is 2 feet.

Check $P = 6 \text{ ft} + 6 \text{ ft} + 2 \text{ ft} + 2 \text{ ft}$
$P = 16 \text{ ft}$

33. $P = 2l + 2w$

$6 \text{ m} = 2l + 2 \cdot 1 \text{ m}$ *Replace w with 1 m and P with 6 m.*

$6 \text{ m} = 2l + 2 \text{ m}$ *Multiply on the right.*

$\underline{-2 \text{ m}} \qquad \underline{-2 \text{ m}}$ *Add −2 m to both sides.*

$4 \text{ m} = 2l + 0 \text{ m}$

$\dfrac{4 \text{ m}}{2} = \dfrac{2}{2}l$ *Divide both sides by 2.*

$2 \text{ m} = l$

The length of the door is 2 meters.

Check $P = 2 \text{ m} + 2 \text{ m} + 1 \text{ m} + 1 \text{ m}$
$P = 6 \text{ m}$

35. Add all four sides to find the perimeter.

$P = 100 \text{ ft} + 60 \text{ ft} + 100 \text{ ft} + 60 \text{ ft}$
$P = 320 \text{ ft}$

The perimeter is 320 feet.

37. Add all three sides to find the perimeter.

$$P = 12 \text{ mm} + 26 \text{ mm} + 16 \text{ mm}$$
$$P = 54 \text{ mm}$$

The perimeter is 54 mm.

39. Add all six sides to find the perimeter.

$$P = 4 \text{ ft} + 12 \text{ ft} + 12 \text{ ft} + 3 \text{ ft} + 8 \text{ ft} + 9 \text{ ft}$$
$$P = 48 \text{ ft}$$

The perimeter is 48 ft.

41. Add all six sides to find the perimeter.

$$P = 13 \text{ in.} + 8 \text{ in.} + 18 \text{ in.} + 13 \text{ in.} + 18 \text{ in.} + 8 \text{ in.}$$
$$P = 78 \text{ in.}$$

The perimeter is 78 in.

43. Add all five sides to find the perimeter.

$$P = 34 \text{ m} + 22 \text{ m} + 20 \text{ m} + 22 \text{ m} + 27 \text{ m}$$
$$P = 125 \text{ m}$$

The perimeter is 125 m.

45. The perimeter is 115 cm.

$$P = 10 \text{ cm} + 10 \text{ cm} + 30 \text{ cm} + 25 \text{ cm} + ?$$
$$P = 75 \text{ cm} + ?$$

Because the perimeter is 115 cm, replace P with 115 cm.

$$115 \text{ cm} = 75 \text{ cm} + ?$$
$$115 \text{ cm} = 75 \text{ cm} + ?$$

$$\underline{-75 \text{ cm} \qquad -75 \text{ cm}} \quad \begin{array}{l} \textit{Add} -75 \textit{ cm to} \\ \textit{both sides.} \end{array}$$

$$40 \text{ cm} = 0 \text{ cm} + ?$$
$$40 \text{ cm} = ?$$

The length of the unknown side is 40 cm.

47. The perimeter is 78 in.

$$P = \text{sum of the lengths of all sides}$$
$$78 \text{ in.} = 15 \text{ in.} + 6 \text{ in.} + 6 \text{ in.} + 6 \text{ in.}$$
$$+ 6 \text{ in.} + 9 \text{ in.} + ? + 6 \text{ in.} + 6 \text{ in.}$$
$$78 \text{ in.} = 66 \text{ in.} + ?$$
$$\underline{-66 \text{ in.} \qquad -66 \text{ in.} \quad \textit{Add} -66 \textit{ in. to both sides.}}$$
$$12 \text{ in.} = ?$$

The length of the unknown side is 12 in.

49. (a) Sketches will vary. Below is a sketch of an equilateral triangle with sides of length 5 in. Its perimeter is 5 in. + 5 in. + 5 in. = 15 in. Note that the perimeter is always 3 times the length of an individual side.

(b) Formula for perimeter of an equilateral triangle is $P = 3s$, where s is the length of one side.

(c) The formula will not work for other kinds of triangles because the sides will have different lengths.

Relating Concepts (Exercises 51–54)

51. Use $d = rt$ with $r = 70$ miles per hour, so that $d = 70t$.

(a) In $t = 2$ hours, you will travel
$$d = 70 \cdot 2 = 140 \text{ miles.}$$

(b) In $t = 5$ hours, you will travel
$$d = 70 \cdot 5 = 350 \text{ miles.}$$

(c) In $t = 8$ hours, you will travel
$$d = 70 \cdot 8 = 560 \text{ miles.}$$

52. Use $d = rt$ with $r = 35$ miles per hour, so that $d = 35t$.

(a) In $t = 2$ hours, you will travel
$$d = 35 \cdot 2 = 70 \text{ miles.}$$

(b) In $t = 5$ hours, you will travel
$$d = 35 \cdot 5 = 175 \text{ miles.}$$

(c) In $t = 8$ hours, you will travel
$$d = 35 \cdot 8 = 280 \text{ miles.}$$

(d) The rate is half of 70 miles per hour, so in each case the distance traveled will be half as far. Divide each result in Exercise 51 by 2.

53. Use $rt = d$ with $d = 3000$ miles and r in miles per hour, so that $rt = 3000$. In each case, divide by the rate.

(a) $60t = 3000$ *Let r = 60.*

$$\frac{60t}{60} = \frac{3000}{60} \qquad \textit{Divide by 60.}$$
$$t = 50 \text{ hours}$$

(b) $50t = 3000$ *Let r = 50.*

$$\frac{50t}{50} = \frac{3000}{50} \qquad \textit{Divide by 50.}$$
$$t = 60 \text{ hours}$$

(c) $20t = 3000$ *Let r = 20.*

$$\frac{20t}{20} = \frac{3000}{20} \qquad \textit{Divide by 20.}$$
$$t = 150 \text{ hours}$$

54. Use $rt = d$. In each case, divide by the time.

(a) $r \cdot 11 = 671$ *Let t = 11, d = 671.*

$$\frac{11r}{11} = \frac{671}{11} \qquad \textit{Divide by 11.}$$
$$r = 61 \text{ miles per hour}$$

(b) $r \cdot 27 = 1539$ *Let t = 27, d = 1539.*

$$\frac{27r}{27} = \frac{1539}{27} \quad \text{Divide by 27.}$$

$$r = 57 \text{ miles per hour}$$

(c) $r \cdot 16 = 1040$ *Let t = 16, d = 1040.*

$$\frac{16r}{16} = \frac{1040}{16} \quad \text{Divide by 16.}$$

$$r = 65 \text{ miles per hour}$$

3.2 Problem Solving: Area

3.2 Margin Exercises

1. **(a)** $A = l \cdot w$

   $A = 9 \text{ ft} \cdot 4 \text{ ft}$ *Replace l with 9 ft and w with 4 ft.*

   $A = \underline{36 \text{ ft}^2}$

   The area of the rectangle is 36 ft^2.

   (b)

   20 yd

   35 yd

   $A = l \cdot w$

   $A = 35 \text{ yd} \cdot 20 \text{ yd}$ *Replace l with 35 yd and w with 20 yd.*

   $A = 700 \text{ yd}^2$

   The area of the rectangle is 700 yd^2.

   (c)

   2 m

   3 m

   $A = l \cdot w$

   $A = 3 \text{ m} \cdot 2 \text{ m}$ *Replace l with 3 m and w with 2 m.*

   $A = 6 \text{ m}^2$

   The area of the patio is 6 m^2.

2. **(a)** $A = l \cdot w$

   $12 \text{ cm}^2 = 6 \text{ cm} \cdot w$ *Replace l with 6 cm and A with 12 cm^2.*

   $$\frac{12 \text{ cm} \cdot \text{cm}}{6 \text{ cm}} = \frac{6 \text{ cm} \cdot w}{6 \text{ cm}}$$ *To get w by itself, divide both sides by 6 cm.*

   $\underline{2 \text{ cm}} = w$

   The width of the slide is 2 cm.

   2 cm

   6 cm

   Check $A = l \cdot w$

   $A = 6 \text{ cm} \cdot 2 \text{ cm}$

   $A = 12 \text{ cm}^2$

   12 cm^2 matches the original area, so 2 cm is the correct width.

(b) $A = l \cdot w$

$160 \text{ ft}^2 = l \cdot 10 \text{ ft}$ *Replace w with 10 ft and A with 160 ft^2.*

$$\frac{160 \text{ ft} \cdot \text{ft}}{10 \text{ ft}} = \frac{l \cdot 10 \text{ ft}}{10 \text{ ft}}$$ *To get l by itself, divide both sides by 10 ft.*

$16 \text{ ft} = l$

The length of the lot is 16 ft.

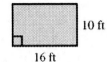

10 ft

16 ft

Check $A = l \cdot w$

$A = 16 \text{ ft} \cdot 10 \text{ ft}$

$A = 160 \text{ ft}^2$

160 ft^2 matches the original area, so 16 ft is the correct length.

(c) $A = l \cdot w$

$93 \text{ m}^2 = 31\text{m} \cdot w$ *Replace A with 93 m^2 and l with 31 m.*

$$\frac{93 \text{ m} \cdot \text{m}}{31 \text{ m}} = \frac{31 \text{ m} \cdot w}{31 \text{ m}}$$ *To get w by itself, divide both sides by 31 m.*

$3 \text{ m} = w$

The width of the floor is 3 m.

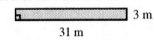

3 m

31 m

Check $A = l \cdot w$

$A = 31 \text{ m} \cdot 3 \text{ m}$

$A = 93 \text{ m}^2$

93 m^2 matches the original area, so 3 m is the correct width.

3. **(a)**

   12 in.

   12 in.

   $A = s^2$

   $A = s \cdot s$

   $A = 12 \text{ in.} \cdot 12 \text{ in.}$

   $A = \underline{144 \text{ in.}^2}$

   The area of the fabric is 144 in.^{2}.

   (b)

   7 mi

   7 mi

   $A = s^2$

   $A = s \cdot s$

   $A = 7 \text{ mi} \cdot 7 \text{ mi}$

   $A = 49 \text{ mi}^2$

   The area of the township is 49 mi^2.

(c) 20 mm

$A = s^2$
$A = s \cdot s$
$A = 20 \text{ mm} \cdot 20 \text{ mm}$
$A = 400 \text{ mm}^2$

The area of the earring is 400 mm².

4. **(a)** $A = s^2$

$16 \text{ mi}^2 = s^2$ *Replace A with 16 mi².*

$16 \text{ mi}^2 = s \cdot s$
$16 \text{ mi}^2 = \underline{4 \text{ mi}} \cdot \underline{4 \text{ mi}}$

The length of one side of the nature center is 4 mi.

(b) $A = s^2$

$100 \text{ m}^2 = s^2$ *Replace A with 100 m².*

$100 \text{ m}^2 = s \cdot s$
$100 \text{ m}^2 = 10 \text{ m} \cdot 10 \text{ m}$

The length of one side of the floor is 10 m.

(c) $A = s^2$

$81 \text{ in.}^2 = s^2$ *Replace A with 81 in.².*

$81 \text{ in.}^2 = s \cdot s$
$81 \text{ in.}^2 = 9 \text{ in.} \cdot 9 \text{ in.}$

The length of one side of the clock face is 9 inches.

5. **(a)** *Note*: The base is 50 ft and the height is 42 ft. The height forms a 90° angle with the base.

$A = b \cdot h$

$A = 50 \text{ ft} \cdot 42 \text{ ft}$ *Replace b with 50 ft and h with 42 ft.*

$A = \underline{2100 \text{ ft}^2}$

(b) *Note*: The base is 18 in. and the height is 10 in. The height forms a 90° angle with the base.

$A = b \cdot h$

$A = 18 \text{ in.} \cdot 10 \text{ in.}$ *Replace b with 18 in. and h with 10 in.*

$A = 180 \text{ in.}^2$

(c) $A = b \cdot h$

$A = 8 \text{ cm} \cdot 1 \text{ cm}$ *Replace b with 8 cm and h with 1 cm.*

$A = 8 \text{ cm}^2$

6. **(a)** $A = b \cdot h$

$140 \text{ in.}^2 = 14 \text{ in.} \cdot h$ *Replace A with 140 in.² and b with 14 in.*

$\dfrac{140 \text{ in.} \cdot \text{in.}}{14 \text{ in.}} = \dfrac{14 \text{ in.} \cdot h}{14 \text{ in.}}$ *Divide both sides by 14 in.*

$10 \text{ in.} = h$

The height is 10 inches.

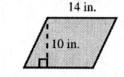

Check $A = b \cdot h$
$A = 14 \text{ in.} \cdot 10 \text{ in.}$
$A = 140 \text{ in.}^2$

140 in.² matches the original area, so 10 inches is the correct height.

(b) $A = b \cdot h$

$4 \text{ yd}^2 = b \cdot 1 \text{ yd}$ *Replace A with 4 yd² and h with 1 yd.*

$\dfrac{4 \text{ yd} \cdot \text{yd}}{1 \text{ yd}} = \dfrac{b \cdot 1 \text{ yd}}{1 \text{ yd}}$ *Divide both sides by 1 yd.*

$4 \text{ yd} = b$

The base is 4 yd.

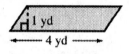

Check $A = b \cdot h$
$A = 4 \text{ yd} \cdot 1 \text{ yd}$
$A = 4 \text{ yd}^2$

4 yd² matches the original area, so 4 yd is the correct base.

7. The sod will *cover* the playground, so we need to find the area of the playground. The playground is rectangular. Use the formula for the area of a rectangle.

$A = l \cdot w$

$A = 22 \text{ yd} \cdot 16 \text{ yd}$ *Replace l with 22 yd and w with 16 yd.*

$A = 352 \text{ yd}^2$

The area is 352 yd², so the neighbors need to buy 352 yd² of sod. The cost of the sod is $3 per square yard, which means $3 for 1 square yard.

Cost $= \$3 \cdot 352$
     $= \$1056$

The neighbors will spend $1056 on sod.

3.2 Section Exercises

1. $A = s^2$ or $A = s \cdot s$; A stands for area; s stands for side (the length of one side).

3. **(a)** $10^2 = 10 \cdot 10 = 100$

(b) $2 \cdot 10 = 20$

(c) $25^2 = 25 \cdot 25 = 625$

(d) $25 \cdot 2 = 50$

5. The figure is a rectangle.

$A = l \cdot w$

$A = 11 \text{ ft} \cdot 7 \text{ ft}$ *Replace l with 11 ft and w with 7 ft.*

$A = \underline{77 \text{ ft}^2}$

The area is 77 ft^2, or 77 square feet.

7. The figure is a square.

$A = s^2$

$A = s \cdot s$ *Remember, s^2 means $s \cdot s$.*

$A = \underline{6 \text{ in.}} \cdot \underline{6 \text{ in.}}$ *Replace s with 6 in.*

$A = \underline{36 \text{ in.}^2}$

The area is 36 in.^{2}, or 36 square inches.

9. The figure is a parallelogram. Turn your book sideways to identify that the height is 25 mm and the base is 31 mm.

$A = b \cdot h$

$A = 31 \text{ mm} \cdot 25 \text{ mm}$ *Replace b with 31 mm and h with 25 mm.*

$A = 775 \text{ mm}^2$

The area is 775 mm^2, or 775 square millimeters.

11.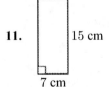

$A = l \cdot w$ *Area of a rectangle*

$A = 15 \text{ cm} \cdot 7 \text{ cm}$ *Replace l with 15 cm and w with 7 cm.*

$A = 105 \text{ cm}^2$

The area of the rectangular calculator is 105 cm^2.

13.

$A = b \cdot h$ *Area of a parallelogram*

$A = 8 \text{ ft} \cdot 9 \text{ ft}$ *Replace b with 8 ft and h with 9 ft.*

$A = 72 \text{ ft}^2$

The area of the parallelogram is 72 ft^2.

15. 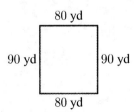 25 mi

25 mi

$A = s^2$ *Area of a square*

$A = s \cdot s$

$A = 25 \text{ mi} \cdot 25 \text{ mi}$ *Replace s with 25 mi.*

$A = 625 \text{ mi}^2$

The square-shaped forest has an area of 625 mi^2.

17. $A = l \cdot w$

$18 \text{ ft}^2 = l \cdot 3 \text{ ft}$ *Replace A with 18 ft^2, and w with 3 ft.*

$\dfrac{18 \text{ ft} \cdot \text{ft}}{3 \text{ ft}} = \dfrac{l \cdot 3 \text{ ft}}{3 \text{ ft}}$ *Divide both sides by 3 ft.*

$6 \text{ ft} = l$

The length of the desk is 6 ft.

6 ft

3 ft   3 ft

6 ft

Check $A = l \cdot w$

$A = 6 \text{ ft} \cdot 3 \text{ ft}$

$A = 18 \text{ ft}^2$

19. $A = l \cdot w$

$7200 \text{ yd}^2 = 90 \text{ yd} \cdot w$ *Replace A with 7200 yd^2, and l with 90 yd.*

$\dfrac{7200 \text{ yd} \cdot \text{yd}}{90 \text{ yd}} = \dfrac{90 \text{ yd} \cdot w}{90 \text{ yd}}$ *Divide both sides by 90 yd.*

$80 \text{ yd} = w$

The width of the parking lot is 80 yd.

80 yd

90 yd   90 yd

80 yd

Check $A = l \cdot w$

$A = 90 \text{ yd} \cdot 80 \text{ yd}$

$A = 7200 \text{ yd}^2$

21.

$$A = l \cdot w$$

$$154 \text{ in.}^2 = l \cdot 11 \text{ in.}$$ *Replace A with 154 in.², and w with 11 in.*

$$\frac{154 \text{ in.} \cdot \cancel{\text{in.}}}{11 \cancel{\text{in.}}} = \frac{l \cdot 11 \text{ in.}}{11 \text{ in.}}$$ *Divide both sides by 11 in.*

$$14 \text{ in.} = l$$

The length of the photo is 14 inches.

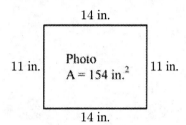

14 in.

11 in.   Photo A = 154 in.²   11 in.

14 in.

Check $A = l \cdot w$

$A = 14 \text{ in.} \cdot 11 \text{ in.}$

$A = 154 \text{ in.}^2$

23.

$$A = s^2$$ *Area formula for a square*

$$A = s \cdot s$$

$$36 \text{ m}^2 = s \cdot s$$ *Replace A with 36 m².*

$$36 \text{ m}^2 = 6 \text{ m} \cdot 6 \text{ m}$$ *By inspection, 6 m · 6 m is 36 m².*

The length of one side of the floor is <u>6 m</u>.

25.

$$A = s^2$$ *Area formula for a square*

$$A = s \cdot s$$

$$4 \text{ ft}^2 = s \cdot s$$ *Replace A with 4 ft².*

$$4 \text{ ft}^2 = 2 \text{ ft} \cdot 2 \text{ ft}$$ *By inspection, 2 ft · 2 ft is 4 ft².*

The length of one side of the sign is 2 ft.

27. Use the area formula for a parallelogram.

$$A = b \cdot h$$

$$500 \text{ cm}^2 = 25 \text{ cm} \cdot h$$ *Replace A with 500 cm², and b with 25 cm.*

$$\frac{500 \text{ cm} \cdot \cancel{\text{cm}}}{25 \cancel{\text{cm}}} = \frac{25 \text{ cm} \cdot h}{25 \text{ cm}}$$ *Divide both sides by 25 cm.*

$$20 \text{ cm} = h$$

The height is 20 cm.

20 cm

25 cm

Check $A = b \cdot h$

$A = 25 \text{ cm} \cdot 20 \text{ cm}$

$A = 500 \text{ cm}^2$

29. Use the area formula for a parallelogram.

$$A = b \cdot h$$

$$221 \text{ in.}^2 = b \cdot 13 \text{ in.}$$ *Replace A with 221 in.², and h with 13 in.*

$$\frac{221 \text{ in.} \cdot \cancel{\text{in.}}}{13 \cancel{\text{in.}}} = \frac{b \cdot 13 \text{ in.}}{13 \text{ in.}}$$ *Divide both sides by 13 in.*

$$17 \text{ in.} = b$$

The base is 17 inches.

13 in.

17 in.

Check $A = b \cdot h$

$A = 17 \text{ in.} \cdot 13 \text{ in.}$

$A = 221 \text{ in.}^2$

31. Use the area formula for a parallelogram.

$$A = b \cdot h$$

$$9 \text{ m}^2 = 9 \text{ m} \cdot h$$ *Replace A with 9 m², and b with 9 m.*

$$\frac{9 \text{ m} \cdot \cancel{\text{m}}}{9 \cancel{\text{m}}} = \frac{9 \text{ m} \cdot h}{9 \text{ m}}$$ *Divide both sides by 9 m.*

$$1 \text{ m} = h$$

The height is 1 m.

1 m

9 m

Check $A = b \cdot h$

$A = 9 \text{ m} \cdot 1 \text{ m}$

$A = 9 \text{ m}^2$

33. The height of the parallelogram is not part of the perimeter. Also, linear units are used for perimeter, so cm² should be cm.

$$P = 25 \text{ cm} + 25 \text{ cm} + 25 \text{ cm} + 25 \text{ cm}$$

$$P = 100 \text{ cm}$$

35. The figure is a square with side 45 in.

$$P = 4s$$

$$P = 4 \cdot 45 \text{ in.}$$

$$P = 180 \text{ in.}$$

$$A = s^2 \qquad\qquad s^2 = s \cdot s$$

$$A = 45 \text{ in.} \cdot 45 \text{ in.}$$

$$A = 2025 \text{ in.}^2$$

37. The figure is a parallelogram with base 18 cm and height 10 cm.

$$P = 18 \text{ cm} + 12 \text{ cm} + 18 \text{ cm} + 12 \text{ cm}$$
$$P = 60 \text{ cm}$$

$$A = b \cdot h$$
$$A = 18 \text{ cm} \cdot 10 \text{ cm}$$
$$A = 180 \text{ cm}^2$$

39. The mat is a square.

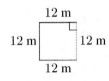

12 m

12 m   12 m

12 m

$$P = 4 \cdot s \qquad A = s \cdot s$$
$$P = 4 \cdot 12 \text{ m} \qquad A = 12 \text{ m} \cdot 12 \text{ m}$$
$$P = 48 \text{ m} \qquad A = 144 \text{ m}^2$$

41. The kitchen floor is a rectangle.

5 m

4 m

Tyra is decorating the top edges of her walls, so find the perimeter of her ceiling.

$$P = 2 \cdot l + 2 \cdot w$$
$$P = 2 \cdot 5 \text{ m} + 2 \cdot 4 \text{ m} \qquad \textit{Replace l with 5 m and w with 4 m.}$$
$$P = 10 \text{ m} + 8 \text{ m}$$
$$P = 18 \text{ m}$$

The strip costs $6 per meter. To find the cost of 18 meters, multiply $6 \cdot 18$ to get $108.

To have the top edges of her walls decorated, Tyra will spend $108.

43. Mr. and Mrs. Gomez are *covering* the bedroom floor, so find the area of the square-shaped floor.

5 yd

5 yd

$$A = s^2 \qquad s^2 = s \cdot s$$
$$A = 5 \text{ yd} \cdot 5 \text{ yd}$$
$$A = 25 \text{ yd}^2$$

The carpet cost is $23 per square yard. To find the cost for 25 square yards, multiply $23 \cdot 25$ to get $575.

The cost of padding and installation is $6 per square yard. To find the cost for 25 square yards, multiply $6 \cdot 25$ to get $150.

To have the bedroom carpeted, Mr. and Mrs. Gomez will spend $575 + $150, or $725 total.

45. The football field is a rectangle with length 100 yards and area 5300 yd^2.

$$A = l \cdot w \qquad \textit{Area formula for a rectangle.}$$

$$5300 \text{ yd}^2 = 100 \text{ yd} \cdot w \qquad \textit{Replace A with 5300 yd}^2 \textit{ and l with 100 yd.}$$

$$\frac{5300 \text{ yd} \cdot \cancel{\text{yd}}}{100 \ \cancel{\text{yd}}} = \frac{100 \text{ yd} \cdot w}{100 \text{ yd}} \qquad \textit{Divide both sides by 100 yd.}$$

$$53 \text{ yd} = w$$

The width of the field is 53 yards.

47.
$$P = 4 \cdot s \qquad\qquad A = s \cdot s$$
$$= 4 \cdot 13 \text{ ft} \qquad\quad = 13 \text{ ft} \cdot 13 \text{ ft}$$
$$= 52 \text{ ft} \qquad\qquad = 169 \text{ ft}^2$$

$$\frac{169 \text{ ft}^2}{8 \text{ campers}} = 21.125 \text{ ft}^2 \text{ for each camper}$$
$$\approx 21 \text{ ft}^2 \text{ for each camper}$$

Relating Concepts (Exercises 49–52)

49. Since perimeter $P = 2 \cdot l + 2 \cdot w$, if $w = 1$ and $P = 12$, then $2 \cdot l = 12 - 2 \cdot 1 = 10$ and $l = 5$. Similarly, if $w = 2$, $l = 4$, and if $w = 3$, $l = 3$. So there are only three possibilities for plots with whole number dimensions, perimeter 12 ft, and $l \geq w$.

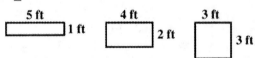

5 ft   1 ft   4 ft   2 ft   3 ft   3 ft

50. **(a)** Use the area formula for a rectangle, $A = l \cdot w$.

l	w	$A = l \cdot w$
5 ft	1 ft	5 ft^2
4 ft	2 ft	8 ft^2
3 ft	3 ft	9 ft^2

(b) The table in part (a) shows that the square plot 3 ft by 3 ft has the greatest area.

51. As in Exercise 49, we let w equal a whole number and find l.

w	$2 \cdot w$	$2l = 16 - 2 \cdot w$	l
1	2	14	7
2	4	12	6
3	6	10	5
4	8	8	4

So there are only four possibilities for plots with whole number dimensions, perimeter 16 ft, and $l \geq w$.

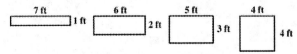

7 ft   1 ft   6 ft   2 ft   5 ft   3 ft   4 ft   4 ft

52. **(a)** Use the area formula for a rectangle, $A = l \cdot w$.

w	l	$A = l \cdot w$
1 ft	7 ft	7 ft^2
2 ft	6 ft	12 ft^2
3 ft	5 ft	15 ft^2
4 ft	4 ft	16 ft^2

(b) Based on Exercises 50 (a) and 52 (a), square plots have the greatest area.

Summary Exercises *Perimeter and Area*

1. The figure is a rectangle with length 13 m and width 3 m.

$$P = 13 \text{ m} + 13 \text{ m} + 3 \text{ m} + 3 \text{ m}$$
$$P = 32 \text{ m}$$

$$A = l \cdot w$$
$$A = 3 \text{ m} \cdot 13 \text{ m}$$
$$A = 39 \text{ m}^2$$

3. The figure is a parallelogram with base 7 yd and height 8 yd.

$$P = 10 \text{ yd} + 10 \text{ yd} + 7 \text{ yd} + 7 \text{ yd}$$
$$P = 34 \text{ yd}$$

$$A = b \cdot h$$
$$A = 7 \text{ yd} \cdot 8 \text{ yd}$$
$$A = 56 \text{ yd}^2$$

5. The figure is a square with side 9 in.

$P = 4 \cdot s$ $A = s \cdot s$
$P = 4 \cdot 9 \text{ in.}$ $A = 9 \text{ in.} \cdot 9 \text{ in.}$
$P = 36 \text{ in.}$ $A = 81 \text{ in.}^2$

7. The figure is a rectangle with length 9 ft and width 4 ft.

$$P = 9 \text{ ft} + 9 \text{ ft} + 4 \text{ ft} + 4 \text{ ft}$$
$$P = 26 \text{ ft}$$

$$A = l \cdot w$$
$$A = 9 \text{ ft} \cdot 4 \text{ ft}$$
$$A = 36 \text{ ft}^2$$

9. $P = 2l + 2w$

$16 \text{ ft} = 2 \cdot 7 \text{ ft} + 2w$ *Replace P with 16 ft and l with 7 ft.*

$16 \text{ ft} = 14 \text{ ft} + 2w$ *Multiply.*

$\underline{-14 \text{ ft}\quad -14 \text{ ft}}$ *Add -14 ft to both sides.*

$2 \text{ ft} = 0 \text{ ft} + 2w$

$\dfrac{2 \text{ ft}}{2} = \dfrac{2w}{2}$ *Divide both sides by 2.*

$1 \text{ ft} = w$

The width is 1 ft.

Check $P = 7 \text{ ft} + 7 \text{ ft} + 1 \text{ ft} + 1 \text{ ft}$
$P = 16 \text{ ft}$

11. $A = s^2$

$36 \text{ in.}^2 = s^2$ *Replace A with 36 in.^{2}.*

$36 \text{ in.}^2 = s \cdot s$

$36 \text{ in.}^2 = 6 \text{ in.} \cdot 6 \text{ in.}$ *By inspection*

The length of one side of the photograph is 6 in.

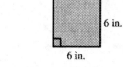

6 in.
6 in.

Check $A = s \cdot s$
$A = 6 \text{ in.} \cdot 6 \text{ in.}$
$A = 36 \text{ in.}^2$

13. $P = 4s$

$64 \text{ cm} = 4s$ *Replace P with 64 cm.*

$\dfrac{64 \text{ cm}}{4} = \dfrac{4s}{4}$ *Divide both sides by 4.*

$16 \text{ cm} = s$

The length of one side of the map is 16 cm.

16 cm
16 cm   16 cm
16 cm

Check $P = 16 \text{ cm} + 16 \text{ cm} + 16 \text{ cm} + 16 \text{ cm}$
$P = 64 \text{ cm}$

15. The figure is a triangle.

$$P = 168 \text{ m} + 168 \text{ m} + 168 \text{ m}$$
$$P = 504 \text{ m}$$

504 meters of fencing are needed.

17. $P = 2l + 2w$

$42 \text{ in.} = 2 \cdot 12 \text{ in.} + 2w$ *Replace l with 12 in. and P with 42 in.*

$42 \text{ in.} = 24 \text{ in.} + 2w$ *Multiply on the right.*

$\underline{-24 \text{ in.}\quad -24 \text{ in.}}$ *Add -24 in. to both sides.*

$18 \text{ in.} = 0 \text{ in.} + 2w$

$\dfrac{18 \text{ in.}}{2} = \dfrac{2w}{2}$ *Divide both sides by 2.*

$9 \text{ in.} = w$

The width of the screen is 9 inches.

19. Each flag is rectangular in shape.

$$P = 3 \text{ ft} + 5 \text{ ft} + 3 \text{ ft} + 5 \text{ ft}$$
$$P = 16 \text{ ft}$$

$$A = l \cdot w$$
$$A = 5 \text{ ft} \cdot 3 \text{ ft}$$
$$A = 15 \text{ ft}^2$$

To make seven flags, complete with binding, we will need $7(15 \text{ ft}^2) = 105 \text{ ft}^2$ of fabric and $7(16 \text{ ft}) = 112 \text{ ft}$ of binding.

3.3 Solving Application Problems with One Unknown Quantity

3.3 Margin Exercises

1. **(a)** 15 less than a number

$$x - 15$$

(b) 12 more than a number

$$x + 12 \quad \text{or} \quad 12 + x$$

(c) A number increased by 13

$$x + 13 \quad \text{or} \quad 13 + x$$

(d) A number minus 8

$$x - 8$$

(e) 10 plus a number

$$10 + x \quad \text{or} \quad x + 10$$

(f) A number subtracted from 6

$$6 - x$$

(g) 6 subtracted from a number

$$x - 6$$

2. **(a)** Double a number

$$2x$$

(b) The product of -8 and a number

$$-8x$$

(c) The quotient of 15 and a number

$$\frac{15}{x}$$

(d) 5 times a number subtracted from 30

$$30 - 5 \cdot x \quad \text{or} \quad 30 - 5x$$

Note: $5x - 30$ would be "30 subtracted from 5 times a number."

3. **(a)** Let x represent the unknown number.

$$\underbrace{3 \text{ times a number}}_{3x} \quad \underbrace{\text{is added to}}_{+} \quad \underset{\downarrow}{4} \quad \underbrace{\text{the result is}}_{=} \quad \underset{\downarrow}{19}$$

$$3x + 4 = 19 \quad \textit{To get 3x by itself,}$$
$$\underline{-4 \qquad -4} \quad \textit{add } -4 \textit{ to both sides.}$$
$$3x + 0 = 15$$
$$\frac{3x}{3} = \frac{15}{3} \quad \textit{Divide both sides by 3.}$$
$$x = 5$$

Check   3 times 5 [= 15] is added to 4 [= 19]. The result is 19, so 5 is the number.

(b) Let x represent the unknown number.

$$\underbrace{6 \text{ times a number}}_{6x} \quad \underbrace{7 \text{ is subtracted from}}_{-} \quad \underbrace{\text{the result is}}_{=} \quad \underset{\downarrow}{-25}$$

$$6x - 7 = -25$$
$$\underline{+7 \qquad +7} \quad \textit{Add 7 to both sides.}$$
$$6x + 0 = -18$$
$$\frac{6x}{6} = \frac{-18}{6} \quad \textit{Divide both sides by 6.}$$
$$x = -3$$

Check   6 times -3 [$= -18$] minus 7 [$= -25$] is -25, so -3 is the number.

4. **Step 1**   The problem is about the number of people on a bus.

Unknown: number of people who got on at the first stop

Known: 3 got on; 5 got on; 10 got off; 4 people still on the bus

Step 2(a)   Let p be the number of people who got on at the first stop.

Step 3 $p + \underline{3} + \underline{5} - \underline{10} = \underline{4}$

Step 4
$$p + 3 + 5 - 10 = 4$$
$$p + 8 + (-10) = 4$$
$$p + (-2) = 4$$
$$\underline{\qquad +2 \qquad +2}$$
$$p + 0 = 6$$
$$p = 6$$

Step 5   6 people got on at the first stop.

Step 6

First stop:	6 got on	
Second stop:	3 got on	$6 + 3 = 9$ on
Third stop:	5 got on	$9 + 5 = 14$ on
Fourth stop:	10 got on	$14 - 10 = 4$ on

4 people are left. **Matches** ⟵⟶

5. **Step 1**   The problem is about money being donated and spent by a college.

Unknown: amount of money given by each donor.

Known: Each donor gave the "same" dollar amount. College gave out \$1250, \$900, and \$850. \$250 was left over.

Step 2(a)   Let m be the amount of money donated by each donor.

Step 3

$$\underbrace{5 \text{ equal donations}}_{5 \cdot m} \quad \underbrace{\text{money spent on scholarships}}_{-(1250 + 900 + 850)} \quad \underbrace{\text{money left over}}_{250}$$

Step 4

$$5m - (1250 + 900 + 850) = 250$$
$$5m - 3000 = 250$$
$$5m + (-3000) = 250$$
$$\underline{+3000 \qquad +3000}$$
$$5m + 0 = 3250$$
$$\frac{5m}{5} = \frac{3250}{5}$$
$$m = 650$$

Step 5   Each donor gave $650.

Step 6   5 donors each gave $650, so
$5 \cdot \$650 = \$3250$ donated

College gave out $1250, $900, $850, so
$\$3250 - \$1250 - \$900 - \$850 = \$250$.
College had $250 left. ***Matches*** ⟵⎯⎤

6.   ***Step 1***   The problem is about money being donated by two people.

Unknown: LuAnn's donation

Known: Susan donated $10 more than twice what LuAnn donated. Susan donated $22.

Step 2(a)   You know the least about LuAnn's donation. Let d represent LuAnn's donation.

Step 3

Susan's donation	is	$10 more than	twice	LuAnn's donation
↓	↓	↓	↓	↓
22	=	10 +	2·	d

Step 4 $22 = 10 + 2d$
$$\underline{-10 \qquad -10}$$
$$12 = 0 + 2d$$
$$\frac{12}{2} = \frac{2d}{2}$$
$$6 = d$$

Step 5   LuAnn donated $6.

Step 6 $10 more than twice $6 is
$\$10 + (2 \cdot \$6) = \$10 + \$12 = \$22$.

Susan donated $22. ***Matches*** ⟵⎯⎤

3.3   Section Exercises

1.   **(a)** $14 - x$ means "14 subtracted from a number" is *false*. $14 - x$ means "a number subtracted from 14" or "14 minus a number."

(b) $x - 7$ means "a number decreased by 7" is *true*.

(c) $\dfrac{-20}{x}$ means "the product of -20 and a number" is *false*. $\dfrac{-20}{x}$ means "the quotient of -20 and a number" or "-20 divided by a number."

3.   14 plus a number
$$14 + x \quad \text{or} \quad x + 14$$

5. -5 added to a number
$$-5 + x \quad \text{or} \quad x + (-5)$$

7.   20 minus a number
$$20 - x$$

9.   9 less than a number
$$x - 9$$

11. -6 times a number
$$-6x$$

13.   Double a number
$$2x$$

15.   A number divided by 2
$$\frac{x}{2}$$

17.   Twice a number added to 8
$$8 + 2x \quad \text{or} \quad 2x + 8$$

19.   10 fewer than seven times a number
$$7x - 10$$

21.   The sum of twice a number and the number
$$2x + x \quad \text{or} \quad x + 2x$$

23.   Let n represent the unknown number.

four times a number	decreased by	2	result is	26.
↓	↓	↓	↓	↓
$4n$	−	2	=	26

$$4n - 2 = 26$$
$$4n - 2 + 2 = 26 + 2$$
$$4n = 28$$
$$\frac{4n}{4} = \frac{28}{4}$$
$$n = 7$$

The number is 7.

Check   Four times 7 [$= 28$] is decreased by 2 [$= 26$]. *True*

25.   Let n represent the unknown number.

Twice a number	added to	the number	is	-15.
↓	↓	↓	↓	↓
$2n$	+	n	=	-15

$$2n + n = -15$$
$$3n = -15$$
$$\frac{3n}{3} = \frac{-15}{3}$$
$$n = -5$$

The number is -5.

Check   Twice -5 [$= -10$] is added to -5 [$= -15$]. *True*

27. Let n represent the unknown number.

Product of a number and 5	increased by	12	the result is	7 times the number.
↓	↓	↓	↓	↓
$5n$	$+$	12	$=$	$7n$

$$5n + 12 = 7n$$
$$5n + 12 + (-5n) = 7n + (-5n)$$
$$12 = 2n$$
$$\frac{12}{2} = \frac{2n}{2}$$
$$6 = n$$

The number is 6.

Check   The product of 6 and 5 [$= 30$] is increased by 12 [$= 42$] is seven times 6 [$= 42$]. *True*

29. Let n represent the unknown number.

30	subtract	3 times a number	is	2	plus	the number
↓	↓	↓	↓	↓	↓	↓
30	$-$	$3n$	$=$	2	$+$	n

$$30 - 3n = 2 + n$$
$$30 - 3n + 3n = 2 + n + 3n$$
$$30 = 2 + 4n$$
$$30 + (-2) = 2 + 4n + (-2)$$
$$28 = 4n$$
$$\frac{28}{4} = \frac{4n}{4}$$
$$7 = n$$

The number is 7.

Check   Three times 7 [$= 21$] is subtracted from 30 [$= 9$] is 2 plus 7 [$= 9$]. *True*

31.   *Step 1* The problem is about Ricardo's weight.

Unknown: his original weight

Known: He gained 15 pounds, lost 28, and then regained 5 pounds to weigh 177.

Step 2(a)   Let w represent Ricardo's original weight.

Step 3

weight at beginning	pounds gained	pounds lost	pounds regained	final weight
↓	↓	↓	↓	↓
w	$+\ 15$	$-\ 28$	$+\ 5$	$=\ 177$

Step 4

$$w + 15 - 28 + 5 = 177$$
$$w - 8 = 177 \qquad \textit{Combine like terms.}$$
$$w - 8 + 8 = 177 + 8 \quad \textit{Add 8 to both sides.}$$
$$w = 185$$

Step 5   He weighed 185 pounds originally.

Step 6 $185 + 15 - 28 + 5 = 177$

The correct answer is 185 pounds.

33.   *Step 1*   The problem is about the number of cookies in the cookie jar.

Unknown: number of cookies the children ate

Known: 18 cookies at the start, three dozen cookies put in the jar, and 49 cookies at the end

Step 2(a)   Let c represent the number of cookies the children ate.

Step 3

cookies at the start	cookies eaten	cookies added	ended up with 49 cookies
↓	↓	↓	↓
18	$-\ c$	$+\ 36$	$=\ 49$

Step 4

$$18 - c + 36 = 49$$

Write the understood 1c;
$$54 + (-1c) = 49 \quad \textit{change subtraction;}$$
combine like terms.
$$\underline{-54 \qquad\qquad -54} \quad \textit{Add } -54 \textit{ to both sides.}$$
$$0 + (-1c) = -5$$
$$0 + (-1c) = -5$$
$$\frac{-1c}{-1} = \frac{-5}{-1} \quad \textit{Divide both sides by } -1.$$
$$c = 5$$

Step 5   The children ate 5 cookies.

Step 6   There were 18 cookies at the start. The children ate 5 cookies, so that left $18 - 5 = 13$ cookies in the jar.

Three dozen cookies were added, so there were $13 + 36 = 49$ cookies.

The correct answer is 5 cookies.

35.   *Step 1*   The problem is about boxes of pens.

Unknown: the number of pens in each box

Known: The bookstore started with 6 boxes of pens. The bookstore sold 32 pens, then 35 pens, and 5 pens were left.

Step 2(a)   Let p represent the number of pens in each box.

Step 3

total number of boxes	number of pens in each box			
↓	↓	↓	↓	↓
6	$\cdot\quad p$	$-\ 32$	$-\ 35$	$=\ 5$

Step 4
$$6p - 32 - 35 = 5$$
$$6p - 67 = 5$$
$$\underline{ +67 \quad +67}$$
$$6p + 0 = 72$$
$$\frac{6p}{6} = \frac{72}{6}$$
$$p = 12$$

Step 5   There were 12 pens in each box.

Step 6   The bookstore ordered 6 boxes of red pens each containing 12 pens. $6(12) = 72$ pens
32 were sold: $72 - 32 = 40$
35 were sold: $40 - 35 = 5$
5 pens were left on the shelf.

The correct answer is 12 pens per box.

37. ***Step 1***   The problem is about the bank account of a music club.

Unknown: the dues paid by each member

Known: 14 total members, earned \$340, spent \$575, and account is overdrawn by \$25

Step 2(a)   Let d represent the dues paid by each member.

Step 3

total dues paid by 14 members	+	earned \$340	−	spent \$575	=	ended up with an overdrawn bank account
↓		↓		↓		↓
$14 \cdot d$	+	340	−	575	=	-25

Step 4
$$14d + 340 - 575 = -25$$
$$14d + 340 + (-575) = -25$$
$$14d + (-235) = -25$$
$$\underline{ +235 \qquad +235}$$
$$14d + 0 = 210$$
$$\frac{14d}{14} = \frac{210}{14}$$
$$d = 15$$

Step 5   Each member paid \$15 in dues.

Step 6   14 members paid dues of \$15 each: $14(\$15) = \$210$ in the bank account.

Earned \$340: $\$210 + \$340 = \$550$ in the account.

Spent \$575: $\$550 - \$575 = -\$25$ in the account.

The account was overdrawn by \$25.

The correct answer is \$15 dues.

39. ***Step 1***   The problem is about Tamu's age.

Unknown: how old Tamu is right now

Known: Tamu's age equals the number you get after multiplying his age by 4 and then subtracting 75.

Step 2(a)   Let a represent Tamu's age.

Step 3

Tamu's age	=	his age multiplied by 4	subtract	75
↓		↓	↓	↓
a	=	$4a$	−	75

Step 4
$$a = 4a - 75$$
$$\underline{-a \qquad -a}$$
$$0 = 3a - 75$$
$$\underline{+75 \qquad +75}$$
$$75 = 3a$$
$$\frac{75}{3} = \frac{3a}{3}$$
$$25 = a$$

Step 5   Tamu is 25 years old.

Step 6   75 subtracted from 4 times 25 :
$4(25) - 75 = 100 - 75 = 25$

The correct solution is 25 years old.

41. ***Step 1***   The problem is about spending money for clothes.

Unknown: amount spent on clothes by Brenda

Known: Consuelo spent \$3 less than twice the amount that Brenda spent, Consuelo spent \$81

Step 2(a)   Let m represent the amount of money that Brenda spent.

Step 3

Consuelo spent	=	\$3 less than twice the amount that Brenda spent
↓		↓
81	=	$2m - 3$

Step 4
$$81 = 2m - 3$$
$$81 = 2m + (-3)$$
$$\underline{+3 \qquad\qquad +3}$$
$$\frac{84}{2} = \frac{2m}{2}$$
$$42 = m$$

Step 5   Brenda spent \$42 on clothes.

Step 6   Consuelo spent \$3 less than twice the amount that Brenda spent, or
$2 \cdot \$42 + (-\$3) = \$84 - (\$3) = \$81$.

The correct answer is \$42.

43. ***Step 1***   The problem concerns bags of candy.

Unknown: the number of pieces of candy in each bag

Known: There were originally 5 bags. She gave 3 pieces of candy each to 48 children. Afterwards 1 bag remained.

Step 2(a) Let b represent the number of pieces in each bag.

Step 3

total amount of candy at the beginning	subtract	total amount of candy given out	results in	amount of candy in one bag
↓	↓	↓	↓	↓
$5 \cdot b$	$-$	$3 \cdot 48$	$=$	b

Step 4
$$5b - 3 \cdot 48 = b$$
$$5b - 144 = b$$
$$\frac{+144 \quad\quad +144}{5b + 0 = b + 144}$$
$$\frac{-b \quad\quad -b}{4b = 144}$$
$$\frac{4b}{4} = \frac{144}{4}$$
$$b = 36$$

Step 5 There were 36 pieces of candy in each bag.

Step 6 5 bags of 36 pieces of candy:
$5 \cdot 36 = 180$
3 pieces were handed out to each of 48 children:
$3 \cdot 48 = 144$
What remained: $180 - 144 = 36$
Equals the amount in one bag: 36

The correct answer is 36.

45. Step 1 The problem is about the recommended daily vitamin C intake.

Unknown: the recommended daily vitamin C intake for a young child

Known: The recommended daily vitamin C intake for an adult female is 8 mg more than four times that for a young child. The amount for an adult female is 60 mg.

Step 2(a) Let d represent the daily amount of vitamin C for a young child.

Step 3

the amount for an adult female	is	8 mg	more than	four times	the amount for a young child
↓	↓	↓	↓	↓	↓
60	$=$	8	$+$	$4 \cdot$	d

Step 4
$$60 = 4d + 8$$
$$\frac{-8 \quad\quad -8}{52 = 4d}$$
$$\frac{52}{4} = \frac{4d}{4}$$
$$13 = d$$

Step 5 The daily recommended vitamin C intake for a young child is 13 mg.

Step 6 8 more than four times 13:
$8 + 4 \cdot 13 = 8 + 52 = 60$
Equals the amount for an adult female: 60

The correct answer is 13 mg.

3.4 Solving Application Problems with Two Unknown Quantities

3.4 Margin Exercises

1. Step 1 The problem is about the amount made by Keonda and her daughter.

Unknowns: how much Keonda made;
how much the daughter made

Known: Keonda made \$12 more than her daughter. Together they made \$182.

Step 2(b) There are two unknowns. You know the least about the money made by the daughter, so let m be the money made by the daughter.

Since Keonda made \$12 more, $\underline{m + 12}$ represents Keonda's earnings.

Step 3

Daughter's money	+	Keonda's money	=	total money
↓	↓	↓	↓	↓
m	$+$	$m + 12$	$=$	182

Step 4
$$m + m + 12 = 182$$
$$2m + 12 = 182$$
$$\frac{-12 \quad\quad -12}{2m + 0 = 170}$$
$$\frac{2m}{2} = \frac{170}{2}$$
$$m = 85$$

Step 5 The daughter made \$85.
Keonda made \$85 + \$12 = \$97.

Step 6 \$97 is \$12 more than \$85 and the sum of their earnings is \$85 + \$97 = \$182 (matches).

2. Step 1 The problem is about putting fishing line on two reels.

Unknowns: lengths of the two lines

Known: The length of one line is 25 yd less than the length of the other line. The total length is 175 yd.

Step 2(b) There are two unknowns. Let r represent the length of the line on the first reel, then $r - 25$ is the length of the line on the second reel.

Step 3

length of first line	+	length of second line	=	total length
↓	↓	↓	↓	↓
r	$+$	$r - 25$	$=$	175

Step 4
$$r + (r - 25) = 175$$
$$2r - 25 = 175$$
$$\underline{+ 25 \quad\quad + 25}$$
$$2r = 200$$
$$\frac{2r}{2} = \frac{200}{2}$$
$$r = 100$$

Step 5   The first reel had 100 yards of line on it and the second reel had $100 - 25 = 75$ yards of line on it.

Step 6   75 yd is 25 yd less than 100 yd and $100 \text{ yd} + 75 \text{ yd} = 175 \text{ yd}$.

3.   Step 1   The problem is about a rectangular garden plot.

Unknowns: garden length and width

Known: The length is 3 yd longer than the width. The perimeter of the garden is 22 yd.

Step 2(b)   There are two unknowns. You know the least about the width, so let $\underline{w}$ represent the width.
The length is 3 yd more than the width, so let $\underline{w} + \underline{3}$ represent the length.

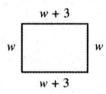

Step 3 $P = 2l + 2w$
$$22 = 2 \cdot (w + 3) + 2w$$

Step 4
$$22 = 2(w + 3) + 2w$$
$$22 = 2 \cdot w + 2 \cdot 3 + 2w$$
$$22 = 2w + 6 + 2w$$
$$22 = 4w + 6$$
$$\underline{-6 \quad\quad\quad -6}$$
$$16 = 4w + 0$$
$$\frac{16}{4} = \frac{4w}{4}$$
$$4 = w$$

Step 5   The width is 4 yd.
The length is $4 + 3 = 7$ yd.

Step 6   7 yd is 3 yd more than 4 yd.
$$P = 2 \cdot 7 \text{ yd} + 2 \cdot 4 \text{ yd}$$
$$P = 14 \text{ yd} + 8 \text{ yd}$$
$$= 22 \text{ yd (matches)}$$

3.4   Section Exercises

1.   (a) As you read the problem, identify the <u>known parts</u> and the <u>unknown parts</u>.

(b) When there is more than one unknown quantity, assign a variable to represent the thing you know the <u>least</u> about.

3.   Step 1   The problem is about someone's age.

Unknowns: how old I am; how old my sister is

Known: My sister is 9 years older than I am. The sum of our ages is 51.

Step 2(b)   Let a represent how old I am. Since my sister is 9 years older, let $a + 9$ represent her age.

Step 3

My age	+	sister's age	=	sum of ages
↓		↓		↓
a	+	$a + 9$	=	51

Step 4
$$a + a + 9 = 51$$
$$2a + 9 = 51$$
$$\underline{-9 \quad\quad -9}$$
$$2a + 0 = 42$$
$$\frac{2a}{2} = \frac{42}{2}$$
$$a = 21$$

Step 5   I am 21 years old, and my sister is $21 + 9 = 30$ years old.

Step 6   Check   30 is 9 more than 21 and the sum of our ages is $21 + 30 = 51$.

5.   Step 1   The problem concerns a couple's earnings.

Unknowns: how much each earned

Known: Lien earned $1500 more than her husband. Together they earned $37,500.

Step 2(b)   Let m represent the husband's earnings. Then Lien earned $m + 1500$.

Step 3

Husband's earnings	+	Lien's earnings	=	amount earned together
↓		↓		↓
m	+	$m + 1500$	=	37,500

Step 4
$$m + m + 1500 = 37,500$$
$$2m + 1500 = 37,500$$
$$\underline{-1500 \quad\quad\quad -1500}$$
$$2m + 0 = 36,000$$
$$\frac{2m}{2} = \frac{36,000}{2}$$
$$m = 18,000$$

Step 5   Her husband earned $18,000 and Lien earned $18,000 + \$1500 = \$19,500$.

Step 6 $19,500 is $1500 more than $18,000 and the sum of $18,000 and $19,500 is $37,500.

7. **Step 1** The problem is about the price of a computer and the price of a printer.

Unknowns: the computer's price and the printer's price

Known: The computer's price is eight times the printer's price. The total paid is $396.

Step 2(b) Let m represent the price of the printer. The price of the computer can be represented by $8 \cdot m$, or $8m$.

Step 3 The total price for both is $396.

$$
\begin{array}{ccccc}
\text{printer} & & \text{computer} & & \text{total} \\
\text{price} & + & \text{price} & = & \text{price} \\
\downarrow & & \downarrow & & \downarrow \\
m & + & 8m & = & 396
\end{array}
$$

Step 4
$$
\begin{aligned}
m + 8m &= 396 \\
9m &= 396 \\
\frac{9m}{9} &= \frac{396}{9} \\
m &= 44
\end{aligned}
$$

Step 5 The printer's price is $44. The computer's price is $8 \cdot \$44 = \$352$.

Step 6 Eight times $44 is $352 and the sum of $44 and $352 is $396.

9. **Step 1** The problem concerns cutting a board into two pieces.

Unknowns: the length of each piece of board

Known: The board had a total length of 78 cm. One piece is 10 cm longer than the other.

Step 2(b) Let x represent the length of the shorter piece. Then $x + 10$ is the length of the longer piece.

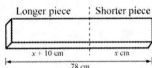

Step 3 The sum of the lengths is 78 cm, so
$$x + (x + 10) = 78.$$

Step 4
$$
\begin{aligned}
x + x + 10 &= 78 \\
2x + 10 &= 78 \\
-10 & -10 \\
\hline
2x + 0 &= 68 \\
\frac{2x}{2} &= \frac{68}{2} \\
x &= 34
\end{aligned}
$$

Step 5 The shorter length is $x = 34$ cm, so the longer length is $34 + 10 = 44$ cm.

Step 6 44 cm is 10 cm longer than 34 cm, and the sum of 34 cm and 44 cm is 78 cm.

11. **Step 1** The problem is about the lengths of two pieces of wire.

Unknowns: the length of each piece

Known: One piece is 7 ft shorter than the other. The wire was 31 ft long before it was cut into two pieces.

Step 2(b) Let x represent the length of the longer piece. Then $x - 7$ represents the piece that is 7 ft shorter.

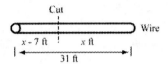

Step 3 The lengths of the two pieces together should be 31, so
$$x + (x - 7) = 31.$$

Step 4
$$
\begin{aligned}
x + x - 7 &= 31 \\
2x + (-7) &= 31 \\
7 & 7 \\
\hline
2x + 0 &= 38 \\
2x &= 38 \\
x &= 19
\end{aligned}
$$

Step 5 The longer piece is 19 ft. The shorter piece is $19 - 7 = 12$ ft.

Step 6 12 ft is 7 ft shorter than 19 ft and the sum of 19 ft and 12 ft is 31 ft.

13. **Step 1** The problem is about the number of Senators and Representatives in Congress.

Unknowns: The number of Senators and the number of Representatives

Known: The number of Representatives is 65 less than 5 times the number of Senators. The total number of Representatives and Senators is 535.

Step 2(b) Let s represent the number of Senators. Then the number of Representatives is $5s - 65$.

Step 3 Since the total number in both houses of Congress equals 535, we have
$$s + (5s - 65) = 535.$$

Step 4
$$
\begin{aligned}
s + 5s - 65 &= 535 \\
6s - 65 &= 535 \\
+ 65 & + 65 \\
\hline
6s + 0 &= 600 \\
\frac{6s}{6} &= \frac{600}{6} \\
s &= 100
\end{aligned}
$$

Step 5 The number of Senators is 100 and the number of Representatives is $5(100) - 65 = 435$.

Step 6 Since $5(100) - 65 = 435$ and $100 + 435 = 535$, this answer is correct.

15.   ***Step 1***   The problem is about cutting a fence into parts.

Unknowns: The length of each part

Known: Two parts are of equal length. The third part is 25 m longer than each of the other parts. The fence is 706 m long.

Step 2(b)   Let x represent the length of one of the two equal parts. Then $x + 25$ is the length of the third part.

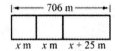

Step 3   Since the sum of the three parts is 706 m,

$$x + x + (x + 25) = 706.$$

Step 4
$$
\begin{array}{rcl}
x + x + x + 25 &=& 706 \\
3x + 25 &=& 706 \\
-25 & & -25 \\
\hline
3x + 0 &=& 681 \\
\dfrac{3x}{3} &=& \dfrac{681}{3} \\
x &=& 227
\end{array}
$$

Step 5   The two equal parts are each 227 m long, so the longer part is $x + 25 = 227 + 25 = 252$ m in length.

Step 6   Since $252 - 227 = 25$ and $227 + 227 + 252 = 706$, this answer is correct.

17.   ***Step 1***   The problem is about the length l of a rectangle.

Unknown: Rectangle's length

Known: Rectangle's width is 5 yd and perimeter is 48 yd

Step 2(a)   Let x represent the length.

$$
\begin{array}{c}
l \\
\text{5 yd} \boxed{} \text{5 yd} \\
l
\end{array}
$$

Step 3 $P = 2l + 2w$
$48 = 2 \cdot l + 2 \cdot 5$ *Perimeter = 48 yd*

Step 4
$$
\begin{array}{rcl}
48 &=& 2l + 10 \\
-10 & & -10 \\
\hline
38 &=& 2l + 0 \\
\dfrac{38}{2} &=& \dfrac{2l}{2} \\
19 &=& l
\end{array}
$$

Step 5   The rectangle's length is 19 yd.

Step 6 $P = 2l + 2w$
$P = 2 \cdot 19 + 2 \cdot 5$
$P = 38 + 10$
$P = 48$

19.   ***Step 1***   The problem is about the dimensions of a rectangular dog pen.

Unknowns: The length and width of the dog pen

Known: The perimeter is 36 ft. The length is twice the width.

Step 2(b)   Let w represent the width. Then the length equals $2w$.

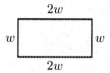

Step 3 $P = 2l + 2w$
$36 = 2 \cdot 2w + 2w$ *Perimeter = 36 ft*

Step 4
$$
\begin{array}{rcl}
36 &=& 4w + 2w \\
36 &=& 6w \\
\dfrac{36}{6} &=& \dfrac{6w}{6} \\
6 &=& w
\end{array}
$$

Step 5   The width is 6 ft, so the length is $2w = 2 \cdot 6 = 12$ ft.

Step 6   Since 12 is twice 6 and $2 \cdot 12 + 2 \cdot 6 = 24 + 12 = 36$, this answer is correct.

21.   ***Step 1***   The problem is about finding the length and the width of a jewelry box.

Unknowns: The length and width of the box

Known: The length is 3 in. more than twice the width. The perimeter is 36 in.

Step 2(b)   Let w represent the width. Then, since the length is 3 in. more than twice the width, use $2w + 3$ to represent the length.

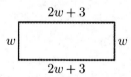

Step 3
$$
\begin{array}{c}
2l + 2w = P \\
2 \cdot (2w + 3) + 2 \cdot w = 36
\end{array}
$$

Step 4
$$
\begin{array}{rcl}
4w + 6 + 2w &=& 36 \\
6w + 6 &=& 36 \\
-6 & & -6 \\
\hline
6w + 0 &=& 30 \\
\dfrac{6w}{6} &=& \dfrac{30}{6} \\
w &=& 5
\end{array}
$$

Step 5 The width is 5 inches. The length is
$2(5) + 3 = 13$ inches.

Step 6 The length of 13 inches is 3 inches longer
than twice the width of 5 inches.

$$P = 2l + 2w$$
$$P = 2 \cdot 13 + 2 \cdot 5$$
$$P = 26 + 10$$
$$P = 36 \text{ in.}$$

23. ***Step 1*** The problem is about a framed
photograph.

Unknowns: The outside perimeter and the total
area of the photograph and frame

Known: The photo inside the frame is a rectangle
with length 10 in. and width 8 in. The frame is 2
in. wide.

Step 2(b) Let l and w represent the length and
width of the photograph. Let P represent the
outside perimeter and A represent the total area of
the photograph and frame.

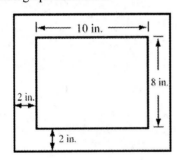

Step 3 $\quad P = 2l + 2w$
$$P = 2(10 + 2 + 2) + 2(8 + 2 + 2)$$

$$A = lw$$
$$A = (10 + 2 + 2) \cdot (8 + 2 + 2)$$

Step 4 $\quad P = 2(14) + 2(12) \qquad A = 14 \cdot 12$
$$ P = 28 + 24 \qquad\qquad A = 168 \text{ in.}^2$$
$$ P = 52 \text{ in.}$$

Step 5 The outside perimeter is 52 in. and the
total area of the photograph and frame is 168 in.^{2}.

Step 6
$$P = 12 \text{ in.} + 14 \text{ in.} + 12 \text{ in.} + 14 \text{ in.} = 52 \text{ in.}$$
$$A = 12 \cdot 14 = 168 \text{ in.}^2.$$

Chapter 3 Review Exercises

1. The figure is a square.

$$P = 4s$$
$$P = 4 \cdot 28 \text{ cm}$$
$$P = 112 \text{ cm}$$

2. The figure is a rectangle.

$$P = 2 \cdot l + 2 \cdot w$$
$$P = 2 \cdot 8 \text{ mi} + 2 \cdot 3 \text{ mi}$$
$$P = 16 \text{ mi} + 6 \text{ mi}$$
$$P = 22 \text{ mi}$$

3. The figure is a parallelogram.

$$P = \text{ sum of all four sides}$$
$$P = 7 \text{ yd} + 14 \text{ yd} + 7 \text{ yd} + 14 \text{ yd}$$
$$P = 42 \text{ yd}$$

4. $\quad P = \text{sum of all sides}$
$$P = 26 \text{ m} + 44 \text{ m} + 14 \text{ m} + 20 \text{ m} + 12 \text{ m} + 24 \text{ m}$$
$$P = 140 \text{ m}$$

5. $\qquad P = 4s$
$$12 \text{ ft} = 4s \qquad \textit{Replace P with 12 ft.}$$
$$\frac{12 \text{ ft}}{4} = \frac{4s}{4} \qquad \textit{Divide both sides by 4.}$$
$$3 \text{ ft} = s$$

One side of the table is 3 ft.

6. $\qquad P = 2l + 2w$
$$128 \text{ yd} = 2l + 2 \cdot 31 \text{ yd} \quad \begin{array}{l}\textit{Replace P with 128 yd} \\ \textit{and w with 31 yd.}\end{array}$$
$$128 \text{ yd} = 2l + 62 \text{ yd}$$
$$\underline{-62 \text{ yd} \qquad\quad -62 \text{ yd}} \quad \begin{array}{l}\textit{Add} -62 \textit{ yd} \\ \textit{to both sides.}\end{array}$$
$$66 \text{ yd} = 2l + 0$$
$$\frac{66 \text{ yd}}{2} = \frac{2l}{2} \qquad \begin{array}{l}\textit{Divide both} \\ \textit{sides by 2.}\end{array}$$
$$33 \text{ yd} = l$$

The length of the playground is 33 yd.

7. $\qquad P = 2l + 2w$
$$72 \text{ in.} = 2 \cdot 21 \text{ in.} + 2w \quad \begin{array}{l}\textit{Replace P with 72 in.} \\ \textit{and l with 21 in.}\end{array}$$
$$72 \text{ in.} = \quad 42 \text{ in.} + 2w$$
$$\underline{-42 \text{ in.} \qquad -42 \text{ in.}} \quad \begin{array}{l}\textit{Add} -42 \textit{ in.} \\ \textit{to both sides.}\end{array}$$
$$30 \text{ in.} = 0 + 2w$$
$$\frac{30 \text{ in.}}{2} = \frac{2w}{2} \qquad \begin{array}{l}\textit{Divide both} \\ \textit{sides by 2.}\end{array}$$
$$15 \text{ in.} = w$$

The width of the painting is 15 inches.

8.

```
       8 ft
     ┌──────┐
5 ft │      │ 5 ft
     └──────┘
       8 ft
```

$$A = l \cdot w \qquad \textit{Rectangle}$$
$$A = 8 \text{ ft} \cdot 5 \text{ ft}$$
$$A = 40 \text{ ft}^2$$

The area of the tablecloth is 40 ft^2.

9.

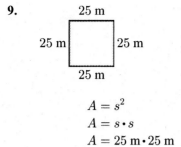

$$A = s^2 \qquad \textit{Square}$$
$$A = s \cdot s$$
$$A = 25 \text{ m} \cdot 25 \text{ m}$$
$$A = 625 \text{ m}^2$$

The area of the square dance floor is 625 m².

10.

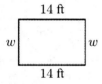

$$A = b \cdot h \qquad \textit{Parallelogram}$$
$$A = 16 \text{ yd} \cdot 13 \text{ yd}$$
$$A = 208 \text{ yd}^2$$

The area of the lot is 208 yd².

11.

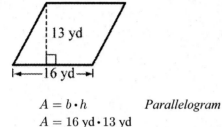

$$A = l \cdot w$$

$$126 \text{ ft}^2 = 14 \text{ ft} \cdot w \qquad \begin{array}{l}\textit{Replace A with 126 ft}^2\\ \textit{and l with 14 ft.}\end{array}$$

$$\frac{126 \text{ ft} \cdot \cancel{\text{ft}}}{14 \, \cancel{\text{ft}}} = \frac{14 \text{ ft} \cdot w}{14 \text{ ft}}$$

$$9 \text{ ft} = w$$

The width of the patio is 9 ft.

12.

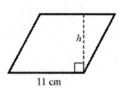

$$A = b \cdot h$$

$$88 \text{ cm}^2 = 11 \text{ cm} \cdot h \qquad \begin{array}{l}\textit{Replace A with 88 cm}^2\\ \textit{and b with 11 cm.}\end{array}$$

$$\frac{88 \text{ cm} \cdot \cancel{\text{cm}}}{11 \, \cancel{\text{cm}}} = \frac{11 \text{ cm} \cdot h}{11 \text{ cm}}$$

$$8 \text{ cm} = h$$

The height is 8 cm.

13.

$$A = s^2$$

$$100 \text{ mi}^2 = s^2 \qquad \begin{array}{l}\textit{Replace A with 100 mi}^2\\ \textit{and solve by inspection.}\end{array}$$

$$100 \text{ mi}^2 = s \cdot s$$
$$100 \text{ mi}^2 = 10 \text{ mi} \cdot 10 \text{ mi}$$

The length of one side of the piece of land is 10 mi.

14.   A number subtracted from 57

$$57 - x$$

(*Note:* $x - 57$ would be the translation of "57 subtracted from a number.")

15.   The sum of 15 and twice a number

$$15 + 2x \quad \text{or} \quad 2x + 15$$

16.   The product of -9 and a number

$$-9x$$

17.   Let n represent the unknown number.

$$\begin{array}{ccccc}\text{The sum of}\\ \text{four times}\\ \text{a number} & \text{and} & 6 & \text{is} & -30.\\ \downarrow & & \downarrow & \downarrow & \downarrow\\ 4n & + & 6 & = & -30\end{array}$$

$$\begin{aligned} 4n + 6 &= -30\\ -6 \quad\;\; &\quad -6\\ \hline 4n + 0 &= -36\\ \frac{4n}{4} &= \frac{-36}{4}\\ n &= -9 \end{aligned}$$

The number is -9.

18.   Let n represent the unknown number.

$$\begin{array}{cccccc}\text{When twice a} & & \text{the}\\ \text{number is} & & \text{result} & & & \text{the}\\ \text{subtracted from 10} & & \text{is} & 4 & \text{plus} & \text{number}\\ \downarrow & & \downarrow & \downarrow & \downarrow & \downarrow\\ 10 - 2n & & = & 4 & + & n\end{array}$$

$$\begin{aligned} 10 - 2n &= 4 + n\\ + 2n \quad\;\; &\quad\;\; + 2n\\ \hline 10 + 0 &= 4 + 3n\\ 10 &= 4 + 3n\\ -4 \quad &\quad -4\\ \hline 6 &= 0 + 3n\\ \frac{6}{3} &= \frac{3n}{3}\\ 2 &= n \end{aligned}$$

The number is 2.

19.   *Step 1*   Unknown: Amount in Grace's account before writing her rent check

Known: $600 check, $750 deposit, $75 deposit, $309 ending balance

Step 2(a)   Let m represent the amount of money originally in the account.

Step 3 $m - 600 + 750 + 75 = 309$

Step 4

$$m + (-600) + 750 + 75 = 309$$
$$m + 225 = 309$$
$$\underline{-225 \qquad -225}$$
$$m + 0 = 84$$
$$m = 84$$

Step 5 $84 was originally in Grace's account.

Step 6 $84 - $600 + $750 + $75
$= -$516 + $750 + 75
$= $234 + 75
$= 309 (ending balance checks)

20.   *Step 1*   Unknown: Number of candles in each box

Known: There were 4 boxes, one candle on each of 25 tables, 23 candles left.

Step 2(a)   Let c represent the number of candles per box.

Step 3

Total number of candles in 4 boxes	−	number of candles used	=	number of candles left
$4 \cdot c$	−	25	=	23

Step 4

$$4c - 25 = 23$$
$$\underline{+25 \qquad +25}$$
$$4c + 0 = 48$$
$$\frac{4c}{4} = \frac{48}{4}$$
$$c = 12$$

Step 5   There were 12 candles in each box.

Step 6   4 boxes with 12 candles per box $= 48$ candles.

25 candles used: $48 - 25 = 23$ left (matches)

21.   *Step 1*   Unknowns: How much Reggie won and how much Donald won

Known: $1000 prize, Donald should get $300 more than Reggie

Step 2(b)   There are two unknowns. You know the least about how much Reggie received in prize money. Let p represent the prize money won by Reggie. Then $p + 300$ represents the prize money won by Donald.

Step 3

Reggie's prize money	+	Donald's prize money	=	Total prize money
p	+	$p + 300$	=	1000

Step 4

$$p + p + 300 = 1000$$
$$2p + 300 = 1000$$
$$\underline{-300 \qquad -300}$$
$$2p + 0 = 700$$
$$\frac{2p}{2} = \frac{700}{2}$$
$$p = 350$$

Step 5   Reggie won $350 and Donald won $350 + $300 = $650.

Step 6 $650 is $300 more than $350 and the sum of $650 and $350 is $1000.

22.   *Step 1*   Unknowns: length and width of photograph

Known: Perimeter is 84 cm, the photo is twice as long as it is wide

Step 2(b)   There are two unknowns. You know the least about the width. Let w represent the width. Then $2w$ represents the length.

Step 3 $2l + 2w = P$

Step 4

$$2 \cdot 2w + 2w = 84 \text{ cm} \qquad \textit{Replace P with 84 cm and l with 2w.}$$
$$4w + 2w = 84 \text{ cm}$$
$$6w = 84 \text{ cm}$$
$$\frac{6w}{6} = \frac{84}{6} \text{ cm}$$
$$w = 14 \text{ cm}$$

Step 5   The width is 14 cm and the length is $2 \cdot 14 = 28$ cm.

Step 6

$$2 \cdot 28 \text{ cm} + 2 \cdot 14 \text{ cm} = P$$
$$56 \text{ cm} + 28 \text{ cm} = P$$
$$84 \text{ cm} = P \quad \text{(matches)}$$

23.   **(a)** **[3.1]** Kit #2 has 36 feet of fencing to make a square dog pen.

$$P = 4s$$
$$36 \text{ ft} = 4s \qquad \textit{Replace P with 36 ft.}$$
$$\frac{36 \text{ ft}}{4} = \frac{4s}{4}$$
$$9 \text{ ft} = s$$

The length of one side of the dog pen is 9 ft.

(b) **[3.2]** Find the area of the square dog pen.

$$A = s^2 \qquad s^2 = s \cdot s$$
$$A = 9 \text{ ft} \cdot 9 \text{ ft} \qquad 9 \text{ ft from part (a)}$$
$$A = 81 \text{ ft}^2$$

The area of the dog pen is 81 ft^2.

24. **[3.2]** Rectangles will vary. To create your rectangles, choose a length and width that add up to 10, which is one-half of the perimeter of 20. Two possibilities are:

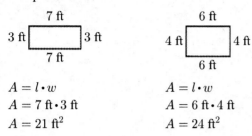

$$A = l \cdot w \qquad\qquad A = l \cdot w$$
$$A = 7 \text{ ft} \cdot 3 \text{ ft} \qquad A = 6 \text{ ft} \cdot 4 \text{ ft}$$
$$A = 21 \text{ ft}^2 \qquad\qquad A = 24 \text{ ft}^2$$

25. **[3.3]** *Step 1* Unknown: Amount of fencing Timotha put around her garden

Known: Kit #1 contains 20 ft, Kit #2 contains 36 ft, 41 ft of fencing available for dog pen

Step 2(a) Let f represent the number of feet of fencing for the garden.

Step 3

Fencing in Kit #2	−	Fencing around garden	+	Fencing in Kit #1	=	Fencing for dog pen
↓		↓		↓		↓
36	−	f	+	20	=	41

Step 4
$$36 - f + 20 = 41$$
$$56 - f = 41$$
$$\underline{-56 \qquad\quad -56}$$
$$-f = -15$$
$$\frac{-1f}{-1} = \frac{-15}{-1}$$
$$f = 15$$

Step 5 15 ft of fencing went around the garden.

Step 6 15 ft + 41 ft = 56 ft matches the total fencing from Kits #1 and #2 [20 + 36].

26. **[3.4]** *Step 1* Unknowns: The width and the length of the dog pen

Known: 36 ft of fencing was used, the length was 2 ft more than the width

Step 2(b) There are two unknowns. You know the least about the width. Let w represent the width. Then $w + 2$ represents the length.

Step 3
$$2l + 2w = P$$
$$2 \cdot (w + 2) + 2w = 36$$

Step 4
$$2(w + 2) + 2w = 36$$
$$2w + 4 + 2w = 36$$
$$4w + 4 = 36$$
$$\underline{-4 \qquad\quad -4}$$
$$4w = 32$$
$$\frac{4w}{4} = \frac{32}{4}$$
$$w = 8$$

Step 5 The width is 8 ft and the length is $8 + 2 = 10$ ft.

Step 6 10 ft is 2 ft longer than 8 ft.
$$P = 2 \cdot 10 \text{ ft} + 2 \cdot 8 \text{ ft}$$
$$P = 20 \text{ ft} + 16 \text{ ft}$$
$$P = 36 \text{ ft} \qquad\qquad \text{(matches)}$$

Chapter 3 Test

1. $P = 59 \text{ m} + 72 \text{ m} + 59 \text{ m} + 72 \text{ m}$
   $P = 262 \text{ m}$

2. $P = 8 \text{ in.} + 4 \text{ in.} + 4 \text{ in.} + 4 \text{ in.} + 8 \text{ in.}$
   $+ 4 \text{ in.} + 4 \text{ in.} + 4 \text{ in.}$
   $P = 40 \text{ in.}$

3. $P = 4s \qquad\qquad Square$
   $P = 4 \cdot 3 \text{ miles}$
   $P = 12 \text{ miles}$

4. $P = 2l + 2w \qquad\qquad Rectangle$
   $P = 2 \cdot 4 \text{ ft} + 2 \cdot 2 \text{ ft}$
   $P = 8 \text{ ft} + 4 \text{ ft}$
   $P = 12 \text{ ft}$

5. $P = 45 \text{ cm} + 50 \text{ cm} + 15 \text{ cm}$
   $P = 110 \text{ cm}$

6. $A = l \cdot w \qquad\qquad Rectangle$
   $A = 27 \text{ mm} \cdot 18 \text{ mm}$
   $A = 486 \text{ mm}^2$

7. *Note:* The base is perpendicular to the height.
   $A = b \cdot h \qquad\qquad Parallelogram$
   $A = 14 \text{ cm} \cdot 10 \text{ cm}$
   $A = 140 \text{ cm}^2$

8. $A = l \cdot w \qquad\qquad Rectangle$
   $A = 68 \text{ mi} \cdot 55 \text{ mi}$
   $A = 3740 \text{ mi}^2$

9. $A = s^2 \qquad\qquad Square$
   $A = s \cdot s$
   $A = 6 \text{ m} \cdot 6 \text{ m}$
   $A = 36 \text{ m}^2$

10.
$$P = 4s \quad Square$$
$$12 \text{ ft} = 4s$$
$$\frac{12 \text{ ft}}{4} = \frac{4s}{4}$$
$$3 \text{ ft} = s$$

The length of one side of the table is 3 ft.

11.
$$P = 2l + 2w$$
$$34 \text{ ft} = 2l + 2(6 \text{ ft})$$
$$34 \text{ ft} = 2l + 12 \text{ ft}$$
$$\underline{-12 \text{ ft} \qquad -12 \text{ ft}}$$
$$22 \text{ ft} = 2l + 0$$
$$\frac{22 \text{ ft}}{2} = \frac{2l}{2}$$
$$11 \text{ ft} = l$$

The length of the plot is 11 ft.

12.
$$A = bh$$
$$65 \text{ in.}^2 = 13 \text{ in.} \cdot h$$
$$\frac{65 \text{ in.} \cdot \cancel{in.}}{13 \cancel{in.}} = \frac{13 \text{ in.} \cdot h}{13 \text{ in.}}$$
$$5 \text{ in.} = h$$

The parallelogram's height is 5 inches.

13.
$$A = l \cdot w$$
$$12 \text{ cm}^2 = 4 \text{ cm} \cdot w$$
$$\frac{12 \text{ cm} \cdot \cancel{cm}}{4 \cancel{cm}} = \frac{4 \text{ cm} \cdot w}{4 \text{ cm}}$$
$$3 \text{ cm} = w$$

The width of the postage stamp is 3 cm.

14.
$$A = s^2$$
$$A = s \cdot s$$
$$16 \text{ ft}^2 = s \cdot s \qquad$$ *Replace A with 16 ft² and solve by inspection.*
$$16 \text{ ft}^2 = 4 \text{ ft} \cdot 4 \text{ ft}$$

Each side of the bulletin board is 4 feet.

15. Answers will vary; here is one possibility. Linear units like ft are used to measure length, width, height, and perimeter. Area is measured in square units like ft² (squares that measure 1 ft on each side).

16. Let n represent the unknown number.

If 40 added to	is	four times a number	the result is	zero.
↓				↓
40 +		4n	=	0

$$40 + 4n = 0$$
$$\underline{-40 \qquad\quad -40}$$
$$0 + 4n = -40$$
$$\frac{4n}{4} = \frac{-40}{4}$$
$$n = -10$$

The number is -10.

17. Let n represent the unknown number.

When 7 times a number is decreased by 23	the result is	the number plus 7.
$7n - 23$	=	$n + 7$

$$7n - 23 = n + 7$$
$$\underline{-n \qquad\quad -n}$$
$$6n - 23 = 7$$
$$\underline{+ 23 \qquad + 23}$$
$$6n + 0 = 30$$
$$6n = 30$$
$$\frac{6n}{6} = \frac{30}{6}$$
$$n = 5$$

The number is 5.

18. **Step 1** Unknown: Amount of money used by son

Known: $43 originally, some spent by son on groceries, $16 put into wallet, and $44 at the end

Step 2(a) Let m represent the money spent on groceries.

Step 3 $43 - m + 16 = 44$

Step 4
$$43 - m + 16 = 44$$
$$59 - m = 44$$
$$\underline{-59 \qquad\quad -59}$$
$$0 - m = -15$$
$$\frac{-1m}{-1} = \frac{-15}{-1}$$
$$m = 15$$

Step 5 Her son spent $15.

Step 6 $43 - $15 + $16 = $44 (matches)

19. **Step 1** Unknown: Daughter's age

Known: Ray is 39 years old and his age is 4 years more than five times his daughter's age.

Step 2(a) Let d represent the daughter's age.

Step 3

Ray's age	is	4 years more than five times his daughter's age.
39	=	$4 + 5d$

Step 4
$$39 = 4 + 5d$$
$$\underline{-4 \qquad -4}$$
$$35 = 0 + 5d$$
$$\frac{35}{5} = \frac{5d}{5}$$
$$7 = d$$

Step 5   The daughter is 7 years old.

Step 6   Four years more than five times 7 years is $4 + 5 \cdot 7 = 4 + 35 = 39$ yrs (matches).

20.   Step 1   Unknowns: The length of each piece

Known: Total board length is 118 cm, one piece is 4 cm longer than the other.

Step 2(b)   There are two unknowns. Let p represent the length of the shorter piece. Then $p + 4$ represents the length of the longer piece.

Step 3

$$\underbrace{\text{long piece}}_{p+4} + \underbrace{\text{short piece}}_{p} = \underbrace{\text{total length}}_{118}$$

Step 4
$$p + 4 + p = 118$$
$$2p + 4 = 118$$
$$\underline{\quad -4 \qquad -4}$$
$$2p + 0 = 114$$
$$\frac{2p}{2} = \frac{114}{2}$$
$$p = 57$$

Step 5   One piece is 57 cm and the other is $57 + 4 = 61$ cm.

Step 6   61 cm is 4 cm longer than 57 cm and the sum of 61 cm and 57 cm is 118 cm.

21.   Step 1   Unknowns: Length and width

Known: Perimeter is 420 ft, length is four times as long as the width

Step 2(b)   Let w represent the width. Then $4w$ represents the length.

Step 3
$$P = 2l + 2w$$
$$420 \text{ ft} = 2 \cdot 4w + 2 \cdot w$$

Step 4
$$420 = 8w + 2w$$
$$420 = 10w$$
$$\frac{420}{10} = \frac{10w}{10}$$
$$42 = w$$

Step 5   The width is 42 ft and the length is $4(42) = 168$ ft.

Step 6   168 ft is four times 42 ft.
$$P = 2 \cdot 168 \text{ ft} + 2 \cdot 42 \text{ ft}$$
$$P = 336 \text{ ft} + 84 \text{ ft}$$
$$P = 420 \text{ ft} \qquad \text{(matches)}$$

22.   Step 1   Unknowns: How many hours each person worked

Known: 19 hours worked by both, Tim worked 3 hours less than Marcella

Step 2(b)   Let h represent the hours worked by Marcella. Then $h - 3$ represents the hours worked by Tim.

Step 3

$$\underbrace{\text{Tim's hours}}_{h-3} + \underbrace{\text{Marcella's hours}}_{h} = \underbrace{\text{Total hours}}_{19}$$

Step 4
$$h - 3 + h = 19$$
$$2h - 3 = 19$$
$$\underline{\qquad +3 \qquad +3}$$
$$2h + 0 = 22$$
$$\frac{2h}{2} = \frac{22}{2}$$
$$h = 11$$

Step 5   Marcella worked 11 hours and Tim worked $11 - 3 = 8$ hours.

Step 6   8 hours is 3 hours less than 11 hours and the sum of 8 and 11 is 19.

CHAPTER 4 RATIONAL NUMBERS: POSITIVE AND NEGATIVE FRACTIONS

4.1 Introduction to Signed Fractions

4.1 Margin Exercises

1. (a) Number of shaded parts $\rightarrow 3$
       Number of equal parts $\rightarrow 5$

   The 3 shaded parts are represented by the fraction $\frac{3}{5}$.

   Number of **un**shaded parts $\rightarrow 2$
   Number of equal parts $\rightarrow 5$

   The 2 shaded parts are represented by the fraction $\frac{2}{5}$.

   (b) The 1 shaded part is represented by the fraction $\frac{1}{6}$; the unshaded parts by $\frac{5}{6}$.

   (c) The 7 shaded parts are represented by the fraction $\frac{7}{8}$; the unshaded part by $\frac{1}{8}$.

2. (a) An area equal to 8 of the $\frac{1}{7}$ parts is shaded, so $\frac{8}{7}$ is shaded.

   (b) An area equal to 7 of the $\frac{1}{4}$ parts is shaded, so $\frac{7}{4}$ is shaded.

3. (a) $\frac{2}{3}$ ← 2 is the *numerator*
       ← 3 is the *denominator*

   (b) $\frac{1}{4}$ ← 1 is the *numerator*
       ← 4 is the *denominator*

   (c) $\frac{8}{5}$ ← 8 is the *numerator*
       ← 5 is the *denominator*

   (d) $\frac{5}{2}$ ← 5 is the *numerator*
       ← 2 is the *denominator*

   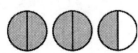

4. (a) Proper fractions have a numerator that is *less* than the denominator.

   $$\frac{3}{4}, \frac{5}{7}, \frac{1}{2}$$

   (b) Improper fractions have a numerator that is *equal to or greater* than the denominator.

   $$\frac{8}{7}, \frac{6}{6}, \frac{2}{1}$$

5. (a) $\frac{2}{4}$ is positive. Divide the space between 0 and 1 into 4 equal parts. Start at 0 and count 2 parts to the right.

   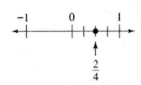

   (b) $\frac{1}{2}$ is positive. Divide the space between 0 and 1 into 2 equal parts. Start at 0 and count 1 part to the right.

   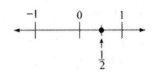

   (c) $-\frac{2}{3}$ is negative. Divide the space between 0 and -1 into 3 equal parts. Start at 0 and count 2 parts to the left.

   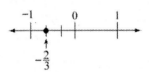

6. (a) $\left|-\frac{3}{4}\right| = \frac{3}{4}$ because the distance from 0 to $-\frac{3}{4}$ on the number line is $\frac{3}{4}$.

   (b) $\left|\frac{5}{8}\right| = \frac{5}{8}$ because the distance from 0 to $\frac{5}{8}$ on the number line is $\frac{5}{8}$.

   (c) $|0| = 0$ because the distance from 0 to 0 on the number line is 0.

7. (a) $\dfrac{2}{5} = \dfrac{?}{20}$

   Multiply 5 by $\underline{4}$ to get the denominator of 20. So multiply both the numerator and denominator by 4.

   $$\frac{2}{5} = \frac{2 \cdot 4}{5 \cdot 4} = \frac{8}{20}$$

   (b) $-\dfrac{21}{28} = \dfrac{?}{4}$

   Divide 28 by 7 to get the denominator of 4. So divide both the numerator and denominator by 7.

   $$-\frac{21}{28} = -\frac{21 \div 7}{28 \div 7} = -\frac{3}{4}$$

8. (a) $\frac{10}{10}$ ▪ Think of $\frac{10}{10}$ as $10 \div 10$.
       The result is $\underline{1}$, so $\frac{10}{10} = \underline{1}$.

   (b) $-\frac{3}{1}$ ▪ Think of $-\frac{3}{1}$ as $-3 \div 1$.
       The result is -3, so $-\frac{3}{1} = -3$.

   (c) $\frac{8}{2}$ ▪ Think of $\frac{8}{2}$ as $8 \div 2$.
       The result is 4, so $\frac{8}{2} = 4$.

(d) $-\frac{25}{5}$ ■   Think of $-\frac{25}{5}$ as $-25 \div 5$.

The result is -5, so $-\frac{25}{5} = -5$.

4.1   Section Exercises

1. (a) When using fractions, the whole must be cut into <u>parts of equal size</u>.

   (b) In a fraction, the <u>denominator</u> tells how many equal parts are in the whole.

3. Number of shaded parts → 5
   Number of equal parts → 8

   The 5 shaded parts are represented by the fraction $\frac{5}{8}$; the 3 unshaded parts by $\frac{3}{8}$.

5. The figure has 3 equal parts. The 2 shaded parts are represented by the fraction $\frac{2}{3}$; the 1 unshaded part by $\frac{1}{3}$.

7. An area equal to 3 of the $\frac{1}{2}$ parts is shaded: $\frac{3}{2}$

   An area equal to 1 of the $\frac{1}{2}$ parts is unshaded: $\frac{1}{2}$

9. An area equal to 11 of the $\frac{1}{6}$ parts is shaded: $\frac{11}{6}$

   An area equal to 1 of the $\frac{1}{6}$ parts is unshaded: $\frac{1}{6}$

11. Two of the 11 coins are dimes: $\frac{2}{11}$

    Three of the 11 coins are pennies: $\frac{3}{11}$

    Four of the 11 coins are nickels: $\frac{4}{11}$

13. (a) 6 of 20 women would like flowers delivered at work: $\frac{6}{20}$

    (b) $13 + 6 = 19$ of 20 women would like flowers delivered either at home or at work: $\frac{19}{20}$

15. $\frac{3}{4}$ ← 3 is the *numerator*
         ← 4 is the *denominator*

    There are 4 equal parts in the whole.

17. $\frac{12}{7}$ ← 12 is the *numerator*
          ← 7 is the *denominator*

    There are 7 equal parts in the whole.

19. Proper fractions have a numerator that is *less* than the denominator.

    $$\frac{1}{3}, \frac{5}{8}, \frac{7}{16}$$

    Improper fractions have a numerator that is *equal to or greater* than the denominator.

    $$\frac{8}{5}, \frac{6}{6}, \frac{12}{2}$$

21. To graph $\frac{1}{4}$ and $-\frac{1}{4}$ on the number line, divide the space between 0 and 1 into 4 equal parts. Start at zero and count 1 part to the right. This spot represents $\frac{1}{4}$. Now divide the space between -1 and 0 into 4 equal parts. Start at zero and count 1 part to the left. This spot represents $-\frac{1}{4}$.

23. Graph $-\frac{3}{5}$ and $\frac{3}{5}$ on the number line.

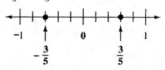

25. Graph $\frac{7}{8}$ and $-\frac{7}{8}$ on the number line.

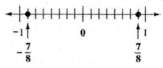

27. "lost" $\frac{3}{4}$ pound: $-\frac{3}{4}$ pound

29. $\frac{3}{10}$ mile long: $+\frac{3}{10}$ mile or $\frac{3}{10}$ mile

31. $\left|-\frac{2}{5}\right| = \frac{2}{5}$ because the distance from 0 to $-\frac{2}{5}$ on the number line is $\frac{2}{5}$.

33. $|0| = 0$ because the distance from 0 to 0 on the number line is 0.

35. (a) $\dfrac{1}{2} = \dfrac{?}{24}$

    Multiply 2 by <u>12</u> to get the denominator of 24. So multiply both the numerator and denominator by 12.

    $$\frac{1}{2} = \frac{1 \cdot 12}{2 \cdot 12} = \frac{12}{24}$$

    (b) $\dfrac{1}{3} = \dfrac{1 \cdot 8}{3 \cdot 8} = \dfrac{8}{24}$

    (c) $\dfrac{2}{3} = \dfrac{2 \cdot 8}{3 \cdot 8} = \dfrac{16}{24}$

    (d) $\dfrac{1}{4} = \dfrac{1 \cdot 6}{4 \cdot 6} = \dfrac{6}{24}$

    (e) $\dfrac{3}{4} = \dfrac{3 \cdot 6}{4 \cdot 6} = \dfrac{18}{24}$

    (f) $\dfrac{1}{6} = \dfrac{1 \cdot 4}{6 \cdot 4} = \dfrac{4}{24}$

    (g) $\dfrac{5}{6} = \dfrac{5 \cdot 4}{6 \cdot 4} = \dfrac{20}{24}$

    (h) $\dfrac{1}{8} = \dfrac{1 \cdot 3}{8 \cdot 3} = \dfrac{3}{24}$

    (i) $\dfrac{3}{8} = \dfrac{3 \cdot 3}{8 \cdot 3} = \dfrac{9}{24}$

    (j) $\dfrac{5}{8} = \dfrac{5 \cdot 3}{8 \cdot 3} = \dfrac{15}{24}$

37. (a) $-\dfrac{2}{6} = -\dfrac{2 \div 2}{6 \div 2} = -\dfrac{1}{3}$

    (b) $-\dfrac{4}{6} = -\dfrac{4 \div 2}{6 \div 2} = -\dfrac{2}{3}$

(c) $-\dfrac{12}{18} = -\dfrac{12 \div 6}{18 \div 6} = -\dfrac{2}{3}$

(d) $-\dfrac{6}{18} = -\dfrac{6 \div 6}{18 \div 6} = -\dfrac{1}{3}$

(e) $-\dfrac{200}{300} = -\dfrac{200 \div 100}{300 \div 100} = -\dfrac{2}{3}$

(f) Some possibilities are: $-\dfrac{4}{12} = -\dfrac{1}{3}$; $-\dfrac{8}{24} = -\dfrac{1}{3}$; $-\dfrac{20}{30} = -\dfrac{2}{3}$; $-\dfrac{24}{36} = -\dfrac{2}{3}$.

39. You cannot do it if you want the numerator to be a whole number, because 5 does not divide into 18 evenly. You could use multiples of 5 as the denominator, such as 10, 15, 20, etc.

41. $\dfrac{10}{1} = 10 \div 1 = \underline{10}$

43. $-\dfrac{16}{16} = -16 \div 16 = -1$

45. $-\dfrac{18}{3} = -18 \div 3 = -6$

47. $\dfrac{24}{8} = 24 \div 8 = 3$

49. $\dfrac{14}{7} = 14 \div 7 = 2$

51. $-\dfrac{90}{10} = -90 \div 10 = -9$

53. $\dfrac{150}{150} = 150 \div 150 = 1$

55. $-\dfrac{32}{4} = -32 \div 4 = -8$

57. $\frac{3}{5}$ is shaded; $\frac{2}{5}$ is unshaded.

59. 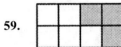 $\frac{3}{8}$ is shaded; $\frac{5}{8}$ is unshaded.

61. One possibility is shown.

63. One possibility is shown.

(!)!!!,...???

Relating Concepts (Exercises 65–68)

65. (a) $\dfrac{3}{8} = \dfrac{3 \cdot 489}{8 \cdot 489} = \dfrac{1467}{3912}$

(b) Divide 3912 by 8 to get 489; multiply 3 by 489 to get 1467.

66. (a) $\dfrac{7}{9} = \dfrac{7 \cdot 608}{9 \cdot 608} = \dfrac{4256}{5472}$

(b) Divide 5472 by 9 to get 608; multiply 7 by 608 to get 4256.

67. (a) $-\dfrac{697}{3485} = -\dfrac{697 \div 697}{3485 \div 697} = -\dfrac{1}{5}$

(b) Divide 3485 by 2, by 3, and by 5 to see that dividing by 5 gives 697. Or divide 3485 by 697 to get 5.

68. (a) $-\dfrac{817}{4902} = -\dfrac{817 \div 817}{4902 \div 817} = -\dfrac{1}{6}$

(b) Divide 4902 by 4, by 6, and by 8 to see that dividing by 6 gives 817. Or divide 4902 by 817 to get 6.

4.2 Writing Fractions in Lowest Terms

4.2 Margin Exercises

1. (a) $\frac{2}{3}$ is in lowest terms since the numerator and denominator have no common factor other than 1.

(b) $-\frac{8}{10}$ is not in lowest terms. A common factor of 8 and 10 is 2.

(c) $-\frac{9}{11}$ is in lowest terms since the numerator and denominator have no common factor other than 1.

(d) $\frac{15}{20}$ is not in lowest terms. A common factor of 15 and 20 is 5.

2. (a) $\dfrac{5}{10}$ ▪ $\underline{5}$ is a common factor of 5 and 10.

$$\dfrac{5}{10} = \dfrac{5 \div 5}{10 \div 5} = \dfrac{1}{2}$$

(b) $\dfrac{9}{12} = \dfrac{9 \div 3}{12 \div 3} = \dfrac{3}{4}$

(c) $-\dfrac{24}{30} = -\dfrac{24 \div 6}{30 \div 6} = -\dfrac{4}{5}$

(d) $\dfrac{15}{40} = \dfrac{15 \div 5}{40 \div 5} = \dfrac{3}{8}$

(e) $-\dfrac{50}{90} = -\dfrac{50 \div 10}{90 \div 10} = -\dfrac{5}{9}$

3. 1 is *neither* prime nor composite.

$2, 3, 7, 13, 19,$ and 29 are *prime* since each of these numbers is divisible only by itself and 1.

$4, 9,$ and 25 are *composite*:

4 because 4 can be divided by 2.
9 because 9 can be divided by 3.
25 because 25 can be divided by 5.

4. (a)

$\begin{array}{l} 1 \quad \textit{Quotient is 1.} \\ 2\lceil 2 \quad \textit{Divide 2 by 2.} \\ 2\lceil 4 \quad \textit{Divide 4 by 2.} \\ 2\lceil 8 \quad \textit{Divide 8 by 2.} \end{array}$

The prime factorization of 8 is $\underline{2} \cdot \underline{2} \cdot \underline{2}$.

(b) $\begin{array}{l} 1 \\ 7\overline{)7} \\ 3\overline{)21} \\ 2\overline{)42} \end{array}$ *Quotient is 1.*
7 is not divisible by 2, 3, or 5; use 7.
21 is not divisible by 2; use 3.
Divide 42 by 2.

The prime factorization of 42 is $2 \cdot 3 \cdot 7$.

(c) $\begin{array}{l} 1 \\ 5\overline{)5} \\ 3\overline{)15} \\ 3\overline{)45} \\ 2\overline{)90} \end{array}$ *Quotient is 1.*
5 is not divisible by 2 or 3; use 5.
Divide 15 by 3.
45 is not divisible by 2; use 3.
Divide 90 by 2.

The prime factorization of 90 is $2 \cdot 3 \cdot 3 \cdot 5$.

(d) $\begin{array}{l} 1 \\ 5\overline{)5} \\ 5\overline{)25} \\ 2\overline{)50} \\ 2\overline{)100} \end{array}$ *Quotient is 1.*
Divide 5 by 5.
25 is not divisible by 2 or 3; use 5.
Divide 50 by 2.
Divide 100 by 2.

The prime factorization of 100 is $2 \cdot 2 \cdot 5 \cdot 5$.

(e) $\begin{array}{l} 1 \\ 3\overline{)3} \\ 3\overline{)9} \\ 3\overline{)27} \\ 3\overline{)81} \end{array}$ *Quotient is 1.*
Divide 3 by 3.
Divide 9 by 3.
Divide 27 by 3.
81 is not divisible by 2; use 3.

The prime factorization of 81 is $3 \cdot 3 \cdot 3 \cdot 3$.

5. (a)

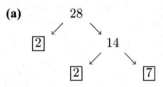

$28 = 2 \cdot 2 \cdot 7$

(b)

$35 = 5 \cdot 7$

(c)

$90 = 2 \cdot 3 \cdot 3 \cdot 5$

6. (a) $\dfrac{16}{48} = \dfrac{\overset{1}{\cancel{2}} \cdot \overset{1}{\cancel{2}} \cdot \overset{1}{\cancel{2}} \cdot \overset{1}{\cancel{2}}}{\underset{1}{\cancel{2}} \cdot \underset{1}{\cancel{2}} \cdot \underset{1}{\cancel{2}} \cdot \underset{1}{\cancel{2}} \cdot 3}$

$= \dfrac{1}{3} \quad\begin{array}{l}\leftarrow \text{Multiply } 1 \cdot 1 \cdot 1 \cdot 1 \\ \leftarrow \text{Multiply } 1 \cdot 1 \cdot 1 \cdot 1 \cdot 3\end{array}$

(b) $\dfrac{28}{60} = \dfrac{\overset{1}{\cancel{2}} \cdot \overset{1}{\cancel{2}} \cdot 7}{\underset{1}{\cancel{2}} \cdot \underset{1}{\cancel{2}} \cdot 3 \cdot 5}$

$= \dfrac{7}{15} \quad\begin{array}{l}\leftarrow \text{Multiply } 1 \cdot 1 \cdot 7 \\ \leftarrow \text{Multiply } 1 \cdot 1 \cdot 3 \cdot 5\end{array}$

(c) $\dfrac{74}{111} = \dfrac{2 \cdot \overset{1}{\cancel{37}}}{3 \cdot \underset{1}{\cancel{37}}}$

$= \dfrac{2}{3} \quad\begin{array}{l}\leftarrow \text{Multiply } 2 \cdot 1 \\ \leftarrow \text{Multiply } 3 \cdot 1\end{array}$

(d) $\dfrac{124}{340} = \dfrac{\overset{1}{\cancel{2}} \cdot \overset{1}{\cancel{2}} \cdot 31}{\underset{1}{\cancel{2}} \cdot \underset{1}{\cancel{2}} \cdot 5 \cdot 17}$

$= \dfrac{31}{85} \quad\begin{array}{l}\leftarrow \text{Multiply } 1 \cdot 1 \cdot 31 \\ \leftarrow \text{Multiply } 1 \cdot 1 \cdot 5 \cdot 17\end{array}$

7. (a) $\dfrac{5c}{15} = \dfrac{\overset{1}{\cancel{5}} \cdot c}{3 \cdot \underset{1}{\cancel{5}}} = \dfrac{1c}{3}, \quad\text{or}\quad \dfrac{c}{3}$

(b) $\dfrac{10x^2}{8x^2} = \dfrac{\overset{1}{\cancel{2}} \cdot 5 \cdot \overset{1}{\cancel{x}} \cdot \overset{1}{\cancel{x}}}{\underset{1}{\cancel{2}} \cdot 2 \cdot 2 \cdot \underset{1}{\cancel{x}} \cdot \underset{1}{\cancel{x}}} = \dfrac{5}{4}$

(c) $\dfrac{9a^3}{11b^3} = \dfrac{9 \cdot a \cdot a \cdot a}{11 \cdot b \cdot b \cdot b}$

There are no common factors, so the fraction is already in lowest terms.

(d) $\dfrac{6m^2n}{9n^2} = \dfrac{2 \cdot \overset{1}{\cancel{3}} \cdot m \cdot m \cdot \overset{1}{\cancel{n}}}{\underset{1}{\cancel{3}} \cdot 3 \cdot \underset{1}{\cancel{n}} \cdot n} = \dfrac{2m^2}{3n}$

4.2   Section Exercises

1. (a) $-\frac{3}{10}$ is in lowest terms since the numerator and denominator have no common factor other than 1.

(b) $\frac{10}{15}$ is not in lowest terms. A common factor of 10 and 15 is 5.

(c) $\frac{9}{16}$ is in lowest terms.

(d) $-\frac{4}{21}$ is in lowest terms.

(e) $\frac{6}{9}$ is not in lowest terms. A common factor of 6 and 9 is 3.

(f) $-\frac{7}{28}$ is not in lowest terms. A common factor of 7 and 28 is 7.

3. (a) $\dfrac{10}{15} = \dfrac{10 \div 5}{15 \div 5} = \dfrac{2}{3}$

(b) $\dfrac{6}{9} = \dfrac{6 \div 3}{9 \div 3} = \dfrac{2}{3}$

(c) $-\dfrac{7}{28} = -\dfrac{7 \div 7}{28 \div 7} = -\dfrac{1}{4}$

(d) $-\dfrac{25}{50} = -\dfrac{25 \div 25}{50 \div 25} = -\dfrac{1}{2}$

(e) $\dfrac{16}{18} = \dfrac{16 \div 2}{18 \div 2} = \dfrac{8}{9}$

(f) $-\dfrac{8}{20} = -\dfrac{8 \div 4}{20 \div 4} = -\dfrac{2}{5}$

5. 1 is *neither* prime nor composite.

2, 5, and 11 are *prime* since each of these numbers is divisible only by itself and 1.

9, 8, 10, and 21 are *composite*:

    9 since 9 is divisible by 3
    8 since 8 is divisible by 2 and 4
    10 since 10 is divisible by 2 and 5
    21 since 21 is divisible by 3 and 7

7.

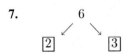

The prime factorization of 6 is $2 \cdot 3$.

9.
$$\begin{array}{ll} 1 & \textit{Quotient is 1.} \\ 5\overline{)5} & \text{5 is not divisible by 2 or 3; use 5.} \\ 2\overline{)10} & \text{Divide 10 by 2.} \\ 2\overline{)20} & \text{Divide 20 by 2.} \end{array}$$

The prime factorization of 20 is $2 \cdot 2 \cdot 5$.

11.

The prime factorization of 25 is $5 \cdot 5$.

13.

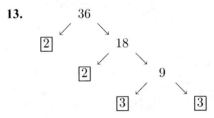

The prime factorization of 36 is $2 \cdot 2 \cdot 3 \cdot 3$.

15. (a)

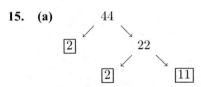

The prime factorization of 44 is $2 \cdot 2 \cdot 11$.

(b) We recognize that 88 is 2 times 44, so using the prime factorization from part (a), we see that the prime factorization of 88 is $2 \cdot 2 \cdot 2 \cdot 11$.

17. (a)

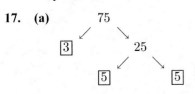

The prime factorization of 75 is $3 \cdot 5 \cdot 5$.

(b)

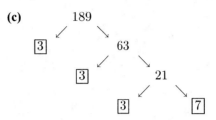

The prime factorization of 68 is $2 \cdot 2 \cdot 17$.

(c)

189
/ \
③ 63
/ \
③ 21
/ \
③ ⑦

The prime factorization of 189 is $3 \cdot 3 \cdot 3 \cdot 7$.

19. $\dfrac{8}{16} = \dfrac{2 \cdot 2 \cdot 2}{2 \cdot 2 \cdot 2 \cdot 2} = \dfrac{\cancel{2} \cdot \cancel{2} \cdot \cancel{2}}{\cancel{2} \cdot \cancel{2} \cdot \cancel{2} \cdot 2} = \dfrac{1}{2}$

21. $\dfrac{32}{48} = \dfrac{2 \cdot 2 \cdot 2 \cdot 2 \cdot 2}{2 \cdot 2 \cdot 2 \cdot 2 \cdot 3} = \dfrac{\cancel{2} \cdot \cancel{2} \cdot \cancel{2} \cdot \cancel{2} \cdot 2}{\cancel{2} \cdot \cancel{2} \cdot \cancel{2} \cdot \cancel{2} \cdot 3} = \dfrac{2}{3}$

23. $\dfrac{14}{21} = \dfrac{2 \cdot 7}{3 \cdot 7} = \dfrac{2 \cdot \cancel{7}}{3 \cdot \cancel{7}} = \dfrac{2}{3}$

25. $\dfrac{36}{42} = \dfrac{\cancel{2} \cdot 2 \cdot \cancel{3} \cdot 3}{\cancel{2} \cdot \cancel{3} \cdot 7} = \dfrac{6}{7}$

27. $\dfrac{50}{63} = \dfrac{2 \cdot 5 \cdot 5}{3 \cdot 3 \cdot 7}$

The fraction is already in lowest terms.

29. $\dfrac{27}{45} = \dfrac{\cancel{3} \cdot \cancel{3} \cdot 3}{\cancel{3} \cdot \cancel{3} \cdot 5} = \dfrac{3}{5}$

31. $\dfrac{12}{18} = \dfrac{\cancel{2} \cdot 2 \cdot \cancel{3}}{\cancel{2} \cdot \cancel{3} \cdot 3} = \dfrac{2}{3}$

33. $\dfrac{35}{40} = \dfrac{\cancel{5} \cdot 7}{2 \cdot 2 \cdot 2 \cdot \cancel{5}} = \dfrac{7}{8}$

35. $\dfrac{90}{180} = \dfrac{\cancel{2} \cdot \cancel{3} \cdot \cancel{3} \cdot \cancel{5}}{\cancel{2} \cdot 2 \cdot \cancel{3} \cdot \cancel{3} \cdot \cancel{5}} = \dfrac{1}{2}$

37. $\dfrac{210}{315} = \dfrac{2 \cdot \cancel{3} \cdot \cancel{5} \cdot \cancel{7}}{3 \cdot \cancel{3} \cdot \cancel{5} \cdot \cancel{7}} = \dfrac{2}{3}$

39. $\dfrac{429}{495} = \dfrac{\cancel{3} \cdot \cancel{11} \cdot 13}{\cancel{3} \cdot 3 \cdot 5 \cdot \cancel{11}} = \dfrac{13}{15}$

41. 60 minutes in an hour ⇒ 60 parts in the whole

(a) $\dfrac{15}{60} = \dfrac{\cancel{3}\cdot\cancel{5}}{2\cdot2\cdot\cancel{3}\cdot\cancel{5}} = \dfrac{1}{4}$ of an hour

(b) $\dfrac{30}{60} = \dfrac{\cancel{2}\cdot\cancel{3}\cdot\cancel{5}}{2\cdot2\cdot\cancel{3}\cdot\cancel{5}} = \dfrac{1}{2}$ of an hour

(c) $\dfrac{6}{60} = \dfrac{\cancel{2}\cdot\cancel{3}}{\cancel{2}\cdot2\cdot\cancel{3}\cdot5} = \dfrac{1}{10}$ of an hour

(d) $\dfrac{60}{60} = 1$ hour

43. (a) $\dfrac{800}{2400} = \dfrac{8\cdot\cancel{100}}{24\cdot\cancel{100}} = \dfrac{8}{24} = \dfrac{\cancel{8}}{3\cdot\cancel{8}} = \dfrac{1}{3}$

(b) $\dfrac{400}{2400} = \dfrac{4\cdot\cancel{100}}{24\cdot\cancel{100}} = \dfrac{4}{24} = \dfrac{\cancel{4}}{\cancel{4}\cdot6} = \dfrac{1}{6}$

(c) $\$2400 - \$800 - \$400 = \$1200$

$\dfrac{1200}{2400} = \dfrac{12\cdot\cancel{100}}{24\cdot\cancel{100}} = \dfrac{12}{24} = \dfrac{\cancel{12}}{2\cdot\cancel{12}} = \dfrac{1}{2}$

45. (a) $\dfrac{10}{25} = \dfrac{2\cdot\cancel{5}}{5\cdot\cancel{5}} = \dfrac{2}{5}$

(b) $\dfrac{15}{25} = \dfrac{3\cdot\cancel{5}}{5\cdot\cancel{5}} = \dfrac{3}{5}$

47. The result of dividing 3 by 3 is 1, so 1 should be written above and below all the slashes. The numerator is $1\cdot1$, so the correct answer is $\frac{1}{4}$.

$\dfrac{9}{36} = \dfrac{\cancel{3}\cdot\cancel{3}}{2\cdot2\cdot\cancel{3}\cdot\cancel{3}} = \dfrac{1}{4}$

49. $\dfrac{16c}{40} = \dfrac{\cancel{2}\cdot\cancel{2}\cdot\cancel{2}\cdot2\cdot c}{\cancel{2}\cdot\cancel{2}\cdot\cancel{2}\cdot5} = \dfrac{2c}{5}$

51. $\dfrac{20x}{35x} = \dfrac{2\cdot2\cdot\cancel{5}\cdot\cancel{x}}{\cancel{5}\cdot7\cdot\cancel{x}} = \dfrac{4}{7}$

53. $\dfrac{18r^2}{15rs} = \dfrac{2\cdot\cancel{3}\cdot3\cdot\cancel{r}\cdot r}{\cancel{3}\cdot5\cdot\cancel{r}\cdot s} = \dfrac{6r}{5s}$

55. $\dfrac{6m}{42mn^2} = \dfrac{\cancel{2}\cdot\cancel{3}\cdot\cancel{m}}{\cancel{2}\cdot\cancel{3}\cdot7\cdot\cancel{m}\cdot n\cdot n} = \dfrac{1}{7n^2}$

57. $\dfrac{9x^2}{16y^2} = \dfrac{3\cdot3\cdot x\cdot x}{2\cdot2\cdot2\cdot2\cdot y\cdot y}$

There are no common factors, so the fraction is already in lowest terms.

59. $\dfrac{7xz}{9xyz} = \dfrac{7\cdot\cancel{x}\cdot\cancel{z}}{3\cdot3\cdot\cancel{x}\cdot y\cdot\cancel{z}} = \dfrac{7}{9y}$

61. $\dfrac{21k^3}{6k^2} = \dfrac{\cancel{3}\cdot7\cdot\cancel{k}\cdot\cancel{k}\cdot k}{2\cdot\cancel{3}\cdot\cancel{k}\cdot\cancel{k}} = \dfrac{7k}{2}$

63. $\dfrac{13a^2bc^3}{39a^2bc^3} = \dfrac{\cancel{13}\cdot\cancel{a}\cdot\cancel{a}\cdot\cancel{b}\cdot\cancel{c}\cdot\cancel{c}\cdot\cancel{c}}{3\cdot\cancel{13}\cdot\cancel{a}\cdot\cancel{a}\cdot\cancel{b}\cdot\cancel{c}\cdot\cancel{c}\cdot\cancel{c}} = \dfrac{1}{3}$

65. $\dfrac{14c^2d}{14cd^2} = \dfrac{\cancel{2}\cdot\cancel{7}\cdot\cancel{c}\cdot c\cdot\cancel{d}}{\cancel{2}\cdot\cancel{7}\cdot\cancel{c}\cdot\cancel{d}\cdot d} = \dfrac{c}{d}$

67. $\dfrac{210ab^3c}{35b^2c^2} = \dfrac{2\cdot3\cdot\cancel{5}\cdot\cancel{7}\cdot a\cdot\cancel{b}\cdot\cancel{b}\cdot b\cdot\cancel{c}}{\cancel{5}\cdot\cancel{7}\cdot\cancel{b}\cdot\cancel{b}\cdot\cancel{c}\cdot c} = \dfrac{6ab}{c}$

69. $\dfrac{25m^3rt^2}{36n^2s^3w^2} = \dfrac{5\cdot5\cdot m\cdot m\cdot m\cdot r\cdot t\cdot t}{2\cdot2\cdot3\cdot3\cdot n\cdot n\cdot s\cdot s\cdot s\cdot w\cdot w}$

There are no common factors, so the fraction is already in lowest terms.

71. $\dfrac{33e^2fg^3}{11efg} = \dfrac{3\cdot\cancel{11}\cdot\cancel{e}\cdot e\cdot\cancel{f}\cdot\cancel{g}\cdot g\cdot g}{\cancel{11}\cdot\cancel{e}\cdot\cancel{f}\cdot\cancel{g}} = 3eg^2$

4.3 Multiplying and Dividing Signed Fractions

4.3 Margin Exercises

1. (a) The product of a negative number and a positive number is negative.

$$-\dfrac{3}{4}\cdot\dfrac{1}{2} = -\dfrac{3\cdot1}{4\cdot2} = -\dfrac{3}{8}$$

(b) The product of two negative numbers is positive.

$$\left(-\dfrac{2}{5}\right)\left(-\dfrac{2}{3}\right) = \dfrac{2\cdot2}{5\cdot3} = \dfrac{4}{15}$$

(c) $\dfrac{3}{4}\left(\dfrac{3}{8}\right) = \dfrac{3\cdot3}{4\cdot8} = \dfrac{9}{32}$

2. (a) $\dfrac{15}{28}\left(-\dfrac{6}{5}\right) = -\dfrac{3\cdot\cancel{5}\cdot\cancel{2}\cdot3}{\cancel{2}\cdot2\cdot7\cdot\cancel{5}} = -\dfrac{9}{14}$

(b) The product will be positive.

$$\dfrac{12}{7}\cdot\dfrac{7}{24} = \dfrac{\cancel{2}\cdot\cancel{2}\cdot\cancel{3}\cdot\cancel{7}}{\cancel{7}\cdot\cancel{2}\cdot\cancel{2}\cdot2\cdot\cancel{3}} = \dfrac{1}{2}$$

(c) The product will be <u>positive</u>.

$$\left(-\frac{11}{18}\right)\left(-\frac{9}{20}\right) = \frac{11 \cdot \overset{1}{\cancel{3}} \cdot \overset{1}{\cancel{3}}}{2 \cdot \cancel{3} \cdot \cancel{3} \cdot 2 \cdot 2 \cdot 5} = \frac{11}{40}$$

3.   (a) $\frac{3}{4}$ of $36 = \frac{3}{4} \cdot \frac{36}{1}$

$$= \frac{3 \cdot \overset{1}{\cancel{2}} \cdot \overset{1}{\cancel{2}} \cdot 3 \cdot 3}{\underset{1}{\cancel{2}} \cdot \underset{1}{\cancel{2}} \cdot 1} = \frac{27}{1} = 27$$

(b) $-10 \cdot \frac{2}{5} = -\frac{10}{1} \cdot \frac{2}{5}$ product is <u>negative</u>

$$= -\frac{2 \cdot \overset{1}{\cancel{5}} \cdot 2}{1 \cdot \underset{1}{\cancel{5}}} = -\frac{4}{1} = -4$$

(c) $\left(-\frac{7}{8}\right)(-24)$

$$= \left(-\frac{7}{8}\right)\left(-\frac{24}{1}\right) \text{ product is } \underline{positive}$$

$$= \frac{7 \cdot \overset{1}{\cancel{2}} \cdot \overset{1}{\cancel{2}} \cdot \overset{1}{\cancel{2}} \cdot 3}{\underset{1}{\cancel{2}} \cdot \underset{1}{\cancel{2}} \cdot \underset{1}{\cancel{2}} \cdot 1} = \frac{21}{1} = 21$$

4.   (a) $\frac{2c}{5} \cdot \frac{c}{4} = \frac{\overset{1}{\cancel{2}} \cdot c \cdot c}{5 \cdot \underset{1}{\cancel{2}} \cdot 2} = \frac{c^2}{10}$

(b) $\left(\frac{m}{6}\right)\left(\frac{9}{m^2}\right) = \frac{\overset{1}{\cancel{m}} \cdot \overset{1}{\cancel{3}} \cdot 3}{2 \cdot \underset{1}{\cancel{3}} \cdot \underset{1}{\cancel{m}} \cdot m} = \frac{3}{2m}$

(c) $\left(\frac{w^2}{y}\right)\left(\frac{x^2 y}{w}\right) = \frac{\overset{1}{\cancel{w}} \cdot w \cdot x \cdot x \cdot \overset{1}{\cancel{y}}}{\underset{1}{\cancel{y}} \cdot \underset{1}{\cancel{w}}}$

$$= \frac{wx^2}{1} = wx^2$$

5.   (a) $-\frac{3}{4} \div \frac{5}{8} = -\frac{3}{4} \cdot \frac{8}{5} = -\frac{3 \cdot 2 \cdot \overset{1}{\cancel{4}}}{\underset{1}{\cancel{4}} \cdot 5} = -\frac{6}{5}$

(b) $0 \div \left(-\frac{7}{12}\right) = 0 \cdot \left(-\frac{12}{7}\right) = 0$

(c) $\frac{5}{6} \div 10 = \frac{5}{6} \cdot \frac{1}{10} = \frac{\overset{1}{\cancel{5}} \cdot 1}{6 \cdot 2 \cdot \underset{1}{\cancel{5}}} = \frac{1}{12}$

(d) $-9 \div \left(-\frac{9}{16}\right) = -\frac{9}{1} \div \left(-\frac{9}{16}\right)$

$$= -\frac{9}{1} \cdot \left(-\frac{16}{9}\right)$$

$$= \frac{\overset{1}{\cancel{9}} \cdot 16}{1 \cdot \underset{1}{\cancel{9}}} = \frac{16}{1} = 16$$

(e) $\frac{2}{5} \div 0$ is *undefined*. This division can't be written as multiplication because 0 does not have a reciprocal.

6.   (a) $\frac{c^2 d^2}{4} \div \frac{c^2 d}{4} = \frac{c^2 d^2}{4} \cdot \frac{4}{c^2 d}$

$$= \frac{\overset{1}{\cancel{c}} \cdot \overset{1}{\cancel{c}} \cdot \overset{1}{\cancel{d}} \cdot d \cdot \overset{1}{\cancel{4}}}{\underset{1}{\cancel{4}} \cdot \underset{1}{\cancel{c}} \cdot \underset{1}{\cancel{c}} \cdot \underset{1}{\cancel{d}}}$$

$$= \frac{d}{1} = d$$

(b) $\frac{20}{7h} \div \frac{5h}{7} = \frac{20}{7h} \cdot \frac{7}{5h} = \frac{4 \cdot \overset{1}{\cancel{5}} \cdot \overset{1}{\cancel{7}}}{\underset{1}{\cancel{7}} \cdot h \cdot \underset{1}{\cancel{5}} \cdot h} = \frac{4}{h^2}$

(c) $\frac{n}{8} \div mn = \frac{n}{8} \div \frac{mn}{1} = \frac{n}{8} \cdot \frac{1}{mn} = \frac{\overset{1}{\cancel{n}} \cdot 1}{8 \cdot m \cdot \underset{1}{\cancel{n}}}$

$$= \frac{1}{8m}$$

7.   (a) Rewording the information might help. 18 quarts will be used up by repeatedly filling a $\frac{2}{3}$ quart spray bottle.

$$18 \div \frac{2}{3} = \frac{18}{1} \cdot \frac{3}{2} = \frac{\overset{1}{\cancel{2}} \cdot 9 \cdot 3}{1 \cdot \underset{1}{\cancel{2}}}$$

$$= \frac{27}{1} = 27$$

The spray bottle can be filled 27 times.

(b) $\frac{5}{8}$ *of* her highest annual salary, which is \$64,000, suggests multiplication.

$$\frac{5}{8} \cdot 64{,}000 = \frac{5}{8} \cdot \frac{64{,}000}{1}$$

$$= \frac{5 \cdot \overset{1}{\cancel{8}} \cdot 8000}{\underset{1}{\cancel{8}} \cdot 1}$$

$$= \frac{40{,}000}{1} = 40{,}000$$

The officer will receive \$40,000.

4.3   Section Exercises

1. To multiply fractions, you should <u>multiply</u> the numerators and multiply the <u>denominators</u>.

3. $-\frac{3}{8} \cdot \frac{1}{2} = -\frac{3 \cdot 1}{8 \cdot 2} = -\frac{3}{16}$

5. $\left(-\frac{3}{8}\right)\left(-\frac{12}{5}\right) = \frac{3 \cdot 3 \cdot \overset{1}{\cancel{4}}}{2 \cdot \underset{1}{\cancel{4}} \cdot 5} = \frac{9}{10}$

7. $\frac{21}{30}\left(\frac{5}{7}\right) = \frac{\overset{1}{\cancel{3}} \cdot \overset{1}{\cancel{7}} \cdot \overset{1}{\cancel{5}}}{2 \cdot \underset{1}{\cancel{3}} \cdot \underset{1}{\cancel{5}} \cdot \underset{1}{\cancel{7}}} = \frac{1}{2}$

9. $10\left(-\dfrac{3}{5}\right) = \dfrac{10}{1}\left(-\dfrac{3}{5}\right) = -\dfrac{2 \cdot \overset{1}{\cancel{5}} \cdot 3}{1 \cdot \cancel{5}} = -\dfrac{6}{1} = -6$

11. $\dfrac{4}{9}$ of $81 = \dfrac{4}{9} \cdot \dfrac{81}{1} = \dfrac{4 \cdot \overset{1}{\cancel{9}} \cdot 9}{\cancel{9} \cdot 1} = \dfrac{36}{1} = 36$

13. $\left(\dfrac{3x}{4}\right)\left(\dfrac{5}{xy}\right) = \dfrac{3 \cdot \overset{1}{\cancel{x}} \cdot 5}{4 \cdot \cancel{x} \cdot y} = \dfrac{15}{4y}$

15. $\dfrac{1}{6} \div \dfrac{1}{3} = \dfrac{1}{6} \cdot \dfrac{3}{1} = \dfrac{1 \cdot \overset{1}{\cancel{3}}}{2 \cdot \cancel{3} \cdot 1} = \dfrac{1}{2}$

17. $-\dfrac{3}{4} \div \left(-\dfrac{5}{8}\right) = -\dfrac{3}{4} \cdot \left(-\dfrac{8}{5}\right) = \dfrac{3 \cdot 2 \cdot \overset{1}{\cancel{4}}}{\cancel{4} \cdot 5} = \dfrac{6}{5}$

19. $6 \div \left(-\dfrac{2}{3}\right) = \dfrac{6}{1} \cdot \left(-\dfrac{3}{2}\right) = -\dfrac{\overset{1}{\cancel{2}} \cdot 3 \cdot 3}{1 \cdot \cancel{2}}$

$\qquad = -\dfrac{9}{1} = -9$

21. $-\dfrac{2}{3} \div 4 = -\dfrac{2}{3} \div \dfrac{4}{1} = -\dfrac{2}{3} \cdot \dfrac{1}{4} = -\dfrac{\overset{1}{\cancel{2}} \cdot 1}{3 \cdot \cancel{2} \cdot 2} = -\dfrac{1}{6}$

23. $\dfrac{11c}{5d} \div 3c = \dfrac{11c}{5d} \div \dfrac{3c}{1} = \dfrac{11c}{5d} \cdot \dfrac{1}{3c}$

$\qquad = \dfrac{11 \cdot \overset{1}{\cancel{c}} \cdot 1}{5 \cdot d \cdot 3 \cdot \cancel{c}} = \dfrac{11}{15d}$

25. $\dfrac{ab^2}{c} \div \dfrac{ab}{c} = \dfrac{ab^2}{c} \cdot \dfrac{c}{ab} = \dfrac{\overset{1}{\cancel{a}} \cdot \overset{1}{\cancel{b}} \cdot b \cdot \overset{1}{\cancel{c}}}{\underset{1}{\cancel{c}} \cdot \underset{1}{\cancel{a}} \cdot \underset{1}{\cancel{b}}} = \dfrac{b}{1} = b$

27. **(a)** Forgot to write 1s in numerator when dividing out common factors. Answer is $\frac{1}{6}$.

(b) Used reciprocal of $\frac{2}{3}$ in multiplication, but the reciprocal is used only in division. Correct answer is $8 \cdot \frac{2}{3} = \frac{8 \cdot 2}{3} = \frac{16}{3}$.

29. **(a)** Forgot to use reciprocal of $\frac{4}{1}$; correct answer is

$$\dfrac{2}{3} \cdot \dfrac{1}{4} = \dfrac{\overset{1}{\cancel{2}} \cdot 1}{3 \cdot 2 \cdot \underset{1}{\cancel{2}}} = \dfrac{1}{6}.$$

(b) Used reciprocal of $\frac{5}{6}$ instead of reciprocal of $\frac{10}{9}$; correct answer is

$$\dfrac{5}{6} \cdot \dfrac{9}{10} = \dfrac{\overset{1}{\cancel{5}} \cdot \overset{1}{\cancel{3}} \cdot 3}{2 \cdot \underset{1}{\cancel{3}} \cdot 2 \cdot \underset{1}{\cancel{5}}} = \dfrac{3}{4}.$$

31. $\dfrac{4}{5} \div 3 = \dfrac{4}{5} \div \dfrac{3}{1} = \dfrac{4}{5} \cdot \dfrac{1}{3} = \dfrac{4 \cdot 1}{5 \cdot 3} = \dfrac{4}{15}$

33. $-\dfrac{3}{8}\left(\dfrac{3}{4}\right) = -\dfrac{3 \cdot 3}{8 \cdot 4} = -\dfrac{9}{32}$

35. $\dfrac{3}{5}$ of $35 = \dfrac{3}{5} \cdot 35 = \dfrac{3}{5} \cdot \dfrac{35}{1} = \dfrac{3 \cdot \overset{1}{\cancel{5}} \cdot 7}{\cancel{5} \cdot 1} = \dfrac{21}{1} = 21$

37. $-9 \div \left(-\dfrac{3}{5}\right) = -\dfrac{9}{1} \cdot \left(-\dfrac{5}{3}\right) = \dfrac{\overset{1}{\cancel{3}} \cdot 3 \cdot 5}{1 \cdot \cancel{3}}$

$\qquad = \dfrac{15}{1} = 15$

39. $\dfrac{12}{7} \div 0$ is *undefined*.

41. $\left(\dfrac{11}{2}\right)\left(-\dfrac{5}{6}\right) = -\dfrac{11 \cdot 5}{2 \cdot 6} = -\dfrac{55}{12}$

43. $\dfrac{4}{7}$ of $14b = \dfrac{4}{7} \cdot 14b = \dfrac{4}{7} \cdot \dfrac{14b}{1} = \dfrac{4 \cdot 2 \cdot \overset{1}{\cancel{7}} \cdot b}{\cancel{7} \cdot 1}$

$\qquad = \dfrac{8b}{1} = 8b$

45. $\dfrac{12}{5} \div 4d = \dfrac{12}{5} \div \dfrac{4d}{1} = \dfrac{12}{5} \cdot \dfrac{1}{4d}$

$\qquad = \dfrac{3 \cdot \overset{1}{\cancel{4}} \cdot 1}{5 \cdot \cancel{4} \cdot d} = \dfrac{3}{5d}$

47. $\dfrac{x^2}{y} \div \dfrac{w}{2y} = \dfrac{x^2}{y} \cdot \dfrac{2y}{w} = \dfrac{x^2 \cdot 2 \cdot \overset{1}{\cancel{y}}}{\cancel{y} \cdot w} = \dfrac{2x^2}{w}$

49. The top of a table is a rectangle.

$A = l \cdot w$

$A = \dfrac{4}{5}$ yd $\cdot \dfrac{3}{8}$ yd

$A = \dfrac{\overset{1}{\cancel{4}} \cdot 3}{5 \cdot 2 \cdot \cancel{4}}$ yd^2

$A = \dfrac{3}{10}$ yd^2 or $\dfrac{3}{10}$ square yard

The area of the table top is $\frac{3}{10}$ yd^2.

51. Splitting 10 ounces into equal size parts indicates division. Each part will contain $\frac{1}{8}$ ounce.

$$10 \div \dfrac{1}{8} = \dfrac{10}{1} \cdot \dfrac{8}{1} = \dfrac{80}{1} = 80$$

80 $\frac{1}{8}$-ounce eyedrop dispensers can be filled.

53. Todd must earn $\frac{3}{4}$ of the cost:

$$\dfrac{3}{4} \cdot \$12{,}400 = \dfrac{3}{4} \cdot \dfrac{12{,}400}{1} = \dfrac{3 \cdot \overset{1}{\cancel{4}} \cdot 3100}{\cancel{4} \cdot 1}$$

$$= \dfrac{9300}{1} = \$9300$$

Todd must borrow the rest:

$$\$12{,}400 - \$9300 = \$3100$$

Thus, he must earn $9300 and borrow $3100.

55. The total number of cords divided by the number of cords per trip will give us the number of trips.

$$6 \div \frac{2}{3} = \frac{6}{1} \div \frac{2}{3} = \frac{6}{1} \cdot \frac{3}{2} = \frac{\overset{1}{\cancel{2}} \cdot 3 \cdot 3}{1 \cdot \underset{1}{\cancel{2}}} = \frac{9}{1} = 9$$

9 trips are needed to deliver 6 cords.

57. $300 \cdot \frac{1}{4} = \frac{300}{1} \cdot \frac{1}{4} = \frac{75 \cdot \overset{1}{\cancel{4}} \cdot 1}{1 \cdot \underset{1}{\cancel{4}}} = 75$

About 75 infield players are in the Hall of Fame.

59. The weight of the adult divided by the weight of the hatchling is

$$400 \div \frac{1}{8} = \frac{400}{1} \cdot \frac{8}{1} = \frac{400 \cdot 8}{1 \cdot 1} = 3200.$$

The adult weighs 3200 times the hatchling.

61. Student borrowing:

$$\frac{3}{20} \text{ of } 1600 = \frac{3}{20} \cdot \frac{1600}{1}$$
$$= \frac{3 \cdot \overset{1}{\cancel{20}} \cdot 80}{\underset{1}{\cancel{20}} \cdot 1} = \frac{240}{1} = 240$$

240 students in the survey borrowed money for college expenses.

63. Student income and savings:

$$\frac{1}{10} \text{ of } 1600 = \frac{1}{10} \cdot \frac{1600}{1}$$
$$= \frac{1 \cdot \overset{1}{\cancel{10}} \cdot 160}{\underset{1}{\cancel{10}} \cdot 1} = \frac{160}{1} = 160$$

Relatives and friends:

$$\frac{1}{25} \text{ of } 1600 = \frac{1}{25} \cdot \frac{1600}{1}$$
$$= \frac{1 \cdot \overset{1}{\cancel{25}} \cdot 64}{\underset{1}{\cancel{25}} \cdot 1} = \frac{64}{1} = 64$$

$160 - 64 = 96$ more students used their own income rather than money from relatives and friends.

65. Horses: $175 \cdot \frac{1}{25} = \frac{175}{1} \cdot \frac{1}{25} = \frac{7 \cdot \overset{1}{\cancel{25}} \cdot 1}{1 \cdot \underset{1}{\cancel{25}}} = 7$

7 million U.S. pets are horses.

67. Dogs: $175 \cdot \frac{2}{5} = \frac{175}{1} \cdot \frac{2}{5} = \frac{\overset{1}{\cancel{5}} \cdot 35 \cdot 2}{1 \cdot \underset{1}{\cancel{5}}} = 70$

Cats: $175 \cdot \frac{12}{25} = \frac{175 \cdot 12}{1 \cdot 25} = \frac{7 \cdot \overset{1}{\cancel{25}} \cdot 12}{\underset{1}{\cancel{25}}} = 84$

$70 + 84 = 154$ million U.S. pets are dogs and cats.

69. Rewrite division as multiplication. Leave the first number (dividend) the same. Change the second number (divisor) to its reciprocal by "flipping" it. Then multiply.

4.4 Adding and Subtracting Signed Fractions

4.4 Margin Exercises

1. **(a)** $\frac{1}{6} + \frac{5}{6} = \frac{1+5}{6} = \frac{6}{6} = 1$

(b) $-\frac{11}{12} + \frac{5}{12} = \frac{-11+5}{12}$
$$= \frac{-6}{12} \text{ or } -\frac{6}{12}$$
$$= -\frac{\overset{1}{\cancel{6}}}{2 \cdot \underset{1}{\cancel{6}}} = -\frac{1}{2} \quad \textit{Reduce to lowest terms}$$

(c) $-\frac{2}{9} - \frac{3}{9} = \frac{-2-3}{9} = \frac{-2+(-3)}{9}$
$$= \frac{-5}{9} \text{ or } -\frac{5}{9}$$

(d) $\frac{8}{ab} + \frac{3}{ab} = \frac{8+3}{ab} = \frac{11}{ab}$

2. **(a)** The LCD for $\frac{3}{5}$ and $\frac{3}{10}$ is $\underline{10}$, since the larger denominator, 10, is divisible by the smaller denominator, 5.

(b) The LCD of $\frac{1}{2}$ and $\frac{2}{5}$ is $\underline{10}$.

Since the larger denominator, 5, is not divisible by the smaller denominator, 2, check multiples of 5: $5, 10, 15, 20$, etc. 10 is the smallest one divisible by 2.

(c) The LCD of $\frac{3}{4}$ and $\frac{1}{6}$ is $\underline{12}$.

Since the larger denominator, 6, is not divisible by the smaller denominator, 4, check multiples of 6: $6, 12, 18, 24, 30$, etc. 12 is the smallest one divisible by 4.

(d) The LCD of $\frac{5}{6}$ and $\frac{7}{18}$ is $\underline{18}$, since the larger denominator, 18, is divisible by the smaller denominator, 6.

3. **(a)** $\frac{1}{10}$ and $\frac{13}{14}$
$$\left.\begin{array}{l} 10 = 2 \cdot 5 \\ 14 = 2 \cdot 7 \end{array}\right\} \text{ LCD} = 2 \cdot 5 \cdot 7 = \underline{70}$$

(b) $\frac{5}{12}$ and $\frac{17}{20}$
$$\left.\begin{array}{l} 12 = 2 \cdot 2 \cdot 3 \\ 20 = 2 \cdot 2 \cdot 5 \end{array}\right\} \text{ LCD} = 2 \cdot 2 \cdot 3 \cdot 5 = 60$$

(c) $\frac{7}{15}$ and $\frac{7}{9}$
$$\left.\begin{array}{l} 15 = 3 \cdot 5 \\ 9 = 3 \cdot 3 \end{array}\right\} \text{ LCD} = 3 \cdot 3 \cdot 5 = 45$$

4. **(a)** $\dfrac{2}{3} + \dfrac{1}{6}$

Step 1   The LCD is 6, the larger denominator.

Step 2 $\dfrac{2}{3} = \dfrac{2\cdot 2}{3\cdot 2} = \dfrac{4}{6}$ and $\dfrac{1}{6}$ already has the LCD.

Step 3 $\dfrac{2}{3} + \dfrac{1}{6} = \dfrac{4}{6} + \dfrac{1}{6} = \dfrac{4+1}{6} = \dfrac{5}{6}$

Step 4 $\dfrac{5}{6}$ is in lowest terms.

(b) $\dfrac{1}{12} - \dfrac{5}{6}$

Step 1   The LCD is 12, the larger denominator.

Step 2 $\dfrac{1}{12}$ already has the LCD and $\dfrac{5}{6} = \dfrac{5\cdot 2}{6\cdot 2} = \dfrac{10}{12}$.

Step 3 $\dfrac{1}{12} - \dfrac{5}{6} = \dfrac{1}{12} - \dfrac{10}{12} = \dfrac{1-10}{12}$

$= \dfrac{1+(-10)}{12} = \dfrac{-9}{12}$ or $-\dfrac{9}{12}$

Step 4 $-\dfrac{9}{12} = -\dfrac{\overset{1}{\cancel{3}}\cdot 3}{\underset{1}{\cancel{3}}\cdot 4} = -\dfrac{3}{4}$

(c) $3 - \dfrac{4}{5}$

Step 1   Think of 3 as $\dfrac{3}{1}$. The LCD for $\dfrac{3}{1}$ and $\dfrac{4}{5}$ is 5, the larger denominator.

Step 2 $\dfrac{3}{1} = \dfrac{3\cdot 5}{1\cdot 5} = \dfrac{15}{5}$ and $\dfrac{4}{5}$ already has the LCD.

Step 3 $3 - \dfrac{4}{5} = \dfrac{15}{5} - \dfrac{4}{5} = \dfrac{15-4}{5} = \dfrac{11}{5}$

Step 4 $\dfrac{11}{5}$ is in lowest terms.

(d) $-\dfrac{5}{12} + \dfrac{9}{16}$

Step 1   Use prime factorization to find the LCD.

$\left.\begin{array}{l} 12 = 2\cdot 2\cdot 3 \\ 16 = 2\cdot 2\cdot 2\cdot 2 \end{array}\right\}$ LCD $= 2\cdot 2\cdot 2\cdot 2\cdot 3 = 48$

Step 2

$-\dfrac{5}{12} = -\dfrac{5\cdot 4}{12\cdot 4} = -\dfrac{20}{48}$ and $\dfrac{9}{16} = \dfrac{9\cdot 3}{16\cdot 3} = \dfrac{27}{48}$

Step 3 $-\dfrac{5}{12} + \dfrac{9}{16} = -\dfrac{20}{48} + \dfrac{27}{48}$

$= \dfrac{-20+27}{48} = \dfrac{7}{48}$

Step 4 $\dfrac{7}{48}$ is in lowest terms.

5. **(a)** $\dfrac{5}{6} - \dfrac{h}{2}$

Step 1   The LCD is 6, the larger denominator.

Step 2

$\dfrac{5}{6}$ already has the LCD and $\dfrac{h}{2} = \dfrac{h\cdot 3}{2\cdot 3} = \dfrac{3h}{6}$.

Step 3 $\dfrac{5}{6} - \dfrac{h}{2} = \dfrac{5}{6} - \dfrac{3h}{6}$

$= \dfrac{5-3h}{6}$ *Lowest terms*

(b) $\dfrac{7}{t} + \dfrac{3}{5}$

Step 1   The LCD is $5\cdot t$ or $5t$.

Step 2 $\dfrac{7}{t} = \dfrac{7\cdot 5}{t\cdot 5} = \dfrac{35}{5t}$ and $\dfrac{3}{5} = \dfrac{3\cdot t}{5\cdot t} = \dfrac{3t}{5t}$

Step 3 $\dfrac{7}{t} + \dfrac{3}{5} = \dfrac{35}{5t} + \dfrac{3t}{5t}$

$= \dfrac{35+3t}{5t}$ *Lowest terms*

(c) $\dfrac{4}{x} - \dfrac{8}{3}$

Step 1   The LCD is $3\cdot x$ or $3x$.

Step 2 $\dfrac{4}{x} = \dfrac{4\cdot 3}{x\cdot 3} = \dfrac{12}{3x}$ and $\dfrac{8}{3} = \dfrac{8\cdot x}{3\cdot x} = \dfrac{8x}{3x}$

Step 3 $\dfrac{4}{x} - \dfrac{8}{3} = \dfrac{12}{3x} - \dfrac{8x}{3x}$

$= \dfrac{12-8x}{3x}$ *Lowest terms*

4.4   Section Exercises

1. **(a)** When two fractions have the same denominator, they are called <u>like fractions</u>.

(b) Two fractions need to have a common denominator before you can <u>add them</u>.

3. $\dfrac{3}{4} + \dfrac{1}{8}$

Step 1   The LCD is 8, the larger denominator.

Step 2 $\dfrac{3}{4} = \dfrac{3\cdot 2}{4\cdot 2} = \dfrac{6}{8}$ and $\dfrac{1}{8}$ has the LCD.

Step 3 $\dfrac{3}{4} + \dfrac{1}{8} = \dfrac{6}{8} + \dfrac{1}{8} = \dfrac{6+1}{8} = \dfrac{7}{8}$

Step 4 $\dfrac{7}{8}$ is already in lowest terms.

5. $-\dfrac{1}{14} + \left(-\dfrac{3}{7}\right)$

Step 1   The LCD is 14.

Step 2 $\dfrac{1}{14}$ has the LCD, $\dfrac{3}{7} = \dfrac{3\cdot 2}{7\cdot 2} = \dfrac{6}{14}$.

Step 3 $-\dfrac{1}{14} + \left(-\dfrac{3}{7}\right) = -\dfrac{1}{14} + \left(-\dfrac{6}{14}\right)$

$= \dfrac{-1+(-6)}{14}$

$= -\dfrac{7}{14}$

Step 4 $-\dfrac{7}{14} = -\dfrac{\overset{1}{\cancel{7}}}{2\cdot \underset{1}{\cancel{7}}} = -\dfrac{1}{2}$

7. $\dfrac{2}{3} - \dfrac{1}{6}$

   Step 1 The LCD is 6.

   Step 2 $\dfrac{2}{3} = \dfrac{2 \cdot 2}{3 \cdot 2} = \dfrac{4}{6}, \dfrac{1}{6}$ has the LCD.

   Step 3 $\dfrac{2}{3} - \dfrac{1}{6} = \dfrac{4}{6} - \dfrac{1}{6} = \dfrac{4-1}{6} = \dfrac{3}{6}$

   Step 4 $\dfrac{3}{6} = \dfrac{\overset{1}{\cancel{3}}}{2 \cdot \underset{1}{\cancel{3}}} = \dfrac{1}{2}$

9. $\dfrac{3}{8} - \dfrac{3}{5}$

   Step 1 The LCD is $8 \cdot 5 = 40$.

   Step 2 $\dfrac{3}{8} = \dfrac{3 \cdot 5}{8 \cdot 5} = \dfrac{15}{40}, \dfrac{3}{5} = \dfrac{3 \cdot 8}{5 \cdot 8} = \dfrac{24}{40}$

   Step 3 $\dfrac{3}{8} - \dfrac{3}{5} = \dfrac{15}{40} - \dfrac{24}{40} = \dfrac{15-24}{40} = -\dfrac{9}{40}$

   Step 4 $-\dfrac{9}{40}$ is already in lowest terms.

11. $-\dfrac{5}{8} + \dfrac{1}{12}$

   Step 1 Use prime factorization to find the LCD.

   $\left. \begin{array}{l} 8 = 2 \cdot 2 \cdot 2 \\ 12 = 2 \cdot 2 \cdot 3 \end{array} \right\}$ LCD $= 2 \cdot 2 \cdot 2 \cdot 3 = 24$

   Step 2 $\dfrac{5}{8} = \dfrac{5 \cdot 3}{8 \cdot 3} = \dfrac{15}{24}, \dfrac{1}{12} = \dfrac{1 \cdot 2}{12 \cdot 2} = \dfrac{2}{24}$

   Step 3 $-\dfrac{5}{8} + \dfrac{1}{12} = -\dfrac{15}{24} + \dfrac{2}{24} = \dfrac{-15+2}{24}$
   $= \dfrac{-13}{24}$ or $-\dfrac{13}{24}$

   Step 4 $-\dfrac{13}{24}$ is already in lowest terms.

13. $-\dfrac{7}{20} - \dfrac{5}{20}$

   Step 1 The LCD is 20.

   Step 2 Each fraction has the LCD.

   Step 3 $-\dfrac{7}{20} - \dfrac{5}{20} = \dfrac{-7-5}{20} = -\dfrac{12}{20}$

   Step 4 $-\dfrac{12}{20} = -\dfrac{\overset{1}{\cancel{4}} \cdot 3}{\underset{1}{\cancel{4}} \cdot 5} = -\dfrac{3}{5}$

15. $0 - \dfrac{7}{18} = 0 + \left(-\dfrac{7}{18}\right) = -\dfrac{7}{18}$

   Addition property of zero

17. $2 - \dfrac{6}{7} = \dfrac{2}{1} - \dfrac{6}{7}$

   Step 1 The LCD is 7.

   Step 2 $2 = \dfrac{2 \cdot 7}{7} = \dfrac{14}{7}, \dfrac{6}{7}$ has the LCD.

   Step 3 $2 - \dfrac{6}{7} = \dfrac{14}{7} - \dfrac{6}{7} = \dfrac{14-6}{7} = \dfrac{8}{7}$

   Step 4 $\dfrac{8}{7}$ is already in lowest terms.

19. $-\dfrac{1}{2} + \dfrac{3}{24}$

   Step 1 The LCD is 24, the larger denominator.

   Step 2 $\dfrac{1}{2} = \dfrac{1 \cdot 12}{2 \cdot 12} = \dfrac{12}{24}, \dfrac{3}{24}$ has the LCD.

   Step 3 $-\dfrac{1}{2} + \dfrac{3}{24} = -\dfrac{12}{24} + \dfrac{3}{24} = \dfrac{-12+3}{24}$
   $= \dfrac{-9}{24}$ or $-\dfrac{9}{24}$

   Step 4 $-\dfrac{9}{24} = -\dfrac{\overset{1}{\cancel{3}} \cdot 3}{\underset{1}{\cancel{3}} \cdot 8} = -\dfrac{3}{8}$

21. $\dfrac{1}{5} + \dfrac{c}{3}$

   Step 1 The LCD is $5 \cdot 3 = 15$.

   Step 2 $\dfrac{1}{5} = \dfrac{1 \cdot 3}{5 \cdot 3} = \dfrac{3}{15}, \dfrac{c}{3} = \dfrac{c \cdot 5}{3 \cdot 5} = \dfrac{5c}{15}$

   Step 3 $\dfrac{1}{5} + \dfrac{c}{3} = \dfrac{3}{15} + \dfrac{5c}{15} = \dfrac{3+5c}{15}$

   Step 4 $\dfrac{3+5c}{15}$ is already in lowest terms.

23. $\dfrac{5}{m} - \dfrac{1}{2}$

   Step 1 The LCD is $2 \cdot m$ or $2m$.

   Step 2 $\dfrac{5}{m} = \dfrac{5 \cdot 2}{m \cdot 2} = \dfrac{10}{2m}, \dfrac{1}{2} = \dfrac{1 \cdot m}{2 \cdot m} = \dfrac{1m}{2m}$

   Step 3 $\dfrac{5}{m} - \dfrac{1}{2} = \dfrac{10}{2m} - \dfrac{1m}{2m}$
   $= \dfrac{10-1m}{2m},$ or $\dfrac{10-m}{2m}$

   Step 4 $\dfrac{10-m}{2m}$ is already in lowest terms.

25. $\dfrac{3}{b^2} + \dfrac{5}{b^2} = \dfrac{3+5}{b^2} = \dfrac{8}{b^2}$, which is in lowest terms.

27. $\dfrac{c}{7} + \dfrac{3}{b}$

   Step 1 The LCD is $7 \cdot b$ or $7b$.

   Step 2 $\dfrac{c}{7} = \dfrac{c \cdot b}{7 \cdot b} = \dfrac{bc}{7b}, \dfrac{3}{b} = \dfrac{3 \cdot 7}{b \cdot 7} = \dfrac{21}{7b}$

   Step 3 $\dfrac{c}{7} + \dfrac{3}{b} = \dfrac{bc}{7b} + \dfrac{21}{7b} = \dfrac{bc+21}{7b}$

   Step 4 $\dfrac{bc+21}{7b}$ is already in lowest terms.

29. $-\dfrac{4}{c^2} - \dfrac{d}{c}$

Step 1   The LCD is c^2.

Step 2 $\dfrac{4}{c^2}$ has the LCD, $\dfrac{d}{c} = \dfrac{d \cdot c}{c \cdot c} = \dfrac{cd}{c^2}$.

Step 3 $-\dfrac{4}{c^2} - \dfrac{d}{c} = -\dfrac{4}{c^2} - \dfrac{cd}{c^2} = \dfrac{-4 - cd}{c^2}$

Step 4 $\dfrac{-4 - cd}{c^2}$ is already in lowest terms.

31. $-\dfrac{11}{42} - \dfrac{11}{70}$

Step 1   Use prime factorization to find the LCD.

$\left.\begin{array}{l} 42 = 2 \cdot 3 \cdot 7 \\ 70 = 2 \cdot 5 \cdot 7 \end{array}\right\}$ LCD $= 2 \cdot 3 \cdot 5 \cdot 7 = 210$

Step 2
$\dfrac{11}{42} = \dfrac{11 \cdot 5}{42 \cdot 5} = \dfrac{55}{210}, \quad \dfrac{11}{70} = \dfrac{11 \cdot 3}{70 \cdot 3} = \dfrac{33}{210}$

Step 3
$-\dfrac{11}{42} - \dfrac{11}{70}$
$= -\dfrac{55}{210} - \dfrac{33}{210}$
$= \dfrac{-55 - 33}{210} = \dfrac{-55 + (-33)}{210}$
$= \dfrac{-88}{210} \text{ or } -\dfrac{88}{210}$

Step 4 $-\dfrac{88}{210} = -\dfrac{2 \cdot 44}{2 \cdot 105} = -\dfrac{44}{105}$

33. $\dfrac{b}{3} + \dfrac{4}{a^2}$

Step 1   The LCD is $3a^2$.

Step 2 $\dfrac{b}{3} = \dfrac{b \cdot a^2}{3 \cdot a^2} = \dfrac{a^2 b}{3a^2}, \dfrac{4}{a^2} = \dfrac{4 \cdot 3}{a^2 \cdot 3} = \dfrac{12}{3a^2}$.

Step 3 $\dfrac{b}{3} + \dfrac{4}{a^2} = \dfrac{a^2 b}{3a^2} + \dfrac{12}{3a^2} = \dfrac{a^2 b + 12}{3a^2}$

Step 4 $\dfrac{a^2 b + 12}{3a^2}$ is already in lowest terms.

35. $-\dfrac{w}{10} + \dfrac{5}{w^2}$

Step 1   The LCD is $10w^2$.

Step 2
$\dfrac{w}{10} = \dfrac{w \cdot w^2}{10 \cdot w^2} = \dfrac{w^3}{10w^2}, \dfrac{5}{w^2} = \dfrac{5 \cdot 10}{w^2 \cdot 10} = \dfrac{50}{10w^2}$.

Step 3
$-\dfrac{w}{10} + \dfrac{5}{w^2} = -\dfrac{w^3}{10w^2} + \dfrac{50}{10w^2} = \dfrac{-w^3 + 50}{10w^2}$

Step 4 $\dfrac{-w^3 + 50}{10w^2}$ is already in lowest terms.

37. $\dfrac{m^4}{n^2} - 0 = \dfrac{m^4}{n^2} + (-0) = \dfrac{m^4}{n^2}$

Addition property of zero

39. **(a)** You cannot add fractions with unlike denominators; use 20 as the LCD.

$$\dfrac{3}{4} + \dfrac{2}{5} = \dfrac{15}{20} + \dfrac{8}{20} = \dfrac{23}{20}$$

(b) When rewriting fractions with 18 as the denominator, you must multiply denominator and numerator by the same number. The correct answer is

$$\dfrac{5}{6} - \dfrac{4}{9} = \dfrac{15}{18} - \dfrac{8}{18} = \dfrac{7}{18}.$$

41. $\dfrac{1}{5} + \dfrac{1}{3} + \dfrac{1}{4} = \dfrac{12}{60} + \dfrac{20}{60} + \dfrac{15}{60}$

$= \dfrac{12 + 20 + 15}{60} = \dfrac{47}{60}$

The total length of the bolt is $\frac{47}{60}$ inches.

43. $\dfrac{1}{8} + \dfrac{1}{2} + \dfrac{1}{3}$

$= \dfrac{3}{24} + \dfrac{12}{24} + \dfrac{8}{24}$ *LCD of 8, 2, and 3 is 24.*

$= \dfrac{3 + 12 + 8}{24} = \dfrac{23}{24}$

She needs $\frac{23}{24}$ cup of milk.

45.   Add the two fractions of land that were planted.

$$\dfrac{5}{12} + \dfrac{11}{12} = \dfrac{16}{12}$$

Then subtract the fraction of land that was destroyed.

$$\dfrac{16}{12} - \dfrac{7}{12} = \dfrac{16 - 7}{12} = \dfrac{9}{12} = \dfrac{3}{4}$$

$\frac{3}{4}$ acre of seedlings remained.

47. $\dfrac{2}{5} + \dfrac{3}{50} = \dfrac{20}{50} + \dfrac{3}{50} = \dfrac{20 + 3}{50} = \dfrac{23}{50}$

$\frac{23}{50}$ of workers are self-taught or learned from friends or family.

49. $\dfrac{2}{5} - \dfrac{6}{25} = \dfrac{10}{25} - \dfrac{6}{25} = \dfrac{10 - 6}{25} = \dfrac{4}{25}$

$\frac{4}{25}$ of workers is the difference.

51.   rightmost size minus leftmost size:

$$\dfrac{1}{2} - \dfrac{3}{16} = \dfrac{8}{16} - \dfrac{3}{16} = \dfrac{8 - 3}{16} = \dfrac{5}{16}$$

The rightmost driver fits a nut that is $\frac{5}{16}$ inch larger than the nut for the leftmost driver.

53. Since the total perimeter is $\frac{7}{8}$ mile, we will use subtraction to find the length of the fourth side.

Length of the fourth side

$$= \frac{7}{8} - \frac{1}{4} - \frac{1}{6} - \frac{3}{8}$$

$$= \frac{7}{8} + \left(-\frac{1}{4}\right) + \left(-\frac{1}{6}\right) + \left(-\frac{3}{8}\right)$$

$$= \frac{21}{24} + \left(-\frac{6}{24}\right) + \left(-\frac{4}{24}\right) + \left(-\frac{9}{24}\right)$$

$$= \frac{21 + (-6) + (-4) + (-9)}{24}$$

$$= \frac{2}{24} = \frac{1}{12} \text{ mile}$$

The fourth side is $\frac{1}{12}$ mile long.

Relating Concepts (Exercises 55–56)

55. (a) $-\frac{2}{3} + \frac{3}{4} = -\frac{8}{12} + \frac{9}{12} = \frac{-8 + 9}{12} = \frac{1}{12}$

$$\frac{3}{4} + \left(-\frac{2}{3}\right) = \frac{9}{12} + \left(-\frac{8}{12}\right) = \frac{9 + (-8)}{12} = \frac{1}{12}$$

The sums are the same because *addition is commutative.*

(b) $\frac{5}{6} - \frac{1}{2} = \frac{5}{6} - \frac{3}{6} = \frac{5 - 3}{6} = \frac{2}{6} = \frac{\overset{1}{\cancel{2}}}{\underset{}{\cancel{2} \cdot 3}} = \frac{1}{3}$

$$\frac{1}{2} - \frac{5}{6} = \frac{3}{6} - \frac{5}{6} = \frac{3 - 5}{6} = \frac{-2}{6} = -\frac{2}{6}$$

$$= \frac{-\overset{1}{\cancel{2}}}{\underset{1}{\cancel{2} \cdot 3}} = -\frac{1}{3}$$

The differences are different because *subtraction is not commutative.*

(c) $\left(-\frac{2}{3}\right)\left(\frac{9}{10}\right) = -\frac{\overset{1}{\cancel{2}} \cdot \overset{1}{\cancel{3}} \cdot 3}{\cancel{3} \cdot \cancel{2} \cdot 5} = -\frac{3}{5}$

$$\left(\frac{9}{10}\right)\left(-\frac{2}{3}\right) = -\frac{\overset{1}{\cancel{3}} \cdot 3 \cdot \overset{1}{\cancel{2}}}{\underset{1}{\cancel{2}} \cdot 5 \cdot \underset{1}{\cancel{3}}} = -\frac{3}{5}$$

The products are the same because *multiplication is commutative.*

(d) $\frac{2}{5} \div \frac{1}{15} = \frac{2}{5} \cdot \frac{15}{1} = \frac{2 \cdot 3 \cdot \overset{1}{\cancel{5}}}{\underset{1}{\cancel{5}} \cdot 1} = 6$

$$\frac{1}{15} \div \frac{2}{5} = \frac{1}{15} \cdot \frac{5}{2} = \frac{1 \cdot \overset{1}{\cancel{5}}}{3 \cdot \underset{1}{\cancel{5}} \cdot 2} = \frac{1}{6}$$

The quotients are different because *division is not commutative.*

56. (a) $-\frac{7}{12} + \frac{7}{12} = \frac{-7 + 7}{12} = \frac{0}{12} = 0$

$$\frac{3}{5} + \left(-\frac{3}{5}\right) = \frac{3 + (-3)}{5} = \frac{0}{5} = 0$$

The sum of a number and its opposite is 0.

(b) $-\frac{13}{16} \div \left(-\frac{13}{16}\right) = -\frac{\overset{1}{\cancel{13}}}{\underset{1}{\cancel{16}}} \cdot \left(-\frac{\overset{1}{\cancel{16}}}{\underset{1}{\cancel{13}}}\right) = 1$

$$\frac{1}{8} \div \frac{1}{8} = \frac{1}{\cancel{8}} \cdot \frac{\overset{1}{\cancel{8}}}{1} = 1$$

When a nonzero number is divided by itself, the quotient is 1.

(c) $\frac{5}{6} \cdot 1 = \frac{5}{6}$

$$1\left(-\frac{17}{20}\right) = -\frac{17}{20}$$

Multiplying by 1 leaves a number unchanged.

(d) $\left(-\frac{4}{5}\right)\left(-\frac{5}{4}\right) = \left(-\frac{\overset{1}{\cancel{4}}}{\underset{1}{\cancel{5}}}\right)\left(-\frac{\overset{1}{\cancel{5}}}{\underset{1}{\cancel{4}}}\right) = 1$

$$7 \cdot \frac{1}{7} = \frac{7}{1} \cdot \frac{1}{\underset{1}{\cancel{7}}} = 1$$

A number times its reciprocal is 1.

4.5 Problem Solving: Mixed Numbers and Estimating

4.5 Margin Exercises

1. (a) There are 5 sections shaded from "whole" parts divided into 3 sections. $1\frac{2}{3}$ written as an improper fraction is $\frac{5}{3}$.

(b) Graph $1\frac{2}{3}$ and $-1\frac{2}{3}$ on a number line.

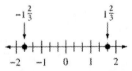

(c) There are 9 sections shaded from "whole" parts divided into 4 sections. $2\frac{1}{4}$ written as an improper fraction is $\frac{9}{4}$.

(d) Graph $2\frac{1}{4}$ and $-2\frac{1}{4}$ on a number line.

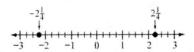

2.   **(a)** $3\frac{2}{3}$ **Step 1** $3 \cdot 3 = 9;\ 9 + 2 = 11$

Step 2 $3\frac{2}{3} = \frac{11}{3}$

(b) $4\frac{7}{10}$ **Step 1** $10 \cdot 4 = 40;\ 40 + 7 = 47$

Step 2 $4\frac{7}{10} = \frac{47}{10}$

(c) $5\frac{3}{4}$ **Step 1** $4 \cdot 5 = 20;\ 20 + 3 = 23$

Step 2 $5\frac{3}{4} = \frac{23}{4}$

(d) $8\frac{5}{6}$ **Step 1** $6 \cdot 8 = 48;\ 48 + 5 = 53$

Step 2 $8\frac{5}{6} = \frac{53}{6}$

3.   **(a)** $\frac{7}{2}$ ▪ Divide 7 by 2.

$$2\overline{)7} \atop \underline{6} \atop 1 \quad \text{so} \quad \frac{7}{2} = 3\frac{1}{2}$$
(quotient 3)

(b) $\frac{14}{4}$ ▪ Divide 14 by 4.

$$4\overline{)14} \atop \underline{12} \atop 2 \quad \text{so} \quad \frac{14}{4} = 3\frac{2}{4} = 3\frac{1}{2}$$
(quotient 3)

You could write $\frac{14}{4}$ in lowest terms first. This would give us the fraction $\frac{7}{2}$, which is the same as part (a).

(c) $\frac{33}{5}$ ▪ Divide 33 by 5.

$$5\overline{)33} \atop \underline{30} \atop 3 \quad \text{so} \quad \frac{33}{5} = 6\frac{3}{5}$$
(quotient 6)

(d) $\frac{58}{10}$ ▪ Reduce first. $\frac{58}{10} = \frac{\cancel{2} \cdot 29}{\cancel{2} \cdot 5} = \frac{29}{5}$

Divide 29 by 5. $5\overline{)29} \atop \underline{25} \atop 4$ so $\frac{29}{5} = 5\frac{4}{5}$
(quotient 5)

4.   **(a)** $7\frac{3}{4}$ ▪ Half of the denominator 4 is 2. Since the numerator, 3, is half or more of the denominator, $7\frac{3}{4}$ rounds up to 8.

(b) $6\frac{3}{8}$ ▪ Half of the denominator 8 is 4. Since the numerator, 3, is less than half of the denominator, $6\frac{3}{8}$ rounds to 6.

(c) $4\frac{2}{3}$ ▪ 2 is more than half of 3. $4\frac{2}{3}$ rounds up to 5.

(d) $1\frac{7}{10}$ ▪ 7 is more than half of 10. $1\frac{7}{10}$ rounds up to 2.

(e) $3\frac{1}{2}$ ▪ 1 is half of 2. $3\frac{1}{2}$ rounds up to 4.

(f) $5\frac{4}{9}$ ▪ 4 is less than half of 9. $5\frac{4}{9}$ rounds to 5.

5.   **(a)** $2\frac{1}{4}$ rounds to 2. $7\frac{1}{3}$ rounds to 7.

Estimate: $2 \cdot 7 = \underline{14}$

Exact: $2\frac{1}{4} \cdot 7\frac{1}{3} = \frac{9}{4} \cdot \frac{22}{3} = \frac{\cancel{3} \cdot 3 \cdot \cancel{2} \cdot 11}{\cancel{2} \cdot 2 \cdot \cancel{3}}$

$$= \frac{33}{2} = 16\frac{1}{2}$$

(b) $4\frac{1}{2}$ rounds to 5. $1\frac{2}{3}$ rounds to 2.

Estimate: $(\underline{5})(\underline{2}) = \underline{10}$

Exact: $\left(4\frac{1}{2}\right)\left(1\frac{2}{3}\right) = \frac{9}{2} \cdot \frac{5}{3} = \frac{3 \cdot \cancel{3} \cdot 5}{2 \cdot \cancel{3}}$

$$= \frac{15}{2} = 7\frac{1}{2}$$

(c) $3\frac{3}{5}$ rounds to 4. $4\frac{4}{9}$ rounds to 4.

Estimate: $\underline{4} \cdot \underline{4} = \underline{16}$

Exact: $3\frac{3}{5} \cdot 4\frac{4}{9} = \frac{18}{5} \cdot \frac{40}{9} = \frac{2 \cdot \cancel{9} \cdot \cancel{5} \cdot 8}{\cancel{5} \cdot \cancel{9}}$

$$= \frac{16}{1} = 16$$

(d) $3\frac{1}{5}$ rounds to 3. $5\frac{3}{8}$ rounds to 5.

Estimate: $(\underline{3})(\underline{5}) = \underline{15}$

Exact: $\left(3\frac{1}{5}\right)\left(5\frac{3}{8}\right) = \frac{16}{5} \cdot \frac{43}{8} = \frac{2 \cdot \cancel{8} \cdot 43}{5 \cdot \cancel{8}}$

$$= \frac{86}{5} = 17\frac{1}{5}$$

6.   **(a)** $6\frac{1}{4}$ rounds to 6. $3\frac{1}{3}$ rounds to 3.

Estimate: $6 \div 3 = \underline{2}$

Exact: $6\frac{1}{4} \div 3\frac{1}{3} = \frac{25}{4} \div \frac{10}{3} = \frac{25}{4} \cdot \frac{3}{10}$

$$= \frac{5 \cdot \cancel{5} \cdot 3}{4 \cdot 2 \cdot \cancel{5}} = \frac{15}{8} = 1\frac{7}{8}$$

(b) $3\frac{3}{8}$ rounds to 3. $2\frac{4}{7}$ rounds to 3.

Estimate: $\underline{3} \div \underline{3} = \underline{1}$

Exact: $3\frac{3}{8} \div 2\frac{4}{7} = \frac{27}{8} \div \frac{18}{7} = \frac{27}{8} \cdot \frac{7}{18}$

$$= \frac{3 \cdot \cancel{9} \cdot 7}{8 \cdot 2 \cdot \cancel{9}} = \frac{21}{16} = 1\frac{5}{16}$$

(c) $5\frac{1}{3}$ rounds to 5.

Estimate: $\underline{8} \div \underline{5} = \dfrac{8}{5} = 1\dfrac{3}{5}$

Exact: $8 \div 5\dfrac{1}{3} = \dfrac{8}{1} \div \dfrac{16}{3} = \dfrac{8}{1} \cdot \dfrac{3}{16}$

$$= \dfrac{\overset{1}{\cancel{8}} \cdot 3}{1 \cdot 2 \cdot \underset{1}{\cancel{8}}} = \dfrac{3}{2} = 1\dfrac{1}{2}$$

(d) $4\frac{1}{2}$ rounds to 5.

Estimate: $\underline{5} \div \underline{6} = \dfrac{5}{6}$

Exact: $4\dfrac{1}{2} \div 6 = \dfrac{9}{2} \div \dfrac{6}{1} = \dfrac{9}{2} \cdot \dfrac{1}{6}$

$$= \dfrac{3 \cdot \overset{1}{\cancel{3}} \cdot 1}{2 \cdot 2 \cdot \underset{1}{\cancel{3}}} = \dfrac{3}{4}$$

7.   (a) $5\frac{1}{3}$ rounds to 5. $2\frac{5}{6}$ rounds to 3.

Estimate: $5 - \underline{3} = \underline{2}$

Exact: $5\dfrac{1}{3} - 2\dfrac{5}{6} = \dfrac{16}{3} - \dfrac{17}{6} = \dfrac{32}{6} - \dfrac{17}{6}$

$$= \dfrac{15}{6} = 2\dfrac{3}{6} = 2\dfrac{1}{2}$$

(b) $\frac{3}{4}$ rounds to 1. $3\frac{1}{8}$ rounds to 3.

Estimate: $\underline{1} + \underline{3} = \underline{4}$

Exact: $\dfrac{3}{4} + 3\dfrac{1}{8} = \dfrac{3}{4} + \dfrac{25}{8} = \dfrac{6}{8} + \dfrac{25}{8}$

$$= \dfrac{31}{8} = 3\dfrac{7}{8}$$

(c) $3\frac{4}{5}$ rounds to 4.

Estimate: $\underline{6} - \underline{4} = \underline{2}$

Exact: $6 - 3\dfrac{4}{5} = \dfrac{6}{1} - \dfrac{19}{5} = \dfrac{30}{5} - \dfrac{19}{5}$

$$= \dfrac{11}{5} = 2\dfrac{1}{5}$$

8.   (a) $3\frac{5}{8}$ inches rounds to $\underline{4}$ inches.
               $2\frac{1}{4}$ inches rounds to $\underline{2}$ inches.

Richard's son grew about 4 inches last year and 2 inches this year. How much has his height increased over the two years? Using rounded numbers makes it easier to see that you need to *add*.

Estimate: $4 + 2 = 6$ inches

Exact: $3\dfrac{5}{8} + 2\dfrac{1}{4} = \dfrac{29}{8} + \dfrac{9}{4} = \dfrac{29}{8} + \dfrac{18}{8}$

$$= \dfrac{47}{8} = 5\dfrac{7}{8} \text{ inches}$$

(b) $2\frac{1}{2}$ packages rounds to 3 packages.
              $5\frac{1}{2}$ ounces rounds to 6 ounces.

Ernestine used about 3 packages of chips and each package contained about 6 ounces. How many ounces of chips did she use in the recipe? Using rounded numbers makes it easier to see that you need to *multiply*.

Estimate: $3 \cdot 6 = 18$ ounces

Exact: $2\dfrac{1}{2} \cdot 5\dfrac{1}{2} = \dfrac{5}{2} \cdot \dfrac{11}{2}$

$$= \dfrac{55}{4} = 13\dfrac{3}{4} \text{ ounces}$$

4.5   Section Exercises

1.   Graph $2\frac{1}{3}$ and $-2\frac{1}{3}$ on a number line.

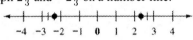

Shade $2\frac{1}{3}$ of the circles.

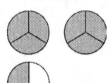

3.   Graph $\frac{3}{2} = 1\frac{1}{2}$ and $-\frac{3}{2} = -1\frac{1}{2}$ on a number line.

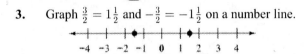

Shade $\frac{3}{2}$ of the squares.

5. $4\frac{1}{2}$ ***Step 1*** $2 \cdot 4 = 8; \ 8 + 1 = 9$

                Step 2 $4\frac{1}{2} = \frac{9}{2}$

7. $-1\frac{3}{5}$ ***Step 1*** $5 \cdot 1 = 5; \ 5 + 3 = 8$

                Step 2 $-1\frac{3}{5} = -\frac{8}{5}$

9. $2\frac{3}{8}$ ***Step 1*** $8 \cdot 2 = 16; \ 16 + 3 = 19$

                Step 2 $2\frac{3}{8} = \frac{19}{8}$

11. $-5\frac{7}{10}$ ***Step 1*** $10 \cdot 5 = 50; \ 50 + 7 = 57$

                Step 2 $-5\frac{7}{10} = -\frac{57}{10}$

13. $10\frac{11}{15}$ ***Step 1*** $15 \cdot 10 = 150; \ 150 + 11 = 161$

                Step 2 $10\frac{11}{15} = \frac{161}{15}$

15. $\frac{13}{3}$ ▪ Divide 13 by 3.

$$3\overline{\smash{\big)}\,1\,3} \qquad \text{so} \quad \dfrac{13}{3} = 4\dfrac{1}{3}$$
$$\underline{1\,2}$$
$$1$$

17. $-\frac{10}{4}$ ■ Divide 10 by 4.

$$
\begin{array}{r}
2 \\
4\overline{)1\,0} \\
8 \\
\hline
2
\end{array}
\quad \text{so} \quad -\frac{10}{4} = -2\frac{2}{4} = -2\frac{1}{2}
$$

19. $\frac{22}{6}$ ■ Divide 22 by 6.

$$
\begin{array}{r}
3 \\
6\overline{)2\,2} \\
1\,8 \\
\hline
4
\end{array}
\quad \text{so} \quad \frac{22}{6} = 3\frac{4}{6} = 3\frac{2}{3}
$$

21. $-\frac{51}{9}$ ■ Divide 51 by 9.

$$
\begin{array}{r}
5 \\
9\overline{)5\,1} \\
4\,5 \\
\hline
6
\end{array}
\quad \text{so} \quad -\frac{51}{9} = -5\frac{6}{9} = -5\frac{2}{3}
$$

23. $\frac{188}{16}$ ■ Divide 188 by 16.

$$
\begin{array}{r}
1\,1 \\
16\overline{)1\,8\,8} \\
1\,6 \\
\hline
2\,8 \\
1\,6 \\
\hline
1\,2
\end{array}
\quad \text{so} \quad \frac{188}{16} = 11\frac{12}{16} = 11\frac{3}{4}
$$

25. $2\frac{1}{4}$ rounds to 2. $3\frac{1}{2}$ rounds to 4.

Estimate: $\underline{2} \cdot \underline{4} = \underline{8}$

Exact: $2\frac{1}{4} \cdot 3\frac{1}{2} = \frac{9}{4} \cdot \frac{7}{2} = \frac{63}{8} = 7\frac{7}{8}$

27. $3\frac{1}{4}$ rounds to 3. $2\frac{5}{8}$ rounds to 3.

Estimate: $\underline{3} \div \underline{3} = \underline{1}$

Exact: $3\frac{1}{4} \div 2\frac{5}{8} = \frac{13}{4} \div \frac{21}{8} = \frac{13}{4} \cdot \frac{8}{21}$

$$
= \frac{13 \cdot 2 \cdot \cancel{4}}{\cancel{4} \cdot 21} = \frac{26}{21} = 1\frac{5}{21}
$$

29. $3\frac{2}{3}$ rounds to 4. $1\frac{5}{6}$ rounds to 2.

Estimate: $\underline{4} + \underline{2} = \underline{6}$

Exact: $3\frac{2}{3} + 1\frac{5}{6} = \frac{11}{3} + \frac{11}{6}$

$$
= \frac{22}{6} + \frac{11}{6} = \frac{22 + 11}{6}
$$

$$
= \frac{33}{6} = 5\frac{3}{6} = 5\frac{1}{2}
$$

31. $4\frac{1}{4}$ rounds to 4. $\frac{7}{12}$ rounds to 1.

Estimate: $\underline{4} - \underline{1} = \underline{3}$

Exact: $4\frac{1}{4} - \frac{7}{12} = \frac{17}{4} - \frac{7}{12}$

$$
= \frac{51}{12} - \frac{7}{12} = \frac{51 - 7}{12}
$$

$$
= \frac{44}{12} = \frac{11}{3} = 3\frac{2}{3}
$$

33. $5\frac{2}{3}$ rounds to 6.

Estimate: $\underline{6} \div \underline{6} = \underline{1}$

Exact: $5\frac{2}{3} \div 6 = \frac{17}{3} \div \frac{6}{1} = \frac{17}{3} \cdot \frac{1}{6} = \frac{17}{18}$

35. $1\frac{4}{5}$ rounds to 2.

Estimate: $\underline{8} - \underline{2} = \underline{6}$

Exact: $8 - 1\frac{4}{5} = \frac{8}{1} - \frac{9}{5}$

$$
= \frac{40}{5} - \frac{9}{5} = \frac{40 - 9}{5}
$$

$$
= \frac{31}{5} = 6\frac{1}{5}
$$

37. The figure is a square.

$$
P = 4s = 4 \cdot 1\frac{3}{4}
$$

$$
= \frac{4}{1} \cdot \frac{7}{4} = \frac{\cancel{4} \cdot 7}{1 \cdot \cancel{4}} = \frac{7}{1} = 7 \text{ in.}
$$

$$
A = s \cdot s = 1\frac{3}{4} \cdot 1\frac{3}{4}
$$

$$
= \frac{7}{4} \cdot \frac{7}{4} = \frac{49}{16}, \quad \text{or} \quad 3\frac{1}{16} \text{ in.}^2
$$

39. The figure is a rectangle.

$$
P = 2l + 2w = 2 \cdot 6\frac{1}{2} + 2 \cdot 3\frac{1}{4}
$$

$$
= \frac{2}{1} \cdot \frac{13}{2} + \frac{2}{1} \cdot \frac{13}{4}
$$

$$
= \frac{26}{2} + \frac{13}{2} = \frac{39}{2}, \quad \text{or} \quad 19\frac{1}{2} \text{ yd}
$$

$$
A = lw = 6\frac{1}{2} \cdot 3\frac{1}{4}
$$

$$
= \frac{13}{2} \cdot \frac{13}{4} = \frac{169}{8}, \quad \text{or} \quad 21\frac{1}{8} \text{ yd}^2
$$

41. The figure is a parallelogram.

$$
P = 3\frac{1}{4} + 2\frac{2}{3} + 3\frac{1}{4} + 2\frac{2}{3} \qquad \textit{add all four sides}
$$

$$
= \frac{13}{4} + \frac{8}{3} + \frac{13}{4} + \frac{8}{3}
$$

$$
= \frac{39}{12} + \frac{32}{12} + \frac{39}{12} + \frac{32}{12}
$$

$$
= \frac{142}{12} = \frac{71}{6}, \quad \text{or} \quad 11\frac{5}{6} \text{ ft}
$$

$$
A = b \cdot h = 3\frac{1}{4} \cdot 2 = \frac{13}{4} \cdot \frac{2}{1}
$$

$$
= \frac{26}{4} = \frac{13}{2}, \quad \text{or} \quad 6\frac{1}{2} \text{ ft}^2
$$

43. $12\frac{1}{2}$ ft rounds to 13 ft. $8\frac{2}{3}$ ft rounds to 9 ft.

"In all" implies addition.

Estimate: $13 + 9 = 22$ ft

Exact: $12\dfrac{1}{2} + 8\dfrac{2}{3} = \dfrac{25}{2} + \dfrac{26}{3} = \dfrac{75}{6} + \dfrac{52}{6}$

$\qquad\qquad\qquad = \dfrac{127}{6} = 21\dfrac{1}{6}$ ft of trim

He has a total of $21\dfrac{1}{6}$ ft of oak trim.

45. $1\dfrac{3}{4}$ ounces/gallon rounds to 2 ounces/gallon.

$5\dfrac{1}{2}$ gallons rounds to 6 gallons.

Estimate: $2 \cdot 6 = 12$ ounces

Exact: $1\dfrac{3}{4} \cdot 5\dfrac{1}{2} = \dfrac{7}{4} \cdot \dfrac{11}{2}$

$\qquad\qquad = \dfrac{77}{8} = 9\dfrac{5}{8}$ ounces

$9\dfrac{5}{8}$ ounces of chemical should be mixed with $5\dfrac{1}{2}$ gallons of water.

47. $1\dfrac{7}{10}$ miles rounds to 2 miles.

Amount left to be picked up implies subtraction.

Estimate: $4 - 2 = 2$ miles

Exact: $4 - 1\dfrac{7}{10} = \dfrac{4}{1} - \dfrac{17}{10}$

$\qquad\qquad = \dfrac{40}{10} - \dfrac{17}{10}$

$\qquad\qquad = \dfrac{23}{10} = 2\dfrac{3}{10}$

The Boy Scout troop needs to pick up trash along the highway for another $2\dfrac{3}{10}$ miles.

49. $3\dfrac{3}{4}$ yd rounds to 4 yd.

Estimate: $4 \cdot 5 = 20$ yd

Exact: $3\dfrac{3}{4} \cdot 5 = \dfrac{15}{4} \cdot \dfrac{5}{1} = \dfrac{75}{4} = 18\dfrac{3}{4}$ yd

$18\dfrac{3}{4}$ yards of material are needed to make dresses for five bridesmaids.

51. The distance from Devils Kitchen to the beach parking is the *difference* in the given distances.

$2\dfrac{3}{8} - 1\dfrac{3}{4} = \dfrac{19}{8} - \dfrac{7}{4} = \dfrac{19}{8} - \dfrac{14}{8}$

$\qquad\qquad = \dfrac{19 - 14}{8} = \dfrac{5}{8}$

The distance is $\dfrac{5}{8}$ mile.

53. *To the picnic parking:* $2\dfrac{1}{4}$ miles

Back to Face Rock: $2\dfrac{1}{4} - \dfrac{1}{4} = 2$ miles

Face Rock to the beach parking:

$2\dfrac{3}{8} - \dfrac{1}{4} = \dfrac{19}{8} - \dfrac{1}{4} = \dfrac{19}{8} - \dfrac{2}{8}$

$\qquad\qquad = \dfrac{19 - 2}{8} = \dfrac{17}{8} = 2\dfrac{1}{8}$

Add the three distances to find the sum.

$2\dfrac{1}{4} + 2 + 2\dfrac{1}{8} = 6 + \dfrac{1}{4} + \dfrac{1}{8}$

$\qquad\qquad\qquad = 6 + \dfrac{2}{8} + \dfrac{1}{8} = 6\dfrac{3}{8}$

You will have traveled $6\dfrac{3}{8}$ miles.

55. $29\dfrac{1}{2} - 6\dfrac{1}{4} - 1\dfrac{7}{8} = \dfrac{59}{2} - \dfrac{25}{4} - \dfrac{15}{8}$

$\qquad\qquad\qquad = \dfrac{236}{8} - \dfrac{50}{8} - \dfrac{15}{8}$

$\qquad\qquad\qquad = \dfrac{236 - 50 - 15}{8}$

$\qquad\qquad\qquad = \dfrac{171}{8} = 21\dfrac{3}{8}$ in.

The length of the arrow shaft is $21\dfrac{3}{8}$ inches.

57. $23\dfrac{3}{4}$ in. rounds to 24 in. $34\dfrac{1}{2}$ in. rounds to 35 in.

Estimate: $24 + 35 + 24 + 35 = 118$ in.

Exact: $23\dfrac{3}{4} + 34\dfrac{1}{2} + 23\dfrac{3}{4} + 34\dfrac{1}{2}$

$\qquad\quad = 116.5$ (by calculator)

The length of lead stripping needed is $116\dfrac{1}{2}$ in.

59. $1\dfrac{1}{2}$ ounces rounds to 2 ounces.

126 ounces of detergent are available and each load requires about 2 ounces. Using estimated numbers, it is easier to see that you need *division*.

Estimate: $126 \div 2 = 63$ loads

Exact: $126 \div 1\dfrac{1}{2} = 126 \div \dfrac{3}{2} = \dfrac{126}{1} \cdot \dfrac{2}{3}$

$\qquad\qquad = \dfrac{42 \cdot \overset{1}{\cancel{3}} \cdot 2}{1 \cdot \underset{1}{\cancel{3}}} = 84$ loads

84 loads of clothes can be washed with 126-ounce container of powdered laundry detergent.

61. Round the finished lengths, $21\dfrac{7}{8}, 22\dfrac{5}{8}$, and $23\dfrac{1}{2}$, to 22, 23, and 24. Round the $\dfrac{3}{4}$ inch seam allowance to 1 inch. There are $4 + 5 + 3 = 12$ total bands.

Estimate:
$(4 \cdot 22) + (5 \cdot 23) + (3 \cdot 24) + (12 \cdot 1) = 287$ in.

Exact:

$\left(4 \cdot 21\dfrac{7}{8}\right) + \left(5 \cdot 22\dfrac{5}{8}\right) + \left(3 \cdot 23\dfrac{1}{2}\right) + \left(12 \cdot \dfrac{3}{4}\right)$

$\qquad = 280.125$ in. (by calculator)

$280\dfrac{1}{8}$ inches of fabric strip are needed to make the bands including the seam allowance.

Summary Exercises
Computation with Fractions

1. **(a)** 3 of 8 equally sized portions are shaded: $\frac{3}{8}$

5 of 8 equally sized portions are unshaded: $\frac{5}{8}$

(b) 4 of 5 equally sized portions are shaded: $\frac{4}{5}$

1 of 5 equally sized portions are unshaded: $\frac{1}{5}$

3. **(a)** $30 \div 5 = 6$

$$-\frac{4}{5} = -\frac{4 \cdot 6}{5 \cdot 6} = -\frac{24}{30}$$

(b) $14 \div 7 = 2$

$$\frac{2}{7} = \frac{2 \cdot 2}{7 \cdot 2} = \frac{4}{14}$$

5. **(a)**

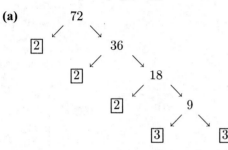

The prime factorization of 72 is $2 \cdot 2 \cdot 2 \cdot 3 \cdot 3$.

(b)

The prime factorization of 105 is $3 \cdot 5 \cdot 7$.

7. $\left(-\frac{3}{4}\right)\left(-\frac{2}{3}\right) = \frac{\overset{1}{\cancel{3}} \cdot \overset{1}{\cancel{2}}}{\underset{1}{\cancel{2}} \cdot 2 \cdot \underset{1}{\cancel{3}}} = \frac{1}{2}$

9. $\frac{7}{16} + \frac{5}{8} = \frac{7}{16} + \frac{10}{16} = \frac{7 + 10}{16} = \frac{17}{16}$

11. $\frac{2}{3} - \frac{4}{5} = \frac{2 \cdot 5}{3 \cdot 5} - \frac{4 \cdot 3}{5 \cdot 3}$

$= \frac{10}{15} - \frac{12}{15} = \frac{10 - 12}{15} = -\frac{2}{15}$

13. $-21 \div \left(-\frac{3}{8}\right) = \frac{21}{1} \cdot \left(\frac{8}{3}\right) = \frac{\overset{1}{\cancel{3}} \cdot 7 \cdot 8}{\underset{1}{\cancel{3}}} = 56$

15. $-\frac{35}{45} \div \frac{10}{15} = -\frac{35}{45} \cdot \frac{15}{10} = -\frac{\overset{1}{\cancel{5}} \cdot 7 \cdot \overset{1}{\cancel{3}} \cdot \overset{1}{\cancel{5}}}{\underset{1}{\cancel{5}} \cdot 3 \cdot \underset{1}{\cancel{3}} \cdot 2 \cdot \underset{1}{\cancel{5}}} = -\frac{7}{6}$

17. $\frac{7}{12} + \frac{5}{6} + \frac{2}{3} = \frac{7}{12} + \frac{10}{12} + \frac{8}{12}$

$= \frac{7 + 10 + 8}{12} = \frac{25}{12}$

19. $4\frac{3}{4}$ rounds to 5. $2\frac{5}{6}$ rounds to 3.

Estimate: $\underline{5} + \underline{3} = \underline{8}$

Exact: $4\frac{3}{4} + 2\frac{5}{6} = \frac{19}{4} + \frac{17}{6}$

$= \frac{57}{12} + \frac{34}{12} = \frac{91}{12} = 7\frac{7}{12}$

21. $2\frac{7}{10}$ rounds to 3.

Estimate: $\underline{6} - \underline{3} = \underline{3}$

Exact: $6 - 2\frac{7}{10} = \frac{60}{10} - \frac{27}{10}$

$= \frac{60 - 27}{10} = \frac{33}{10} = 3\frac{3}{10}$

23. $4\frac{2}{3}$ rounds to 5. $1\frac{1}{6}$ rounds to 1.

Estimate: $\underline{5} \div \underline{1} = \underline{5}$

Exact: $4\frac{2}{3} \div 1\frac{1}{6} = \frac{14}{3} \div \frac{7}{6}$

$= \frac{14}{3} \cdot \frac{6}{7} = \frac{2 \cdot \overset{1}{\cancel{7}} \cdot 2 \cdot \overset{1}{\cancel{3}}}{\underset{1}{\cancel{3}} \cdot \underset{1}{\cancel{7}}} = \frac{4}{1} = 4$

25. **(a)** $\frac{1}{4} + \frac{11}{16} + 1\frac{1}{8} = \frac{4}{16} + \frac{11}{16} + \frac{18}{16}$

$= \frac{33}{16} = 2\frac{1}{16}$

The total length is $2\frac{1}{16}$ inches.

(b) $2\frac{1}{16} - 1\frac{3}{4} = \frac{33}{16} - \frac{7}{4}$

$= \frac{33}{16} - \frac{28}{16} = \frac{33 - 28}{16} = \frac{5}{16}$

$\frac{5}{16}$ in. will stick out the back of the board.

27. $9 \div \frac{3}{4} = \frac{9}{1} \cdot \frac{4}{3} = \frac{3 \cdot \overset{1}{\cancel{3}} \cdot 4}{1 \cdot \underset{1}{\cancel{3}}} = 12$

You can make 12 batches of cookies.

29. Not sure: $\frac{3}{20} \cdot 1500 = \frac{3 \cdot 75 \cdot \overset{1}{\cancel{20}}}{\underset{1}{\cancel{20}}} = 225$ adults

Real: $\frac{9}{20} \cdot 1500 = \frac{9 \cdot 75 \cdot \overset{1}{\cancel{20}}}{\underset{1}{\cancel{20}}} = 675$ adults

Imaginary: $\frac{2}{5} \cdot 1500 = \frac{2 \cdot 300 \cdot \overset{1}{\cancel{5}}}{\underset{1}{\cancel{5}}} = 600$ adults

31. $15\frac{1}{3} \div \frac{1}{3} = \frac{46}{3} \cdot \frac{3}{1} = \frac{46 \cdot \overset{1}{\cancel{3}}}{\underset{1}{\cancel{3}} \cdot 1} = 46$

46 bottles can be filled.

4.6 Exponents, Order of Operations, and Complex Fractions

4.6 Margin Exercises

1. **(a)** $\left(-\dfrac{3}{5}\right)^2 = \left(-\dfrac{3}{5}\right)\left(-\dfrac{3}{5}\right) = \dfrac{3\cdot 3}{5\cdot 5} = \dfrac{9}{25}$

(b) $\left(\dfrac{1}{3}\right)^4 = \dfrac{1}{3}\cdot\dfrac{1}{3}\cdot\dfrac{1}{3}\cdot\dfrac{1}{3} = \dfrac{1\cdot 1\cdot 1\cdot 1}{3\cdot 3\cdot 3\cdot 3} = \dfrac{1}{81}$

(c) $\left(-\dfrac{2}{3}\right)^3\left(\dfrac{1}{2}\right)^2$

$= \left(-\dfrac{2}{3}\right)\left(-\dfrac{2}{3}\right)\left(-\dfrac{2}{3}\right)\left(\dfrac{1}{2}\right)\left(\dfrac{1}{2}\right)$

$= -\dfrac{2\cdot\overset{1}{\cancel{2}}\cdot\overset{1}{\cancel{2}}\cdot 1\cdot 1}{3\cdot 3\cdot 3\cdot\underset{1}{\cancel{2}}\cdot\underset{1}{\cancel{2}}} = -\dfrac{2}{27}$

(d) $\left(-\dfrac{1}{2}\right)^2\left(\dfrac{1}{4}\right)^2 = \left(-\dfrac{1}{2}\right)\left(-\dfrac{1}{2}\right)\left(\dfrac{1}{4}\right)\left(\dfrac{1}{4}\right)$

$= \dfrac{1\cdot 1\cdot 1\cdot 1}{2\cdot 2\cdot 4\cdot 4} = \dfrac{1}{64}$

2. **(a)** $\dfrac{1}{3} - \dfrac{5}{9}\left(\dfrac{3}{4}\right)$

$= \dfrac{1}{3} - \dfrac{15}{36}$ *Multiply first.*

$= \dfrac{12}{36} - \dfrac{15}{36}$ *The LCD is 36.*

$= \dfrac{12-15}{36}$ *Now subtract.*

$= -\dfrac{3}{36}$

$= -\dfrac{1}{12}$ *Lowest terms*

(b) $-\dfrac{3}{4} + \left(-\dfrac{1}{2}\right)^2 \div \dfrac{2}{3}$

$= -\dfrac{3}{4} + \dfrac{1}{4} \div \dfrac{2}{3}$ *Exponent first*

$= -\dfrac{3}{4} + \dfrac{1}{4}\cdot\dfrac{3}{2}$ *Change division to multiplication.*

$= -\dfrac{3}{4} + \dfrac{3}{8}$ *Multiply.*

$= -\dfrac{6}{8} + \dfrac{3}{8}$ *The LCD is 8.*

$= \dfrac{-6+3}{8}$ *Add.*

$= -\dfrac{3}{8}$

(c) $\dfrac{12}{5} - \dfrac{1}{6}\left(3 - \dfrac{3}{5}\right)$

$= \dfrac{12}{5} - \dfrac{1}{6}\left(\dfrac{15}{5} - \dfrac{3}{5}\right)$ *Parentheses first*

$= \dfrac{12}{5} - \dfrac{1}{6}\left(\dfrac{12}{5}\right)$ *Subtract.*

$= \dfrac{12}{5} - \dfrac{1\cdot\overset{2}{\cancel{12}}}{\underset{1}{\cancel{6}}\cdot 5}$ *Multiply.*

$= \dfrac{12}{5} - \dfrac{2}{5}$

$= \dfrac{12-2}{5}$ *Subtract.*

$= \dfrac{10}{5}$

$= 2$

3. **(a)** $\dfrac{-\frac{3}{5}}{\frac{9}{10}} = -\dfrac{3}{5} \div \dfrac{9}{10}$

$= -\dfrac{3}{5}\cdot\dfrac{10}{9}$ *Change division to multiplication.*

$= -\dfrac{\overset{1}{\cancel{3}}\cdot 2\cdot\overset{1}{\cancel{5}}}{\underset{1}{\cancel{5}}\cdot\underset{1}{\cancel{3}}\cdot 3}$ *Divide out common factors.*

$= -\dfrac{2}{3}$

(b) $\dfrac{6}{\frac{3}{4}} = 6 \div \dfrac{3}{4}$

$= \dfrac{6}{1}\cdot\dfrac{4}{3} = \dfrac{2\cdot\overset{1}{\cancel{3}}\cdot 4}{1\cdot\underset{1}{\cancel{3}}} = \dfrac{8}{1} = 8$

(c) $\dfrac{-\frac{15}{16}}{-5} = -\dfrac{15}{16} \div \left(-\dfrac{5}{1}\right)$

$= -\dfrac{15}{16}\cdot\left(-\dfrac{1}{5}\right) = \dfrac{3\cdot\overset{1}{\cancel{5}}\cdot 1}{16\cdot\underset{1}{\cancel{5}}} = \dfrac{3}{16}$

(d) $\dfrac{-3}{\left(-\frac{3}{4}\right)^2} = -3 \div \left(-\dfrac{3}{4}\right)^2 = -\dfrac{3}{1} \div \dfrac{9}{16}$

$= -\dfrac{3}{1}\cdot\dfrac{16}{9} = -\dfrac{\overset{1}{\cancel{3}}\cdot 16}{1\cdot\underset{3}{\cancel{9}}}$

$= -\dfrac{16}{3}, \text{ or } -5\dfrac{1}{3}$

4.6 Section Exercises

1. **(a)** In the expression $\left(-\dfrac{3}{8}\right)^2$, $-\dfrac{3}{8}$ is called the base.

(b) The small raised "2" is called the <u>exponent</u>.

(c) $\left(-\dfrac{3}{8}\right)^2 = \left(-\dfrac{3}{8}\right)\left(-\dfrac{3}{8}\right)$

3. $\left(-\dfrac{3}{4}\right)^2 = \left(-\dfrac{3}{4}\right)\left(-\dfrac{3}{4}\right) = \dfrac{3\cdot 3}{4\cdot 4} = \dfrac{9}{16}$

Signs match so the product is <u>positive</u>.

5. $\left(\dfrac{2}{5}\right)^3 = \dfrac{2}{5}\cdot\dfrac{2}{5}\cdot\dfrac{2}{5}$

$= \dfrac{4}{25}\cdot\dfrac{2}{5} = \dfrac{8}{125}$

7. $\left(-\dfrac{1}{3}\right)^3 = \left(-\dfrac{1}{3}\right)\left(-\dfrac{1}{3}\right)\left(-\dfrac{1}{3}\right)$

$\qquad = \dfrac{1}{9}\left(-\dfrac{1}{3}\right) = -\dfrac{1}{27}$

9. $\left(\dfrac{1}{2}\right)^5 = \dfrac{1}{2}\cdot\dfrac{1}{2}\cdot\dfrac{1}{2}\cdot\dfrac{1}{2}\cdot\dfrac{1}{2}$

$\qquad = \dfrac{1}{4}\cdot\dfrac{1}{2}\cdot\dfrac{1}{2}\cdot\dfrac{1}{2}$

$\qquad = \dfrac{1}{8}\cdot\dfrac{1}{2}\cdot\dfrac{1}{2}$

$\qquad = \dfrac{1}{16}\cdot\dfrac{1}{2}$

$\qquad = \dfrac{1}{32}$

11. $\left(\dfrac{7}{10}\right)^2 = \dfrac{7}{10}\cdot\dfrac{7}{10} = \dfrac{49}{100}$

13. $\left(-\dfrac{6}{5}\right)^2 = \left(-\dfrac{6}{5}\right)\left(-\dfrac{6}{5}\right)$

$\qquad = \dfrac{36}{25},\ \text{ or }\ 1\dfrac{11}{25}$

15. $\dfrac{15}{16}\left(\dfrac{4}{5}\right)^3$

$\qquad = \dfrac{15}{16}\left(\dfrac{4}{5}\cdot\dfrac{4}{5}\cdot\dfrac{4}{5}\right)$ *Exponent first*

$\qquad = \dfrac{15}{16}\left(\dfrac{64}{125}\right)$ *Multiply.*

$\qquad = \dfrac{3\cdot\overset{1}{\cancel{5}}\cdot\overset{1}{\cancel{16}}\cdot 4}{\underset{1}{\cancel{16}}\cdot\underset{1}{\cancel{5}}\cdot 25}$ *Divide out common factors.*

$\qquad = \dfrac{12}{25}$

17. $\left(\dfrac{1}{3}\right)^4\left(\dfrac{9}{10}\right)^2 = \left(\dfrac{1}{3}\cdot\dfrac{1}{3}\cdot\dfrac{1}{3}\cdot\dfrac{1}{3}\right)\left(\dfrac{9}{10}\cdot\dfrac{9}{10}\right)$

$\qquad = \dfrac{1\cdot 1\cdot 1\cdot 1\cdot\overset{1}{\cancel{3}}\cdot\overset{1}{\cancel{3}}\cdot\overset{1}{\cancel{3}}\cdot\overset{1}{\cancel{3}}}{\underset{1}{\cancel{3}}\cdot\underset{1}{\cancel{3}}\cdot\underset{1}{\cancel{3}}\cdot\underset{1}{\cancel{3}}\cdot 10\cdot 10}$

$\qquad = \dfrac{1}{100}$

19. $\left(-\dfrac{3}{2}\right)^3\left(-\dfrac{2}{3}\right)^2$

$\qquad = \left(-\dfrac{3}{2}\right)\left(-\dfrac{3}{2}\right)\left(-\dfrac{3}{2}\right)\left(-\dfrac{2}{3}\right)\left(-\dfrac{2}{3}\right)$

$\qquad = -\dfrac{\overset{1}{\cancel{3}}\cdot\overset{1}{\cancel{3}}\cdot 3\cdot\overset{1}{\cancel{2}}\cdot\overset{1}{\cancel{2}}}{\underset{1}{\cancel{2}}\cdot\underset{1}{\cancel{2}}\cdot 2\cdot\underset{1}{\cancel{3}}\cdot\underset{1}{\cancel{3}}}$

$\qquad = -\dfrac{3}{2},\ \text{ or }\ -1\dfrac{1}{2}$

21. (a) $\left(-\dfrac{1}{2}\right)^2 = \left(-\dfrac{1}{2}\right)\left(-\dfrac{1}{2}\right) = \dfrac{1}{4}$

$\left(-\dfrac{1}{2}\right)^3 = \left(-\dfrac{1}{2}\right)^2\left(-\dfrac{1}{2}\right)$

$\qquad = \left(\dfrac{1}{4}\right)\left(-\dfrac{1}{2}\right) = -\dfrac{1}{8}$

$\left(-\dfrac{1}{2}\right)^4 = \left(-\dfrac{1}{2}\right)^3\left(-\dfrac{1}{2}\right)$

$\qquad = \left(-\dfrac{1}{8}\right)\left(-\dfrac{1}{2}\right) = \dfrac{1}{16}$

$\left(-\dfrac{1}{2}\right)^5 = \left(-\dfrac{1}{2}\right)^4\left(-\dfrac{1}{2}\right)$

$\qquad = \left(\dfrac{1}{16}\right)\left(-\dfrac{1}{2}\right) = -\dfrac{1}{32}$

$\left(-\dfrac{1}{2}\right)^6 = \left(-\dfrac{1}{2}\right)^5\left(-\dfrac{1}{2}\right)$

$\qquad = \left(-\dfrac{1}{32}\right)\left(-\dfrac{1}{2}\right) = \dfrac{1}{64}$

$\left(-\dfrac{1}{2}\right)^7 = \left(-\dfrac{1}{2}\right)^6\left(-\dfrac{1}{2}\right)$

$\qquad = \left(\dfrac{1}{64}\right)\left(-\dfrac{1}{2}\right) = -\dfrac{1}{128}$

$\left(-\dfrac{1}{2}\right)^8 = \left(-\dfrac{1}{2}\right)^7\left(-\dfrac{1}{2}\right)$

$\qquad = \left(-\dfrac{1}{128}\right)\left(-\dfrac{1}{2}\right) = \dfrac{1}{256}$

$\left(-\dfrac{1}{2}\right)^9 = \left(-\dfrac{1}{2}\right)^8\left(-\dfrac{1}{2}\right)$

$\qquad = \left(\dfrac{1}{256}\right)\left(-\dfrac{1}{2}\right) = -\dfrac{1}{512}$

(b) When a negative number is raised to an even power, the answer is positive. When a negative number is raised to an odd power, the answer is negative.

23. $\dfrac{1}{5} - 6\left(\dfrac{7}{10}\right)$

$\qquad = \dfrac{1}{5} - \dfrac{6}{1}\left(\dfrac{7}{10}\right)$ *Rewrite 6 as $\dfrac{6}{1}$.*

$\qquad = \dfrac{1}{5} - \dfrac{42}{10}$ *Multiply.*

$\qquad = \dfrac{2}{10} - \dfrac{42}{10}$ *The LCD is 10.*

$\qquad = \dfrac{2-42}{10}$ *Subtract.*

$\qquad = -\dfrac{40}{10}$

$\qquad = -4$ *Lowest terms*

25. $\left(\dfrac{4}{3} \div \dfrac{8}{3}\right) + \left(-\dfrac{3}{4} \cdot \dfrac{1}{4}\right)$

$= \left(\dfrac{4}{3} \cdot \dfrac{3}{8}\right) + \left(-\dfrac{3}{4} \cdot \dfrac{1}{4}\right)$ *Change division to multiplication.*

$= \left(\dfrac{\overset{1}{\cancel{4}} \cdot \overset{1}{\cancel{3}}}{\underset{1}{\cancel{3}} \cdot 2 \cdot \underset{1}{\cancel{4}}}\right) + \left(-\dfrac{3}{16}\right)$ *Parentheses*

$= \dfrac{1}{2} - \dfrac{3}{16}$

$= \dfrac{8}{16} - \dfrac{3}{16}$ *LCD is 16.*

$= \dfrac{8-3}{16} = \dfrac{5}{16}$ *Subtract.*

27. $-\dfrac{3}{10} \div \dfrac{3}{5}\left(-\dfrac{2}{3}\right)$

$= -\dfrac{3}{10} \cdot \dfrac{5}{3}\left(-\dfrac{2}{3}\right)$ *Change division to multiplication.*

$= -\dfrac{\overset{1}{\cancel{3}} \cdot \overset{1}{\cancel{5}}}{2 \cdot \underset{1}{\cancel{5}} \cdot \underset{1}{\cancel{3}}}\left(-\dfrac{2}{3}\right)$ *Multiply.*

$= -\dfrac{1}{2}\left(-\dfrac{2}{3}\right)$

$= \dfrac{1 \cdot \overset{1}{\cancel{2}}}{\underset{1}{\cancel{2}} \cdot 3}$ *Multiply.*

$= \dfrac{1}{3}$

29. $\dfrac{8}{3}\left(\dfrac{1}{4} - \dfrac{1}{2}\right)^2$

$= \dfrac{8}{3}\left(\dfrac{1}{4} - \dfrac{2}{4}\right)^2$ *LCD inside parentheses is 4.*

$= \dfrac{8}{3}\left(-\dfrac{1}{4}\right)^2$ *Subtract.*

$= \dfrac{8}{3}\left(\dfrac{1}{16}\right)$ *Exponent*

$= \dfrac{\overset{1}{\cancel{8}} \cdot 1}{3 \cdot 2 \cdot \underset{1}{\cancel{8}}}$ *Multiply.*

$= \dfrac{1}{6}$

31. $-\dfrac{3}{8} + \dfrac{2}{3}\left(-\dfrac{2}{3} + \dfrac{1}{6}\right)$

$= -\dfrac{3}{8} + \dfrac{2}{3}\left(-\dfrac{4}{6} + \dfrac{1}{6}\right)$ *LCD inside parentheses is 6.*

$= -\dfrac{3}{8} + \dfrac{2}{3}\left(-\dfrac{3}{6}\right)$ *Add.*

$= -\dfrac{3}{8} + \dfrac{2}{3}\left(-\dfrac{1}{2}\right)$ *Reduce.*

$= -\dfrac{3}{8} + \left(-\dfrac{1}{3}\right)$ *Multiply.*

$= -\dfrac{9}{24} + \left(-\dfrac{8}{24}\right)$ *LCD is 24.*

$= -\dfrac{17}{24}$ *Add.*

33. $2\left(\dfrac{1}{3}\right)^3 - \dfrac{2}{9}$

$= \dfrac{2}{1}\left(\dfrac{1}{27}\right) - \dfrac{2}{9}$ *Exponent*

$= \dfrac{2}{27} - \dfrac{2}{9}$ *Multiply.*

$= \dfrac{2}{27} - \dfrac{6}{27}$ *LCD is 27.*

$= \dfrac{2-6}{27} = -\dfrac{4}{27}$ *Subtract.*

35. $\left(-\dfrac{2}{3}\right)^3\left(\dfrac{1}{8} - \dfrac{1}{2}\right) - \dfrac{2}{3}\left(\dfrac{1}{8}\right)$

$= \left(-\dfrac{2}{3}\right)^3\left(\dfrac{1}{8} - \dfrac{4}{8}\right) - \dfrac{2}{3}\left(\dfrac{1}{8}\right)$ *LCD inside paren. is 8.*

$= \left(-\dfrac{2}{3}\right)^3\left(-\dfrac{3}{8}\right) - \dfrac{2}{3}\left(\dfrac{1}{8}\right)$ *Subtract.*

$= \left(-\dfrac{8}{27}\right)\left(-\dfrac{3}{8}\right) - \dfrac{2}{3}\left(\dfrac{1}{8}\right)$ *Exponent*

$= \dfrac{\overset{1}{\cancel{8}} \cdot \overset{1}{\cancel{3}}}{\underset{1}{\cancel{3}} \cdot 9 \cdot \underset{1}{\cancel{8}}} - \dfrac{\overset{1}{\cancel{2}} \cdot 1}{3 \cdot \underset{1}{\cancel{2}} \cdot 4}$ *Multiply.*

$= \dfrac{1}{9} - \dfrac{1}{12}$

$= \dfrac{4}{36} - \dfrac{3}{36}$ *LCD is 36.*

$= \dfrac{4-3}{36}$ *Subtract.*

$= \dfrac{1}{36}$

37. $A = s^2 = \left(\dfrac{3}{8}\right)^2$

$= \left(\dfrac{3}{8}\right)\left(\dfrac{3}{8}\right) = \dfrac{3 \cdot 3}{8 \cdot 8} = \dfrac{9}{64}$ in.^{2}

39. $P = 2l + 2w$

$= 2\left(\dfrac{7}{10}\right) + 2\left(\dfrac{1}{4}\right)$

$= \dfrac{2}{1}\left(\dfrac{7}{10}\right) + \dfrac{2}{1}\left(\dfrac{1}{4}\right)$ *Rewrite as $\dfrac{2}{1}$.*

$= \dfrac{14}{10} + \dfrac{2}{4}$ *Multiply.*

$= \dfrac{7}{5} + \dfrac{1}{2}$ *Reduce.*

$= \dfrac{14}{10} + \dfrac{5}{10}$ *LCD is 10.*

$= \dfrac{19}{10}$, or $1\dfrac{9}{10}$ miles *Add.*

41. $\dfrac{-\frac{7}{9}}{-\frac{7}{36}}$

$$= -\frac{7}{9} \div \left(-\frac{7}{36}\right) \qquad \textit{Rewrite using the} \div \textit{symbol.}$$

$$= -\frac{7}{9} \cdot \left(-\frac{36}{7}\right) \qquad \textit{Change division to multiplication.}$$

$$= \frac{\overset{1}{\cancel{7}} \cdot \overset{1}{\cancel{9}} \cdot 4}{\underset{1}{\cancel{9}} \cdot \underset{1}{\cancel{7}}} \qquad \textit{Multiply and reduce.}$$

$$= \frac{4}{1} = 4$$

43. $\dfrac{-15}{\frac{6}{5}}$

$$= -15 \div \frac{6}{5} \qquad \textit{Rewrite using the} \div \textit{symbol.}$$

$$= -\frac{15}{1} \cdot \left(\frac{5}{6}\right) \qquad \textit{Change division to multiplication.}$$

$$= -\frac{\overset{1}{\cancel{3}} \cdot 5 \cdot 5}{1 \cdot 2 \cdot \underset{1}{\cancel{3}}} \qquad \textit{Multiply and reduce.}$$

$$= -\frac{25}{2}, \quad \text{or} \quad -12\frac{1}{2}$$

45. $\dfrac{\frac{4}{7}}{8}$

$$= \frac{4}{7} \div 8 \qquad \textit{Rewrite using the} \div \textit{symbol.}$$

$$= \frac{4}{7} \cdot \left(\frac{1}{8}\right) \qquad \textit{Change division to multiplication.}$$

$$= \frac{\overset{1}{\cancel{4}} \cdot 1}{7 \cdot \underset{1}{\cancel{4}} \cdot 2} \qquad \textit{Multiply and reduce.}$$

$$= \frac{1}{14}$$

47. $\dfrac{-\frac{2}{3}}{-2\frac{2}{5}}$

$$= \frac{-\frac{2}{3}}{-\frac{12}{5}} \qquad \textit{Change to improper fraction.}$$

$$= \frac{2}{3} \div \frac{12}{5} \qquad \textit{Rewrite using the} \div \textit{symbol.}$$

$$= \frac{2}{3} \cdot \frac{5}{12} \qquad \textit{Change division to multiplication.}$$

$$= \frac{\overset{1}{\cancel{2}} \cdot 5}{3 \cdot \underset{1}{\cancel{2}} \cdot 6} \qquad \textit{Multiply and reduce.}$$

$$= \frac{5}{18}$$

49. $\dfrac{-4\frac{1}{2}}{\left(\frac{3}{4}\right)^2}$

$$= \frac{-4\frac{1}{2}}{\frac{9}{16}} \qquad \textit{Exponent first}$$

$$= \frac{-\frac{9}{2}}{\frac{9}{16}} \qquad \textit{Change to improper fraction.}$$

$$= -\frac{9}{2} \div \frac{9}{16} \qquad \textit{Rewrite using the} \div \textit{symbol.}$$

$$= -\frac{9}{2} \cdot \frac{16}{9} \qquad \textit{Change division to multiplication.}$$

$$= -\frac{\overset{1}{\cancel{9}} \cdot \overset{1}{\cancel{2}} \cdot 8}{\underset{1}{\cancel{2}} \cdot \underset{1}{\cancel{9}}} \qquad \textit{Multiply and reduce.}$$

$$= -\frac{8}{1} = -8$$

51. $\dfrac{\left(\frac{2}{5}\right)^2}{\left(-\frac{1}{3}\right)^2}$

$$= \frac{\frac{4}{25}}{\frac{16}{9}} \qquad \textit{Exponents first}$$

$$= \frac{4}{25} \div \frac{16}{9} \qquad \textit{Rewrite using the} \div \textit{symbol.}$$

$$= \frac{4}{25} \cdot \frac{9}{16} \qquad \textit{Change division to multiplication.}$$

$$= \frac{\overset{1}{\cancel{4}} \cdot 9}{25 \cdot \underset{1}{\cancel{4}} \cdot 4} \qquad \textit{Multiply and reduce.}$$

$$= \frac{9}{100}$$

4.7   Problem Solving: Equations Containing Fractions

4.7   Margin Exercises

1.   (a) $\dfrac{1}{6}m = 3$

$$\frac{\overset{1}{\cancel{6}}}{1}\left(\frac{1}{\underset{1}{\cancel{6}}}m\right) = \frac{6}{1}\left(\frac{3}{1}\right) \qquad \begin{array}{l}\textit{Multiply both} \\ \textit{sides by } \frac{6}{1}, \textit{ the} \\ \textit{reciprocal of } \frac{1}{6}.\end{array}$$

$$m = \underline{18}$$

Check $\frac{1}{6}(\underline{18}) = 3$ *Replace m with $\underline{18}$.*

$$\frac{18}{6} = 3$$

$$\underline{3} = 3 \quad \text{Balances}$$

The solution is 18.

(b) $\dfrac{3}{2}a = -9$

$$\dfrac{\cancel{2}^{1}}{\cancel{3}_{1}}\left(\dfrac{\cancel{3}^{1}}{\cancel{2}_{1}}a\right) = \dfrac{2}{3}\left(-\dfrac{9}{1}\right)$$ *Multiply both sides by $\frac{2}{3}$, the reciprocal of $\frac{3}{2}$.*

$$a = -\dfrac{18}{3}$$

$$a = -6$$

Check $\frac{3}{2}(-6) = -9$ *Replace a with −6.*

$$\dfrac{-18}{2} = -9$$

$$-9 = -9$$ Balances

The solution is −6.

(c) $\dfrac{3}{14} = -\dfrac{2}{7}x$

$$-\dfrac{7}{2}\left(\dfrac{3}{14}\right) = -\dfrac{\cancel{7}^{1}}{\cancel{2}_{1}}\left(-\dfrac{\cancel{2}^{1}}{\cancel{7}_{1}}x\right)$$ *Multiply both sides by $-\frac{7}{2}$.*

$$-\dfrac{\cancel{7}^{1}\cdot 3}{2\cdot 2\cdot\cancel{7}_{1}} = x$$

$$-\dfrac{3}{4} = x$$

Check $\dfrac{3}{14} = -\dfrac{2}{7}\left(-\dfrac{3}{4}\right)$ *Replace x with $-\frac{3}{4}$.*

$$\dfrac{3}{14} = \dfrac{\cancel{2}^{1}\cdot 3}{7\cdot\cancel{2}\cdot 2_{1}}$$

$$\dfrac{3}{14} = \dfrac{3}{14}$$ Balances

The solution is $-\frac{3}{4}$.

2. (a) $18 = \dfrac{4}{5}x + 2$

$$\underline{\quad -2 \qquad\qquad -2\quad}$$ *Add −2 to both sides.*

$$16 = \dfrac{4}{5}x + 0$$

$$\dfrac{5}{4}(16) = \dfrac{\cancel{5}^{1}}{\cancel{4}_{1}}\left(\dfrac{\cancel{4}^{1}}{\cancel{5}_{1}}x\right)$$ *Multiply both sides by $\frac{5}{4}$, the reciprocal of $\frac{4}{5}$.*

$$\dfrac{5\cdot\cancel{4}^{1}\cdot 4}{\cancel{4}_{1}} = x$$

$$20 = x$$

Check $18 = \dfrac{4}{5}(20) + 2$ *Replace x with 20.*

$$18 = \dfrac{4\cdot\cancel{5}^{1}\cdot 4}{\cancel{5}_{1}} + 2$$

$$18 = 16 + 2$$

$$18 = 18$$ Balances

The solution is 20.

(b) $\dfrac{1}{4}h - 5 = 1$

$$\underline{\qquad 5 \qquad\qquad 5\qquad}$$ *Add 5 to both sides.*

$$\dfrac{1}{4}h + 0 = 6$$

$$\dfrac{\cancel{4}^{1}}{1}\left(\dfrac{1}{\cancel{4}_{1}}h\right) = \dfrac{4}{1}(6)$$ *Multiply both sides by $\frac{4}{1}$, the reciprocal of $\frac{1}{4}$.*

$$h = 24$$

Check $\frac{1}{4}(24) - 5 = 1$ *Replace h with 24.*

$$6 - 5 = 1$$

$$1 = 1$$ Balances

The solution is 24.

(c) $\dfrac{4}{3}r + 4 = -8$

$$\underline{\qquad -4 \qquad\qquad -4\qquad}$$ *Add −4 to both sides.*

$$\dfrac{4}{3}r + 0 = -12$$

$$\dfrac{\cancel{3}^{1}}{\cancel{4}_{1}}\left(\dfrac{\cancel{4}^{1}}{\cancel{3}_{1}}r\right) = \dfrac{3}{4}\left(-\dfrac{12}{1}\right)$$ *Multiply both sides by $\frac{3}{4}$, the reciprocal of $\frac{4}{3}$.*

$$r = -\dfrac{3\cdot 3\cdot\cancel{4}^{1}}{\cancel{4}_{1}}$$

$$r = -9$$

Check $\frac{4}{3}(-9) + 4 = -8$ *Replace r with −9.*

$$-\dfrac{36}{3} + 4 = -8$$

$$-12 + 4 = -8$$

$$-8 = -8$$ Balances

The solution is −9.

3. ***Step 1***   Unknown: woman's age
Known: blood pressure expression is $100 + \frac{\text{age}}{2}$; woman's pressure is 111

Step 2(a)   Let a represent the woman's age.

Step 3 $100 + \dfrac{a}{2} = $ blood pressure

Step 4 $100 + \dfrac{a}{2} = \underline{111}$

$$\underline{\,-100 \qquad\qquad -100\,}$$ *Add −100 to both sides.*

$$\dfrac{a}{2} = 11$$

$$\dfrac{\cancel{2}^{1}}{1}\left(\dfrac{1}{\cancel{2}_{1}}a\right) = \dfrac{2}{1}(11)$$ *Multiply both sides by $\frac{2}{1}$, the reciprocal of $\frac{1}{2}$.*

$$a = 22$$

Step 5   The woman is 22 years old.

Step 6 $100 + \frac{22}{2} = 111$ *Replace a with 22.*

$\qquad\quad 100 + 11 = 111$

$\qquad\qquad\qquad 111 = 111$ Balances

4.7 Section Exercises

1. **(a)** The coefficient of the variable c is $\frac{2}{5}$. Multiply both sides by the reciprocal, that is, $\frac{5}{2}$.

(b) $\frac{5}{2}$ is the reciprocal of $\frac{2}{5}$. Multiplying $\frac{5}{2}$ times $\frac{2}{5}c$ gives $1c$, or just c, on the left side of the equal sign.

3. $\qquad \frac{1}{3}a = 10$

$\frac{1}{\cancel{3}} \left(\frac{1}{\cancel{3}}a \right) = \frac{3}{1} \left(\frac{10}{1} \right)$ *Multiply both sides by $\frac{3}{1}$, the reciprocal of $\frac{1}{3}$.*

$\qquad\quad a = \frac{3}{1} \cdot \frac{10}{1}$

$\qquad\quad a = \underline{30}$

Check $\frac{1}{3}(30) = 10$ *Replace a with 30.*

$\qquad\quad \frac{30}{3} = 10$

$\qquad\quad 10 = 10$ Balances

The solution is 30.

5. $\qquad -20 = \frac{5}{6}b$

$\frac{6}{5} \left(-\frac{20}{1} \right) = \frac{1}{\cancel{6}} \left(\frac{1}{\cancel{5}}b \right)$ *Multiply both sides by $\frac{6}{5}$, the reciprocal of $\frac{5}{6}$.*

$-\frac{6 \cdot 4 \cdot \cancel{5}}{\cancel{5}} = b$

$\qquad -24 = b$

Check $-20 = \frac{5}{6}(-24)$ *Replace b with -24.*

$\qquad -20 = -20$ Balances

The solution is -24.

7. $\qquad -\frac{7}{2}c = -21$

$-\frac{\cancel{2}}{7} \left(-\frac{1}{\cancel{2}}c \right) = -\frac{2}{7} \left(-\frac{\cancel{21}}{1} \right)$ *Multiply both sides by $-\frac{2}{7}$, the reciprocal of $-\frac{7}{2}$.*

$\qquad\quad c = \frac{6}{1}$

$\qquad\quad c = 6$

Check $-\frac{7}{2}(6) = -21$ *Replace c with 6.*

$\qquad\quad -\frac{42}{2} = -21$

$\qquad\quad -21 = -21$ Balances

The solution is 6.

9. $\qquad \frac{3}{10} = -\frac{1}{4}d$

$-\frac{4}{1} \left(\frac{3}{10} \right) = -\frac{4}{1} \left(\frac{-1}{\cancel{4}}d \right)$ *Multiply both sides by $-\frac{4}{1}$, the reciprocal of $-\frac{1}{4}$.*

$\qquad -\frac{12}{10} = d$

$\qquad -\frac{6}{5} = d$

Check $\frac{3}{10} = -\frac{1}{4} \left(-\frac{6}{5} \right)$ *Replace d with $-\frac{6}{5}$.*

$\qquad \frac{3}{10} = \frac{1 \cdot \cancel{2} \cdot 3}{\cancel{2} \cdot 2 \cdot 5}$

$\qquad \frac{3}{10} = \frac{3}{10}$ Balances

The solution is $-\frac{6}{5}$.

11. $\frac{1}{6}n + 7 = 9$

$\qquad \underline{-7 \qquad -7}$ *Add -7 to both sides.*

$\qquad \frac{1}{6}n = 2$

$\frac{\cancel{6}}{1} \cdot \frac{1}{\cancel{6}}n = \frac{6}{1} \cdot 2$ *Multiply both sides by $\frac{6}{1}$, the reciprocal of $\frac{1}{6}$.*

$\qquad n = 12$

Check $\frac{1}{6}(12) + 7 = 9$ *Replace n with 12.*

$\qquad\quad 2 + 7 = 9$

$\qquad\qquad 9 = 9$ Balances

The solution is 12.

13. $\qquad -10 = \frac{5}{3}r + 5$

$\qquad \underline{-5 \qquad\qquad -5}$ *Add -5 to both sides.*

$\qquad -15 = \frac{5}{3}r + 0$

$\frac{3}{5} \left(-\frac{15}{1} \right) = \frac{\cancel{3}}{\cancel{5}} \left(\frac{\cancel{5}}{\cancel{3}}r \right)$ *Multiply both sides by $\frac{3}{5}$, the reciprocal of $\frac{5}{3}$.*

$-\frac{3 \cdot 3 \cdot \cancel{5}}{\cancel{5}} = r$

$\qquad -9 = r$

Check $-10 = \frac{5}{3}(-9) + 5$ *Replace r with -9.*

$\qquad -10 = -\frac{45}{3} + 5$

$\qquad -10 = -15 + 5$

$\qquad -10 = -10$ Balances

The solution is -9.

15. $\frac{3}{8}x - 9 = 0$

$$\underline{99}$$ *Add 9 to both sides.*

$$\frac{3}{8}x = 9$$

$$\frac{\cancel{8}^{1}}{\cancel{3}_{1}} \cdot \frac{\cancel{3}^{1}}{\cancel{8}_{1}}x = \frac{8}{\cancel{3}_{1}} \cdot \frac{\cancel{9}^{3}}{1}$$ *Multiply both sides by $\frac{8}{3}$, the reciprocal of $\frac{3}{8}$.*

$$x = 24$$

Check $\frac{3}{8}(24) - 9 = 0$ *Replace x with 24.*

$$9 - 9 = 0$$

$$0 = 0 \quad \text{Balances}$$

The solution is 24.

17. $\underbrace{7 - 2}_{} = \frac{1}{5}y - 4$

$$5 = \frac{1}{5}y + (-4)$$

$$\underline{44}$$ *Add 4 to both sides.*

$$9 = \frac{1}{5}y + 0$$

$$\frac{5}{1}\left(\frac{9}{1}\right) = \frac{\cancel{5}}{1}\left(\frac{1}{\cancel{5}_{1}}y\right)$$ *Multiply both sides by $\frac{5}{1}$, the reciprocal of $\frac{1}{5}$.*

$$45 = y$$

The solution is 45.

19. $4 + \frac{2}{3}n = -10 + 2$

$$4 + \frac{2}{3}n = -8$$

$$\underline{-4 -4}$$ *Add -4 to both sides.*

$$\frac{2}{3}n = -12$$

$$\frac{\cancel{3}^{1}}{\cancel{2}_{1}}\left(\frac{\cancel{2}^{1}}{\cancel{3}_{1}}n\right) = \frac{3}{2}(-12)$$ *Multiply both sides by $\frac{3}{2}$, the reciprocal of $\frac{2}{3}$.*

$$n = -\frac{36}{2}$$

$$n = -18$$

The solution is -18.

21. $3x + \frac{1}{2} = \frac{3}{4}$

$$\underline{-\frac{1}{2}-\frac{1}{2}}$$ *Add $-\frac{1}{2}$ to both sides.*

$$3x + 0 = \frac{3}{4} + \left(-\frac{1}{2}\right)$$

$$3x = \frac{3}{4} + \left(-\frac{2}{4}\right)$$

$$3x = \frac{1}{4}$$

$$\frac{1}{3}(3x) = \frac{1}{3}\left(\frac{1}{4}\right)$$ *Multiply both sides by $\frac{1}{3}$, the reciprocal of 3.*

$$x = \frac{1}{12}$$

The solution is $\frac{1}{12}$.

23. $\frac{3}{10} = -4b - \frac{1}{5}$

$$\underline{\frac{1}{5}\frac{1}{5}}$$ *Add $\frac{1}{5}$ to both sides.*

$$\frac{3}{10} + \frac{1}{5} = -4b$$

$$\frac{3}{10} + \frac{2}{10} = -4b$$

$$\frac{5}{10} = -4b$$

$$-\frac{1}{4} \cdot \frac{5}{10} = -\frac{1}{4}(-4b)$$ *Multiply both sides by $-\frac{1}{4}$, the reciprocal of -4.*

$$-\frac{5}{40} = b$$

$$-\frac{1}{8} = b$$

The solution is $-\frac{1}{8}$.

25. $-1 - \frac{3}{8}n = 5 - 6$

$$-1 - \frac{3}{8}n = -1$$

$$\underline{11}$$ *Add 1 to both sides.*

$$-\frac{3}{8}n = 0$$

$$-\frac{8}{3}\left(-\frac{3}{8}n\right) = -\frac{8}{3}(0)$$ *Multiply both sides by $-\frac{8}{3}$, the reciprocal of $-\frac{3}{8}$.*

$$n = 0$$

The solution is 0.

27. (a) $\frac{1}{6}x + 1 = -2$

$$\frac{1}{6}\left(\frac{18}{1}\right) + 1 = -2$$ *Replace x with 18.*

$$\frac{18}{6} + 1 = -2$$

$$3 + 1 = -2$$

$$4 \neq -2$$

No, it does not balance. $x = 18$ is not the solution, so we'll solve the equation.

$$\frac{1}{6}x + 1 = -2$$

$$\underline{\quad -1 \qquad -1 \quad}$$ *Add -1 to both sides.*

$$\frac{1}{6}x + 0 = -3$$

$$\frac{1}{6}x = -3$$

$$\frac{\cancel{6}}{1}\left(\frac{1}{\cancel{6}}x\right) = \frac{6}{1}(-3)$$ *Multiply both sides by $\frac{6}{1}$, the reciprocal of $\frac{1}{6}$.*

$$x = -18$$

The correct solution is -18.

(b) $-\dfrac{3}{2} = \dfrac{9}{4}k$

$$-\frac{3}{2} = \frac{9}{4}\left(-\frac{2}{3}\right)$$ *Replace k with $-\frac{2}{3}$.*

$$-\frac{3}{2} = -\frac{18}{12}$$

$$-\frac{3}{2} = -\frac{6 \cdot 3}{6 \cdot 2}$$

$$-\frac{3}{2} = -\frac{3}{2}$$ *Balances*

Yes, $k = -\frac{2}{3}$ is the correct solution because the equation balances.

29.   *Step 1*   Unknown: man's age
Known: blood pressure expression is $100 + \frac{\text{age}}{2}$; man's pressure is 109

Step 2(a)   Let a represent the man's age.

Step 3 $100 + \frac{a}{2} = $ blood pressure

Step 4 $100 + \dfrac{a}{2} = 109$

$$\underline{-100 \qquad\qquad -100}$$ *Add -100 to both sides.*

$$\frac{a}{2} = 9$$

$$\frac{\cancel{2}}{1}\left(\frac{1}{\cancel{2}}a\right) = \frac{2}{1}(9)$$ *Multiply both sides by $\frac{2}{1}$, the reciprocal of $\frac{1}{2}$.*

$$a = 18$$

Step 5   The man is 18 years old.

Step 6 $100 + \frac{18}{2} = 109$ *Replace a with 18.*
$100 + 9 = 109$
$109 = 109$ *Balances*

31.   *Step 1*   Unknown: woman's age
Known: blood pressure expression is $100 + \frac{\text{age}}{2}$; woman's pressure is 122

Step 2(a)   Let a represent the woman's age.

Step 3 $100 + \frac{a}{2} = $ blood pressure

Step 4 $100 + \dfrac{a}{2} = 122$

$$\underline{-100 \qquad\qquad -100}$$ *Add -100 to both sides.*

$$\frac{a}{2} = 22$$

$$\frac{\cancel{2}}{1}\left(\frac{1}{\cancel{2}}a\right) = \frac{2}{1}(22)$$ *Multiply both sides by $\frac{2}{1}$, the reciprocal of $\frac{1}{2}$.*

$$a = 44$$

Step 5   The woman is 44 years old.

Step 6 $100 + \frac{44}{2} = 122$ *Replace a with 44.*
$100 + 22 = 122$
$122 = 122$ *Balances*

33.   *Step 1*   Unknown: woman's age
Known: blood pressure expression is $100 + \frac{\text{age}}{2}$; woman's pressure is 128

Step 2(a)   Let a represent the woman's age.

Step 3 $100 + \frac{a}{2} = $ blood pressure

Step 4 $100 + \dfrac{a}{2} = 128$

$$\underline{-100 \qquad\qquad -100}$$ *Add -100 to both sides.*

$$\frac{a}{2} = 28$$

$$\frac{\cancel{2}}{1}\left(\frac{1}{\cancel{2}}a\right) = \frac{2}{1}(28)$$ *Multiply both sides by $\frac{2}{1}$, the reciprocal of $\frac{1}{2}$.*

$$a = 56$$

Step 5   The woman is 56 years old.

Step 6 $100 + \frac{56}{2} = 128$ *Replace a with 56.*
$100 + 28 = 128$
$128 = 128$ *Balances*

35.   *Step 1*   Unknown: the penny size
Known: penny size equation is

$$\frac{\text{penny size}}{4} + \frac{1}{2} \text{ inch} = \text{length of nail};$$

length of a common nail is 3 inches;

Step 2(a)   Let p be the penny size.

Step 3 $\frac{\text{penny size}}{4} + \frac{1}{2} = $ length of nail

Step 4 $\dfrac{p}{4} + \dfrac{1}{2} = 3$

$$\underline{\quad -\frac{1}{2} \qquad -\frac{1}{2} \quad}$$ *Add $-\frac{1}{2}$ to both sides.*

$$\frac{p}{4} = 3 - \frac{1}{2}$$

$$\frac{p}{4} = \frac{6}{2} - \frac{1}{2}$$

$$\frac{1}{4}p = \frac{5}{2}$$

$$\frac{1}{\cancel{4}}\left(\frac{1}{\cancel{4}}p\right) = \frac{\cancel{4}}{1} \cdot \frac{5}{\cancel{2}}\ \ \begin{array}{l}\textit{Multiply both}\\ \textit{sides by } \frac{4}{1}, \textit{ the}\\ \textit{reciprocal of } \frac{1}{4}.\end{array}$$

$$p = 10$$

Step 5 The penny size is 10.

Step 6 $\frac{10}{4} + \frac{1}{2} = 3$ *Replace p with 10.*
  $\frac{5}{2} + \frac{1}{2} = 3$
    $3 = 3$ Balances

37. Step 1 Unknown: the penny size
Known: penny size equation is

$$\frac{\text{penny size}}{4} + \frac{1}{2} \text{ inch} = \text{length of nail};$$

length of a box nail is $2\frac{1}{2}$ inches;

Step 2(a) Let p be the penny size.

Step 3 $\frac{\text{penny size}}{4} + \frac{1}{2} = \text{length of nail}$

Step 4 $\frac{p}{4} + \frac{1}{2} = 2\frac{1}{2}$

$$\underline{\phantom{\frac{p}{4}+}-\frac{1}{2}-\frac{1}{2}}\ \ \begin{array}{l}\textit{Add } -\frac{1}{2} \textit{ to}\\ \textit{both sides.}\end{array}$$

$$\frac{p}{4} = 2$$

$$\frac{1}{\cancel{4}}\left(\frac{1}{\cancel{4}}p\right) = 4 \cdot 2\ \ \begin{array}{l}\textit{Multiply both}\\ \textit{sides by } \frac{4}{1}, \textit{ the}\\ \textit{reciprocal of } \frac{1}{4}.\end{array}$$

$$p = 8$$

Step 5 The penny size is 8.

Step 6 $\frac{8}{4} + \frac{1}{2} = 2\frac{1}{2}$ *Replace p with 8.*
  $2 + \frac{1}{2} = 2\frac{1}{2}$
    $2\frac{1}{2} = 2\frac{1}{2}$ Balances

Relating Concepts (Exercises 39–40)

39. Start with the equation $x = 8$. Now multiply both sides by any fraction, say $\frac{1}{2}$. This gives us the equation $\frac{1}{2}x = 4$.

Some possibilities are:

$$\frac{1}{2}x = 4;\ -\frac{1}{4}a = -2;\ \frac{3}{4}b = 6.$$

40. Start with the equation $y = -12$. Now multiply both sides by any fraction, say $\frac{1}{2}$. This gives us the equation $\frac{1}{2}y = -6$.

Some possibilities are:

$$\frac{1}{2}y = -6;\ \frac{2}{3}w = -8;\ -\frac{1}{12}d = 1.$$

4.8 Geometry Applications: Area and Volume

4.8 Margin Exercises

1. **(a)** $A = \frac{1}{2} \cdot b \cdot h$

$A = \frac{1}{2} \cdot 26 \text{ m} \cdot 20 \text{ m}$

$A = \frac{1}{2} \cdot \frac{26 \text{ m}}{1} \cdot \frac{20 \text{ m}}{1}$

$A = \dfrac{1 \cdot \overset{1}{\cancel{2}} \cdot 13 \text{ m} \cdot 20 \text{ m}}{\underset{1}{\cancel{2}} \cdot 1 \cdot 1}$

$A = \underline{260 \text{ m}^2}$

(b) $A = \frac{1}{2}bh$

$A = \frac{1}{2} \cdot 6 \text{ yd} \cdot 5 \text{ yd}$

$A = \frac{1}{2} \cdot \frac{6 \text{ yd}}{1} \cdot \frac{5 \text{ yd}}{1}$

$A = \dfrac{1 \cdot \overset{1}{\cancel{2}} \cdot 3 \text{ yd} \cdot 5 \text{ yd}}{\underset{1}{\cancel{2}} \cdot 1 \cdot 1}$

$A = \underline{15 \text{ yd}^2}$

(c) $A = \frac{1}{2}bh$

$A = \frac{1}{2} \cdot 9\frac{1}{2} \text{ ft} \cdot 7 \text{ ft}$

$A = \frac{1}{2} \cdot \frac{19}{2} \text{ ft} \cdot \frac{7}{1} \text{ ft}$

$A = \dfrac{1 \cdot 19 \text{ ft} \cdot 7 \text{ ft}}{2 \cdot 2 \cdot 1}$

$A = \underline{\frac{133}{4}} \text{ ft}^2, \text{ or } \underline{33\frac{1}{4}} \text{ ft}^2$

2. The entire figure is a square. Find its area.

$$A = s^2$$
$$A = (25 \text{ m})^2$$
$$A = 25 \text{ m} \cdot 25 \text{ m}$$
$$A = \underline{625 \text{ m}^2}$$

The unshaded parts are triangles of equal size. Find the area of one of them.

$$A = \frac{1}{2}bh$$
$$A = \frac{1}{2} \cdot 25 \text{ m} \cdot 10 \text{ m}$$
$$A = \frac{1}{2} \cdot \frac{25 \text{ m}}{1} \cdot \frac{10 \text{ m}}{1}$$
$$A = \dfrac{1 \cdot 25 \text{ m} \cdot \overset{1}{\cancel{2}} \cdot 5 \text{ m}}{\underset{1}{\cancel{2}} \cdot 1 \cdot 1}$$
$$A = 125 \text{ m}^2$$

So for both triangles the area is

$$2(125 \text{ m}^2) = 250 \text{ m}^2.$$

Subtract to find the area of the shaded part.

$$\begin{array}{ccc} \text{entire} & \text{unshaded} & \text{shaded} \\ \text{area} & - \text{ parts} & = \text{ part} \\ \downarrow & \downarrow & \downarrow \end{array}$$

$$A = 625 \text{ m}^2 - 250 \text{ m}^2 = 375 \text{ m}^2$$

The shaded area has area 375 m².

3. (a) $V = l \cdot w \cdot h$
       $V = 8 \text{ m} \cdot 3 \text{ m} \cdot 3 \text{ m}$
       $V = \underline{72 \text{ m}^3}$

    (b) $V = l \cdot w \cdot h$
       $V = 6\frac{1}{4} \text{ ft} \cdot 3\frac{1}{2} \text{ ft} \cdot 2 \text{ ft}$
       $V = \dfrac{25 \text{ ft}}{4} \cdot \dfrac{7 \text{ ft}}{2} \cdot \dfrac{2 \text{ ft}}{1}$
       $V = \dfrac{25 \text{ ft} \cdot 7 \text{ ft} \cdot \overset{1}{\cancel{2}} \text{ ft}}{4 \cdot \underset{1}{\cancel{2}} \cdot 1}$
       $V = \dfrac{175}{4} \text{ ft}^3, \quad \text{or} \quad 43\frac{3}{4} \text{ ft}^3$

4. The base is square and measures 10 ft by 10 ft.

    First find the area of the square base.

    $$\begin{aligned} B &= s^2 = s \cdot s \\ B &= (10 \text{ ft})^2 \\ B &= 10 \text{ ft} \cdot 10 \text{ ft} \\ B &= \underline{100 \text{ ft}^2} \end{aligned}$$

    Now find the volume of the pyramid.

    $$\begin{aligned} V &= \tfrac{1}{3} \cdot B \cdot h \\ V &= \tfrac{1}{3} \cdot 100 \text{ ft}^2 \cdot 6 \text{ ft} \\ V &= \dfrac{1}{3} \cdot \dfrac{100 \text{ ft}^2}{1} \cdot \dfrac{6 \text{ ft}}{1} \\ V &= \dfrac{1 \cdot 100 \text{ ft}^2 \cdot 2 \cdot \overset{1}{\cancel{3}} \text{ ft}}{\underset{1}{\cancel{3}} \cdot 1 \cdot 1} \\ V &= 200 \text{ ft}^3 \end{aligned}$$

4.8 Section Exercises

1. Multiply $\frac{1}{2}$ times 58 m times 66 m; the units in the answer will be m².

3. $P = 58 \text{ m} + 72 \text{ m} + 72 \text{ m}$
    $P = 202 \text{ m}$

    $A = \frac{1}{2}bh$
    $A = \frac{1}{2}(58 \text{ m})(66 \text{ m})$
    $A = \dfrac{1}{2} \cdot \dfrac{58 \text{ m}}{1} \cdot \dfrac{66 \text{ m}}{1}$
    $A = \dfrac{1 \cdot \overset{1}{\cancel{2}} \cdot 29 \text{ m} \cdot 66 \text{ m}}{\underset{1}{\cancel{2}} \cdot 1 \cdot 1}$
    $A = 1914 \text{ m}^2$

5. $P = 2\frac{1}{4} \text{ ft} + 1\frac{1}{2} \text{ ft} + 1\frac{1}{4} \text{ ft}$
    $P = \dfrac{9}{4} \text{ ft} + \dfrac{3}{2} \text{ ft} + \dfrac{5}{4} \text{ ft}$
    $P = \dfrac{9}{4} \text{ ft} + \dfrac{6}{4} \text{ ft} + \dfrac{5}{4} \text{ ft}$
    $P = \dfrac{9 + 6 + 5}{4} \text{ ft}$
    $P = \dfrac{20}{4} \text{ ft} = 5 \text{ ft}$

    $A = \frac{1}{2}bh$
    $A = \dfrac{1}{2}\left(2\frac{1}{4} \text{ ft}\right)\left(\dfrac{3}{4} \text{ ft}\right)$
    $A = \dfrac{1}{2}\left(\dfrac{9}{4} \text{ ft}\right)\left(\dfrac{3}{4} \text{ ft}\right)$
    $A = \dfrac{1 \cdot 9 \cdot 3}{2 \cdot 4 \cdot 4} \text{ ft}^2 = \dfrac{27}{32} \text{ ft}^2$

7. $P = 9 \text{ yd} + 7 \text{ yd} + 10\frac{1}{4} \text{ yd}$
    $P = 16 \text{ yd} + 10\frac{1}{4} \text{ yd}$
    $P = 26\frac{1}{4} \text{ yd}$

    $A = \frac{1}{2}bh$
    $A = \dfrac{1}{2}\left(10\frac{1}{4} \text{ yd}\right)(6 \text{ yd})$
    $A = \dfrac{1}{2}\left(\dfrac{41}{4} \text{ yd}\right)\left(\dfrac{6}{1} \text{ yd}\right)$
    $A = \dfrac{1 \cdot 41 \cdot \overset{1}{\cancel{2}} \cdot 3}{\underset{1}{\cancel{2}} \cdot 4 \cdot 1} \text{ yd}^2$
    $A = \dfrac{123}{4}, \quad \text{or} \quad 30\frac{3}{4} \text{ yd}^2$

9. $P = 12\frac{3}{5} \text{ yd} + 7\frac{2}{3} \text{ yd} + 10 \text{ yd}$
    $P = \dfrac{63}{5} \text{ yd} + \dfrac{23}{3} \text{ yd} + 10 \text{ yd}$
    $P = \dfrac{189}{15} \text{ yd} + \dfrac{115}{15} \text{ yd} + \dfrac{150}{15} \text{ yd}$
    $P = \dfrac{454}{15}, \quad \text{or} \quad 30\frac{4}{15} \text{ yd}$

    $A = \frac{1}{2}bh$
    $A = \dfrac{1}{2}(10 \text{ yd})\left(7\frac{2}{3} \text{ yd}\right)$
    $A = (5 \text{ yd})\left(7\frac{2}{3} \text{ yd}\right)$
    $A = \left(\dfrac{5}{1} \text{ yd}\right)\left(\dfrac{23}{3} \text{ yd}\right)$
    $A = \dfrac{115}{3}, \quad \text{or} \quad 38\frac{1}{3} \text{ yd}^2$

11. To find the shaded area, find the area of the entire rectangle and then subtract the area of the triangle.

$$\begin{array}{c} \text{Shaded} \\ \text{area } A \end{array} = \begin{array}{c} \text{Area of} \\ \text{rectangle} \end{array} - \begin{array}{c} \text{Area of} \\ \text{triangle} \end{array}$$

$$A = lw - \tfrac{1}{2}bh$$
$$A = (52 \text{ m})(\underline{37 \text{ m}}) - \tfrac{1}{2}(52 \text{ m})(8 \text{ m})$$
$$A = \underline{1924 \text{ m}^2} - 208 \text{ m}^2$$
$$A = 1716 \text{ m}^2$$

The shaded area is 1716 m^2.

13. $A = \tfrac{1}{2}bh$

$$A = \frac{1}{2}\left(3\frac{1}{2} \text{ ft}\right)\left(4\frac{1}{2} \text{ ft}\right)$$
$$A = \frac{1}{2}\left(\frac{7}{2} \text{ ft}\right)\left(\frac{9}{2} \text{ ft}\right)$$
$$A = \frac{63}{8}, \quad \text{or} \quad 7\frac{7}{8} \text{ ft}^2$$

$7\frac{7}{8}$ ft^2 of material is needed.

15. Amount of curbing to go "around" the space implies perimeter.

$$P = 33 \text{ m} + 55 \text{ m} + 44 \text{ m}$$
$$P = 132 \text{ m}$$

132 m of curbing will be needed.

Amount of sod to "cover" the space implies area.

$$A = \tfrac{1}{2}bh \qquad \textit{triangular area}$$
$$A = \tfrac{1}{2} \cdot 44 \text{ m} \cdot 33 \text{ m}$$
$$A = \frac{1}{2} \cdot \frac{44 \text{ m}}{1} \cdot \frac{33 \text{ m}}{1}$$
$$A = \frac{1 \cdot \overset{1}{\cancel{2}} \cdot 22 \text{ m} \cdot 33 \text{ m}}{\underset{1}{\cancel{2}} \cdot 1 \cdot 1}$$
$$A = 726 \text{ m}^2$$

726 m^2 of sod will be needed.

17. The figure is a rectangular solid.

$$V = lwh$$
$$V = (12 \text{ cm})(4 \text{ cm})(11 \text{ cm})$$
$$V = 528 \text{ cm}^3$$

19. The figure is a rectangular solid or cube.

$$V = lwh$$
$$V = 2\frac{1}{2} \text{ in.} \cdot 2\frac{1}{2} \text{ in.} \cdot 2\frac{1}{2} \text{ in.}$$
$$V = \frac{5}{2} \text{ in.} \cdot \frac{5}{2} \text{ in.} \cdot \frac{5}{2} \text{ in.}$$
$$V = \frac{125}{8}, \quad \text{or} \quad 15\frac{5}{8} \text{ in.}^3$$

21. The figure is a pyramid.

First find the area of the rectangular base.

$$B = l \cdot w$$
$$B = 15 \text{ cm} \cdot 8 \text{ cm}$$
$$B = 120 \text{ cm}^2$$

Now find the volume of the pyramid.

$$V = \frac{1}{3} \cdot B \cdot h$$
$$V = \frac{1}{3} \cdot 120 \text{ cm}^2 \cdot 20 \text{ cm}$$
$$V = 800 \text{ cm}^3$$

23. First find the area of the square base.

$$B = l \cdot w$$
$$B = 8 \text{ ft} \cdot 8 \text{ ft}$$
$$B = 64 \text{ ft}^2$$

Now find the volume of the pyramid.

$$V = \frac{1}{3} \cdot B \cdot h$$
$$V = \frac{1}{3} \cdot 64 \text{ ft}^2 \cdot 5 \text{ ft}$$
$$V = \frac{320}{3}, \quad \text{or} \quad 106\frac{2}{3} \text{ ft}^3$$

25. The figure is a rectangular box.

$$V = lwh$$
$$V = 8 \text{ in.} \cdot 3 \text{ in.} \cdot \frac{3}{4} \text{ in.}$$
$$V = \frac{8 \text{ in.}}{1} \cdot \frac{3 \text{ in.}}{1} \cdot \frac{3 \text{ in.}}{4}$$
$$V = \frac{2 \cdot \overset{2}{\cancel{4}} \text{ in.} \cdot 3 \text{ in.} \cdot 3 \text{ in.}}{1 \cdot 1 \cdot \underset{1}{\cancel{4}}}$$
$$V = 18 \text{ in.}^3$$

27. Use the formula for the volume of a pyramid.

First find the area of the square base.

$$B = s^2 = s \cdot s$$
$$B = 145 \text{ m} \cdot 145 \text{ m}$$
$$B = 21{,}025 \text{ m}^2$$

Now find the volume of the pyramid.

$$V = \frac{1}{3} \cdot B \cdot h$$
$$V = \frac{1}{3} \cdot 21{,}025 \text{ m}^2 \cdot 93 \text{ m}$$
$$V = 651{,}775 \text{ m}^3$$

The volume of the ancient stone pyramid is $651{,}775 \text{ m}^3$.

Chapter 4 Review Exercises

1. 2 of the 5 figures are squares: $\frac{2}{5}$

1 of the 5 figures is a circle: $\frac{1}{5}$

2. 3 of the 10 squares are shaded: $\frac{3}{10}$

7 of the 10 squares are unshaded: $\frac{7}{10}$

3. Graph $-\frac{1}{2}$ and $1\frac{1}{2}$ on the number line.

4. **(a)** $-\dfrac{20}{5} = -\dfrac{20 \div 5}{5 \div 5} = -\dfrac{4}{1} = -4$

(b) $\dfrac{8}{1} = 8 \div 1 = 8$

(c) $-\dfrac{3}{3} = -3 \div 3 = -1$

5. $\dfrac{28}{32} = \dfrac{\overset{1}{\cancel{2}} \cdot \overset{1}{\cancel{2}} \cdot 7}{\underset{1}{\cancel{2}} \cdot \underset{1}{\cancel{2}} \cdot 2 \cdot 2 \cdot 2} = \dfrac{7}{8}$

6. $\dfrac{54}{90} = \dfrac{\overset{1}{\cancel{2}} \cdot \overset{1}{\cancel{3}} \cdot \overset{1}{\cancel{3}} \cdot 3}{\underset{1}{\cancel{2}} \cdot \underset{1}{\cancel{3}} \cdot \underset{1}{\cancel{3}} \cdot 5} = \dfrac{3}{5}$

7. $\dfrac{16}{25} = \dfrac{2 \cdot 2 \cdot 2 \cdot 2}{5 \cdot 5}$ is already in lowest terms.

8. $\dfrac{15x^2}{40x} = \dfrac{3 \cdot \overset{1}{\cancel{5}} \cdot \overset{1}{\cancel{x}} \cdot x}{2 \cdot 2 \cdot 2 \cdot \underset{1}{\cancel{5}} \cdot \underset{1}{\cancel{x}}} = \dfrac{3x}{8}$

9. $\dfrac{7a^3}{35a^3b} = \dfrac{\overset{1}{\cancel{7}} \cdot \overset{1}{\cancel{a}} \cdot \overset{1}{\cancel{a}} \cdot \overset{1}{\cancel{a}}}{5 \cdot \underset{1}{\cancel{7}} \cdot \underset{1}{\cancel{a}} \cdot \underset{1}{\cancel{a}} \cdot \underset{1}{\cancel{a}} \cdot b} = \dfrac{1}{5b}$

10. $\dfrac{12mn^2}{21m^3n} = \dfrac{2 \cdot 2 \cdot \overset{1}{\cancel{3}} \cdot \overset{1}{\cancel{m}} \cdot \overset{1}{\cancel{n}} \cdot n}{\underset{1}{\cancel{3}} \cdot 7 \cdot \underset{1}{\cancel{m}} \cdot m \cdot m \cdot \underset{1}{\cancel{n}}} = \dfrac{4n}{7m^2}$

11. $-\dfrac{3}{8} \div (-6) = \dfrac{3}{8} \div \dfrac{6}{1}$

$\qquad = \dfrac{3}{8} \cdot \dfrac{1}{6} = \dfrac{3}{8}\left(\dfrac{1}{6}\right)$

$\qquad = \dfrac{\overset{1}{\cancel{3}} \cdot 1}{8 \cdot 2 \cdot \underset{1}{\cancel{3}}} = \dfrac{1}{16}$

12. $\dfrac{2}{5}$ of (-30)

$\dfrac{2}{5} \cdot (-30) = \dfrac{2}{5} \cdot -\dfrac{30}{1} = -\dfrac{2 \cdot 30}{5 \cdot 1}$

$\qquad = -\dfrac{2 \cdot \overset{1}{\cancel{5}} \cdot 6}{\underset{1}{\cancel{5}} \cdot 1} = -\dfrac{12}{1} = -12$

13. $\dfrac{4}{9}\left(\dfrac{2}{3}\right) = \dfrac{4 \cdot 2}{9 \cdot 3} = \dfrac{8}{27}$

14. $\left(\dfrac{7}{3x^3}\right)\left(\dfrac{x^2}{14}\right) = \dfrac{\overset{1}{\cancel{7}} \cdot \overset{1}{\cancel{x}} \cdot \overset{1}{\cancel{x}}}{3 \cdot \underset{1}{\cancel{x}} \cdot \underset{1}{\cancel{x}} \cdot x \cdot 2 \cdot \underset{1}{\cancel{7}}} = \dfrac{1}{6x}$

15. $\dfrac{ab}{5} \div \dfrac{b}{10a} = \dfrac{ab}{5} \cdot \dfrac{10a}{b}$

$\qquad = \dfrac{a \cdot \overset{1}{\cancel{b}} \cdot 2 \cdot \overset{1}{\cancel{5}} \cdot a}{\underset{1}{\cancel{5}} \cdot \underset{1}{\cancel{b}}} = 2a^2$

16. $\dfrac{18}{7} \div 3k = \dfrac{18}{7} \div \dfrac{3k}{1}$

$\qquad = \dfrac{18}{7} \cdot \dfrac{1}{3k}$

$\qquad = \dfrac{6 \cdot \overset{1}{\cancel{3}} \cdot 1}{7 \cdot \underset{1}{\cancel{3}} \cdot k} = \dfrac{6}{7k}$

17. $-\dfrac{5}{12} + \dfrac{5}{8}$

Step 1   The LCD is 24.

Step 2 $-\dfrac{5}{12} = -\dfrac{5 \cdot 2}{12 \cdot 2} = -\dfrac{10}{24}$

$\qquad\qquad \dfrac{5}{8} = \dfrac{5 \cdot 3}{8 \cdot 3} = \dfrac{15}{24}$

Step 3 $-\dfrac{5}{12} + \dfrac{5}{8} = -\dfrac{10}{24} + \dfrac{15}{24}$

$\qquad\qquad\qquad = \dfrac{-10 + 15}{24} = \dfrac{5}{24}$

Step 4 $\frac{5}{24}$ is in lowest terms.

18. $\dfrac{2}{3} - \dfrac{4}{5}$

Step 1   The LCD is 15.

Step 2 $\dfrac{2}{3} = \dfrac{2 \cdot 5}{3 \cdot 5} = \dfrac{10}{15}$

$\qquad\qquad \dfrac{4}{5} = \dfrac{4 \cdot 3}{5 \cdot 3} = \dfrac{12}{15}$

Step 3 $\dfrac{2}{3} - \dfrac{4}{5} = \dfrac{10}{15} - \dfrac{12}{15}$

$\qquad\qquad\qquad = \dfrac{10 - 12}{15} = -\dfrac{2}{15}$

Step 4 $-\frac{2}{15}$ is in lowest terms.

19. $4 - \dfrac{5}{6} = \dfrac{4}{1} - \dfrac{5}{6}$

Step 1   The LCD is 6.

Step 2 $\dfrac{4}{1} = \dfrac{4 \cdot 6}{1 \cdot 6} = \dfrac{24}{6}$

Step 3 $\dfrac{4}{1} - \dfrac{5}{6} = \dfrac{24}{6} - \dfrac{5}{6}$

$\qquad\qquad\qquad = \dfrac{24 - 5}{6} = \dfrac{19}{6},$ or $3\dfrac{1}{6}$

Step 4 $\frac{19}{6}$ is in lowest terms.

20. $\dfrac{7}{9} + \dfrac{13}{18}$

Step 1   The LCD is 18.

Step 2 $\dfrac{7}{9} = \dfrac{7 \cdot 2}{9 \cdot 2} = \dfrac{14}{18}$

Step 3 $\dfrac{7}{9} + \dfrac{13}{18} = \dfrac{14}{18} + \dfrac{13}{18} = \dfrac{14 + 13}{18} = \dfrac{27}{18}$

Step 4 $\dfrac{27}{18} = \dfrac{3 \cdot \overset{1}{\cancel{9}}}{2 \cdot \underset{1}{\cancel{9}}} = \dfrac{3}{2}, \quad \text{or} \quad 1\dfrac{1}{2}$

21. $\dfrac{n}{5} + \dfrac{3}{4}$

Step 1   The LCD is 20.

Step 2 $\dfrac{n}{5} = \dfrac{n \cdot 4}{5 \cdot 4} = \dfrac{4n}{20}, \dfrac{3}{4} = \dfrac{3 \cdot 5}{4 \cdot 5} = \dfrac{15}{20}$

Step 3 $\dfrac{n}{5} + \dfrac{3}{4} = \dfrac{4n}{20} + \dfrac{15}{20} = \dfrac{4n + 15}{20}$

Step 4 $\dfrac{4n + 15}{20}$ is in lowest terms.

22. $\dfrac{3}{10} - \dfrac{7}{y}$

Step 1   The LCD is $10 \cdot y$ or $10y$.

Step 2 $\dfrac{3}{10} = \dfrac{3 \cdot y}{10 \cdot y} = \dfrac{3y}{10y}, \dfrac{7}{y} = \dfrac{7 \cdot 10}{y \cdot 10} = \dfrac{70}{10y}$

Step 3 $\dfrac{3}{10} - \dfrac{7}{y} = \dfrac{3y}{10y} - \dfrac{70}{10y} = \dfrac{3y - 70}{10y}$

Step 4 $\dfrac{3y - 70}{10y}$ is in lowest terms.

23. $2\dfrac{1}{4}$ rounds to 2. $1\dfrac{5}{8}$ rounds to 2.

Estimate: $\underline{2} \div \underline{2} = \underline{1}$

Exact: $2\dfrac{1}{4} \div 1\dfrac{5}{8} = \dfrac{9}{4} \div \dfrac{13}{8} = \dfrac{9}{4} \cdot \dfrac{8}{13}$

$= \dfrac{9 \cdot 2 \cdot \overset{1}{\cancel{4}}}{\underset{1}{\cancel{4}} \cdot 13} = \dfrac{18}{13}, \quad \text{or} \quad 1\dfrac{5}{13}$

24. $7\dfrac{1}{3}$ rounds to 7. $4\dfrac{5}{6}$ rounds to 5.

Estimate: $\underline{7} - \underline{5} = \underline{2}$

Exact: $7\dfrac{1}{3} - 4\dfrac{5}{6} = \dfrac{22}{3} - \dfrac{29}{6} = \dfrac{44}{6} - \dfrac{29}{6}$

$= \dfrac{15}{6} = \dfrac{\overset{1}{\cancel{3}} \cdot 5}{\underset{1}{\cancel{3}} \cdot 2} = \dfrac{5}{2}, \quad \text{or} \quad 2\dfrac{1}{2}$

25. $1\dfrac{3}{4}$ rounds to 2. $2\dfrac{3}{10}$ rounds to 2.

Estimate: $\underline{2} + \underline{2} = \underline{4}$

Exact: $1\dfrac{3}{4} + 2\dfrac{3}{10} = \dfrac{7}{4} + \dfrac{23}{10} = \dfrac{35}{20} + \dfrac{46}{20}$

$= \dfrac{81}{20}, \quad \text{or} \quad 4\dfrac{1}{20}$

26. $\left(-\dfrac{3}{4}\right)^3 = -\dfrac{3}{4}\left(-\dfrac{3}{4}\right)\left(-\dfrac{3}{4}\right)$

$= \dfrac{9}{16}\left(-\dfrac{3}{4}\right) = -\dfrac{27}{64}$

27. $\left(\dfrac{2}{3}\right)^2\left(-\dfrac{1}{2}\right)^4$

$= \dfrac{2}{3} \cdot \dfrac{2}{3}\left(-\dfrac{1}{2}\right)\left(-\dfrac{1}{2}\right)\left(-\dfrac{1}{2}\right)\left(-\dfrac{1}{2}\right)$

$= \dfrac{\overset{1}{\cancel{2}} \cdot \overset{1}{\cancel{2}}}{3 \cdot 3 \cdot \underset{1}{\cancel{2}} \cdot \underset{1}{\cancel{2}} \cdot 2 \cdot 2} = \dfrac{1}{36}$

28. $\dfrac{2}{5} + \dfrac{3}{10}(-4)$

$= \dfrac{2}{5} + \dfrac{3}{10} \cdot \left(\dfrac{-4}{1}\right)$

$= \dfrac{2}{5} + \left(-\dfrac{12}{10}\right)$

$= \dfrac{2}{5} + \left(-\dfrac{6}{5}\right)$

$= \dfrac{2 + (-6)}{5} = \dfrac{-4}{5} \quad \text{or} \quad -\dfrac{4}{5}$

29. $-\dfrac{5}{8} \div \left(-\dfrac{1}{2}\right)\left(\dfrac{14}{15}\right)$

$= -\dfrac{5}{8} \div \left(-\dfrac{1}{2}\right) \cdot \left(\dfrac{14}{15}\right)$

$= -\dfrac{5}{8}\left(-\dfrac{2}{1}\right)\left(\dfrac{14}{15}\right)$ *Change division to multiplication.*

$= \dfrac{\overset{1}{\cancel{5}} \cdot \overset{1}{\cancel{2}} \cdot \overset{1}{\cancel{2}} \cdot 7}{\underset{1}{\cancel{2}} \cdot \underset{1}{\cancel{2}} \cdot 2 \cdot 3 \cdot \underset{1}{\cancel{5}}}$ *Multiply.*

$= \dfrac{7}{6}, \quad \text{or} \quad 1\dfrac{1}{6}$

30. $\dfrac{\frac{5}{8}}{\frac{1}{16}} = \dfrac{5}{8} \div \dfrac{1}{16}$

$= \dfrac{5}{8} \cdot \dfrac{16}{1} = \dfrac{5 \cdot 2 \cdot \overset{1}{\cancel{8}}}{\underset{1}{\cancel{8}} \cdot 1}$

$= \dfrac{10}{1} = 10$

31. $\dfrac{\frac{8}{9}}{-6} = \dfrac{8}{9} \div (-6)$

$= \dfrac{8}{9} \div \left(-\dfrac{6}{1}\right) = \dfrac{8}{9}\left(-\dfrac{1}{6}\right)$

$= -\dfrac{\overset{1}{\cancel{2}} \cdot 4 \cdot 1}{9 \cdot \underset{1}{\cancel{2}} \cdot 3} = -\dfrac{4}{27}$

32.
$$-12 = -\frac{3}{5}w$$

$$-\frac{5}{3}\left(-\frac{12}{1}\right) = -\frac{5}{3}\left(-\frac{3}{5}w\right) \quad \text{Multiply both sides by } -\frac{5}{3}.$$

$$-\frac{5}{\underset{1}{\cancel{3}}}\left(-\frac{\overset{1}{\cancel{3}}\cdot 4}{1}\right) = -\frac{\overset{1}{\cancel{5}}}{\underset{1}{\cancel{3}}}\left(-\frac{\overset{1}{\cancel{3}}}{\underset{1}{\cancel{5}}}\right)w$$

$$20 = w$$

The solution is 20.

33.
$$18 + \frac{6}{5}r = 0$$

$$\underline{-18 \qquad\qquad -18} \quad \begin{array}{l} Add -18\ to \\ both\ sides. \end{array}$$

$$0 + \frac{6}{5}r = -18$$

$$\frac{5}{6}\cdot\left(\frac{6}{5}r\right) = \frac{5}{6}(-18) \quad \begin{array}{l} Multiply\ both \\ sides\ by\ \frac{5}{6}. \end{array}$$

$$r = -15$$

The solution is -15.

34.
$$3x - \frac{2}{3} = \frac{5}{6}$$

$$\underline{\frac{2}{3}\qquad\quad \frac{2}{3}} \quad \begin{array}{l} Add\ \frac{2}{3}\ to \\ both\ sides. \end{array}$$

$$3x + 0 = \frac{5}{6} + \frac{2}{3} \qquad \frac{5}{6} + \frac{2}{3} = \frac{5}{6} + \frac{4}{6} = \frac{9}{6}$$

$$3x = \frac{9}{6} \qquad \frac{9}{6} = \frac{\overset{1}{\cancel{3}}\cdot 3}{\underset{1}{\cancel{3}}\cdot 2} = \frac{3}{2}$$

$$\frac{1}{3}(3x) = \frac{1}{3}\left(\frac{3}{2}\right) \quad \begin{array}{l} Multiply\ both \\ sides\ by\ \frac{1}{3}. \end{array}$$

$$x = \frac{1}{2}$$

The solution is $\frac{1}{2}$.

35. $A = \frac{1}{2}bh$

$$A = \frac{1}{2}\cdot 8\ \text{ft}\cdot 3\frac{1}{2}\ \text{ft}$$

$$A = \frac{1}{2}\cdot\frac{8}{1}\ \text{ft}\cdot\frac{7}{2}\ \text{ft}$$

$$A = \frac{1\cdot\overset{1}{\cancel{2}}\cdot\overset{1}{\cancel{2}}\cdot 2\cdot 7}{\underset{1}{\cancel{2}}\cdot 1\cdot\underset{1}{\cancel{2}}}\ \text{ft}^2$$

$$A = 14\ \text{ft}^2$$

36. The figure is a rectangular solid.
$$V = l\cdot w\cdot h$$

$$V = 4\ \text{in.}\cdot 3\frac{1}{4}\ \text{in.}\cdot 2\frac{1}{2}\ \text{in.}$$

$$V = \frac{4}{1}\ \text{in.}\cdot\frac{13}{4}\ \text{in.}\cdot\frac{5}{2}\ \text{in.}$$

$$V = \frac{\overset{1}{\cancel{4}}\cdot 13\cdot 5}{1\cdot\underset{1}{\cancel{4}}\cdot 2}\ \text{in.}^3$$

$$V = \frac{65}{2},\quad \text{or}\quad 32\frac{1}{2}\ \text{in.}^3$$

37. The figure is a pyramid.

First find the area of the rectangular base.
$$B = l\cdot w$$
$$B = 8\ \text{yd}\cdot 5\ \text{yd}$$
$$B = 40\ \text{yd}^2$$

Now find the volume of the pyramid.
$$V = \frac{1}{3}\cdot B\cdot h$$

$$V = \frac{1}{3}\cdot\frac{40\ \text{yd}^2}{1}\cdot\frac{7\ \text{yd}}{1}$$

$$V = \frac{280}{3},\quad \text{or}\quad 93\frac{1}{3}\ \text{yd}^3$$

38. **[4.3]** The $2\frac{1}{2}$ pounds of meat will be used to make 10 servings:

"$2\frac{1}{2}$ pounds divided or separated into 10 servings"

$$2\frac{1}{2} \div 10 = \frac{5}{2} \div \frac{10}{1}$$

$$= \frac{5}{2}\cdot\frac{1}{10} = \frac{\overset{1}{\cancel{5}}\cdot 1}{2\cdot 2\cdot\underset{1}{\cancel{5}}} = \frac{1}{4}$$

$\frac{1}{4}$ pound will be used in each serving.

For 30 servings:

$$30\left(\frac{1}{4}\right) = \frac{30}{1}\cdot\frac{1}{4}$$

$$= \frac{\overset{1}{\cancel{2}}\cdot 15\cdot 1}{1\cdot\underset{1}{\cancel{2}}\cdot 2} = \frac{15}{2} = 7\frac{1}{2}$$

You would need $7\frac{1}{2}$ pounds of meat.

39. **[4.4]** To find out how much longer she worked on Monday than on Friday, use subtraction.

$$4\frac{1}{2} - 3\frac{2}{3} = \frac{9}{2} - \frac{11}{3}$$

$$= \frac{27}{6} - \frac{22}{6} \quad LCD\ is\ 6.$$

$$= \frac{5}{6}\ \text{hour}$$

She worked $\frac{5}{6}$ hour longer on Monday than on Friday.

To find how many hours she worked in all, use addition.

$$4\frac{1}{2} + 2\frac{3}{4} + 3\frac{2}{3}$$

$$= \frac{9}{2} + \frac{11}{4} + \frac{11}{3}$$

$$= \frac{54}{12} + \frac{33}{12} + \frac{44}{12} \quad LCD\ is\ 12.$$

$$= \frac{54 + 33 + 44}{12}$$

$$= \frac{131}{12}\ \text{or}\ 10\frac{11}{12}\ \text{hours}$$

She worked a total of $10\frac{11}{12}$ hours.

40. [4.3] 60 total children is the "whole."

$\frac{1}{5}$ of the "whole" are preschoolers:

$$\frac{1}{5} \cdot 60 = \frac{1}{5} \cdot \frac{60}{1} = \frac{60}{5} = 12 \text{ preschoolers}$$

$\frac{2}{3}$ of the "whole" are toddlers:

$$\frac{2}{3} \cdot 60 = \frac{2}{3} \cdot \frac{60}{1} = \frac{120}{3} = 40 \text{ toddlers}$$

"The rest" are infants; use subtraction:

$$60 - 12 - 40 = 8 \text{ infants}$$

41. [4.5] $P = 2l + 2w$

$$P = \frac{2}{1} \cdot \frac{3}{4} \text{ mi} + \frac{2}{1} \cdot \frac{3}{10} \text{ mi}$$

$$P = \frac{3}{2} \text{ mi} + \frac{3}{5} \text{ mi}$$

$$P = \frac{15}{10} \text{ mi} + \frac{6}{10} \text{ mi}$$

$$P = \frac{21}{10}, \quad \text{or} \quad 2\frac{1}{10} \text{ miles}$$

The perimeter of the park is $2\frac{1}{10}$ miles.

$$A = l \cdot w$$

$$A = \frac{3}{4} \text{ mi} \cdot \frac{3}{10} \text{ mi}$$

$$A = \frac{9}{40} \text{ mi}^2$$

The area of the park is $\frac{9}{40}$ mi².

Chapter 4 Test

1. 5 of the 6 squares are shaded: $\frac{5}{6}$

1 of the 6 squares are unshaded: $\frac{1}{6}$

2. Graph $-\frac{2}{3}$ and $2\frac{1}{3}$ on the number line.

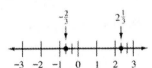

3. $\frac{21}{84} = \frac{\overset{1}{\cancel{3}} \cdot \overset{1}{\cancel{7}}}{2 \cdot 2 \cdot \underset{1}{\cancel{3}} \cdot \underset{1}{\cancel{7}}} = \frac{1}{4}$

4. $\frac{25}{54} = \frac{5 \cdot 5}{2 \cdot 3 \cdot 3 \cdot 3}$ is already in lowest terms.

5. $\frac{6a^2b}{9b^2} = \frac{2 \cdot \overset{1}{\cancel{3}} \cdot a \cdot a \cdot \overset{1}{\cancel{b}}}{\underset{1}{\cancel{3}} \cdot 3 \cdot \underset{1}{\cancel{b}} \cdot b} = \frac{2a^2}{3b}$

6. $\frac{1}{6} + \frac{7}{10}$

$= \frac{5}{30} + \frac{21}{30}$ *LCD is 30.*

$= \frac{5 + 21}{30} = \frac{26}{30} = \frac{13}{15}$

7. $-\frac{3}{4} \div \frac{3}{8} = -\frac{3}{4} \cdot \frac{8}{3} = -\frac{\overset{1}{\cancel{3}} \cdot 2 \cdot \overset{1}{\cancel{4}}}{\underset{1}{\cancel{4}} \cdot \underset{1}{\cancel{3}}}$

$= -\frac{2}{1} = -2$

8. $\frac{5}{8} - \frac{4}{5} = \frac{25}{40} - \frac{32}{40}$

$= \frac{25 - 32}{40} = \frac{-7}{40}$ or $-\frac{7}{40}$

9. $(-20)\left(-\frac{7}{10}\right) = -\frac{20}{1}\left(-\frac{7}{10}\right)$

$= \frac{2 \cdot \overset{1}{\cancel{10}} \cdot 7}{1 \cdot \underset{1}{\cancel{10}}} = \frac{14}{1} = 14$

10. $\frac{\frac{4}{9}}{-6} = \frac{4}{9} \div (-6)$

$= \frac{4}{9} \div \left(-\frac{6}{1}\right) = \frac{4}{9}\left(-\frac{1}{6}\right)$

$= -\frac{\overset{1}{\cancel{2}} \cdot 2 \cdot 1}{9 \cdot \underset{1}{\cancel{2}} \cdot 3} = -\frac{2}{27}$

11. $4 - \frac{7}{8} = \frac{4}{1} - \frac{7}{8}$

$= \frac{32}{8} - \frac{7}{8} = \frac{32 - 7}{8} = \frac{25}{8}$,   or $3\frac{1}{8}$

12. $-\frac{2}{9} + \frac{2}{3} = -\frac{2}{9} + \frac{6}{9} = \frac{-2 + 6}{9} = \frac{4}{9}$

13. $\frac{21}{24}\left(\frac{9}{14}\right) = \frac{\overset{1}{\cancel{3}} \cdot \overset{1}{\cancel{7}} \cdot 3 \cdot 3}{\underset{1}{\cancel{3}} \cdot 8 \cdot 2 \cdot \underset{1}{\cancel{7}}} = \frac{9}{16}$

14. $\frac{12x}{7y} \div 3x = \frac{12x}{7y} \div \frac{3x}{1}$

$= \frac{12x}{7y} \cdot \frac{1}{3x} = \frac{\overset{1}{\cancel{3}} \cdot 4 \cdot \overset{1}{\cancel{x}} \cdot 1}{7 \cdot y \cdot \underset{1}{\cancel{3}} \cdot \underset{1}{\cancel{x}}} = \frac{4}{7y}$

15. $\frac{6}{n} - \frac{1}{4} = \frac{24}{4n} - \frac{n}{4n} = \frac{24 - n}{4n}$

16. $\frac{2}{3} + \frac{a}{5} = \frac{10}{15} + \frac{3a}{15} = \frac{10 + 3a}{15}$

17. $\left(\frac{5}{9b^2}\right)\left(\frac{b}{10}\right) = \frac{\overset{1}{\cancel{5}} \cdot \overset{1}{\cancel{b}}}{3 \cdot 3 \cdot \underset{1}{\cancel{b}} \cdot b \cdot 2 \cdot \underset{1}{\cancel{5}}} = \frac{1}{18b}$

18. $\left(-\frac{1}{2}\right)^3\left(\frac{2}{3}\right)^2 = -\frac{1}{2}\left(-\frac{1}{\cancel{2}}\right)\left(-\frac{1}{\cancel{2}}\right) \cdot \frac{\overset{1}{\cancel{2}}}{3} \cdot \frac{\overset{1}{\cancel{2}}}{3}$

$= -\frac{1}{18}$

19. $\dfrac{1}{6} + 4\left(\dfrac{2}{5} - \dfrac{7}{10}\right)$

$= \dfrac{1}{6} + 4\left(\dfrac{4}{10} - \dfrac{7}{10}\right)$

$= \dfrac{1}{6} + 4\left(\dfrac{-3}{10}\right)$

$= \dfrac{1}{6} + \dfrac{4}{1} \cdot \dfrac{-3}{10}$

Note that $\dfrac{4}{1} \cdot \dfrac{-3}{10} = \dfrac{\overset{1}{\cancel{2}} \cdot 2 \cdot (-3)}{1 \cdot \cancel{2} \cdot 5} = \dfrac{-6}{5}$.

$= \dfrac{1}{6} + \dfrac{-6}{5}$

$= \dfrac{5}{30} + \dfrac{-36}{30}$

$= \dfrac{5 + (-36)}{30}$

$= -\dfrac{31}{30},$ or $-1\dfrac{1}{30}$

20. $4\dfrac{4}{5}$ rounds to 5. $1\dfrac{1}{8}$ rounds to 1.

Estimate: $5 \div 1 = 5$

Exact: $4\dfrac{4}{5} \div 1\dfrac{1}{8} = \dfrac{24}{5} \div \dfrac{9}{8} = \dfrac{24}{5} \cdot \dfrac{8}{9}$

$= \dfrac{\overset{1}{\cancel{3}} \cdot 8 \cdot 8}{5 \cdot \cancel{3} \cdot 3}$

$= \dfrac{64}{15},$ or $4\dfrac{4}{15}$

21. $3\dfrac{2}{5}$ rounds to 3. $1\dfrac{9}{10}$ rounds to 2.

Estimate: $3 - 2 = 1$

Exact: $3\dfrac{2}{5} - 1\dfrac{9}{10} = \dfrac{17}{5} - \dfrac{19}{10} = \dfrac{34}{10} - \dfrac{19}{10}$

$= \dfrac{15}{10} = \dfrac{3}{2},$ or $1\dfrac{1}{2}$

22. $7 = \dfrac{1}{5}d$

$\dfrac{5}{1}(7) = \dfrac{5}{1}\left(\dfrac{1}{5}d\right)$ *Multiply both sides by $\frac{5}{1}$.*

$35 = d$

The solution is 35.

23. $-\dfrac{3}{10}t = \dfrac{9}{14}$

$-\dfrac{10}{3}\left(-\dfrac{3}{10}t\right) = -\dfrac{10}{3}\left(\dfrac{9}{14}\right)$ *Multiply both sides by $-\frac{10}{3}$.*

$t = -\dfrac{\overset{1}{\cancel{2}} \cdot 5 \cdot \overset{1}{\cancel{3}} \cdot 3}{\underset{1}{\cancel{3}} \cdot \underset{1}{\cancel{2}} \cdot 7}$

$t = -\dfrac{15}{7},$ or $-2\dfrac{1}{7}$

The solution is $-\frac{15}{7}$.

24. $0 = \dfrac{1}{4}b - 2$

$\underline{2}\;\underline{2}$ *Add 2 to both sides.*

$2 = \dfrac{1}{4}b + 0$

$\dfrac{4}{1}(2) = \dfrac{4}{1}\left(\dfrac{1}{4}b\right)$ *Multiply both sides by $\frac{4}{1}$.*

$8 = b$

The solution is 8.

25. $\dfrac{4}{3}x + 7 = -13$

$\underline{-7}\;\underline{-7}$ *Add -7 to both sides.*

$\dfrac{4}{3}x + 0 = -20$

$\dfrac{3}{4}\left(\dfrac{4}{3}x\right) = \dfrac{3}{4}(-20)$ *Multiply both sides by $\frac{3}{4}$.*

$x = -15$

The solution is -15.

26. $A = \frac{1}{2}bh$

$A = \frac{1}{2} \cdot 13 \text{ m} \cdot 8 \text{ m}$

$A = \dfrac{1}{2} \cdot \dfrac{13 \text{ m}}{1} \cdot \dfrac{8 \text{ m}}{1}$

$A = \dfrac{13 \cdot \overset{1}{\cancel{2}} \cdot 4}{\underset{1}{\cancel{2}}} \text{ m}^2$

$A = 52 \text{ m}^2$

27. $A = \frac{1}{2}bh$

$A = \frac{1}{2} \cdot 9 \text{ yd} \cdot 13 \text{ yd}$

$A = \dfrac{1}{2} \cdot \dfrac{9 \text{ yd}}{1} \cdot \dfrac{13 \text{ yd}}{1}$

$A = \dfrac{117}{2},$ or $58\dfrac{1}{2} \text{ yd}^2$

28. The figure is a rectangular solid.

$V = l \cdot w \cdot h$

$V = 30 \text{ m} \cdot 18 \text{ m} \cdot 12 \text{ m}$

$V = 6480 \text{ m}^3$

29. The figure is a pyramid. The base is a rectangle. The area of the base is:

$B = l \cdot w$

$B = 4 \text{ yd} \cdot 3 \text{ yd}$

$B = 12 \text{ yd}^2$

The volume of the pyramid is:

$V = \frac{1}{3} \cdot B \cdot h$

$V = \frac{1}{3} \cdot 12 \text{ yd}^2 \cdot 4 \text{ yd}$

$V = \dfrac{48}{3} \text{ yd}^3$

$V = 16 \text{ yd}^3$

30. To find her total training hours, use addition.

$$4\frac{5}{6} + 6\frac{2}{3} + 3\frac{1}{4}$$

$$= \frac{29}{6} + \frac{20}{3} + \frac{13}{4}$$

$$= \frac{58}{12} + \frac{80}{12} + \frac{39}{12} \qquad \text{\textit{LCD is 12.}}$$

$$= \frac{177}{12} = \frac{59}{4}, \quad \text{or} \quad 14\frac{3}{4}$$

She spent $14\frac{3}{4}$ hours training.

Use subtraction to find how many more hours she trained on Tuesday than on Monday.

$$6\frac{2}{3} - 4\frac{5}{6}$$

$$= \frac{20}{3} - \frac{29}{6}$$

$$= \frac{40}{6} - \frac{29}{6} \qquad \text{\textit{LCD is 6.}}$$

$$= \frac{11}{6}, \quad \text{or} \quad 1\frac{5}{6}$$

She spent $1\frac{5}{6}$ hours more on Tuesday than on Monday.

31. $8\frac{3}{4}$ ounces can be synthesized in how many days if $2\frac{1}{2}$ ounces can be synthesized in a day?

Estimation will help. 9 ounces is needed.
3 ounces can be synthesized per day.

Estimate: $9 \div 3 = 3$ days

Exact: $8\dfrac{3}{4} \div 2\dfrac{1}{2}$

$$= \frac{35}{4} \div \frac{5}{2} = \frac{35}{4} \cdot \frac{2}{5}$$

$$= \frac{\overset{1}{\cancel{5}} \cdot 7 \cdot \overset{1}{\cancel{2}}}{\underset{1}{\cancel{2}} \cdot 2 \cdot \underset{1}{\cancel{5}}} = \frac{7}{2}, \quad \text{or} \quad 3\frac{1}{2} \text{ days}$$

32. $\frac{7}{8}$ of the 8448 students work:

$$\frac{7}{8} \cdot 8448 = \frac{7 \cdot \overset{1}{\cancel{8}} \cdot 1056}{\underset{1}{\cancel{8}}} = 7392$$

7392 students work.

CHAPTER 5   RATIONAL NUMBERS: POSITIVE AND NEGATIVE DECIMALS

5.1   Reading and Writing Decimal Numbers

5.1   Margin Exercises

1.   **(a)** The figure has 10 equal parts; 1 part is yellow shaded. The fraction is $\frac{1}{10}$, the decimal is 0.1, and the words that name the yellow shaded portion are one tenth.

   (b) The figure has 10 equal parts; 3 parts are shaded.

$$\frac{3}{10}; \ 0.3; \ \text{three tenths}$$

   (c) The figure has 10 equal parts; 9 parts are shaded.

$$\frac{9}{10}; \ 0.9; \ \text{nine tenths}$$

2.   **(a)** The figure has 10 equal parts; 3 parts are shaded.

$$\frac{3}{10}; \ 0.\underline{3}; \ \underline{\text{three}} \text{ tenths}$$

   (b) The figure has 100 equal parts; 41 parts are shaded.

$$\frac{41}{100}; \ 0.41; \ \text{forty-one hundredths}$$

3.   **(a)** $-0.7 = -\frac{7}{10}$

   (b) $0.2 = \frac{2}{10}$

   (c) $-0.03 = -\frac{3}{100}$

   (d) $0.69 = \frac{69}{100}$

   (e) $0.047 = \frac{47}{1000}$

   (f) $-0.351 = -\frac{351}{1000}$

4.   **(a)**

hundreds	tens	ones		tenths	hundredths
9	7	1	.	5	4

   (b)

ones		tenths
0	.	4

   (c)

ones		tenths	hundredths
5	.	6	0

   (d)

ones		tenths	hundredths	thousandths	ten-thousandths
0	.	0	8	3	5

5.   **(a)** 0.6 is six <u>tenths</u>.

   (b) 0.46 is forty-six <u>hundredths</u>.

   (c) 0.05 is five hundredths.

   (d) 0.409 is four hundred nine thousandths.

   (e) 0.0003 is three ten-thousandths.

   (f) 0.2703 is two thousand seven hundred three ten-thousandths.

   (g) 0.088 is eighty-eight thousandths.

6.   **(a)** 3.8 is read three **and** eight <u>tenths</u>.

   (b) 15.001 is fifteen and one thousandth.

   (c) 0.0073 is seventy-three ten-thousandths.

   (d) 764.309 is seven hundred sixty-four and three hundred nine thousandths.

7.   **(a)** $0.7 = \frac{7}{10}$

   (b) $12.21 = 12\frac{21}{100}$

   (c) $0.101 = \frac{101}{1000}$

   (d) $0.007 = \frac{7}{1000}$

   (e) $1.3717 = 1\frac{3717}{10,000}$

8.   **(a)** $0.5 = \dfrac{5}{10} = \dfrac{5 \div 5}{10 \div 5} = \dfrac{1}{2}$

   (b) $12.6 = 12\dfrac{6}{10} = 12\dfrac{6 \div 2}{10 \div 2} = 12\dfrac{3}{5}$

   (c) $0.85 = \dfrac{85}{100} = \dfrac{85 \div 5}{100 \div 5} = \dfrac{17}{20}$

   (d) $3.05 = 3\dfrac{5}{100} = 3\dfrac{5 \div 5}{100 \div 5} = 3\dfrac{1}{20}$

   (e) $0.225 = \dfrac{225}{1000} = \dfrac{225 \div 25}{1000 \div 25} = \dfrac{9}{40}$

   (f) $420.0802 = 420\dfrac{802}{10,000} = 420\dfrac{802 \div 2}{10,000 \div 2}$

$$= 420\dfrac{401}{5000}$$

5.1   Section Exercises

1.   The number 11.084 has 3 decimal places (three numbers *after* the decimal point).

3.   In 22.85, the 8 means 8 tenths.

5.
	tens	ones		tenths		
	7	0	.	4	8	9

7.
	tenths		thousandths	ten-thousandths	
0	. 8	3	4	7	2

9.
hundreds		ones		hundredths		
1	4	9	. 0	8	3	2

11.
thousands				hundredths	thousandths		
6	2	8	5 . 7	1	2	5

13. 0 ones, 5 hundredths, 1 ten, 4 hundreds, 2 tenths:

$$\underline{4}\underline{1}\underline{0}.\underline{2}\underline{5}$$

15. 3 thousandths, 4 hundredths, 6 ones,
2 ten-thousandths, 5 tenths: **6.5432**

17. 4 hundredths, 4 hundreds, 0 tens, 0 tenths,
5 thousandths, 5 thousands, 6 ones: **5406.045**

19. $0.7 = \frac{7}{10}$

21. $13.4 = 13\frac{4}{10} = 13\frac{4 \div 2}{10 \div 2} = 13\frac{2}{5}$

23. $0.35 = \frac{35}{100} = \frac{35 \div 5}{100 \div 5} = \frac{7}{20}$

25. $0.66 = \frac{66}{100} = \frac{66 \div 2}{100 \div 2} = \frac{33}{50}$

27. $10.17 = 10\frac{17}{100}$

29. $0.06 = \frac{6}{100} = \frac{6 \div 2}{100 \div 2} = \frac{3}{50}$

31. $0.205 = \frac{205}{1000} = \frac{205 \div 5}{1000 \div 5} = \frac{41}{200}$

33. $5.002 = 5\frac{2}{1000} = 5\frac{2 \div 2}{1000 \div 2} = 5\frac{1}{500}$

35. $0.686 = \frac{686}{1000} = \frac{686 \div 2}{1000 \div 2} = \frac{343}{500}$

37. 0.5 is five tenths.

39. 0.78 is seventy-eight hundredths.

41. 0.105 is one hundred five thousandths.

43. 12.04 is twelve and four hundredths.

45. 1.075 is one and seventy-five thousandths.

47. Six and seven tenths:

$$6\frac{7}{10} = \underline{6}.\underline{7}$$

49. Thirty-two hundredths:

$$\frac{32}{100} = 0.32$$

51. Four hundred twenty and eight thousandths:

$$420\frac{8}{1000} = 420.008$$

53. Seven hundred three ten-thousandths:

$$\frac{703}{10,000} = 0.0703$$

55. Seventy-five and thirty thousandths:

$$75\frac{30}{1000} = 75.030$$

57. Anne should not say "and" because that denotes a decimal point.

59. 8-pound test line has a diameter of 0.010 inch.
0.010 inch is read ten thousandths of an inch.

$$0.010 = \frac{10}{1000} = \frac{10 \div 10}{1000 \div 10} = \frac{1}{100} \text{ inch}$$

61. $\frac{13}{1000}$ inch $= 0.013$ inch

A diameter of 0.013 inch corresponds to a test strength of 12 pounds.

63. Six tenths is 0.6, so the correct part number is 3-C.

65. One and six thousandths is 1.006, which is part number 4-A.

67. The size of part number 4-E is 1.602 centimeters, which in words is one and six hundred two thousandths centimeters.

Relating Concepts (Exercises 69–76)

69. Use the Whole Number Place Value Chart on text page 4 to find the names after hundred thousands. Then change "s" to "*ths*" in those names. So millions becomes million*ths* when used on the right side of the decimal point, and so on for ten-millionths, hundred-millionths, and billionths.

70. The first place to the left of the decimal point is ones. "Oneths" would mean a fraction with a denominator of 1, which would equal 1 or more. Anything that is 1 or more is to the *left* of the decimal point.

71. The eighth place value to the right of the decimal point is hundred-millionths, so 0.72436955 is seventy-two million four hundred thirty-six thousand nine hundred fifty-five hundred-millionths.

72. The ninth place value to the right of the decimal point is billionths, so 0.000678554 is six hundred seventy-eight thousand five hundred fifty-four billionths.

73. 8006.500001 is eight thousand six and five hundred thousand one millionths.

74. 20,060.000505 is twenty thousand sixty and five hundred five millionths.

75. Three hundred two thousand forty ten-millionths:

0.0302040

76. Nine billion, eight hundred seventy-six million, five hundred forty-three thousand, two hundred ten and one hundred million two hundred thousand three hundred billionths:

9,876,543,210.100200300

5.2 Rounding Decimal Numbers

5.2 Margin Exercises

1. Round to the nearest thousandth.

(a) Draw a cut-off line: 0.334|92
The first digit cut is 9, which is 5 or more, so round up.

$$\begin{array}{r} 0.334 \\ + 0.001 \\ \hline 0.335 \end{array}$$

Answer: ≈ 0.335

(b) Draw a cut-off line: 8.008|51
The first digit cut is 5, which is 5 or more, so round up.

$$\begin{array}{r} 8.008 \\ + 0.001 \\ \hline 8.009 \end{array}$$

Answer: ≈ 8.009

(c) Draw a cut-off line: 265.420|68
The first digit cut is 6, which is 5 or more, so round up.

$$\begin{array}{r} 265.420 \\ + 0.001 \\ \hline 265.421 \end{array}$$

Answer: ≈ 265.421

(d) Draw a cut-off line: 10.701|80
The first digit cut is 8, which is 5 or more, so round up.

$$\begin{array}{r} 10.701 \\ + 0.001 \\ \hline 10.702 \end{array}$$

Answer: ≈ 10.702

2. **(a)** 0.8988 to the nearest hundredth

Draw a cut-off line: 0.89|88
The first digit cut is 8, which is 5 or more, so round up.

$$\begin{array}{r} 0.89 \\ + 0.01 \\ \hline 0.90 \end{array}$$

Answer: ≈ 0.90

(b) 5.8903 to the nearest hundredth

Draw a cut-off line: 5.89|03
The first digit cut is 0, which is 4 or less. The part you keep stays the same.

Answer: ≈ 5.89

(c) 11.0299 to the nearest thousandth

Draw a cut-off line: 11.029|9
The first digit cut is 9, which is 5 or more, so round up.

$$\begin{array}{r} 11.029 \\ + 0.001 \\ \hline 11.030 \end{array}$$

Answer: ≈ 11.030

(d) 0.545 to the nearest tenth

Draw a cut-off line: 0.5|45
The first digit cut is 4, which is 4 or less. The part you keep stays the same.

Answer: ≈ 0.5

3. Round to the nearest cent (hundredths).

(a) Draw a cut-off line: $14.59|5
The first digit cut is 5, which is 5 or more, so round up.

$$\begin{array}{r} \$14.59 \\ + 0.01 \\ \hline \$14.60 \end{array}$$

Answer: $\approx \$14.60$

(b) Draw a cut-off line: $578.06|63
The first digit cut is 6, which is 5 or more, so round up.

$$\begin{array}{r} \$578.06 \\ + 0.01 \\ \hline \$578.07 \end{array}$$

Answer: $\approx \$578.07$

(c) Draw a cut-off line: $0.84|9
The first digit cut is 9, which is 5 or more, so round up.

$$\begin{array}{r} \$0.84 \\ + 0.01 \\ \hline \$0.85 \end{array}$$

Answer: $\approx \$0.85$

(d) Draw a cut-off line: $0.05|48
The first digit cut is 4, which is 4 or less. The part you keep stays the same.

Answer: ≈ $0.05

4.   Round to the nearest dollar.

(a) Draw a cut-off line: $29|.10
The first digit cut is 1, which is 4 or less. The part you keep stays the same.

Answer: ≈ $29

(b) Draw a cut-off line: $136|.49
The first digit cut is 4, which is 4 or less. The part you keep stays the same.

Answer: ≈ $136

(c) Draw a cut-off line: $990|.91
The first digit cut is 9, which is 5 or more, so round up by adding $1.

$$\begin{array}{r} \$990 \\ +\ 1 \\ \hline \$991 \end{array}$$

Answer: ≈ $991

(d) Draw a cut-off line: $5999|.88
The first digit cut is 8, which is 5 or more, so round up by adding $1.

$$\begin{array}{r} \$5999 \\ +\ \ 1 \\ \hline \$6000 \end{array}$$

Answer: ≈ $6000

(e) Draw a cut-off line: $49|.60
The first digit cut is 6, which is 5 or more, so round up by adding $1.

$$\begin{array}{r} \$49 \\ +1 \\ \hline \$50 \end{array}$$

Answer: ≈ $50

(f) Draw a cut-off line: $0|.55
The first digit cut is 5, which is 5 or more, so round up by adding $1.

$$\begin{array}{r} \$0 \\ +1 \\ \hline \$1 \end{array}$$

Answer: ≈ $1

(g) Draw a cut-off line: $1|.08
The first digit cut is 0, which is 4 or less. The part you keep stays the same.

Answer: ≈ $1

5.2   Section Exercises

1.   When rounding 4.8073 to the nearest tenth, you would look at the first digit *after* the tenths place, that is, <u>0</u>.

3.   To round 5.70961 to the nearest thousandth, look only at the 6, which is *5 or more*. So round up by adding 1 thousandth to 5.709 to get 5.710.

5.   16.8974 to the nearest tenth   ▪ Draw a cut-off line after the tenths place:

$$16.8|974$$

The first digit cut is 9, which is 5 or more, so round up the tenths place.

$$\begin{array}{r} 16.8 \\ +\ 0.1 \\ \hline 16.9 \quad \textit{Answer} \end{array}$$

7.   0.95647 to the nearest thousandth   ▪ Draw a cut-off line after the thousandths place:

$$0.956|47$$

The first digit cut is 4, which is 4 or less. The part you keep stays the same. Answer is 0.956.

9.   0.799 to the nearest hundredth   ▪ Draw a cut-off line after the hundredths place:

$$0.79|9$$

The first digit cut is 9, which is 5 or more, so round up the hundredths place.

$$\begin{array}{r} 0.79 \\ +\ 0.01 \\ \hline 0.80 \quad \textit{Answer} \end{array}$$

11.   3.66062 to the nearest thousandth   ▪ Draw a cut-off line after the thousandths place:

$$3.660|62$$

The first digit cut is 6, which is 5 or more, so round up the thousandths place.

$$\begin{array}{r} 3.660 \\ +\ 0.001 \\ \hline 3.661 \quad \textit{Answer} \end{array}$$

13.   793.988 to the nearest tenth   ▪ Draw a cut-off line after the tenths place:

$$793.9|88$$

The first digit cut is 8, which is 5 or more, so round up the tenths place.

$$\begin{array}{r} 793.9 \\ +\ 0.1 \\ \hline 794.0 \quad \textit{Answer} \end{array}$$

15. 0.09804 to the nearest ten-thousandth ▪ Draw a cut-off line after the ten-thousandths place:

$$0.0980|4$$

The first digit cut is 4, which is 4 or less. The part you keep stays the same. Answer is 0.0980.

17. 9.0906 to the nearest hundredth ▪ Draw a cut-off line after the hundredths place:

$$9.09|06$$

The first digit cut is 0, which is 4 or less. The part you keep stays the same. Answer is 9.09.

19. 82.000151 to the nearest ten-thousandth ▪ Draw a cut-off line after the ten-thousandths place:

$$82.0001|51$$

The first digit cut is 5, which is 5 or more, so round up the ten-thousandths place.

$$
\begin{array}{r}
82.0001 \\
+\,0.0001 \\
\hline
82.0002 \quad \textit{Answer}
\end{array}
$$

21. Round $0.81666 to the nearest cent.
Draw a cut-off line: $0.81|666
The first digit cut is 6, which is 5 or more, so round up.

$$
\begin{array}{r}
\$0.81 \\
+\,0.01 \\
\hline
\$0.82
\end{array}
$$

Nardos pays <u>$0.82</u>.

23. Round $1.2225 to the nearest cent.
Draw a cut-off line: $1.22|25
The first digit cut is 2, which is 4 or less. The part you keep stays the same.

Nardos pays <u>$1.22</u>.

25. Round $0.6983 to the nearest cent.
Draw a cut-off line: $0.69|83
The first digit cut is 8, which is 5 or more, so round up.

$$
\begin{array}{r}
\$0.69 \\
+\,0.01 \\
\hline
\$0.70
\end{array}
$$

Nardos pays <u>$0.70</u>.

27. Round $48,649.60 to the nearest dollar.
Draw a cut-off line: $48,649|.60
The first digit cut is 6, which is 5 or more, so round up.

$$
\begin{array}{r}
\$48,649 \\
+\qquad 1 \\
\hline
\$48,650
\end{array}
$$

Income from job: $\approx \$48,650$

29. Round $840.08 to the nearest dollar.
Draw a cut-off line: $840|.08
The first digit cut is 0, which is 4 or less. The part you keep stays the same.

Donations to charity: $\approx \$840$

31. $499.98 to the nearest dollar ▪ Draw a cut-off line after the ones place:

$$\$499|.98$$

The first digit cut is 9, which is 5 or more, so round up the ones place.

$$
\begin{array}{r}
\$499 \\
+\quad 1 \\
\hline
\$500 \quad \textit{Answer}
\end{array}
$$

33. $0.996 to the nearest cent ▪ Draw a cut-off line after the hundredths place:

$$\$0.99|6$$

The first digit cut is 6, which is 5 or more, so round up the hundredths place.

$$
\begin{array}{r}
0.99 \\
+\,0.01 \\
\hline
\$1.00 \quad \textit{Answer}
\end{array}
$$

35. $999.73 to the nearest dollar ▪ Draw a cut-off line after the ones place:

$$\$999|.73$$

The first digit cut is 7, which is 5 or more, so round up the ones place.

$$
\begin{array}{r}
\$999 \\
+\quad 1 \\
\hline
\$1000 \quad \textit{Answer}
\end{array}
$$

37. **(a)** Round 252.662 to the nearest whole number.
Draw a cut-off line: 252|.6
The first digit cut is 6, which is 5 or more, so round up.

$$
\begin{array}{r}
252 \\
+1 \\
\hline
253
\end{array}
$$

The record speed for a motorcycle is about 253 miles per hour.

(b) Round 134.8 to the nearest whole number.
Draw a cut-off line: 134|.8
The first digit cut is 8, which is 5 or more, so round up.

$$
\begin{array}{r}
134 \\
+1 \\
\hline
135
\end{array}
$$

The fastest roller-coaster speed record is about 135 miles per hour.

39. **(a)** Round 185.981 to the nearest tenth.
Draw a cut-off line: 185.9|81
The first digit cut is 8, which is 5 or more, so round up.

$$
\begin{array}{r}
185.9 \\
+\ 0.1 \\
\hline
186.0
\end{array}
$$

The Indianapolis 500 fastest average winning speed is about 186.0 miles per hour.

(b) Round 763.04 to the nearest tenth.
Draw a cut-off line: 763.0|4
The first digit cut is 4, which is 4 or less. The part you keep stays the same.

The land speed record is about 763.0 miles per hour.

Relating Concepts (Exercises 41–44)

41. If you round $0.499 to the nearest dollar, it will round to $0 (zero dollars) because $0.499 is closer to $0 than to $1.

42. Rounding $0.499 to the nearest cent to get $0.50 is more helpful. Guideline: Round amounts less than $1.00 to the nearest cent instead of the nearest dollar.

43. If you round $0.0015 to the nearest cent, it will round to $0.00 (zero cents) because $0.0015 is closer to $0.00 than to $0.01.

44. Both $0.5968 and $0.6014 round to $0.60. Rounding to nearest thousandth (tenth of a cent) would allow you to identify $0.597 as less than $0.601.

5.3 Adding and Subtracting Signed Decimal Numbers

5.3 Margin Exercises

1. **(a)** 2.86 + 7.09

$$
\begin{array}{r}
\overset{1}{2.86} \\
+\ 7.09 \quad \text{\textit{Line up decimal points.}} \\
\hline
9.95
\end{array}
$$

(b) 13.761 + 8.325

$$
\begin{array}{r}
\overset{11}{13.761} \\
+\ 8.325 \quad \text{\textit{Line up decimal points.}} \\
\hline
22.086
\end{array}
$$

(c) 0.319 + 56.007 + 8.252

$$
\begin{array}{r}
\overset{1}{0.3}\overset{1}{1}9 \\
56.007 \\
+\ 8.252 \quad \text{\textit{Line up decimal points.}} \\
\hline
64.578
\end{array}
$$

(d) 39.4 + 0.4 + 177.2

$$
\begin{array}{r}
\overset{1}{\ }\overset{11}{\ } \\
39.4 \\
0.4 \\
+\ 177.2 \quad \text{\textit{Line up decimal points.}} \\
\hline
217.0
\end{array}
$$

2. **(a)** 6.54 + 9.8

$$
\begin{array}{r}
\overset{1}{6.54} \quad \text{\textit{Line up decimal points.}} \\
+\ 9.80 \quad \text{\textit{Write in one 0.}} \\
\hline
16.\underline{34}
\end{array}
$$

(b) 0.831 + 222.2 + 10

$$
\begin{array}{r}
\overset{1}{0.831} \quad \text{\textit{Line up decimal points.}} \\
222.200 \quad \text{\textit{Write in zeros.}} \\
+\ 10.000 \\
\hline
233.031
\end{array}
$$

(c) 8.64 + 39.115 + 3.0076

$$
\begin{array}{r}
\overset{2}{\ }\overset{1}{8.6400} \quad \text{\textit{Line up decimal points.}} \\
39.1150 \quad \text{\textit{Write in zeros.}} \\
+\ 3.0076 \\
\hline
50.7626
\end{array}
$$

(d) 5 + 429.823 + 0.76

$$
\begin{array}{r}
\overset{11}{5.000} \quad \text{\textit{Line up decimal points.}} \\
429.823 \quad \text{\textit{Write in zeros.}} \\
+\ 0.760 \\
\hline
435.583
\end{array}
$$

3. **(a)** Subtract 22.7 from 72.9.

$$
\begin{array}{r}
72.9 \\
-\ 22.7 \\
\hline
50.\underline{2}
\end{array}
\qquad
\textbf{Check}
\begin{array}{r}
22.7 \\
+\ 50.2 \\
\hline
72.9
\end{array}
$$

(b) Subtract 6.425 from 11.813.

$$
\begin{array}{r}
{}^{71013} \\
11.\cancel{8}\cancel{1}\cancel{3} \\
-\ 6.425 \\
\hline
5.388
\end{array}
\qquad
\textbf{Check}
\begin{array}{r}
6.425 \\
+\ 5.388 \\
\hline
11.813
\end{array}
$$

(c) $20.15 − $19.67

$$
\begin{array}{r}
{}^{1\ 9\ 1015} \\
\$\cancel{2}\cancel{0}.\cancel{1}\cancel{5} \\
-\ 19.67 \\
\hline
\$0.48
\end{array}
\qquad
\textbf{Check}
\begin{array}{r}
\$19.67 \\
+\ 0.48 \\
\hline
\$20.15
\end{array}
$$

4. **(a)** Subtract 18.651 from 25.3.

$$
\begin{array}{r}
{}^{1\ 14\ 12\ 9\ 10} \\
\cancel{2}\cancel{5}.\cancel{3}\cancel{0}\cancel{0} \quad \text{\textit{Write two zeros.}} \\
-\ 18.651 \\
\hline
6.649
\end{array}
$$

$$
\textbf{Check}
\begin{array}{r}
18.651 \\
+\ 6.649 \\
\hline
25.300
\end{array}
$$

(b) $5.816 - 4.98$

$$
\begin{array}{r}
\scriptstyle 4\ 1711 \\
\cancel{5}.\cancel{8}\cancel{1}6 \\
- 4.9\ 8\ 0 \qquad \textit{Write one zero.} \\
\hline
0.8\ 3\ 6
\end{array}
$$

Check $\quad \begin{array}{r} 4.980 \\ + 0.836 \\ \hline 5.816 \end{array}$

(c) 40 less 3.66

$$
\begin{array}{r}
\scriptstyle 3\ 9\ 910 \\
\cancel{4}\cancel{0}.\cancel{0}\cancel{0} \\
- 3.6\ 6 \\
\hline
3\ 6.3\ 4
\end{array}
$$

Check $\quad \begin{array}{r} 36.34 \\ + 3.66 \\ \hline 40.00 \end{array}$

(d) $1 - 0.325$

$$
\begin{array}{r}
\scriptstyle 0\ 9\ 910 \\
\cancel{1}.\cancel{0}\cancel{0}\cancel{0} \\
- 0.3\ 2\ 5 \\
\hline
0.6\ 7\ 5
\end{array}
$$

Check $\quad \begin{array}{r} 0.325 \\ + 0.675 \\ \hline 1.000 \end{array}$

5. **(a)** $13.245 + (-18)$ ▪ The addends have unlike signs.

$$|13.245| \text{ is } \underline{13.245}, \quad |-18| \text{ is } \underline{18}$$

-18 has the larger absolute value, so the sum will be negative. Subtract the absolute values.

$$
\begin{array}{r}
18.000 \\
- 13.245 \\
\hline
4.\underline{755}
\end{array}
$$

$13.245 + (-18) = \underline{-4.755}$

(b) $-0.7 + (-0.33)$ ▪ Both addends are negative, so the sum will be negative.

$$|-0.7| = 0.7, \quad |-0.33| = 0.33$$

$$
\begin{array}{r}
0.70 \\
+ 0.33 \\
\hline
1.03
\end{array}
$$

$-0.7 + (-0.33) = -1.03$

(c) $-6.02 + 100.5$ ▪ The addends have unlike signs.

$$|-6.02| = 6.02, \quad |100.5| = 100.5$$

100.5 has the larger absolute value, so the sum will be positive. Subtract the absolute values.

$$
\begin{array}{r}
100.50 \\
- 6.02 \\
\hline
94.48
\end{array}
$$

$-6.02 + 100.5 = 94.48$

6. **(a)** $\quad -0.37 \quad - \quad (-6)$

$\qquad\qquad \downarrow \qquad\quad \downarrow \qquad$ *Rewrite subtraction as*

$\quad -0.37 \quad + \quad 6 \qquad$ *adding the opposite.*

6 has the larger absolute value, so the sum will be positive. Subtract the absolute values.

$$
\begin{array}{r}
6.00 \\
- 0.37 \\
\hline
5.63
\end{array}
$$

Answer: 5.63

(b) $\quad 5.8 \quad - \quad\quad 10.03$

$\qquad\quad \downarrow \qquad\quad\quad \downarrow \qquad$ *Rewrite subtraction as*

$\quad 5.8 \quad + \quad (-10.03) \qquad$ *adding the opposite.*

-10.03 has the larger absolute value, so the sum will be negative. Subtract the absolute values.

$$
\begin{array}{r}
10.03 \\
- 5.80 \\
\hline
4.23
\end{array}
$$

Answer: -4.23

(c) $\quad -312.72 \quad - \quad\quad 65.7$

$\qquad\qquad\quad \downarrow \qquad\quad\quad \downarrow \qquad$ *Rewrite subt. as*

$\quad -312.72 \quad + \quad (-65.7) \qquad$ *adding the opp.*

Both addends are negative, so the sum will be negative.

$$
\begin{array}{r}
312.72 \\
+ 65.70 \\
\hline
378.42
\end{array}
$$

Answer: -378.42

(d) $0.8 - (6 - 7.2)$

$\qquad\qquad\qquad\qquad$ *Work inside*
$\qquad\qquad\qquad\qquad$ *parentheses first.*

$= 0.8 - [6 + (-7.2)] \quad$ *Change subtraction*
$\qquad\qquad\qquad\qquad$ *to adding the*
$\qquad\qquad\qquad\qquad$ *opposite.*

$\qquad\qquad\qquad\qquad$ *-7.2 has the larger*

$= 0.8 - (-1.2) \qquad$ *absolute value, so*
$\qquad\qquad\qquad\qquad$ *the sum will be*
$\qquad\qquad\qquad\qquad$ *negative.*

$\qquad\qquad\qquad\qquad$ *Change subtraction*

$= 0.8 + 1.2 \qquad\quad$ *to adding the*
$\qquad\qquad\qquad\qquad$ *opposite.*

$= 2.0 \ \text{ or } \ 2$

7. **(a)** $2.83 + 5.009 + 76.1$

$$
\begin{array}{ll}
\textit{Estimate:} & \textit{Exact:} \\
\quad 3 \leftarrow & \quad 2.830 \\
\quad 5 \leftarrow & \quad 5.009 \\
+ 80 \leftarrow & + 76.100 \\
\hline
\quad 88 & \quad 83.939
\end{array}
$$

(b) \$19.28 less \$1.53

$$
\begin{array}{ll}
\textit{Estimate:} & \textit{Exact:} \\
\$20 \leftarrow & \$19.28 \\
- 2 \leftarrow & - 1.53 \\
\hline
\$18 & \$17.75
\end{array}
$$

(c) $11.365 - 38$

Estimate: $10 - 40 = 10 + (-40) = -30$

Exact: $11.365 - 38 = 11.365 + (-38)$

-38 has the larger absolute value and is negative, so the sum is negative.

$$\begin{array}{r} 38.000 \\ -\ 11.365 \\ \hline 26.635 \end{array}$$

Answer: -26.635

(d) $-214.6 + 300.72$

Estimate: $-200 + 300 = 100$

Exact: $-214.6 + 300.72$

300.72 has the larger absolute value, so the sum is positive.

$$\begin{array}{r} 300.72 \\ -\ 214.60 \\ \hline 86.12 \end{array}$$

Answer: 86.12

5.3 Section Exercises

1. $6.42 + 10.163$ ▪ Line up decimal points.

$$\begin{array}{r} 6.420 \quad \textit{Write in one 0.} \\ +\ 10.163 \\ \hline \end{array}$$

3. $20 - 9.1263$ ▪ Line up decimal points.

$$\begin{array}{r} 20.0000 \quad \textit{Write in 4 zeros.} \\ -\ 9.1263 \\ \hline \end{array}$$

5.
$$\begin{array}{r} \overset{1\ 2}{5.69} \quad \textit{Line up decimal points.} \\ 0.24 \\ +\ 11.79 \\ \hline 17.72 \end{array}$$

7. $0.38 + 7 + 4.6$ ▪ Line up decimal points.

$$\begin{array}{r} 0.38 \\ 7.00 \quad \textit{Write in 2 zeros.} \\ +\ 4.60 \quad \textit{Write in one 0.} \\ \hline 11.98 \end{array}$$

9. $14.23 + 8 + 74.63 + 18.715 + 0.286$

$$\begin{array}{r} \overset{2\ 1\ 1\ 1}{14.230} \quad \textit{Write in one 0.} \\ 8.000 \quad \textit{Write in 3 zeros.} \\ 74.630 \quad \textit{Write in one 0.} \\ 18.715 \\ +\ 0.286 \\ \hline 115.861 \end{array}$$

11. $27.65 + 18.714 + 9.749 + 3.21$

$$\begin{array}{r} \overset{2\ 2\ 1\ 1}{27.650} \quad \textit{Write in one 0.} \\ 18.714 \\ 9.749 \\ +\ 3.210 \quad \textit{Write in one 0.} \\ \hline 59.323 \end{array}$$

13. The decimal points were not lined up. 6 should be written as 6.00.

$$\begin{array}{r} 0.72 \\ 6.00 \\ +\ 39.50 \\ \hline 46.22 \end{array}$$

15. 0.3000 written as a fraction in lowest terms is $\frac{3}{10}$, which is equivalent to 0.3.

$$0.3000 = \frac{3000}{10{,}000} = \frac{3000 \div 1000}{10{,}000 \div 1000} = \frac{3}{10} = 0.3$$

17. $90.5 - 0.8$ ▪ Line up decimal points.

$$\begin{array}{r} \overset{9}{8\,\overset{10}{\cancel{1}}5} \\ \cancel{9}\,\cancel{0}.\cancel{5} \\ -\ 0.8 \\ \hline 89.7 \quad \textit{Subtract as usual.} \end{array}$$

19. 0.4 less 0.291 ▪ Line up decimal points.

$$\begin{array}{r} \overset{9}{3\,\overset{10}{\cancel{1}}0} \\ 0.\cancel{4}\,\cancel{0}\,\cancel{0} \quad \textit{Write in 2 zeros.} \\ -\ 0.291 \\ \hline 0.109 \quad \textit{Subtract as usual.} \end{array}$$

21. 6 minus 5.09 ▪ Line up decimal points.

$$\begin{array}{r} \overset{9}{5\,\overset{10}{\cancel{1}}0} \\ \cancel{6}.\cancel{0}\,\cancel{0} \quad \textit{Write in 2 zeros.} \\ -\ 5.09 \\ \hline 0.91 \quad \textit{Subtract as usual.} \end{array}$$

23. Subtract 8.339 from 15 ▪ Line up decimal points.

$$\begin{array}{r} \overset{9\ \ 9}{14\,\overset{10}{\cancel{1}}\overset{10}{\cancel{0}}\overset{10}{\cancel{1}}0} \\ \cancel{1}\cancel{5}.\cancel{0}\,\cancel{0}\,\cancel{0} \quad \textit{Write in 3 zeros.} \\ -\ 8.339 \\ \hline 6.661 \quad \textit{Subtract as usual.} \end{array}$$

25. "Subtract 7.45 from 15.32" requires 15.32 to be on top.

$$\begin{array}{r} 15.32 \quad \textit{Line up decimal points.} \\ -\ 7.45 \\ \hline 7.87 \quad \textit{Subtract as usual.} \end{array}$$

27. **(a)** Humerus 14.35 in., radius 10.4 in.

$$\begin{array}{r} 14.35 \quad \textit{Line up decimal points.} \\ +\ 10.40 \quad \textit{Write in zero.} \\ \hline 24.75 \end{array}$$

The combined length is 24.75 inches.

(b)
$$\begin{array}{r} 14.35 \\ -\ 10.40 \\ \hline 3.95 \end{array}$$

The difference is 3.95 inches.

29. (a) Humerus 14.35 in., ulna 11.1 in., femur 19.88 in., tibia 16.94 in.

$$\begin{array}{r} 14.35 \\ 11.10 \\ 19.88 \\ +\ 16.94 \\ \hline 62.27 \end{array}$$ *Line up decimal points.*
Write in zero.

The sum of the lengths is 62.27 inches.

(b) 8th rib 9.06 in., 7th rib 9.45 in.

$$\begin{array}{r} 9.45 \\ -\ 9.06 \\ \hline 0.39 \end{array}$$

The 8th rib is 0.39 in. shorter than the 7th rib.

31. $24.008 + (-0.995)$ ■ The addends have unlike signs.

$$|24.008| = 24.008, \quad |-0.995| = 0.995$$

24.008 has the larger absolute value, so the sum will be positive. Subtract the absolute values.

$$\begin{array}{r} 24.008 \\ -\ 0.995 \\ \hline 23.013 \end{array}$$

33. $-6.05 + (-39.7)$ ■ Both addends are negative, so the sum will be negative.

$$|-6.05| = 6.05, \quad |-39.7| = 39.7$$

$$\begin{array}{r} 6.05 \\ +\ 39.70 \\ \hline 45.75 \end{array}$$

$$-6.05 + (-39.7) = -45.75$$

35. $0.9 - 7.59 = 0.9 + (-7.59)$ ■ The addends have unlike signs.

$$|0.9| = 0.9, \quad |-7.59| = 7.59$$

-7.59 has the larger absolute value, so the sum will be negative. Subtract the absolute values.

$$\begin{array}{r} 7.59 \\ -\ 0.90 \\ \hline 6.69 \end{array}$$

$$0.9 - 7.59 = -6.69$$

37. $-2 - 4.99 = -2 + (-4.99)$ ■ Both addends are negative, so the sum will be negative.

$$|-2| = 2, \quad |-4.99| = 4.99$$

$$\begin{array}{r} 2.00 \\ +\ 4.99 \\ \hline 6.99 \end{array}$$

$$-2 - 4.99 = -6.99$$

39. $-5.009 + 0.73$ ■ The addends have unlike signs.

$$|-5.009| = 5.009, \quad |-0.73| = 0.73$$

-5.009 has the larger absolute value, so the sum will be negative. Subtract the absolute values.

$$\begin{array}{r} 5.009 \\ -\ 0.730 \\ \hline 4.279 \end{array}$$

$$-5.009 + 0.73 = -4.279$$

41. $-1.7035 - (5 - 6.7)$ ■ Work inside parentheses first.

$$5 - 6.7 = 5 + (-6.7) = -1.7$$

Now the problem becomes:

$$= -1.7035 - (-1.7)$$
$$= -1.7035 + 1.7$$
$$= -0.0035$$

43. $8000 - (8002.63 - 8)$ ■ Work inside parentheses first.

$$8002.63 - 8 = 8002.63 + (-8) = 7994.63$$

Now the problem becomes:

$$= 8000 - (7994.63)$$
$$= 8000 + (-7994.63)$$
$$= 5.37$$

45. We can estimate $18 - 11.725$ as $20 - 10 = 10$, so 6.275 is the most reasonable answer.

47. We can estimate $-6.5 + 0.7$ as $-7 + 1 = -6$, so -5.8 is the most reasonable answer.

49. We can estimate $-0.671 - 9$ as $-1 + (-9) = -10$, so -9.671 is the most reasonable answer.

51. We can estimate $8.4 - (-50.83)$ as $8 + (+50) = 58$, so 59.23 is the most reasonable answer.

53. Mexico: 30.6 million; India: 81 million

Estimate:		*Exact*:
80	⟵	81.0
− 30	⟵	− 30.6
50		50.4

There are 50.4 million fewer Internet users in Mexico than there are in India.

55. Add the number of users from all the countries.

Estimate:		Exact:
400	←	420.00
200	←	239.90
100	←	99.14
80	←	81.00
40	←	43.98
30	←	30.60
+ 30	←	+ 26.20
880		940.82

There are 940.82 million Internet users in all the countries listed in the table.

57. Add the three heights.

Estimate:	Exact:
2	2.13
2	1.98
+ 2	+ 2.08
6 meters	6.19 meters

The NBA stars' combined height, 6.19 meters, is *less than* the rhino's height of 6.4 meters. Subtract to determine by how much.

6.40 *rhino's height*
− 6.19 *players' combined height*
0.21 meter

59. Subtract $9.12 from $20.

Estimate:	Exact:
$20	$20.00
− 9	− 9.12
$11	$10.88

He received $10.88 in change.

61. Subtract the price of the regular fishing line, $4.84, from the price of the fluorescent fishing line, $5.14.

Estimate:	Exact:
$5	$5.14
− 5	− 4.84
$0	$0.30

The fluorescent fishing line costs $0.30 more than the regular fishing line.

63. Add the price of the four items and the sales tax.

Estimate:	Exact:	
$19	$18.84	*spinning reel*
2	2.07	*tin split shot*
2	2.07	*tin split shot*
10	9.96	*tackle box*
+ 2	+ 2.31	*sales tax*
$35	$35.25	

The total cost of the four items and the sales tax is $35.25.

65. Add the monthly expenses.

$994.00
290.78
205.00
49.95
29.95
40.80
57.32
186.81
97.75
+ 107.00
$2059.36

Olivia's total expenses were $2059.36 per month.

67. Subtract to find the difference.

$290.78
− 186.81
$103.97

The difference in the amounts spent for groceries and for the car payment is $103.97.

69. Subtract the two inside measurements from the total length.

3.00 *Total length*
− 0.91 *Leftmost measurement*
2.09
− 0.70 *Middle measurement*
1.39

$b = 1.39$ cm

71. Add the given lengths and subtract the sum from the total length.

2.981		13.905
+ 2.981		− 5.962
5.962 feet		7.943 feet

$q = 7.943$ feet

5.4 Multiplying Signed Decimal Numbers

5.4 Margin Exercises

1.   **(a)** $-2.6(0.4)$

2.6	←	*1 decimal place*
× 0.4	←	*1 decimal place*
−1.04	←	*2 decimal places*

The factors have *different* signs, so the product is *negative*.

(b) (45.2)(0.25)

$$
\begin{array}{r}
45.2 \quad \leftarrow \textit{1 decimal place} \\
\times\, 0.25 \quad \leftarrow \textit{2 decimal places} \\
\hline
2\,26\,0 \\
9\,04 \\
\hline
11.30\,0 \quad \leftarrow \textit{3 decimal places}
\end{array}
$$

or 11.3

The factors have the *same* sign, so the product is *positive*.

(c)
$$
\begin{array}{r}
0.104 \quad \leftarrow \textit{3 decimal places} \\
\times\, 7 \quad \leftarrow \textit{0 decimal places} \\
\hline
0.728 \quad \leftarrow \textit{3 decimal places}
\end{array}
$$

(d) $(-3.18)^2$ means $(-3.18)(-3.18)$.

$$
\begin{array}{r}
3.18 \quad \leftarrow \textit{2 decimal places} \\
\times\, 3.18 \quad \leftarrow \textit{2 decimal places} \\
\hline
2544 \\
318 \\
9\,54 \\
\hline
10.1124 \quad \leftarrow \textit{4 decimal places}
\end{array}
$$

Now count over 4 places and write in the decimal point. The factors have the *same* sign, so the product is *positive*.

2. (a) 0.04(−0.09)

$$
\begin{array}{r}
0.04 \quad \leftarrow \textit{2 decimal places} \\
\times\, 0.09 \quad \leftarrow \textit{2 decimal places} \\
\hline
-0.0036 \quad \leftarrow \textit{4 decimal places}
\end{array}
$$

Count 4 places. Write in the decimal point and zeros. The factors have *different* signs, so the product is *negative*.

(b) (0.2)(0.008)

$$
\begin{array}{r}
0.008 \quad \leftarrow \textit{3 decimal places} \\
\times\, 0.2 \quad \leftarrow \textit{1 decimal place} \\
\hline
0.0016 \quad \leftarrow \textit{4 decimal places}
\end{array}
$$

Count 4 places. Write in the decimal point and zeros. The factors have the *same* sign, so the product is *positive*.

(c) (−0.063)(−0.04)

$$
\begin{array}{r}
0.063 \quad \leftarrow \textit{3 decimal places} \\
\times\, 0.04 \quad \leftarrow \textit{2 decimal places} \\
\hline
0.00252 \quad \leftarrow \textit{5 decimal places}
\end{array}
$$

Count 5 places. Write in the decimal point and zeros. The factors have the *same* sign, so the product is *positive*.

(d) $(0.003)^2$ means $(0.003)(0.003)$.

$$
\begin{array}{r}
0.003 \quad \leftarrow \textit{3 decimal places} \\
\times\, 0.003 \quad \leftarrow \textit{3 decimal places} \\
\hline
0.000009 \quad \leftarrow \textit{6 decimal places}
\end{array}
$$

Count 6 places. Write in the decimal point and zeros. The factors have the *same* sign, so the product is *positive*.

3. (a) (11.62)(4.01)

Estimate: *Exact:*

$$
\begin{array}{r}
10 \\
\times\, 4 \\
\hline
40
\end{array}
\qquad
\begin{array}{r}
11.62 \quad \leftarrow \textit{2 decimal places} \\
\times\, 4.01 \quad \leftarrow \textit{2 decimal places} \\
\hline
1162 \\
46\,480 \\
\hline
46.5962 \quad \leftarrow \textit{4 decimal places}
\end{array}
$$

The factors have the *same* sign, so the product is *positive*.

(b) (−5.986)(−33)

Estimate: *Exact:*

$$
\begin{array}{r}
6 \\
\times\, 30 \\
\hline
180
\end{array}
\qquad
\begin{array}{r}
5.986 \quad \leftarrow \textit{3 decimal places} \\
\times\, 33 \quad \leftarrow \textit{0 decimal places} \\
\hline
17\,958 \\
179\,58 \\
\hline
197.538 \quad \leftarrow \textit{3 decimal places}
\end{array}
$$

The factors have the *same* sign, so the product is *positive*.

(c) 8($4.35)

Estimate: *Exact:*

$$
\begin{array}{r}
4 \\
\times\, 8 \\
\hline
32
\end{array}
\qquad
\begin{array}{r}
\$4.35 \quad \leftarrow \textit{2 decimal places} \\
\times\, 8 \quad \leftarrow \textit{0 decimal places} \\
\hline
\$34.80 \quad \leftarrow \textit{2 decimal places}
\end{array}
$$

The factors have the *same* sign, so the product is *positive*.

(d) 58.6(−17.4)

Estimate: *Exact:*

$$
\begin{array}{r}
60 \\
\times\, 20 \\
\hline
-1200
\end{array}
\qquad
\begin{array}{r}
58.6 \quad \leftarrow \textit{1 decimal place} \\
\times\, 17.4 \quad \leftarrow \textit{1 decimal place} \\
\hline
23\,44 \\
410\,2 \\
586 \\
\hline
-1019.64 \quad \leftarrow \textit{2 decimal places}
\end{array}
$$

The factors have *different* signs, so the product is *negative*.

5.4 Section Exercises

1. In 4.2×3.46, the product will have 1 decimal place from 4.2 and 2 decimal places from 3.46 for a total of <u>3</u> decimal places.

3. In (0.12)(0.03), we will need 4 decimal places. Multiplying 12×3 gives us 36, so we need to move the decimal point 4 places to the left. Thus, we need to write in <u>2</u> zeros as placeholders.

5. Multiply the numbers as if they were whole numbers.

$$
\begin{array}{r}
0.042 \quad \leftarrow \textit{3 decimal places} \\
\times \ 3.2 \quad \leftarrow \textit{1 decimal place} \\
\hline
84 \\
126 \\
\hline
0.1344 \quad \leftarrow \textit{4 decimal places}
\end{array}
$$

Count 4 places. Write in the decimal point and zero. The factors have the *same* sign, so the product is *positive*.

7. $-21.5(7.4)$

$$
\begin{array}{r}
21.5 \quad \leftarrow \textit{1 decimal place} \\
\times \ 7.4 \quad \leftarrow \textit{1 decimal place} \\
\hline
8\ 60 \\
150\ 5 \\
\hline
-159.10 \quad \leftarrow \textit{2 decimal places}
\end{array}
$$

The factors have *different* signs, so the product is *negative*.

9. $(-23.4)(-0.66)$

$$
\begin{array}{r}
23.4 \quad \leftarrow \textit{1 decimal place} \\
\times \ 0.66 \quad \leftarrow \textit{2 decimal places} \\
\hline
1\ 404 \\
14\ 04 \\
\hline
15.444 \quad \leftarrow \textit{3 decimal places}
\end{array}
$$

The factors have the *same* sign, so the product is *positive*.

11. Use a calculator.

$$
\begin{array}{r}
\$51.88 \\
\times \ \ \ 665 \\
\hline
\$34,500.20
\end{array}
$$

13. $72(-0.6)$ ∎ The factors have *different* signs, so the product is *negative*.

72 has 0 decimal places. ⎫ Answer has 1
0.6 has 1 decimal place. ⎭→ decimal place.

$$72(-0.6) = -43.2$$

15. $(7.2)(0.06)$ ∎ The factors have the *same* sign, so the product is *positive*.

7.2 has 1 decimal place. ⎫ Answer has 3
0.06 has 2 decimal places. ⎭→ decimal places.

$$(7.2)(0.06) = 0.432$$

17. $-0.72(-0.06)$ ∎ The factors have the *same* sign, so the product is *positive*.

0.72 has 2 decimal places. ⎫ Answer has 4
0.06 has 2 decimal places. ⎭→ decimal places.

$$-0.72(-0.06) = 0.0432$$

19. $(0.006)(0.0052)$

$$
\begin{array}{r}
0.0052 \quad \leftarrow \textit{4 decimal places} \\
\times \ 0.006 \quad \leftarrow \textit{3 decimal places} \\
\hline
0.0000312 \quad \leftarrow \textit{7 decimal places}
\end{array}
$$

Write in 4 zeros to get 7 decimal places.

21. $(-0.003)^2$ means $(-0.003)(-0.003)$.

$$
\begin{array}{r}
0.003 \quad \leftarrow \textit{3 decimal places} \\
\times \ 0.003 \quad \leftarrow \textit{3 decimal places} \\
\hline
0.000009 \quad \leftarrow \textit{6 decimal places}
\end{array}
$$

Write in 5 zeros to get 6 decimal places. The factors have the *same* sign, so the product is *positive*.

23. *Estimate*:   *Exact*:

$$
\begin{array}{rr}
40 \longleftarrow & 39.6 \quad \leftarrow \textit{1 decimal place} \\
\times 5 \longleftarrow & \times \ 4.8 \quad \leftarrow \textit{1 decimal place} \\
\hline
200 & 31\ 68 \\
& 158\ 4 \\
& \hline
& 190.08 \quad \leftarrow \textit{2 decimal places}
\end{array}
$$

25. *Estimate*:   *Exact*:

$$
\begin{array}{rr}
40 \longleftarrow & 37.1 \quad \leftarrow \textit{1 decimal place} \\
\times 40 \longleftarrow & \times \ \ 42 \quad \leftarrow \textit{0 decimal places} \\
\hline
1600 & 74\ 2 \\
& 1484 \\
& \hline
& 1558.2 \quad \leftarrow \textit{1 decimal place}
\end{array}
$$

27. *Estimate*:   *Exact*:

$$
\begin{array}{rr}
7 \longleftarrow & 6.53 \quad \leftarrow \textit{2 decimal places} \\
\times 5 \longleftarrow & \times \ 4.6 \quad \leftarrow \textit{1 decimal place} \\
\hline
35 & 3\ 918 \\
& 26\ 12 \\
& \hline
& 30.038 \quad \leftarrow \textit{3 decimal places}
\end{array}
$$

29. Use a calculator.

Estimate: *Exact*:

$$
\begin{array}{rr}
3 \longleftarrow & 2.809 \quad \leftarrow \textit{3 decimal places} \\
\times 7 \longleftarrow & \times \ 6.85 \quad \leftarrow \textit{2 decimal places} \\
\hline
21 & 19.24165 \quad \leftarrow \textit{5 decimal places}
\end{array}
$$

31. An $28.90 car payment is *unreasonable*. A reasonable answer would be $289.00.

33. A height of 60.5 inches (about 5 feet) is *reasonable*.

35. A gallon of milk for $419 is *unreasonable*. A reasonable answer would be $4.19.

37. 0.095 pound for a baby's weight is *unreasonable*. A reasonable answer would be 9.5 pounds.

39. Multiply her pay per hour times the hours she worked.

$$
\begin{array}{r}
\$18.73 \quad \leftarrow \textit{2 decimal places} \\
\times \quad 50.5 \quad \leftarrow \textit{1 decimal place} \\
\hline
9\ 365 \\
936\ 50 \\
\hline
\$945.865 \quad \leftarrow \textit{3 decimal places}
\end{array}
$$

Round $945.865 to the nearest cent. LaTasha made $945.87 (rounded).

41. Multiply the cost of one meter of canvas by the number of meters needed.

$$
\begin{array}{r}
\$4.09 \\
\times\ 0.6 \\
\hline
\$2.454
\end{array}
$$

$2.454 rounds to $2.45. Sid will spend $2.45 on the canvas.

43. Multiply the number of gallons that she pumped into her pickup truck by the price per gallon.

$$
\begin{array}{r}
20.510 \\
\times\ 3.759 \\
\hline
77.09709
\end{array}
$$

Round 77.09709 to the nearest cent. Michelle paid $77.10 for the gas.

45. Multiply the cost of the home by 0.07.

$$
\begin{array}{r}
\$289{,}500 \\
\times\quad 0.07 \\
\hline
\$20{,}265.00
\end{array}
$$

Ms. Rolack's fee was $20,265.

47. **(a) Area before 1929:**

$$
\begin{array}{r}
7.4218 \\
\times\quad 3.125 \\
\hline
23.1931250
\end{array}
$$

Rounding to the nearest tenth gives us 23.2 in.^{2}.

Area after 1929:

$$
\begin{array}{r}
6.14 \\
\times\ 2.61 \\
\hline
16.0254
\end{array}
$$

Rounding to the nearest tenth gives us 16.0 in.^{2}.

(b) Subtract to find the difference.

$$
\begin{array}{r}
23.2 \\
-\ 16.0 \\
\hline
7.2
\end{array}
$$

The difference in rounded areas is 7.2 in.^{2}.

49. (a) Multiply the thickness of one bill times the number of bills.

$$
\begin{array}{r}
0.0043 \\
\times\ 100 \\
\hline
0.4300
\end{array}
$$

A pile of 100 bills would be 0.43 inch high.

(b)
$$
\begin{array}{r}
0.0043 \\
\times\ 1000 \\
\hline
4.3000
\end{array}
$$

A pile of 1000 bills would be 4.3 inches high.

51. Multiply the monthly cost of cable times 24 months (two years).

$$
\begin{array}{r}
\$48.96 \quad \textit{basic per month} \\
\times\quad 24 \\
\hline
195\ 84 \\
979\ 2 \\
\hline
\$1175.04 \quad \textit{monthly total} \\
+\ 89.00 \quad \textit{one-time installation} \\
\hline
\$1264.04 \quad \textit{two-year total}
\end{array}
$$

The total cost for basic cable is $1264.04. *cont.*

$$
\begin{array}{r}
\$109.78 \quad \textit{deluxe per month} \\
\times\quad 24 \\
\hline
439\ 12 \\
2195\ 6 \\
\hline
\$2634.72 \quad \textit{monthly total} \\
+\ 89.00 \quad \textit{one-time installation} \\
\hline
\$2723.72 \quad \textit{two-year total}
\end{array}
$$

The total cost for deluxe cable is $2723.72.

53. Multiply to find the cost of the rope, then multiply to find the cost of the wire. Add the results to find Barry's total purchases. Subtract the purchases from $15 (three $5 bills is $15).

Cost of rope	Cost of wire
16.5	$1.05
× $0.47	× 3
$7.755	$3.15

The cost of the rope rounds to $7.76.

Purchases	Change
$7.76 *rope*	$15.00
+ 3.15 *wire*	− 10.91
$10.91	$4.09

Barry received $4.09 in change.

55. **(a)** Find the cost for 3 short-sleeved, solid-color shirts.

$$\begin{array}{r} \$14.75 \\ \times\ \ \ \ 3 \\ \hline \$44.25 \end{array}$$

Based on this subtotal, shipping is $7.95. Find the cost of the 3 monograms.

$$\begin{array}{r} \$4.95 \\ \times\ \ \ 3 \\ \hline \$14.85 \end{array}$$

Add these amounts, plus $5.00 for a gift box.

$$\begin{array}{r} \$44.25 \\ 7.95 \\ 14.85 \\ +\ \ 5.00 \\ \hline \$72.05 \end{array}$$

The total cost is $72.05.

(b) Subtract the cost of the shirts to find the difference.

$$\begin{array}{r} \$72.05 \\ -\ 44.25 \\ \hline \$27.80 \end{array}$$

The monograms, gift box, and shipping added $27.80 to the cost of the gift.

Relating Concepts (Exercises 57–58)

57. $(5.96)(10) = \underline{59.6}$ $(3.2)(10) = \underline{32}$
$(0.476)(10) = \underline{4.76}$ $(80.35)(10) = \underline{803.5}$
$(722.6)(10) = \underline{7226}$ $(0.9)(10) = \underline{9}$

Multiplying by 10, decimal point moves one place to the right; by 100, two places to the right; by 1000, three places to the right.

58. $(59.6)(0.1) = \underline{5.96}$ $(3.2)(0.1) = \underline{0.32}$
$(0.476)(0.1) = \underline{0.0476}$ $(80.35)(0.1) = \underline{8.035}$
$(65)(0.1) = \underline{6.5}$ $(523)(0.1) = \underline{52.3}$

Multiplying by 0.1, decimal point moves one place to the left; by 0.01, two places to the left; by 0.001, three places to the left.

5.5 Dividing Signed Decimal Numbers

5.5 Margin Exercises

1. **(a)**
$$\begin{array}{r} 2\,3.\underline{4} \\ 4\overline{\smash{)}9\,3.\,6} \\ \underline{8} \\ 1\,3 \\ \underline{1\,2} \\ 1\ 6 \\ \underline{1\ 6} \\ 0 \end{array}$$

Check $\begin{array}{r} 23.4 \\ \times\ \ \ 4 \\ \hline 93.6 \end{array}$

(b)
$$\begin{array}{r} 1.\,1\,3\,4 \\ 6\overline{\smash{)}6.\,8\,0\,4} \\ \underline{6} \\ 0\ 8 \\ \underline{6} \\ 2\ 0 \\ \underline{1\ 8} \\ 2\ 4 \\ \underline{2\ 4} \\ 0 \end{array}$$

Check $\begin{array}{r} 1.134 \\ \times\ \ \ \ 6 \\ \hline 6.804 \end{array}$

(c) $\dfrac{278.3}{11}$

$$\begin{array}{r} 2\,5.\,3 \\ 11\overline{\smash{)}2\,7\,8.\,3} \\ \underline{2\,2} \\ 5\ 8 \\ \underline{5\ 5} \\ 3\ 3 \\ \underline{3\ 3} \\ 0 \end{array}$$

Check $\begin{array}{r} 25.3 \\ \times\ \ 11 \\ \hline 253 \\ 253\ \\ \hline 278.3 \end{array}$

(d) $-0.51835 \div 5$

Different signs; *negative* quotient

$$\begin{array}{r} -0.\,1\,0\,3\,6\,7 \\ 5\overline{\smash{)}\ \ 0.\,5\,1\,8\,3\,5} \\ \underline{5} \\ 1 \\ \underline{0} \\ 1\ 8 \\ \underline{1\ 5} \\ 3\ 3 \\ \underline{3\ 0} \\ 3\ 5 \\ \underline{3\ 5} \\ 0 \end{array}$$

Check $\begin{array}{r} -0.10367 \\ \times\ \ \ \ \ \ \ 5 \\ \hline -0.51835 \end{array}$

(e) $-213.45 \div (-15)$

Same signs; *positive* quotient

$$\begin{array}{r} 1\,4.\,2\,3 \\ 15\overline{\smash{)}2\,1\,3.\,4\,5} \\ \underline{1\ 5} \\ 6\ 3 \\ \underline{6\ 0} \\ 3\ 4 \\ \underline{3\ 0} \\ 4\ 5 \\ \underline{4\ 5} \\ 0 \end{array}$$

Check $\begin{array}{r} 14.23 \\ \times\ \ -15 \\ \hline 71\,15 \\ 142\,3\ \ \\ \hline -213.45 \end{array}$

2. (a)

$$
\begin{array}{r}
1.\,2\,8 \\
5\overline{)6.\,4\,0} \quad \leftarrow \textit{Write one zero.} \\
\underline{5} \\
1\;4 \\
\underline{1\;0} \\
4\;0 \\
\underline{4\;0} \\
0
\end{array}
$$

Check
$$
\begin{array}{r}
1.28 \\
\times\;5 \\
\hline
6.40 \\
\text{or} \\
6.4
\end{array}
$$

(b) $30.87 \div (-14)$

Different signs; *negative* quotient

$$
\begin{array}{r}
-\,2.\,2\,0\,5 \\
14\overline{)3\;0.\,8\,7\,0} \quad \leftarrow \textit{Write one zero.} \\
2\;8 \\
\hline
2\;8 \\
\underline{2\;8} \\
0\;7 \\
\underline{0} \\
7\;0 \\
\underline{7\;0} \\
0
\end{array}
$$

Check
$$
\begin{array}{r}
2.205 \\
\times\;-14 \\
\hline
8\,820 \\
22\,05 \\
\hline
-30.870 \text{ or } -30.87
\end{array}
$$

(c) $\dfrac{-259.5}{-30}$

Same signs; *positive* quotient

$$
\begin{array}{r}
8.\,6\,5 \\
30\overline{)2\;5\;9.\,5\,0} \quad \leftarrow \textit{Write one zero.} \\
2\;4\;0 \\
\hline
1\;9\;5 \\
\underline{1\;8\;0} \\
1\;5\;0 \\
\underline{1\;5\;0} \\
0
\end{array}
$$

Check
$$
\begin{array}{r}
8.65 \\
\times\;-30 \\
\hline
-259.50 \text{ or } -259.5
\end{array}
$$

(d) $0.3 \div 8$

$$
\begin{array}{r}
0.\,0\,3\,7\,5 \\
8\overline{)0.\,3\,0\,0\,0} \quad \leftarrow \textit{Write three zeros.} \\
0\;0 \\
\hline
3\;0 \\
\underline{2\;4} \\
6\;0 \\
\underline{5\;6} \\
4\;0 \\
\underline{4\;0} \\
0
\end{array}
$$

Check 0.0375
$$
\begin{array}{r}
0.0375 \\
\times\;\;8 \\
\hline
0.3000 \text{ or } 0.3
\end{array}
$$

3. (a) $13\overline{)2\;6\;7.\,0\,1}$

From a calculator, $267.01 \div 13 \approx 20.539231.$ There are no repeating digits visible on the calculator.

$20.539|231$ rounds to $20.539.$

Check $(20.539)(13) = 267.007 \approx 267.01$

(b) $6\overline{)2\;0.\,5}$

$20.5 \div 6 \approx 3.416666$

There is a repeating decimal: $3.41\overline{6}.$

$3.416|666$ rounds to $3.417.$

Check $(3.417)(6) = 20.502 \approx 20.5$

(c) $\dfrac{10.22}{9} = 10.22 \div 9 \approx 1.135555$

There is a repeating decimal: $1.13\overline{5}.$

$1.135|555$ rounds to $1.136.$

Check $(1.136)(9) = 10.224 \approx 10.22$

(d) $16.15 \div 3 \approx 5.383333$

There is a repeating decimal: $5.38\overline{3}.$

$5.383|333$ rounds to $5.383.$

Check $(5.383)(3) = 16.149 \approx 16.15$

(e) $116.3 \div 11 \approx 10.5727272$

The answer has a repeating decimal that starts repeating in the fourth place: $10.5\overline{72}.$

$10.572|72$ rounds to $10.573.$

Check $(10.573)(11) = 116.303 \approx 116.3$

4. (a)

$$
\begin{array}{r}
5.\,2 \\
0.2_\wedge\overline{)1.\,0_\wedge\,4} \\
1\;0 \\
\hline
0\;4 \\
\underline{4} \\
0
\end{array}
$$

(b)

$$
\begin{array}{r}
3\,0.\,1\,2 \\
0.06_\wedge\overline{)1.\,8\,0_\wedge\,7\,2} \\
1\;8 \\
\hline
0\;0 \\
\underline{0} \\
0\;7 \\
\underline{6} \\
1\;2 \\
\underline{1\;2} \\
0
\end{array}
$$

(c)

```
                6 4 0 0
0.005∧ ⟌ 3 2. 0 0 0∧
                3 0
                ───
                  2 0
                  2 0
                  ───
                    0 0
                     0
                    ───
                     0 0
                      0
                     ───
                      0
```

(d) $-8.1 \div 0.025$

Different signs; negative quotient

```
                 − 3 2 4
0.025∧ ⟌ 8. 1 0 0∧
                 7 5
                 ───
                   6 0
                   5 0
                   ───
                     1 0 0
                     1 0 0
                     ─────
                         0
```

(e) $\dfrac{7}{1.3}$

```
                 5. 3 8 4   ≈ 5.38
1.3∧ ⟌ 7. 0∧ 0 0 0
                 6 5
                 ───
                   5 0
                   3 9
                   ───
                     1 1 0
                     1 0 4
                     ─────
                         6 0
                         5 2
                         ───
                           8
```

(f) $-5.3091 \div (-6.2)$

Same signs; positive quotient

```
                 0. 8 5 6   ≈ 0.86
6.2∧ ⟌ 5. 3∧ 0 9 1
                 4 9 6
                 ─────
                   3 4 9
                   3 1 0
                   ─────
                     3 9 1
                     3 7 2
                     ─────
                       1 9
```

5. **(a)** $42.75 \div 3.8 = 1.125$

Estimate: $40 \div 4 = 10$

The answer is **not** reasonable.

```
                 1 1. 2 5
3.8∧ ⟌ 4 2. 7∧ 5 0
                 3 8
                 ───
                   4 7
                   3 8
                   ───
                     9 5
                     7 6
                     ───
                       1 9 0
                       1 9 0
                       ─────
                           0
```

The answer should be 11.25

(b) $807.1 \div 1.76 = 458.580$ to nearest thousandth

Estimate: $800 \div 2 = 400$

The answer is reasonable.

(c) $48.63 \div 52 = 93.519$ to nearest thousandth

Estimate: $50 \div 50 = 1$

The answer is **not** reasonable.

```
              0. 9 3 5 1   ≈ 0.935
52 ⟌ 4 8. 6 3 0 0
              4 6 8
              ─────
                1 8 3
                1 5 6
                ─────
                  2 7 0
                  2 6 0
                  ─────
                    1 0 0
                      5 2
                    ─────
                      4 8
```

The answer should be 0.935.

(d) $9.0584 \div 2.68 = 0.338$

Estimate: $9 \div 3 = 3$

The answer is **not** reasonable.

```
                 3. 3 8
2.68∧ ⟌ 9. 0 5∧ 8 4
                 8 0 4
                 ─────
                   1 0 1 8
                     8 0 4
                   ─────
                     2 1 4 4
                     2 1 4 4
                     ───────
                           0
```

The answer should be 3.38.

6. **(a)** $-4.6 - 0.79 \; + \; \underbrace{1.5^2}$

$= -4.6 - 0.79 + 2.25$ *Exponent first*

$= -4.6 + (-0.79) + 2.25$ *Add the opposite.*

$= -5.39 + 2.25$ *Add.*

$= -3.14$ *Add.*

(b) $\underbrace{3.64 \div 1.3}(3.6)$

$= 2.8(3.6)$ *Divide first.*

$= 10.08$ *Multiply.*

(c) $0.08 + 0.6(\underbrace{2.99 - 3})$

$= 0.08 + 0.6(-0.01)$ *Parentheses first*

$= 0.08 + (-0.006)$ *Multiply.*

$= 0.074$ *Add.*

(d) $10.85 - \underbrace{2.3 \cdot (5.2)} \div 3.2$

$= 10.85 - 11.96 \div 3.2$ *Multiply first.*

$= 10.85 - 3.7375$ *Divide.*

$= 7.1125$ *Subtract.*

5.5 Section Exercises

1. $25.5 \div 5$ has a whole number for the divisor and can be written as $5\overline{)25.5}$.

3. You make 2.2 a whole number by moving the decimal point one place to the right, and you must do the same to 8.24.

$$2.2\overline{)8.24}$$

5. $27.3 \div (-7)$ ∎ The numbers have different signs, so the quotient is negative. Divide as if both numbers are whole numbers.

```
        3. 9   Line up decimal points.
    7 ) 2 7. 3
        2 1
        6 3
        6 3
          0
```

$$27.3 \div (-7) = -3.9$$

7. $\dfrac{4.23}{9}$ ∎ The numbers have the same sign, so the quotient is positive. Divide as if both numbers are whole numbers.

```
      0. 4 7   Line up decimal points.
    9 ) 4. 2 3
        3 6
        6 3
        6 3
          0
```

$$\frac{4.23}{9} = 0.47$$

9. $-20.01 \div (-0.05)$ ∎ The numbers have the same sign, so the quotient is positive.

```
            4 0 0. 2   Line up decimal points.
  0.05ʌ) 2 0. 0 1ʌ 0   Move decimal point
         2 0            2 places in dividend
         0 0 1 0        and divisor.
             1  0       Write 0 in dividend.
               0
```

$$-20.01 \div (-0.05) = 400.2$$

11. $1.5\overline{)54}$ ∎ The numbers have the same sign, so the quotient is positive.

```
            3  6.   Line up decimal points.
  1.5ʌ) 5 4. 0ʌ    Move decimal point
        4 5         1 place in dividend
        9  0        and divisor; write 0.
        9  0
          0
```

$$1.5\overline{)54} = 36$$

13. Given: $108 \div 18 = 6$

Find: $1.8\overline{)0.108}$ or $0.108 \div 1.8$

The new dividend, 0.108, has the effect of moving the decimal place in the answer *left* three places. The new divisor, 1.8, has the effect of moving the decimal place in the answer *right* one place. So the new answer has the decimal point moved to the left two places.

$$0.108 \div 1.8 = 0.06$$

15. Given: $108 \div 18 = 6$

Find: $0.018\overline{)108}$ or $108 \div 0.018$

The new divisor, 0.018, has the effect of moving the decimal place in the answer *right* three places.

$$108 \div 0.018 = 6000$$

17. Given: $108 \div 18 = 6$

Find: $0.18\overline{)10.8}$ or $10.8 \div 0.18$

The new dividend, 10.8, has the effect of moving the decimal place in the answer *left* one place. The new divisor, 0.18, has the effect of moving the decimal place in the answer *right* two places. So the new answer has the decimal point moved to the right one place.

$$10.8 \div 0.18 = 60$$

19. Given: $108 \div 18 = 6$

Find: $18\overline{)0.0108}$ or $0.0108 \div 18$

The new dividend, 0.0108, has the effect of moving the decimal place in the answer *left* four places. So the new answer has the decimal point moved to the left four places.

$$0.0108 \div 18 = 0.0006$$

21. $4.6\overline{)116.38}$

```
            2 5. 3   Line up decimal points.
  4.6ʌ) 1 1 6. 3ʌ 8  Move decimal point 1
         9 2          place in dividend and
         2 4 3        divisor.
         2 3 0
           1 3 8
           1 3 8
               0
```

23. $\dfrac{-3.1}{-0.006}$ ■ The numbers have the same sign, so the quotient is positive.

```
              5 1 6. 6 6 6   Line up decimal points.
  0.006∧ 3. 1 0 0∧ 0 0 0    Move decimal point 3
        3 0                 places in dividend and
        1 0                 divisor; write 2 zeros.
          6
          4 0
          3 6
            4 0    Write 0 in dividend.
            3 6
              4 0    Write 0 in dividend.
              3 6
                4 0   Write 0 in dividend.
                3 6
                  4   Stop and round answer
                      to the nearest hundredth.
```

516.666 rounds to 516.67.

25. $-240.8 \div 9$

```
        2 6. 7 5 5 5  Line up decimal points.
  9 240. 8 0 0 0
    1 8
      6 0
      5 4
        6 8
        6 3
          5 0    Write 0 in dividend.
          4 5
            5 0    Write 0 in dividend.
            4 5
              5 0  Write 0 in dividend.
              4 5
                5   Stop and round answer
                    to the nearest thousandth.
```

The numbers have different signs, so the quotient is negative.
Round -26.7555 to -26.756.

27. $0.034 \overline{)342.81}$

    Enter on calculator: $342.81 \boxed{\div} 0.034 \boxed{=}$

Round 10,082.64706 to 10,082.647.

29. $37.8 \div 8 = 47.25$

Estimate: $40 \div 8 = 5$

The answer 47.25 is *unreasonable.*

```
          4. 7 2 5
  8 3 7. 8 0 0
    3 2
      5 8
      5 6
        2 0
        1 6
          4 0
          4 0
            0
```

The correct answer is 4.725.

31. $54.6 \div 48.1 \approx 1.135$

Estimate: $50 \div 50 = 1$

The answer 1.135 is *reasonable.*

33. $307.02 \div 5.1 = 6.2$

Estimate: $300 \div 5 = 60$

The answer 6.2 is *unreasonable.*

```
                6 0. 2
  5.1∧ 3 0 7. 0∧ 2
      3 0 6
          1 0
            0
          1 0 2
          1 0 2
              0
```

The correct answer is 60.2.

35. Divide the cost by the number of pairs of tights.

```
          3. 9 9 6
  6 2 3. 9 8 0
    1 8
      5 9
      5 4
        5 8
        5 4
          4 0
          3 6
            4
```

$3.996 rounds to $4.00.

One pair costs $4.00 (rounded).

37. Divide the balance by the number of months.

```
          6 7. 0 8
  21 1 4 0 8. 6 8
     1 2 6
       1 4 8
       1 4 7
           1 6 8
           1 6 8
               0
```

Aimee is paying $67.08 per month.

39. Divide the total earnings by the number of hours.

```
          1 1. 9 2
  40 4 7 6. 8 0
     4 0
       7 6
       4 0
         3 6 8
         3 6 0
             8 0
             8 0
               0
```

Darren earns $11.92 per hour.

41. Divide the miles driven by the gallons of gas purchased.

$$344.1 \div 12.3 \approx 28.0$$

She got 28.0 miles per gallon (rounded).

43. Multiply the price per can by the number of cans.

$$\begin{array}{r} \$0.57 \\ \times\ \ 6 \\ \hline \$3.42 \end{array}$$

Subtract to find total savings.

$$\begin{array}{r} \$3.42 \\ -\ 3.25 \\ \hline \$0.17 \end{array}$$

There are six cans, so divide by 6 to find the savings per can.

$$\begin{array}{r} 0.\,0\,2\,8 \\ 6\overline{)0.\,1\,7\,0} \\ \underline{1\,2} \\ 5\,0 \\ \underline{4\,8} \\ 2 \end{array}$$

$0.028 rounds to $0.03.

You will save $0.03 per can (rounded).

45. First find the sum of lengths for Jackie Joyner-Kersee.

$$\begin{array}{r} 7.49 \\ 7.45 \\ 7.40 \\ 7.32 \\ +\ 7.20 \\ \hline 36.86 \end{array}$$

Then divide by the number of jumps, namely 5.

$$\begin{array}{r} 7.\,3\,7\,2 \\ 5\overline{)3\,6.\,8\,6\,0} \\ \underline{3\,5} \\ 1\ 8 \\ \underline{1\ 5} \\ 3\,6 \\ \underline{3\,5} \\ 1\,0 \\ \underline{1\,0} \\ 0 \end{array}$$

7.372 rounds to 7.37.

The average length of the long jumps made by Jackie Joyner-Kersee is 7.37 meters (rounded).

47. Subtract to find the difference.

$$\begin{array}{rl} 7.40 & \textit{fifth longest jump} \\ -\ 7.32 & \textit{sixth longest jump} \\ \hline 0.08 & \textit{difference} \end{array}$$

The fifth longest jump was 0.08 meter longer than the sixth longest jump.

49. Add the lengths of the jumps of the top three athletes.

$$\begin{array}{r} 7.52 \\ 7.49 \\ +\ 7.48 \\ \hline 22.49 \end{array}$$

The total length jumped by the top three athletes was 22.49 meters.

51. (a) Work inside parentheses; subtract $9.5 - 3.1$ to get 6.4

(b) Apply the exponent; multiply $(2.2)(2.2)$ to get 4.84

(c) Add 4.84 plus 6.4 to get 11.24

53.
$$\begin{aligned} & 7.2 - 5.2 + 3.5^2 \\ &= 7.2 - 5.2 + 12.25 && \textit{Exponent} \\ &= 2 + 12.25 && \textit{Subtract.} \\ &= 14.25 && \textit{Add.} \end{aligned}$$

55.
$$\begin{aligned} & 38.6 + 11.6(10.4 - 13.4) \\ &= 38.6 + 11.6[10.4 + (-13.4)] && \textit{Change to addition.} \\ &= 38.6 + 11.6(-3) && \textit{Brackets} \\ &= 38.6 + (-34.8) && \textit{Multiply.} \\ &= 3.8 && \textit{Add.} \end{aligned}$$

57.
$$\begin{aligned} & -8.68 - 4.6(10.4) \div 6.4 \\ &= -8.68 - 47.84 \div 6.4 && \textit{Multiply.} \\ &= -8.68 - 7.475 && \textit{Divide.} \\ &= -16.155 && \textit{Subtract.} \end{aligned}$$

59.
$$\begin{aligned} & 33 - 3.2(0.68 + 9) + (-1.3)^2 \\ &= 33 - 3.2(9.68) + 1.69 && \textit{Parentheses; Exponent} \\ &= 33 - 30.976 + 1.69 && \textit{Multiply.} \\ &= 2.024 + 1.69 && \textit{Subtract.} \\ &= 3.714 && \textit{Add.} \end{aligned}$$

61. (a) $\frac{26{,}000{,}000}{24} \approx 1{,}083{,}333$ pieces each hour

(b) There are $24 \times 60 = 1440$ minutes in a day.

$\frac{26{,}000{,}000}{1440} \approx 18{,}056$ pieces each minute

(c) There are $24 \times 60 \times 60 = 86{,}400$ seconds in a day.

$\frac{26{,}000{,}000}{86{,}400} \approx 301$ pieces each second

63. Divide $10,000 by 10¢.

$$
\begin{array}{r}
1\ 0\ 0,0\ 0\ 0 \\
0.10_\wedge\overline{)1\ 0,0\ 0\ 0.\ 0\ 0_\wedge} \\
\underline{1\ 0} \\
0\ 0\ 0\ 0\ 0
\end{array}
$$

The school would need to collect 100,000 box tops.

65. Divide 100,000 (the answer from Exercise 63) by 38.

$$
\begin{array}{r}
2\ 6\ 3\ 1.\ 5 \\
38\overline{)1\ 0\ 0,0\ 0\ 0.\ 0} \\
\underline{7\ 6} \\
2\ 4\ 0 \\
\underline{2\ 2\ 8} \\
1\ 2\ 0 \\
\underline{1\ 1\ 4} \\
6\ 0 \\
\underline{3\ 8} \\
2\ 2\ 0 \\
\underline{1\ 9\ 0} \\
3\ 0
\end{array}
$$

2631.5 rounds to 2632. The school needs to collect 2632 box tops (rounded) during each of the 38 weeks.

Relating Concepts (Exercises 67–68)

67. $3.77 \div 10 = \underline{0.377}$ $9.1 \div 10 = \underline{0.91}$

$0.886 \div 10 = \underline{0.0886}$ $30.19 \div 10 = \underline{3.019}$

$406.5 \div 10 = \underline{40.65}$ $6625.7 \div 10 = \underline{662.57}$

(a) Dividing by 10, decimal point moves one place to the left; by 100, two places to the left; by 1000, three places to the left.

(b) The decimal point moved to the *right* when multiplying by 10, by 100, or by 1000; here it moves to the *left* when dividing by 10, by 100, or by 1000.

68. $40.2 \div 0.1 = \underline{402}$ $7.1 \div 0.1 = \underline{71}$

$0.339 \div 0.1 = \underline{3.39}$ $15.77 \div 0.1 = \underline{157.7}$

$46 \div 0.1 = \underline{460}$ $873 \div 0.1 = \underline{8730}$

(a) Dividing by 0.1, decimal point moves one place to the right; by 0.01, two places to the right; by 0.001, three places to the right.

(b) The decimal point moved to the *left* when multiplying by 0.1, 0.01, or 0.001; here it moves to the *right* when dividing by 0.1, 0.01, or 0.001.

Summary Exercises
Computation wth Decimal Numbers

1. $0.8 = \dfrac{8}{10} = \dfrac{8 \div 2}{10 \div 2} = \dfrac{4}{5}$

3. $0.35 = \dfrac{35}{100} = \dfrac{35 \div 5}{100 \div 5} = \dfrac{7}{20}$

5. 2.0003 is two and three ten-thousandths.

7. five hundredths:

$$\frac{5}{100} = 0.05$$

9. ten and seven tenths:

$$10\frac{7}{10} = 10.7$$

11. 0.95 to the nearest tenth

Draw a cut-off line: 0.9|5

The first digit cut is 5, which is 5 or more, so round up.

$$
\begin{array}{r}
0.9 \\
+\ 0.1 \\
\hline
1.0
\end{array}
$$

Answer: ≈ 1.0

13. $0.893 to the nearest cent

Draw a cut-off line: $0.89|3

The first digit cut is 3, which is 4 or less. The part you keep stays the same.

Answer: $\approx \$0.89$

15. $99.64 to the nearest dollar

Draw a cut-off line: $99|.64

The first digit cut is 6, which is 5 or more, so round up.

$$
\begin{array}{r}
\$99 \\
+\ 1 \\
\hline
\$100
\end{array}
$$

Answer: $\approx \$100$

17. $50 - 0.3801$

$$
\begin{array}{r}
{\scriptstyle 4\ 9\ 9\ 9\ 9\ 10} \\
\cancel{5}\ \cancel{0}.\cancel{0}\ \cancel{0}\ \cancel{0}\ \cancel{0} \quad \text{\textit{Write four zeros.}} \\
-\ 0.3\ 8\ 0\ 1 \\
\hline
4\ 9\ .6\ 1\ 9\ 9
\end{array}
$$

Check
$$
\begin{array}{r}
0.3801 \\
+\ 49.6199 \\
\hline
50.0000
\end{array}
$$

19. $\dfrac{-90.18}{-6}$

Same signs; positive quotient

$$
\begin{array}{r}
1\ 5.\ 0\ 3 \\
6\overline{)9\ 0.\ 1\ 8} \\
\underline{6} \\
3\ 0 \\
\underline{3\ 0} \\
0\ 1 \\
\underline{0} \\
1\ 8 \\
\underline{1\ 8} \\
0
\end{array}
$$

21. $1.55 - 3.7 = 1.55 + (-3.7)$

Subtract absolute values.

$$\begin{array}{r} {\scriptstyle 6\ 10} \\ 3.\not{7}\,\not{0} \\ -\ 1.5\,5 \\ \hline 2.1\,5 \end{array}$$

Since -3.7 has the larger absolute value, the sum is negative.

$$1.55 - 3.7 = -2.15$$

23. $3.6 + 0.718 + 9 + 5.0829$

$$\begin{array}{l} {\scriptstyle 1\ 11} \\ 3.6000 \quad \textit{Line up decimal points.} \\ 0.7180 \quad \textit{Write in zeros.} \\ 9.0000 \\ +\ 5.0829 \\ \hline 18.4009 \end{array}$$

25. $-8.9 + 4^2 \div (-0.02)$

$$\begin{array}{ll} = -8.9 + 16 \div (-0.02) & \textit{Exponent} \\ = -8.9 - 800 & \textit{Divide.} \\ = -808.9 & \textit{Subtract.} \end{array}$$

27. $0.64 \div 16.3 \approx 0.039 \approx 0.04$

29. To find the perimeter, add the lengths of all the sides.

$$\begin{array}{r} 2.000 \\ 1.000 \\ 1.700 \\ 0.860 \\ 2.095 \\ 1.180 \\ +\ 0.900 \\ \hline 9.735\ \text{meters} \end{array}$$

The perimeter is 9.735 meters.

31. Find their total revenue.

$$\begin{array}{r} {\scriptstyle 77\ 4} \\ \$18.95 \quad \textit{price per blanket} \\ \times\quad 8 \quad \textit{number of blankets} \\ \hline \$151.60 \quad \textit{total revenue} \end{array}$$

Subtract their cost from the revenue.

$$\begin{array}{r} {\scriptstyle 510} \\ \$151.\not{6}\,\not{0} \quad \textit{total revenue} \\ -\ 60.3\,2 \quad \textit{cost} \\ \hline \$91.2\,8 \quad \textit{profit} \end{array}$$

They made a profit of $91.28.

33. (a) $A = s^2$

$$= (12\ \text{ft})(12\ \text{ft})$$
$$= 144\ \text{ft}^2$$

(b) Cost per square foot:

$$\frac{\$99}{144\ \text{ft}^2} = \$0.6875 \approx \$0.69\ (\text{rounded})$$

35. Multiply 0.004 and 80.

$$\begin{array}{r} 0.004 \\ \times\ 80 \\ \hline 0.320 \end{array}$$

The average weight of food eaten each day by a queen bee is 0.32 ounce.

5.6 Fractions and Decimals

5.6 Margin Exercises

1. (a) $\frac{1}{9}$ is written $9\overline{)1}$.

(b) $\frac{2}{3}$ is written $3\overline{)2}$.

(c) $\frac{5}{4}$ is written $4\overline{)5}$.

(d) $\frac{3}{10}$ is written $10\overline{)3}$.

(e) $\frac{21}{16}$ is written $16\overline{)21}$.

(f) $\frac{1}{50}$ is written $50\overline{)1}$.

2. (a) $\frac{1}{4}$

$$\begin{array}{r} 0.2\,5 \\ 4\overline{)1.0\,0} \\ \underline{8} \\ 2\,0 \\ \underline{2\,0} \\ 0 \end{array}$$

$\frac{1}{4} = 0.25$

(b) $2\frac{1}{2} = \frac{5}{2}$

$$\begin{array}{r} 2.5 \\ 2\overline{)5.0} \\ \underline{4} \\ 1\,0 \\ \underline{1\,0} \\ 0 \end{array}$$

$2\frac{1}{2} = 2.5$

(c) $\frac{5}{8}$

$$\begin{array}{r} 0.6\,2\,5 \\ 8\overline{)5.0\,0\,0} \\ \underline{4\,8} \\ 2\,0 \\ \underline{1\,6} \\ 4\,0 \\ \underline{4\,0} \\ 0 \end{array}$$

$\frac{5}{8} = 0.625$

(d) $4\frac{3}{5}$

$$\begin{array}{r} 0.6 \\ 5\overline{)3.0} \\ \underline{3\,0} \\ 0 \end{array}$$

$4 + 0.6 = 4.6$

$4\frac{3}{5} = 4.6$

(e) $\frac{7}{8}$

$$\begin{array}{r} 0.\,8\,7\,5 \\ 8\overline{\smash)7.\,0\,0\,0} \\ \underline{6\,4} \\ 6\,0 \\ \underline{5\,6} \\ 4\,0 \\ \underline{4\,0} \\ 0 \end{array}$$

$\frac{7}{8} = 0.875$

3.   (a) $\frac{1}{3} = 1 \div 3$ Use a calculator.
Rounded to the nearest thousandth, $\frac{1}{3} \approx 0.333$.

(b) $2\frac{7}{9} = \frac{25}{9} = 25 \div 9$ Use a calculator.
Rounded to the nearest thousandth, $2\frac{7}{9} \approx 2.778$.

(c) $\frac{10}{11} = 10 \div 11$ Use a calculator.
Rounded to the nearest thousandth, $\frac{10}{11} \approx 0.909$.

(d) $\frac{3}{7} = 3 \div 7$ Use a calculator.
Rounded to the nearest thousandth, $\frac{3}{7} \approx 0.429$.

(e) $3\frac{5}{6} = \frac{23}{6} = 23 \div 6$ Use a calculator.
Rounded to the nearest thousandth, $3\frac{5}{6} \approx 3.833$.

4.   (a) On the first number line, 0.4375 is to the *left*
of 0.5, so use the $\boxed{<}$ symbol: $0.4375 < 0.5$

(b) On the first number line, 0.75 is to the *right* of
0.6875, so use the $\boxed{>}$ symbol: $0.75 > 0.6875$

(c) $0.625 > 0.0625$

(d) $\frac{2}{8} = \frac{1}{4} = 0.25$

$\qquad 0.25 < 0.375$, so $\frac{2}{8} < 0.375$.

(e) On the second number line, $0.8\overline{3}$ and $\frac{5}{6}$ are at
the *same point*, so use the $\boxed{=}$ symbol: $0.8\overline{3} = \frac{5}{6}$

(f) $\frac{1}{2} < 0.\overline{5}$ $\left(\text{because } \frac{1}{2} = 0.5\right)$

(g) $0.\overline{1} < 0.1\overline{6}$

(h) $\frac{8}{9} = 0.\overline{8}$

(i) $\frac{4}{6} = \frac{2}{3} = 0.\overline{6}$

$\qquad 0.\overline{7} > 0.\overline{6}$, so $0.\overline{7} > \frac{4}{6}$.

(j) $\frac{1}{4} = 0.25$

5.   (a)   0.7   0.703   0.7029
$\qquad\quad\downarrow\qquad\quad\downarrow\qquad\quad\downarrow$
$\qquad$ 0.7000   0.7030   0.7029

From least to greatest:
0.7000, 0.7029, 0.7030
or
0.7, 0.7029, 0.703

(b)   6.39   6.309   6.401   6.4
$\qquad\downarrow\qquad\quad\downarrow\qquad\quad\downarrow\qquad\quad\downarrow$
$\qquad$ 6.390   6.309   6.401   6.400

From least to greatest:
6.309, 6.390, 6.400, 6.401
or
6.309, 6.39, 6.4, 6.401

(c)   1.085 $1\frac{3}{4}$ 0.9
$\qquad\downarrow\qquad\quad\downarrow\qquad\quad\downarrow$
$\qquad$ 1.085   1.750   0.900

From least to greatest:
0.900, 1.085, 1.750
or
0.9, 1.085, $1\frac{3}{4}$

(d) $\frac{1}{4}, \frac{2}{5}, \frac{3}{7}, 0.428$

To compare, change fractions to decimals.

$$\frac{1}{4} = 0.250 \qquad \frac{2}{5} = 0.400 \qquad \frac{3}{7} \approx 0.429$$

From least to greatest:
0.250, 0.400, 0.428, 0.429
or
$\frac{1}{4}, \frac{2}{5}, 0.428, \frac{3}{7}$

5.6   Section Exercises

1.   (a) incorrect, $\frac{2}{5}$ is written $5\overline{\smash)2}$.

(b) correct

(c) correct

3. $\dfrac{3}{4} \rightarrow \begin{array}{r} 0. \\ 4\overline{\smash)3.} \end{array}$

To continue dividing, write zeros in the dividend.

5.   (a) $\frac{2}{5}$ ($= 0.4$) is <u>less than</u> 0.5.

(b) $\frac{3}{4}$ ($= 0.75$) is <u>greater than</u> 0.6.

(c) 0.2 is <u>less than</u> $\frac{5}{8}$ ($= 0.625$).

7. $\dfrac{1}{2} = 0.5$ $\qquad \begin{array}{r} 0.\,5 \\ 2\overline{\smash)1.\,0} \\ \underline{1\,0} \\ 0 \end{array}$

9. $\dfrac{3}{4} = 0.75$ $\qquad \begin{array}{r} 0.\,7\,5 \\ 4\overline{\smash)3.\,0\,0} \\ \underline{2\,8} \\ 2\,0 \\ \underline{2\,0} \\ 0 \end{array}$

11. $\dfrac{3}{10} = 0.3$ $\qquad \begin{array}{r} 0.\,3 \\ 10\overline{\smash)3.\,0} \\ \underline{3\,0} \\ 0 \end{array}$

13. $\frac{9}{10} = 0.9$

$$10\overline{)9.0}$$
$$\underline{9\ 0}$$
$$0$$

15. $\frac{3}{5} = 0.6$

$$5\overline{)3.0}$$
$$\underline{3\ 0}$$
$$0$$

17. $\frac{7}{8} = 0.875$

$$8\overline{)7.000}$$
$$\underline{6\ 4}$$
$$6\ 0$$
$$\underline{5\ 6}$$
$$4\ 0$$
$$\underline{4\ 0}$$
$$0$$

19. $2\frac{1}{4} = \frac{9}{4}$

$$4\overline{)9.00}$$
$$\underline{8}$$
$$1\ 0$$
$$\underline{8}$$
$$2\ 0$$
$$\underline{2\ 0}$$
$$0$$

21. $14\frac{7}{10} = 14 + \frac{7}{10} = 14 + 0.7 = 14.7$

23. $3\frac{5}{8} = \frac{29}{8} = 3.625$

$$8\overline{)29.000}$$
$$\underline{2\ 4}$$
$$5\ 0$$
$$\underline{4\ 8}$$
$$2\ 0$$
$$\underline{1\ 6}$$
$$4\ 0$$
$$\underline{4\ 0}$$
$$0$$

25. $6\frac{1}{3} = \frac{19}{3} \approx 6.3333333 \approx 6.333$

27. $\frac{5}{6} \approx 0.8333333 \approx 0.833$

29. $1\frac{8}{9} = \frac{17}{9} \approx 1.8888889 \approx 1.889$

31. $0.4 = \frac{4}{10} = \frac{4 \div 2}{10 \div 2} = \frac{2}{5}$

33. $0.625 = \frac{625}{1000} = \frac{625 \div 125}{1000 \div 125} = \frac{5}{8}$

35. $0.35 = \frac{35}{100} = \frac{35 \div 5}{100 \div 5} = \frac{7}{20}$

37. $\frac{7}{20} = 0.35$

$$20\overline{)7.00}$$
$$\underline{6\ 0}$$
$$1\ 00$$
$$\underline{1\ 00}$$
$$0$$

39. $0.04 = \frac{4}{100} = \frac{4 \div 4}{100 \div 4} = \frac{1}{25}$

41. $\frac{1}{5} = 0.2$

$$5\overline{)1.0}$$
$$\underline{1\ 0}$$
$$0$$

43. $0.09 = \frac{9}{100}$

45. Compare the two lengths.
average length $\rightarrow$ 20.80 *longer*
Charlene's baby $\rightarrow$ 20.08 *shorter*

$$\begin{array}{r} 20.80 \\ -\ 20.08 \\ \hline 0.72 \end{array}$$

Her baby is 0.72 inch *shorter* than the average length.

47. Compare the two amounts.

$3\frac{3}{4} \rightarrow 3.75$ *less*
$3.8 \rightarrow 3.80$ *more*

$$\begin{array}{r} 3.80 \\ -\ 3.75 \\ \hline 0.05 \end{array}$$

3.8 inches is 0.05 inch *more* than Ginny had hoped for.

49. Write two zeros to the right of 0.5 so it has the same number of decimal places as 0.505. Then you can compare the numbers: $0.505 > 0.500$. There was too much calcium in each capsule. Subtract to find the difference.

$$\begin{array}{r} 0.505 \\ -\ 0.500 \\ \hline 0.005 \end{array}$$

There was 0.005 gram *too much*.

51. Write two zeros so that all the numbers have four decimal places.

$1.0100 > 1.0020$ *unacceptable*
$0.9991 > 0.9980$ and $0.9991 < 1.0020$ *acceptable*
$1.0007 > 0.9980$ and $1.0007 < 1.0020$ *acceptable*
$0.9900 < 0.9980$ *unacceptable*

The lengths of 0.9991 cm and 1.0007 cm are acceptable.

53. **(a)** On the first number line, 0.3125 is to the *left* of 0.375, so use the $\boxed{<}$ symbol: $0.3125 < 0.375$

(b) Write $\frac{6}{8}$ in lowest terms as $\frac{3}{4}$. On the first number line, $\frac{3}{4}$ and 0.75 are at the *same point*, so use the $\boxed{=}$ symbol: $\frac{3}{4} = 0.75$

(c) On the second number line, $0.\overline{8}$ is to the *right* of $0.8\overline{3}$, so use the $\boxed{>}$ symbol: $0.\overline{8} > 0.8\overline{3}$

(d) On the second number line, 0.5 is to the *left* of $\frac{5}{9}$, so use the $\boxed{<}$ symbol: $0.5 < \frac{5}{9}$

55. $0.54, 0.5455, 0.5399$

$$0.54 = 0.5400$$
$$0.5455 = 0.5455 \quad \textit{greatest}$$
$$0.5399 = 0.5399 \quad \textit{least}$$

From least to greatest: $0.5399, 0.54, 0.5455$

57. $5.8, 5.79, 5.0079, 5.804$

$$5.8 = 5.8000$$
$$5.79 = 5.7900$$
$$5.0079 = 5.0079 \quad \textit{least}$$
$$5.804 = 5.8040 \quad \textit{greatest}$$

From least to greatest: $5.0079, 5.79, 5.8, 5.804$

59. $0.628, 0.62812, 0.609, 0.6009$

$$0.628 = 0.62800$$
$$0.62812 = 0.62812 \quad \textit{greatest}$$
$$0.609 = 0.60900$$
$$0.6009 = 0.60090 \quad \textit{least}$$

From least to greatest:
$0.6009, 0.609, 0.628, 0.62812$

61. $5.8751, 4.876, 2.8902, 3.88$

The numbers after the decimal places are irrelevant since the whole number parts are all different.

From least to greatest:
$2.8902, 3.88, 4.876, 5.8751$

63. $0.043, 0.051, 0.006, \frac{1}{20}$

$$0.043 = 0.043$$
$$0.051 = 0.051 \quad \textit{greatest}$$
$$0.006 = 0.006 \quad \textit{least}$$
$$\frac{1}{20} = 0.050$$

From least to greatest: $0.006, 0.043, \frac{1}{20}, 0.051$

65. $\frac{3}{8}, \frac{2}{5}, 0.37, 0.4001$

$$\frac{3}{8} = 0.3750$$
$$\frac{2}{5} = 0.4000$$
$$0.37 = 0.3700 \quad \textit{least}$$
$$0.4001 = 0.4001 \quad \textit{greatest}$$

From least to greatest: $0.37, \frac{3}{8}, \frac{2}{5}, 0.4001$

67. **(a)** Find the greatest of:
$0.018, 0.01, 0.008, 0.010$

$$0.018 = 0.018 \quad \textit{greatest}$$
$$0.01 = 0.010$$
$$0.008 = 0.008 \quad \textit{least}$$
$$0.010 = 0.010$$

List from least to greatest:

$$0.008, 0.01 = 0.010, 0.018$$

The red box, labeled 0.018 in. diameter, has the strongest line.

The green box, labeled 0.008 in. diameter, has the line with the least strength.

(b) Subtract 0.008 from 0.018.

$$\begin{array}{r} 0.018 \\ -\ 0.008 \\ \hline 0.010 \end{array}$$

The difference in diameter is 0.01 inch.

69. $1\frac{7}{16} = \frac{23}{16} \approx 1.43$. Length (a) is 1.4 inches (rounded to the nearest tenth).

71. $\frac{1}{4} = 0.25 \approx 0.3$. Length (c) is 0.3 inch (rounded to the nearest tenth).

73. $\frac{3}{8} = 0.375 \approx 0.4$. Length (e) is 0.4 inch (rounded to the nearest tenth).

Relating Concepts (Exercises 75–78)

75. **(a)** A proper fraction like $\frac{5}{9}$ is less than 1, so $\frac{5}{9}$ cannot be equivalent to a decimal number that is greater than 1.

(b) $\frac{5}{9}$ means $5 \div 9$ or $9\overline{)5}$, so the correct answer is 0.556 (rounded). This makes sense because both the fraction and decimal are less than 1.

$$\begin{array}{r} 0.\ 5\ 5\ 5\ 5 \\ 9\overline{)5.\ 0\ 0\ 0\ 0} \\ \underline{4\ 5} \\ 5\ 0 \\ \underline{4\ 5} \\ 5\ 0 \\ \underline{4\ 5} \\ 5\ 0 \\ \underline{4\ 5} \\ 5 \end{array}$$

76. **(a)** $2.035 = 2\frac{35}{1000} = 2\frac{7}{200}$, not $2\frac{7}{20}$.

(b) Adding the whole number part gives $2 + 0.35$, which is 2.35, not 2.035. To check, $2.35 = 2\frac{35}{100} = 2\frac{7}{20}$ but $2.035 = 2\frac{35}{1000} = 2\frac{7}{200}$.

77. Just add the whole number part to 0.375. So $1\frac{3}{8} = 1.375$; $3\frac{3}{8} = 3.375$; $295\frac{3}{8} = 295.375$.

78. It works only when the fraction part has a one-digit numerator and a denominator of 10, or a two-digit numerator and a denominator of 100, and so on.

5.7 Problem Solving with Statistics: Mean, Median, and Mode

5.7 Margin Exercises

1. $$\text{mean} = \frac{\text{sum of all values}}{\text{number of values}}$$

    $$\text{mean} = \frac{95 + 91 + 81 + 78 + 81 + 90}{6}$$

    $$\text{mean} = \frac{516}{6}$$

    $$\text{mean} = 86$$

2. **(a)** sum of all values

    $= \$25.12 + \$42.58 + \$76.19 + \$32$
    $+ \$81.11 + \$26.41 + \$49.76 + \$59.32$
    $+ \$71.18 + \$30.09 + \$60.50 + \$79.84$
    $= \$634.10$

    $$\text{mean} = \frac{\$634.10}{12} \approx \$52.84 \text{ (rounded)}$$

    His average monthly cell phone bill was $52.84.

    (b) sum of all values

    $= \$749,820 + \$765,480 + \$643,744$
    $+ \$824,222 + \$485,886 + \$668,178$
    $+ \$702,294 + \$525,800$
    $= \$5,365,424$

    $$\text{mean} = \frac{\$5,365,424}{8} = \$670,678$$

    The average sales for each office supply store were $670,678.

3.
Parking Fee	Frequency	Product
$6	2	$12
$7	6	$42
$8	3	$24
$9	4	$36
$10	6	$60
	21	$174

    $$\text{weighted mean} = \frac{\$174}{21} \approx \$8.29$$

    Her average daily parking cost was about $8.29.

4.
Course	Credits	Grade	Credits · Grade
Math	5	A (= 4)	5·4 = 20
English	3	C (= 2)	3·2 = 6
Biology	4	B (= 3)	4·3 = 12
History	3	B (= 3)	3·3 = 9
Totals	15		47

    $$\text{GPA} = \frac{\text{sum of Credits} \cdot \text{Grade}}{\text{total number of credits}}$$

    $$= \frac{47}{15} \approx 3.13 \text{ (rounded)}$$

5. Arrange the 9 numbers in numerical order from least to greatest.

    25, 27, 30, 30, 33, 35, 39, 50, 59

    The middle number is the fifth number, which is 33. The median is 33 students.

6. Arrange in numerical order.

    121, 121, 126, 178, 189, 195, 200, 261

    The middle values are 178 and 189. The median is the mean of the two middle values.

    $$\text{median} = \frac{178 \text{ ft} + 189 \text{ ft}}{2} = 183.5 \text{ feet}$$

7. **(a)** 312, 219, 782, 312, 219, 426, 507, 600

    Because both 219 and 312 occur two times, which is more than any other number, each is a mode. This list is *bimodal.*

    (b) 28, 21, 16, 22, 28, 34, 22, 28, 19, 18

    The only number that occurs three times, which is more than any other number, is 28, so the mode is 28 years.

    (c) $1706, $1989, $1653, $1892, $2001, $1782, $1450, $1566

    No number occurs more than once. This list has *no mode.*

5.7 Section Exercises

1. **(a)** Another name for "mean" is <u>average</u>.

    (b) To find the mean, add all the values and then divide by <u>the number of values</u>.

3. $$\text{mean} = \frac{\text{sum of all values}}{\text{number of values}}$$

    $$\text{mean} = \frac{92 + 51 + 59 + 86 + 68 + 73 + 49 + 80}{8}$$

    $$\text{mean} = \frac{558}{8} = 69.75$$

    The mean (average) final exam score was <u>69.8</u> (rounded).

5. $$\text{mean} = \frac{\text{sum of all values}}{\text{number of values}}$$

    $$\text{mean} = \frac{(\$31,900 + 32,850 + 34,930 + 39,712 + 38,340 + 60,000)}{6}$$

    $$\text{mean} = \frac{\$237,732}{6} = \$39,622$$

    The mean (average) annual salary was $39,622.

7.

Quiz Score	Frequency	Product
3	4	12
5	2	10
6	5	30
8	5	40
9	2	18
Totals	18	110

$$\text{weighted mean} = \frac{\text{sum of products}}{\text{total number of quizzes}}$$

$$= \frac{110}{18} = 6.\overline{1}$$

The mean (average) quiz score was 6.1 (rounded).

9.

Course	Credits	Grade	Credits·Grade
Biology	4	B (= 3)	$4 \cdot 3 = 12$
Biology Lab	2	A (= 4)	$2 \cdot 4 = 8$
Mathematics	5	C (= 2)	$5 \cdot 2 = 10$
Health	1	F (= 0)	$1 \cdot 0 = 0$
Psychology	3	B (= 3)	$3 \cdot 3 = 9$
Totals	15		39

$$\text{GPA} = \frac{\text{sum of Credits} \cdot \text{Grade}}{\text{total number of credits}}$$

$$= \frac{39}{15} = 2.60$$

11. **(a)** In Exercise 9, replace $1 \cdot 0$ with $1 \cdot 3$ to get

$$\text{GPA} = \frac{42}{15} = 2.80.$$

(b) In Exercise 9, replace $5 \cdot 2$ with $5 \cdot 3$ to get

$$\text{GPA} = \frac{44}{15} = 2.9\overline{3} = 2.93 \text{ (rounded)}.$$

(c) Making both of those changes gives us

$$\text{GPA} = \frac{47}{15} = 3.1\overline{3} = 3.13 \text{ (rounded)}.$$

13. The numbers are already arranged in numerical order from least to greatest.

$$9, 12, 14, \underline{15}, 23, 24, 28$$

The list has 7 numbers. The middle number is the 4th number, so the median is $\underline{15}$ text messages.

15. Arrange the numbers in numerical order from least to greatest.

$$328, 420, \underline{483}, \underline{549}, 592, 715$$

The list has 6 numbers. The middle numbers are the 3rd and 4th numbers, so the median is

$$\frac{483 + 549}{2} = 516 \text{ students.}$$

17. Arrange the numbers in numerical order from least to greatest.

$$34, 40, 40, 47, \underline{48}, \underline{49}, 51, 56, 95, 96$$

The list has 10 numbers. The middle numbers are the 5th and 6th numbers, so the median is

$$\frac{48 + 49}{2} = 48.5 \text{ pounds of shrimp.}$$

19. mean

$$= \frac{\text{sum of all values}}{\text{number of values}}$$

$$= \frac{2500 + 3009 + 7767 + 8357 + 1740 + 2400}{6}$$

$$= \frac{25{,}773}{6} = 4295.5 \text{ miles}$$

The mean distance flown without refueling was 4296 miles (rounded).

21. $1740, 2400, \underline{2500}, \underline{3009}, 7767, 8357$

There are 6 distances. The middle numbers are the 3rd and 4th numbers, so the median is

$$\frac{2500 + 3009}{2} = 2754.5 \text{ miles.}$$

The median distance flown without refueling was 2754.5 miles.

23. $3, \underline{8}, 5, 1, 7, 6, \underline{8}, 4, 5, \underline{8}$

The number 8 occurs three times, which is more often than any other number. Therefore, 8 samples is the mode.

25. $\underline{74}, \underline{68}, \underline{68}, \underline{68}, 75, 75, \underline{74}, \underline{74}, 70, 77$

Because both 68 and 74 years occur three times, which is more often than any other values, each is a mode. This list is *bimodal.*

27. $5, 9, 17, 3, 2, 8, 19, 1, 4, 20, 10, 6$

No number occurs more than once. This list has *no mode.*

29. **(i)** Barrow's

$$\text{mean} = \frac{\begin{array}{c}-2 + (-11) + (-13) \\ + (-18) + (-15) + (-2)\end{array}}{6}$$

$$= \frac{-61}{6} \approx -10°F$$

Fairbanks'

$$\text{mean} = \frac{3 + (-7) + (-10) + (-4) + 11 + 31}{6}$$

$$= \frac{24}{6} = 4°F$$

(ii) Find the difference: $4 - (-10) = 14$

Fairbanks' mean is 14 degrees warmer than Barrow's mean.

5.8   Geometry Applications: Pythagorean Theorem and Square Roots

5.8   Margin Exercises

1.   (a) $\sqrt{36} = \underline{6}$ because $6 \cdot 6 = 36$.

   (b) $\sqrt{25} = 5$ because $5 \cdot 5 = 25$.

   (c) $\sqrt{9} = 3$ because $3 \cdot 3 = 9$.

   (d) $\sqrt{100} = 10$ because $10 \cdot 10 = 100$.

   (e) $\sqrt{121} = 11$ because $11 \cdot 11 = 121$.

2.   (a) $\sqrt{11}$ ▪   Calculator shows 3.31662479; round to 3.317.

   (b) $\sqrt{40}$ ▪   Calculator shows 6.32455532; round to 6.325.

   (c) $\sqrt{56}$ ▪   Calculator shows 7.48331477; round to 7.483.

   (d) $\sqrt{196}$ ▪   Calculator shows 14; $\sqrt{196} = 14$ because $14 \cdot 14 = 196$.

   (e) $\sqrt{147}$ ▪   Calculator shows 12.12435565; round to 12.124.

3.   (a) legs: 5 in. and 12 in.

$$\text{hypotenuse} = \sqrt{(\text{leg})^2 + (\text{leg})^2}$$
$$\text{hypotenuse} = \sqrt{(5)^2 + (12)^2}$$
$$= \sqrt{25 + \underline{144}}$$
$$= \sqrt{\underline{169}}$$
$$= \underline{13 \text{ in.}}$$

   (b) hypotenuse: 25 cm, leg: 7 cm

$$\text{leg} = \sqrt{(\text{hypotenuse})^2 - (\text{leg})^2}$$
$$\text{leg} = \sqrt{(25)^2 - (7)^2}$$
$$= \sqrt{625 - \underline{49}}$$
$$= \sqrt{\underline{576}}$$
$$= \underline{24 \text{ cm}}$$

   (c) legs: 17 m and 13 m

$$\text{hypotenuse} = \sqrt{(\text{leg})^2 + (\text{leg})^2}$$
$$\text{hypotenuse} = \sqrt{(17)^2 + (13)^2}$$
$$= \sqrt{289 + 169}$$
$$= \sqrt{458}$$
$$\approx 21.4 \text{ m}$$

   (d) hypotenuse: 20 ft, leg: 18 ft

$$\text{leg} = \sqrt{(\text{hypotenuse})^2 - (\text{leg})^2}$$
$$\text{leg} = \sqrt{(20)^2 - (18)^2}$$
$$= \sqrt{400 - 324}$$
$$= \sqrt{76}$$
$$\approx 8.7 \text{ ft}$$

4.   (a) hypotenuse: 25 ft, leg: 20 ft

$$\text{leg} = \sqrt{(\text{hypotenuse})^2 - (\text{leg})^2}$$
$$\text{leg} = \sqrt{(25)^2 - (20)^2}$$
$$= \sqrt{625 - 400}$$
$$= \sqrt{225}$$
$$= 15$$

The bottom of the ladder is 15 ft from the building.

   (b) legs: 11 ft and 8 ft

$$\text{hypotenuse} = \sqrt{(\text{leg})^2 + (\text{leg})^2}$$
$$\text{hypotenuse} = \sqrt{(11)^2 + (8)^2}$$
$$= \sqrt{121 + 64}$$
$$= \sqrt{185}$$
$$\approx 13.6 \text{ ft}$$

The ladder is about 13.6 ft long.

   (c) hypotenuse: 17 ft, leg: 10 ft

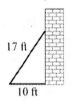

$$\text{leg} = \sqrt{(\text{hypotenuse})^2 - (\text{leg})^2}$$
$$\text{leg} = \sqrt{(17)^2 - (10)^2}$$
$$= \sqrt{289 - 100}$$
$$= \sqrt{189}$$
$$\approx 13.7 \text{ ft}$$

The ladder will reach about 13.7 ft high up on the building.

5.8   Section Exercises

1. $\sqrt{16} = 4$ because $4 \cdot 4 = 16$.

3. $\sqrt{64} = 8$ because $8 \cdot 8 = 64$.

5. $\sqrt{11}$ ▪   Calculator shows 3.31662479; rounds to 3.317.

7. $\sqrt{5}$ ▪   Calculator shows 2.236067977; rounds to 2.236.

9. $\sqrt{73}$ ▪   Calculator shows 8.544003745; rounds to 8.544.

11. $\sqrt{101}$ ▪   Calculator shows 10.04987562; rounds to 10.050.

13. On a calculator, $\sqrt{361} = 19$.

15. $\sqrt{1000}$ ▪   Calculator shows 31.6227766; rounds to 31.623.

17. 30 is about halfway between 25 and 36, so $\sqrt{30}$ should be about halfway between 5 and 6, or about 5.5. Using a calculator, $\sqrt{30} \approx 5.477$. Similarly, $\sqrt{26}$ should be a little more than $\sqrt{25}$; by calculator, $\sqrt{26} \approx 5.099$. And $\sqrt{35}$ should be a little less than $\sqrt{36}$; by calculator, $\sqrt{35} \approx 5.916$.

19. legs: 15 ft and 36 ft

$$\text{hypotenuse} = \sqrt{(\text{leg})^2 + (\text{leg})^2}$$
$$\text{hypotenuse} = \sqrt{(15)^2 + (36)^2}$$
$$= \sqrt{225 + 1296}$$
$$= \sqrt{1521}$$
$$= 39 \text{ ft}$$

21. legs: 8 in. and 15 in.

$$\text{hypotenuse} = \sqrt{(\text{leg})^2 + (\text{leg})^2}$$
$$\text{hypotenuse} = \sqrt{(8)^2 + (15)^2}$$
$$= \sqrt{64 + 225}$$
$$= \sqrt{289}$$
$$= 17 \text{ in.}$$

23. hypotenuse: 20 mm, leg: 16 mm

$$\text{leg} = \sqrt{(\text{hypotenuse})^2 - (\text{leg})^2}$$
$$\text{leg} = \sqrt{(20)^2 - (16)^2}$$
$$= \sqrt{400 - 256}$$
$$= \sqrt{144}$$
$$= 12 \text{ mm}$$

25. legs: 8 in. and 3 in.

$$\text{hypotenuse} = \sqrt{(\text{leg})^2 + (\text{leg})^2}$$
$$\text{hypotenuse} = \sqrt{(8)^2 + (3)^2}$$
$$= \sqrt{64 + 9}$$
$$= \sqrt{73}$$
$$\approx 8.5 \text{ in.}$$

27. legs: 7 yd and 4 yd

$$\text{hypotenuse} = \sqrt{(\text{leg})^2 + (\text{leg})^2}$$
$$\text{hypotenuse} = \sqrt{(7)^2 + (4)^2}$$
$$= \sqrt{49 + 16}$$
$$= \sqrt{65}$$
$$\approx 8.1 \text{ yd}$$

29. hypotenuse: 22 cm, leg: 17 cm

$$\text{leg} = \sqrt{(\text{hypotenuse})^2 - (\text{leg})^2}$$
$$\text{leg} = \sqrt{(22)^2 - (17)^2}$$
$$= \sqrt{484 - 289}$$
$$= \sqrt{195}$$
$$\approx 14.0 \text{ cm}$$

31. legs: 1.3 m and 2.5 m

$$\text{hypotenuse} = \sqrt{(\text{leg})^2 + (\text{leg})^2}$$
$$\text{hypotenuse} = \sqrt{(1.3)^2 + (2.5)^2}$$
$$= \sqrt{1.69 + 6.25}$$
$$= \sqrt{7.94}$$
$$\approx 2.8 \text{ m}$$

33. hypotenuse: 11.5 cm, leg: 8.2 cm

$$\text{leg} = \sqrt{(\text{hypotenuse})^2 - (\text{leg})^2}$$
$$\text{leg} = \sqrt{(11.5)^2 - (8.2)^2}$$
$$= \sqrt{132.25 - 67.24}$$
$$= \sqrt{65.01}$$
$$\approx 8.1 \text{ cm}$$

35. hypotenuse: 21.6 km, leg: 13.2 km

$$\text{leg} = \sqrt{(\text{hypotenuse})^2 - (\text{leg})^2}$$
$$\text{leg} = \sqrt{(21.6)^2 - (13.2)^2}$$
$$= \sqrt{466.56 - 174.24}$$
$$= \sqrt{292.32}$$
$$\approx 17.1 \text{ km}$$

37. The student used the formula for finding the hypotenuse but the unknown side is a leg, so $\text{leg} = \sqrt{(20)^2 - (13)^2}$. Also, the final answer should be m, not m^2. The correct answer is $\sqrt{231} \approx 15.2$ m.

39. legs: 4 ft and 7 ft

$$\text{hypotenuse} = \sqrt{(\text{leg})^2 + (\text{leg})^2}$$
$$\text{hypotenuse} = \sqrt{(4)^2 + (7)^2}$$
$$= \sqrt{16 + 49}$$
$$= \sqrt{65}$$
$$\approx 8.1 \text{ ft}$$

The length of the loading ramp is about 8.1 ft.

41. hypotenuse: 1000 m, leg: 800 m

$$\text{leg} = \sqrt{(\text{hypotenuse})^2 - (\text{leg})^2}$$
$$\text{leg} = \sqrt{(1000)^2 - (800)^2}$$
$$= \sqrt{1,000,000 - 640,000}$$
$$= \sqrt{360,000}$$
$$= 600 \text{ m}$$

The airplane is 600 meters above the ground.

43. legs: 6.5 ft and 2.5 ft

$$\text{hypotenuse} = \sqrt{(\text{leg})^2 + (\text{leg})^2}$$
$$\text{hypotenuse} = \sqrt{(6.5)^2 + (2.5)^2}$$
$$= \sqrt{42.25 + 6.25}$$
$$= \sqrt{48.5}$$
$$\approx 6.96 \text{ ft}$$

The diagonal brace is about 7.0 ft long.

45.

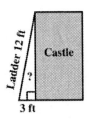

3 ft

hypotenuse: 12 ft, leg: 3 ft

$$\text{leg} = \sqrt{(\text{hypotenuse})^2 - (\text{leg})^2}$$
$$\text{leg} = \sqrt{(12)^2 - (3)^2}$$
$$= \sqrt{144 - 9}$$
$$= \sqrt{135}$$
$$\approx 11.6 \text{ ft}$$

The ladder will reach about 11.6 ft high on the castle.

Relating Concepts (Exercises 47–50)

47.   legs: 90 ft and 90 ft

$$\text{hypotenuse} = \sqrt{(\text{leg})^2 + (\text{leg})^2}$$
$$\text{hypotenuse} = \sqrt{(90)^2 + (90)^2}$$
$$= \sqrt{8100 + 8100}$$
$$= \sqrt{16{,}200}$$
$$\approx 127.3 \text{ ft}$$

The distance from home plate to second base is about 127.3 ft.

48. (a)

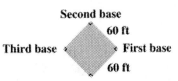

Second base
60 ft
Third base   First base
60 ft
Home plate

(b) legs: 60 ft and 60 ft

$$\text{hypotenuse} = \sqrt{(\text{leg})^2 + (\text{leg})^2}$$
$$\text{hypotenuse} = \sqrt{(60)^2 + (60)^2}$$
$$= \sqrt{3600 + 3600}$$
$$= \sqrt{7200}$$
$$\approx 84.9 \text{ ft}$$

The distance from home plate to second base is about 84.9 ft.

49.   The distance from third to first is the same as the distance from home to second because the baseball diamond is a square.

50. (a) Since 80 ft < 84.9 ft, then the side length is less than 60 ft.

(b) $80^2 = 6400$

$$\frac{6400}{2} = 3200$$

$$\sqrt{3200} \approx 56.6$$

The side length of each side is about 56.6 ft.

5.9 Problem Solving: Equations Containing Decimals

5.9 Margin Exercises

1.   (a) $8.1 = h + 9$ *To get h by itself,*
$$\underline{ -9 \qquad\qquad -9}\ \ \textit{add −9 to both sides.}$$
$$-0.9 = \underbrace{h + 0}$$
$$-0.9 = \quad h$$

Check $8.1 = h + 9$
$$8.1 = -0.9 + 9 \quad \textit{Replace h with −0.9.}$$
$$8.1 = 8.1 \qquad\quad \text{Balances}$$

The solution is −0.9.

(b) $-0.75 + y = 0$
$$\underline{+0.75 \qquad\qquad +0.75}\ \ \textit{Add 0.75 to both sides.}$$
$$0 + y = 0.75$$
$$y = 0.75$$

Check $-0.75 + y = 0$
$$-0.75 + 0.75 = 0 \quad \textit{Replace y with 0.75.}$$
$$0 = 0 \quad \text{Balances}$$

The solution is 0.75.

(c) $c - 6.8 = -4.8$
$$c + (-6.8) = -4.8$$
$$\underline{+6.8 \qquad\quad +6.8}\ \ \textit{Add 6.8 to both sides.}$$
$$c + 0 = 2$$
$$c = 2$$

Check $c - 6.8 = -4.8$
$$2 - 6.8 = -4.8 \quad \textit{Replace c with 2.}$$
$$2 + (-6.8) = -4.8$$
$$-4.8 = -4.8 \quad \text{Balances}$$

The solution is 2.

2.   (a) $-3y = -0.63$
$$\frac{-3y}{-3} = \frac{-0.63}{-3} \quad \textit{Divide both sides by the coefficient, −3.}$$
$$\frac{\overset{1}{-\cancel{3}} \cdot y}{\underset{1}{-\cancel{3}}} = 0.21 \quad \textit{On the left, divide out the common factor of −3.}$$
$$y = \underline{0.21}$$

Check $-3y = -0.63$
$$-3(\underline{0.21}) = -0.63 \quad \textit{Replace y with 0.21.}$$
$$\underline{-0.63 = -0.63} \quad \text{Balances}$$

The solution is 0.21.

(b) $2.25r = -18$

$\dfrac{2.25r}{2.25} = \dfrac{-18}{2.25}$ *Divide both sides by 2.25.*

$r = -8$

Check $2.25r = -18$

$2.25(-8) = -18$ *Replace r with −8.*

$-18 = -18$ Balances

The solution is −8.

(c) $1.7 = 0.5n$

$\dfrac{1.7}{0.5} = \dfrac{0.5n}{0.5}$ *Divide both sides by 0.5.*

$3.4 = n$

Check $1.7 = 0.5n$

$1.7 = 0.5(3.4)$ *Replace n with 3.4.*

$1.7 = 1.7$ Balances

The solution is 3.4.

3.   (a) $4 = 0.2c - 2.6$

$4 = 0.2c + (-2.6)$

$\underline{+2.6 \qquad\qquad +2.6}$ *Add 2.6 to both sides.*

$6.6 = 0.2c + 0$

$\dfrac{6.6}{0.2} = \dfrac{0.2c}{0.2}$ *Divide both sides by 0.2.*

$33 = c$

Check $4 = 0.2c - 2.6$

$4 = 0.2(33) - 2.6$ *Replace c with 33.*

$4 = 6.6 - 2.6$

$4 = 4$ Balances

The solution is 33.

(b) $3.1k - 4 = 0.5k + 13.42$

$\underline{-0.5k \qquad\quad -0.5k}$ *Add −0.5k to both sides.*

$2.6k - 4 = 0 + 13.42$

$2.6k + (-4) = 13.42$

$\underline{+4 \qquad +4}$ *Add 4 to both sides.*

$2.6k + 0 = 17.42$

$2.6k = 17.42$

$\dfrac{2.6k}{2.6} = \dfrac{17.42}{2.6}$ *Divide both sides by 2.6.*

$k = 6.7$

Check $3.1k - 4 = 0.5k + 13.42$

$3.1(6.7) - 4 = 0.5(6.7) + 13.42$

$16.77 = 16.77$ Balances

The solution is 6.7.

(c) $-2y + 3 = 3y - 6$

$\underline{+2y \qquad\qquad +2y}$ *Add 2y to both sides.*

$0 + 3 = 5y - 6$

$3 = 5y + (-6)$

$\underline{+6 \qquad\qquad +6}$ *Add 6 to both sides.*

$9 = 5y + 0$

$\dfrac{9}{5} = \dfrac{5y}{5}$ *Divide both sides by 5.*

$1.8 = y$

Check $-2y + 3 = 3y - 6$

$-2(1.8) + 3 = 3(1.8) - 6$

$-3.6 + 3 = 5.4 - 6$

$-0.6 = -0.6$ Balances

The solution is 1.8.

4.   *Step 1*   It is about the cost of a telephone call.

Unknown: number of minutes the call lasted

Known: Costs are $1 connection fee plus $0.06 per minute; total cost was $4.

Step 2(a)   There is only one unknown; so let m be the number of minutes.

Step 3

cost per minute		number of minutes		connection fee		total cost
0.06	·	m	+	1	=	4

Step 4 $0.06m + 1 = 4$

$\underline{\qquad -1 \qquad -1}$

$0.06m + 0 = 3$

$\dfrac{0.06m}{0.06} = \dfrac{3}{0.06}$

$m = 50$

Step 5   The call lasted 50 minutes.

Step 6 $0.06 per minute times 50 minutes = $3

$3 plus $1 connection fee = $4 (matches)

5.9   Section Exercises

1.   Add 6.2 to both sides because $-6.2 + 6.2$ gives $x + 0$ on the right side.

3. $-20.6 + n = -22$

$\underline{+20.6 \qquad\qquad +20.6}$ *Add 20.6 to both sides.*

$0 + n \qquad -1.4$

$\underline{n = -1.4}$

Check $-20.6 + n = -22$

$-20.6 + (-1.4) = -22$

$-22 = -22$ Balances

The solution is −1.4.

5.
$$0 = b - 0.008$$

$\underline{+0.008 \qquad +0.008}$ *Add 0.008 to both sides.*

$$0.008 = b + 0$$
$$0.008 = b$$

Check $0 = b - 0.008$
$$0 = 0.008 - 0.008$$
$$0 = 0 \qquad \text{Balances}$$

The solution is 0.008.

7. $2.03 = 7a$

$\dfrac{2.03}{7} = \dfrac{7a}{7}$ *Divide both sides by 7.*

$0.29 = a$

Check $2.03 = 7a$
$$2.03 = 7(0.29)$$
$$2.03 = 2.03 \qquad \text{Balances}$$

The solution is 0.29.

9. $0.8p = -96$

$\dfrac{0.8p}{0.8} = \dfrac{-96}{0.8}$ *Divide both sides by 0.8.*

$p = -120$

Check $0.8p = -96$
$$0.8(-120) = -96$$
$$-96 = -96 \qquad \text{Balances}$$

The solution is -120.

11. $-3.3t = -2.31$

$\dfrac{-3.3t}{-3.3} = \dfrac{-2.31}{-3.3}$ *Divide both sides by -3.3.*

$t = 0.7$

Check $-3.3t = -2.31$
$$-3.3(0.7) = -2.31$$
$$-2.31 = -2.31 \qquad \text{Balances}$$

The solution is 0.7.

13. $7.5x + 0.15 = -6$

$\underline{\qquad -0.15 \qquad -0.15}$ *Add -0.15 to both sides.*

$$7.5x + 0 = -6.15$$
$$7.5x = -6.15$$

$\dfrac{7.5x}{7.5} = \dfrac{-6.15}{7.5}$ *Divide both sides by 7.5.*

$x = -0.82$

Check $7.5x + 0.15 = -6$
$$7.5(-0.82) + 0.15 = -6$$
$$-6.15 + 0.15 = -6$$
$$-6 = -6 \qquad \text{Balances}$$

The solution is -0.82.

15. $-7.38 = 2.05z - 7.38$

$\underline{+7.38 \qquad \quad + 7.38}$ *Add 7.38 to both sides.*

$$0 = 2.05z + 0$$
$$0 = 2.05z$$

$\dfrac{0}{2.05} = \dfrac{2.05z}{2.05}$ *Divide both sides by 2.05.*

$0 = z$

Check $-7.38 = 2.05z - 7.38$
$$-7.38 = 2.05(0) - 7.38$$
$$-7.38 = 0 - 7.38$$
$$-7.38 = -7.38 \qquad \text{Balances}$$

The solution is 0.

17. $-0.9 = 0.2 - 0.01h$

$\underline{-0.2 \qquad -0.2}$ *Add -0.2 to both sides.*

$$-1.1 = 0 - 0.01h$$

$\dfrac{-1.1}{-0.01} = \dfrac{-0.01h}{-0.01}$ *Divide both sides by -0.01.*

$110 = h$

Check $-0.9 = 0.2 - 0.01h$
$$-0.9 = 0.2 - 0.01(110)$$
$$-0.9 = 0.2 - 1.1$$
$$-0.9 = -0.9 \qquad \text{Balances}$$

The solution is 110.

19. $3c + 10 = 6c + 8.65$

$\underline{-3c \qquad \qquad -3c}$ *Add $-3c$ to both sides.*

$$10 = 3c + 8.65$$

$\underline{-8.65 \qquad \qquad -8.65}$ *Add -8.65 to both sides.*

$$1.35 = 3c$$

$\dfrac{1.35}{3} = \dfrac{3c}{3}$ *Divide both sides by 3.*

$0.45 = c$

Check
$$3c + 10 = 6c + 8.65$$
$$3(0.45) + 10 = 6(0.45) + 8.65$$
$$1.35 + 10 = 2.7 + 8.65$$
$$11.35 = 11.35 \qquad \text{Balances}$$

The solution is 0.45.

21.

$$0.8w - 0.4 = -6 + w$$

$$\underline{-0.8w \qquad\qquad -0.8w}$$ *Add $-0.8w$ to both sides.*

$$-0.4 = -6 + 0.2w$$

$$\underline{+6 \qquad +6}$$ *Add 6 to both sides.*

$$5.6 = 0.2w$$

$$\frac{5.6}{0.2} = \frac{0.2w}{0.2}$$ *Divide both sides by 0.2.*

$$28 = w$$

Check $0.8w - 0.4 = -6 + w$

$$0.8(28) - 0.4 = -6 + 28$$

$$22.4 - 0.4 = 22$$

$$22 = 22 \qquad \text{Balances}$$

The solution is 28.

23.

$$-10.9 + 0.5p = 0.9p + 5.3$$

$$\underline{-0.5p \qquad -0.5p}$$ *Add $-0.5p$ to both sides.*

$$-10.9 = 0.4p + 5.3$$

$$\underline{-5.3 \qquad\qquad -5.3}$$ *Add -5.3 to both sides.*

$$-16.2 = 0.4p$$

$$\frac{-16.2}{0.4} = \frac{0.4p}{0.4}$$ *Divide both sides by 0.4.*

$$-40.5 = p$$

Check

$$-10.9 + 0.5p = 0.9p + 5.3$$

$$-10.9 + 0.5(-40.5) = 0.9(-40.5) + 5.3$$

$$-10.9 - 20.25 = -36.45 + 5.3$$

$$-31.15 = -31.15 \qquad \text{Balances}$$

The solution is -40.5.

25. *Step 1* Unknown: the adult dose

Known: child's dose is 0.3 times the adult dose; child's dose is 9 mg

Step 2(a) Let d be the adult dose.

Step 3

adult dose multiplied by 0.3	is	child's dose
↓	↓	↓
$0.3d$	$=$	9

Step 4 $0.3d = 9$

$$\frac{0.3d}{0.3} = \frac{9}{0.3}$$ *Divide both sides by 0.3.*

$$d = 30$$

Step 5 The adult dose is 30 milligrams.

Step 6 $0.3(30) = 9$ (matches)

27. *Step 1* Unknown: number of days the saw was rented

Known: $65.95 charge per day, $12 sharpening fee, $275.80 total

Step 2(a) Let d be the number of days.

Step 3

65.95 per day	plus	sharpening fee	equals	total bill
↓	↓	↓	↓	↓
$65.95 \cdot d$	$+$	12	$=$	275.80

Step 4

$$65.95d + 12 = 275.80$$

$$\underline{-12 \qquad -12}$$

$$65.95d = 263.80$$

$$\frac{65.95d}{65.95} = \frac{263.80}{65.95}$$

$$d = 4$$

Step 5 The saw was rented for 4 days.

Step 6 Charge for 4 days: $4(\$65.95) = \$263.80$
Add sharpening fee: $\$263.80 + \$12 = \$275.80$
This value matches the total charge.

Relating Concepts (Exercises 29–32)

29.

$$0.7(220 - a) = 140$$

$$154 - 0.7a = 140$$ *Distributive property*

$$\underline{-154 \qquad\qquad -154}$$ *Add -154 to both sides.*

$$-0.7a = -14$$

$$\frac{-0.7a}{-0.7} = \frac{-14}{-0.7}$$ *Divide both sides by -0.7.*

$$a = 20$$

The person is 20 years old.

30.

$$0.7(220 - a) = 126$$

$$154 - 0.7a = 126$$ *Distributive property*

$$\underline{-154 \qquad\qquad -154}$$ *Add -154 to both sides.*

$$-0.7a = -28$$

$$\frac{-0.7a}{-0.7} = \frac{-28}{-0.7}$$ *Divide both sides by -0.7.*

$$a = 40$$

The person is 40 years old.

31.

$$0.7(220 - a) = 134$$

$$154 - 0.7a = 134$$ *Distributive property*

$$\underline{-154 \qquad\qquad -154}$$ *Add -154 to both sides.*

$$-0.7a = -20$$

$$\frac{-0.7a}{-0.7} = \frac{-20}{-0.7}$$ *Divide both sides by -0.7.*

$$a \approx 28.57$$

The person is about 29 years old.

32.

$$0.7(220 - a) = 117$$

$$154 - 0.7a = 117 \quad \textit{Distributive property}$$

$$\underline{-154 \qquad\qquad -154} \quad \textit{Add } -154 \textit{ to both sides.}$$

$$-0.7a = -37$$

$$\frac{-0.7a}{-0.7} = \frac{-37}{-0.7} \quad \textit{Divide both sides by } -0.7.$$

$$a \approx 52.86$$

The person is about 53 years old.

5.10 Geometry Applications: Circles, Cylinders, and Surface Area

5.10 Margin Exercises

1.   (a) diameter: 40 ft

$$r = \frac{d}{2} = \frac{40 \text{ ft}}{2} = 20 \text{ ft}$$

(b) diameter: 11 cm

$$r = \frac{d}{2} = \frac{11 \text{ cm}}{2} = 5.5 \text{ cm}$$

(c) radius: 32 yd

$$d = 2 \cdot r = 2 \cdot 32 \text{ yd} = 64 \text{ yd}$$

(d) radius: 9.5 m

$$d = 2 \cdot r = 2 \cdot 9.5 \text{ m} = 19 \text{ m}$$

2.   (a) diameter: 150 ft

$$C = \pi \cdot d$$
$$\approx 3.14 \cdot 150 \text{ ft}$$
$$\approx \underline{471 \text{ ft}}$$

(b) radius: 7 in.

$$C = 2 \cdot \pi \cdot r$$
$$\approx 2 \cdot 3.14 \cdot 7 \text{ in.}$$
$$\approx 44.0 \text{ in.}$$

(c) diameter: 0.9 km

$$C = \pi \cdot d$$
$$\approx 3.14 \cdot 0.9 \text{ km}$$
$$\approx 2.8 \text{ km}$$

(d) radius: 4.6 m

$$C = 2 \cdot \pi \cdot r$$
$$\approx 2 \cdot 3.14 \cdot 4.6 \text{ m}$$
$$\approx 28.9 \text{ m}$$

3.   (a) radius: 4 ft

$$A = \pi \cdot r \cdot r$$
$$\approx 3.14 \cdot 4 \text{ ft} \cdot 4 \text{ ft}$$
$$\approx \underline{50.2 \text{ ft}^2}$$

(b) diameter: 12 in., so $r = \underline{6}$ in.

$$A = \pi \cdot r \cdot r$$
$$\approx 3.14 \cdot 6 \text{ in.} \cdot 6 \text{ in.}$$
$$\approx 113.0 \text{ in.}^2$$

(c) radius: 1.8 km

$$A = \pi \cdot r \cdot r$$
$$\approx 3.14 \cdot 1.8 \text{ km} \cdot 1.8 \text{ km}$$
$$\approx 10.2 \text{ km}^2$$

(d) diameter: 8.4 cm, so $r = 4.2$ cm

$$A = \pi \cdot r \cdot r$$
$$\approx 3.14 \cdot 4.2 \text{ cm} \cdot 4.2 \text{ cm}$$
$$\approx 55.4 \text{ cm}^2$$

4.   (a) Area of the circle (radius: 24 m):

$$A = \pi \cdot r \cdot r$$
$$\approx 3.14 \cdot 24 \text{ m} \cdot 24 \text{ m}$$
$$\approx 1808.64 \text{ m}^2$$

Area of the semicircle:

$$\frac{1808.64 \text{ m}^2}{2} \approx 904.3 \text{ m}^2$$

(b) Area of the circle (diameter: 35.4 ft, so $r = 17.7$ ft):

$$A = \pi \cdot r \cdot r$$
$$\approx 3.14 \cdot 17.7 \text{ ft} \cdot 17.7 \text{ ft}$$
$$\approx 983.7306 \text{ ft}^2$$

Area of the semicircle:

$$\frac{983.7306 \text{ ft}^2}{2} \approx 491.9 \text{ ft}^2$$

(c) Area of the circle (radius: 9.8 m):

$$A = \pi \cdot r \cdot r$$
$$\approx 3.14 \cdot 9.8 \text{ m} \cdot 9.8 \text{ m}$$
$$\approx 301.5656 \text{ m}^2$$

Area of the semicircle:

$$\frac{301.5656 \text{ m}^2}{2} \approx 150.8 \text{ m}^2$$

5.

$$C = \pi \cdot d$$
$$\approx 3.14 \cdot 3 \text{ m}$$
$$\approx 9.42 \text{ m}$$

$$\text{Cost of binding} = \frac{9.42 \text{ m}}{1} \cdot \frac{\$4.50}{1 \text{ m}} = \$42.39$$

6.   diameter: 3 m, so $r = 1.5$ m

$$A = \pi \cdot r \cdot r$$
$$\approx 3.14 \cdot 1.5 \text{ m} \cdot 1.5 \text{ m}$$
$$\approx 7.065 \text{ m}^2$$

$$\text{Total cost} = \frac{7.065 \text{ m}^2}{1} \cdot \frac{\$3.89}{\text{m}^2} \approx \$27.48$$

7. **(a)** $r = 4$ ft, $h = 12$ ft

$$V = \pi \cdot r \cdot r \cdot h$$
$$V \approx 3.14 \cdot 4 \text{ ft} \cdot 4 \text{ ft} \cdot 12 \text{ ft}$$
$$V \approx 602.9 \text{ ft}^3 \text{ (rounded)}$$

(b) $r = \frac{7 \text{ cm}}{2} = 3.5$ cm, $h = 6$ cm

$$V = \pi \cdot r \cdot r \cdot h$$
$$V \approx 3.14 \cdot 3.5 \text{ cm} \cdot 3.5 \text{ cm} \cdot 6 \text{ cm}$$
$$V \approx 230.8 \text{ cm}^3 \text{ (rounded)}$$

(c) $r = 14.5$ yd, $h = 3.2$ yd

$$V = \pi \cdot r \cdot r \cdot h$$
$$V \approx 3.14 \cdot 14.5 \text{ yd} \cdot 14.5 \text{ yd} \cdot 3.2 \text{ yd}$$
$$V \approx 2112.592 \text{ yd}^3$$
$$V \approx 2112.6 \text{ yd}^3 \text{ (rounded)}$$

8. **(a)** $V = lwh$

$$V = \underline{10 \text{ yd}} \cdot \underline{6 \text{ yd}} \cdot \underline{9 \text{ yd}}$$
$$V = \underline{540 \text{ yd}^3} \text{ (cubic yd for volume)}$$

$$S = 2lw + 2lh + 2wh$$
$$S = (2 \cdot 10 \text{ yd} \cdot 6 \text{ yd}) + (2 \cdot 10 \text{ yd} \cdot 9 \text{ yd})$$
$$+ (2 \cdot 6 \text{ yd} \cdot 9 \text{ yd})$$
$$S = 120 \text{ yd}^2 + 180 \text{ yd}^2 + 108 \text{ yd}^2$$
$$S = 408 \text{ yd}^2 \text{ (square yd for area)}$$

(b) $V = lwh$

$$V = (16 \text{ m})(7 \text{ m})(7 \text{ m})$$
$$V = 784 \text{ m}^3$$

$$S = 2lw + 2lh + 2wh$$
$$S = (2 \cdot 16 \text{ m} \cdot 7 \text{ m}) + (2 \cdot 16 \text{ m} \cdot 7 \text{ m})$$
$$+ (2 \cdot 7 \text{ m} \cdot 7 \text{ m})$$
$$S = 224 \text{ m}^2 + 224 \text{ m}^2 + 98 \text{ m}^2$$
$$S = 546 \text{ m}^2$$

9. **(a)** $V = \pi r^2 h$

$$V \approx 3.14 \cdot \underline{5 \text{ cm}} \cdot \underline{5 \text{ cm}} \cdot \underline{15 \text{ cm}}$$
$$V \approx \underline{1177.5 \text{ cm}^3} \text{ (cubic units for volume)}$$

$$S = 2\pi rh + 2\pi r^2$$
$$S \approx (2 \cdot 3.14 \cdot 5 \text{ cm} \cdot 15 \text{ cm}) + (2 \cdot 3.14 \cdot 5 \text{ cm} \cdot 5 \text{ cm})$$
$$S \approx 471 \text{ cm}^2 + 157 \text{ cm}^2$$
$$S \approx 628 \text{ cm}^2 \text{ (square units for area)}$$

(b) $r = \dfrac{d}{2} = \dfrac{17 \text{ in.}}{2} = 8.5$ in.

$$V = \pi r^2 h$$
$$V \approx 3.14 \cdot 8.5 \text{ in.} \cdot 8.5 \text{ in.} \cdot 8 \text{ in.}$$
$$V \approx 1814.92 \text{ in.}^3$$
$$V \approx 1814.9 \text{ in.}^3$$
$$S = 2\pi rh + 2\pi r^2$$
$$S \approx (2 \cdot 3.14 \cdot 8.5 \text{ in.} \cdot 8 \text{ in.})$$
$$+ (2 \cdot 3.14 \cdot 8.5 \text{ in.} \cdot 8.5 \text{ in.})$$
$$S \approx 427.04 \text{ in.}^2 + 453.73 \text{ in.}^2$$
$$S \approx 880.77 \text{ in.}^2$$
$$S \approx 880.8 \text{ in.}^2$$

5.10   Section Exercises

1. Finding the circumference of a circle is like finding the <u>perimeter</u> of a rectangle. If the circle's radius and diameter are measured in feet, then the units for the circumference will be <u>ft</u>.

3. The radius r is 9 mm, so the diameter d is

$$d = 2 \cdot r = 2 \cdot 9 \text{ mm} = 18 \text{ mm}.$$

5. The diameter d is 0.7 km, so the radius r is

$$r = \frac{d}{2} = \frac{0.7 \text{ km}}{2} = 0.35 \text{ km}.$$

7. The radius r is 11 ft.

$$C = 2 \cdot \pi \cdot r$$
$$\approx 2 \cdot 3.14 \cdot 11 \text{ ft}$$
$$\approx 69.1 \text{ ft}$$
$$A = \pi \cdot r \cdot r$$
$$\approx 3.14 \cdot 11 \text{ ft} \cdot 11 \text{ ft}$$
$$\approx 379.9 \text{ ft}^2$$

9. The diameter d is 2.6 m, so the radius r is

$$r = \tfrac{1}{2}d = \tfrac{1}{2}(2.6 \text{ m}) = 1.3 \text{ m}.$$

$$C = \pi \cdot d$$
$$\approx 3.14 \cdot 2.6 \text{ m}$$
$$\approx 8.2 \text{ m}$$
$$A = \pi \cdot r \cdot r$$
$$\approx 3.14 \cdot 1.3 \text{ m} \cdot 1.3 \text{ m}$$
$$\approx 5.3 \text{ m}^2$$

11. The diameter d is 15 cm, so the radius r is

$$r = \tfrac{1}{2}d = \tfrac{1}{2}(15 \text{ cm}) = 7.5 \text{ cm}.$$

$$C = \pi \cdot d$$
$$\approx 3.14 \cdot 15 \text{ cm}$$
$$\approx \underline{47.1 \text{ cm}}$$
$$A = \pi \cdot r \cdot r$$
$$\approx 3.14 \cdot 7.5 \text{ cm} \cdot 7.5 \text{ cm}$$
$$\approx 176.6 \text{ cm}^2$$

13. The diameter d is $7\frac{1}{2}$ ft, so the radius r is

$$r = \tfrac{1}{2}d = \tfrac{1}{2}(7.5 \text{ ft}) = 3.75 \text{ ft}.$$

$$C = \pi \cdot d$$
$$\approx 3.14 \cdot 7.5 \text{ ft}$$
$$\approx 23.6 \text{ ft}$$
$$A = \pi \cdot r \cdot r$$
$$\approx 3.14 \cdot 3.75 \text{ ft} \cdot 3.75 \text{ ft}$$
$$\approx 44.2 \text{ ft}^2$$

15. Area of the circle (radius: 7 in.):

$$A = \pi \cdot r \cdot r$$
$$\approx 3.14 \cdot 7 \text{ in.} \cdot 7 \text{ in.}$$
$$\approx 153.86 \text{ in.}^2$$

Area of the semicircle:

$$\frac{153.86 \text{ in.}^2}{2} \approx 76.9 \text{ in.}^2$$

17. Area of the circle:

$$A = \pi \cdot r \cdot r$$
$$\approx 3.14 \cdot 10 \text{ cm} \cdot 10 \text{ cm}$$
$$\approx 314 \text{ cm}^2$$

Area of the semicircle:

$$\frac{314 \text{ cm}^2}{2} = 157 \text{ cm}^2$$

Area of the triangle:

$$A = \tfrac{1}{2} \cdot b \cdot h$$
$$= \tfrac{1}{2} \cdot 20 \text{ cm} \cdot 10 \text{ cm}$$
$$= 100 \text{ cm}^2$$

The shaded area is about

$$157 \text{ cm}^2 - 100 \text{ cm}^2 = 57 \text{ cm}^2.$$

19. Find the area of the circular region shown in the figure.

$$A = \pi \cdot r \cdot r$$
$$\approx 3.14 \cdot 50 \text{ yd} \cdot 50 \text{ yd}$$
$$\approx 7850 \text{ yd}^2$$

The watered area is about 7850 yd^2.

21. A point on the tire tread moves the length of the circumference in one complete turn.

$$C = \pi \cdot d$$
$$\approx 3.14 \cdot 29.10 \text{ in.}$$
$$\approx 91.4 \text{ in.}$$

Bonus question:

$$\frac{1 \text{ revolution}}{91.4 \text{ inches}} \cdot \frac{12 \text{ inches}}{1 \text{ foot}} \cdot \frac{5280 \text{ feet}}{1 \text{ mile}}$$
$$\approx 693 \text{ revolutions/mile}$$

23.

$$A = \pi \cdot r \cdot r$$
$$\approx 3.14 \cdot 150 \text{ mi} \cdot 150 \text{ mi}$$
$$\approx 70,650 \text{ mi}^2$$

There are about $70,650 \text{ mi}^2$ in the broadcast area.

Watch

25.   Watch: $d = 1$ in., $r = 0.5$ in.

$$
\begin{array}{ll}
C = \pi \cdot 1 \text{ in.} & A = \pi \cdot r \cdot r \\
\approx 3.14 \cdot 1 \text{ in.} & \approx 3.14 \cdot 0.5 \text{ in.} \cdot 0.5 \text{ in.} \\
\approx 3.1 \text{ in.} & \approx 0.8 \text{ in.}^2
\end{array}
$$

Wall Clock

    Wall clock: $r = 3$ in.

$$
\begin{array}{ll}
C = 2 \cdot \pi \cdot 3 \text{ in.} & A = \pi \cdot r \cdot r \\
\approx 2 \cdot 3.14 \cdot 3 \text{ in.} & \approx 3.14 \cdot 3 \text{ in.} \cdot 3 \text{ in.} \\
\approx 18.8 \text{ in.} & \approx 28.3 \text{ in.}^2
\end{array}
$$

27.

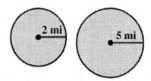

Area of larger circle:

$$A = \pi \cdot r \cdot r$$
$$\approx 3.14 \cdot 5 \text{ mi} \cdot 5 \text{ mi}$$
$$\approx 78.5 \text{ mi}^2$$

Area of smaller circle:

$$A = \pi \cdot r \cdot r$$
$$\approx 3.14 \cdot 2 \text{ mi} \cdot 2 \text{ mi}$$
$$\approx 12.6 \text{ mi}^2$$

The difference in the area covered is about

$$78.5 \text{ mi}^2 - 12.6 \text{ mi}^2 = 65.9 \text{ mi}^2.$$

29. (a)
$$
\begin{array}{ll}
C = \pi \cdot d & \\
144 \text{ cm} = \pi \cdot d & \textit{Replace C with 144 cm.} \\
144 \text{ cm} \approx 3.14 \cdot d & \textit{Replace } \pi \textit{ with 3.14.} \\
\dfrac{144 \text{ cm}}{3.14} \approx d & \textit{Divide by 3.14.} \\
d \approx 45.9 \text{ cm} &
\end{array}
$$

The diameter is about 45.9 cm.

(b) Divide the circumference by π
($144 \text{ cm} \div 3.14$).

31. $C = 2 \cdot \pi \cdot r$
$$\approx 2 \cdot 3.14 \cdot 2.33 \text{ ft}$$
$$\approx 14.6 \text{ ft}$$

The rag traveled about 14.6 ft with each revolution of the wheel.

33. Area of larger circle:
$$A = \pi \cdot r \cdot r$$
$$\approx 3.14 \cdot 12 \text{ cm} \cdot 12 \text{ cm}$$
$$\approx 452.16 \text{ cm}^2$$

Area of smaller circle:
$$A = \pi \cdot r \cdot r$$
$$\approx 3.14 \cdot 9 \text{ cm} \cdot 9 \text{ cm}$$
$$\approx 254.34 \text{ cm}^2$$
$$452.16 \text{ cm}^2 - 254.34 \text{ cm}^2 = 197.82 \text{ cm}^2$$

The shaded area is about 197.8 cm^2.

35. $$V = \pi \cdot r^2 \cdot h$$
$$\approx 3.14 \cdot 5 \text{ ft} \cdot 5 \text{ ft} \cdot 6 \text{ ft}$$
$$\approx 471 \text{ ft}^3$$
$$S = 2\pi rh + 2\pi r^2$$
$$\approx 2 \cdot 3.14 \cdot 5 \text{ ft} \cdot 6 \text{ ft} + 2 \cdot 3.14 \cdot 5 \text{ ft} \cdot 5 \text{ ft}$$
$$\approx 345.4 \text{ ft}^2$$

37. $$V = lwh$$
$$= 16.5 \text{ m} \cdot 9.8 \text{ m} \cdot 10 \text{ m}$$
$$= 1617 \text{ m}^3$$
$$S = 2lw + 2lh + 2wh$$
$$= 2 \cdot 16.5 \text{ m} \cdot 9.8 \text{ m} + 2 \cdot 16.5 \text{ m} \cdot 10 \text{ m}$$
$$+ 2 \cdot 9.8 \text{ m} \cdot 10 \text{ m}$$
$$= 849.4 \text{ m}^2$$

39. $d = 18$ in., so $r = \frac{d}{2} = \frac{18 \text{ in.}}{2} = 9$ in.
$$V = \pi r^2 h$$
$$\approx 3.14 \cdot 9 \text{ in.} \cdot 9 \text{ in.} \cdot 3 \text{ in.}$$
$$\approx 763.0 \text{ in.}^3 \text{ (rounded)}$$
$$S = 2\pi rh + 2\pi r^2$$
$$\approx 2 \cdot 3.14 \cdot 9 \text{ in.} \cdot 3 \text{ in.} + 2 \cdot 3.14 \cdot 9 \text{ in.} \cdot 9 \text{ in.}$$
$$\approx 678.2 \text{ in.}^2 \text{ (rounded)}$$

41. $$V = lwh$$
$$= 15 \text{ mm} \cdot 10 \text{ mm} \cdot 37 \text{ mm}$$
$$= 5550 \text{ mm}^3$$
$$S = 2lw + 2lh + 2wh$$
$$= 2 \cdot 15 \text{ mm} \cdot 10 \text{ mm} + 2 \cdot 15 \text{ mm} \cdot 37 \text{ mm}$$
$$+ 2 \cdot 10 \text{ mm} \cdot 37 \text{ mm}$$
$$= 2150 \text{ mm}^2$$

43. Student should use radius of 3.5 cm instead of diameter of 7 cm in the formula; units for volume are cm^3, not cm^2. Correct answer is $V \approx 192.3 \text{ cm}^3$.

45. Use the formula for the volume of a cylinder.
$d = 5$ ft, so $r = \frac{d}{2} = 2.5$ ft.
$$V = \pi \cdot r^2 \cdot h$$
$$\approx 3.14 \cdot 2.5 \text{ ft} \cdot 2.5 \text{ ft} \cdot 200 \text{ ft}$$
$$\approx 3925 \text{ ft}^3$$

The volume of the city sewer pipe is about 3925 ft^3.

47. Use the formula for the surface area of a rectangular solid.
$$S = 2lw + 2lh + 2wh$$
$$= 2 \cdot 5.5 \text{ in.} \cdot 2.8 \text{ in.} + 2 \cdot 5.5 \text{ in.} \cdot 8 \text{ in.}$$
$$+ 2 \cdot 2.8 \text{ in.} \cdot 8 \text{ in.}$$
$$= 163.6 \text{ in.}^2$$

The amount of material needed is 163.6 in.^2.

Chapter 5 Review Exercises

1.

			tenths	hundredths
2	4	3 . 0	5	9

2.

ones		tenths		
0 .	6	8	1	7

3.

	hundreds			hundredths
$5	8	2	4 . 3	9

4.

	tens		tenths	
8	9	6 .	5 0	3

5.

	tenths			ten-thousandths
2	0 . 7	3	8 6	1

6. $0.5 = \dfrac{5}{10} = \dfrac{1}{2}$

7. $0.75 = \dfrac{75}{100} = \dfrac{75 \div 25}{100 \div 25} = \dfrac{3}{4}$

8. $4.05 = 4\dfrac{5}{100} = 4\dfrac{5 \div 5}{100 \div 5} = 4\dfrac{1}{20}$

9. $0.875 = \dfrac{875}{1000} = \dfrac{875 \div 125}{1000 \div 125} = \dfrac{7}{8}$

10. $0.027 = \dfrac{27}{1000}$

11. $27.8 = 27\dfrac{8}{10} = 27\dfrac{8 \div 2}{10 \div 2} = 27\dfrac{4}{5}$

12. 0.8 is eight tenths.

13. 400.29 is four hundred and twenty-nine hundredths.

14. 12.007 is twelve and seven thousandths.

15. 0.0306 is three hundred six ten-thousandths.

16. eight and three tenths:

$$8\frac{3}{10} = 8.3$$

17. two hundred five thousandths:

$$\frac{205}{1000} = 0.205$$

18. seventy and sixty-six ten-thousandths:

$$70\frac{66}{10,000} = 70.0066$$

19. thirty hundredths:

$$\frac{30}{100} = 0.30$$

20. 275.635 to the nearest tenth ■ Draw a cut-off line after the tenths place:

$$275.6|35$$

The first digit cut is 3, which is 4 or less. The part you keep stays the same. Answer is 275.6.

21. 72.789 to the nearest hundredth ■ Draw a cut-off line after the hundredths place:

$$72.78|9$$

The first digit cut is 9, which is 5 or more, so round up the hundredths place.

$$\begin{array}{r} 72.78 \\ + \ 0.01 \\ \hline 72.79 \end{array} \quad \textit{Answer}$$

22. 0.1604 to the nearest thousandth ■ Draw a cut-off line after the thousandths place:

$$0.160|4$$

The first digit cut is 4, which is 4 or less. The part you keep stays the same. Answer is 0.160.

23. 0.0905 to the nearest thousandth ■ Draw a cut-off line after the thousandths place:

$$0.090|5$$

The first digit cut is 5, which is 5 or more, so round up the thousandths place.

$$\begin{array}{r} 0.090 \\ + \ 0.001 \\ \hline 0.091 \end{array} \quad \textit{Answer}$$

24. 0.98 to the nearest tenth ■ Draw a cut-off line after the tenths place:

$$0.9|8$$

The first digit cut is 8, which is 5 or more, so round up the tenths place.

$$\begin{array}{r} 0.9 \\ + \ 0.1 \\ \hline 1.0 \end{array} \quad \textit{Answer}$$

25. \$15.8333 to the nearest cent ■ Draw a cut-off line after the hundredths place:

$$\$15.83|33$$

The first digit cut is 3, which is 4 or less. The part you keep stays the same. Answer is \$15.83.

26. \$0.698 to the nearest cent ■ Draw a cut-off line after the hundredths place:

$$\$0.69|8$$

The first digit cut is 8, which is 5 or more, so round up the hundredths place.

$$\begin{array}{r} \$0.69 \\ + \ 0.01 \\ \hline \$0.70 \end{array} \quad \textit{Answer}$$

27. \$17,625.7906 to the nearest cent ■ Draw a cut-off line after the hundredths place:

$$\$17,625.79|06$$

The first digit cut is 0, which is 4 or less. The part you keep stays the same. Answer is \$17,625.79.

28. \$350.48 to the nearest dollar ■ Draw a cut-off line after the ones place:

$$\$350|.48$$

The first digit cut is 4, which is 4 or less. The part you keep stays the same. Answer is \$350.

29. \$129.50 to the nearest dollar ■ Draw a cut-off line after the ones place:

$$\$129|.50$$

The first digit cut is 5, which is 5 or more, so round up the ones place.

$$\begin{array}{r} \$129 \\ + \ 1 \\ \hline \$130 \end{array} \quad \textit{Answer}$$

30. \$99.61 to the nearest dollar ■ Draw a cut-off line after the ones place:

$$\$99|.61$$

The first digit cut is 6, which is 5 or more, so round up the ones place.

$$\begin{array}{r} \$99 \\ + \ 1 \\ \hline \$100 \end{array} \quad \textit{Answer}$$

31. $29.37 to the nearest dollar   ▪   Draw a cut-off line after the ones place:

$$\$29|.37$$

The first digit cut is 3, which is 4 or less. The part you keep stays the same. Answer is $29.

32. $0.4 - 6.07 = 0.4 + (-6.07)$

$$|-6.07| = 6.07, \quad |0.4| = 0.4$$

-6.07 has a larger absolute value, so the sum will be negative. Subtract absolute values.

$$\begin{array}{r} 6.07 \\ -\ 0.40 \\ \hline 5.67 \end{array}$$

$0.4 - 6.07 = \ = -5.67$

33. $-20 + 19.97$

$$|-20| = 20, \quad |19.97| = 19.97$$

-20 has the larger absolute value, so the sum will be negative. Subtract absolute values.

$$\begin{array}{r} 20.00 \\ -\ 19.97 \\ \hline 0.03 \end{array}$$

$-20 + 19.97 = -0.03$

34. $-1.35 + 7.229$

$$|-1.35| = 1.35, \quad |7.229| = 7.229$$

7.229 has the larger absolute value, so the sum will be positive. Subtract absolute values.

$$\begin{array}{r} {}^{6}\ {}^{11\,12} \\ 7.\not{2}\,\not{2}\,9 \\ -\ 1.3\ 5\ 0 \\ \hline 5.8\ 7\ 9 \end{array}$$

$-1.35 + 7.229 = 5.879$

35. $0.005 + (3 - 9.44) = 0.005 + [3 + (-9.44)]$
$$= 0.005 + (-6.44)$$
$$= -6.435$$

36.

Estimate:		Exact:	
100	walking	95.8	walking
− 40	camping	− 44.7	camping
60	(million)	51.1	(million)

51.1 million more people go walking than camping.

37.

Estimate:		Exact:	
400	start amt.	406.00	start amt.
− 300	day care	− 315.53	day care
100		90.47	
− 70	groceries	− 74.67	groceries
30	end amt.	15.80	end amt.

The new amount in her account is $15.80.

38.   First total the money that Joey spent.

Estimate:		Exact:	
$2	toothpaste	$1.59	toothpaste
5	vitamins	5.33	vitamins
+ 20	toaster	+ 18.94	toaster
$27		$25.86	

Then subtract to find the change.

Estimate:		Exact:	
$30	three $10s	$30.00	three $10s
− 27	estimate	− 25.86	exact
$3		$4.14	

Joey's change was $4.14.

39.   Add the kilometers that she raced each day.

Estimate:		Exact:	
2	Monday	2.30	Monday
4	Wednesday	4.00	Wednesday
+ 5	Friday	+ 5.25	Friday
11	(km)	11.55	(km)

Altogether, Roseanne raced 11.55 kilometers.

40.

Estimate:	Exact:
6	6.138
× 4	× 3.7
24	4 2966
	18 414
	22.7106

41.

Estimate:	Exact:
40	42.9
× 3	× 3.3
120	12 87
	128 7
	141.57

42. $(-5.6)(-0.002)$ ▪   The signs are the *same*, so the product will be *positive*.

$$\begin{array}{rll} 5.6 & \leftarrow & \text{1 decimal place} \\ \times\ 0.002 & \leftarrow & \text{3 decimal places} \\ \hline 0.0112 & \leftarrow & \text{4 decimal places} \end{array}$$

43. $(0.071)(-0.005)$ ▪   The signs are *different*, so the product will be *negative*.

$$\begin{array}{rll} 0.071 & \leftarrow & \text{3 decimal places} \\ \times\ 0.005 & \leftarrow & \text{3 decimal places} \\ \hline -0.000355 & \leftarrow & \text{6 decimal places} \end{array}$$

44. $706.2 \div 12 = 58.85$

Estimate: $700 \div 10 = 70$

58.85 is *reasonable*.

45. $26.6 \div 2.8 = 0.95$

Estimate: $30 \div 3 = 10$

0.95 is *not reasonable*.

```
            9. 5
2.8∧⟌2 6. 6∧0   Move decimal point 1
    2 5 2        place in divisor and
      1 4  0     dividend; write one zero.
      1 4  0
           0
```

The correct answer is 9.5.

46.
```
     1 4. 4 6 6 6
  3⟌4 3. 4 0 0 0
     3
     1 3
     1 2
       1 4
       1 2
         2 0   Write one zero.
         1 8
           2 0   Write one zero.
           1 8
             2 0  Write one zero.
             1 8
               2
```

14.4666 rounds to 14.467.

47. $\dfrac{-72}{-0.06}$

$= -72.00 \div (-0.06)$

$\qquad\qquad$ *Move both*
$\qquad\qquad$ *decimal points*
$= -7200 \div (-6)$ $\quad$ *two places*
$\qquad\qquad$ *to the right.*

$= 1200$

48. $-0.00048 \div 0.0012$

$\qquad\qquad$ *Move both*
$\qquad\qquad$ *decimal points*
$= -4.8 \div 12$ $\quad$ *four places*
$\qquad\qquad$ *to the right.*
$= -0.4$ $\qquad$ *See below.*

```
       0. 4
  12⟌4. 8
     4 8
       0
```

49. Multiply the hourly wage times the hours worked.

Pay for first 40 hours	Pay rate after 40 hours	Pay for over 40 hours
14.24	14.24	21.36
× 40	× 1.5	× 6.5
569.60	7 120	10 680
	14 24	128 16
	21.360	138.840

Add the two pay amounts.

$569.60
138.84
$708.44

Adrienne's total earnings were $708 (rounded).

50. Divide the cost of the book by the number of tickets in the book.

```
         4. 9 9 0
  12⟌5 9. 8 9 0
     4 8
     1 1 8
     1 0 8
       1 0 9
       1 0 8
           1 0
```

Round to the nearest cent.
Draw a cut-off line: $4.99|0
The first digit cut is 0, which is 4 or less. The part you keep stays the same.

Each ticket costs $4.99 (rounded).

51. Divide the amount of the investment by the price per share.

```
            1 3 3. 3
  3.75∧⟌5 0 0. 0 0∧0
       3 7 5
       1 2 5 0
       1 1 2 5
         1 2 5 0
         1 1 2 5
           1 2 5 0
           1 1 2 5
             1 2 5
```

Round 133.3 down to the nearest whole number.
Kenneth could buy 133 shares (rounded).

52. Multiply the price per pound by the amount to be purchased.

$$
\begin{array}{r}
\$2.29 \\
\times\ 3.5 \\
\hline
1\,1\,4\,5 \\
6\,8\,7\ \ \\
\hline
\$8.0\,1\,5
\end{array}
$$

Ms. Lee will pay $8.02 (rounded).

53. $3.5^2 + 8.7 \cdot (-1.95)$

$\quad = 12.25 + 8.7(-1.95)\quad$ *Exponent*

$\quad = 12.25 + (-16.965)\quad$ *Multiply.*

$\quad = -4.715\quad$ *Add.*

54. $11 - 3.06 \div (3.95 - 0.35)$

$\quad = 11 - 3.06 \div 3.6\quad$ *Parentheses*

$\quad = 11 - 0.85\quad$ *Divide.*

$\quad = 10.15\quad$ *Subtract.*

55. $3\dfrac{4}{5} = \dfrac{19}{5}$ ▪

$$
\begin{array}{r}
3.\,8 \\
5\overline{)1\,9.\,0} \\
1\,5\ \ \\
\hline
4\,0 \\
4\,0 \\
\hline
0
\end{array}
$$

$$3\dfrac{4}{5} = 3.8$$

56. $\dfrac{16}{25}$ ▪

$$
\begin{array}{r}
0.\,6\,4 \\
25\overline{)1\,6.\,0\,0} \\
1\,5\,0\ \ \\
\hline
1\,0\,0 \\
1\,0\,0 \\
\hline
0
\end{array}
$$

$$\dfrac{16}{25} = 0.64$$

57. $1\dfrac{7}{8} = \dfrac{15}{8}$ ▪

$$
\begin{array}{r}
1.\,8\,7\,5 \\
8\overline{)1\,5.\,0\,0\,0} \\
8\ \ \\
\hline
7\,0 \\
6\,4 \\
\hline
6\,0 \\
5\,6 \\
\hline
4\,0 \\
4\,0 \\
\hline
0
\end{array}
$$

$$1\dfrac{7}{8} = 1.875$$

58. $\dfrac{1}{9}$ ▪

$$
\begin{array}{r}
0.\,1\,1\,1\,1 \\
9\overline{)1.\,0\,0\,0\,0} \\
9\ \ \\
\hline
1\,0 \\
9\ \ \\
\hline
1\,0 \\
9\ \ \\
\hline
1\,0 \\
9\ \ \\
\hline
1
\end{array}
$$

0.1111 rounds to 0.111. $\dfrac{1}{9} \approx 0.111$

59. 3.68, 3.806, 3.6008

$\quad 3.68 = 3.6800$

$\quad 3.806 = 3.8060\quad$ *greatest*

$\quad 3.6008 = 3.6008\quad$ *least*

From least to greatest:

$\quad\quad$ 3.6008, 3.68, 3.806

60. 0.215, 0.22, 0.209, 0.2102

$\quad 0.215 = 0.2150$

$\quad 0.22 = 0.2200\quad$ *greatest*

$\quad 0.209 = 0.2090\quad$ *least*

$\quad 0.2102 = 0.2102$

From least to greatest:

$\quad\quad$ 0.209, 0.2102, 0.215, 0.22

61. $0.17, \dfrac{3}{20}, \dfrac{1}{8}, 0.159$

$\quad 0.17 = 0.170\quad$ *greatest*

$\quad \dfrac{3}{20} = 0.150$

$\quad \dfrac{1}{8} = 0.125\quad$ *least*

$\quad 0.159 = 0.159$

From least to greatest:

$\quad\quad \dfrac{1}{8}, \dfrac{3}{20}, 0.159, 0.17$

62. $\text{mean} = \dfrac{\text{sum of all values}}{\text{number of values}}$

$$\text{mean} = \dfrac{\begin{array}{c}(18 + 12 + 15 + 24 + 9 \\ + 42 + 54 + 87 + 21 + 3)\end{array}}{10}$$

$$\text{mean} = \dfrac{285}{10} = 28.5 \text{ phones}$$

To find the median, arrange the data from high to low or low to high.

$\quad\quad$ 3, 9, 12, 15, $\underline{18}$, $\underline{21}$, 24, 42, 54, 87

Since the number of items is even, the median is the average of the middle items.

$$\text{median} = \dfrac{18 + 21}{2} = 19.5 \text{ phones}$$

63.
$$\text{mean} = \frac{\text{sum of all values}}{\text{number of values}}$$
$$\text{mean} = \frac{(54 + 28 + 35 + 43 + 17}{9}$$
$$\frac{+ 37 + 68 + 75 + 39)}{9}$$
$$\text{mean} = \frac{396}{9} = 44 \text{ loans}$$

To find the median, arrange the data from high to low or low to high.

$$17, 28, 35, 37, \underline{39}, 43, 54, 68, 75$$

Since the number of items is odd, the median is the middle item.

$$\text{median} = 39 \text{ loans}$$

64.

Dollar Value	Frequency	Product
$42	3	$126
$47	7	$329
$53	2	$106
$55	3	$165
$59	5	$295
	20	$1021

$$\text{weighted mean} = \frac{\$1021}{20} = \$51.05$$

65. Arrange the data in order to more easily identify the most often used value.

Store J: $69, $84, $107, $107, $139, $160, $160
There are two modes: $107 and $160 (bimodal).

Store K: $95, $99, $119, $119, $119, $136, $139
The mode is $119.

66.
$$\text{hypotenuse} = \sqrt{(\text{leg})^2 + (\text{leg})^2}$$
$$= \sqrt{(15)^2 + (8)^2}$$
$$= \sqrt{225 + 64}$$
$$= \sqrt{289}$$
$$= 17 \text{ in.}$$

67.
$$\text{leg} = \sqrt{(\text{hypotenuse})^2 - (\text{leg})^2}$$
$$= \sqrt{(25)^2 - (24)^2}$$
$$= \sqrt{625 - 576}$$
$$= \sqrt{49}$$
$$= 7 \text{ cm}$$

68.
$$\text{leg} = \sqrt{(\text{hypotenuse})^2 - (\text{leg})^2}$$
$$= \sqrt{(15)^2 - (11)^2}$$
$$= \sqrt{225 - 121}$$
$$= \sqrt{104}$$
$$\approx 10.2 \text{ cm (rounded)}$$

69.
$$\text{hypotenuse} = \sqrt{(\text{leg})^2 + (\text{leg})^2}$$
$$= \sqrt{(6)^2 + (4)^2}$$
$$= \sqrt{36 + 16}$$
$$= \sqrt{52}$$
$$\approx 7.2 \text{ in. (rounded)}$$

70.
$$\text{hypotenuse} = \sqrt{(\text{leg})^2 + (\text{leg})^2}$$
$$= \sqrt{(2.2)^2 + (1.3)^2}$$
$$= \sqrt{4.84 + 1.69}$$
$$= \sqrt{6.53}$$
$$\approx 2.6 \text{ m (rounded)}$$

71.
$$\text{leg} = \sqrt{(\text{hypotenuse})^2 - (\text{leg})^2}$$
$$= \sqrt{(12)^2 - (8.5)^2}$$
$$= \sqrt{144 - 72.25}$$
$$= \sqrt{71.75}$$
$$\approx 8.5 \text{ km (rounded)}$$

72.
$$-0.1 = b - 0.35$$
$$-0.1 = b + (-0.35) \quad \textit{Add the opposite.}$$
$$+0.35 \qquad +0.35 \quad \textit{Add 0.35 to both sides.}$$
$$\overline{0.25 = b + 0}$$
$$0.25 = b$$

The solution is 0.25.

73.
$$-3.8x = 0$$
$$\frac{-3.8x}{-3.8} = \frac{0}{-3.8} \quad \textit{Divide both sides by −3.8.}$$
$$x = 0$$

The solution is 0.

74.
$$6.8 + 0.4n = 1.6$$
$$-6.8 \qquad -6.8 \quad \textit{Add −6.8 to both sides.}$$
$$\overline{0 + 0.4n = -5.2}$$
$$\frac{0.4n}{0.4} = \frac{-5.2}{0.4} \quad \textit{Divide both sides by 0.4.}$$
$$n = -13$$

The solution is −13.

75.
$$-0.375 + 1.75a = 2a$$
$$-1.75a \qquad -1.75a \quad \textit{Add −1.75a to both sides.}$$
$$\overline{-0.375 + 0 = 0.25a}$$
$$-0.375 = 0.25a$$
$$\frac{-0.375}{0.25} = \frac{0.25a}{0.25} \quad \textit{Divide both sides by 0.25.}$$
$$-1.5 = a$$

The solution is −1.5.

76. $0.3y - 5.4 = 2.7 + 0.8y$

$$\begin{array}{ll} \underline{-0.3y \qquad\qquad -0.3y} & \textit{Add } -0.3y \textit{ to} \\ & \textit{both sides.} \\ 0 - 5.4 = 2.7 + 0.5y & \\ -5.4 = \quad 2.7 + 0.5y & \\ \underline{-2.7 \qquad -2.7} & \textit{Add } -2.7 \textit{ to} \\ & \textit{both sides.} \\ -8.1 = 0 + 0.5y & \\ \dfrac{-8.1}{0.5} = \dfrac{0.5y}{0.5} & \textit{Divide both} \\ & \textit{sides by 0.5.} \\ -16.2 = y & \end{array}$$

The solution is -16.2.

77. $d = 2 \cdot r = 2 \cdot 68.9 \text{ m} = 137.8 \text{ m}$

The diameter of the field is 137.8 m.

78. $d = 3$ in., so $r = \dfrac{d}{2} = \dfrac{3 \text{ in.}}{2} = 1\dfrac{1}{2}$ in. or 1.5 in.

The radius of the juice can is $1\dfrac{1}{2}$ in. or 1.5 in.

79.   radius: 1 cm, $d = 2$ cm

$$\begin{aligned} C &= \pi \cdot d \\ &\approx 3.14 \cdot 2 \text{ cm} \\ &\approx 6.3 \text{ cm (rounded)} \\ A &= \pi \cdot r \cdot r \\ &\approx 3.14 \cdot 1 \text{ cm} \cdot 1 \text{ cm} \\ &\approx 3.1 \text{ cm}^2 \text{ (rounded)} \end{aligned}$$

80.   radius: 17.4 m, $d = 34.8$ m

$$\begin{aligned} C &= \pi \cdot d \\ &\approx 3.14 \cdot 34.8 \text{ m} \\ &\approx 109.3 \text{ m (rounded)} \\ A &= \pi \cdot r \cdot r \\ &\approx 3.14 \cdot 17.4 \text{ m} \cdot 17.4 \text{ m} \\ &\approx 950.7 \text{ m}^2 \text{ (rounded)} \end{aligned}$$

81.   diameter: 12 in., radius: 6 in.

$$\begin{aligned} C &= \pi \cdot d \\ &\approx 3.14 \cdot 12 \text{ in.} \\ &\approx 37.7 \text{ in. (rounded)} \\ A &= \pi \cdot r \cdot r \\ &\approx 3.14 \cdot 6 \text{ in.} \cdot 6 \text{ in.} \\ &\approx 113.0 \text{ in.}^2 \text{ (rounded)} \end{aligned}$$

82.   The figure is a cylinder.

$$\begin{aligned} V &= \pi \cdot r^2 \cdot h \\ &\approx 3.14 \cdot 5 \text{ cm} \cdot 5 \text{ cm} \cdot 7 \text{ cm} \\ &\approx 549.5 \text{ cm}^3 \\ S &= 2\pi rh + 2\pi r^2 \\ &\approx 2 \cdot 3.14 \cdot 5 \text{ cm} \cdot 7 \text{ cm} + 2 \cdot 3.14 \cdot 5 \text{ cm} \cdot 5 \text{ cm} \\ &\approx 376.8 \text{ cm}^2 \end{aligned}$$

83.   The figure is a cylinder.

$$\begin{aligned} V &= \pi \cdot r^2 \cdot h \quad (d = 24 \text{ m}; \ r = 12 \text{ m}) \\ &\approx 3.14 \cdot 12 \text{ m} \cdot 12 \text{ m} \cdot 4 \text{ m} \\ &\approx 1808.6 \text{ m}^3 \text{ (rounded)} \end{aligned}$$

$$\begin{aligned} S &= 2\pi rh + 2\pi r^2 \\ &\approx 2 \cdot 3.14 \cdot 12 \text{ m} \cdot 4 \text{ m} + 2 \cdot 3.14 \cdot 12 \text{ m} \cdot 12 \text{ m} \\ &\approx 1205.8 \text{ m}^2 \text{ (rounded)} \end{aligned}$$

84. $l = 3.5$ ft, $w = 1.5$ ft, $h = 1.5$ ft

$$\begin{aligned} V &= lwh \\ &= 3.5 \text{ ft} \cdot 1.5 \text{ ft} \cdot 1.5 \text{ ft} \\ &= 7.875 \text{ ft}^3 \approx 7.9 \text{ ft}^3 \text{ (rounded)} \end{aligned}$$

$$\begin{aligned} S &= 2lw + 2lh + 2wh \\ &= 2 \cdot 3.5 \text{ ft} \cdot 1.5 \text{ ft} + 2 \cdot 3.5 \text{ ft} \cdot 1.5 \text{ ft} \\ &\quad + 2 \cdot 1.5 \text{ ft} \cdot 1.5 \text{ ft} \\ &= 25.5 \text{ ft}^2 \end{aligned}$$

85.   **[5.3]** $89.19 + 0.075 + 310.6 + 5$

$$\begin{array}{rl} \overset{11}{8}\overset{1}{9}.190 & \textit{Line up decimal points.} \\ 0.075 & \textit{Write in zeros.} \\ 310.600 & \\ \underline{+ \ 5.000} & \\ 404.865 & \end{array}$$

86.   **[5.4]** $72.8(-3.5)$ ■ The signs are *different*, so the product will be *negative*.

$$\begin{array}{rl} 72.8 & \leftarrow \quad \textit{1 decimal place} \\ \underline{\times \ 3.5} & \leftarrow \quad \textit{1 decimal place} \\ 36\ 40 & \\ \underline{218\ 4} & \\ 254.80 & \leftarrow \quad \textit{2 decimal places} \end{array}$$

$72.8(-3.5) = -254.80$

87.   **[5.5]** $1648.3 \div 0.46 \approx 3583.2609$

3583.2609 rounds to 3583.261.

88.   **[5.3]** $30 - 0.9102$

$$\begin{array}{r} {\overset{2}{\cancel{3}}}{\overset{9}{\cancel{0}}}.{\overset{9}{\cancel{0}}}{\overset{9}{\cancel{0}}}{\overset{9}{\cancel{0}}}{\overset{10}{\cancel{0}}} \\ - \ 0 \ . \ 9 \ 1 \ 0 \ 2 \\ \hline 2 \ 9 \ . \ 0 \ 8 \ 9 \ 8 \end{array}$$

$30 - 0.9102 = 29.0898$

89.   **[5.4]** $(4.38)(0.007)$

$$\begin{array}{rl} 4.38 & \leftarrow \quad \textit{2 decimal places} \\ \underline{\times \ 0.007} & \leftarrow \quad \textit{3 decimal places} \\ 0.03066 & \leftarrow \quad \textit{5 decimal places} \end{array}$$

$(4.38)(0.007) = 0.03066$

90. [5.5]

$$0.005_\wedge \overline{)0.047_\wedge 0}$$

Move decimal point 3 places in divisor and dividend; write 0

$$
\begin{array}{r}
9.\,4 \\
\hline
4\,5 \\
\hline
2\ 0 \\
2\ 0 \\
\hline
0
\end{array}
$$

91. [5.3] $72.105 + 8.2 - 95.37$

$= 72.105 + 8.2 + (-95.37)$ *Change.*

$= 80.305 + (-95.37)$ *Add.*

$= -15.065$ *Add.*

92. [5.5] $\dfrac{-81.36}{9}$ ▪ The signs are different, so the quotient is negative.

$$
\begin{array}{r}
9.04 \\
9\overline{)81.36} \\
81 \\
\hline
0\ 3 \\
0 \\
\hline
36 \\
36 \\
\hline
0
\end{array}
$$

$\dfrac{-81.36}{9} = -9.04$

93. [5.5] $(0.6 - 1.22) + 4.8(-3.15)$

$= [0.6 + (-1.22)] + 4.8(-3.15)$ *Change.*

$= (-0.62) + 4.8(-3.15)$ *Brackets*

$= -0.62 + (-15.12)$ *Multiply.*

$= -15.74$ *Add.*

94. [5.4] $0.455(18)$

$$
\begin{array}{r}
0.455 \quad \leftarrow \textit{3 decimal places} \\
\times\ 18 \quad \leftarrow \textit{0 decimal places} \\
\hline
3\,640 \\
4\,55 \\
\hline
8.190 \quad \leftarrow \textit{3 decimal places}
\end{array}
$$

$0.455(18) = 8.19$

95. [5.4] $(-1.6)(-0.58)$ ▪ The signs are the same, so the product is positive.

$$
\begin{array}{r}
0.58 \quad \leftarrow \textit{2 decimal places} \\
\times\ 1.6 \quad \leftarrow \textit{1 decimal place} \\
\hline
348 \\
58 \\
\hline
0.928 \quad \leftarrow \textit{3 decimal places}
\end{array}
$$

$(-1.6)(-0.58) = 0.928$

96. [5.5] $0.218\overline{)7.63}$

$7.63 \div 0.218 = 35$

97. [5.3] $-21.059 - 20.8$

$= -21.059 + (-20.8)$ *Change subt.*

$= -41.859$ *Add.*

98. [5.5] $18.3 - 3^2 \div 0.5$

$= 18.3 - 9 \div 0.5$ *Exponent*

$= 18.3 - 18$ *Divide.*

$= 0.3$ *Subtract.*

99. [5.5] Men's socks are 3 pairs for $8.99. Divide the price by 3 to find the cost of one pair.

$$
\begin{array}{r}
2.\,9\,9\,6 \\
3\overline{)8.\,9\,9\,0} \\
6 \\
\hline
2\ 9 \\
2\ 7 \\
\hline
2\ 9 \\
2\ 7 \\
\hline
2\ 0 \\
1\ 8 \\
\hline
2
\end{array}
$$

$2.996 rounds to $3.00.

One pair of men's socks costs $3.00 (rounded).

100. [5.5] Children's socks cost 6 pairs for $5.00. Find the cost for one pair.

$$
\begin{array}{r}
0.\,8\,3\,3 \\
6\overline{)5.\,0\,0\,0} \\
4\ 8 \\
\hline
2\ 0 \\
1\ 8 \\
\hline
2\ 0 \\
1\ 8 \\
\hline
2
\end{array}
$$

$0.833 rounds to $0.83.

Subtract to find how much more men's socks cost.

$$
\begin{array}{r}
\$3.00 \\
-\ 0.83 \\
\hline
\$2.17
\end{array}
$$

Men's socks cost $2.17 (rounded) more.

101. [5.4] A dozen pair of socks is $4 \cdot 3 = 12$ pairs of socks. So multiply 4 times $8.99.

$$
\begin{array}{r}
\$8.99 \\
\times\ 4 \\
\hline
\$35.96
\end{array}
$$

The cost for a dozen pair of men's socks is $35.96.

102. [5.4] Five pairs of teen jeans cost $19.95 times 5.

$$
\begin{array}{r}
\$19.95 \\
\times\ 5 \\
\hline
\$99.75
\end{array}
$$

Four pairs of women's jeans cost $24.99 times 4.

$$
\begin{array}{r}
\$24.99 \\
\times\ 4 \\
\hline
\$99.96
\end{array}
$$

Add the two amounts.

$$\begin{array}{r} \$99.75 \\ +\ 99.96 \\ \hline \$199.71 \end{array}$$

Akiko would pay $199.71.

103. **[5.3]** The highest regular price for athletic shoes is $149.50. The cheapest sale price is $71.

Subtract to find the difference.

$$\begin{array}{r} \$149.50 \\ -\ 71.00 \\ \hline \$78.50 \end{array}$$

The difference in price is $78.50.

104. **[5.9]** $4.62 = -6.6y$

$\dfrac{4.62}{-6.6} = \dfrac{-6.6y}{-6.6}$ *Divide both sides by −6.6.*

$-0.7 = y$

The solution is -0.7.

105. **[5.9]** $1.05x - 2.5 = 0.8x + 5$

$$\begin{array}{ccc} -0.8x & & -0.8x \end{array}$$ *Add −0.8x to both sides.*

$\dfrac{}{0.25x - 2.5 = 0 + 5}$

$0.25x - 2.5 = 5$

$$\begin{array}{ccc} +\,2.5 & & +\,2.5 \end{array}$$ *Add 2.5 to both sides.*

$\dfrac{}{0.25x + 0 = 7.5}$

$\dfrac{0.25x}{0.25} = \dfrac{7.5}{0.25}$ *Divide both sides by 0.25.*

$x = 30$

The solution is 30.

106. **[5.10]** Rubber strip to go "around" the edge of the table indicates circumference of the table.

$C = \pi \cdot d$

$C = \pi \cdot 5 \text{ ft}$

$C \approx 3.14 \cdot 5 \text{ ft}$

$C \approx 15.7 \text{ ft long rubber strip}$

Since the diameter is 5, the radius is $\frac{5}{2}$ or 2.5.

$A = \pi \cdot r^2$

$A = \pi \cdot (2.5 \text{ ft})^2$

$A \approx 3.14 \cdot 2.5 \text{ ft} \cdot 2.5 \text{ ft}$

$A \approx 19.6 \text{ ft}^2 \text{ (rounded)}$

107. **[5.7]** $\text{mean} = \dfrac{\text{sum of all values}}{\text{number of values}}$

$\text{mean} = \dfrac{82 + 0 + 78 + 93 + 85}{5}$

$\text{mean} = \dfrac{338}{5} = 67.6$

Order the scores:

0, 78, <u>82</u>, 85, 93

The median score for 5 scores is the third score, that is, 82.

108. **[5.8]** $\text{hypotenuse} = \sqrt{(\text{leg})^2 + (\text{leg})^2}$

$= \sqrt{(16)^2 + (12)^2}$

$= \sqrt{256 + 144}$

$= \sqrt{400}$

$= 20 \text{ miles}$

She is 20 miles from her starting point.

109. **[5.10]** A "can" is geometrically a cylinder.

$d = 3$ in., so $r = 1.5$ in.

$V = \pi \cdot r^2 \cdot h$

$V = \pi \cdot (1.5 \text{ in.})^2 \cdot 7 \text{ in.}$

$V \approx 3.14 \cdot 1.5 \text{ in.} \cdot 1.5 \text{ in.} \cdot 7 \text{ in.}$

$V \approx 49.5 \text{ in.}^3 \text{ (rounded)}$

The volume of the can is about 49.5 in.^{3}.

Chapter 5 Test

1. $18.4 = 18\dfrac{4}{10} = 18\dfrac{4 \div 2}{10 \div 2} = 18\dfrac{2}{5}$

2. $0.075 = \dfrac{75}{1000} = \dfrac{75 \div 25}{1000 \div 25} = \dfrac{3}{40}$

3.   60.007 is sixty and seven thousandths.

4.   0.0208 is two hundred eight ten-thousandths.

5. $7.6 + 82.0128 + 39.59$

Estimate:	*Exact:*	
8	7.6000	*Line up decimals.*
80	82.0128	*Write in zeros.*
+ 40	+ 39.5900	
128	129.2028	

6. $-5.79(1.2)$ ▪ The signs are *different*, so the product is *negative*.

Estimate:	*Exact:*
-6	-5.79
$\times\ 1$	$\times\ 1.2$
-6	$1\ 158$
	$5\ 79$
	-6.948

$-5.79(1.2) = -6.948$

7. $-79.1 - 3.602 = -79.1 + (-3.602)$

Both addends have the same sign, so the sum will be negative.

Estimate:	*Exact:*
-80	-79.100
$+\ -4$	$+\ -3.602$
-84	-82.702

$-79.1 - 3.602 = -82.702$

8. $-20.04 \div (-4.8)$ ■ The signs are the *same*, so the quotient is *positive*.

Estimate: *Exact*:

$$\begin{array}{r} 4 \\ 5\overline{)20} \end{array}$$

$$\begin{array}{r} 4.\ 1\ 7\ 5 \\ 4.8\overline{)2\ 0.\ 0_{\wedge}\ 4\ 0\ 0} \\ \underline{1\ 9\ 2} \\ 8\ 4 \\ \underline{4\ 8} \\ 3\ 6\ 0 \\ \underline{3\ 3\ 6} \\ 2\ 4\ 0 \\ \underline{2\ 4\ 0} \\ 0 \end{array}$$

$-20.04 \div (-4.8) = 4.175$

9. $670 - 0.996$

$$\begin{array}{r} 670.000 \\ -\ 0.996 \\ \hline 669.004 \end{array}$$ *Line up decimals.*
Write in zeros.

$670 - 0.996 = 669.004$

10.

$$\begin{array}{r} 4\ 8\ 0. \\ 0.15_{\wedge}\overline{)7\ 2.\ 0\ 0_{\wedge}} \\ \underline{6\ 0} \\ 1\ 2\ 0 \\ \underline{1\ 2\ 0} \\ 0\ 0 \\ \underline{0} \\ 0 \end{array}$$ *Move decimal point 2 places in dividend and divisor; write two zeros.*

11. $(-0.006)(-0.007)$ ■ The signs are the *same*, so the product is *positive*.

$$\begin{array}{r} 0.006 \\ \times\ 0.007 \\ \hline 0.000042 \end{array}$$ ← *3 decimal places*
← *3 decimal places*
← *6 decimal places*

$(-0.006)(-0.007) = 0.000042$

12. Divide to find the cost per foot.

$$\begin{array}{r} 3.\ 7\ 9 \\ 6.5_{\wedge}\overline{)\$2\ 4.\ 6_{\wedge}\ 4\ 0} \\ \underline{1\ 9\ 5} \\ 5\ 1\ 4 \\ \underline{4\ 5\ 5} \\ 5\ 9\ 0 \\ \underline{5\ 8\ 5} \\ 5 \end{array}$$ *Move decimal point 1 place in dividend and divisor; write one zero.*

The chain costs $3.79 per foot (rounded).

13. Compare the two times.

$$\begin{array}{r} 3.500 \\ -\ 3.059 \\ \hline 0.441 \end{array}$$ *Angela's time (more)*
Davida's time (less)
(in minutes)

Davida won by 0.441 minute.

14. Multiply the price per pound by the amount purchased.

$$\begin{array}{r} \$2.89 \\ \times\ 1.85 \\ \hline 1445 \\ 2\ 312 \\ 2\ 89 \\ \hline \$5.3465 \end{array}$$

Round $5.3465 to the nearest cent.
Draw a cut-off line: $5.34|65
The first digit cut is 6, which is 5 or more, so round up.

Mr. Yamamoto paid $5.35 (rounded) for the cheese.

15.

$$\begin{array}{rcl} -5.9 & = & y + 0.25 \\ -0.25 & & -0.25 \\ \hline -6.15 & = & y + 0 \\ -6.15 & = & y \end{array}$$ *Add -0.25 to both sides.*

The solution is -6.15.

16.

$$\begin{array}{rcl} -4.2x & = & 1.47 \\ \dfrac{-4.2x}{-4.2} & = & \dfrac{1.47}{-4.2} \\ x & = & -0.35 \end{array}$$ *Divide both sides by -4.2.*

The solution is -0.35.

17.

$$\begin{array}{rcl} 3a - 22.7 & = & 10 \\ 3a + (-22.7) & = & 10 \\ +\ 22.7 & & +\ 22.7 \\ \hline 3a + 0 & = & 32.7 \\ \dfrac{3a}{3} & = & \dfrac{32.7}{3} \\ a & = & 10.9 \end{array}$$ *Add 22.7 to both sides.*
Divide both sides by 3.

The solution is 10.9.

18.

$$\begin{array}{rcl} -0.8n + 1.88 & = & 2n - 6.1 \\ +0.8n & & +0.8n \\ \hline 0 + 1.88 & = & 2.8n - 6.1 \\ 1.88 & = & 2.8n - 6.1 \\ +\ 6.1 & & +\ 6.1 \\ \hline 7.98 & = & 2.8n + 0 \\ \dfrac{7.98}{2.8} & = & \dfrac{2.8n}{2.8} \\ 2.85 & = & n \end{array}$$ *Add 0.8n to both sides.*
Add 6.1 to both sides.
Divide both sides by 2.8.

The solution is 2.85.

19. $0.44, 0.451, \frac{9}{20}, 0.4506$

Change to ten-thousandths, if necessary, and compare.

$$0.44 = 0.4400 \quad \textit{least}$$
$$0.451 = 0.4510 \quad \textit{greatest}$$
$$\frac{9}{20} = 0.4500$$
$$0.4506 = 0.4506$$

From least to greatest: $0.44, \frac{9}{20}, 0.4506, 0.451$

20. $6.3^2 - 5.9 + 3.4(-0.5)$

$$\begin{aligned}
&= 39.69 - 5.9 + 3.4(-0.5) &&\textit{Exponent} \\
&= 39.69 - 5.9 + (-1.7) &&\textit{Multiply.} \\
&= 39.69 + (-5.9) + (-1.7) &&\textit{Add the oppos.} \\
&= 33.79 + (-1.7) &&\textit{Add.} \\
&= 32.09 &&\textit{Add.}
\end{aligned}$$

21. $\text{mean} = \dfrac{\text{sum of all values}}{\text{number of values}}$

$$\text{mean} = \frac{\begin{array}{c}(52 + 61 + 68 + 69 + 73 \\ + 75 + 79 + 84 + 91 + 98)\end{array}}{10}$$

$$\text{mean} = \frac{750}{10} = 75 \text{ books}$$

22. $96°, \underline{104°}, \underline{103°}, \underline{104°}, \underline{103°}, \underline{104°}, 91°, 74°, \underline{103°}$

The mode is the value that occurs most often. The modes are 103° and 104° (bimodal).

23.

Cost	Frequency	Product
$6	7	$42
$10	3	$30
$11	4	$44
$14	2	$28
$19	3	$57
$24	1	$24
	20	$225

$$\text{weighted mean} = \frac{\$225}{20} = \$11.25$$

24. Arrange the prices in numerical order.

$39.75, $46.90, $48, $\underline{49.30}, $\underline{51.80}, $54.50, $56.25, $89

Because there is an even number of items, find the middle two numbers (4th and 5th of 8):

$$\text{median} = \frac{\$49.30 + \$51.80}{2} = \frac{\$101.10}{2} = \$50.55$$

25. $\text{hypotenuse} = \sqrt{(\text{leg})^2 + (\text{leg})^2}$

$$\begin{aligned}
&= \sqrt{(7)^2 + (6)^2} \\
&= \sqrt{49 + 36} \\
&= \sqrt{85} \\
&\approx 9.2 \text{ cm}
\end{aligned}$$

26. The unknown side is a leg of the right triangle.

$$\begin{aligned}
\text{leg} &= \sqrt{(\text{hypotenuse})^2 - (\text{leg})^2} \\
&= \sqrt{(20)^2 - (11)^2} \\
&= \sqrt{400 - 121} \\
&= \sqrt{279} \\
&\approx 16.7 \text{ ft (nearest tenth)}
\end{aligned}$$

27. $d = 25$ in., so $r = \frac{d}{2} = \frac{25 \text{ in.}}{2} = 12.5$ in.

28. $C = 2 \cdot \pi \cdot r$

$$\begin{aligned}
&\approx 2 \cdot 3.14 \cdot 0.9 \text{ km} \\
&\approx 5.7 \text{ km (rounded)}
\end{aligned}$$

29. $d = 16.2$ cm, so $r = \frac{d}{2} = \frac{16.2 \text{ cm}}{2} = 8.1$ cm.

$$\begin{aligned}
A &= \pi \cdot r^2 \\
&\approx 3.14 \cdot 8.1 \text{ cm} \cdot 8.1 \text{ cm} \\
&\approx 206.0 \text{ cm}^2 \text{ (rounded)}
\end{aligned}$$

30. The figure is a cylinder.

$$\begin{aligned}
V &= \pi \cdot r^2 \cdot h \\
&\approx 3.14 \cdot 18 \text{ ft} \cdot 18 \text{ ft} \cdot 5 \text{ ft} \\
&\approx 5086.8 \text{ ft}^3
\end{aligned}$$

31. $S = 2\pi rh + 2\pi r^2$

$$\begin{aligned}
&\approx 2 \cdot 3.14 \cdot 18 \text{ ft} \cdot 5 \text{ ft} + 2 \cdot 3.14 \cdot 18 \text{ ft} \cdot 18 \text{ ft} \\
&\approx 2599.9 \text{ ft}^2 \text{ (rounded)}
\end{aligned}$$

CHAPTER 6 RATIO, PROPORTION, AND LINE/ANGLE/ TRIANGLE RELATIONSHIPS

6.1 Ratios

6.1 Margin Exercises

1. **(a)** $7 spent on fresh fruit to $5 spent on milk

$$\text{Spent on fruit} \rightarrow \quad \frac{7}{5}$$
$$\text{Spent on milk} \rightarrow$$

(b) $5 spent on milk to $14 spent on meat

$$\text{Spent on milk} \rightarrow \quad \frac{5}{14}$$
$$\text{Spent on meat} \rightarrow$$

(c) $14 spent on meat to $5 spent on milk

$$\text{Spent on meat} \rightarrow \quad \frac{14}{5}$$
$$\text{Spent on milk} \rightarrow$$

2. **(a)** 9 hours to 12 hours

$$\frac{9 \text{ hours}}{12 \text{ hours}} = \frac{9}{12} = \frac{9 \div 3}{12 \div 3} = \frac{3}{4}$$

(b) 100 meters to 50 meters

$$\frac{100 \text{ meters}}{50 \text{ meters}} = \frac{100}{50} = \frac{100 \div 50}{50 \div 50} = \frac{2}{1}$$

(c) width to length or 24 feet to 48 feet

$$\frac{\text{width}}{\text{length}} = \frac{24 \text{ feet}}{48 \text{ feet}} = \frac{24}{48} = \frac{24 \div 24}{48 \div 24} = \frac{1}{2}$$

3. **(a)** The increase in price is

$$\$9.00 - \$5.50 = \$3.50.$$

The ratio of increase in price to original price is

$$\frac{3.50}{5.50}.$$

Rewrite as two whole numbers.

$$\frac{3.50}{5.50} = \frac{3.50 \times 100}{5.50 \times 100} = \frac{350}{550}$$

Write in lowest terms.

$$\frac{350 \div 50}{550 \div 50} = \frac{7}{11}$$

(b) The decrease in hours is

$$4.5 \text{ hours} - 3 \text{ hours} = 1.5 \text{ hours}.$$

The ratio of decrease in hours to the original number of hours is

$$\frac{1.5}{4.5}.$$

Rewrite as two whole numbers.

$$\frac{1.5 \cdot 10}{4.5 \cdot 10} = \frac{15}{45}$$

Write in lowest terms.

$$\frac{15 \div 15}{45 \div 15} = \frac{1}{3}$$

4. **(a)** $3\frac{1}{2}$ to 4

$$\frac{3\frac{1}{2}}{4} = \frac{\frac{7}{2}}{\frac{4}{1}} = \frac{7}{2} \div \frac{4}{1} = \frac{7}{2} \cdot \frac{1}{4} = \frac{7}{8}$$

(b) $5\frac{5}{8}$ pounds to $3\frac{3}{4}$ pounds

$$\frac{5\frac{5}{8}}{3\frac{3}{4}} = \frac{\frac{45}{8}}{\frac{15}{4}} = \frac{45}{8} \div \frac{15}{4} = \frac{\overset{3}{\cancel{45}}}{\underset{2}{\cancel{8}}} \cdot \frac{\overset{1}{\cancel{4}}}{\underset{1}{\cancel{15}}} = \frac{3}{2}$$

(c) $3\frac{1}{3}$ inches to $\frac{5}{6}$ inch

$$\frac{3\frac{1}{3}}{\frac{5}{6}} = \frac{\frac{10}{3}}{\frac{5}{6}}$$

$$= \frac{10}{3} \div \frac{5}{6} = \frac{10}{3} \cdot \frac{6}{5} = \frac{\overset{2}{\cancel{10}} \cdot \overset{2}{\cancel{6}}}{\underset{1}{\cancel{3}} \cdot \underset{1}{\cancel{5}}} = \frac{4}{1}$$

5. **(a)** 9 inches to 6 feet

Change 6 feet to inches.

$$6 \text{ feet} = 6 \cdot 12 \text{ inches} = \underline{72} \text{ inches}$$

$$\frac{9 \text{ in.}}{6 \text{ ft}} = \frac{9 \text{ in.}}{\underline{72} \text{ in.}} = \frac{9}{72} = \frac{9 \div 9}{\underline{72 \div 9}} = \frac{1}{8}$$

(b) 2 days to 8 hours

$$2 \text{ days} = 2 \cdot 24 \text{ hours} = 48 \text{ hours}$$

$$\frac{2 \text{ days}}{8 \text{ hours}} = \frac{48 \text{ hours}}{8 \text{ hours}} = \frac{48}{8} = \frac{48 \div 8}{8 \div 8} = \frac{6}{1}$$

(c) 7 yards to 14 feet

$$7 \text{ yards} = 7 \cdot 3 \text{ feet} = 21 \text{ feet}$$

$$\frac{7 \text{ yards}}{14 \text{ feet}} = \frac{21 \text{ feet}}{14 \text{ feet}} = \frac{21}{14} = \frac{21 \div 7}{14 \div 7} = \frac{3}{2}$$

(d) 3 quarts to 3 gallons

$$3 \text{ gallons} = 3 \cdot 4 = 12 \text{ quarts}$$

$$\frac{3 \text{ quarts}}{3 \text{ gallons}} = \frac{3 \text{ quarts}}{12 \text{ quarts}} = \frac{3}{12} = \frac{3 \div 3}{12 \div 3} = \frac{1}{4}$$

(e) 25 minutes to 2 hours

$$2 \text{ hours} = 2 \cdot 60 \text{ minutes} = 120 \text{ minutes}$$

$$\frac{25 \text{ minutes}}{2 \text{ hours}} = \frac{25 \text{ minutes}}{120 \text{ minutes}}$$

$$= \frac{25}{120} = \frac{25 \div 5}{120 \div 5} = \frac{5}{24}$$

(f) 4 pounds to 12 ounces

4 pounds $= 4 \cdot 16$ ounces $= 64$ ounces

$$\frac{4 \text{ pounds}}{12 \text{ ounces}} = \frac{64 \text{ ounces}}{12 \text{ ounces}} = \frac{64}{12} = \frac{64 \div 4}{12 \div 4} = \frac{16}{3}$$

6.1   Section Exercises

1. Answers will vary. One possibility is: A ratio compares two quantities with the same units. Examples will vary.

3. To rewrite the ratio $\frac{80}{20}$ in lowest terms, divide both the numerator and the denominator by $\underline{20}$. The ratio in lowest terms is

$$\frac{80}{20} = \frac{80 \div 20}{120 \div 20} = \frac{4}{1}.$$

5. 8 days to 9 days: $\dfrac{8 \text{ days}}{9 \text{ days}} = \dfrac{8}{9}$

7. $100 to $50

$$\frac{\$100}{\$50} = \frac{100}{50} = \frac{100 \div 50}{50 \div 50} = \frac{2}{1}$$

9. 30 minutes to 90 minutes

$$\frac{30 \text{ minutes}}{90 \text{ minutes}} = \frac{30}{90} = \frac{30 \div 30}{90 \div 30} = \frac{1}{3}$$

11. $4.50 to $3.50

$$\frac{\$4.50}{\$3.50} = \frac{4.50}{3.50} = \frac{4.50 \cdot 10}{3.50 \cdot 10} = \frac{45}{35}$$
$$= \frac{45 \div 5}{35 \div 5} = \frac{9}{7}$$

13. $1\frac{1}{4}$ to $1\frac{1}{2}$

$$\frac{1\frac{1}{4}}{1\frac{1}{2}} = \frac{\frac{5}{4}}{\frac{3}{2}} = \frac{5}{4} \div \frac{3}{2} = \frac{5}{\overset{2}{\underset{2}{4}}} \cdot \frac{\overset{1}{2}}{3} = \frac{5}{6}$$

15. When writing ratios with ounces and pounds, it is easier to use $\underline{\text{ounces}}$ because the conversion is easier to work with. That is, 1 pound $= 16$ ounces is easier to work with than 1 ounce $= \frac{1}{16}$ pound.

17. 4 feet to 30 inches

4 feet $= 4 \cdot 12$ inches $= \underline{48}$ inches

$$\frac{4 \text{ feet}}{30 \text{ inches}} = \frac{48 \text{ inches}}{30 \text{ inches}} = \frac{48}{30} = \frac{48 \div 6}{30 \div 6} = \frac{8}{5}$$

(Do not write the ratio as $1\frac{3}{5}$.)

19. 5 minutes to 1 hour

1 hour $= 60$ minutes

$$\frac{5 \text{ minutes}}{1 \text{ hour}} = \frac{5 \text{ minutes}}{60 \text{ minutes}} = \frac{5}{60} = \frac{5 \div 5}{60 \div 5} = \frac{1}{12}$$

21. 5 gallons to 5 quarts

5 gallons $= 5 \cdot 4$ quarts $= 20$ quarts

$$\frac{5 \text{ gallons}}{5 \text{ quarts}} = \frac{20 \text{ quarts}}{5 \text{ quarts}} = \frac{20}{5} = \frac{20 \div 5}{5 \div 5} = \frac{4}{1}$$

23. Thanksgiving cards to graduation cards

$$\frac{10 \text{ million}}{60 \text{ million}} = \frac{10}{60} = \frac{10 \div 10}{60 \div 10} = \frac{1}{6}$$

25. Valentine's Day cards to Halloween cards

$$\frac{150 \text{ million}}{10 \text{ million}} = \frac{150}{10} = \frac{150 \div 10}{10 \div 10} = \frac{15}{1}$$

27. To get a ratio of $\frac{3}{1}$, we can start with the smallest value in the table and see if there is a value that is three times the smallest. $3 \times 9 = 27$ and 27 million is not in the table. $3 \times 10 = 30$, so there are at least 2 pairs of songs that give a ratio of $\frac{3}{1}$: *White Christmas* to *It's Now or Never* and *White Christmas* to *I Will Always Love You*.
$3 \times 12 = 36$, so there is another pair: *Candle in the Wind* to *I Want to Hold Your Hand*.
$3 \times 17 = 51$, which is greater than the largest value in the table, so we do not need to look for any more pairs.

29. $\dfrac{\text{longest side}}{\text{shortest side}} = \dfrac{1.8 \text{ meters}}{0.3 \text{ meters}} = \dfrac{1.8}{0.3}$

$$= \frac{1.8 \cdot 10}{0.3 \cdot 10} = \frac{18}{3} = \frac{18 \div 3}{3 \div 3} = \frac{6}{1}$$

The ratio of the length of the longest side to the length of the shortest side is $\frac{6}{1}$.

31. $\dfrac{\text{longest side}}{\text{shortest side}} = \dfrac{9\frac{1}{2} \text{ inches}}{4\frac{1}{4} \text{ inches}} = \dfrac{9\frac{1}{2}}{4\frac{1}{4}} = \dfrac{\frac{19}{2}}{\frac{17}{4}}$

$$= \frac{19}{2} \div \frac{17}{4} = \frac{19}{\underset{1}{2}} \cdot \frac{\overset{2}{4}}{17} = \frac{38}{17}$$

The ratio of the length of the longest side to the length of the shortest side is $\frac{38}{17}$.

33. The increase in price is

$$\$12.50 - \$10.00 = \$2.50.$$
$$\frac{\text{increase}}{\text{original}} = \frac{\$2.50}{\$10.00} = \frac{2.50}{10.00}$$
$$= \frac{2.50 \cdot 10}{10.00 \cdot 10} = \frac{25}{100}$$
$$= \frac{25 \div 25}{100 \div 25} = \frac{1}{4}$$

The ratio of the increase in price to the original price is $\frac{1}{4}$.

35. $59\frac{1}{2}$ days to $8\frac{3}{4}$ weeks

$$59\frac{1}{2} \text{ days} \div 7 = \frac{\overset{17}{\cancel{119}}}{2} \cdot \frac{1}{\underset{1}{7}} = \frac{17}{2} \text{ weeks}$$

$$\frac{59\frac{1}{2} \text{ days}}{8\frac{3}{4} \text{ weeks}} = \frac{\frac{17}{2} \text{ weeks}}{8\frac{3}{4} \text{ weeks}} = \frac{\frac{17}{2}}{\frac{35}{4}}$$

$$= \frac{17}{2} \div \frac{35}{4} = \frac{17}{\underset{1}{2}} \cdot \frac{\overset{2}{4}}{35} = \frac{34}{35}$$

The ratio of the first movie's filming time to the second movie's filming time is $\frac{34}{35}$.

6.2   Rates

6.2   Margin Exercises

1.   **(a)** $6 for 30 packets

To write the rate in lowest terms, divide both numerator and denominator by <u>6</u>.

$$\frac{\$6 \div 6}{30 \text{ packets} \div 6} = \frac{\$\underline{1}}{\underline{5} \text{ packets}}$$

(b) 500 miles in 10 hours

$$\frac{500 \text{ miles} \div 10}{10 \text{ hours} \div 10} = \frac{50 \text{ miles}}{1 \text{ hour}}$$

(c) 4 teachers for 90 students

$$\frac{4 \text{ teachers} \div 2}{90 \text{ students} \div 2} = \frac{2 \text{ teachers}}{45 \text{ students}}$$

(d) 1270 bushels from 30 acres

$$\frac{1270 \text{ bushels} \div 10}{30 \text{ acres} \div 10} = \frac{127 \text{ bushels}}{3 \text{ acres}}$$

2.   **(a)** $4.35 for 3 pounds of cheese

$$\frac{\$4.35 \div 3}{3 \text{ pounds} \div 3} = \frac{\$1.45}{1 \text{ pound}}$$

The unit rate is $1.45/pound.

(b) 304 miles on 9.5 gallons of gas

$$\frac{304 \text{ miles} \div 9.5}{9.5 \text{ gallons} \div 9.5} = \frac{32 \text{ miles}}{1 \text{ gallon}}$$

The unit rate is 32 miles/gallon.

(c) $850 in 5 days

$$\frac{\$850 \div 5}{5 \text{ days} \div 5} = \frac{\$170}{1 \text{ day}}$$

The unit rate is $170/day.

(d) 24-pound turkey for 15 people

$$\frac{24 \text{ pounds} \div 15}{15 \text{ people} \div 15} = \frac{1.6 \text{ pounds}}{1 \text{ person}}$$

The unit rate is 1.6 pounds/person.

3.   **(a)** Divide to find each unit cost.

Size	Cost per Unit
2 quarts	cost→ per→ quart→ $\dfrac{\$3.25}{2 \text{ quarts}}$

The unit cost is $<u>1.625</u> per quart. Similarly,

| 3 quarts | $\dfrac{\$4.95}{3 \text{ quarts}} = \$1.65/\text{qt}$ |

The unit cost is $1.65 per quart.

| 4 quarts | $\dfrac{\$6.48}{4 \text{ quarts}} = \$1.62/\text{qt}$ |

The unit cost is $1.62 per quart.

The lowest cost per quart is $1.62, so the best buy is 4 quarts for $6.48.

(b)

Size	Cost per Unit
6 cans	$\dfrac{\$1.99}{6 \text{ cans}} \approx \$0.332/\text{can}$
12 cans	$\dfrac{\$3.49}{12 \text{ cans}} \approx \$0.291/\text{can}$
24 cans	$\dfrac{\$7}{24 \text{ cans}} \approx \$0.292/\text{can}$

The lowest price per can is about $0.291, so the best buy is 12 cans of cola for $3.49.

4.   **(a)** One battery that lasts twice as long is like getting two batteries.

$$\frac{\$1.19 \div 2}{2 \text{ batteries} \div 2} = \$0.595/\text{battery. } (*)$$

The cost of the four-pack is

$$\frac{\$2.79 \div 4}{4 \text{ batteries} \div 4} = \$0.6975/\text{battery.}$$

The best buy is the single AA-size battery.

(b)   Brand C: $3.89 − $0.85 = $3.04

$$\frac{\$3.04 \div 6}{6 \text{ ounces} \div 6} \approx \$0.507/\text{ounce } (*)$$

Brand D: $1.89 − $0.50 = $1.39

$$\frac{\$1.39 \div 2.5}{2.5 \text{ ounces} \div 2.5} = \$0.556/\text{ounce}$$

Brand C with the 85¢ coupon is the best buy at $0.507 per ounce (rounded).

6.2   Section Exercises

1.   15 feet in 35 seconds

$$\frac{15 \text{ feet} \div 5}{35 \text{ seconds} \div 5} = \frac{3 \text{ feet}}{7 \text{ seconds}}$$

3.   25 letters in 5 minutes

$$\frac{25 \text{ letters} \div 5}{5 \text{ minutes} \div 5} = \frac{5 \text{ letters}}{1 \text{ minute}}$$

5.   72 miles on 4 gallons

$$\frac{72 \text{ miles} \div 4}{4 \text{ gallons} \div 4} = \frac{18 \text{ miles}}{1 \text{ gallon}}$$

7.   To find a unit rate, you <u>divide</u>. The correct set-up to find the unit rate of $5.85 for 3 boxes is

$$\frac{\$5.85}{3 \text{ boxes}}, \text{ or, equivalently, } 3\overline{)5\,.\,8\,5}$$

9. $60 in 5 hours

$$\frac{\$60 \div 5}{5 \text{ hours} \div 5} = \frac{\$12}{1 \text{ hour}}$$

The unit rate is $12 per hour or $12/hour.

11. 7.5 pounds for 6 people

$$\frac{7.5 \text{ pounds} \div 6}{6 \text{ people} \div 6} = \frac{1.25 \text{ pounds}}{1 \text{ person}}$$

The unit rate is 1.25 pounds/person.

13. $413.20 for 4 days

$$\frac{\$413.20 \div 4}{4 \text{ days} \div 4} = \frac{\$103.30}{1 \text{ day}}$$

The unit rate is $103.30/day.

15. Miles traveled:

$$27{,}758.2 - 27{,}432.3 = 325.9$$

Miles per gallon:

$$\frac{325.9 \text{ miles} \div 15.5}{15.5 \text{ gallons} \div 15.5} \approx 21.03 \approx 21.0$$

17. Miles traveled:

$$28{,}396.7 - 28{,}058.1 = 338.6$$

Miles per gallon:

$$\frac{338.6 \text{ miles} \div 16.2}{16.2 \text{ gallons} \div 16.2} \approx 20.90 \approx 20.9$$

19.

Size	Cost per Unit
2 ounces	$\dfrac{\$2.25}{2 \text{ ounces}} = \$1.125$
4 ounces	$\dfrac{\$3.65}{4 \text{ ounces}} = \$0.9125 \ (*)$

The best buy is 4 ounces for $3.65, about $0.913/ounce.

21.

Size	Cost per Unit
12 ounces	$\dfrac{\$2.49}{12 \text{ ounces}} = \$0.2075$
14 ounces	$\dfrac{\$2.89}{14 \text{ ounces}} \approx \$0.2064 \ (*)$
18 ounces	$\dfrac{\$3.96}{18 \text{ ounces}} = \$0.22$

The best buy is 14 ounces for $2.89, about $0.206/ounce.

23.

Size	Cost per Unit
12 ounces	$\dfrac{\$1.29}{12 \text{ ounces}} \approx \$0.108$
18 ounces	$\dfrac{\$1.79}{18 \text{ ounces}} \approx \$0.099 \ (*)$
28 ounces	$\dfrac{\$3.39}{28 \text{ ounces}} \approx \$0.121$
40 ounces	$\dfrac{\$4.39}{40 \text{ ounces}} \approx \$0.110$

The best buy is 18 ounces for $1.79, about $0.099/ounce.

25. Answers will vary. For example, you might choose Brand B because you like more chicken, so the cost per chicken chunk may actually be the same as or less than Brand A.

27. 10.5 pounds in 6 weeks

$$\frac{10.5 \text{ pounds} \div 6}{6 \text{ weeks} \div 6} = \frac{1.75 \text{ pounds}}{1 \text{ week}}$$

Her rate of loss was 1.75 pounds/week.

29. 7 hours to earn $85.82

$$\frac{\$85.82 \div 7}{7 \text{ hours} \div 7} = \frac{\$12.26}{1 \text{ hour}}$$

His pay rate is $12.26/hour.

31. (a) Add the connection fee and five times the cost per minute.

Penny Saver:
$0.39 + 5($0.005) = $0.39 + $0.025 = $0.415

Most Minutes: $0.49 + 5($0.0025) =
$0.49 + $0.0125 = $0.5025 ≈ $0.503

USA Card:
$0.25 + 5($0.01) = $0.30

(b) Divide the total cost for each card by five minutes.

Penny Saver: $\dfrac{\$0.415 \div 5}{5 \text{ minutes} \div 5} = \$0.083/\text{minute}$

Most Minutes: $\dfrac{\$0.503 \div 5}{5 \text{ minutes} \div 5} \approx \$0.101/\text{minute}$

USA Card: $\dfrac{\$0.30 \div 5}{5 \text{ minutes} \div 5} = \$0.06/\text{minute}$

USA Card is the best buy.

33. Add the connection fee and thirty times the cost per minute.

Penny Saver:
$0.39 + 30($0.005) = $0.39 + $0.15 = $0.54

$$\frac{\$0.54 \div 30}{30 \text{ minutes} \div 30} \approx \$0.018/\text{minute}$$

Most Minutes: $0.49 + 30($0.0025) =
$0.49 + $0.075 = $0.565

$$\frac{\$0.565 \div 30}{30 \text{ minutes} \div 30} \approx \$0.019/\text{minute}$$

USA Card:
$0.25 + 30($0.01) = $0.25 + $0.3 = $0.55

$$\frac{\$0.55 \div 30}{30 \text{ minutes} \div 30} \approx \$0.018/\text{minute}$$

For the 30-minute call, all three unit rates are very similar.

35. One battery for $1.79 is like getting 3 batteries for $1.79 ÷ 3 ≈ $0.597 per battery.

An eight-pack of AAA batteries for $4.99 is $4.99 ÷ 8 ≈ $0.624 per battery.

The best buy is the one battery package.

37. Brand G: $2.39 − $0.60 coupon = $1.79

$$\frac{\$1.79}{10 \text{ ounces}} = \$0.179 \text{ per ounce}$$

Brand K: $3.99, no coupon

$$\frac{\$3.99}{20.3 \text{ ounces}} \approx \$0.197 \text{ per ounce}$$

Brand P: $3.39 − $0.50 coupon = $2.89

$$\frac{\$2.89}{16.5 \text{ ounces}} \approx \$0.175 \text{ per ounce } (*)$$

Brand P with the 50¢ coupon has the lowest cost per ounce and is the best buy.

Relating Concepts (Exercises 39–43)

39. Plan A: $\dfrac{\$39.95}{450 \text{ anytime minutes}} \approx \$0.09/\text{minute}$

Plan B: $\dfrac{\$59.95}{900 \text{ anytime minutes}} \approx \$0.07/\text{minute}$

Plan B is the better buy.

40. Divide the number of "anytime minutes" by the number of days in a month.

(a) Plan A: $\dfrac{450 \text{ anytime minutes}}{30 \text{ days}} = 15 \text{ min/day}$

(b) Plan B: $\dfrac{900 \text{ anytime minutes}}{30 \text{ days}} = 30 \text{ min/day}$

41. 2 extra minutes per day times 30 days gives us 60 extra minutes. 60 extra minutes times 45¢ per extra minute gives us $60 \times \$0.45 = \$27$ in overage charges.

$$\frac{\text{Total charge}}{\text{Total minutes}} = \frac{\$39.95 + \$27}{450 + 60}$$

$$= \frac{\$66.95}{510 \text{ min}} \approx \$0.13/\text{min}$$

42. 3 extra minutes per day times 30 days gives us 90 extra minutes. 90 extra minutes times 40¢ per extra minute gives us $90 \times \$0.40 = \$36$ in overage charges.

$$\frac{\text{Total charge}}{\text{Total minutes}} = \frac{\$59.95 + \$36}{900 + 90}$$

$$= \frac{\$95.95}{990 \text{ min}} \approx \$0.10/\text{min}$$

43. Use $69.95 as the total charge.

From Exercise 41: $\dfrac{\$69.95}{510 \text{ min}} \approx \$0.14/\text{min}$

From Exercise 42: $\dfrac{\$69.95}{990 \text{ min}} \approx \$0.07/\text{min}$

6.3 Proportions

6.3 Margin Exercises

1. **(a)** $7 is to 3 cans as $28 is to 12 cans.

$$\frac{\$7}{3 \text{ cans}} = \frac{\$28}{12 \text{ cans}}$$

(b) 9 meters is to 16 meters as 18 meters is to 32 meters

$$\frac{9 \text{ m}}{16 \text{ m}} = \frac{18 \text{ m}}{32 \text{ m}}, \quad \text{so} \quad \frac{9}{16} = \frac{18}{32} \quad \textit{Common units cancel}$$

(c) 5 is to 7 as 35 is to 49.

$$\frac{5}{7} = \frac{35}{49}$$

(d) 10 is to 30 as 60 is to 180.

$$\frac{10}{30} = \frac{60}{180}$$

2. **(a)** $\dfrac{6}{12} = \dfrac{15}{30}$

$$\frac{6 \div 6}{12 \div 6} = \frac{1}{2} \text{ and } \frac{15 \div 15}{30 \div 15} = \frac{1}{2}$$

Both ratios are equivalent to $\frac{1}{2}$, so the proportion is *true.*

(b) $\dfrac{20}{24} = \dfrac{3}{4}$

$$\frac{20}{24} = \frac{20 \div 4}{24 \div 4} = \frac{5}{6} \text{ and } \frac{3}{4}$$

Because $\frac{5}{6}$ is *not* equivalent to $\frac{3}{4}$, the proportion is *false.*

(c) $\dfrac{25}{40} = \dfrac{30}{48}$

$$\frac{25 \div 5}{40 \div 5} = \frac{5}{8} \text{ and } \frac{30 \div 6}{48 \div 6} = \frac{5}{8}$$

Both ratios are equivalent to $\frac{5}{8}$, so the proportion is *true.*

(d) $\dfrac{35}{45} = \dfrac{12}{18}$

$$\frac{35 \div 5}{45 \div 5} = \frac{7}{9} \text{ and } \frac{12 \div 6}{18 \div 6} = \frac{2}{3}$$

Because $\frac{7}{9}$ is *not* equivalent to $\frac{2}{3}$, the proportion is *false.*

3. **(a)** $\dfrac{5}{9} = \dfrac{10}{18}$ Cross products: $\begin{aligned}(9)(10) &= \underline{90}\\(5)(18) &= \underline{90}\end{aligned}$

The cross products are *equal,* so the proportion is *true.*

(b) $\dfrac{32}{15} = \dfrac{16}{8}$ Cross products: $\begin{aligned}(15)(16) &= 240\\(32)(8) &= 256\end{aligned}$

The cross products are *unequal,* so the proportion is *false.*

(c) $\dfrac{10}{17} = \dfrac{20}{34}$ Cross products: $\begin{array}{l}(17)(20) = 340\\(10)(34) = 340\end{array}$

The cross products are *equal*, so the proportion is *true*.

(d) $\dfrac{2.4}{6} = \dfrac{5}{12}$

Cross products: $\begin{array}{l}(6)(5) = \underline{30}\\(2.4)(12) = \underline{28.8}\end{array}$

The cross products are *unequal*, so the proportion is *false*.

(e) $\dfrac{3}{4.25} = \dfrac{24}{34}$

Cross products: $\begin{array}{l}(4.25)(24) = 102\\(3)(34) = 102\end{array}$

The cross products are *equal*, so the proportion is *true*.

(f) $\dfrac{1\frac{1}{6}}{2\frac{1}{3}} = \dfrac{4}{8}$

Cross products: $2\dfrac{1}{3} \cdot 4 = \dfrac{7}{3} \cdot \dfrac{4}{1} = \dfrac{28}{3}$

$1\dfrac{1}{6} \cdot 8 = \dfrac{7}{\cancel{6}} \cdot \dfrac{\cancel{8}}{1} = \dfrac{28}{3}$

The cross products are *equal*, so the proportion is *true*.

4. (a) $\dfrac{1}{2} = \dfrac{x}{12}$

$2 \cdot x = 1 \cdot 12$ *Show that cross products are equal.*

$\dfrac{\cancel{2} \cdot x}{\cancel{2}} = \dfrac{1 \cdot \cancel{12}^{6}}{\cancel{2}}$ *Divide both sides by 2.*

$x = 6$

We will check our answers by showing that the cross products are equal.

Check $1 \cdot 12 = 12 = 2 \cdot 6$

(b) $\dfrac{6}{10} = \dfrac{15}{x}$

$6 \cdot x = 10 \cdot 15$ *Show that cross products are equal.*

$\dfrac{\cancel{6} \cdot x}{\cancel{6}} = \dfrac{\cancel{150}^{25}}{\cancel{6}}$ *Divide both sides by 6.*

$x = 25$

Check $6 \cdot 25 = 150 = 10 \cdot 15$

(c) $\dfrac{28}{x} = \dfrac{21}{9}$

$x \cdot 21 = 28 \cdot 9$ *Show that cross products are equal.*

$\dfrac{x \cdot \cancel{21}}{\cancel{21}} = \dfrac{\cancel{28}^{4} \cdot 9}{\cancel{21}_{3}}$ *Divide both sides by 21.*

$x = 12$

Check $28 \cdot 9 = 252 = 12 \cdot 21$

(d) $\dfrac{x}{8} = \dfrac{3}{5}$

$x \cdot 5 = 8 \cdot 3$ *Show that cross products are equal.*

$\dfrac{x \cdot \cancel{5}^{1}}{\cancel{5}_{1}} = \dfrac{24}{5}$ *Divide both sides by 5.*

$x = \dfrac{24}{5} = 4.8$

Check $(4.8)(5) = 24 = 8 \cdot 3$

(e) $\dfrac{14}{11} = \dfrac{x}{3}$

$11 \cdot x = 14 \cdot 3$ *Show that cross products are equal.*

$\dfrac{\cancel{11}^{1} \cdot x}{\cancel{11}_{1}} = \dfrac{14 \cdot 3}{11}$ *Divide both sides by 11.*

$x = \dfrac{42}{11} \approx 3.82$ *(rounded to the nearest hundredth)*

Check $14 \cdot 3 = 42 = 11 \cdot \frac{42}{11}$

5. (a) $\dfrac{3\frac{1}{4}}{2} = \dfrac{x}{8}$

$2 \cdot x = 3\dfrac{1}{4} \cdot 8$ *Show that cross products are equal.*

$2 \cdot x = \dfrac{13}{\cancel{4}_{1}} \cdot \dfrac{\cancel{8}^{2}}{1}$

$\dfrac{\cancel{2}^{1} \cdot x}{\cancel{2}_{1}} = \dfrac{\cancel{26}^{13}}{\cancel{2}_{1}}$ *Divide both sides by 2.*

$x = 13$

Check $3\frac{1}{4} \cdot 8 = 26 = 2 \cdot 13$

(b) $\dfrac{x}{3} = \dfrac{1\frac{2}{3}}{5}$

$5 \cdot x = 3 \cdot 1\dfrac{2}{3}$ *Show that cross products are equal.*

$5 \cdot x = \cancel{3}^{1} \cdot \dfrac{5}{\cancel{3}_{1}}$

$\dfrac{\cancel{5}^{1} \cdot x}{\cancel{5}_{1}} = \dfrac{1 \cdot \cancel{5}^{1}}{\cancel{5}_{1}}$ *Divide both sides by 5.*

$x = 1$

Check $1 \cdot 5 = 5 = 3 \cdot 1\frac{2}{3}$

(c) $\dfrac{0.06}{x} = \dfrac{0.3}{0.4}$

$0.3 \cdot x = 0.06 \cdot 0.4$ *Show that cross products are equal.*

$\dfrac{\cancel{0.3}^{\,1} \cdot x}{\cancel{0.3}_{\,1}} = \dfrac{\overset{0.08}{\cancel{0.024}}}{\cancel{0.3}_{\,1}}$ *Divide both sides by 0.3.*

$x = 0.08$

Check $(0.06)(0.4) = 0.024 = (0.08)(0.3)$

(d) $\dfrac{2.2}{5} = \dfrac{13}{x}$

$2.2 \cdot x = 5 \cdot 13$ *Show that cross products are equal.*

$\dfrac{\cancel{2.2}^{\,1} \cdot x}{\cancel{2.2}_{\,1}} = \dfrac{65}{2.2}$ *Divide both sides by 2.2.*

$x = \dfrac{65}{2.2} \approx 29.55$ *(rounded to the nearest hundredth)*

Check $2.2\left(\dfrac{65}{2.2}\right) = 65 = 5 \cdot 13$

(e) $\dfrac{x}{6} = \dfrac{0.5}{1.2}$

$1.2 \cdot x = 6(0.5)$ *Show that cross products are equal.*

$\dfrac{\cancel{1.2}^{\,1} \cdot x}{\cancel{1.2}_{\,1}} = \dfrac{3}{1.2} = \dfrac{30}{12} = \dfrac{5}{2}$ *Divide both sides by 1.2.*

$x = 2.5$

Check $(2.5)(1.2) = 3 = 6(0.5)$

(f) $\dfrac{0}{2} = \dfrac{x}{7.092}$ Since $\dfrac{0}{2} = 0$, x *must* equal 0.

Note: The denominator 7.092 could be any number (except 0).

6.3 Section Exercises

1. $9 is to 12 cans as $18 is to 24 cans.

$$\dfrac{\$9}{12 \text{ cans}} = \dfrac{\$18}{24 \text{ cans}}$$

3. 200 adults is to 450 children as 4 adults is to 9 children.

$$\dfrac{200 \text{ adults}}{450 \text{ children}} = \dfrac{4 \text{ adults}}{9 \text{ children}}$$

5. 120 feet is to 150 feet as 8 feet is to 10 feet.

$\dfrac{120}{150} = \dfrac{8}{10}$ *The common units (feet) cancel.*

7. $\dfrac{6}{10} = \dfrac{3}{5}$

$$\dfrac{6 \div 2}{10 \div 2} = \dfrac{3}{5} \text{ and } \dfrac{3}{5}$$

Both ratios are equivalent to $\frac{3}{5}$, so the proportion is *true.*

9. $\dfrac{5}{8} = \dfrac{25}{40}$

$$\dfrac{5}{8} \text{ and } \dfrac{25 \div 5}{40 \div 5} = \dfrac{5}{8}$$

Both ratios are equivalent to $\frac{5}{8}$, so the proportion is *true.*

11. $\dfrac{150}{200} = \dfrac{200}{300}$

$$\dfrac{150 \div 50}{200 \div 50} = \dfrac{3}{4} \text{ and } \dfrac{200 \div 100}{300 \div 100} = \dfrac{2}{3}$$

Because $\frac{3}{4}$ is *not* equivalent to $\frac{2}{3}$, the proportion is *false.*

13. Answers may vary. One example: A proportion states that two ratios (or rates) are equal. The multiplications on the proportion should show that $6 \cdot 45 = 270$ and $30 \cdot 9 = 270$.

15. $\dfrac{2}{9} = \dfrac{6}{27}$ Cross products: $\begin{matrix} 9 \cdot 6 = 54 \\ 2 \cdot 27 = 54 \end{matrix}$

The cross products are *equal,* so the proportion is *true.*

17. $\dfrac{20}{28} = \dfrac{12}{16}$ Cross products: $\begin{matrix} 28 \cdot 12 = 336 \\ 20 \cdot 16 = 320 \end{matrix}$

The cross products are *unequal,* so the proportion is *false.*

19. $\dfrac{110}{18} = \dfrac{160}{27}$ Cross products: $\begin{matrix} 18 \cdot 160 = 2880 \\ 110 \cdot 27 = 2970 \end{matrix}$

The cross products are *unequal,* so the proportion is *false.*

21. $\dfrac{3.5}{4} = \dfrac{7}{8}$ Cross products: $\begin{matrix} 4 \cdot 7 = 28 \\ (3.5)(8) = 28 \end{matrix}$

The cross products are *equal,* so the proportion is *true.*

23. $\dfrac{18}{15} = \dfrac{2\frac{5}{6}}{2\frac{1}{2}}$ Cross products:

$$15 \cdot 2\frac{5}{6} = \dfrac{\cancel{15}^{\,5}}{1} \cdot \dfrac{17}{\cancel{6}_{\,2}} = \dfrac{85}{2} = 42\frac{1}{2}$$

$$18 \cdot 2\frac{1}{2} = \dfrac{\cancel{18}^{\,9}}{1} \cdot \dfrac{5}{\cancel{2}_{\,1}} = 45$$

The cross products are *unequal,* so the proportion is *false.*

25. $\dfrac{6}{3\frac{2}{3}} = \dfrac{18}{11}$

Cross products: $3\frac{2}{3} \cdot 18 = \dfrac{11}{\cancel{3}_{\,1}} \cdot \dfrac{\cancel{18}^{\,6}}{1} = 66$

$$6 \cdot 11 = 66$$

The cross products are *equal,* so the proportion is *true.*

27. (Answers may vary: proportion may also be set up with "at bats" in both numerators and "hits" in both denominators. Cross product will be the same.)

$$\frac{\overset{\text{Joe Mauer}}{68 \text{ hits}}}{200 \text{ at bats}} = \frac{\overset{\text{Joey Votto}}{153 \text{ hits}}}{450 \text{ at bats}}$$

Cross products: $\begin{array}{l} 200 \cdot 153 = 30{,}600 \\ 68 \cdot 450 = 30{,}600 \end{array}$

The cross products are *equal,* so the proportion is *true.* Paula is correct; the two men hit equally well.

29. $\dfrac{1}{3} = \dfrac{x}{12}$

$3 \cdot x = 1 \cdot 12$ *Show that cross products are equal.*

$3 \cdot x = \underline{12}$

$\dfrac{\overset{1}{\cancel{3}} \cdot x}{\underset{1}{\cancel{3}}} = \dfrac{12}{3}$ *Divide both sides by 3.*

$x = \underline{4}$

We will check our answers by showing that the cross products are equal.

Check $1 \cdot 12 = 12 = 3 \cdot 4$

31. It is a good idea to first reduce any fractions. We will do this when possible, and work with the reduced fraction.

$$\frac{15}{10} = \frac{3}{x} \quad \textbf{OR} \quad \frac{3}{2} = \frac{3}{x}$$

Since the numerators are equal, the denominators must be equal, so $x = 2$.

33. $\dfrac{x}{11} = \dfrac{32}{4} \quad \textbf{OR} \quad \dfrac{x}{11} = \dfrac{8}{1}$

$1 \cdot x = 11 \cdot 8$ *Show that cross products are equal.*

$x = 88$

Check $88 \cdot 4 = 352 = 11 \cdot 32$

35. $\dfrac{42}{x} = \dfrac{18}{39}$

$x \cdot 18 = 42 \cdot 39$ *Show that cross products are equal.*

$\dfrac{x \cdot \overset{1}{\cancel{18}}}{\underset{1}{\cancel{18}}} = \dfrac{\overset{7}{\cancel{42}} \cdot \overset{13}{\cancel{39}}}{\underset{\underset{1}{\cancel{6}}}{\cancel{18}}}$ *Divide both sides by 18.*

$x = 91$

Check $42 \cdot 39 = 1638 = 91 \cdot 18$

37. $\dfrac{x}{25} = \dfrac{4}{20} \quad \textbf{OR} \quad \dfrac{x}{25} = \dfrac{1}{5}$

$5 \cdot x = 25 \cdot 1$ *Show that cross products are equal.*

$\dfrac{\overset{1}{\cancel{5}} \cdot x}{\underset{1}{\cancel{5}}} = \dfrac{\overset{5}{\cancel{25}}}{\underset{1}{\cancel{5}}}$ *Divide both sides by 5.*

$x = 5$

Check $5 \cdot 20 = 100 = 25 \cdot 4$

39. $\dfrac{8}{x} = \dfrac{24}{30} \quad \textbf{OR} \quad \dfrac{8}{x} = \dfrac{4}{5}$

$4 \cdot x = 8 \cdot 5$ *Show that cross products are equal.*

$\dfrac{\overset{1}{\cancel{4}} \cdot x}{\underset{1}{\cancel{4}}} = \dfrac{\overset{2}{\cancel{8}} \cdot 5}{\underset{1}{\cancel{4}}}$ *Divide both sides by 4.*

$x = 10$

Check $8 \cdot 30 = 240 = 10 \cdot 24$

41. $\dfrac{99}{55} = \dfrac{44}{x} \quad \textbf{OR} \quad \dfrac{9}{5} = \dfrac{44}{x}$

$9 \cdot x = 5 \cdot 44$ *Show that cross products are equal.*

$\dfrac{\overset{1}{\cancel{9}} \cdot x}{\underset{1}{\cancel{9}}} = \dfrac{220}{9}$ *Divide both sides by 9.*

$x = \dfrac{220}{9}$

≈ 24.44 (rounded)

Check $99 \cdot \frac{220}{9} = 2420 = 55 \cdot 44$

43. $\dfrac{0.7}{9.8} = \dfrac{3.6}{x}$

$0.7 \cdot x = 9.8\,(3.6)$ *Show that cross products are equal.*

$\dfrac{\overset{1}{\cancel{0.7}} \cdot x}{\underset{1}{\cancel{0.7}}} = \dfrac{35.28}{0.7}$ *Divide both sides by 0.7.*

$x = 50.4$

Check $0.7(50.4) = 35.28 = 9.8(3.6)$

45. $\dfrac{250}{24.8} = \dfrac{x}{1.75}$

$24.8 \cdot x = 250(1.75)$ *Show that cross products are equal.*

$\dfrac{\overset{1}{\cancel{24.8}} \cdot x}{\underset{1}{\cancel{24.8}}} = \dfrac{437.5}{24.8}$ *Divide both sides by 24.8.*

≈ 17.64 (rounded)

Check $250(1.75) = 437.5 = 24.8 \cdot \frac{437.5}{24.8}$

47.
$$\frac{15}{1\frac{2}{3}} = \frac{9}{x}$$

$15 \cdot x = 1\frac{2}{3} \cdot 9$ *Show that cross products are equal.*

$15 \cdot x = \frac{5}{\cancel{3}} \cdot \frac{\cancel{9}^{3}}{1}$

$\frac{\cancel{15}^{1} \cdot x}{\cancel{15}_{1}} = \frac{15}{15}$ *Divide both sides by 15.*

$x = 1$

49.
$$\frac{2\frac{1}{3}}{1\frac{1}{2}} = \frac{x}{2\frac{1}{4}}$$

$1\frac{1}{2} \cdot x = 2\frac{1}{3} \cdot 2\frac{1}{4}$ *Show that cross products are equal.*

$\frac{3}{2} \cdot x = \frac{7}{\cancel{3}} \cdot \frac{\cancel{9}^{3}}{4}$

$\frac{3}{2} \cdot x = \frac{21}{4}$

$\frac{\frac{3}{2} \cdot x}{\frac{3}{2}} = \frac{\frac{21}{4}}{\frac{3}{2}}$ *Divide both sides by 3/2.*

$x = 3\frac{1}{2}$ $\frac{21}{4} \div \frac{3}{2} = \frac{\cancel{21}^{7}}{\cancel{4}_{2}} \cdot \frac{\cancel{2}^{1}}{\cancel{3}_{1}} = \frac{7}{2}$

51. Change $\frac{1}{2}$ to a decimal by dividing: $1 \div 2 = 0.5$

$$\frac{\frac{1}{2}}{x} = \frac{2}{0.8}$$

$$\frac{0.5}{x} = \frac{2}{0.8}$$

$x \cdot 2 = 0.5(0.8)$

$x \cdot 2 = 0.4$

$\frac{x \cdot \cancel{2}^{1}}{\cancel{2}_{1}} = \frac{0.4}{2}$

$x = 0.2$

Now change 0.8 to a fraction and write it in lowest terms.

$$0.8 = \frac{8 \div 2}{10 \div 2} = \frac{4}{5}$$

$$\frac{\frac{1}{2}}{x} = \frac{2}{\frac{4}{5}}$$

$x \cdot 2 = \frac{1}{\cancel{2}} \cdot \frac{\cancel{4}^{2}}{5}$

$x \cdot 2 = \frac{2}{5}$

$$\frac{x \cdot \cancel{2}}{\cancel{2}_{1}} = \frac{\frac{2}{5}}{2}$$

$x = \frac{2}{5} \div 2 = \frac{\cancel{2}}{5} \cdot \frac{1}{\cancel{2}_{1}} = \frac{1}{5}$

53.
$$\frac{x}{\frac{3}{50}} = \frac{0.15}{1\frac{4}{5}}$$

Change to decimals:

$$\frac{x}{0.06} = \frac{0.15}{1.8}$$

$x \cdot 1.8 = 0.15(0.06)$

$\frac{x \cdot 1.8}{1.8} = \frac{0.009}{1.8}$

$x = 0.005$

Change to fractions:

$$\frac{x}{\frac{3}{50}} = \frac{\frac{3}{20}}{1\frac{4}{5}}$$

$1\frac{4}{5} \cdot x = \frac{3}{50} \cdot \frac{3}{20}$

$\frac{1\frac{4}{5} \cdot x}{1\frac{4}{5}} = \frac{\frac{9}{1000}}{1\frac{4}{5}}$

$x = \frac{9}{1000} \div 1\frac{4}{5} = \frac{\cancel{9}^{1}}{\cancel{1000}_{200}} \cdot \frac{\cancel{5}}{\cancel{9}_{1}} = \frac{1}{200}$

Relating Concepts (Exercises 55–56)

55. $\frac{10}{4} = \frac{5}{3}$ ▪ Find the cross products.

$$4 \cdot 5 = 20$$
$$10 \cdot 3 = 30$$

The cross products are *unequal,* so the proportion is *not* true.

Replace 10 with x.

$$\frac{x}{4} = \frac{5}{3}$$

$3 \cdot x = 4 \cdot 5$

$x = \frac{20}{3} = 6\frac{2}{3}$, so

$\frac{6\frac{2}{3}}{4} = \frac{5}{3}$ is a true proportion.

Replace 4 with x.

$$\frac{10}{x} = \frac{5}{3}$$

$5 \cdot x = 10 \cdot 3$

$x = \frac{30}{5} = 6$, so

$\frac{10}{6} = \frac{5}{3}$ is a true proportion.

Replace 5 with x.

$$\frac{10}{4} = \frac{x}{3}$$
$$4 \cdot x = 10 \cdot 3$$
$$x = \frac{30}{4} = \frac{15}{2} = 7.5, \text{ so}$$
$$\frac{10}{4} = \frac{7.5}{3} \text{ is a true proportion.}$$

Replace 3 with x.

$$\frac{10}{4} = \frac{5}{x}$$
$$10 \cdot x = 4 \cdot 5$$
$$x = \frac{20}{10} = 2, \text{ so}$$
$$\frac{10}{4} = \frac{5}{2} \text{ is a true proportion.}$$

56. $\dfrac{6}{8} = \dfrac{24}{30}$ ■ Find the cross products.

$$8 \cdot 24 = 192$$
$$6 \cdot 30 = 180$$

The cross products are *unequal,* so the proportion is *not* true.

Replace 6 with x.

$$\frac{x}{8} = \frac{24}{30}$$
$$30 \cdot x = 8 \cdot 24$$
$$x = \frac{8 \cdot \overset{4}{\cancel{24}}}{\underset{5}{\cancel{30}}} = \frac{32}{5} = 6.4, \text{ so}$$
$$\frac{6.4}{8} = \frac{24}{30} \text{ is a true proportion.}$$

Replace 8 with x.

$$\frac{6}{x} = \frac{24}{30}$$
$$24 \cdot x = 6 \cdot 30$$
$$x = \frac{6 \cdot \overset{5}{\cancel{30}}}{\underset{4}{\cancel{24}}} = \frac{30}{4} = 7.5, \text{ so}$$
$$\frac{6}{7.5} = \frac{24}{30} \text{ is a true proportion.}$$

Replace 24 with x.

$$\frac{6}{8} = \frac{x}{30}$$
$$8 \cdot x = 6 \cdot 30$$
$$x = \frac{\overset{3}{\cancel{6}} \cdot \overset{15}{\cancel{30}}}{\underset{\underset{2}{\cancel{4}}}{\cancel{8}}} = \frac{45}{2} = 22.5, \text{ so}$$
$$\frac{6}{8} = \frac{22.5}{30} \text{ is a true proportion.}$$

Replace 30 with x.

$$\frac{6}{8} = \frac{24}{x}$$
$$6 \cdot x = 8 \cdot 24$$
$$x = \frac{8 \cdot \overset{4}{\cancel{24}}}{\underset{1}{\cancel{6}}} = 32, \text{ so}$$
$$\frac{6}{8} = \frac{24}{32} \text{ is a true proportion.}$$

Summary Exercises
Ratios, Rates, and Proportions

1. 50 million websites in Japan to 25 million websites in Germany

$$\frac{50 \text{ million}}{25 \text{ million}} = \frac{50 \div 25}{25 \div 25} = \frac{2}{1}$$

3. combined number of websites in Italy and Germany (20 million + 25 million = 45 million) to 50 million websites in Japan

$$\frac{45 \text{ million}}{50 \text{ million}} = \frac{45 \div 5}{50 \div 5} = \frac{9}{10}$$

5. **(i)** 2 million violin players to 22 million piano players

$$\frac{2 \text{ million}}{22 \text{ million}} = \frac{2 \div 2}{22 \div 2} = \frac{1}{11}$$

(ii) 2 million violin players to 20 million guitar players

$$\frac{2 \text{ million}}{20 \text{ million}} = \frac{2 \div 2}{20 \div 2} = \frac{1}{10}$$

(iii) 2 million violin players to 6 million organ players

$$\frac{2 \text{ million}}{6 \text{ million}} = \frac{2 \div 2}{6 \div 2} = \frac{1}{3}$$

(iv) 2 million violin players to 4 million clarinet players

$$\frac{2 \text{ million}}{4 \text{ million}} = \frac{2 \div 2}{4 \div 2} = \frac{1}{2}$$

(v) 2 million violin players to 3 million drum players

$$\frac{2 \text{ million}}{3 \text{ million}} = \frac{2}{3}$$

7. $\dfrac{100 \text{ points}}{48 \text{ minutes}} = \dfrac{100 \div 48 \text{ points}}{48 \div 48 \text{ minutes}}$
$$= 2.08\overline{3} \approx 2.1 \text{ points/minute}$$

$\dfrac{48 \text{ minutes}}{100 \text{ points}} = \dfrac{48 \div 100 \text{ minutes}}{100 \div 100 \text{ points}}$
$$= 0.48 \approx 0.5 \text{ minute/point}$$

9. $\dfrac{\$652.80}{40 \text{ hours}} = \dfrac{\$652.80 \div 40}{40 \div 40 \text{ hours}} = \$16.32/\text{hour}$

Her regular hourly pay rate is $\$16.32/\text{hour}$.

$\dfrac{\$195.84}{8 \text{ hours}} = \dfrac{\$195.84 \div 8}{8 \div 8 \text{ hours}} = \$24.48/\text{hour}$

Her overtime hourly pay rate is $\$24.48/\text{hour}$.

11.

Size	Cost per Unit
11 ounces	$\dfrac{\$6.79}{11 \text{ ounces}} \approx \$0.62$
12 ounces	$\dfrac{\$7.24}{12 \text{ ounces}} \approx \$0.60 \; (*)$
16 ounces	$\dfrac{\$10.99}{16 \text{ ounces}} \approx \$0.69$

The best buy is 12 ounces for $7.24, about $0.60 per ounce.

13. $\dfrac{28}{21} = \dfrac{44}{33}$

$\dfrac{28 \div 7}{21 \div 7} = \dfrac{4}{3}$ and $\dfrac{44 \div 11}{33 \div 11} = \dfrac{4}{3}$

Both ratios are equivalent to $\frac{4}{3}$, so the proportion is *true*.

Note: We could also use the method of finding cross products and show that both cross products equal 924.

15. $\dfrac{2\frac{5}{8}}{3\frac{1}{4}} = \dfrac{21}{26}$

$3\dfrac{1}{4} \cdot 21 = \dfrac{13}{4} \cdot \dfrac{21}{1} = \dfrac{273}{4}$

Cross products:

$2\dfrac{5}{8} \cdot 26 = \dfrac{21}{\cancel{8}} \cdot \dfrac{\overset{13}{\cancel{26}}}{1} = \dfrac{273}{4}$

The cross products are *equal*, so the proportion is *true*.

17. $\dfrac{15}{8} = \dfrac{6}{x}$

$15 \cdot x = 8 \cdot 6$ *Show that cross products are equal.*

$\dfrac{\overset{1}{\cancel{15}} \cdot x}{\cancel{15}} = \dfrac{8 \cdot \overset{2}{\cancel{6}}}{\underset{5}{\cancel{15}}}$ *Divide both sides by 15.*

$x = \dfrac{16}{5} = 3.2$

Check $15(3.2) = 48 = 8 \cdot 6$

19. $\dfrac{10}{11} = \dfrac{x}{4}$

$11 \cdot x = 10 \cdot 4$ *Show that cross products are equal.*

$\dfrac{\overset{1}{\cancel{11}} \cdot x}{\cancel{11}} = \dfrac{40}{11}$ *Divide both sides by 11.*

$x \approx 3.64$ (rounded)

Check $10 \cdot 4 = 40 = 11 \cdot \frac{40}{11}$

21. $\dfrac{2.6}{x} = \dfrac{13}{7.8}$

$13 \cdot x = 2.6(7.8)$ *Show that cross products are equal.*

$\dfrac{\overset{1}{\cancel{13}} \cdot x}{\underset{1}{\cancel{13}}} = \dfrac{20.28}{13}$ *Divide both sides by 13.*

$x = 1.56$

Check $2.6(7.8) = 20.28 = 1.56(13)$

23. $\dfrac{\frac{1}{3}}{8} = \dfrac{x}{24}$

$8 \cdot x = \dfrac{1}{3} \cdot 24$ *Show that cross products are equal.*

$\dfrac{\overset{1}{\cancel{8}} \cdot x}{\underset{1}{\cancel{8}}} = \dfrac{8}{8}$ *Divide both sides by 8.*

$x = 1$

Check $\frac{1}{3}(24) = 8 = 8 \cdot 1$

6.4 Problem Solving with Proportions

6.4 Margin Exercises

1. **(a)** $\dfrac{2 \text{ pounds}}{50 \text{ square feet}} = \dfrac{x \text{ pounds}}{225 \text{ square feet}}$

$50 \cdot x = 2 \cdot 225$ *Show that cross products are equal.*

$\dfrac{50 \cdot x}{50} = \dfrac{450}{50}$ *Divide both sides by 50.*

$x = 9$

9 pounds of fertilizer are needed for 225 square feet.

(b) $\dfrac{1 \text{ inch}}{75 \text{ miles}} = \dfrac{4.75 \text{ inches}}{x \text{ miles}}$

$1 \cdot x = 75(4.75)$ *Show that cross products are equal.*

$x = 356.25 \text{ miles}$

The lake's actual length is about 356 miles.

(c) $\dfrac{30 \text{ milliliters}}{100 \text{ pounds}} = \dfrac{x \text{ milliliters}}{34 \text{ pounds}}$

$100 \cdot x = 30 \cdot 34$ *Show that cross products are equal.*

$\dfrac{100 \cdot x}{100} = \dfrac{1020}{100}$ *Divide both sides by 100.*

$x = 10.2 \text{ milliliters}$

A 34-pound child should be given about 10 milliliters of cough syrup.

2.   (a) $\dfrac{\text{(lose weight) 2 people}}{\text{(surveyed) 3 people}} = \dfrac{x \text{ (lose weight)}}{150 \text{ people (in group)}}$

$3 \cdot x = 2 \cdot 150$ *Show that cross products are equal.*

$\dfrac{3 \cdot x}{3} = \dfrac{300}{3}$ *Divide both sides by 3.*

$x = 100$

In a group of 150, 100 people want to lose weight. This is a reasonable answer because it's more than half the people, but less than all the people.

Incorrect setup

$$\dfrac{3 \text{ surveyed}}{2 \text{ want to lose}} = \dfrac{x}{150 \text{ total}}$$
$$2 \cdot x = 3 \cdot 150$$
$$\dfrac{2 \cdot x}{2} = \dfrac{450}{2}$$
$$x = 225$$

The incorrect setup gives an answer of 225 people, which is unreasonable since there are only 150 people in the group.

(b) $\dfrac{3 \text{ FA students}}{5 \text{ students}} = \dfrac{x}{4500 \text{ students}}$

$5 \cdot x = 3 \cdot 4500$ *Show that cross products are equal.*

$\dfrac{5 \cdot x}{5} = \dfrac{13,500}{5}$ *Divide both sides by 5.*

$x = 2700$

At Central Community College 2700 students receive financial aid. This is a reasonable answer because it's more than half the people, but less than all the people.

Incorrect setup

$$\dfrac{5 \text{ students}}{3 \text{ FA students}} = \dfrac{x}{4500 \text{ students}}$$
$$3 \cdot x = 5 \cdot 4500$$
$$\dfrac{3 \cdot x}{3} = \dfrac{22,500}{3}$$
$$x = 7500$$

The incorrect setup gives an answer of 7500 students, which is unreasonable since there are only 4500 students at the college.

6.4   Section Exercises

1. $\dfrac{6 \text{ potatoes}}{4 \text{ eggs}} = \dfrac{12 \text{ potatoes}}{x \text{ eggs}}$

3. Here's one set-up:

$$\dfrac{4 \text{ cartoon strips}}{5 \text{ hours}} = \dfrac{18 \text{ cartoon strips}}{x \text{ hours}}$$

Here's another set-up and solution:

$\dfrac{5 \text{ hours}}{4 \text{ cartoon strips}} = \dfrac{x \text{ hours}}{18 \text{ cartoon strips}}$

$$\dfrac{5}{4} = \dfrac{x}{18}$$
$$4 \cdot x = 5 \cdot 18$$
$$4 \cdot x = 90$$
$$\dfrac{\overset{1}{\cancel{4}} \cdot x}{\underset{1}{\cancel{4}}} = \dfrac{90}{4}$$
$$x = 22.5$$

It will take 22.5 hours to sketch 18 cartoon strips.

5. $\dfrac{60 \text{ newspapers}}{\$27} = \dfrac{16 \text{ newspapers}}{x}$

$$60 \cdot x = 27 \cdot 16$$
$$\dfrac{60 \cdot x}{60} = \dfrac{432}{60}$$
$$x = 7.2$$

The cost of 16 newspapers is \$7.20.

7. $\dfrac{3 \text{ pounds}}{350 \text{ square feet}} = \dfrac{x}{4900 \text{ square feet}}$

$$350 \cdot x = 3 \cdot 4900$$
$$\dfrac{350 \cdot x}{350} = \dfrac{14,700}{350}$$
$$x = 42$$

42 pounds are needed to cover 4900 square feet.

9. $\dfrac{\$672.80}{5 \text{ days}} = \dfrac{x}{3 \text{ days}}$

$$5 \cdot x = (672.80)(3)$$
$$\dfrac{5 \cdot x}{5} = \dfrac{2018.40}{5}$$
$$x = 403.68$$

In 3 days Tom makes \$403.68.

11. $\dfrac{6 \text{ ounces}}{7 \text{ servings}} = \dfrac{x}{12 \text{ servings}}$

$$7 \cdot x = 6 \cdot 12$$
$$\dfrac{7 \cdot x}{7} = \dfrac{72}{7}$$
$$x = 10\dfrac{2}{7} \approx 10$$

You need about 10 ounces for 12 servings.

13. $\dfrac{3 \text{ quarts}}{270 \text{ square feet}} = \dfrac{x}{(350 + 100) \text{ square feet}}$

$$\dfrac{3}{270} = \dfrac{x}{450}$$
$$270 \cdot x = 3 \cdot 450$$
$$\dfrac{270 \cdot x}{270} = \dfrac{1350}{270}$$
$$x = 5$$

You will need 5 quarts.

15. First find the length.

$$\frac{1 \text{ inch}}{4 \text{ feet}} = \frac{3.5 \text{ inches}}{x}$$
$$1 \cdot x = 4(3.5)$$
$$x = 14$$

The kitchen is 14 feet long.

Then find the width.

$$\frac{1 \text{ inch}}{4 \text{ feet}} = \frac{2.5 \text{ inches}}{x \text{ feet}}$$
$$1 \cdot x = 4(2.5)$$
$$x = 10$$

The kitchen is 10 feet wide.

17. The length of the dining area is the same as the length of the kitchen, which is 14 feet by Exercise 15.

Find the width of the dining area. It is $4.5 - 2.5$ inches $= 2$ inches on the floor plan.

$$\frac{1 \text{ inch}}{4 \text{ feet}} = \frac{2 \text{ inches}}{x}$$
$$1 \cdot x = 4 \cdot 2$$
$$x = 8$$

The dining area is 8 feet wide.

19. Set up and solve a proportion with pieces of chicken and number of guests.

$$\frac{40 \text{ pieces}}{25 \text{ guests}} = \frac{x}{60 \text{ guests}}$$
$$25 \cdot x = 40 \cdot 60$$
$$\frac{25 \cdot x}{25} = \frac{40 \cdot 60}{25}$$
$$x = \frac{40 \cdot 60}{25} = 96 \text{ pieces of chicken}$$

For the other food items, the proportions are similar, so we simply replace "40" in the last step with the appropriate value.

$$\frac{14 \cdot 60}{25} = 33.6 \text{ pounds of lasagna}$$

$$\frac{4.5 \cdot 60}{25} = 10.8 \text{ pounds of deli meats}$$

$$\frac{\frac{7}{3} \cdot 60}{25} = \frac{28}{5} = 5\frac{3}{5} \text{ pounds of cheese}$$

$$\frac{3 \text{ dozen} \cdot 60}{25} = 7.2 \text{ dozen (about 86) buns}$$

$$\frac{6 \cdot 60}{25} = 14.4 \text{ pounds of salad}$$

21. 44 students is an *unreasonable* answer because there are only 35 students in the class.

23.
$$\frac{7 \text{ refresher}}{10 \text{ entering}} = \frac{x}{2950 \text{ entering}}$$
$$10 \cdot x = 7 \cdot 2950$$
$$\frac{10 \cdot x}{10} = \frac{20{,}650}{10}$$
$$x = 2065$$

2065 students will probably need a refresher course. This is a reasonable answer because it's more than half the students, but not all the students.

Incorrect setup
$$\frac{10 \text{ entering}}{7 \text{ refresher}} = \frac{x}{2950 \text{ entering}}$$
$$7 \cdot x = 10 \cdot 2950$$
$$\frac{7 \cdot x}{7} = \frac{29{,}500}{7}$$
$$x \approx 4214$$

The incorrect setup gives an estimate of 4214 entering students, which is unreasonable since there are only 2950 entering students.

25.
$$\frac{1 \text{ chooses vanilla}}{3 \text{ people}} = \frac{x \text{ choose vanilla}}{238 \text{ people}}$$
$$3 \cdot x = 1 \cdot 238$$
$$\frac{3 \cdot x}{3} = \frac{238}{3}$$
$$x \approx 79.3 \approx 79$$

You would expect about 79 people to choose vanilla ice cream. This is a reasonable answer.

Incorrect setup
$$\frac{3 \text{ people}}{1 \text{ chooses vanilla}} = \frac{x \text{ choose vanilla}}{238 \text{ people}}$$
$$1 \cdot x = 3 \cdot 238$$
$$x = 714$$

With an incorrect setup, 714 people choose vanilla ice cream. This is unreasonable because only 238 people attended the ice cream social.

27.
$$\frac{5 \text{ stocks up}}{6 \text{ stocks down}} = \frac{x \text{ stocks up}}{750 \text{ stocks down}}$$
$$\frac{5}{6} = \frac{x}{750}$$
$$6 \cdot x = 5 \cdot 750$$
$$\frac{6 \cdot x}{6} = \frac{3750}{6}$$
$$x = 625$$

625 stocks went up.

29.
$$\frac{8 \text{ length}}{1 \text{ width}} = \frac{32.5 \text{ meters length}}{x \text{ meters width}}$$
$$8 \cdot x = 1 \cdot 32.5$$
$$\frac{8 \cdot x}{8} = \frac{32.5}{8}$$
$$x = 4.0625 \approx 4.06$$

The wing must be about 4.06 meters wide.

31. $\dfrac{150 \text{ pounds}}{222 \text{ calories}} = \dfrac{210 \text{ pounds}}{x}$

$$150 \cdot x = 222 \cdot 210$$
$$\dfrac{150 \cdot x}{150} = \dfrac{46{,}620}{150}$$
$$x = 310.8 \approx 311$$

A 210-pound person would burn about 311 calories.

33. $\dfrac{1.05 \text{ meters}}{1.68 \text{ meters}} = \dfrac{6.58 \text{ meters}}{x}$

$$1.05 \cdot x = 1.68(6.58)$$
$$\dfrac{1.05 \cdot x}{1.05} = \dfrac{11.0544}{1.05}$$
$$x = 10.528 \approx 10.53$$

The height of the tree is about 10.53 meters.

35. You cannot solve this problem using a proportion because the ratio of age to weight is not constant. As Jim's age increases from 25 to 50 years old, his weight may decrease, stay the same, or increase.

37. First find the number of coffee drinkers.

$$\dfrac{4 \text{ coffee drinkers}}{5 \text{ students}} = \dfrac{x}{56{,}100 \text{ students}}$$
$$5 \cdot x = 4 \cdot 56{,}100$$
$$\dfrac{5 \cdot x}{5} = \dfrac{224{,}400}{5}$$
$$x = 44{,}880$$

The survey showed that 44,880 students drink coffee. Now find the number of coffee drinkers who use cream.

$$\dfrac{1 \text{ uses cream}}{8 \text{ coffee drinkers}} = \dfrac{x}{44{,}880 \text{ coffee drinkers}}$$
$$8 \cdot x = 44{,}880$$
$$\dfrac{8 \cdot x}{8} = \dfrac{44{,}880}{8}$$
$$x = 5610$$

According to the survey, 5610 students use cream.

39. First find the number of calories in a $\frac{1}{2}$-cup serving of bran cereal.

$$\dfrac{\frac{1}{3} \text{ cup}}{80 \text{ calories}} = \dfrac{\frac{1}{2} \text{ cup}}{x}$$
$$\dfrac{1}{3} \cdot x = 80 \cdot \dfrac{1}{2}$$
$$\dfrac{\frac{1}{3} \cdot x}{\frac{1}{3}} = \dfrac{40}{\frac{1}{3}} = \dfrac{40}{1} \cdot \dfrac{3}{1}$$
$$x = 120$$

Then find the number of grams of fiber in a $\frac{1}{2}$-cup serving of bran cereal.

$$\dfrac{\frac{1}{3} \text{ cup}}{8 \text{ grams of fiber}} = \dfrac{\frac{1}{2} \text{ cup}}{x}$$
$$\dfrac{1}{3} \cdot x = 8 \cdot \dfrac{1}{2}$$
$$\dfrac{\frac{1}{3} \cdot x}{\frac{1}{3}} = \dfrac{4}{\frac{1}{3}} = \dfrac{4}{1} \cdot \dfrac{3}{1}$$
$$x = 12$$

A $\frac{1}{2}$-cup serving of bran cereal provides 120 calories and 12 grams of fiber.

Relating Concepts (Exercises 41–42)

41.   *Use proportions.*

Water: $\dfrac{3\frac{1}{2} \text{ cups}}{12 \text{ servings}} = \dfrac{x}{6 \text{ servings}}$

$$12 \cdot x = 3\dfrac{1}{2} \cdot 6$$
$$12 \cdot x = \dfrac{7}{2} \cdot 6$$
$$\dfrac{12 \cdot x}{12} = \dfrac{21}{12}$$
$$x = \dfrac{7}{4} = 1\dfrac{3}{4}$$

Margarine: $\dfrac{6 \text{ Tbsp}}{12 \text{ servings}} = \dfrac{x}{6 \text{ servings}}$

$$12 \cdot x = 6 \cdot 6$$
$$\dfrac{12 \cdot x}{12} = \dfrac{36}{12}$$
$$x = 3$$

Milk: $\dfrac{1\frac{1}{2} \text{ cups}}{12 \text{ servings}} = \dfrac{x}{6 \text{ servings}}$

$$12 \cdot x = 1\dfrac{1}{2} \cdot 6$$
$$12 \cdot x = \dfrac{3}{2} \cdot 6$$
$$\dfrac{12 \cdot x}{12} = \dfrac{9}{12}$$
$$x = \dfrac{3}{4}$$

Potato flakes: $\dfrac{4 \text{ cups}}{12 \text{ servings}} = \dfrac{x}{6 \text{ servings}}$

$$12 \cdot x = 4 \cdot 6$$
$$\dfrac{12 \cdot x}{12} = \dfrac{24}{12}$$
$$x = 2$$

Multiply the quantities by $\frac{1}{2}$ (or divide by 2), since *6 servings is $\frac{1}{2}$ of 12 servings.*

Water: $\dfrac{3\frac{1}{2}}{2} = \dfrac{\frac{7}{2}}{2} = \dfrac{7}{2} \div 2 = \dfrac{7}{2} \cdot \dfrac{1}{2}$

$$= \dfrac{7}{4} = 1\dfrac{3}{4}$$

Margarine: $\dfrac{6}{2} = 3$

Milk:
$$\frac{1\frac{1}{2}}{2} = \frac{\frac{3}{2}}{2} = \frac{3}{2} \div 2$$
$$= \frac{3}{2} \cdot \frac{1}{2} = \frac{3}{4}$$

Potato flakes:
$$\frac{4}{2} = 2$$

For 6 servings, use $1\frac{3}{4}$ cups water, 3 Tbsp margarine, $\frac{3}{4}$ cup milk, and 2 cups potato flakes.

42. *Use proportions.*

Water:
$$\frac{3\frac{1}{2} \text{ cups}}{12 \text{ servings}} = \frac{x}{18 \text{ servings}}$$
$$12 \cdot x = 3\frac{1}{2} \cdot 18$$
$$12 \cdot x = \frac{7}{2} \cdot 18$$
$$\frac{12 \cdot x}{12} = \frac{63}{12}$$
$$x = 5\frac{1}{4}$$

Margarine:
$$\frac{6 \text{ Tbsp}}{12 \text{ servings}} = \frac{x}{18 \text{ servings}}$$
$$12 \cdot x = 6 \cdot 18$$
$$\frac{12 \cdot x}{12} = \frac{108}{12}$$
$$x = 9$$

Milk:
$$\frac{1\frac{1}{2} \text{ cups}}{12 \text{ servings}} = \frac{x}{18 \text{ servings}}$$
$$12 \cdot x = 1\frac{1}{2} \cdot 18$$
$$12 \cdot x = \frac{3}{2} \cdot 18$$
$$\frac{12 \cdot x}{12} = \frac{27}{12}$$
$$x = 2\frac{1}{4}$$

Potato flakes:
$$\frac{4 \text{ cups}}{12 \text{ servings}} = \frac{x}{18 \text{ servings}}$$
$$12 \cdot x = 4 \cdot 18$$
$$\frac{12 \cdot x}{12} = \frac{72}{12}$$
$$x = 6$$

Multiply the quantities in Exercise 41 by 3, since 18 servings is 3 times 6 servings.

Water: $1\frac{3}{4} \cdot 3 = \frac{7}{4} \cdot \frac{3}{1} = \frac{21}{4} = 5\frac{1}{4}$

Margarine: $3 \cdot 3 = 9$

Milk: $\frac{3}{4} \cdot 3 = \frac{3}{4} \cdot \frac{3}{1} = \frac{9}{4} = 2\frac{1}{4}$

Potato flakes: $2 \cdot 3 = 6$

For 18 servings, use $5\frac{1}{4}$ cups water, 9 Tbsp margarine, $2\frac{1}{4}$ cups milk, and 6 cups potato flakes.

6.5 Geometry: Lines and Angles

6.5 Margin Exercises

1. **(a)** The figure has two endpoints, so it is a *line segment* named $\overline{EF}$ or $\overline{FE}$.

(b) The figure starts at point S and goes on forever in one direction, so it is a *ray* named $\overrightarrow{SR}$.

(c) The figure goes on forever in both directions, so it is a *line* named $\overleftrightarrow{WX}$ or $\overleftrightarrow{XW}$.

(d) The figure has two endpoints, so it is a *line segment* named $\overline{CD}$ or $\overline{DC}$.

2. **(a)** The lines cross at E, so they are <u>intersecting</u> lines.

(b) The lines never intersect (cross), so they appear to be <u>parallel</u> lines.

(c) The lines never intersect (cross), so they appear to be *parallel lines.*

3. **(a)** The angle is named $\angle 3$, $\angle CQD$, or $\angle DQC$.

(b) Darken rays $\overrightarrow{TW}$ and $\overrightarrow{TZ}$.

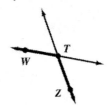

(c) The angle can be named $\angle 1$, $\angle R$, $\angle MRN$, and $\angle NRM$.

4. **(a)** Since there is a small square at the vertex, the angle measures exactly 90°, so it is a *right angle.*

(b) The angle measures exactly 180°, so it is a *straight angle.*

(c) The angle measures between 90° and 180°, so it is *obtuse.*

(d) The angle measures between 0° and 90°, so it is *acute.*

5. The lines in (b) show perpendicular lines, because they intersect at right angles. The lines in (a) show intersecting lines. They cross, but not at right angles.

6. $\angle COD$ and $\angle DOE$ are complementary angles because $70° + 20° = 90°$.
$\angle RST$ and $\angle XPY$ are complementary angles because $45° + 45° = 90°$.

7. **(a)** The complement of 35° is 55°, because $90° - 35° = \underline{55°}$.

(b) The complement of 80° is 10°, because $90° - 80° = 10°$.

8. ∠*CRF* and ∠*BRF* are supplementary angles because 127° + 53° = 180°.
    ∠*CRE* and ∠*ERB* are supplementary angles because 53° + 127° = 180°.

    ∠*BRF* and ∠*BRE* are supplementary angles because 53° + 127° = 180°.
    ∠*CRE* and ∠*CRF* are supplementary angles because 53° + 127° = 180°.

9. **(a)** The supplement of 175° is 5°, because 180° − 175° = 5°.

    (b) The supplement of 30° is 150°, because 180° − 30° = 150°.

10. ∠*BOC* ≅ ∠*AOD* because they each measure 150°.
    ∠*AOB* ≅ ∠*DOC* because they each measure 30°.

11. ∠*SPB* and ∠*MPD* are vertical angles because they do *not* share a common side and the angles are formed by intersecting lines.
    Similarly, ∠*BPD* and ∠*SPM* are vertical angles. Vertical angles are congruent (they measure the same number of degrees).

12. **(a)** ∠*TOS*
    ∠*TOS* is a right angle, so its measure is 90°.

    (b) ∠*QOR*
    ∠*QOR* and ∠*SOT* are vertical angles, so they are also congruent and have the same measure. The measure of ∠*SOT* is 90°, so the measure of ∠*QOR* is 90°.

    (c) ∠*VOR*
    ∠*VOR* and ∠*SOP* are vertical angles, so they are also congruent and have the same measure. The measure of ∠*SOP* is 38°, so the measure of ∠*VOR* is 38°.

    (d) ∠*POQ*
    ∠*POQ* and ∠*SOP* are complementary angles, so the sum of their angle measures is 90°. Since the degree measure of ∠*SOP* is 38°, the measure of ∠*POQ* equals 90° − 38° = 52°.

    (e) ∠*TOV*
    ∠*TOV* and ∠*POQ* are vertical angles, so they are congruent. The measure of ∠*POQ* is 52° (from part d), so the measure of ∠*TOV* is 52°.

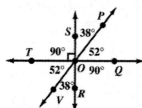

13. **(a)** There are four pairs of corresponding angles:

    ∠1 and ∠5,   ∠2 and ∠6,

    ∠7 and ∠3,   ∠8 and ∠4

There are two pairs of alternate interior angles:

∠7 and ∠6,   ∠8 and ∠5

(b) There are four pairs of corresponding angles:

∠5 and ∠7,   ∠6 and ∠8,

∠1 and ∠3,   ∠2 and ∠4

There are two pairs of alternate interior angles:

∠6 and ∠3, ∠2 and ∠7

14. **(a)** ∠6 ≅ ∠7 ≅ ∠2 ≅ ∠3, so each measures 150°.

    ∠8 measures 180° − 150° = 30°.

    ∠8 ≅ ∠5 ≅ ∠4 ≅ ∠1, so each measures 30°.

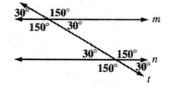

(b) ∠1 ≅ ∠6 ≅ ∠3 ≅ ∠8, so each measures 45°.

∠2 measures 180° − 45° = 135°.

∠2 ≅ ∠5 ≅ ∠4 ≅ ∠7, so each measures 135°.

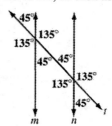

6.5   Section Exercises

1. Answer wording may vary slightly: a **line** is a row of points continuing in both directions forever; the drawing should have arrows on both ends. A **line segment** is a piece of a line; the drawing should have an endpoint (dot) on each end. A **ray** has one endpoint and goes on forever in one direction; the drawing should have one endpoint (dot) and an arrowhead on the other end.

3. This is a *line* named $\overleftrightarrow{CD}$ or $\overleftrightarrow{DC}$. A line is a straight row of points that goes on forever in both directions.

5. The figure is a piece of a line that has two endpoints, so it is a *line segment* named $\overline{GF}$ or $\overline{FG}$.

7. This is a *ray* named $\overrightarrow{PQ}$. A ray is a part of a line that has only one endpoint and goes on forever in one direction.

9. The lines are *perpendicular* because they intersect at right angles.

11. These lines appear to be *parallel* lines. Parallel lines are lines in the same plane that never intersect (cross).

13. The lines intersect so they are *not* parallel. At their intersection they *do not* form a right angle so they are not perpendicular. The lines are *intersecting*.

15. The angle can be named ∠*AOS* or ∠*SOA*. The middle letter, *O*, identifies the vertex.

17. The angle can be named ∠*AQC* or ∠*CQA*. The middle letter, *Q*, identifies the vertex.

19. The angle is a *right angle*, as indicated by the small square at the vertex. Right angles measure exactly 90°.

21. The measure of the angle is between 0° and 90°, so it is an *acute angle*.

23. Two rays in a straight line pointing opposite directions measure 180°. An angle that measures 180° is called a *straight angle*.

25. The pairs of complementary angles are:
∠*EOD* and ∠*COD* because 75° + 15° = 90°;
∠*AOB* and ∠*BOC* because 25° + 65° = 90°.

27. The pairs of supplementary angles are:
∠*HNE* and ∠*ENF* because 77° + 103° = 180°;
∠*ACB* and ∠*KOL* because 120° + 60° = 180°.

29. The complement of 40° is 50°, because
90° − 40° = <u>50°</u>.

31. The complement of 86° is 4°, because
90° − 86° = 4°.

33. The supplement of 130° is 50°, because
180° − 130° = <u>50°</u>.

35. The supplement of 90° is 90°, because
180° − 90° = 90°.

37. ∠*SON* ≅ ∠*TOM* because they are vertical angles.
∠*TOS* ≅ ∠*MON* because they are vertical angles.

39. Because ∠*COE* and ∠*GOH* are vertical angles, they are also congruent. This means they have the same measure. ∠*COE* measures 63°, so ∠*GOH* measures 63°.

The sum of the measures of ∠*COE*, ∠*AOC*, and ∠*AOH* equals 180°. Therefore, ∠*AOC* measures 180° − (63° + 37°) = 180° − 100° = 80°. Since ∠*AOC* and ∠*GOF* are vertical, they are congruent, so ∠*GOF* measures 80°. Since ∠*AOH* and ∠*EOF* are vertical, they are congruent, so ∠*EOF* measures 37°.

41. "∠*UST* is 90°" is *true* because $\overleftrightarrow{UQ}$ is perpendicular to $\overleftrightarrow{ST}$.

43. "The measure of ∠*USQ* is less than the measure of ∠*PQR*" is *false*. ∠*USQ* is a straight angle and so is ∠*PQR*, therefore each measures 180°. A true statement would be: ∠*USQ* has the same measure as ∠*PQR*.

45. "$\overleftrightarrow{QU}$ and $\overleftrightarrow{TS}$ are parallel" is *false*. $\overleftrightarrow{QU}$ and $\overleftrightarrow{TS}$ are perpendicular because they intersect at right angles.

47. There are four pairs of corresponding angles:

∠1 and ∠8,   ∠2 and ∠5,

∠3 and ∠6,   ∠4 and ∠7

There are two pairs of alternate interior angles:

∠4 and ∠5,   ∠3 and ∠8

49. ∠8 ≅ ∠6 ≅ ∠2 ≅ ∠4, so each measures 130°.

∠5 measures 180° − 130° = 50°.

∠5 ≅ ∠7 ≅ ∠1 ≅ ∠3, so each measures 50°.

51. ∠6 ≅ ∠1 ≅ ∠8 ≅ ∠3, so each measures 47°.

∠5 measures 180° − 47° = 133°.

∠5 ≅ ∠2 ≅ ∠7 ≅ ∠4, so each measures 133°.

53. ∠6 ≅ ∠8 ≅ ∠4 ≅ ∠2, so each measures 114°.

∠7 measures 180° − 114° = 66°.

∠7 ≅ ∠5 ≅ ∠3 ≅ ∠1, so each measures 66°.

55. ∠2 and ∠*ABC* are alternate interior angles, so they have the same measure, 42°.
∠1 and ∠*ABC* are supplementary angles, so ∠1 = 180° − 42° = 138°.
∠1 and ∠3 are supplements of alternate interior angles, so they have the same measure, 138°.

6.6 Geometry Applications: Congruent and Similar Triangles

6.6 Margin Exercises

1. (a) If you picked up △*ABC* and slid it over on top of △*DEF*, the two triangles would match.

The corresponding parts are congruent, so:

∠1 and ∠4	$\overline{AC}$ and $\overline{DF}$
∠2 and ∠5	$\overline{AB}$ and $\overline{DE}$
∠3 and ∠6	$\overline{BC}$ and $\overline{EF}$

(b) If you rotate △*FGH*, then slide it on top of △*JLK*, the two triangles would match.

∠1 and ∠6	$\overline{GF}$ and $\overline{KL}$
∠2 and ∠5	$\overline{FH}$ and $\overline{LJ}$
∠3 and ∠4	$\overline{GH}$ and $\overline{KJ}$

(c) If you flipped $\triangle RST$ over, then slid it on top of $\triangle VWX$, the two triangles would match.

$\angle 1$ and $\angle 5$	$\overline{RS}$ and $\overline{XV}$
$\angle 2$ and $\angle 4$	$\overline{RT}$ and $\overline{XW}$
$\angle 3$ and $\angle 6$	$\overline{ST}$ and $\overline{VW}$

2. **(a)** On both triangles, two corresponding angles and the side that connects them measure the same, so the Angle-Side-Angle (ASA) method can be used to prove that the triangles are congruent.

(b) On both triangles, two corresponding sides and the angle between them measure the same, so the Side-Angle-Side (SAS) method can be used to prove that the triangles are congruent.

(c) Each pair of corresponding sides has the same length, so the Side-Side-Side (SSS) method can be used to prove that the triangles are congruent.

3. **(a)** Corresponding angles have the same measure.

The corresponding angles are
1 and $\underline{4}$, 2 and $\underline{5}$, 3 and $\underline{6}$.
$\overline{PN}$ and $\overline{ZX}$ are opposite corresponding angles 3 and 6.

$\overline{PM}$ and $\overline{ZY}$ are opposite corresponding angles 1 and 4.
$\overline{NM}$ and $\overline{XY}$ are opposite corresponding angles 2 and 5.
Thus, the corresponding sides are
$\overline{PN}$ and $\overline{ZX}$, $\overline{PM}$ and $\overline{ZY}$, $\overline{NM}$ and $\overline{XY}$.

(b) The corresponding angles are
1 and $\underline{6}$, 2 and $\underline{4}$, 3 and $\underline{5}$.
The corresponding sides are
$\overline{AB}$ and $\overline{EF}$, $\overline{BC}$ and $\overline{FG}$, $\overline{AC}$ and $\overline{EG}$.

4. Find the length of $\overline{EF}$.

$$\frac{EF}{CB} = \frac{ED}{CA} \qquad \textit{Corresponding sides}$$

$$\frac{EF}{CB} = \frac{5}{15} \qquad \textit{Replace ED with 5 and CA with 15.}$$

$$\frac{x}{33} = \frac{1}{3} \qquad \textit{Replace EF with x and CB with 33.}$$

$$3 \cdot x = 33 \cdot 1$$

$$\frac{3 \cdot x}{3} = \frac{33}{3}$$

$$x = 11$$

$\overline{EF}$ has a length of 11 m.

5. **(a)** Find the length of $\overline{AB}$.
From Example 4 in the text,

$$\frac{PR}{AC} = \frac{7}{14} = \frac{1}{2}, \text{ so } \frac{PQ}{AB} = \frac{1}{2}.$$

Replace PQ with 3 and AB with y.

$$\frac{3}{y} = \frac{1}{2}$$

$$y \cdot 1 = 3 \cdot 2$$

$$y = 6$$

$\overline{AB}$ is 6 ft.
Find the perimeter.
Perimeter $= 14 \text{ ft} + 10 \text{ ft} + 6 \text{ ft} = 30 \text{ ft}$

(b) Set up ratios.

$$\frac{PQ}{AB} = \frac{10 \text{ m}}{30 \text{ m}} = \frac{1}{3}, \text{ so } \frac{QR}{BC} = \frac{1}{3} \text{ and } \frac{PR}{AC} = \frac{1}{3}.$$

Replace QR with x and BC with 18.

$$\frac{x}{18} = \frac{1}{3}$$

$$3 \cdot x = 18 \cdot 1$$

$$\frac{3 \cdot x}{3} = \frac{18}{3}$$

$$x = 6$$

$\overline{QR}$ is 6 m.
The perimeter of triangle PQR is

$$10 \text{ m} + 8 \text{ m} + 6 \text{ m} = 24 \text{ m}.$$

Next replace PR with 8 and AC with y.

$$\frac{8}{y} = \frac{1}{3}$$

$$y \cdot 1 = 8 \cdot 3$$

$$y = 24$$

$\overline{AC}$ is 24 m.
The perimeter of triangle ABC is

$$30 \text{ m} + 18 \text{ m} + 24 \text{ m} = 72 \text{ m}.$$

6. **(a)** Write a proportion.

$$\frac{\text{longer height}}{\text{shorter height}} = \frac{\text{longer shadow}}{\text{shorter shadow}}$$

$$\frac{h}{5} = \frac{48}{12}$$

$$\frac{h}{5} = \frac{4}{1} \qquad \textit{Lowest terms}$$

$$1 \cdot h = 5 \cdot 4$$

$$h = 20$$

The flagpole is 20 ft high.

(b) Write a proportion.

$$\frac{\text{longer height}}{\text{shorter height}} = \frac{\text{longer base}}{\text{shorter base}}$$

$$\frac{h}{7.2} = \frac{12.5}{5}$$

$$5 \cdot h = 7.2 \cdot 12.5$$

$$\frac{5 \cdot h}{5} = \frac{90}{5}$$

$$h = 18$$

The flagpole is 18 m high.

6.6 Section Exercises

1. One dictionary definition is "conforming; agreeing." Examples of congruent objects include two matching chairs, and two pieces of paper from the same notebook.

3. If you picked up $\triangle ABC$ and slid it over on top of $\triangle DEF$, the two triangles would match.

The corresponding angles are

$\angle 1$ and $\angle 4$, $\angle 2$ and $\angle 5$, $\angle 3$ and $\angle 6$.

The corresponding sides are

$\overline{AB}$ and $\overline{DE}$, $\overline{BC}$ and $\overline{EF}$, $\overline{AC}$ and $\overline{DF}$.

5. If you rotate $\triangle TUS$, then slide it on top of $\triangle WXY$, the two triangles would match.

The corresponding angles are

$\angle 1$ and $\angle 6$, $\angle 2$ and $\angle 4$, $\angle 3$ and $\angle 5$.

The corresponding sides are

$\overline{ST}$ and $\overline{YW}$, $\overline{TU}$ and $\overline{WX}$, $\overline{SU}$ and $\overline{YX}$.

7. If you flipped $\triangle MNL$ over, then slid it on top of $\triangle SRT$, the two triangles would match.

The corresponding angles are

$\angle 1$ and $\angle 6$, $\angle 2$ and $\angle 5$, $\angle 3$ and $\angle 4$.

The corresponding sides are

$\overline{LM}$ and $\overline{TS}$, $\overline{LN}$ and $\overline{TR}$, $\overline{MN}$ and $\overline{SR}$.

9. On both triangles, two corresponding sides and the angle between them measure the same, so the Side-Angle-Side (SAS) method can be used to prove that the triangles are congruent.

11. Each pair of corresponding sides has the same length, so the Side-Side-Side (SSS) method can be used to prove that the triangles are congruent.

13. On both triangles, two corresponding angles and the side that connects them measure the same, so the Angle-Side-Angle (ASA) method can be used to prove that the triangles are congruent.

15. use SAS: $BC = CE$, $\angle ABC \cong \angle DCE$, $BA = CD$

17. use SAS: $PS = SR$, $m\angle QSP = m\angle QSR = 90°$, $QS = QS$ (common side)

19. Write a proportion to find a.

$$\frac{a}{12 \text{ cm}} = \frac{6 \text{ cm}}{12 \text{ cm}} \quad \text{OR} \quad \frac{a}{12} = \frac{1}{2}$$
$$2 \cdot a = 12 \cdot 1$$
$$\frac{2 \cdot a}{2} = \frac{12}{2}$$
$$a = 6 \text{ cm}$$

Write a proportion to find b.

$$\frac{7.5 \text{ cm}}{b} = \frac{6 \text{ cm}}{12 \text{ cm}} \quad \text{OR} \quad \frac{7.5}{b} = \frac{1}{2}$$
$$1 \cdot b = 2 \cdot 7.5$$
$$b = 15 \text{ cm}$$

21. Write a proportion to find a.

$$\frac{a}{10 \text{ mm}} = \frac{6 \text{ mm}}{12 \text{ mm}} \quad \text{OR} \quad \frac{a}{10} = \frac{1}{2}$$
$$a \cdot 2 = 10 \cdot 1$$
$$\frac{a \cdot 2}{2} = \frac{10}{2}$$
$$a = 5 \text{ mm}$$

Write a proportion to find b.

$$\frac{b}{6 \text{ mm}} = \frac{6 \text{ mm}}{12 \text{ mm}} \quad \text{OR} \quad \frac{b}{6} = \frac{1}{2}$$
$$b \cdot 2 = 6 \cdot 1$$
$$\frac{b \cdot 2}{2} = \frac{6}{2}$$
$$b = 3 \text{ mm}$$

23. Write a proportion to find a.

$$\frac{a}{16 \text{ in.}} = \frac{18 \text{ in.}}{12 \text{ in.}} \quad \text{OR} \quad \frac{a}{16} = \frac{3}{2}$$
$$a \cdot 2 = 16 \cdot 3$$
$$\frac{a \cdot 2}{2} = \frac{48}{2}$$
$$a = 24 \text{ inches}$$

Write a proportion to find b.

$$\frac{30 \text{ in.}}{b} = \frac{18 \text{ in.}}{12 \text{ in.}} \quad \text{OR} \quad \frac{30}{b} = \frac{3}{2}$$
$$b \cdot 3 = 30 \cdot 2$$
$$\frac{b \cdot 3}{3} = \frac{60}{3}$$
$$b = 20 \text{ inches}$$

25. Write a proportion to find x.

$$\frac{x}{18.6} = \frac{28}{21} \quad \text{OR} \quad \frac{x}{18.6} = \frac{4}{3}$$
$$3 \cdot x = 4 \cdot 18.6$$
$$\frac{3 \cdot x}{3} = \frac{74.4}{3}$$
$$x = 24.8 \text{ m}$$

$P = 24.8 \text{ m} + 28 \text{ m} + 20 \text{ m} = 72.8 \text{ m}$

Write a proportion to find y.

$$\frac{y}{20} = \frac{21}{28} \quad \text{OR} \quad \frac{y}{20} = \frac{3}{4}$$
$$4 \cdot y = 20 \cdot 3$$
$$\frac{4 \cdot y}{4} = \frac{60}{4}$$
$$y = 15 \text{ m}$$

$P = 15 \text{ m} + 21 \text{ m} + 18.6 \text{ m} = 54.6 \text{ m}$

27. Set up a ratio of corresponding sides to find the length of the missing sides in triangle *FGH*.

$$\frac{DE}{GH} = \frac{CE}{FH}$$

$$\frac{12}{8} = \frac{12}{x} \quad \text{Let } x = \overline{FH}.$$

$$12 \cdot x = 8 \cdot 12$$

$$\frac{12 \cdot x}{12} = \frac{96}{12}$$

$$x = 8$$

Each missing side of triangle *FGH* is 8 cm.

Perimeter of triangle FGH
$$= 8 \text{ cm} + 8 \text{ cm} + 8 \text{ cm}$$
$$= 24 \text{ cm}$$

Set up a ratio of corresponding sides to find the height *h* of triangle *FGH*.

$$\frac{10.4}{12} = \frac{h}{8}$$

$$12 \cdot h = 8 \cdot 10.4$$

$$\frac{12 \cdot h}{12} = \frac{83.2}{12}$$

$$h \approx 6.9 \text{ cm}$$

Area of triangle FGH $= 0.5 \cdot b \cdot h$
$$\approx 0.5 \cdot 8 \text{ cm} \cdot 6.9 \text{ cm}$$
$$\approx 27.6 \text{ cm}^2$$

The area is approximate because the height was rounded to the nearest tenth.

29. Write a proportion to find *h*.

$$\frac{2}{16} = \frac{3}{h}$$

$$2 \cdot h = 3 \cdot 16$$

$$\frac{2 \cdot h}{2} = \frac{48}{2}$$

$$h = 24 \text{ ft}$$

The height of the house is 24 ft.

31. Using the hint, we can write a proportion to find *x*.

$$\frac{x}{120} = \frac{100}{100 + 140} \quad \text{OR} \quad \frac{x}{120} = \frac{5}{12}$$

$$x \cdot 12 = 120 \cdot 5$$

$$\frac{x \cdot 12}{12} = \frac{600}{12}$$

$$x = 50 \text{ m}$$

33. Write a proportion to find *n*.

$$\frac{50}{n} = \frac{100}{100 + 120} \quad \text{OR} \quad \frac{50}{n} = \frac{5}{11}$$

$$5 \cdot n = 50 \cdot 11$$

$$\frac{5 \cdot n}{5} = \frac{550}{5}$$

$$n = 110 \text{ m}$$

Chapter 6 Review Exercises

1. orca whale's length to whale shark's length

$$\frac{30 \text{ ft}}{40 \text{ ft}} = \frac{30}{40} = \frac{30 \div 10}{40 \div 10} = \frac{3}{4}$$

2. blue whale's length to great white shark's length

$$\frac{80 \text{ ft}}{20 \text{ ft}} = \frac{80}{20} = \frac{80 \div 20}{20 \div 20} = \frac{4}{1}$$

3. To get a ratio of $\frac{1}{2}$, we can start with the smallest value in the table and see if there is a value that is two times the smallest. $2 \times 20 = 40$, so the ratio of the *great white shark's* length to the *whale shark's* length is $\frac{20}{40} = \frac{1}{2}$. The length of the orca whale is 30 ft, but $2 \times 30 = 60$ is not in the table. $2 \times 40 = 80$, so the ratio of the *whale shark's* length to the *blue whale's* length is $\frac{40}{80} = \frac{1}{2}$.

4. $2.50 to $1.25

$$\frac{\$2.50}{\$1.25} = \frac{2.50}{1.25} = \frac{2.50 \div 1.25}{1.25 \div 1.25} = \frac{2}{1}$$

5. $0.30 to $0.45

$$\frac{\$0.30}{\$0.45} = \frac{0.30}{0.45} = \frac{0.30 \div 0.15}{0.45 \div 0.15} = \frac{2}{3}$$

6. $1\frac{2}{3}$ cups to $\frac{2}{3}$ cup

$$\frac{1\frac{2}{3} \text{ cups}}{\frac{2}{3} \text{ cup}} = \frac{1\frac{2}{3}}{\frac{2}{3}} = \frac{5}{3} \div \frac{2}{3}$$

$$= \frac{5}{\cancel{3}} \cdot \frac{\cancel{3}^{\,1}}{2} = \frac{5}{2}$$

7. $2\frac{3}{4}$ miles to $16\frac{1}{2}$ miles

$$\frac{2\frac{3}{4} \text{ miles}}{16\frac{1}{2} \text{ miles}} = \frac{2\frac{3}{4}}{16\frac{1}{2}} = \frac{\frac{11}{4}}{\frac{33}{2}} = \frac{11}{4} \div \frac{33}{2}$$

$$= \frac{\cancel{11}^{\,1}}{\cancel{4}_{\,2}} \cdot \frac{\cancel{2}^{\,1}}{\cancel{33}_{\,3}} = \frac{1}{6}$$

8. 5 hours to 100 minutes

5 hours $= 5 \cdot 60$ minutes $= 300$ minutes

$$\frac{5 \text{ hours}}{100 \text{ minutes}} = \frac{300 \text{ minutes}}{100 \text{ minutes}}$$

$$= \frac{300}{100} = \frac{300 \div 100}{100 \div 100} = \frac{3}{1}$$

9. 9 inches to 2 feet

$$\frac{9 \text{ inches}}{2 \text{ feet}} = \frac{9 \text{ inches}}{24 \text{ inches}} = \frac{9}{24} = \frac{9 \div 3}{24 \div 3} = \frac{3}{8}$$

10. 1 ton to 1500 pounds

$$\frac{1 \text{ ton}}{1500 \text{ pounds}} = \frac{2000 \text{ pounds}}{1500 \text{ pounds}}$$

$$= \frac{2000}{1500} = \frac{2000 \div 500}{1500 \div 500} = \frac{4}{3}$$

11. 8 hours to 3 days

3 days $= 3 \cdot 24$ hours $= 72$ hours

$$\frac{8 \text{ hours}}{3 \text{ days}} = \frac{8 \text{ hours}}{72 \text{ hours}} = \frac{8}{72} = \frac{8 \div 8}{72 \div 8} = \frac{1}{9}$$

12. Ramona's \$500 to Jake's \$350

$$\frac{\$500}{\$350} = \frac{500 \div 50}{350 \div 50} = \frac{10}{7}$$

The ratio of Ramona's sales to Jake's sales is $\frac{10}{7}$.

13. $\frac{35 \text{ miles per gallon}}{25 \text{ miles per gallon}} = \frac{35}{25} = \frac{35 \div 5}{25 \div 5} = \frac{7}{5}$

The ratio of the new car's mileage to the old car's mileage is $\frac{7}{5}$.

14. $\frac{6000 \text{ students}}{7200 \text{ students}} = \frac{6000}{7200} = \frac{6000 \div 1200}{7200 \div 1200} = \frac{5}{6}$

The ratio of the math students to the English students is $\frac{5}{6}$.

15. \$88 for 8 dozen

$$\frac{\$88 \div 8}{8 \text{ dozen} \div 8} = \frac{\$11}{1 \text{ dozen}}$$

16. 96 children in 40 families

$$\frac{96 \text{ children} \div 8}{40 \text{ families} \div 8} = \frac{12 \text{ children}}{5 \text{ families}}$$

17. 4 pages in 20 minutes

(i) $\frac{4 \text{ pages} \div 20}{20 \text{ minutes} \div 20} = \frac{0.2 \text{ page}}{1 \text{ minute}}$
$= 0.2$ page/minute
or $\frac{1}{5}$ page/minute

(ii) $\frac{20 \text{ minutes} \div 4}{4 \text{ pages} \div 4} = \frac{5 \text{ minutes}}{1 \text{ page}}$
$= 5$ minutes/page

His rate is $\frac{1}{5}$ page/minute or 5 minutes/page.

18. \$60 in 3 hours

(i) $\frac{\$60 \div 3}{3 \text{ hours} \div 3} = \frac{\$20}{1 \text{ hour}} = \$20/\text{hour}$

(ii) $\frac{3 \text{ hours} \div 60}{\$60 \div 60} = \frac{0.05 \text{ hour}}{\$1}$
$= 0.05$ hour/dollar
or $\frac{1}{20}$ hour/dollar

Her earnings are \$20/hour or $\frac{1}{20}$ hour/dollar.

19.

Size	Cost per Unit
8 ounces	$\frac{\$4.98}{8 \text{ ounces}} \approx \$0.623 \ (*)$
3 ounces	$\frac{\$2.49}{3 \text{ ounces}} = \$0.83$
2 ounces	$\frac{\$1.89}{2 \text{ ounces}} = \$0.945$

The best buy is 8 ounces for \$4.98, about \$0.623/ounce.

20. 35.2 pounds for \$36.96 − \$1 (coupon) = \$35.96

$$\frac{\$35.96}{35.2 \text{ pounds}} \approx \$1.022$$

17.6 pounds for \$18.69 − \$1 (coupon) = \$17.69

$$\frac{\$17.69}{17.6 \text{ pounds}} \approx \$1.005 \ (*)$$

3.5 pounds for \$4.25

$$\frac{\$4.25}{3.5 \text{ pounds}} \approx \$1.214$$

The best buy is 17.6 pounds for \$18.69 with the \$1 coupon, about \$1.005/pound.

21. $\frac{6}{10} = \frac{9}{15}$

$$\frac{6 \div 2}{10 \div 2} = \frac{3}{5} \text{ and } \frac{9 \div 3}{15 \div 3} = \frac{3}{5}$$

Both ratios are equal to $\frac{3}{5}$, so the proportion is *true*.

22. $\frac{6}{48} = \frac{9}{36}$ Cross products: $48 \cdot 9 = 432$, $6 \cdot 36 = 216$

The cross products are *unequal*, so the proportion is *false*.

23. $\frac{47}{10} = \frac{98}{20}$ Cross products: $10 \cdot 98 = 980$, $47 \cdot 20 = 940$

The cross products are *unequal*, so the proportion is *false*.

24. $\frac{1.5}{2.4} = \frac{2}{3.2}$ Cross products: $2.4 \cdot 2 = 4.8$, $1.5(3.2) = 4.8$

The cross products are *equal*, so the proportion is *true*.

25. $\frac{3\frac{1}{2}}{2\frac{1}{3}} = \frac{6}{4}$

Cross products:
$$2\frac{1}{3} \cdot 6 = \frac{7}{3} \cdot \frac{\overset{2}{\cancel{6}}}{1} = 14$$
$$3\frac{1}{2} \cdot 4 = \frac{7}{2} \cdot \frac{\overset{2}{\cancel{4}}}{1} = 14$$

The cross products are *equal*, so the proportion is *true*.

26. $\dfrac{4}{42} = \dfrac{150}{x}$ **OR** $\dfrac{2}{21} = \dfrac{150}{x}$

$2 \cdot x = 21 \cdot 150$ *Show that cross*
products are equal.

$\dfrac{2 \cdot x}{2} = \dfrac{3150}{2}$ *Divide both*
sides by 2.

$x = 1575$

Check $4 \cdot 1575 = 6300 = 42 \cdot 150$

27. $\dfrac{16}{x} = \dfrac{12}{15}$ **OR** $\dfrac{16}{x} = \dfrac{4}{5}$

$4 \cdot x = 5 \cdot 16$ *Show that cross*
products are equal.

$\dfrac{4 \cdot x}{4} = \dfrac{80}{4}$ *Divide both*
sides by 4.

$x = 20$

Check $16 \cdot 15 = 240 = 20 \cdot 12$

28. $\dfrac{100}{14} = \dfrac{x}{56}$ **OR** $\dfrac{50}{7} = \dfrac{x}{56}$

$7 \cdot x = 50 \cdot 56$ *Show that cross*
products are equal.

$\dfrac{7 \cdot x}{7} = \dfrac{2800}{7}$ *Divide both*
sides by 7.

$x = 400$

Check $100 \cdot 56 = 5600 = 14 \cdot 400$

29. $\dfrac{5}{8} = \dfrac{x}{20}$

$8 \cdot x = 5 \cdot 20$ *Show that cross*
products are equal.

$\dfrac{8 \cdot x}{8} = \dfrac{100}{8}$ *Divide both*
sides by 8.

$x = 12.5$

Check $5 \cdot 20 = 100 = 8(12.5)$

30. $\dfrac{x}{24} = \dfrac{11}{18}$

$18 \cdot x = 24 \cdot 11$ *Show that cross*
products are equal.

$\dfrac{18 \cdot x}{18} = \dfrac{264}{18}$ *Divide both*
sides by 18.

$x = \dfrac{44}{3} \approx 14.67$ (rounded)

Check $\frac{44}{3} \cdot 18 = 264 = 24 \cdot 11$

31. $\dfrac{7}{x} = \dfrac{18}{21}$ **OR** $\dfrac{7}{x} = \dfrac{6}{7}$

$6 \cdot x = 7 \cdot 7$ *Show that cross*
products are equal.

$\dfrac{6 \cdot x}{6} = \dfrac{49}{6}$ *Divide both*
sides by 6.

$x = \dfrac{49}{6} \approx 8.17$ (rounded)

Check $7 \cdot 21 = 147 = \frac{49}{6} \cdot 18$

32. $\dfrac{x}{3.6} = \dfrac{9.8}{0.7}$

$0.7 \cdot x = 9.8(3.6)$ *Show that cross*
products are equal.

$\dfrac{0.7 \cdot x}{0.7} = \dfrac{35.28}{0.7}$ *Divide both*
sides by 0.7.

$x = 50.4$

Check $50.4(0.7) = 35.28 = 3.6(9.8)$

33. $\dfrac{13.5}{1.7} = \dfrac{4.5}{x}$

$13.5 \cdot x = 1.7(4.5)$ *Show that cross*
products are equal.

$\dfrac{13.5 \cdot x}{13.5} = \dfrac{7.65}{13.5}$ *Divide both*
sides by 13.5.

$x \approx 0.57$ (rounded)

Check $13.5\left(\frac{7.65}{13.5}\right) = 7.65 = 1.7(4.5)$

34. $\dfrac{0.82}{1.89} = \dfrac{x}{5.7}$

$1.89 \cdot x = 0.82(5.7)$ *Show that cross*
products are equal.

$\dfrac{1.89 \cdot x}{1.89} = \dfrac{4.674}{1.89}$ *Divide both*
sides by 1.89.

$x \approx 2.47$ (rounded)

Check $0.82(5.7) = 4.674 = 1.89\left(\frac{4.674}{1.89}\right)$

35. $\dfrac{3 \text{ cats}}{5 \text{ dogs}} = \dfrac{x}{45 \text{ dogs}}$

$5 \cdot x = 3 \cdot 45$

$\dfrac{5 \cdot x}{5} = \dfrac{135}{5}$

$x = 27$

There are 27 cats.

36. $\dfrac{8 \text{ hits}}{28 \text{ at bats}} = \dfrac{x}{161 \text{ at bats}}$

$28 \cdot x = 8 \cdot 161$

$\dfrac{28 \cdot x}{28} = \dfrac{1288}{28}$

$x = 46$

She will get 46 hits.

37. $\dfrac{3.5 \text{ pounds}}{\$9.77} = \dfrac{5.6 \text{ pounds}}{x}$

$3.5 \cdot x = 9.77(5.6)$

$\dfrac{3.5 \cdot x}{3.5} = \dfrac{54.712}{3.5}$

$x \approx 15.63$

The cost for 5.6 pounds of ground beef is $15.63 (rounded).

38. $\dfrac{4 \text{ voting students}}{10 \text{ students}} = \dfrac{x}{8247 \text{ students}}$

$$10 \cdot x = 4 \cdot 8247$$
$$\frac{10 \cdot x}{10} = \frac{32,988}{10}$$
$$x \approx 3299$$

They should expect about 3299 students to vote.

39. $\dfrac{1 \text{ inch}}{16 \text{ feet}} = \dfrac{4.25 \text{ inches}}{x}$

$$1 \cdot x = 16(4.25)$$
$$x = 68$$

The length of the real boxcar is 68 feet.

40. 2 dozen necklaces $= 2 \cdot 12 = 24$ necklaces

$$\frac{24 \text{ necklaces}}{16\frac{1}{2} \text{ hours}} = \frac{40 \text{ necklaces}}{x}$$
$$24 \cdot x = 16\frac{1}{2} \cdot 40 = \frac{33}{2} \cdot \frac{40}{1}$$
$$\frac{24 \cdot x}{24} = \frac{660}{24}$$
$$x = 27.5 = 27\frac{1}{2}$$

It will take Marvette $27\frac{1}{2}$ hours or 27.5 hours to make 40 necklaces.

41. $\dfrac{284 \text{ calories}}{25 \text{ minutes}} = \dfrac{x}{45 \text{ minutes}}$

$$25 \cdot x = 284 \cdot 45$$
$$\frac{25 \cdot x}{25} = \frac{12,780}{25}$$
$$x = 511.2 \approx 511$$

A 180-pound person would burn about 511 calories in 45 minutes.

42. $\dfrac{3.5 \text{ milligrams}}{50 \text{ pounds}} = \dfrac{x}{210 \text{ pounds}}$

$$50 \cdot x = 3.5 \cdot 210$$
$$\frac{50 \cdot x}{50} = \frac{735}{50}$$
$$x = 14.7$$

A patient who weighs 210 pounds should be given 14.7 milligrams of the medicine.

43. The figure is a piece of a line that has two endpoints, so it is a *line segment* named $\overline{AB}$ or $\overline{BA}$.

44. This is a *line* named $\overleftrightarrow{CD}$ or $\overleftrightarrow{DC}$. A line is a straight row of points that goes on forever in both directions.

45. This is a *ray* named $\overrightarrow{OP}$. A ray is a part of a line that has only one endpoint and goes on forever in one direction.

46. These lines appear to be *parallel* lines. Parallel lines are lines in the same plane that never intersect (cross).

47. The lines are *perpendicular* because they intersect at right angles.

48. The lines intersect so they are *not* parallel. At their intersection they *do not* form a right angle so they are not perpendicular. The lines are *intersecting*.

49. The measure of the angle is between $0°$ and $90°$, so it is an *acute angle*.

50. The measure of the angle is between $90°$ and $180°$, so it is an *obtuse angle*.

51. Two rays in a straight line pointing opposite directions measure $180°$. An angle that measures $180°$ is called a *straight angle*.

52. The angle is a *right angle*, as indicated by the small square at the vertex. Right angles measure exactly $90°$.

53. (a) The complement of $80°$ is $10°$, because $90° - 80° = 10°$.

(b) The complement of $45°$ is $45°$, because $90° - 45° = 45°$.

(c) The complement of $7°$ is $83°$, because $90° - 7° = 83°$.

54. (a) The supplement of $155°$ is $25°$, because $180° - 155° = 25°$.

(b) The supplement of $90°$ is $90°$, because $180° - 90° = 90°$.

(c) The supplement of $33°$ is $147°$, because $180° - 33° = 147°$.

55. $\angle 2 \cong \angle 5$, so $\angle 5$ measures $60°$.
$\angle 6 \cong \angle 3$, so $\angle 6$ and $\angle 3$ measure $90°$.
$\angle 1$ measures $90° - 60° = 30°$.
$\angle 1 \cong \angle 4$, so $\angle 4$ measures $30°$.

56. $\angle 8 \cong \angle 3 \cong \angle 6 \cong \angle 1$, so each measures $160°$.
$\angle 4$ measures $180° - 160° = 20°$.
$\angle 4 \cong \angle 7 \cong \angle 2 \cong \angle 5$, so each measures $20°$.

57. Each pair of corresponding sides has the same length, so the Side-Side-Side (SSS) method can be used to prove that the triangles are congruent.

58. On both triangles, two corresponding sides and the angle between them measure the same, so the Side-Angle-Side (SAS) method can be used to prove that the triangles are congruent.

59. On both triangles, two corresponding angles and the side that connects them measure the same, so the Angle-Side-Angle (ASA) method can be used to prove that the triangles are congruent.

60. Write a proportion to find y.

$$\frac{y}{15 \text{ ft}} = \frac{40 \text{ ft}}{20 \text{ ft}} \quad \textbf{OR} \quad \frac{y}{15} = \frac{2}{1}$$
$$y \cdot 1 = 2 \cdot 15$$
$$y = 30 \text{ ft}$$

Write a proportion to find x.

$$\frac{x}{17 \text{ ft}} = \frac{40 \text{ ft}}{20 \text{ ft}} \quad \textbf{OR} \quad \frac{x}{17} = \frac{2}{1}$$
$$x \cdot 1 = 17 \cdot 2$$
$$x = 34 \text{ ft}$$

$P = 34 \text{ ft} + 30 \text{ ft} + 40 \text{ ft}$
$P = 104 \text{ ft}$

61. Write a proportion to find x.

$$\frac{6 \text{ m}}{x} = \frac{4 \text{ m}}{6 \text{ m}} \quad \textbf{OR} \quad \frac{6}{x} = \frac{2}{3}$$
$$2 \cdot x = 3 \cdot 6$$
$$\frac{2 \cdot x}{2} = \frac{18}{2}$$
$$x = 9 \text{ m}$$

Write a proportion to find y.

$$\frac{5 \text{ m}}{y} = \frac{4 \text{ m}}{6 \text{ m}} \quad \textbf{OR} \quad \frac{5}{y} = \frac{2}{3}$$
$$2 \cdot y = 5 \cdot 3$$
$$\frac{2 \cdot y}{2} = \frac{15}{2}$$
$$y = 7.5 \text{ m}$$

$P = 9 \text{ m} + 7.5 \text{ m} + 6 \text{ m}$
$P = 22.5 \text{ m}$

62. Write a proportion to find x.

$$\frac{x}{9 \text{ mm}} = \frac{16 \text{ mm}}{12 \text{ mm}} \quad \textbf{OR} \quad \frac{x}{9} = \frac{4}{3}$$
$$3 \cdot x = 9 \cdot 4$$
$$\frac{3 \cdot x}{3} = \frac{36}{3}$$
$$x = 12 \text{ mm}$$

Write a proportion to find y.

$$\frac{10 \text{ mm}}{y} = \frac{16 \text{ mm}}{12 \text{ mm}} \quad \textbf{OR} \quad \frac{10}{y} = \frac{4}{3}$$
$$y \cdot 4 = 10 \cdot 3$$
$$\frac{y \cdot 4}{4} = \frac{30}{4}$$
$$y = 7.5 \text{ mm}$$

$P = 12 \text{ mm} + 10 \text{ mm} + 16 \text{ mm}$
$P = 38 \text{ mm}$

63. **[6.3]** $\dfrac{x}{45} = \dfrac{70}{30} \quad \textbf{OR} \quad \dfrac{x}{45} = \dfrac{7}{3}$

Show that cross
$$3 \cdot x = 7 \cdot 45 \qquad \textit{products are equal.}$$
$$\frac{3 \cdot x}{3} = \frac{315}{3} \qquad \textit{Divide both}$$
$$\qquad\qquad\qquad \textit{sides by 3.}$$
$$x = 105$$

Check $105 \cdot 30 = 3150 = 45 \cdot 70$

64. **[6.3]** $\dfrac{x}{52} = \dfrac{0}{20}$

Since $\frac{0}{20} = 0$, x *must* equal 0. The denominator 52 could be any number (except 0).

65. **[6.3]** $\dfrac{64}{10} = \dfrac{x}{20} \quad \textbf{OR} \quad \dfrac{32}{5} = \dfrac{x}{20}$

$$5 \cdot x = 32 \cdot 20 \qquad \begin{array}{l}\textit{Show that cross} \\ \textit{products are equal.}\end{array}$$
$$\frac{5 \cdot x}{5} = \frac{640}{5} \qquad \begin{array}{l}\textit{Divide both} \\ \textit{sides by 5.}\end{array}$$
$$x = 128$$

Check $64 \cdot 20 = 1280 = 10 \cdot 128$

66. **[6.3]** $\dfrac{15}{x} = \dfrac{65}{100} \quad \textbf{OR} \quad \dfrac{15}{x} = \dfrac{13}{20}$

$$13 \cdot x = 15 \cdot 20 \qquad \begin{array}{l}\textit{Show that cross} \\ \textit{products are equal.}\end{array}$$
$$\frac{13 \cdot x}{13} = \frac{300}{13} \qquad \begin{array}{l}\textit{Divide both} \\ \textit{sides by 13.}\end{array}$$
$$x = \frac{300}{13} \approx 23.08 \text{ (rounded)}$$

Check $15 \cdot 100 = 1500 = \frac{300}{13} \cdot 65$

67. **[6.3]** $\dfrac{7.8}{3.9} = \dfrac{13}{x} \quad \textbf{OR} \quad \dfrac{2}{1} = \dfrac{13}{x}$

$$2 \cdot x = 1 \cdot 13 \qquad \begin{array}{l}\textit{Show that cross} \\ \textit{products are equal.}\end{array}$$
$$\frac{2 \cdot x}{2} = \frac{13}{2} \qquad \begin{array}{l}\textit{Divide both} \\ \textit{sides by 2.}\end{array}$$
$$x = 6.5$$

Check $7.8(6.5) = 50.7 = 3.9 \cdot 13$

68. **[6.3]** $\dfrac{34.1}{x} = \dfrac{0.77}{2.65}$

$$0.77 \cdot x = 34.1(2.65) \qquad \begin{array}{l}\textit{Show that cross} \\ \textit{products are equal.}\end{array}$$
$$\frac{0.77 \cdot x}{0.77} = \frac{90.365}{0.77} \qquad \begin{array}{l}\textit{Divide both} \\ \textit{sides by 0.77.}\end{array}$$
$$x \approx 117.36 \qquad \textit{(rounded)}$$

Check $34.1(2.65) = 90.365 = \left(\frac{90.365}{0.77}\right)(0.77)$

69. **[6.1]** 4 dollars to 10 quarters

4 dollars $= 4 \cdot 4 = 16$ quarters

$$\frac{4 \text{ dollars}}{10 \text{ quarters}} = \frac{16 \text{ quarters}}{10 \text{ quarters}} = \frac{16 \div 2}{10 \div 2} = \frac{8}{5}$$

70. **[6.1]** $4\frac{1}{8}$ inches to 10 inches

$$\frac{4\frac{1}{8} \text{ inches}}{10 \text{ inches}} = \frac{\frac{33}{8}}{10} = \frac{33}{8} \div 10$$
$$= \frac{33}{8} \cdot \frac{1}{10} = \frac{33}{80}$$

71. **[6.1]** 10 yards to 8 feet

10 yards $= 10 \cdot 3 = 30$ feet

$$\frac{10 \text{ yards}}{8 \text{ feet}} = \frac{30 \text{ feet}}{8 \text{ feet}} = \frac{30 \div 2}{8 \div 2} = \frac{15}{4}$$

72. **[6.1]** $3.60 to $0.90

$$\frac{\$3.60}{\$0.90} = \frac{3.60 \div 0.90}{0.90 \div 0.90} = \frac{4}{1}$$

73. **[6.1]** 12 eggs to 15 eggs

$$\frac{12 \text{ eggs}}{15 \text{ eggs}} = \frac{12 \div 3}{15 \div 3} = \frac{4}{5}$$

74. **[6.1]** 37 meters to 7 meters

$$\frac{37 \text{ meters}}{7 \text{ meters}} = \frac{37}{7}$$

75. **[6.1]** 3 pints to 4 quarts

4 quarts $= 4 \cdot 2 = 8$ pints

$$\frac{3 \text{ pints}}{4 \text{ quarts}} = \frac{3 \text{ pints}}{8 \text{ pints}} = \frac{3}{8}$$

76. **[6.1]** 15 minutes to 3 hours

3 hours $= 3 \cdot 60 = 180$ minutes

$$\frac{15 \text{ minutes}}{3 \text{ hours}} = \frac{15 \text{ minutes}}{180 \text{ minutes}}$$
$$= \frac{15 \div 15}{180 \div 15} = \frac{1}{12}$$

77. **[6.1]** $4\frac{1}{2}$ miles to $1\frac{3}{10}$ miles

$$\frac{4\frac{1}{2} \text{ miles}}{1\frac{3}{10} \text{ miles}} = \frac{4\frac{1}{2}}{1\frac{3}{10}} = 4\frac{1}{2} \div 1\frac{3}{10}$$
$$= \frac{9}{2} \div \frac{13}{10} = \frac{9}{\overset{1}{\underset{1}{2}}} \cdot \frac{\overset{5}{\cancel{10}}}{13}$$
$$= \frac{45}{13}$$

78. **[6.4]**

$$\frac{7 \text{ buying fans}}{8 \text{ fans}} = \frac{x}{28{,}500 \text{ fans}}$$
$$8 \cdot x = 7 \cdot 28{,}500$$
$$\frac{8 \cdot x}{8} = \frac{199{,}500}{8} \quad \textit{Divide both}$$
$$\textit{sides by 8.}$$
$$x \approx 24{,}937.5$$

At today's concert, about 24,900 (rounded to the nearest hundred) fans can be expected to buy a beverage.

79. **[6.1]**

$$\frac{\$400 \text{ spent on car insurance}}{\$150 \text{ spent on repairs}} = \frac{400 \div 50}{150 \div 50}$$
$$= \frac{8}{3}$$

The ratio of the amount spent on insurance to the amount spent on repairs is $\frac{8}{3}$.

80. **[6.2]** 25 feet for $0.78

$$\frac{\$0.78}{25 \text{ feet}} \approx \$0.031 \text{ per foot}$$

75 feet for $1.99 - $0.50 (coupon) $=$ $1.49

$$\frac{\$1.49}{75 \text{ feet}} \approx \$0.020 \text{ per foot } (*)$$

100 feet for $2.59 - $0.50 (coupon) $=$ $2.09

$$\frac{\$2.09}{100 \text{ feet}} \approx \$0.021 \text{ per foot}$$

The best buy is 75 feet for $1.99 with a 50¢ coupon.

81. **[6.4]** First find the length.

$$\frac{0.5 \text{ inch}}{6 \text{ feet}} = \frac{1.75 \text{ inches}}{x}$$
$$0.5 \cdot x = 6(1.75)$$
$$\frac{0.5 \cdot x}{0.5} = \frac{10.5}{0.5}$$
$$x = 21$$

When it is built, the actual length of the patio will be 21 feet.

Then find the width.

$$\frac{0.5 \text{ inch}}{6 \text{ feet}} = \frac{1.25 \text{ inches}}{x}$$
$$0.5 \cdot x = 6(1.25)$$
$$\frac{0.5 \cdot x}{0.5} = \frac{7.5}{0.5}$$
$$x = 15$$

When it is built, the actual width of the patio will be 15 feet.

82. **[6.4]** **(a)** $\dfrac{1000 \text{ milligrams}}{5 \text{ pounds}} = \dfrac{x}{7 \text{ pounds}}$

$$5 \cdot x = 1000 \cdot 7$$
$$\frac{5 \cdot x}{5} = \frac{7000}{5}$$
$$x = 1400$$

A 7-pound cat should be given 1400 milligrams.

(b) 8 ounces $= \frac{8}{16}$ pound $= 0.5$ pound

$$\frac{1000 \text{ milligrams}}{5 \text{ pounds}} = \frac{x}{0.5 \text{ pounds}}$$
$$5 \cdot x = 1000(0.5)$$
$$\frac{5 \cdot x}{5} = \frac{500}{5}$$
$$x = 100$$

An 8-ounce kitten should be given 100 milligrams.

83. **[6.4]** $\dfrac{251 \text{ points}}{169 \text{ minutes}} = \dfrac{x}{14 \text{ minutes}}$

$$169 \cdot x = 251 \cdot 14$$
$$\frac{169 \cdot x}{169} = \frac{3514}{169}$$
$$x \approx 20.792899$$

Charles should score 21 points (rounded).

84. **[6.4]**

$$\frac{1\frac{1}{2}\text{ teaspoons}}{24\text{ pounds}} = \frac{x}{8\text{ pounds}}$$

$$24 \cdot x = 1\frac{1}{2} \cdot 8 = \frac{3}{2} \cdot \frac{8}{1} = \frac{24}{2} = 12$$

$$\frac{24 \cdot x}{24} = \frac{12}{24}$$

$$x = \frac{1}{2}\text{ or }0.5$$

The infant should be given $\frac{1}{2}$ or 0.5 teaspoon.

85. **[6.5]** $\overleftrightarrow{WX}$ and $\overleftrightarrow{YZ}$ are parallel lines.

86. **[6.5]** $\overline{QR}$ is a line segment.

87. **[6.5]** $\angle CQD$ is an acute angle.

88. **[6.5]** $\overleftrightarrow{PQ}$ and $\overleftrightarrow{NO}$ are intersecting lines.

89. **[6.5]** $\angle APB$ is a right angle measuring $90°$.

90. **[6.5]** $\overrightarrow{AB}$ is a ray.

91. **[6.5]** $\overleftrightarrow{T}$ is a straight angle measuring $180°$.

92. **[6.5]** $\angle FEG$ is an obtuse angle.

93. **[6.5]** $\overleftrightarrow{LM}$ and $\overrightarrow{JK}$ are perpendicular lines.

94. **[6.5] (a)** If your car "did a 360," the car turned around in a complete circle.

(b) If the governor's view on taxes "took a 180° turn," he or she took the opposite view. For example, he or she may have opposed taxes but now supports them.

95. **[6.5] (a)** No; obtuse angles are $> 90°$, so their sum would be $> 180°$.

(b) Yes; acute angles are $< 90°$, so their sum could equal $90°$.

96. **[6.5]** $\angle 1$ measures $90° - 45° = 45°$.
$\angle 7 \cong \angle 4$, so each measures $55°$.
$\angle 3$ measures $90° - 55° = 35°$.
$\angle 3 \cong \angle 6$, so $\angle 6$ measures $35°$.
$\angle 5$ measures $90°$.

97. **[6.5]** $\angle 5 \cong \angle 2 \cong \angle 7 \cong \angle 4$, so each measures $75°$.
$\angle 6$ measures $180° - 75° = 105°$.
$\angle 6 \cong \angle 1 \cong \angle 8 \cong \angle 3$, so each measures $105°$.

Chapter 6 Test

1. $15 for 75 minutes

$$\frac{\$15 \div 15}{75\text{ minutes} \div 15} = \frac{\$1}{5\text{ minutes}}$$

2. 3 hours to 40 minutes

3 hours $= 3 \cdot 60$ minutes $= 180$ minutes

$$\frac{3\text{ hours}}{40\text{ minutes}} = \frac{180\text{ minutes}}{40\text{ minutes}} = \frac{180 \div 20}{40 \div 20} = \frac{9}{2}$$

3. $$\frac{1200\text{ seats}}{320\text{ seats}} = \frac{1200 \div 80}{320 \div 80} = \frac{15}{4}$$

4. 26 ounces of Brand X for
$3.89 − $0.75 (coupon) = $3.14

$$\frac{\$3.14}{26\text{ ounces}} \approx \$0.121\text{ per ounce}$$

16 ounces of Brand Y for
$1.89 − $0.50 (coupon) = $1.39

$$\frac{\$1.39}{16\text{ ounces}} \approx \$0.087\text{ per ounce }(*)$$

14 ounces of Brand Z for $1.29.

$$\frac{\$1.29}{14\text{ ounces}} \approx \$0.092\text{ per ounce}$$

The best buy is 16 ounces of Brand Y for $1.89 with a 50¢ coupon.

5. $$\frac{5}{9} = \frac{x}{45}$$

$9 \cdot x = 5 \cdot 45$ *Show that cross products are equal.*

$\dfrac{9 \cdot x}{9} = \dfrac{225}{9}$ *Divide both sides by 9.*

$x = 25$

Check $5 \cdot 45 = 225 = 9 \cdot 25$

6. $$\frac{3}{1} = \frac{8}{x}$$

$3 \cdot x = 8 \cdot 1$ *Show that cross products are equal.*

$\dfrac{3 \cdot x}{3} = \dfrac{8}{3}$ *Divide both sides by 3.*

$x \approx 2.67$ (rounded)

Check $3\left(\frac{8}{3}\right) = 8 = 1 \cdot 8$

7. $$\frac{x}{20} = \frac{6.5}{0.4}$$

$0.4 \cdot x = 20(6.5)$ *Show that cross products are equal.*

$\dfrac{0.4 \cdot x}{0.4} = \dfrac{130}{0.4}$ *Divide both sides by 0.4.*

$x = 325$

Check $325(0.4) = 130 = 20(6.5)$

8. $$\frac{2\frac{1}{3}}{x} = \frac{\frac{8}{9}}{4}$$ *Show that cross products are equal.*

$\dfrac{8}{9} \cdot x = 2\dfrac{1}{3} \cdot 4$ $\dfrac{7}{3} \cdot \dfrac{4}{1} = \dfrac{28}{3}$

$\dfrac{\frac{8}{9} \cdot x}{\frac{8}{9}} = \dfrac{\frac{28}{3}}{\frac{8}{9}}$ *Divide both sides by 8/9.*

$x = 10\dfrac{1}{2}$ $\dfrac{28}{3} \div \dfrac{8}{9} = \dfrac{\overset{7}{\cancel{28}}}{\underset{1}{\cancel{3}}} \cdot \dfrac{\overset{3}{\cancel{9}}}{\underset{2}{\cancel{8}}} = \dfrac{21}{2}$

Check $2\frac{1}{3} \cdot 4 = \frac{28}{3} = \frac{21}{2} \cdot \frac{8}{9}$

9.
$$\frac{18 \text{ orders}}{30 \text{ minutes}} = \frac{x}{40 \text{ minutes}}$$

$$30 \cdot x = 18 \cdot 40 \qquad \textit{Show that cross}$$
$$\qquad\qquad\qquad\qquad \textit{products are equal.}$$
$$\frac{30 \cdot x}{30} = \frac{720}{30} \qquad \textit{Divide both}$$
$$\qquad\qquad\qquad\qquad \textit{sides by 30.}$$
$$x = 24$$

Pedro could enter 24 orders in 40 minutes.

10.
$$\frac{2 \text{ left-handed people}}{15 \text{ people}} = \frac{x}{650 \text{ students}}$$

$$15 \cdot x = 2 \cdot 650$$
$$\frac{15 \cdot x}{15} = \frac{1300}{15}$$
$$x = \frac{260}{3} \approx 87$$

You could expect about 87 students to be left-handed.

11.
$$\frac{8.2 \text{ grams}}{50 \text{ pounds}} = \frac{x}{145 \text{ pounds}}$$

$$50 \cdot x = 8.2(145) \qquad \textit{Show that cross}$$
$$\qquad\qquad\qquad\qquad \textit{products are equal.}$$
$$\frac{50 \cdot x}{50} = \frac{1189}{50} \qquad \textit{Divide both}$$
$$\qquad\qquad\qquad\qquad \textit{sides by 50.}$$
$$x = 23.78 \approx 23.8$$

A 145-pound person should be given 23.8 grams (rounded).

12.
$$\frac{1 \text{ inch}}{8 \text{ feet}} = \frac{7.5 \text{ inches}}{x}$$
$$1 \cdot x = 8(7.5)$$
$$x = 60$$

The actual height of the building is 60 feet.

13. $\angle LOM$ is an acute angle, so the answer is (e).

14. $\angle YOX$ is a right angle, so the answer is (a). Its measure is $90°$.

15. $\overrightarrow{GH}$ is a ray, so the answer is (d).

16. $\overleftrightarrow{W}$ is a straight angle, so the answer is (g). Its measure is $180°$.

17. **Parallel lines** are lines in the same plane that never intersect.
Perpendicular lines intersect to form a right angle.

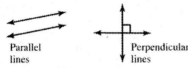

Parallel lines   Perpendicular lines

18. The complement of an angle measuring $81°$ is $90° - 81° = 9°$.

19. The supplement of an angle measuring $20°$ is $180° - 20° = 160°$.

20. $\angle 4 \cong \angle 1$, so each measures $50°$.
$\angle 6 \cong \angle 3$, so each measures $95°$.
$\angle 2$ measures $180° - 50° - 95° = 35°$.
$\angle 2 \cong \angle 5$, so each measures $35°$.

21. $\angle 3 \cong \angle 1 \cong \angle 5 \cong \angle 7$, so each measures $65°$.
$\angle 4$ measures $180° - 65° = 115°$.
$\angle 4 \cong \angle 2 \cong \angle 6 \cong \angle 8$, so each measures $115°$.

22. On both triangles, two corresponding angles and the side that connects them measure the same, so the Angle-Side-Angle (ASA) method can be used to prove that the triangles are congruent.

23. On both triangles, two corresponding sides and the angle between them measure the same, so the Side-Angle-Side (SAS) method can be used to prove that the triangles are congruent.

24. Write a proportion to find y.

$$\frac{y}{18 \text{ cm}} = \frac{10 \text{ cm}}{15 \text{ cm}} \quad \textbf{OR} \quad \frac{y}{18} = \frac{2}{3}$$
$$y \cdot 3 = 18 \cdot 2$$
$$\frac{y \cdot 3}{3} = \frac{36}{3}$$
$$y = 12 \text{ cm}$$

Write a proportion to find z.

(Note that the third side of the first triangle is 9 cm, which was covered in the first printing of the text.)

$$\frac{z}{9 \text{ cm}} = \frac{10 \text{ cm}}{15 \text{ cm}} \quad \textbf{OR} \quad \frac{z}{9} = \frac{2}{3}$$
$$z \cdot 3 = 9 \cdot 2$$
$$\frac{z \cdot 3}{3} = \frac{18}{3}$$
$$z = 6 \text{ cm}$$

25. Write a proportion to find x.

$$\frac{x}{10 \text{ mm}} = \frac{18 \text{ mm}}{15 \text{ mm}} \quad \textbf{OR} \quad \frac{x}{10} = \frac{6}{5}$$
$$x \cdot 5 = 10 \cdot 6$$
$$\frac{x \cdot 5}{5} = \frac{60}{5}$$
$$x = 12 \text{ mm}$$

Write a proportion to find y.

$$\frac{16.8 \text{ mm}}{y} = \frac{18 \text{ mm}}{15 \text{ mm}} \quad \textbf{OR} \quad \frac{16.8}{y} = \frac{6}{5}$$
$$y \cdot 6 = 16.8 \cdot 5$$
$$\frac{y \cdot 6}{6} = \frac{84}{6}$$
$$y = 14 \text{ mm}$$

The perimeter of the larger triangle:

$$P = 18 \text{ mm} + 12 \text{ mm} + 16.8 \text{ mm}$$
$$P = 46.8 \text{ mm}$$

The perimeter of the smaller triangle:

$$P = 15 \text{ mm} + 10 \text{ mm} + 14 \text{ mm}$$
$$P = 39 \text{ mm}$$

CHAPTER 7   PERCENT

7.1   The Basics of Percent

7.1   Margin Exercises

1.   (a) The tip is \$20 per \$100 or $\frac{20}{100}$ or 20%.

   (b) The tax rate is \$5 per \$100 or $\frac{5}{100}$ or 5%.

   (c) 94 points out of 100 points or $\frac{94}{100}$ or 94%

2.   (a) $68\% = 68 \div \underline{100} = 0.\underline{68}$

   (b) $5\% = 5 \div 100 = 0.05$

   (c) $40.6\% = 40.6 \div 100 = 0.406$

   (d) $200\% = 200 \div \underline{100} = 2.00,$ or $\underline{2}$

   (e) $350\% = 350 \div 100 = 3.50,$ or 3.5

3.   (a) 90%   ▪   Drop the percent sign and move the decimal point two places to the left.

$$90.\% = 0.90, \quad \text{or} \quad \underline{0.9}$$

   (b) $9\% = 09.\% = 0.09$

   (c) $900\% = 900.\% = 9.00,$ or 9

   (d) $9.9\% = 09.9\% = \underline{0.099}$

   (e) $0.9\% = 00.9\% = 0.009$

4.   (a) 0.95   ▪   Move the decimal point two places to the right. (The decimal point is not written with whole number percents.) Attach a percent sign.

$$0.95 = \underline{95\%}$$

   (b) $0.16 = 16\%$

   (c) $0.09 = 9\%$

   (d) $0.617 = 61.7\%$

   (e) $0.4 = 0.40 = 40\%$

   (f) $5.34 = 534\%$

   (g) $2.8 = 2.80 = 280\%$ ▪   One zero is attached so the decimal point can be moved two places to the right. Attach a percent sign.

   (h) $4 = 4.00 = 400\%$ ▪   Two zeros are attached so the decimal point can be moved two places to the right. Attach a percent sign.

5.   (a) $50\% = \frac{50}{100} = \frac{50 \div 50}{100 \div 50} = \frac{1}{2}$

   (b) $19\% = \frac{19}{100}$

   (c) $80\% = \frac{80}{100} = \frac{80 \div 20}{100 \div 20} = \frac{4}{5}$

   (d) $6\% = \frac{6}{100} = \frac{6 \div 2}{100 \div 2} = \frac{3}{50}$

   (e) $125\% = \frac{125}{100} = \frac{125 \div 25}{100 \div 25} = \frac{5}{4} = 1\frac{1}{4}$

   (f) $300\% = \frac{300}{100} = \frac{300 \div 100}{100 \div 100} = \frac{3}{1} = 3$

6.   (a) $18.5\% = \frac{18.5}{100} = \frac{(18.5)(10)}{(100)(10)} = \frac{185}{1000}$

$$= \frac{185 \div 5}{1000 \div 5} = \frac{37}{200}$$

   (b) $87.5\% = \frac{87.5}{100} = \frac{(87.5)(10)}{(100)(10)} = \frac{875}{1000}$

$$= \frac{875 \div 125}{1000 \div 125} = \frac{7}{8}$$

   (c) $6.5\% = \frac{6.5}{100} = \frac{(6.5)(10)}{(100)(10)} = \frac{65}{1000}$

$$= \frac{65 \div 5}{1000 \div 5} = \frac{13}{200}$$

   (d) $66\frac{2}{3}\% = \frac{66\frac{2}{3}}{100} = \frac{\frac{200}{3}}{100}$

$$= \frac{200}{3} \div 100 = \frac{200}{3} \div \frac{100}{1}$$

$$= \frac{200}{3} \cdot \frac{1}{100} = \frac{2 \cdot \overset{1}{\cancel{100}} \cdot 1}{3 \cdot \underset{1}{\cancel{100}}} = \frac{2}{3}$$

   (e) $12\frac{1}{3}\% = \frac{12\frac{1}{3}}{100} = \frac{\frac{37}{3}}{100} = \frac{37}{3} \div 100$

$$= \frac{37}{3} \div \frac{100}{1} = \frac{37}{3} \cdot \frac{1}{100} = \frac{37}{300}$$

   (f) $62\frac{1}{2}\% = \frac{62\frac{1}{2}}{100} = \frac{\frac{125}{2}}{100} = \frac{125}{2} \div 100$

$$= \frac{125}{2} \div \frac{100}{1} = \frac{125}{2} \cdot \frac{1}{100}$$

$$= \frac{5 \cdot \overset{1}{\cancel{25}} \cdot 1}{2 \cdot 4 \cdot \underset{1}{\cancel{25}}} = \frac{5}{8}$$

OR

$$62\frac{1}{2}\% = 62.5\% = \frac{62.5}{100} = \frac{(62.5)(10)}{(100)(10)}$$

$$= \frac{625}{1000} = \frac{625 \div 125}{1000 \div 125} = \frac{5}{8}$$

7.   (a) $\frac{1}{2} = \left(\frac{1}{2}\right)(100\%) = \left(\frac{1}{2}\right)\left(\frac{100}{1}\%\right)$

$$= \frac{1 \cdot \overset{1}{\cancel{2}} \cdot 50}{\underset{1}{\cancel{2}} \cdot 1}\% = \frac{50}{1}\% = 50\%$$

(b) $\frac{3}{4} = \left(\frac{3}{4}\right)(100\%) = \left(\frac{3}{4}\right)\left(\frac{100}{1}\%\right)$

$= \frac{3 \cdot \overset{1}{\cancel{4}} \cdot 25}{\underset{1}{\cancel{4}} \cdot 1}\% = \frac{75}{1}\% = 75\%$

(c) $\frac{1}{10} = \left(\frac{1}{10}\right)(100\%) = \left(\frac{1}{10}\right)\left(\frac{100}{1}\%\right)$

$= \frac{1 \cdot \overset{1}{\cancel{10}} \cdot 10}{\underset{1}{\cancel{10}} \cdot 1}\% = \frac{10}{1}\% = 10\%$

(d) $\frac{7}{8} = \left(\frac{7}{8}\right)(100\%) = \left(\frac{7}{8}\right)\left(\frac{100}{1}\%\right)$

$= \frac{7 \cdot \overset{1}{\cancel{4}} \cdot 25}{2 \cdot \underset{1}{\cancel{4}}}\% = \frac{175}{2}\% = 87\frac{1}{2}\%,$

or   87.5% (Both are exact answers.)

(e) $\frac{5}{6} = \left(\frac{5}{6}\right)(100\%) = \left(\frac{5}{6}\right)\left(\frac{100}{1}\%\right)$

$= \frac{5 \cdot \overset{1}{\cancel{2}} \cdot 50}{\underset{1}{\cancel{2}} \cdot 3 \cdot 1}\% = \frac{250}{3}\% = 83\frac{1}{3}\% \text{ (exact)},$

or   ≈ 83.3% (rounded)

(f) $\frac{2}{3} = \left(\frac{2}{3}\right)(100\%) = \left(\frac{2}{3}\right)\left(\frac{100}{1}\%\right)$

$= \frac{200}{3}\% = 66\frac{2}{3}\% \text{ (exact)}, \quad \text{or}$

≈ 66.7% (rounded)

8.   **(a)** 100% of \$4.60 is <u>\$4.60</u>. 100% is *all* of the money.

(b) 100% of 3000 students is <u>3000 students</u>.

(c) 100% of 7 pages is <u>7 pages</u>.

(d) 100% of 272 miles is <u>272 miles</u>.

(e) 100% of $10\frac{1}{2}$ hours is <u>$10\frac{1}{2}$ hours</u>.

9.   **(a)** 50% of \$4.60 is <u>\$2.30</u>. 50% is *half* of the money.

(b) 50% of 3000 students is *half* of 3000 students, or <u>1500 students</u>.

(c) 50% of 7 pages is *half* of 7 pages, or <u>$3\frac{1}{2}$ pages</u>.

(d) 50% of 272 miles is *half* of 272 miles, or <u>136 miles</u>.

(e) 50% of $10\frac{1}{2}$ hours is *half* of $10\frac{1}{2}$ hours, or <u>$5\frac{1}{4}$ hours</u>.

7.1 Section Exercises

1.   **(a)** A shortcut for dividing by 100 is to move the decimal point <u>2 places to the left</u>.

(b) $48.\% = 0.48$ $1\,20.\% = 1.20$ or 1.2

$03.5\% = 0.035$ $06.\% = 0.06$

3. $25\% = 25.\% = 0.25$

Drop the percent sign and move the decimal point two places to the left.

5. $30\% = 30.\% = 0.30,$ or 0.3

Drop the percent sign and move the decimal point two places to the left.

7. $6\% = 06.\% = 0.\underline{06}$

0 is attached so the decimal point can be moved two places to the left.

9. $140\% = 140.\% = 1.40,$ or 1.4

Drop the percent sign and move the decimal point two places to the left.

11. $7.8\% = 07.8\% = 0.078$

0 is attached so the decimal point can be moved two places to the left.

13. $100\% = 100.\% = 1.00,$ or 1

15. $0.5\% = 00.5\% = 0.005$

0 is attached so the decimal point can be moved two places to the left.

17. $0.35\% = 00.35\% = 0.0035$

0 is attached so the decimal point can be moved two places to the left.

19. $0.5 = 0.50 = 50\%$

0 is attached so the decimal point can be moved two places to the right. Attach a percent sign. (The decimal point is not written with whole numbers.)

21. $0.62 = 62\%$

Move the decimal point two places to the right and attach a percent sign.

23. $0.03 = 3\%$

Move the decimal point two places to the right and attach a percent sign.

25. $0.125 = 12.5\%$

Move the decimal point two places to the right and attach a percent sign.

27. $2 = 2.00 = 200\%$

Two zeros are attached so the decimal point can be moved two places to the right. Attach a percent sign.

29. $2.6 = 2.60 = 260\%$

One zero is attached so the decimal point can be moved two places to the right. Attach a percent sign.

31. $0.0312 = 3.12\%$

Move the decimal point two places to the right and attach a percent sign.

33. $20\% = \dfrac{20}{100} = \dfrac{20 \div 20}{100 \div 20} = \dfrac{1}{5}$

35. $50\% = \dfrac{50}{100} = \dfrac{50 \div 50}{100 \div 50} = \dfrac{1}{2}$

37. $55\% = \dfrac{55}{100} = \dfrac{55 \div 5}{100 \div 5} = \dfrac{11}{20}$

39. $37.5\% = \dfrac{37.5}{100} = \dfrac{37.5(10)}{100(10)} = \dfrac{375 \div 125}{1000 \div 125} = \dfrac{3}{8}$

41. $6.25\% = \dfrac{6.25}{100} = \dfrac{6.25(100)}{100(100)}$

$= \dfrac{625 \div 625}{10{,}000 \div 625} = \dfrac{1}{16}$

43. $16\dfrac{2}{3}\% = \dfrac{16\frac{2}{3}}{100} = 16\dfrac{2}{3} \div \dfrac{100}{1} = \dfrac{\overset{1}{\cancel{50}}}{3} \cdot \dfrac{1}{\underset{2}{\cancel{100}}} = \dfrac{1}{6}$

45. $130\% = \dfrac{130}{100} = \dfrac{130 \div 10}{100 \div 10} = \dfrac{13}{10} = 1\dfrac{3}{10}$

47. $250\% = \dfrac{250}{100} = \dfrac{250 \div 50}{100 \div 50} = \dfrac{5}{2} = 2\dfrac{1}{2}$

49. $\dfrac{1}{4} = \left(\dfrac{1}{4}\right)(100\%) = \left(\dfrac{1}{4}\right)\left(\dfrac{100}{1}\%\right)$

$= \dfrac{1 \cdot \overset{1}{\cancel{4}} \cdot 25}{\underset{1}{\cancel{4}} \cdot 1}\% = \dfrac{25}{1}\% = 25\%$

51. $\dfrac{3}{10} = \left(\dfrac{3}{10}\right)(100\%) = \left(\dfrac{3}{10}\right)\left(\dfrac{100}{1}\%\right)$

$= \dfrac{3 \cdot \overset{1}{\cancel{10}} \cdot 10}{\underset{1}{\cancel{10}} \cdot 1}\% = \dfrac{30}{1}\% = 30\%$

53. $\dfrac{3}{5} = \left(\dfrac{3}{5}\right)(100\%) = \left(\dfrac{3}{5}\right)\left(\dfrac{100}{1}\%\right)$

$= \dfrac{3 \cdot \overset{1}{\cancel{5}} \cdot 20}{\underset{1}{\cancel{5}} \cdot 1}\% = \dfrac{60}{1}\% = 60\%$

55. The denominator is already 100.

$$\dfrac{37}{100} = 37\%$$

57. $\dfrac{3}{8} = \left(\dfrac{3}{8}\right)(100\%) = \left(\dfrac{3}{8}\right)\left(\dfrac{100}{1}\%\right)$

$= \dfrac{3 \cdot \overset{1}{\cancel{4}} \cdot 25}{2 \cdot \underset{1}{\cancel{4}} \cdot 1}\% = \dfrac{75}{2}\% = 37\dfrac{1}{2}\%, \quad \text{or} \quad 37.5\%$

59. $\dfrac{1}{20} = \left(\dfrac{1}{20}\right)(100\%) = \left(\dfrac{1}{20}\right)\left(\dfrac{100}{1}\%\right)$

$= \dfrac{1 \cdot \overset{1}{\cancel{20}} \cdot 5}{\underset{1}{\cancel{20}} \cdot 1}\% = \dfrac{5}{1}\% = 5\%$

61. $\dfrac{5}{9} = \left(\dfrac{5}{9}\right)(100\%) = \left(\dfrac{5}{9}\right)\left(\dfrac{100}{1}\%\right)$

$= \dfrac{500}{9}\% = 55\dfrac{5}{9}\%, \quad \text{or} \quad 55.6\% \text{ (rounded)}$

63. $\dfrac{1}{7} = \left(\dfrac{1}{7}\right)(100\%) = \left(\dfrac{1}{7}\right)\left(\dfrac{100}{1}\%\right)$

$= \dfrac{100}{7}\% = 14\dfrac{2}{7}\%, \quad \text{or} \quad 14.3\% \text{ (rounded)}$

65. Drop the percent sign and move the decimal point two places to the left.

$$8\% \text{ of U.S. homes} = 0.08$$

67. Drop the percent sign and move the decimal point two places to the left.

$$49\% \text{ of tornadoes} = 0.49$$

69. Move the decimal point two places to the right and attach a percent sign.

$$0.035 \text{ property tax rate} = 3.5\%$$

71. Attach two 0's so that the decimal point can be moved two places to the right. Attach a percent sign.

$$2 \text{ times that of the last session} = 200\%$$

73. Ninety-five parts of the one hundred parts are shaded.

$$\dfrac{95}{100} = 0.95 = 95\%$$

Five parts of the one hundred parts are unshaded.

$$\dfrac{5}{100} = 0.05 = 5\%$$

75. Three parts of the four parts are shaded.

$$\dfrac{3}{4} = \dfrac{3(25)}{4(25)} = \dfrac{75}{100} = 75\%$$

One part of the four parts is unshaded.

$$\dfrac{1}{4} = \dfrac{1(25)}{4(25)} = \dfrac{25}{100} = 25\%$$

77. $50\% = 50.\% = 0.50$ or 0.5.

$$50\% = \dfrac{50}{100} = \dfrac{50 \div 50}{100 \div 50} = \dfrac{1}{2}$$

So 50%, 0.5, and $\dfrac{1}{2}$ are equivalent.

79. $\dfrac{1}{100} = 0.01$ ←   Decimal

$= 1\%$ ←   Percent

81. $0.2 = 0.20 = \dfrac{20}{100} = \dfrac{1}{5}$ ←   Fraction

$= 20\%$ ←   Percent

83. $30\% = \dfrac{30}{100} = \dfrac{3}{10}$ ←   Fraction

$= 0.30 = 0.3$ ←   Decimal

85. $\dfrac{1}{2} = \dfrac{50}{100} = 0.50 = 0.5$ ←   Decimal

$= 50\%$ ←   Percent

87. $90\% = \dfrac{90}{100} = \dfrac{9}{10}$ ←   Fraction

$= 0.90 = 0.9$ ←   Decimal

89. $1.5 = \dfrac{150}{100} = \dfrac{3}{2} = 1\dfrac{1}{2}$ ←   Fraction

$= 150\%$ ←   Percent

91. $8\% = \dfrac{8}{100} = \dfrac{2}{25}$ ←   Fraction

$= 0.08$ ←   Decimal

93. $\dfrac{5 \text{ ft for}}{\text{every } 100 \text{ ft}} = \dfrac{5}{100} = \dfrac{1}{20}$ ←   Fraction

$= 0.05$ ←   Decimal

$= 5\%$ ←   Percent

95. There are 8 incisors and 32 total teeth.

$$\dfrac{8}{32} = \dfrac{8 \div 8}{32 \div 8} = \dfrac{1}{4}$$

$$\dfrac{1}{4} = \dfrac{1(25)}{4(25)} = \dfrac{25}{100} = 0.25$$

$$= 25\%$$

97. There are 12 molars and 32 total teeth.

$$\dfrac{12}{32} = \dfrac{12 \div 4}{32 \div 4} = \dfrac{3}{8}$$

$$\dfrac{3}{8} = \dfrac{3(125)}{8(125)} = \dfrac{375}{1000} = 0.375$$

$$= 37.5\%$$

99. There are 4 canines and 28 total teeth.

$$\dfrac{4}{28} = \dfrac{4 \div 4}{28 \div 4} = \dfrac{1}{7}$$

$$\dfrac{1}{7} = \dfrac{1}{7} \cdot 100\% = \dfrac{100}{7}\% = 14\dfrac{2}{7}\% \text{ (exactly)}$$

$$\approx 14.3\% \text{ (rounded)}$$

101. (a) The student forgot to move the decimal point in 0.35 two places to the right. So $\dfrac{7}{20} = 35\%$.

(b) The student did the division in the wrong order. Enter $16 \div 25$ to get 0.64 and then move the decimal point two places to the right. So $\dfrac{16}{25} = 0.64 = 64\%$.

103. (a) 100% of $78 is $\underline{\$78}$.

(b) 50% of $78 is half of $78, or $\underline{\$39}$.

105. (a) 100% of 15 inches is $\underline{15 \text{ inches}}$.

(b) 50% of 15 inches is half of 15 inches, or $\underline{7\dfrac{1}{2} \text{ inches}}$.

107. (a) 100% of 2.8 miles is $\underline{2.8 \text{ miles}}$.

(b) 50% of 2.8 miles is half of 2.8 miles, or $\underline{1.4 \text{ miles}}$.

109. 100% of 20 children is 20 children. $\underline{20 \text{ children}}$ are served both meals.

111. 50% of 120 credits is half of 120 credits, or $\underline{60 \text{ credits}}$.

113. (a) 50% of $285 is half of $285, or $\underline{\$142.50}$.

(b) John will have to pay the remaining $\underline{50\%}$ of the tuition ($100\% - 50\% = 50\%$).

(c) John will have to pay the same amount that financial aid will cover, that is, $\underline{\$142.50}$.

115. (a) 50% of 8200 is 4100, so the number of students who work more than 20 hours per week is *about* $\underline{4100}$.

(b) The percent of students who work 20 hours or less per week is $100\% - 50\% = \underline{50\%}$.

117. (a) 100% of 35 problems is $\underline{35 \text{ problems}}$.

(b) He correctly worked all 35 problems, so he missed $\underline{0 \text{ problems}}$.

119. 50% means 50 parts out of 100 parts. That's half of the number. A shortcut for finding 50% of a number is to divide the number by 2. Examples will vary.

7.2   The Percent Proportion

7.2 Margin Exercises

1. **(a)** The percent is 15. The number 15 appears with the symbol %.

(b) The word *percent* has no number with it, so the percent is the *unknown* part of the problem.

(c) The percent is $6\dfrac{1}{2}$ because $6\dfrac{1}{2}$ appears with the word *percent*.

(d) The percent is 48. The number 48 appears with the symbol %.

(e) The word *percent* has no number with it, so the percent is the *unknown* part of the problem.

2. **(a)** The whole is $2000. The number $2000 appears after the word *of*.

(b) The whole is 750 employees. The number 750 appears after the word *of*.

(c) The whole is $590. The number 590 appears after the word *of*.

(d) The whole is *unknown*. The word *of* appears before "what amount."

(e) The whole is 110 rental cars. The number 110 appears after the word *of*.

3. **(a)** The part of the $2000 that will be spent on a washing machine is *unknown*.

(b) The part of the 750 employees is 60 employees.

(c) The percent is $6\frac{1}{2}$. The whole price is $590. The sales tax would be part of the whole price of $590. The sales tax (part) is *unknown*.

(d) The part is $30. $30 is part of some unknown dollar value.

(e) The part is 75 rental cars. 75 rental cars is part of the 110 rental cars.

4. **(a)** Part: unknown (n)

Percent → $\dfrac{9}{100}$ = $\dfrac{n}{3250}$ ← Part
Always 100 → ← Whole

Step 1 $\dfrac{9}{100} = \dfrac{n}{3250}$

Step 2 $100 \cdot n = 9 \cdot 3250$

Step 3 $\dfrac{100n}{100} = \dfrac{29{,}250}{100}$

$n = 292.5$

The part is 292.5 miles. 292.5 miles is 9% of 3250 miles.

(b) Part: unknown (n)

Percent → $\dfrac{78}{100}$ = $\dfrac{n}{5.50}$ ← Part
Always 100 → ← Whole

Step 1 $\dfrac{78}{100} = \dfrac{n}{5.50}$

Step 2 $100 \cdot n = 78(5.50)$

Step 3 $\dfrac{100n}{100} = \dfrac{429}{100}$

$n = 4.29$

The part is $4.29. $4.29 is 78% of $5.50.

(c) Part: unknown (n)

Percent → $\dfrac{12.5}{100}$ = $\dfrac{n}{400}$ ← Part
Always 100 → ← Whole

Step 1 $\dfrac{12.5}{100} = \dfrac{n}{400}$

Step 2 $100 \cdot n = 12.5(400)$

Step 3 $\dfrac{100n}{100} = \dfrac{5000}{100}$

$n = 50$

The part is 50 homes. 50 homes are $12\frac{1}{2}$% of 400.

5. **(a)** Percent: unknown (p)

Percent → $\dfrac{p}{100}$ = $\dfrac{1200}{5000}$ ← Part
Always 100 → ← Whole

Step 1 $\dfrac{p}{100} = \dfrac{1200}{5000}$

Step 2 $p \cdot 5000 = 100(1200)$

Step 3 $\dfrac{p \cdot 5000}{5000} = \dfrac{120{,}000}{5000}$

$p = 24$

The percent is 24%. 1200 is 24% of 5000.

(b) Percent: unknown (p)

Percent → $\dfrac{p}{100}$ = $\dfrac{0.52}{6.50}$ ← Part
Always 100 → ← Whole

Step 1 $\dfrac{p}{100} = \dfrac{0.52}{6.50}$

Step 2 $6.50 \cdot p = 100(0.52)$

Step 3 $\dfrac{6.50p}{6.50} = \dfrac{52}{6.50}$

$p = 8$

The percent is 8%. $0.52 is 8% of $6.50.

(c) Percent: unknown (p)

Percent → $\dfrac{p}{100}$ = $\dfrac{20}{32}$ ← Part
Always 100 → ← Whole

Step 1 $\dfrac{p}{100} = \dfrac{20}{32}$

Step 2 $32 \cdot p = 100(20)$

Step 3 $\dfrac{32p}{32} = \dfrac{2000}{32}$

$p = 62.5$

The percent is 62.5% or $62\frac{1}{2}$%. 20 is 62.5% of 32.

6. **(a)** Whole: unknown (n)

Percent → $\dfrac{74}{100}$ = $\dfrac{37}{n}$ ← Part
Always 100 → ← Whole

Step 1 $\dfrac{74}{100} = \dfrac{37}{n}$

Step 2 $74 \cdot n = 100 \cdot 37$

Step 3 $\dfrac{74n}{74} = \dfrac{3700}{74}$

$n = 50$

The whole is 50 cars. 37 cars is 74% of 50 cars.

(b) Whole: unknown (n)

Percent → $\dfrac{45}{100}$ = $\dfrac{139.59}{n}$ ← Part
Always 100 → ← Whole

Step 1 $\dfrac{45}{100} = \dfrac{139.59}{n}$

Step 2 $45 \cdot n = 100(139.59)$

Step 3 $\dfrac{45n}{45} = \dfrac{13{,}959}{45}$

$n = 310.20$

The whole is $310.20. $139.59 is 45% of $310.20.

(c) Whole: unknown (n)

$$\text{Percent} \to \frac{2.5}{100} = \frac{1.2}{n} \leftarrow \text{Part}$$
$$\text{Always } 100 \to \qquad\qquad \leftarrow \text{Whole}$$

Step 1 $\dfrac{2.5}{100} = \dfrac{1.2}{n}$

Step 2 $2.5 \cdot n = 100(1.2)$

Step 3 $\dfrac{2.5n}{2.5} = \dfrac{120}{2.5}$

$\qquad\qquad n = 48$

The whole is 48 tons. 1.2 tons is $2\frac{1}{2}\%$ of 48 tons.

7.   (a) Part: unknown (n)

$$\text{Percent} \to \frac{350}{100} = \frac{n}{6} \leftarrow \text{Part}$$
$$\text{Always } 100 \to \qquad\qquad \leftarrow \text{Whole}$$

Step 1 $\dfrac{350}{100} = \dfrac{n}{6}$

Step 2 $100 \cdot n = 350 \cdot 6$

Step 3 $\dfrac{100n}{100} = \dfrac{2100}{100}$

$\qquad\qquad n = 21$

The part is \$21. 350% of \$6 is \$21.

(b) Percent: unknown (p)

$$\text{Percent} \to \frac{p}{100} = \frac{23}{20} \leftarrow \text{Part}$$
$$\text{Always } 100 \to \qquad\qquad \leftarrow \text{Whole}$$

Step 1 $\dfrac{p}{100} = \dfrac{23}{20}$

Step 2 $20 \cdot p = 100 \cdot 23$

Step 3 $\dfrac{20p}{20} = \dfrac{2300}{20}$

$\qquad\qquad p = 115$

The percent is 115%.
23 hours is 115% of 20 hours.

(c) Whole: unknown (n)

$$\text{Percent} \to \frac{225}{100} = \frac{106.47}{n} \leftarrow \text{Part}$$
$$\text{Always } 100 \to \qquad\qquad\qquad \leftarrow \text{Whole}$$

Step 1 $\dfrac{225}{100} = \dfrac{106.47}{n}$

Step 2 $225 \cdot n = 100(106.47)$

Step 3 $\dfrac{225n}{225} = \dfrac{10{,}647}{225}$

$\qquad\qquad n = 47.32$

The whole is \$47.32. \$106.47 is 225% of \$47.32.

7.2 Section Exercises

1.   (a) The percent is 10%.

(b) The whole is 3000 runners.

(c) The part is unknown (n).

(d)
$$\text{Percent} \to \frac{10}{100} = \frac{n}{3000} \leftarrow \text{Part}$$
$$\text{Always } 100 \to \qquad\qquad \leftarrow \text{Whole}$$

3.   (a) The percent is unknown (p).

(b) The whole is 32 pizzas.

(c) The part is 16 pizzas.

(d)
$$\text{Percent} \to \frac{p}{100} = \frac{16}{32} \leftarrow \text{Part}$$
$$\text{Always } 100 \to \qquad\qquad \leftarrow \text{Whole}$$

5.   (a) The percent is 90%.

(b) The whole is unknown (n).

(c) The part is 495 students.

(d)
$$\text{Percent} \to \frac{90}{100} = \frac{495}{n} \leftarrow \text{Part}$$
$$\text{Always } 100 \to \qquad\qquad \leftarrow \text{Whole}$$

7. Part: unknown (n)

$$\text{Percent} \to \frac{4}{100} = \frac{n}{120} \leftarrow \text{Part}$$
$$\text{Always } 100 \to \qquad\qquad \leftarrow \text{Whole}$$

Step 1 $\dfrac{4}{100} = \dfrac{n}{120}$

Step 2 $100 \cdot n = 4 \cdot 120$

Step 3 $\dfrac{100n}{100} = \dfrac{480}{100}$

$\qquad\qquad n = 4.8$

4% of 120 feet is 4.8 feet.

9. Percent: unknown (p)

$$\text{Percent} \to \frac{p}{100} = \frac{16}{200} \leftarrow \text{Part}$$
$$\text{Always } 100 \to \qquad\qquad \leftarrow \text{Whole}$$

Step 1 $\dfrac{p}{100} = \dfrac{16}{200}$

Step 2 $200 \cdot p = 100 \cdot 16$

Step 3 $\dfrac{200p}{200} = \dfrac{1600}{200}$

$\qquad\qquad p = 8$

8% of 200 calories is 16 calories.

11. Whole: unknown (n)

$$\text{Percent} \to \frac{12.5}{100} = \frac{3.50}{n} \leftarrow \text{Part}$$
$$\text{Always } 100 \to \qquad\qquad\qquad \leftarrow \text{Whole}$$

Step 1 $\dfrac{12.5}{100} = \dfrac{3.50}{n}$

Step 2 $12.5 \cdot n = 100 \cdot 3.50$

Step 3 $\dfrac{12.5n}{12.5} = \dfrac{350}{12.5}$

$\qquad\qquad n = 28$

$12\frac{1}{2}\%$ of \$28 is \$3.50.

13. Part: unknown (n)

$$\text{Percent} \to \frac{250}{100} = \frac{n}{7} \leftarrow \text{Part}$$
$$\text{Always } 100 \to \qquad\qquad \leftarrow \text{Whole}$$

Step 1 $\dfrac{250}{100} = \dfrac{n}{7}$

Step 2 $100 \cdot n = 250 \cdot 7$

Step 3 $\dfrac{100n}{100} = \dfrac{1750}{100}$

$\qquad\qquad n = 17.5$

250% of 7 hours is 17.5 hours.

15. Percent: unknown (p)

Percent $\rightarrow \dfrac{p}{100} = \dfrac{32}{172} \leftarrow$ Part
Always 100 $\rightarrow$ $\phantom{\dfrac{p}{100}}$ $\phantom{\dfrac{32}{172}}$ $\leftarrow$ Whole

Step 1 $\quad \dfrac{p}{100} = \dfrac{32}{172}$

Step 2 $\quad 172 \cdot p = 100 \cdot 32$

Step 3 $\quad \dfrac{172p}{172} = \dfrac{3200}{172}$

$\quad\quad\quad\quad p \approx 18.6$

$32 is 18.6% (rounded) of $172.

17. Whole: unknown (n)

Percent $\rightarrow \dfrac{110}{100} = \dfrac{748}{n} \leftarrow$ Part
Always 100 $\rightarrow$ $\phantom{\dfrac{110}{100}}$ $\phantom{\dfrac{748}{n}}$ $\leftarrow$ Whole

Step 1 $\quad \dfrac{110}{100} = \dfrac{748}{n}$

Step 2 $\quad 110 \cdot n = 100 \cdot 748$

Step 3 $\quad \dfrac{110n}{110} = \dfrac{74,800}{110}$

$\quad\quad\quad\quad n = 680$

748 e-books is 110% of 680 e-books.

19. Part: unknown (n)

Percent $\rightarrow \dfrac{14.7}{100} = \dfrac{n}{274} \leftarrow$ Part
Always 100 $\rightarrow$ $\phantom{\dfrac{14.7}{100}}$ $\phantom{\dfrac{n}{274}}$ $\leftarrow$ Whole

Step 1 $\quad \dfrac{14.7}{100} = \dfrac{n}{274}$

Step 2 $\quad 100 \cdot n = 14.7 \cdot 274$

Step 3 $\quad \dfrac{100n}{100} = \dfrac{4027.8}{100}$

$\quad\quad\quad\quad n \approx 40.28$

14.7% of $274 is $40.28 (rounded).

21. Percent: unknown (p)

Percent $\rightarrow \dfrac{p}{100} = \dfrac{105}{54} \leftarrow$ Part
Always 100 $\rightarrow$ $\phantom{\dfrac{p}{100}}$ $\phantom{\dfrac{105}{54}}$ $\leftarrow$ Whole

Step 1 $\quad \dfrac{p}{100} = \dfrac{105}{54}$

Step 2 $\quad 54 \cdot p = 100 \cdot 105$

Step 3 $\quad \dfrac{54p}{54} = \dfrac{10,500}{54}$

$\quad\quad\quad\quad p \approx 194.4$

105 employees is 194.4% (rounded) of 54 employees.

23. Whole: unknown (n)

Percent $\rightarrow \dfrac{5}{100} = \dfrac{0.33}{n} \leftarrow$ Part
Always 100 $\rightarrow$ $\phantom{\dfrac{5}{100}}$ $\phantom{\dfrac{0.33}{n}}$ $\leftarrow$ Whole

Step 1 $\quad \dfrac{5}{100} = \dfrac{0.33}{n}$

Step 2 $\quad 5 \cdot n = 100 \cdot 0.33$

Step 3 $\quad \dfrac{5n}{5} = \dfrac{33}{5}$

$\quad\quad\quad\quad n = 6.6$

$0.33 is 5% of $6.60.

25. 150% of $30 cannot be *less* than $30 because 150% is greater than 100%. The answer must be greater than $30.

25% of $16 cannot be *greater* than $16 because 25% is less than 100%. The answer must be less than $16.

27. The correct proportion and solution follow.

$$\dfrac{p}{100} = \dfrac{14}{8}$$
$$8 \cdot p = 100 \cdot 14$$
$$\dfrac{8p}{8} = \dfrac{1400}{8}$$
$$p = 175$$

The answer should be labeled with the % symbol. Correct answer is 175%.

7.3 The Percent Equation

7.3 Margin Exercises

1. **(a)** Round 110.38 to $100. 25% is $\frac{1}{4}$, so divide $100 by 4. The estimate is $25.

(b) Round 7.6 hours to 8 hours. 25% is $\frac{1}{4}$, so divide 8 hours by 4. The estimate is 2 hours.

(c) Round 34 pounds to 30 pounds. 25% is $\frac{1}{4}$, so divide 30 pounds by 4. The estimate is 7.5 pounds. Or, 34 pounds rounds to 32 pounds (a multiple of 4). Divide 32 pounds by 4. The estimate is 8 pounds.

2. **(a)** 10% of $110.38 = $11.038 or $11.04 to the nearest cent.

(b) 10% of 7.6 hours = 0.76 hours

(c) 10% of 34. pounds = 3.4 pounds

3. **(a)** 1% of $110.38 = $1.1038 or $1.10 to the nearest cent.

(b) 1% of 07.6 hours = 0.076 hour

(c) 1% of 34. pounds = 0.34 pound

4. **(a)** 9% of 3250 miles is how many miles?

$$\text{percent} \cdot \text{whole} = \text{part}$$
$$(0.09)(3250) = n$$
$$292.5 = n$$

9% of 3250 is 292.5 miles.

(b) 78% of $5.50 is how much?

$$\text{percent} \cdot \text{whole} = \text{part}$$
$$(0.78)(5.50) = n$$
$$4.29 = n$$

78% of $5.50 is $4.29.

(c) What is 12.5% of 400 homes?

$$\text{part} = \text{percent} \cdot \text{whole}$$
$$n = (0.125)(400)$$
$$n = 50$$

50 homes is 12.5% of 400 homes.

(d) How much is 350% of $6?

$$\text{part} = \text{percent} \cdot \text{whole}$$
$$n = (3.5)(6)$$
$$n = 21$$

$21 is 350% of $6.

5.   (a) 1200 books is what percent of 5000 books?

$$\text{part} = \text{percent} \cdot \text{whole}$$
$$1200 = p \cdot 5000$$
$$\frac{1200}{5000} = \frac{5000p}{5000}$$
$$0.24 = p$$

1200 books is 24% of 5000 books.

(b) 23 hours is what percent of 20 hours?

$$\text{part} = \text{percent} \cdot \text{whole}$$
$$23 = p \cdot 20$$
$$\frac{23}{20} = \frac{20p}{20}$$
$$1.15 = p$$

23 hours is 115% of 20 hours.

(c) What percent of $6.50 is $0.52?

$$\text{percent} \cdot \text{whole} = \text{part}$$
$$p \cdot 6.50 = 0.52$$
$$\frac{6.50p}{6.50} = \frac{0.52}{6.50}$$
$$p = 0.08$$

8% of $6.50 is $0.52.

6. **(a)** 74% of how many cars is 37 cars?

$$\text{percent} \cdot \text{whole} = \text{part}$$
$$0.74 \cdot n = 37$$
$$\frac{0.74n}{0.74} = \frac{37}{0.74}$$
$$n = 50$$

74% of 50 cars is 37 cars.

(b) 1.2 tons is $2\frac{1}{2}$% of how many tons?

Note that $2\frac{1}{2}\% = 2.5\% = 0.025.$

$$\text{part} = \text{percent} \cdot \text{whole}$$
$$1.2 = 0.025 \cdot n$$
$$\frac{1.2}{0.025} = \frac{0.025n}{0.025}$$
$$48 = n$$

1.2 tons is $2\frac{1}{2}$% of 48 tons.

(c) 216 calculators is 160% of how many calculators?

$$\text{part} = \text{percent} \cdot \text{whole}$$
$$216 = 1.60 \cdot n$$
$$\frac{216}{1.60} = \frac{1.60n}{1.60}$$
$$135 = n$$

216 calculators is 160% of 135 calculators.

7.3   Section Exercises

1. 50% of 3000 is the same as half of 3000. Choose 1500 patients.

3. 25% of $60 is the same as dividing $60 by 4. 25% of $60 = $\frac{\$60}{4}$ = $15. Choose $15.

5. 10% of 45 pounds is the same as dividing by 10. Move the decimal one place to the left.

10% of 45. = 4.5 pounds. Choose 4.5 pounds.

7. 200% of $3.50 is the same as multiplying $3.50 by 2. 200% of $3.50 = 2 · $3.50 = $7.00. Choose $7.00.

9. 1% of 5200 students is the same as dividing by 100. Move the decimal two places to the left.

1% of 5200. = 52 students. Choose 52 students.

11. 10% of 8700 smartphones is the same as dividing by 10. Move the decimal one place to the left.

10% of 8700. = 870 smartphones. Choose 870 phones.

13. 25% of 19 hours is the same as dividing 19 by 4. 25% of 19 hours = $\frac{19}{4} = 4\frac{3}{4}$ = 4.75 hours. Choose 4.75 hours.

15. **(a)** 10% means $\frac{10}{100}$ or $\frac{1}{10}$. The denominator tells you to divide the whole by 10. The shortcut for dividing by 10 is to move the decimal point one place to the left.

(b) Once you find 10% of a number, multiply the result by 2 for 20% and by 3 for 30%.

17.

35% of	660	is	how many
	programs		programs?
↓ ↓	↓	↓	↓
0.35 ·	660	=	n
	231	=	n

35% of 660 programs is 231 programs.

19.

70	is	what	of	140
coffee mugs		percent		mugs?
↓	↓	↓	↓	↓
70	=	p	·	140

$$\frac{70}{140} = \frac{140p}{140}$$
$$0.5 = p \quad (50\%)$$

70 coffee mugs is 50% of 140 mugs.

21.

476	is	70%	of	what number
circuits				of circuits?

$$476 = 0.70 \cdot n$$
$$\frac{476}{0.70} = \frac{0.70n}{0.70}$$
$$680 = n$$

476 circuits is 70% of 680 circuits.

23.

$12\frac{1}{2}\%$	of	what number	is	135
		of people		people?

$$0.125 \cdot n = 135$$
$$\frac{0.125n}{0.125} = \frac{135}{0.125}$$
$$n = 1080$$

$12\frac{1}{2}\%$ of 1080 people is 135 people.

25.

What	is	65%	of	1300
				blog posts?

$$n = 0.65 \cdot 1300$$
$$n = 845$$

845 blog posts is 65% of 1300 blog posts.

27.

4%	of	$520	is	how much?

$$0.04 \cdot 520 = n$$
$$20.8 = n$$

4% of $520 is $20.80.

29.

38	is	what	of	50 styles?
styles		percent		

$$38 = p \cdot 50$$
$$\frac{38}{50} = \frac{50p}{50}$$
$$0.76 = p \quad (76\%)$$

38 styles is 76% of 50 styles.

31.

What	of	$264	is	$330?
percent				

$$p \cdot 264 = 330$$
$$\frac{264p}{264} = \frac{330}{264}$$
$$p = 1.25 \quad (125\%)$$

125% of $264 is $330.

33.

141	is	3%	of	what number
employees				of employees?

$$141 = 0.03 \cdot n$$
$$\frac{141}{0.03} = \frac{0.03n}{0.03}$$
$$4700 = n$$

141 employees is 3% of 4700 employees.

35.

32%	of	260	is	how many
		quarts		quarts?

$$0.32 \cdot 260 = n$$
$$83.2 = n$$

32% of 260 quarts is 83.2 quarts.

37.

$1.48	is	what	of	$74?
		percent		

$$1.48 = p \cdot 74$$
$$\frac{1.48}{74} = \frac{74p}{74}$$
$$0.02 = p \quad (2\%)$$

$1.48 is 2% of $74.

39.

How many	is	140%	of	500
tablets				tablets?

$$n = 1.40 \cdot 500$$
$$n = 700$$

700 tablets is 140% of 500 tablets.

41.

40%	of	what number	is	130
		of babies		babies?

$$0.40 \cdot n = 130$$
$$\frac{0.40n}{0.40} = \frac{130}{0.40}$$
$$n = 325$$

40% of 325 babies is 130 babies.

43.

What	of	160	is	2.4
percent		liters		liters?

$$p \cdot 160 = 2.4$$
$$\frac{160p}{160} = \frac{2.4}{160}$$
$$p = 0.015 \quad (1.5\%)$$

1.5% of 160 liters is 2.4 liters.

45.

$$\underbrace{225.\%}_{}\quad \text{of}\quad \underbrace{\text{what number}}_{\text{of gallons}}\ \text{is}\ \underbrace{11.25}_{\text{gallons}}?$$

$$\begin{array}{ccccc}
\downarrow & \downarrow & \downarrow & \downarrow & \downarrow \\
2.25 & \cdot & n & = & 11.25
\end{array}$$

$$\frac{2.25n}{2.25} = \frac{11.25}{2.25}$$

$$n = 5$$

225% of 5 gallons is 11.25 gallons.

47.

$$\underbrace{\text{What}}\ \text{is}\ \underbrace{12.4\%}\ \text{of}\ \underbrace{\begin{array}{c}8300 \\ \text{meters}?\end{array}}$$

$$\begin{array}{ccccc}
\downarrow & \downarrow & \downarrow & \downarrow & \downarrow \\
n & = & 0.124 & \cdot & 8300
\end{array}$$

$$n = 1029.2$$

1029.2 meters is 12.4% of 8300 meters.

49. Multiply 0.2 by 100% to change it from a decimal to a percent. So, $0.20 = 20\%$.

51. The correct equation is $50 = p \cdot 20$.

$$\frac{50}{20} = \frac{20p}{20}$$

$$2.5 = p$$

The solution is 250%.

Relating Concepts (Exercises 53–56)

53. 25% of 80 is the same as dividing 80 by 4.
25% of $80 = \frac{80}{4} = 20$. There are 20 boys.

54. 100% is *all* the children, so $100\% - 25\%$ who are boys $= 75\%$ who are girls.

55. **Method 1:** 80 children $-$ 20 boys $=$ 60 girls.

Method 2: From Exercise 53, 25% of the children is 20 boys. So, 75% is 3 times $20 = 60$ girls.

56. There are 20 boys (from Exercise 53). The shortcut for finding 10% is dividing by 10, or moving the decimal point one place to the left. 10% of 20. is 2 boys who are left handed. There are no left-handed girls, so the total for left-handed children is 2.

Summary Exercises *Percent Computation*

1. **(a)** $\frac{3}{100} = 0.03 \leftarrow$ Decimal
$\phantom{\frac{3}{100}} = 3\% \leftarrow$ Percent

(b) $30\% = \frac{30}{100} = \frac{3}{10} \leftarrow$ Fraction
$ = 0.30 = 0.3 \leftarrow$ Decimal

(c) $0.375 = \frac{375}{1000} = \frac{375 \div 125}{1000 \div 125} = \frac{3}{8} \leftarrow$ Fraction
$ = 37.5\% \leftarrow$ Percent

(d) $160\% = \frac{160}{100} = \frac{8}{5} = 1\frac{3}{5} \leftarrow$ Fraction
$ = 1.6 \leftarrow$ Decimal

(e) Since 16 doesn't divide evenly into 100 or 1000, but does divide evenly into 10,000, we divide 10,000 by 16 to get 625.

$$\frac{1}{16} = \frac{625}{10,000} = 0.0625 \leftarrow \text{Decimal}$$
$$\phantom{\frac{1}{16}} = 6.25\% \leftarrow \text{Percent}$$

(f) $5\% = \frac{5}{100} = \frac{1}{20} \leftarrow$ Fraction
$ = 0.05 \leftarrow$ Decimal

(g) $2.0 = \frac{200}{100} = \frac{2}{1} = 2 \leftarrow$ Fraction
$ = 200\% \leftarrow$ Percent

(h) $\frac{4}{5} = \frac{4(20)}{5(20)} = \frac{80}{100} = 0.8 \leftarrow$ Decimal
$\phantom{\frac{4}{5}} = 80\% \leftarrow$ Percent

(i) $0.072 = \frac{72}{1000} = \frac{72 \div 8}{1000 \div 8} = \frac{9}{125} \leftarrow$ Fraction
$ = 7.2\% \leftarrow$ Percent

3. Part: 9
Whole: 72
Percent: unknown (p)

Step 1 $\dfrac{p}{100} = \dfrac{9}{72}$

Step 2 $72 \cdot p = 100 \cdot 9$

Step 3 $\dfrac{72p}{72} = \dfrac{900}{72}$

$ \ p = 12.5$

9 websites is 12.5% of 72 websites.

5. Part: unknown (n)
Whole: $8.79
Percent: 6%

Step 1 $\dfrac{6}{100} = \dfrac{n}{8.79}$

Step 2 $100 \cdot n = 6 \cdot 8.79$

Step 3 $\dfrac{100n}{100} = \dfrac{52.74}{100}$

$ \ n \approx 0.53$

6% of $8.79 is $0.53 (rounded).

7. Part: unknown (n)
Whole: 168
Percent: $3\frac{1}{2}\%$

Step 1 $\dfrac{3.5}{100} = \dfrac{n}{168}$

Step 2 $100 \cdot n = 3.5 \cdot 168$

Step 3 $\dfrac{100n}{100} = \dfrac{588}{100}$

$ \ n = 5.88$

$3\frac{1}{2}\%$ of 168 pounds is 5.9 pounds (rounded).

9. Part: 40,000

Whole: 80,000

Percent: unknown (p)

Step 1 $\dfrac{p}{100} = \dfrac{40,000}{80,000}$

Step 2 $80,000 \cdot p = 100 \cdot 40,000$

Step 3 $\dfrac{80,000p}{80,000} = \dfrac{4,000,000}{80,000}$

$p = 50$

50% of 80,000 deer is 40,000 deer.

11. Part: unknown (n)

Whole: 35

Percent: 280%

Step 1 $\dfrac{280}{100} = \dfrac{n}{35}$

Step 2 $100 \cdot n = 280 \cdot 35$

Step 3 $\dfrac{100n}{100} = \dfrac{9800}{100}$

$n = 98$

98 golf balls is 280% of 35 golf balls.

13.

9% of what number is 207
          of tweets   tweets?

$0.09 \cdot n = 207$

$\dfrac{0.09n}{0.09} = \dfrac{207}{0.09}$

$n = 2300$

9% of 2300 tweets is 207 tweets.

15.

$1160 is what of $800?
              percent

$1160 = p \cdot 800$

$\dfrac{1160}{800} = \dfrac{800p}{800}$

$1.45 = p \quad (145\%)$

$1160 is 145% of $800.

17.

What is 300% of 0.007
                        inch?

$n = 3 \cdot 0.007$

$n = 0.021$

0.021 inch is 300% of 0.007 inch.

19.

What of 60 is 4.8
percent yards   yards?

$p \cdot 60 = 4.8$

$\dfrac{60p}{60} = \dfrac{4.8}{60}$

$p = 0.08 \quad (8\%)$

8% of 60 yards is 4.8 yards.

21. The smallest percentage is $2\frac{3}{4}\%$ for invitations.
$2\frac{3}{4}\%$ of $30,800 = 0.0275(\$30,800) = \$847$.

23. The cost of photography and videography is
12% of $30,800 = 0.12(\$30,800) = \$3696$.

25. **(a)** The amount spent on the reception is
48% of $30,800 = 0.48(\$30,800) = \$14,784$.

(b) On average, the cost for each guest at a reception is

$\dfrac{\$14,784}{178 \text{ guests}} \approx \$83.06/\text{guest}.$

7.4 Problem Solving with Percent

7.4 Margin Exercises

1. **Step 1** Unknown: number of students receiving aid
Known: 65% receive aid; 9280 students

Step 2 Let n be the number of students receiving aid.

Step 3 percent $\cdot$ whole $=$ part
$65\% \cdot 9280 = n$

Step 4 $(0.65)(9280) = n$
$6032 = n$

Step 5 6032 students receive financial aid.

Step 6 10% of $928\overset{\vee}{0}$. students is 928, which

rounds to 900 students. So 60% would be 6 times 900 students $= 5400$ students, and 70% would be $7 \cdot 900 = 6300$. The solution falls between 5400 and 6300 students, so it is reasonable.

2. **Step 1** Unknown: number of points earned by Hue
Known: 50 total points; 83% correct

Step 2 Let n be the number of points earned.

Step 3 percent $\cdot$ whole $=$ part
$83\% \cdot 50 = n$

Step 4 $(0.83)(50) = n$
$41.5 = n$

Step 5 Hue earned 41.5 points.

Step 6 10% of $5\overset{\vee}{0}$. points is 5 points, so 80%

would be 8 times 5 points $= 40$ points. Hue earned a little more than 80%, so 41.5 points is reasonable.

3. **Step 1** Unknown: percentage of made field goals
Known: 47 made; 80 attempted

Step 2 Let p be the unknown percent.

Step 3   percent • whole = part
$$p \cdot 80 = 47$$

Step 4 $$\frac{80p}{80} = \frac{47}{80}$$
$$p = 0.5875 = 58.75\% \approx 59\%$$

Step 5   The Lakers made about 59% of their field goals.

Step 6   The Lakers made a little more than half of their field goals. $\frac{1}{2} = 50\%$, so the solution of 59% is reasonable.

4.   ***Step 1***   Unknown: percent of the predicted enrollment that the actual enrollment is
Known: 1200 students predicted; 1620 students actual enrollment

Step 2   Let p be the unknown percent.

Step 3

actual enrollment	is	what percent	of	predicted number
↓	↓	↓	↓	↓
1620	=	p	•	1200

Step 4 $$\frac{1620}{1200} = \frac{1200p}{1200}$$
$$p = 1.35 = 135\%$$

Step 5   Enrollment is 135% of the predicted number.

Step 6   More than 1200 students enrolled (more than 100%), so 1620 students must be more than 100% of the predicted number. The solution of 135% is reasonable.

5.   **(a)** ***Step 1***   Unknown: number of problems on the test
Known: 15 correct; $62\frac{1}{2}\%$ correct

Step 2   Let n be the number of problems on the test.

Step 3   percent • whole = part
$$62\frac{1}{2}\% \cdot n = 15$$

Step 4 $$(0.625)(n) = 15$$
$$\frac{0.625n}{0.625} = \frac{15}{0.625}$$
$$n = 24$$

Step 5   There were 24 problems on the test.

Step 6   50% of 24 problems is $24 \div 2 = 12$ problems correct, so it is reasonable that $62\frac{1}{2}\%$ would be 15 problems correct.

(b) ***Step 1***   Unknown: total number of calories
Known: 55 calories from fat; 18% of the calories are from fat

Step 2   Let n be the total number of calories

Step 3   percent • whole = part
$$18\% \cdot n = 55$$

Step 4 $$(0.18)(n) = 55$$
$$\frac{0.18n}{0.18} = \frac{55}{0.18}$$
$$n \approx 306$$

Step 5   There were 306 calories (rounded) in the dinner.

Step 6   18% is close to 20%. 10% of 306 calories is about 30 calories, so 20% would be $2 \cdot 30 = 60$ calories, which is close to the number given (55 calories).

6.   **(a)** ***Step 1***   Unknown: percent of increase
Known: original rent was $650; new rent is $767

Step 2   Let p be the percent increase.

Step 3
Amount of increase = $767 - $650 = $117

percent	of	original rent	=	amount of increase
↓	↓	↓	↓	↓
p	•	$650	=	$117

Step 4 $$\frac{650p}{650} = \frac{117}{650}$$
$$p = 0.18 = 18\%$$

Step 5   Duyen's rent increased 18%.

Step 6   10% increase would be $650. = $65; 20% increase would be $2 \cdot \$65 = \$130$; so an 18% increase is reasonable.

(b) ***Step 1***   Unknown: percent of increase
Known: number of spaces changed from 8 to 20

Step 2   Let p be the percent of increase.

Step 3   Amount of increase = $20 - 8 = 12$

percent	of	original # of spaces	=	amount of increase
↓	↓	↓	↓	↓
p	•	8	=	12

Step 4 $$\frac{8p}{8} = \frac{12}{8}$$
$$p = 1.5 = 150\%$$

Step 5   The number of accessible parking spaces increased 150%.

Step 6   150% is $1\frac{1}{2}$, and $1\frac{1}{2} \cdot 8 = 12$ space increase, so it checks.

7. **(a)** *Step 1*   Unknown: percent of decrease
Known: number of students changed from
425 to 200

Step 2   Let p be the percent of decrease.

Step 3   Amount of decrease $= 425 - 200 = 225$

$$\underbrace{\text{percent of}\quad \text{original}}_{} = \underbrace{\text{amount of}}_{}$$
$$\text{attendance} \qquad \text{decrease}$$
$$\downarrow \quad \downarrow \qquad \downarrow \qquad \downarrow \qquad \downarrow$$
$$p \quad \cdot \qquad 425 \qquad = \qquad 225$$

Step 4 $\dfrac{425p}{425} = \dfrac{225}{425}$
$$p \approx 0.529 \approx 53\%$$

Step 5   Daily attendance decreased 53%
(rounded).

Step 6   A 50% decrease would be
$425 \div 2 \approx 212$, so a 53% decrease is reasonable.

(b) *Step 1*   Unknown: percent of decrease
Known: number of calories from fat dropped from
70 to 60

Step 2   Let p be the percent of decrease.

Step 3   Amount of decrease $= 70 - 60 = 10$

$$\underbrace{\text{percent of}\quad \text{original}}_{} = \underbrace{\text{amount of}}_{}$$
$$\text{\# of calories} \qquad \text{decrease}$$
$$\downarrow \quad \downarrow \qquad \downarrow \qquad \downarrow \qquad \downarrow$$
$$p \quad \cdot \qquad 70 \qquad = \qquad 10$$

Step 4 $\dfrac{70p}{70} = \dfrac{10}{70}$
$$p \approx 0.143 \approx 14\%$$

Step 5   The claim of a 20% decrease is not true;
the decrease in calories is about 14%.

Step 6   A 10% decrease would be
$70. = 7$ calories; a 20% decrease would be
$2 \cdot 7 = 14$ calories, so a 14% decrease is
reasonable.

7.4   Section Exercises

1. **(a)** *Step 1*   Unknown: percent
Known: $20 withdrawal; $2 fee

Step 2   Let p be the percent.

Step 3   percent $\cdot$ whole $=$ part
$$p \cdot 20 = 2$$

Step 4 $\dfrac{20p}{20} = \dfrac{2}{20}$
$$p = 0.10 = 10\%$$

Step 5   The $2 fee is 10% of the $20 withdrawal.

Step 6   10% of $20. is $2.

Use the same method for parts (b) and (c).

(b) $p = \dfrac{2}{40} = 0.05 = 5\%$

(c) $p = \dfrac{2}{200} = 0.01 = 1\%$

3. *Step 1*   Unknown: amount withheld
Known: withhold 18% of $210

Step 2   Let n be the amount withheld.

Step 3 $\underset{\curvearrowright}{18.\%}$ of   pay   is   withheld
$$\downarrow \qquad \downarrow \qquad \downarrow \quad \downarrow \qquad \downarrow$$
$$0.18 \quad \cdot \quad \$210 \quad = \qquad n$$

Step 4 $(0.18)(\$210) = n$
$$\$37.80 = n$$

Step 5   The amount withheld is $37.80.

Step 6   10% of $210 is $21, so 20% would be
$2 \cdot \$21 = \$42$. $37.80 is slightly less than $42, so
it is reasonable.

5. **(a)** *Step 1*   Unknown: number of pounds of water
Known: water weight is 61.6% of 165 pounds

Step 2   Let n be the number of pounds of water.

Step 3   percent $\cdot$ whole $=$ part
$$61.6\% \cdot 165 = n$$

Step 4 $(0.616)(165) = n$
$$101.64 = n$$

Step 5   101.6 pounds (rounded) of the 165
pounds is water.

Step 6   50% of 165 pounds is 82.5 pounds, so
101.6 pounds is reasonable.

(b) (minerals) $n = 6.1\%$ of 165
$$= (0.061)(165)$$
$$= 10.065$$
$$\approx 10.1 \text{ pounds (rounded)}$$

(c) (proteins) $n = 17\%$ of 165
$$= (0.17)(165)$$
$$= 28.05$$
$$\approx 28.1 \text{ pounds (rounded)}$$

7. *Step 1*   Unknown: percent of people that gave
each answer
Known: 480 people were asked; 283 said they
would pay off debt (we'll solve this part and
summarize the other parts at the end of this
solution)

Step 2   Let p be the percent that said they would
pay off debt.

Step 3   percent · whole = part
$$p \cdot 480 = 283$$
Step 4 $$\frac{480p}{480} = \frac{283}{480}$$
$$p \approx 0.59 \text{ (rounded)}$$

Step 5   About 59.0% said they would pay off debt.

Step 6   50% of 480 is 240, so 283 being about 59% is reasonable.

Similarly, $\frac{317}{480} \approx 66.0\%$ said they would save or invest the money, and $\frac{110}{480} \approx 22.9\%$ said they would buy something.

9.   Whole: Total U.S. population is unknown (n)
Part: 40.3 million are 65 years of age or older
Percent: 13.0% (13.0% = 0.13)

$$\text{Percent} \cdot \text{Whole} = \text{Part}$$
$$0.13 \cdot n = 40.3$$
$$\frac{0.13n}{0.13} = \frac{40.3}{0.13}$$
$$n = 310$$

There were 310 million people in the U.S. at the time of the 2010 census.

11.   Whole: $50,000 is the goal
Percent: unknown (p)
Part: $69,000 was actually raised

$$\text{Percent} \cdot \text{Whole} = \text{Part}$$
$$p \cdot 50,000 = 69,000$$
$$\frac{50,000p}{50,000} = \frac{69,000}{50,000}$$
$$p = 1.38 \qquad (138\%)$$

The society raised 138% of their goal.

13.   Whole: Number of problems on the test is
          unknown (n)
Part: 34 problems done correctly
Percent: 85%

$$\text{Percent} \cdot \text{Whole} = \text{Part}$$
$$0.85 \cdot n = 34$$
$$\frac{0.85n}{0.85} = \frac{34}{0.85}$$
$$n = 40$$

There were 40 problems on the test.

15.   Whole: number of shots tried is
          unknown (n)
Percent: 52.7%
Part: 307 shots made

$$\text{Percent} \cdot \text{Whole} = \text{Part}$$
$$0.527 \cdot n = 307$$
$$\frac{0.527n}{0.527} = \frac{307}{0.527}$$
$$n \approx 583$$

He tried 583 shots (rounded).

17.   First calculate her increase in mileage.

Percent ·	Original mileage	=	Amount of increase
0.15 ·	20.6	=	n
	3.09	=	n

The new tires should increase her mileage by 3.09 miles per gallon.

Her new mileage should be
$20.6 + 3.09 = 23.69 \approx 23.7$ miles per gallon (rounded)

19.   Whole: weekly earnings of a person with an
          associate degree is $767
Percent: unknown (p)
Part: weekly earnings of a person with less than a
      high school diploma is $444

$$\text{Percent} \cdot \text{Whole} = \text{Part}$$
$$p \cdot 767 = 444$$
$$\frac{767p}{767} = \frac{444}{767}$$
$$p \approx 0.58 \quad (58\%)$$

The weekly earnings of a person with less than a high school diploma is about 58% of the weekly earnings of a person with an associate degree.

In one week, the difference in earnings is
$$\$767 - \$444 = \$323.$$

For a 52-week year, the difference in earnings is
$$52(\$323) = \$16,796.$$

21.   Whole: weekly earnings of a person with an
          associate degree is $767
Percent: unknown (p)
Part: weekly earnings of a person with a bachelor's
      degree is $1038

$$\text{Percent} \cdot \text{Whole} = \text{Part}$$
$$p \cdot 767 = 1038$$
$$\frac{767p}{767} = \frac{1038}{767}$$
$$p \approx 1.35 \quad (135\%)$$

The weekly earnings of a person with a bachelor's degree is about 135% of the weekly earnings of a person with an associate degree.

23.   Amount of decrease: $825 - $290 = $535

Let p be the percent of decrease.

percent	of	original cost	=	amount of decrease
↓	↓	↓ ↓		↓
p	·	$825	=	$535
		$\frac{825p}{825}$	=	$\frac{535}{825}$
		p	$\approx$	0.648 (64.8%)

The percent of decrease was 64.8% (rounded).

25. Original tuition last semester: $1328
New tuition this semester: $1449
Amount of increase: $1449 − $1328 = $121

Let p be the percent of increase.

$$\underbrace{\text{percent}}_{\downarrow} \; \text{of} \; \underbrace{\begin{array}{c}\text{original} \\ \text{tuition}\end{array}}_{\downarrow} = \underbrace{\begin{array}{c}\text{amount of} \\ \text{increase}\end{array}}_{\downarrow}$$

$$p \cdot 1328 = 121$$
$$\frac{1328p}{1328} = \frac{121}{1328}$$
$$p \approx 0.091 \quad (9.1\%)$$

The percent increase is 9.1% (rounded).

27. Original hours: 30 hours
New hours: 18 hours
Amount of decrease: 30 − 18 = 12

Let p be the percent of decrease.

$$\underbrace{\text{percent}}_{\downarrow} \; \text{of} \; \underbrace{\begin{array}{c}\text{original} \\ \text{amount}\end{array}}_{\downarrow} = \underbrace{\begin{array}{c}\text{amount of} \\ \text{decrease}\end{array}}_{\downarrow}$$

$$p \cdot 30 = 12$$
$$\frac{30p}{30} = \frac{12}{30}$$
$$p = 0.4 \quad (40\%)$$

The percent decrease is 40%.

29. Original number: 78
New number: 588
Amount of increase: 588 − 78 = 510

Let p be the percent of increase.

$$\underbrace{\text{percent}}_{\downarrow} \; \text{of} \; \underbrace{\begin{array}{c}\text{original} \\ \text{number}\end{array}}_{\downarrow} = \underbrace{\begin{array}{c}\text{amount of} \\ \text{increase}\end{array}}_{\downarrow}$$

$$p \cdot 78 = 510$$
$$\frac{78p}{78} = \frac{510}{78}$$
$$p \approx 6.54 \quad (654\%)$$

The percent of increase is 654% (rounded).

31. No. 100% is the entire price, so a decrease of 100% would take the price down to 0. Therefore, 100% is the maximum possible decrease in the price of something.

Relating Concepts (Exercises 33–36)

33. George ate more than 65 grams, so the percent must be $> 100\%$. Use $p \cdot 65 = 78$ to get 120%.

34. The team won more than half the games, so the percent must be $> 50\%$. Correct solution is $0.72 = 72\%$.

35. The brain could not weigh 375 pounds, which is more than the person weighs.
$2\frac{1}{2}\% = 2.5\% = 0.025$, so $(0.025)(150) = n$ and $n = 3.75$ pounds.

36. If 80% were absent, then only 20% made it to class. $800 − 640 = 160$ students, or use $n = (0.20)(800) = 160$ students.

7.5 Consumer Applications: Sales Tax, Tips, Discounts, and Simple Interest

7.5 Margin Exercises

1. (a) Percent: tax rate $= 5\frac{1}{2}\%$ or 5.5%
Whole: cost of item $=$ $495
Part: sales tax $=$ unknown (n)

$$\underbrace{\begin{array}{c}\text{Tax} \\ \text{Rate}\end{array}}_{\downarrow} \cdot \underbrace{\begin{array}{c}\text{Cost of} \\ \text{Item}\end{array}}_{\downarrow} = \underbrace{\begin{array}{c}\text{Sales} \\ \text{Tax}\end{array}}_{\downarrow}$$

$$5.5\% \cdot \$495 = n$$
$$0.055 \cdot \$495 = n$$
$$\$27.225 = n$$

The sales tax is $27.23 (rounded).

Total cost = $495 + $27.23 = $522.23

Check 1% of $500. is $5;
6% is 6 times $5 = $30

(b) Percent: tax rate $= 4\%$
Whole: cost of item $=$ $1.19
Part: sales tax $=$ unknown (n)

$$\underbrace{\begin{array}{c}\text{Tax} \\ \text{Rate}\end{array}}_{\downarrow} \cdot \underbrace{\begin{array}{c}\text{Cost of} \\ \text{Item}\end{array}}_{\downarrow} = \underbrace{\begin{array}{c}\text{Sales} \\ \text{Tax}\end{array}}_{\downarrow}$$

$$4\% \cdot \$1.19 = n$$
$$0.04 \cdot \$1.19 = n$$
$$\$0.0476 = n$$

The sales tax is $0.05 (rounded).

Total cost = $1.19 + $0.05 = $1.24

Check 1% of $1 is $0.01;
4% is 4 times $0.01 = $0.04

2. (a) Percent: tax rate $=$ unknown (p)
Whole: cost of item $=$ $97
Part: sales tax $=$ $5.82

$$\underbrace{\begin{array}{c}\text{Tax} \\ \text{Rate}\end{array}}_{\downarrow} \cdot \underbrace{\begin{array}{c}\text{Cost of} \\ \text{Item}\end{array}}_{\downarrow} = \underbrace{\begin{array}{c}\text{Sales} \\ \text{Tax}\end{array}}_{\downarrow}$$

$$p \cdot \$97 = \$5.82$$
$$\frac{97p}{97} = \frac{5.82}{97}$$
$$p = 0.06 \quad (6\%)$$

The tax rate is 6%.

Check 1% of $100. is $1; 6% would be 6 times $1 or $6 (which is close to $5.82 given in the problem).

(b) Percent: tax rate = unknown (p)
Whole: cost of item = $4
Part: sales tax = $0.18

$$\underset{\downarrow}{\underset{\text{Rate}}{\text{Tax}}} \cdot \underset{\downarrow}{\underset{\text{Item}}{\text{Cost of}}} = \underset{\downarrow}{\underset{\text{Tax}}{\text{Sales}}}$$

$$p \ \cdot \ \$4 \ = \ \$0.18$$
$$\frac{4p}{4} = \frac{0.18}{4}$$
$$p = 0.045 \quad (4.5\%)$$

The tax rate is 4.5%.

Check 1% of $4 is $0.04; round 4.5% to 5%; 5% of $4 is 5 times $0.04, or $0.20 (which is close to $0.18 given in the problem).

3. **(a)** *Estimate:* Round $58.37 to $60. Then 10% of $60 is $6. 20% is 2 times $6, or $\underline{\$12}$.

Exact: Percent $\cdot$ Whole = Part
$$0.20 \ \cdot \ \$58.37 \ = \ \text{tip}$$
$$\$11.674 \ = \ \text{tip}$$

Exact tip amount = $11.67 (rounded).

(b) *Estimate:* Round $11.93 to $12. Then 10% of $12 is $1.20. 5% is half of $1.20, or $0.60. So 15% is $1.20 + $0.60 = $1.80.

Exact: Percent $\cdot$ Whole = Part
$$0.15 \ \cdot \ \$11.93 \ = \ \text{tip}$$
$$\$1.7895 \ = \ \text{tip}$$

Exact tip amount = $1.79 (rounded).

(c) *Estimate:* Round $89.02 to $90. Then 10% of $90 is $9. 5% is half of $9, or $4.50. So 15% is $9 + $4.50 = $13.50.

Exact: Percent $\cdot$ Whole = Part
$$0.15 \ \cdot \ \$89.02 \ = \ \text{tip}$$
$$\$13.353 \ = \ \text{tip}$$

The tip amount is about $13.35, which is close to the estimate of $13.50.

Total cost = $89.02 + $13.35 = $102.37

Amount paid by each friend:

$$\$102.37 \div 4 = \$25.59 \text{ (rounded)}$$

4. **(a)** Percent: rate of discount = 35%
Whole: original price = $950
Part: amount of discount = unknown (n)

$$\underset{\downarrow}{\underset{\text{discount}}{\text{rate of}}} \cdot \underset{\downarrow}{\underset{\text{price}}{\text{original}}} = \underset{\downarrow}{\underset{\text{discount}}{\text{amount of}}}$$

$$3\,5.\% \ \cdot \ \$950 \ = \ n$$
$$0.35 \ \cdot \ 950 \ = \ n$$
$$332.50 \ = \ n$$

The amount of discount is $332.50.

Sale price = original price − amount of discount
$$= \$950 - \$332.50 = \$617.50$$

(b) Percent: rate of discount = 40%
Whole: original price = $68 and $97
Part: amount of discount = unknown (n)

Calculations for the $68 suit:

$$\text{Percent} \cdot \text{Whole} = \text{Part}$$
$$0.40 \ \cdot \ 68 \ = \ n$$
$$27.20 \ = \ n$$

The amount of discount is $27.20.

Sale price = $68 − $27.20 = $40.80

Calculations for the $97 suit:

$$\text{Percent} \cdot \text{Whole} = \text{Part}$$
$$0.40 \ \cdot \ 97 \ = \ n$$
$$38.80 \ = \ n$$

The amount of discount is $38.80.

Sale price = $97 − $38.80 = $58.20

5. **(a)** $I = p \cdot r \cdot t$ $4\% = 0.04$
$$I = (500)(0.04)(1)$$
$$I = 20$$

The interest is $20.

amount due = principal + interest
$$= \$500 + \$20$$
$$= \$520$$

The total amount due is $520.

(b) $I = p \cdot r \cdot t$ $9\frac{1}{2}\% = 0.095$
$$I = (1850)(0.095)(1)$$
$$I = 175.75$$

The interest is $175.75.

amount due = principal + interest
$$= \$1850 + \$175.75$$
$$= \$2025.75$$

The total amount due is $2025.75.

6. (a) $I = p \cdot r \cdot t$ $5\% = 0.05$
$I = (340)(0.05)(3.5)$
$I = 59.50$

The interest is $59.50.

amount due = principal + interest
= $340 + $59.50
= $399.50

The total amount due is $399.50.

(b) $I = p \cdot r \cdot t$ $8\% = 0.08$
$I = (2450)(0.08)(3.25)$
$I = 637$

The interest is $637.

amount due = principal + interest
= $2450 + $637
= $3087

The total amount due is $3087.

(c) $7\frac{1}{2}\% = 7.5\%$ and $2\frac{3}{4}$ yrs $= 2.75$ yrs

$I = p \cdot r \cdot t$ $7.5\% = 0.075$
$I = (14{,}200)(0.075)(2.75)$
$I = 2928.75$

The interest is $2928.75.

amount due = principal + interest
= $14,200 + $2928.75
= $17,128.75

The total amount due is $17,128.75.

7. (a) $I = p \cdot r \cdot t$
$I = (1600)(0.07)(\frac{1}{12})$ $\frac{4}{12} = \frac{1}{3}$
$I = 112(\frac{1}{3})$
$I = \frac{112}{1} \cdot \frac{1}{3} = \frac{112}{3} = 37\frac{1}{3} \approx 37.33$

The interest is $37.33.

amount due = principal + interest
= $1600 + $37.33
= $1637.33

The total amount due is $1637.33.

(b) $I = p \cdot r \cdot t$ $10\frac{1}{2}\% = 0.105$
$I = (25{,}000)(0.105)(\frac{3}{12})$
$I = (2625)(\frac{1}{4})$
$I = \frac{2625}{1} \cdot \frac{1}{4} = \frac{2625}{4} = 656.25$

The interest is $656.25.

amount due = principal + interest
= $25,000 + $656.25
= $25,656.25

The total amount due is $25,656.25.

7.5 Section Exercises

1. (a) $6\frac{1}{2}\% = 6.5\% = 0.065$ ←   Decimal

(b) $1\frac{1}{4}\% = 1.25\% = 0.0125$ ←   Decimal

(c) $10\frac{1}{4}\% = 10.25\% = 0.1025$ ←   Decimal

(d) $5\frac{3}{4}\% = 5.75\% = 0.0575$ ←   Decimal

3. Cost of Item = $365.98
Tax Rate = $8\% = 0.08$

Tax Rate	$\cdot$	Cost of Item	=	Amount of Tax
↓	↓	↓	↓	↓
0.08	$\cdot$	$365.98	=	n
		$29.28	$\approx$	n

Total Cost = $365.98 + $29.28 = $395.26

5. Cost of Item = $68
Tax Rate = unknown (p)
Amount of Tax = $2.04
Total Cost = $68 + $2.04 = $70.04

Tax Rate	$\cdot$	Cost of Item	=	Amount of Tax
↓	↓	↓	↓	↓
p	$\cdot$	68	=	2.04
		$\frac{68p}{68}$	=	$\frac{2.04}{68}$
		p	=	0.03 (3%)

The tax rate is 3%.

7. Cost of Item = $2.10
Tax Rate = $5\frac{1}{2}\% = 0.055$

Tax Rate	$\cdot$	Cost of Item	=	Amount of Tax
↓	↓	↓	↓	↓
0.055	$\cdot$	$2.10	=	n
		$0.12	$\approx$	n

Total Cost = $2.10 + $0.12 = $2.22

9. Cost of Item = $12,600
Tax Rate = unknown (p)
Amount of Tax = $567
Total Cost = $12,600 + $567 = $13,167

Tax Rate	$\cdot$	Cost of Item	=	Amount of Tax
↓	↓	↓	↓	↓
p	$\cdot$	12,600	=	567
		$\frac{12{,}600p}{12{,}600}$	=	$\frac{567}{12{,}600}$
		p	=	0.045 (4.5%)

The tax rate is 4.5% or $4\frac{1}{2}\%$.

11. The bill of $32.17 rounds to $30.

Estimate of 15% tip:

10% of $30 is $3. 5% of $30 is half of $3, or $1.50. An estimate is $3 + $1.50 = $4.50.

Exact 15% tip:

$0.15(\$32.17) = \$4.8255 = \$4.83$ (rounded)

Estimate of 20% tip:

10% of $30 is $3. 20% is 2 times $3, or $6. An estimate is $6.

Exact 20% tip:

$0.20(\$32.17) = \$6.434 = \$6.43$ (rounded)

13. The bill of $78.33 rounds to $80.

Estimate of 15% tip:

10% of $80 is $8. 5% of $80 is half of $8, or $4. An estimate is $8 + $4 = $12.

Exact 15% tip:

$0.15(\$78.33) = \$11.7495 = \$11.75$ (rounded)

Estimate of 20% tip:

10% of $80 is $8. 20% is 2 times $8, or $16. An estimate is $16.

Exact 20% tip:

$0.20(\$78.33) = \$15.666 = \$15.67$ (rounded)

15. The bill of $9.55 rounds to $10.

Estimate of 15% tip:

10% of $10 is $1. 5% of $10 is half of $1, or $0.50. An estimate is $1 + $0.50 = $1.50.

Exact 15% tip:

$0.15(\$9.55) = \$1.4325 = \$1.43$ (rounded)

Estimate of 20% tip:

10% of $10 is $1. 20% is 2 times $1, or $2. An estimate is $2.

Exact 20% tip:

$0.20(\$9.55) = \$1.91$

17. Original Price = $100
Rate of Discount = 15% = 0.15

$$\underset{\underset{0.15}{\downarrow}}{\underset{\text{Discount}}{\text{Rate of}}} \cdot \underset{\underset{\$100}{\downarrow}}{\underset{\text{Price}}{\text{Original}}} = \underset{\underset{n}{\downarrow}}{\underset{\text{Discount}}{\text{Amount of}}}$$
$$\$15 = n$$

Sale Price = $100 − $15 = $85

19. Original Price = $180
Rate of Discount = unknown (p)
Amount of Discount = $54
Sale Price = $180 − $54 = $126

$$\underset{\underset{p}{\downarrow}}{\underset{\text{Discount}}{\text{Rate of}}} \cdot \underset{\underset{\$180}{\downarrow}}{\underset{\text{Price}}{\text{Original}}} = \underset{\underset{\$54}{\downarrow}}{\underset{\text{Discount}}{\text{Amount of}}}$$
$$\frac{180p}{180} = \frac{54}{180}$$
$$p = 0.3 \quad (30\%)$$

The rate of discount is 30%.

21. Original Price = $17.50
Rate of Discount = 25% = 0.25

$$\underset{\underset{0.25}{\downarrow}}{\underset{\text{Discount}}{\text{Rate of}}} \cdot \underset{\underset{\$17.50}{\downarrow}}{\underset{\text{Price}}{\text{Original}}} = \underset{\underset{n}{\downarrow}}{\underset{\text{Discount}}{\text{Amount of}}}$$
$$\$4.38 \approx n$$

Sale Price = $17.50 − $4.38 = $13.12

23. Original Price = $37.88
Rate of Discount = 10% = 0.10

$$\underset{\underset{0.10}{\downarrow}}{\underset{\text{Discount}}{\text{Rate of}}} \cdot \underset{\underset{\$37.88}{\downarrow}}{\underset{\text{Price}}{\text{Original}}} = \underset{\underset{n}{\downarrow}}{\underset{\text{Discount}}{\text{Amount of}}}$$
$$\$3.79 \approx n$$

Sale Price = $37.88 − $3.79 = $34.09

25. $300 at 14% for 1 year

$$\begin{aligned} I &= p \cdot r \cdot t \\ &= (300)(0.14)(1) \\ &= 42 \end{aligned}$$

The interest is $42.

$$\begin{aligned} \text{amount due} &= \text{principal} + \text{interest} \\ &= \$300 + \$42 \\ &= \$342 \end{aligned}$$

The total amount due is $342.

27. $740 at 6% for 9 months

$$\begin{aligned} I &= p \cdot r \cdot t \\ &= (740)(0.06)\left(\tfrac{9}{12}\right) \\ &= 33.30 \end{aligned}$$

The interest is $33.30.

$$\begin{aligned} \text{amount due} &= \text{principal} + \text{interest} \\ &= \$740 + \$33.30 \\ &= \$773.30 \end{aligned}$$

The total amount due is $773.30.

29. $1500 at $2\frac{1}{2}$% for $1\frac{1}{2}$ years

$$I = p \cdot r \cdot t$$
$$= (1500)(0.025)(1.5)$$
$$= 56.25$$

The interest is $56.25.

$$\text{amount due} = \text{principal} + \text{interest}$$
$$= \$1500 + \$56.25$$
$$= \$1556.25$$

The total amount due is $1556.25.

31. $17,800 at $7\frac{3}{4}$% for 8 months

$$I = p \cdot r \cdot t$$
$$= (17,800)(0.0775)\left(\tfrac{8}{12}\right)$$
$$= 919.67 \text{ (rounded)}$$

The interest is $919.67.

$$\text{amount due} = \text{principal} + \text{interest}$$
$$= \$17,800 + \$919.67$$
$$= \$18,719.67$$

The total amount due is $18,719.67.

33. Original Price = $1950
Rate of Discount = 40% = 0.40
Amount of Discount = 0.40(1950) = $780
Sale Price = $1950 − $780 = $1170

The sale price of the ring is $1170.

35. $7500 at $8\frac{1}{2}$% for 9 months

$$I = p \cdot r \cdot t$$
$$= (7500)(0.085)\left(\tfrac{9}{12}\right)$$
$$= 478.13 \text{ (rounded)}$$

The interest is $478.13.

$$\text{amount due} = \text{principal} + \text{interest}$$
$$= \$7500 + \$478.13$$
$$= \$7978.13$$

Rick will owe his mother $7978.13 (rounded).

37. Cost of Item = $24.99
Tax Rate = $6\frac{1}{2}$% = 0.065
Amount of Tax = 0.065(24.99) = $1.62 (rounded)
Total Cost = $24.99 + $1.62 = $26.61

The total cost of the headset is $26.61 (rounded).

39. Cost of Item = $99
Tax Rate = unknown(p)
Amount of Tax = $7.92

Tax Rate	·	Cost of Item	=	Amount of Tax
↓	↓	↓	↓	↓
p	·	99	=	7.92

$$\frac{99p}{99} = \frac{7.92}{99}$$
$$p = 0.08 \quad (8\%)$$

The sales tax rate is 8%.

41. Original Price = $135
Rate of Discount = 45%
Amount of Discount = 0.45(135) = $60.75
Sale Price = $135 − $60.75 = $74.25

The sale price of the parka is $74.25.

43.

Tip rate	·	Original price	=	Tip amount
↓	↓	↓	↓	↓
0.15	·	$43.70	=	n
		$6.555	=	n

The tip amount is $6.56 (rounded).
Total = $43.70 + $6.56 = $50.26
Amount paid by each = $50.26 ÷ 2 = $25.13

45. Original Price = $1199
Rate of Discount = 33%
Amount of Discount = 0.33(1199)
$$= \$395.67$$
Sale Price = $1199 − $395.67 = $803.33

The discount is $395.67 and the sale price is $803.33.

47. $1900 at $12\frac{1}{4}$% for 6 months

$$I = p \cdot r \cdot t$$
$$= (1900)(0.1225)\left(\tfrac{6}{12}\right)$$
$$= 116.375$$

The interest is $116.38 (rounded).

$$\text{amount due} = \text{principal} + \text{interest}$$
$$= \$1900 + \$116.38$$
$$= \$2016.38$$

She must pay a total amount of $2016.38 (rounded).

49.

Tip rate	·	Original price	=	Tip amount
↓	↓	↓	↓	↓
0.15	·	$17.98	=	n
		$2.697	=	n

The 15% tip would be $2.70 (rounded).
Total = $17.98 + $2.70 = $20.68

They would give the delivery person $21 (rounded to the nearest dollar).

51. Computer modem: ($129, 65% off)

Discount amount = 0.65($129) = $83.85
Sale price = $129 − $83.85 = $45.15
Tax amount = 0.06($45.15) = $2.71 (rounded)
Total = $45.15 + $2.71 = **$47.86**

Earrings: ($60, 30% off)

Discount amount = 0.30($60) = $18
Sale price = $60 − $18 = $42
Tax amount = 0.06($42) = $2.52
Total = $42 + $2.52 = **$44.52**

Bill = $47.86 + $44.52 = **$92.38**

53. Camcorder: ($287.95, 65% off)

Discount amount = 0.65($287.95) = $187.17
(rounded)
Sale price = $287.95 − $187.17 = $100.78
Tax amount = 0.06($100.78) = $6.05 (rounded)
Total = $100.78 + $6.05 = **$106.83**

Jeans: (2 @ $48, 45% off)

Discount amount = 0.45($48) = $21.60
Sale price = $48 − $21.60 = $26.40
Tax amount: no tax on clothing
Total for 2 pairs = 2($26.40) = **$52.80**

Ring: ($95, 30% off)

Discount amount = 0.30($95) = $28.50
Sale price = $95 − $28.50 = $66.50
Tax amount = 0.06($66.50) = $3.99
Total = $66.50 + $3.99 = **$70.49**

Bill = $106.83 + $52.80 + $70.49 = **$230.12**

Relating Concepts (Exercises 55–57)

55. (a) $10,000 at 5% for 10 years

$$I = p \cdot r \cdot t$$
$$= (10,000)(0.05)(10)$$
$$= 5000$$

The interest is $5000. The amount you would
have is $10,000 + $5000 = $15,000.

(b) From the table, $10,000 invested at 5% for 10
years would be worth $16,289, which is $1289
more than the amount in part (a).

56. (a) $10,000 at 8% for 30 years

$$I = p \cdot r \cdot t$$
$$= (10,000)(0.08)(30)$$
$$= 24,000$$

The interest is $24,000. The amount you would
have is $10,000 + $24,000 = $34,000.

(b) From the table, $10,000 invested at 8% for 30
years would be worth $100,627, which is $66,627
more than the amount in part (a).

57. (a) $10,000 at 10% for 50 years gives
($10,000)(0.10)(50) = $50,000 in interest for a
total of $60,000.

From the table, $10,000 invested at 10% for 50
years would be worth $1,173,909, which is
$1,113,909 more than $60,000.

(b) $\dfrac{\$1,173,909}{\$60,000} \approx 19.565$, so the compounded
amount is more than 19 times greater than the
simple interest amount.

Chapter 7 Review Exercises

1. $25\% = 25 \div 100 = 0.25$

2. $180\% = 180 \div 100 = 1.80$

3. $12.5\% = 12.5 \div 100 = 0.125$

4. $7\% = 7 \div 100 = 0.07$

5. $2.65 = 2.65 \cdot 100\% = 265\%$

6. $0.02 = 0.02 \cdot 100\% = 2\%$

7. $0.3 = 0.3 \cdot 100\% = 30\%$

8. $0.002 = 0.002 \cdot 100\% = 0.2\%$

9. $12\% = \dfrac{12}{100} = \dfrac{12 \div 4}{100 \div 4} = \dfrac{3}{25}$

10. $37.5\% = \dfrac{37.5}{100} = \dfrac{37.5 \cdot 10}{100 \cdot 10} = \dfrac{375}{1000}$
$$= \dfrac{375 \div 125}{1000 \div 125} = \dfrac{3}{8}$$

11. $250\% = \dfrac{250}{100} = \dfrac{250 \div 50}{100 \div 50} = \dfrac{5}{2} = 2\dfrac{1}{2}$

12. $5\% = \dfrac{5}{100} = \dfrac{5 \div 5}{100 \div 5} = \dfrac{1}{20}$

13. $\dfrac{3}{4} = \left(\dfrac{3}{4}\right)(100\%) = \left(\dfrac{3}{4}\right)\left(\dfrac{100}{1}\%\right)$
$$= \dfrac{3 \cdot \overset{1}{\cancel{4}} \cdot 25}{\underset{1}{\cancel{4}} \cdot 1}\% = \dfrac{75}{1}\% = 75\%$$

14. $\dfrac{5}{8} = \left(\dfrac{5}{8}\right)(100\%) = \left(\dfrac{5}{8}\right)\left(\dfrac{100}{1}\%\right)$
$$= \dfrac{5 \cdot \overset{1}{\cancel{4}} \cdot 25}{2 \cdot \underset{1}{\cancel{4}} \cdot 1}\% = \dfrac{125}{2}\% = 62\dfrac{1}{2}\%, \quad \text{or} \quad 62.5\%$$

15. $3\dfrac{1}{4} = \dfrac{13}{4} = \left(\dfrac{13}{4}\right)(100\%) = \left(\dfrac{13}{4}\right)\left(\dfrac{100}{1}\%\right)$
$$= \dfrac{13 \cdot \overset{1}{\cancel{4}} \cdot 25}{\underset{1}{\cancel{4}} \cdot 1}\% = \dfrac{325}{1}\% = 325\%$$

16. $\dfrac{3}{50} = \left(\dfrac{3}{50}\right)(100\%) = \left(\dfrac{3}{50}\right)\left(\dfrac{100}{1}\%\right)$
$$= \dfrac{3 \cdot 2 \cdot \overset{1}{\cancel{50}}}{\underset{1}{\cancel{50}} \cdot 1}\% = \dfrac{6}{1}\% = 6\%$$

17. $\dfrac{1}{8} = 1 \div 8 = 0.125$

18. $0.125 = 0.125 \cdot 100\% = 12.5\%$

19. $0.15 = \dfrac{15}{100} = \dfrac{15 \div 5}{100 \div 5} = \dfrac{3}{20}$

20. $0.15 = 0.15 \cdot 100\% = 15\%$

21. $180\% = \dfrac{180}{100} = \dfrac{18}{10} = \dfrac{9}{5} = 1\dfrac{4}{5}$

22. $180\% = 180 \div 100 = 1.80 = 1.8$

23. 100% of $46 is all of the money, or $46.

24. 50% of $46 is half of the money, or $23.

25. 100% of 9 hours is all of the hours, or 9 hours.

26. 50% of 9 hours is half of the hours, which is $4\frac{1}{2}$ or 4.5 hours.

27. Whole: unknown (n)

$$\text{Percent} \rightarrow \dfrac{140}{100} = \dfrac{338.8}{n} \begin{array}{l} \leftarrow \text{Part} \\ \leftarrow \text{Whole} \end{array}$$
$$\text{Always 100} \rightarrow$$

Step 1 $\dfrac{140}{100} = \dfrac{338.8}{n}$

Step 2 $140 \cdot n = 100 \cdot 338.8$

Step 3 $\dfrac{140n}{140} = \dfrac{33{,}880}{140}$

$n = 242$

338.8 meters is 140% of 242 meters.

28. Whole: unknown (n)

$$\text{Percent} \rightarrow \dfrac{2.5}{100} = \dfrac{425}{n} \begin{array}{l} \leftarrow \text{Part} \\ \leftarrow \text{Whole} \end{array}$$
$$\text{Always 100} \rightarrow$$

Step 1 $\dfrac{2.5}{100} = \dfrac{425}{n}$

Step 2 $2.5 \cdot n = 100 \cdot 425$

Step 3 $\dfrac{2.5n}{2.5} = \dfrac{42{,}500}{2.5}$

$n = 17{,}000$

2.5% of 17,000 cases is 425 cases.

29. Part: unknown (n)

$$\text{Percent} \rightarrow \dfrac{6}{100} = \dfrac{n}{450} \begin{array}{l} \leftarrow \text{Part} \\ \leftarrow \text{Whole} \end{array}$$
$$\text{Always 100} \rightarrow$$

Step 1 $\dfrac{6}{100} = \dfrac{n}{450}$

Step 2 $100 \cdot n = 6 \cdot 450$

Step 3 $\dfrac{100n}{100} = \dfrac{2700}{100}$

$n = 27$

6% of 450 smartphones is 27 smartphones.

30. Part: unknown (n)

$$\text{Percent} \rightarrow \dfrac{60}{100} = \dfrac{n}{1450} \begin{array}{l} \leftarrow \text{Part} \\ \leftarrow \text{Whole} \end{array}$$
$$\text{Always 100} \rightarrow$$

Step 1 $\dfrac{60}{100} = \dfrac{n}{1450}$

Step 2 $100 \cdot n = 60 \cdot 1450$

Step 3 $\dfrac{100n}{100} = \dfrac{87{,}000}{100}$

$n = 870$

60% of 1450 math books is 870 math books.

31. Percent: unknown (p)

$$\text{Percent} \rightarrow \dfrac{p}{100} = \dfrac{36}{380} \begin{array}{l} \leftarrow \text{Part} \\ \leftarrow \text{Whole} \end{array}$$
$$\text{Always 100} \rightarrow$$

Step 1 $\dfrac{p}{100} = \dfrac{36}{380}$

Step 2 $380 \cdot p = 100 \cdot 36$

Step 3 $\dfrac{380p}{380} = \dfrac{3600}{380}$

$p \approx 9.5$

36 pairs is 9.5% (rounded) of 380 pairs.

32. Percent: unknown (p)

$$\text{Percent} \rightarrow \dfrac{p}{100} = \dfrac{1440}{640} \begin{array}{l} \leftarrow \text{Part} \\ \leftarrow \text{Whole} \end{array}$$
$$\text{Always 100} \rightarrow$$

Step 1 $\dfrac{p}{100} = \dfrac{1440}{640}$

Step 2 $640 \cdot p = 100 \cdot 1440$

Step 3 $\dfrac{640p}{640} = \dfrac{144{,}000}{640}$

$p = 225$

1440 cans is 225% of 640 cans.

33. $\underline{11\%}$ of $\underline{\$23.60}$ is $\underline{\text{how much?}}$

$0.11 \cdot 23.60 = n$

$2.596 = n$

11% of $23.60 is $2.60 (rounded).

34. $\underline{\text{What}}$ is $\underline{125\%}$ of $\underline{64 \text{ days?}}$

$n = 1.25 \cdot 64$

$n = 80$ days

80 days is 125% of 64 days.

35. $\underline{1.28 \text{ ounces}}$ is $\underline{\text{what percent}}$ of $\underline{32 \text{ ounces?}}$

$1.28 = p \cdot 32$

$1.28 = 32p$

$\dfrac{1.28}{32} = \dfrac{32p}{32}$

$0.04 = p$

$4\% = p$

1.28 ounces is 4% of 32 ounces.

36. $\underbrace{\$46}$ is $\underbrace{8\%}$ of $\underbrace{\text{what number of dollars?}}$

$\downarrow \quad \downarrow \quad \downarrow \quad \downarrow \qquad\qquad \downarrow$

$46 \;=\; 0.08 \;\cdot\qquad\qquad n$

$46 = 0.08n$

$\dfrac{46}{0.08} = \dfrac{0.08n}{0.08}$

$\$575 \;=\; n$

$\$46$ is 8% of $\$575$.

37. $\underbrace{\text{8 people}}$ is $\underbrace{40\%}$ of $\underbrace{\text{what number of people?}}$

$\downarrow \qquad \downarrow \quad \downarrow \quad \downarrow \qquad\qquad \downarrow$

$8 \;=\; 0.40 \;\cdot\qquad\qquad n$

$8 = 0.40n$

$\dfrac{8}{0.40} = \dfrac{0.40n}{0.40}$

$20 \text{ people} \;=\; n$

8 people is 40% of 20 people.

38. $\underbrace{\text{What percent}}$ of $\underbrace{\text{174 ft}}$ is $\underbrace{\text{304.5 ft?}}$

$\downarrow \qquad\quad \downarrow \quad \downarrow \quad \downarrow \qquad \downarrow$

$p \quad\cdot\quad 174 \;=\; 304.5$

$174p \;=\; 304.5$

$\dfrac{174p}{174} = \dfrac{304.5}{174}$

$p \;=\; 1.75 \;(175\%)$

175% of 174 ft is 304.5 ft.

39. **(a)** Part: 504 late patients
Whole: total number of patients, unknown (n)
Percent: 16.8%

$$\text{Percent} \cdot \text{Whole} = \text{Part}$$
$$0.168 \cdot n = 504$$
$$\dfrac{0.168n}{0.168} = \dfrac{504}{0.168}$$
$$n = 3000$$

There were 3000 patients in January.

(b) Amount of decrease $= 504 - 345 = 159$

Let p be the percent of decrease.

$\underbrace{\text{percent}}_{}$ of $\underbrace{\text{original}}_{\text{value}}$ $=$ $\underbrace{\text{amount of}}_{\text{decrease}}$

$\downarrow \quad \downarrow \quad\;\; \downarrow \qquad \downarrow \qquad \downarrow$

$p \quad\cdot\quad 504 \;=\; 159$

$\dfrac{504p}{504} = \dfrac{159}{504}$

$p \;\approx\; 0.315 \quad (31.5\%)$

The percent of decrease was 31.5% (rounded).

40. **(a)** Part: actual amount spent, unknown (n)
Whole: $280 budgeted for food
Percent: 130%

$$\text{Percent} \cdot \text{Whole} = \text{Part}$$
$$1.30 \cdot 280 = n$$
$$364 = n$$

$364 was actually spent on food.

(b) Part: amount spent, $112
Whole: amount budgeted, $50
Percent: unknown (p)

$$\text{Percent} \cdot \text{Whole} = \text{Part}$$
$$p \cdot 50 = 112$$
$$\dfrac{50p}{50} = \dfrac{112}{50}$$
$$p = 2.24 \quad (224\%)$$

She spent 224% of the budgeted amount.

41. **(a)** Part: 640 trees still living after 1 year
Whole: 800 trees planted
Percent: unknown (p)

$$\text{Percent} \cdot \text{Whole} = \text{Part}$$
$$p \cdot 800 = 640$$
$$\dfrac{800p}{800} = \dfrac{640}{800}$$
$$p = 0.8 \quad (80\%)$$

80% of the trees were still living.

(b) Amount of increase $= 850 - 800 = 50$

Let p be the percent of increase.

$\underbrace{\text{percent}}_{}$ of $\underbrace{\text{original}}_{\text{value}}$ $=$ $\underbrace{\text{amount of}}_{\text{increase}}$

$\downarrow \quad \downarrow \quad\;\; \downarrow \qquad \downarrow \qquad \downarrow$

$p \quad\cdot\quad 800 \;=\; 50$

$\dfrac{800p}{800} = \dfrac{50}{800}$

$p \;=\; 0.0625 \quad (6.25\%)$

The percent of increase was 6.3% (rounded).

42. Amount of increase $= 13 - 7 = 6$

Let p be the percent of increase.

$\underbrace{\text{percent}}_{}$ of $\underbrace{\text{original}}_{\text{value}}$ $=$ $\underbrace{\text{amount of}}_{\text{increase}}$

$\downarrow \quad \downarrow \quad\;\; \downarrow \qquad \downarrow \qquad \downarrow$

$p \quad\cdot\quad 7 \;=\; 6$

$\dfrac{7p}{7} = \dfrac{6}{7}$

$p \;=\; 0.857 \quad (85.7\%)$

The percent of increase was 85.7% (rounded).

43. Part: Sales tax = unknown (n)
Whole: Cost of item = $2.79
Percent: Tax rate = 4%

$$\text{Percent} \cdot \text{Whole} = \text{Part}$$
$$0.04 \cdot 2.79 = n$$
$$0.1116 = n$$

Amount of tax = $0.11 (rounded)
Total cost = $2.79 + $0.11 = $2.90

44. Part: Sales tax = $58.50
Whole: Cost of item = $780
Percent: Tax rate = unknown (p)

$$\text{Percent} \cdot \text{Whole} = \text{Part}$$
$$p \cdot 780 = 58.50$$
$$\frac{780p}{780} = \frac{58.50}{780}$$
$$p = 0.075 \quad (7.5\%)$$

The tax rate is 7.5% or $7\frac{1}{2}$%.
Total cost = $780 + $58.50 = $838.50

45. The bill of $42.73 rounds to $40.

Estimate of 15% tip:

10% of $40 is $4. 5% of $40 is half of $4, or $2. An estimate is $4 + $2 = $6.

Exact 15% tip:

0.15($42.73) = $6.4095 = $6.41 (rounded)

Estimate of 20% tip:

10% of $40 is $4. 20% is 2 times $4, or $8. An estimate is $8.

Exact 20% tip:

0.20($42.73) = $8.546 = $8.55 (rounded)

46. The bill of $8.05 rounds to $8.

Estimate of 15% tip:

10% of $8 is $0.80. 5% of $8 is half of $0.80, or $0.40. An estimate is $0.80 + $0.40 = $1.20.

Exact 15% tip:

0.15($8.05) = $1.2075 = $1.21 (rounded)

Estimate of 20% tip:

10% of $8 is $0.80. 20% is 2 times $0.80, or $1.60. An estimate is $1.60.

Exact 20% tip:

0.20($8.05) = $1.61

47. Original Price = $37.50
Rate of Discount = 10% = 0.10
Amount of Discount = 0.10(37.50) = $3.75
Sale Price = $37.50 − $3.75 = $33.75

48. Original Price = $252
Rate of Discount = unknown (p)
Amount of Discount = $63
Sale Price = $252 − $63 = $189

$$\frac{\text{amount of}}{\text{discount}} = \frac{\text{rate of}}{\text{discount}} \cdot \frac{\text{original}}{\text{price}}$$
$$63 = p \cdot 252$$
$$\frac{63}{252} = \frac{252p}{252}$$
$$0.25 = p \quad (25\%)$$

The rate of discount is 25%.

49. $350 at $3\frac{1}{4}$% for 3 years

$$I = p \cdot r \cdot t$$
$$= (350)(0.0325)(3)$$
$$= 34.125 = 34.13 \text{ (rounded)}$$

The interest is $34.13.

$$\text{amount due} = \text{principal} + \text{interest}$$
$$= \$350 + \$34.13$$
$$= \$384.13$$

The total amount due is $384.13.

50. $1530 at 16% for 9 months

$$I = p \cdot r \cdot t$$
$$= (1530)(0.16)\left(\frac{9}{12}\right)$$
$$= 183.60$$

The interest is $183.60.

$$\text{amount due} = \text{principal} + \text{interest}$$
$$= \$1530 + \$183.60$$
$$= \$1713.60$$

The total amount due is $1713.60.

51. **[7.1]** $\frac{1}{3} = \frac{1}{3} \cdot 100\% = \frac{100}{3}\% = 33\frac{1}{3}\%$ (exact)

$33\frac{1}{3}\% \approx 0.333 = 33.3\%$ (rounded)

52. **[7.1]** The largest percent is 68%, which represents cards, so cards is the most popular game.

$$68\% = 0.68 = \frac{68}{100} = \frac{68 \div 4}{100 \div 4} = \frac{17}{25}$$

53. **[7.3]** $\frac{1}{3} \cdot 830 \approx 276.7 \approx 277$

277 adults (rounded) play electronic/computer games.

54. **[7.3]** Most popular: cards, 68%

0.68(277) = 188.36 ≈ 188 adults (rounded)

Least popular: sci-fi/simulation, 37%

0.37(277) = 102.49 ≈ 102 adults (rounded)

55. [7.4] Part: 3440 dogs placed
Whole: 4629 dogs received
Percent: unknown (p)

$$\text{Percent} \cdot \text{Whole} = \text{Part}$$
$$p \cdot 4629 = 3440$$
$$\frac{4629p}{4629} = \frac{3440}{4629}$$
$$p \approx 0.743 \quad (74.3\%)$$

74.3% (rounded) of the dogs were placed in new homes.

56. [7.4] Part: 4629 received so far this year
Whole: number expected, unknown (n)
Percent: 81%

$$\text{Percent} \cdot \text{Whole} = \text{Part}$$
$$0.81 \cdot n = 4629$$
$$\frac{0.81n}{0.81} = \frac{4629}{0.81}$$
$$n \approx 5715$$

They expect to receive 5715 dogs (rounded).

57. [7.4] Part: 3891 placed in new homes
Whole: 6230 received
Percent: unknown (p)

$$\text{Percent} \cdot \text{Whole} = \text{Part}$$
$$p \cdot 6230 = 3891$$
$$\frac{6230p}{6230} = \frac{3891}{6230}$$
$$p \approx 0.625 \quad (62.5\%)$$

About 62.5% of the cats were placed in new homes.

58. [7.4] During the first 6 months:
Last year = 8748 cats received
This year = 6230 cats received
Amount of decrease = 8748 − 6230
= 2518 cats

$$\frac{\text{Percent}}{\text{decrease}} \cdot \frac{\text{Original}}{\text{amount}} = \frac{\text{Amount of}}{\text{decrease}}$$
$$p \cdot 8748 = 2518$$
$$\frac{8748p}{8748} = \frac{2518}{8748}$$
$$p \approx 0.288 \quad (28.8\%)$$

The percent decrease is 28.8% (rounded).

59. [7.4] Total number of animals received
$$= 4629 + 6230 + 944 = 11{,}803$$

Total number of animals placed
$$= 3440 + 3891 + 852 = 8183$$

$$\text{Percent} \cdot \text{Whole} = \text{Part}$$
$$p \cdot 11{,}803 = 8183$$
$$\frac{11{,}803p}{11{,}803} = \frac{8183}{11{,}803}$$
$$p \approx 0.693 \quad (69.3\%)$$

About 69.3% of the animals received were placed in new homes.

60. [7.4] 70% of all animals received:
$$0.70(11{,}803) \approx 8262$$

Since they placed 8183 animals,
$8262 - 8183 = 79$ (rounded) animals need to be placed in a home in order to reach their goal.

Chapter 7 Test

1. $75\% = 75 \div 100 = 0.75$

2. $0.6 = 0.60 = 60\%$

3. $1.8 = 1.80 = 180\%$

4. $0.075 = 7.5\%, \quad \text{or} \quad 7\frac{1}{2}\%$

5. $300\% = 300 \div 100 = 3.00, \quad \text{or} \quad 3$

6. $2\% = 2 \div 100 = 0.02$

7. $62.5\% = \frac{62.5}{100} = \frac{62.5 \cdot 10}{100 \cdot 10} = \frac{625}{1000}$
$$= \frac{625 \div 125}{1000 \div 125} = \frac{5}{8}$$

8. $240\% = \frac{240}{100} = \frac{24}{10} = \frac{12}{5} = 2\frac{2}{5}$

9. $\frac{1}{20} = \left(\frac{1}{20}\right)(100\%) = \left(\frac{1}{20}\right)\left(\frac{100}{1}\%\right)$
$$= \frac{1 \cdot \overset{1}{\cancel{20}} \cdot 5}{\underset{1}{\cancel{20}} \cdot 1}\% = \frac{5}{1}\% = 5\%$$

10. $\frac{7}{8} = \left(\frac{7}{8}\right)(100\%) = \left(\frac{7}{8}\right)\left(\frac{100}{1}\%\right)$
$$= \frac{7 \cdot \overset{1}{\cancel{4}} \cdot 25}{2 \cdot \underset{1}{\cancel{4}}}\% = \frac{175}{2}\% = 87\frac{1}{2}\%, \quad \text{or} \quad 87.5\%$$

11. $1\frac{3}{4} = \frac{7}{4} = \left(\frac{7}{4}\right)(100\%) = \left(\frac{7}{4}\right)\left(\frac{100}{1}\%\right)$
$$= \frac{7 \cdot \overset{1}{\cancel{4}} \cdot 25}{\underset{1}{\cancel{4}} \cdot 1}\% = \frac{175}{1}\% = 175\%$$

12. 16 laptops is 5% of what number of laptops?
$$16 = 0.05 \cdot n$$
$$\frac{16}{0.05} = \frac{0.05n}{0.05}$$
$$320 = n$$

16 laptops is 5% of 320 laptops.

13. $192 is what percent of $48?
$$192 = p \cdot 48$$
$$\frac{192}{48} = \frac{48p}{48}$$
$$4 = p \quad (400\%)$$

$192 is 400% of $48.

14. Part: $14,625
Whole: Amount needed = unknown (n)
Percent: 75%

$$\text{Percent} \cdot \text{Whole} = \text{Part}$$
$$0.75 \cdot n = 14,625$$
$$\frac{0.75n}{0.75} = \frac{14,625}{0.75}$$
$$n = 19,500$$

$19,500 is needed for a down payment.

15. Part: Sales tax = unknown (n)
Whole: Cost of item = $7950
Percent: Tax rate = $6\frac{1}{2}\% = 6.5\% = 0.065$

$$\text{Percent} \cdot \text{Whole} = \text{Part}$$
$$0.065 \cdot 7950 = n$$
$$516.75 = n$$

The sales tax is $516.75.

$$\text{Total cost of the car} = \$7950 + \$516.75$$
$$= \$8466.75$$

16. Last semester = 1440 students
Current semester = 1925 students
Amount of increase
$$= 1925 - 1440 = 485 \text{ students}$$

$$
\begin{array}{ccccc}
\text{Percent} & \cdot & \text{Original} & = & \text{Amount of} \\
 & & & & \text{increase} \\
p & \cdot & 1440 & = & 485 \\
 & & \dfrac{1440p}{1440} & = & \dfrac{485}{1440} \\
 & & p & \approx & 0.34 \quad (34\%)
\end{array}
$$

The percent of increase is 34% (rounded).

17. To find 50% of a number, divide the number by 2. To find 25% of a number, divide the number by 4. Examples will vary.

18. Round $31.94 to $30. Then 10% of $30. is $3 and 5% of $30 is half of $3 or $1.50, so a 15% tip estimate is $3 + $1.50 = $4.50. A 20% tip estimate is 2($3) = $6.

19. $\text{Percent} \cdot \text{Whole} = \text{Part}$
$0.15 \cdot \$31.94 = \text{Tip amount}$
$\$4.791 = \text{Tip amount}$

The tip amount would be $4.79 (rounded).

$$\text{Total expense} = \$31.94 + \$4.79$$
$$= \$36.73$$

If the total expense is shared by 3 friends, each person will pay $36.73 ÷ 3 = $12.24 (rounded).

20. Discount rate · Original price = Discount amount
$0.08 \cdot \$48 = \text{Discount amount}$
$\$3.84 = \text{Discount amount}$

Sale price = Original price − Discount amount
$= \$48 - \$3.84$
$= \$44.16$

21. Discount rate · Original price = Discount amount
$0.18 \cdot \$229.95 = \text{Discount amount}$
$\$41.391 = \text{Discount amount}$

Rounded to the nearest cent, the discount amount is $41.39.

Sale price = Original price − Discount amount
$= \$229.95 - \$41.39$
$= \$188.56$

22. Discount rate · Original price = Discount amount
$0.30 \cdot \$1089 = \text{Discount amount}$
$\$326.70 = \text{Discount amount}$

Discounted price = $1089 − $326.70 = $762.30

$$\text{Tax rate} \cdot \text{Price} = \text{Sales tax}$$
$$0.07 \cdot \$762.30 = \text{Sales tax}$$
$$\$53.361 = \text{Sales tax}$$
$$\$53.36 \text{ (rounded)} = \text{Sales tax}$$

Total bill = $762.30 + $53.36
$= \$815.66$

Since the payments will be spread out over 6 months, Jamal will need to pay
$815.66 ÷ 6 = $135.94 per month (rounded).

23. $5000 at $8\frac{1}{4}\%$ for 4 years ($8\frac{1}{4}\% = 0.0825$)

$$I = p \cdot r \cdot t$$
$$= (5000)(0.0825)(4)$$
$$= 1650$$

The simple interest due on the loan is $1650.

Total amount due = principal + interest
$= \$5000 + \$1650$
$= \$6650$

The total amount due is $6650.

24. $860 at 12% for 6 months

$$I = p \cdot r \cdot t$$
$$= (860)(0.12)\left(\tfrac{6}{12}\right)$$
$$= 51.60$$

The simple interest due on the loan is $51.60.

Total amount due = principal + interest
$= \$860 + \$51.60$
$= \$911.60$

The total amount due is $911.60.

CHAPTER 8   MEASUREMENT

8.1   Problem Solving with U.S. Measurement Units

8.1 Margin Exercises

1. (a) 1 c = $\underline{8}$ fl oz

(b) $\underline{4}$ qt = 1 gal

(c) 1 wk = $\underline{7}$ days

(d) $\underline{3}$ ft = 1 yd

(e) 1 ft = $\underline{12}$ in.

(f) $\underline{16}$ oz = 1 lb

(g) 1 ton = $\underline{2000}$ lb

(h) $\underline{60}$ min = 1 hr

(i) 1 pt = $\underline{2}$ c

(j) $\underline{24}$ hr = 1 day

(k) 1 min = $\underline{60}$ sec

(l) 1 qt = $\underline{2}$ pt

(m) $\underline{5280}$ ft = 1 mi

2. (a) $5\frac{1}{2}$ ft to inches   ▪   You are converting from a *larger* unit to a *smaller* unit, so $\underline{multiply}$.

Because 1 ft = 12 in., multiply by 12.

$$5\frac{1}{2} \text{ ft} = 5\frac{1}{2} \cdot 12 = \frac{11}{\underset{1}{\cancel{2}}} \cdot \frac{\overset{6}{\cancel{12}}}{1} = \frac{66}{1} = 66 \text{ in.}$$

(b) 64 oz to pounds   ▪   You are converting from a *smaller* unit to a *larger* unit, so $\underline{divide}$.

Because 16 oz = 1 lb, divide by 16.

$$64 \text{ oz} = \frac{64}{16} = 4 \text{ lb}$$

(c) 6 yd to feet   ▪   You are converting from a *larger* unit to a *smaller* unit, so multiply.

Because 1 yd = 3 ft, multiply by 3.

$$6 \text{ yd} = 6 \cdot 3 = 18 \text{ ft}$$

(d) 2 tons to pounds   ▪   You are converting from a *larger* unit to a *smaller* unit, so multiply.

Because 1 ton = 2000 lb, multiply by 2000.

$$2 \text{ tons} = 2 \cdot 2000 = 4000 \text{ lb}$$

(e) 35 pt to quarts   ▪   You are converting from a *smaller* unit to a *larger* unit, so divide.

Because 1 qt = 2 pt, divide by 2.

$$35 \text{ pt} = \frac{35}{2} = 17\frac{1}{2} \text{ qt}$$

(f) 20 min to hours   ▪   You are converting from a *smaller* unit to a *larger* unit, so divide.

Because 60 min = 1 hr, divide by 60.

$$20 \text{ min} = \frac{20}{60} = \frac{20 \div 20}{60 \div 20} = \frac{1}{3} \text{ hr}$$

3. (a) 36 in. to feet   ▪   unit fraction $\left.\right\}$ $\dfrac{1 \text{ ft}}{12 \text{ in.}}$

$$36 \text{ in.} = \frac{\overset{3}{\cancel{36}} \text{ in.}}{1} \cdot \frac{1 \text{ ft}}{\underset{1}{\cancel{12}} \text{ in.}} = \frac{3 \cdot 1 \text{ ft}}{1} = 3 \text{ ft}$$

(b) 14 ft to inches   ▪   unit fraction $\left.\right\}$ $\dfrac{12 \text{ in.}}{1 \text{ ft}}$

$$14 \text{ ft} = \frac{14 \cancel{\text{ft}}}{1} \cdot \frac{12 \text{ in.}}{1 \cancel{\text{ft}}} = \frac{14 \cdot 12 \text{ in.}}{1} = 168 \text{ in.}$$

(c) 60 in. to feet   ▪   unit fraction $\left.\right\}$ $\dfrac{1 \text{ ft}}{12 \text{ in.}}$

$$60 \text{ in.} = \frac{\overset{5}{\cancel{60}} \text{ in.}}{1} \cdot \frac{1 \text{ ft}}{\underset{1}{\cancel{12}} \text{ in.}} = \frac{5 \cdot 1 \text{ ft}}{1} = 5 \text{ ft}$$

(d) 4 yd to feet   ▪   unit fraction $\left.\right\}$ $\dfrac{3 \text{ ft}}{1 \text{ yd}}$

$$4 \text{ yd} = \frac{4 \cancel{\text{yd}}}{1} \cdot \frac{3 \text{ ft}}{1 \cancel{\text{yd}}} = \frac{4 \cdot 3 \text{ ft}}{1} = 12 \text{ ft}$$

(e) 39 ft to yards   ▪   unit fraction $\left.\right\}$ $\dfrac{1 \text{ yd}}{3 \text{ ft}}$

$$39 \text{ ft} = \frac{\overset{13}{\cancel{39}} \cancel{\text{ft}}}{1} \cdot \frac{1 \text{ yd}}{\underset{1}{\cancel{3}} \cancel{\text{ft}}} = \frac{13 \cdot 1 \text{ yd}}{1} = 13 \text{ yd}$$

(f) 2 mi to feet   ▪   unit fraction $\left.\right\}$ $\dfrac{5280 \text{ ft}}{1 \text{ mi}}$

$$2 \text{ mi} = \frac{2 \cancel{\text{mi}}}{1} \cdot \frac{5280 \text{ ft}}{1 \cancel{\text{mi}}} = \frac{2 \cdot 5280 \text{ ft}}{1} = 10{,}560 \text{ ft}$$

4. (a) 16 qt to gallons   ▪   unit fraction $\left.\right\}$ $\dfrac{1 \text{ gallon}}{4 \text{ quarts}}$

$$16 \text{ qt} = \frac{\overset{4}{\cancel{16}} \cancel{\text{qt}}}{1} \cdot \frac{1 \text{ gal}}{\underset{1}{\cancel{4}} \cancel{\text{qt}}} = \frac{4 \cdot 1 \text{ gal}}{1} = \underline{4} \text{ gal}$$

(b) 3 c to pints   ▪   unit fraction $\left.\right\}$ $\dfrac{1 \text{ pt}}{2 \text{ c}}$

$$3 \text{ c} = \frac{3 \cancel{\text{c}}}{1} \cdot \frac{1 \text{ pt}}{2 \cancel{\text{c}}} = \frac{3 \cdot 1 \text{ pt}}{2} = 1\frac{1}{2} \text{ pt or } 1.5 \text{ pt}$$

(c) $3\frac{1}{2}$ tons to pounds   ▪   unit fraction $\left.\right\}$ $\dfrac{2000 \text{ lb}}{1 \text{ ton}}$

$$3\frac{1}{2} \text{ tons} = \frac{3\frac{1}{2} \cancel{\text{tons}}}{1} \cdot \frac{2000 \text{ lb}}{1 \cancel{\text{ton}}} = \frac{7}{\underset{1}{\cancel{2}}} \cdot \frac{\overset{1000}{\cancel{2000}}}{1}$$

$$= 7000 \text{ lb}$$

(d) $1\frac{3}{4}$ lb to ounces ∙ unit fraction $\Big\}$ $\dfrac{16 \text{ oz}}{1 \text{ lb}}$

$$1\frac{3}{4}\text{ lb} = \frac{1\frac{3}{4}\cancel{\text{lb}}}{1} \cdot \frac{16 \text{ oz}}{1 \cancel{\text{lb}}} = \frac{7}{\underset{1}{\cancel{4}}} \cdot \frac{\overset{4}{\cancel{16}}}{1} = 28 \text{ oz}$$

(e) 4 oz to pounds ∙ unit fraction $\Big\}$ $\dfrac{1 \text{ lb}}{16 \text{ oz}}$

$$4 \text{ oz} = \frac{\overset{1}{\cancel{4}}\cancel{\text{oz}}}{1} \cdot \frac{1 \text{ lb}}{\underset{4}{\cancel{16}}\cancel{\text{oz}}} = \frac{1}{4} \text{ lb or } 0.25 \text{ lb}$$

5. **(a)** 4 tons to ounces

unit fractions $\Big\}$ $\dfrac{2000 \text{ lb}}{1 \text{ ton}}, \dfrac{16 \text{ oz}}{1 \text{ lb}}$

$$\frac{4 \cancel{\text{tons}}}{1} \cdot \frac{2000 \cancel{\text{lb}}}{1 \cancel{\text{ton}}} \cdot \frac{16 \text{ oz}}{1 \cancel{\text{lb}}} = 4 \cdot 2000 \cdot 16 \text{ oz}$$
$$= 128{,}000 \text{ oz}$$

(b) 3 mi to inches

unit fractions $\Big\}$ $\dfrac{5280 \text{ ft}}{1 \text{ mi}}, \dfrac{12 \text{ in.}}{1 \text{ ft}}$

$$\frac{3 \cancel{\text{mi}}}{1} \cdot \frac{5280 \cancel{\text{ft}}}{1 \cancel{\text{mi}}} \cdot \frac{12 \text{ in.}}{1 \cancel{\text{ft}}} = 3 \cdot 5280 \cdot 12 \text{ in.}$$
$$= 190{,}080 \text{ in.}$$

(c) 36 pt to gallons

unit fractions $\Big\}$ $\dfrac{1 \text{ qt}}{2 \text{ pt}}, \dfrac{1 \text{ gal}}{4 \text{ qt}}$

$$\frac{36 \cancel{\text{pt}}}{1} \cdot \frac{1 \cancel{\text{qt}}}{2 \cancel{\text{pt}}} \cdot \frac{1 \text{ gal}}{4 \cancel{\text{qt}}} = \frac{\overset{9}{\cancel{36}}}{1} \cdot \frac{1}{2} \cdot \frac{1}{\underset{1}{\cancel{4}}} \text{ gal} = \frac{9}{2} \text{ gal}$$
$$= 4\frac{1}{2} \text{ gal or } 4.5 \text{ gal}$$

(d) 2 wk to minutes

unit fractions $\Big\}$ $\dfrac{7 \text{ days}}{1 \text{ wk}}, \dfrac{24 \text{ hr}}{1 \text{ day}}, \dfrac{60 \text{ min}}{1 \text{ hr}}$

$$\frac{2 \cancel{\text{wk}}}{1} \cdot \frac{7 \cancel{\text{days}}}{1 \cancel{\text{wk}}} \cdot \frac{24 \cancel{\text{hr}}}{1 \cancel{\text{day}}} \cdot \frac{60 \text{ min}}{1 \cancel{\text{hr}}} = 2 \cdot 7 \cdot 24 \cdot 60 \text{ min}$$
$$= 20{,}160 \text{ min}$$

6. **(a)** *Step 1* The problem asks for the price per pound.

Step 2 Convert ounces to pounds. Then divide the cost by the pounds.

Step 3 To estimate, round $3.29 to $3. There are 16 oz in a pound, so 12 oz is about 1 lb. Thus, $3 ÷ 1 pound = $3 per pound is our estimate.

Step 4 $\dfrac{\overset{3}{\cancel{12}}\cancel{\text{oz}}}{1} \cdot \dfrac{1 \text{ lb}}{\underset{4}{\cancel{16}}\cancel{\text{oz}}} = \dfrac{3}{4} \text{ lb} = 0.75 \text{ lb}$

Then divide: $\dfrac{\$3.29}{0.75 \text{ lb}} = \$4.38\overline{6} \approx \$4.39$ per lb

Step 5 The cheese costs $4.39 per pound (to the nearest cent).

Step 6 The answer, $4.39, is close to our estimate of $3.

(b) *Step 1* The problem asks for the number of tons of furnishings they moved.

Step 2 First convert pounds to tons. Then multiply to find the number of tons of furnishings for 5 houses.

Step 3 To estimate, round 11,000 lb to 10,000 lb. There are 2000 lb in a ton, so $10{,}000 \div 2000 = 5$ tons. Multiply by 5 to get 25 tons as our estimate.

Step 4 $\dfrac{\overset{11}{\cancel{11{,}000}}\cancel{\text{lb}}}{1} \cdot \dfrac{1 \text{ ton}}{\underset{2}{\cancel{2000}}\cancel{\text{lb}}} = \dfrac{11}{2} \text{ tons}$
$$= 5.5 \text{ tons}$$
$$5(5.5 \text{ tons}) = 27.5 \text{ tons}$$

Step 5 The company moved 27.5 tons or $27\frac{1}{2}$ tons of furnishings.

Step 6 The answer, $27\frac{1}{2}$ tons, is close to our estimate of 25 tons.

8.1 Section Exercises

1. **(a)** foot, mile, inch, yard arranged from smallest to largest is inch, foot, yard, mile

(b) pint, fluid ounce, gallon, quart, cup arranged from smallest to largest is fluid ounce, cup, pint, quart, gallon

3. **(a)** 1 yd = $\underline{3}$ ft; **(b)** $\underline{12}$ in. = 1 ft

5. **(a)** $\underline{8}$ fl oz = 1 c; **(b)** 1 qt = $\underline{2}$ pt

7. **(a)** $\underline{2000}$ lb = 1 ton; **(b)** 1 lb = $\underline{16}$ oz

9. **(a)** 1 min = $\underline{60}$ sec; **(b)** $\underline{60}$ min = 1 hr

11. **(a)** 120 sec to minutes ∙ You are converting from a *smaller* unit to a *larger* unit, so divide. Because 60 sec = 1 min, divide by 60.

$$120 \text{ sec} = \frac{120}{60} = \underline{2} \text{ min}$$

(b) 4 hr to minutes ∙ You are converting from a *larger* unit to a *smaller* unit, so multiply. Because 1 hr = 60 min, multiply by 60.

$$4 \text{ hr} = 4 \cdot 60 = \underline{240} \text{ min}$$

13. **(a)** 2 qt to gallons ∙ You are converting from a *smaller* unit to a *larger* unit, so divide. Because 4 qt = 1 gal, divide by 4.

$$2 \text{ qt} = \frac{2}{4} = \frac{2 \div 2}{4 \div 2} = \frac{1}{2} \text{ or } 0.5 \text{ gal}$$

(b) $6\frac{1}{2}$ ft to inches ▪ You are converting from a *larger* unit to a *smaller* unit, so multiply. Because 1 ft = 12 inches, multiply by 12.

$$6\frac{1}{2} \text{ ft} = 6\frac{1}{2} \cdot 12 = \frac{13}{\underset{1}{\cancel{2}}} \cdot \frac{\overset{6}{\cancel{12}}}{1} = \frac{78}{1} = \underline{78} \text{ in.}$$

15. 7 to 8 tons to pounds ▪ You are converting from a *larger* unit to a *smaller* unit, so multiply. Because 1 ton = 2000 pounds, multiply by 2000.

$$7 \text{ tons} = 7 \cdot 2000 = 14{,}000 \text{ lb}$$

$$8 \text{ tons} = 8 \cdot 2000 = 16{,}000 \text{ lb}$$

An adult African elephant may weigh 14,000 to 16,000 lb.

17. 9 yd to feet ▪ $\left.\begin{array}{c}\text{unit}\\\text{fraction}\end{array}\right\}$ $\dfrac{3\text{ ft}}{1\text{ yd}}$

$$\frac{9 \text{ y\cancel{d}}}{1} \cdot \frac{3 \text{ ft}}{1 \text{ y\cancel{d}}} = 9 \cdot 3 \text{ ft} = \underline{27} \text{ ft}$$

19. 7 lb to ounces

$$\frac{7 \text{ \cancel{lb}}}{1} \cdot \frac{16 \text{ oz}}{1 \text{ \cancel{lb}}} = 7 \cdot 16 \text{ oz} = \underline{112} \text{ oz}$$

21. 5 qt to pints

$$\frac{5 \text{ \cancel{qt}}}{1} \cdot \frac{2 \text{ pt}}{1 \text{ \cancel{qt}}} = 5 \cdot 2 \text{ pt} = \underline{10} \text{ pt}$$

23. 90 min to hours

$$\frac{90 \text{ \cancel{min}}}{1} \cdot \frac{1 \text{ hr}}{60 \text{ \cancel{min}}} = \frac{90}{60} \text{ hr} = 1\frac{1}{2} \text{ or } \underline{1.5 \text{ hr}}$$

25. 3 in. to feet

$$\frac{\overset{1}{\cancel{3}} \text{ \cancel{in}}}{1} \cdot \frac{1 \text{ ft}}{\underset{4}{\cancel{12}} \text{ \cancel{in.}}} = \frac{1}{4} \text{ or } \underline{0.25 \text{ ft}}$$

27. 24 oz to pounds

$$\frac{24 \text{ \cancel{oz}}}{1} \cdot \frac{1 \text{ lb}}{16 \text{ \cancel{oz}}} = \frac{24}{16} \text{ lb} = 1\frac{1}{2} \text{ or } \underline{1.5 \text{ lb}}$$

29. 5 c to pints

$$\frac{5 \text{ \cancel{c}}}{1} \cdot \frac{1 \text{ pt}}{2 \text{ \cancel{c}}} = \frac{5}{2} \text{ pt} = 2\frac{1}{2} \text{ or } \underline{2.5 \text{ pt}}$$

31. $\frac{1}{2}$ ft $= \dfrac{\frac{1}{2} \text{ \cancel{ft}}}{1} \cdot \dfrac{12 \text{ in.}}{1 \text{ \cancel{ft}}} = \dfrac{1}{2} \cdot 12 \text{ in.} = 6 \text{ in.}$

The ice will safely support a snowmobile or ATV or a person walking.

33. $2\frac{1}{2}$ tons to lb

$$\frac{2\frac{1}{2} \text{ \cancel{tons}}}{1} \cdot \frac{2000 \text{ lb}}{1 \text{ \cancel{ton}}} = \frac{5}{\underset{1}{\cancel{2}}} \cdot \frac{\overset{1000}{\cancel{2000}}}{1} \text{ lb} = \underline{5000} \text{ lb}$$

35. $4\frac{1}{4}$ gal to quarts

$$\frac{4\frac{1}{4} \text{ \cancel{gal}}}{1} \cdot \frac{4 \text{ qt}}{1 \text{ \cancel{gal}}} = \frac{17}{\cancel{4}} \cdot \frac{\overset{1}{\cancel{4}}}{1} \text{ qt} = \underline{17} \text{ qt}$$

37. $\frac{1}{3}$ ft $= \dfrac{\frac{1}{3} \text{ \cancel{ft}}}{1} \cdot \dfrac{12 \text{ in.}}{1 \text{ \cancel{ft}}} = \dfrac{1}{\underset{1}{\cancel{3}}} \cdot \dfrac{\overset{4}{\cancel{12}}}{1} \text{ in.} = 4 \text{ in.}$

Two-thirds of a foot would be twice as high; that is, $2(4 \text{ in.}) = 8 \text{ in.}$ The cactus could be 4 to 8 in. tall.

39. 6 yd to inches

$$\frac{6 \text{ y\cancel{d}}}{1} \cdot \frac{3 \text{ \cancel{ft}}}{1 \text{ y\cancel{d}}} \cdot \frac{12 \text{ in.}}{1 \text{ \cancel{ft}}} = 6 \cdot 3 \cdot 12 \text{ in.} = \underline{216} \text{ in.}$$

41. 112 c to quarts

$$\frac{\overset{\overset{28}{\cancel{56}}}{\cancel{112}} \text{ \cancel{c}}}{1} \cdot \frac{1 \text{ \cancel{pt}}}{\underset{1}{\cancel{2}} \text{ \cancel{c}}} \cdot \frac{1 \text{ qt}}{\underset{1}{\cancel{2}} \text{ \cancel{pt}}} = \underline{28} \text{ qt}$$

43. 6 days to seconds

$$\frac{6 \text{ days}}{1} \cdot \frac{24 \text{ \cancel{hr}}}{1 \text{ \cancel{day}}} \cdot \frac{60 \text{ \cancel{min}}}{1 \text{ \cancel{hr}}} \cdot \frac{60 \text{ sec}}{1 \text{ \cancel{min}}}$$
$$= 6 \cdot 24 \cdot 60 \cdot 60 \text{ sec}$$
$$= \underline{518{,}400} \text{ sec}$$

45. $1\frac{1}{2}$ tons to ounces

$$\frac{1\frac{1}{2} \text{ \cancel{tons}}}{1} \cdot \frac{2000 \text{ \cancel{lb}}}{1 \text{ \cancel{ton}}} \cdot \frac{16 \text{ oz}}{1 \text{ \cancel{lb}}} = \frac{3}{\underset{1}{\cancel{2}}} \cdot \frac{\overset{1000}{\cancel{2000}}}{1} \cdot \frac{16}{1} \text{ oz}$$
$$= \underline{48{,}000} \text{ oz}$$

47. **(a)** There is only one relationship that uses 1 to 16: 1 <u>pound</u> = 16 <u>ounces</u>

(b) 10 to 20 is the same as 1 to 2. There are 2 relationships like this: 1 pint = 2 cups and 1 quart = 2 pints.
10 <u>quarts</u> = 20 <u>pints</u> or 10 <u>pints</u> = 20 <u>cups</u>

(c) 120 to 2 is the same as 60 to 1. There are 2 relationships like this: 60 minutes = 1 hour and 60 seconds = 1 minute.
120 <u>minutes</u> = 2 <u>hours</u> or 120 <u>seconds</u> = 2 <u>minutes</u>

(d) 2 to 24 is the same as 1 to 12. Use 1 foot = 12 inches.
2 <u>feet</u> = 24 <u>inches</u>

(e) 6000 to 3 is the same as 2000 to 1. Use 2000 pounds = 1 ton.
6000 <u>pounds</u> = 3 <u>tons</u>

(f) 35 to 5 is the same as 7 to 1. Use 7 days = 1 week.
35 <u>days</u> = 5 <u>weeks</u>

49. $2\frac{3}{4}$ miles to inches

$$\frac{2\frac{3}{4} \text{ mi}}{1} \cdot \frac{5280 \text{ ft}}{1 \text{ mi}} \cdot \frac{12 \text{ in.}}{1 \text{ ft}} = \frac{11}{4} \cdot \frac{5280}{1} \cdot \frac{\overset{3}{\cancel{12}}}{1} \text{ in.}$$

$$= \underline{174{,}240} \text{ in.}$$

51. $6\frac{1}{4}$ gal to fluid ounces

$$\frac{6\frac{1}{4} \text{ gal}}{1} \cdot \frac{4 \text{ qt}}{1 \text{ gal}} \cdot \frac{32 \text{ fl oz}}{1 \text{ qt}} = \frac{25}{\cancel{4}} \cdot \frac{1}{\cancel{4}} \cdot 32 \text{ fl oz}$$

$$= \underline{800} \text{ fl oz}$$

53. 24,000 oz to tons

$$\frac{24{,}000 \text{ oz}}{1} \cdot \frac{1 \text{ lb}}{16 \text{ oz}} \cdot \frac{1 \text{ ton}}{2000 \text{ lb}}$$

$$= \frac{\overset{3}{\cancel{24{,}000}}}{1} \cdot \frac{1}{\underset{4}{\cancel{16}}} \cdot \frac{1}{\underset{1}{\cancel{2000}}} \text{ ton}$$

$$= \frac{3}{4} \text{ or } 0.75 \text{ ton}$$

For Exercises 55–62, the six problem-solving steps should be used, but are only shown for Exercise 55.

55. *Step 1* The problem asks for the price per pound of strawberries.

Step 2 Convert ounces to pounds. Then divide the cost by the pounds.

Step 3 To estimate, round $2.29 to $2. Then, there are 16 oz in a pound, so 20 oz is a little more than 1 lb. Thus, $2 ÷ 1 = $2 per pound is our estimate.

Step 4 $\dfrac{\overset{5}{\cancel{20}} \text{ oz}}{1} \cdot \dfrac{1 \text{ lb}}{\underset{4}{\cancel{16}} \text{ oz}} = \dfrac{5}{4} \text{ lb} = 1.25 \text{ lb}$

$$\frac{\$2.29}{1.25 \text{ lb}} = 1.832 \approx \$1.83/\text{lb (rounded)}$$

Step 5 The strawberries are $1.83 per pound (to the nearest cent).

Step 6 The answer, $1.83, is close to our estimate of $2.

57. Find the total number of feet needed.

$$24 \cdot 2 = 48 \text{ ft}$$

Then convert feet to yards.

$$\frac{\overset{16}{\cancel{48}} \text{ ft}}{1} \cdot \frac{1 \text{ yd}}{\underset{1}{\cancel{3}} \text{ ft}} = 16 \text{ yd}$$

Then multiply to find the cost.

$$\frac{16 \text{ yd}}{1} \cdot \frac{\$8.75}{\text{yd}} = \$140$$

It will cost $140 to equip all the stations.

59. **(a)** Convert seconds per foot to seconds per mile.

$$\frac{1 \text{ sec}}{5 \text{ ft}} \cdot \frac{5280 \text{ ft}}{1 \text{ mi}} = \frac{\overset{1056}{\cancel{5280}}}{\underset{1}{\cancel{5}}} \text{ sec/mi} = 1056 \text{ sec/mi}$$

It would take the cockroach 1056 seconds to travel 1 mile.

(b) $\dfrac{1056 \text{ sec}}{1 \text{ mi}} \cdot \dfrac{1 \text{ min}}{60 \text{ sec}} = \dfrac{\overset{88}{\cancel{1056}}}{\underset{5}{\cancel{60}}} \text{ min/mi}$

$$= 17.6 \text{ min/mi}$$

It would take the cockroach 17.6 minutes to travel 1 mile.

61. **(a)** Find the total number of cups per week. Then convert cups to quarts.

$$\frac{2}{3} \cdot 15 \cdot 5 = \frac{2}{\cancel{3}} \cdot \frac{\overset{5}{\cancel{15}}}{1} \cdot \frac{5}{1} = 50 \text{ c}$$

$$\frac{\overset{25}{\cancel{50}} \text{ c}}{1} \cdot \frac{1 \text{ pt}}{\underset{1}{\cancel{2}} \text{ c}} \cdot \frac{1 \text{ qt}}{2 \text{ pt}} = \frac{25}{2} \text{ qt} = 12\frac{1}{2} \text{ qt}$$

The center needs $12\frac{1}{2}$ qt of milk per week.

(b) Convert quarts to gallons.

$$\frac{12\frac{1}{2} \text{ qt}}{1} \cdot \frac{1 \text{ gal}}{4 \text{ qt}} = \frac{25}{2} \cdot \frac{1}{4} \text{ gal} = 3.125 \text{ gal}$$

The center should order 4 containers, because you can't buy part of a container.

Relating Concepts (Exercises 63–66)

63. **(a)** Convert 4150 feet to miles.

$$\frac{4150 \text{ ft}}{1} \cdot \frac{1 \text{ mi}}{5280 \text{ ft}} = \frac{4150}{5280} \text{ mi} \approx 0.8 \text{ mi (rnd)}$$

(b) Multiply the distance across the crater, 4150 feet, by 3.14 to get 13,031 ft. Since 3.14 is an approximation, so is 13,031.

(c) Convert the answer in part (b) to miles.

$$\frac{13{,}031 \text{ ft}}{1} \cdot \frac{1 \text{ mi}}{5280 \text{ ft}} = \frac{13{,}031}{5280} \text{ mi} \approx 2.5 \text{ mi (rnd)}$$

64. **(a)** Convert 550 feet to yards.

$$\frac{550 \text{ ft}}{1} \cdot \frac{1 \text{ yd}}{3 \text{ ft}} = \frac{550}{3} \text{ yd} \approx 183 \text{ yd (rnd)}$$

(b) Convert 550 feet to inches.

$$\frac{550 \text{ ft}}{1} \cdot \frac{12 \text{ in.}}{1 \text{ ft}} = 550 \cdot 12 \text{ in.} = 6600 \text{ in.}$$

(c) Convert 550 feet to miles.

$$\frac{550 \text{ ft}}{1} \cdot \frac{1 \text{ mi}}{5280 \text{ ft}} = \frac{550}{5280} \text{ mi} \approx 0.1 \text{ mi (rnd)}$$

(d) Divide the depth of the crater, 550 feet, by 60 to get about 9.2 feet per story, or 9 ft (rounded to the nearest foot).

65. **(a)** Convert 18 and 30 inches to feet.

$$\frac{\overset{3}{\cancel{18}}\ \cancel{\text{in.}}}{1} \cdot \frac{1\ \text{ft}}{\underset{2}{\cancel{12}}\ \cancel{\text{in.}}} = \frac{3}{2} \text{ or } 1\frac{1}{2} \text{ ft}$$

$$\frac{\overset{5}{\cancel{30}}\ \cancel{\text{in.}}}{1} \cdot \frac{1\ \text{ft}}{\underset{2}{\cancel{12}}\ \cancel{\text{in.}}} = \frac{5}{2} \text{ or } 2\frac{1}{2} \text{ ft}$$

The trees are $1\frac{1}{2}$ to $2\frac{1}{2}$ ft, or 1.5 to 2.5 ft tall.

(b) Convert 700 years to months.

$$\frac{700\ \cancel{\text{yr}}}{1} \cdot \frac{12\ \text{mo}}{1\ \cancel{\text{yr}}} = 700 \cdot 12\ \text{mo} = 8400 \text{ months}$$

(c) Three feet is thirty-six inches, which is six inches more than 30 inches. Set up and solve a proportion with age (in years) and tree height (in inches).

$$\frac{700\ \text{years}}{30\ \text{inches}} = \frac{x}{6\ \text{inches}}$$
$$30 \cdot x = 700 \cdot 6$$
$$\frac{30 \cdot x}{30} = \frac{700 \cdot 6}{30}$$
$$x = \frac{700 \cdot 6}{30} = 140 \text{ years}$$

It will take 140 more years to grow 6 inches, for a total of 840 years in all.

66. **(a)** Multiply the distance across the smaller crater, 90 miles, by 3.14 to get 282.6, or about 283 mi (rnd). Since 3.14 is an approximation, so is 283.

(b) Multiply the distance across the larger crater, 120 miles, by 3.14 to get 376.8, or about 377 mi (rnd).

8.2   The Metric System—Length

8.2   Margin Exercises

1. The length of a baseball bat, the height of a doorknob from the floor, and a basketball player's arm length are each about 1 meter in length.

2. **(a)** The woman's height is 168 <u>cm</u>.

(b) The man's waist is 90 <u>cm</u> around.

(c) Louise ran the 100 <u>m</u> dash in the track meet.

(d) A postage stamp is 22 <u>mm</u> wide.

(e) Michael paddled his canoe 2 <u>km</u> down the river.

(f) The pencil lead is 1 <u>mm</u> thick.

(g) A stick of gum is 7 <u>cm</u> long.

(h) The highway speed limit is 90 <u>km</u> per hour.

(i) The classroom was 12 <u>m</u> long.

(j) A penny is about 18 <u>mm</u> across.

3. **(a)** 3.67 m to cm   ▪   unit fraction } $\frac{100\ \text{cm}}{1\ \text{m}}$

$$\frac{3.67\ \cancel{\text{m}}}{1} \cdot \frac{100\ \text{cm}}{1\ \cancel{\text{m}}} = \frac{3.67 \cdot 100\ \text{cm}}{1} = 367 \text{ cm}$$

(b) 92 cm to m   ▪   unit fraction } $\frac{1\ \text{m}}{100\ \text{cm}}$

$$\frac{92\ \cancel{\text{cm}}}{1} \cdot \frac{1\ \text{m}}{100\ \cancel{\text{cm}}} = \frac{92}{100} \text{ m} = 0.92 \text{ m}$$

(c) 432.7 cm to m   ▪   unit fraction } $\frac{1\ \text{m}}{100\ \text{cm}}$

$$\frac{432.7\ \cancel{\text{cm}}}{1} \cdot \frac{1\ \text{m}}{100\ \cancel{\text{cm}}} = \frac{432.7}{100} \text{ m} = 4.327 \text{ m}$$

(d) 65 mm to cm   ▪   unit fraction } $\frac{1\ \text{cm}}{10\ \text{mm}}$

$$\frac{65\ \cancel{\text{mm}}}{1} \cdot \frac{1\ \text{cm}}{10\ \cancel{\text{mm}}} = \frac{65}{10} \text{ cm} = 6.5 \text{ cm}$$

(e) 0.9 m to mm   ▪   unit fraction } $\frac{1000\ \text{mm}}{1\ \text{m}}$

$$\frac{0.9\ \cancel{\text{m}}}{1} \cdot \frac{1000\ \text{mm}}{1\ \cancel{\text{m}}} = (0.9)(1000) \text{ mm} = 900 \text{ mm}$$

(f) 2.5 cm to mm   ▪   unit fraction } $\frac{10\ \text{mm}}{1\ \text{cm}}$

$$\frac{2.5\ \cancel{\text{cm}}}{1} \cdot \frac{10\ \text{mm}}{1\ \cancel{\text{cm}}} = (2.5)(10) \text{ mm} = 25 \text{ mm}$$

4. **(a)** $(43.5)(10) = \underline{435}$ ▪   43.5∧ gives 435.

(b) $43.5 \div 10 = \underline{4.35}$ ▪   4∧3.5 gives $\underline{4.35}$.

(c) $(28)(100) = \underline{2800}$ ▪   28.00∧ gives $\underline{2800}$.

(d) $28 \div 100 = \underline{0.28}$ ▪   ∧28. gives $\underline{0.28}$.

(e) $(0.7)(1000) = \underline{700}$ ▪   0.700∧ gives $\underline{700}$.

(f) $0.7 \div 1000 = \underline{0.0007}$ ▪   ∧000.7 gives $\underline{0.0007}$.

5. **(a)** 12.008 km to m   ▪   Count <u>three</u> places to the <u>right</u> on the conversion line.

$$12.008\text{∧ km} = \underline{12,008} \text{ m}$$

(b) 561.4 m to km ▪ Count <u>three</u> places to the <u>left</u> on the conversion line.

$$\wedge561.4 \text{ m} = \underline{0.5614} \text{ km}$$

(c) 20.7 cm to m ▪ Count 2 places to the *left* on the conversion line.

$$\wedge20.7 \text{ cm} = 0.207 \text{ m}$$

(d) 20.7 cm to mm ▪ Count 1 place to the *right* on the conversion line.

$$20.7\wedge \text{ cm} = 207 \text{ mm}$$

(e) 4.66 m to cm ▪ Count 2 places to the *right* on the conversion line.

$$4.66\wedge \text{ m} = 466 \text{ cm}$$

6. **(a)** <u>9 m to mm</u> ▪ Count <u>three</u> places to the <u>right</u> on the conversion line. Three zeros are written in as placeholders.

$$9.000\wedge \text{ m} = \underline{9000} \text{ mm}$$

(b) 3 cm to m ▪ Count <u>two</u> places to the <u>left</u> on the conversion line. One zero is written in as a placeholder.

$$\wedge03. \text{ cm} = \underline{0.03} \text{ m}$$

(c) 5 mm to cm ▪ Count 1 place to the *left* on the conversion line.

$$\wedge5. \text{ mm} = 0.5 \text{ cm}$$

(d) 70 m to km ▪ Count 3 places to the *left* on the conversion line. One zero is written in as a placeholder.

$$\wedge070. \text{ m} = 0.07 \text{ km}$$

(e) 0.8 m to cm ▪ Count 2 places to the *right* on the conversion line. One zero is written in as a placeholder.

$$0.80\wedge \text{ m} = 80 \text{ cm}$$

8.2 Section Exercises

1. *Kilo* means 1000, so 1 km $= \underline{1000}$ m.

3. *Milli* means $\frac{1}{1000}$ or 0.001, so

$$1 \text{ mm} = \frac{1}{1000} \text{ or } 0.001 \text{ m.}$$

5. *Centi* means $\frac{1}{100}$ or 0.01, so 1 cm $= \frac{1}{100}$ or 0.01 m.

7. **(a)** <u>cm</u> is shorter than m

(b) <u>mm</u> is shorter than m

(c) <u>cm</u> is shorter than km

9. The width of your hand in centimeters ▪ Answers will vary; about 8 to 10 cm.

11. The width of your thumb in millimeters ▪ Answers will vary; about 20 to 25 mm.

13. The child was 91 <u>cm</u> tall.

15. Ming-Na swam in the 200 <u>m</u> backstroke race.

17. Adriana drove 400 <u>km</u> on her vacation.

19. An aspirin tablet is 10 <u>mm</u> across.

21. A paper clip is about 3 <u>cm</u> long.

23. Dave's truck is 5 <u>m</u> long.

25. 7 m to cm

$$\frac{7 \text{ m}}{1} \cdot \frac{100 \text{ cm}}{1 \text{ m}} = \frac{7 \cdot 100 \text{ cm}}{1} = 700 \text{ cm}$$

27. 40 mm to m

$$\frac{40 \text{ mm}}{1} \cdot \frac{1 \text{ m}}{1000 \text{ mm}} = \frac{40}{1000} \text{ m} = 0.040 \text{ m or } 0.04 \text{ m}$$

29. 9.4 km to m

$$\frac{9.4 \text{ km}}{1} \cdot \frac{1000 \text{ m}}{1 \text{ km}} = (9.4)(1000) \text{ m} = 9400 \text{ m}$$

31. 509 cm to m

$$\frac{509 \text{ cm}}{1} \cdot \frac{1 \text{ m}}{100 \text{ cm}} = \frac{509}{100} \text{ m} = 5.09 \text{ m}$$

33. 400 mm to cm ▪ Count 1 place to the *left* on the conversion line.

$$40\wedge0. \text{ mm} = 40.0 \text{ cm} = 40 \text{ cm}$$

35. 0.91 m to mm ▪ Count 3 places to the *right* on the conversion line. One zero is written in as a placeholder.

$$0.910\wedge \text{ m} = 910 \text{ mm}$$

37. 82 cm to m ▪ Count 2 places to the *left* on the conversion line.

$$\wedge82. \text{ cm} = 0.82 \text{ m}$$

0.82 m is less than 1 m, so 82 cm is **less than** 1 m. The difference in length is 1 m $-$ 0.82 m $=$ 0.18 m or 100 cm $-$ 82 cm $=$ 18 cm.

39.

5 mm to centimeters   ▪   Count one place to the *left* on the conversion line.

$$_\wedge 5. \text{ mm} = 0.5 \text{ cm}$$

Similarly, 1 mm = 0.1 cm.

41.   61 m to km   ▪   Count 3 places to the *left* on the conversion line. One zero is written in as a placeholder.

$$_\wedge 061. \text{ m} = 0.061 \text{ km}$$

The Roe River is just under 0.061 km long.

43.   1.64 m to centimeters and millimeters   ▪   Count two (three) places to the *right* on the conversion line.

$$1.64_\wedge \text{ m} = 164 \text{ cm}; \qquad 1.640_\wedge \text{ m} = 1640 \text{ mm}$$

The median height for U.S. females who are 20 to 29 years old is about 164 cm, or equivalently, 1640 mm.

45.   5.6 mm to km   ▪

$$\frac{5.6 \text{ mm}}{1} \cdot \frac{1 \text{ m}}{1000 \text{ mm}} \cdot \frac{1 \text{ km}}{1000 \text{ m}}$$

$$= \frac{5.6}{1,000,000} \text{ km}$$

$$= 0.0000056 \text{ km}$$

8.3   The Metric System—Capacity and Weight (Mass)

8.3   Margin Exercises

1.   The liter would be used to measure the amount of water in the bathtub, the amount of gasoline you buy for your car, and the amount of water in a pail.

2.   **(a)** I bought 8 <u>L</u> of soda at the store.

(b) The nurse gave me 10 <u>mL</u> of cough syrup.

(c) This is a 100 <u>L</u> garbage can.

(d) It took 10 <u>L</u> of paint to cover the bedroom walls.

(e) My car's gas tank holds 50 <u>L</u>.

(f) I added 15 <u>mL</u> of oil to the pancake mix.

(g) The can of orange soda holds 350 <u>mL</u>.

(h) My friend gave me a 30 <u>mL</u> bottle of expensive perfume.

3.   **(a)** 9 L to mL   ▪   On the (metric capacity) conversion line, L to mL is three places to the

right, which gives us 9000 mL. Alternatively, we could use a unit fraction as follows:

$$\frac{9 \text{ L}}{1} \cdot \frac{1000 \text{ mL}}{1 \text{ L}} = \underline{9000} \text{ mL}$$

(b)   0.75 L to mL   ▪   Count 3 places to the *right* on the (metric capacity) conversion line.

$$0.750_\wedge \text{ L} = 750 \text{ mL}$$

(c)   500 mL to L

$$\frac{500 \text{ mL}}{1} \cdot \frac{1 \text{ L}}{1000 \text{ mL}} = \frac{500}{1000} \text{ L} = 0.5 \text{ L}$$

(d)   5 mL to L   ▪   Count 3 places to the *left* on the (metric capacity) conversion line.

$$_\wedge 005. \text{ mL} = 0.005 \text{ L}$$

(e)   2.07 L to mL

$$\frac{2.07 \text{ L}}{1} \cdot \frac{1000 \text{ mL}}{1 \text{ L}} = 2070 \text{ mL}$$

(f)   3275 mL to L   ▪   Count 3 places to the *left* on the (metric capacity) conversion line.

$$3_\wedge 275. \text{ mL} = 3.275 \text{ L}$$

4.   A small paper clip, one playing card from a deck of cards, and the check you wrote to the cable company would each weigh about 1 gram.

5.   **(a)** A thumbtack weights 800 <u>mg</u>.

(b) A teenager weighs 50 <u>kg</u>.

(c) This large cast-iron frying pan weighs 1 <u>kg</u>.

(d) Jerry's basketball weighed 600 <u>g</u>.

(e) Tamlyn takes a 500 <u>mg</u> calcium tablet every morning.

(f) On his diet, Greg can eat 90 <u>g</u> of meat for lunch.

(g) One strand of hair weighs 2 <u>mg</u>.

(h) One banana might weigh 150 <u>g</u>.

6.   **(a)** 10 kg to g   ▪   On the [metric weight (mass)] conversion line, kg to g is <u>three</u> places to the <u>right</u>, which gives us 10,000 g. Alternatively, we could use a unit fraction as follows:

$$\frac{10 \text{ kg}}{1} \cdot \frac{1000 \text{ g}}{1 \text{ kg}} = \underline{10,000} \text{ g}$$

(b)   45 mg to g

$$\frac{45 \text{ mg}}{1} \cdot \frac{1 \text{ g}}{1000 \text{ mg}} = \frac{45}{1000} \text{ g} = 0.045 \text{ g}$$

(c) 6.3 kg to g ▪ Count 3 places to the *right* on the [metric weight (mass)] conversion line.

$$6.300_\wedge \text{ kg} = 6300 \text{ g}$$

(d) 0.077 g to mg ▪ Count 3 places to the *right* on the [metric weight (mass)] conversion line.

$$0.077_\wedge \text{ g} = 77 \text{ mg}$$

(e) 5630 g to kg ▪ Count 3 places to the *left* on the [metric weight (mass)] conversion line.

$$5_\wedge 630. \text{ g} = 5.63 \text{ kg}$$

(f) 90 g to kg

$$\frac{90 \text{ g}}{1} \cdot \frac{1 \text{ kg}}{1000 \text{ g}} = \frac{9}{100} \text{ kg} = 0.09 \text{ kg}$$

7. (a) Gail bought a 4 L can of paint. Use <u>capacity</u> units.

(b) The bag of chips weighed 450 g. Use <u>weight</u> units.

(c) Give the child 5 <u>mL</u> of cough syrup. Use <u>capacity</u> units.

(d) The width of the window is 55 <u>cm</u>. Use <u>length</u> units.

(e) Akbar drives 18 <u>km</u> to work. Use <u>length</u> units.

(f) The laptop computer weighs 2 <u>kg</u>. Use <u>weight</u> units.

(g) A credit card is 55 <u>mm</u> wide. Use <u>length</u> units.

8.3 Section Exercises

1. (a) For a dose of cough syrup, use <u>milliliters</u>.

(b) For a large carton of milk, use <u>liters</u>.

(c) For a vitamin pill, use <u>milligrams</u>.

(d) For a heavyweight wrestler, use <u>kilograms</u>.

3. The glass held 250 <u>mL</u> of water. (Liquids are measured in mL or L.)

5. Dolores can make 10 <u>L</u> of soup in that pot. (Liquids are measured in mL or L.)

7. Our yellow Labrador dog grew up to weigh 40 <u>kg</u>. (Weight is measured in mg, g, or kg.)

9. Lori caught a small sunfish weighing 150 <u>g</u>. (Weight is measured in mg, g, or kg.)

11. Andre donated 500 <u>mL</u> of blood today. (Blood is a liquid, and liquids are measured in mL or L.)

13. The patient received a 250 <u>mg</u> tablet of medication each hour. (Weight is measured in mg, g, or kg.)

15. The gas can for the lawnmower holds 4 <u>L</u>. (Gasoline is a liquid, and liquids are measured in mL or L.)

17. Pam's backpack weighs 5 <u>kg</u> when it is full of books. (Weight is measured in mg, g, or kg.)

19. This is unreasonable (too much) since 4.1 liters would be about 4 quarts.

21. This is unreasonable (too much) since 5 kilograms of Epsom salts would be about 11 pounds with approximately a quart of water.

23. This is reasonable because 15 milliliters would be 3 teaspoons.

25. This is reasonable because 350 milligrams would be a little more than 1 tablet, which is about 325 milligrams.

27. The unit for your answer (g) is in the numerator. The unit being changed (kg) is in denominator, so it will divide out. The unit fraction is $\frac{1000 \text{ g}}{1 \text{ kg}}$.

29. 15 L to mL ▪ unit fraction $\Big\}$ $\frac{1000 \text{ mL}}{1 \text{ L}}$

$$\frac{15 \text{ L}}{1} \cdot \frac{1000 \text{ mL}}{1 \text{ L}} = \underline{15,000} \text{ mL}$$

OR
Count <u>three</u> places to the <u>right</u> on the conversion line.

$$15.000_\wedge \text{ L} = \underline{15,000} \text{ mL}$$

31. In Exercises 31–54, we use a unit fraction or a conversion line, but not both.

3000 mL to L ▪ Count 3 places to the *left* on the conversion line.

$$3_\wedge 000. \text{ mL} = 3 \text{ L}$$

33. 925 mL to L

$$\frac{925 \text{ mL}}{1} \cdot \frac{1 \text{ L}}{1000 \text{ mL}} = \frac{925}{1000} \text{ L} = 0.925 \text{ L}$$

35. 8 mL to L

$$\frac{8 \text{ mL}}{1} \cdot \frac{1 \text{ L}}{1000 \text{ mL}} = \frac{8}{1000} \text{ L} = 0.008 \text{ L}$$

37. 4.15 L to mL ▪ Count 3 places to the *right* on the conversion line.

$$4.150_\wedge \text{ L} = 4150 \text{ mL}$$

39. 8000 g to kg ■ Count 3 places to the *left* on the conversion line.

$$8{\wedge}000. \text{ g} = 8 \text{ kg}$$

41. 5.2 kg to g ■ Count 3 places to the *right* on the conversion line.

$$5.200{\wedge} \text{ kg} = 5200 \text{ g}$$

43. 0.85 g to mg ■ Count 3 places to the *right* on the conversion line.

$$0.850{\wedge} \text{ g} = 850 \text{ mg}$$

45. 30,000 mg to g ■ Count 3 places to the *left* on the conversion line.

$$30{\wedge}000. \text{ mg} = 30 \text{ g}$$

47. 598 mg to g ■ Count 3 places to the *left* on the conversion line.

$${\wedge}598. \text{ mg} = 0.598 \text{ g}$$

49. 60 mL to L

$$\frac{60 \text{ mL}}{1} \cdot \frac{1 \text{ L}}{1000 \text{ mL}} = \frac{60}{1000} \text{ L} = 0.06 \text{ L}$$

51. 3 g to kg

$$\frac{3 \text{ g}}{1} \cdot \frac{1 \text{ kg}}{1000 \text{ g}} = \frac{3}{1000} \text{ kg} = 0.003 \text{ kg}$$

53. 0.99 L to mL

$$\frac{0.99 \text{ L}}{1} \cdot \frac{1000 \text{ mL}}{1 \text{ L}} = 990 \text{ mL}$$

55. The masking tape is 19 <u>mm</u> wide. (Length is measured in mm, cm, m, or km.)

57. Buy a 60 <u>mL</u> jar of acrylic paint for art class. (Paint is a liquid and liquids are measured in mL or L.)

59. My waist measurement is 65 <u>cm</u>. (Length is measured in mm, cm, m, or km.)

61. A single postage stamp weighs 90 <u>mg</u>. (Weight is measured in mg, g, or kg.)

63. Convert 300 mL to L. Count 3 places to the *left* on the conversion line.

$${\wedge}300. \text{ mL} = 0.3 \text{ L}$$

Each day 0.3 L of sweat is released.

65. Convert 1.34 kg to g. Count 3 places to the *right* on the conversion line.

$$1.340{\wedge} \text{ kg} = 1340 \text{ g}$$

The average weight of a human brain is 1340 g.

Convert 5 kg to g. Count 3 places to the *right* on the conversion line.

$$5.000{\wedge} \text{ kg} = 5000 \text{ g}$$

An adult human's skin weighs about 5000 g.

67. Convert 900 mL to L.

$$\frac{900 \text{ mL}}{1} \cdot \frac{1 \text{ L}}{1000 \text{ mL}} = \frac{900}{1000} \text{ L} = 0.9 \text{ L}$$

On average, we breathe in and out roughly 0.9 L of air every 10 seconds.

69. Convert 3000 g to kg and 4000 g to kg.

$$\frac{3000 \text{ g}}{1} \cdot \frac{1 \text{ kg}}{1000 \text{ g}} = \frac{3000}{1000} \text{ kg} = 3 \text{ kg}$$

$$\frac{4000 \text{ g}}{1} \cdot \frac{1 \text{ kg}}{1000 \text{ g}} = \frac{4000}{1000} \text{ kg} = 4 \text{ kg}$$

A small adult cat weighs from 3 kg to 4 kg.

71. There are 1000 milligrams in a gram so 1005 mg is *greater* than 1 g. The difference in weight is 1005 mg − 1000 mg = 5 mg or 1.005 g − 1 g = 0.005 g.

73. Convert 1 kg to g.

$$\frac{1 \text{ kg}}{1} \cdot \frac{1000 \text{ g}}{1 \text{ kg}} = 1000 \text{ g}$$

Divide the 1000 g by the weight of 1 nickel.

$$\frac{1000 \text{ g}}{5 \text{ g}} = 200$$

There are 200 nickels in 1 kg of nickels.

Relating Concepts (Exercises 75–78)

75. **(a)** 1 Mm = <u>1,000,000</u> m

(b) 3.5 Mm to m

$$\frac{3.5 \text{ Mm}}{1} \cdot \frac{1,000,000 \text{ m}}{1 \text{ Mm}} = 3,500,000 \text{ m}$$

76. **(a)** 1 Gm = <u>1,000,000,000</u> m

(b) 2500 m to Gm

$$\frac{2500 \text{ m}}{1} \cdot \frac{1 \text{ Gm}}{1,000,000,000 \text{ m}} = 0.0000025 \text{ Gm}$$

77. (a) 1 Tm = 1,000,000,000,000 m

(b) $\dfrac{1 \text{ Tm}}{1} \cdot \dfrac{1,000,000,000,000 \text{ m}}{1 \text{ Tm}} \cdot \dfrac{1 \text{ Gm}}{1,000,000,000 \text{ m}}$

$= \dfrac{1,000,000,000,000}{1,000,000,000} \text{ Gm} = 1000 \text{ Gm}$

So 1 Tm = 1000 Gm.

$\dfrac{1 \text{ Tm}}{1} \cdot \dfrac{1,000,000,000,000 \text{ m}}{1 \text{ Tm}} \cdot \dfrac{1 \text{ Mm}}{1,000,000 \text{ m}}$

$= \dfrac{1,000,000,000,000}{1,000,000} \text{ Mm} = 1,000,000 \text{ Mm}$

So 1 Tm = 1,000,000 Mm.

78. 1 MB = 1,000,000 bytes

1 GB = 1,000,000,000 bytes

$2^{20} = 1,048,576$

$2^{30} = 1,073,741,824$

Summary Exercises
U.S. and Metric Measurement Units

1. **U.S. Measurement Units and Abbreviations**

Length	Weight	Capacity
inch (in.)	ounce (oz)	fluid ounce (fl oz)
foot (ft)	pound (lb)	cup (c)
yard (yd)	ton (ton)	pint (pt)
mile (mi)		quart (qt)
		gallon (gal)

3. (a) 1 ft = 12 in.

(b) 3 ft = 1 yd

(c) 1 mi = 5280 ft

5. (a) 1 cup = 8 fl oz

(b) 4 qt = 1 gal

(c) 2 pt = 1 qt

7. My water bottle holds 450 mL.

9. The child weighed 23 kg.

11. The red pen is 14 cm long.

13. Merlene made 12 L of fruit punch for her daughter's birthday party.

15. 45 cm to meters

$\dfrac{45 \text{ cm}}{1} \cdot \dfrac{1 \text{ m}}{100 \text{ cm}} = \dfrac{45}{100} \text{ m} = 0.45 \text{ m}$

17. 0.6 L to milliliters

$\dfrac{0.6 \text{ L}}{1} \cdot \dfrac{1000 \text{ mL}}{1 \text{ L}} = 600 \text{ mL}$

19. 300 mm to centimeters

$\dfrac{300 \text{ mm}}{1} \cdot \dfrac{1 \text{ cm}}{10 \text{ mm}} = \dfrac{300}{10} \text{ cm} = 30 \text{ cm}$

21. 50 mL to liters ▪ Count 3 places to the *left* on the conversion line.

$_\wedge 050. \text{ mL} = 0.050 \text{ or } 0.05 \text{ L}$

Note: You could also use unit fractions.

23. 7.28 kg to grams ▪ Count 3 places to the *right* on the conversion line.

$7.280_\wedge \text{ kg} = 7280 \text{ g}$

25. 9 g to kilograms

$\dfrac{9 \text{ g}}{1} \cdot \dfrac{1 \text{ kg}}{1000 \text{ g}} = \dfrac{9}{1000} \text{ kg} = 0.009 \text{ kg}$

27. 2.72 m to centimeters ▪ Count two places to the *right* on the conversion line.

$2.72_\wedge \text{ m} = 272 \text{ cm};$

Now convert 2.72 m to millimeters. Count three places to the *right* on the conversion line.

$2.720_\wedge \text{ m} = 2720 \text{ mm}$

29. The tallest person is 2.72 m tall and the second tallest person is 268 cm tall. Since

$2_\wedge 68. \text{ cm} = 2.68 \text{ m},$

the difference is $2.72 - 2.68 = 0.04$ m.

Now convert 0.04 m to cm. Count two places to the *right* on the conversion line.

$0.04_\wedge \text{ m} = 4 \text{ cm};$

Finally, convert 0.04 m to mm. Count three places to the *right* on the conversion line.

$0.040_\wedge \text{ m} = 40 \text{ mm}$

31. We are given dollars and ounces, so we need to use a unit fraction involving ounces and pounds to get an answer in dollars and pounds.

$\dfrac{\$3.89}{\underset{3}{12} \text{ oz}} \cdot \dfrac{\overset{4}{16} \text{ oz}}{1 \text{ lb}} = \dfrac{\$15.56}{3 \text{ lb}} \approx \$5.19/\text{lb (rounded)}$

The price per pound to the nearest cent was $5.19 (rounded).

8.4   Problem Solving with Metric Measurement

8.4   Margin Exercises

1. ***Step 1***   The problem asks for the cost of 75 <u>cm</u> of ribbon.

    Step 2   The price is $0.89 per <u>meter</u>, but the length wanted is cm. Convert cm to <u>m</u>.

    Step 3   There are 100 cm in a meter, so 75 cm is $\frac{3}{4}$ of a meter. Round the cost of 1 m from $0.89 to $0.88 so that it is divisible by 4.

    An estimate is: $\frac{3}{4} \cdot \$0.88 = \$0.66$

    Step 4   Convert 75 cm to m: 75 cm = <u>0.75 m</u>

    Find the cost.

    $$\frac{0.75 \; \cancel{m}}{1} \cdot \frac{\$0.89}{1 \; \cancel{m}} = \$0.6675 \approx \underline{\$0.67} \text{ (rounded)}$$

    Step 5   The cost for 75 cm of ribbon is <u>$0.67</u> (rounded).

    Step 6   The rounded answer, $0.67, is close to our estimate of $0.66.

2. Convert 1.2 grams to milligrams.

    $$1.200_\wedge \; g = 1200 \text{ mg}$$

    Divide by the number of doses.

    $$\frac{1200 \text{ mg}}{3 \text{ doses}} = 400 \text{ mg/dose}$$

    Each dose should be 400 mg.

3. Convert centimeters to meters.

    $$2 \text{ m } 35 \text{ cm} = 2.35 \text{ m}$$
    $$1 \text{ m } 85 \text{ cm} = 1.85 \text{ m}$$

    Add the measurements of the fabric.

one piece	2.35 m
other piece	+ 1.85 m
	4.20 m

    Andrea has 4.2 m of fabric.

8.4   Section Exercises

1. An important thing to include in the answer is the units, choice **(c)**.

3. Convert 850 g to <u>kg</u>.

    $$\frac{850 \; \cancel{g}}{1} \cdot \frac{1 \text{ kg}}{1000 \; \cancel{g}} = 0.85 \text{ kg}$$

    $$\frac{\$0.98}{1 \; \cancel{kg}} \cdot \frac{0.85 \; \cancel{kg}}{1} = \$0.833 \approx \$0.83$$

    She will pay $0.83 (rounded) for the rice.

5. Convert 500 g to kg.

    $$\frac{500 \; \cancel{g}}{1} \cdot \frac{1 \text{ kg}}{1000 \; \cancel{g}} = 0.5 \text{ kg}$$

    Find the difference.

    $$90 \text{ kg} - 0.5 \text{ kg} = 89.5 \text{ kg}$$

    The difference in the weights of the two dogs is 89.5 kg.

7. Write 5 L in terms of mL.

    $$5 \text{ L} = \frac{5 \; \cancel{L}}{1} \cdot \frac{1000 \text{ mL}}{1 \; \cancel{L}} = 5000 \text{ mL}$$

    $$\frac{5000 \; \cancel{mL}}{1} \cdot \frac{1 \text{ beat}}{70 \; \cancel{mL}} \approx 71.42857 \text{ beats} \approx 71 \text{ beats}$$

    It takes 71 beats (rounded) to pass all the blood through the heart.

9. Multiply to find the total length in mm, then convert mm to cm.

    $$60 \text{ mm} \cdot 30 = 1800 \text{ mm}$$

    $$\frac{\overset{180}{\cancel{1800}} \; \cancel{mm}}{1} \cdot \frac{1 \text{ cm}}{\underset{1}{\cancel{10}} \; \cancel{mm}} = 180 \text{ cm}$$

    The total length is 180 cm.

    Divide to find the cost per cm.

    $$\frac{\$3.29}{180 \text{ cm}} \approx \$0.0183/\text{cm} \approx \$0.02/\text{cm}$$

    The cost is $0.02/cm (rounded).

11. Convert centimeters to meters, then add the lengths of the two pieces.

2 m 8 cm	=	2.08 m
2 m 95 cm	=	+ 2.95 m
		5.03 m

    The total length of the two boards is 5.03 m.

13. Convert milliliters to liters.

    $$85 \text{ mL} = 0.085 \text{ L}$$

    Multiply by the number of students.

    $$(0.085 \text{ L})(45) = 3.825 \text{ L}$$

    <u>Four</u> one-liter bottles should be ordered.

    To determine how much acid will be left over, subtract.

    $$4 \text{ L} - 3.825 \text{ L} = 0.175 \text{ L, or } 175 \text{ mL}.$$

15. Three cups a day for one week is $3 \cdot 7 = 21$ cups in one week.

    $$\frac{21 \; \cancel{cups}}{1} \cdot \frac{90 \text{ mg}}{1 \; \cancel{cup}} = 1890 \text{ mg}$$

Convert mg to g.

$$\frac{1890 \text{ mg}}{1} \cdot \frac{1 \text{ g}}{1000 \text{ mg}} = 1.89 \text{ g}$$

She will consume 1.89 grams of caffeine in one week.

17. Boston: 106 cm = 1.06 m (212 cm in 2 years)
Lander: 2.60 m = 260 cm (520 cm in 2 years)

(a) In one year, the difference in meters is

$$2.60 - 1.06 = 1.54.$$

(b) In two years, the difference in centimeters is

$$520 - 212 = 308.$$

Relating Concepts (Exercises 19–22)

19. A box of 50 envelopes weighs 255 g. Subtract the packaging weight to find the net weight.

$$\begin{array}{r} 255 \text{ g} \\ - 40 \text{ g} \\ \hline 215 \text{ g} \end{array}$$

There are 50 envelopes. Divide to find the weight of one envelope.

$$\frac{215 \text{ g}}{50 \text{ envelopes}} = \frac{4.3 \text{ g}}{\text{envelope}}$$

One envelope weighs 4.3 grams.
Convert grams to milligrams.

$$\frac{4.3 \text{ g}}{1} \cdot \frac{1000 \text{ mg}}{1 \text{ g}} = 4300 \text{ mg}$$

20. Box of 1000 staples ■ Subtract the weight of the packaging to find the net weight.

$$\begin{array}{r} 350 \text{ g} \\ - 20 \text{ g} \\ \hline 330 \text{ g} \end{array}$$

There are 1000 staples, so divide the net weight by 1000 or move the decimal point 3 places to the left to find the weight of one staple.

$$330 \text{ g} \div 1000 = 0.33 \text{ g}$$

The weight of one staple is 0.33 g or 330 mg.

21. Given that the weight of one sheet of paper is 3000 mg, find the weight of one sheet in grams.

$$3000 \text{ mg} = 3{\scriptstyle\wedge}000. \text{ g} = 3 \text{ g}$$

To find the net weight of the ream of paper, multiply 500 sheets by 3 g per sheet.

$$\frac{500 \text{ sheets}}{1} \cdot \frac{3 \text{ g}}{1 \text{ sheet}} = 1500 \text{ g}$$

The net weight of the paper is 1500 g.
Since the weight of the packaging is 50 g, the total weight is 1500 g + 50 g = 1550 g.

22. Box of 100 small paper clips ■ Given that the weight of the box of clips is 500 mg, use the metric conversion line to find the weight of one paper clip in grams.

$$500 \text{ mg} = {\scriptstyle\wedge}500. \text{ g} = 0.5 \text{ g}$$

The net weight is (0.5 g)(100) = 50 g.
The total weight is 50 g + 5 g = 55 g.

8.5 Metric–U.S. Measurement Conversions and Temperature

8.5 Margin Exercises

1. **(a)** 23 m to yards ■ Look at the "Metric to U.S. Units" side of the table, where you will see that

$$1 \text{ meter} \approx 1.09 \text{ yards}.$$

$$\frac{23 \text{ m}}{1} \cdot \frac{1.09 \text{ yd}}{1 \text{ m}} \approx 25.1 \text{ yd}$$

(b) 40 cm to inches

$$\frac{40 \text{ cm}}{1} \cdot \frac{0.39 \text{ in.}}{1 \text{ cm}} \approx 15.6 \text{ in.}$$

(c) 5 mi to kilometers ■ Look at the "U.S. to Metric Units" side of the table, where you will see that

$$1 \text{ mile} \approx 1.61 \text{ kilometers}.$$

$$\frac{5 \text{ mi}}{1} \cdot \frac{1.61 \text{ km}}{1 \text{ mi}} \approx 8.1 \text{ km}$$

(d) 12 in. to centimeters

$$\frac{12 \text{ in.}}{1} \cdot \frac{2.54 \text{ cm}}{1 \text{ in.}} \approx 30.5 \text{ cm}$$

2. **(a)** 17 kg to pounds

From the table, 1 kg ≈ 2.20 lb.

$$\frac{17 \text{ kg}}{1} \cdot \frac{2.20 \text{ lb}}{1 \text{ kg}} = \frac{(17)(2.20 \text{ lb})}{1} \approx 37.4 \text{ lb}$$

(b) 5 L to quarts

$$\frac{5 \text{ L}}{1} \cdot \frac{1.06 \text{ qt}}{1 \text{ L}} \approx 5.3 \text{ qt}$$

(c) 90 g to ounces

$$\frac{90 \text{ g}}{1} \cdot \frac{0.035 \text{ oz}}{1 \text{ g}} \approx 3.2 \text{ oz}$$

(d) 3.5 gal to liters

$$\frac{3.5 \text{ gal}}{1} \cdot \frac{3.79 \text{ L}}{1 \text{ gal}} \approx 13.3 \text{ L}$$

(e) 145 lb to kilograms

$$\frac{145 \text{ lb}}{1} \cdot \frac{0.45 \text{ kg}}{1 \text{ lb}} \approx 65.3 \text{ kg}$$

(f) 8 oz to grams

$$\frac{8 \cancel{oz}}{1} \cdot \frac{28.35 \text{ g}}{1 \cancel{oz}} \approx 226.8 \text{ g}$$

3. **(a)** Set the living room thermostat at: 21 °C

(b) The baby has a fever of: 39 °C

(c) Wear a sweater outside because it's: 15 °C

(d) My iced tea is: 5 °C

(e) Let's go swimming! It's: 35 °C

(f) Inside a refrigerator (not the freezer) it's: 3 °C

(g) There is a blizzard outside. It's: −20 °C

(h) I need hot water to get these clothes clean. The water should be: 55 °C

4. Use $C = \dfrac{5(F - 32)}{9}$.

(a) 59 °F

$$C = \frac{5(59 - 32)}{9} = \frac{5 \cdot \overset{3}{\cancel{27}}}{\underset{1}{\cancel{9}}} = 15$$

59 °F = 15 °C

(b) 20 °F

$$C = \frac{5(20 - 32)}{9} = \frac{5 \cdot (-12)}{9}$$

$$= \frac{-5 \cdot \overset{1}{\cancel{3}} \cdot 4}{\underset{1}{\cancel{3}} \cdot 3} = \frac{-20}{3} \approx -7$$

20 °F ≈ −7 °C (rounded)

(c) 212 °F

$$C = \frac{5(212 - 32)}{9} = \frac{5 \cdot \overset{20}{\cancel{180}}}{\underset{1}{\cancel{9}}} = 100$$

212 °F = 100 °C

(d) 98.6 °F

$$C = \frac{5(98.6 - 32)}{9} = \frac{5 \cdot \overset{7.4}{\cancel{66.6}}}{\underset{1}{\cancel{9}}} = 37$$

98.6 °F = 37 °C

5. Use $F = \dfrac{9C}{5} + 32$.

(a) 100 °C

$$F = \frac{9 \cdot \overset{20}{\cancel{100}}}{\underset{1}{\cancel{5}}} + 32 = 180 + 32 = 212$$

100 °C = 212 °F

(b) −25 °C

$$F = \frac{-9 \cdot \overset{5}{\cancel{25}}}{\underset{1}{\cancel{5}}} + 32 = -45 + 32 = -13$$

−25 °C = −13 °F

(c) 32 °C

$$F = \frac{9 \cdot 32}{5} + 32 = 57.6 + 32 = 89.6$$

32 °C ≈ 90 °F (rounded)

(d) −18 °C

$$F = \frac{-9 \cdot 18}{5} + 32 = -32.4 + 32 = -0.4$$

−18 °C ≈ 0 °F (rounded)

8.5 Section Exercises

1. **(a)** On the "Metric to U.S." side of the table, we see that 1 kilogram ≈ 2.20 pounds. So two unit fractions using that information are

$$\frac{1 \text{ kg}}{2.20 \text{ lb}} \quad \text{and} \quad \frac{2.20 \text{ lb}}{1 \text{ kg}}.$$

(b) When converting 5 kg to pounds, we want to eliminate <u>kg</u>, so use the unit fraction with kg in the <u>denominator</u>.

3. 20 m to yards

$$\frac{20 \cancel{m}}{1} \cdot \frac{1.09 \text{ yd}}{1 \cancel{m}} \approx 21.8 \text{ yd}$$

5. 80 m to feet

$$\frac{80 \cancel{m}}{1} \cdot \frac{3.28 \text{ ft}}{1 \cancel{m}} \approx 262.4 \text{ ft}$$

7. 16 ft to meters

$$\frac{16 \cancel{ft}}{1} \cdot \frac{0.30 \text{ m}}{1 \cancel{ft}} \approx 4.8 \text{ m}$$

9. 150 g to ounces

$$\frac{150 \cancel{g}}{1} \cdot \frac{0.035 \text{ oz}}{1 \cancel{g}} \approx 5.3 \text{ oz}$$

11. 248 lb to kilograms

$$\frac{248 \cancel{lb}}{1} \cdot \frac{0.45 \text{ kg}}{1 \cancel{lb}} \approx 111.6 \text{ kg}$$

13. 28.6 L to quarts

$$\frac{28.6 \cancel{L}}{1} \cdot \frac{1.06 \text{ qt}}{1 \cancel{L}} \approx 30.3 \text{ qt}$$

15. **(a)** Convert 5 g to ounces.

$$\frac{5 \cancel{g}}{1} \cdot \frac{0.035 \text{ oz}}{1 \cancel{g}} \approx 0.2 \text{ oz}$$

About 0.2 oz (rounded) of gold was used.

(b) The extra weight probably did not slow him down.

17. Convert 8.4 gal to liters.

$$\frac{8.4 \text{ gal}}{1} \cdot \frac{3.79 \text{ L}}{1 \text{ gal}} \approx 31.8 \text{ L}$$

The dishwasher uses about 31.8 L (rounded).

19. Convert 0.5 in. to centimeters.

$$\frac{0.5 \text{ in.}}{1} \cdot \frac{2.54 \text{ cm}}{1 \text{ in.}} \approx 1.3$$

The dwarf gobie is about 1.3 cm (rounded) long.

21. A snowy day   ■ $-8 \,°\text{C}$ is the most reasonable temperature. $12 \,°\text{C}$ and $28 \,°\text{C}$ are temperatures well above freezing.

23. A high fever   ■ $40 \,°\text{C}$ is the most reasonable temperature because normal body temperature is about $37 \,°\text{C}$.

25. Oven temperature   ■ $150 \,°\text{C}$ is the most reasonable temperature. $50 \,°\text{C}$ is the temperature of hot bath water, which is clearly not hot enough for an oven temperature.

27. $60 \,°\text{F}$ ■   Use $C = \dfrac{5(F - 32)}{9}$.

$$C = \frac{5(60 - 32)}{9} = \frac{5 \cdot 28}{9} = \frac{140}{9} \approx 15.6 \approx 16$$

$60 \,°\text{F} \approx 16 \,°\text{C}$ (rounded)

29. $-4 \,°\text{F}$ ■   Use $C = \dfrac{5(F - 32)}{9}$.

$$C = \frac{5(-4 - 32)}{9} = \frac{5(-36)}{9}$$

$$= \frac{-5 \cdot 4 \cdot \overset{1}{\cancel{9}}}{\underset{1}{\cancel{9}}} = -20$$

$-4 \,°\text{F} = -20 \,°\text{C}$

31. $8 \,°\text{C}$ ■   Use $F = \dfrac{9C}{5} + 32$.

$$F = \frac{9 \cdot 8}{5} + 32 = \frac{72}{5} + 32 = 14.4 + 32 = 46.4$$

$8 \,°\text{C} \approx 46 \,°\text{F}$ (rounded)

33. $-5 \,°\text{C}$ ■   Use $F = \dfrac{9C}{5} + 32$.

$$F = \frac{9(-5)}{5} + 32 = \frac{-9 \cdot \overset{1}{\cancel{5}}}{\underset{1}{\cancel{5}}} + 32 = -9 + 32 = 23$$

$-5 \,°\text{C} = 23 \,°\text{F}$

35. $136 \,°\text{F}$ (highest)

$$C = \frac{5(136 - 32)}{9} = \frac{5 \cdot 104}{9} = \frac{520}{9} \approx 57.8$$

$136 \,°\text{F} \approx 58 \,°\text{C}$

$-129 \,°\text{F}$ (lowest)

$$C = \frac{5(-129 - 32)}{9} = \frac{5(-161)}{9} \approx -89$$

$-129 \,°\text{F} \approx -89 \,°\text{C}$

37. (a) Since the comfort range of the boots is from $24 \,°\text{C}$ to $4 \,°\text{C}$, you would wear these boots in pleasant weather—above freezing, but not hot.

(b) Change $24 \,°\text{C}$ to Fahrenheit.

$$F = \frac{9 \cdot C}{5} + 32 = \frac{9 \cdot 24}{5} + 32 = \frac{216}{5} + 32$$
$$= 43.2 + 32 = 75.2$$

Thus, $24 \,°\text{C} \approx 75 \,°\text{F}$ (rounded).

Change $4 \,°\text{C}$ to Fahrenheit.

$$F = \frac{9 \cdot C}{5} + 32 = \frac{9 \cdot 4}{5} + 32 = \frac{36}{5} + 32$$
$$= 7.2 + 32 = 39.2$$

Thus, $4 \,°\text{C} \approx 39 \,°\text{F}$ (rounded).

The boots are designed for Fahrenheit temperatures of about $75 \,°\text{F}$ to about $39 \,°\text{F}$.

(c) The range of metric temperatures in January would depend on where you live. In Minnesota, it's $0 \,°\text{C}$ to $-40 \,°\text{C}$, and in California it's $24 \,°\text{C}$ to $0 \,°\text{C}$.

39. $10 \,°\text{C}$ (low)

$$F = \frac{9 \cdot \overset{2}{\cancel{10}}}{\underset{1}{\cancel{5}}} + 32 = 18 + 32 = 50$$

$40 \,°\text{C}$ (high)

$$F = \frac{9 \cdot \overset{8}{\cancel{40}}}{\underset{1}{\cancel{5}}} + 32 = 72 + 32 = 104$$

These temperatures are $50 \,°\text{F}$ and $104 \,°\text{F}$ in the U.S. system.

Relating Concepts (Exercises 41–48)

41. Length of model plane

$$\frac{6 \text{ ft}}{1} \cdot \frac{0.30 \text{ m}}{1 \text{ ft}} \approx 6(0.30 \text{ m}) = 1.8 \text{ m (rnd)}$$

42. Weight of plane

$$\frac{11 \text{ lb}}{1} \cdot \frac{0.45 \text{ kg}}{1 \text{ lb}} \approx 11(0.45 \text{ kg})$$
$$= 4.95 \text{ kg} \approx 5.0 \text{ kg (rnd)}$$

43. Length of flight path

$$\frac{1888.3 \text{ mi}}{1} \cdot \frac{1.61 \text{ km}}{1 \text{ mi}} \approx 1888.3(1.61 \text{ km})$$
$$\approx 3040.2 \text{ km (rnd)}$$

44. Time of flight = 38 hr 23 min

45. Cruising altitude

$$\frac{1000 \text{ ft}}{1} \cdot \frac{0.30 \text{ m}}{1 \text{ ft}} \approx 1000(0.30 \text{ m}) = 300 \text{ m (rnd)}$$

46. Fuel at the start

$$\frac{1 \text{ gal}}{1} \cdot \frac{3.79 \text{ L}}{1 \text{ gal}} \cdot \frac{1000 \text{ mL}}{1 \text{ L}} \approx (3.79)(1000 \text{ mL})$$
$$= 3790 \text{ mL}$$

Since the plane carried "less than a gallon of fuel," there would be fewer than 3790 mL.

47. Fuel left after landing

$$\frac{\overset{1}{2 \text{ fl oz}}}{1} \cdot \frac{1 \text{ qt}}{\underset{16}{32 \text{ fl oz}}} \cdot \frac{0.95 \text{ L}}{1 \text{ qt}} \cdot \frac{1000 \text{ mL}}{1 \text{ L}}$$
$$= \frac{950}{16} = 59.375 \approx 59.4 \text{ mL (rnd)}$$

48. The plane started with less than a gallon of fuel. Convert that amount to fluid ounces.

$$\frac{1 \text{ gal}}{1} \cdot \frac{4 \text{ qt}}{1 \text{ gal}} \cdot \frac{32 \text{ fl oz}}{1 \text{ qt}} = 128 \text{ fl oz}$$

It had less than 2 fluid ounces of fuel left. Divide to determine the *percent* of the fuel that was left at the end of the flight.

$$\frac{2 \text{ fl oz}}{128 \text{ fl oz}} = \frac{1}{64} = 0.015625 \approx 1.6\% \text{ (rnd)}$$

Chapter 8 Review Exercises

1. 1 lb = <u>16</u> oz

2. <u>3</u> ft = 1 yd

3. 1 ton = <u>2000</u> lb

4. <u>4</u> qt = 1 gal

5. 1 hr = <u>60</u> min

6. 1 c = <u>8</u> fl oz

7. <u>60</u> sec = 1 min

8. <u>5280</u> ft = 1 mi

9. <u>12</u> in. = 1 ft

10. 4 ft to inches

$$\frac{4 \text{ ft}}{1} \cdot \frac{12 \text{ in.}}{1 \text{ ft}} = 4 \cdot 12 \text{ in.} = \underline{48} \text{ in.}$$

11. 6000 lb to tons

$$\frac{\overset{3}{6000 \text{ lb}}}{1} \cdot \frac{1 \text{ ton}}{\underset{1}{2000 \text{ lb}}} = \underline{3} \text{ tons}$$

12. 64 oz to pounds

$$\frac{\overset{4}{64 \text{ oz}}}{1} \cdot \frac{1 \text{ lb}}{\underset{1}{16 \text{ oz}}} = \underline{4} \text{ lb}$$

13. 18 hr to days

$$\frac{\overset{3}{18 \text{ hr}}}{1} \cdot \frac{1 \text{ day}}{\underset{4}{24 \text{ hr}}} = \frac{3}{4} \text{ or 0.75 day}$$

14. 150 min to hours

$$\frac{\overset{5}{150 \text{ min}}}{1} \cdot \frac{1 \text{ hr}}{\underset{2}{60 \text{ min}}} = \frac{5}{2} \text{ hr} = 2\frac{1}{2} \text{ or 2.5 hr}$$

15. $1\frac{3}{4}$ lb to ounces

$$\frac{1\frac{3}{4} \text{ lb}}{1} \cdot \frac{16 \text{ oz}}{1 \text{ lb}} = \frac{7}{4} \cdot \frac{\overset{4}{16}}{1} \text{ oz} = \underline{28} \text{ oz}$$

16. $6\frac{1}{2}$ ft to inches

$$\frac{6\frac{1}{2} \text{ ft}}{1} \cdot \frac{12 \text{ in.}}{1 \text{ ft}} = \frac{13}{\underset{1}{2}} \cdot \frac{\overset{6}{12}}{1} \text{ in.} = \underline{78} \text{ in.}$$

17. 7 gal to cups

$$\frac{7 \text{ gal}}{1} \cdot \frac{4 \text{ qt}}{1 \text{ gal}} \cdot \frac{2 \text{ pt}}{1 \text{ qt}} \cdot \frac{2 \text{ c}}{1 \text{ pt}} = 7 \cdot 4 \cdot 2 \cdot 2 \text{ c} = \underline{112} \text{ c}$$

18. 4 days to seconds

$$\frac{4 \text{ days}}{1} \cdot \frac{24 \text{ hr}}{1 \text{ day}} \cdot \frac{60 \text{ min}}{1 \text{ hr}} \cdot \frac{60 \text{ sec}}{1 \text{ min}}$$
$$= 4 \cdot 24 \cdot 60 \cdot 60 \text{ sec}$$
$$= \underline{345,600} \text{ sec}$$

19. (a) 12,460 ft to yards

$$\frac{12,460 \text{ ft}}{1} \cdot \frac{1 \text{ yd}}{3 \text{ ft}} = \frac{12,460}{3} \text{ yd}$$
$$= 4153\frac{1}{3} \text{ yd (exact)}$$
$$\text{or } 4153.3 \text{ yd (rounded)}$$

The average depth is $4153\frac{1}{3}$ yd (exact) or 4153.3 yd (rounded).

(b) 12,460 ft to miles

$$\frac{12,460 \text{ ft}}{1} \cdot \frac{1 \text{ mi}}{5280 \text{ ft}} \approx 2.36 \text{ mi} \approx 2.4 \text{ mi}$$

The average depth is 2.4 mi (rounded).

20. *Step 1* The problem asks for the amount of money the company made.

Step 2 Convert pounds to tons. Then multiply to find the total amount.

Step 3   To estimate, round 123,260 lb to 100,000 lb. Then, there are 2000 lb in a ton, so 100,000 lb is 50 tons. So $50 \cdot \$40 = \$2000$ is our estimate.

Step 4 $\dfrac{123,260 \text{ lb}}{1} \cdot \dfrac{1 \text{ ton}}{2000 \text{ lb}} = 61.63$ tons

$61.63 \cdot \$40 = \$2465.20$

Step 5   The company made $2465.20.

Step 6   The answer, $2465.20, is close to our estimate of $2000.

21. My thumb is 20 <u>mm</u> wide.

22. Her waist measurement is 66 <u>cm</u>.

23. The two towns are 40 <u>km</u> apart.

24. A basketball court is 30 <u>m</u> long.

25. The height of the picnic bench is 45 <u>cm</u>.

26. The eraser on the end of my pencil is 5 <u>mm</u> long.

27. 5 m to cm ▪ Count 2 places to the *right* on the metric conversion line.

$$5.00_\wedge \text{ m} = 500 \text{ cm}$$

28. 8.5 km to m

$$\dfrac{8.5 \text{ km}}{1} \cdot \dfrac{1000 \text{ m}}{1 \text{ km}} = 8500 \text{ m}$$

29. 85 mm to cm ▪ Count 1 place to the *left* on the metric conversion line.

$$8_\wedge 5. \text{ mm} = 8.5 \text{ cm}$$

30. 370 cm to m

$$\dfrac{370 \text{ cm}}{1} \cdot \dfrac{1 \text{ m}}{100 \text{ cm}} = 3.7 \text{ m}$$

31. 70 m to km
Count 3 places to the *left* on the metric conversion line.

$$_\wedge 070. \text{ m} = 0.07 \text{ km}$$

32. 0.93 m to mm

$$\dfrac{0.93 \text{ m}}{1} \cdot \dfrac{1000 \text{ mm}}{1 \text{ m}} = 930 \text{ mm}$$

33. The eyedropper holds 1 <u>mL</u>.

34. I can heat 3 <u>L</u> of water in this pan.

35. Loretta's hammer weighed 650 g.

36. Yongshu's large suitcase weighed 20 <u>kg</u> when it was packed.

37. My fish tank holds 80 <u>L</u> of water.

38. I'll buy the 500 <u>mL</u> bottle of mouthwash.

39. Mara took a 200 <u>mg</u> antibiotic pill.

40. This piece of chicken weighs 100 g.

41. 5000 mL to L ▪ Count 3 places to the *left* on the metric conversion line.

$$5_\wedge 000. \text{ mL} = 5 \text{ L}$$

42. 8 L to mL ▪ Count 3 places to the *right* on the metric conversion line.

$$8.000_\wedge \text{ L} = 8000 \text{ mL}$$

43. 4.58 g to mg

$$\dfrac{4.58 \text{ g}}{1} \cdot \dfrac{1000 \text{ mg}}{1 \text{ g}} = 4580 \text{ mg}$$

44. 0.7 kg to g

$$\dfrac{0.7 \text{ kg}}{1} \cdot \dfrac{1000 \text{ g}}{1 \text{ kg}} = 700 \text{ g}$$

45. 6 mg to g ▪ Count 3 places to the *left* on the metric conversion line.

$$_\wedge 006. \text{ mg} = 0.006 \text{ g}$$

46. 35 mL to L ▪ Count 3 places to the *left* on the metric conversion line.

$$_\wedge 035. \text{ mL} = 0.035 \text{ L}$$

47. Convert milliliters to liters.

$$\dfrac{180 \text{ mL}}{1} \cdot \dfrac{1 \text{ L}}{1000 \text{ mL}} = \dfrac{180}{1000} \text{ L} = 0.18 \text{ L}$$

Multiply 0.18 L by the number of servings.
$(0.18 \text{ L})(175) = 31.5 \text{ L}$

For 175 servings, 31.5 L of punch are needed.

48. Convert kilograms to grams
$10 \text{ kg} = 10{,}000 \text{ g}$
Divide by the number of people.

$$\dfrac{10{,}000 \text{ g}}{28 \text{ people}} \approx 357$$

Jason is allowing 357 g (rounded) of turkey for each person.

49. Convert grams to kilograms.
Using the metric conversion line,

$$4 \text{ kg } 750 \text{ g} = 4.75 \text{ kg.}$$

Subtract the weight loss from his original weight.

$$\begin{array}{r} 92.00 \text{ kg} \\ - \ 4.75 \text{ kg} \\ \hline 87.25 \text{ kg} \end{array}$$

Yerald weighs 87.25 kg.

50. Convert 950 g to kilograms.
Using the metric conversion line,

$$950 \text{ g} = 0.95 \text{ kg}.$$

Multiply by the price per kilogram.

$$
\begin{array}{r}
\$1.49 \quad \leftarrow \textit{2 decimal places} \\
\times\, 0.95 \quad \leftarrow \textit{2 decimal places} \\
\hline
745 \\
1\,341 \\
\hline
\$1.4155 \quad \leftarrow \textit{4 decimal places}
\end{array}
$$

Young-Mi paid $1.42 (rounded) for the onions.

51. 6 m to yards

$$\frac{6 \text{ m}}{1} \cdot \frac{1.09 \text{ yd}}{1 \text{ m}} \approx 6.5 \text{ yd (rounded)}$$

52. 30 cm to inches

$$\frac{30 \text{ cm}}{1} \cdot \frac{0.39 \text{ in.}}{1 \text{ cm}} \approx 11.7 \text{ in. (rounded)}$$

53. 108 km to miles

$$\frac{108 \text{ km}}{1} \cdot \frac{0.62 \text{ mi}}{1 \text{ km}} \approx 67.0 \text{ mi (rounded)}$$

54. 800 mi to kilometers

$$\frac{800 \text{ mi}}{1} \cdot \frac{1.61 \text{ km}}{1 \text{ mi}} \approx 1288 \text{ km}$$

55. 23 qt to liters

$$\frac{23 \text{ qt}}{1} \cdot \frac{0.95 \text{ L}}{1 \text{ qt}} \approx 21.9 \text{ L (rounded)}$$

56. 41.5 L to quarts

$$\frac{41.5 \text{ L}}{1} \cdot \frac{1.06 \text{ qt}}{1 \text{ L}} \approx 44.0 \text{ qt (rounded)}$$

57. Water freezes at 0 °C.

58. Water boils at 100 °C.

59. Normal body temperature is about 37 °C.

60. Comfortable room temperature is about 20 °C.

61. 77 °F ▪ Use $C = \frac{5(F-32)}{9}$.

$$C = \frac{5(77-32)}{9} = \frac{5(\overset{5}{\cancel{45}})}{\underset{1}{\cancel{9}}} = 25$$

77 °F = 25 °C

62. 5 °F ▪ Use $C = \frac{5(F-32)}{9}$.

$$C = \frac{5(5-32)}{9} = \frac{5(-27)}{9} = \frac{-5\cdot3\cdot\overset{1}{\cancel{9}}}{\underset{1}{\cancel{9}}} = -15$$

5 °F = −15 °C

63. −2 °C ▪ Use $F = \frac{9C}{5} + 32$.

$$F = \frac{9(-2)}{5} + 32 = \frac{-18}{5} + 32 = -3.6 + 32$$
$$= 28.4$$

−2 °C ≈ 28 °F (rounded)

64. 49 °C ▪ Use $F = \frac{9C}{5} + 32$.

$$F = \frac{9\cdot49}{5} + 32 = 88.2 + 32 = 120.2 \approx 120$$

49 °C ≈ 120 °F (rounded)

65. [8.3] I added 1 L of oil to my car.

66. [8.3] The box of books weighed 15 kg.

67. [8.2] Larry's shoe is 30 cm long.

68. [8.3] Jan used 15 mL of shampoo on her hair.

69. [8.2] My fingernail is 10 mm wide.

70. [8.2] I walked 2 km to school.

71. [8.3] The tiny bird weighed 15 g.

72. [8.2] The new library building is 18 m wide.

73. [8.3] The cookie recipe uses 250 mL of milk.

74. [8.3] Renee's pet mouse weighs 30 g.

75. [8.3] One postage stamp weighs 90 mg.

76. [8.3] I bought 30 L of gas for my car.

77. [8.2] 10.5 cm to millimeters ▪ Count 1 place to the *right* on the metric conversion line.

$$10.5_\wedge \text{ cm} = 105 \text{ mm}$$

78. [8.1] 45 min to hours

$$\frac{\overset{3}{\cancel{45}} \text{ min}}{1} \cdot \frac{1 \text{ hr}}{\underset{4}{\cancel{60}} \text{ min}} = \frac{3}{4} \text{ hr or } 0.75 \text{ hr}$$

79. [8.1] 90 in. to feet

$$\frac{\overset{15}{\cancel{90}} \text{ in.}}{1} \cdot \frac{1 \text{ ft}}{\underset{2}{\cancel{12}} \text{ in.}} = \frac{15}{2} \text{ ft} = 7\frac{1}{2} \text{ ft or } 7.5 \text{ ft}$$

80. [8.2] 1.3 m to centimeters ▪ Count 2 places to the *right* on the metric conversion line.

$$1.30_\wedge \text{ m} = 130 \text{ cm}$$

81. [8.5] 25 °C to Fahrenheit

$$F = \frac{9\cdot\overset{5}{\cancel{25}}}{\underset{1}{\cancel{5}}} + 32 = 45 + 32 = 77$$

25 °C = 77 °F

82. **[8.1]** $3\frac{1}{2}$ gal to quarts

$$\frac{3\frac{1}{2}\text{ gal}}{1}\cdot\frac{4\text{ qt}}{1\text{ gal}}=\frac{7}{\underset{1}{2}}\cdot\frac{\overset{2}{\cancel{4}}}{1}\text{ qt}=14\text{ qt}$$

83. **[8.3]** 700 mg to grams   ▪ Count 3 places to the *left* on the metric conversion line.

$$_{\wedge}700.\text{ mg}=0.7\text{ g}$$

84. **[8.3]** 0.81 L to milliliters   ▪ Count 3 places to the *right* on the metric conversion line.

$$0.810_{\wedge}\text{ L}=810\text{ mL}$$

85. **[8.1]** 5 lb to ounces

$$\frac{5\text{ lb}}{1}\cdot\frac{16\text{ oz}}{1\text{ lb}}=5\cdot16\text{ oz}=80\text{ oz}$$

86. **[8.3]** 60 kg to g

$$\frac{60\text{ kg}}{1}\cdot\frac{1000\text{ g}}{1\text{ kg}}=60{,}000\text{ g}$$

87. **[8.3]** 1.8 L to milliliters   ▪ Count 3 places to the *right* on the metric conversion line.

$$1.800_{\wedge}\text{ L}=1800\text{ mL}$$

88. **[8.5]** 86 °F to Celsius

$$C=\frac{5(86-32)}{9}=\frac{5(\overset{6}{\cancel{54}})}{\underset{1}{\cancel{9}}}=30$$

86 °F = 30 °C

89. **[8.2]** 0.36 m to centimeters   ▪ Count 2 places to the *right* on the metric conversion line.

$$0.36_{\wedge}\text{ m}=36\text{ cm}$$

90. **[8.3]** 55 mL to liters   ▪ Count 3 places to the *left* on the metric conversion line.

$$_{\wedge}055.\text{ mL}=0.055\text{ L}$$

91. **[8.4]** Convert centimeters to meters.

$$\begin{aligned}2\text{ m }4\text{ cm}&=2+0.04\text{ m}=&2.04\text{ m}\\78\text{ cm}&=&\underline{-0.78\text{ m}}\\&&1.26\text{ m}\end{aligned}$$

The board is 1.26 m long.

92. **[8.1]** 3000 lb to tons

$$\frac{3000\text{ lb}}{1}\cdot\frac{1\text{ ton}}{2000\text{ lb}}=\frac{3}{2}\text{ T}=1.5\text{ tons}$$

Multiply by 12 days.

$$(1.5\text{ tons})(12)=18\text{ tons}$$

18 tons of cookies are sold in all.

93. **[8.5]** Convert ounces to grams.

$$\frac{4\text{ oz}}{1}\cdot\frac{28.35\text{ g}}{1\text{ oz}}\approx113\text{ g (rounded)}$$

Convert 350 °F to Celsius.

$$\begin{aligned}C&=\frac{5(350-32)}{9}\\&=\frac{5(318)}{9}=\frac{1590}{9}\approx177\text{ °C (rounded)}\end{aligned}$$

94. **[8.5]** Convert kilograms to pounds.

$$\frac{80.9\text{ kg}}{1}\cdot\frac{2.20\text{ lb}}{1\text{ kg}}\approx178.0\text{ lb}$$

Convert meters to feet.

$$\frac{1.83\text{ m}}{1}\cdot\frac{3.28\text{ ft}}{1\text{ m}}\approx6.0\text{ ft}$$

Jalo weighs about 178.0 lb and is about 6.0 ft.

95. **[8.5]** 44 ft to meters

$$\frac{44\text{ ft}}{1}\cdot\frac{0.30\text{ m}}{1\text{ ft}}\approx13.2\text{ m}$$

96. **[8.5]** 6.7 m to feet

$$\frac{6.7\text{ m}}{1}\cdot\frac{3.28\text{ ft}}{1\text{ m}}\approx22.0\text{ ft}$$

97. **[8.5]** 4 yd to meters

$$\frac{4\text{ yd}}{1}\cdot\frac{0.91\text{ m}}{1\text{ yd}}\approx3.6\text{ m}$$

98. **[8.5]** 102 cm to inches

$$\frac{102\text{ cm}}{1}\cdot\frac{0.39\text{ in.}}{1\text{ cm}}\approx39.8\text{ in.}$$

99. **[8.5]** 220,000 kg to pounds

$$\frac{220{,}000\text{ kg}}{1}\cdot\frac{2.20\text{ lb}}{1\text{ kg}}\approx484{,}000\text{ lb}$$

100. **[8.5]** 5 gal to liters

$$\frac{5\text{ gal}}{1}\cdot\frac{3.79\text{ L}}{1\text{ gal}}=18.95\approx19.0\text{ L}$$

6 gal to liters

$$\frac{6\text{ gal}}{1}\cdot\frac{3.79\text{ L}}{1\text{ gal}}\approx22.7\text{ L}$$

101. **[8.1]** **(a)** From Exercise 99:

$$\frac{484{,}000\text{ lb}}{1}\cdot\frac{1\text{ ton}}{2000\text{ lb}}=242\text{ tons}$$

(b) From Exercise 97:

$$\frac{4\text{ yd}}{1}\cdot\frac{3\text{ ft}}{1\text{ yd}}\cdot\frac{12\text{ in.}}{1\text{ ft}}=144\text{ in.}$$

102. **[8.2]** **(a)** From Exercise 98:

$$\frac{102 \cancel{\text{cm}}}{1} \cdot \frac{1 \text{ m}}{100 \cancel{\text{cm}}} = 1.02 \text{ m}$$

(b) From Exercise 96:

$$\frac{6.7 \cancel{\text{m}}}{1} \cdot \frac{100 \text{ cm}}{1 \cancel{\text{m}}} = 670 \text{ cm}$$

Chapter 8 Test

1. 9 gal to quarts

$$\frac{9 \cancel{\text{gal}}}{1} \cdot \frac{4 \text{ qt}}{1 \cancel{\text{gal}}} = \underline{36} \text{ qt}$$

2. 45 ft to yards

$$\frac{45 \cancel{\text{ft}}}{1} \cdot \frac{1 \text{ yd}}{3 \cancel{\text{ft}}} = \frac{45}{3} \text{ yd} = \underline{15} \text{ yd}$$

3. 135 min to hours

$$\frac{\overset{27}{\cancel{135}} \text{ min}}{1} \cdot \frac{1 \text{ hr}}{\underset{12}{\cancel{60}} \text{ min}} = \frac{27}{12} \text{ hr} = 2.25 \text{ or } 2\frac{1}{4} \text{ hr}$$

4. 9 in. to feet

$$\frac{\overset{3}{\cancel{9}} \text{ in.}}{1} \cdot \frac{1 \text{ ft}}{\underset{4}{\cancel{12}} \text{ in.}} = \frac{3}{4} \text{ or } 0.75 \text{ ft}$$

5. $3\frac{1}{2}$ lb to ounces

$$\frac{3\frac{1}{2} \cancel{\text{lb}}}{1} \cdot \frac{16 \text{ oz}}{1 \cancel{\text{lb}}} = \frac{7}{\underset{1}{\cancel{2}}} \cdot \frac{\overset{8}{\cancel{16}}}{1} \text{ oz} = \underline{56} \text{ oz}$$

6. 5 days to minutes

$$\frac{5 \cancel{\text{days}}}{1} \cdot \frac{24 \cancel{\text{hr}}}{1 \cancel{\text{day}}} \cdot \frac{60 \text{ min}}{1 \cancel{\text{hr}}} = \frac{5 \cdot 24 \cdot 60}{1} \text{ min}$$
$$= \underline{7200} \text{ min}$$

7. My husband weighs 75 <u>kg</u>.

8. I hiked 5 <u>km</u> this morning.

9. She bought 125 <u>mL</u> of cough syrup.

10. This apple weighs 180 <u>g</u>.

11. This page is about 21 <u>cm</u> wide.

12. My watch band is 10 <u>mm</u> wide.

13. I bought 10 <u>L</u> of soda for the picnic.

14. The bracelet is 16 <u>cm</u> long.

15. 250 cm to meters ▪ Count 2 places to the *left* on the metric conversion line.

$$2_\wedge 50. \text{ cm} = 2.5 \text{ m}$$

16. 4.6 km to meters

$$\frac{4.6 \cancel{\text{km}}}{1} \cdot \frac{1000 \text{ m}}{1 \cancel{\text{km}}} = 4600 \text{ m}$$

17. 5 mm to centimeters ▪ Count 1 place to the *left* on the metric conversion line.

$$_\wedge 5. \text{ mm} = 0.5 \text{ cm}$$

18. 325 mg to grams

$$\frac{325 \cancel{\text{mg}}}{1} \cdot \frac{1 \text{ g}}{1000 \cancel{\text{mg}}} = \frac{325}{1000} \text{ g} = 0.325 \text{ g}$$

19. 16 L to milliliters ▪ Count 3 places to the *right* on the metric conversion line.

$$16.000_\wedge \text{ L} = 16,000 \text{ mL}$$

20. 0.4 kg to grams

$$\frac{0.4 \cancel{\text{kg}}}{1} \cdot \frac{1000 \text{ g}}{1 \cancel{\text{kg}}} = 400 \text{ g}$$

21. 10.55 m to centimeters ▪ Count 2 places to the *right* on the metric conversion line.

$$10.55_\wedge \text{ m} = 1055 \text{ cm}$$

22. 95 mL to liters ▪ Count 3 places to the *left* on the metric conversion line.

$$_\wedge 095. \text{ mL} = 0.095 \text{ L}$$

23. Convert 460 in. to feet.

$$\frac{460 \cancel{\text{in.}}}{1} \cdot \frac{1 \text{ ft}}{12 \cancel{\text{in.}}} = \frac{460}{12} \text{ ft} = \frac{115}{3} \text{ ft}$$

Divide by 12 months

$$\frac{\frac{115}{3} \text{ ft}}{12 \text{ months}} = \frac{115}{3} \cdot \frac{1}{12} \approx 3.2 \text{ ft/month (rounded)}$$

It rains about 3.2 ft per month.

24. **(a)** Convert 1.7 g to mg.

$$\frac{1.7 \cancel{\text{g}}}{1} \cdot \frac{1000 \text{ mg}}{1 \cancel{\text{g}}} = 1700 \text{ mg}$$

Find the difference.

$$1700 \text{ mg} - 310 \text{ mg} = 1390 \text{ mg}$$

It has 1390 mg more sodium.

(b) 1700 mg is more than 1500 mg
1700 mg − 1500 mg = 200 mg

$$\frac{200 \cancel{\text{mg}}}{1} \cdot \frac{1 \text{ g}}{1000 \cancel{\text{mg}}} = 0.2 \text{ g}$$

The Spicy Italian has 200 mg or 0.2 g more sodium than the 1500 mg recommended daily amount.

25. The water is almost boiling. On the Celsius scale, water boils at 100°, so 95 °C would be almost boiling.

26. The tomato plants may freeze tonight. Water freezes at 0 °C, so tomatoes would also likely freeze at 0 °C, so choose 0 °C.

27. 6 ft to meters

$$\frac{6 \cancel{ft}}{1} \cdot \frac{0.30 \text{ m}}{1 \cancel{ft}} \approx 1.8 \text{ m}$$

28. 125 lb to kilograms

$$\frac{125 \cancel{lb}}{1} \cdot \frac{0.45 \text{ kg}}{1 \cancel{lb}} \approx 56.3 \text{ kg (rounded)}$$

29. 50 L to gallons

$$\frac{50 \cancel{L}}{1} \cdot \frac{0.26 \text{ gal}}{1 \cancel{L}} \approx 13 \text{ gal}$$

30. 8.1 km to miles

$$\frac{8.1 \cancel{km}}{1} \cdot \frac{0.62 \text{ mi}}{1 \cancel{km}} \approx 5.0 \text{ mi (rounded)}$$

31. 74 °F to Celsius

$$C = \frac{5(74 - 32)}{9} = \frac{5(\overset{14}{\cancel{42}})}{\underset{3}{\cancel{9}}} = \frac{70}{3} \approx 23 \text{ °C}$$

74 °F ≈ 23 °C (rounded)

32. −12 °C to Fahrenheit

$$F = \frac{9(-12)}{5} + 32 = \frac{-108}{5} + 32$$
$$= -21.6 + 32 = 10.4$$

−12 °C ≈ 10 °F (rounded)

33. Convert 120 cm to meters. Count 2 places to the *left* on the metric conversion line.

$$1 \wedge 20. \text{ cm} = 1.2 \text{ m}$$

Multiply to find the number of meters for 5 pillows.

$$5(1.2 \text{ m}) = 6 \text{ m}$$

Convert 6 m to yards.

$$\frac{6 \cancel{m}}{1} \cdot \frac{1.09 \text{ yd}}{1 \cancel{m}} \approx 6.54 \text{ yd}$$

Multiply by \$3.98 to find the cost.

$$\frac{6.54 \cancel{yd}}{1} \cdot \frac{\$3.98}{\cancel{yd}} \approx \$26.03$$

It will cost about \$26.03.

34. Possible answers: Use same system as rest of the world; easier system for children to learn; less use of fractional numbers; compete internationally.

CHAPTER 9 GRAPHS AND GRAPHING

9.1 Problem Solving with Tables and Pictographs

9.1 Margin Exercises

1. **(a)** Look down the column headed On-Time Performance.

   To find the best performance, look for the highest percent. The highest percent is 93%. Then look to the left to find the name of the airline: Hawaiian.

   (b) Look down the column headed Luggage Handling.

   To find the best record, look for the lowest number. The lowest number is 2.2. Then look to the left to find the airline, which is JetBlue.

   (c) Look down the column headed Luggage Handling and find all the numbers less than or equal to 3. Then look to the left to find the airlines, which are Delta, Hawaiian, and JetBlue.

   (d) Add the values in the Luggage Handling column and divide by 8.

   $$\frac{3.6 + 5.5 + 3.4 + 2.7 + 2.6 + 2.2 + 3.7 + 3.7}{8}$$

   $$= \frac{27.4}{8} \approx 3.4$$

   The average number of luggage problems was 3.4 (rounded) per 1000 passengers.

2. **(a)** The table shows that the price per mile in Chicago is $1.80, so the cost for 6.5 miles is $6.5(\$1.80) = \$11.70$. Then add the flag drop charge: $\$11.70 + \$2.25 = \$13.95$.

   The table shows that the price per mile in San Francisco is $2.75, so the cost for 6.5 miles is $6.5(\$2.75) = \$17.88$ (rounded). Then add the flag drop charge: $\$17.88 + \$3.50 = \$21.38$ (rounded).

   The difference in the fares is

   $$\$21.38 - \$13.95 = \$7.43.$$

   The San Francisco fare is $7.43 higher.

   (b) Price per mile for Denver = $2.25
   Price for 4.5 miles = $4.5(\$2.25) = \$10.13$ (rounded)
   Figure out the cost of the wait time.

   $$\frac{\$24}{60 \text{ min}} = \frac{\$x}{30 \text{ min}}$$
   $$60 \cdot x = 24 \cdot 30$$
   $$\frac{60x}{60} = \frac{720}{60}$$
   $$x = 12$$

Now add the cost of the wait time charge of $12 and the flag drop charge of $2.50.

Total fare = $\$10.13 + \$12 + \$2.50 = \$24.63$

Use the percent equation to find the tip.

$$\text{percent} \cdot \text{whole} = \text{part}$$
$$(0.15)(\$24.63) = n$$
$$\$3.69 \approx n$$

Rounded to the nearest $0.25, the tip is $3.75. The total cost is $\$24.63 + \$3.75 = \$28.38$.

3. **(a)** The population of Chicago is represented by 5 whole symbols and each whole symbol represents 2 million. The approximate population of Chicago is

   $5 \cdot 2$ million = 10 million people (or 10,000,000).

   (b) The population of Atlanta is represented by 2 whole symbols $(2 \cdot 2$ million = 4 million$)$ plus half of a symbol.

   ($\frac{1}{2}$ of 2 million is 1 million) for a total of 5 million people (or 5,000,000).

   (c) Los Angeles has $6\frac{1}{2}$ symbols and New York has $9\frac{1}{2}$ symbols, so Los Angeles has 3 fewer symbols than New York. This is $3 \cdot 2$ million = 6 million people, so Los Angeles' population is 6 million or 6,000,000 less than that of New York.

   (d) Dallas has 3 symbols and Atlanta has $2\frac{1}{2}$ symbols, so Dallas has $\frac{1}{2}$ more symbol than Atlanta. This is $\frac{1}{2} \cdot 2$ million = 1 million people, so Dallas' population is 1 million or 1,000,000 greater than that of Atlanta.

9.1 Section Exercises

1. A table presents data organized into **(b) rows and columns**.

3. An advantage of a pictograph is that you can **(b) easily make comparisons**.

5. **(a)** Look in the points column for Wilt Chamberlain. He scored 31,419 points.

   (b) Look down the points column and find the number(s) greater than 31,419, then read across to the player's name.

   We see that Kareem Abdul-Jabbar, Karl Malone, and Michael Jordan scored more points than Wilt Chamberlain.

7. **(a)** Look down the games column and find the largest number, which is 1560. Then read across to find the player's name.

   Kareem Abdul-Jabbar has been in the greatest number of games.

(b) The smallest number in the games column is 1045.

Wilt Chamberlain has been in the fewest number of games.

9. Look down the points column and find the largest and smallest number. Then find the difference.

$$38{,}387 - 28{,}316 = 10{,}071$$

The difference is 10,071 points.

11. Round answers to the nearest tenth to match other numbers in column.

For Shaquille O'Neal:

$$\begin{array}{l} \text{Total points} \\ \text{Total games} \end{array} \rightarrow \frac{28{,}596}{1207} \approx 23.69 \approx 23.7$$

For Kobe Bryant:

$$\begin{array}{l} \text{Total points} \\ \text{Total games} \end{array} \rightarrow \frac{28{,}316}{1117} \approx 25.35 \approx 25.4$$

13. (a) Look down the 140 Pounds column and across the aerobic dance activity row.

The person will burn 255 calories.

(b) Look down the 140 Pounds column for the largest number. Then read across to the activity. Moderate jogging burns the most calories.

15. (a) Look down the 110 Pounds column for numbers greater than or equal to 200. Then read across to the activities. Moderate jogging, aerobic dance, and racquetball are activities that burn at least 200 calories.

(b) Use the same method as in part (a), except use the 170 Pound column. Moderate jogging, moderate bicycling, aerobic dance, racquetball, and tennis are activities that burn at least 200 calories.

17. 15 minutes is $\frac{1}{2}$ of 30 minutes.
60 minutes is 2 times 30 minutes.
Look down the 140 Pound column.

$$\frac{1}{2} \cdot 180 + 2 \cdot 140 = 90 + 280 = 370$$

The person will burn 370 calories.

For Exercises 19 and 21, other proportions are possible.

19. (a) From the table, a 110-pound person burns 322 calories during 30 minutes of moderate jogging.

$$\frac{322 \text{ calories}}{110\text{-pound person}} = \frac{x \text{ calories}}{125\text{-pound person}}$$
$$110 \cdot x = 322 \cdot 125$$
$$\frac{110x}{110} = \frac{40{,}250}{110}$$
$$x \approx 366$$

The person would burn approximately 366 calories.

(b) $\dfrac{210 \text{ calories}}{110\text{-pound person}} = \dfrac{x \text{ calories}}{125\text{-pound person}}$
$$110 \cdot x = 210 \cdot 125$$
$$\frac{110x}{110} = \frac{26{,}250}{110}$$
$$x \approx 239$$

The person would burn approximately 239 calories.

21. Aerobic dance:

$$\frac{255 \text{ calories}}{140\text{-pound person}} = \frac{x \text{ calories}}{158\text{-pound person}}$$
$$140 \cdot x = 255 \cdot 158$$
$$\frac{140x}{140} = \frac{40{,}290}{140}$$
$$x \approx 288$$

15 min is $\frac{1}{2}$ of 30 min: $\frac{1}{2} \cdot 288 = 144$

Walking:

$$\frac{140 \text{ calories}}{140\text{-pound person}} = \frac{x \text{ calories}}{158\text{-pound person}}$$
$$140 \cdot x = 140 \cdot 158$$
$$x = 158$$

20 min is $\frac{2}{3}$ of 30 min: $\frac{2}{3} \cdot 158 \approx 105$

The difference is about $144 - 105 = 39$ calories.

For Exercises 23–30, each symbol represents 10 million passenger arrivals and departures.

23. (a) Atlanta is represented by 9 symbols.

$9(10 \text{ million}) = 90$ million or $90{,}000{,}000$

(b) New York (JFK) is represented by $4\frac{1}{2}$ or 4.5 symbols.

$4.5(10 \text{ million}) = 45$ million or $45{,}000{,}000$

25. Atlanta: $9(10 \text{ million}) = 90$ million

Chicago: $6.5(10 \text{ million}) = 65$ million

The total number is
90 million $+ 65$ million $= 155$ million or $155{,}000{,}000$.

27. Las Vegas: $4(10 \text{ million}) = 40$ million

New York (JFK): $4.5(10 \text{ million}) = 45$ million

The difference is $45 - 40 = 5$ million or $5{,}000{,}000$.

29. There are $9 + 5.5 + 6.5 + 4.5 + 4 = 29.5$ symbols.

$$29.5(10 \text{ million}) = 295 \text{ million}$$

The total number is 295 million or $295{,}000{,}000$.

9.2   Reading and Constructing Circle Graphs

9.2   Margin Exercises

1.   **(a)** The circle graph shows that the greatest number of hours is spent sleeping, which is 7 hours.

   (b) 4 hours studying
   6 hours working
   $\underline{+\,3}$ hours attending class
   13 hours

   Thirteen hours are spent studying, working, and attending classes.

2.   **(a)** $\dfrac{2 \text{ hours (driving)}}{24 \text{ hours (whole day)}} = \dfrac{2 \text{ hours}}{24 \text{ hours}}$

   $= \dfrac{\overset{1}{\cancel{2}}}{\underset{1}{\cancel{2}} \cdot 12} = \dfrac{1}{12}$

   (b) $\dfrac{4 \text{ hours (studying)}}{24 \text{ hours (whole day)}} = \dfrac{4 \text{ hours}}{24 \text{ hours}}$

   $= \dfrac{\overset{1}{\cancel{4}}}{\underset{1}{\cancel{4}} \cdot 6} = \dfrac{1}{6}$

3.   **(a)** $\dfrac{4 \text{ hours (studying)}}{6 \text{ hours (working)}} = \dfrac{4 \text{ hours}}{6 \text{ hours}}$

   $= \dfrac{\overset{1}{\cancel{2}} \cdot 2}{\underset{1}{\cancel{2}} \cdot 3} = \dfrac{2}{3}$

   (b) $\dfrac{4 \text{ hours (studying)}}{2 \text{ hours (driving)}} = \dfrac{4 \text{ hours}}{2 \text{ hours}}$

   $= \dfrac{\overset{1}{\cancel{2}} \cdot 2}{\underset{1}{\cancel{2}} \cdot 1} = \dfrac{2}{1}$

4.   **(a)** Of the 5 million barrels, the percent that is still at sea or on shores is 26%.

   $$\text{percent} \cdot \text{whole} = \text{part}$$
   $$26.\% \cdot 5 \text{ million} = n$$
   $$\rotatebox{0}{$\curvearrowright$}$$
   $$(0.26)(5 \text{ million}) = n$$
   $$1.3 \text{ million} = n$$

   1.3 million, or 1,300,000, barrels are still at sea or on shores.

   (b) The percent that has been skimmed is 3%.

   $$(0.03)(5 \text{ million}) = n$$
   $$0.15 \text{ million} = n$$

   0.15 million, or 150,000, barrels have been skimmed.

(c) The percent that has evaporated or dissolved is 25%.

   $$(0.25)(5 \text{ million}) = n$$
   $$1.25 \text{ million} = n$$

1.25 million, or 1,250,000, barrels have evaporated or dissolved.

(d) The percent that has been chemically dispersed is 8%.

   $$(0.08)(5 \text{ million}) = n$$
   $$0.4 \text{ million} = n$$

0.4 million, or 400,000, barrels have been chemically dispersed.

5.   **(a)** "Ages 10–11" makes up 25% of the total number of players, so that sector needs 25% of the circle or 25% of 360°.

   $$\text{percent} \cdot \text{whole} = \text{part}$$
   $$25.\% \cdot 360° = n$$
   $$\rotatebox{0}{$\curvearrowright$}$$
   $$\underline{(0.25)(360°)} = n$$
   $$90° = n$$

   $(360°)(25\%) = (360°)(0.25) = 90°$

   (b) "Ages 12–13" sector:

   $(360°)(25\%) = (360°)(0.25) = 90°$

   (c) "Ages 14–15" sector:

   $(360°)(15\%) = (360°)(0.15) = 54°$

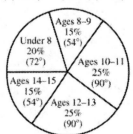

9.2   Section Exercises

1.   A circle graph represents **(b) how a total is divided into parts**.

3.   The tool used to draw circle graphs is called a **(c) protractor**.

5.   The total number of people in the survey is $50 + 180 + 120 + 35 + 20 + 30 + 65 = 500$.

7.   Friday is the most popular day of the week to order food. 180 people picked it.

9.   From Exercise 5, the total number of people in the survey is 500.

   $\begin{array}{l} \text{Picked Thursday} \rightarrow \\ \text{Total people} \rightarrow \end{array} \dfrac{50}{500} = \dfrac{50 \div 50}{500 \div 50} = \dfrac{1}{10}$

11. Picked Friday $\rightarrow$ $\dfrac{180}{120} = \dfrac{180 \div 60}{120 \div 60} = \dfrac{3}{2}$
Picked Saturday $\rightarrow$

13. The number of people who picked *either* Friday or Saturday was $180 + 120 = 300$.

Picked Either $\rightarrow$ $\dfrac{300}{20} = \dfrac{300 \div 20}{20 \div 20} = \dfrac{15}{1}$
Picked Monday $\rightarrow$

15. There are 400 people in the survey. 35% said their cat mingles with the guests.

$$\text{percent} \cdot \text{whole} = \text{part}$$
$$35.\% \cdot 400 = n$$
$$\underbrace{(0.35)(400)}_{140} = n$$
$$140 = n$$

140 people said their cat mingles with the guests.

17. The response "Raids the buffet table" was given least often at 4%.

$$(0.04)(400) = n$$
$$16 = n$$

16 people said their cat raids the buffet table.

19. Watches from a safe perch: $0.19(400) = 76$
Runs and hides: $0.35(400) = 140$

Difference: $140 - 76 = 64$

64 fewer people said "watches from a safe perch" than said "runs and hides."

21. To determine how many people prefer onions for their hot dog topping, use the percent equation.

$$\text{percent} \cdot \text{whole} = \text{part}$$
$$(0.05)(3200) = n$$
$$160 = n$$

160 people favored onions.

23. The most popular topping was mustard (30%).

$$(0.30)(3200) = n$$
$$960 = n$$

960 people favored mustard.

25. 12% of 3200 = chili   10% of 3200 = relish
$(0.12)(3200) = n$ $(0.10)(3200) = n$
$384 = n$ $320 = n$

There were $384 - 320 = 64$ more people who chose chili than chose relish.

27. First, find the percent of the total that is represented by each item. Next, multiply the percent by 360° to find the size of each sector. Finally, use a protractor to draw each sector.

29. (a) 50% of total is "chapter tests."

Degrees of a circle = 50% of 360°
$$= (0.50)(360)$$
$$= 180°$$

(b) Percent for final exam $= \dfrac{160}{800} = \dfrac{1}{5}$
$$= 0.20 = 20\%$$

Degrees of a circle = 20% of 360°
$$= (0.20)(360)$$
$$= 72°$$

(c) Percent for homework $= \dfrac{40}{800} = \dfrac{1}{20}$
$$= 0.05 = 5\%$$

Degrees of a circle = 5% of 360°
$$= (0.05)(360)$$
$$= 18°$$

(d) Percent for quizzes $= \dfrac{120}{800} = \dfrac{3}{20}$
$$= 0.15 = 15\%$$

Degrees of a circle = 15% of 360°
$$= (0.15)(360)$$
$$= 54°$$

(e) Percent for project $= \dfrac{80}{800} = \dfrac{1}{10}$
$$= 0.10 = 10\%$$

Degrees of a circle = 10% of 360°
$$= (0.10)(360)$$
$$= 36°$$

(f)

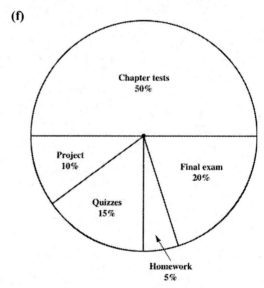

31. (a) Total sales = \$12,500 + \$40,000 + \$60,000 + \$50,000 + \$37,500 = \$200,000

(b) Adventure classes = \$12,500

percent of total $= \dfrac{12,500}{200,000} = 0.0625 = 6.25\%$

Grocery/provision sales = \$40,000

percent of total $= \dfrac{40,000}{200,000} = 0.2 = 20\%$

Equipment rentals = $60,000

$$\text{percent of total} = \frac{60,000}{200,000} = 0.3 = 30\%$$

Rafting tours = $50,000

$$\text{percent of total} = \frac{50,000}{200,000} = 0.25 = 25\%$$

Equipment sales = $37,500

$$\text{percent of total} = \frac{37,500}{200,000} = 0.1875 = 18.75\%$$

(c) Adventure classes:

number of degrees $= (0.0625)(360°) = 22.5°$

Grocery/provision sales:

number of degrees $= (0.2)(360°) = 72°$

Equipment rentals:

number of degrees $= (0.3)(360°) = 108°$

Rafting tours:

number of degrees $= (0.25)(360°) = 90°$

Equipment sales:

number of degrees $= (0.1875)(360°) = 67.5°$

(d)

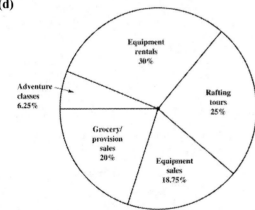

9.3 Bar Graphs and Line Graphs

9.3 Margin Exercises

1. **(a)** The bar for 2011 rises to 30, showing that the number of K–12 students in the school district was

$$30 \times \underline{1000} = \underline{30,000}.$$

(b) 2009: $20 \times 1000 = 20,000$ students

(c) 2008: $25 \times 1000 = 25,000$ students

(d) 2012: $35 \times 1000 = 35,000$ students

2. **(a)** The first red bar represents the 1st quarter in 2011. It rises to 4, and 4 times $\underline{1000}$ is $\underline{4000}$ new connections.

The first green bar represents the 1st quarter in 2012. It rises to $\underline{3}$, and $\underline{3}$ times $\underline{1000}$ is $\underline{3000}$ new connections.

(b) 3rd quarter, 2011:

$$7 \times 1000 = 7000 \text{ connections}$$

3rd quarter, 2012:

$$8 \times 1000 = 8000 \text{ connections}$$

(c) 4th quarter, 2011:

$$5 \times 1000 = 5000 \text{ connections}$$

4th quarter, 2012:

$$4 \times 1000 = 4000 \text{ connections}$$

3. **(a)** June: The dot is at 5.5 on the vertical axis.

$$5.5 \times 10,000 = 55,000$$

There were 55,000 trout stocked in June.

(b) May: $3 \times 10,000 = 30,000$
April: $4 \times 10,000 = 40,000$

There were $40,000 - 30,000 = 10,000$ fewer trout stocked in May than in April.

(c) July: $6 \times 10,000 = 60,000$
August: $2.5 \times 10,000 = 25,000$

Total (in thousands) =

$$40 + 30 + 55 + 60 + 25 = 210$$

There were 210,000 trout stocked during the five months.

(d) The highest point on the graph corresponds to July. From part (c), 60,000 trout were stocked in July.

4. **(a)** The dot on the blue line above 2007 lines up with 14 on the left edge. Then multiply 14 by 1,000,000 to get sales of 14,000,000 cars. Find the following in a similar fashion.

2009: $10 \times 1,000,000 = 10,000,000$
2010: $12 \times 1,000,000 = 12,000,000$
2011: $12 \times 1,000,000 = 12,000,000$

(b) The number of pickup trucks sold:

2007: $3 \times 1,000,000 = 3,000,000$
2008: $2 \times 1,000,000 = 2,000,000$
2009: $1 \times 1,000,000 = 1,000,000$
2011: $2 \times 1,000,000 = 2,000,000$

(c) The graph is flat from 2010 to 2011, so 2011 is the year in which sales of pickup trucks were unchanged from the previous year.

(d) From part (a), the amount of increase in car sales from 2009 to 2010 is

$$12,000,000 - 10,000,000 = 2,000,000 \text{ cars.}$$

percent **of** 2009 sales = amount of increase

$$p \cdot 10,000,000 = 2,000,000$$
$$\frac{p \cdot 10,000,000}{10,000,000} = \frac{2,000,000}{10,000,000}$$
$$p = \tfrac{1}{5} = 20\%$$

The percent of increase is 20%.

9.3 Section Exercises

1. A comparison of two sets of data can be easily shown in a **(c) double-bar graph**.

3. If a dot on a line graph lines up with "3" on the left side, and the label on the left side says, "in ten-thousands," then the dot represents **(b) 30,000**. Note that $3 \times 10,000 = 30,000$.

5. Look for the longest bar. "Music and videos" is tied with "Electronics" for the highest percent, which is 74%.

7. The portion of the total that was spent online for toys and games was 61% of $4.4 billion.

$$\underset{\curvearrowright}{61.\%} \cdot 4.4 = n$$
$$\underbrace{(0.61)(4.4)}_{} = n$$
$$2.684 \quad = n$$

About $2.7 billion was spent online for toys and games.

9. $\frac{2}{3} \approx 0.67 = 67\%$

"Clothing" (at 66%) is the category that had about $\frac{2}{3}$ of its total sales online.

11. Look for the longest green bar. July's bar goes to 12, which corresponds to

$$12 \times 10,000 = 120,000 \text{ unemployed.}$$

13. Unemployed workers in Dec. of 2011 = 90,000
Unemployed workers in Dec. of 2010 = 80,000

$$90,000 - 80,000 = 10,000$$

There were 10,000 more workers unemployed in December of 2011 than in December 2010.

15. The number of unemployed workers decreased from 120,000 in July of 2010 to 80,000 in December of 2010. The decrease was $120,000 - \underline{80,000} = \underline{40,000}$ workers.

$$
\begin{array}{ccc}
\text{July 2010} & = & \text{amount} \\
\text{percent } \mathbf{of} \text{ unemployed} & & \text{of} \\
\text{workers} & & \text{decrease} \\
p \quad \cdot \quad 120,000 & = & 40,000 \\
\dfrac{p \cdot 120,000}{120,000} & = & \dfrac{40,000}{120,000} \\
p & = & \dfrac{1}{3} \approx 0.33 = 33\%
\end{array}
$$

The percent of decrease was 33% (rounded).

17. Look at the purple bar over 2009–10. It goes to <u>15%</u>, so that is the percent of employers expected to hire more graduates in 2009–10.

19. By visual inspection, the largest difference in heights between a purple bar and a pink bar occurs over <u>2011-12</u>, so that is the year that had the greatest difference between the percent of employers expecting to hire more graduates and the percent expecting to hire fewer graduates.

21. The number of employers that expected to hire fewer graduates in 2011–12 is 10% of 172.

$$\underset{\curvearrowright}{10.\%} \cdot 172 = n$$
$$\underbrace{(0.10)(172)}_{} = n$$
$$17.2 \quad = n$$

About 17 employers (rounded) expected to hire fewer graduates in 2011–12.

23. The number of PCs sold in 1990 was 24.1 million or 24,100,000.

25. The increase in the number of PCs sold in 2005 from the number sold in 1985 was $197.4 - 11.8 = 185.6$ million or 185,600,000.

27. The amount of increase in sales from 1995 to 2000 was $144.6 - 62.3 = 82.3$ million or 82,300,000 PCs.

$$
\begin{array}{ccc}
\text{percent of } 1995 & = & \text{amount} \\
\text{sales} & & \text{of increase} \\
p \quad \cdot \quad 62.3 & = & 82.3 \\
\dfrac{p \cdot 62.3}{62.3} & = & \dfrac{82.3}{62.3} \\
p & \approx & 1.32 = 132\%
\end{array}
$$

The percent of increase was 132% (rounded).

29. **(a)** Chain Store A sold $3000 \cdot 1000 = 3,000,000$ DVDs in 2008.

(b) Chain Store B sold $1500 \cdot 1000 = 1,500,000$ DVDs in 2008.

31. **(a)** Chain Store A sold $2500 \cdot 1000 = 2,500,000$ DVDs in 2011.

(b) Chain Store A sold $3000 \cdot 1000 = 3,000,000$ DVDs in 2012.

33. Answers will vary. Possibilities include: Both stores had decreased sales from 2008 to 2009 and increased sales from 2010 to 2012; Store B had lower sales than Store A in 2008-2009 but higher sales than Store A in 2010-2012.

35. On the blue line graph (Sales), the lowest point corresponds to the year 2010 and the amount $25,000.

37. **For 2009:** Profit = $10,000, Sales = $35,000

Percent the profit is of sales
$$= \frac{\$10,000}{\$35,000} \approx 0.29 = 29\% \text{ (rounded)}$$

For 2010: Profit = $5,000, Sales = $25,000

Percent the profit is of sales

$$= \frac{\$5,000}{\$25,000} = 0.20 = 20\%$$

For 2011: Profit = $5,000, Sales = $30,000

Percent the profit is of sales

$$= \frac{\$5,000}{\$30,000} \approx 0.17 = 17\% \text{ (rounded)}$$

For 2012: Profit = $15,000, Sales = $40,000

Percent the profit is of sales

$$= \frac{\$15,000}{\$40,000} = 0.375 \approx 38\% \text{ (rounded)}$$

39. Answers will vary. Possibilities include: The decrease in sales may have resulted from poor service or greater competition; the increase in sales may have been a result of more advertising or better service.

Relating Concepts (Exercises 41–46)

41. Sales have increased at a rapid rate since 1985.

42. Answers will vary. Some possibilities are: lower prices; more uses and applications for students, home use, and businesses; improved technology.

43. See Exercises 27 and 28 for more detail.

(a) 1985 to 1990:

$$11.8p = 24.1 - 11.8$$
$$p = \frac{12.3}{11.8} \approx 1.04 = \textbf{104\%} \text{ (rounded)}$$

(b) 1990 to 1995:

$$24.1p = 62.3 - 24.1$$
$$p = \frac{38.2}{24.1} \approx 1.59 = \textbf{159\%} \text{ (rounded)}$$

(c) 1995 to 2000:

$$62.3p = 144.6 - 62.3$$
$$p = \frac{82.3}{62.3} \approx 1.32 = \textbf{132\%} \text{ (rounded)}$$

(d) 2000 to 2005:

$$144.6p = 197.4 - 144.6$$
$$p = \frac{52.8}{144.6} \approx 0.37 = \textbf{37\%} \text{ (rounded)}$$

(e) 2005 to 2010:

$$197.4p = 304.9 - 197.4$$
$$p = \frac{107.5}{197.4} \approx 0.54 = \textbf{54\%} \text{ (rounded)}$$

44. Since 2000, the percent of increase for each 5-year period has been much lower than in earlier periods.

45. Answers will vary. One possibility is that people will buy new devices such as smartphones and tablets instead of PCs.

46. No. The line graph shows the amount of sales in previous years but cannot accurately predict future sales; we can only make a guess based on the trend in the graph.

9.4 The Rectangular Coordinate System

9.4 Margin Exercises

1. **(a)** To plot the point $(1, 4)$ on the grid, start at 0. Move *to the right* along the horizontal axis until you reach 1. Then move *up* <u>4</u> units so that you are aligned with 4 on the vertical axis. Make a dot. This is the plot, or graph, of the point $(1, 4)$. Write $(\underline{1}, \underline{4})$ next to the dot. See the graph after part (d).

(b) $(5, 2)$ ■ Move right 5, then up 2.

(c) $(4, 1)$ ■ Move right 4, then up 1.

(d) $(3, 3)$ ■ Move right 3, then up 3.

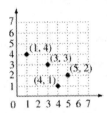

2. **(a)** $(5, -3)$ ■ Start at 0. Move to the <u>right</u> until you reach <u>5</u>. Then move <u>down</u> 3 units. Make a dot and label it $(\underline{5}, \underline{-3})$.

(b) $(-5, 3)$ ■ Move left 5, then up 3.

(c) $(0, 3)$ ■ Start at 0. Do ***not*** move left or right. Move <u>up</u> 3 units. Make a dot and label it $(\underline{0}, \underline{3})$. This point is on the y-axis.

(d) $(-4, -4)$ ■ Move left 4, then down 4.

(e) $(-2, 0)$ ■ Start at 0. Move to the <u>left</u> until you reach <u>−2</u>. Do ***not*** move up or down. Make a dot and label it $(\underline{-2}, \underline{0})$. This point is on the $\underline{x}$-axis.

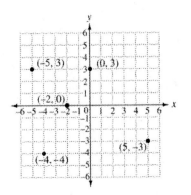

3. A is $(-4, -1)$.
B is $(3, -3)$.
C is $(5, 0)$.
D is $(0, -2)$.
E is approximately $\left(-1\frac{1}{2}, 5\right)$.

4. **(a)** The pattern for all points in Quadrant IV is
$(+, -)$.

Examples: $(3, -10), (1, -4), (12, -21)$

(b) For $(-2, -6)$, the pattern is $(-, -)$, so the
point is in **quadrant III**.

The point corresponding to $(0, 5)$ is on the y-axis,
so it isn't in any quadrant. **no quadrant**

For $(-3, 1)$, the pattern is $(-, +)$, so the point is
in **quadrant II**.

For $(4, -1)$, the pattern is $(+, -)$, so the point is
in **quadrant IV**.

9.4 Section Exercises

1. **(a)** The origin is labeled in the following figure.

(b) The x-axis and the y-axis are labeled in the
following figure.

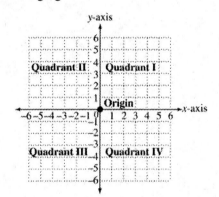

3.

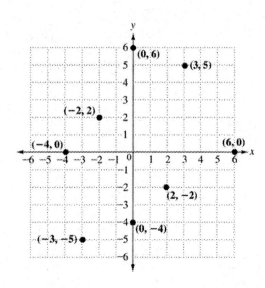

5.

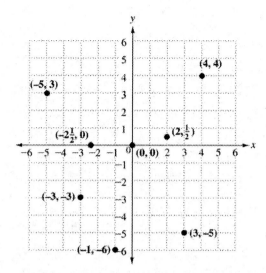

7. A is $(3, 4)$; B is $(5, -5)$; C is $(-4, -2)$;
D is approximately $\left(4, \frac{1}{2}\right)$; E is $(0, -6)$;
F is $(-5, 5)$; G is $(-2, 0)$; H is $(0, 0)$.

9. $(-3, -7)$ is in Quadrant III.
$(0, 4)$ is on the y-axis, so it is not in a quadrant.
$(10, -16)$ is in Quadrant IV.
$(-9, 5)$ is in Quadrant II.

11. **(a)** Any *positive* number, because points in
Quadrant II have the pattern $(-, +)$.

(b) Any *negative* number, because points in
Quadrant IV have the pattern $(+, -)$.

(c) 0, because points not in a quadrant have the
form $(0, \pm)$ or $(\pm, 0)$.

(d) Any *negative* number, because points in
Quadrant III have the pattern $(-, -)$.

(e) Any *positive* number, because points in
Quadrant I have the pattern $(+, +)$.

13. Starting at the origin, move left or right along the
x-axis to the number a; then move up if b is
positive or move down if b is negative.

9.5 Introduction to Graphing Linear Equations

9.5 Margin Exercises

1. $x + y = 5$

x	y	Check that $x + y = 5$	Ordered Pair (x, y)
0	5	$0 + 5 = \underline{5}$	$(0, \underline{5})$
1	4	$1 + 4 = \underline{5}$	$(\underline{1}, \underline{4})$
2	$\underline{3}$	$2 + \underline{3} = \underline{5}$	$(2, \underline{3})$

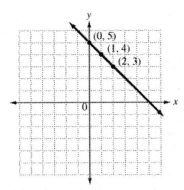

Some other possible solutions are $(-1, 6)$, $(3, 2)$, $(4, 1)$, $(5, 0)$, and $(6, -1)$.

2. $y = 2x$

x	$2 \cdot x = y$	Ordered Pair (x, y)
0	$2 \cdot 0 = 0$	$(0, \underline{0})$
1	$2 \cdot 1 = \underline{2}$	$(1, \underline{2})$
2	$2 \cdot 2 = \underline{4}$	$(\underline{2}, \underline{4})$

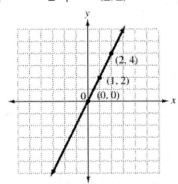

Some other possible solutions are $(3, 6)$, $(-1, -2)$, $(-2, -4)$, and $(-3, -6)$.

3. $y = -\frac{1}{2}x$

x	$-\frac{1}{2} \cdot x = y$	Ordered Pair (x, y)
2	$-\frac{1}{2} \cdot 2 = -\frac{2}{2} = -1$	$(2, \underline{-1})$
4	$-\frac{1}{2} \cdot 4 = -\frac{4}{2} = \underline{-2}$	$(4, \underline{-2})$
6	$-\frac{1}{2} \cdot \underline{6} = -\frac{6}{2} = \underline{-3}$	$(\underline{6}, \underline{-3})$

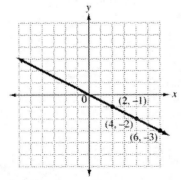

Some other possible solutions are $(0, 0)$, $(-2, 1)$, $(-4, 2)$, $(-6, 3)$, $(1, -\frac{1}{2})$, $(3, -1\frac{1}{2})$, and $(5, -2\frac{1}{2})$.

4. $y = x - 5$

x	$x - 5 = y$	Ordered Pair (x, y)
1	$1 - 5 = -4$	$(1, \underline{-4})$
2	$2 - 5 = \underline{-3}$	$(2, \underline{-3})$
3	$\underline{3} - 5 = \underline{-2}$	$(\underline{3}, \underline{-2})$

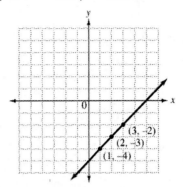

Some other possible solutions are $(-1, -6)$, $(0, -5)$, $(4\frac{1}{2}, -\frac{1}{2})$, $(5, 0)$, and $(5\frac{1}{2}, \frac{1}{2})$.

5. **(a)** The graph of $y = 2x$ has a <u>positive</u> slope. As the value of x increases, the value of y <u>increases</u>.

(b) The graph of $y = -\frac{1}{2}x$ has a <u>negative</u> slope. As the value of x increases, the value of y <u>decreases</u>.

9.5 **Section Exercises**

1. The graph of a linear equation is a **(c) straight line**.

3. $x + y = 4$

x	y	Ordered Pair (x, y)
0	4	$(0, 4)$
1	3	$(1, \underline{3})$
2	2	$(2, \underline{2})$

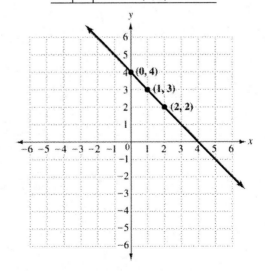

Two other possible solutions are $(3, 1)$ and $(4, 0)$. All points on the line are solutions.

5. $x + y = -1$

x	y	Ordered Pair (x, y)
0	-1	$(0, -1)$
1	-2	$(1, -2)$
2	-3	$(2, -3)$

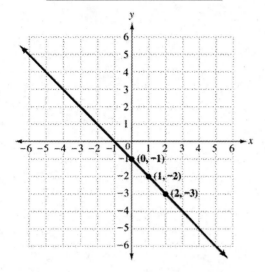

Two other possible solutions are $(-1, 0)$ and $(-2, 1)$. All points on the line are solutions.

7.   The line in Exercise 3, $x + y = 4$, crosses the y-axis at <u>4</u> or $(0, 4)$.

The line in Exercise 5, $x + y = -1$, crosses the y-axis at <u>−1</u> or $(0, -1)$.

Based on these examples:

The line $x + y = -6$ will cross the y-axis at <u>−6</u> or $(0, -6)$.

The line $x + y = 99$ will cross the y-axis at <u>99</u> or $(0, 99)$.

9. $y = x - 2$

x	$x - 2 = y$	Ordered Pair (x, y)
1	$1 - 2 = -1$	$(1, -1)$
2	$2 - 2 = 0$	$(2, 0)$
3	$3 - 2 = 1$	$(3, 1)$

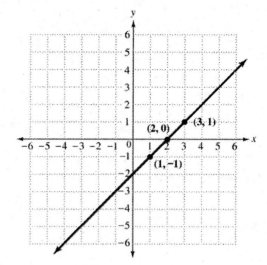

11. $y = x + 2$

x	$x + 2 = y$	(x, y)
0	$0 + 2 = 2$	$(0, 2)$
-1	$-1 + 2 = 1$	$(-1, 1)$
-2	$-2 + 2 = 0$	$(-2, 0)$

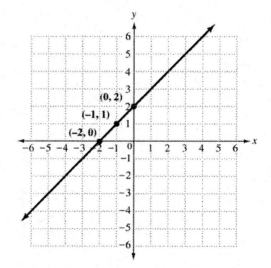

13. $y = -3x$

x	$-3 \cdot x = y$	(x, y)
0	$-3(0) = 0$	$(0, 0)$
1	$-3(1) = -3$	$(1, -3)$
2	$-3(2) = -6$	$(2, -6)$

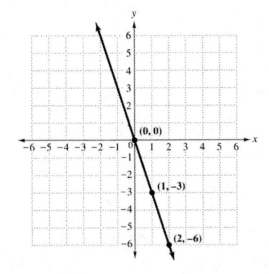

19. $y = x$

x	$y = x$	(x, y)
-1	-1	$(-1, -1)$
-2	-2	$(-2, -2)$
-3	-3	$(-3, -3)$

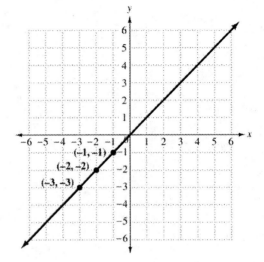

15. The lines in Exercises 9 and 11 have a positive slope. The lines in Exercises 3, 5, and 13 have a negative slope.

17. $y = \frac{1}{3}x$

x	$\frac{1}{3} \cdot x = y$	(x, y)
0	$\frac{1}{3}(0) = 0$	$(0, 0)$
3	$\frac{1}{3}(3) = 1$	$(3, 1)$
6	$\frac{1}{3}(6) = 2$	$(6, 2)$

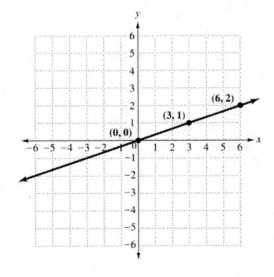

21. $y = -2x + 3$

x	$-2 \cdot x + 3 = y$	(x, y)
0	$-2(0) + 3 = 3$	$(0, 3)$
1	$-2(1) + 3 = 1$	$(1, 1)$
2	$-2(2) + 3 = -1$	$(2, -1)$

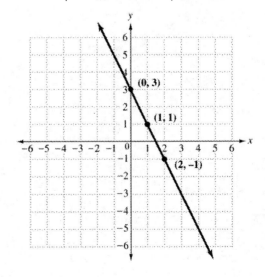

23. $x + y = -3$ ■ One possible table:

x	y	(x, y)
0	-3	$(0, -3)$
1	-4	$(1, -4)$
2	-5	$(2, -5)$

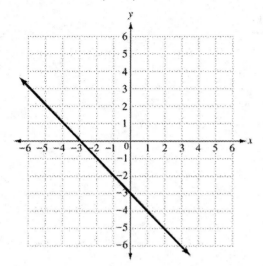

The line has a negative slope (falls).

27. $y = x - 5$ ■ One possible table:

x	$x - 5 = y$	(x, y)
2	$2 - 5 = -3$	$(2, -3)$
3	$3 - 5 = -2$	$(3, -2)$
4	$4 - 5 = -1$	$(4, -1)$

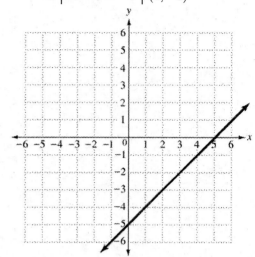

The line has a positive slope (rises).

25. $y = \frac{1}{4}x$ ■ One possible table:

x	$\frac{1}{4} \cdot x = y$	(x, y)
0	$\frac{1}{4}(0) = 0$	$(0, 0)$
4	$\frac{1}{4}(4) = 1$	$(4, 1)$
8	$\frac{1}{4}(8) = 2$	$(8, 2)$

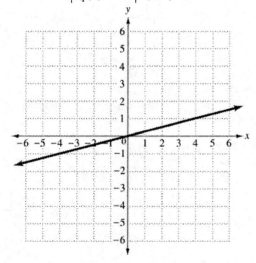

The line has a positive slope (rises).

29. $y = -3x + 1$ ■ One possible table:

x	$-3 \cdot x + 1 = y$	(x, y)
0	$-3(0) + 1 = 1$	$(0, 1)$
1	$-3(1) + 1 = -2$	$(1, -2)$
2	$-3(2) + 1 = -5$	$(2, -5)$

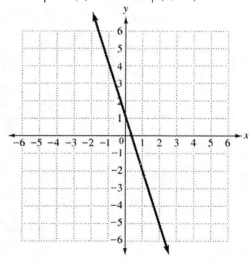

The line has a negative slope (falls).

Chapter 9 Review Exercises

1. **(a)** Look down the column for male teams to find the smallest number (17). Then look across to find the sport. Gymnastics had the fewest men's teams.

(b) Look down the column for female teams and find the second largest number (1039). Then look across to find the sport. Volleyball had the second-greatest number of women's teams.

2. **(a)** Look down the column for female athletes for the number closest to 16,000 (15,708). Then look across to find the sport. Basketball had about 16,000 female athletes.

(b) Look down the column for male athletes for the number closest to 14,000 (13,976). Then look across to find the sport. Cross country had about 14,000 male athletes.

3. **(a)** From the table, basketball had 17,500 male athletes and cross country had 13,976 male athletes.

$$17,500 - 13,976 = 3524$$

3524 more men participated in basketball than cross country.

(b) From the table, gymnastics had 1463 female athletes and volleyball had 15,597 female athletes.

$$15,597 - 1463 = 14,134$$

14,134 fewer women participated in gymnastics than volleyball.

4. The average squad size for men's volleyball is

$$\frac{1456}{95} \approx 15.3 \text{ (rounded)}.$$

The average squad size for women's volleyball is

$$\frac{15,597}{1039} \approx 15.0 \text{ (rounded)}.$$

For Exercises 5–8, each symbol represents 10 inches of snowfall.

5. **(a)** Juneau is represented by 10 symbols.

$$10(10 \text{ inches}) = 100 \text{ inches}$$

(b) Washington D.C. is represented by $1\frac{1}{2}$ or 1.5 symbols.

$$1.5(10 \text{ inches}) = 15 \text{ inches}$$

6. **(a)** Minneapolis is represented by 5 symbols.

$$5(10 \text{ inches}) = 50 \text{ inches}$$

(b) Cleveland is represented by $5\frac{1}{2}$ or 5.5 symbols.

$$5.5(10 \text{ inches}) = 55 \text{ inches}$$

7. **(a)** Buffalo: $9(10 \text{ inches}) = 90 \text{ inches}$
Cleveland: $5.5(10 \text{ inches}) = 55 \text{ inches}$

The difference in average yearly snowfall is

$$90 \text{ inches} - 55 \text{ inches} = 35 \text{ inches}.$$

(b) Memphis: $\frac{1}{2}(10 \text{ inches}) = 5 \text{ inches}$
Minneapolis: $5(10 \text{ inches}) = 50 \text{ inches}$

The difference in average yearly snowfall is

$$50 \text{ inches} - 5 \text{ inches} = 45 \text{ inches}.$$

8. greatest amount: Juneau, 100 inches
least amount: Memphis, 5 inches

The difference in average yearly snowfall between Juneau and Memphis is

$$100 \text{ inches} - 5 \text{ inches} = 95 \text{ inches}.$$

9. **(a)** Lodging is the largest sector in the circle graph: $560

(b) Food is the second largest sector in the circle graph: $400

(c) Total cost
$$= \$560 + \$400 + \$300 + \$280 + \$160$$
$$= \$1700$$

10. $\dfrac{\text{lodging}}{\text{food}} = \dfrac{\$560}{\$400} = \dfrac{560 \div 80}{400 \div 80} = \dfrac{7}{5}$

11. $\dfrac{\text{gasoline}}{\text{total cost}} = \dfrac{\$300}{\$1700} = \dfrac{300}{1700} = \dfrac{3}{17}$

12. $\dfrac{\text{sightseeing}}{\text{total cost}} = \dfrac{\$280}{\$1700} = \dfrac{280 \div 20}{1700 \div 20} = \dfrac{14}{85}$

13. $\dfrac{\text{gasoline}}{\text{other}} = \dfrac{\$300}{\$160} = \dfrac{300 \div 20}{160 \div 20} = \dfrac{15}{8}$

14. The most popular project is painting and wallpapering since it has the longest bar. The percent is 63%.

15. The project selected the least is construction work at 33%.

16. 43% of the homeowners in the survey selected carpentry projects.

$$\text{percent} \cdot \text{whole} = \text{part}$$
$$(0.43)(341) = n$$
$$n = 146.63 \approx 147$$

About 147 homeowners selected carpentry.

17. 54% of the homeowners in the survey selected landscaping or gardening projects.

$$\text{percent} \cdot \text{whole} = \text{part}$$
$$(0.54)(341) = n$$
$$n = 184.14 \approx 184$$

About 184 homeowners selected landscaping or gardening projects.

18.   **(a)** $\frac{1}{2} = 0.5 = 50\%$

51% (about $\frac{1}{2}$) of the homeowners selected interior decorating. 54% (about $\frac{1}{2}$) of the homeowners selected landscaping/gardening.

    (b) $\frac{1}{3} = 0.\overline{3} = 33.\overline{3}\%$

33% (about $\frac{1}{3}$) of the homeowners selected construction work and 34% (about $\frac{1}{3}$) of the homeowners selected window treatments.

19.   Answers will vary. Possibilities include: painting and wallpapering are easier to do, take less time, or cost less than construction work.

20.   In 2012, the greatest amount of water in the lake occurred in March when there were 8,000,000 acre-feet of water.

21.   In 2011, the least amount of water in the lake occurred in June when there were 2,000,000 acre-feet of water.

22.   In June of 2012, there were 5,000,000 acre-feet of water in the lake.

23.   In January of 2011, there were 6,000,000 acre-feet of water in the lake.

24.   March 2012: 8,000,000 acre-feet
June 2012: 5,000,000 acre-feet

This is a $8,000,000 - 5,000,000 = 3,000,000$ acre-feet decrease.

$$
\begin{array}{ccc}
\text{percent of} & \text{March 2012} & \text{amount} \\
& \text{amount} & \text{of decrease} \\
p \quad \cdot & 8,000,000 & = \quad 3,000,000 \\
\dfrac{p \cdot 8,000,000}{8,000,000} & = & \dfrac{3,000,000}{8,000,000} \\
p & = & \tfrac{3}{8} = 0.375 = 37.5\%
\end{array}
$$

In 2012, the percent of decrease from March to June was 37.5%.

25.   April 2011: 5,000,000 acre-feet
June 2011: 2,000,000 acre-feet

This is a $5,000,000 - 2,000,000 = 3,000,000$ acre-feet decrease.

$$
\begin{array}{ccc}
\text{percent of} & \text{April 2011} & \text{amount} \\
& \text{amount} & \text{of decrease} \\
p \quad \cdot & 5,000,000 & = \quad 3,000,000 \\
\dfrac{p \cdot 5,000,000}{5,000,000} & = & \dfrac{3,000,000}{5,000,000} \\
p & = & \tfrac{3}{5} = 0.6 = 60\%
\end{array}
$$

In 2011, the percent of decrease from April to June was 60%.

26.   In 2009, Center A sold $50,000,000 worth of floor covering.

27.   In 2011, Center A sold $20,000,000 worth of floor covering.

28.   In 2010, Center B sold $20,000,000 worth of floor covering.

29.   In 2012, Center B sold $40,000,000 worth of floor covering.

30.   Sales decreased for two years and then moved up slightly. Answers will vary. Perhaps there is less new construction, remodeling, and home improvement in the area near Center A. Or, better product selection and service may have reversed the decline in sales.

31.   Sales are increasing. Answers will vary. New construction may have increased in the area near Center B, or greater advertising may attract more attention.

32.   Plumbing and electrical changes

$$
\begin{aligned}
\text{Degrees} &= 10\% \text{ of } 360° \\
&= (0.10)(360°) = 36°
\end{aligned}
$$

33.   Work stations (total = $22,400)

$$
\text{Percent} = \frac{\$7840}{\$22,400} = 0.35 = 35\%
$$

$$
\begin{aligned}
\text{Degrees} &= (0.35)(360°) \\
&= 126°
\end{aligned}
$$

34.   Small appliances

$$
\text{Percent} = \frac{\$4480}{\$22,400} = 0.20 = 20\%
$$

$$
\begin{aligned}
\text{Degrees} &= (0.20)(360°) \\
&= 72°
\end{aligned}
$$

35.   Interior decoration

$$
\text{Percent} = \frac{\$5600}{\$22,400} = 0.25 = 25\%
$$

$$
\begin{aligned}
\text{Degrees} &= (0.25)(360°) \\
&= 90°
\end{aligned}
$$

36.   Supplies

$$
\text{Percent} = \frac{36°}{360°} = 0.10 \text{ or } 10\%
$$

Dollar amount $= (0.10)(\$22,400) = \$2240$

37.

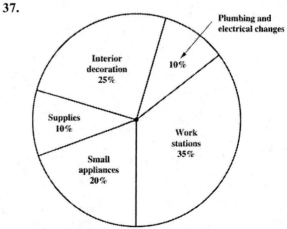

38.

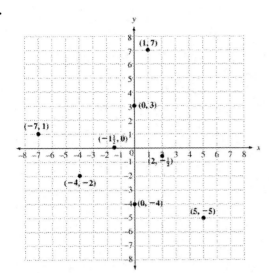

39. *A* is $(0, 6)$; *B* is approximately $\left(-2, 2\frac{1}{2}\right)$;
C is $(0, 0)$; *D* is $(-6, -6)$; *E* is $(4, 3)$;
F is approximately $\left(3\frac{1}{2}, 0\right)$; *G* is $(2, -4)$.

40. $x + y = -2$

x	y	(x, y)
0	-2	$(0, -2)$
1	-3	$(1, -3)$
2	-4	$(2, -4)$

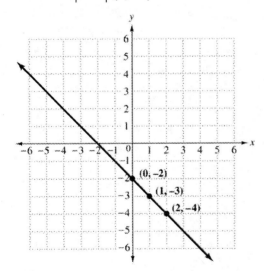

The graph of $x + y = -2$ has a *negative* slope.
Two other solutions of $x + y = -2$ are
$(-2, 0)$ and $(-1, -1)$.
All points on the line are solutions.

41. $y = x + 3$

x	$x + 3 =$	y	(x, y)
0	$0 + 3 =$	3	$(0, 3)$
1	$1 + 3 =$	4	$(1, 4)$
2	$2 + 3 =$	5	$(2, 5)$

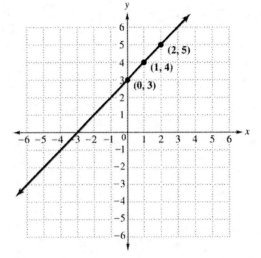

The graph of $y = x + 3$ has a *positive* slope.
Two other solutions for $y = x + 3$ are
$(-1, 2)$ and $(-2, 1)$.
All points on the line are solutions.

42. $y = -4x$

x	$-4 \cdot x = y$	(x, y)
-1	$-4(-1) = 4$	$(-1, 4)$
0	$-4(0) = 0$	$(0, 0)$
1	$-4(1) = -4$	$(1, -4)$

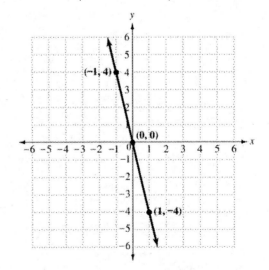

The graph of $y = -4x$ has a *negative* slope.
Two other solutions of $y = -4x$ are
$\left(-\frac{1}{2}, 2\right)$ and $\left(\frac{1}{2}, -2\right)$.
All points on the line are solutions.

Chapter 9 Test

1. **(a)** Look down the calcium column and find the largest number (371). Then look across to find the food. Sardines have the greatest amount of calcium.

(b) Look down the calcium column and find the smallest number (23). Then look across to find the food. Cream cheese has the least amount of calcium.

2. Set up a proportion.

$$\frac{95 \text{ calories}}{1 \text{ oz}} = \frac{x \text{ calories}}{1.75 \text{ oz}}$$
$$1 \cdot x = 95 \cdot 1.75$$
$$x \approx 166$$

There are approximately 166 calories in 1.75 ounces of Swiss cheese.

3. Set up a proportion.

$$\frac{345 \text{ mg}}{8 \text{ oz}} = \frac{x \text{ mg}}{6 \text{ oz}}$$
$$8 \cdot x = 345 \cdot 6$$
$$\frac{8x}{8} = \frac{2070}{8}$$
$$x \approx 259$$

There are approximately 259 mg of calcium in 6 oz of fruit-flavored yogurt.

For Exercises 4–6, each symbol represents 10 species.

4. **(a)** Birds and fish are represented by $7\frac{1}{2}$ symbols, more than any other group. Thus, birds and fish are tied for the greatest number of endangered species.

(b) $7\frac{1}{2}(10 \text{ species}) = 75$ species, so there are 75 bird and 75 fish species in the top groups.

5. Mammals: $7(10 \text{ species}) = 70$ species
Reptiles: $1.5(10 \text{ species}) = 15$ species

There are $70 - 15 = 55$ more species of endangered mammals than endangered reptiles.

6. There is a total of $7 + 7.5 + 1.5 + 7.5 + 3.5 = 27$ symbols shown in the pictograph.

There are $27(10) = 270$ endangered species shown in the pictograph.

7. Look for the largest sector (or largest percent).

29% of $2,800,000
$$= 0.29(\$2,800,000) = \$812,000$$

Television has the largest budget of $812,000.

8. Look for the smallest sector (or smallest percent).

3% of $2,800,000
$$= 0.03(\$2,800,000) = \$84,000$$

Miscellaneous has the smallest budget of $84,000.

9. 11% of $2,800,000
$$= 0.11(\$2,800,000) = \$308,000$$

$308,000 is budgeted for internet advertising.

10. 23% of $2,800,000
$$= 0.23(\$2,800,000) = \$644,000$$

$644,000 is budgeted for newspaper ads.

11. In 2011, expenses exceeded income by

$$\$21,000 - \$17,000 = \$4000.$$

12. From 2009 to 2010, expenses increased by

$$\$18,000 - \$13,000 = \$5000$$

$$\begin{array}{ccc} \text{percent of} & 2009 & = & \text{amount} \\ & \text{expenses} & & \text{of increase} \\ p & \cdot \quad 13,000 & = & 5,000 \end{array}$$
$$\frac{p \cdot 13,000}{13,000} = \frac{5,000}{13,000}$$
$$p = \frac{5}{13} \approx 0.38 = 38\%$$

The amount of increase in the student's expenses was $5000 and the percent of increase was 38%.

13. In 2011, the student's income declined. Explanations will vary. Some possibilities are: laid off from work, changed jobs, was ill, cut down on hours worked.

14. 5500 students were enrolled at College A in 2009.

3000 students were enrolled at College B in 2010.

15. College B had a higher enrollment in 2012.

College A = 4500 students

College B = 5500 students

College B had $5500 - 4500 = 1000$ more students.

16. Explanations will vary. For example, College B may have added new courses or lowered tuition or added child care.

17. $\text{Percent} = \dfrac{\$168,000}{\$480,000} = 0.35 = 35\%$

35% of $360° = 0.35(360°) = 126°$

18. $\text{Percent} = \dfrac{\$24{,}000}{\$480{,}000} = 0.05 = 5\%$

5% of $360° = 0.05(360°) = 18°$

19. $\text{Percent} = \dfrac{\$96{,}000}{\$480{,}000} = 0.20 = 20\%$

20% of $360° = 0.20(360°) = 72°$

20. $\text{Percent} = \dfrac{\$144{,}000}{\$480{,}000} = 0.30 = 30\%$

30% of $360° = 0.30(360°) = 108°$

21. $\text{Percent} = \dfrac{36°}{360°} = 0.10 = 10\%$

10% of $\$480{,}000 = 0.10(\$480{,}000) = \$48{,}000$

22.

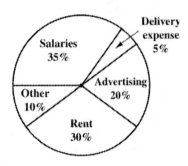

23. $(-5, 3)$ ▪ Move left 5, then up 3.

24. $(1, -4)$ ▪ Move right 1, then down 4.

25. $(0, 6)$ ▪ Move up 6 (on the y-axis).

26. $(2, 0)$ ▪ Move right 2 (on the x-axis).

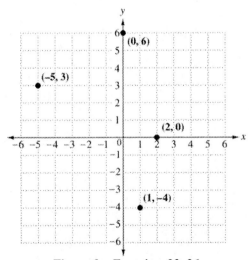

Figure for Exercises 23–26

27. Point A is $(0, 0)$; no quadrant

28. Point B is $(-5, -4)$; Quadrant III

29. Point C is $(3, 3)$; Quadrant I

30. Point D is $(-2, 4)$; Quadrant II

31. $y = x - 4$

x	$x - 4 = y$	(x, y)
0	$0 - 4 = -4$	$(0, -4)$
1	$1 - 4 = -3$	$(1, -3)$
2	$2 - 4 = -2$	$(2, -2)$

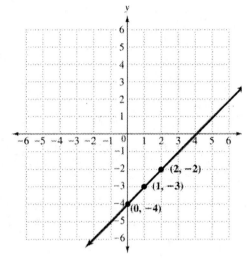

32. **(a)** Answers will vary; all points on the line are solutions. Some possibilities are $(3, -1)$ and $(4, 0)$.

(b) For $y = x - 4$, as the value of x increases, the value of y *increases*. Thus, the graph of $y = x - 4$ has a *positive* slope.

CHAPTER 10 REAL NUMBERS, EQUATIONS, AND INEQUALITIES

10.1 Real Numbers and Expressions

10.1 Margin Exercises

1. To compare these numbers and locate them on the number line, rewrite -2.75 and $\frac{17}{8}$ as mixed numbers (or change to decimals).

$$-2.75 = -2\frac{3}{4} \quad \text{and} \quad \frac{17}{8} = 2\frac{1}{8}$$

The order of the numbers from smallest to largest is:

$$-3, \quad -2.75, \quad -\frac{3}{4}, \quad 1\frac{1}{2}, \quad \frac{17}{8}$$

2. **(a)** $\sqrt{36}$ is rational because $\sqrt{36} = 6$.

(b) $0.6666\ldots$ is rational because the digit 6 repeats.

(c) $\sqrt{13} \approx 3.605551275$ is irrational because the decimal value never ends or repeats in a fixed block.

(d) $0.454545\ldots$ is rational because the decimal digits repeat in a fixed block.

(e) $0.131131113\ldots$ is irrational because the decimal digits do not repeat in a fixed block.

(f) 9.4375 is rational because the decimal terminates.

(g) $0.\overline{27}$ is rational because the decimal digits repeat in a fixed block.

3. **(a)** $0 \geq -\frac{3}{4}$ is *true* because $0 > -\frac{3}{4}$.

(b) $\sqrt{25} \leq \sqrt{25}$ is *true* because $\sqrt{25} = \sqrt{25}$.

(c) $3.06 \geq 3.6$ is *false* because $3.06 < 3.6$ and $3.06 \neq 3.6$.

(d) $\frac{2}{3}(10) \neq \frac{3}{2}(10)$ is *true* because $6\frac{2}{3} \neq 15$.

(e) $-15 \leq -16$ is *false* because $-15 > -16$ and $-15 \neq -16$.

(f) $\frac{1}{4} \neq \frac{12}{48}$ is *false* because $\frac{12}{48}$ in lowest terms is $\frac{1}{4}$.

(g) $0.5 \geq \frac{1}{2}$ is *true* because $0.5 = \frac{1}{2}$.

4. **(a)** $0.3 \leq 0.33$ becomes $0.33 \geq 0.3$.

(b) $\frac{2}{3} > \frac{2}{9}$ becomes $\frac{2}{9} < \frac{2}{3}$.

(c) $-5 < -1$ becomes $-1 > -5$.

(d) $\frac{9}{10} \geq 0.7$ becomes $0.7 \leq \frac{9}{10}$.

5. **(a)** $4\left[7 + 3(\underline{6 + 1})\right]$

$4\left[7 + 3(\underline{7})\right]$ *Work inside parentheses.*

$4\left[7 + \underline{21}\right]$ *Multiply.*

$4\underline{[28]}$ *Add.*

112 *Multiply.*

(b) $3[(-20 \div 5) - 7]$

$3[(-4) - 7]$ *Work inside parentheses.*

$3[-11]$ *Subtract.*

-33 *Multiply.*

6. **(a)** $-3(4b + 1)$

$\underbrace{(-3 \cdot 4b)} + \underbrace{(-3 \cdot 1)}$

$\underline{-12b} + \underline{(-3)}$

$\underline{-12b - 3}$ *Change to opposite.*

(b) $-7(2x - 3)$

$(-7 \cdot 2x) - (-7 \cdot 3)$

$-14x - (-21)$

$-14x + 21$ *Change to opposite.*

(c) $-4(h - 5)$

$(-4 \cdot h) - (-4 \cdot 5)$

$-4h - (-20)$

$-4h + 20$ *Change to opposite.*

(d) $-6(-2y + 4)$

$(-6 \cdot -2y) + (-6 \cdot 4)$

$12y + (-24)$

$12y - 24$ *Change to opposite.*

7. **(a)** $-(6k - 5)$ can be written $-1(6k - 5)$

$(-1 \cdot 6k) - (-1 \cdot 5)$

$\underline{-6k} - \underline{(-5)}$

$\underline{-6k + 5}$

The simplified expression is $-6k + 5$.

(b) $-(-2 - r)$ can be written $-1(-2 - r)$

$(-1 \cdot -2) - (-1 \cdot r)$

$2 - (-r)$

$2 + r$

The simplified expression is $2 + r$.

(c) $-(-5y + 8)$ can be written $-1(-5y + 8)$

$(-1 \cdot -5y) + (-1 \cdot 8)$

$5y + (-8)$

$5y - 8$

The simplified expression is $5y - 8$.

(d) $-(z + 4)$ can be written $-1(z + 4)$

$(-1 \cdot z) + (-1 \cdot 4)$

$-z + (-4)$

$-z - 4$

The simplified expression is $-z - 4$.

8.   (a) $10p + 3(5 + 2p)$
$10p + 15 + 6p$
$16p + 15$

(b) $-7x - 2 - (1 + x)$
$-7x - 2 - 1(1 + x)$ *Rewrite with −1.*
$-7x - 2 - 1 - x$ *Distributive property*
$-8x - 3$ *Combine like terms.*

(c) $-(3k^2 + 5k) + 7(k^2 - 4k)$
$-1(3k^2 + 5k) + 7(k^2 - 4k)$
$-3k^2 - 5k + 7k^2 - 28k$
$4k^2 - 33k$

(d) $-2(4b - 3) - 5(2b^2 + 1)$
$-8b + 6 - 10b^2 - 5$
$-10b^2 - 8b + 1$

10.1   Section Exercises

1. -0.0625 is rational because the decimal terminates.

3. π is irrational because the decimal form never ends or repeats in a fixed block.

5. $\frac{3}{4}$ is rational because it is the quotient of two integers (denominator not equal to 0).

7. $0.636363\ldots$ is rational because the decimal digits repeat in a fixed block.

9. $\sqrt{2} \approx 1.41421356$ is irrational because the decimal form never ends or repeats in a fixed block.

11. $\left\{-9, -\sqrt{7}, -1\frac{1}{4}, -\frac{3}{5}, 0, \sqrt{5}, 3, 5.9, 7\right\}$

(a) The natural numbers in the given set are 3 and 7, since they are in the natural number set $\{1, 2, 3, \ldots\}$.

(b) The set of whole numbers includes the natural numbers and 0. The whole numbers in the given set are 0, 3, and 7.

(c) The integers are the set of numbers $\{\ldots, -3, -2, -1, 0, 1, 2, 3, \ldots\}$. The integers in the given set are $-9, 0, 3,$ and 7.

(d) Rational numbers are the numbers which can be expressed as the quotient of two integers, with denominators not equal to 0.

We can write numbers from the given set in this form as follows:

$$-9 = \frac{-9}{1}, \quad -1\frac{1}{4} = \frac{-5}{4}, \quad -\frac{3}{5} = \frac{-3}{5}, \quad 0 = \frac{0}{1},$$

$$3 = \frac{3}{1}, \quad 5.9 = \frac{59}{10}, \quad \text{and} \quad 7 = \frac{7}{1}.$$

Thus, the rational numbers in the given set are

$$-9, \quad -1\frac{1}{4}, \quad -\frac{3}{5}, \quad 0, \quad 3, \quad 5.9, \quad \text{and} \quad 7.$$

(e) Irrational numbers are real numbers that are not rational. $-\sqrt{7}$ and $\sqrt{5}$ can be represented by points on the number line but cannot be written as a quotient of integers. Thus, the irrational numbers in the given set are $-\sqrt{7}$ and $\sqrt{5}$.

(f) Real numbers are all numbers that can be represented on the number line. All the numbers in the given set are real.

13.   The statement $0.75 \neq \frac{3}{4}$ can be written as 75 hundredths does not equal three-fourths. It is *false* because $0.75 = \frac{3}{4}$.

15.   The statement $-4 \leq -5$ can be written as negative 4 is less than or equal to negative 5. It is *false* because $-4 > -5$ and $-4 \neq -5$.

17.   The statement $0 \geq 0$ can be written as zero is greater than or equal to zero. It is *true* because $0 = 0$.

19. $12 < 19$ becomes $19 > 12$.

21. $\frac{4}{5} \geq \frac{1}{2}$ becomes $\frac{1}{2} \leq \frac{4}{5}$.

23. $-17 \leq 1 - 18$
$-17 \leq -17$ *True*

25. $-6(8) + 9(5) \geq 0$
$-48 + 45 \geq 0$
$-3 \geq 0$ *False*

27. $6[5 - 3(4 - 2)] \neq 6$
$6[5 - 3(2)] \neq 6$
$6[5 - 6] \neq 6$
$6[-1] \neq 6$
$-6 \neq 6$ *True*

29. $3^2[(-10 \div 5) + 7] \leq 40$
$9[(-10 \div 5) + 7] \leq 40$
$9[-2 + 7] \leq 40$
$9[5] \leq 40$
$45 \leq 40$ *False*

31. $[6 - 4(4)] \div (-10) \geq 1$
$[6 - 16] \div (-10) \geq 1$
$[-10] \div (-10) \geq 1$
$1 \geq 1$ *True*

33. $0 \neq 12 \div [3(2) - 6]$
$0 \neq 12 \div [6 - 6]$
$0 \neq 12 \div [0]$
$0 \neq$ undefined   *True*

35. **(a)** The film that had gross receipts greater than 658.7 million dollars is *Avatar*. (Note that the sales of *Titanic* were *equal* to 658.7 million dollars, not *greater than*.)

(b) Since *Titanic* had gross receipts of 658.7 million dollars, the films that had gross receipts greater than or equal to 658.7 million dollars were *Avatar* and *Titanic*.

37. Answers will vary. One possibility is:

gross receipts ≤ 617.8 million dollars.

39. $-(6h - n)$
$-1(6h - n)$ *Rewrite with* -1.
$(-1 \cdot 6h) - (-1 \cdot n)$ *Distributive property*
$-6h - (-n)$
$-6h + n$ *Change to opposite.*

41. $-(-3q + 5r - 8s)$
$-1(-3q + 5r - 8s)$
$(-1 \cdot -3q) + (-1 \cdot 5r) - (-1 \cdot 8s)$
$3q + (-5r) - (-8s)$
$3q - 5r + 8s$

43. $13p + 4(4 - 8p)$
$13p + 16 - 32p$
$-19p + 16$

45. $-4(y - 7) - 6$
$-4y + 28 - 6$
$-4y + 22$

47. $-(6 - y) + y^2 - 6$
$-1(6 - y) + y^2 - 6$
$-6 + y + y^2 - 6$
$y^2 + y - 12$

49. $2(3b^2 - b) - 4(b - 2)$
$6b^2 - 2b - 4b + 8$
$6b^2 - 6b + 8$

51. $-3(-a + 1) - (2a - 4)$
$3a - 3 - 2a + 4$
$a + 1$

53. $-10(-3k - 2) + 6(-4 - k)$
$30k + 20 - 24 - 6k$
$24k - 4$

10.2 More on Solving Linear Equations

10.2 Margin Exercises

1. **(a)** *Step 1*   *Distributive property*
$2p + 4 = 7(p - 2) + p$
$2p + 4 = 7p - 14 + p$

Step 2   *Combine like terms.*
$2p + 4 = \underline{8p} - 14$

Step 3   *Add or subtract.*
$2p + 4 - 4 = 8p - 14 - 4$ *Subtract 4.*
$2p = 8p - 18$
$2p - 8p = 8p - 18 - 8p$ *Subtract 8p.*
$-6p = -18$

Step 4   *Multiply or divide.*
$\dfrac{-6p}{-6} = \dfrac{-18}{-6}$ *Divide by* -6.
$p = 3$

Step 5   **Check** $p = 3$.
$2p + 4 = 7(p - 2) + p$
$2(3) + 4 = 7(3 - 2) + 3$
$6 + 4 = 7(1) + 3$
$10 = 7 + 3$
$10 = 10$ Balances

The solution is 3.

(b) *Step 2*   *Combine like terms.*
$-5y + 7y - 6y - 9 = 3 + 2y$
$-4y - 9 = 3 + 2y$

Step 3   *Add or subtract.*
$-4y - 9 + 9 = 3 + 2y + 9$ *Add 9.*
$-4y = 12 + 2y$
$-4y - 2y = 12 + 2y - 2y$ *Subtract 2y.*
$-6y = 12$

Step 4   *Multiply or divide.*
$\dfrac{-6y}{-6} = \dfrac{12}{-6}$ *Divide by* -6.
$y = -2$

Step 5   **Check** $y = -2$.
$-5y + 7y - 6y - 9 = 3 + 2y$
$-5(-2) + 7(-2) - 6(-2) - 9 = 3 + 2(-2)$
$10 - 14 + 12 - 9 = 3 - 4$
$-4 + 12 - 9 = -1$
$8 - 9 = -1$
$-1 = -1$
       Balances

The solution is -2.

2. **(a)** *Step 1*   *Distributive property*
$2m - (7m + 6) = 39$
$2m - 1(7m + 6) = 39$
$2m - \underline{7m} - \underline{6} = 39$

Step 2   *Combine like terms.*
$-5m - 6 = 39$

Step 3   *Add or subtract.*
$$-5m - 6 + 6 = 39 + 6 \quad Add\ 6.$$
$$-5m = 45$$

Step 4   *Multiply or divide.*
$$\frac{-5m}{-5} = \frac{45}{-5} \qquad Divide\ by\ -5.$$
$$m = -9$$

Step 5   **Check** $m = -9.$
$$2m - (7m + 6) = 39$$
$$2(-9) - [7(-9) + 6] = 39$$
$$-18 - (-63 + 6) = 39$$
$$-18 - (-57) = 39$$
$$-18 + 57 = 39$$
$$39 = 39 \quad Balances$$

The solution is $-9.$

(b) ***Step 1***   *Distributive property*
$$6 = 4x - (3 - 2x)$$
$$6 = 4x - 3 + 2x$$

Step 2   *Combine like terms.*
$$6 = 6x - 3$$

Step 3   *Add or subtract.*
$$6 + 3 = 6x - 3 + 3 \qquad Add\ 3.$$
$$9 = 6x$$

Step 4   *Multiply or divide.*
$$\frac{9}{6} = \frac{6x}{6} \qquad Divide\ by\ 6.$$
$$\frac{3}{2} = x$$

Step 5   **Check** $x = \frac{3}{2}.$
$$6 = 4x - (3 - 2x)$$
$$6 = 4\left(\tfrac{3}{2}\right) - [3 - 2\left(\tfrac{3}{2}\right)]$$
$$6 = 6 - (3 - 3)$$
$$6 = 6 - 0$$
$$6 = 6 \qquad\qquad Balances$$

The solution is $\frac{3}{2}.$

3.   (a) $2(4 - 3r) = 3(r + 1) + 14$

$$\underline{8} - \underline{6r} = \underline{3r} + \underline{3} + 14 \qquad \begin{array}{l}Distributive\\property\end{array}$$
$$8 - 6r = 3r + 17 \qquad Combine\ terms.$$
$$\underline{-8 \qquad\qquad -8} \quad Subtract\ 8.$$
$$-6r = 3r + 9$$
$$\underline{-3r \qquad -3r} \qquad Subtract\ 3r.$$
$$-9r = 9$$
$$\frac{-9r}{-9} = \frac{9}{-9} \qquad Divide\ by\ -9.$$
$$r = -1$$

Check $r = -1:$ $14 = 14$ Balances

The solution is $-1.$

(b) $2 - 3(2 + 6z) = 4(z + 1)$

$$2 - 6 - 18z = 4z + 4 \qquad \begin{array}{l}Distributive\\property\end{array}$$
$$-4 - 18z = 4z + 4 \quad Combine\ terms.$$
$$\underline{-4z \qquad -4z} \quad Subtract\ 4z.$$
$$-4 - 22z = 4$$
$$\underline{+4 \qquad\qquad +4} \qquad Add\ 4.$$
$$-22z = 8$$
$$\frac{-22z}{-22} = \frac{8}{-22} \qquad Divide\ by\ -22.$$
$$z = -\frac{4}{11}$$

Check $z = -\frac{4}{11}:$ $\frac{28}{11} = \frac{28}{11}$ Balances

The solution is $-\frac{4}{11}.$

4.   (a) $2(x - 6) = 3 + 2x - 15$

$$2x - 12 = 3 + 2x - 15 \qquad \begin{array}{l}Distributive\\property\end{array}$$
$$2x - 12 = 2x - 12 \qquad Combine\ terms.$$
$$\underline{+12 \qquad\quad +12} \qquad Add\ 12.$$
$$2x = 2x$$
$$\underline{-2x \qquad -2x} \qquad Subtract\ 2x.$$
$$0 = 0 \qquad\qquad True\ statement$$

The solution is *all real numbers.*

(b) $4b + 2(3 - 2b) = 6$

$$4b + 6 - 4b = 6 \qquad \begin{array}{l}Distributive\\property\end{array}$$
$$6 = 6 \qquad \begin{array}{l}Combine\ terms.\\True\ statement\end{array}$$

The solution is *all real numbers.*

5.   (a) $9y - 4 = 3y + 6(y + 1)$

$$9y - 4 = 3y + 6y + 6 \qquad \begin{array}{l}Distributive\\property\end{array}$$
$$9y - 4 = 9y + 6 \qquad Combine\ terms.$$
$$\underline{-9y \qquad\quad -9y} \qquad Subtract\ 9y.$$
$$-4 = 6 \qquad\qquad False\ statement$$

The equation has *no solution.*

(b) $2m - 5(m + 2) = -3(m - 5)$

$$2m - 5m - 10 = -3m + 15 \qquad \begin{array}{l}Distributive\\property\end{array}$$
$$-3m - 10 = -3m + 15 \quad Combine\ terms.$$
$$\underline{+3m \qquad\qquad +3m} \qquad Add\ 3m.$$
$$-10 = 15 \qquad\qquad False\ statement$$

The equation has *no solution.*

6. $\dfrac{1}{4}x - 4 = \dfrac{3}{2}x + \dfrac{3}{4}x$

The LCD of all the fractions in the equation is $\underline{4}$, so multiply *both* sides by 4 to clear the fractions.

$$4\left(\dfrac{1}{4}x - 4\right) = 4\left(\dfrac{3}{2}x + \dfrac{3}{4}x\right)$$

$$4\left(\dfrac{1}{4}x\right) - 4(4) = 4\left(\dfrac{3}{2}x\right) + 4\left(\dfrac{3}{4}x\right)$$

$$\begin{aligned} x - 16 &= 6x + 3x && \textit{Dist. property} \\ x - 16 &= 9x && \textit{Combine terms.} \\ x - 16 - x &= 9x - x && \textit{Subtract x.} \\ -16 &= 8x \\ \dfrac{-16}{8} &= \dfrac{8x}{8} && \textit{Divide by 8.} \\ -2 &= x \end{aligned}$$

Check $x = -2$: $-\dfrac{9}{2} = -\dfrac{9}{2}$ Balances

The solution is -2.

7. $0.06(100 - y) + 0.4y = 0.05(86)$

To clear decimals, multiply both sides by $\underline{100}$. Move the decimal point $\underline{2}$ places to the $\underline{right}$.

$$100[0.06(100 - y) + 0.4y] = 100[0.05(86)]$$

$$\begin{aligned} 6(100 - y) + 40y &= 5(86) && \textit{Dist. prop.} \\ 600 - 6y + 40y &= 430 && \textit{Dist. prop.} \\ 600 + 34y &= 430 && \textit{Comb. terms.} \\ -600 && -600 && \textit{Subtract 600.} \\ \hline 34y &= -170 \\ \dfrac{34y}{34} &= \dfrac{-170}{34} && \textit{Divide by 34.} \\ y &= -5 \end{aligned}$$

Check $y = -5$: $4.3 = 4.3$ Balances

The solution is -5.

10.2   Section Exercises

1. **(a)** $\dfrac{3}{2}x + 1 = \dfrac{2}{3}x - \dfrac{1}{6}$ ▪ Clear <u>fractions</u> by multiplying both sides by $\underline{6}$, the LCD of 2, 3, and 6.

(b) $6 = 0.09m + 0.3$ ▪ Clear <u>decimals</u> by multiplying both sides by $\underline{100}$ since the decimal needs to be moved 2 places to the right in 0.09.

Use the five-step method for solving linear equations as given in the text. The details of these steps will only be shown for a few of the exercises.

3. $5m + 8 = 7 + 4m$

For this equation, steps 1 and 2 are not needed.

Step 3
$$\begin{aligned} 5m + 8 - 8 &= 7 + 4m - 8 && \textit{Subtract 8.} \\ 5m &= 4m - 1 \\ 5m - 4m &= 4m - 1 - 4m && \textit{Subtract 4m.} \\ m &= -1 \end{aligned}$$

For this equation, step 4 is not needed since the coefficient of m is already 1.

Step 5
Substitute -1 for m in the original equation.

$$\begin{aligned} 5m + 8 &= 7 + 4m \\ 5(-1) + 8 &= 7 + 4(-1) && \textit{Let m = -1.} \\ -5 + 8 &= 7 + (-4) \\ 3 &= 3 && \textit{True} \end{aligned}$$

The solution is -1.

5. ***Step 1***
$$\begin{aligned} 10p + 6 &= 4(3p - 1) \\ 10p + 6 &= 12p - 4 \end{aligned}$$

For this equation, step 2 is not needed.

Step 3
$$\begin{aligned} 10p + 6 - 12p &= 12p - 4 - 12p && \textit{Subtract 12p.} \\ -2p + 6 &= -4 \\ -2p + 6 - 6 &= -4 - 6 && \textit{Subtract 6.} \\ -2p &= -10 \end{aligned}$$

Step 4
$$\begin{aligned} \dfrac{-2p}{-2} &= \dfrac{-10}{-2} && \textit{Divide by } -2. \\ p &= 5 \end{aligned}$$

Step 5 **Check** $p = 5$: $56 = 56$ Balances

The solution is 5.

7. ***Step 2***
$$\begin{aligned} 7r - 5r + 2 &= 5r - r \\ 2r + 2 &= 4r && \textit{Combine terms.} \end{aligned}$$

Step 3
$$\begin{aligned} 2r + 2 - 2 &= 4r - 2 && \textit{Subtract 2.} \\ 2r &= 4r - 2 \\ 2r - 4r &= 4r - 2 - 4r && \textit{Subtract 4r.} \\ -2r &= -2 \end{aligned}$$

Step 4
$$\begin{aligned} \dfrac{-2r}{-2} &= \dfrac{-2}{-2} && \textit{Divide by } -2. \\ r &= 1 \end{aligned}$$

Step 5 **Check** $r = 1$: $4 = 4$ Balances

The solution is 1.

9. ***Step 1***
$$\begin{aligned} x + 3 &= -(2x + 2) \\ x + 3 &= -1(2x + 2) \\ x + 3 &= -2x - 2 && \textit{Distributive property} \end{aligned}$$

Step 3
$$\begin{aligned} x + 3 + 2x &= -2x - 2 + 2x && \textit{Add 2x.} \\ 3x + 3 &= -2 \\ 3x + 3 - 3 &= -2 - 3 && \textit{Subtract 3.} \\ 3x &= -5 \end{aligned}$$

Step 4

$$\frac{3x}{3} = \frac{-5}{3} \quad \textit{Divide by 3.}$$

$$x = -\frac{5}{3}$$

Step 5 **Check** $x = -\frac{5}{3}$: $\quad \frac{4}{3} = \frac{4}{3} \quad$ Balances

The solution is $-\frac{5}{3}$.

11. $\quad 4(2x - 1) = -6(x + 3)$

$\qquad 8x - 4 = -6x - 18$

$\qquad 14x - 4 = -18 \qquad \textit{Add 6x.}$

$\qquad 14x = -14 \qquad \textit{Add 4.}$

$\qquad x = -1 \qquad \textit{Divide by 14.}$

Check $x = -1$: $\quad -12 = -12 \quad$ Balances

The solution is -1.

13. $\quad 2(p + 5) - (9 + p) = -3$

$\qquad 2p + 10 - 9 - p = -3$

$\qquad p + 1 = -3$

$\qquad p = -4 \quad \textit{Subtract 1.}$

Check $p = -4$: $\quad 2 - 5 = -3 \quad$ Balances

The solution is -4.

15. $\quad -6(2b + 1) + (13b - 7) = 0$

$\qquad -12b - 6 + 13b - 7 = 0$

$\qquad b - 13 = 0$

$\qquad b = 13 \quad \textit{Add 13.}$

Check $b = 13$: $\quad -162 + 162 = 0 \quad$ Balances

The solution is 13.

17. $\quad -2(8p + 2) - 3(2 - 7p) = 2(4 + 2p)$

$\qquad -16p - 4 - 6 + 21p = 8 + 4p$

$\qquad 5p - 10 = 4p + 8$

$\qquad 5p = 4p + 18 \quad \textit{Add 10.}$

$\qquad p = 18 \quad \textit{Subt. 4p.}$

Check $p = 18$: $\quad -292 + 372 = 80 \quad$ Balances

The solution is 18.

19. $\quad 4(7x - 1) + 3(2 - 5x) = 4(3x + 5) - 6$

$\qquad 28x - 4 + 6 - 15x = 12x + 20 - 6$

$\qquad 13x + 2 = 12x + 14$

$\qquad x + 2 = 14 \quad \textit{Subt. 12x.}$

$\qquad x = 12 \quad \textit{Subt. 2.}$

Check $x = 12$: $\quad 332 - 174 = 158 \quad$ Balances

The solution is 12.

21. $\quad 3x + 9 = 3x + 8$

$\qquad 3x + 1 = 3x \quad \textit{Subtract 8.}$

$\qquad 1 = 0 \quad \textit{Subtract 3x.}$

Because $1 = 0$ is a *false* statement, the equation has *no solution*.

23. $\quad 8x + 1 = 1 + 8x$

$\qquad 8x = 8x \quad \textit{Subtract 1.}$

$\qquad 0 = 0 \quad \textit{Subtract 8x.}$

Since $0 = 0$ is a *true* statement, *all real numbers* are solutions of the equation.

25. $\quad 6x + 5 - 7x + 3 = 5x - 6x - 4$

$\qquad -x + 8 = -x - 4$

$\qquad 8 = -4 \quad \textit{Add x.}$

Because $8 = -4$ is a *false* statement, the equation has *no solution.*

27. $\quad 6(4x - 1) = 12(2x + 3)$

$\qquad 24x - 6 = 24x + 36$

$\qquad -6 = 36 \quad \textit{Subtract 24x.}$

Because $-6 = 36$ is a *false* statement, the equation has *no solution.*

29. $\quad 3(2x - 4) = 6(x - 2)$

$\qquad 6x - 12 = 6x - 12$

$\qquad -12 = -12 \quad \textit{Subtract 6x.}$

Since $-12 = -12$ is a *true* statement, *all real numbers* are solutions of the equation.

31. $\quad 10(-2x + 1) = -14(x + 2) + 38 - 6x$

$\qquad -20x + 10 = -14x - 28 + 38 - 6x$

$\qquad -20x + 10 = -20x + 10$

$\qquad 10 = 10 \quad \textit{Add 20x.}$

Since $10 = 10$ is a *true* statement, *all real numbers* are solutions of the equation.

33. No, it is incorrect to divide both sides by a variable. If $-3x$ is added to both sides, the equation becomes $4x = 0$, so $x = 0$ and 0 is the correct solution.

35. $\quad \frac{3}{5}t - \frac{1}{10}t = t - \frac{5}{2} \quad$ ■ The least common denominator of all the fractions in the equation is 10.

$$10\left(\frac{3}{5}t - \frac{1}{10}t\right) = 10\left(t - \frac{5}{2}\right)$$

$$\textit{Multiply both sides by 10.}$$

$$10\left(\frac{3}{5}t\right) + 10\left(-\frac{1}{10}t\right) = 10t + 10\left(-\frac{5}{2}\right)$$

$$\textit{Distributive property}$$

$$6t - t = 10t - 25$$

$$5t = 10t - 25$$

$$-5t = -25 \quad \textit{Subtract 10t.}$$

$$\frac{-5t}{-5} = \frac{-25}{-5} \quad \textit{Divide by -5.}$$

$$t = 5$$

Check $t = 5$: $\quad \frac{5}{2} = \frac{5}{2} \quad$ Balances

The solution is 5.

37. $-\dfrac{1}{4}(x-12)+\dfrac{1}{2}(x+2)=x+4$ ▪ The LCD of all the fractions is 4.

$$4\left[-\dfrac{1}{4}(x-12)+\dfrac{1}{2}(x+2)\right]$$

$$=4(x+4)\qquad \text{\textit{Multiply by 4.}}$$

$$4\left(-\dfrac{1}{4}\right)(x-12)+4\left(\dfrac{1}{2}\right)(x+2)$$

$$=4x+16\qquad \text{\textit{Dist. prop.}}$$

$$(-1)(x-12)+2(x+2)=4x+16\qquad \text{\textit{Multiply.}}$$

$$-x+12+2x+4=4x+16\qquad \text{\textit{Dist. prop.}}$$

$$x+16=4x+16$$
$$-3x+16=16$$
$$-3x=0$$
$$\dfrac{-3x}{-3}=\dfrac{0}{-3}\qquad \text{\textit{Divide by }-3.}$$
$$x=0$$

Check $x=0$: $4=4$ Balances

The solution is 0.

39. $\dfrac{2}{3}k-\left(k+\dfrac{1}{4}\right)=\dfrac{1}{12}(k+4)$ ▪ The least common denominator of all the fractions in the equation is 12, so multiply both sides by 12 and solve for k.

$$12\left[\tfrac{2}{3}k-\left(k+\tfrac{1}{4}\right)\right]=12\left[\tfrac{1}{12}(k+4)\right]$$
$$12\left(\tfrac{2}{3}k\right)-12\left(k+\tfrac{1}{4}\right)=12\left[\tfrac{1}{12}(k+4)\right]$$
$$\text{\textit{Distributive property}}$$
$$8k-12k-12\left(\tfrac{1}{4}\right)=1(k+4)$$
$$8k-12k-3=k+4$$
$$-4k-3=k+4$$
$$-5k-3=4$$
$$-5k=7$$
$$k=-\tfrac{7}{5}$$

Check $k=-\tfrac{7}{5}$: $\tfrac{13}{60}=\tfrac{13}{60}$ Balances

The solution is $-\tfrac{7}{5}$.

41. $\qquad 5.6x+2=4.6x$

$$10(5.6x+2)=10(4.6x)\qquad \text{\textit{Multiply by \underline{10}.}}$$
$$56x+20=46x\qquad \text{\textit{Dist. property}}$$
$$10x+20=0\qquad \text{\textit{Subtract 46x.}}$$
$$10x=-20\qquad \text{\textit{Subtract 20.}}$$
$$x=-2\qquad \text{\textit{Divide by 10.}}$$

Check $x=-2$: $-11.2+2=-9.2$ Balances

The solution is -2.

43. $5.2q-4.6-7.1q=-2.1-1.9q-2.5$
$$52q-46-71q=-21-19q-25$$
$$\text{\textit{Multiply both sides by 10.}}$$
$$-19q-46=-19q-46$$
$$-46=-46\qquad \text{\textit{Add 19q.}}$$

Since $-46=-46$ is a *true* statement, *all real numbers* are solutions of the equation.

45. $0.2(60)+0.05x=0.1(60+x)$ ▪ To clear the decimal in 0.2 and 0.1, we need to multiply the equation by 10. But to clear the decimal in 0.05, we need to multiply by 100, so we choose <u>100</u>.

$$100[0.2(60)+0.05x]=100[0.1(60+x)]$$
$$\text{\textit{Multiply by 100.}}$$
$$100[0.2(60)]+100(0.05x)=100[0.1(60+x)]$$
$$\text{\textit{Distributive property}}$$
$$20(60)+5x=10(60+x)$$
$$\text{\textit{Multiply.}}$$
$$1200+5x=600+10x$$
$$1200-5x=600$$
$$-5x=-600$$
$$x=\dfrac{-600}{-5}=120$$

Check $x=120$: $18=18$ Balances

The solution is 120.

47. $x+0.05(12-x)=0.1(63)$ ▪ To clear the equation of decimals, we multiply both sides by 100.

$$100[x+0.05(12-x)]=100[0.1(63)]$$
$$100(x)+100[0.05(12-x)]=(100)(0.1)(63)$$
$$100x+5(12-x)=10(63)$$
$$100x+60-5x=630$$
$$95x+60=630$$
$$95x=570$$
$$x=\dfrac{570}{95}=6$$

Check $x=6$: $6.3=6.3$ Balances

The solution is 6.

49. $0.06(10,000)+0.08x=0.072(10,000+x)$
$$1000[0.06(10,000)]+1000(0.08x)=$$
$$1000[0.072(10,000+x)]$$
$$\text{\textit{Multiply both sides by 1000, not 100.}}$$
$$60(10,000)+80x=72(10,000+x)$$
$$600,000+80x=720,000+72x$$
$$600,000+8x=720,000$$
$$8x=120,000$$
$$x=\dfrac{120,000}{8}=15,000$$

Check $x=15,000$: $1800=1800$ Balances

The solution is 15,000.

51. $9(v+1) - 3v = 2(3v+1) - 8$
$9v + 9 - 3v = 6v + 2 - 8$
$6v + 9 = 6v - 6$
$9 = -6$

Because $9 = -6$ is a *false* statement, the equation has *no solution*.

53. $\dfrac{1}{2}(x+2) + \dfrac{3}{4}(x+4) = x+5$ ■ To clear fractions, multiply both sides by the LCD, which is 4.

$$4\left[\tfrac{1}{2}(x+2) + \tfrac{3}{4}(x+4)\right] = 4(x+5)$$
$$4\left(\tfrac{1}{2}\right)(x+2) + 4\left(\tfrac{3}{4}\right)(x+4) = 4x+20$$
$$2(x+2) + 3(x+4) = 4x+20$$
$$2x + 4 + 3x + 12 = 4x+20$$
$$5x + 16 = 4x+20$$
$$x + 16 = 20$$
$$x = 4$$

Check $x = 4$: $9 = 9$ Balances

The solution is 4.

55. $-(4y+2) - (-3y-5) = 3$
$-1(4y+2) - 1(-3y-5) = 3$
$-4y - 2 + 3y + 5 = 3$
$-y + 3 = 3$
$-y = 0$
$y = 0$

Check $y = 0$: $3 = 3$ Balances

The solution is 0.

57. $0.10(x+80) + 0.20x = 14$ ■ To clear the decimals, multiply both sides by 10.

$$10[0.10(x+80) + 0.20x] = 10(14)$$
$$1(x+80) + 2x = 140$$
$$x + 80 + 2x = 140$$
$$3x + 80 = 140$$
$$3x = 60$$
$$x = 20$$

Check $x = 20$: $14 = 14$ Balances

The solution is 20.

59. $4(x+8) = 2(2x+6) + 20$
$4x + 32 = 4x + 12 + 20$
$4x + 32 = 4x + 32$
$32 = 32$

Since $32 = 32$ is a *true* statement, *all real numbers* are solutions of the equation.

61. $-2(2s-4) - 8 = -3(4s+4) - 1$
$-4s + 8 - 8 = -12s - 12 - 1$
$-4s = -12s - 13$
$8s = -13$ *Add 12s.*
$\dfrac{8s}{8} = \dfrac{-13}{8}$ *Divide by 8.*
$s = -\dfrac{13}{8}$

Check $s = -\dfrac{13}{8}$: $\dfrac{13}{2} = \dfrac{13}{2}$ Balances

The solution is $-\dfrac{13}{8}$.

10.3 Formulas and Solving for a Specified Variable

10.3 Margin Exercises

1. **(a)** Given the formula $I = prt$, the problem is to find p when $I = \$246$, $r = 0.06$, and $t = 2$. Substitute these values into the formula, and solve for p.

$$I = prt$$
$$\$246 = p(0.06)(2)$$
$$\$246 = 0.12p \quad \textit{Multiply.}$$
$$\dfrac{\$246}{0.12} = \dfrac{0.12p}{0.12} \quad \textit{Divide by 0.12.}$$
$$\$2050 = p$$

(b) Using $P = 2l + 2w$ as the formula, find l when $P = 126$ and $w = 25$. Substitute these values into the formula, and solve for l.

$$P = 2l + 2w$$
$$126 = 2l + 2(25)$$
$$126 = 2l + 50$$
$$126 - 50 = 2l + 50 - 50 \; \textit{Subtract 50.}$$
$$76 = 2l$$
$$\dfrac{76}{2} = \dfrac{2l}{2} \quad \textit{Divide by 2.}$$
$$38 = l$$

2. The fence will enclose the perimeter of the rectangular field, so use the formula for the perimeter of a rectangle. Find the length of the field by substituting $P = 800$ and $w = 175$ into the formula and solving for l.

$$P = 2l + 2w$$
$$800 = 2l + 2(175)$$
$$800 = 2l + 350$$
$$800 - 350 = 2l + 350 - 350 \; \textit{Subtract 350.}$$
$$450 = 2l$$
$$\dfrac{450}{2} = \dfrac{2l}{2} \quad \textit{Divide by 2.}$$
$$225 = l$$

The length of the field is 225 meters.

3. **(a)** Solve $I = prt$ for t. Our goal is to isolate $\underline{t}$.

$$I = prt$$
$$\frac{I}{pr} = \frac{prt}{pr} \qquad \textit{Divide by pr.}$$
$$\frac{I}{pr} = t, \ \text{ or } \ t = \frac{I}{pr}$$

(b) Solve $P = a + b + c$ for a.

$$P = a + b + c$$
$$P - b - c = a + b + c - b - c$$
$$\qquad\qquad\qquad \textit{Subtract b and c.}$$
$$P - b - c = a, \ \text{ or } \ a = P - b - c$$

4. **(a)** Solve $A = p + prt$ for t.

$$A = p + prt$$
$$A - p = p + prt - p \qquad \textit{Subtract p.}$$
$$A - p = prt$$
$$\frac{A - p}{pr} = \frac{prt}{pr} \qquad \textit{Divide by pr.}$$
$$\frac{A - p}{pr} = t, \ \text{ or } \ t = \frac{A - p}{pr}$$

(b) Solve $Ax + By = C$ for y.

$$Ax + By = C$$
$$Ax + By - Ax = C - Ax \quad \textit{Subtract Ax.}$$
$$By = C - Ax$$
$$\frac{By}{B} = \frac{C - Ax}{B} \qquad \textit{Divide by B.}$$
$$y = \frac{C - Ax}{B}$$

10.3 Section Exercises

1. **(a)** Perimeter is the distance around the outside edges of the figure.

(b) You would find the perimeter for "fencing around a park" and "lace trim for the edges of a scarf."

In Exercises 3–14, substitute the given values into the formula and then solve for the remaining variable.

3. $P = 2l + 2w$; $P = 20$, $w = 4$

$$P = 2l + 2w$$
$$20 = 2l + 2(4)$$
$$20 = 2l + 8$$
$$12 = 2l$$
$$6 = l$$

5. $A = \frac{1}{2}bh$; $A = 70, b = 10$

$$A = \frac{1}{2}bh$$
$$70 = \frac{1}{2}(10)h$$
$$70 = 5h$$
$$\frac{70}{5} = \frac{5h}{5}$$
$$14 = h$$

7. $P = a + b + c$; $P = 15, a = 3, b = 7$

$$P = a + b + c$$
$$15 = 3 + 7 + c$$
$$15 = 10 + c$$
$$5 = c$$

9. $d = rt$; $d = 100$, $t = 2.5$

$$d = rt$$
$$100 = r(2.5)$$
$$100 = 2.5r$$
$$\frac{100}{2.5} = \frac{2.5r}{2.5}$$
$$40 = r$$

11. $C = 2\pi r$; $C = 8.164$, $\pi \approx 3.14$

$$C = 2\pi r$$
$$8.164 \approx 2(3.14)r$$
$$8.164 \approx 6.28r$$
$$\frac{8.164}{6.28} \approx \frac{6.28r}{6.28}$$
$$1.3 \approx r$$

13. $V = lwh$; $V = 384, l = 12, h = 4$

$$V = lwh$$
$$384 = (12)w(4)$$
$$384 = 48w$$
$$\frac{384}{48} = \frac{48w}{48}$$
$$8 = w$$

15. A page of the newspaper is a rectangle with length 51 inches and perimeter 172 square inches, so use the formulas for the perimeter and area of a rectangle.

(a) $P = 2l + 2w$
$$172 = 2(51) + 2w$$
$$172 = 102 + 2w$$
$$70 = 2w$$
$$\frac{70}{2} = \frac{2w}{2}$$
$$w = 35 \text{ inches}$$

(b) $A = lw$
$$A = (51)(35)$$
$$A = 1785 \text{ square inches}$$

17. The circumference is 1978 feet. Use the circumference of a circle formula.

$$C = 2\pi r$$
$$C = \pi d \qquad 2r = d$$
$$1978 \approx (3.14)d$$
$$d \approx 630 \text{ ft} \qquad \textit{Divide by 3.14.}$$

19. $d = rt$ for r

$$d = rt$$

$$\frac{d}{t} = \frac{rt}{t} \qquad \textit{Divide by t.}$$

$$\frac{d}{t} = r, \text{ or } r = \frac{d}{t}$$

21. $A = lw$ for l

$$A = lw$$

$$\frac{A}{w} = \frac{lw}{w} \qquad \textit{Divide by w.}$$

$$\frac{A}{w} = l, \text{ or } l = \frac{A}{w}$$

23. $P = a + b + c$ for a

$$P = a + b + c$$

$$P - b - c = a + b + c - b - c$$

$$\textit{Subtract b and c.}$$

$$P - b - c = a, \text{ or } a = P - b - c$$

25. $I = prt$ for p

$$I = prt$$

$$\frac{I}{rt} = \frac{prt}{rt} \qquad \textit{Divide by rt.}$$

$$\frac{I}{rt} = p \text{ or } p = \frac{I}{rt}$$

27. $A = \frac{1}{2}bh$ for b

$$A = \tfrac{1}{2}bh$$

$$2A = 2\left(\tfrac{1}{2}bh\right) \qquad \textit{Multiply by 2.}$$

$$2A = bh$$

$$\frac{2A}{h} = \frac{bh}{h} \qquad \textit{Divide by h.}$$

$$\frac{2A}{h} = b, \text{ or } b = \frac{2A}{h}$$

29. $A = p + prt$ for r

$$A = p + prt$$

$$A - p = p + prt - p \quad \textit{Subtract p.}$$

$$A - p = prt$$

$$\frac{A - p}{pt} = \frac{prt}{pt} \qquad \textit{Divide by pt.}$$

$$\frac{A - p}{pt} = r, \text{ or } r = \frac{A - p}{pt}$$

31. $V = \pi r^2 h$ for h

$$V = \pi r^2 h$$

$$\frac{V}{\pi r^2} = \frac{\pi r^2 h}{\pi r^2} \qquad \textit{Divide by } \pi r^2.$$

$$\frac{V}{\pi r^2} = h, \text{ or } h = \frac{V}{\pi r^2}$$

33. $V = \frac{1}{3}Bh$ for B

$$V = \tfrac{1}{3}Bh$$

$$3V = 3\left(\tfrac{1}{3}Bh\right) \qquad \textit{Multiply by 3.}$$

$$3V = Bh$$

$$\frac{3V}{h} = \frac{Bh}{h} \qquad \textit{Divide by h.}$$

$$\frac{3V}{h} = B, \text{ or } B = \frac{3V}{h}$$

10.4   Solving Linear Inequalities

10.4   Margin Exercises

1.   **(a)** $x \le 3$ ▪ The statement $x \le 3$ says that x can be any real number less than or equal to 3. To graph this inequality, place a closed circle at $\underline{3}$ on a number line and draw an arrow extending from the circle to the <u>left</u>.

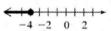

   (b) $x > -4$ ▪ x can be any real number greater than -4, but not equal to -4. To graph this inequality, place an open circle at -4 on a number line and draw an arrow to the right.

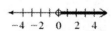

   (c) $-4 \ge x$ ▪ $-4 \ge x$ is the same as $x \le -4$. Graph this inequality by placing a closed circle at -4 on a number line and drawing an arrow to the left.

   (d) $0 < x$ ▪ $0 < x$ is the same as $x > 0$. Graph this inequality by placing an open circle at 0 on a number line and drawing an arrow to the right.

2.   **(a)** $-7 < x < -2$ ▪ To graph this inequality, place open circles at -7 and -2 on a number line. Then draw a line segment between the two circles.

   (b) $-6 < x \le -4$ ▪ To graph this inequality, draw an open circle at -6 and a closed circle at -4 on a number line. Then draw a line segment between the open circle and the closed circle.

3. **(a)**
$$-1 + 8r < 7r + 2$$
$$-1 + 8r - 7r < 7r + 2 - 7r \quad \textit{Subtract 7r.}$$
$$-1 + r < 2$$
$$-1 + r + 1 < 2 + 1 \quad \textit{Add 1.}$$
$$r < 3$$

To graph this inequality, place an open circle at 3 on a number line and draw an arrow to the left.

(b) $4m \geq 3m - 1$
$$m \geq -1 \quad \textit{Subtract 3m.}$$

To graph this inequality, place a closed circle at -1 on a number line and draw an arrow to the right.

4. **(a) (i)**
$$-2 < 8$$
$$6(-2) < 6(8) \quad \textit{Multiply by 6.}$$
$$-12 < 48 \quad \textit{True}$$

(ii)
$$-2 < 8$$
$$-5(-2) > -5(8) \quad \textit{Multiply by } -5 \textit{ and reverse the symbol.}$$
$$10 > -40 \quad \textit{True}$$

(b) (i)
$$-4 > -9$$
$$2(-4) > 2(-9) \quad \textit{Multiply by 2.}$$
$$-8 > -18 \quad \textit{True}$$

(ii)
$$-4 > -9$$
$$-8(-4) < -8(-9) \quad \textit{Multiply by } -8 \textit{ and reverse the symbol.}$$
$$32 < 72 \quad \textit{True}$$

5. **(a)** $9y < -18$
$$\frac{9y}{9} < \frac{-18}{9} \quad \textit{Divide both sides by 9.}$$
$$y < -2$$

(b) $-2r > -12$
$$\frac{-2r}{-2} < \frac{-12}{-2} \quad \textit{Divide by } -2 \textit{ and reverse the symbol.}$$
$$r < 6$$

(c) $-5p \leq 0$
$$\frac{-5p}{-5} \geq \frac{0}{-5} \quad \textit{Divide by } -5; \textit{ reverse the symbol.}$$
$$p \geq 0$$

6. **(a)**
$$\underline{5r - r} + 2 < 7r - 5$$
$$4r + 2 < 7r - 5$$
$$4r + 2 - 7r < 7r - 5 - 7r \quad \textit{Subtract 7r.}$$
$$-3r + 2 < \underline{-5}$$
$$-3r + 2 - 2 < -5 - 2 \quad \textit{Subtract 2.}$$
$$-3r < -7$$
$$\frac{-3r}{-3} > \frac{-7}{-3} \quad \textit{Divide by } -3 \textit{ and reverse the symbol.}$$
$$r > \frac{7}{3}$$

(b) $4(y - 1) - 3y > -15 - (2y + 1)$
$$4y - 4 - 3y > -15 - 2y - 1 \quad \textit{Distributive property}$$
$$y - 4 > -16 - 2y$$
$$y - 4 + 2y > -16 - 2y + 2y \quad \textit{Add 2y.}$$
$$3y - 4 > -16$$
$$3y - 4 + 4 > -16 + 4 \quad \textit{Add 4.}$$
$$3y > -12$$
$$\frac{3y}{3} > \frac{-12}{3} \quad \textit{Divide by 3.}$$
$$y > -4$$

7. Let x represent Maggie's score on the fourth test.

The average	is at least	90.
↓	↓	↓
$\dfrac{98 + 86 + 88 + x}{4}$	$\geq$	90

Solve the inequality.

$$4\left(\frac{272 + x}{4}\right) \geq 4(90) \quad \textit{Add in the numerator; multiply by 4.}$$
$$272 + x \geq 360$$
$$272 + x - 272 \geq 360 - 272 \quad \textit{Subtract 272.}$$
$$x \geq 88 \quad \textit{Combine terms.}$$

She must score 88 or more on the fourth test to have an average of *at least* 90.

10.4 Section Exercises

1. Use an open circle if the symbol is $>$ or $<$. Use a closed circle if the symbol is $\geq$ or $\leq$.

3. Every real number less than 2 is a solution of the inequality $x < 2$. There are an *infinite* number of solutions, so it is not possible to list all of them.

5. $k \le 4$ ■ k can be any real number less than or equal to 4. To graph this inequality, place a closed circle at 4 on a number line (because 4 is part of the graph) and draw an arrow to the left.

7. $x < -3$ ■ x can be any real number less than -3, but not equal to -3. To graph this inequality, place an open circle at -3 on a number line (because -3 is not part of the graph) and draw an arrow to the left.

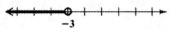

9. $8 \le x \le 10$ ■ On a number line, place a closed circle at 8 and a closed circle at 10. Then draw a line segment between the two circles.

11. $0 < y \le 10$ ■ On a number line, place an open circle at 0 (because 0 is not part of the graph) and a closed circle at 10 (because 10 is part of the graph). Then draw a line segment between the two circles.

13.
$$z - 8 \ge -7$$
$$z - 8 + 8 \ge -7 + 8 \quad \textit{Add 8.}$$
$$z \ge 1$$

15.
$$2k + 3 \ge k + 8$$
$$2k + 3 - k \ge k + 8 - k \quad \textit{Subtract k.}$$
$$k + 3 \ge 8$$
$$k + 3 - 3 \ge 8 - 3 \quad \textit{Subtract 3.}$$
$$k \ge 5$$

17.
$$3n + 5 < 2n - 6$$
$$3n + 5 - 2n < 2n - 6 - 2n \quad \textit{Subtract 2n.}$$
$$n + 5 < -6$$
$$n + 5 - 5 < -6 - 5 \quad \textit{Subtract 5.}$$
$$n < -11$$

19. When solving an inequality, the inequality symbol must be reversed when multiplying or dividing by a negative number.

21.
$$3x < 18$$
$$\frac{3x}{3} < \frac{18}{3} \quad \textit{Divide both sides by 3.}$$
$$x < 6$$

23.
$$2y \ge -20$$
$$\frac{2y}{2} \ge \frac{-20}{2} \quad \textit{Divide both sides by 2.}$$
$$y \ge -10$$

25.
$$-8t > 24$$
$$\frac{-8t}{-8} < \frac{24}{-8} \quad \textit{Divide by } -8; \textit{ reverse the symbol from } > \textit{ to } <.$$
$$t < -3$$

27.
$$-x \ge 0$$
$$-1x \ge 0$$
$$\frac{-1x}{-1} \le \frac{0}{-1} \quad \textit{Divide by } -1; \textit{ reverse the symbol from } \ge \textit{ to } \le.$$
$$x \le 0$$

29.
$$-\frac{3}{4}r < -15$$
$$\left(-\frac{4}{3}\right)\left(-\frac{3}{4}r\right) > \left(-\frac{4}{3}\right)(-15)$$

Multiply by $-\frac{4}{3}$ (the reciprocal of $-\frac{3}{4}$); reverse the symbol from $<$ to $>$.
$$r > 20$$

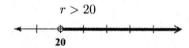

31.
$$-0.02x \le 0.06$$
$$\frac{-0.02x}{-0.02} \ge \frac{0.06}{-0.02} \quad \textit{Divide by } -0.02; \textit{ reverse the symbol from } \le \textit{ to } \ge.$$
$$x \ge -3$$

33.
$$5r + 1 \ge 3r - 9$$
$$2r + 1 \ge -9 \quad \textit{Subtract 3r.}$$
$$2r \ge -10 \quad \textit{Subtract 1.}$$
$$r \ge -5 \quad \textit{Divide by 2.}$$

35.
$$6x + 3 + x < 2 + 4x + 4$$
$$7x + 3 < 4x + 6 \quad \textit{Combine like terms.}$$
$$3x + 3 < 6 \quad \textit{Subtract 4x.}$$
$$3x < 3 \quad \textit{Subtract 3.}$$
$$x < 1 \quad \textit{Divide by 3.}$$

37. $-x + 4 + 7x \leq -2 + 3x + 6$

$\qquad 6x + 4 \leq 4 + 3x \qquad$ *Combine terms.*

$\qquad 3x + 4 \leq 4 \qquad$ *Subtract 3x.*

$\qquad 3x \leq 0 \qquad$ *Subtract 4.*

$\qquad x \leq 0 \qquad$ *Divide by 3.*

39. $5(x + 3) - 6x \leq 3(2x + 1) - 4x$

$\qquad 5x + 15 - 6x \leq 6x + 3 - 4x$

$\qquad -x + 15 \leq 2x + 3$

$\qquad -3x + 15 \leq 3$

$\qquad -3x \leq -12$

$\qquad \dfrac{-3x}{-3} \geq \dfrac{-12}{-3} \qquad$ *Divide by –3 and reverse the symbol.*

$\qquad x \geq 4$

41. $\dfrac{2}{3}(p + 3) > \dfrac{5}{6}(p - 4)$

$\qquad 6\left(\dfrac{2}{3}\right)(p + 3) > 6\left(\dfrac{5}{6}\right)(p - 4) \qquad$ *Multiply by 6, the LCD.*

$\qquad 4(p + 3) > 5(p - 4)$

$\qquad 4p + 12 > 5p - 20$

$\qquad -p + 12 > -20$

$\qquad -p > -32$

$\qquad \dfrac{-p}{-1} < \dfrac{-32}{-1} \qquad$ *Divide by –1 and reverse the symbol.*

$\qquad p < 32$

43. $4x - (6x + 1) \leq 8x + 2(x - 3)$

$\qquad 4x - 6x - 1 \leq 8x + 2x - 6$

$\qquad -2x - 1 \leq 10x - 6$

$\qquad -12x - 1 \leq -6$

$\qquad -12x \leq -5$

$\qquad \dfrac{-12x}{-12} \geq \dfrac{-5}{-12} \qquad$ *Divide by –12 and reverse the symbol.*

$\qquad x \geq \dfrac{5}{12}$

45. $5(2k + 3) - 2(k - 8) > 3(2k + 4) + k - 2$

$\qquad 10k + 15 - 2k + 16 > 6k + 12 + k - 2$

$\qquad 8k + 31 > 7k + 10$

$\qquad k + 31 > 10$

$\qquad k > -21$

47. Let x represent Tara's score on the third test.

The average	is at least	85.
↓	↓	↓
$\dfrac{78 + 83 + x}{3}$	$\geq$	85

Solve the inequality.

$3\left(\dfrac{161 + x}{3}\right) \geq 3(85) \qquad$ *Add in the numerator; multiply by 3.*

$\qquad 161 + x \geq 255$

$161 + x - 161 \geq 255 - 161 \qquad$ *Subtract 161.*

$\qquad x \geq 94 \qquad$ *Combine terms.*

Tara must score 94 or more on the third test to have an average of *at least* 85.

49. Let x represent Twylene's score on the fifth test.

The average	is at least	80.
↓	↓	↓
$\dfrac{89 + 78 + 73 + 81 + x}{5}$	$\geq$	80

Solve the inequality.

$5\left(\dfrac{321 + x}{5}\right) \geq 5(80) \qquad$ *Add in the numerator; multiply by 5.*

$\qquad 321 + x \geq 400$

$321 + x - 321 \geq 400 - 321 \qquad$ *Subtract 321.*

$\qquad x \geq 79 \qquad$ *Combine terms.*

Twylene must score 79 or more on the fifth test to have an average of *at least* 80.

51. The Fahrenheit temperature must correspond to a Celsius temperature that is less than or equal to 30 degrees.

$$\text{C} \leq 30$$

$$\dfrac{5}{9}(\text{F} - 32) \leq 30$$

$$\dfrac{9}{5}\left[\dfrac{5}{9}(\text{F} - 32)\right] \leq \dfrac{9}{5}(30)$$

$$\text{F} - 32 \leq 54$$

$$\text{F} \leq 86$$

The temperature in Houston on a certain day is never more than 86 degrees Fahrenheit.

53. $P = 2l + 2w;\ P \geq 400$

From the figure, we have $l = 4x + 3$ and $w = x + 37$. Thus, we have the inequality

$$2(4x + 3) + 2(x + 37) \geq 400.$$

Solve this inequality.

$$8x + 6 + 2x + 74 \geq 400$$
$$10x + 80 \geq 400$$
$$10x \geq 320$$
$$x \geq 32$$

The rectangle will have a perimeter of at least 400 if the value of x is 32 or greater.

55. Let $x =$ the amount she must earn in October.

$$\frac{200 + 300 + 225 + x}{4} \geq 250$$
$$200 + 300 + 225 + x \geq 1000 \quad \textit{Multiply by 4.}$$
$$725 + x \geq 1000$$
$$x \geq 275 \quad \textit{Subtract 725.}$$

She must earn at least \$275 in October to average \$250 for July–October.

Chapter 10 Review Exercises

1. 24.625 is rational because the decimal terminates.

2. 0.363636... is rational because the decimal digits repeat in a fixed block.

3. $\sqrt{144}$ is rational because $\sqrt{144} = 12$.

4. $\sqrt{23} \approx 4.79583152$ is irrational because the decimal form never ends or repeats in a fixed block.

5. $\frac{2}{3} \neq 0.6$ is *true* because $\frac{2}{3}$ does not equal $\frac{6}{10}$.

6. $-\frac{5}{8} \leq -\frac{5}{8}$ is *true* because $-\frac{5}{8} = -\frac{5}{8}$.

7. $2 - 10 \geq 10 - 2$ is *false* because $-8 < 8$ and $-8 \neq 8$.

8. $-75 \geq -50$ is *false* because $-75 < -50$ and $-75 \neq -50$.

9. $6[2 + 8(3^3)]$
$6[2 + 8(27)]$ *Exponent*
$6[2 + 216]$ *Multiply.*
$6[218]$ *Add.*
1308 *Multiply.*

10. $3^2[(11 + 3) - 4]$
$9[(11 + 3) - 4]$ *Exponent*
$9[14 - 4]$ *Add within parentheses.*
$9[10]$ *Subtract.*
90 *Multiply.*

11. $-8 + [(-4 + 17) - (-3 - 3)]$
$-8 + [13 - (-6)]$ *Parentheses*
$-8 + [13 + 6]$ *Change to add.*
$-8 + [19]$ *Add.*
11 *Add.*

12. $[-9 - (2 \cdot 1) - (-3)] + [8 + (-13 + 13)]$
$[-9 - 2 - (-3)] + [8 + 0]$ *Parentheses*
$[-9 - 2 + 3] + [8 + 0]$ *Change to add.*
$[-11 + 3] + [8]$ *Subtract; add.*
$[-8] + 8$ *Add.*
0 *Add.*

13. $-5(2x - 4)$
$(-5 \cdot 2x) - (-5 \cdot 4)$
$-10x - (-20)$
$-10x + 20$

14. $-(-5 + 3p)$
$-1(-5 + 3p)$
$(-1 \cdot -5) + (-1 \cdot 3p)$
$5 + (-3p)$
$5 - 3p$

15. $-(-17c - 6)$
$-1(-17c - 6)$
$(-1 \cdot -17c) - (-1 \cdot 6)$
$17c - (-6)$
$17c + 6$

16. $-2 - (8 - 10y)$
$-2 - 1(8 - 10y)$
$-2 - 8 + 10y$
$-10 + 10y$

17. $-10 - (7 + 14r)$
$-10 - 1(7 + 14r)$
$-10 - 7 - 14r$
$-17 - 14r$

18. $-8 - (-3r - 6)$
$-8 - 1(-3r - 6)$
$-8 + 3r + 6$
$-2 + 3r$

19. $-8(5k - 6) + 3(7k + 2)$
$-40k + 48 + 21k + 6$
$-19k + 54$

20. $-7(2t - 4) - 4(3t + 8) + 27t$
$-14t + 28 - 12t - 32 + 27t$
$-26t - 4 + 27t$
$t - 4$

21. $5x + 8 = -2(-2x - 1)$
$5x + 8 = 4x + 2$
$x + 8 = 2$ *Subtract 4x.*
$x = -6$ *Subtract 8.*

Check $x = -6$: $-22 = -22$ Balances

The solution is -6.

22. $8t = 7t + \frac{3}{2}$

$t = \frac{3}{2}$ *Subtract 7t.*

Check $t = \frac{3}{2}$: $12 = 12$ Balances

The solution is $\frac{3}{2}$.

23. $(4r - 8) - (3r + 12) = 0$

$1(4r - 8) - 1(3r + 12) = 0$

$4r - 8 - 3r - 12 = 0$

$r - 20 = 0$

$r = 20$

Check $r = 20$: $0 = 0$ Balances

The solution is 20.

24. $7(2x + 1) = 6(2x - 9)$

$14x + 7 = 12x - 54$

$2x + 7 = -54$ *Subtract 12x.*

$2x = -61$ *Subtract 7.*

$x = -\frac{61}{2}$ *Divide by 2.*

Check $x = -\frac{61}{2}$: $-420 = -420$ Balances

The solution is $-\frac{61}{2}$.

25. $-\frac{6}{5}y = -18$

$\left(-\frac{5}{6}\right)\left(-\frac{6}{5}y\right) = \left(-\frac{5}{6}\right)(-18)$

$y = 15$

Check $y = 15$: $-18 = -18$ Balances

The solution is 15.

26. $\frac{1}{2}r - \frac{1}{6}r + 3 = 2 + \frac{1}{6}r + 1$ ■ Multiply both sides by 6, the LCD.

$6\left(\frac{1}{2}r - \frac{1}{6}r + 3\right) = 6\left(2 + \frac{1}{6}r + 1\right)$

$3r - r + 18 = 12 + r + 6$

$2r + 18 = 18 + r$

$r + 18 = 18$

$r = 0$

Check $r = 0$: $3 = 3$ Balances

The solution is 0.

27. $3x - (-2x + 6) = 4(x - 4) + x$

$3x + 2x - 6 = 4x - 16 + x$

$5x - 6 = 5x - 16$

$-6 = -16$

This statement is *false*, so there is *no solution*.

28. $0.10(x + 80) + 0.20x = 14$ ■ To clear decimals, multiply both sides by 10.

$10[0.10(x + 80) + 0.20x] = 10(14)$

$1(x + 80) + 2x = 140$

$x + 80 + 2x = 140$

$3x + 80 = 140$

$3x = 60$

$x = 20$

Check $x = 20$: $10 + 4 = 14$ Balances

The solution is 20.

In Exercises 29–30, substitute the given values into the given formula and then solve for the remaining variable.

29. $A = \frac{1}{2}bh$; $A = 44$, $b = 8$

$A = \frac{1}{2}bh$

$44 = \frac{1}{2}(8)h$

$44 = 4h$

$11 = h$ *Divide by 4.*

30. $C = 2\pi r$; $C = 29.83$, $\pi \approx 3.14$

$C = 2\pi r$

$29.83 \approx 2(3.14)r$

$29.83 \approx 6.28r$

$\frac{29.83}{6.28} \approx \frac{6.28r}{6.28}$

$4.75 \approx r$

31. $V = lwh$ for w

$V = lwh$

$\frac{V}{lh} = \frac{lwh}{lh}$ *Divide by lh.*

$\frac{V}{lh} = w$,   or $w = \frac{V}{lh}$

32. $A = \frac{1}{2}h(b + B)$ for h

$A = \frac{1}{2}h(b + B)$

$2A = 2\left[\frac{1}{2}h(b + B)\right]$ *Multiply by 2.*

$2A = h(b + B)$

$\frac{2A}{(b + B)} = \frac{h(b + B)}{(b + B)}$ *Divide by b + B.*

$\frac{2A}{b + B} = h$,   or $h = \frac{2A}{b + B}$

33. $P = 2l + 2w$

$326.5 = 2(92.75) + 2w$ *Substitute given values.*

$326.5 = 185.5 + 2w$ *Multiply.*

$141 = 2w$ *Subtract 185.5.*

$70.5 = w$ *Divide by 2.*

The screen's width is 70.5 feet.

34.
$$V = lwh$$
$$6624 = (40)(20.7)h \quad \textit{Substitute given values.}$$
$$6624 = 828h \quad \textit{Multiply.}$$
$$8 = h \quad \textit{Divide by 828.}$$

The height of the box was 8 feet.

35. $p \geq -4$ ▪ Place a closed circle at -4 (to show that -4 is part of the graph) and draw an arrow extending to the right.

36. $x < 7$ ▪ Place an open circle at 7 (to show that 7 is not part of the graph) and draw an arrow extending to the left.

37. $-5 \leq y < 6$ ▪ Place a closed circle at -5 and an open circle at 6. Then draw a line segment between them.

38. $r \geq \frac{1}{2}$ ▪ Place a closed circle at $\frac{1}{2}$ and draw an arrow extending to the right.

39.
$$y + 6 \geq 3$$
$$y + 6 - 6 \geq 3 - 6 \quad \textit{Subtract 6.}$$
$$y \geq -3$$

40.
$$5t < 4t + 2$$
$$5t - 4t < 4t + 2 - 4t \quad \textit{Subtract 4t.}$$
$$t < 2$$

41.
$$-6x \leq -18$$
$$\frac{-6x}{-6} \geq \frac{-18}{-6} \quad \textit{Divide by } -6 \textit{ and reverse the symbol.}$$
$$x \geq 3$$

42.
$$8(k - 5) - (2 + 7k) \geq 4$$
$$8k - 40 - 2 - 7k \geq 4 \quad \textit{Distributive property}$$
$$k - 42 \geq 4 \quad \textit{Combine like terms.}$$
$$k \geq 46 \quad \textit{Add 42.}$$

43.
$$4x - 3x > 10 - 4x + 7x$$
$$x > 10 + 3x \quad \textit{Combine like terms.}$$
$$-2x > 10 \quad \textit{Subtract 3x.}$$
$$\frac{-2x}{-2} < \frac{10}{-2} \quad \textit{Divide by } -2 \textit{ and reverse the symbol.}$$
$$x < -5$$

44.
$$3(2w + 5) + 4(8 + 3w) < 5(3w + 2) + 2w$$
$$6w + 15 + 32 + 12w < 15w + 10 + 2w$$
$$18w + 47 < 17w + 10$$
$$w + 47 < 10$$
$$w < -37$$

45. Let x represent Carlotta's score on the fourth test.

The average	is at least	85.
↓	↓	↓
$\dfrac{81 + 77 + 88 + x}{4}$	$\geq$	85

Solve the inequality.

$$4\left(\frac{246 + x}{4}\right) \geq 4(85) \quad \textit{Add in the numerator; multiply by 4.}$$
$$246 + x \geq 340$$
$$246 + x - 246 \geq 340 - 246 \quad \textit{Subtract 246.}$$
$$x \geq 94 \quad \textit{Combine terms.}$$

Carlotta must score 94 or more on the fourth test to have an average of *at least* 85.

46. Let $n =$ the number. "If nine times a number is added to 6, the result is at most 3" can be translated as

$$9n + 6 \leq 3.$$

Solve the inequality.

$$9n \leq -3 \quad \textit{Subtract 6.}$$
$$\frac{9n}{9} \leq \frac{-3}{9} \quad \textit{Divide by 9.}$$
$$n \leq -\frac{1}{3} \quad \textit{Reduce.}$$

All numbers less than or equal to $-\frac{1}{3}$ satisfy the given condition.

47. [10.2] $\dfrac{y}{7} = \dfrac{y-5}{2}$ ■ The LCD of all the fractions is 14.

$$14\left(\frac{y}{7}\right) = 14\left(\frac{y-5}{2}\right) \quad \textit{Multiply by 14.}$$
$$2y = 7(y-5)$$
$$2y = 7y - 35$$
$$-5y = -35 \qquad \textit{Subtract 7y.}$$
$$y = 7 \qquad \textit{Divide by } -5.$$

Check $y = 7$: $1 = 1$ Balances

The solution is 7.

48. [10.3] $I = prt$ for r

$$I = prt$$
$$\frac{I}{pt} = \frac{prt}{pt} \quad \textit{Divide by pt.}$$
$$\frac{I}{pt} = r, \text{ or } r = \frac{I}{pt}$$

49. [10.4] $-2x > -4$

$$\frac{-2x}{-2} < \frac{-4}{-2} \quad \begin{array}{l}\textit{Divide by } -2 \textit{ and}\\ \textit{reverse the symbol.}\end{array}$$
$$x < 2$$

50. [10.2] $\quad 2k - 5 = 4k + 13$

$$-2k - 5 = 13 \qquad \textit{Subtract 4k.}$$
$$-2k = 18 \qquad \textit{Add 5.}$$
$$k = -9 \qquad \textit{Divide by } -2.$$

Check $k = -9$: $-23 = -23$ Balances

The solution is -9.

51. [10.2] $0.05x + 0.02x = 4.9$ ■ To clear decimals, multiply both sides by 100.

$$100(0.05x + 0.02x) = 100(4.9)$$
$$5x + 2x = 490$$
$$7x = 490$$
$$x = 70$$

Check $x = 70$: $3.5 + 1.4 = 4.9$ Balances

The solution is 70.

52. [10.2] $\quad 2 - 3(y-5) = 4 + y$

$$2 - 3y + 15 = 4 + y$$
$$17 - 3y = 4 + y$$
$$17 - 4y = 4$$
$$-4y = -13$$
$$y = \frac{-13}{-4} = \frac{13}{4}$$

Check $y = \frac{13}{4}$: $\frac{8}{4} + \frac{21}{4} = \frac{29}{4}$ Balances

The solution is $\frac{13}{4}$.

53. [10.2] $\quad 9x - (7x+2) = 3x + (2-x)$

$$9x - 7x - 2 = 3x + 2 - x$$
$$2x - 2 = 2x + 2$$
$$-2 = 2$$

Because $-2 = 2$ is a *false* statement, the given equation has *no solution*.

54. [10.2] $\quad \dfrac{1}{3}s + \dfrac{1}{2}s + 7 = \dfrac{5}{6}s + 5 + 2$

$$\frac{1}{3}s + \frac{1}{2}s = \frac{5}{6}s \qquad \textit{Subtract 7.}$$

The least common denominator is 6.

$$6\left(\frac{1}{3}s + \frac{1}{2}s\right) = 6\left(\frac{5}{6}s\right)$$
$$2s + 3s = 5s$$
$$5s = 5s$$
$$0 = 0$$

Because $0 = 0$ is a *true* statement, the solution is *all real numbers*.

55. [10.2]
$$4 - 5(a+2) = 3(a+1) - 1$$
$$4 - 5a - 10 = 3a + 3 - 1$$
$$-5a - 6 = 3a + 2$$
$$-8a - 6 = 2 \qquad \textit{Subtract 3a.}$$
$$-8a = 8 \qquad \textit{Add 6.}$$
$$a = -1 \qquad \textit{Divide by } -8.$$

Check $a = -1$: $4 - 5 = -1$ Balances

The solution is -1.

56. [10.3] $P = a + b + c$ for a

$$P = a + b + c$$
$$P - b - c = a + b + c - b - c$$
$$\textit{Subtract b and c.}$$
$$P - b - c = a, \text{ or } a = P - b - c$$

57. [10.2] $\quad \dfrac{2}{3}y + \dfrac{3}{4}y = -17$

The least common denominator is 12.

$$12\left(\frac{2}{3}y + \frac{3}{4}y\right) = 12(-17)$$
$$8y + 9y = -204$$
$$17y = -204$$
$$y = -12$$

Check $y = -12$: $-8 + (-9) = -17$ Balances

The solution is -12.

58. [10.2] $\quad 2 - 6(z+1) = 4(z-2) + 10$

$$2 - 6z - 6 = 4z - 8 + 10$$
$$-6z - 4 = 4z + 2$$
$$-10z - 4 = 2$$
$$-10z = 6$$
$$z = -\frac{6}{10} = -\frac{3}{5}$$

Check $z = -\frac{3}{5}$: $2 - \frac{12}{5} = -\frac{52}{5} + 10$ Balances

The solution is $-\frac{3}{5}$.

59. **[10.3]** $A = \frac{1}{2}bh$
$$182 = \tfrac{1}{2}b(14)$$
$$182 = 7b$$
$$26 = b \qquad \textit{Divide by 7.}$$

The length of the base is 26 inches.

60. **[10.3]** $P = 2l + 2w$
$$75 = 2l + 2(17)$$
$$75 = 2l + 34$$
$$41 = 2l$$
$$l = \frac{41}{2} \quad \text{or} \quad 20\frac{1}{2} \quad \text{or} \quad 20.5$$

The length of the rectangle is 20.5 inches.

61. **[10.4]** Let $x =$ the distance they must drive on the third day.

$$\frac{430 + 470 + x}{3} \geq 450$$
$$430 + 470 + x \geq 1350 \quad \textit{Multiply by 3.}$$
$$900 + x \geq 1350$$
$$x \geq 450 \quad \textit{Subtract 900.}$$

They must drive at least 450 miles on the third day in order to average 450 miles per day.

62. **[10.4]** Let x represent Nalima's score on the fifth test.

The average	is at least	90.
↓	↓	↓
$\dfrac{82 + 91 + 97 + 96 + x}{5}$	$\geq$	90

Solve the inequality.

$$5\left(\frac{366 + x}{5}\right) \geq 5(90) \qquad \begin{array}{l}\textit{Add in the}\\ \textit{numerator;}\\ \textit{multiply by 5.}\end{array}$$
$$366 + x \geq 450$$
$$366 + x - 366 \geq 450 - 366 \quad \textit{Subtract 366.}$$
$$x \geq 84 \quad \textit{Combine terms.}$$

Nalima must score 84 or more on the fifth test to have an average of *at least* 90.

63. **[10.3]** The 4.7 meters represents the circumference of a circle.

$$C = 2\pi r$$
$$C = \pi d \qquad 2r = d$$
$$4.7 \approx (3.14)d$$
$$1.5 \approx d \qquad \textit{Divide by 3.14.}$$

The diameter of the cloth, to the nearest tenth of a meter, is 1.5 meters.

64. **[10.3]** $A = lw$
$$93.5 = l(8\tfrac{1}{2})$$
$$11 = l \qquad \textit{Divide by 8.5 (or } 8\tfrac{1}{2}\textit{).}$$

The length of the rug is 11 feet.

Chapter 10 Test

1. π is irrational because the decimal form never ends or repeats in a fixed block.

2. $3.909090\ldots$ is rational because the decimal digits repeat in a fixed block.

3. $-6 \leq -8$ is *false* because $-6 > -8$ and $-6 \neq -8$.

4. $\frac{3}{4} \neq 0.75$ is *false* because $\frac{3}{4}$ in decimal form *is* 0.75.

5. $1\frac{3}{10} \geq 0.95$ is *true* because $1.3 > 0.95$.

6. $4[-20 + 7(-2)]$
$$4[-20 + (-14)] \quad \textit{Multiply.}$$
$$4[-34] \qquad\qquad \textit{Add.}$$
$$-136 \qquad\qquad \textit{Multiply.}$$

7. $-6 - [-7 + (2 - 3)]$
$$-6 - [-7 + (-1)] \quad \textit{Subtract.}$$
$$-6 - [-8] \qquad\quad \textit{Add.}$$
$$-6 + 8 \qquad\qquad \textit{Change to addition.}$$
$$2 \qquad\qquad\qquad \textit{Add.}$$

8. $-(-5a + 14)$
$$-1(-5a + 14)$$
$$(-1 \cdot -5a) + (-1 \cdot 14)$$
$$5a + (-14)$$
$$5a - 14$$

9. $-2(3x^2 + 4) - 3(x^2 + 2x)$
$$-6x^2 - 8 - 3x^2 - 6x$$
$$-9x^2 - 6x - 8$$

10. $5x + 9 = 7x + 21$
$$-2x + 9 = 21 \qquad \textit{Subtract 7x.}$$
$$-2x = 12 \qquad\quad \textit{Subtract 9.}$$
$$x = -6 \qquad\quad \textit{Divide by –2.}$$

Check $x = -6$: $-21 = -21$ Balances

The solution is -6.

11. $2 - 3(y - 5) = 3 + (y + 1)$
$$2 - 3y + 15 = 3 + y + 1$$
$$-3y + 17 = y + 4$$
$$-4y + 17 = 4 \qquad\qquad \textit{Subtract y.}$$
$$-4y = -13 \qquad\quad \textit{Subtract 17.}$$
$$y = \tfrac{13}{4} \qquad\qquad \textit{Divide by –4.}$$

Check $x = \frac{13}{4}$: $\frac{29}{4} = \frac{29}{4}$ Balances

The solution is $\frac{13}{4}$.

12. $2.3x + 13.7 = 1.3x + 2.9$

$x + 13.7 = 2.9$ *Subtract 1.3x.*

$ x = -10.8$ *Subtract 13.7.*

Check $x = -10.8$: $-11.14 = -11.14$
Balances

The solution is -10.8.

13. $7 - (m - 4) = -3m + 2(m + 1)$

$ 7 - m + 4 = -3m + 2m + 2$

$ -m + 11 = -m + 2$

$ 11 = 2$

Because the last statement is *false*, the equation has *no solution*.

14. $-\dfrac{4}{7}x = -1 - \dfrac{2}{3}x$ ■   Multiply both sides by 21, the LCD.

$$21\left(-\tfrac{4}{7}x\right) = 21\left(-1 - \tfrac{2}{3}x\right)$$

$$-12x = -21 - 14x$$

$$2x = -21 \qquad \textit{Add 14x.}$$

$$x = -\frac{21}{2} \ \text{ or } \ -10\frac{1}{2} \text{ or } -10.5$$

Check $x = -\frac{21}{2}$: $6 = -1 + 7$ Balances

The solution is $-\frac{21}{2}$.

15. $0.06(x + 20) + 0.08(x - 10) = 4.6$ ■   To clear decimals, multiply both sides by 100.

$$100[0.06(x + 20) + 0.08(x - 10)] = 100(4.6)$$

$$6(x + 20) + 8(x - 10) = 460$$

$$6x + 120 + 8x - 80 = 460$$

$$14x + 40 = 460$$

$$14x = 420$$

$$x = 30$$

Check $x = 30$: $4.6 = 4.6$ Balances

The solution is 30.

16. $-8(2x + 4) = -4(4x + 8)$

$ -16x - 32 = -16x - 32$

$ -32 = -32$

Because the last statement is *true*, the solution is *all real numbers*.

17. $I = prt$

$600 = p(0.08)(3)$ *Substitute given values.*

$600 = 0.24p$ *Multiply.*

$p = \dfrac{600}{0.24} = 2500$ *Divide by 0.24.*

The principal is $2500.

18. $C = 2\pi r$

$ C = \pi d \qquad\qquad 2r = d$

$62.5 \approx (3.14)d$

$19.9 \approx d \qquad\qquad \textit{Divide by 3.14.}$

The diameter of the turntable, to the nearest tenth of a foot, is 19.9 feet.

19. Solve $d = 2r$ for r.

$$\frac{d}{2} = r \qquad\qquad \textit{Divide by 2.}$$

20. Solve $A = \dfrac{1}{2}bh$ for h.

$$2A = bh \qquad\qquad \textit{Multiply by 2.}$$

$$\frac{2A}{b} = h \qquad\qquad \textit{Divide by b.}$$

21. Solve $A = p + prt$ for r.

$$A - p = prt \qquad\qquad \textit{Subtract p.}$$

$$\frac{A - p}{pt} = r \qquad\qquad \textit{Divide by pt.}$$

22. $-3x > -33$

$$\frac{-3x}{-3} < \frac{-33}{-3} \qquad \begin{array}{l}\textit{Divide by } -3 \textit{ and} \\ \textit{reverse the symbol.}\end{array}$$

$$x < 11$$

23. $-4x + 2(x - 3) \geq 4x - (3 + 5x) - 7$

$ -4x + 2x - 6 \geq 4x - 3 - 5x - 7$

$ -2x - 6 \geq -x - 10$

$ -x - 6 \geq -10$

$ -x \geq -4$

$$\frac{-1x}{-1} \leq \frac{-4}{-1} \qquad \begin{array}{l}\textit{Divide by } -1 \textit{ and} \\ \textit{reverse the symbol.}\end{array}$$

$$x \leq 4$$

24. Let x represent Paula's score on the fourth test.

The average	is at least	90.
↓	↓	↓
$\dfrac{95 + 84 + 100 + x}{4}$	$\geq$	90

Solve the inequality.

$$4\left(\frac{279 + x}{4}\right) \geq 4(90) \qquad \begin{array}{l}\textit{Add in the} \\ \textit{numerator;} \\ \textit{multiply by 4.}\end{array}$$

$$279 + x \geq 360$$

$$279 + x - 279 \geq 360 - 279 \qquad \textit{Subtract 279.}$$

$$x \geq 81 \qquad\qquad\qquad\quad \textit{Combine terms.}$$

Paula must score 81 or more on the fourth test to have an average of *at least* 90.

CHAPTER 11 GRAPHS OF LINEAR EQUATIONS AND INEQUALITIES IN TWO VARIABLES

11.1 Linear Equations in Two Variables; The Rectangular Coordinate System

11.1 Margin Exercises

1. **(a)** Locate 7000 on the vertical axis and follow the line across to the right. Three years—2004, 2005, and 2006—have bars that extend below the line for 7000, so per-capita health care spending was less than $7000 in those years.

    (b) Locate the top of the bar for 2006 and move horizontally across to the vertical scale to see that it is about 6750. Per-capita health care spending for 2006 was about $6750.

    Similarly, follow the top of the bar for 2007 across to the vertical scale to see that it is about 7000, so per-capita health care spending in 2007 was about $7000.

    (c) Per-capita health care spending increased by about $250 ($7000 − $6750) from 2006 to 2007.

2. **(a)** The dot on the graph over '05 is slightly higher than 2.25, so the average price of a gallon of regular unleaded gasoline in 2005 was about $2.30.

    (b) The dot on the graph over '08 corresponds to 3.25, so the average price of a gallon of regular unleaded gasoline increased about $3.25 − $2.30 = $0.95 from 2005 to 2008.

    (c) The dot on the graph over '09 corresponds to 2.35, so the average price of a gallon of regular unleaded gasoline decreased about $3.25 − $2.35 = $0.90 from 2008 to 2009.

3. **(a)** $x = 5$ and $y = 7$ is written as the ordered pair $(\underline{5}, \underline{7})$ [always (x-value, y-value)].

    (b) $y = 6$ and $x = -1$ is written as the ordered pair $(\underline{-1}, \underline{6})$ [always (x-value, y-value)].

    (c) $y = 4$ and $x = -3$ is written as the ordered pair $(-3, 4)$.

    (d) $x = \frac{2}{3}$ and $y = -12$ is written as the ordered pair $\left(\frac{2}{3}, -12\right)$.

    (e) $y = 1.5$ and $x = -2.4$ is written as the ordered pair $(-2.4, 1.5)$.

    (f) $x = 0$ and $y = 0$ is written as the ordered pair $(0, 0)$.

4. **(a)** $(0, 10)$
    $$5x + 2y = 20$$
    $$5(\underline{0}) + 2(\underline{10}) \stackrel{?}{=} 20 \quad \textit{Let x = 0, y = 10.}$$
    $$\underline{0} + 20 \stackrel{?}{=} 20$$
    $$\underline{20} = 20 \quad \textit{True}$$
    Yes, $(0, 10)$ is a solution.

    (b) $(2, -5)$
    $$5x + 2y = 20$$
    $$5(2) + 2(-5) \stackrel{?}{=} 20 \quad \textit{Let x = 2, y = −5.}$$
    $$10 + (-10) \stackrel{?}{=} 20$$
    $$0 = 20 \quad \textit{False}$$
    No, $(2, -5)$ is not a solution.

    (c) $(3, 2)$
    $$5x + 2y = 20$$
    $$5(3) + 2(2) \stackrel{?}{=} 20 \quad \textit{Let x = 3, y = 2.}$$
    $$15 + 4 \stackrel{?}{=} 20$$
    $$19 = 20 \quad \textit{False}$$
    No, $(3, 2)$ is not a solution.

    (d) $(-4, 20)$
    $$5x + 2y = 20$$
    $$5(-4) + 2(20) \stackrel{?}{=} 20 \quad \textit{Let x = −4, y = 20.}$$
    $$-20 + 40 \stackrel{?}{=} 20$$
    $$20 = 20 \quad \textit{True}$$
    Yes, $(-4, 20)$ is a solution.

5. **(a)** In the ordered pair $(5, \_\_)$, $x = \underline{5}$. Find the corresponding value of y by replacing x with 5 in the given equation.
    $$y = 2x - 9$$
    $$y = 2(\underline{5}) - 9 \quad \textit{Let x = 5.}$$
    $$y = \underline{10} - 9$$
    $$y = \underline{1}$$
    The ordered pair is $\underline{(5, 1)}$.

    (b) In the ordered pair $(2, \_\_)$, $x = 2$.
    $$y = 2x - 9$$
    $$y = 2(2) - 9 \quad \textit{Let x = 2.}$$
    $$y = 4 - 9$$
    $$y = -5$$
    The ordered pair is $(2, -5)$.

    (c) In the ordered pair $(\_\_, 7)$, $y = 7$. Find the corresponding value of x by replacing y with 7 in the given equation.
    $$y = 2x - 9$$
    $$7 = 2x - 9 \quad \textit{Let y = 7.}$$
    $$16 = 2x \quad \textit{Add 9.}$$
    $$8 = x \quad \textit{Divide by 2.}$$
    The ordered pair is $(8, 7)$.

(d) In the ordered pair $(\_\_, -13)$, $y = -13$.

$$y = 2x - 9$$
$$-13 = 2x - 9 \quad \textit{Let } y = -13.$$
$$-4 = 2x \quad \textit{Add 9.}$$
$$-2 = x \quad \textit{Divide by 2.}$$

The ordered pair is $(-2, -13)$.

6.   (a) To complete the first ordered pair, let $x = \underline{0}$.

$$2x - 3y = 12$$
$$2(\underline{0}) - 3y = 12 \quad \textit{Let } x = 0.$$
$$\underline{0} - 3y = 12$$
$$-3y = 12$$
$$\frac{-3y}{-3} = \frac{12}{-3}$$
$$y = \underline{-4}$$

The first ordered pair is $(0, -4)$.

To complete the second ordered pair, let $y = 0$.

$$2x - 3y = 12$$
$$2x - 3(0) = 12 \quad \textit{Let } y = 0.$$
$$2x = 12$$
$$x = 6$$

The second ordered pair is $(6, 0)$.

To complete the third ordered pair, let $x = 3$.

$$2x - 3y = 12$$
$$2(3) - 3y = 12 \quad \textit{Let } x = 3.$$
$$6 - 3y = 12$$
$$-3y = 6$$
$$y = -2$$

The third ordered pair is $(3, -2)$.

To complete the last ordered pair, let $y = -3$.

$$2x - 3y = 12$$
$$2x - 3(-3) = 12 \quad \textit{Let } y = -3.$$
$$2x + 9 = 12$$
$$2x = 3$$
$$x = \frac{3}{2}$$

The last ordered pair is $(\frac{3}{2}, -3)$.

The completed table of ordered pairs follows.

x	y
0	-4
6	0
3	-2
$\frac{3}{2}$	-3

(b) The given equation is $x = -1$. No matter which value of y might be chosen, the value of x is always the same, -1. Each ordered pair can be completed by placing -1 in the first position.

x	y
-1	-4
-1	0
-1	2

(c) The given equation is $y = 4$. No matter which value of x might be chosen, the value of y is always the same, 4. Each ordered pair can be completed by placing 4 in the second position.

x	y
-3	4
2	4
5	4

7. A is in quadrant II, B is in quadrant IV, C is in quadrant I, D is in quadrant II, and E is in quadrant III.

8.   To plot the ordered pairs, start at the origin in each case. See the art that follows part (h).

(a) $(3, 5)$ ▪   Go 3 units to the right along the x-axis; then go up 5 units, parallel to the y-axis.

(b) $(-2, 6)$ ▪   Go 2 units to the left along the x-axis; then go up 6 units.

(c) $(-4.5, 0)$ ▪   Go 4.5 units to the left. This point is on the x-axis since the y-coordinate is 0.

(d) $(-5, -2)$ ▪   Go 5 units to the left; then go down 2 units.

(e) $(6, -2)$ ▪   Go 6 units to the right; then go down 2 units.

(f) $(0, -6)$ ▪   Go down 6 units. This point is on the y-axis since the x-coordinate is 0.

(g) $(0, 0)$ ▪   This point is the origin. Plot it.

(h) $(-3, \frac{5}{2})$ ▪   Go 3 units to the left along the x-axis; then go up $2\frac{1}{2}$ units.

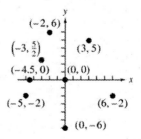

quadrant I: $(3, 5)$

quadrant II: $(-2, 6)$, $(-3, \frac{5}{2})$

quadrant III: $(-5, -2)$

quadrant IV: $(6, -2)$

no quadrant: $(-4.5, 0)$, $(0, 0)$, $(0, -6)$

9. **(a)** To find y when $x = 2004$, substitute 2004 for x into the equation and use a calculator.

$$y = 2.929x - 5738$$
$$y = 2.929(\underline{2004}) - \underline{5738}$$
$$y = 131.716 \approx \underline{132}$$

This means that in 2004, there were about 132 thousand (or 132,000) twin births in the United States.

(b) To find y when $x = 2007$, substitute 2007 for x into the equation and use a calculator.

$$y = 2.929x - 5738$$
$$y = 2.929(2007) - 5738$$
$$y = 140.503 \approx 141$$

This means that in 2007, there were about 141 thousand (or 141,000) twin births in the United States.

11.1 Section Exercises

1. The symbol (x, y) <u>does</u> represent an ordered pair, while the symbols $[x, y]$ and $\{x, y\}$ <u>do not</u> represent ordered pairs. (Note that only parentheses are used to write ordered pairs.)

3. All points having x-coordinate 0 lie on the y-axis, so the point whose graph has coordinates $(0, 5)$ lies on the <u>y</u>-axis.

5. All ordered pairs that are solutions of the equation $x = 6$ have x-coordinates equal to 6, so the ordered pair $(\underline{\;6\;}, -2)$ is a solution of the equation $x = 6$.

7. No. The ordered pairs $(4, -1)$ and $(-1, 4)$ are different ordered pairs because the order of the x- and y- coordinates has been reversed. For two ordered pairs to be equal, their x-values must be equal and their y-values must be equal. Here we have $4 \neq -1$ for the x-values and $-1 \neq 4$ for the y-values.

9. The point with coordinates (x, y) is in quadrant III if x is <u>negative</u> and y is <u>negative</u>.

11. The point with coordinates (x, y) is in quadrant IV if x is <u>positive</u> and y is <u>negative</u>.

13. The bars are above 185, and hence, the U.S. milk production was greater than 185 billion pounds, in 2007, 2008, 2009, and 2010.

15. For 2004, the U.S. milk production was about 171 billion pounds. For 2010, the U.S. milk production was about 193 billion pounds.

17. The line between 2009 and 2010 shows the steepest rise, so from 2009 to 2010 the greatest increase in the number of imported cars occurred.

Subtract the number of imported cars for 2009 from the number of imported cars for 2010.

$$5.7 - 4.3 = 1.4$$

The increase was about 1.4 or 1.5 million.

19. The line between 2006 and 2009 falls, so the number of cars imported each year was decreasing from 2006 to 2009.

21. $x + y = 9$; $(0, 9)$ ∎ To determine whether $(0, 9)$ is a solution of the given equation, substitute 0 for x and 9 for y.

$$x + y = 9$$
$$0 + 9 \overset{?}{=} 9 \quad \textit{Let x = 0, y = 9.}$$
$$9 = 9 \quad \textit{True}$$

The result is true, so $(0, 9)$ is a solution of the given equation $x + y = 9$.

23. $2x - y = 6$; $(4, 2)$ ∎ Substitute 4 for x and 2 for y.

$$2x - y = 6$$
$$2(4) - 2 \overset{?}{=} 6 \quad \textit{Let x = 4, y = 2.}$$
$$8 - 2 \overset{?}{=} 6$$
$$6 = 6 \quad \textit{True}$$

The result is true, so $(4, 2)$ is a solution of $2x - y = 6$.

25. $4x - 3y = 6$; $(2, 1)$ ∎ Substitute 2 for x and 1 for y.

$$4x - 3y = 6$$
$$4(2) - 3(1) \overset{?}{=} 6 \quad \textit{Let x = 2, y = 1.}$$
$$8 - 3 \overset{?}{=} 6$$
$$5 = 6 \quad \textit{False}$$

The result is false, so $(2, 1)$ is not a solution of $4x - 3y = 6$.

27. $y = \frac{2}{3}x$; $(-6, -4)$ ∎ Substitute -6 for x and -4 for y.

$$y = \frac{2}{3}x$$
$$-4 \overset{?}{=} \frac{2}{3}(-6) \quad \textit{Let x = -6, y = -4.}$$
$$-4 = -4 \quad \textit{True}$$

The result is true, so $(-6, -4)$ is a solution of $y = \frac{2}{3}x$.

29. $x = -6$; $(5, -6)$ ∎ Since y does not appear in the equation, we just substitute 5 for x.

$$x = -6$$
$$5 = -6 \quad \textit{Let x = 5.} \quad \textit{False}$$

The result is false, so $(5, -6)$ is not a solution of $x = -6$.

31. $y = 2$; $(2, 4)$ ▪ Since x does not appear in the equation, we just substitute 4 for y.

$$y = 2$$
$$4 = 2 \quad \textit{Let y = 4.} \quad \textit{False}$$

The result is false, so $(2, 4)$ is not a solution of $y = 2$.

33. $y = 2x + 7$; $(2, \_\_)$ ▪ In this ordered pair, $x = 2$. Find the corresponding value of y by replacing x with 2 in the given equation.

$$y = 2x + 7$$
$$y = 2(2) + 7 \quad \textit{Let x = 2.}$$
$$y = 4 + 7$$
$$y = 11$$

The ordered pair is $(2, 11)$.

35. $y = 2x + 7$; $(\_\_, 0)$ ▪ In this ordered pair, $y = 0$. Find the corresponding value of x by replacing y with 0 in the given equation.

$$y = 2x + 7$$
$$0 = 2x + 7 \quad \textit{Let y = 0.}$$
$$-7 = 2x$$
$$\frac{-7}{2} = x$$

The ordered pair is $\left(-\frac{7}{2}, 0\right)$.

37. $y = -4x - 4$; $(0, \_\_)$

$$y = -4x - 4$$
$$y = -4(0) - 4 \quad \textit{Let x = 0.}$$
$$y = 0 - 4$$
$$y = -4$$

The ordered pair is $(0, -4)$.

39. $y = -4x - 4$; $(\_\_, 16)$

$$y = -4x - 4$$
$$16 = -4x - 4 \quad \textit{Let y = 16.}$$
$$20 = -4x$$
$$-5 = x$$

The ordered pair is $(-5, 16)$.

41. $2x + 3y = 12$ ▪ If $x = 0$,

$$2(0) + 3y = 12$$
$$0 + 3y = 12$$
$$3y = 12$$
$$y = 4. \quad (0, 4)$$

If $y = 0$,

$$2x + 3(0) = 12$$
$$2x + 0 = 12$$
$$2x = 12$$
$$x = 6. \quad (6, 0)$$

If $y = 8$,

$$2x + 3(8) = 12$$
$$2x + 24 = 12$$
$$2x = -12$$
$$x = -6. \quad (-6, 8)$$

The completed table of values is shown below.

x	y
0	4
6	0
-6	8

43. $3x - 5y = -15$ ▪ If $x = 0$,

$$3(0) - 5y = -15$$
$$0 - 5y = -15$$
$$-5y = -15$$
$$y = 3. \quad (0, 3)$$

If $y = 0$,

$$3x - 5(0) = -15$$
$$3x - 0 = -15$$
$$3x = -15$$
$$x = -5. \quad (-5, 0)$$

If $y = -6$,

$$3x - 5(-6) = -15$$
$$3x + 30 = -15$$
$$3x = -45$$
$$x = -15. \quad (-15, -6)$$

The completed table of values is shown below.

x	y
0	3
-5	0
-15	-6

45. $x = -9$ ▪ No matter which value of y is chosen, the value of x will always be -9. Each ordered pair can be completed by placing -9 in the first position.

x	y	Ordered Pair
-9	6	$(-9, 6)$
-9	2	$(-9, 2)$
-9	-3	$(-9, -3)$

47. $y = -6$ ▪ No matter which value of x is chosen, the value of y will always be -6. Each ordered pair can be completed by placing -6 in the second position.

x	y	Ordered Pair
8	-6	$(8, -6)$
4	-6	$(4, -6)$
-2	-6	$(-2, -6)$

49. $x - 8 = 0$ ∎ The given equation may be written $x = 8$. No matter which value of y is chosen, the value of x will always be 8. Each ordered pair can be completed by placing 8 in the first position.

x	y	Ordered Pair
8	8	$(8, 8)$
8	3	$(8, 3)$
8	0	$(8, 0)$

51. $y + 2 = 0$ ∎ The given equation may be written $y = -2$. No matter which value of x is chosen, the value of y will always be -2. Each ordered pair can be completed by placing -2 in the second position.

x	y	Ordered Pair
9	-2	$(9, -2)$
2	-2	$(2, -2)$
0	-2	$(0, -2)$

53. Point A was plotted by starting at the origin, going 2 units to the right along the x-axis, and then going up 4 units on a line parallel to the y-axis. The ordered pair for this point is $(2, 4)$, which is in quadrant I.

55. Point C was plotted by starting at the origin, going 5 units to the left along the x-axis, and then going up 4 units on a line parallel to the y-axis. The ordered pair for this point is $(-5, 4)$, which is in quadrant II.

57. Point E was plotted by starting at the origin, going 3 units to the right along the x-axis. The ordered pair for this point is $(3, 0)$, which is not in any quadrant.

59. If $xy < 0$, then either $x < 0$ and $y > 0$ or $x > 0$ and $y < 0$. If $x < 0$ and $y > 0$, then the point lies in quadrant II. If $x > 0$ and $y < 0$, then the point lies in quadrant IV.

For Exercises 61–72, the ordered pairs are plotted on the graph following the solution for Exercise 71.

61. To plot $(6, 2)$, start at the origin, go 6 units to the right, and then go up 2 units.

63. To plot $(-4, 2)$, start at the origin, go 4 units to the left, and then go up 2 units.

65. To plot $\left(-\frac{4}{5}, -1\right)$, start at the origin, go $\frac{4}{5}$ unit to the left, and then go down 1 unit.

67. To plot $(3, -1.75)$, start at the origin, go 3 units to the right, and then go down 1.75 units.

69. To plot $(0, 4)$, start at the origin and go up 4 units. The point lies on the y-axis.

71. To plot $(4, 0)$, start at the origin and go 4 units to the right along the x-axis. The point lies on the x-axis.

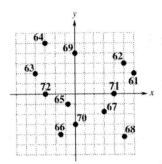

73. $x - 2y = 6$

x	y
0	
	0
2	
	-1

Substitute the given values to complete the ordered pairs.

$$x - 2y = 6$$
$$0 - 2y = 6 \quad \text{Let } x = 0.$$
$$-2y = 6$$
$$y = -3$$

$$x - 2y = 6$$
$$x - 2(0) = 6 \quad \text{Let } y = 0.$$
$$x - 0 = 6$$
$$x = 6$$

$$x - 2y = 6$$
$$2 - 2y = 6 \quad \text{Let } x = 2.$$
$$-2y = 4$$
$$y = -2$$

$$x - 2y = 6$$
$$x - 2(-1) = 6 \quad \text{Let } y = -1.$$
$$x + 2 = 6$$
$$x = 4$$

The completed table of values follows.

x	y
0	-3
6	0
2	-2
4	-1

Plot the points $(0, -3)$, $(6, 0)$, $(2, -2)$, and $(4, -1)$ on a coordinate system.

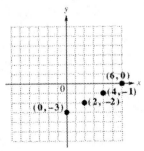

75. $3x - 4y = 12$

x	y
0	
	0
−4	
	−4

Substitute the given values to complete the ordered pairs.

$3x - 4y = 12$
$3(0) - 4y = 12$ *Let x = 0.*
$0 - 4y = 12$
$-4y = 12$
$y = -3$

$3x - 4y = 12$
$3x - 4(0) = 12$ *Let y = 0.*
$3x - 0 = 12$
$3x = 12$
$x = 4$

$3x - 4y = 12$
$3(-4) - 4y = 12$ *Let x = −4.*
$-12 - 4y = 12$
$-4y = 24$
$y = -6$

$3x - 4y = 12$
$3x - 4(-4) = 12$ *Let y = −4.*
$3x + 16 = 12$
$3x = -4$
$x = -\frac{4}{3}$

The completed table is as follows.

x	y
0	−3
4	0
−4	−6
$-\frac{4}{3}$	−4

Plot the points $(0, -3)$, $(4, 0)$, $(-4, -6)$, and $(-\frac{4}{3}, -4)$ on a coordinate system.

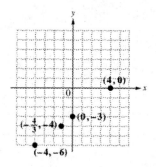

77. The given equation, $y + 4 = 0$, can be written as $y = -4$. So regardless of the value of x, the value of y is -4.

x	y
0	−4
5	−4
−2	−4
−3	−4

Plot the points $(0, -4)$, $(5, -4)$, $(-2, -4)$, and $(-3, -4)$ on a coordinate system.

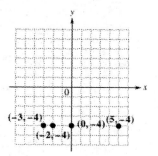

79. The points in each graph appear to lie on a straight line.

81. (a) When $x = 5$ (the number of days), $y = 45$ (the cost in dollars). Thus, the ordered pair is $(5, 45)$.

(b) When $y = 50$, $x = 6$. Thus, the ordered pair is $(6, 50)$.

83. (a) We can write the results from the table as ordered pairs (x, y).

$(2007, 27.1)$, $(2008, 29.3)$, $(2009, 28.3)$, $(2010, 28.0)$, $(2011, 26.9)$

(b) $(2000, 32.4)$ means that 32.4 percent of 2-year college students in 2000 received a degree within 3 years.

(c)

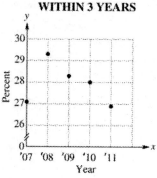

2-YEAR COLLEGE STUDENTS COMPLETING A DEGREE WITHIN 3 YEARS

(d) With the exception of the point for 2007, the points lie approximately on a straight line. Rates at which 2-year college students complete a degree within 3 years were generally decreasing.

11.2 Graphing Linear Equations in Two Variables

11.2 Margin Exercises

1. **(a)** $(-3, \underline{\quad})$ ▪ Substitute -3 for x in the equation and solve for y.

$$x + 2y = 7 \quad \textit{Given equation}$$
$$-3 + 2y = 7 \quad \textit{Let x = -3.}$$
$$2y = 10 \quad \textit{Add 3.}$$
$$y = 5 \quad \textit{Divide by 2.}$$

The ordered pair is $(-3, 5)$ is a solution of the equation.

(b) $(-1, \underline{\quad})$ ▪
$$x + 2y = 7$$
$$-1 + 2y = 7 \quad \textit{Let x = -1.}$$
$$2y = 8$$
$$y = 4$$

The ordered pair is $(-1, 4)$ is a solution of the equation.

(c) $(3, \underline{\quad})$ ▪
$$x + 2y = 7$$
$$3 + 2y = 7 \quad \textit{Let x = 3.}$$
$$2y = 4$$
$$y = 2$$

The ordered pair is $(3, 2)$ is a solution of the equation.

(d) $(5, \underline{\quad})$ ▪
$$x + 2y = 7$$
$$5 + 2y = 7 \quad \textit{Let x = 5.}$$
$$2y = 2$$
$$y = 1$$

The ordered pair is $(5, 1)$ is a solution of the equation.

(e) $(7, \underline{\quad})$ ▪
$$x + 2y = 7$$
$$7 + 2y = 7 \quad \textit{Let x = 7.}$$
$$2y = 0$$
$$y = 0$$

The ordered pair is $(7, 0)$ is a solution of the equation.

2. $x + y = 6$

x	y
0	
	0
2	

Find the first missing value in the table by substituting $\underline{0}$ for x in the equation and solving for y.

$$x + y = 6$$
$$\underline{0} + y = 6 \quad \textit{Let x = \underline{0}.}$$
$$y = \underline{6}$$

The first ordered pair is $\underline{(0, 6)}$.

Find the second missing value.

$$x + y = 6$$
$$x + 0 = 6 \quad \textit{Let y = 0.}$$
$$x = 6$$

The second ordered pair is $(6, 0)$.

Find the third missing value.

$$x + y = 6$$
$$2 + y = 6 \quad \textit{Let x = 2.}$$
$$y = 4$$

The last ordered pair is $(2, 4)$.

The completed table of ordered pairs follows.

x	y
0	6
6	0
2	4

To graph the line, plot the corresponding points and draw a <u>line</u> through them.

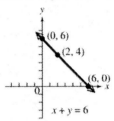

3. $5x + 2y = 10$

Find the y-intercept by letting $x = \underline{0}$.

$$5x + 2y = 10$$
$$5(0) + 2y = 10 \quad \textit{Let x = 0.}$$
$$0 + 2y = 10$$
$$2y = 10$$
$$y = 5$$

The y-intercept is $\underline{(0, 5)}$.

Find the x-intercept by letting $y = \underline{0}$.

$$5x + 2y = 10$$
$$5x + 2(0) = 10 \quad \textit{Let y = 0.}$$
$$5x + 0 = 10$$
$$5x = 10$$
$$x = 2$$

The x-intercept is $\underline{(2, 0)}$.

Get a third point as a check. For example, choosing $x = 4$ gives $y = -5$. Plot $(2, 0)$, $(0, 5)$, and $(4, -5)$ and draw a line through them.

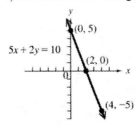

4. Find the y-intercept by letting $x = 0$.

$$y = \tfrac{2}{3}x - 2$$
$$y = \tfrac{2}{3}(0) - 2 \quad \textit{Let } x = 0.$$
$$y = 0 - 2$$
$$y = -2$$

This gives the y-intercept $(0, -2)$.

Find the x-intercept by letting $y = 0$.

$$y = \tfrac{2}{3}x - 2$$
$$0 = \tfrac{2}{3}x - 2 \quad \textit{Let } y = 0.$$
$$2 = \tfrac{2}{3}x$$
$$3 = x \qquad \textit{Multiply by } \tfrac{3}{2}.$$

This gives the x-intercept $(3, 0)$.

Ordered pairs may vary. Since the denominator of the coefficient of x is 3, choosing multiples of 3 for x leads to integer values of y, which makes it easier to plot points.

$$y = \tfrac{2}{3}x - 2$$
$$y = \tfrac{2}{3}(6) - 2 \quad \textit{Let } x = 6.$$
$$y = 4 - 2$$
$$y = 2$$

This gives the ordered pair $(6, 2)$.

Graph the equation by plotting these three points and drawing a line through them.

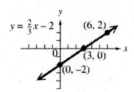

5. Graph $2x - y = 0$. ■ Three ordered pairs that can be used are shown in the following table.

x	y
0	0
2	4
-2	-4

Graph the equation by plotting these points and drawing a line through them.

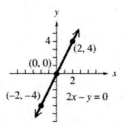

6. Graph $y = -5$. ■ No matter what value we choose for x, y is always -5, as shown in the table of ordered pairs.

x	y
0	-5
-5	-5
3	-5

The graph is a horizontal line through $(0, -5)$. There is no x-intercept.

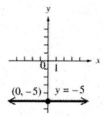

7. Graph $x + 4 = 6$. ■ Subtracting $\underline{4}$ from each side gives us the equivalent equation, $x = \underline{2}$. Thus, x is always 2 regardless of the value of y, as shown in the table of ordered pairs.

x	y
2	0
2	-5
2	5

The graph is a <u>vertical</u> line through $(2, 0)$. There is no y-intercept.

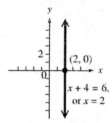

8. See the table on this page in the text.

(a) $x = 5$ is of the form $x = k$, and its graph is a vertical line. Choice **C**

(b) $2x - 5y = 8$ is of the form $Ax + By = C$. Choice **D** appears to be the most likely answer. To determine if the graph passes through the point $(9, 2)$, substitute $(9, 2)$ in $2x - 5y = 8$.

$$2(9) - 5(2) = 8$$
$$18 - 10 = 8$$
$$8 = 8 \quad \textit{True}$$

Choice **D** is the correct answer.

(c) $y - 2 = 3$, or $y = 5$, is of the form $y = k$, and its graph is a horizontal line. Choice **A**

(d) $x + 4y = 0$ is of the form $Ax + By = 0$, and its graph passes through the origin, $(0, 0)$. Choice **B**

9. **(a)** The year 2000 corresponds to $x = 0$, so the year 2006 corresponds to $x = 2006 - 2000 = 6$. On the graph, find 6 on the horizontal axis, move up to the graphed line and then across to the vertical axis. It appears that credit card debt in 2006 was about 875 billion dollars.

(b) Substitute $x = 6$ into the equation.
$$y = 32.0x + 684$$
$$y = 32.0(6) + 684 \quad \textit{Let } x = 6.$$
$$y = 876$$

From the equation, it is 876 billion dollars.

11.2 Section Exercises

1. A linear equation in two variables can be written in the form $Ax + \underline{By} = C$, where A, B, and C are real numbers and A and B are not both $\underline{0}$.

3. **(a)** To determine which equation has y-intercept $(0, -4)$, set x equal to 0 (if there is an x) and see which equation is equivalent to $y = -4$. Choice **A** is correct since
$$3x + y = -4$$
$$3(0) + y = -4$$
$$y = -4$$

(b) If the graph of the equation goes through the origin, then substituting 0 for x and 0 for y will result in a true statement. Choice **C** is correct since
$$y = 4x$$
$$0 = 4(0)$$
$$0 = 0$$
is a true statement.

(c) Choice **D** is correct since the graph of $y = 4$ is a horizontal line.

(d) To determine which equation has x-intercept $(4, 0)$, set y equal to 0 (if there is a y) and see which equation is equivalent to $x = 4$. Choice **B** is correct since
$$x - 4 = 0$$
$$x = 4$$

5. The dot at 4 on the x-axis is the x-intercept, that is, $(4, 0)$. The dot at -4 on the y-axis is the y-intercept, that is, $(0, -4)$.

7. The dot at -2 on the x-axis is the x-intercept, that is, $(-2, 0)$. The dot at -3 on the y-axis is the y-intercept, that is, $(0, -3)$.

9. $x + y = 5$

$(0, \underline{\quad})$, $\quad$ $(\underline{\quad}, 0)$, $\quad$ $(2, \underline{\quad})$

If $x = 0$, $\quad$ If $y = 0$, $\quad$ If $x = 2$,

$0 + y = 5$ $\quad$ $x + 0 = 5$ $\quad$ $2 + y = 5$

$y = 5$. $\quad\quad$ $x = 5$. $\quad\quad$ $y = 3$.

The ordered pairs are $(0, 5)$, $(5, 0)$, and $(2, 3)$. Plot the corresponding points and draw a line through them.

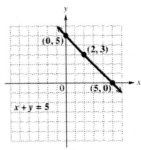

11. $y = \frac{2}{3}x + 1$ $\;(0, \underline{\quad})$, $\;(3, \underline{\quad})$, $\;(-3, \underline{\quad})$

If $x = 0$, $\quad\quad$ If $x = 3$,

$y = \frac{2}{3}(0) + 1$ $\quad\quad$ $y = \frac{2}{3}(3) + 1$

$y = 0 + 1$ $\quad\quad\quad$ $y = 2 + 1$

$y = 1$. $\quad\quad\quad\quad$ $y = 3$.

If $x = -3$,

$y = \frac{2}{3}(-3) + 1$

$y = -2 + 1$

$y = -1$.

The ordered pairs are $(0, 1)$, $(3, 3)$, and $(-3, -1)$. Plot the corresponding points and draw a line through them.

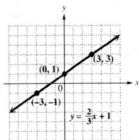

13. $3x = -y - 6$ $\;(0, \underline{\quad})$, $\;(\underline{\quad}, 0)$, $\;(-\frac{1}{3}, \underline{\quad})$

If $x = 0$, $\quad\quad\quad$ If $y = 0$,

$3(0) = -y - 6$ $\quad\quad$ $3x = -0 - 6$

$0 = -y - 6$ $\quad\quad\;$ $3x = -6$

$y = -6$. $\quad\quad\quad\quad$ $x = -2$.

If $x = -\frac{1}{3}$,

$3(-\frac{1}{3}) = -y - 6$

$-1 = -y - 6$

$y - 1 = -6$

$y = -5$.

The ordered pairs are $(0, -6)$, $(-2, 0)$, and $(-\frac{1}{3}, -5)$. Plot the corresponding points and draw a line through them.

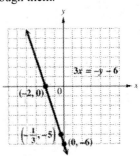

15. To find the y-intercept, let $x = \underline{0}$.

$$x - y = 8$$
$$0 - y = 8$$
$$-y = 8$$
$$y = -8$$

The y-intercept is $\underline{(0, -8)}$.

To find the x-intercept, let $y = \underline{0}$.

$$x - y = 8$$
$$x - 0 = 8$$
$$x = 8$$

The x-intercept is $\underline{(8, 0)}$.

17. To find the x-intercept, let $y = 0$.

$$2x - 3y = 24$$
$$2x - 3(0) = 24$$
$$2x - 0 = 24$$
$$2x = 24$$
$$x = 12$$

The x-intercept is $(12, 0)$.

To find the y-intercept, let $x = 0$.

$$2x - 3y = 24$$
$$2(0) - 3y = 24$$
$$0 - 3y = 24$$
$$-3y = 24$$
$$y = -8$$

The y-intercept is $(0, -8)$.

19. To find the x-intercept, let $y = 0$.

$$x + 6y = 0$$
$$x + 6(0) = 0$$
$$x + 0 = 0$$
$$x = 0$$

The x-intercept is $(0, 0)$. Since we have found the point with x equal to 0, this is also the y-intercept.

21. Begin by finding the intercepts.

$$y = x - 2$$
$$y = 0 - 2 \quad \textit{Let x = 0.}$$
$$y = -2$$

$$y = x - 2$$
$$0 = x - 2 \quad \textit{Let y = 0.}$$
$$2 = x$$

The x-intercept is $(2, 0)$ and the y-intercept is $(0, -2)$. To find a third point, choose $x = 4$.

$$y = x - 2$$
$$y = 4 - 2 \quad \textit{Let x = 4.}$$
$$y = 2$$

This gives the ordered pair $(4, 2)$. Plot $(2, 0)$, $(0, -2)$, and $(4, 2)$ and draw a line through them.

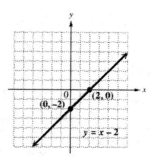

23. Find the intercepts.

$$x - y = 4$$
$$x - 0 = 4 \quad \textit{Let y = 0.}$$
$$x = 4$$

$$x - y = 4$$
$$0 - y = 4 \quad \textit{Let x = 0.}$$
$$y = -4$$

The x-intercept is $(4, 0)$ and the y-intercept is $(0, -4)$. To find a third point, choose $y = 1$.

$$x - y = 4$$
$$x - 1 = 4 \quad \textit{Let y = 1.}$$
$$x = 5$$

This gives the ordered pair $(5, 1)$. Plot $(4, 0)$, $(0, -4)$, and $(5, 1)$ and draw a line through them.

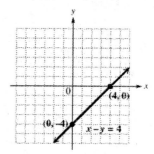

25. Find the intercepts.

$$2x + y = 6$$
$$2x + 0 = 6 \quad \textit{Let y = 0.}$$
$$2x = 6$$
$$x = 3$$

$$2x + y = 6$$
$$2(0) + y = 6 \quad \textit{Let x = 0.}$$
$$0 + y = 6$$
$$y = 6$$

The x-intercept is $(3, 0)$ and the y-intercept is $(0, 6)$. To find a third point, choose $x = 1$.

$$2x + y = 6$$
$$2(1) + y = 6 \quad \textit{Let x = 1.}$$
$$2 + y = 6$$
$$y = 4$$

This gives the ordered pair $(1, 4)$. Plot $(3, 0)$, $(0, 6)$, and $(1, 4)$ and draw a line through them.

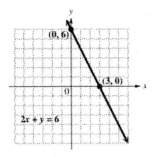

27. Find the intercepts.

$y = 2x - 5$ $y = 2x - 5$

$0 = 2x - 5$ *Let y = 0.* $y = 2(0) - 5$ *Let x = 0.*

$5 = 2x$ $y = 0 - 5$

$\frac{5}{2} = x$ $y = -5$

The x-intercept is $\left(\frac{5}{2}, 0\right)$ and the y-intercept is $(0, -5)$. To find a third point, choose $x = 1$.

$y = 2x - 5$

$y = 2(1) - 5$ *Let x = 1.*

$y = 2 - 5$

$y = -3$

This gives the ordered pair $(1, -3)$. Plot $\left(\frac{5}{2}, 0\right)$, $(0, -5)$, and $(1, -3)$ and draw a line through them.

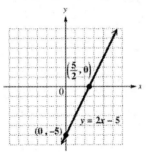

29. Find the intercepts.

$3x + 7y = 14$

$3x + 7(0) = 14$ *Let y = 0.*

$3x + 0 = 14$

$3x = 14$

$x = \frac{14}{3}$

$3x + 7y = 14$

$3(0) + 7y = 14$ *Let x = 0.*

$0 + 7y = 14$

$7y = 14$

$y = 2$

The x-intercept is $\left(\frac{14}{3}, 0\right)$ and the y-intercept is $(0, 2)$. To find a third point, choose $x = 2$.

$3x + 7y = 14$

$3(2) + 7y = 14$ *Let x = 2.*

$6 + 7y = 14$

$7y = 8$

$y = \frac{8}{7}$

This gives the ordered pair $\left(2, \frac{8}{7}\right)$. Plot $\left(\frac{14}{3}, 0\right)$, $(0, 2)$, and $\left(2, \frac{8}{7}\right)$. Writing $\frac{14}{3}$ as the mixed number $4\frac{2}{3}$ and $\frac{8}{7}$ as $1\frac{1}{7}$ will be helpful for plotting. Draw a line through these three points.

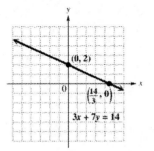

31. $y - 2x = 0$ ■ If $y = 0$, $x = 0$. Both intercepts are the origin, $(0, 0)$. Find two additional points.

$y - 2x = 0$

$y - 2(2) = 0$ *Let x = 2.*

$y - 4 = 0$

$y = 4$

$y - 2x = 0$

$y - 2(-3) = 0$ *Let x = -3.*

$y + 6 = 0$

$y = -6$

Plot $(0, 0)$, $(2, 4)$, and $(-3, -6)$ and draw a line through them.

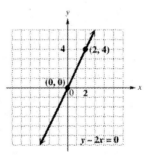

33. $y = -6x$ ■ Find three points on the line.

If $x = 0$, $y = -6(0) = 0$.

If $x = 1$, $y = -6(1) = -6$.

If $x = -1$, $y = -6(-1) = 6$.

Plot $(0, 0)$, $(1, -6)$, and $(-1, 6)$ and draw a line through these points.

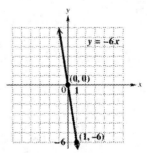

35. $x = -2$ ■ For any value of y, the value of x is -2. Three ordered pairs are $(-2, 3)$, $(-2, 0)$, and $(-2, -4)$. Plot these points and draw a line through them. The graph is a vertical line.

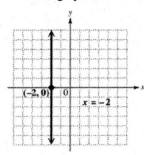

37. $y - 3 = 0$
      $y = 3$

For any value of x, the value of y is 3.
Three ordered pairs are $(-2, 3)$, $(0, 3)$, and $(4, 3)$.
Plot these points and draw a line through them.
The graph is a horizontal line.

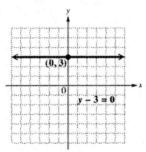

39. $-3y = 15$ *Divide by −3.*
      $y = -5$

For any value of x, the value of y is -5. Three ordered pairs are $(-2, -5)$, $(0, -5)$, and $(4, -5)$. Plot these points and draw a line through them. The graph is a horizontal line.

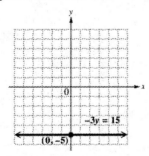

41. (a) The graph of $x = -2$ is a vertical line with x-intercept $(-2, 0)$, which is choice **D**.

(b) The graph of $y = -2$ is a horizontal line with y-intercept $(0, -2)$, which is choice **C**.

(c) The graph of $x = 2$ is a vertical line with x-intercept $(2, 0)$, which is choice **B**.

(d) The graph of $y = 2$ is a horizontal line with y-intercept $(0, 2)$, which is choice **A**.

43. $3x = y - 9$

   If $x = 0$,   If $y = 0$,
   $3(0) = y - 9$ $3x = 0 - 9$
   $0 = y - 9$ $3x = -9$
   $9 = y.$ $x = -3.$

The graph of this equation is a line with x-intercept $(-3, 0)$ and y-intercept $(0, 9)$.

45. $3y = -6$
      $y = -2$ *Divide by 3.*

The graph of this equation is a horizontal line with y-intercept $(0, -2)$.

47. (a) $y = 3.9x + 73.5$

Let $x = 20$. $y = 3.9(20) + 73.5 = 151.5$

Let $x = 22$. $y = 3.9(22) + 73.5 = 159.3$

Let $x = 26$. $y = 3.9(26) + 73.5 = 174.9$

The approximate heights of women with radius bones of lengths 20 cm, 22 cm, and 26 cm are 151.5 cm, 159.3 cm, and 174.9 cm, respectively.

(b) The three ordered pairs from part (a) are $(20, 151.5)$, $(22, 159.3)$, and $(26, 174.9)$.

(c) Plot the points in part (b) and connect them with a smooth line.

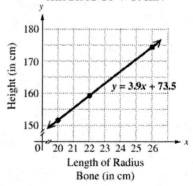

HEIGHTS OF WOMEN

(d) Locate 167 on the vertical scale, then move across to the line, then down to the horizontal scale. From the graph, the radius bone in a woman who is 167 cm tall is about 24 cm.

Now substitute 167 for y in the equation.

$$y = 3.9x + 73.5$$
$$167 = 3.9x + 73.5$$
$$93.5 = 3.9x$$
$$x = \frac{93.5}{3.9} \approx 23.97$$

From the equation, the length of the radius bone is 24 cm to the nearest centimeter.

49. **(a)** Let $x = 50$.　　Let $x = 100$.

$y = 0.75x + 25$　　$y = 0.75x + 25$
$y = 0.75(50) + 25$　$y = 0.75(100) + 25$
$y = 62.50$　　　$y = 100$

Thus, 50 posters cost \$62.50 and 100 posters cost \$100.

(b) Let $y = 175$.　　$175 = 0.75x + 25$
$150 = 0.75x$
$x = \dfrac{150}{0.75} = 200$

Thus, 200 posters cost \$175.

(c) The three ordered pairs from parts (a) and (b) are $(50, 62.50)$, $(100, 100)$, and $(200, 175)$.

(d) Plot the points in part (c) and connect them with a smooth line.

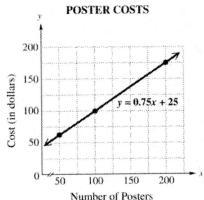

POSTER COSTS

$y = 0.75x + 25$

51. **(a)** At $x = 0$, $y = 30{,}000$.
The initial value of the SUV is \$30,000.

(b) At $x = 3$, $y = 15{,}000$.
$30{,}000 - 15{,}000 = 15{,}000$
The depreciation after the first 3 years is \$15,000.

(c) The line connecting consecutive years from 0 to 1, 1 to 2, 2 to 3, and so on, drops 5000 for each segment. Therefore, the annual depreciation in each of the first 5 years is \$5000.

(d) $(5, 5000)$ means after 5 years the SUV has a value of \$5000.

53. **(a)** $y = 0.5249x + 18.54$

For 1990, let $x = 10$.
$y = 0.5249(10) + 18.54 = 23.789$

For 2000, let $x = 20$.
$y = 0.5249(20) + 18.54 = 29.038$

For 2009, let $x = 29$.
$y = 0.5249(29) + 18.54 = 33.7621$

The approximate per capita consumptions for 1990, 2000, and 2009 are 23.8 lb, 29.0 lb, and 33.8 lb, respectively.

(b) Locate 10, 20, and 29 on the horizontal scale, then find the corresponding value on the vertical scale.

The approximate consumptions for 1990, 2000, and 2009 are 25 lb, 30 lb, and 33 lb, respectively.

(c) The values are quite close.

11.3 The Slope of a Line

11.3 Margin Exercises

1. **(a)** The indicated points have coordinates $(-1, -4)$ and $(1, -1)$. Begin at $(-1, -4)$. To get to $(1, -1)$, count up $\underline{3}$ units. Then count to the right $\underline{2}$ units.

$$\text{slope} = \frac{\text{vertical change in } y \text{ (rise)}}{\text{horizontal change in } x \text{ (run)}}$$

$$= \frac{3}{2}$$

The slope is $\frac{3}{2}$.

(b) The indicated points have coordinates $(2, 2)$ and $(4, -2)$.

$$\text{slope} = \frac{\text{vertical change in } y \text{ (rise)}}{\text{horizontal change in } x \text{ (run)}}$$

$$= \frac{-2 - 2}{4 - 2} = \frac{-4}{2} = -2$$

2. **(a)** Use the slope formula with $(6, -2) = (x_1, y_1)$ and $(5, 4) = (x_2, y_2)$.

$$\text{slope } m = \frac{\text{change in } y}{\text{change in } x} = \frac{y_2 - y_1}{x_2 - x_1}$$

$$= \frac{4 - (-2)}{5 - 6} = \frac{6}{-1} = -6$$

(b) Use the slope formula with $(-3, 5) = (x_1, y_1)$ and $(-4, -7) = (x_2, y_2)$.

$$\text{slope } m = \frac{\text{change in } y}{\text{change in } x} = \frac{y_2 - y_1}{x_2 - x_1}$$

$$= \frac{-7 - 5}{-4 - (-3)} = \frac{-12}{-1} = 12$$

(c) Use the slope formula with $(6, -8) = (x_1, y_1)$ and $(-2, 4) = (x_2, y_2)$.

$$\text{slope } m = \frac{\text{change in } y}{\text{change in } x} = \frac{y_2 - y_1}{x_2 - x_1}$$

$$= \frac{4 - (-8)}{-2 - 6} = \frac{12}{-8} = -\frac{3}{2}$$

Now subtract in reverse order:

$$m = \frac{-8 - 4}{6 - (-2)} = \frac{-12}{8} = -\frac{3}{2}$$

3. **(a)** Use the slope formula with $(2, 5)$ and $(-1, 5)$.

$$\text{slope } m = \frac{\text{change in } y}{\text{change in } x} = \frac{5 - 5}{-1 - 2} = \frac{0}{-3} = 0$$

(b) Use the slope formula with $(3, 1)$ and $(3, -4)$.

$$\text{slope } m = \frac{\text{change in } y}{\text{change in } x} = \frac{-4 - 1}{3 - 3} = \frac{-5}{0},$$

which is undefined.

4. **(a)** The line with equation $y = -1$ ■ All lines with an equation of the form $y = k$ are horizontal and have a slope of 0.

(b) The line with equation $x - 4 = 0$ ■ The equation $x - 4 = 0$ is equivalent to the equation $x = \underline{4}$. All lines with equation $x = k$ are vertical and have undefined slope.

5. **(a)** $y = -\frac{7}{2}x + 1$ ■ Since this equation is already solved for y, Step 1 is not needed. The slope, which is given by the <u>coefficient</u> of $\underline{x}$, can be read directly from the equation. The slope is $-\frac{7}{2}$.

(b) $4y = 4x - 3$ ■ Solve for y.

$$4y = 4x - 3$$
$$y = x - \frac{3}{4} \quad \textit{Divide by 4.}$$

The slope is the coefficient of x, 1.

(c) $3x + 2y = 9$ ■ Solve for y.

$$3x + 2y = 9$$
$$2y = -3x + 9 \quad \textit{Subtract 3x.}$$
$$y = -\frac{3}{2}x + \frac{9}{2} \quad \textit{Divide by 2.}$$

The slope is the coefficient of x, $-\frac{3}{2}$.

(d) $5y - x = 10$ ■ Solve for y.

$$5y - x = 10$$
$$5y = x + 10 \quad \textit{Add x.}$$
$$y = \frac{1}{5}x + 2 \quad \textit{Divide by 5.}$$

The slope is the coefficient of x, $\frac{1}{5}$.

6. Find the slope of each line by first solving each equation for y.

(a) $x + y = 6$ $x + y = 1$
     $y = -x + 6$ $y = -x + 1$
The slope is -1.   The slope is -1.

The lines have the same slope, so they are parallel.

(b) $3x - y = 4$ $x + 3y = 9$
     $-y = -3x + 4$ $3y = -x + 9$
     $y = 3x - 4$ $y = -\frac{1}{3}x + 3$
The slope is 3.   The slope is $-\frac{1}{3}$.

The product of the slopes is $3(-\frac{1}{3}) = -1$, so the lines are perpendicular.

(c) $2x - y = 5$ $2x + y = 3$
     $-y = -2x + 5$ $y = -2x + 3$
     $y = 2x - 5$ The slope is -2.
The slope is 2.

The slopes are not equal, and the product of the slopes is $2(-2) = -4$, not -1. The lines are neither parallel nor perpendicular.

(d) $3x - 7y = 35$ $7x - 3y = -6$
     $-7y = -3x + 35$ $-3y = -7x - 6$
     $y = \frac{3}{7}x - 5$ $y = \frac{7}{3}x + 2$
The slope is $\frac{3}{7}$.   The slope is $\frac{7}{3}$.

The slopes are not equal, and the product of the slopes is $(\frac{3}{7})(\frac{7}{3}) = 1$, not -1. The lines are neither parallel nor perpendicular.

11.3 Section Exercises

1. Slope is used to measure the <u>steepness</u> of a line. Slope is the <u>vertical</u> change compared to the <u>horizontal</u> change while moving along the line from left to right from one point to another.

3. **(a)** Count up 6 units. The vertical change is $\underline{6}$.

(b) Count 4 units to the right. The horizontal change is $\underline{4}$.

(c) The quotient of the numbers found in parts (a) and (b) is $\frac{6}{4}$, or $\frac{3}{2}$. We call this number the <u>*slope of the line*</u>.

(d) Yes, the slope will be the same. If we *start* at $(-1, -4)$ and *end* at $(3, 2)$, the vertical change will be 6 (6 units up) and the horizontal change will be 4 (4 units to the right), giving a slope of

$$m = \frac{6}{4} = \frac{3}{2}.$$

If we *start* at $(3, 2)$ and *end* at $(-1, -4)$, the vertical change will be -6 (6 units down) and the horizontal change will be -4 (4 units to the left), giving a slope of

$$m = \frac{-6}{-4} = \frac{3}{2}.$$

5. The indicated points have coordinates $(-1, -4)$ and $(1, 4)$.

$$\text{slope} = \frac{\text{change in } y \text{ (rise)}}{\text{change in } x \text{ (run)}}$$
$$= \frac{4 - (-4)}{1 - (-1)} = \frac{8}{2} = 4$$

7. The indicated points have coordinates $(-3, 2)$ and $(5, -2)$.

$$\text{slope} = \frac{\text{change in } y \text{ (rise)}}{\text{change in } x \text{ (run)}}$$
$$= \frac{-2 - 2}{5 - (-3)} = \frac{-4}{8} = -\frac{1}{2}$$

9. **(a)** Negative slope ▪ Sketches will vary. The line must fall from left to right. One such line is shown in the graph following part (d).

(b) Positive slope ▪ Sketches will vary. The line must rise from left to right. One such line is shown in the graph following part (d).

(c) Undefined slope ▪ Sketches will vary. The line must be vertical. One such line is shown in the graph following part (d).

(d) Zero slope ▪ Sketches will vary. The line must be horizontal. One such line is shown in the following graph.

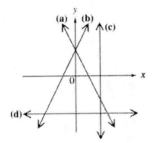

11. **(a)** Because the line *falls* from left to right, its slope is *negative*.

(b) Because the line intersects the y-axis *at* the origin, the y-value of its y-intercept is *zero*.

13. **(a)** Because the line *rises* from left to right, its slope is *positive*.

(b) Because the line intersects the y-axis *below* the origin, the y-value of its y-intercept is *negative*.

15. **(a)** The line is *horizontal*, so its slope is *zero*.

(b) The line intersects the y-axis *below* the origin, so the y-value of its y-intercept is *negative*.

17. The slope (or grade) of the hill is the ratio of the rise to the run, or the ratio of the vertical change to the horizontal change. Since the rise is 32 and the run is 108, the slope is

$$\frac{32}{108} = \frac{8 \cdot 4}{27 \cdot 4} = \frac{8}{27}.$$

19. $\text{slope} = \dfrac{\text{vertical change (rise)}}{\text{horizontal change (run)}}$

$= \frac{-8}{12}$ ("drops" indicates negative)

$= -\frac{2}{3}$

21. Because the student found the difference $3 - 5 = -2$ in the numerator, he or she should have subtracted in the same order in the denominator to get $-1 - 2 = -3$. The correct slope is $\frac{-2}{-3} = \frac{2}{3}$. Note that the student's slope and the correct slope are opposites of one another.

23. Use the slope formula with $(1, -2) = (x_1, y_1)$ and $(-3, -7) = (x_2, y_2)$.

$$\text{slope } m = \frac{\text{change in } y}{\text{change in } x} = \frac{y_2 - y_1}{x_2 - x_1}$$

$$= \frac{-7 - (-2)}{-3 - 1} = \frac{-5}{-4} = \frac{5}{4}$$

25. Use the slope formula with $(0, 3) = (x_1, y_1)$ and $(-2, 0) = (x_2, y_2)$.

$$\text{slope } m = \frac{\text{change in } y}{\text{change in } x} = \frac{y_2 - y_1}{x_2 - x_1}$$

$$= \frac{0 - 3}{-2 - 0} = \frac{-3}{-2} = \frac{3}{2}$$

27. Use the slope formula with $(-2, 4) = (x_1, y_1)$ and $(-3, 7) = (x_2, y_2)$.

$$\text{slope } m = \frac{\text{change in } y}{\text{change in } x} = \frac{y_2 - y_1}{x_2 - x_1}$$

$$= \frac{7 - 4}{-3 - (-2)} = \frac{3}{-1} = -3$$

29. Use the slope formula with $(4, 3) = (x_1, y_1)$ and $(-6, 3) = (x_2, y_2)$.

$$\text{slope } m = \frac{\text{change in } y}{\text{change in } x} = \frac{y_2 - y_1}{x_2 - x_1}$$

$$= \frac{3 - 3}{-6 - 4} = \frac{0}{-10} = 0$$

31. Use the slope formula with $(-12, 3) = (x_1, y_1)$ and $(-12, -7) = (x_2, y_2)$.

$$\text{slope } m = \frac{\text{change in } y}{\text{change in } x} = \frac{y_2 - y_1}{x_2 - x_1}$$

$$= \frac{-7 - 3}{-12 - (-12)} = \frac{-10}{0},$$

which is undefined.

33. Use the slope formula with $(4.8, 2.5) = (x_1, y_1)$ and $(3.6, 2.2) = (x_2, y_2)$.

$$\text{slope } m = \frac{\text{change in } y}{\text{change in } x} = \frac{y_2 - y_1}{x_2 - x_1}$$

$$= \frac{2.2 - 2.5}{3.6 - 4.8} = \frac{-0.3}{-1.2} = \frac{1}{4}$$

35. Use the slope formula with $\left(-\frac{7}{5}, \frac{3}{10}\right) = (x_1, y_1)$ and $\left(\frac{1}{5}, -\frac{1}{2}\right) = (x_2, y_2)$.

$$\text{slope } m = \frac{\text{change in } y}{\text{change in } x} = \frac{y_2 - y_1}{x_2 - x_1}$$

$$= \frac{-\frac{1}{2} - \frac{3}{10}}{\frac{1}{5} - \left(-\frac{7}{5}\right)} = \frac{-\frac{5}{10} - \frac{3}{10}}{\frac{1}{5} + \frac{7}{5}} = \frac{-\frac{8}{10}}{\frac{8}{5}}$$

$$= \left(-\frac{8}{10}\right)\left(\frac{5}{8}\right) = -\frac{5}{10} = -\frac{1}{2}$$

37. $y = 5x + 12$ ■ Since the equation is already solved for y, the slope is given by the coefficient of x, which is 5. Thus, the slope of the line is 5.

39. Solve the equation for y.

$$4y = x + 1$$
$$y = \tfrac{1}{4}x + \tfrac{1}{4} \quad \textit{Divide by 4.}$$

The slope of the line is given by the coefficient of x, so the slope is $\tfrac{1}{4}$.

41. Solve the equation for y.

$$3x - 2y = 3$$
$$-2y = -3x + 3 \quad \textit{Subtract 3x.}$$
$$y = \tfrac{3}{2}x - \tfrac{3}{2} \quad \textit{Divide by −2.}$$

The slope of the line is given by the coefficient of x, so the slope is $\tfrac{3}{2}$.

43. Solve the equation for y.

$$-2y - 3x = 5$$
$$-2y = 3x + 5 \quad \textit{Add 3x.}$$
$$y = -\tfrac{3}{2}x - \tfrac{5}{2} \quad \textit{Divide by −2.}$$

The slope of the line is given by the coefficient of x, so the slope is $-\tfrac{3}{2}$.

45. $y = 6$ ■ This is an equation of a horizontal line. Its slope is 0. (This equation may be rewritten in the form $y = 0x + 6$, where the coefficient of x gives the slope.)

47. $x = -2$ ■ This is an equation of a vertical line. Its slope is *undefined.*

49. Solve the equation for y.

$$x - y = 0$$
$$x = y \quad \textit{Add y.}$$

The slope of the line is given by the coefficient of x, so the slope is 1.

51. Find the slope of each line by solving the equations for y.

$$-4x + 3y = 4$$
$$3y = 4x + 4 \quad \textit{Add 4x.}$$
$$y = \tfrac{4}{3}x + \tfrac{4}{3} \quad \textit{Divide by 3.}$$

The slope of the first line is $\tfrac{4}{3}$.

$$-8x + 6y = 0$$
$$6y = 8x \quad \textit{Add 8x.}$$
$$y = \tfrac{8}{6}x \quad \textit{Divide by 6.}$$
$$y = \tfrac{4}{3}x \quad \textit{Reduce.}$$

The slope of the second line is $\tfrac{4}{3}$.

The slopes are equal, so the lines are *parallel.*

53. Find the slope of each line by solving the equations for y.

$$5x - 3y = -2$$
$$-3y = -5x - 2 \quad \textit{Subtract 5x.}$$
$$y = \tfrac{5}{3}x + \tfrac{2}{3} \quad \textit{Divide by −3.}$$

The slope of the first line is $\tfrac{5}{3}$.

$$3x - 5y = -8$$
$$-5y = -3x - 8 \quad \textit{Subtract 3x.}$$
$$y = \tfrac{3}{5}x + \tfrac{8}{5} \quad \textit{Divide by −5.}$$

The slope of the second line is $\tfrac{3}{5}$.

The slopes are not equal, so the lines are not parallel. The product of the slopes of the (nonvertical) lines is not -1, so the lines are not perpendicular. Thus, the lines are *neither* parallel nor perpendicular.

55. Find the slope of each line by solving the equations for y.

$$3x - 5y = -1$$
$$-5y = -3x - 1 \quad \textit{Subtract 3x.}$$
$$y = \tfrac{3}{5}x + \tfrac{1}{5} \quad \textit{Divide by −5.}$$

The slope of the first line is $\tfrac{3}{5}$.

$$5x + 3y = 2$$
$$3y = -5x + 2 \quad \textit{Subtract 5x.}$$
$$y = -\tfrac{5}{3}x + \tfrac{2}{3} \quad \textit{Divide by 3.}$$

The slope of the second line is $-\tfrac{5}{3}$.

The product of the slopes is

$$\tfrac{3}{5}\left(-\tfrac{5}{3}\right) = -1,$$

so the lines are *perpendicular.*

Relating Concepts (Exercises 57–62)

57. We use the points with coordinates $(2000, 6.6)$ and $(2010, 10.8)$.

$$m = \frac{10.8 - 6.6}{2010 - 2000} = \frac{4.2}{10} = 0.42 \,\%/\text{yr}$$

58. The slope of the line in Figure A is <u>positive</u>. This means that during the period represented, the percent of freshmen planning to major in the Biological Sciences <u>increased</u>.

59. The increase is 0.42% per year.

60. We use the points with coordinates $(2000, 16.7)$ and $(2010, 13.7)$.

$$m = \frac{13.7 - 16.7}{2010 - 2000} = \frac{-3}{10}$$
$$= -0.3 \text{ percent per year}$$

61. The slope of the line in Figure B is <u>negative</u>. This means that during the period represented, the percent of freshmen planning to major in Business <u>decreased</u>.

62. The decrease is 0.3% per year.

11.4   Writing and Graphing Equations of Lines

11.4   Margin Exercises

1. **(a)** $y = 2x - 6$ ∎ The slope is the coefficient of x, that is, 2. The y-intercept has x-coordinate 0 and y-coordinate -6, that is, $(0, -6)$.

(b) $y = -\frac{3}{5}x - 9$ ∎ The slope is the coefficient of x, that is, $-\frac{3}{5}$. The y-intercept has x-coordinate 0 and y-coordinate -9, that is, $(0, -9)$.

(c) $y = -\frac{x}{3} + \frac{7}{3}$ can be written as $y = -\frac{1}{3}x + \frac{7}{3}$.

The slope is $-\frac{1}{3}$ and the y-intercept is $(0, \frac{7}{3})$.

(d) $y = -x$ can be written as $y = -1x + 0$.

The slope is -1 and the y-intercept is $(0, 0)$.

2. **(a)** slope $\frac{1}{2}$; y-intercept $(0, -4)$ ∎ Use the slope-intercept form of a line with $m = \frac{1}{2}$ and $b = -4$.

$$y = mx + b$$
$$y = \frac{1}{2}x - 4$$

(b) slope -1; y-intercept $(0, 8)$

$$y = (-1)x + 8 \quad \text{or} \quad y = -x + 8$$

(c) slope 3; y-intercept $(0, 0)$

$$y = (3)x + 0 \quad \text{or} \quad y = 3x$$

(d) slope 0; y-intercept $(0, 2)$

$$y = (0)x + 2 \quad \text{or} \quad y = 2$$

(e) slope 1; y-intercept $(0, 0.75)$

$$y = (1)x + 0.75 \quad \text{or} \quad y = x + 0.75$$

3. ***Step 1*** Begin by solving for y.

$$3x - 4y = 8 \qquad \textit{Given equation}$$
$$-4y = -3x + 8 \qquad \textit{Subtract 3x.}$$
$$y = \frac{3}{4}x - 2 \qquad \textit{Divide by } -4.$$

Step 2 The y-intercept is $(0, -2)$. Graph this point.

Step 3 The slope is $\dfrac{3}{4} = \dfrac{\text{change in } y}{\text{change in } x}$.

Starting at the y-intercept, we count 3 units up and 4 units right to obtain another point on the graph, $(4, 1)$.

Step 4 Draw the line through the points $(0, -2)$ and $(4, 1)$ to obtain the graph of $3x - 4y = 8$.

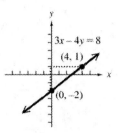

4. Through $(2, -3)$, with slope $-\frac{1}{3}$

To graph the line, write the slope as

$$m = \frac{\text{change in } y}{\text{change in } x} = \frac{-1}{3}.$$

$\left(\text{This could also be } \frac{1}{-3}.\right)$

Locate $(2, -3)$. Count 1 unit down and 3 units to the right. Draw a line through this point, $(5, -4)$, and $(2, -3)$.

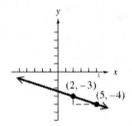

5. **(a)** The line having slope -2 passing through the point $(-1, 4)$

$$
\begin{array}{ll}
y = mx + b & \textit{Slope-intercept form} \\
\underline{4} = \underline{-2}(\underline{-1}) + b & \textit{Let } x = -1, y = 4, \\
& \textit{and } m = -2. \\
4 = \underline{2} + b & \textit{Multiply.} \\
\underline{2} = b & \textit{Subtract 2.}
\end{array}
$$

The y-intercept is $(0, 2)$ and the slope-intercept form of the equation of the line is

$$y = \underline{-2x + 2}.$$

(b) The line having slope 3 passing through the point $(-2, 1)$

$$
\begin{array}{ll}
y = mx + b & \textit{Slope-intercept form} \\
1 = 3(-2) + b & \textit{Let } x = -2, y = 1, \\
& \textit{and } m = 3. \\
1 = -6 + b & \textit{Multiply.} \\
7 = b & \textit{Add 6.}
\end{array}
$$

The y-intercept is $(0, 7)$ and the slope-intercept form of the equation of the line is

$$y = 3x + 7.$$

6.   **(a)** The line passing through $(-1, 3)$, with slope -2

Use the point-slope form of the equation of a line with $x_1 = -1$, $y_1 = 3$, and $m = -2$.

$$y - y_1 = m(x - x_1)$$
$$y - \underline{3} = \underline{-2}\,[x - (\underline{-1})]$$
$$y - 3 = -2(x + \underline{1})$$
$$y - 3 = -2x - \underline{2}$$
$$y = \underline{-2x + 1}$$

(b) The line passing through $(5, 2)$, with slope $-\frac{1}{3}$

Use the point-slope form of the equation of a line with $x_1 = 5$, $y_1 = 2$, and $m = -\frac{1}{3}$.

$$y - y_1 = m(x - x_1)$$
$$y - 2 = -\frac{1}{3}(x - 5)$$
$$y - 2 = -\frac{1}{3}x + \frac{5}{3}$$
$$y = -\frac{1}{3}x + \frac{5}{3} + \frac{6}{3}$$
$$y = -\frac{1}{3}x + \frac{11}{3}$$

7.   **(a)** $(2, 5)$ and $(-1, 6)$ ▪   First, find the slope of the line.

$$m = \frac{6 - 5}{-1 - 2} = \frac{1}{-3} = -\frac{1}{3}$$

Now use the point $(2, 5)$ for (x_1, y_1) and $m = -\frac{1}{3}$ in the point-slope form.

$$y - y_1 = m(x - x_1)$$
$$y - 5 = -\frac{1}{3}(x - 2)$$
$$y - 5 = -\frac{1}{3}x + \frac{2}{3}$$
$$y = -\frac{1}{3}x + \frac{2}{3} + \frac{15}{3}$$
$$y = -\frac{1}{3}x + \frac{17}{3}$$

$$y = -\frac{1}{3}x + \frac{17}{3}\quad \textit{Slope-intercept form}$$
$$3y = -x + 17\quad \textit{Multiply by 3.}$$
$$x + 3y = 17\quad \textit{Add x.}$$

The standard form is $x + 3y = 17$.

(b) $(-3, 1)$ and $(2, 4)$ ▪   First, find the slope of the line.

$$m = \frac{4 - 1}{2 - (-3)} = \frac{3}{5}$$

Now use the point $(-3, 1)$ for (x_1, y_1) and $m = \frac{3}{5}$ in the point-slope form.

$$y - y_1 = m(x - x_1)$$
$$y - 1 = \frac{3}{5}[x - (-3)]$$
$$y - 1 = \frac{3}{5}(x + 3)$$
$$y - 1 = \frac{3}{5}x + \frac{9}{5}$$
$$y = \frac{3}{5}x + \frac{9}{5} + \frac{5}{5}$$
$$y = \frac{3}{5}x + \frac{14}{5}$$

$$y = \frac{3}{5}x + \frac{14}{5}\quad \textit{Slope-intercept form}$$
$$5y = 3x + 14\quad \textit{Multiply by 5.}$$
$$-3x + 5y = 14\quad \textit{Subtract 3x.}$$
$$3x - 5y = -14\quad \textit{Multiply by -1.}$$

The standard form is $3x - 5y = -14$.

8.   Let $(x_1, y_1) = (1, 3766)$ and $(x_2, y_2) = (9, 7050)$.

$$m = \frac{y_2 - y_1}{x_2 - x_1} = \frac{7050 - 3766}{9 - 1}$$
$$= \frac{3284}{8} = 410.5$$

Now use this slope and the point $(1, 3766)$ in the point-slope form to find an equation of the line.

$$y - y_1 = m(x - x_1)$$
$$y - 3766 = 410.5(x - 1)$$
$$y - 3766 = 410.5x - 410.5$$
$$y = 410.5x + 3355.5$$

For 2007, let $x = 7$.

$$y = 410.5(7) + 3355.5$$
$$y = 6229$$

The equation gives $y = 6229$ when $x = 7$, which is a little high of the cost given in the table, 6185, but a reasonable approximation.

11.4   Section Exercises

1.   **(a)** The point-slope form of the equation of a line with slope -2 passing through the point $(4, 1)$ is

$$y - 1 = -2(x - 4).$$

Choice **D** is correct.

(b) The slope-intercept form of the equation of a line with slope -2 and y-intercept $(0, 1)$ is

$$y = -2x + 1.$$

Choice **C** is correct.

(c) A line that passes through the points $(0, 0)$ [its y-intercept] and $(4, 1)$ has slope

$$m = \frac{\text{change in } y}{\text{change in } x} = \frac{1 - 0}{4 - 0} = \frac{1}{4}.$$

Its slope-intercept form is

$$y = \frac{1}{4}x + 0\quad \text{or}\quad y = \frac{1}{4}x.$$

Choice **B** is correct.

(d) A line that passes through the points $(0, 0)$ [its y-intercept] and $(1, 4)$ has slope

$$m = \frac{\text{change in } y}{\text{change in } x} = \frac{4 - 0}{1 - 0} = \frac{4}{1} = 4.$$

Its slope-intercept form is

$$y = 4x + 0\quad \text{or}\quad y = 4x.$$

Choice **A** is correct.

3. **(a)** $y = x + 3 = 1x + 3$

The graph of this equation is a line with slope 1 and y-intercept $(0, 3)$. The only graph which has a *positive* slope and intersects the y-axis *above* the origin is **C**.

(b) $y = -x + 3 = -1x + 3$

The graph of this equation is a line with slope -1 and y-intercept $(0, 3)$. The only graph which has a *negative* slope and intersects the y-axis *above* the origin is **B**.

(c) $y = x - 3 = 1x - 3$

The graph of this equation is a line with slope 1 and y-intercept $(0, -3)$. The only graph which has a *positive* slope and intersects the y-axis *below* the origin is **A**.

(d) $y = -x - 3 = -1x - 3$

The graph of this equation is a line with slope -1 and y-intercept $(0, -3)$. The only graph which has a *negative* slope and intersects the y-axis *below* the origin is **D**.

5. $y = \frac{5}{2}x - 4$ ■ The slope is $\frac{5}{2}$ and the y-intercept is $(0, -4)$.

7. $y = -x + 9$ can be written as $y = -1x + 9$.

The slope is -1 and the y-intercept is $(0, 9)$.

9. $y = \frac{x}{5} - \frac{3}{10}$ can be written as $y = \frac{1}{5}x - \frac{3}{10}$.

The slope is $\frac{1}{5}$ and the y-intercept is $\left(0, -\frac{3}{10}\right)$.

11. slope 4, y-intercept $(0, -3)$ ■ Since the y-intercept is $(0, -3)$, we have $b = -3$. Use the slope-intercept form.

$$y = mx + b$$
$$y = 4x + (-3)$$
$$y = 4x - 3$$

13. $m = -1$, $(0, -7)$ ■ Since the y-intercept is $(0, -7)$, we have $b = -7$. Use the slope-intercept form.

$$y = mx + b$$
$$y = -1x - 7$$
$$y = -x - 7$$

15. slope 0, y-intercept $(0, 3)$ ■ Since the y-intercept is $(0, 3)$, we have $b = 3$. Use the slope-intercept form.

$$y = mx + b$$
$$y = 0x + 3$$
$$y = 3$$

17. undefined slope, $(0, -2)$ ■ Since the slope is undefined, the line is vertical and has equation $x = k$. Because the line goes through a point $(x, y) = (0, -2)$, we must have $x = 0$.

19. The rise is 3 and the run is 1, so the slope is given by

$$m = \frac{\text{rise}}{\text{run}} = \frac{3}{1} = 3.$$

The y-intercept is $(0, -3)$, so $b = -3$. The equation of the line, written in slope-intercept form, is

$$y = 3x - 3.$$

21. Since the line falls from left to right, the "rise" is negative. For this line, the rise is -3 and the run is 3, so the slope is

$$m = \frac{\text{rise}}{\text{run}} = \frac{-3}{3} = -1.$$

The y-intercept is $(0, 3)$, so $b = 3$. The equation of the line, written in slope-intercept form, is

$$y = -1x + 3$$
$$y = -x + 3.$$

23. Since the line falls from left to right, the "rise" is negative. For this line, the rise is -2 and the run is 4, so the slope is

$$m = \frac{\text{rise}}{\text{run}} = \frac{-2}{4} = -\frac{1}{2}.$$

The y-intercept is $(0, 2)$, so $b = 2$. The equation of the line, written in slope-intercept form, is

$$y = -\frac{1}{2}x + 2.$$

25. $y = 3x + 2$ ■ The slope is $3 = \frac{3}{1} = \frac{\text{change in } y}{\text{change in } x}$, and the y-intercept is $(0, 2)$. Graph that point and count up 3 units and right 1 unit to get to the point $(1, 5)$. Draw a line through the points.

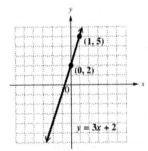

27. $y = \frac{3}{4}x - 1$ ■ The slope is $\frac{3}{4} = \frac{\text{change in } y}{\text{change in } x}$, and the y-intercept is $(0, -1)$. Graph that point and count up 3 units and right 4 units to get to the point $(4, 2)$. Draw a line through the points.

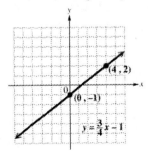

29. $2x + y = -5$ can be written as $y = -2x - 5$.

The slope is $-2 = \frac{-2}{1} = \frac{\text{change in } y}{\text{change in } x}$, and the y-intercept is $(0, -5)$. Graph that point and count down 2 units and right 1 unit to get to the point $(1, -7)$. Draw a line through the points.

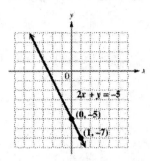

31. $x + 2y = 4$ ■   Solve the given equation for y.

$$2y = -x + 4 \quad \textit{Subtract x.}$$
$$y = -\tfrac{1}{2}x + 2 \quad \textit{Divide by 2.}$$

The slope is $-\frac{1}{2} = \frac{-1}{2} = \frac{\text{change in } y}{\text{change in } x}$, and the y-intercept is $(0, 2)$. Graph that point and count down 1 unit and right 2 units to get to the point $(2, 1)$. Draw a line through the points.

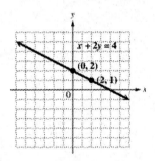

33. $(-2, 3), m = \frac{1}{2}$ ■   First, locate the point $(-2, 3)$. Write the slope as

$$m = \frac{\text{rise}}{\text{run}} = \frac{1}{2}.$$

Locate another point by counting 1 unit up and then 2 units to the right. Draw a line through this new point, $(0, 4)$, and the given point $(-2, 3)$. The slope-intercept form of the equation of this line is $y = \frac{1}{2}x + 4$.

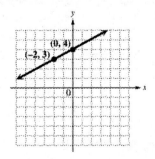

35. $(1, -5), m = -\frac{2}{5}$ ■   First, locate the point $(1, -5)$. Write the slope as

$$m = \frac{\text{rise}}{\text{run}} = \frac{-2}{5}.$$

Locate another point by counting 2 units down (because of the negative sign) and then 5 units to the right. Draw a line through this new point, $(6, -7)$, and the given point, $(1, -5)$.

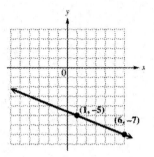

To find the slope-intercept form of the equation of this line, we will use the point-slope form and solve for y.

$$y - y_1 = m(x - x_1)$$
$$y - (-5) = -\tfrac{2}{5}(x - 1) \qquad (x_1, y_1) = (1, -5),\ m = -\tfrac{2}{5}$$
$$y + 5 = -\tfrac{2}{5}x + \tfrac{2}{5}$$
$$y = -\tfrac{2}{5}x + \tfrac{2}{5} - \tfrac{25}{5} \quad \textit{Subtract } 5 = \tfrac{25}{5}.$$
$$y = -\tfrac{2}{5}x - \tfrac{23}{5}$$

37. $(3, 2), m = 0$ ■   First, locate the point $(3, 2)$. Since the slope is 0, the line will be horizontal. Draw a horizontal line through the point $(3, 2)$.

The slope-intercept form of the equation of this line is $y = 0x + 2$, or $y = 2$.

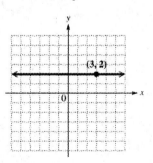

39. $(3, -2)$, undefined slope   ■   First, locate the point $(3, -2)$. Since the slope is undefined, the line will be vertical. Draw a vertical line through the point $(3, -2)$.

An equation of this line is $x = 3$. There is no slope-intercept form.

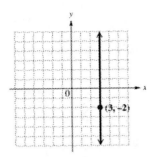

41. $(0,0), m = \frac{2}{3}$ ▪ First, locate the point $(0,0)$. Write the slope as

$$m = \frac{\text{rise}}{\text{run}} = \frac{2}{3}.$$

Locate another point by counting 2 units up and then 3 units to the right. Draw a line through this new point, $(3,2)$, and the given point $(0,0)$. The slope-intercept form of the equation of this line is $y = \frac{2}{3}x$.

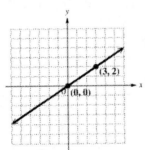

43. $(4,1), m = 2$ ▪ The given point is $(4,1)$, so $x_1 = 4$ and $y_1 = 1$. Also, $m = 2$. Substitute these values into the point-slope form. Then solve for y to obtain the slope-intercept form.

$$y - y_1 = m(x - x_1)$$
$$y - 1 = 2(x - 4)$$
$$y - 1 = 2x - 8 \qquad \text{Dist. prop.}$$
$$y = 2x - 7 \qquad \text{Add 1.}$$

45. $(3,-10), m = -2$ ▪ Use the values $x_1 = 3$, $y_1 = -10$, and $m = -2$ in the point-slope form.

$$y - y_1 = m(x - x_1)$$
$$y - (-10) = -2(x - 3)$$
$$y + 10 = -2x + 6 \qquad \text{Dist. prop.}$$
$$y = -2x - 4 \qquad \text{Subtract 10.}$$

47. $(-2,5), m = \frac{2}{3}$ ▪ Use the values $x_1 = -2$, $y_1 = 5$, and $m = \frac{2}{3}$ in the point-slope form.

$$y - y_1 = m(x - x_1)$$
$$y - 5 = \frac{2}{3}[x - (-2)]$$
$$y - 5 = \frac{2}{3}(x + 2)$$
$$y - 5 = \frac{2}{3}x + \frac{4}{3}$$
$$y = \frac{2}{3}x + \frac{19}{3} \qquad \text{Add } 5 = \frac{15}{3}.$$

49. **(a)** $(8,5)$ and $(9,6)$ ▪ First, find the slope of the line.

$$m = \frac{6 - 5}{9 - 8} = \frac{1}{1} = 1$$

Now use the point $(8,5)$ for (x_1, y_1) and $m = 1$ in the point-slope form.

$$y - y_1 = m(x - x_1)$$
$$y - 5 = 1(x - 8)$$
$$y - 5 = x - 8$$
$$y = x - 3$$

The same result would be found by using $(9,6)$ for (x_1, y_1).

(b)
$$y = x - 3$$
$$-x + y = -3 \qquad \text{Subtract } x.$$
$$x - y = 3 \qquad \text{Multiply by } -1.$$

51. **(a)** $(-1,-7)$ and $(-8,-2)$ ▪ First, find the slope of the line.

$$m = \frac{-2 - (-7)}{-8 - (-1)} = \frac{5}{-7} = -\frac{5}{7}$$

Now use the point $(-1,-7)$ for (x_1, y_1) and $m = -\frac{5}{7}$ in the point-slope form.

$$y - y_1 = m(x - x_1)$$
$$y - (-7) = -\frac{5}{7}[x - (-1)]$$
$$y + 7 = -\frac{5}{7}(x + 1)$$
$$y + 7 = -\frac{5}{7}x - \frac{5}{7}$$
$$y = -\frac{5}{7}x - \frac{54}{7} \qquad \text{Subtract } 7 = \frac{49}{7}.$$

(b)
$$y = -\frac{5}{7}x - \frac{54}{7}$$
$$7y = -5x - 54 \qquad \text{Multiply by } 7.$$
$$5x + 7y = -54 \qquad \text{Add } 5x.$$

53. **(a)** $(0,-2)$ and $(-3,0)$ ▪ First, find the slope of the line.

$$m = \frac{0 - (-2)}{-3 - 0} = \frac{2}{-3} = -\frac{2}{3}$$

Now use the point $(-3,0)$ for (x_1, y_1) and $m = -\frac{2}{3}$ in the point-slope form.

$$y - y_1 = m(x - x_1)$$
$$y - 0 = -\frac{2}{3}[x - (-3)]$$
$$y = -\frac{2}{3}(x + 3)$$
$$y = -\frac{2}{3}x - 2$$

(b)
$$y = -\frac{2}{3}x - 2$$
$$3y = -2x - 6 \qquad \text{Multiply by } 3.$$
$$2x + 3y = -6 \qquad \text{Add } 2x.$$

55. (a) $\left(\frac{1}{2}, \frac{3}{2}\right)$ and $\left(-\frac{1}{4}, \frac{5}{4}\right)$ ▪ First, find the slope of the line.

$$m = \frac{\frac{5}{4} - \frac{3}{2}}{-\frac{1}{4} - \frac{1}{2}} = \frac{\frac{5}{4} - \frac{6}{4}}{-\frac{1}{4} - \frac{2}{4}} = \frac{-\frac{1}{4}}{-\frac{3}{4}}$$

$$= \left(-\frac{1}{4}\right)\left(-\frac{4}{3}\right) = \frac{1}{3}$$

Now use the point $\left(\frac{1}{2}, \frac{3}{2}\right)$ for (x_1, y_1) and $m = \frac{1}{3}$ in the point-slope form.

$$y - y_1 = m(x - x_1)$$
$$y - \frac{3}{2} = \frac{1}{3}\left(x - \frac{1}{2}\right)$$
$$y - \frac{3}{2} = \frac{1}{3}x - \frac{1}{6}$$
$$y = \frac{1}{3}x - \frac{1}{6} + \frac{9}{6} \quad \left(\frac{3}{2} = \frac{9}{6}\right)$$
$$y = \frac{1}{3}x + \frac{4}{3} \quad \left(\frac{8}{6} = \frac{4}{3}\right)$$

(b) $\qquad y = \frac{1}{3}x + \frac{4}{3}$
$\qquad\qquad 3y = x + 4 \qquad$ *Multiply by 3.*
$\quad -x + 3y = 4 \qquad$ *Subtract x.*
$\qquad x - 3y = -4 \qquad$ *Multiply by −1.*

57. Solve the equation for y.

$$x - 2y = 7$$
$$-2y = -x + 7$$
$$y = \frac{1}{2}x - \frac{7}{2}$$

The slope is $\frac{1}{2}$. A line perpendicular to this line has slope -2, since $\frac{1}{2}(-2) = -1$.
Now use the slope-intercept form with $m = -2$ and y-intercept $(0, -3)$.

$$y = mx + b$$
$$y = -2x - 3$$

59. Solve the equation for y.

$$4x - y = -2$$
$$4x + 2 = y$$

The slope is 4. A line parallel to this line has the same slope. Now use the point-slope form with $m = 4$ and $(x_1, y_1) = (2, 3)$.

$$y - y_1 = m(x - x_1)$$
$$y - 3 = 4(x - 2)$$
$$y - 3 = 4x - 8$$
$$y = 4x - 5$$

61. Solve the equation for y.

$$3x = 4y + 5$$
$$-4y = -3x + 5$$
$$y = \frac{3}{4}x - \frac{5}{4}$$

The slope is $\frac{3}{4}$. A line parallel to this line has the same slope. Now use the point-slope form with $m = \frac{3}{4}$ and $(x_1, y_1) = (2, -3)$.

$$y - y_1 = m(x - x_1)$$
$$y - (-3) = \frac{3}{4}(x - 2)$$
$$y + 3 = \frac{3}{4}x - \frac{3}{2}$$
$$y = \frac{3}{4}x - \frac{3}{2} - \frac{6}{2}$$
$$y = \frac{3}{4}x - \frac{9}{2}$$

63. (a) The fixed cost is $400.

(b) The variable cost is $0.25.

(c) Substitute $m = 0.25$ and $b = 400$ into $y = mx + b$ to get the cost equation

$$y = 0.25x + 400.$$

(d) Let $x = 100$ in the cost equation.
$$y = 0.25(100) + 400$$
$$y = 25 + 400$$
$$y = 425$$

The cost to produce 100 campaign buttons will be $425.

(e) Let $y = 775$ in the cost equation.

$$775 = 0.25x + 400$$
$$375 = 0.25x \qquad\qquad \textit{Subtract 400.}$$
$$x = \frac{375}{0.25} = 1500 \quad \textit{Divide by 0.25.}$$

If the total cost is $775, 1500 campaign buttons will be produced.

65. (a) x represents the year and y represents the cost in the ordered pairs (x, y). The ordered pairs are $(1, 2294), (2, 2372), (3, 2558), (4, 2727),$ and $(5, 2963)$.

(b)

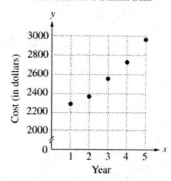

AVERAGE ANNUAL COSTS AT 2-YEAR COLLEGES

Yes, the points lie approximately in a straight line.

(c) Find the slope using $(x_1, y_1) = (2, 2372)$ and $(x_2, y_2) = (5, 2963)$.

$$m = \frac{y_2 - y_1}{x_2 - x_1} = \frac{2963 - 2372}{5 - 2} = \frac{591}{3} = 197$$

Now use the point-slope form with $m = 197$ and $(x_1, y_1) = (2, 2372)$.

$$y - y_1 = m(x - x_1)$$
$$y - 2372 = 197(x - 2)$$
$$y - 2372 = 197x - 394$$
$$y = 197x + 1978$$

(d) Since year 1 represents 2007, year 0 represents 2006.

For 2012, $x = 2012 - 2006 = 6$.

$$y = 197x + 1978$$
$$y = 197(6) + 1978$$
$$y = 3160$$

In 2012, the estimate of the average annual cost at 2-year colleges is $3160.

67. Find the slope using
$(x_1, y_1) = (2002, 52)$ and $(x_2, y_2) = (2007, 127)$.

$$m = \frac{y_2 - y_1}{x_2 - x_1} = \frac{127 - 52}{2007 - 2002} = \frac{75}{5} = 15$$

Now use the point-slope form with $m = 15$ and $(x_2, y_2) = (2007, 127)$.

$$y - y_1 = m(x - x_1)$$
$$y - 127 = 15(x - 2007)$$
$$y - 127 = 15x - 30,105$$
$$y = 15x - 29,978$$

Relating Concepts (Exercises 69–76)

69. When $C = 0$, $F = 32$. This gives the ordered pair $(0, 32)$. When $C = 100$, $F = 212$. This gives the ordered pair $(100, 212)$.

70. Use the two points $(0, 32)$ and $(100, 212)$.

$$m = \frac{212 - 32}{100 - 0} = \frac{180}{100} = \frac{9}{5}$$

71. Write the point-slope form as
$$F - F_1 = m(C - C_1).$$

We may use either the point $(0, 32)$ or the point $(100, 212)$ with the slope $\frac{9}{5}$, which we found in Exercise 70. Using $F_1 = 32$, $C_1 = 0$, and $m = \frac{9}{5}$, we obtain

$$F - 32 = \frac{9}{5}(C - 0).$$

If we use $F_1 = 212$, $C_1 = 100$, and $m = \frac{9}{5}$, we obtain

$$F - 212 = \frac{9}{5}(C - 100).$$

72. We want to write an equation in the form
$$F = mC + b.$$

From Exercise 70, we have $m = \frac{9}{5}$. The point $(0, 32)$ is the y-intercept of the graph of the desired equation, so we have $b = 32$. Thus, an equation for F in terms of C is

$$F = \frac{9}{5}C + 32.$$

73. The expression for F in terms of C obtained in Exercise 72 is

$$F = \frac{9}{5}C + 32.$$

To obtain an expression for C in terms of F, solve this equation for C.

$$F - 32 = \frac{9}{5}C \qquad \textit{Subtract 32.}$$

$$\frac{5}{9}(F - 32) = \frac{5}{9} \cdot \frac{9}{5}C \qquad \textit{Multiply by } \frac{5}{9}, \textit{ the reciprocal of } \frac{9}{5}.$$

$$\frac{5}{9}(F - 32) = C \quad \text{or} \quad C = \frac{5}{9}F - \frac{160}{9}$$

74. $F = \frac{9}{5}C + 32$

$$\text{If } C = 30, \qquad F = \frac{9}{5}(30) + 32$$
$$= 54 + 32 = 86.$$

Thus, when $C = 30°$, $F = 86°$.

75. $C = \frac{5}{9}(F - 32)$

$$\text{When } F = 50, \qquad C = \frac{5}{9}(50 - 32)$$
$$= \frac{5}{9}(18) = 10.$$

Thus, when $F = 50°$, $C = 10°$.

76. Let $F = C$ in the equation obtained in Exercise 72.

$$F = \frac{9}{5}C + 32$$
$$C = \frac{9}{5}C + 32 \qquad \textit{Let F = C.}$$
$$5C = 5\left(\frac{9}{5}C + 32\right) \quad \textit{Multiply by 5.}$$
$$5C = 9C + 160$$
$$-4C = 160 \qquad \textit{Subtract 9C.}$$
$$C = -40 \qquad \textit{Divide by } -4.$$

(The same result may be found by using either form of the equation obtained in Exercise 73.) The Celsius and Fahrenheit temperatures are equal $(F = C)$ at -40 degrees.

Summary Exercises
Applying Graphing and Equation-Writing Techniques for Lines

1. **(a)** "Slope -0.5" indicates that $m = -\frac{1}{2}$ and "$b = -2$" indicates that the y-intercept is $(0, -2)$. Given the slope and y-intercept, we know that one form of the equation of the line is $y = -\frac{1}{2}x - 2$, choice **B**.

(b) The slope of the line passing through the x-intercept of $(4, 0)$ and the y-intercept of $(0, 2)$ is

$$m = \frac{2 - 0}{0 - 4} = \frac{2}{-4} = -\frac{1}{2}.$$

The slope-intercept form is $y = -\frac{1}{2}x + 2$. None of the choices has this equation, so change the form of $y = -\frac{1}{2}x + 2$ by multiplying each side by 2 to clear fractions.

$$2(y) = 2(-\tfrac{1}{2}x + 2)$$
$$2y = -x + 4$$

Add x to both sides to get $x + 2y = 4$, choice **D**.

(c) The slope of the line passing through $(4, -2)$ and $(0, 0)$ is

$$m = \frac{0 - (-2)}{0 - 4} = \frac{2}{-4} = -\frac{1}{2}.$$

The slope-intercept form is $y = -\frac{1}{2}x$, choice **A**.

(d) $m = \frac{1}{2}$; passes through $(-2, -2)$

$$
\begin{array}{ll}
y = mx + b & \textit{Slope-intercept form} \\
-2 = \frac{1}{2}(-2) + b & \textit{Let } x = -2, y = -2, m = \frac{1}{2}. \\
-2 = -1 + b & \textit{Multiply.} \\
-1 = b & \textit{Add 1.}
\end{array}
$$

The line has equation $y = \frac{1}{2}x - 1$. Multiply each side by 2 to get $2y = x - 2$, which is the same as $2 = x - 2y$, choice **C**.

3. $m = 1, b = -2$ ▪ The slope is $1 = \frac{1}{1} = \frac{\text{change in } y}{\text{change in } x}$, and the y-intercept is $(0, -2)$. Graph that point and count up 1 unit and right 1 unit to get to the point $(1, -1)$. Draw a line through the points.

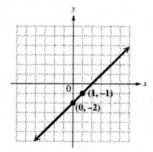

5. $y = -2x + 6$ ▪ The slope is $-2 = \frac{-2}{1} = \frac{\text{change in } y}{\text{change in } x}$, and the y-intercept is $(0, 6)$. Graph that point and count down 2 units and right 1 unit to get to the point $(1, 4)$. Draw a line through the points.

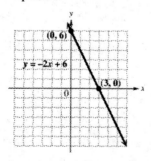

7. $m = -\frac{2}{3}$, passes through $(3, -4)$ ▪ The slope is $-\frac{2}{3} = \frac{2}{-3} = \frac{\text{change in } y}{\text{change in } x}$, and the line passes through $(3, -4)$. Graph that point and count up 2 units and left 3 units to get to the point $(0, -2)$. Draw a line through the points.

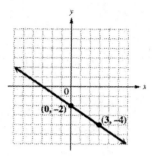

9. $y - 4 = -9$, or equivalently, $y = -5$, is an equation of a horizontal line with y-intercept $(0, -5)$.

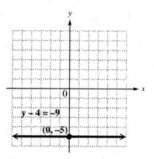

11. Undefined slope, passes through $(3.5, 0)$

Since the slope is undefined, the line is vertical. Draw a vertical line through the point $(3.5, 0)$.

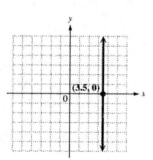

13. $4x - 5y = 20$ ▪ Solve the equation for y.

$$
\begin{array}{ll}
4x - 5y = 20 & \textit{Given equation} \\
-5y = -4x + 20 & \textit{Subtract 4x.} \\
y = \frac{4}{5}x - 4 & \textit{Divide by } -5.
\end{array}
$$

The slope is $\frac{4}{5} = \frac{\text{change in } y}{\text{change in } x}$, and the y-intercept is $(0, -4)$. Graph that point and count up 4 units and right 5 units to get to the point $(5, 0)$. Draw a line through the points.

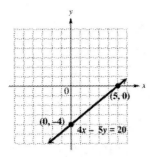

15. $x - 4y = 0$ ■ Solve the equation for y.

$$
\begin{aligned}
x - 4y &= 0 && \textit{Given equation} \\
-4y &= -x && \textit{Subtract x.} \\
y &= \tfrac{1}{4}x && \textit{Divide by } -4.
\end{aligned}
$$

The slope is $\frac{1}{4} = \frac{\text{change in } y}{\text{change in } x}$, and the y-intercept is $(0, 0)$. Graph that point and count up 1 unit and right 4 units to get to the point $(4, 1)$. Draw a line through the points.

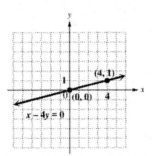

17. $3y = 12 - 2x$ ■ Solve the equation for y.

$$
\begin{aligned}
3y &= 12 - 2x && \textit{Given equation} \\
y &= 4 - \tfrac{2}{3}x && \textit{Divide by 3.}
\end{aligned}
$$

The slope is $-\frac{2}{3} = \frac{-2}{3} = \frac{\text{change in } y}{\text{change in } x}$, and the y-intercept is $(0, 4)$. Graph that point and count down 2 units and right 3 units to get to the point $(3, 2)$. Draw a line through the points.

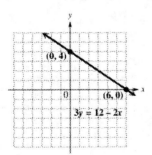

19. $m = -3, b = -6$ ■ Use the slope-intercept form of a line.

$$
\begin{aligned}
y &= mx + b \\
y &= -3x - 6
\end{aligned}
$$

21. Passes through $(0, 0)$ and $(5, 3)$ ■ First, find the slope of the line.

$$
m = \frac{3 - 0}{5 - 0} = \frac{3}{5}
$$

Now use the slope-intercept form of a line.

$$
\begin{aligned}
y &= mx + b \\
y &= \tfrac{3}{5}x + 0 \\
y &= \tfrac{3}{5}x
\end{aligned}
$$

23. Passes through $(0, 0)$, $m = 0$

A line with slope 0 has equation
$y = y$-intercept value; in this case, $y = 0$.

25. $m = \frac{5}{3}$, through $(-3, 0)$ ■ Use the point-slope form of a line.

$$
\begin{aligned}
y - y_1 &= m(x - x_1) \\
y - 0 &= \tfrac{5}{3}[x - (-3)] \\
y &= \tfrac{5}{3}(x + 3) \\
y &= \tfrac{5}{3}x + 5
\end{aligned}
$$

11.5 Graphing Linear Inequalities in Two Variables

11.5 Margin Exercises

1.
$$
\begin{aligned}
3x + 4y &\le 12 && \textit{Original inequality} \\
3(\underline{0}) + 4(\underline{0}) &\overset{?}{\le} 12 && \textit{Let x = 0 and y = 0.} \\
0 + 0 &\overset{?}{\le} 12 && \textit{Multiply.} \\
\underline{0} &\le 12 && \textit{True}
\end{aligned}
$$

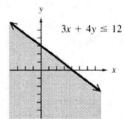

2. Graph $4x - 5y \le 20$.

Start by graphing the equation

$$
4x - 5y = 20.
$$

The intercepts are $(5, 0)$ and $(0, -4)$. Draw a solid line through these points to show that the points on the line are solutions to the inequality

$$
4x - 5y \le 20.
$$

Choose a test point not on the line, such as $(0, 0)$.

$$
\begin{aligned}
4x - 5y &\le 20 && \textit{Original inequality} \\
4(0) - 5(0) &\overset{?}{\le} 20 && \textit{Let x = 0, y = 0.} \\
0 - 0 &\overset{?}{\le} 20 && \\
0 &\le 20 && \textit{True}
\end{aligned}
$$

Since the last statement is true, shade the region that includes the test point $(0, 0)$, that is, the region above the line.

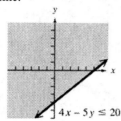

$4x - 5y \le 20$

3.

$$3x + 5y > 15 \quad \text{Original inequality}$$

$$3(0) + 5(0) \overset{?}{>} 15 \quad \text{Let } x = 0 \text{ and } y = 0.$$

$$0 + 0 \overset{?}{>} 15$$

$$0 > 15 \quad \text{False}$$

Since the last statement is false, shade the region that does *not* include the test point, $(0, 0)$, that is, shade the region above the line.

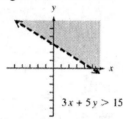

$3x + 5y > 15$

4. Graph $x + 2y > 6$.

Start by graphing the equation

$$x + 2y = 6.$$

The intercepts are $(6, 0)$ and $(0, 3)$. Draw a dashed line through these points to show that the points on the line are *not* solutions to the inequality

$$x + 2y > 6.$$

Choose a test point not on the line, such as $(0, 0)$.

$$x + 2y > 6 \quad \text{Original inequality}$$

$$(0) + 2(0) \overset{?}{>} 6 \quad \text{Let } x = 0 \text{ and } y = 0.$$

$$0 > 6 \quad \text{False}$$

Since the last statement is false, shade the region that does *not* include the test point, $(0, 0)$, that is, shade the region above the line.

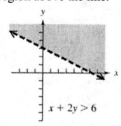

$x + 2y > 6$

5. $y < 4$ ▪ First, graph $y = 4$, a horizontal line through the point $(0, 4)$. Use a dashed line because of the $<$ symbol. Choose $(0, 0)$ as a test point.

$$y < 4 \quad \text{Original inequality}$$

$$0 < 4 \quad \text{Let } y = 0. \quad \text{True}$$

Since the last statement is true, shade the region that includes the test point $(0, 0)$, that is, the region below the line.

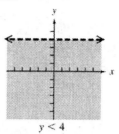

$y < 4$

6. $x \ge -3y$ ▪ Graph $x = -3y$ as a solid line through $(0, 0)$ and $(-3, 1)$. We cannot use $(0, 0)$ as a test point because $(0, 0)$ is on the line $x = -3y$. Instead, we choose a test point off the line, say $(0, 1)$.

$$x \ge -3y \quad \text{Original inequality}$$

$$0 \overset{?}{\ge} -3(1) \quad \text{Let } x = 0 \text{ and } y = 1.$$

$$0 \ge -3 \quad \text{True}$$

Since the last statement is true, shade the region that includes the test point $(0, 1)$, that is, the region above the line.

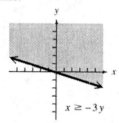

$x \ge -3y$

11.5 Section Exercises

1. The key phrase is "more than" *(occurs twice), so use the symbol $>$ twice.

3. The key phrase is "at most," so use the symbol $\le$.

5. For the point $(4, 0)$, substitute 4 for x and 0 for y in the inequality.

$$3x - 4y < 12$$

$$3(4) - 4(0) \overset{?}{<} 12$$

$$12 - 0 \overset{?}{<} 12$$

$$12 < 12 \quad \text{False}$$

The *false* result shows that $(4, 0)$ *is not* a solution of the given inequality. The point $(4, 0)$ lies on the boundary line $3x - 4y = 12$, which is *not* part of the graph because the symbol $<$ does not involve equality.

7. False; because $(0, 0)$ is on the boundary line $x + 4y = 0$, it cannot be used as a test point. Use a test point *off* the line.

9. $3x - 2y \ge 0$

(i) For the point $(4, 1)$, substitute 4 for x and 1 for y in the given inequality.

$$3x - 2y \geq 0$$
$$3(4) - 2(1) \overset{?}{\geq} 0$$
$$12 - 2 \overset{?}{\geq} 0$$
$$10 \geq 0 \quad \textit{True}$$

The *true* result shows that $(4, 1)$ *is* a solution of the given inequality.

(ii) For the point $(0, 0)$, substitute 0 for x and 0 for y in the given inequality.

$$3x - 2y \geq 0$$
$$3(0) - 2(0) \overset{?}{\geq} 0$$
$$0 \geq 0 \quad \textit{True}$$

The *true* result shows that $(0, 0)$ *is* a solution of the given inequality.

Since (i) and (ii) are true, the given statement is *true*.

11. $x + y \geq 4$ ▪ Use $(0, 0)$ as a test point.

$$x + y \geq 4$$
$$0 + 0 \overset{?}{\geq} 4 \quad \textit{Let x = 0, y = 0.}$$
$$0 \geq 4 \quad \textit{False}$$

Because the last statement is *false*, we shade the region that *does not* include the test point $(0, 0)$. This is the region above the line.

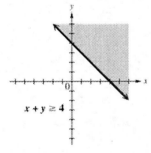

13. $x + 2y \geq 7$ ▪ Use $(0, 0)$ as a test point.

$$x + 2y \geq 7$$
$$0 + 2(0) \overset{?}{\geq} 7 \quad \textit{Let x = 0, y = 0.}$$
$$0 \geq 7 \quad \textit{False}$$

Because the last statement is *false*, we shade the region that *does not* include the test point $(0, 0)$. This is the region above the line.

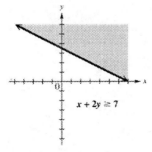

15. $-3x + 4y > 12$ ▪ Use $(0, 0)$ as a test point.

$$-3x + 4y > 12$$
$$-3(0) + 4(0) \overset{?}{>} 12 \quad \textit{Let x = 0, y = 0.}$$
$$0 > 12 \quad \textit{False}$$

Because the last statement is *false*, we shade the region that *does not* include the test point $(0, 0)$. This is the region above the line.

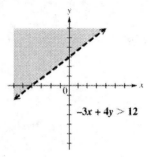

17. $x > 4$ ▪ Use $(0, 0)$ as a test point.

$$x > 4$$
$$0 \overset{?}{>} 4 \quad \textit{Let x = 0. False}$$

Because $0 > 4$ is *false*, shade the region *not* containing $(0, 0)$. This is the region to the right of the line.

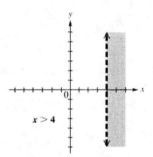

19. $x \geq -y$ ▪ Since $(0, 0)$ is on the boundary line $x = -y$, we will use $(0, 1)$ as a test point.

$$x \geq -y$$
$$0 \overset{?}{\geq} -1 \quad \textit{Let x = 0, y = 1.}$$
$$0 \geq -1 \quad \textit{True}$$

Because the last statement is *true*, we shade the region including the test point $(0, 1)$. This is the region above the line.

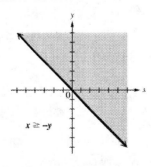

21. Use a dashed line if the symbol is $<$ or $>$. Use a solid line if the symbol is $\leq$ or $\geq$.

23. $x + y \leq 5$

Step 1 Graph the boundary of the region, the line with equation $x + y = 5$.

If $y = 0$, $x = 5$, so the x-intercept is $(5, 0)$.
If $x = 0$, $y = 5$, so the y-intercept is $(0, 5)$.

Draw the line through these intercepts. Make the line solid because of the $\leq$ sign.

Step 2 Choose the point $(0, 0)$ as a test point.

$$x + y \leq 5$$
$$0 + 0 \overset{?}{\leq} 5 \quad \text{Let } x = 0, y = 0.$$
$$0 \leq 5 \quad \text{True}$$

Because $0 \leq 5$ is true, shade the region containing the origin. The shaded region, along with the boundary, is the desired graph.

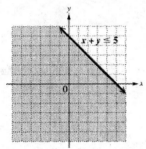

25. $x + 2y < 4$ ▪ Start by graphing the boundary, the line with equation $x + 2y = 4$, through its intercepts $(0, 2)$ and $(4, 0)$. Make the line dashed because of the $<$ sign. Choose $(0, 0)$ as a test point.

$$x + 2y < 4$$
$$0 + 2(0) \overset{?}{<} 4 \quad \text{Let } x = 0, y = 0.$$
$$0 < 4 \quad \text{True}$$

Because the last statement is *true*, shade the region containing the origin. The dashed line indicates that the boundary is not part of the graph.

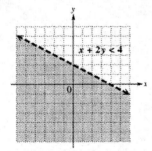

27. $2x + 6 > -3y$ ▪ Start by graphing the boundary, the line with equation $2x + 6 = -3y$, through its intercepts $(-3, 0)$ and $(0, -2)$. Make the line dashed because of the $>$ sign.

Choose $(0, 0)$ as a test point.

$$2x + 6 > -3y$$
$$2(0) + 6 \overset{?}{>} -3(0) \quad \text{Let } x = 0, y = 0.$$
$$6 > 0 \quad \text{True}$$

Because the last statement is *true*, shade the region containing the origin. The dashed line indicates that the boundary is not part of the graph.

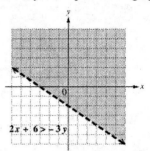

29. $y \geq 2x + 1$ ▪ The boundary is the line with equation $y = 2x + 1$. This line has slope 2 and y-intercept $(0, 1)$. It may be graphed by starting at $(0, 1)$ and going 2 units up and then 1 unit to the right to reach the point $(1, 3)$. Draw a solid line through $(0, 1)$ and $(1, 3)$.

The ordered pairs for which y is *greater than* $2x + 1$ are *above* this line. Indicate the solution by shading the region above the line. The solid line shows that the boundary is part of the graph.

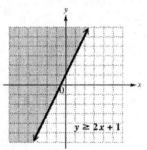

31. $x \leq -2$ ▪ The boundary is the line with equation $x = -2$. This is a vertical line through $(-2, 0)$. Make this line solid because of the $\leq$ sign.

Using $(0, 0)$ as a test point will result in the inequality $0 \leq -2$, which is false. Shade the region not containing the origin. This is the region to the left of the boundary. The solid line shows that the boundary is part of the graph.

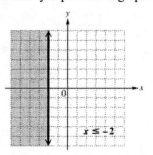

33. $y < 5$ ▪ The boundary is the line with equation $y = 5$. This is the horizontal line through $(0, 5)$. Make this line dashed because of the $<$ sign.

The ordered pairs for which y is *less than* 5 are *below* this line. Indicate the solution by shading the region below the line. The dashed line shows that the boundary is not part of the graph.

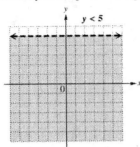

35. $y \geq 4x$ ▪ The boundary has the equation $y = 4x$. This line goes through the points $(0, 0)$ and $(1, 4)$. Make the line solid because of the $\geq$ sign.

The ordered pairs for which y is *greater than* $4x$ are *above* this line. Indicate the solution by shading the region above the line. The solid line shows that the boundary is part of the graph.

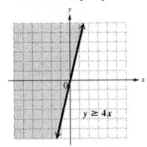

37. $x < -2y$ ▪ The boundary has the equation $x = -2y$. This line goes through the points $(0, 0)$ and $(-2, 1)$. Make the line dashed because of the $>$ sign. Because the boundary passes through the origin, we cannot use $(0, 0)$ as a test point. Choose $(0, 1)$ as a test point.

$$x > -2y$$
$$0 \overset{?}{>} -2(1) \quad \textit{Let } x = 0, y = 1.$$
$$0 > -2 \quad \textit{True}$$

Because the last statement is *true*, shade the region containing the test point $(0, 1)$.

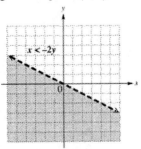

Chapter 11 Review Exercises

1. In this case, x represents the year and y represents the percent. Ordered pairs that reflect the data shown in the graph are $(2006, 52.5)$ and $(2011, 55.4)$.

2. The ordered pair $(2000, 52.0)$ means that in the year 2000, 52.0% of the first-year college students at two-year public institutions returned for a second year.

3. The graph is flat between the years 2008 and 2009. The percent is about 53.7%.

4. The percent increased over the years 2007 to 2008 and 2009 to 2010. It was about a 2% increase each time.

5. $y = 3x + 2;\ (-1, \_\_),\ (0, \_\_),\ (\_\_, 5)$

$$y = 3x + 2$$
$$y = 3(-1) + 2 \quad \textit{Let x = -1.}$$
$$y = -3 + 2$$
$$y = -1$$

$$y = 3x + 2$$
$$y = 3(0) + 2 \quad \textit{Let x = 0.}$$
$$y = 0 + 2$$
$$y = 2$$

$$y = 3x + 2$$
$$5 = 3x + 2 \quad \textit{Let y = 5.}$$
$$3 = 3x$$
$$1 = x$$

The ordered pairs are $(-1, -1)$, $(0, 2)$, and $(1, 5)$.

6. $4x + 3y = 6;\ (0, \_\_),\ (\_\_, 0),\ (-2, \_\_)$

$$4x + 3y = 6$$
$$4(0) + 3y = 6 \quad \textit{Let x = 0.}$$
$$3y = 6$$
$$y = 2$$

$$4x + 3y = 6$$
$$4x + 3(0) = 6 \quad \textit{Let y = 0.}$$
$$4x = 6$$
$$x = \frac{6}{4} = \frac{3}{2}$$

$$4x + 3y = 6$$
$$4(-2) + 3y = 6 \quad \textit{Let x = -2.}$$
$$-8 + 3y = 6$$
$$3y = 14$$
$$y = \frac{14}{3}$$

The ordered pairs are $(0, 2)$, $(\frac{3}{2}, 0)$, and $(-2, \frac{14}{3})$.

7. $x = 3y$; $(0, \_\_)$, $(8, \_\_)$, $(\_\_, -3)$

$$x = 3y$$
$$0 = 3y \quad \textit{Let x = 0.}$$
$$0 = y$$

$$x = 3y$$
$$8 = 3y \quad \textit{Let x = 8.}$$
$$\frac{8}{3} = y$$

$$x = 3y$$
$$x = 3(-3) \quad \textit{Let y = -3.}$$
$$x = -9$$

The ordered pairs are $(0, 0)$, $(8, \frac{8}{3})$, and $(-9, -3)$.

8. $x - 7 = 0$; $(\_\_, -3)$, $(\_\_, 0)$, $(\_\_, 5)$

The given equation may be written $x = 7$. For any value of y, the value of x will always be 7. The ordered pairs are $(7, -3)$, $(7, 0)$, and $(7, 5)$.

9. $x + y = 7$; $(2, 5)$ ▪ Substitute 2 for x and 5 for y in the given equation.

$$x + y = 7$$
$$2 + 5 \stackrel{?}{=} 7$$
$$7 = 7 \quad \textit{True}$$

Yes, $(2, 5)$ is a solution of the given equation.

10. $2x + y = 5$; $(-1, 3)$ ▪ Substitute -1 for x and 3 for y in the given equation.

$$2x + y = 5$$
$$2(-1) + 3 \stackrel{?}{=} 5$$
$$-2 + 3 \stackrel{?}{=} 5$$
$$1 = 5 \quad \textit{False}$$

No, $(-1, 3)$ is not a solution of the given equation.

11. $3x - y = 4$; $(\frac{1}{3}, -3)$ ▪ Substitute $\frac{1}{3}$ for x and -3 for y in the given equation.

$$3x - y = 4$$
$$3(\tfrac{1}{3}) - (-3) \stackrel{?}{=} 4$$
$$1 + 3 \stackrel{?}{=} 4$$
$$4 = 4 \quad \textit{True}$$

Yes, $(\frac{1}{3}, -3)$ is a solution of the given equation.

12. $x = -1$; $(0, -1)$ ▪ For the equation $x = -1$, the x-coordinate of any solution must be -1. So no, $(0, -1)$ is not a solution of the given equation.

For Exercises 13–16, the ordered pairs are plotted on the graph following the solution for Exercise 16.

13. To plot $(2, 3)$, start at the origin, go 2 units to the right, and then go up 3 units. The point lies in quadrant I.

14. To plot $(-4, 2)$, start at the origin, go 4 units to the left, and then go up 2 units. The point lies in quadrant II.

15. To plot $(3, 0)$, start at the origin, go 3 units to the right. The point lies on the x-axis (not in any quadrant).

16. To plot $(0, -6)$, start at the origin, go down 6 units. The point lies on the y-axis (not in any quadrant).

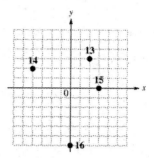

17. x is positive in quadrants I and IV; y is negative in quadrants III and IV. Thus, if x is positive and y is negative, (x, y) must lie in quadrant IV.

18. In the ordered pair $(k, 0)$, the y-value is 0, so the point lies on the x-axis. In the ordered pair $(0, k)$, the x-value is 0, so the point lies on the y-axis.

19. Begin by finding the intercepts.

$$2x - y = 3$$
$$2x - 0 = 3 \quad \textit{Let y = 0.}$$
$$2x = 3$$
$$x = \tfrac{3}{2}$$

The x-intercept is $(\frac{3}{2}, 0)$.

$$2x - y = 3$$
$$2(0) - y = 3 \quad \textit{Let x = 0.}$$
$$0 - y = 3$$
$$y = -3$$

The y-intercept is $(0, -3)$.
To find a third point, choose $y = 3$.

$$2x - y = 3$$
$$2x - 3 = 3 \quad \textit{Let y = 3.}$$
$$2x = 6$$
$$x = 3$$

This gives the ordered pair $(3, 3)$. Plot $(\frac{3}{2}, 0)$, $(0, -3)$, and $(3, 3)$ and draw a line through them.

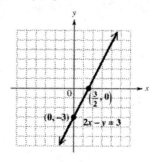

20. $x + 2y = -4$ ▪ Find the intercepts.

If $y = 0$, $x = -4$, so the x-intercept is $(-4, 0)$.
If $x = 0$, $y = -2$, so the y-intercept is $(0, -2)$.

To find a third point, choose $x = 2$.

$$x + 2y = -4$$
$$2 + 2y = -4 \quad \text{Let } x = 2.$$
$$2y = -6$$
$$y = -3$$

This gives the ordered pair $(2, -3)$. Plot $(-4, 0)$, $(0, -2)$, and $(2, -3)$ and draw a line through them.

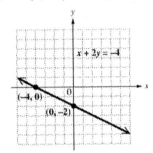

21. $x + y = 0$ ▪ This line goes through the origin, so both intercepts are $(0, 0)$. Two other points on the line are $(2, -2)$ and $(-2, 2)$.

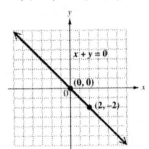

22. $y = 3x + 4$ ▪ Find the intercepts.

If $y = 0$, $x = -\frac{4}{3}$, so the x-intercept is $\left(-\frac{4}{3}, 0\right)$.
If $x = 0$, $y = 4$, so the y-intercept is $(0, 4)$.

To find a third point, choose $x = 1$.

$$y = 3x + 4$$
$$y = 3(1) + 4 \quad \text{Let } x = 1.$$
$$y = 7$$

This gives the ordered pair $(1, 7)$. Plot $\left(-\frac{4}{3}, 0\right)$, $(0, 4)$, and $(1, 7)$ and draw a line through them.

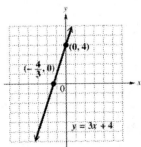

23. The line passing through $(2, 3)$ and $(-4, 6)$

Let $(2, 3) = (x_1, y_1)$ and $(-4, 6) = (x_2, y_2)$.

$$\text{slope } m = \frac{y_2 - y_1}{x_2 - x_1} = \frac{6 - 3}{-4 - 2} = \frac{3}{-6} = -\frac{1}{2}$$

24. The line passing through $(0, 0)$ and $(-3, 2)$

Let $(0, 0) = (x_1, y_1)$ and $(-3, 2) = (x_2, y_2)$.

$$\text{slope } m = \frac{2 - 0}{-3 - 0} = \frac{2}{-3} = -\frac{2}{3}$$

25. The line passing through $(0, 6)$ and $(1, 6)$

Let $(0, 6) = (x_1, y_1)$ and $(1, 6) = (x_2, y_2)$.

$$\text{slope } m = \frac{6 - 6}{1 - 0} = \frac{0}{1} = 0$$

26. The line passing through $(2, 5)$ and $(2, 8)$

Let $(2, 5) = (x_1, y_1)$ and $(2, 8) = (x_2, y_2)$.

$$\text{slope } m = \frac{8 - 5}{2 - 2} = \frac{3}{0}, \text{ which is } \textit{undefined.}$$

27. $y = 3x - 4$ ▪ The equation is already solved for y, so the slope of the line is given by the coefficient of x. Thus, the slope is 3.

28. $y = \frac{2}{3}x + 1$ ▪ The equation is already solved for y, so the slope is given by the coefficient of x. Thus, the slope is $\frac{2}{3}$.

29. The indicated points have coordinates $(0, -2)$ and $(2, 1)$. Use the definition of slope with $(0, -2) = (x_1, y_1)$ and $(2, 1) = (x_2, y_2)$.

$$m = \frac{y_2 - y_1}{x_2 - x_1} = \frac{1 - (-2)}{2 - 0} = \frac{3}{2}$$

30. The indicated points have coordinates $(0, 1)$ and $(3, 0)$. Use the definition of slope with $(0, 1) = (x_1, y_1)$ and $(3, 0) = (x_2, y_2)$.

$$m = \frac{y_2 - y_1}{x_2 - x_1} = \frac{0 - 1}{3 - 0} = \frac{-1}{3} = -\frac{1}{3}$$

31. From the table, we choose two points $(0, 1)$ and $(2, 4)$. Therefore,

$$\text{slope } m = \frac{4 - 1}{2 - 0} = \frac{3}{2}.$$

Any points from the table will give the same result.

32. $x = 0$ is the equation of the y-axis, which is a vertical line. Its slope is undefined.

33. $y = 4$ is the equation of a horizontal line. Its slope is 0.

34. **(a)** Because parallel lines have equal slopes and the slope of the graph of $y = 2x + 3$ is 2, the slope of a line parallel to it will also be 2.

(b) Because perpendicular lines have slopes whose product is -1 and the slope of the graph of $y = -3x + 3$ is -3, the slope of a line perpendicular to it will be $\frac{1}{3}$, since $-3\left(\frac{1}{3}\right) = -1$.

35. Find the slope of each line by solving the equations for y.

$$3x + 2y = 6$$
$$2y = -3x + 6 \quad \textit{Subtract 3x.}$$
$$y = -\tfrac{3}{2}x + 3 \quad \textit{Divide by 2.}$$

The slope of the first line is $-\tfrac{3}{2}$.

$$6x + 4y = 8$$
$$4y = -6x + 8 \quad \textit{Subtract 6x.}$$
$$y = -\tfrac{6}{4}x + 2 \quad \textit{Divide by 4.}$$
$$y = -\tfrac{3}{2}x + 2 \quad \textit{Reduce.}$$

The slope of the second line is $-\tfrac{3}{2}$. The slopes are equal so the lines are *parallel*.

36. Find the slope of each line by solving the equations for y.

$$x - 3y = 1$$
$$-3y = -x + 1 \quad \textit{Subtract x.}$$
$$y = \tfrac{1}{3}x - \tfrac{1}{3} \quad \textit{Divide by –3.}$$

The slope of the first line is $\tfrac{1}{3}$.

$$3x + y = 4$$
$$y = -3x + 4 \quad \textit{Subtract 3x.}$$

The slope of the second line is -3.

The product of the slopes is

$$\tfrac{1}{3}(-3) = -1,$$

so the lines are *perpendicular*.

37. Find the slope of each line by solving the equations for y.

$$x - 2y = 8$$
$$-2y = -x + 8 \quad \textit{Subtract x.}$$
$$y = \tfrac{1}{2}x - 4 \quad \textit{Divide by –2.}$$

The slope of the first line is $\tfrac{1}{2}$.

$$x + 2y = 8$$
$$2y = -x + 8 \quad \textit{Subtract x.}$$
$$y = -\tfrac{1}{2}x + 4 \quad \textit{Divide by 2.}$$

The slope of the second line is $-\tfrac{1}{2}$.

The slopes are not equal and their product is

$$\left(\tfrac{1}{2}\right)\left(-\tfrac{1}{2}\right) = -\tfrac{1}{4} \neq -1,$$

so the lines are *neither* parallel nor perpendicular.

38. A line with undefined slope is vertical. Any line perpendicular to a vertical line is a horizontal line, which has a slope of 0.

39. $m = -1, b = \tfrac{2}{3}$ ▪ Use the slope-intercept form.

$$y = mx + b$$
$$y = -1 \cdot x + \tfrac{2}{3}$$
$$y = -x + \tfrac{2}{3}$$

40. The line in Exercise 30 has slope $-\tfrac{1}{3}$ and y-intercept $(0, 1)$, so we may write its equation in slope-intercept form as

$$y = -\tfrac{1}{3}x + 1.$$

41. The line passing through $(4, -3)$, $m = 1$

Use the point-slope form.

$$y - y_1 = m(x - x_1)$$
$$y - (-3) = 1(x - 4)$$
$$y + 3 = x - 4$$
$$y = x - 7$$

42. The line passing through $(-1, 4)$, $m = \tfrac{2}{3}$

Use the point-slope form.

$$y - y_1 = m(x - x_1)$$
$$y - 4 = \tfrac{2}{3}[x - (-1)]$$
$$y - 4 = \tfrac{2}{3}(x + 1)$$
$$y - 4 = \tfrac{2}{3}x + \tfrac{2}{3}$$
$$y = \tfrac{2}{3}x + \tfrac{14}{3}$$

43. The line passing through $(1, -1)$, $m = -\tfrac{3}{4}$

Use the point-slope form.

$$y - (-1) = -\tfrac{3}{4}(x - 1)$$
$$y + 1 = -\tfrac{3}{4}x + \tfrac{3}{4}$$
$$y = -\tfrac{3}{4}x - \tfrac{1}{4}$$

44. The line passing through $(2, 1)$ and $(-2, 2)$

First, find the slope of the line.

$$m = \frac{2 - 1}{-2 - 2} = \frac{1}{-4} = -\frac{1}{4}$$

Using $(2, 1)$ and $m = -\tfrac{1}{4}$, substitute into the point-slope form.

$$y - y_1 = m(x - x_1)$$
$$y - 1 = -\tfrac{1}{4}(x - 2)$$
$$y - 1 = -\tfrac{1}{4}x + \tfrac{1}{2}$$
$$y = -\tfrac{1}{4}x + \tfrac{1}{2} + \tfrac{2}{2}$$
$$y = -\tfrac{1}{4}x + \tfrac{3}{2}$$

45. The line passing through $(-4, 1)$ with slope 0

Horizontal lines have 0 slope and equations of the form $y = k$. In this case, k must equal 1 since the line goes through $(-4, 1)$, so the equation is

$$y = 1.$$

46. The line passing through $\left(\frac{1}{3}, -\frac{3}{1}\right)$ with undefined slope

Vertical lines have undefined slope and equations of the form $x = k$. In this case, k must equal $\frac{1}{3}$ since the line goes through $\left(\frac{1}{3}, -\frac{3}{1}\right)$, so the equation is

$$x = \tfrac{1}{3}.$$

It is not possible to express this equation as $y = mx + b$.

47. **(a)** Solve the equation for y.

$$x + 3y = 15 \qquad \textit{Given equation}$$
$$3y = -x + 15 \qquad \textit{Subtract x.}$$
$$y = -\tfrac{1}{3}x + 5 \qquad \textit{Divide by 3.}$$

(b) The slope is $-\frac{1}{3}$ and the y-intercept is $(0, 5)$.

(c) The slope is $-\dfrac{1}{3} = \dfrac{-1}{3} = \dfrac{\text{change in } y}{\text{change in } x}$, and the y-intercept is $(0, 5)$. Graph that point and count down 1 unit and right 3 units to get to the point $(3, 4)$. Draw a line through the points.

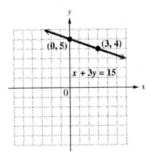

48. $3x + 5y > 9$ ■ To graph the boundary, which is the line $3x + 5y = 9$, find its intercepts.

$$3x + 5y = 9 \qquad\qquad 3x + 5y = 9$$
$$3x + 5(0) = 9 \qquad\qquad 3(0) + 5y = 9$$
$$\textit{Let y = 0.} \qquad\qquad\quad \textit{Let x = 0.}$$
$$3x = 9 \qquad\qquad\qquad 5y = 9$$
$$x = 3 \qquad\qquad\qquad y = \tfrac{9}{5}$$

The x-intercept is $(3, 0)$ and the y-intercept is $\left(0, \frac{9}{5}\right)$. (A third point may be used as a check.) Draw a dashed line through these points. In order to determine which side of the line should be shaded, use $(0, 0)$ as a test point. Substituting 0 for x and 0 for y will result in the inequality $0 > 9$,

which is false. Shade the region *not* containing the origin. This is the region above the line. The dashed line shows that the boundary is not part of the graph.

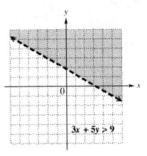

49. $2x - 3y > -6$ ■ Use intercepts to graph the boundary, $2x - 3y = -6$.

If $y = 0$, $x = -3$, so the x-intercept is $(-3, 0)$.

If $x = 0$, $y = 2$, so the y-intercept is $(0, 2)$.

Draw a dashed line through $(-3, 0)$ and $(0, 2)$.

Using $(0, 0)$ as a test point will result in the inequality $0 > -6$, which is true. Shade the region containing the origin. This is the region below the line. The dashed line shows that the boundary is not part of the graph.

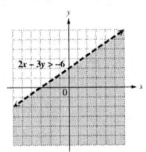

50. $x \geq -4$ ■ The boundary line has the equation $x = -4$. This is the vertical line through $(-4, 0)$. Make this line solid because of the $\geq$ sign. Using $(0, 0)$ as a test point will result in the inequality $0 \geq -4$, which is true. Shade the region containing the origin. The solid line shows that the boundary is part of the graph.

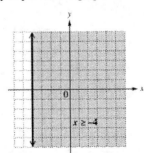

51. **[11.3]** Vertical lines have undefined slopes. The answer is **A**.

52. **[11.2]** Two graphs pass through $(0, -3)$. **C** and **D** are the answers.

53. **[11.2]** Three graphs pass through the point $(-3, 0)$. **A**, **B**, and **D** are the answers.

54. **[11.3]** Lines that fall from left to right have negative slope. The answer is **D**.

55. **[11.2]** $y = -3$ is a horizontal line passing through $(0, -3)$. **C** is the answer.

56. **[11.3]** **B** is the only graph that has a positive slope, so it is the only one we need to investigate. **B** passes through the points $(0, 3)$ and $(-3, 0)$. Find the slope.

$$m = \frac{0 - 3}{-3 - 0} = \frac{-3}{-3} = 1$$

B is the answer.

57. **[11.3]** $y = -2x - 5$ ▪   The equation is in the slope-intercept form, so the slope is -2 and the y-intercept is $(0, -5)$. To find the x-intercept, let $y = 0$.

$$0 = -2x - 5 \quad \textit{Let y = 0.}$$
$$2x = -5$$
$$x = -\tfrac{5}{2}$$

The x-intercept is $\left(-\tfrac{5}{2}, 0\right)$.

Graph the line using the intercepts.

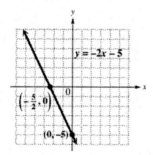

58. **[11.3]** $x + 3y = 0$ ▪   Solve the equation for y.

$$x + 3y = 0$$
$$3y = -x \quad \textit{Subtract x.}$$
$$y = -\tfrac{1}{3}x \quad \textit{Divide by 3.}$$

From this slope-intercept form, we see that the slope is $-\tfrac{1}{3}$ and the y-intercept is $(0, 0)$, which is also the x-intercept. To find another point, let $y = 1$.

$$x + 3(1) = 0 \quad \textit{Let y = 1.}$$
$$x = -3$$

So the point $(-3, 1)$ is on the graph. Graph the line through $(0, 0)$ and $(-3, 1)$.

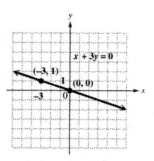

59. **[11.3]** $y - 5 = 0$ or $y = 5$.   ▪   This is a horizontal line passing through the point $(0, 5)$, which is the y-intercept.

There is no x-intercept.
Horizontal lines have slopes of 0.

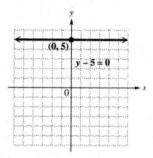

60. **[11.3]** $x + 4 = 3$ or $x = -1$.   ▪   This is a vertical line passing through the point $(-1, 0)$, which is the x-intercept.

There is no y-intercept.
Vertical lines have undefined slopes.

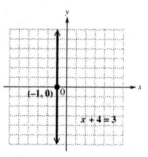

61. **[11.4]** $m = -\tfrac{1}{4}$, $b = -\tfrac{5}{4}$

(a) Substitute the values $m = -\tfrac{1}{4}$ and $b = -\tfrac{5}{4}$ into the slope-intercept form.

$$y = mx + b$$
$$y = -\tfrac{1}{4}x - \tfrac{5}{4}$$

(b)
$$y = -\tfrac{1}{4}x - \tfrac{5}{4}$$
$$4y = -x - 5 \quad \textit{Multiply by 4.}$$
$$x + 4y = -5 \quad \textit{Add x.}$$

62. **[11.4]** The line passing through $(8, 6)$; $m = -3$

(a) Use the point-slope form with $(x_1, y_1) = (8, 6)$ and $m = -3$.

$$y - y_1 = m(x - x_1)$$
$$y - 6 = -3(x - 8)$$
$$y - 6 = -3x + 24$$
$$y = -3x + 30$$

(b) $\quad y = -3x + 30$

$3x + y = 30 \qquad$ *Add 3x.*

63. **[11.4]** The line passing through $(3, -5)$ and $(-4, -1)$

(a) First, find the slope of the line.

$$m = \frac{-1 - (-5)}{-4 - 3} = \frac{4}{-7} = -\frac{4}{7}$$

Now use either point and the slope in the point-slope form. If we use $(3, -5)$, we get the following.

$$y - y_1 = m(x - x_1)$$
$$y - (-5) = -\frac{4}{7}(x - 3)$$
$$y + 5 = -\frac{4}{7}x + \frac{12}{7}$$
$$y = -\frac{4}{7}x - \frac{23}{7} \qquad \text{Subtract } 5 = \frac{35}{7}.$$

(b) $\quad y = -\frac{4}{7}x - \frac{23}{7}$

$7y = -4x - 23 \qquad$ *Multiply by 7.*

$4x + 7y = -23 \qquad$ *Add 4x.*

64. **[11.5]** $y < -4x$ ∎ This is a linear inequality, so its graph will be a shaded region. Graph the boundary, $y = -4x$, as a dashed line through $(0, 0)$, $(1, -4)$, and $(-1, 4)$.

The ordered pairs for which y is *less than* $-4x$ are *below* this line. Indicate the solution by shading the region below the line. The dashed line shows that the boundary is not part of the graph.

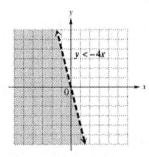

65. **[11.5]** $x - 2y \leq 6$ ∎ This is a linear inequality, so its graph will be a shaded region. Graph the boundary, $x - 2y = 6$, as a solid line through the intercepts $(0, -3)$ and $(6, 0)$.

Using $(0, 0)$ as a test point results in the true statement $0 \leq 6$, so shade the region containing the origin. This is the region above the line. The solid line shows that the boundary is part of the graph.

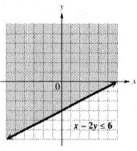

Relating Concepts (Exercises 66–72)

66. In 2004, 42.3% of four-year college students in public institutions earned a degree within five years. In 2009, it was 44.0%. The total increase in percent was

$$44.0\% - 42.3\% = 1.7\%.$$

67. Since the graph rises from left to right, the slope is positive.

68. Let x represent the year and y represent the percent in the ordered pairs (x, y). The ordered pairs are $(2004, 42.3)$ and $(2009, 44.0)$.

69. Find the slope using $(x_1, y_1) = (2004, 42.3)$ and $(x_2, y_2) = (2009, 44.0)$.

$$m = \frac{y_2 - y_1}{x_2 - x_1} = \frac{44.0 - 42.3}{2009 - 2004}$$
$$= \frac{1.7}{5} = 0.34$$

Now use the point-slope form with $m = 0.34$ and $(x_1, y_1) = (2004, 42.3)$.

$$y - y_1 = m(x - x_1)$$
$$y - 42.3 = 0.34(x - 2004)$$
$$y - 42.3 = 0.34x - 681.36$$
$$y = 0.34x - 639.06$$

70. From the slope-intercept form, $y = mx + b$, the slope of the line is 0.34. Yes, the slope agrees with the answer in Exercise 67 because the slope is positive.

71. Substitute 2005, 2006, 2007, and 2008 for x in $y = 0.34x - 639.06$ to complete the table (percents rounded to the nearest tenth).

Year (x)	Percent (y)
2005	42.6
2006	43.0
2007	43.3
2008	43.7

72. For 2010, let $x = 2010$.

$$y = 0.34(2010) - 639.06$$
$$= 44.34$$

The estimate is 44.3% for 2010. We cannot be sure that this estimate is accurate since the equation is based on data only from 2004 through 2009.

Chapter 11 Test

1.   The line segments between 2003 and 2004, 2004 and 2005, and 2005 and 2006 fall, so these are the pairs of consecutive years in which the unemployment rate decreased.

2.   The line segments rise for each pair of consecutive years from 2007 to 2010, so the unemployment rate was increasing.

3.   For 2008, the unemployment rate was about 5.8%. For 2009, the unemployment rate was about 9.3%. The increase is $9.3\% - 5.8\% = 3.5\%$.

4. $3x + y = 6$

If $x = 0$, $y = 6$, so the y-intercept is $(0, 6)$.
If $y = 0$, $x = 2$, so the x-intercept is $(2, 0)$.

A third point, such as $(1, 3)$, can be used as a check. Draw a line through $(0, 6)$, $(1, 3)$, and $(2, 0)$.

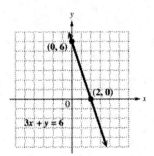

5. $y - 2x = 0$ ∎ Solving for y gives us the slope-intercept form of the line, $y = 2x$. We see that the y-intercept is $(0, 0)$ and so the x-intercept is also $(0, 0)$. The slope is 2 and can be written as

$$m = \frac{\text{rise}}{\text{run}} = \frac{2}{1}.$$

Starting at the origin and moving to the right 1 unit and then up 2 units gives us the point $(1, 2)$. Draw a line through $(0, 0)$ and $(1, 2)$.

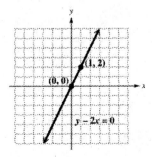

6. $x + 3 = 0$ can also be written as $x = -3$. Its graph is a vertical line with x-intercept $(-3, 0)$. There is no y-intercept.

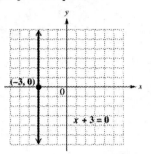

7. $y = 1$ is the graph of a horizontal line with y-intercept $(0, 1)$. There is no x-intercept.

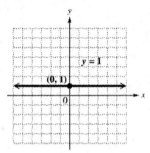

8. $x - y = 4$

If $x = 0$, $y = -4$, so the y-intercept is $(0, -4)$.
If $y = 0$, $x = 4$, so the x-intercept is $(4, 0)$.

A third point, such as $(2, -2)$, can be used as a check. Draw a line through $(0, -4)$, $(2, -2)$, and $(4, 0)$.

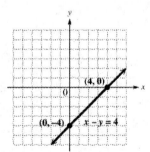

9.   The line passing through $(-4, 6)$ and $(-1, -2)$

Use the definition of slope with $(x_1, y_1) = (-4, 6)$ and $(x_2, y_2) = (-1, -2)$.

$$\text{slope } m = \frac{y_2 - y_1}{x_2 - x_1} = \frac{-2 - 6}{-1 - (-4)} = \frac{-8}{3} = -\frac{8}{3}$$

10. $2x + y = 10$ ∎ To find the slope, solve the given equation for y.

$$2x + y = 10$$
$$y = -2x + 10$$

The equation is now written in $y = mx + b$ form, so the slope is given by the coefficient of x, which is -2.

11. $x + 12 = 0$ can also be written as $x = -12$. Its graph is a vertical line with x-intercept $(-12, 0)$. The slope is undefined.

12. The indicated points are $(0, -4)$ and $(2, 1)$. Use the definition of slope with $(x_1, y_1) = (0, -4)$ and $(x_2, y_2) = (2, 1)$.

$$\text{slope } m = \frac{y_2 - y_1}{x_2 - x_1} = \frac{1 - (-4)}{2 - 0} = \frac{5}{2}$$

13. $y - 4 = 6$ can also be written as $y = 10$. Its graph is a horizontal line with y-intercept $(0, 10)$. Its slope is 0, as is the slope of any line parallel to it.

14. The line passing through $(-1, 4)$, $m = 2$

Let $x_1 = -1$, $y_1 = 4$, and $m = 2$ in the point-slope form.

$$\begin{aligned}
y - y_1 &= m(x - x_1) \\
y - 4 &= 2[x - (-1)] \\
y - 4 &= 2(x + 1) \\
y - 4 &= 2x + 2 \\
y &= 2x + 6
\end{aligned}$$

15. The indicated points are $(0, -4)$ and $(2, 1)$. The slope of the line through those points is

$$m = \frac{1 - (-4)}{2 - 0} = \frac{5}{2}.$$

The y-intercept is $(0, -4)$, so the slope-intercept form is

$$y = \frac{5}{2}x - 4.$$

16. The line passing through $(2, -6)$ and $(1, 3)$

The slope of the line through these points is

$$m = \frac{3 - (-6)}{1 - 2} = \frac{9}{-1} = -9.$$

Use the point-slope form of a line with $(1, 3) = (x_1, y_1)$ and $m = -9$.

$$\begin{aligned}
y - y_1 &= m(x - x_1) \\
y - 3 &= -9(x - 1) \\
y - 3 &= -9x + 9 \\
y &= -9x + 12
\end{aligned}$$

17. x-intercept: $(3, 0)$, y-intercept: $\left(0, \frac{9}{2}\right)$

First, find the slope of the line.

$$m = \frac{\frac{9}{2} - 0}{0 - 3} = \frac{\frac{9}{2}}{-3} = \frac{9}{2} \cdot \left(-\frac{1}{3}\right) = -\frac{3}{2}$$

Since one of the given points is the y-intercept, use the slope-intercept form of a line.

$$y = mx + b$$

$$y = -\frac{3}{2}x + \frac{9}{2}$$

18. $x + y \leq 3$ ▪ Graph the boundary, $x + y = 3$, as a solid line through the intercepts $(3, 0)$ and $(0, 3)$.

Using $(0, 0)$ as a test point results in the true statement $0 \leq 3$, so shade the region containing the origin. This is the region below the line. The solid line shows that the boundary is part of the graph.

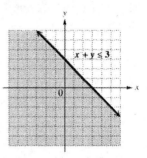

19. $3x - y > 0$ ▪ The boundary, $3x - y = 0$, goes through the origin, so both intercepts are $(0, 0)$. Two other points on this line are $(1, 3)$ and $(-1, -3)$. Draw the boundary as a dashed line.

Choose a test point which is not on the boundary. Using $(3, 0)$ results in the true statement $9 > 0$, so shade the region containing $(3, 0)$. This is the region below the line. The dashed line shows that the boundary is not part of the graph.

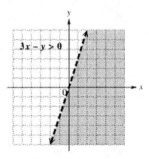

20. The slope is negative since worldwide snowmobile sales are decreasing, indicated by the line which falls from left to right.

21. Two ordered pairs are $(0, 209)$ and $(11, 123)$. Use these points to find the slope.

$$m = \frac{y_2 - y_1}{x_2 - x_1} = \frac{123 - 209}{11 - 0} = \frac{-86}{11} \approx -7.8$$

The slope is about -7.8.

22. Using the ordered pair $(0, 209)$, which is the y-intercept, and the slope of -7.8, we get the slope-intercept form

$$y = -7.8x + 209.$$

23.   For 2007, $x = 2007 - 2000 = 7$.

$$y = -7.8x + 209$$
$$y = -7.8(7) + 209$$
$$y = 154.4$$

This represents worldwide snowmobile sales of 154.4 thousand in 2007. The equation gives a number that is lower than the actual sales, which are 160.3 thousand.

24.   The ordered pair $(11, 123)$ means that in the year 2011, worldwide snowmobile sales were 123 thousand.

CHAPTER 12 SYSTEMS OF LINEAR EQUATIONS AND INEQUALITIES

12.1 Solving Systems of Linear Equations by Graphing

12.1 Margin Exercises

1.   **(a)** $(2,5)$ $3x - 2y = -4$
$$5x + y = 15$$

To decide whether $(2,5)$ is a solution, substitute 2 for x and 5 for y in each equation.

$$3x - 2y = -4$$
$$3(\underline{2}) - 2(\underline{5}) \overset{?}{=} -4$$
$$6 - 10 \overset{?}{=} -4$$
$$-4 = -4 \quad \textit{True}$$

$$5x + y = 15$$
$$5(2) + \underline{5} \overset{?}{=} \underline{15}$$
$$10 + 5 \overset{?}{=} 15$$
$$15 = 15 \quad \textit{True}$$

Since $(2,5)$ satisfies both equations, $(2,5)$ <u>is</u> a solution of the system.

(b) $(1,-2)$ $x - 3y = 7$
$$4x + y = 5$$

To decide whether $(1,-2)$ is a solution, substitute 1 for x and -2 for y in each equation.

$$x - 3y = -7$$
$$1 - 3(-2) \overset{?}{=} 7$$
$$1 + 6 \overset{?}{=} 7$$
$$7 = 7 \quad \textit{True}$$

$$4x + y = 5$$
$$4(1) + (-2) \overset{?}{=} 5$$
$$4 - 2 \overset{?}{=} 5$$
$$2 = 5 \quad \textit{False}$$

Since $(1,-2)$ does not satisfy the second equation, $(1,-2)$ <u>is not</u> a solution of the system.

2.   **(a)** $5x - 3y = 9$
$$x + 2y = 7$$

Graph each equation by plotting several points for each line. The intercepts are good choices.

Graph $5x - 3y = 9$.

x	y
0	-3
$\frac{9}{5}$	0
3	2

Graph $x + 2y = 7$.

x	y
0	$\frac{7}{2}$
7	0
-1	4

(The line $5x - 3y = 9$ is already graphed in the textbook.)

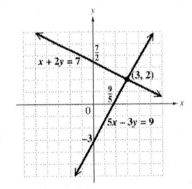

As suggested by the figure, $\underline{(3,2)}$ is the point at which the graphs of the two lines intersect, so the solution set of the system of equations is $\underline{\{(3,2)\}}$. Check by substituting 3 for x and 2 for y in both equations of the system.

(b) $x + y = 4$
$$2x - y = -1$$

Graph $x + y = 4$.

x	y
4	0
0	4
3	1

Graph $2x - y = -1$.

x	y
$-\frac{1}{2}$	0
0	1
2	5

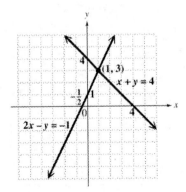

The point at which the graphs of the two lines intersect is $(1,3)$, and the solution set is $\{(1,3)\}$. Check by substituting 1 for x and 3 for y in both equations of the system.

3.   (a) $3x - y = 4$
$6x - 2y = 12$

Graph $3x - y = 4$.

x	y
$\frac{4}{3}$	0
0	-4
1	-1

Graph $6x - 2y = 12$.

x	y
2	0
0	-6
1	-3

(The line $3x - y = 4$ is already graphed in the text.)

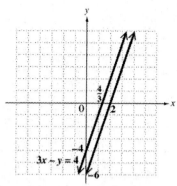

Solving the equations for y gives us $y = 3x - 4$ and $y = 3x - 6$. Since the slope of each line is 3, the two lines are parallel. Since they have no points in common, there is no solution, and we write the solution set as $\emptyset$. The **system is inconsistent** and the **equations are independent**.

(b) $x - 3y = -2$
$2x - 6y = -4$

Graph $x - 3y = -2$.

x	y
-2	0
0	$\frac{2}{3}$
1	1

Graph $2x - 6y = -4$.

x	y
-2	0
0	$\frac{2}{3}$
1	1

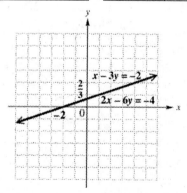

The graphs of these two equations are the same line. There are an infinite number of solutions, and we write the solution set as

$$\{(x, y) \mid x - 3y = -2\}.$$

The **system is consistent** and the **equations are dependent**.

4.   (a) Solve each equation for y.

$$2x - 3y = 5 \qquad\qquad 3y = 2x - 7$$
$$-3y = -2x + 5 \qquad\; y = \tfrac{2}{3}x - \tfrac{7}{3}$$
$$y = \tfrac{2}{3}x - \tfrac{5}{3}$$

The equations have the same slope, $\frac{2}{3}$, but different y-intercepts, $(0, -\frac{5}{3})$ and $(0, -\frac{7}{3})$. The equations represent parallel lines. The system has no solution.

(b) Solve each equation for y.

$$-x + 3y = 2 \qquad\qquad 2x - 6y = -4$$
$$3y = x + 2 \qquad\qquad -6y = -2x - 4$$
$$y = \tfrac{1}{3}x + \tfrac{2}{3} \qquad\qquad y = \tfrac{1}{3}x + \tfrac{2}{3}$$

The slope-intercept forms are the same, so the equations represent the same line. The system has an infinite number of solutions.

(c) Solve each equation for y.

$$6x + y = 3 \qquad\qquad 2x - y = -11$$
$$y = -6x + 3 \qquad\qquad -y = -2x - 11$$
$$y = 2x + 11$$

The slopes, -6 and 2, are different, so the graphs of these equations are neither parallel nor the same line. The system has exactly one solution.

12.1   Section Exercises

1.   A <u>system of linear equations</u> consists of two or more linear equations with the <u>same</u> variables.

3.   The equations of two parallel lines form an <u>inconsistent</u> system that has <u>no</u> solution. The equations are <u>independent</u> because their graphs are different.

5.   If two equations of a linear system have the same graph, the equations are <u>dependent</u>. The system is <u>consistent</u> and has <u>infinitely many</u> solutions.

7.   The ordered pair $(1, -2)$ does make the first equation true, but if we replace x with 1 and y with -2 in the second equation, $2x + y = 4$, we get $2(1) + (-2) = 4$, or $0 = 4$, which is false. Because the ordered pair does not satisfy *both* equations in the system, it is not a solution of the system.

9.   From the graph, the ordered pair that is a solution of the system is in the second quadrant. Choice **B**, $(-2, 2)$, is the only ordered pair given that is in quadrant II, so it is the only valid choice.

11. $(2, -3)$ $\qquad\; x + y = -1$
$2x + 5y = 19$

To decide whether $(2, -3)$ is a solution of the system, substitute 2 for x and -3 for y in each equation.

$$x + y = -1$$
$$2 + (-3) \stackrel{?}{=} -1$$
$$-1 = -1 \quad \textit{True}$$

$$2x + 5y = 19$$
$$2(2) + 5(-3) \stackrel{?}{=} 19$$
$$4 + (-15) \stackrel{?}{=} 19$$
$$-11 = 19 \quad \textit{False}$$

The ordered pair $(2, -3)$ satisfies the first equation but not the second. Because it does not satisfy *both* equations, it is not a solution of the system.

13. $(-1, -3)$ $\quad 3x + 5y = -18$
$\quad\quad\quad\quad\quad\quad 4x + 2y = -10$

Substitute -1 for x and -3 for y in each equation.

$$3x + 5y = -18$$
$$3(-1) + 5(-3) \stackrel{?}{=} -18$$
$$-3 - 15 \stackrel{?}{=} -18$$
$$-18 = -18 \quad \textit{True}$$

$$4x + 2y = -10$$
$$4(-1) + 2(-3) \stackrel{?}{=} -10$$
$$-4 - 6 \stackrel{?}{=} -10$$
$$-10 = -10 \quad \textit{True}$$

Since $(-1, -3)$ satisfies both equations, it is a solution of the system.

15. $(7, -2)$ $\quad 4x = 26 - y$
$\quad\quad\quad\quad\quad\quad 3x = 29 + 4y$

Substitute 7 for x and -2 for y in each equation.

$$4x = 26 - y$$
$$4(7) \stackrel{?}{=} 26 - (-2)$$
$$28 \stackrel{?}{=} 26 + 2$$
$$28 = 28 \quad\quad \textit{True}$$

$$3x = 29 + 4y$$
$$3(7) \stackrel{?}{=} 29 + 4(-2)$$
$$21 \stackrel{?}{=} 29 - 8$$
$$21 = 21 \quad\quad \textit{True}$$

Since $(7, -2)$ satisfies both equations, it is a solution of the system.

17. $(6, -8)$ $\quad -2y = x + 10$
$\quad\quad\quad\quad\quad\quad 3y = 2x + 30$

Substitute 6 for x and -8 for y in each equation.

$$-2y = x + 10$$
$$-2(-8) \stackrel{?}{=} 6 + 10$$
$$16 = 16 \quad\quad \textit{True}$$

$$3y = 2x + 30$$
$$3(-8) \stackrel{?}{=} 2(6) + 30$$
$$-24 \stackrel{?}{=} 12 + 30$$
$$-24 = 42 \quad\quad \textit{False}$$

The ordered pair $(6, -8)$ satisfies the first equation but not the second. Because it does not satisfy *both* equations, it is not a solution of the system.

19. $\quad x - y = 2$
$\quad\quad x + y = 6$

To graph the equations, find the intercepts.

$x - y = 2$: Let $y = 0$; then $x = 2$.
$\quad\quad\quad\quad\quad$ Let $x = 0$; then $y = -2$.

Plot the intercepts, $(2, 0)$ and $(0, -2)$, and draw the line through them.

$x + y = 6$: Let $y = 0$; then $x = 6$.
$\quad\quad\quad\quad\quad$ Let $x = 0$; then $y = 6$.

Plot the intercepts, $(6, 0)$ and $(0, 6)$, and draw the line through them.

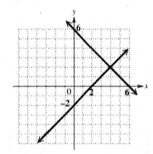

It appears that the lines intersect at the point $(4, 2)$. Check this by substituting 4 for x and 2 for y in both equations. Since $(4, 2)$ satisfies both equations, the solution set of this system is $\{(4, 2)\}$.

21. $\quad x + y = 4$
$\quad\quad y - x = 4$

To graph the equations, find the intercepts.

$x + y = 4$: Let $y = 0$; then $x = 4$.
$\quad\quad\quad\quad\quad$ Let $x = 0$; then $y = 4$.

Plot the intercepts, $(0, 4)$ and $(4, 0)$, and draw the line through them.

$y - x = 4$: Let $y = 0$; then $x = -4$.
$\quad\quad\quad\quad\quad$ Let $x = 0$; then $y = 4$.

Plot the intercepts, $(-4, 0)$ and $(0, 4)$, and draw the line through them. *graph follows*

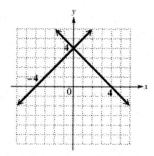

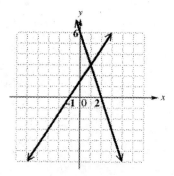

The lines intersect at their common y-intercept, $(0, 4)$, so $\{(0, 4)\}$ is the solution set of the system.

23. $x - 2y = 6$
$x + 2y = 2$

To graph the equations, find the intercepts.

$\boldsymbol{x - 2y = 6}$: Let $y = 0$; then $x = 6$.
Let $x = 0$; then $y = -3$.

Plot the intercepts, $(6, 0)$ and $(0, -3)$, and draw the line through them.

$\boldsymbol{x + 2y = 2}$: Let $y = 0$; then $x = 2$.
Let $x = 0$; then $y = 1$.

Plot the intercepts, $(2, 0)$ and $(0, 1)$, and draw the line through them.

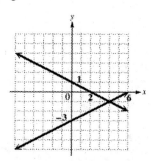

It appears that the lines intersect at the point $(4, -1)$. Since $(4, -1)$ satisfies both equations, the solution set of this system is $\{(4, -1)\}$.

25. $3x - 2y = -3$
$-3x - y = -6$

To graph the equations, find the intercepts.

$\boldsymbol{3x - 2y = -3}$: Let $y = 0$; then $x = -1$.
Let $x = 0$; then $y = \frac{3}{2}$.

Plot the intercepts, $(-1, 0)$ and $(0, \frac{3}{2})$ and draw the line through them.

$\boldsymbol{-3x - y = -6}$: Let $y = 0$; then $x = 2$.
Let $x = 0$; then $y = 6$.

Plot the intercepts, $(2, 0)$ and $(0, 6)$, and draw the line through them.

It appears that the lines intersect at the point $(1, 3)$. Since $(1, 3)$ satisfies both equations, the solution set of this system is $\{(1, 3)\}$.

27. $2x - 3y = -6$
$y = -3x + 2$

To graph the first line, find the intercepts.

$\boldsymbol{2x - 3y = -6}$: Let $y = 0$; then $x = -3$.
Let $x = 0$; then $y = 2$.

Plot the intercepts, $(-3, 0)$ and $(0, 2)$, and draw the line through them.

To graph the second line, start by plotting the y-intercept, $(0, 2)$. From this point, go 3 units down and 1 unit to the right (because the slope is -3) to reach the point $(1, -1)$. Draw the line through $(0, 2)$ and $(1, -1)$.

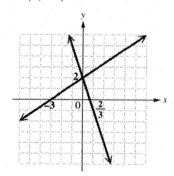

The lines intersect at their common y-intercept, $(0, 2)$, so $\{(0, 2)\}$ is the solution set of the system.

29. $x + 2y = 6$
$2x + 4y = 8$

To graph the lines, find the intercepts.

$\boldsymbol{x + 2y = 6}$: Let $y = 0$; then $x = 6$.
Let $x = 0$; then $y = 3$.

Draw the line through $(6, 0)$ and $(0, 3)$.

$\boldsymbol{2x + 4y = 8}$: Let $y = 0$; then $x = 4$.
Let $x = 0$; then $y = 2$.

Draw the line through $(4, 0)$ and $(0, 2)$.

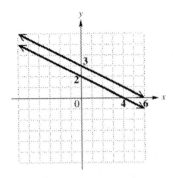

The lines appear to be parallel. By finding the slope of the line through the intercepts, we see that each line has slope $-\frac{1}{2}$. Thus, the lines are parallel, and the system has no solution. This is an inconsistent system and the solution set is $\emptyset$.

31. $5x - 3y = 2$
   $10x - 6y = 4$

Notice that if we multiply each side of the first equation by 2, we obtain the second equation. Thus, the graphs of these two equations are the same line.

To graph the first line, find the intercepts.

   $5x - 3y = 2$:   Let $y = 0$; then $x = \frac{2}{5}$.
   Let $x = 0$; then $y = -\frac{2}{3}$.

Plot the intercepts, $\left(\frac{2}{5}, 0\right)$ and $\left(0, -\frac{2}{3}\right)$, and draw the line through them.

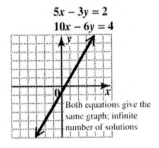

In this case, every point on the line is a solution of the system, and the solution set contains an infinite number of ordered pairs, each of which satisfies both equations of the system. We write the solution set as $\{(x, y) \mid 5x - 3y = 2\}$ (dependent equations).

33. $3x - 4y = 24$
   $y = -\frac{3}{2}x + 3$

Graph the line $3x - 4y = 24$ through its intercepts, $(8, 0)$ and $(0, -6)$.

To graph the line $y = -\frac{3}{2}x + 3$, plot the y-intercept $(0, 3)$ and then go 3 units down and 2 units to the right (because the slope is $-\frac{3}{2}$) to

reach the point $(2, 0)$. Draw the line through $(0, 3)$ and $(2, 0)$.

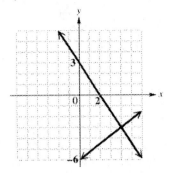

It appears that the lines intersect at the point $(4, -3)$. Since $(4, -3)$ satisfies both equations, the solution set of this system is $\{(4, -3)\}$.

35. $4x - 2y = 8$
   $2x = y + 4$

Graph the line $4x - 2y = 8$, or equivalently, $2x - y = 4$, through its intercepts, $(2, 0)$ and $(0, -4)$.

Graph the line $2x = y + 4$ through its intercepts, $(2, 0)$ and $(0, -4)$.

Since both equations have the same intercepts, they are equations of the same line.

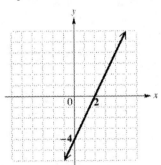

There are an infinite number of solutions. The equations are dependent equations and the solution contains an infinite number of ordered pairs. The solution set is $\{(x, y) \mid 2x - y = 4\}$.

37. $3x = y + 5$
   $6x - 5 = 2y$

To graph the lines, find the intercepts.

   $3x = y + 5$:   Let $y = 0$; then $x = \frac{5}{3}$.
   Let $x = 0$; then $y = -5$.

Draw the line through $\left(\frac{5}{3}, 0\right)$ and $(0, -5)$.

   $6x - 5 = 2y$:   Let $y = 0$; then $x = \frac{5}{6}$.
   Let $x = 0$; then $y = -\frac{5}{2}$.

Draw the line through $\left(\frac{5}{6}, 0\right)$ and $\left(0, -\frac{5}{2}\right)$.

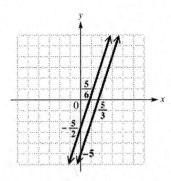

The lines appear to be parallel. By finding the slope of the line through the intercepts, we see that each line has slope 3. Thus, the lines are parallel, and the system has no solution. This is an inconsistent system and the solution set is $\emptyset$.

39. $y - x = -5$
$x + y = 1$

Write the equations in slope-intercept form.

$y - x = -5$	$x + y = 1$
$y = x - 5$	$y = -x + 1$
$m = 1$	$m = -1$

The lines have different slopes.

(a) The system is consistent because it has a solution. The equations are independent because they have different graphs. Therefore, the answer is "neither."

(b) The graph is a pair of intersecting lines.

(c) The system has one solution.

41. $x + 2y = 0$
$4y = -2x$

Write the equations in slope-intercept form.

$x + 2y = 0$	$4y = -2x$
$2y = -x$	$y = -\frac{1}{2}x$
$y = -\frac{1}{2}x$	

For both lines, $m = -\frac{1}{2}$ and $b = 0$.

(a) Since the equations have the same slope and y-intercept, they are dependent.

(b) The graph is one line.

(c) The system has an infinite number of solutions.

43. $5x + 4y = 7$

$10x + 8y = 4$

Write the equations in slope-intercept form.

$5x + 4y = 7$	$10x + 8y = 4$
$4y = -5x + 7$	$8y = -10x + 4$
$y = -\frac{5}{4}x + \frac{7}{4}$	$y = -\frac{10}{8}x + \frac{4}{8}$
$m = -\frac{5}{4}, b = \frac{7}{4}$	$y = -\frac{5}{4}x + \frac{1}{2}$
	$m = -\frac{5}{4}, b = \frac{1}{2}$

The lines have the same slope but different y-intercepts.

(a) The system is inconsistent because it has no solution.

(b) The graph is a pair of parallel lines.

(c) The system has no solution.

45. $x - 3y = 5$
$2x + y = 8$

Write the equations in slope-intercept form.

$x - 3y = 5$	$2x + y = 8$
$-3y = -x + 5$	$y = -2x + 8$
$y = \frac{1}{3}x - \frac{5}{3}$	$m = -2$
$m = \frac{1}{3}$	

The lines have different slopes.

(a) The system is consistent because it has a solution. The equations are independent because they have different graphs. Therefore, the answer is "neither."

(b) The graph is a pair of intersecting lines.

(c) The system has one solution.

47. **(a)** Supply equals demand at the point where the two lines intersect, or when $x = 40$.

(b) Supply equals demand at the point where the two lines intersect, or when $p = 30$.

49. The red line (beef) is above the blue line (poultry) for the years 1980–2000, so that's when per-capita consumption of beef exceeded that of poultry.

51. The graphs intersect when the year is about 2006. The intersection point is at about 600 on the vertical axis, so there were about 600 million units sold of each type in 2006.

53. The slope would be negative. Sales of CDs were decreasing during this period.

12.2   Solving Systems of Linear Equations by Substitution

12.2   Margin Exercises

1. $2x + 3y = 6$
$x - 3y = 5$

To graph the equations, find the intercepts.

$2x + 3y = 6$: Let $y = 0$; then $x = 3$.
Let $x = 0$; then $y = 2$.

Plot the intercepts, $(3, 0)$ and $(0, 2)$, and draw the line through them.

$x - 3y = 5$: Let $y = 0$; then $x = 5$.
Let $x = 0$; then $y = -\frac{5}{3}$.

Plot the intercepts, $(5, 0)$ and $(0, -\frac{5}{3})$, and draw the line through them.

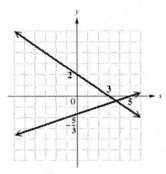

It appears that the lines intersect near the point $(3.5, -0.5)$, but the answer cannot be determined from the graph because it is too difficult to determine the exact coordinates.

2.　　$3x + 5y = 69$　　(1)
　　　　　　$y = 4x$　　　(2)

Substitute $4x$ for y in the first equation, and solve for x.

$$3x + 5y = 69 \quad (1)$$
$$3x + 5(\underline{4x}) = 69 \quad Let\ y = 4x.$$
$$3x + \underline{20x} = 69 \quad Multiply.$$
$$\underline{23x} = 69 \quad Combine\ terms.$$
$$x = \underline{3} \quad Divide\ by\ 23.$$

To find y, substitute 3 for x in $y = 4x$.

$$y = 4x = 4(\underline{3}) = \underline{12}$$

Check that $(3, 12)$ is the solution.

$$3x + 5y = 69 \quad (1) \quad \bigg| \quad y = 4x \quad (2)$$
$$3(3) + 5(12) \overset{?}{=} 69 \quad \bigg| \quad 12 \overset{?}{=} 4(3)$$
$$9 + 60 \overset{?}{=} 69 \quad \bigg| \quad 12 = 12 \quad True$$
$$69 = 69 \quad True \quad \bigg|$$

The solution set is $\underline{\{(3, 12)\}}$.

3.　　**(a)** $2x + 7y = -12$　　　(1)
　　　　　　　$x = 3 - 2y$　　　(2)

Substitute $3 - 2y$ for x in the first equation, and solve for y.

$$2x + 7y = -12$$
$$2(3 - 2y) + 7y = -12 \quad Let\ x = 3 - 2y.$$
$$6 - 4y + 7y = -12$$
$$6 + 3y = -12$$
$$3y = -18$$
$$y = -6$$

To find x, use $x = 3 - 2y$ and $y = -6$.

$$x = 3 - 2y = 3 - 2(-6) = 3 + 12 = 15$$

Check that $(15, -6)$ is the solution.

$$2x + 7y = -12 \ (1) \quad \bigg| \quad x = 3 - 2y \quad (2)$$
$$2(15) + 7(-6) \overset{?}{=} -12 \quad \bigg| \quad 15 \overset{?}{=} 3 - 2(-6)$$
$$30 - 42 \overset{?}{=} -12 \quad \bigg| \quad 15 \overset{?}{=} 3 + 12$$
$$-12 = -12 \ True \quad \bigg| \quad 15 = 15 \qquad True$$

The solution set is $\{(15, -6)\}$.

(b)　　　　$x = y - 3$
　　　　　$4x + 9y = 1$

Substitute $y - 3$ for x in the second equation, and solve for y.

$$4x + 9y = 1$$
$$4(y - 3) + 9y = 1 \quad Let\ x = y - 3.$$
$$4y - 12 + 9y = 1$$
$$13y - 12 = 1$$
$$13y = 13$$
$$y = 1$$

To find x, use $x = y - 3$ and $y = 1$.

$$x = y - 3 = 1 - 3 = -2$$

The solution set is $\{(-2, 1)\}$.

4.　　**(a)** $x + 4y = -1$　　(1)
　　　　　　$2x - 5y = 11$　　(2)

Solve (1) for x.

$$x + 4y = -1$$
$$x = -1 - \underline{4y} \quad (3)$$

Substitute $-1 - 4y$ for x in equation (2) and solve for y.

$$2(\underline{-1 - 4y}) - 5y = 11 \quad Let\ x = -1 - 4y.$$
$$-2 - 8y - 5y = 11$$
$$-2 - \underline{13}y = 11$$
$$\underline{-13}y = 13$$
$$y = \underline{-1}$$

To find x, let $y = -1$ in equation (3).

$$x = -1 - 4y = -1 - 4(-1)$$
$$= -1 - (\underline{-4}) = -1 + 4 = \underline{3}$$

Check that $(3, -1)$ is the solution.

$$x + 4y = -1 \ (1) \quad \bigg| \quad 2x - 5y = 11 \ (2)$$
$$3 + 4(-1) \overset{?}{=} -1 \quad \bigg| \quad 2(3) - 5(-1) \overset{?}{=} 11$$
$$-1 = -1 \ True \quad \bigg| \quad 11 = 11 \ True$$

The solution set of the system is $\underline{\{(3, -1)\}}$.

(b) $2x + 5y = 4$　　(1)
　　　$x + y = -1$　　(2)

Solve equation (2) for x.

$$x + y = -1$$
$$x = -1 - y \quad (3)$$

Substitute $-1 - y$ for x in equation (1) and solve for y.

$$2x + 5y = 4$$
$$2(-1 - y) + 5y = 4 \quad \textit{Let } x = -1 - y.$$
$$-2 - 2y + 5y = 4$$
$$-2 + 3y = 4$$
$$3y = 6$$
$$y = 2$$

To find x, let $y = 2$ in equation (3).

$$x = -1 - y = -1 - 2 = -3$$

The solution set is $\{(-3, 2)\}$. Check this solution in both of the original equations.

5. $8x - 2y = 1$ (1)
$y = 4x - 8$ (2)

Substitute $4x - 8$ for y in equation (1).

$$8x - 2y = 1$$
$$8x - 2(4x - 8) = 1 \quad \textit{Let } y = 4x - 8.$$
$$8x - 8x + 16 = 8$$
$$16 = 8 \quad \textit{False}$$

This false result indicates that the system has no solution, so the solution set is $\emptyset$.

6.   **(a)** $7x - 6y = 10$ (1)
$-14x + 20 = -12y$ (2)

Solve equation (1) for x.

$$7x - 6y = 10$$
$$7x = 6y + 10$$
$$x = \frac{6y + 10}{7}$$

Substitute $\frac{6y + 10}{7}$ for x in equation (2).

$$-14x + 20 = -12y$$
$$-14\left(\frac{6y + 10}{7}\right) + 20 = -12y \quad \textit{Let } x = \frac{6y + 10}{7}.$$
$$-2(6y + 10) + 20 = -12y$$
$$-12y - 20 + 20 = -12y$$
$$-12y = -12y$$
$$0 = 0 \quad \textit{True}$$

This true result indicates that the system has an infinite number of solutions. The solution set is

$$\{(x, y) \mid 7x - 6y = 10\}.$$

(b) $8x - y = 4$ (1)
$y = 8x + 4$ (2)

Substitute $8x + 4$ for y in equation (1).

$$8x - y = 4$$
$$8x - (8x + 4) = 4 \quad \textit{Let } y = 8x + 4.$$
$$8x - 8x - 4 = 4$$
$$-4 = 4 \quad \textit{False}$$

This false result indicates that the system has no solution, and the solution set is $\emptyset$.

7.   **(a)** $\frac{2}{3}x + \frac{1}{2}y = 6$ (1)
$\frac{1}{2}x - \frac{3}{4}y = 0$ (2)

First, clear all fractions.

Equation (1):

$$\underline{6}\left(\tfrac{2}{3}x + \tfrac{1}{2}y\right) = \underline{6}(6) \quad \begin{array}{l}\textit{Multiply by}\\ \textit{the LCD, } \underline{6}.\end{array}$$

$$\underline{6}\left(\tfrac{2}{3}x\right) + 6\left(\tfrac{1}{2}y\right) = 36 \quad \begin{array}{l}\textit{Distributive}\\ \textit{property}\end{array}$$

$$\underline{4x} + 3y = 36 \quad (3)$$

Equation (2):

$$4\left(\tfrac{1}{2}x - \tfrac{3}{4}y\right) = 4(0) \quad \begin{array}{l}\textit{Multiply by}\\ \textit{the LCD, } \underline{4}.\end{array}$$

$$4\left(\tfrac{1}{2}x\right) - 4\left(\tfrac{3}{4}y\right) = 0 \quad \begin{array}{l}\textit{Distributive}\\ \textit{property}\end{array}$$

$$2x - 3y = 0 \quad (4)$$

The system has been simplified to

$$4x + 3y = 36 \quad (3)$$
$$2x - 3y = 0 \quad (4)$$

Solve equation (4) for x.

$$2x = 3y$$
$$x = \tfrac{3}{2}y \quad (5)$$

Substitute $\frac{3}{2}y$ for x in equation (3).

$$4x + 3y = 36$$
$$4\left(\tfrac{3}{2}y\right) + 3y = 36 \quad \textit{Let } x = \tfrac{3}{2}y.$$
$$6y + 3y = 36$$
$$9y = 36$$
$$y = 4$$

To find x, let $y = 4$ in equation (5).

$$x = \tfrac{3}{2}y = \tfrac{3}{2}(4) = 6$$

The solution set is $\underline{\{(6, 4)\}}$. Check in both of the original equations.

(b) $x + \tfrac{1}{2}y = \tfrac{1}{2}$ (1)
$\tfrac{1}{6}x - \tfrac{1}{3}y = \tfrac{4}{3}$ (2)

First, clear all fractions.

Equation (1):

$$2\left(x + \tfrac{1}{2}y\right) = 2\left(\tfrac{1}{2}\right) \quad \begin{array}{l}\textit{Multiply by}\\ \textit{the LCD, 2.}\end{array}$$

$$2(x) + 2\left(\tfrac{1}{2}y\right) = 1 \quad \begin{array}{l}\textit{Distributive}\\ \textit{property}\end{array}$$

$$2x + y = 1 \quad (3)$$

Equation (2):

$$6(\tfrac{1}{6}x - \tfrac{1}{3}y) = 6(\tfrac{4}{3}) \quad \textit{Multiply by the LCD, 6.}$$

$$6(\tfrac{1}{6}x) - 6(\tfrac{1}{3}y) = 8 \quad \textit{Distributive property}$$

$$x - 2y = 8 \quad (4)$$

The system has been simplified to

$$2x + y = 1 \quad (3)$$
$$x - 2y = 8 \quad (4)$$

Solve this system by the substitution method.

$$x = 2y + 8 \quad (5) \quad \textit{Solve (4) for x.}$$

$$2(2y + 8) + y = 1 \quad \textit{Substitute for x in (3).}$$
$$4y + 16 + y = 1$$
$$5y + 16 = 1$$
$$5y = -15$$
$$y = -3$$

To find x, let $y = -3$ in equation (5).

$$x = 2(-3) + 8 = -6 + 8 = 2$$

The solution set is $\{(2, -3)\}$.

8. $0.2x + 0.3y = 0.5 \quad (1)$

$0.3x - 0.1y = 1.3 \quad (2)$

First, clear all decimals.

Equation (1):

$$10(0.2x + 0.3y) = 10(0.5) \quad \textit{Multiply by 10.}$$
$$10(0.2x) + 10(0.3y) = 10(0.5) \quad \textit{Dist. property}$$
$$2x + 3y = 5 \quad (3)$$

Equation (2):

$$10(0.3x - 0.1y) = 10(1.3) \quad \textit{Multiply by 10.}$$
$$10(0.3x) - 10(0.1y) = 10(1.3) \quad \textit{Dist. property}$$
$$3x - y = 13 \quad (4)$$

The system has been simplified to

$$2x + 3y = 5 \quad (3)$$
$$3x - y = 13 \quad (4)$$

Solve this system by the substitution method.

$$y = 3x - 13 \quad (5) \quad \textit{Solve (4) for y.}$$

$$2x + 3(3x - 13) = 5 \quad \textit{Substitute for y in (3).}$$
$$2x + 9x - 39 = 5$$
$$11x - 39 = 5$$
$$11x = 44$$
$$x = 4$$

To find y, let $x = 4$ in equation (5).

$$y = 3(4) - 13 = 12 - 13 = -1$$

The solution set is $\{(4, -1)\}$.

12.2 Section Exercises

1. The student must find the value of y and write the solution as an ordered pair. Substituting 3 for x in the first equation, $5x - y = 15$, gives us $15 - y = 15$, so $y = 0$. The correct solution set is $\{(3, 0)\}$.

3. A false statement, such as $0 = 3$, occurs when there is no solution.

In this section, all solutions should be checked by substituting the proposed solution in *both* equations of the original system. Checks will not be shown here.

5. $x + y = 12 \quad (1)$

$y = 3x \quad (2)$

Equation (2) is already solved for y. Substitute $3x$ for y in equation (1) and solve the resulting equation for x.

$$x + y = 12$$
$$x + 3x = 12 \quad \textit{Let y = 3x.}$$
$$4x = 12$$
$$x = 3$$

To find the y-value of the solution, substitute 3 for x in equation (2).

$$y = 3x$$
$$y = 3(3) \quad \textit{Let x = 3.}$$
$$= 9$$

The solution set is $\{(3, 9)\}$.

To check this solution, substitute 3 for x and 9 for y in both equations of the given system.

7. $3x + 2y = 27 \quad (1)$

$x = y + 4 \quad (2)$

Equation (2) is already solved for x. Substitute $y + 4$ for x in equation (1).

$$3x + 2y = 27$$
$$3(y + 4) + 2y = 27$$
$$3y + 12 + 2y = 27$$
$$5y = 15$$
$$y = 3$$

To find x, substitute 3 for y in equation (2).

$$x = y + 4 = 3 + 4 = 7$$

The solution set is $\{(7, 3)\}$. Check in both of the original equations.

9. $3x + 5y = 14 \quad (1)$

$x - 2y = -10 \quad (2)$

Solve equation (2) for x since its coefficient is 1.

$$x - 2y = -10$$
$$x = 2y - 10 \quad (3)$$

Substitute $2y - 10$ for x in equation (1) and solve for y.

$$3x + 5y = 14$$
$$3(2y - 10) + 5y = 14$$
$$6y - 30 + 5y = 14$$
$$11y = 44$$
$$y = 4$$

To find x, substitute 4 for y in equation (3).

$$x = 2y - 10 = 2(4) - 10 = -2$$

The solution set is $\{(-2, 4)\}$. Check in both of the original equations.

11. $3x + 4 = -y$ (1)
  $2x + y = 0$ (2)

Solve equation (1) for y.

$$3x + 4 = -y$$
$$y = -3x - 4 \quad (3)$$

Substitute $-3x - 4$ for y in equation (2) and solve for x.

$$2x + y = 0$$
$$2x + (-3x - 4) = 0$$
$$-x - 4 = 0$$
$$-x = 4$$
$$x = -4$$

To find y, substitute -4 for x in equation (3).

$$y = -3x - 4 = -3(-4) - 4 = 8$$

The solution set is $\{(-4, 8)\}$. Check in both of the original equations.

13. $7x + 4y = 13$ (1)
  $x + y = 1$ (2)

Solve equation (2) for y.

$$x + y = 1$$
$$y = 1 - x \quad (3)$$

Substitute $1 - x$ for y in equation (1).

$$7x + 4y = 13$$
$$7x + 4(1 - x) = 13$$
$$7x + 4 - 4x = 13$$
$$3x + 4 = 13$$
$$3x = 9$$
$$x = 3$$

To find y, substitute 3 for x in equation (3).

$$y = 1 - x = 1 - 3 = -2$$

The solution set is $\{(3, -2)\}$. Check in both of the original equations.

15. $3x - y = 5$ (1)
  $y = 3x - 5$ (2)

Equation (2) is already solved for y, so we substitute $3x - 5$ for y in equation (1).

$$3x - (3x - 5) = 5$$
$$3x - 3x + 5 = 5$$
$$5 = 5 \quad \textit{True}$$

This true result means that every solution of one equation is also a solution of the other, so the system has an infinite number of solutions. The solution set is $\{(x, y) \mid 3x - y = 5\}$.

17. $2x + 8y = 3$ (1)
  $x = 8 - 4y$ (2)

Equation (2) is already solved for x, so substitute $8 - 4y$ for x in equation (1).

$$2(8 - 4y) + 8y = 3$$
$$16 - 8y + 8y = 3$$
$$16 = 3 \quad \textit{False}$$

This false statement means that the system has no solution. The equations of the system represent parallel lines. The solution set is $\emptyset$.

19. $2y = 4x + 24$ (1)
  $2x - y = -12$ (2)

Solve equation (2) for y.

$$2x - y = -12$$
$$2x = y - 12$$
$$2x + 12 = y$$

Substitute $2x + 12$ for y in (1) and solve for x.

$$2(2x + 12) = 4x + 24$$
$$4x + 24 = 4x + 24 \quad \textit{True}$$

This true result means that every solution of one equation is also a solution of the other, so the system has an infinite number of solutions. The solution set is $\{(x, y) \mid 2x - y = -12\}$.

21. $6x - 8y = 6$ (1)
  $2y = -2 + 3x$ (2)

Solve equation (2) for y.

$$2y = -2 + 3x$$
$$y = \frac{3x - 2}{2} \quad (3)$$

Substitute $\dfrac{3x - 2}{2}$ for y in equation (1).

$$6x - 8y = 6$$
$$6x - 8\left(\frac{3x - 2}{2}\right) = 6$$
$$6x - 4(3x - 2) = 6$$
$$6x - 12x + 8 = 6$$
$$-6x + 8 = 6$$
$$-6x = -2$$
$$x = \frac{-2}{-6} = \frac{1}{3}$$

To find y, let $x = \frac{1}{3}$ in equation (3).

$$y = \frac{3x - 2}{2} = \frac{3\left(\frac{1}{3}\right) - 2}{2} = \frac{1 - 2}{2} = -\frac{1}{2}$$

The solution set is $\left\{\left(\frac{1}{3}, -\frac{1}{2}\right)\right\}$. Check in both of the original equations.

23. $\frac{1}{5}x + \frac{2}{3}y = -\frac{8}{5}$ (1)
    $3x - y = 9$ (2)

Multiply each side of equation (1) by 15 to clear fractions.

$$15\left(\frac{1}{5}x + \frac{2}{3}y\right) = 15\left(-\frac{8}{5}\right)$$
$$15\left(\frac{1}{5}x\right) + 15\left(\frac{2}{3}y\right) = -24$$
$$3x + 10y = -24 \quad\quad (3)$$

Solve equation (2) for y.

$$3x - y = 9$$
$$-y = -3x + 9 \quad\quad \textit{Subtract } 3x.$$
$$y = 3x - 9 \quad (4) \quad \textit{Divide by } -1.$$

Substitute $3x - 9$ for y in equation (3).

$$3x + 10y = -24 \quad (3)$$
$$3x + 10(3x - 9) = -24$$
$$3x + 30x - 90 = -24$$
$$33x = 66 \quad\quad \textit{Add } 90.$$
$$x = 2 \quad\quad \textit{Divide by } 33.$$

To find y, let $x = 2$ in equation (4).

$$y = 3(2) - 9 = 6 - 9 = -3$$

The solution set is $\{(2, -3)\}$. Check in both of the original equations.

25. $\frac{1}{2}x + \frac{1}{3}y = -\frac{1}{3}$ (1)
    $\frac{1}{2}x + 2y = -7$ (2)

First, clear all fractions.

Equation (1):

$$6\left(\frac{1}{2}x + \frac{1}{3}y\right) = 6\left(-\frac{1}{3}\right) \quad \begin{array}{l}\textit{Multiply by}\\\textit{the LCD, 6.}\end{array}$$

$$6\left(\frac{1}{2}x\right) + 6\left(\frac{1}{3}y\right) = -2 \quad \begin{array}{l}\textit{Distributive}\\\textit{property}\end{array}$$

$$3x + 2y = -2 \quad (3)$$

Equation (2):

$$2\left(\frac{1}{2}x + 2y\right) = 2(-7) \quad \textit{Multiply by 2.}$$
$$x + 4y = -14 \quad (4)$$

The system has been simplified to

$$3x + 2y = -2 \quad\quad (3)$$
$$x + 4y = -14 \quad\quad (4)$$

Solve this system by the substitution method.

$$x = -4y - 14 \quad (5) \quad \textit{Solve (4) for x.}$$

$$3(-4y - 14) + 2y = -2 \quad \textit{Substitute for x in (3).}$$
$$-12y - 42 + 2y = -2$$
$$-10y - 42 = -2$$
$$-10y = 40$$
$$y = -4$$

To find x, let $y = -4$ in equation (5).

$$x = -4(-4) - 14 = 16 - 14 = 2$$

The solution set is $\{(2, -4)\}$. Check in both of the original equations.

27. $\dfrac{x}{5} + 2y = \dfrac{16}{5}$ (1)

    $\dfrac{3x}{5} + \dfrac{y}{2} = -\dfrac{7}{5}$ (2)

Multiply each side of equation (1) by 5.

$$5\left(\frac{x}{5} + 2y\right) = 5\left(\frac{16}{5}\right)$$
$$x + 10y = 16 \quad\quad (3)$$

Multiply each side of equation (2) by 10.

$$10\left(\frac{3x}{5} + \frac{y}{2}\right) = 10\left(-\frac{7}{5}\right)$$
$$6x + 5y = -14 \quad\quad (4)$$

We now have the simplified system

$$x + 10y = 16 \quad\quad (3)$$
$$6x + 5y = -14. \quad\quad (4)$$

To solve this system by the substitution method, solve equation (3) for x.

$$x = 16 - 10y \quad (5)$$

Substitute $16 - 10y$ for x in equation (4).

$$6x + 5y = -14$$
$$6(16 - 10y) + 5y = -14$$
$$96 - 60y + 5y = -14$$
$$-55y = -110$$
$$y = 2$$

To find x, let $y = 2$ in equation (5).

$$x = 16 - 10y$$
$$= 16 - 10(2) = -4$$

The solution set is $\{(-4, 2)\}$. Check in both of the original equations.

29. $0.1x + 0.9y = -2$ (1)
    $0.5x - 0.2y = 4.1$ (2)

Multiply equation (1) and equation (2) by 10 to eliminate the decimals. Then solve the resulting system of equations by the substitution method.

$$x + 9y = -20 \quad (3)$$
$$5x - 2y = 41 \quad (4)$$

Solve equation (3) for x.

$$x = -9y - 20 \quad (5)$$

Substitute $-9y - 20$ for x in equation (4).

$$5x - 2y = 41$$
$$5(-9y - 20) - 2y = 41 \quad \textit{Let } x = -9y - 20.$$
$$-45y - 100 - 2y = 41$$
$$-47y = 141$$
$$y = -3$$

To find x, let $y = -3$ in equation (5).

$$x = -9(-3) - 20 = 7$$

The solution set is $\{(7, -3)\}$. Check in both of the original equations.

31. $0.8x - 0.1y = 1.3 \quad (1)$
$2.2x + 1.5y = 8.9 \quad (2)$

Multiply equation (1) and equation (2) by 10 to eliminate the decimals. Then solve the resulting system of equations by the substitution method.

$$8x - y = 13 \quad (3)$$
$$22x + 15y = 89 \quad (4)$$

Solve equation (3) for y.

$$y = 8x - 13 \quad (5)$$

Substitute $8x - 13$ for y in equation (4).

$$22x + 15y = 89$$
$$22x + 15(8x - 13) = 89 \quad \textit{Let } x = 8x - 13.$$
$$22x + 120x - 195 = 89$$
$$142x = 284$$
$$x = 2$$

To find y, let $x = 2$ in equation (5).

$$y = 8(2) - 13 = 3$$

The solution set is $\{(2, 3)\}$. Check in both of the original equations.

Relating Concepts (Exercises 33–36)

33. To find the total cost, multiply the number of bicycles (x) by the cost per bicycle ($400), and add the fixed cost ($5000). Thus, $y_1 = 400x + 5000$ gives the total cost (in dollars).

34. Since each bicycle sells for $600, the total revenue for selling x bicycles is $600x$ (in dollars). Thus, $y_2 = 600x$ gives the total revenue.

35. $y_1 = 400x + 5000 \quad (1)$
$y_2 = 600x \quad (2)$

To solve this system by the substitution method, substitute $600x$ for y_1 in equation (1).

$$600x = 400x + 5000$$
$$200x = 5000$$
$$x = 25$$

If $x = 25$, $y_2 = 600(25) = 15{,}000$.

The solution set is $\{(25, 15{,}000)\}$.

36. The value of x from Exercise 35 is the number of bikes it takes to break even. When <u>25</u> bikes are sold, the break-even point is reached. At that point, you have spent <u>15,000</u> dollars and taken in <u>15,000</u> dollars.

12.3 Solving Systems of Linear Equations by Elimination

12.3 Margin Exercises

1. **(a)** $\begin{aligned} x + y &= 8 \quad (1) \\ x - y &= 2 \quad (2) \\ \hline 2x + 0 &= 10 \quad \textit{Add (1) and (2).} \\ x &= 5 \quad \textit{Divide by 2.} \end{aligned}$

To find y, substitute 5 for x in either of the original equations.

$$x - y = 2 \quad \textit{Equation (2)}$$
$$5 - y = 2 \quad \textit{Let } x = 5.$$
$$-y = -3$$
$$y = 3$$

Check by substituting 5 for x and 3 for y in both equations of the system.

$x + y = 8 \quad (1)$	$x - y = 2 \quad (2)$
$5 + 3 \stackrel{?}{=} 8$	$5 - 3 \stackrel{?}{=} 2$
$8 = 8$ *True*	$2 = 2$ *True*

The solution set is $\underline{\{(5, 3)\}}$.

(b) $\begin{aligned} 3x - y &= 7 \quad (1) \\ 2x + y &= 3 \quad (2) \\ \hline 5x &= 10 \quad \textit{Add (1) and (2).} \\ x &= 2 \quad \textit{Divide by 5.} \end{aligned}$

To find y, substitute 2 for x in either of the original equations.

$$2x + y = 3 \quad \textit{Equation (2)}$$
$$2(2) + y = 3 \quad \textit{Let } x = 2.$$
$$4 + y = 3$$
$$y = -1$$

Check $x = 2$, $y = -1$: $7 = 7$, $3 = 3$

The solution set is $\{(2, -1)\}$.

2. **(a)** $\begin{aligned} 2x - y &= 2 \quad (1) \\ 4x + y &= 10 \quad (2) \\ \hline 6x &= 12 \quad \textit{Add (1) and (2).} \\ x &= 2 \quad \textit{Divide by 6.} \end{aligned}$

To find y, substitute 2 for x in either of the original equations.

$$4x + y = 10 \quad \textit{Equation (2)}$$
$$4(2) + y = 10 \quad \textit{Let x = 2.}$$
$$8 + y = 10$$
$$y = 2$$

Check $x = 2, y = 2$: $2 = 2, 10 = 10$

The solution set is $\{(2, 2)\}$.

(b)
$$\begin{array}{rl} 8x - 5y = 32 & (1) \\ 4x + 5y = 4 & (2) \\ \hline 12x = 36 & \textit{Add (1) and (2).} \\ x = 3 & \textit{Divide by 12.} \end{array}$$

To find y, substitute 3 for x in either of the original equations.

$$4x + 5y = 4 \quad \textit{Equation (2)}$$
$$4(3) + 5y = 4 \quad \textit{Let x = 3.}$$
$$12 + 5y = 4$$
$$5y = -8$$
$$y = -\tfrac{8}{5}$$

Check $x = 3, y = -\tfrac{8}{5}$: $32 = 32, 4 = 4$

The solution set is $\{(3, -\tfrac{8}{5})\}$.

3. (a)
$$\begin{array}{rl} 2x + 3y = -15 & (1) \\ 5x + 2y = 1 & (2) \end{array}$$

To eliminate y, multiply equation (1) by 2 and equation (2) by -3. Notice that we multiplied by numbers that will cause the coefficients of y to be opposites of each other ($+6$ and -6).

$$\begin{array}{rl} 4x + 6y = -30 & (3) \\ -15x - 6y = -3 & (4) \\ \hline -11x = -33 & \textit{Add (3) and (4).} \\ x = 3 & \textit{Divide by } -11. \end{array}$$

Substitute 3 for x in equation (1).

$$2x + 3y = -15$$
$$2(3) + 3y = -15$$
$$6 + 3y = -15$$
$$3y = -21$$
$$y = -7$$

Check $x = 3, y = -7$: $-15 = -15, 1 = 1$

The solution set is $\{(3, -7)\}$.

(b)
$$\begin{array}{rl} 6x + 7y = 4 & (1) \\ 5x + 8y = -1 & (2) \end{array}$$

To eliminate x, multiply equation (1) by -5 and equation (2) by 6.

$$\begin{array}{rl} -30x - 35y = -20 & (3) \\ 30x + 48y = -6 & (4) \\ \hline 13y = -26 & \textit{Add (3) and (4).} \\ y = -2 & \textit{Divide by 13.} \end{array}$$

To find x, substitute -2 for y in equation (1).

$$6x + 7y = 4$$
$$6x + 7(-2) = 4 \quad \textit{Let y = -2.}$$
$$6x - 14 = 4$$
$$6x = 18$$
$$x = 3$$

Check $x = 3, y = -2$: $4 = 4, -1 = -1$

The solution set is $\{(3, -2)\}$.

4. (a)
$$\begin{array}{rl} 5x = 7 + 2y & (1) \\ 5y = 5 - 3x & (2) \end{array}$$

Rearrange the terms in both equations as follows so that like terms can be aligned.

$$5x - 2y = 7 \quad (3)$$
$$3x + 5y = 5 \quad (4)$$

To eliminate y, multiply equation (3) by 5 and equation (4) by 2.

$$\begin{array}{rl} 25x - 10y = 35 & (5) \\ 6x + 10y = 10 & (6) \\ \hline 31x = 45 & \textit{Add (5) and (6).} \\ x = \tfrac{45}{31} & \textit{Divide by 31.} \end{array}$$

Substituting $\tfrac{45}{31}$ for x to find y in one of the original equations would be messy. Instead, solve for y by starting with equations (3) and (4) and eliminating x. Multiply equation (3) by -3 and equation (4) by 5.

$$\begin{array}{rl} -15x + 6y = -21 & (7) \\ 15x + 25y = 25 & (8) \\ \hline 31y = 4 & \textit{Add (7) and (8).} \\ y = \tfrac{4}{31} & \textit{Divide by 31.} \end{array}$$

(See the calculator note at the end of the solution for Exercise 31 in this section.)

Check $x = \tfrac{45}{31}, y = \tfrac{4}{31}$: $\tfrac{225}{31} = \tfrac{225}{31}, \tfrac{20}{31} = \tfrac{20}{31}$

The solution set is $\{(\tfrac{45}{31}, \tfrac{4}{31})\}$.

(b)
$$\begin{array}{rl} 3y = 8 + 4x & (1) \\ 6x = 9 - 2y & (2) \end{array}$$

Rearrange the terms in both equations.

$$-4x + 3y = 8 \quad (3)$$
$$6x + 2y = 9 \quad (4)$$

To eliminate x, multiply equation (3) by 3 and equation (4) by 2.

$$\begin{array}{rl} -12x + 9y = 24 & (5) \\ 12x + 4y = 18 & (6) \\ \hline 13y = 42 & \textit{Add (5) and (6).} \\ y = \tfrac{42}{13} & \textit{Divide by 13.} \end{array}$$

To eliminate y, multiply equation (3) by 2 and equation (4) by -3.

$$
\begin{array}{rcll}
-8x + 6y &=& 16 & (7) \\
-18x - 6y &=& -27 & (8) \\
\hline
-26x &=& -11 & \text{Add (7) and (8).} \\
x &=& \frac{11}{26} & \text{Divide by } -26.
\end{array}
$$

Check $x = \frac{11}{26}$, $y = \frac{42}{13}$: $\frac{126}{13} = \frac{126}{13}$, $\frac{33}{13} = \frac{33}{13}$

The solution set is $\{(\frac{11}{26}, \frac{42}{13})\}$.

5. **(a)** $4x + 3y = 10$ (1)

$2x + \frac{3}{2}y = 12$ (2)

To eliminate x, multiply equation (2) by -2.

$$
\begin{array}{rcll}
4x + 3y &=& 10 & (1) \\
-4x - 3y &=& -24 & (3) \\
\hline
0 &=& -14 & \text{False}
\end{array}
$$

A false statement results. The graphs of these equations are parallel lines, so there is no solution. The solution set is $\emptyset$.

(b) $4x - 6y = 10$ (1)

$-10x + 15y = -25$ (2)

To eliminate x, multiply equation (1) by 5 and equation (2) by 2.

$$
\begin{array}{rcll}
20x - 30y &=& 50 & (3) \\
-20x + 30y &=& -50 & (4) \\
\hline
0 &=& 0 & \text{True}
\end{array}
$$

A true statement occurs when the equations are equivalent. This indicates that every solution of one equation is also a solution of the other. The system has an infinite number of solutions. The solution set is $\{(x, y) \mid 2x - 3y = 5\}$.

(*Note:* To write the solution set, we divided each term of the equation $4x - 6y = 10$ by 2 so that the coefficients would have greatest common factor 1.)

12.3 Section Exercises

1. The statement is *true*. Both lines would have x- and y-intercepts at $(0, 0)$. Since the point $(0, 0)$ satisfies both equations, it is a solution of the system.

3. It is impossible to have two numbers whose sum is both 1 and 2, so the given statement is *true*.

For all systems in this section, solutions should be checked by substituting the proposed solution in *both* equations of the original system. A detailed check will only be shown here for Exercises 5 and 7.

5. $$
\begin{array}{rcll}
x + y &=& 2 & (1) \\
2x - y &=& -5 & (2) \\
\hline
3x &=& -3 & \text{Add (1) and (2).} \\
x &=& -1 & \text{Divide by 3.}
\end{array}
$$

Replace x with -1 in either equation.

If we use equation (1), $x + y = 2$, we get $-1 + y = 2$ or $y = 3$.

If we use equation (2), $2x - y = -5$, we get $-2 - y = -5$ or $y = 3$.

Check $x = -1$, $y = 3$:

$$
\begin{array}{rcl}
x + y &=& 2 \qquad (1) \\
-1 + 3 &\overset{?}{=}& 2 \\
2 &=& 2 \qquad \text{True}
\end{array}
$$

$$
\begin{array}{rcl}
2x - y &=& -5 \qquad (2) \\
2(-1) - 3 &\overset{?}{=}& -5 \\
-2 - 3 &\overset{?}{=}& -5 \\
-5 &=& -5 \qquad \text{True}
\end{array}
$$

The solution set is $\{(-1, 3)\}$.

7. $$
\begin{array}{rcll}
2x + y &=& -5 & (1) \\
x - y &=& 2 & (2) \\
\hline
3x &=& -3 & \text{Add (1) and (2).} \\
x &=& -1 &
\end{array}
$$

Substitute -1 for x in equation (1) to find the y-value of the solution.

$$
\begin{array}{rcl}
2x + y &=& -5 \\
2(-1) + y &=& -5 \qquad \text{Let } x = -1. \\
-2 + y &=& -5 \\
y &=& -3
\end{array}
$$

Check $x = -1$, $y = -3$:

$$
\begin{array}{rcl}
2x + y &=& -5 \qquad (1) \\
2(-1) + (-3) &\overset{?}{=}& -5 \\
-2 + (-3) &\overset{?}{=}& -5 \\
-5 &=& -5 \qquad \text{True}
\end{array}
$$

$$
\begin{array}{rcl}
x - y &=& 2 \qquad (2) \\
-1 - (-3) &\overset{?}{=}& 2 \\
-1 + 3 &\overset{?}{=}& 2 \\
2 &=& 2 \qquad \text{True}
\end{array}
$$

The solution set is $\{(-1, -3)\}$.

9. $$
\begin{array}{rcll}
3x + 2y &=& 0 & (1) \\
-3x - y &=& 3 & (2) \\
\hline
y &=& 3 & \text{Add (1) and (2).}
\end{array}
$$

Substitute 3 for y in equation (1).

$$
\begin{array}{rcl}
3x + 2y &=& 0 \\
3x + 2(3) &=& 0 \\
3x + 6 &=& 0 \\
3x &=& -6 \\
x &=& -2
\end{array}
$$

The solution set is $\{(-2, 3)\}$.

11. $6x - y = -1$
       $5y = 17 + 6x$

Rewrite in standard form.

$$\begin{array}{rcrcrl}
6x & - & y & = & -1 & \quad (1)\\
-6x & + & 5y & = & 17 & \quad (2)\\
\hline
 & & 4y & = & 16 & \quad \textit{Add (1) and (2).}\\
 & & y & = & 4 & \quad \textit{Divide by 4.}
\end{array}$$

Substitute 4 for y in equation (1).

$$6x - y = -1$$
$$6x - 4 = -1$$
$$6x = 3$$
$$x = \frac{3}{6} = \frac{1}{2}$$

The solution set is $\{(\frac{1}{2}, 4)\}$.

13. $2x - y = 12$ (1)
       $3x + 2y = -3$ (2)

If we simply add the equations, we will not eliminate either variable. To eliminate y, multiply equation (1) by 2 and add the result to equation (2).

$$\begin{array}{rcrcrl}
4x & - & 2y & = & 24 & \quad (3)\\
3x & + & 2y & = & -3 & \quad (2)\\
\hline
7x & & & = & 21 & \quad \textit{Add (3) and (2).}\\
 & & x & = & 3 &
\end{array}$$

Substitute 3 for x in equation (1).

$$2x - y = 12$$
$$2(3) - y = 12$$
$$-y = 6$$
$$y = -6$$

The solution set is $\{(3, -6)\}$.

15. $x + 3y = 19$ (1)
       $2x - y = 10$ (2)

If we simply add the equations, we will not eliminate either variable. To eliminate y, multiply equation (2) by 3 and add the result to equation (1).

$$\begin{array}{rcrcrl}
x & + & 3y & = & 19 & \quad (1)\\
6x & - & 3y & = & 30 & \quad (3)\\
\hline
7x & & & = & 49 & \quad \textit{Add (1) and (3).}\\
 & & x & = & 7 &
\end{array}$$

Substitute 7 for x in equation (1) to find the y-value of the solution.

$$x + 3y = 19$$
$$7 + 3y = 19$$
$$3y = 12$$
$$y = 4$$

The solution set is $\{(7, 4)\}$.

17. $x + 4y = 16$ (1)
       $3x + 5y = 20$ (2)

To eliminate x, multiply equation (1) by -3 and add the result to equation (2).

$$\begin{array}{rcrcrl}
-3x & - & 12y & = & -48 & \quad (3)\\
3x & + & 5y & = & 20 & \quad (2)\\
\hline
 & & -7y & = & -28 & \quad \textit{Add (3) and (2).}\\
 & & y & = & 4 &
\end{array}$$

Substitute 4 for y in equation (1).

$$x + 4y = 16$$
$$x + 4(4) = 16$$
$$x + 16 = 16$$
$$x = 0$$

The solution set is $\{(0, 4)\}$.

19. $5x - 3y = -20$ (1)
       $-3x + 6y = 12$ (2)

To eliminate y, multiply equation (1) by 2 and add the result to equation (2).

$$\begin{array}{rcrcrl}
10x & - & 6y & = & -40 & \quad (3)\\
-3x & + & 6y & = & 12 & \quad (2)\\
\hline
7x & & & = & -28 & \quad \textit{Add (3) and (2).}\\
 & & x & = & -4 &
\end{array}$$

Substitute -4 for x in equation (2).

$$-3x + 6y = 12$$
$$-3(-4) + 6y = 12$$
$$12 + 6y = 12$$
$$6y = 0$$
$$y = 0$$

The solution set is $\{(-4, 0)\}$.

21. $2x - 8y = 0$ (1)
       $4x + 5y = 0$ (2)

To eliminate x, multiply equation (1) by -2 and add the result to equation (2).

$$\begin{array}{rcrcrl}
-4x & + & 16y & = & 0 & \quad (3)\\
4x & + & 5y & = & 0 & \quad (2)\\
\hline
 & & 21y & = & 0 & \quad \textit{Add (3) and (2).}\\
 & & y & = & 0 &
\end{array}$$

Substitute 0 for y in equation (1).

$$2x - 8y = 0$$
$$2x - 8(0) = 0$$
$$2x = 0$$
$$x = 0$$

The solution set is $\{(0, 0)\}$.

23. $x + y = 7$ (1)
$x + y = -3$ (2)

Multiply equation (2) by -1 and add the result to equation (1).

$$
\begin{array}{rcll}
x + y &=& 7 & (1) \\
-x - y &=& 3 & (3) \\
\hline
0 &=& 10 & \textit{Add (1) and (3).}
\end{array}
$$

The false statement $0 = 10$ shows that the given system has no solution. The solution set is $\emptyset$.

25. $-x + 3y = 4$ (1)
$-2x + 6y = 8$ (2)

$$
\begin{array}{rcll}
2x - 6y &=& -8 & (3) \quad -2 \times \text{Eq.(1)} \\
-2x + 6y &=& 8 & (2) \\
\hline
0 &=& 0 & \textit{Add (3) and (2).}
\end{array}
$$

Since $0 = 0$ is a *true* statement, the equations are equivalent. This result indicates that every solution of one equation is also a solution of the other; there are an *infinite number of solutions.* The solution set is $\{(x, y) \mid x - 3y = -4\}$.

27. $2x + 3y = 21$ (1)
$5x - 2y = -14$ (2)

To eliminate y, we could multiply equation (1) by $\frac{2}{3}$, but that would introduce fractions and make the solution more complicated. Instead, we will work with the least common multiple of the coefficients of y, which is 6, and choose suitable multipliers of these coefficients so that the new coefficients are opposites.

In this case, we could pick 2 times equation (1) and 3 times equation (2). If we wanted to eliminate x, we could multiply equation (1) by 5 and equation (2) by -2 *or* equation (1) by -5 and equation (2) by 2.

$$
\begin{array}{rcll}
4x + 6y &=& 42 & (3) \quad 2 \times \text{Eq.(1)} \\
15x - 6y &=& -42 & (4) \quad 3 \times \text{Eq.(2)} \\
\hline
19x &=& 0 & \textit{Add (3) and (4).} \\
x &=& 0 &
\end{array}
$$

Substitute 0 for x in equation (1).

$$
\begin{aligned}
2x + 3y &= 21 \\
2(0) + 3y &= 21 \\
3y &= 21 \\
y &= 7
\end{aligned}
$$

The solution set is $\{(0, 7)\}$.

29. $3x - 7 = -5y$
$5x + 4y = -10$

Rewrite in standard form.

$3x + 5y = 7$ (1)
$5x + 4y = -10$ (2)

To eliminate x, multiply equation (1) by 5 and equation (2) by -3.

$$
\begin{array}{rcll}
15x + 25y &=& 35 & (3) \\
-15x - 12y &=& 30 & (4) \\
\hline
13y &=& 65 & \textit{Add (3) and (4).} \\
y &=& 5 &
\end{array}
$$

Substitute 5 for y in equation (2).

$$
\begin{aligned}
5x + 4y &= -10 \\
5x + 4(5) &= -10 \\
5x + 20 &= -10 \\
5x &= -30 \\
x &= -6
\end{aligned}
$$

The solution set is $\{(-6, 5)\}$.

31. $2x + 3y = 0$
$4x + 12 = 9y$

Rewrite in standard form.

$2x + 3y = 0$ (1)
$4x - 9y = -12$ (2)

$$
\begin{array}{rcll}
6x + 9y &=& 0 & (3) \quad 3 \times \text{Eq.(1)} \\
4x - 9y &=& -12 & (2) \\
\hline
10x &=& -12 & \textit{Add (3) and (2).} \\
x &=& \frac{-12}{10} = -\frac{6}{5} &
\end{array}
$$

$$
\begin{array}{rcll}
-4x - 6y &=& 0 & (4) \quad -2 \times \text{Eq.(1)} \\
4x - 9y &=& -12 & (2) \\
\hline
-15y &=& -12 & \textit{Add (4) and (2).} \\
y &=& \frac{-12}{-15} = \frac{4}{5} &
\end{array}
$$

The solution set is $\{(-\frac{6}{5}, \frac{4}{5})\}$.

When you get a solution that has non-integer coordinates, it is sometimes more difficult to check the problem than it was to solve it. A graphing calculator can be very helpful in this case. Just store the values for x and y in their respective memory locations, and then type the expressions as shown in the following screen. The results 0 and -12 (the right sides of the equations) indicate that we have found the correct solution.

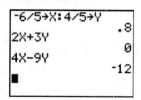

33. $24x + 12y = -7$
$16x - 17 = 18y$

Rewrite in standard form.

$$24x + 12y = -7 \qquad (1)$$
$$16x - 18y = 17 \qquad (2)$$

$$
\begin{array}{rll}
48x \; + \; 24y \; = \; -14 & (3) & 2 \times \text{Eq.}(1) \\
-48x \; + \; 54y \; = \; -51 & (4) & -3 \times \text{Eq.}(2) \\
\hline
78y \; = \; -65 & & \text{Add } (3) \text{ and } (4). \\
y \; = \; \frac{-65}{78} = -\frac{5}{6} & &
\end{array}
$$

$$
\begin{array}{rll}
72x \; + \; 36y \; = \; -21 & (5) & 3 \times \text{Eq.}(1) \\
32x \; - \; 36y \; = \; 34 & (6) & 2 \times \text{Eq.}(2) \\
\hline
104x \; = \; 13 & & \text{Add } (5) \text{ and } (6). \\
x \; = \; \frac{13}{104} = \frac{1}{8} & &
\end{array}
$$

The solution set is $\left\{ \left(\frac{1}{8}, -\frac{5}{6} \right) \right\}$.

35. $5x - 2y = 3 \qquad (1)$
$10x - 4y = 5 \qquad (2)$

$$
\begin{array}{rll}
-10x \; + \; 4y \; = \; -6 & (3) & -2 \times \text{Eq.}(1) \\
10x \; - \; 4y \; = \; 5 & (2) & \\
\hline
0 \; = \; -1 & & \text{Add } (3) \text{ and } (2).
\end{array}
$$

Since $0 = -1$ is a *false* statement, there are no solutions of the system. The solution set is $\emptyset$.

37. $6x + 3y = 0 \qquad (1)$
$-18x - 9y = 0 \qquad (2)$

Multiply equation (1) by 3 and add the result to equation (2).

$$
\begin{array}{rll}
18x \; + \; 9y \; = \; 0 & (3) & \\
-18x \; - \; 9y \; = \; 0 & (2) & \\
\hline
0 \; = \; 0 & & \text{Add } (3) \text{ and } (2).
\end{array}
$$

This true result, $0 = 0$, means that the system has an infinite number of solutions. The equations of the system represent the same line. The solution set is $\{ (x, y) \mid 2x + y = 0 \}$.

39. $3x = 3 + 2y \qquad (1)$
$-\frac{4}{3}x + y = \frac{1}{3} \qquad (2)$

Rewrite equation (1) in standard form and multiply equation (2) by 3 to clear fractions.

$$3x - 2y = 3 \qquad (3)$$
$$-4x + 3y = 1 \qquad (4)$$

$$
\begin{array}{rll}
12x \; - \; 8y \; = \; 12 & (5) & 4 \times \text{Eq.}(3) \\
-12x \; + \; 9y \; = \; 3 & (6) & 3 \times \text{Eq.}(4) \\
\hline
y \; = \; 15 & & \text{Add } (5) \text{ and } (6).
\end{array}
$$

$$
\begin{array}{rll}
9x \; - \; 6y \; = \; 9 & (7) & 3 \times \text{Eq.}(3) \\
-8x \; + \; 6y \; = \; 2 & (8) & 2 \times \text{Eq.}(4) \\
\hline
x \; = \; 11 & & \text{Add } (7) \text{ and } (8).
\end{array}
$$

The solution set is $\{ (11, 15) \}$.

41. $\frac{1}{5}x + y = \frac{6}{5} \qquad (1)$
$\frac{1}{10}x + \frac{1}{3}y = \frac{5}{6} \qquad (2)$

First, clear all fractions.

Equation (1):

$$5\left(\tfrac{1}{5}x + y \right) = 5\left(\tfrac{6}{5} \right) \quad \textit{Multiply by LCD, 5.}$$
$$x + 5y = 6 \qquad (3)$$

Equation (2):

$$30\left(\tfrac{1}{10}x + \tfrac{1}{3}y \right) = 30\left(\tfrac{5}{6} \right) \quad \textit{Multiply by LCD, 30.}$$
$$3x + 10y = 25 \qquad (4)$$

The system has been simplified to

$$x + 5y = 6 \qquad (3)$$
$$3x + 10y = 25 \qquad (4)$$

To eliminate y, add -2 times equation (3) to (4).

$$
\begin{array}{rll}
-2x \; - \; 10y \; = \; -12 & (5) & -2 \times \text{Eq.}(3) \\
3x \; + \; 10y \; = \; 25 & (4) & \\
\hline
x \; = \; 13 & & \text{Add } (5) \text{ and } (4).
\end{array}
$$

Substitute 13 for x in equation (3).

$$
\begin{array}{rl}
x + 5y = 6 & (3) \\
13 + 5y = 6 & \\
5y = -7 & \\
y = -\frac{7}{5} &
\end{array}
$$

The solution set is $\left\{ \left(13, -\frac{7}{5} \right) \right\}$.

43. $2.4x + 1.7y = 7.6 \quad (1)$
$1.2x - 0.5y = 9.2 \quad (2)$

Multiply equation (1) and equation (2) by 10 to eliminate the decimals. Then solve the resulting system of equations by the elimination method.

$$24x + 17y = 76 \quad (3)$$
$$12x - 5y = 92 \quad (4)$$

Multiply equation (2) by -2.

$$
\begin{array}{rll}
24x \; + \; 17y \; = \; 76 & (3) \\
-24x \; + \; 10y \; = \; -184 & (5) \\
\hline
27y \; = \; -108 & \text{Add .} \\
y \; = \; -4 &
\end{array}
$$

Substitute -4 for y in equation (4).

$$
\begin{array}{rl}
12x - 5y = 92 & \\
12x - 5(-4) = 92 & \textit{Let y = -4.} \\
12x + 20 = 92 & \\
12x = 72 & \\
x = 6 & \textit{Divide by 12.}
\end{array}
$$

The solution set is $\{ (6, -4) \}$.

Relating Concepts (Exercises 45–48)

45.
$$y = ax + b$$
$$5.66 = a(2001) + b \quad \textit{Let x = 2001, y = 5.66.}$$
$$5.66 = 2001a + b$$

46. As in Exercise 45,
$$7.89 = 2010a + b.$$

47.
$$2001a + b = 5.66 \quad (1)$$
$$2010a + b = 7.89 \quad (2)$$

Multiply equation (1) by -1 and add the result to equation (2) to eliminate b,

$$
\begin{array}{rcr}
-2001a - b & = & -5.66 \\
2010a + b & = & 7.89 \\
\hline
9a & = & 2.23 \\
a & \approx & 0.248
\end{array}
$$

Substitute 0.248 for a in equation (2).

$$2010(0.248) + b = 7.89$$
$$498.48 + b = 7.89$$
$$b = -490.590$$

The solution set is $\{(0.248, -490.590)\}$. Note that results can vary slightly depending on method of solution.

48. **(a)** An equation of the segment PQ is

$$y = 0.248x - 490.590$$

for $2001 \le x \le 2010$.

(b)
$$y = 0.248x - 490.590$$
$$y = 0.248(2008) - 490.590 \quad \textit{Let x = 2008.}$$
$$= 497.984 - 490.590$$
$$= 7.394 \approx \$7.39$$

This is a bit more than the actual figure of $7.18.

Summary Exercises *Applying Techniques for Solving Systems of Linear Equations*

1. **(a)** $3x + 2y = 18$
$$y = 3x$$

Use substitution since the second equation is solved for y.

(b) $3x + y = -7$
$$x - y = -5$$

Use elimination since the coefficients of the y-terms are opposites.

(c) $3x - 2y = 0$
$$9x + 8y = 7$$

Use elimination since the equations are in standard form with no coefficients of 1 or -1. Solving by substitution would involve fractions.

3.
$$4x - 3y = -8 \quad (1)$$
$$x + 3y = 13 \quad (2)$$

(a) Solve the system by the elimination method.

$$
\begin{array}{rcrcl}
4x & - & 3y & = & -8 \quad (1) \\
x & + & 3y & = & 13 \quad (2) \\
\hline
5x & & & = & 5 \quad \textit{Add (1) and (2).} \\
& & x & = & 1
\end{array}
$$

To find y, let $x = 1$ in equation (2).

$$x + 3y = 13$$
$$1 + 3y = 13$$
$$3y = 12$$
$$y = 4$$

The solution set is $\{(1, 4)\}$.

(b) To solve this system by the substitution method, begin by solving equation (2) for x.

$$x + 3y = 13$$
$$x = -3y + 13$$

Substitute $-3y + 13$ for x in equation (1).

$$4(-3y + 13) - 3y = -8$$
$$-12y + 52 - 3y = -8$$
$$-15y = -60$$
$$y = 4$$

To find x, let $y = 4$ in equation (2).

$$x + 3y = 13$$
$$x + 3(4) = 13$$
$$x + 12 = 13$$
$$x = 1$$

The solution set is $\{(1, 4)\}$.

(c) For this particular system, the elimination method is preferable because both equations are already written in the form $Ax + By = C$, and the equations can be added without multiplying either by a constant. Comparing the two methods, we see that the elimination method requires fewer steps than the substitution method for this system.

5.
$$3x + 2y = 18 \quad (1)$$
$$y = 3x \quad (2)$$

Equation (2) is already solved for y, so we will use the substitution method. Substitute $3x$ for y in equation (1) and solve the resulting equation for x.

$$3x + 2y = 18$$
$$3x + 2(3x) = 18 \quad \textit{Let y = 3x.}$$
$$3x + 6x = 18$$
$$9x = 18$$
$$x = \frac{18}{9} = 2$$

To find the y-value of the solution, substitute 2 for x in equation (2).

$$y = 3x = 3(2) = 6$$

The solution set is $\{(2, 6)\}$.

7. $3x - 2y = 0$ (1)
$9x + 8y = 7$ (2)

$$\begin{array}{rcl} 12x & - & 8y & = & 0 \\ 9x & + & 8y & = & 7 \\ \hline 21x & & & = & 7 \end{array}$$ (3) $4 \times$ Eq. (1)
(2)
Add (3) *and* (2).

$$x = \frac{7}{21} = \frac{1}{3}$$

$$\begin{array}{rcl} -9x & + & 6y & = & 0 \\ 9x & + & 8y & = & 7 \\ \hline & & 14y & = & 7 \end{array}$$ (4) $-3 \times$ Eq.(1)
(2)
Add (4) *and* (2).

$$y = \frac{7}{14} = \frac{1}{2}$$

The solution set is $\left\{ \left(\frac{1}{3}, \frac{1}{2} \right) \right\}$.

9. $5x - 4y = 15$ (1)
$-3x + 6y = -9$ (2)

$$\begin{array}{rcl} 15x & - & 12y & = & 45 \\ -15x & + & 30y & = & -45 \\ \hline & & 18y & = & 0 \\ & & y & = & 0 \end{array}$$ (3) $3 \times$ Eq.(1)
(4) $5 \times$ Eq.(2)
Add (3) *and* (4).

Substitute 0 for y in equation (1).

$$5x - 4y = 15$$
$$5x - 4(0) = 15$$
$$5x = 15$$
$$x = 3$$

The solution set is $\{(3, 0)\}$.

11. $3x = 7 - y$ (1)
$2y = 14 - 6x$ (2)

Solve equation (1) for y.

$$3x + y = 7$$
$$y = 7 - 3x$$

Substitute $7 - 3x$ for y in equation (2).

$$2(7 - 3x) = 14 - 6x$$
$$14 - 6x = 14 - 6x$$
$$14 = 14$$

The last equation is true, so there are an infinite number of solutions. The solution set is $\{(x, y) \mid 3x + y = 7\}$.

13. $5x = 7 + 2y$
$5y = 5 - 3x$

Rewrite in standard form.

$5x - 2y = 7$ (1)
$3x + 5y = 5$ (2)

$$\begin{array}{rcl} 25x & - & 10y & = & 35 \\ 6x & + & 10y & = & 10 \\ \hline 31x & & & = & 45 \end{array}$$ (3) $5 \times$ Eq.(1)
(4) $2 \times$ Eq.(2)
Add (3) *and* (4).

$$x = \frac{45}{31}$$

$$\begin{array}{rcl} -15x & + & 6y & = & -21 \\ 15x & + & 25y & = & 25 \\ \hline & & 31y & = & 4 \end{array}$$ (5) $-3 \times$ Eq.(1)
(6) $5 \times$ Eq.(2)
Add (5) *and* (6).

$$y = \frac{1}{31}$$

The solution set is $\left\{ \left(\frac{45}{31}, \frac{4}{31} \right) \right\}$.

15. $2x - 3y = 7$ (1)
$-4x + 6y = 14$ (2)

$$\begin{array}{rcl} 4x & - & 6y & = & 14 \\ -4x & + & 6y & = & 14 \\ \hline & & 0 & = & 28 \end{array}$$ (3) $2 \times$ Eq.(1)
(2)
Add (3) *and* (2).

Since $0 = 28$ is *false*, there is no solution. The solution set is $\emptyset$.

17. $6x + 5y = 13$ (1)
$3x + 3y = 4$ (2)

Multiply equation (2) by -2.

$$\begin{array}{rcl} 6x & + & 5y & = & 13 \\ -6x & - & 6y & = & -8 \\ \hline & & -y & = & 5 \\ & & y & = & -5 \end{array}$$ (1)
(3)
Add (1) *and* (3).

Substitute -5 for y in equation (2).

$$3x + 3y = 4$$
$$3x + 3(-5) = 4 \quad \textit{Let } y = -5.$$
$$3x - 15 = 4$$
$$3x = 19$$
$$x = \frac{19}{3}$$

The solution set is $\left\{ \left(\frac{19}{3}, -5 \right) \right\}$.

19. $\frac{1}{4}x - \frac{1}{5}y = 9$ (1)
$y = 5x$ (2)

First, clear all fractions in equation (1).

$$20\left(\frac{1}{4}x - \frac{1}{5}y \right) = 20(9) \qquad \textit{Multiply by the LCD, 20.}$$

$$20\left(\frac{1}{4}x \right) - 20\left(\frac{1}{5}y \right) = 180 \qquad \textit{Distributive property}$$

$$5x - 4y = 180 \qquad (3)$$

From equation (2), substitute $5x$ for y in equation (3).

$$5x - 4(5x) = 180$$
$$5x - 20x = 180$$
$$-15x = 180$$
$$x = -12$$

To find y, let $x = -12$ in equation (2).

$$y = 5(-12) = -60$$

The solution set is $\{(-12, -60)\}$.

21. $\frac{1}{6}x + \frac{1}{6}y = 2$ (1)

$-\frac{1}{2}x - \frac{1}{3}y = -8$ (2)

Multiply each side of equation (1) by 6 to clear fractions.

$$6\left(\frac{1}{6}x + \frac{1}{6}y\right) = 6(2)$$
$$6\left(\frac{1}{6}x\right) + 6\left(\frac{1}{6}y\right) = 6(2)$$
$$x + y = 12 \qquad (3)$$

Multiply each side of equation (2) by the LCD, 6, to clear fractions.

$$6\left(-\frac{1}{2}x - \frac{1}{3}y\right) = 6(-8)$$
$$6\left(-\frac{1}{2}x\right) + 6\left(-\frac{1}{3}y\right) = 6(-8)$$
$$-3x - 2y = -48 \qquad (4)$$

The given system of equations has been simplified as follows.

$$x + y = 12 \qquad (3)$$
$$-3x - 2y = -48 \qquad (4)$$

Multiply equation (3) by 3 and add the result to equation (4).

$$
\begin{array}{rcrcr}
3x & + & 3y & = & 36 \\
-3x & - & 2y & = & -48 \\
\hline
 & & y & = & -12
\end{array}
$$

To find x, let $y = -12$ in equation (3).

$$x + y = 12$$
$$x + (-12) = 12$$
$$x - 12 = 12$$
$$x = 24$$

The solution set is $\{(24, -12)\}$.

23. $\frac{x}{5} + y = 6$ (1)

$\frac{x}{10} + \frac{y}{3} = \frac{5}{6}$ (2)

First, clear all fractions.

Equation (1):

$$5\left(\frac{x}{5} + y\right) = 5(6) \quad \textit{Multiply by 5.}$$
$$x + 5y = 30 \qquad (3)$$

Equation (2):

$$30\left(\frac{x}{10} + \frac{y}{3}\right) = 30\left(\frac{5}{6}\right) \quad \textit{Multiply by the LCD, 30.}$$
$$3x + 10y = 25 \qquad (4)$$

The system has been simplified to

$$x + 5y = 30 \qquad (3)$$
$$3x + 10y = 25. \qquad (4)$$

Solve this system by the substitution method.

$x = -5y + 30$ (5)   *Solve (3) for x.*

$3(-5y + 30) + 10y = 25$ *Substitute for x in (4).*

$$-15y + 90 + 10y = 25$$
$$-5y + 90 = 25$$
$$-5y = -65$$
$$y = 13$$

To find x, let $y = 13$ in equation (5).

$$x = -5(13) + 30 = -65 + 30 = -35$$

The solution set is $\{(-35, 13)\}$.

25. $0.2x + 0.3y = 1.0$ (1)

$-0.3x + 0.1y = 1.8$ (2)

Multiply equation (1) and equation (2) by 10 to eliminate the decimals. Then solve the resulting system of equations by the substitution method.

$$2x + 3y = 10 \qquad (3)$$
$$-3x + y = 18 \qquad (4)$$

Solve equation (4) for y.

$$y = 3x + 18 \quad (5)$$

Substitute $3x + 18$ for y in equation (3).

$$2x + 3y = 10 \quad \textit{Equation (3)}$$
$$2x + 3(3x + 18) = 10 \quad \textit{Let y = 3x + 18.}$$
$$2x + 9x + 54 = 10$$
$$11x + 54 = 10$$
$$11x = -44$$
$$x = -4$$

Substitute -4 for x in equation (5).

$$y = 3(-4) + 18 = -12 + 18 = 6$$

The solution is $\{(-4, 6)\}$.

12.4 Applications of Linear Systems

12.4 Margin Exercises

1. **(a)** *Step 1* We must find <u>equipment sales for each sport</u>.

Step 2 Let $x = $ equipment sales for tennis (in millions of dollars), and $y = $ equipment sales for <u>archery</u> (in millions of dollars).

Step 3 The sales were $26 million less for tennis than for archery, so one equation is

$$x = \underline{y - 26}. \quad (1)$$

Together, total equipment sales for these two sports were $876 million, so another equation is

$$x + y = 876. \quad (2)$$

Step 4 Substitute $y - 26$ for x in the second equation, and solve for y.

$$x + y = 876 \quad (2)$$
$$(y - 26) + y = 876 \quad \textit{Let x = y - 26.}$$
$$2y - 26 = 876$$
$$2y = 902$$
$$y = 451$$

To find x, use $x = y - 26$ and $y = 451$.

$$x = y - 26 = 451 - 26 = 425$$

Step 5 The equipment sales for tennis and archery were $425 million and $451 million, respectively.

Step 6 The sum of 425 and 451 is 876. Also, 425 is 26 less than 451. The solution satisfies the conditions of the problem.

(b) Step 1 We must find the amount that each movie grossed.

Step 2 Let $x =$ the amount that *Fast Five* grossed, and $y =$ the amount that *Cars 2* grossed. (Both are in millions of dollars.)

Step 3 Together, their gross was $401 million, so one equation is

$$x + y = 401. \quad (1)$$

Cars 2 grossed $18 million less than *Fast Five*, so another equation is

$$y = x - 18. \quad (2)$$

Step 4 Substitute $x - 18$ for y in the first equation, and solve for x.

$$x + y = 401 \quad (1)$$
$$x + (x - 18) = 401 \quad \textit{Let y = x - 18.}$$
$$2x - 18 = 401$$
$$2x = 419$$
$$x = 209.5$$

To find y, use $y = x - 18$ and $x = 209.5$.

$$y = x - 18 = 209.5 - 18 = 191.5$$

Step 5 *Fast Five* grossed $209.5 million and *Cars 2* grossed $191.5 million.

Step 6 The sum of 209.5 and 191.5 is 401. Also, 191.5 is 18 less than 209.5. The solution satisfies the conditions of the problem.

2. (a)

	Number of Tickets	Price per Ticket (in $)	Total Value (in $)
Orchestra	x	129	$129x$
Mezzanine	y	87	$87y$
Total	18	XXXXXX	1776

(b) From the second column in part (a),

$$x + y = 18.$$

From the fourth column in part (a),

$$129x + 87y = 1776.$$

(c) Multiply the first equation by -87 and add to the second equation.

$$
\begin{array}{rcrcr}
-87x & - & 87y & = & -1566 \\
129x & + & 87y & = & 1776 \\
\hline
42x & & & = & 210 \\
& & x & = & 5
\end{array}
$$

Substitute 5 for x in the first equation.

$$5 + y = 18$$
$$y = 13$$

There were 5 Orchestra tickets sold and 13 Mezzanine tickets sold.

Check: The sum of 5 and 13 is 18, so the number of tickets is correct. Since 5 Orchestra tickets were sold for $129 each and 13 Mezzanine tickets were sold for $87 each, the total value of the tickets is

$$\$129(5) + \$87(13) = \$1776,$$

which agrees with the total spent stated in the problem.

3. (a)

Liters	Percent (as a decimal)	Liters of Pure Alcohol
x	0.25	$0.25x$
y	0.12	$0.12y$
13	0.15	$0.15(13)$

(b)

$$x + y = 13 \quad \text{\textit{From first column}}$$
$$0.25x + 0.12y = 0.15(13) \quad \text{\textit{From third column}}$$

To eliminate the x-terms, multiply the first equation by -0.25. Then add the result to the second equation.

$$
\begin{array}{rcrcrl}
-0.25x & - & 0.25y & = & -3.25 & \\
0.25x & + & 0.12y & = & 1.95 & \\
\hline
& & -0.13y & = & -1.3 & \textit{Add.} \\
& & y & = & 10 &
\end{array}
$$

To find x, substitute 10 for y in the first equation of the original system.

$$x + y = 13$$
$$x + 10 = 13 \quad \textit{Let y = 10.}$$
$$x = 3$$

The solution is $x = 3$, $y = 10$. Mix 3 L of 25% solution with 10 L of 12% solution.

4. Let $x =$ the amount of 10% solution needed, and $y =$ the amount of 25% solution needed.

Make a table.

Milliliters	Percent	Pure Acid
x	10	$0.10x$
y	25	$0.25y$
60	20	$0.20(60)$

Set up a system of equations.

$$x + y = 60$$
$$0.10x + 0.25y = 0.20(60)$$

To eliminate the x-terms, multiply the first equation by -0.10. Then add the result to the second equation.

$$
\begin{array}{rcrcl}
-0.10x & - & 0.10y & = & -6 \\
0.10x & + & 0.25y & = & 12 \\
\hline
 & & 0.15y & = & 6 \\
 & & y & = & \dfrac{6}{0.15} = 40
\end{array}
$$

To find x, substitute 40 for y in the first equation of the original system.

$$x + y = 60$$
$$x + 40 = 60 \quad \text{Let } y = 40.$$
$$x = 20$$

20 mL of 10% solution and 40 mL of 25% solution must be mixed.

5. Use the relationship

$$\text{distance} = \text{rate} \times \text{time}.$$

The rate is 244 km per hr and the time is 1.7 hr.

$$\text{distance} = (244 \text{ km per hr}) \times (1.7 \text{ hr})$$
$$= 414.8 \text{ km}$$

The distance, to the nearest kilometer, is 415 km.

6. **(a)** Let $x =$ the rate of the faster car, and $y =$ the rate of the slower car.

	r	t	d
Faster Car	x	5	$5x$
Slower Car	y	5	$5y$

(b) Write a system of equations.

$$5x + 5y = 450 \quad \textit{Total distance}$$
$$x = 2y \quad \begin{array}{l}\textit{Faster car is} \\ \textit{twice as fast.}\end{array}$$

Substitute $2y$ for x in the first equation, and solve for y.

$$5x + 5y = 450$$
$$5(2y) + 5y = 450 \quad \textit{Let x = 2y.}$$
$$10y + 5y = 450$$
$$15y = 450$$
$$y = 30$$

To find x, use $x = 2y$ and $y = 30$.

$$x = 2y = 2(30) = 60$$

The faster car's rate is 60 miles per hour and the slower car's rate is 30 miles per hour.

7. Let $x =$ the rate of the faster truck, and $y =$ the rate of the slower truck.

Summarize the information in a table.

	r	t	d
Faster Truck	x	3	$3x$
Slower Truck	y	3	$3y$

Write a system of equations.

$$3x + 3y = 405 \quad \textit{Total distance}$$
$$x = y + 5 \quad \begin{array}{l}\textit{Faster truck is} \\ \textit{5 mph faster.}\end{array}$$

Substitute $y + 5$ for x in the first equation, and solve for y.

$$3x + 3y = 405$$
$$3(y + 5) + 3y = 405 \quad \textit{Let x = y + 5.}$$
$$3y + 15 + 3y = 405$$
$$6y + 15 = 405$$
$$6y = 390$$
$$y = 65$$

To find x, use $x = y + 5$ and $y = 65$.

$$x = y + 5 = 65 + 5 = 70$$

The faster truck's rate is 70 miles per hour and the slower truck's rate is 65 miles per hour.

8. **(a)** Let $x =$ the rate of the current, and $y = \underline{\text{Gigi's rate}}$ in still water.

When Gigi rows against the current, the current works against her, so the rate of the current is subtracted from her rate. When she rows with the current, the rate of the current is added to her rate.

Make a table.

	r	t	d
Against the Current	$y - x$	1	2
With the Current	$y + x$	1	10

Use $d = rt$ to get each equation of the system.

$$
\begin{array}{rcl}
(y - x)(1) = 2 & \rightarrow & -x + y = 2 \\
(y + x)(1) = 10 & \rightarrow & x + y = 10 \\
\hline
 & & 2y = 12 \quad \textit{Add.} \\
 & & y = 6
\end{array}
$$

Substitute 6 for y in the second equation of the system.

$$x + y = 10$$
$$x + 6 = 10 \quad \text{Let } y = 6.$$
$$x = 4$$

The rate of the current is 4 miles per hour. Gigi's rate in still water is 6 miles per hour.

(b) Let x = the rate of the current, and y = the boat's rate in still water.

When the boat travels against the current, the current works against the boat, so the rate of the current is subtracted from the rate of the boat. When the boat travels with the current, the rate of the current is added to the rate of the boat.

Make a table.

	r	t	d
Against the Current	$y - x$	1	9
With the Current	$y + x$	1	15

Use $d = rt$ to get each equation of the system.

$$
\begin{aligned}
(y - x)(1) = 9 &\rightarrow -x + y = 9 \\
(y + x)(1) = 15 &\rightarrow x + y = 15 \\
\hline
& 2y = 24 \quad \text{Add.} \\
& y = 12
\end{aligned}
$$

Substitute 12 for y in the second equation of the system.

$$x + y = 15$$
$$x + 12 = 15 \quad \text{Let } y = 12.$$
$$x = 3$$

The rate of the current is 3 miles per hour. The boat's rate in still water is 12 miles per hour.

12.4   Section Exercises

1. To represent the monetary value of x 20-dollar bills, multiply 20 times x. The answer is **D**, $20x$ dollars.

3. The amount of interest earned on d dollars at an interest rate of 2% (0.02) for 1 yr is choice **B**, $0.02d$ dollars.

5. If a cheetah's rate is 70 mph and it runs at that rate for x hours, then the distance covered is choice **D**, $70x$ miles.

7. Since the plane is traveling *with* the wind, add the rate of the plane, 560 miles per hour, to the rate of the wind, r miles per hour. The answer is **C**, $(560 + r)$ mph.

9. **Step 2**   Let x = the first number and y = _the second number_ .

Step 3   First equation: $x + y = 98$
Second equation: $\underline{x - y = 48}$

Step 4   Add the two equations.

$$
\begin{aligned}
x &+ y &= 98 \\
x &- y &= 48 \\
\hline
2x & &= 146 \\
 & x &= 73
\end{aligned}
$$

Substitute 73 for x in either equation to find that $y = 25$.

Step 5   The two numbers are 73 and 25.

Step 6   The sum of 73 and 25 is 98. The difference between 73 and 25 is 48. The solution satisfies the conditions of the problem.

11. **Step 2**   Let x = the number of performances of *Cats* and y = the number of performances of *Phantom of the Opera*.

Step 3   The total number of performances was 17,288, so one equation is

$$x + y = 17{,}288. \quad (1)$$

Phantom of the Opera had 2318 more performances than *Cats*, so another equation is

$$y = x + 2318. \quad (2)$$

Step 4   Substitute $x + 2318$ for y in equation (1).

$$x + y = 17{,}288$$
$$x + (x + 2318) = 17{,}288$$
$$2x + 2318 = 17{,}288$$
$$2x = 14{,}970$$
$$x = 7485$$

Substitute 7485 for x in (2) to find $y = 7485 + 2318 = 9803$.

Step 5   *Cats* had 7485 performances and *Phantom of the Opera* had 9803 performances.

Step 6   The sum of 7485 and 9803 is 17,288 and 9803 is 2318 more than 7485.

13. **Step 2**   Let x = the amount grossed by *Harry Potter and the Deathly Hallows Part 2* and y = the amount grossed by *Transformers: Dark of the Moon*.

Step 3   *Transformers: Dark of the Moon* grossed $29 million less than *Harry Potter and the Deathly Hallows Part 2*, so

$$y = x - 29. \quad (1)$$

The total grossed by these two films was $733 million, so

$$x + y = 733 \quad (2)$$

Step 4 Substitute $x - 29$ for y in equation (2).

$$x + (x - 29) = 733$$
$$2x - 29 = 733$$
$$2x = 762$$
$$x = 381$$

To find y, let $x = 381$ in equation (1).

$$y = x - 29$$
$$y = 381 - 29 = 352$$

Step 5 *Harry Potter and the Deathly Hallows Part 2* grossed \$381 million and *Transformers: Dark of the Moon* grossed \$352 million.

Step 6 The sum of 381 and 352 is 733 and 352 is 29 less than 381.

15. **Step 2** Let $t =$ the height of the Terminal Tower and $k =$ the height of the Key Tower.

Step 3 The Terminal Tower is 239 feet shorter than the Key Tower, so

$$t = k - 239. \quad (1)$$

The total of the two heights is 1655 feet, so

$$t + k = 1655. \quad (2)$$

Step 4 Substitute $k - 239$ for t in equation (2).

$$(k - 239) + k = 1655$$
$$2k - 239 = 1655$$
$$2k = 1894$$
$$k = 947$$

To find t, let $k = 947$ in equation (1).

$$t = k - 239$$
$$t = 947 - 239 = 708$$

Step 5 The height of the Terminal Tower is 708 feet and the height of the Key Tower is 947 feet.

Step 6 The sum of 708 and 947 is 1655 and 708 is 239 less than 947.

17. **Step 2** Let $l =$ the length of the field (in yards) and $w =$ the width of the field (in yards).

Step 3 The length is 38 yards longer than the width, so

$$l = 38 + w. \quad (1)$$

The perimeter is 188 yards, so

$$2l + 2w = 188. \quad (2)$$

Step 4 Substitute $38 + w$ for l in equation (2).

$$2(38 + w) + 2w = 188$$
$$76 + 2w + 2w = 188$$
$$4w + 76 = 188$$
$$4w = 112$$
$$w = 28$$

From (1), $l = 38 + 28 = 66$.

Step 5 The length is 66 yards and the width is 28 yards.

Step 6 66 is 38 more than 28 and the perimeter is $2(66) + 2(28) = 188$ yards.

19. **(a)** $C = 85x + 900$; $R = 105x$; no more than 38 units can be sold.

To find the break-even quantity, let $C = R$.

$$85x + 900 = 105x$$
$$900 = 20x$$
$$x = \frac{900}{20} = 45$$

The break-even quantity is 45 units.

(b) Since no more than 38 units can be sold, do not produce the product (since $38 < 45$). The product will lead to a loss.

21. **Step 2** Let $x =$ the number of \$1 bills and $y =$ the number of \$10 bills.

Complete the table in the text, realizing that the entries in the "total value" column were found by multiplying the denomination of the bill by the number of bills.

Number of Bills	Denomination of Bill (in dollars)	Total Value (in dollars)
x	1	$1x$
y	10	$10y$
74	XXXXXX	326

Step 3 The total number of bills is 74, so

$$x + y = 74. \quad (1)$$

Since the total value is \$326, the third column leads to

$$x + 10y = 326. \quad (2)$$

These two equations give the system:

$$x + y = 74 \quad\quad (1)$$
$$x + 10y = 326 \quad (2)$$

Step 4 To solve this system by the elimination method, multiply equation (1) by -1 and add this result to equation (2).

$$\begin{array}{rrrcr} -x & - & y & = & -74 \\ x & + & 10y & = & 326 \\ \hline & & 9y & = & 252 \\ & & y & = & 28 \end{array}$$

Substitute 28 for y in equation (2).

$$x + 10y = 326$$
$$x + 10(28) = 326$$
$$x + 280 = 326$$
$$x = 46$$

Step 5 The clerk has 46 ones and 28 tens.

Step 6 46 ones and 28 tens give us 74 bills worth $326.

23. ***Step 2*** Let $x =$ the number of *Moneyball* DVDs and $y =$ the number of *The Concert for George* Blu-ray discs.

Type of Gift	Number Bought	Cost of each (in dollars)	Total Value (in $)
DVD	x	16.99	$16.99x$
Blu-ray	y	22.42	$22.42y$
Totals	7	—	129.79

Step 3 From the second and fourth columns of the table, we obtain the system

$$x + y = 7 \qquad (1)$$
$$16.99x + 22.42y = 129.79 \qquad (2)$$

Step 4 Multiply equation (1) by -16.99 and add the result to equation (2).

$$
\begin{array}{rrrrr}
-16.99x & - & 16.99y & = & -118.93 \\
16.99x & + & 22.42y & = & 129.79 \\
\hline
 & & 5.43y & = & 10.86 \\
 & & y & = & 2
\end{array}
$$

From (1) with $y = 2$, we get $x = 5$.

Step 5 She bought 5 DVDs and 2 Blu-ray discs.

Step 6 Five $16.99 DVDs and two $22.42 Blu-ray discs give us 7 gifts worth $129.79.

25. ***Step 2*** Let $x =$ the amount invested at 5% and $y =$ the amount invested at 4%.

Amount Invested (in dollars)	Rate of Interest	Interest for One Year (in dollars)
x	5%, or 0.05	$0.05x$
y	4%, or 0.04	$0.04y$
XXXXXX	XXXXXX	350

Step 3 Maria has invested twice as much at 5% as at 4%, so

$$x = 2y. \qquad (1)$$

Her total interest income is $350, so

$$0.05x + 0.04y = 350. \qquad (2)$$

Step 4 Solve the system by substitution. Substitute $2y$ for x in equation (2).

$$0.05(2y) + 0.04y = 350$$
$$0.14y = 350$$
$$y = \frac{350}{0.14} = 2500$$

Substitute 2500 for y in equation (1).

$$x = 2y = 2(2500) = 5000$$

Step 5 Maria has $5000 invested at 5% and $2500 invested at 4%.

Step 6 $5000 is twice as much as $2500. 5% of $5000 is $250 and 4% of $2500 is $100, which is a total of $350 in interest each year.

27. ***Step 2*** Let $x =$ the average ticket cost for U2 and $y =$ the average ticket cost for Taylor Swift.

Step 3 Six tickets for U2 plus five tickets for Taylor Swift cost $912, so

$$6x + 5y = 912. \qquad (1)$$

Three tickets for U2 plus four tickets for Taylor Swift cost $564, so

$$3x + 4y = 564. \qquad (2)$$

Step 4 Multiply (2) by -2 and add the results.

$$
\begin{array}{rrrrr}
6x & + & 5y & = & 912 \\
-6x & - & 8y & = & -1128 \\
\hline
 & & -3y & = & -216 \\
 & & y & = & 72
\end{array}
$$

Substitute $y = 72$ in (1).

$$6x + 5(72) = 912$$
$$6x + 360 = 912$$
$$6x = 552$$
$$x = 92$$

Step 5 The average ticket cost for U2 is $92 and the average ticket cost for Taylor Swift is $72.

Step 6 Six tickets (on average) for U2 cost $552 and five tickets for Taylor Swift cost $360; a sum of $912. Three tickets for U2 cost $276 and four tickets for Taylor Swift cost $288; a sum of $564.

29. ***Step 2*** Let $x =$ the amount of 40% solution and $y =$ the amount of 70% solution.

Liters of Solution	Percent (as a decimal)	Liters of Pure Dye
x	0.40	$0.40x$
y	0.70	$0.70y$
120	0.50	$0.50(120) = 60$

Step 3 The total number of liters in the final mixture is 120, so

$$x + y = 120. \qquad (1)$$

The amount of pure dye in the 40% solution added to the amount of pure dye in the 70% solution is equal to the amount of pure dye in the 50% mixture, so

$$0.40x + 0.70y = 60. \qquad (2)$$

Multiply equation (2) by 10 to clear decimals.

$$4x + 7y = 600 \qquad (3)$$

We now have the system

$$x + y = 120 \qquad (1)$$
$$4x + 7y = 600. \qquad (3)$$

Step 4 Solve this system by the elimination method.

$$\begin{array}{rl} -4x - 4y = -480 & \textit{Multiply (1) by } -4. \\ \underline{4x + 7y = 600} & \\ 3y = 120 & \\ y = 40 & \end{array}$$

From (1) with $y = 40$, $x = 80$.

Step 5 80 liters of 40% solution should be mixed with 40 liters of 70% solution.

Step 6 Since $80 + 40 = 120$ and $0.40(80) + 0.70(40) = 60$, this mixture will give the 120 liters of 50 percent solution, as required in the original problem.

31. ***Step 2*** Let $x =$ the number of pounds of coffee worth $6 per pound and $y =$ the number of pounds of coffee worth $3 per pound. Complete the table in the text.

Number of Pounds	Dollars per Pound	Cost (in dollars)
x	6	$6x$
y	3	$3y$
90	4	$90(4) = 360$

Step 3 The mixture contains 90 pounds, so

$$x + y = 90. \quad (1)$$

The cost of the mixture is $360, so

$$6x + 3y = 360. \quad (2)$$

Equation (2) may be simplified by dividing each side by 3.

$$2x + y = 120 \quad (3)$$

We now have the system

$$x + y = 90 \qquad (1)$$
$$2x + y = 120. \qquad (3)$$

Step 4 To solve this system by the elimination method, multiply equation (1) by -1 and add the result to equation (3).

$$\begin{array}{rcrcr} -x & - & y & = & -90 \\ \underline{2x} & + & y & = & 120 \\ x & & & = & 30 \end{array}$$

From (1) with $x = 30$, $y = 60$.

Step 5 Ahmad will need to mix 30 pounds of coffee at $6 per pound with 60 pounds at $3 per pound.

Step 6 Since $30 + 60 = 90$ and $\$6(30) + \$3(60) = \$360$, this mixture will give the 90 pounds worth $4 per pound, as required in the original problem.

33. ***Step 2*** Let $x =$ the number of pounds of nuts selling for $6 per pound and $y =$ the number of pounds of raisins selling for $3 per pound. Complete a table.

Number of Pounds	Dollars per Pound	Cost (in dollars)
x	6	$6x$
y	3	$3y$
60	5	$60(5) = 300$

Step 3 The mixture contains 60 pounds, so

$$x + y = 60. \quad (1)$$

The cost of the mixture is $300, so

$$6x + 3y = 300. \quad (2)$$

Equation (2) may be simplified by dividing each side by 3.

$$2x + y = 100 \quad (3)$$

We now have the system

$$x + y = 60 \qquad (1)$$
$$2x + y = 100. \qquad (3)$$

Step 4 To solve this system by the elimination method, multiply equation (1) by -1 and add the result to equation (3).

$$\begin{array}{rcrcr} -x & - & y & = & -60 \\ \underline{2x} & + & y & = & 100 \\ x & & & = & 40 \end{array}$$

From (1), $y = 20$.

Step 5 Kelli will need to mix 40 pounds of nuts at $6 per pound with 20 pounds of raisins at $3 per pound.

Step 6 Since $40 + 20 = 60$ and $\$6(40) + \$3(20) = \$300$, this mixture will give the 60 pounds worth $5 per pound, as required in the original problem.

35. ***Step 2*** Let $x =$ the average rate of the slower train and $y =$ the average rate of the faster train.

	r	t	d
Slower train	x	4.5	$4.5x$
Faster train	y	4.5	$4.5y$

Step 3 The total distance is 495 miles, so

$$4.5x + 4.5y = 495 \quad (1)$$

or, dividing equation (1) by 4.5,

$$x + y = 110. \quad (2)$$

The faster train traveled 10 mph faster than the slower train, so

$$y = x + 10. \quad (3)$$

Step 4 From (3), substitute $x + 10$ for y in (2).

$$x + (x + 10) = 110$$
$$2x + 10 = 110$$
$$2x = 100$$
$$x = 50$$

From (3) with $x = 50$, $y = 50 + 10 = 60$.

Step 5 The average rate of the slower train was 50 mph and the average rate of the faster train was 60 mph.

Step 6 60 mph is 10 mph faster than 50 mph. The slower train traveled $4.5(50) = 225$ miles and the faster train traveled $4.5(60) = 270$ miles. The sum is $225 + 270 = 495$, as required.

37. Step 2 Let $x =$ the average rate of the bicycle and $y =$ the average rate of the car.

	r	t	d
Bicycle	x	7.5	$7.5x$
Car	y	7.5	$7.5y$

Step 3 The total distance is 471 miles, so

$$7.5x + 7.5y = 471 \quad (1)$$

or, dividing equation (1) by 7.5,

$$x + y = 62.8. \quad (2)$$

The car traveled 35.8 mph faster than the bicycle, so

$$y = x + 35.8. \quad (3)$$

Step 4 From (3), substitute $x + 35.8$ for y in (2).

$$x + (x + 35.8) = 62.8$$
$$2x + 35.8 = 62.8$$
$$2x = 27.0$$
$$x = 13.5$$

From (3) with $x = 13.5$, $y = 13.5 + 35.8 = 49.3$.

Step 5 The average rate of the bicycle was 13.5 mph and the average rate of the car was 49.3 mph.

Step 6 49.3 mph is 35.8 mph faster than 13.5 mph. The bicycle traveled $7.5(13.5) = 101.25$ miles and the car traveled $7.5(49.3) = 369.75$ miles. The sum is $101.25 + 369.75 = 471$, as required.

39. Step 2 Let $x =$ the average rate of the car leaving Cincinnati and $y =$ the average rate of the car leaving Toledo.

Convert the total time of 1 hour and 36 minutes to hours.

$$1 + \frac{36}{60} = \frac{96}{60} = 1.6$$

	r	t	d
Cincinnati to Toledo	x	1.6	$1.6x$
Toledo to Cincinnati	y	1.6	$1.6y$

Step 3 The total distance is 200 miles, so

$$1.6x + 1.6y = 200 \quad (1)$$

or, dividing equation (1) by 1.6,

$$x + y = 125. \quad (2)$$

The car leaving Toledo averages 15 miles per hour more than the other car, so

$$y = x + 15. \quad (3)$$

Step 4 From (3), substitute $x + 15$ for y in (2).

$$x + (x + 15) = 125$$
$$2x = 110$$
$$x = 55$$

From (3) with $x = 55$, $y = 70$.

Step 5 The average rate of the car leaving Cincinnati was 55 miles per hour and the average rate of the car leaving Toledo was 70 miles per hour.

Step 6 70 miles per hour is 15 miles per hour more than 55 miles per hour. The car leaving Cincinnati travels $1.6(55) = 88$ miles and the car leaving Toledo travels $1.6(70) = 112$ miles. The total distance traveled is $88 + 112 = 200$, as required.

41. Step 2 Let $x =$ Roberto's rate and $y =$ Juana's rate.

Step 3 Use the formula $d = rt$ to complete two tables.

Riding in same direction

	r	t	d
Roberto	x	6	$6x$
Juana	y	6	$6y$

Roberto rode 30 miles farther than Juana, so

$$6x = 6y + 30$$
$$\text{or} \quad x - y = 5. \quad (1)$$

Riding toward each other

	r	t	d
Roberto	x	1	$1x$
Juana	y	1	$1y$

Roberto and Juana rode a total of 30 miles, so

$$x + y = 30 \quad (2)$$

We have the system

$$x - y = 5 \quad (1)$$
$$x + y = 30. \quad (2)$$

Step 4 To solve the system by the elimination method, add equations (1) and (2).

$$
\begin{array}{rcrcr}
x & - & y & = & 5 \\
x & + & y & = & 30 \\
\hline
2x & & & = & 35 \\
& & x & = & 17.5
\end{array}
$$

From (2) with $x = 17.5$, $y = 12.5$.

Step 5 Roberto's rate is 17.5 miles per hour and Juana's rate is 12.5 miles per hour.

Step 6 Riding in the same direction, Roberto rides $6(17.5) = 105$ miles and Juana rides $6(12.5) = 75$ miles. The difference is 30 miles, as required.

Riding toward each other, Roberto rides $1(17.5) = 17.5$ miles and Juana rides $1(12.5) = 12.5$ miles. The sum is 30 miles, as required.

43. **Step 2** Let $x =$ the rate of the boat in still water and $y =$ the rate of the current.

	r	t	d
Downstream	$x + y$	3	36
Upstream	$x - y$	3	24

Step 3 Use the formula $d = rt$ and the completed table to write the system of equations.

$$(x + y)(3) = 36 \quad (1) \quad \textit{Distance downstream}$$

$$(x - y)(3) = 24 \quad (2) \quad \textit{Distance upstream}$$

Step 4 Divide equations (1) and (2) by 3.

$$x + y = 12 \quad (3)$$
$$x - y = 8 \quad (4)$$

Now add equations (3) and (4).

$$
\begin{array}{rcrcrl}
x & + & y & = & 12 & (3) \\
x & - & y & = & 8 & (4) \\
\hline
2x & & & = & 20 & \textit{Add (3) and (4).} \\
& & x & = & 10 &
\end{array}
$$

From (3) with $x = 10$, $y = 2$.

Step 5 The rate of the current is 2 miles per hour; the rate of the boat in still water is 10 miles per hour.

Step 6 Traveling downstream for 3 hours at $10 + 2 = 12$ miles per hour gives us a 36-mile trip. Traveling upstream for 3 hours at $10 - 2 = 8$ miles per hour gives us a 24-mile trip, as required.

45. **Step 2** Let $x =$ the rate of the plane in still air and $y =$ the rate of the wind.

Step 3 The rate of the plane with the wind is $x + y$, so

$$x + y = 500. \quad (1)$$

The rate of the plane into the wind is $x - y$, so

$$x - y = 440. \quad (2)$$

Step 4 To solve the system by the elimination method, add equations (1) and (2).

$$
\begin{array}{rcrcr}
x & + & y & = & 500 \\
x & - & y & = & 440 \\
\hline
2x & & & = & 940 \\
& & x & = & 470
\end{array}
$$

From (1) with $x = 470$, $y = 30$.

Step 5 The rate of the wind is 30 miles per hour; the rate of the plane in still air is 470 miles per hour.

Step 6 The plane travels $470 + 30 = 500$ miles per hour with the wind and $470 - 30 = 440$ miles per hour into the wind, as required.

12.5 Solving Systems of Linear Inequalities

12.5 Margin Exercises

1. **(a)** $x - 2y \leq 8$
$\qquad 3x + y \geq 6$

The graph of $x - 2y = 8$ has intercepts $(8, 0)$ and $(0, -4)$. The graph of $3x + y = 6$ has intercepts $(2, 0)$ and $(0, 6)$. Both are graphed as solid lines because of the $\leq$ and $\geq$ signs. Use $(0, 0)$ as a test point in each case.

$$
\begin{aligned}
x - 2y &\leq 8 \\
0 - 2(0) &\leq 8 \quad \textit{Let x = 0, y = 0.} \\
0 &\leq 8 \quad \textit{True}
\end{aligned}
$$

Shade the side of the graph for $x - 2y = 8$ that contains $(0, 0)$.

$$
\begin{aligned}
3x + y &\geq 6 \\
3(0) + 0 &\geq 6 \quad \textit{Let x = 0, y = 0.} \\
0 &\geq 6 \quad \textit{False}
\end{aligned}
$$

Shade the side of the graph for $3x + y = 6$ that does not contain $(0, 0)$. The graph of the solution set is the overlap of the two shaded regions and portions of the two boundary lines that bound this region.

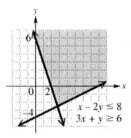

(b) $4x - 2y \leq 8$
$x + 3y \geq 3$

The graph of $4x - 2y = 8$ has intercepts $(2, 0)$ and $(0, -4)$. The graph of $x + 3y = 3$ has intercepts $(3, 0)$ and $(0, 1)$. Both are graphed as solid lines because of the $\leq$ and $\geq$ signs. Use $(0, 0)$ as a test point in each case.

$$4x - 2y \leq 8$$
$$4(0) - 2(0) \leq 8 \quad \text{Let } x = 0, \ y = 0.$$
$$0 \leq 8 \quad \text{True}$$

Shade the side of the graph for $4x - 2y = 8$ that contains $(0, 0)$.

$$x + 3y \geq 3$$
$$0 + 3(0) \geq 3 \quad \text{Let } x = 0, \ y = 0.$$
$$0 \geq 3 \quad \text{False}$$

Shade the side of the graph for $x + 3y = 3$ that does not contain $(0, 0)$.

The graph of the solution set is the intersection (overlap) of the two shaded regions, and includes the portions of the boundary lines that bound this region.

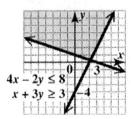

2. $x + 2y < 0$
$3x - 4y < 12$

Graph $x + 2y = 0$ as a dashed line through the points $(0, 0)$ and $(2, -1)$. Use $(-4, 0)$ as a test point.

$$x + 2y < 0$$
$$-4 + 2(0) < 0 \quad \text{Let } x = -4, \ y = 0.$$
$$-4 < 0 \quad \text{True}$$

Shade the side of the graph for $x + 2y = 0$ that contains $(-4, 0)$.

Then graph $3x - 4y = 12$ as a dashed line through its intercepts, $(4, 0)$ and $(0, -3)$. Use $(0, 0)$ as a test point.

$$3x - 4y < 12$$
$$3(0) - 4(0) < 12 \quad \text{Let } x = 0, \ y = 0.$$
$$0 < 12 \quad \text{True}$$

Shade the side of the graph for $3x - 4y = 12$ that contains $(0, 0)$.

The graph of the solution set is the overlap of the two shaded regions. The solution set does not include the boundary lines of this region.

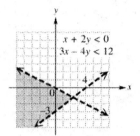

3. $3x + 2y \leq 12$
$x \leq 2$
$y \leq 4$

Graph $3x + 2y = 12$ as a solid line through its intercepts, $(4, 0)$ and $(0, 6)$. Use $(0, 0)$ as a test point.

$$3x + 2y < 12$$
$$3(0) + 2(0) < 12 \quad \text{Let } x = 0, \ y = 0.$$
$$0 < 12 \quad \text{True}$$

Shade the region that includes $(0, 0)$. Recall that $x = 2$ is a vertical line through the point $(2, 0)$, and $y = 4$ is a horizontal line through $(0, 4)$. Shade to the left of $x = 2$ and below $y = 4$. The graph of the solution set is the overlap of the three shaded regions and includes portions of the two boundary lines.

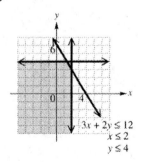

12.5 Section Exercises

1. $x \geq 5$ is the region to the right of the vertical line $x = 5$ and includes the line. $y \leq -3$ is the region below the horizontal line $y = -3$ and includes the line. The correct choice is **C**.

3. $x > 5$ is the region to the right of the vertical line $x = 5$. $y < -3$ is the region below the horizontal line $y = -3$. The correct choice is **B**.

5. $x - 3y \leq 6$

$ x \geq -4$

For $x - 3y \leq 6$: Use $(0, 0)$ as a test point. This results in the true statement $0 \leq 6$, so shade the region containing the origin.

For $x \geq -4$: Use $(0, 0)$ as a test point. This results in the true statement $0 \geq -4$, so shade the region containing the origin.

The solution set of the system is the intersection of the two shaded regions, and includes the portions of the two lines that bound the region.

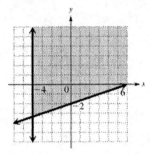

7. $x + y > 4$

$5x - 3y < 15$

For $x + y > 4$: Use $(0, 0)$ as a test point. This results in the false statement $0 > 4$, so shade the region that does *not* contain the origin.

For $5x - 3y < 15$: Use $(0, 0)$ as a test point. This results in the true statement $0 < 15$, so shade the region containing the origin.

The solution set of the system is the intersection of the two shaded regions. It does not include the boundary lines.

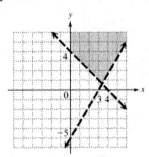

9. $x + y \leq 6$

$x - y \geq 1$

Graph the boundary $x + y = 6$ as a solid line through its intercepts, $(6, 0)$ and $(0, 6)$. Using $(0, 0)$ as a test point will result in the true statement $0 \leq 6$, so shade the region containing the origin.

Graph the boundary $x - y = 1$ as a solid line through its intercepts, $(1, 0)$ and $(0, -1)$. Using $(0, 0)$ as a test point will result in the false

statement $0 \geq 1$, so shade the region *not* containing the origin.

The solution set of this system is the intersection (overlap) of the two shaded regions, and includes the portions of the boundary lines that bound this region.

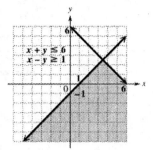

11. $4x + 5y \geq 20$

$x - 2y \leq 5$

Graph the boundary $4x + 5y = 20$ as a solid line through its intercepts, $(0, 4)$ and $(5, 0)$. Using $(0, 0)$ as a test point will result in the false statement $0 \geq 20$, so shade the region *not* containing the origin.

Graph the boundary $x - 2y = 5$ as a solid line through $(5, 0)$ and $(1, -2)$. Using $(0, 0)$ as a test point will result in the true statement $0 \leq 5$, so shade the region containing the origin.

The solution set of this system is the intersection of the two shaded regions, and includes the portions of the boundary lines that bound the region.

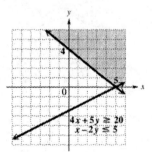

13. $2x + 3y < 6$

$x - y < 5$

Graph $2x + 3y = 6$ as a dashed line through $(3, 0)$ and $(0, 2)$. Using $(0, 0)$ as a test point will result in the true statement $0 < 6$, so shade the region containing the origin.

Now graph $x - y = 5$ as a dashed line through $(5, 0)$ and $(0, -5)$. Using $(0, 0)$ as a test point will result in the true statement $0 < 5$, so shade the region containing the origin.

The solution set of the system is the intersection of the two shaded regions. Because the inequality signs are both $<$, the solution set does not include the boundary lines.

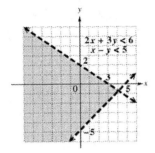

15. $y \leq 2x - 5$
$x < 3y + 2$

Graph $y = 2x - 5$ as a solid line through $(0, -5)$ and $(3, 1)$. Using $(0, 0)$ as a test point will result in the false statement $0 \leq -5$, so shade the region *not* containing the origin.

Now graph $x = 3y + 2$ as a dashed line through $(2, 0)$ and $(-1, -1)$. Using $(0, 0)$ as a test point will result in the true statement $0 < 2$, so shade the region containing the origin.

The solution set of the system is the intersection of the two shaded regions. It includes the portion of the line $y = 2x - 5$ that bounds the region, but not the portion of the line $x = 3y + 2$.

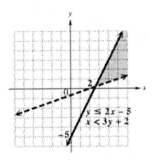

17. $4x + 3y < 6$
$x - 2y > 4$

Graph $4x + 3y = 6$ as a dashed line through $\left(\frac{3}{2}, 0\right)$ and $(0, 2)$. Using $(0, 0)$ as a test point will result in the true statement $0 < 6$, so shade the region containing the origin.

Now graph $x - 2y = 4$ as a dashed line through $(4, 0)$ and $(0, -2)$. Using $(0, 0)$ as a test point will result in the false statement $0 > 4$, so shade the region *not* containing the origin.

The solution set of the system is the intersection of the two shaded regions. It does not include the boundary lines.

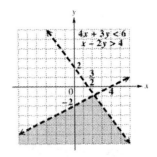

19. $x \leq 2y + 3$
$x + y < 0$

Graph $x = 2y + 3$ as a solid line through $(3, 0)$ and $(5, 1)$. Using $(0, 0)$ as a test point will result in the true statement $0 \leq 3$, so shade the region containing the origin.

Now graph $x + y = 0$ as a dashed line through $(0, 0)$ and $(1, -1)$. Using $(1, 0)$ as a test point will result in the false statement $1 < 0$, so shade the region *not* containing $(1, 0)$.

The solution set of the system is the intersection of the two shaded regions. It includes the portion of the line $x = 2y + 3$ that bounds the region, but not the portion of the line $x + y = 0$.

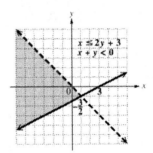

21. $4x + 5y < 8$
$y > -2$
$x > -4$

Graph $4x + 5y = 8$, $y = -2$, and $x = -4$ as dashed lines. All three inequalities are true for $(0, 0)$. Shade the region bounded by the three lines, which contains the test point $(0, 0)$. The solution set is the shaded region. It does not include the boundary lines.

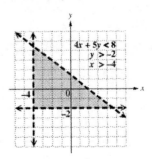

23. $3x - 2y \geq 6$ (1)
$x + y \leq 4$ (2)
$x \geq 0$ (3)
$y \geq -4$ (4)

Graph $3x - 2y = 6$, $x + y = 4$, $x = 0$, and $y = -4$ as solid lines. Inequality (1) is *false* for the test point $(0, 0)$, and the other three inequalities are true for $(0, 0)$. Shade the region bounded by the four lines, which *does not* contain the test point $(0, 0)$. The solution set is the shaded region and includes the portions of the boundary lines that bound this region.

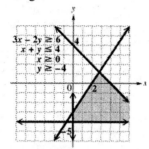

25. There are infinitely many points in any of the shaded regions. There are, however, points that are not in the shaded regions, and thus there are ordered pairs that are not solutions.

Chapter 12 Review Exercises

1. $(3, 4)$ $4x - 2y = 4$
$5x + y = 19$

To decide whether $(3, 4)$ is a solution of the system, substitute 3 for x and 4 for y in each equation.

$$4x - 2y = 4$$
$$4(3) - 2(4) \overset{?}{=} 4$$
$$12 - 8 \overset{?}{=} 4$$
$$4 = 4 \quad \textit{True}$$

$$5x + y = 19$$
$$5(3) + 4 \overset{?}{=} 19$$
$$15 + 4 \overset{?}{=} 19$$
$$19 = 19 \quad \textit{True}$$

Since $(3, 4)$ satisfies both equations, it is a solution of the system.

2. $(-5, 2)$ $x - 4y = -13$ (1)
$2x + 3y = 4$ (2)

Substitute -5 for x and 2 for y in equation (2).

$$2x + 3y = 4$$
$$2(-5) + 3(2) \overset{?}{=} 4$$
$$-10 + 6 \overset{?}{=} 4$$
$$-4 = 4 \quad \textit{False}$$

Since $(-5, 2)$ is not a solution of the second equation, it cannot be a solution of the system.

3. $x + y = 4$
$2x - y = 5$

To graph the equations, find the intercepts.

$x + y = 4$: Let $y = 0$; then $x = 4$.
Let $x = 0$; then $y = 4$.

Plot the intercepts, $(4, 0)$ and $(0, 4)$, and draw the line through them.

$2x - y = 5$: Let $y = 0$; then $x = \frac{5}{2}$.
Let $x = 0$; then $y = -5$.

Plot the intercepts, $(\frac{5}{2}, 0)$ and $(0, -5)$, and draw the line through them.

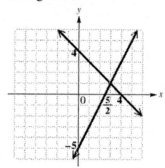

It appears that the lines intersect at the point $(3, 1)$. Check this by substituting 3 for x and 1 for y in both equations. Since $(3, 1)$ satisfies both equations, the solution set of this system is $\{(3, 1)\}$.

4. $x - 2y = 4$
$2x + y = -2$

To graph the equations, find the intercepts.

$x - 2y = 4$: Let $y = 0$; then $x = 4$.
Let $x = 0$; then $y = -2$.

Plot the intercepts, $(4, 0)$ and $(0, -2)$, and draw the line through them.

$2x + y = -2$: Let $y = 0$; then $x = -1$.
Let $x = 0$; then $y = -2$.

Plot the intercepts, $(-1, 0)$ and $(0, -2)$, and draw the line through them.

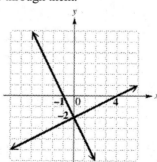

The lines intersect at their common y-intercept, $(0, -2)$, so $\{(0, -2)\}$ is the solution set of the system.

5. $x - 2 = 2y$
         $2x - 4y = 4$

The first equation may be written as $x - 2y = 2$. Find the intercepts for this line.

$x - 2y = 2$: Let $y = 0$; then $x = 2$.
             Let $x = 0$; then $y = -1$.

Draw the line through $(2, 0)$ and $(0, -1)$.

Now find the intercepts for the second line.

$2x - 4y = 4$: Let $y = 0$; then $x = 2$.
             Let $x = 0$; then $y = -1$.

Again, the intercepts are $(2, 0)$ and $(0, -1)$, so the graph is the same line.

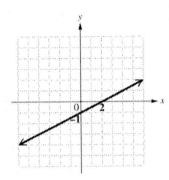

Because the graphs of both equations are the same line, the system has an infinite number of solutions. The solution set is $\{(x, y) \mid x - 2y = 2\}$.

6. $2x + 4 = 2y$
         $y - x = -3$

Graph the line $2x + 4 = 2y$ through its intercepts, $(-2, 0)$ and $(0, 2)$. Graph the line $y - x = -3$ through its intercepts $(3, 0)$ and $(0, -3)$.

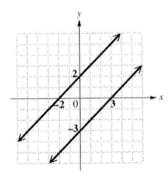

The lines appear to be parallel. By finding the slope of the line through the intercepts, we see that each line has slope 1. Thus, the lines are parallel, and the system has no solution. This is an inconsistent system and the solution set is $\emptyset$.

7. $3x + y = 7$ (1)
         $x = 2y$ (2)

Substitute $2y$ for x in equation (1) and solve the resulting equation for y.

$$3x + y = 7$$
$$3(2y) + y = 7$$
$$6y + y = 7$$
$$7y = 7$$
$$y = 1$$

To find x, let $y = 1$ in equation (2).

$$x = 2y = 2(1) = 2$$

The solution set is $\{(2, 1)\}$.

8. $2x - 5y = -19$ (1)
         $y = x + 2$ (2)

Substitute $x + 2$ for y in equation (1).

$$2x - 5y = -19$$
$$2x - 5(x + 2) = -19$$
$$2x - 5x - 10 = -19$$
$$-3x - 10 = -19$$
$$-3x = -9$$
$$x = 3$$

To find y, let $x = 3$ in equation (2).

$$y = x + 2 = 3 + 2 = 5$$

The solution set is $\{(3, 5)\}$.

9. $4x + 5y = 44$ (1)
         $x + 2 = 2y$ (2)

Solve equation (2) for x.

$$x + 2 = 2y$$
$$x = 2y - 2 \quad (3)$$

Substitute $2y - 2$ for x in equation (1).

$$4x + 5y = 44$$
$$4(2y - 2) + 5y = 44$$
$$8y - 8 + 5y = 44$$
$$13y - 8 = 44$$
$$13y = 52$$
$$y = 4$$

To find x, let $y = 4$ in equation (3).

$$x = 2y - 2 = 2(4) - 2 = 8 - 2 = 6$$

The solution set is $\{(6, 4)\}$.

10.
$$5x + 15y = 3 \quad (1)$$
$$x + 3y = 2 \quad (2)$$

Solve equation (2) for x.

$$x + 3y = 2$$
$$x = 2 - 3y \quad (3)$$

Substitute $2 - 3y$ for x in equation (1).

$$5x + 15y = 3$$
$$5(2 - 3y) + 15y = 3$$
$$10 - 15y + 15y = 3$$
$$10 = 3 \quad \textit{False}$$

This false statement means that the system has no solution. The equations of the system represent parallel lines. The solution set is $\emptyset$.

11.
$$\begin{array}{rcll} 2x & - & y & = & 13 & (1) \\ x & + & y & = & 8 & (2) \\ \hline 3x & & & = & 21 & \textit{Add } (1) \textit{ and } (2). \\ & & x & = & 7 \end{array}$$

From (2) with $x = 7$, $y = 1$.

The solution set is $\{(7, 1)\}$.

12.
$$3x - y = -13 \quad (1)$$
$$x - 2y = -1 \quad (2)$$

Multiply equation (2) by -3 and add the result to equation (1).

$$\begin{array}{rcll} 3x & - & y & = & -13 & (1) \\ -3x & + & 6y & = & 3 & (3) \\ \hline & & 5y & = & -10 & \textit{Add } (1) \textit{ and } (3). \\ & & y & = & -2 \end{array}$$

To find x, let $y = -2$ in equation (2).

$$x - 2y = -1$$
$$x - 2(-2) = -1 \quad \textit{Let } y = -2.$$
$$x + 4 = -1$$
$$x = -5$$

The solution set is $\{(-5, -2)\}$.

13.
$$-4x + 3y = 25 \quad (1)$$
$$6x - 5y = -39 \quad (2)$$

Multiply equation (1) by 3 and equation (2) by 2; then add the results.

$$\begin{array}{rcll} -12x & + & 9y & = & 75 \\ 12x & - & 10y & = & -78 \\ \hline & & -y & = & -3 \\ & & y & = & 3 \end{array}$$

To find x, let $y = 3$ in equation (1).

$$-4x + 3y = 25$$
$$-4x + 3(3) = 25$$
$$-4x + 9 = 25$$
$$-4x = 16$$
$$x = -4$$

The solution set is $\{(-4, 3)\}$.

14.
$$3x - 4y = 9 \quad (1)$$
$$6x - 8y = 18 \quad (2)$$

Multiply equation (1) by -2 and add the result to equation (2).

$$\begin{array}{rcll} -6x & + & 8y & = & -18 \\ 6x & - & 8y & = & 18 \\ \hline & & 0 & = & 0 & \textit{True} \end{array}$$

This result indicates that all solutions of equation (1) are also solutions of equation (2). The given system has an infinite number of solutions. The solution set is $\{(x, y) \mid 3x - 4y = 9\}$.

15.
$$x - 2y = 5 \quad (1)$$
$$y = x - 7 \quad (2)$$

From (2), substitute $x - 7$ for y in equation (1).

$$x - 2y = 5$$
$$x - 2(x - 7) = 5$$
$$x - 2x + 14 = 5$$
$$-x = -9$$
$$x = 9$$

Let $x = 9$ in equation (2) to find y.

$$y = 9 - 7 = 2$$

The solution set is $\{(9, 2)\}$.

16.
$$5x - 3y = 11 \quad (1)$$
$$2y = x - 4 \quad (2)$$

Solve for x in equation (2).

$$2y = x - 4$$
$$2y + 4 = x \quad (3)$$

Substitute $2y + 4$ for x in equation (1) and then solve for y.

$$5(2y + 4) - 3y = 11$$
$$10y + 20 - 3y = 11$$
$$7y + 20 = 11$$
$$7y = -9$$
$$y = -\tfrac{9}{7}$$

To find x, let $y = -\tfrac{9}{7}$ in equation (3).

$$x = 2\left(-\tfrac{9}{7}\right) + 4 = -\tfrac{18}{7} + \tfrac{28}{7} = \tfrac{10}{7}$$

The solution set is $\left\{\left(\tfrac{10}{7}, -\tfrac{9}{7}\right)\right\}$.

17.
$$\frac{x}{2} + \frac{y}{3} = 7 \quad (1)$$
$$\frac{x}{4} + \frac{2y}{3} = 8 \quad (2)$$

Multiply equation (1) by 6 to clear fractions.

$$6\left(\frac{x}{2} + \frac{y}{3}\right) = 6(7)$$
$$3x + 2y = 42 \qquad (3)$$

Multiply equation (2) by 12 to clear fractions.

$$12\left(\frac{x}{4} + \frac{2y}{3}\right) = 12(8)$$
$$3x + 8y = 96 \qquad (4)$$

To solve this system by the elimination method, multiply equation (3) by -1 and add the result to equation (4).

$$
\begin{array}{rcr}
-3x - 2y &=& -42 \\
3x + 8y &=& 96 \\
\hline
6y &=& 54 \\
y &=& 9
\end{array}
$$

To find x, let $y = 9$ in equation (3).

$$3x + 2y = 42$$
$$3x + 2(9) = 42$$
$$3x + 18 = 42$$
$$3x = 24$$
$$x = 8$$

The solution set is $\{(8, 9)\}$.

18. $\dfrac{3x}{4} - \dfrac{y}{3} = \dfrac{7}{6}$ (1)

$\dfrac{x}{2} + \dfrac{2y}{3} = \dfrac{5}{3}$ (2)

Multiply equation (1) by 12 to clear fractions.

$$12\left(\frac{3x}{4}\right) - 12\left(\frac{y}{3}\right) = 12\left(\frac{7}{6}\right)$$
$$9x - 4y = 14 \qquad (3)$$

Multiply equation (2) by 6 to clear fractions.

$$6\left(\frac{x}{2}\right) + 6\left(\frac{2y}{3}\right) = 6\left(\frac{5}{3}\right)$$
$$3x + 4y = 10 \qquad (4)$$

Add equations (3) and (4) to eliminate y.

$$
\begin{array}{rcr}
9x - 4y &=& 14 \\
3x + 4y &=& 10 \\
\hline
12x &=& 24 \\
x &=& 2
\end{array}
$$

To find y, let $x = 2$ in equation (4).

$$3(2) + 4y = 10$$
$$6 + 4y = 10$$
$$4y = 4$$
$$y = 1$$

The solution set is $\{(2, 1)\}$.

19. $0.2x + 1.2y = -1$ (1)
$0.1x + 0.3y = 0.1$ (2)

First, clear all decimals.

Equation (1):

$$10(0.2x + 1.2y) = 10(-1) \quad \textit{Multiply by 10.}$$
$$10(0.2x) + 10(1.2y) = 10(-1) \quad \textit{Dist. property}$$
$$2x + 12y = -10 \qquad (3)$$

Equation (2):

$$10(0.1x + 0.3y) = 10(0.1) \quad \textit{Multiply by 10.}$$
$$10(0.1x) + 10(0.3y) = 10(0.1) \quad \textit{Dist. property}$$
$$x + 3y = 1 \qquad (4)$$

The system has been simplified to

$$2x + 12y = -10 \qquad (3)$$
$$x + 3y = 1 \qquad (4)$$

Solve this system by the substitution method.

$$x = 1 - 3y \quad (5) \quad \textit{Solve (4) for x.}$$
$$2(1 - 3y) + 12y = -10 \quad \textit{Substitute for x in (3).}$$
$$2 - 6y + 12y = -10$$
$$2 + 6y = -10$$
$$6y = -12$$
$$y = -2$$

To find x, let $y = -2$ in equation (5).

$$x = 1 - 3(-2) = 1 + 6 = 7$$

The solution set is $\{(7, -2)\}$.

20. $0.1x + y = 1.6$ (1)
$0.6x + 0.5y = -1.4$ (2)

Multiply equation (1) and equation (2) by 10 to eliminate the decimals. Then solve the resulting system of equations by the elimination method.

$$x + 10y = 16 \quad (3)$$
$$6x + 5y = -14 \quad (4)$$

Multiply equation (1) by -6.

$$
\begin{array}{rcrl}
-6x - 60y &=& -96 & (5) \\
6x + 5y &=& -14 & (4) \\
\hline
-55y &=& -110 & \textit{Add.} \\
y &=& 2 &
\end{array}
$$

Substitute 2 for y in equation (4).

$$6x + 5y = -14 \quad (4)$$
$$6x + 5(2) = -14 \quad \textit{Let y = 2.}$$
$$6x + 10 = -14$$
$$6x = -24$$
$$x = -4 \quad \textit{Divide by 6.}$$

The solution is $\{(-4, 2)\}$. Check in both of the original equations.

21. *Step 2* Let x = the number of Domino's restaurants and y = the number of Pizza Hut restaurants.

Step 3 Pizza Hut had 2613 more locations than Domino's, so

$$y = x + 2613. \quad (1)$$

Together, the two chains had 12,471 locations, so

$$x + y = 12{,}471. \quad (2)$$

Step 4 Substitute $x + 2613$ for y in (2).

$$x + (x + 2613) = 12{,}471$$
$$2x + 2613 = 12{,}471$$
$$2x = 9858$$
$$x = 4929$$

From (1), $y = 4929 + 2613 = 7542$.

Step 5 In 2010, Pizza Hut had 7542 locations and Domino's had 4929 locations.

Step 6 7542 is 2613 more than 4929 and the sum of 7542 and 4929 is 12,471.

22. *Step 2* Let x = the circulation figure for *Reader's Digest* and y = the circulation figure for *People*.

Step 3 The total circulation was 9.3 million, so

$$x + y = 9.3. \quad (1)$$

People's circulation was 2.1 million less than that of *Reader's Digest*, so

$$y = x - 2.1. \quad (2)$$

Step 4 Substitute $x - 2.1$ for y in (1).

$$x + (x - 2.1) = 9.3$$
$$2x - 2.1 = 9.3$$
$$2x = 11.4$$
$$x = 5.7$$

From (2), $y = 5.7 - 2.1 = 3.6$.

Step 5 The average circulation for *Reader's Digest* was 5.7 million and for *People* it was 3.6 million.

Step 6 3.6 is 2.1 less than 5.7 and the sum of 3.6 and 5.7 is 9.3.

23. *Step 2* Let x = the number of pounds of $1.30 per pound candy and y = the number of pounds of $0.90 per pound candy.

Number of Pounds	Cost per Pound (in dollars)	Total Value (in dollars)
x	1.30	$1.30x$
y	0.90	$0.90y$
100	1.00	$100(1) = 100$

Step 3 From the first and third columns of the table, we obtain the system

$$x + y = 100 \quad (1)$$
$$1.30x + 0.90y = 100. \quad (2)$$

Step 4 Multiply equation (2) by 10 (to clear decimals) and equation (1) by -9.

$$
\begin{array}{rrrr}
-9x & - & 9y & = & -900 \\
13x & + & 9y & = & 1000 \\
\hline
4x & & & = & 100 \\
& & x & = & 25
\end{array}
$$

From (1), $y = 100 - 25 = 75$.

Step 5 25 pounds of candy at $1.30 per pound and 75 pounds of candy at $0.90 per pound should be used.

Step 6 The value of the mixture is $25(1.30) + 75(0.90) = 100$, giving us 100 pounds of candy that can sell for $1 per pound.

24. *Step 2* Let x = the number of $10 bills and y = the number of $20 bills.

Number of Bills	Denomination of Bill (in dollars)	Total Value (in dollars)
x	10	$10x$
y	20	$20y$
20	XXXXXX	330

Step 3 The total number of bills is 20, so

$$x + y = 20. \quad (1)$$

The total value of the money is $330, so

$$10x + 20y = 330. \quad (2)$$

We may simplify equation (2) by dividing each side by 10.

$$x + 2y = 33 \quad (3)$$

Step 4 To solve this equation by the elimination method, multiply equation (1) by -1 and add the result to equation (3).

$$
\begin{array}{rrrr}
-x & - & y & = & -20 \\
x & + & 2y & = & 33 \\
\hline
& & y & = & 13
\end{array}
$$

From (1), $x = 20 - 13 = 7$.

Step 5 The cashier has 13 twenties and 7 tens.

Step 6 There are $13 + 7 = 20$ bills worth $13(\$20) + 7(\$10) = \$330$.

25. *Step 2* Let l = the length of the rectangle and w = the width of the rectangle.

Step 3 The perimeter is 90 meters, so

$$2l + 2w = 90. \quad (1)$$

The length is $1\frac{1}{2}$ (or $\frac{3}{2}$) times the width, so

$$l = \frac{3}{2}w. \quad (2)$$

Step 4 Substitute $\frac{3}{2}w$ for l in equation (1).

$$2l + 2w = 90$$
$$2(\tfrac{3}{2}w) + 2w = 90$$
$$3w + 2w = 90$$
$$5w = 90$$
$$w = 18$$

From (2), $l = \frac{3}{2}(18) = 27$.

Step 5 The length is 27 meters and the width is 18 meters.

Step 6 27 is $1\frac{1}{2}$ times 18 and the perimeter is $2(27) + 2(18) = 90$ meters.

26. ***Step 2*** Let x = the rate of the plane in still air and y = the rate of the wind.

	r	t	d
With Wind	$x + y$	2	540
Against Wind	$x - y$	3	690

Step 3 Use the formula $d = rt$.

$540 = (x + y)(2)$	*With wind*
$270 = x + y$ (1)	*Divide by 2.*
$690 = (x - y)(3)$	*Against wind*
$230 = x - y$ (2)	*Divide by 3.*

Step 4 Solve the system of equations (1) and (2) by the elimination method.

$$270 = x + y \quad (1)$$
$$230 = x - y \quad (2)$$
$$\overline{500 = 2x} \qquad \textit{Add (1) and (2).}$$
$$250 = x$$

From (1) with $x = 250$, $y = 20$.

Step 5 The rate of the plane in still air is 250 miles per hour and the rate of the wind is 20 miles per hour.

Step 6 The plane travels $(250 + 20)2 = 540$ miles per hour with the wind and $(250 - 20)3 = 690$ miles per hour into the wind, as required.

27. ***Step 2*** Let x = the amount invested at 3% and y = the amount invested at 4%.

Amount Invested (in dollars)	Percent (as a decimal)	Interest (in dollars)
x	0.03	$0.03x$
y	0.04	$0.04y$
18,000	XXXXXX	650

Step 3 From the chart, we obtain the equations

$$x + y = 18{,}000 \qquad (1)$$
$$0.03x + 0.04y = 650. \qquad (2)$$

Step 4 To clear decimals, multiply each side of equation (2) by 100.

$$100(0.03x + 0.04y) = 100(650)$$
$$3x + 4y = 65{,}000 \qquad (3)$$

Multiply equation (1) by -3 and add the result to equation (3).

$$-3x \quad - \quad 3y \quad = \quad -54{,}000$$
$$\underline{3x \quad + \quad 4y \quad = \quad 65{,}000}$$
$$y \quad = \quad 11{,}000$$

From (1) with $y = 11{,}000$, $x = 7000$.

Step 5 She invested $7000 at 3% and $11,000 at 4%.

Step 6 The sum of $7000 and $11,000 is $18,000. 3% of $7000 is $210 and 4% of $11,000 is $440. This gives us $210 + \$440 = \$650$ in interest, as required.

28. ***Step 2*** Let x = the number of liters of 40% solution and y = the number of liters of 70% solution.

Number of Liters	Percent (as a decimal)	Amount of Pure Antifreeze
x	0.40	$0.40x$
y	0.70	$0.70y$
90	0.50	$0.50(90) = 45$

Step 3 From the first and third columns of the table, we obtain the equations

$$x + y = 90 \qquad (1)$$
$$0.40x + 0.70y = 45. \qquad (2)$$

To clear decimals, multiply both sides of equation (2) by 10.

$$4x + 7y = 450 \quad (3)$$

Step 4 To solve this system by the elimination method, multiply equation (1) by -4 and add the result to equation (3).

$$-4x - 4y = -360 \quad (4)$$
$$\underline{4x + 7y = 450 \quad (3)}$$
$$3y = 90 \quad \textit{Add (4) and (3).}$$
$$y = 30$$

To find x, let $y = 30$ in equation (1).

$$x = 90 - y = 90 - 30 = 60$$

Step 5 To get 90 liters of a 50% antifreeze solution, 60 liters of a 40% solution and 30 liters of a 70% solution will be needed.

Step 6 Since $60 + 30 = 90$ and $0.40(60) + 0.70(30) = 45$, this mixture will give the 90 liters of 50 percent solution, as required in the original problem.

29. The shaded region is to the left of the vertical line $x = 3$, above the horizontal line $y = 1$, and includes both lines. This is the graph of the system
$$x \le 3$$
$$y \ge 1,$$
so the answer is **B**.

30. It is impossible for the sum of any two numbers to be both greater than 4 and less than 3. Therefore, system **B** has no solution.

31. $x + y \ge 2$
    $x - y \le 4$

Graph $x + y = 2$ as a solid line through its intercepts, $(2, 0)$ and $(0, 2)$. Using $(0, 0)$ as a test point will result in the false statement $0 \ge 2$, so shade the region *not* containing the origin.

Graph $x - y = 4$ as a solid line through its intercepts, $(4, 0)$ and $(0, -4)$. Using $(0, 0)$ as a test point will result in the true statement $0 \le 4$, so shade the region containing the origin.

The solution set of this system is the intersection of the two shaded regions, and includes the portions of the two lines that bound this region.

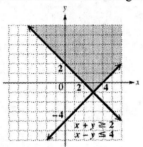

32. $y \ge 2x$
    $2x + 3y \le 6$

Graph $y = 2x$ as a solid line through $(0, 0)$ and $(1, 2)$. This line goes through the origin, so a different test point must be used. Choosing $(-4, 0)$ as a test point will result in the true statement $0 \ge -8$, so shade the region containing $(-4, 0)$.

Graph $2x + 3y = 6$ as a solid line through its intercepts, $(3, 0)$ and $(0, 2)$. Choosing $(0, 0)$ as a test point will result in the true statement $0 \le 6$, so shade the region containing the origin.

The solution set of this system is the intersection of the two shaded regions, and includes the portions of the two lines that bound this region.

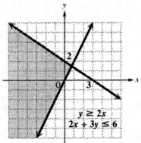

33. $x + y < 3$
    $2x > y$

Graph $x + y = 3$ as a dashed line through $(3, 0)$ and $(0, 3)$. Using $(0, 0)$ as a test point will result in the true statement $0 < 3$, so shade the region containing the origin.

Graph $2x = y$ as a dashed line through $(0, 0)$ and $(1, 2)$. Choosing $(0, -3)$ as a test point will result in the true statement $0 \ge -3$, so shade the region containing $(0, -3)$. The solution set of this system is the intersection of the two shaded regions. It does not contain the boundary lines.

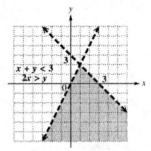

34. **[12.3]** $3x + 4y = 6$ (1)
        $4x - 5y = 8$ (2)

To solve this system by the elimination method, multiply equation (1) by -4 and equation (2) by 3; then add the results.

$$
\begin{array}{rrrcr}
-12x & - & 16y & = & -24 \\
12x & - & 15y & = & 24 \\
\hline
 & & -31y & = & 0 \\
 & & y & = & 0
\end{array}
$$

To find x, substitute 0 for y in equation (1).
$$3x + 4(0) = 6$$
$$3x = 6$$
$$x = 2$$

The solution set is $\{(2, 0)\}$.

35. **[12.2]** $\dfrac{3x}{2} + \dfrac{y}{5} = -3$ (1)

        $4x + \dfrac{y}{3} = -11$ (2)

To clear fractions, multiply each side of equation (1) by 10 and each side of equation (2) by 3.

$$10\left(\frac{3x}{2} + \frac{y}{5}\right) = 10(-3)$$
$$15x + 2y = -30 \quad (3)$$

$$3\left(4x + \frac{y}{3}\right) = 3(-11)$$
$$12x + y = -33 \quad (4)$$

Solve equation (4) for y.
$$y = -33 - 12x \quad (5)$$

Substitute $-33 - 12x$ for y in equation (3) and solve for x.

$$15x + 2y = -30$$
$$15x + 2(-33 - 12x) = -30$$
$$15x - 66 - 24x = -30$$
$$-9x = 36$$
$$x = -4$$

To find y, let $x = -4$ in equation (5).

$$y = -33 - 12(-4)$$
$$= -33 + 48 = 15$$

The solution set is $\{(-4, 15)\}$.

36. **[12.3]** $x + 6y = 3$ (1)
              $2x + 12y = 2$ (2)

To solve this system by the elimination method, multiply equation (1) by -2 and add the result to equation (2).

$$-2x - 12y = -6 \quad (3)$$
$$\underline{2x + 12y = 2 \quad (2)}$$
$$0 = -4 \quad \text{Add (3) and (2).}$$

The last statement, $0 = -4$, is false, which indicates that the system has no solution. The solution set is $\emptyset$.

37. **[12.5]** $x + y < 5$
              $x - y \geq 2$

Graph $x + y = 5$ as a dashed line through its intercepts, $(5, 0)$ and $(0, 5)$. Using $(0, 0)$ as a test point will result in the true statement $0 < 5$, so shade the region containing the origin.

Graph $x - y = 2$ as a solid line through its intercepts, $(2, 0)$ and $(0, -2)$. Using $(0, 0)$ as a test point will result in the false statement $0 \geq 2$, so shade the region *not* containing the origin.

The solution set of this system is the intersection of the two shaded regions. It includes the portion of the line $x - y = 2$ that bounds this region, but not the line $x + y = 5$.

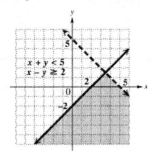

38. **[12.5]** $y \leq 2x$
              $x + 2y > 4$

Graph $y = 2x$ as a solid line through $(0, 0)$ and $(1, 2)$. Shade the region below the line.

Graph $x + 2y = 4$ as a dashed line through its intercepts, $(4, 0)$ and $(0, 2)$. Since $x + 2y > 4$ is equivalent to $y > -\frac{1}{2}x + 2$, shade the region above this line.

The solution set of this system is the intersection of the shaded regions. It includes the portion of the line $y = 2x$ that bounds this region, but not the line $x + 2y = 4$.

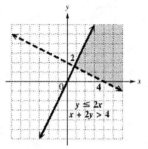

39. **[12.5]** $y < -4x$
              $y < -2$

The graph of $y < -4x$ is the region below the graph of the dashed line $y = -4x$. This line goes through the points $(0, 0)$ and $(1, -4)$.

The graph of $y < -2$ is the region below the graph of the dashed horizontal line $y = -2$. The intersection of these two regions is graphed below. The solution set is the shaded region.

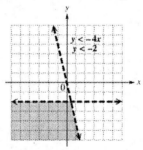

40. **[12.4]** Let $x =$ the length of each of the two equal sides and $y =$ the length of the longer third side.

The perimeter is 29 inches, so

$$x + x + y = 29$$
$$\text{or} \quad 2x + y = 29. \quad (1)$$

The third side is 5 inches longer than each of the two equal sides, so

$$y = x + 5. \quad (2)$$

Substitute $x + 5$ for y in (1).

$$2x + y = 29$$
$$2x + (x + 5) = 29$$
$$3x + 5 = 29$$
$$3x = 24$$
$$x = 8$$

From (2), $y = 8 + 5 = 13$.

The lengths of the sides of the triangle are 8 inches, 8 inches, and 13 inches.

41. **[12.4]** Let $x =$ the Giants' score and $y =$ the Patriots' score.

The Giants beat the Patriots by 4, so

$$x = y + 4. \quad (1)$$

The winning score was 13 less than twice the losing score, so

$$x = 2y - 13. \quad (2)$$

Substitute $2y - 13$ for x in equation (1).

$$2y - 13 = y + 4$$
$$2y = y + 17$$
$$y = 17$$

From (1), $x = 17 + 4 = 21$.

The final score of the game was
New York Giants 21, New England Patriots 17.

42. **[12.4]** **(a)** The graph of the fixed-rate mortgage is above the graph of the variable-rate mortgage for years 0 to 6, so that is when the monthly payment for the fixed-rate mortgage is more than the monthly payment for the variable-rate mortgage.

(b) The graphs intersect at year 6, so that is when the payments would be the same. The monthly payment at that time appears to be about $650.

Chapter 12 Test

1. $2x + y = -3 \quad (1)$
$\quad\ \ x - y = -9 \quad (2)$

(a) $(1, -5)$

Substitute 1 for x and -5 for y in (1) and (2).

(1) $2(1) + (-5) \stackrel{?}{=} -3$
$\qquad\qquad -3 = -3$ *True*

(2) $(1) - (-5) \stackrel{?}{=} -9$
$\qquad\qquad 6 = -9$ *False*

Since $(1, -5)$ does not satisfy both equations, it *is not* a solution of the system.

(b) $(1, 10)$

(1) $2(1) + (10) \stackrel{?}{=} -3$
$\qquad\qquad 12 = -3$ *False*

The false result indicates that $(1, 10)$ *is not* a solution of the system.

(c) $(-4, 5)$

(1) $2(-4) + (5) \stackrel{?}{=} -3$
$\qquad\qquad -3 = -3$ *True*

(2) $(-4) - (5) \stackrel{?}{=} -9$
$\qquad\qquad -9 = -9$ *True*

Since $(-4, 5)$ satisfies both equations, it *is* a solution of the system.

2. $2x + y = 1$
$\quad 3x - y = 9$

The intercepts for the line $2x + y = 1$ are $(\frac{1}{2}, 0)$ and $(0, 1)$. Because one of the coordinates is a fraction and the intercepts are close together, it may be difficult to graph the line accurately using only these two points. A third point, such as $(-1, 3)$ may be helpful. Graph the line $3x - y = 9$ through its intercepts, $(3, 0)$ and $(0, -9)$.

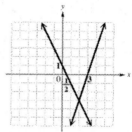

The two lines appear to intersect at the point $(2, -3)$. Checking by substituting 2 for x and -3 for y in the original equations will show that this ordered pair is the solution of the system. The solution set is $\{(2, -3)\}$.

3. Two lines which have the same slope but different y-intercepts are parallel, so the system has no solution.

4. $2x + y = -4 \quad (1)$
$\quad x = y + 7 \quad\ \ (2)$

Substitute $y + 7$ for x in equation (1), and solve for y.

$$2x + y = -4$$
$$2(y + 7) + y = -4$$
$$2y + 14 + y = -4$$
$$3y = -18$$
$$y = -6$$

From (2), $x = -6 + 7 = 1$.

The solution set is $\{(1, -6)\}$.

5. $4x + 3y = -35 \quad (1)$
$\quad\ \ x + y = 0 \quad\quad (2)$

Solve equation (2) for y.

$$y = -x \quad (3)$$

Substitute $-x$ for y in equation (1) and solve for x.

$$4x + 3y = -35$$
$$4x + 3(-x) = -35$$
$$4x - 3x = -35$$
$$x = -35$$

From (3), $y = -(-35) = 35$.

The solution set is $\{(-35, 35)\}$.

6.
$$\begin{array}{rcl} 2x & - & y & = & 4 & (1) \\ 3x & + & y & = & 21 & (2) \\ \hline 5x & & & = & 25 & \textit{Add } (1) \textit{ and } (2). \\ & & x & = & 5 \end{array}$$

To find y, let $x = 5$ in equation (2).
$$3x + y = 21$$
$$3(5) + y = 21$$
$$15 + y = 21$$
$$y = 6$$

The solution set is $\{(5, 6)\}$.

7. $\quad 4x + 2y = 2 \qquad (1)$
$\qquad 5x + 4y = 7 \qquad (2)$

Multiply equation (1) by -2 and add the result to equation (2).
$$\begin{array}{rcl} -8x & - & 4y & = & -4 \\ 5x & + & 4y & = & 7 \\ \hline -3x & & & = & 3 \\ & & x & = & -1 \end{array}$$

To find y, let $x = -1$ in equation (1).
$$4x + 2y = 2$$
$$4(-1) + 2y = 2$$
$$-4 + 2y = 2$$
$$2y = 6$$
$$y = 3$$

The solution set is $\{(-1, 3)\}$.

8. $\quad 3x + 4y = 9 \qquad (1)$
$\qquad 2x + 5y = 13 \qquad (2)$

$$\begin{array}{rcll} 6x & + & 8y & = & 18 & (3) \ 2 \times \text{Eq.}(1) \\ -6x & - & 15y & = & -39 & (4) -3 \times \text{Eq.}(2) \\ \hline & & -7y & = & -21 & \textit{Add } (3) \textit{ and } (4). \\ & & y & = & 3 \end{array}$$

Substitute 3 for y in (1).
$$3x + 4y = 9$$
$$3x + 4(3) = 9$$
$$3x + 12 = 9$$
$$3x = -3$$
$$x = -1$$

The solution set is $\{(-1, 3)\}$.

9. $\quad 6x - 5y = 0 \qquad (1)$
$\qquad -2x + 3y = 0 \qquad (2)$

Multiply equation (2) by 3 and add the result to equation (1).
$$\begin{array}{rcl} 6x & - & 5y & = & 0 \\ -6x & + & 9y & = & 0 \\ \hline & & 4y & = & 0 \\ & & y & = & 0 \end{array}$$

To find x, let $y = 0$ in equation (1).

$$6x - 5y = 0$$
$$6x - 5(0) = 0$$
$$6x = 0$$
$$x = 0$$

The solution set is $\{(0, 0)\}$.

10. $\qquad 4x + 5y = 2 \qquad (1)$
$\qquad -8x - 10y = 6 \qquad (2)$

Multiply equation (1) by 2 and add the result to equation (2).
$$\begin{array}{rcl} 8x & + & 10y & = & 4 \\ -8x & - & 10y & = & 6 \\ \hline & & 0 & = & 10 \quad \textit{False} \end{array}$$

This result indicates that the system has no solution. The solution set is $\emptyset$.

11. $\quad 3x = 6 + y \qquad (1)$
$\qquad 6x - 2y = 12 \qquad (2)$

Write equation (1) in the form $Ax + By = C$.
$$3x - y = 6 \qquad (3)$$
$$6x - 2y = 12 \qquad (2)$$

Multiply equation (3) by -2 and add the result to equation (2).
$$\begin{array}{rcll} -6x & + & 2y & = & -12 & (4) \\ 6x & - & 2y & = & 12 & (2) \\ \hline & & 0 & = & 0 & \textit{Add } (4) \textit{ and } (2). \end{array}$$

This result indicates that the system has an infinite number of solutions. The solution set is $\{(x, y) \mid 3x - y = 6\}$.

12. $\qquad \frac{6}{5}x - \frac{1}{3}y = -20 \qquad (1)$
$\qquad -\frac{2}{3}x + \frac{1}{6}y = 11 \qquad (2)$

To clear fractions, multiply by the LCD for each equation. Multiply equation (1) by 15.
$$15\left(\frac{6}{5}x - \frac{1}{3}y\right) = 15(-20)$$
$$18x - 5y = -300 \qquad (3)$$

Multiply equation (2) by 6.
$$6\left(-\frac{2}{3}x + \frac{1}{6}y\right) = 6(11)$$
$$-4x + y = 66 \qquad (4)$$

To solve this system by the elimination method, multiply equation (4) by 5 and add the result to equation (3).
$$\begin{array}{rcl} 18x & - & 5y & = & -300 \\ -20x & + & 5y & = & 330 \\ \hline -2x & & & = & 30 \\ & & x & = & -15 \end{array}$$

To find y, let $x = -15$ in equation (4).
$$-4x + y = 66$$
$$-4(-15) + y = 66$$
$$60 + y = 66$$
$$y = 6$$

The solution set is $\{(-15, 6)\}$.

13. *Step 2* Let $x =$ the distance between Memphis and Atlanta (in miles), and let $y =$ the distance between Minneapolis and Houston (in miles).

Step 3 Since the distance between Memphis and Atlanta is 782 miles less than the distance between Minneapolis and Houston,

$$x = y - 782. \quad (1)$$

Together the two distances total 1570 miles, so

$$x + y = 1570. \quad (2)$$

Step 4 Substitute $y - 782$ for x in equation (2).

$$(y - 782) + y = 1570$$
$$2y - 782 = 1570$$
$$2y = 2352$$
$$y = 1176$$

From (1), $x = 1176 - 782 = 394$.

Step 5 The distance between Memphis and Atlanta is 394 miles, while the distance between Minneapolis and Houston is 1176 miles.

Step 6 394 is 782 less than 1176 and the sum of 394 and 1176 is 1570, as required.

14. *Step 2* Let $x =$ the number of people who visited the Statue of Liberty and $y =$ the number of people who visited the National World War II Memorial.

Step 3 The total was 7.8 million, so

$$x + y = 7.8. \quad (1)$$

The Statue of Liberty had 0.2 million fewer visitors than the National World War II Memorial, so

$$x = y - 0.2. \quad (2)$$

Step 4 Substitute $y - 0.2$ for x in (1).

$$(y - 0.2) + y = 7.8$$
$$2y - 0.2 = 7.8$$
$$2y = 8.0$$
$$y = 4.0$$

From (2), $x = 4.0 - 0.2 = 3.8$.

Step 5 In 2010, 4.0 million people visited the National World War II Memorial and 3.8 million people visited the Statue of Liberty.

Step 6 The sum of 4.0 (all numbers in millions) and 3.8 is 7.8 and 3.8 is 0.2 less than 4.0.

15. *Step 2* Let $x =$ the number of liters of 15% alcohol solution and $y =$ the number of liters of 40% alcohol solution.

Liters of Solution	Percent (as a decimal)	Liters of Pure Alcohol
x	0.15	$0.15x$
y	0.40	$0.40y$
50	0.30	$0.30(50) = 15$

Step 3 From the first and third columns of the table, we obtain the equations

$$x + y = 50 \qquad (1)$$
$$0.15x + 0.40y = 15. \qquad (2)$$

To clear decimals, multiply both sides of equation (2) by 100.

$$15x + 40y = 1500 \quad (3)$$

Step 4 To solve this system by the elimination method, multiply equation (1) by -15 and add the result to equation (3).

$$\begin{array}{rcr} -15x - 15y &=& -750 \\ 15x + 40y &=& 1500 \\ \hline 25y &=& 750 \\ y &=& \frac{750}{25} = 30 \end{array}$$

From (1), $x = 20$.

Step 5 To get 50 liters of a 30% alcohol solution, 20 liters of a 15% solution and 30 liters of a 40% solution should be used.

Step 6 Since $20 + 30 = 50$ and $0.15(20) + 0.40(30) = 3 + 12 = 15$, this mixture will give the 50 liters of 30% solution, as required.

16. *Step 2* Let $x =$ the rate of the faster car and $y =$ the rate of the slower car.

Use $d = rt$ in constructing the table.

	r	t	d
Faster car	x	3	$3x$
Slower car	y	3	$3y$

Step 3 The faster car travels $1\frac{1}{3}$ times as fast as the other car, so

$$x = 1\tfrac{1}{3}y$$
$$\text{or} \quad x = \tfrac{4}{3}y. \qquad (1)$$

After 3 hours, they are 45 miles apart, that is, the difference between their distances is 45.

$$3x - 3y = 45 \quad (2)$$

Step 4 To solve the system by substitution, substitute $\frac{4}{3}y$ for x in equation (2).

$$3(\tfrac{4}{3}y) - 3y = 45$$
$$4y - 3y = 45$$
$$y = 45$$

From (1), $x = \frac{4}{3}(45) = 60$.

Step 5 The rate of the faster car is 60 miles per hour, and the rate of the slower car is 45 miles per hour.

Step 6 60 is one and one-third times 45. In three hours, the faster car travels $3(60) = 180$ miles and the slower car travels $3(45) = 135$ miles. The cars are $180 - 135 = 45$ miles apart, as required.

17. $2x + 7y \leq 14$
$\quad\ x - y \geq 1$

Graph $2x + 7y = 14$ as a solid line through its intercepts, $(7, 0)$ and $(0, 2)$. Choosing $(0, 0)$ as a test point will result in the true statement $0 \leq 14$, so shade the side of the line containing the origin.

Graph $x - y = 1$ as a solid line through its intercepts, $(1, 0)$ and $(0, -1)$. Choosing $(0, 0)$ as a test point will result in the false statement $0 \geq 1$, so shade the side of the line *not* containing the origin.

The solution set of the given system is the intersection of the two shaded regions, and includes the portions of the two lines that bound this region.

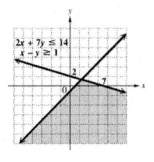

18. $2x - y > 6$
$\quad 4y + 12 \geq -3x$

Graph $2x - y = 6$ as a dashed line through its intercepts, $(3, 0)$ and $(0, -6)$. Choosing $(0, 0)$ as a test point will result in the false statement $0 > 6$, so shade the side of the line *not* containing the origin.

Graph $4y + 12 = -3x$ as a solid line through its intercepts, $(-4, 0)$ and $(0, -3)$. Choosing $(0, 0)$ as a test point will result in the true statement $12 \geq 0$, so shade the side of the line containing the origin.

The solution set of the given system is the intersection of the two shaded regions. It includes the portion of the line $4y + 12 = -3x$ that bounds this region, but not the line $2x - y = 6$.

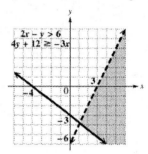

CHAPTER 13 EXPONENTS AND POLYNOMIALS

13.1 Adding and Subtracting Polynomials

13.1 Margin Exercises

1. **(a)** $5x^4 + 7x^4$
    $= (5+7)x^4$ *Distributive property*
    $= 12x^4$

    (b) $9pq + 3pq - 2pq$
    $= (9+3-2)pq$ *Distributive property*
    $= 10pq$

    (c) $r^2 + 3r + 5r^2$
    $= (1r^2 + 5r^2) + 3r$ *Commutative and Associative properties*
    $= (1+5)r^2 + 3r$ *Distributive property*
    $= 6r^2 + 3r$

    (d) $x + \frac{1}{2}x$
    $= 1x + \frac{1}{2}x$ $x = 1x$
    $= (1 + \frac{1}{2})x$ *Distributive property*
    $= \frac{3}{2}x$

    (e) $8t + 6w$
    These are unlike terms. They cannot be combined.

    (f) $3x^4 - 3x^2$
    These are unlike terms. They cannot be combined.

2. **(a)** $3m^5 + 5m^2 - 2m + 1$ is a polynomial written in descending powers, since $m = m^1$ and $1 = m^0$. So **A** and **B** apply.

    (b) $2p^4 + p^6$ is a polynomial, but is not written in descending powers. So **A** applies.

    (c) $\frac{1}{x} + 2x^2 + 3$ is not a polynomial, because $\frac{1}{x}$ has a variable in the denominator. So **C** applies.

    (d) $x - 3$ is a polynomial written in descending powers, since $x = x^1$ and $3 = 3x^0$. So **A** and **B** apply.

3. **(a)** $3x^2 + 2x - 4$ cannot be simplified, because there are no like terms. There are three terms, and the largest exponent on the variable x is 2. So the polynomial has degree 2 and is a trinomial.

    (b) $x^3 + 4x^3 = 1x^3 + 4x^3$ simplifies to $(1+4)x^3 = 5x^3$. This polynomial has one term, and the exponent on the variable x is 3. So the polynomial has degree 3 and is a monomial.

    (c) $x^8 - x^7 + 2x^8 = 1x^8 + 2x^8 - x^7$ simplifies to $(1+2)x^8 - x^7 = 3x^8 - x^7$. This polynomial has two terms, and the largest exponent on the variable x is 8. So the polynomial has degree 8 and is a binomial.

4. **(a)** $2x^3 + 8x - 6$
    $= 2(-1)^3 + 8(-1) - 6$ *Let $x = -1$.*
    $= 2(-1) + 8(-1) - 6$
    $= -2 - 8 - 6$
    $= -16$

    (b) $2x^3 + 8x - 6$
    $= 2(4)^3 + 8(4) - 6$ *Let $x = 4$.*
    $= 2(64) + 8(4) - 6$
    $= 128 + 32 - 6$
    $= 154$

    (c) $2x^3 + 8x - 6$
    $= 2(-2)^3 + 8(-2) - 6$ *Let $x = -2$.*
    $= 2(-8) + 8(-2) - 6$
    $= -16 - 16 - 6$
    $= -38$

5. **(a)** Write like terms in columns and add column by column.

    $$\begin{array}{r} 4x^3 - 3x^2 + 2x \\ 6x^3 + 2x^2 - 3x \\ \hline 10x^3 - x^2 - x \end{array}$$

    (b) Be sure to leave a space for the missing x-term in $4x^2 - 2$.

    $$\begin{array}{r} x^2 - 2x + 5 \\ 4x^2 \qquad - 2 \\ \hline 5x^2 - 2x + 3 \end{array}$$

6. **(a)** Combine like terms.
    $(2x^4 - 6x^2 + 7) + (-3x^4 + 5x^2 + 2)$
    $= (2x^4 - 3x^4) + (-6x^2 + 5x^2) + (7+2)$
    $= -x^4 - x^2 + 9$

    (b) Combine like terms.
    $(3x^2 + 4x + 2) + (6x^3 - 5x - 7)$
    $= 6x^3 + 3x^2 + (4x - 5x) + (2-7)$
    $= 6x^3 + 3x^2 - x - 5$

7. **(a)** Change subtraction to addition.
    $(14y^3 - 6y^2 + 2y - 5) - (2y^3 - 7y^2 - 4y + 6)$
    $= (14y^3 - 6y^2 + 2y - 5)$
       $+ (-2y^3 + 7y^2 + 4y - 6)$
                         Combine like terms.
    $= (14y^3 - 2y^3) + (-6y^2 + 7y^2)$
       $+ (2y + 4y) + (-5 - 6)$
    $= \underline{12y^3 + y^2 + 6y - 11}$

    Check by adding.
    $$\begin{array}{l} 2y^3 - 7y^2 - 4y + 6 \quad \textit{Second polynomial} \\ \underline{12y^3 + y^2 + 6y - 11} \quad \textit{Answer} \\ 14y^3 - 6y^2 + 2y - 5 \quad \textit{First polynomial} \end{array}$$

(b) Subtract $\left(-\frac{3}{2}y^2 + \frac{4}{3}y + 6\right)$
from $\left(\frac{7}{2}y^2 - \frac{11}{3}y + 8\right)$.

$$\left(\frac{7}{2}y^2 - \frac{11}{3}y + 8\right) - \left(-\frac{3}{2}y^2 + \frac{4}{3}y + 6\right)$$

$$= \left(\frac{7}{2}y^2 - \frac{11}{3}y + 8\right) + \left(\frac{3}{2}y^2 - \frac{4}{3}y - 6\right)$$

Definition of subtraction

$$= \left(\frac{7}{2}y^2 + \frac{3}{2}y^2\right) + \left[\left(-\frac{11}{3}y\right) + \left(-\frac{4}{3}y\right)\right]$$
$$+ \left[8 + (-6)\right] \quad \text{\textit{Group like terms.}}$$

$$= \frac{10}{2}y^2 - \frac{15}{3}y + 2 \quad \text{\textit{Add like terms.}}$$

$$= 5y^2 - 5y + 2$$

Check by adding.

$$\left(-\frac{3}{2}y^2 + \frac{4}{3}y + 6\right) + \left(5y^2 - 5y + 2\right)$$

$$= \left(-\frac{3}{2}y^2 + 5y^2\right) + \left(\frac{4}{3}y - 5y\right) + (6 + 2)$$

$$= \left(-\frac{3}{2}y^2 + \frac{10}{2}y^2\right) + \left(\frac{4}{3}y - \frac{15}{3}y\right) + (6 + 2)$$

$$= \frac{7}{2}y^2 - \frac{11}{3}y + 8$$

The answer is correct.

8. $(4y^3 - 16y^2 + 2y) - (12y^3 - 9y^2 + 16)$

Arrange like terms in columns. Insert zeros for missing terms.

$$\begin{array}{r} 4y^3 - 16y^2 + 2y + 0 \\ 12y^3 - 9y^2 + 0y + 16 \end{array}$$

Change all signs in the second row, and then add.

$$\begin{array}{r} 4y^3 - 16y^2 + 2y + 0 \\ -12y^3 + 9y^2 - 0y - 16 \\ \hline -8y^3 - 7y^2 + 2y - 16 \end{array}$$

9. $(6p^4 - 8p^3 + 2p - 1) - (-7p^4 + 6p^2 - 12)$
$\quad + (p^4 - 3p + 8)$
$= (6p^4 - 8p^3 + 2p - 1) + (7p^4 - 6p^2 + 12)$
$\quad + (p^4 - 3p + 8)$
$= (13p^4 - 8p^3 - 6p^2 + 2p + 11) + (p^4 - 3p + 8)$
$= 14p^4 - 8p^3 - 6p^2 - p + 19$

10. (a) $(3mn + 2m - 4n) + (-mn + 4m + n)$
$\quad = [3mn + (-mn)] + (2m + 4m) + [(-4n) + n]$
$\quad = 2mn + 6m - 3n$

(b) $(5p^2q^2 - 4p^2 + 2q) - (2p^2q^2 - p^2 - 3q)$
$\quad = (5p^2q^2 - 4p^2 + 2q) + (-2p^2q^2 + p^2 + 3q)$
$\quad = [5p^2q^2 + (-2p^2q^2)] + [(-4p^2) + p^2]$
$\quad\quad + (2q + 3q)$
$\quad = 3p^2q^2 - 3p^2 + 5q$

13.1 Section Exercises

1. In the term $7x^5$, the coefficient is __7__ and the exponent is __5__.

3. The degree of the term $-4x^8$ is __8__, the exponent.

5. When $x^2 + 10$ is evaluated for $x = 4$, the result is
$$4^2 + 10 = 16 + 10 = \underline{26}.$$

7. The polynomial $6x^4$ has one term. The coefficient of this term is 6.

9. The polynomial t^4 has one term. Since $t^4 = 1 \cdot t^4$, the coefficient of this term is 1.

11. The polynomial $\frac{x}{5}$ has one term. Since $\frac{x}{5} = \frac{1}{5}x$, the coefficient of this term is $\frac{1}{5}$.

13. The polynomial $-19r^2 - r$ has two terms. The coefficient of r^2 is -19 and the coefficient of r is -1.

15. The polynomial $x - 8x^2 + \frac{2}{3}x^3$ has three terms. The coefficient of x is 1, the coefficient of x^2 is -8, and the coefficient of x^3 is $\frac{2}{3}$.

In Exercises 17–30, use the distributive property to add like terms.

17. $-3m^5 + 5m^5 = (-3 + 5)m^5$
$$= 2m^5$$

19. $2r^5 + (-3r^5) = [2 + (-3)]r^5$
$$= -1r^5, \text{ which equals } -r^5.$$

21. $\frac{1}{2}x^4 + \frac{1}{6}x^4 = \left(\frac{1}{2} + \frac{1}{6}\right)x^4$
$$= \left(\frac{3}{6} + \frac{1}{6}\right)x^4$$
$$= \frac{4}{6}x^4 = \frac{2}{3}x^4$$

23. The polynomial $-0.5m^2 + 0.2m^5$ cannot be simplified. The two terms are unlike because the exponents on the variables are different, so they cannot be combined. In descending powers of the variable, the polynomial is written $0.2m^5 - 0.5m^2$.

25. $-3x^5 + 2x^5 - 4x^5 = (-3 + 2 - 4)x^5$
$$= -5x^5$$

27. $-4p^7 + 8p^7 + 5p^9 = (-4 + 8)p^7 + 5p^9$
$$= 4p^7 + 5p^9$$

In descending powers of the variable, this polynomial is written $5p^9 + 4p^7$.

29. $-4y^2 + 3y^2 - 2y^2 + y^2 = (-4 + 3 - 2 + 1)y^2$
$$= -2y^2$$

31. $6x^4 - 9x$ ■ The polynomial has no like terms, so it is already simplified. It is already written in descending powers of the variable x. The highest degree of any nonzero term is 4, so the degree of the polynomial is 4. There are two terms, so this is a *binomial*.

33. $5m^4 - 3m^2 + 6m^5 - 7m^3$ ■ The polynomial has no like terms, so it is already simplified. We can write it in descending powers of the variable m as
$$6m^5 + 5m^4 - 7m^3 - 3m^2.$$

The highest degree of any nonzero term is 5, so the degree of the polynomial is 5. The polynomial is not a monomial, binomial, or trinomial since it has 4 terms.

35. $\frac{5}{3}x^4 - \frac{2}{3}x^4 + \frac{1}{3}x^2 - 4 = \frac{3}{3}x^4 + \frac{1}{3}x^2 - 4$

$\qquad\qquad\qquad\qquad = x^4 + \frac{1}{3}x^2 - 4$

The resulting polynomial is a *trinomial* of degree 4.

37. $0.8x^4 - 0.3x^4 - 0.5x^4 + 7$

$\qquad = (0.8 - 0.3 - 0.5)x^4 + 7$

$\qquad = 0x^4 + 7 = 7$

Since 7 can be written as $7x^0$, the degree of the polynomial is 0. The simplified polynomial has one term, so it is a *monomial*.

39. $2.5x^2 + 0.5x + x^2 - x - 2x^2$

$\qquad = (2.5 + 1 - 2)x^2 + (0.5 - 1)x$

$\qquad = 1.5x^2 - 0.5x$

The resulting polynomial is a *binomial* of degree 2.

41.   **(a)** $-2x + 3 = -2(2) + 3$ *Let x = 2.*

$\qquad\qquad\qquad = -4 + 3$

$\qquad\qquad\qquad = -1$

      (b) $-2x + 3 = -2(-1) + 3$ *Let x = -1.*

$\qquad\qquad\qquad = 2 + 3$

$\qquad\qquad\qquad = 5$

43.   **(a)** $2x^2 + 5x + 1$

$\qquad = 2(2)^2 + 5(2) + 1$ *Let x = 2.*

$\qquad = 2(4) + 10 + 1$

$\qquad = 8 + 10 + 1$

$\qquad = 18 + 1$

$\qquad = 19$

      (b) $2x^2 + 5x + 1$

$\qquad = 2(-1)^2 + 5(-1) + 1$ *Let x = -1.*

$\qquad = 2(1) - 5 + 1$

$\qquad = 2 - 5 + 1$

$\qquad = -3 + 1$

$\qquad = -2$

45.   **(a)** $2x^5 - 4x^4 + 5x^3 - x^2$

$\qquad = 2(2)^5 - 4(2)^4 + 5(2)^3 - (2)^2$ *Let x = 2.*

$\qquad = 2(32) - 4(16) + 5(8) - 4$

$\qquad = 64 - 64 + 40 - 4$

$\qquad = 36$

      (b) $2x^5 - 4x^4 + 5x^3 - x^2$

$\qquad = 2(-1)^5 - 4(-1)^4 + 5(-1)^3 - (-1)^2$

$\qquad\qquad\qquad\qquad\qquad$ *Let x = -1.*

$\qquad = 2(-1) - 4(1) + 5(-1) - 1$

$\qquad = -2 - 4 - 5 - 1$

$\qquad = -12$

47.   **(a)** $-4x^5 + x^2$

$\qquad = -4(2)^5 + (2)^2$ *Let x = 2.*

$\qquad = -4(32) + 4$

$\qquad = -128 + 4$

$\qquad = -124$

      (b) $-4x^5 + x^2$

$\qquad = -4(-1)^5 + (-1)^2$ *Let x = -1.*

$\qquad = -4(-1) + 1$

$\qquad = 4 + 1$

$\qquad = 5$

49.   Add. $3m^2 + 5m$

$\qquad\qquad\quad \underline{2m^2 - 2m}$

$\qquad\qquad\quad\ 5m^2 + 3m$

51.   Subtract. $12x^4 - \ \ x^2$

$\qquad\qquad\qquad\ \underline{8x^4 + 3x^2}$

Change all signs in the second row, and then add.

$$12x^4 - \ \ x^2$$
$$\underline{-8x^4 - 3x^2}$$
$$\ \ 4x^4 - 4x^2$$

53.   Add. $\frac{2}{3}x^2 + \frac{1}{5}x + \frac{1}{6}$

$\qquad\qquad\ \underline{\frac{1}{2}x^2 - \frac{1}{3}x + \frac{2}{3}}$

Rewrite the fractions so that the fractions in each column have a common denominator; then add column by column.

$$\frac{4}{6}x^2 + \frac{3}{15}x + \frac{1}{6}$$
$$\underline{\frac{3}{6}x^2 - \frac{5}{15}x + \frac{4}{6}}$$
$$\frac{7}{6}x^2 - \frac{2}{15}x + \frac{5}{6}$$

55.   Subtract. $12m^3 - 8m^2 + 6m + 7$

$\qquad\qquad\qquad\qquad\quad \underline{5m^2 \qquad\quad - 4}$

Change all signs in the second row, and then add.

$$12m^3 - \ \ 8m^2 + 6m + \ \ 7$$
$$\underline{\ \ \ - \ \ 5m^2 \qquad\quad + \ \ 4}$$
$$12m^3 - 13m^2 + 6m + 11$$

57.   Subtract. $4.3x^3 - 6.1x^2 - 3.0x - 5$

$\qquad\qquad\qquad\qquad \underline{1.4x^3 - 2.6x^2 - 1.5x + 4}$

Change all signs in the second row, and then add.

$$4.3x^3 - 6.1x^2 - 3.0x - 5$$
$$\underline{-1.4x^3 + 2.6x^2 + 1.5x - 4}$$
$$\ \ 2.9x^3 - 3.5x^2 - 1.5x - 9$$

59. $(2r^2 + 3r - 12) + (6r^2 + 2r)$
$$= (2r^2 + 6r^2) + (3r + 2r) - 12$$
$$= 8r^2 + 5r - 12$$

61. $(8m^2 - 7m) - (3m^2 + 7m - 6)$
$$= (8m^2 - 7m) + (-3m^2 - 7m + 6)$$
$$= (8 - 3)m^2 + (-7 - 7)m + 6$$
$$= 5m^2 - 14m + 6$$

63. $(16x^3 - x^2 + 3x) + (-12x^3 + 3x^2 + 2x)$
$$= 16x^3 - x^2 + 3x - 12x^3 + 3x^2 + 2x$$
$$= (16 - 12)x^3 + (-1 + 3)x^2 + (3 + 2)x$$
$$= 4x^3 + 2x^2 + 5x$$

65. $(7y^4 + 3y^2 + 2y) - (18y^5 - 5y^3 + y)$
$$= (7y^4 + 3y^2 + 2y) + (-18y^5 + 5y^3 - y)$$
$$= (-18y^5) + 7y^4 + 5y^3 + 3y^2 + [2y + (-y)]$$
$$= -18y^5 + 7y^4 + 5y^3 + 3y^2 + y$$

67. $[(8m^2 + 4m - 7) - (2m^3 - 5m + 2)] - (m^2 + m)$
$$= (8m^2 + 4m - 7) + (-2m^3 + 5m - 2)$$
$$\quad + (-m^2 - m)$$
$$= 8m^2 + 4m - 7 - 2m^3 + 5m - 2 - m^2 - m$$
$$= -2m^3 + (8m^2 - m^2) + (4m + 5m - m)$$
$$\quad + (-7 - 2)$$
$$= -2m^3 + 7m^2 + 8m - 9$$

69. Subtract $9x^2 - 3x + 7$ from $-2x^2 - 6x + 4$.

$(-2x^2 - 6x + 4) - (9x^2 - 3x + 7)$
$$= (-2x^2 - 6x + 4) + (-9x^2 + 3x - 7)$$
$$= (-2x^2 - 9x^2) + (-6x + 3x) + (4 - 7)$$
$$= -11x^2 - 3x - 3$$

71. Use the formula for the perimeter of a square, $P = 4s$, with $s = \frac{1}{2}x^2 + 2x$.

$$P = 4s$$
$$= 4\left(\tfrac{1}{2}x^2 + 2x\right)$$
$$= 4\left(\tfrac{1}{2}x^2\right) + 4(2x)$$
$$= 2x^2 + 8x$$

A polynomial that represents the perimeter of the square is $2x^2 + 8x$.

73. Use the formula for the perimeter of a rectangle, $P = 2l + 2w$, with length $l = 4x^2 + 3x + 1$ and width $w = x + 2$.

$$P = 2l + 2w$$
$$= 2(4x^2 + 3x + 1) + 2(x + 2)$$
$$= 8x^2 + 6x + 2 + 2x + 4$$
$$= 8x^2 + 8x + 6$$

A polynomial that represents the perimeter of the rectangle is $8x^2 + 8x + 6$.

75. Use the formula for the perimeter of a triangle, $P = a + b + c$, with $a = 3t^2 + 2t + 7$, $b = 5t^2 + 2$, and $c = 6t + 4$.

$$P = (3t^2 + 2t + 7) + (5t^2 + 2) + (6t + 4)$$
$$= (3t^2 + 5t^2) + (2t + 6t) + (7 + 2 + 4)$$
$$= 8t^2 + 8t + 13$$

A polynomial that represents the perimeter of the triangle is $8t^2 + 8t + 13$.

77. $(9a^2b - 3a^2 + 2b) + (4a^2b - 4a^2 - 3b)$
$$= (9a^2b + 4a^2b) + [(-3a^2) + (-4a^2)]$$
$$\quad + [2b + (-3b)]$$
$$= 13a^2b + (-7a^2) + (-b)$$
$$= 13a^2b - 7a^2 - b$$

79. $(2c^4d + 3c^2d^2 - 4d^2) - (c^4d + 8c^2d^2 - 5d^2)$
$$= (2c^4d + 3c^2d^2 - 4d^2) + (-c^4d - 8c^2d^2 + 5d^2)$$
$$= (2c^4d - c^4d) + (3c^2d^2 - 8c^2d^2)$$
$$\quad + (-4d^2 + 5d^2)$$
$$= c^4d - 5c^2d^2 + d^2$$

81. Subtract.
$$9m^3n - 5m^2n^2 + 4mn^2$$
$$-3m^3n + 6m^2n^2 + 8mn^2$$

Change all signs in the second row, and then add.

$$9m^3n - 5m^2n^2 + 4mn^2$$
$$3m^3n - 6m^2n^2 - 8mn^2$$
$$\overline{12m^3n - 11m^2n^2 - 4mn^2}$$

83. **(a)** Use the formula for the perimeter of a triangle, $P = a + b + c$, with $a = 2y - 3t$, $b = 5y + 3t$, and $c = 16y + 5t$.

$$P = (2y - 3t) + (5y + 3t) + (16y + 5t)$$
$$= (2y + 5y + 16y) + (-3t + 3t + 5t)$$
$$= 23y + 5t$$

The perimeter of the triangle is $23y + 5t$.

(b) Use the fact that the sum of the measures of the angles of any triangle is $180°$.

$$(7x - 3)° + (5x + 2)° + (2x - 1)° = 180°$$
$$(7x + 5x + 2x) + (-3 + 2 - 1) = 180$$
$$14x - 2 = 180$$
$$14x = 182$$
$$x = \tfrac{182}{14} = 13$$

If $x = 13$,

$$7x - 3 = 7(13) - 3 = 88,$$
$$5x + 2 = 5(13) + 2 = 67,$$
$$\text{and} \quad 2x - 1 = 2(13) - 1 = 25.$$

The measures of the angles are $25°$, $67°$, and $88°$.

Relating Concepts (Exercises 85–88)

85. $D = 100t - 13t^2$
$= 100(5) - 13(5)^2$ *Let t = 5.*
$= 500 - 13(25)$
$= 500 - 325$
$= 175$

In _5_ seconds, the car will skid _175_ feet.

86. $D = 100t - 13t^2$
$= 100(1) - 13(1)^2$ *Let t = 1.*
$= 100 - 13$
$= 87$

The skidding distance is 87 feet for 1 second.
The ordered pair is $(t, D) = (1, 87)$.

87. If $x = 6$,
$2x + 15 = 2(6) + 15 = 12 + 15 = 27.$

If the saw is rented for _6_ days, the cost is _$27_.

88. $-16t^2 + 60t + 80$
$= -16(2.5)^2 + 60(2.5) + 80$ *Let t = 2.5.*
$= -16(6.25) + 150 + 80$
$= -100 + 150 + 80$
$= 130$

If _2.5_ seconds have elapsed, the height of the object is _130_ feet.

13.2 The Product Rule and Power Rules for Exponents

13.2 Margin Exercises

1. **(a)** $2 \cdot 2 \cdot 2 \cdot 2$
$= 2^4$ 4 factors of 2
$= 16$

(b) $(-3)(-3)(-3)$
$= (-3)^3$ 3 factors of -3
$= -27$

2. **(a)** $2^5 = 2 \cdot 2 \cdot 2 \cdot 2 \cdot 2$
$= 32$

Base: 2; exponent: 5

(b) $-2^5 = -(2 \cdot 2 \cdot 2 \cdot 2 \cdot 2)$
$= -32$

Base: 2; exponent: 5

(c) $(-2)^5 = (-2)(-2)(-2)(-2)(-2)$
$= -32$

Base: -2; exponent: 5

(d) $4^2 = 4 \cdot 4$
$= 16$

Base: 4; exponent: 2

(e) $-4^2 = -(4 \cdot 4)$
$= -16$

Base: 4; exponent: 2

(f) $(-4)^2 = (-4)(-4)$
$= 16$

Base: -4; exponent: 2

3. **(a)** $8^2 \cdot 8^5 = 8^{2+5}$ *Product rule*
$= 8^7$

(b) $(-7)^5 \cdot (-7)^3 = (-7)^{5+3}$ *Product rule*
$= (-7)^8$

(c) $y^3 \cdot y = y^3 \cdot y^1 = y^{3+1}$ *Product rule*
$= y^4$

(d) $z^2 \cdot z^5 \cdot z^6 = z^{2+5+6} = z^{13}$

(e) $4^2 \cdot 3^5$ ▪ The product rule does not apply because the bases are different. We can evaluate the product as follows:
$$4^2 \cdot 3^5 = 16 \cdot 243 = 3888$$

(f) $6^4 + 6^2$ ▪ The product rule does not apply because it is a sum, not a product. We can evaluate the sum as follows:
$$6^4 + 6^2 = 1296 + 36 = 1332$$

4. **(a)** $5m^2 \cdot 2m^6 = (5 \cdot 2) \cdot (m^2 \cdot m^6)$
$= 10m^{2+6} = 10m^8$

(b) $3p^5 \cdot 9p^4 = (3 \cdot 9) \cdot (p^5 \cdot p^4)$
$= 27p^{5+4} = 27p^9$

(c) $-7p^5 \cdot (3p^8) = (-7 \cdot 3) \cdot (p^5 \cdot p^8)$
$= -21p^{5+8} = -21p^{13}$

5. **(a)** $(5^3)^4 = 5^{3 \cdot 4}$ *Power rule (a)*
$= 5^{12}$

(b) $(6^2)^5 = 6^{2 \cdot 5}$ *Power rule (a)*
$= 6^{10}$

(c) $(3^2)^4 = 3^{2 \cdot 4}$ *Power rule (a)*
$= 3^8$

(d) $(a^6)^5 = a^{6 \cdot 5}$ *Power rule (a)*
$= a^{30}$

6. **(a)** $(2ab)^4 = 2^4 a^4 b^4$ *Power rule (b)*
$= 16a^4 b^4$

(b) $5(mn)^3 = 5(m^3 n^3)$ *Power rule (b)*
$= 5m^3 n^3$

(c) $(3a^2 b^4)^5 = 3^5 (a^2)^5 (b^4)^5$ *Power rule (b)*
$= 3^5 a^{10} b^{20}$ *Power rule (a)*
$= 243 a^{10} b^{20}$

(d) $(-5m^2)^3$

$$= (-1 \cdot 5 \cdot m^2)^3 \qquad -a = -1 \cdot a$$
$$= (-1)^3 \cdot 5^3 \cdot (m^2)^3 \quad \textit{Power rule (b)}$$
$$= -1 \cdot 125 \cdot m^{2 \cdot 3} \quad \textit{Power rule (a)}$$
$$= -125 m^6$$

7. (a) $\left(\dfrac{5}{2}\right)^4 = \dfrac{5^4}{2^4} \quad \textit{Power rule (c)}$

$$= \dfrac{625}{16}$$

(b) $\left(\dfrac{p}{q}\right)^2 = \dfrac{p^2}{q^2} \quad \textit{Power rule (c)}$

(c) $\left(\dfrac{r}{t}\right)^3 = \dfrac{r^3}{t^3} \quad \textit{Power rule (c)}$

(d) $\left(\dfrac{1}{3}\right)^5 = \dfrac{1^5}{3^5} \quad \textit{Power rule (c)}$

$$= \dfrac{1}{243}$$

(e) $\left(\dfrac{1}{x}\right)^{10} = \dfrac{1^{10}}{x^{10}} \quad \textit{Power rule (c)}$

$$= \dfrac{1}{x^{10}}$$

8. (a) $(2m)^5 (2m)^3$

$$= (2m)^{5+3} \qquad\qquad \textit{Product rule}$$
$$= (2m)^8$$
$$= 2^8 m^8, \quad \text{or} \quad 256m^8 \quad \textit{Power rule (b)}$$

(b) $\left(\dfrac{5k^3}{3}\right)^2$

$$= \dfrac{(5k^3)^2}{3^2} \qquad\qquad \textit{Power rule (c)}$$
$$= \dfrac{5^2 (k^3)^2}{3^2} \qquad\quad \textit{Power rule (b)}$$
$$= \dfrac{5^2 k^{3 \cdot 2}}{3^2} \qquad\quad \textit{Power rule (a)}$$
$$= \dfrac{5^2 k^6}{3^2}, \quad \text{or} \quad \dfrac{25k^6}{9}$$

(c) $\left(\dfrac{1}{5}\right)^4 (2x)^2$

$$= \dfrac{1^4}{5^4}(2x)^2 \qquad\quad \textit{Power rule (c)}$$
$$= \dfrac{1}{5^4}(2^2 x^2) \qquad\quad \textit{Power rule (b)}$$
$$= \dfrac{2^2 x^2}{5^4}, \quad \text{or} \quad \dfrac{4x^2}{625}$$

(d) $(-3xy^2)^3 (x^2 y)^4$

$$= (-1)^3 3^3 x^3 (y^2)^3 (x^2)^4 y^4 \quad \textit{Power rule (b)}$$
$$= (-1)^3 3^3 x^3 y^{2 \cdot 3} x^{2 \cdot 4} y^4 \quad \textit{Power rule (a)}$$
$$= -3^3 x^3 y^6 x^8 y^4$$
$$= -3^3 x^{3+8} y^{6+4} \qquad\qquad \textit{Product rule}$$
$$= -3^3 x^{11} y^{10}, \quad \text{or} \quad -27x^{11} y^{10}$$

9. (a) Use the formula for the area of a rectangle, $A = lw$, with $l = 8x^4$ and $w = 4x^2$.

$$A = (8x^4)(4x^2) \quad \textit{Area formula}$$
$$= 8 \cdot 4 \cdot x^{4+2} \quad \textit{Product rule}$$
$$= 32x^6$$

(b) Use the formula for the area of a triangle, $A = \frac{1}{2}bh$, with $b = 10x^6$ and $h = 5x^4$.

$$A = \tfrac{1}{2}(10x^6)(5x^4) \qquad \textit{Area formula}$$
$$= \tfrac{1}{2}(10 \cdot 5 \cdot x^{6+4}) \quad \textit{Product rule}$$
$$= 25x^{10}$$

13.2   Section Exercises

1. $3^3 = 3 \cdot 3 \cdot 3 = 27$, so the statement $3^3 = 9$ is *false*.

3. $(a^2)^3 = a^{2(3)} = a^6$, so the statement $(a^2)^3 = a^5$ is *false*.

5. xy^2 can be written as $x^1 y^2$. The exponent on the base x is understood to be 1.

7. $\underbrace{t \cdot t \cdot t \cdot t \cdot t \cdot t \cdot t}_{\text{7 factors of } t} = t^7$

9. $\underbrace{\left(\dfrac{1}{2}\right)\left(\dfrac{1}{2}\right)\left(\dfrac{1}{2}\right)\left(\dfrac{1}{2}\right)\left(\dfrac{1}{2}\right)}_{\text{5 factors of } \left(\frac{1}{2}\right)} = \left(\dfrac{1}{2}\right)^5$

11. $\underbrace{(-8p)(-8p)}_{\text{2 factors of } (-8p)} = (-8p)^2$

13. In the exponential expression 3^5, the base is 3 and the exponent is 5.

$$3^5 = 3 \cdot 3 \cdot 3 \cdot 3 \cdot 3 = 243$$

15. In the exponential expression $(-3)^5$, the base is -3 and the exponent is 5.

$$(-3)^5 = (-3)(-3)(-3)(-3)(-3) = -243$$

17. In the exponential expression $(-6x)^4$, the base is $-6x$ and the exponent is 4.

19. In the exponential expression $-6x^4$, -6 is not part of the base. The base is x and the exponent is 4.

21. $5^2 \cdot 5^6 = 5^{2+6} = 5^8$

23. $4^2 \cdot 4^7 \cdot 4^3 = 4^{2+7+3} = 4^{12}$

25. $(-7)^3 (-7)^6 = (-7)^{3+6} = (-7)^9$

27. $t^3 t^8 t^{13} = t^{3+8+13} = t^{24}$

29. $(-8r^4)(7r^3) = -8 \cdot 7 \cdot r^4 \cdot r^3$
$$= -56 r^{4+3}$$
$$= -56 r^7$$

31. $(-6p^5)(-7p^5) = (-6)(-7)p^5 \cdot p^5$
$$= 42p^{5+5}$$
$$= 42p^{10}$$

33. $3^8 + 3^9$ ■ The product rule does not apply because it is a sum, not a product.

35. $5^8 \cdot 3^8$ ■ The product rule does not apply because the bases are different.

37. $(4^3)^2 = 4^{3 \cdot 2}$ *Power rule (a)*
$$= 4^6$$

39. $(t^4)^5 = t^{4 \cdot 5}$ *Power rule (a)*
$$= t^{20}$$

41. $(7r)^3 = 7^3 r^3$ *Power rule (b)*
$$= 343r^3$$

43. $(-5^2)^6$
$$= (-1 \cdot 5^2)^6 \qquad -a = -1 \cdot a$$
$$= (-1)^6 \cdot (5^2)^6 \quad \text{\emph{Power rule (b)}}$$
$$= 1 \cdot 5^{2 \cdot 6} \qquad \text{\emph{Power rule (a)}}$$
$$= 5^{12}$$

45. $(-8^3)^5$
$$= (-1 \cdot 8^3)^5 \qquad -a = -1 \cdot a$$
$$= (-1)^5 \cdot (8^3)^5 \quad \text{\emph{Power rule (b)}}$$
$$= -1 \cdot 8^{3 \cdot 5} \qquad \text{\emph{Power rule (a)}}$$
$$= -8^{15}$$

47. $(5xy)^5 = 5^5 x^5 y^5$ *Power rule (b)*

49. $8(qr)^3 = 8q^3 r^3$ *Power rule (b)*

51. $\left(\dfrac{1}{2}\right)^3 = \dfrac{1^3}{2^3}$ *Power rule (c)*
$$= \dfrac{1}{8}$$

53. $\left(\dfrac{a}{b}\right)^3 (b \neq 0) = \dfrac{a^3}{b^3}$ *Power rule (c)*

55. $\left(\dfrac{9}{5}\right)^8 = \dfrac{9^8}{5^8}$ *Power rule (c)*

57. $(-2x^2 y)^3 = (-2)^3 \cdot (x^2)^3 \cdot y^3$
$$= (-2)^3 \cdot x^{2 \cdot 3} \cdot y^3$$
$$= -8x^6 y^3$$

59. $(3a^3 b^2)^2 = 3^2 \cdot (a^3)^2 \cdot (b^2)^2$
$$= 3^2 \cdot a^{3 \cdot 2} \cdot b^{2 \cdot 2}$$
$$= 9a^6 b^4$$

61. $\left(\dfrac{5}{2}\right)^3 \cdot \left(\dfrac{5}{2}\right)^2 = \left(\dfrac{5}{2}\right)^{3+2}$ *Product rule*
$$= \left(\dfrac{5}{2}\right)^5$$
$$= \dfrac{5^5}{2^5} \qquad \text{\emph{Power rule (c)}}$$

63. $\left(\dfrac{9}{8}\right)^3 \cdot 9^2 = \dfrac{9^3}{8^3} \cdot \dfrac{9^2}{1}$ *Power rule (c)*
$$= \dfrac{9^3 \cdot 9^2}{8^3 \cdot 1} \qquad \text{\emph{Multiply fractions}}$$
$$= \dfrac{9^{3+2}}{8^3} \qquad \text{\emph{Product rule}}$$
$$= \dfrac{9^5}{8^3}$$

65. $(2x)^9 (2x)^3 = (2x)^{9+3}$ *Product rule*
$$= (2x)^{12}$$
$$= 2^{12} x^{12} \qquad \text{\emph{Power rule (b)}}$$

67. $(-6p)^4 (-6p)$
$$= (-6p)^4 (-6p)^1$$
$$= (-6p)^5 \qquad \text{\emph{Product rule}}$$
$$= (-1 \cdot 6p)^5 \qquad -a = -1 \cdot a$$
$$= (-1)^5 6^5 p^5 \quad \text{\emph{Power rule (b)}}$$
$$= -1 \cdot 6^5 p^5$$
$$= -6^5 p^5$$

69. $(6x^2 y^3)^5 = 6^5 (x^2)^5 (y^3)^5$ *Power rule (b)*
$$= 6^5 x^{2 \cdot 5} y^{3 \cdot 5} \qquad \text{\emph{Power rule (a)}}$$
$$= 6^5 x^{10} y^{15}$$

71. $(x^2)^3 (x^3)^5 = x^6 \cdot x^{15}$ *Power rule (a)*
$$= x^{21} \qquad \text{\emph{Product rule}}$$

73. $(2w^2 x^3 y)^2 (x^4 y)^5$
$$= [2^2 (w^2)^2 (x^3)^2 y^2] [(x^4)^5 y^5] \quad \substack{\text{\emph{Power}} \\ \text{\emph{rule (b)}}}$$
$$= (2^2 w^4 x^6 y^2)(x^{20} y^5) \qquad \substack{\text{\emph{Power}} \\ \text{\emph{rule (a)}}}$$
$$= 2^2 w^4 (x^6 x^{20})(y^2 y^5)$$
$$\text{\emph{Commutative and associative properties}}$$
$$= 4w^4 x^{26} y^7$$

75. $(-r^4 s)^2 (-r^2 s^3)^5$
$$= [(-1)r^4 s]^2 [(-1)r^2 s^3]^5$$
$$= [(-1)^2 (r^4)^2 s^2] [(-1)^5 (r^2)^5 (s^3)^5]$$
$$\qquad\qquad \text{\emph{Power rule (b)}}$$
$$= [(-1)^2 r^8 s^2] [(-1)^5 r^{10} s^{15}] \quad \substack{\text{\emph{Power}} \\ \text{\emph{rule (a)}}}$$
$$= (-1)^7 r^{18} s^{17} \qquad \text{\emph{Product rule}}$$
$$= -r^{18} s^{17}$$

77. $\left(\dfrac{5a^2b^5}{c^6}\right)^3$ $(c \neq 0)$

$= \dfrac{(5a^2b^5)^3}{(c^6)^3}$ *Power rule (c)*

$= \dfrac{5^3(a^2)^3(b^5)^3}{(c^6)^3}$ *Power rule (b)*

$= \dfrac{125a^6b^{15}}{c^{18}}$ *Power rule (a)*

79. $(-5m^3p^4q)^2(p^2q)^3$

$= (-1 \cdot 5m^3p^4q)^2(p^2q)^3$

$= (-1)^2 \cdot 5^2 \cdot (m^3)^2 \cdot (p^4)^2 \cdot q^2 \cdot (p^2)^3 \cdot q^3$

$= 1 \cdot 25 \cdot m^{3 \cdot 2} \cdot p^{4 \cdot 2} \cdot q^2 \cdot p^{2 \cdot 3} \cdot q^3$

$= 25m^6p^8q^2p^6q^3$

$= 25m^6p^{8+6}q^{2+3}$

$= 25m^6p^{14}q^5$

81. $(2x^2y^3z)^4(xy^2z^3)^2$

$= 2^4 \cdot (x^2)^4 \cdot (y^3)^4 \cdot z^4 \cdot x^2 \cdot (y^2)^2 \cdot (z^3)^2$

$= 16x^{2 \cdot 4} \cdot y^{3 \cdot 4} \cdot z^4 \cdot x^2 \cdot y^{2 \cdot 2} \cdot z^{3 \cdot 2}$

$= 16x^8y^{12}z^4x^2y^4z^6$

$= 16x^{8+2}y^{12+4}z^{4+6}$

$= 16x^{10}y^{16}z^{10}$

83. To simplify $(10^2)^3$ as 1000^6 is not correct. Using power rule (a) to simplify $(10^2)^3$, we obtain

$(10^2)^3 = 10^{2 \cdot 3}$

$= 10^6$

$= 10 \cdot 10 \cdot 10 \cdot 10 \cdot 10 \cdot 10$

$= 1,000,000.$

85. Use the formula for the area of a rectangle, $A = lw$, with $l = 10x^5$ and $w = 3x^2$.

$A = (10x^5)(3x^2)$

$= 10 \cdot 3 \cdot x^5 \cdot x^2$

$= 30x^7$

87. Use the formula for the area of a parallelogram, $A = bh$, with $b = 2p^5$ and $h = 3p^2$.

$A = (2p^5)(3p^2)$

$= 2 \cdot 3 \cdot p^5 \cdot p^2$

$= 6p^7$

13.3 Multiplying Polynomials

13.3 Margin Exercises

1. **(a)** $5m^3(2m + 7)$

$= 5m^3(2m) + 5m^3(7)$ *Distributive property*

$= 10m^4 + 35m^3$ *Multiply monomials.*

(b) $2x^4(3x^2 + 2x - 5)$

$= 2x^4(3x^2) + 2x^4(2x) + 2x^4(-5)$
     Distributive property

$= 6x^6 + 4x^5 + (-10x^4)$
     Multiply monomials.

$= 6x^6 + 4x^5 - 10x^4$

(c) $-4y^2(3y^3 + 2y^2 - 4y + 8)$

$= -4y^2(3y^3) + (-4y^2)(2y^2)$
$\quad + (-4y^2)(-4y) + (-4y^2)(8)$
     Distributive property

$= -12y^5 + (-8y^4) + (16y^3) + (-32y^2)$
     Multiply monomials.

$= -12y^5 - 8y^4 + 16y^3 - 32y^2$

2. **(a)** $(m + 3)(m^2 - 2m + 1)$

$= m(m^2) + m(-2m) + m(1)$
$\quad + 3(m^2) + 3(-2m) + 3(1)$

$= m^3 - 2m^2 + m + 3m^2 - 6m + 3$

$= m^3 + m^2 - 5m + 3$

(b) $(6p^2 + 2p - 4)(3p^2 - 5)$

$= 6p^2(3p^2) + 6p^2(-5) + (2p)(3p^2)$
$\quad + (2p)(-5) + (-4)(3p^2) + (-4)(-5)$

$= 18p^4 - 30p^2 + 6p^3 - 10p - 12p^2 + 20$

$= 18p^4 + 6p^3 - 42p^2 - 10p + 20$

3.

$$
\begin{array}{r}
3x^2 + 4x - 5 \\
x + 4 \\
\hline
12x^2 + 16x - 20 \quad \leftarrow \quad 4(3x^2 + 4x - 5) \\
3x^3 + 4x^2 - 5x \quad\quad\quad \leftarrow \quad x(3x^2 + 4x - 5) \\
\hline
3x^3 + 16x^2 + 11x - 20 \leftarrow \textit{Add like terms.}
\end{array}
$$

4. **(a)**

	x	2
$4x$	$4x^2$	$8x$
3	$3x$	6

$(4x + 3)(x + 2) = 4x^2 + 8x + 3x + 6$

$\qquad\qquad\qquad\quad = 4x^2 + 11x + 6$

(b)

	x^2	$3x$	1
x	x^3	$3x^2$	x
5	$5x^2$	$15x$	5

$(x + 5)(x^2 + 3x + 1)$

$= x^3 + 3x^2 + x + 5x^2 + 15x + 5$

$= x^3 + 8x^2 + 16x + 5$

5. $(2p - 5)(3p + 7)$

(a) The product of the first terms is $\underline{2p}(\underline{3p}) = \underline{6p^2}$.

(b) The outer product is $\underline{2p}(\underline{7}) = \underline{14p}$.

(c) The inner product is $-5(3p) = -15p$.

(d) The product of the last terms is $-5(7) = -35$.

(e) The complete product is the sum of the terms in parts (a)–(d), that is,

$$6p^2 + 14p + (-15p) + (-35).$$

Simplified, the complete product is

$$\underline{6p^2 - p - 35.}$$

6. (a) $(m + 4)(m - 3)$

$$
\begin{aligned}
\text{F:} \quad & m(\underline{m}) = m^2 \\
\text{O:} \quad & m(\underline{-3}) = -3m \\
\text{I:} \quad & 4(\underline{m}) = 4m \\
\text{L:} \quad & 4(\underline{-3}) = -12
\end{aligned}
$$

$$
\begin{aligned}
(m + 4)(m - 3) &= m^2 - 3m + 4m - 12 \\
&= \underline{m^2 + m - 12}
\end{aligned}
$$

(b) $(y + 7)(y + 2)$

$$
\begin{aligned}
\text{F:} \quad & y(y) = y^2 \\
\text{O:} \quad & y(2) = 2y \\
\text{I:} \quad & 7(y) = 7y \\
\text{L:} \quad & 7(2) = 14
\end{aligned}
$$

$$
\begin{aligned}
(y + 7)(y + 2) &= y^2 + 2y + 7y + 14 \\
&= y^2 + 9y + 14
\end{aligned}
$$

(c) $(r - 8)(r - 5)$

$$
\begin{aligned}
\text{F:} \quad & r(r) = r^2 \\
\text{O:} \quad & r(-5) = -5r \\
\text{I:} \quad & -8(r) = -8r \\
\text{L:} \quad & -8(-5) = 40
\end{aligned}
$$

$$
\begin{aligned}
(r - 8)(r - 5) &= r^2 - 5r - 8r + 40 \\
&= r^2 - 13r + 40
\end{aligned}
$$

7. $(4x - 3)(2y + 5)$

$$
\begin{aligned}
\text{F:} \quad & 4x(2y) = 8xy \\
\text{O:} \quad & 4x(5) = 20x \\
\text{I:} \quad & -3(2y) = -6y \\
\text{L:} \quad & -3(5) = -15
\end{aligned}
$$

$$(4x - 3)(2y + 5) = 8xy + 20x - 6y - 15$$

8. (a) $(6m + 5)(m - 4)$

$$
\begin{aligned}
\text{F:} \quad & 6m(\underline{m}) = 6m^2 \\
\text{O:} \quad & 6m(\underline{-4}) = -24m \\
\text{I:} \quad & 5(\underline{m}) = 5m \\
\text{L:} \quad & 5(\underline{-4}) = -20
\end{aligned}
$$

$$
\begin{aligned}
(6m + 5)(m - 4) &= 6m^2 - 24m + 5m - 20 \\
&= \underline{6m^2 - 19m - 20}
\end{aligned}
$$

(b) $(3r + 2t)(3r + 4t)$

$$
\begin{aligned}
\text{F:} \quad & 3r(3r) = 9r^2 \\
\text{O:} \quad & 3r(4t) = 12rt \\
\text{I:} \quad & 2t(3r) = 6rt \\
\text{L:} \quad & 2t(4t) = 8t^2
\end{aligned}
$$

$$
\begin{aligned}
(3r + 2t)(3r + 4t) &= 9r^2 + 12rt + 6rt + 8t^2 \\
&= 9r^2 + 18rt + 8t^2
\end{aligned}
$$

(c) $y^2(8y + 3)(2y + 1)$

First multiply the binomials and then multiply that result by y^2.

$$
\begin{aligned}
\text{F:} \quad & 8y(2y) = 16y^2 \\
\text{O:} \quad & 8y(1) = 8y \\
\text{I:} \quad & 3(2y) = 6y \\
\text{L:} \quad & 3(1) = 3
\end{aligned}
$$

$$
\begin{aligned}
y^2(8y + 3)(2y + 1) &= y^2(16y^2 + 8y + 6y + 3) \\
&= y^2(\underline{16y^2 + 14y + 3}) \\
&= \underline{16y^4 + 14y^3 + 3y^2}
\end{aligned}
$$

13.3 Section Exercises

1. (a) $5x^3(6x^7)$

$$
\begin{aligned}
&= 5 \cdot 6x^{3+7} \\
&= 30x^{10}
\end{aligned}
$$

Choice **B** is correct.

(b) $-5x^7(6x^3)$

$$
\begin{aligned}
&= -5 \cdot 6x^{7+3} \\
&= -30x^{10}
\end{aligned}
$$

Choice **D** is correct.

(c) $(5x^7)^3 = (5)^3(x^7)^3$

$$
\begin{aligned}
&= 125x^{7 \cdot 3} \\
&= 125x^{21}
\end{aligned}
$$

Choice **A** is correct.

(d) $(-6x^3)^3 = (-6)^3(x^3)^3$

$$
\begin{aligned}
&= -216x^{3 \cdot 3} \\
&= -216x^9
\end{aligned}
$$

Choice **C** is correct.

3. The first property that is used is the <u>distributive</u> property. $4x$ is being distributed over $(3x^2 + 7x^3)$.

5. $5p(3q^2) = 5(3)pq^2$

$$= 15pq^2$$

7. $-6m^3(3n^2) = -6(3)m^3n^2$

$$= -18m^3n^2$$

9. $y^5 \cdot 9y \cdot y^4 = 9y^{5+1+4}$
$$= 9y^{10}$$

11. $(4x^3)(2x^2)(-x^5) = (-4 \cdot 2)x^{3+2+5}$
$$= -8x^{10}$$

13. $-2m(3m + 2)$

$= -2m(3m) + (-2m)(2)$ *Distributive property*

$= -6m^2 - 4m$ *Multiply monomials.*

15. $\frac{3}{4}p(8 - 6p + 12p^3)$
$$= \frac{3}{4}p(8) + \frac{3}{4}p(-6p) + \frac{3}{4}p(12p^3)$$
$$= 6p - \frac{9}{2}p^2 + 9p^4$$

17. $2y^5(5y^4 + 2y + 3)$
$$= 2y^5(5y^4) + 2y^5(2y) + 2y^5(3)$$
$$= 10y^9 + 4y^6 + 6y^5$$

19. $2y^3(3y^3 + 2y + 1)$
$$= 2y^3(3y^3) + 2y^3(2y) + 2y^3(1)$$
$$= 6y^6 + 4y^4 + 2y^3$$

21. $-4r^3(-7r^2 + 8r - 9)$
$$= -4r^3(-7r^2) + (-4r^3)(8r) + (-4r^3)(-9)$$
$$= 28r^5 - 32r^4 + 36r^3$$

23. $3a^2(2a^2 - 4ab + 5b^2)$
$$= 3a^2(2a^2) + 3a^2(-4ab) + 3a^2(5b^2)$$
$$= 6a^4 - 12a^3b + 15a^2b^2$$

25. $7m^3n^2(3m^2 + 2mn - n^3)$
$$= 7m^3n^2(\underline{3m^2}) + 7m^3n^2(\underline{2mn})$$
$$+ 7m^3n^2(\underline{-n^3})$$
$$= \underline{21m^5n^2 + 14m^4n^3 - 7m^3n^5}$$

In Exercises 27–36, we can multiply the polynomials horizontally or vertically. The following solutions illustrate these two methods.

27. $(6x + 1)(2x^2 + 4x + 1)$
$$= (6x)(2x^2) + (6x)(4x) + (6x)(1)$$
$$+ (1)(2x^2) + (1)(4x) + (1)(1)$$
$$= 12x^3 + 24x^2 + 6x + 2x^2 + 4x + 1$$
$$= 12x^3 + 26x^2 + 10x + 1$$

29. $(2r - 1)(3r^2 + 4r - 4)$

Multiply vertically.

$$
\begin{array}{rrrr}
3r^2 & + & 4r & - & 4 \\
 & & 2r & - & 1 \\
\hline
-3r^2 & - & 4r & + & 4 \\
6r^3 + 8r^2 & - & 8r & & \\
\hline
6r^3 + 5r^2 & - & 12r & + & 4
\end{array}
$$

31. $(4m + 3)(5m^3 - 4m^2 + m - 5)$

Multiply vertically.

$$
\begin{array}{rrrrr}
5m^3 & - & 4m^2 & + & m & - & 5 \\
 & & & & 4m & + & 3 \\
\hline
15m^3 & - & 12m^2 & + & 3m & - & 15 \\
20m^4 - 16m^3 & + & 4m^2 & - & 20m & & \\
\hline
20m^4 - & m^3 & - & 8m^2 & - & 17m & - & 15
\end{array}
$$

33. $(5x^2 + 2x + 1)(x^2 - 3x + 5)$

Multiply vertically.

$$
\begin{array}{rrrrr}
5x^2 & + & 2x & + & 1 \\
x^2 & - & 3x & + & 5 \\
\hline
25x^2 & + & 10x & + & 5 \\
-15x^3 - & 6x^2 & - & 3x & & \\
5x^4 + 2x^3 & + & x^2 & & & \\
\hline
5x^4 - 13x^3 & + & 20x^2 & + & 7x & + & 5
\end{array}
$$

35. $(6x^4 - 4x^2 + 8x)(\frac{1}{2}x + 3)$
$$= 6x^4(\tfrac{1}{2}x) + -4x^2(\tfrac{1}{2}x) + 8x(\tfrac{1}{2}x)$$
$$+ 6x^4(3) + -4x^2(3) + 8x(3)$$
$$= 3x^5 - 2x^3 + 4x^2 + 18x^4 - 12x^2 + 24x$$
$$= 3x^5 + 18x^4 - 2x^3 - 8x^2 + 24x$$

37. $(x + 3)(x + 4)$

	x	4
x	x^2	$4x$
3	$3x$	12

$(x + 3)(x + 4) = x^2 + 4x + 3x + 12$
$$= x^2 + 7x + 12$$

39. $(2x + 1)(x^2 + 3x + 2)$

	x^2	$3x$	2
$2x$	$2x^3$	$6x^2$	$4x$
1	x^2	$3x$	2

$(2x + 1)(x^2 + 3x + 2)$
$$= 2x^3 + 6x^2 + 4x + x^2 + 3x + 2$$
$$= 2x^3 + 7x^2 + 7x + 2$$

41. Multiply each term of the second polynomial by each term of the first. Then combine like terms.

$(m + 7)(m + 5)$

  F   **O**   **I**   **L**

$= m(m) + m(5) + 7(m) + 7(5)$
$$= m^2 + 5m + 7m + 35$$
$$= m^2 + 12m + 35$$

43. Multiply each term of the second polynomial by each term of the first. Then combine like terms.

$(n - 2)(n + 3)$

  F   **O**   **I**   **L**

$= n(n) + n(3) + (-2)(n) + (-2)(3)$
$$= n^2 + 3n + (-2n) + (-6)$$
$$= n^2 + n - 6$$

45. Multiply each term of the second polynomial by each term of the first. Then combine like terms.

$$(4r + 1)(2r - 3)$$
$$\quad\text{F}\qquad\text{O}\qquad\text{I}\qquad\text{L}$$
$$= 4r(2r) + 4r(-3) + 1(2r) + 1(-3)$$
$$= 8r^2 + (-12r) + 2r + (-3)$$
$$= 8r^2 - 10r - 3$$

47. $(3x + 2)(3x - 2)$
$$\quad\text{F}\qquad\text{O}\qquad\text{I}\qquad\text{L}$$
$$= 3x(3x) + 3x(-2) + 2(3x) + 2(-2)$$
$$= 9x^2 - 6x + 6x - 4$$
$$= 9x^2 - 4$$

49. $(3q + 1)(3q + 1)$
$$\quad\text{F}\qquad\text{O}\qquad\text{I}\qquad\text{L}$$
$$= 3q(3q) + 3q(1) + 1(3q) + 1(1)$$
$$= 9q^2 + 3q + 3q + 1$$
$$= 9q^2 + 6q + 1$$

51. $(5x + 7)(3y - 8)$
$$\quad\text{F}\qquad\text{O}\qquad\text{I}\qquad\text{L}$$
$$= 5x(3y) + 5x(-8) + 7(3y) + 7(-8)$$
$$= 15xy - 40x + 21y - 56$$

53. $(3t + 4s)(2t + 5s)$
$$\quad\text{F}\qquad\text{O}\qquad\text{I}\qquad\text{L}$$
$$= 3t(2t) + 3t(5s) + 4s(2t) + 4s(5s)$$
$$= 6t^2 + 15st + 8st + 20s^2$$
$$= 6t^2 + 23st + 20s^2$$

55. $(-0.3t + 0.4)(t + 0.6)$
$$\quad\text{F}\qquad\text{O}\qquad\text{I}\qquad\text{L}$$
$$= -0.3t(t) + (-0.3t)(0.6) + 0.4(t) + 0.4(0.6)$$
$$= -0.3t^2 - 0.18t + 0.4t + 0.24$$
$$= -0.3t^2 + 0.22t + 0.24$$

57. $\left(x - \frac{2}{3}\right)\left(x + \frac{1}{4}\right)$
$$\quad\text{F}\qquad\text{O}\qquad\text{I}\qquad\text{L}$$
$$= (x)(x) + (x)\left(\tfrac{1}{4}\right) + \left(-\tfrac{2}{3}\right)(x) + \left(-\tfrac{2}{3}\right)\left(\tfrac{1}{4}\right)$$
$$= x^2 + \tfrac{1}{4}x - \tfrac{2}{3}x - \tfrac{1}{6}$$
$$= x^2 + \left(\tfrac{3}{12}x - \tfrac{8}{12}x\right) - \tfrac{1}{6}$$
$$= x^2 - \tfrac{5}{12}x - \tfrac{1}{6}$$

59. $\left(-\frac{5}{4} + 2r\right)\left(-\frac{3}{4} - r\right)$
$$\qquad\text{F}\qquad\qquad\text{O}$$
$$= \left(-\tfrac{5}{4}\right)\left(-\tfrac{3}{4}\right) + \left(-\tfrac{5}{4}\right)(-r)$$
$$\qquad\text{I}\qquad\qquad\text{L}$$
$$+ (2r)\left(-\tfrac{3}{4}\right) + (2r)(-r)$$
$$= \tfrac{15}{16} + \tfrac{5}{4}r - \tfrac{6}{4}r - 2r^2$$
$$= \tfrac{15}{16} - \tfrac{1}{4}r - 2r^2$$

61. $x(2x - 5)(x + 3)$
Use FOIL to multiply the binomials.

$$(2x - 5)(x + 3)$$
$$\quad\text{F}\qquad\text{O}\qquad\text{I}\qquad\text{L}$$
$$= 2x(x) + 2x(3) + (-5)(x) + (-5)(3)$$
$$= 2x^2 + 6x - 5x - 15$$
$$= 2x^2 + x - 15$$

Now multiply this result by x.

$$x(2x^2 + x - 15)$$
$$= x(2x^2) + x(x) + x(-15)$$
$$= 2x^3 + x^2 - 15x$$

63. $3y^3(2y + 3)(y - 5)$
Use FOIL to multiply the binomials.

$$(2y + 3)(y - 5)$$
$$\quad\text{F}\qquad\text{O}\qquad\text{I}\qquad\text{L}$$
$$= 2y(y) + 2y(-5) + 3y + 3(-5)$$
$$= 2y^2 - 10y + 3y - 15$$
$$= 2y^2 - 7y - 15$$

Now multiply this result by $3y^3$.

$$3y^3(2y^2 - 7y - 15)$$
$$= 3y^3(2y^2) + 3y^3(-7y) + 3y^3(-15)$$
$$= 6y^5 - 21y^4 - 45y^3$$

65. $-8r^3(5r^2 + 2)(5r^2 - 2)$

$$(5r^2 + 2)(5r^2 - 2)$$
$$\quad\text{F}\qquad\text{O}\qquad\text{I}\qquad\text{L}$$
$$= 5r^2(5r^2) + 5r^2(-2) + 2(5r^2) + 2(-2)$$
$$= 25r^4 - 10r^2 + 10r^2 - 4$$
$$= 25r^4 - 4$$

Now multiply this result by $-8r^3$.

$$-8r^3(25r^4 - 4)$$
$$= -8r^3(25r^4) + (-8r^3)(-4)$$
$$= -200r^7 + 32r^3$$

67. **(a)** Use the formula for the area of a rectangle, $A = lw$, with $l = 3y + 7$ and $w = y + 1$.

$$A = (3y + 7)(y + 1)$$
$$= (3y + 7)(y) + (3y + 7)(1)$$
$$= 3y^2 + 7y + 3y + 7$$
$$= 3y^2 + 10y + 7$$

(b) Use the formula for the perimeter of a rectangle, $P = 2l + 2w$, with $l = 3y + 7$ and $w = y + 1$.

$$P = 2(3y + 7) + 2(y + 1)$$
$$= 2(3y) + 2(7) + 2(y) + 2(1)$$
$$= 6y + 14 + 2y + 2$$
$$= 8y + 16$$

Relating Concepts (Exercises 69–74)

69. Area = width · length
$$= 10(3x + 6)$$
$$= 10(3x) + 10(6)$$
$$= 30x + 60$$

70. Area = 600
$$30x + 60 = 600 \quad \textit{Subtract 60.}$$
$$30x = 540 \quad \textit{Divide by 30.}$$
$$\frac{30x}{30} = \frac{540}{30}$$
$$x = 18$$

The solution set for the equation is {18}.

71. With $x = 18$, $3x + 6 = 3(18) + 6 = 60$.

The rectangle measures 10 yd by 60 yd.

72. Perimeter = 2(length) + 2(width)
$$= 2(60) + 2(10)$$
$$= 120 + 20$$
$$= 140 \text{ yd}$$

The perimeter of the rectangle is 140 yd.

73. ($3.50 per yd^2)(600 yd^2) = $2100

It will cost $2100 to sod the entire lawn.

74. ($9.00 per yd)(140 yd) = $1260

It will cost $1260 to fence the entire lawn.

13.4 Special Products

13.4 Margin Exercises

1. **(a)** $\underline{x}$ is the first term of the binomial $x + 4$.
$$(x)^2 = \underline{x^2}$$

(b) $\underline{4}$ is the last term of the binomial.
$$4^2 = \underline{16}$$

(c) Twice the product of the two terms of the binomial is
$$2 \cdot \underline{x} \cdot \underline{4} = \underline{8x}.$$

(d) Adding the terms from parts (a), (b), and (c) gives us
$$(x + 4)^2 = \underline{x^2 + 8x + 16}.$$

2. **(a)** $(t - 6)^2 = \underline{t}^2 - 2(\underline{t})(\underline{6}) + \underline{6}^2$
$$= \underline{t^2 - 12t + 36}$$

(b) $(2m - p)^2 = (2m)^2 - 2(2m)(p) + p^2$
$$= 4m^2 - 4mp + p^2$$

(c) $(4p + 3q)^2 = (\underline{4p})^2 + \underline{2(4p)(3q)} + (\underline{3q})^2$
$$= \underline{16p^2 + 24pq + 9q^2}$$

(d) $(5r - 6s)^2 = (5r)^2 - 2(5r)(6s) + (6s)^2$
$$= 25r^2 - 60rs + 36s^2$$

(e) $(3k - \frac{1}{2})^2 = (3k)^2 - 2(3k)(\frac{1}{2}) + (\frac{1}{2})^2$
$$= 9k^2 - 3k + \frac{1}{4}$$

(f) $x(2x + 7)^2 = x(4x^2 + 28x + 49)$
$$= 4x^3 + 28x^2 + 49x$$

3. **(a)** $(y + 3)(y - 3) = \underline{y}^2 - \underline{3}^2$
$$= \underline{y^2 - 9}$$

(b) $(10m + 7)(10m - 7) = (10m)^2 - 7^2$
$$= 100m^2 - 49$$

(c) $(7p + 2q)(7p - 2q) = (\underline{7p})^2 - (\underline{2q})^2$
$$= \underline{49p^2 - 4q^2}$$

(d) $(3r - \frac{1}{2})(3r + \frac{1}{2}) = (3r)^2 - (\frac{1}{2})^2$
$$= 9r^2 - \frac{1}{4}$$

(e) $3x(x^3 - 4)(x^3 + 4)$

First use the rule for the product of the sum and difference of two terms.
$$(x^3 - 4)(x^3 + 4) = (x^3)^2 - (4)^2$$
$$= x^6 - 16$$

Now multiply this result by $3x$.
$$3x(x^6 - 16) = 3x(x^6) + 3x(-16)$$
$$= 3x^7 - 48x$$

4. **(a)** $(m + 1)^3$

Since $(m + 1)^3 = \underline{(m + 1)^2(m + 1)}$, the first step is to find the product $(m + 1)^2$.
$$(m + 1)^2 = m^2 + 2 \cdot m \cdot 1 + 1$$
$$= \underline{m^2 + 2m + 1}$$

Now multiply this result by $m + 1$.
$$(m + 1)^3 = (m + 1)(m^2 + 2m + 1)$$
$$= m^3 + 2m^2 + m + m^2 + 2m + 1$$
$$= \underline{m^3 + 3m^2 + 3m + 1}$$

(b) Note that $(3k - 2)^4 = [(3k - 2)^2]^2$.

Since $(3k - 2)^2 = (3k)^2 - 2(3k)(2) + 2^2$
$$= 9k^2 - 12k + 4,$$

we have $(3k - 2)^4 = (9k^2 - 12k + 4)^2$.

Multiply vertically.

$$
\begin{array}{r}
9k^2 - 12k + 4 \\
9k^2 - 12k + 4 \\
\hline
36k^2 - 48k + 16 \\
-108k^3 + 144k^2 - 48k \\
81k^4 - 108k^3 + 36k^2 \\
\hline
81k^4 - 216k^3 + 216k^2 - 96k + 16
\end{array}
$$

(c) $-3x(x-4)^3$

$$= -3x(x-4)(x-4)^2$$
$$= -3x(x-4)(x^2-8x+16)$$
$$= -3x(x^3-8x^2+16x-4x^2+32x-64)$$
$$= -3x(x^3-12x^2+48x-64)$$
$$= -3x^4+36x^3-144x^2+192x$$

13.4 Section Exercises

1. The square of a binomial is a trinomial consisting of the <u>square</u> of the first term + <u>twice</u> the <u>product</u> of the two terms + the <u>square</u> of the last term.

3. The product of the sum and difference of two terms is the <u>difference</u> of the <u>squares</u> of the two terms.

5. $(2x+3)^2$

  (a) The square of the first term is
  $$(2x)^2 = (2x)(2x) = 4x^2.$$

  (b) Twice the product of the two terms is
  $$2(2x)(3) = 12x.$$

  (c) The square of the last term is
  $$3^2 = 9.$$

  (d) The final product is the trinomial
  $$4x^2 + 12x + 9.$$

In Exercises 7–30, use one of the following formulas for the square of a binomial:

$$(a+b)^2 = a^2 + 2ab + b^2$$
$$(a-b)^2 = a^2 - 2ab + b^2$$

7. $(p+2)^2 = p^2 + 2(p)(2) + 2^2$
$$= p^2 + 4p + 4$$

9. $(z-5)^2 = (z)^2 - 2(z)(5) + 5^2$
$$= z^2 - 10z + 25$$

11. $\left(x - \frac{3}{4}\right)^2 = (x)^2 - 2(x)\left(\frac{3}{4}\right) + \left(\frac{3}{4}\right)^2$
$$= x^2 - \frac{3}{2}x + \frac{9}{16}$$

13. $(v+0.4)^2 = (v)^2 + 2(v)(0.4) + (0.4)^2$
$$= v^2 + 0.8v + 0.16$$

15. $(4x-3)^2 = (4x)^2 - 2(4x)(3) + 3^2$
$$= 16x^2 - 24x + 9$$

17. $(10z+6)^2 = (10z)^2 + 2(10z)(6) + 6^2$
$$= 100z^2 + 120z + 36$$

19. $(2p+5q)^2 = (2p)^2 + 2(2p)(5q) + (5q)^2$
$$= 4p^2 + 20pq + 25q^2$$

21. $(0.8t+0.7s)^2$
$$= (0.8t)^2 + 2(0.8t)(0.7s) + (0.7s)^2$$
$$= 0.64t^2 + 1.12ts + 0.49s^2$$

23. $\left(5x + \frac{2}{5}y\right)^2 = (5x)^2 + 2(5x)\left(\frac{2}{5}y\right) + \left(\frac{2}{5}y\right)^2$
$$= 25x^2 + 4xy + \frac{4}{25}y^2$$

25. $t(3t-1)^2$

First square the binomial.

$$(3t-1)^2 = (3t)^2 - 2(3t)(1) + 1^2$$
$$= 9t^2 - 6t + 1$$

Now multiply by t.

$$t(9t^2 - 6t + 1) = 9t^3 - 6t^2 + t$$

27. $3t(4t+1)^2$

First square the binomial.

$$(4t+1)^2 = (4t)^2 + 2(4t)(1) + 1^2$$
$$= 16t^2 + 8t + 1$$

Now multiply by $3t$.

$$3t(16t^2 + 8t + 1) = 48t^3 + 24t^2 + 3t$$

29. $-(4r-2)^2$

First square the binomial.

$$(4r-2)^2 = (4r)^2 - 2(4r)(2) + 2^2$$
$$= 16r^2 - 16r + 4$$

Now multiply by -1.

$$-1(16r^2 - 16r + 4) = -16r^2 + 16r - 4$$

31. $(7x+3y)(7x-3y)$

  (a) $7x(7x) = 49x^2$

  (b) $7x(-3y) + 3y(7x)$
  $$= -21xy + 21xy = 0$$

  (c) $3y(-3y) = -9y^2$

  (d) $49x^2 - 9y^2$

The sum found in part (b) is omitted because it is 0. Adding 0, the identity element for addition, would not change the answer.

In Exercises 33–48, use the following formula for the product of the sum and the difference of two terms.

$$(a+b)(a-b) = a^2 - b^2$$

33. $(q+2)(q-2) = q^2 - 2^2$
$$= q^2 - 4$$

35. $\left(r - \frac{3}{4}\right)\left(r + \frac{3}{4}\right) = r^2 - \left(\frac{3}{4}\right)^2$
$$= r^2 - \frac{9}{16}$$

37. $(s + 2.5)(s - 2.5) = s^2 - (2.5)^2$
$$= s^2 - 6.25$$

39. $(2w + 5)(2w - 5) = (2w)^2 - 5^2$
$$= 4w^2 - 25$$

41. $(10x + 3y)(10x - 3y) = (10x)^2 - (3y)^2$
$$= 100x^2 - 9y^2$$

43. $(2x^2 - 5)(2x^2 + 5) = (2x^2)^2 - 5^2$
$$= 4x^4 - 25$$

45. $\left(7x + \frac{3}{7}\right)\left(7x - \frac{3}{7}\right) = (7x)^2 - \left(\frac{3}{7}\right)^2$
$$= 49x^2 - \frac{9}{49}$$

47. $p(3p + 7)(3p - 7)$

First use the rule for the product of the sum and difference of two terms.

$$(3p + 7)(3p - 7) = (3p)^2 - 7^2$$
$$= 9p^2 - 49$$

Now multiply by p.

$$p(9p^2 - 49) = 9p^3 - 49p$$

49. $(m - 5)^3$

$= (m - 5)^2(m - 5)$ $a^3 = a^2 \cdot a$

$= (m^2 - 10m + 25)(m - 5)$ *Square the binomial.*

$= m^3 - 10m^2 + 25m$

$\quad - 5m^2 + 50m - 125$ *Multiply polynomials.*

$= m^3 - 15m^2 + 75m - 125$ *Combine like terms.*

51. $(y + 2)^3$

$= (y + 2)^2(y + 2)$ $a^3 = a^2 \cdot a$

$= (y^2 + 4y + 4)(y + 2)$ *Square the binomial.*

$= y^3 + 4y^2 + 4y$

$\quad + 2y^2 + 8y + 8$ *Multiply polynomials.*

$= y^3 + 6y^2 + 12y + 8$ *Combine like terms.*

53. $(2a + 1)^3$

$= (2a + 1)^2(2a + 1)$ $a^3 = a^2 \cdot a$

$= (4a^2 + 4a + 1)(2a + 1)$ *Square the binomial.*

$= 8a^3 + 8a^2 + 2a$

$\quad + 4a^2 + 4a + 1$ *Multiply polynomials.*

$= 8a^3 + 12a^2 + 6a + 1$ *Combine like terms.*

55. $(3r - 2t)^4$

$= (3r - 2t)^2(3r - 2t)^2$ $a^4 = a^2 \cdot a^2$

$= (9r^2 - 12rt + 4t^2)(9r^2 - 12rt + 4t^2)$

Square each binomial.

$= 81r^4 - 108r^3t + 36r^2t^2 - 108r^3t$

$\quad + 144r^2t^2 - 48rt^3 + 36r^2t^2 - 48rt^3 + 16t^4$

Multiply polynomials.

$= 81r^4 - 216r^3t + 216r^2t^2 - 96rt^3 + 16t^4$

Combine like terms.

57. $3x^2(x - 3)^3$

$= 3x^2(x - 3)(x - 3)^2$

$= 3x^2(x - 3)(x^2 - 6x + 9)$

$= 3x^2(x^3 - 6x^2 + 9x - 3x^2 + 18x - 27)$

$= 3x^2(x^3 - 9x^2 + 27x - 27)$

$= 3x^5 - 27x^4 + 81x^3 - 81x^2$

59. $-8x^2y(x + y)^4$ ■   First expand $(x + y)^4$.

$(x + y)^4$

$= (x + y)^2(x + y)^2$

$= (x^2 + 2xy + y^2)(x^2 + 2xy + y^2)$

$= x^2(x^2 + 2xy + y^2)$

$\quad + 2xy(x^2 + 2xy + y^2)$

$\quad + y^2(x^2 + 2xy + y^2)$

$= x^4 + 2x^3y + x^2y^2$

$\quad + 2x^3y + 4x^2y^2 + 2xy^3$

$\quad + x^2y^2 + 2xy^3 + y^4$

$= x^4 + 4x^3y + 6x^2y^2 + 4xy^3 + y^4$

Now multiply this result by $-8x^2y$.

$-8x^2y(x^4 + 4x^3y + 6x^2y^2 + 4xy^3 + y^4)$

$= -8x^6y - 32x^5y^2 - 48x^4y^3 - 32x^3y^4 - 8x^2y^5$

61. Use the formula for area the of a triangle, $A = \frac{1}{2}bh$, with $b = m + 2n$ and $h = m - 2n$.

$$A = \frac{1}{2}(m + 2n)(m - 2n)$$
$$= \frac{1}{2}[m^2 - (2n)^2]$$
$$= \frac{1}{2}(m^2 - 4n^2)$$
$$= \frac{1}{2}m^2 - 2n^2$$

63. Use the formula for the area of a parallelogram, $A = bh$, with $b = 3a + 2$ and $h = 3a - 2$.

$$A = (3a + 2)(3a - 2)$$
$$= (3a)^2 - 2^2$$
$$= 9a^2 - 4$$

65. Use the formula for the area of a circle, $A = \pi r^2$, with $r = x + 2$.

$$A = \pi(x + 2)^2$$
$$= \pi(x^2 + 4x + 4)$$
$$= \pi x^2 + 4\pi x + 4\pi$$

67. Use the formula for the volume of a cube, $V = e^3$, with $e = x + 2$.

$$\begin{aligned} V &= (x+2)^3 \\ &= (x+2)^2(x+2) \\ &= (x^2 + 4x + 4)(x+2) \\ &= x^3 + 4x^2 + 4x + 2x^2 + 8x + 8 \\ &= x^3 + 6x^2 + 12x + 8 \end{aligned}$$

Relating Concepts (Exercises 69–78)

69. The large square has sides of length $a + b$, so its area is $(a+b)^2$.

70. The red square has sides of length a, so its area is a^2.

71. Each blue rectangle has length a and width b, so each has an area of ab. Thus, the sum of the areas of the blue rectangles is

$$ab + ab = 2ab.$$

72. The yellow square has sides of length b, so its area is b^2.

73. Sum $= a^2 + 2ab + b^2$

74. The area of the largest square equals the sum of the areas of the two smaller squares and the two rectangles. Therefore, $(a+b)^2$ must equal $a^2 + 2ab + b^2$.

75. $35^2 = (35)(35)$

$$\begin{array}{r} 35 \\ 35 \\ \hline 175 \\ 105 \\ \hline 1225 \end{array}$$

76. $(a+b)^2 = a^2 + 2ab + b^2$
$(30+5)^2 = 30^2 + 2(30)(5) + 5^2$

77. $30^2 + 2(30)(5) + 5^2$
$= 900 + 60(5) + 25$
$= 900 + 300 + 25$
$= 1225$

78. The answers are equal.

13.5 Integer Exponents and the Quotient Rule

13.5 Margin Exercises

1. $2^{-2} = \dfrac{1}{4} \quad \left(\dfrac{1}{2} \div 2 = \dfrac{1}{4}\right)$

$2^{-3} = \dfrac{1}{8} \quad \left(\dfrac{1}{4} \div 2 = \dfrac{1}{8}\right)$

$2^{-4} = \dfrac{1}{16} \quad \left(\dfrac{1}{8} \div 2 = \dfrac{1}{16}\right)$

2. (a) $28^0 = 1$

(b) $(-16)^0 = 1$

(c) $-7^0 = -(7^0) = -1$

(d) $m^0 = 1$, when $m \neq 0$

(e) $-2p^0 = -2(p^0) = -2(1) = -2$, when $p \neq 0$

(f) $(5r)^0 = 1$, when $r \neq 0$

3. (a) $4^{-3} = \dfrac{1}{4^3}$, or $\dfrac{1}{64}$ *Definition of negative exponent*

(b) $6^{-2} = \dfrac{1}{6^2}$, or $\dfrac{1}{36}$ *Definition of negative exponent*

(c) $\left(\dfrac{1}{4}\right)^{-2} = 4^2$, or 16 *$\frac{1}{4}$ and 4 are reciprocals.*

(d) $\left(\dfrac{2}{3}\right)^{-2} = \left(\dfrac{3}{2}\right)^2$, or $\dfrac{9}{4}$ *$\frac{2}{3}$ and $\frac{3}{2}$ are reciprocals.*

(e) $2^{-1} + 5^{-1} = \dfrac{1}{2^1} + \dfrac{1}{5^1}$
$= \dfrac{1}{2} + \dfrac{1}{5}$
$= \dfrac{5}{10} + \dfrac{2}{10} = \dfrac{7}{10}$

(f) $7m^{-5} = \dfrac{7}{m^5}$, $m \neq 0$

(g) $\dfrac{1}{z^{-6}} = \dfrac{1^{-6}}{z^{-6}}$ $1^{-6} = 1, z \neq 0$
$= \left(\dfrac{1}{z}\right)^{-6}$ *Power rule (c)*
$= z^6$ *$\frac{1}{z}$ and z are reciprocals.*

(h) $p^2 q^{-5} = p^2\left(\dfrac{1}{q^5}\right) = \dfrac{p^2}{q^5}$

4. (a) $\dfrac{7^{-1}}{5^{-4}} = \dfrac{5^4}{7^1} = \dfrac{625}{7}$

(b) $\dfrac{x^{-3}}{y^{-2}} = \dfrac{y^2}{x^3}$

(c) $\dfrac{4h^{-5}}{m^{-2}k} = \dfrac{4}{k} \cdot \dfrac{h^{-5}}{m^{-2}} = \dfrac{4}{k} \cdot \dfrac{m^2}{h^5} = \dfrac{4m^2}{h^5 k}$

(d) $\left(\dfrac{3m}{p}\right)^{-2} = \left(\dfrac{p}{3m}\right)^2 = \dfrac{p^2}{9m^2}$

5. (a) $\dfrac{5^{11}}{5^8} = 5^{11-8} = 5^3 = 125$

(b) $\dfrac{4^7}{4^{10}} = 4^{7-10} = 4^{-3} = \dfrac{1}{4^3} = \dfrac{1}{64}$

(c) $\dfrac{6^{-5}}{6^{-2}} = 6^{-5-(-2)}$
$= 6^{-5+2} = 6^{-3} = \dfrac{1}{6^3} = \dfrac{1}{216}$

(d) $\dfrac{8^4 m^9}{8^5 m^{10}} = \dfrac{8^4}{8^5} \cdot \dfrac{m^9}{m^{10}}$

$= 8^{4-5} \cdot m^{9-10}$

$= 8^{-1} \cdot m^{-1}$

$= \dfrac{1}{8} \cdot \dfrac{1}{m}$

$= \dfrac{1}{8m}$

(e) $\dfrac{3^{-1}(x+y)^{-3}}{2^{-2}(x+y)^{-4}} = \dfrac{2^2}{3^1}(x+y)^{-3-(-4)}$

$= \dfrac{4}{3}(x+y)^{-3+4}$

$= \dfrac{4}{3}(x+y) \qquad (x \neq -y)$

Note that x cannot equal $-y$ because then $x + y$ would equal 0, and the original expression would be undefined.

6. **(a)** $\dfrac{(3^4)^2}{3^3} = \dfrac{3^8}{3^3} = 3^{8-3} = 3^5 = 243$

(b) $(4x)^2 (4x)^4 = (4x)^{2+4}$

$= (4x)^6$

$= 4^6 x^6, \quad \text{or} \quad 4096 x^6$

(c) $\left(\dfrac{5}{2z^4}\right)^{-3} = \left(\dfrac{2z^4}{5}\right)^3$

$= \dfrac{2^3 (z^4)^3}{5^3}$

$= \dfrac{8z^{12}}{125}$

(d) $\left(\dfrac{6y^{-4}}{7^{-1}z^5}\right)^{-2} = \dfrac{(6y^{-4})^{-2}}{(7^{-1}z^5)^{-2}}$

$= \dfrac{6^{-2}y^8}{7^2 z^{-10}}$

$= \dfrac{y^8 z^{10}}{6^2 \cdot 7^2}$

$= \dfrac{y^8 z^{10}}{1764}$

(e) $\dfrac{(6x)^{-1}}{(3x^2)^{-2}} = \dfrac{(3x^2)^2}{(6x)^1}$

$= \dfrac{9x^4}{6x}$

$= \dfrac{3}{2}(x^{4-1})$

$= \dfrac{3x^3}{2}$

13.5 Section Exercises

1. $(-2)^{-3} = \dfrac{1}{(-2)^3}$ is negative, because $(-2)^3$ is a negative number raised to an odd exponent, which is negative, and the quotient of a positive number and a negative number is a negative number.

3. $-2^4 = -(2^4)$ is negative, because 2^4 is positive.

5. $\left(\frac{1}{4}\right)^{-2}$ is positive. A positive base to any power will have a positive result.

7. $1 - 5^0 = 1 - 1 = 0$

9. **(a)** $x^0 = 1$ (Choice **B**)

(b) $-x^0 = -1 \cdot x^0 = -1 \cdot 1 = -1$ (Choice **C**)

(c) $7x^0 = 7 \cdot x^0 = 7 \cdot 1 = 7$ (Choice **D**)

(d) $(7x)^0 = 1$ (Choice **B**)

(e) $-7x^0 = -7 \cdot x^0 = -7 \cdot 1 = -7$ (Choice **E**)

(f) $(-7x)^0 = 1$ (Choice **B**)

11. By definition, $a^0 = 1$ $(a \neq 0)$, so $9^0 = 1$.

13. $(-4)^0 = 1$ *Definition of zero exponent*

15. $-9^0 = -(9^0) = -(1) = -1$

17. $(-2)^0 - 2^0 = 1 - 1 = 0$

19. $\dfrac{0^{10}}{10^0} = \dfrac{0}{1} = 0$

21. $7^0 + 9^0 = 1 + 1 = 2$

23. $4^{-3} = \dfrac{1}{4^3}$ *Definition of negative exponent*

$= \frac{1}{64}$

25. When we evaluate a fraction raised to a negative exponent, we can use a shortcut (the second negative-to-positive rule in the text). Note that

$$\left(\dfrac{a}{b}\right)^{-n} = \dfrac{1}{\left(\dfrac{a}{b}\right)^n} = \dfrac{1}{\dfrac{a^n}{b^n}} = \dfrac{b^n}{a^n} = \left(\dfrac{b}{a}\right)^n.$$

In words, a fraction raised to the negative of a number is equal to its reciprocal raised to the number. We will use the simple phrase "$\frac{a}{b}$ and $\frac{b}{a}$ are reciprocals" to indicate our use of this evaluation shortcut.

$\left(\dfrac{1}{2}\right)^{-4} = 2^4 = 16$ *$\frac{1}{2}$ and 2 are reciprocals.*

27. $\left(\dfrac{6}{7}\right)^{-2} = \left(\dfrac{7}{6}\right)^2$ *$\frac{6}{7}$ and $\frac{7}{6}$ are reciprocals.*

$= \dfrac{7^2}{6^2}$ *Power rule (c)*

$= \frac{49}{36}$

29. $(-3)^{-4} = \dfrac{1}{(-3)^4}$

$= \frac{1}{81}$

31. $5^{-1} + 3^{-1} = \dfrac{1}{5^1} + \dfrac{1}{3^1}$

$\qquad = \dfrac{1}{5} + \dfrac{1}{3}$

$\qquad = \dfrac{3}{15} + \dfrac{5}{15} = \dfrac{8}{15}$

33. $-2^{-1} + 3^{-2} = -(2^{-1}) + 3^{-2}$

$\qquad = -\dfrac{1}{2^1} + \dfrac{1}{3^2}$

$\qquad = -\dfrac{1}{2} + \dfrac{1}{9}$

$\qquad = -\dfrac{9}{18} + \dfrac{2}{18}$

$\qquad = -\dfrac{7}{18}$

35. $\dfrac{9^1}{9^5} = 9^{1-5}$

$\qquad = 9^{-1}$

$\qquad = \dfrac{1}{9}$

37. $\dfrac{6^{-3}}{6^2} = 6^{-3-2}$

$\qquad = 6^{-5}$

$\qquad = \dfrac{1}{6^5}, \quad \text{or} \quad \dfrac{1}{7776}$

39. $\dfrac{1}{6^{-3}} = 6^3 = 216$ *Changing from negative to positive exponents*

41. $\dfrac{2}{r^{-4}} = 2r^4$ *Changing from negative to positive exponents*

43. $\dfrac{4^{-3}}{5^{-2}} = \dfrac{5^2}{4^3} = \dfrac{25}{64}$ *Changing from negative to positive exponents*

45. $p^5 q^{-8} = \dfrac{p^5}{q^8}$ *Changing from negative to positive exponents*

47. $\dfrac{r^5}{r^{-4}} = r^5 \cdot r^4 = r^{5+4} = r^9$

Or we can use the quotient rule:

$\qquad \dfrac{r^5}{r^{-4}} = r^{5-(-4)} = r^{5+4} = r^9$

49. $\dfrac{6^4 x^8}{6^5 x^3} = 6^{4-5} \cdot x^{8-3}$

$\qquad = 6^{-1} x^5$

$\qquad = \dfrac{x^5}{6^1} = \dfrac{x^5}{6}$

51. $\dfrac{6y^3}{2y} = \dfrac{6}{2} y^{3-1} = 3y^2$

53. $\dfrac{3x^5}{3x^2} = 3^{1-1} x^{5-2} = 3^0 x^3 = 1x^3 = x^3$

55. $\dfrac{x^{-3} y}{4z^{-2}} = \dfrac{yz^2}{4x^3}$

57. Treat the expression in parentheses as a single variable; that is, treat $(a+b)$ as you would treat x.

$$\dfrac{(a+b)^{-3}}{(a+b)^{-4}} = (a+b)^{-3-(-4)}$$

$$= (a+b)^{-3+4}$$

$$= (a+b)^1 = a+b$$

Another Method:

$$\dfrac{(a+b)^{-3}}{(a+b)^{-4}} = \dfrac{(a+b)^4}{(a+b)^3}$$

$$= (a+b)^{4-3}$$

$$= (a+b)^1 = a+b$$

59. $\dfrac{(7^4)^3}{7^9} = \dfrac{7^{4\cdot3}}{7^9}$ *Power rule (a)*

$\qquad = \dfrac{7^{12}}{7^9}$

$\qquad = 7^{12-9}$ *Quotient rule*

$\qquad = 7^3 = 343$

61. $x^{-3} \cdot x^5 \cdot x^{-4}$

$\qquad = x^{-3+5+(-4)}$ *Product rule*

$\qquad = x^{-2}$

$\qquad = \dfrac{1}{x^2}$ *Definition of negative exponent*

63. $\dfrac{(3x)^{-2}}{(4x)^{-3}} = \dfrac{(4x)^3}{(3x)^2}$ *Changing from negative to positive exponents.*

$\qquad = \dfrac{4^3 x^3}{3^2 x^2}$ *Power rule (b)*

$\qquad = \dfrac{4^3 x^{3-2}}{3^2}$ *Quotient rule*

$\qquad = \dfrac{4^3 x}{3^2} = \dfrac{64x}{9}$

65. $\left(\dfrac{x^{-1} y}{z^2}\right)^{-2}$

$\qquad = \dfrac{(x^{-1} y)^{-2}}{(z^2)^{-2}}$ *Power rule (c)*

$\qquad = \dfrac{(x^{-1})^{-2} y^{-2}}{(z^2)^{-2}}$ *Power rule (b)*

$\qquad = \dfrac{x^2 y^{-2}}{z^{-4}}$ *Power rule (a)*

$\qquad = \dfrac{x^2 z^4}{y^2}$ *Definition of negative exponent*

67. $(6x)^4 (6x)^{-3} = (6x)^{4+(-3)}$ *Product rule*

$\qquad = (6x)^1 = 6x$

69. $\dfrac{(m^7n)^{-2}}{m^{-4}n^3} = \dfrac{(m^7)^{-2}n^{-2}}{m^{-4}n^3}$

$= \dfrac{m^{7(-2)}n^{-2}}{m^{-4}n^3}$

$= \dfrac{m^{-14}n^{-2}}{m^{-4}n^3}$

$= m^{-14-(-4)}n^{-2-3}$

$= m^{-10}n^{-5}$

$= \dfrac{1}{m^{10}n^5}$

71. $\dfrac{5x^{-3}}{(4x)^2} = \dfrac{5x^{-3}}{4^2x^2}$

$= \dfrac{5}{16x^2x^3}$

$= \dfrac{5}{16x^5}$

73. $\left(\dfrac{2p^{-1}q}{3^{-1}m^2}\right)^2 = \dfrac{2^2(p^{-1})^2q^2}{(3^{-1})^2(m^2)^2}$

$= \dfrac{2^2p^{-2}q^2}{3^{-2}m^4}$

$= \dfrac{2^2 \cdot 3^2q^2}{m^4p^2}$

$= \dfrac{4 \cdot 9q^2}{m^4p^2}$

$= \dfrac{36q^2}{m^4p^2}$

75. The student attempted to use the quotient rule with unequal bases. The correct way to simplify this expression is

$\dfrac{16^3}{2^2} = \dfrac{(2^4)^3}{2^2} = \dfrac{2^{12}}{2^2} = 2^{12-2} = 2^{10} = 1024.$

Summary Exercises
Applying the Rules for Exponents

1. $(-5)^4$ is *positive* since a negative number raised to an even exponent is positive.

3. $(-2)^5$ is *negative* since a negative number raised to an odd exponent is negative.

5. $(-\frac{3}{7})^2$ is *positive* since a negative number raised to an even exponent is positive.

7. $(-\frac{3}{7})^{-2} = (-\frac{7}{3})^2$ is *positive* since a negative number raised to an even exponent is positive.

9. $\left(\dfrac{6x^2}{5}\right)^{12} = \dfrac{(6x^2)^{12}}{5^{12}}$

$= \dfrac{6^{12}(x^2)^{12}}{5^{12}}$

$= \dfrac{6^{12}x^{24}}{5^{12}}$

11. $(10x^2y^4)^2(10xy^2)^3$

$= 10^2(x^2)^2(y^4)^2 \cdot 10^3x^3(y^2)^3$

$= 10^2x^4y^8 10^3x^3y^6$

$= 10^5x^7y^{14} = 100,000x^7y^{14}$

13. $\left(\dfrac{9wx^3}{y^4}\right)^3 = \dfrac{(9wx^3)^3}{(y^4)^3}$

$= \dfrac{9^3w^3(x^3)^3}{y^{12}}$

$= \dfrac{729w^3x^9}{y^{12}}$

15. $\dfrac{c^{11}(c^2)^4}{(c^3)^3(c^2)^{-6}} = \dfrac{c^{11}c^8}{c^9c^{-12}}$

$= \dfrac{c^{19}}{c^{-3}}$

$= c^{19-(-3)} = c^{22}$

17. $5^{-1} + 6^{-1} = \dfrac{1}{5^1} + \dfrac{1}{6^1}$

$= \dfrac{6}{30} + \dfrac{5}{30}$

$= \dfrac{11}{30}$

19. $\dfrac{(2xy^{-1})^3}{2^3x^{-3}y^2} = \dfrac{2^3x^3y^{-3}}{2^3x^{-3}y^2}$

$= x^6y^{-5} = \dfrac{x^6}{y^5}$

21. $(z^4)^{-3}(z^{-2})^{-5}$

$= z^{-12}z^{10} = z^{-2} = \dfrac{1}{z^2}$

23. $\dfrac{(3^{-1}x^{-3}y)^{-1}(2x^2y^{-3})^2}{(5x^{-2}y^2)^{-2}}$

$= \dfrac{(5x^{-2}y^2)^2(2x^2y^{-3})^2}{(3^{-1}x^{-3}y)^1}$

$= \dfrac{5^2x^{-4}y^42^2x^4y^{-6}}{3^{-1}x^{-3}y} = \dfrac{25x^0y^{-2} \cdot 4}{3^{-1}x^{-3}y}$

$= 100x^3y^{-3} \cdot 3 = \dfrac{300x^3}{y^3}$

25. $\left(\dfrac{-2x^{-2}}{2x^2}\right)^{-2} = \left(\dfrac{2x^2}{-2x^{-2}}\right)^2$

$= \left(\dfrac{x^4}{-1}\right)^2 = \dfrac{(x^4)^2}{(-1)^2}$

$= \dfrac{x^8}{1} = x^8$

27. $\dfrac{(a^{-2}b^3)^{-4}}{(a^{-3}b^2)^{-2}(ab)^{-4}} = \dfrac{a^8b^{-12}}{a^6b^{-4}a^{-4}b^{-4}}$

$\qquad\qquad = \dfrac{a^8b^{-12}}{a^2b^{-8}}$

$\qquad\qquad = a^6b^{-4}$

$\qquad\qquad = \dfrac{a^6}{b^4}$

29. $5^{-2} + 6^{-2} = \dfrac{1}{5^2} + \dfrac{1}{6^2}$

$\qquad\qquad = \dfrac{1}{25} + \dfrac{1}{36}$

$\qquad\qquad = \dfrac{36}{25 \cdot 36} + \dfrac{25}{25 \cdot 36}$

$\qquad\qquad = \dfrac{36 + 25}{900} = \dfrac{61}{900}$

31. $\left(\dfrac{7a^2b^3}{2}\right)^3 = \dfrac{(7a^2b^3)^3}{2^3}$

$\qquad\qquad = \dfrac{7^3a^6b^9}{8} = \dfrac{343a^6b^9}{8}$

33. $-(-12)^0 = -1$

35. $\dfrac{(2xy^{-3})^{-2}}{(3x^{-2}y^4)^{-3}} = \dfrac{(3x^{-2}y^4)^3}{(2xy^{-3})^2}$

$\qquad\qquad = \dfrac{3^3x^{-6}y^{12}}{2^2x^2y^{-6}}$

$\qquad\qquad = \dfrac{27x^{-8}y^{18}}{4}$

$\qquad\qquad = \dfrac{27y^{18}}{4x^8}$

37. $(6x^{-5}z^3)^{-3} = 6^{-3}x^{15}z^{-9}$

$\qquad\qquad = \dfrac{x^{15}}{6^3z^9}$

$\qquad\qquad = \dfrac{x^{15}}{216z^9}$

39. $\dfrac{(xy)^{-3}(xy)^5}{(xy)^{-4}} = (xy)^{-3+5-(-4)}$

$\qquad\qquad = (xy)^6 = x^6y^6$

41. $\dfrac{(7^{-1}x^{-3})^{-2}(x^4)^{-6}}{7^{-1}x^{-3}}$

$\qquad = \dfrac{7^2x^6x^{-24}}{7^{-1}x^{-3}}$

$\qquad = 7^{2-(-1)}x^{6-24-(-3)}$

$\qquad = 7^3x^{-15} = \dfrac{343}{x^{15}}$

43. $(5p^{-2}q)^{-3}(5pq^3)^4$

$\qquad = 5^{-3}p^6q^{-3}5^4p^4q^{12}$

$\qquad = 5p^{10}q^9$

45. $\left(\dfrac{4r^{-6}s^{-2}t}{2r^8s^{-4}t^2}\right)^{-1}$

$\qquad = \left(2r^{-6-8}s^{-2-(-4)}t^{1-2}\right)^{-1}$

$\qquad = (2r^{-14}s^2t^{-1})^{-1}$

$\qquad = \left(\dfrac{2s^2}{r^{14}t}\right)^{-1}$

$\qquad = \dfrac{r^{14}t}{2s^2}$

47. $\dfrac{(8pq^{-2})^4}{(8p^{-2}q^{-3})^3}$

$\qquad = \dfrac{8^4p^4q^{-8}}{8^3p^{-6}q^{-9}}$

$\qquad = 8^{4-3}p^{4-(-6)}q^{-8-(-9)}$

$\qquad = 8p^{10}q$

49. $-(-3^0)^0 = -(1) = -1$

13.6 Dividing a Polynomial by a Monomial

13.6 Margin Exercises

1. **(a)** $\dfrac{6p^4 + 18p^7}{3p^2} = \dfrac{6p^4}{3p^2} + \dfrac{18p^7}{3p^2}$

$\qquad\qquad = 2p^2 + 6p^5$

(b) $\dfrac{12m^6 + 18m^5 + 30m^4}{6m^2}$

$\qquad = \dfrac{12m^6}{6m^2} + \dfrac{18m^5}{6m^2} + \dfrac{30m^4}{6m^2}$

$\qquad = 2m^4 + 3m^3 + 5m^2$

(c) $(18r^7 - 9r^2) \div (3r) = \dfrac{18r^7 - 9r^2}{3r}$

$\qquad\qquad\qquad = \dfrac{18r^7}{3r} - \dfrac{9r^2}{3r}$

$\qquad\qquad\qquad = 6r^6 - 3r$

2. **(a)** $\dfrac{20x^4 - 25x^3 + 5x}{5x^2}$

$\qquad = \dfrac{20x^4}{5x^2} - \dfrac{25x^3}{5x^2} + \dfrac{5x}{5x^2}$

$\qquad = 4x^2 - 5x + \dfrac{1}{x}$

(b) $\dfrac{50m^4 - 30m^3 + 20m}{10m^3}$

$\qquad = \dfrac{50m^4}{10m^3} - \dfrac{30m^3}{10m^3} + \dfrac{20m}{10m^3}$

$\qquad = 5m - 3 + \dfrac{2}{m^2}$

3. **(a)** $\dfrac{-9y^6 + 8y^7 - 11y - 4}{y^2}$

$\qquad = -\dfrac{9y^6}{y^2} + \dfrac{8y^7}{y^2} - \dfrac{11y}{y^2} - \dfrac{4}{y^2}$

$\qquad = -9y^4 + 8y^5 - \dfrac{11}{y} - \dfrac{4}{y^2}$

$\qquad = 8y^5 - 9y^4 - \dfrac{11}{y} - \dfrac{4}{y^2} \qquad$ *Descending powers*

(b) $\dfrac{-8p^4 - 6p^3 - 12p^5}{-3p^3}$

$= -\dfrac{8p^4}{-3p^3} - \dfrac{6p^3}{-3p^3} - \dfrac{12p^5}{-3p^3}$

$= \dfrac{8p}{3} + 2 + 4p^2$

$= 4p^2 + \dfrac{8p}{3} + 2$ *Descending powers*

4. $\dfrac{45x^4y^3 + 30x^3y^2 - 60x^2y}{-15x^2y}$

$= \dfrac{45x^4y^3}{-15x^2y} + \dfrac{30x^3y^2}{-15x^2y} - \dfrac{60x^2y}{-15x^2y}$

$= -3x^2y^2 - 2xy + 4$

13.6 Section Exercises

1. In the statement $\dfrac{6x^2 + 8}{2} = 3x^2 + 4$, $\underline{6x^2 + 8}$ is the dividend, $\underline{2}$ is the divisor, and $\underline{3x^2 + 4}$ is the quotient.

3. To check the division shown in Exercise 1, multiply $\underline{3x^2 + 4}$ by $\underline{2}$ (or 2 by $3x^2 + 4$) and show that the product is $\underline{6x^2 + 8}$.

5. To use the method of this section, the divisor (denominator) must be a monomial (one term). This is true of

$\dfrac{16m^3 - 12m^2}{4m}$, but not of $\dfrac{4m}{16m^3 - 12m^2}$.

7. The quotient would be a polynomial in the variable x having degree $6 - 4 = 2$.

9. $\dfrac{12m^4 - 6m^3}{6m^2}$

$= \dfrac{12m^4}{6m^2} - \dfrac{6m^3}{6m^2}$

$= 2m^2 - m$

11. $\dfrac{60x^4 - 20x^2 + 10x}{2x}$

$= \dfrac{60x^4}{2x} - \dfrac{20x^2}{2x} + \dfrac{10x}{2x}$

$= \dfrac{60}{2}x^{4-1} - \dfrac{20}{2}x^{2-1} + \dfrac{10}{2}$

$= 30x^3 - 10x + 5$

13. $\dfrac{20m^5 - 10m^4 + 5m^2}{-5m^2}$

$= \dfrac{20m^5}{-5m^2} - \dfrac{10m^4}{-5m^2} + \dfrac{5m^2}{-5m^2}$

$= -4m^3 + 2m^2 - 1$

15. $\dfrac{8t^5 - 4t^3 + 4t^2}{2t} = \dfrac{8t^5}{2t} - \dfrac{4t^3}{2t} + \dfrac{4t^2}{2t}$

$= 4t^4 - 2t^2 + 2t$

17. $\dfrac{4a^5 - 4a^2 + 8}{4a} = \dfrac{4a^5}{4a} - \dfrac{4a^2}{4a} + \dfrac{8}{4a}$

$= a^4 - a + \dfrac{2}{a}$

19. $\dfrac{12x^5 - 4x^4 + 6x^3}{-6x^2}$

$= \dfrac{12x^5}{-6x^2} - \dfrac{4x^4}{-6x^2} + \dfrac{6x^3}{-6x^2}$

$= -2x^3 + \dfrac{2x^2}{3} - x$

21. $\dfrac{4x^2 + 20x^3 - 36x^4}{4x^2}$

$= \dfrac{4x^2}{4x^2} + \dfrac{20x^3}{4x^2} - \dfrac{36x^4}{4x^2}$

$= 1 + 5x - 9x^2$

$= -9x^2 + 5x + 1$ *Descending powers*

23. $\dfrac{-3x^3 - 4x^4 + 2x}{-3x^2}$

$= -\dfrac{3x^3}{-3x^2} - \dfrac{4x^4}{-3x^2} + \dfrac{2x}{-3x^2}$

$= x + \dfrac{4x^2}{3} - \dfrac{2}{3x}$

$= \dfrac{4x^2}{3} + x - \dfrac{2}{3x}$ *Descending powers*

25. $\dfrac{27r^4 - 36r^3 - 6r^2 + 3r - 2}{3r}$

$= \dfrac{27r^4}{3r} - \dfrac{36r^3}{3r} - \dfrac{6r^2}{3r} + \dfrac{3r}{3r} - \dfrac{2}{3r}$

$= 9r^3 - 12r^2 - 2r + 1 - \dfrac{2}{3r}$

27. $\dfrac{2m^5 - 6m^4 + 8m^2}{-2m^3}$

$= \dfrac{2m^5}{-2m^3} - \dfrac{6m^4}{-2m^3} + \dfrac{8m^2}{-2m^3}$

$= -m^2 + 3m - \dfrac{4}{m}$

29. $(120x^{11} - 60x^{10} + 140x^9 - 100x^8) \div (10x^{12})$

$= \dfrac{120x^{11} - 60x^{10} + 140x^9 - 100x^8}{10x^{12}}$

$= \dfrac{120x^{11}}{10x^{12}} - \dfrac{60x^{10}}{10x^{12}} + \dfrac{140x^9}{10x^{12}} - \dfrac{100x^8}{10x^{12}}$

$= \dfrac{12}{x} - \dfrac{6}{x^2} + \dfrac{14}{x^3} - \dfrac{10}{x^4}$

31. $(20a^4b^3 - 15a^5b^2 + 25a^3b) \div (-5a^4b)$

$$= \frac{20a^4b^3}{-5a^4b} - \frac{15a^5b^2}{-5a^4b} + \frac{25a^3b}{-5a^4b}$$

$$= -4b^2 + 3ab - \frac{5}{a}$$

33. Use the formula for the area of a rectangle,
$A = lw$, with $A = 12x^2 - 4x + 2$ and $w = 2x$.

$$12x^2 - 4x + 2 = l(2x)$$

$$\frac{12x^2 - 4x + 2}{2x} = l \qquad \textit{Divide by } 2x.$$

$$\frac{12x^2}{2x} - \frac{4x}{2x} + \frac{2}{2x} = l$$

$$6x - 2 + \frac{1}{x} = l$$

35. $\dfrac{(\ \ )}{5x^3} = 3x^2 - 7x + 7$

Multiply the quotient and the divisor to find the missing polynomial.

$$(5x^3)(3x^2 - 7x + 7)$$
$$= 5x^3(3x^2) + 5x^3(-7x) + 5x^3(7)$$
$$= 15x^5 - 35x^4 + 35x^3$$

13.7 Dividing a Polynomial by a Polynomial

13.7 Margin Exercises

1. **(a)** $(x^3 + x^2 + 4x - 6) \div (x - 1)$

$$
\begin{array}{r}
x^2 + 2x + 6 \\
x - 1 \overline{\smash{\big)}\ x^3 + x^2 + 4x - 6} \\
\underline{x^3 - x^2} \\
2x^2 + 4x \\
\underline{2x^2 - 2x} \\
6x - 6 \\
\underline{6x - 6} \\
0
\end{array}
$$

$(x^3 + x^2 + 4x - 6) \div (x - 1) = x^2 + 2x + 6$

(b) $\dfrac{p^3 - 2p^2 - 5p + 9}{p + 2}$

$$
\begin{array}{r}
p^2 - 4p + 3 \\
p + 2 \overline{\smash{\big)}\ p^3 - 2p^2 - 5p + 9} \\
\underline{p^3 + 2p^2} \\
-4p^2 - 5p \\
\underline{-4p^2 - 8p} \\
3p + 9 \\
\underline{3p + 6} \\
3
\end{array}
$$

Write the remainder, 3, in the numerator of a fraction with the divisor as the denominator. Add this fraction to the quotient to get the answer.

$$\frac{p^3 - 2p^2 - 5p + 9}{p + 2} = p^2 - 4p + 3 + \frac{3}{p + 2}$$

2. **(a)** $\dfrac{r^2 - 5}{r + 4}$

Use 0 as the coefficient of the missing r-term.

$$
\begin{array}{r}
r - 4 \\
r + 4 \overline{\smash{\big)}\ r^2 + 0r - 5} \\
\underline{r^2 + 4r} \\
-4r - 5 \\
\underline{-4r - 16} \\
11 \leftarrow \text{Remainder}
\end{array}
$$

$$\frac{r^2 - 5}{r + 4} = r - 4 + \frac{11}{r + 4}$$

(b) $(x^3 - 8) \div (x - 2)$

Use 0 as the coefficient of the missing x^2- and x-terms.

$$
\begin{array}{r}
x^2 + 2x + 4 \\
x - 2 \overline{\smash{\big)}\ x^3 + 0x^2 + 0x - 8} \\
\underline{x^3 - 2x^2} \\
2x^2 + 0x \\
\underline{2x^2 - 4x} \\
4x - 8 \\
\underline{4x - 8} \\
0
\end{array}
$$

$(x^3 - 8) \div (x - 2) = x^2 + 2x + 4$

3. **(a)** $(2x^4 + 3x^3 - x^2 + 6x + 5) \div (x^2 - 1)$

$$
\begin{array}{r}
2x^2 + 3x + 1 \\
x^2 + 0x - 1 \overline{\smash{\big)}\ 2x^4 + 3x^3 - x^2 + 6x + 5} \\
\underline{2x^4 + 0x^3 - 2x^2} \\
3x^3 + x^2 + 6x \\
\underline{3x^3 + 0x^2 - 3x} \\
x^2 + 9x + 5 \\
\underline{x^2 + 0x - 1} \\
9x + 6
\end{array}
$$

The remainder is $9x + 6$.

The answer is $2x^2 + 3x + 1 + \dfrac{9x + 6}{x^2 - 1}$.

(b) $\dfrac{2m^5 + m^4 + 6m^3 - 3m^2 - 18}{m^2 + 3}$

$$
\begin{array}{r}
2m^3 + m^2 - 6 \\
m^2 + 0m + 3 \overline{\smash{\big)}\ 2m^5 + m^4 + 6m^3 - 3m^2 + 0m - 18} \\
\underline{2m^5 + 0m^4 + 6m^3} \\
m^4 + 0m^3 - 3m^2 \\
\underline{m^4 + 0m^3 + 3m^2} \\
-6m^2 + 0m - 18 \\
\underline{-6m^2 + 0m - 18} \\
0
\end{array}
$$

$$\frac{2m^5 + m^4 + 6m^3 - 3m^2 - 18}{m^2 + 3} = 2m^3 + m^2 - 6$$

4.

$$x^2 + \tfrac{1}{3}x + \tfrac{5}{3}$$

$$3x + 6 \,\overline{\smash{\big)}\, 3x^3 + 7x^2 + 7x + 10}$$
$$\underline{3x^3 + 6x^2}$$
$$x^2 + 7x \qquad \tfrac{x^2}{3x} = \tfrac{1}{3}x$$
$$\underline{x^2 + 2x}$$
$$5x + 10 \quad \tfrac{5x}{3x} = \tfrac{5}{3}$$
$$\underline{5x + 10}$$
$$0$$

$(3x^3 + 7x^2 + 7x + 10)$ divided by $(3x + 6)$ is $x^2 + \tfrac{1}{3}x + \tfrac{5}{3}$.

5.

$$x^2 + 2x + 4$$

$$x + 2 \,\overline{\smash{\big)}\, x^3 + 4x^2 + 8x + 8}$$
$$\underline{x^3 + 2x^2}$$
$$2x^2 + 8x$$
$$\underline{2x^2 + 4x}$$
$$4x + 8$$
$$\underline{4x + 8}$$
$$0$$

$(x^3 + 4x^2 + 8x + 8)$ divided by $(x + 2)$ is $x^2 + 2x + 4$.

13.7 Section Exercises

1. In the division problem, the divisor is $2x + 5$ and the quotient is $2x^3 - 4x^2 + 3x + 2$.

3. In dividing $12m^2 - 20m + 3$ by $2m - 3$, the first step is to divide $12m^2$ by $2m$ to get $6m$.

5. $\dfrac{x^2 - x - 6}{x - 3}$

$$x + 2$$
$$x - 3 \,\overline{\smash{\big)}\, x^2 - x - 6}$$
$$\underline{x^2 - 3x}$$
$$2x - 6$$
$$\underline{2x - 6}$$
$$0$$

The remainder is 0. The answer is the quotient, $x + 2$.

7. $\dfrac{2y^2 + 9y - 35}{y + 7}$

$$2y - 5$$
$$y + 7 \,\overline{\smash{\big)}\, 2y^2 + 9y - 35}$$
$$\underline{2y^2 + 14y}$$
$$-5y - 35$$
$$\underline{-5y - 35}$$
$$0$$

The remainder is 0. The answer is the quotient, $2y - 5$.

9. $\dfrac{p^2 + 2p + 20}{p + 6}$

$$p - 4$$
$$p + 6 \,\overline{\smash{\big)}\, p^2 + 2p + 20}$$
$$\underline{p^2 + 6p}$$
$$-4p + 20$$
$$\underline{-4p - 24}$$
$$44$$

The remainder is 44. Write the remainder as the numerator of a fraction that has the divisor $p + 6$ as its denominator. The answer is

$$p - 4 + \dfrac{44}{p + 6}.$$

11. $(r^2 - 8r + 15) \div (r - 3)$

$$r - 5$$
$$r - 3 \,\overline{\smash{\big)}\, r^2 - 8r + 15}$$
$$\underline{r^2 - 3r}$$
$$-5r + 15$$
$$\underline{-5r + 15}$$
$$0$$

The remainder is 0. The answer is the quotient, $r - 5$.

13. $\dfrac{4a^2 - 22a + 32}{2a + 3}$

$$2a - 14$$
$$2a + 3 \,\overline{\smash{\big)}\, 4a^2 - 22a + 32}$$
$$\underline{4a^2 + 6a}$$
$$-28a + 32$$
$$\underline{-28a - 42}$$
$$74$$

The remainder is 74. The answer is

$$2a - 14 + \dfrac{74}{2a + 3}.$$

15. $\dfrac{8x^3 - 10x^2 - x + 3}{2x + 1}$

$$4x^2 - 7x + 3$$
$$2x + 1 \,\overline{\smash{\big)}\, 8x^3 - 10x^2 - x + 3}$$
$$\underline{8x^3 + 4x^2}$$
$$-14x^2 - x$$
$$\underline{-14x^2 - 7x}$$
$$6x + 3$$
$$\underline{6x + 3}$$
$$0$$

The remainder is 0. The answer is the quotient, $4x^2 - 7x + 3$.

17. $\dfrac{3y^3 + y^2 + 2}{y + 1}$

Use 0 as the coefficient of the missing y-term.

$$
\begin{array}{r}
3y^2 \quad - 2y \quad + 2 \\
y + 1 \enclose{longdiv}{3y^3 \ + \ y^2 \ + \ 0y \ + 2} \\
\underline{3y^3 \ + \ 3y^2} \\
-2y^2 \ + \ 0y \\
\underline{-2y^2 \ - \ 2y} \\
2y \ + 2 \\
\underline{2y \ + 2} \\
0
\end{array}
$$

The remainder is 0. The answer is the quotient, $3y^2 - 2y + 2$.

19. $\dfrac{2x^3 + x + 2}{x + 1}$

Use 0 as the coefficient of the missing x^2-term.

$$
\begin{array}{r}
2x^2 \ - \ 2x \ + \ 3 \\
x + 1 \enclose{longdiv}{2x^3 \ + \ 0x^2 \ + \ x \ + \ 2} \\
\underline{2x^3 \ + \ 2x^2} \\
-2x^2 \ + \ x \\
\underline{-2x^2 \ - \ 2x} \\
3x \ + \ 2 \\
\underline{3x \ + \ 3} \\
-1
\end{array}
$$

$$\frac{2x^3 + x + 2}{x + 1} = 2x^2 - 2x + 3 + \frac{-1}{x + 1}$$

21. $\dfrac{3k^3 - 4k^2 - 6k + 10}{k^2 - 2}$

Use 0 as the coefficient of the missing k-term in the divisor.

$$
\begin{array}{r}
3k \quad - \ 4 \\
k^2 + 0k - 2 \enclose{longdiv}{3k^3 \quad - 4k^2 \quad - 6k \ + 10} \\
\underline{3k^3 \quad + 0k^2 \quad - 6k} \\
-4k^2 \ + 0k \ + 10 \\
\underline{-4k^2 \ + 0k \ + \ 8} \\
2
\end{array}
$$

The remainder is 2. The quotient is $3k - 4$.

The answer is $3k - 4 + \dfrac{2}{k^2 - 2}$.

23. $(x^4 - x^2 - 2) \div (x^2 - 2)$

Use 0 as the coefficient of the missing x^3- and x-terms.

$$
\begin{array}{r}
x^2 \qquad + 1 \\
x^2 + 0x - 2 \enclose{longdiv}{x^4 \ + 0x^3 \ - \ x^2 \ + 0x \ - \ 2} \\
\underline{x^4 \ + 0x^3 \ - 2x^2} \\
x^2 \ + 0x \ - \ 2 \\
\underline{x^2 \ + 0x \ - \ 2} \\
0
\end{array}
$$

The remainder is 0. The answer is the quotient, $x^2 + 1$.

25. $\dfrac{x^4 - 1}{x^2 - 1}$

$$
\begin{array}{r}
x^2 \qquad\qquad + 1 \\
x^2 + 0x - 1 \enclose{longdiv}{x^4 \ + 0x^3 \ + 0x^2 \ + 0x \ - 1} \\
\underline{x^4 \ + 0x^3 \ - \ x^2} \\
x^2 \ + 0x \ - 1 \\
\underline{x^2 \ + 0x \ - 1} \\
0
\end{array}
$$

The remainder is 0. The answer is the quotient, $x^2 + 1$.

27. $\dfrac{6p^4 - 15p^3 + 14p^2 - 5p + 10}{3p^2 + 1}$

$$
\begin{array}{r}
2p^2 \ - 5p \ + \ 4 \\
3p^2 + 0p + 1 \enclose{longdiv}{6p^4 \ - 15p^3 \ + 14p^2 \ - 5p \ + 10} \\
\underline{6p^4 \ + \ 0p^3 \ + \ 2p^2} \\
-15p^3 \ + 12p^2 \ - 5p \\
\underline{-15p^3 \ + \ 0p^2 \ - 5p} \\
12p^2 \ + 0p \ + 10 \\
\underline{12p^2 \ + 0p \ + \ 4} \\
6
\end{array}
$$

The remainder is 6. The quotient is $2p^2 - 5p + 4$.

The answer is $2p^2 - 5p + 4 + \dfrac{6}{3p^2 + 1}$.

29. $\dfrac{2x^5 + x^4 + 11x^3 - 8x^2 - 13x + 7}{2x^2 + x - 1}$

$$
\begin{array}{r}
x^3 \qquad\quad + \ 6x \ - 7 \\
2x^2 + x - 1 \enclose{longdiv}{2x^5 + x^4 + 11x^3 - \ 8x^2 - 13x + 7} \\
\underline{2x^5 + x^4 - \ x^3} \\
12x^3 - \ 8x^2 - 13x \\
\underline{12x^3 + \ 6x^2 - \ 6x} \\
-14x^2 - \ 7x + 7 \\
\underline{-14x^2 - \ 7x + 7} \\
0
\end{array}
$$

The remainder is 0. The answer is the quotient, $x^3 + 6x - 7$.

31. $(10x^3 + 13x^2 + 4x + 1) \div (5x + 5)$

$$
\begin{array}{r}
2x^2 \ + \frac{3}{5}x \ + \frac{1}{5} \\
5x + 5 \enclose{longdiv}{10x^3 \ + 13x^2 \ + 4x \ + 1} \\
\underline{10x^3 \ + 10x^2} \\
3x^2 \ + 4x \\
\underline{3x^2 \ + 3x} \\
x \ + 1 \\
\underline{x \ + 1} \\
0
\end{array}
$$

The remainder is 0. The answer is the quotient, $2x^2 + \frac{3}{5}x + \frac{1}{5}$.

33. Use $A = lw$ with
$$A = 5x^3 + 7x^2 - 13x - 6$$
and $w = 5x + 2$.
$$5x^3 + 7x^2 - 13x - 6 = l(5x + 2)$$
$$\frac{5x^3 + 7x^2 - 13x - 6}{5x + 2} = l$$

$$
\begin{array}{r}
x^2 + x - 3 \\
5x + 2 \overline{\smash{\big)}\, 5x^3 + 7x^2 - 13x - 6} \\
\underline{5x^3 + 2x^2} \\
5x^2 - 13x \\
\underline{5x^2 + 2x} \\
-15x - 6 \\
\underline{-15x - 6} \\
0
\end{array}
$$

The length is $(x^2 + x - 3)$ units.

Relating Concepts (Exercises 35–38)

35. $2x^2 - 4x + 3$
$= 2(-3)^2 - 4(-3) + 3$ *Let x = −3.*
$= 18 + 12 + 3$
$= 33$

36.
$$
\begin{array}{r}
2x - 10 \\
x + 3 \overline{\smash{\big)}\, 2x^2 - 4x + 3} \\
\underline{2x^2 + 6x} \\
-10x + 3 \\
\underline{-10x - 30} \\
33
\end{array}
$$

The remainder is 33.

37. The answers to Exercises 35 and 36 are the same.

38. We will choose the polynomial
$5x^3 - 12x^2 - 25x - 20$ and the value $x = 4$.

Substitution:

$5x^3 - 12x^2 - 25x - 20$
$= 5(4)^3 - 12(4)^2 - 25(4) - 20$ *Let x = 4.*
$= 320 - 192 - 100 - 20$
$= 8$

Division:

$$
\begin{array}{r}
5x^2 + 8x + 7 \\
x - 4 \overline{\smash{\big)}\, 5x^3 - 12x^2 - 25x - 20} \\
\underline{5x^3 - 20x^2} \\
8x^2 - 25x \\
\underline{8x^2 - 32x} \\
7x - 20 \\
\underline{7x - 28} \\
8
\end{array}
$$

The remainder is 8. The answers agree.

13.8 An Application of Exponents: Scientific Notation

13.8 Margin Exercises

1. **(a)** $63,000 = 6.3 \times 10^4$

The decimal point has been moved $\underline{4}$ places to put it after the first nonzero digit (the $\underline{6}$). Since the original number, 63,000, is *greater* than 6.3, we must multiply by a *positive* power of 10 so that the product 6.3×10^4 will equal the larger number.

(b) $5,870,000 = 5.87 \times 10^6$ *6 places*

Since 5,870,000 is *greater* than 5.87, we must multiply by a *positive* power of 10 so that the product 5.87×10^6 will equal the larger number.

(c) For 7.0065, the decimal point is already to the right of the first nonzero digit, so the exponent on 10 is 0.
$$7.0065 = 7.0065 \times 10^0$$

(d) $0.0571 = 5.71 \times 10^{-2}$

The decimal point has been moved $\underline{2}$ places to put it after the first nonzero digit (the $\underline{5}$). Since the original number, 0.0571, is *less* than 5.71, we must multiply by a *negative* power of 10 so that the product 5.71×10^{-2} will equal the smaller number.

(e) $-0.00062. = -6.2 \times 10^{-4}$ *4 places*

When changing a negative number to scientific notation, use the absolute value of the number when moving the decimal point or choosing the exponent. Since 0.00062 is *less* than 6.2, we must multiply by a *negative* power of 10 so that the product 6.2×10^{-4} will equal the smaller number. The leading negative sign belongs in the final answer.

2. **(a)** 4.2×10^3 ▪ Since the exponent is positive, move the decimal point $\underline{3}$ places to the <u>right</u>.
$$4.2 \times 10^3 = \underline{4200}$$

(b) 8.7×10^5 ▪ Since the exponent is positive, move the decimal point 5 places to the *right*.
$$8.7 \times 10^5 = 870,000$$

(c) 6.42×10^{-3} ▪ Since the exponent is negative, move the decimal point 3 places to the left.
$$6.42 \times 10^{-3} = \underline{0.00642}$$

(d) -5.27×10^{-1} ▪ Since the exponent is negative, move the decimal point 1 place to the *left*.
$$-5.27 \times 10^{-1} = -0.527$$

3. **(a)** $(2.6 \times 10^1)(2 \times 10^{-6})$

$= (2.6 \times 2)(10^1 \times 10^{-6})$

$= 5.2 \times 10^{4+(-6)}$

$= 5.2 \times 10^{-2}$ *Scientific notation*

$= 0.052$ *Without exponents*

(b) $(3 \times 10^5)(5 \times 10^{-2})$

$= (3 \times 5)(10^5 \times 10^{-2})$

$= 15 \times 10^3$

$= (1.5 \times 10^1) \times 10^3$

$= 1.5 \times (10^1 \times 10^3)$

$= 1.5 \times 10^4$ *Scientific notation*

$= 15,000$ *Without exponents*

(c) $\dfrac{4.8 \times 10^2}{2.4 \times 10^{-3}}$

$= \dfrac{4.8}{2.4} \times \dfrac{10^2}{10^{-3}}$

$= 2 \times 10^{2-(-3)}$

$= 2 \times 10^{2+3}$

$= 2 \times 10^5$ *Scientific notation*

$= 200,000$ *Without exponents*

4. Use the formula $d = rt$.

$d = \left(3.0 \times 10^5 \dfrac{\text{km}}{\text{sec}}\right)(6.0 \times 10^1 \text{ sec})$

$= (3.0 \times 6.0) \times (10^5 \times 10^1)$

$= 18.0 \times 10^{5+1}$

$= 18.0 \times 10^6$

$= 1.8 \times 10^7$ *Scientific notation*

$= 18,000,000$ *Without exponents*

Light travels 1.8×10^7 km or 18,000,000 km in 6.0×10^1 sec.

5. Use the formula $t = \dfrac{d}{r}$.

$t = \dfrac{1.5 \times 10^8}{3.0 \times 10^5}$

$= \dfrac{1.5}{3.0} \times \dfrac{10^8}{10^5}$

$= 0.5 \times 10^{8-5}$

$= 0.5 \times 10^3$

$= 5.0 \times 10^2$ *Scientific notation*

$= 500$ *Without exponents*

It takes light 500 sec to travel approximately 1.5×10^8 km from the Sun to Earth.

13.8 Section Exercises

1. **(a)** Move the decimal point to the left 4 places due to the exponent, -4.

$$4.6 \times 10^{-4} = 0.000\,46$$

Choice **C** is correct.

(b) $4.6 \times 10^4 = 46,000$

Choice **A** is correct.

(c) Move the decimal point to the right 5 places due to the exponent, 5.

$$4.6 \times 10^5 = 460,000$$

Choice **B** is correct.

(d) $4.6 \times 10^{-5} = 0.000\,046$

Choice **D** is correct.

3. 4.56×10^3 is written in scientific notation because 4.56 is between 1 and 10, and 10^3 is an integer power of 10.

5. 5,600,000 is not written in scientific notation. It is a "large" number and can be written in scientific notation as 5.6×10^6.

7. 0.004 is not written in scientific notation because $|0.004| = 0.004$ is not between 1 and 10. It is a "small" number and can be written in scientific notation as 4×10^{-3}.

9. 0.8×10^2 is not written in scientific notation because $|0.8| = 0.8$ is not between 1 and 10.

$0.8 \times 10^2 = 80$ (move the decimal point 2 places to the right), and $80 = 8 \times 10^1$ in scientific notation.

11. A number is written in scientific notation if it is the product of a number whose absolute value is between 1 and 10 (inclusive of 1) and an integer power of 10.

13. 5,876,000,000 ▪ Move the decimal point to the right of the first nonzero digit and count the number of places the decimal point was moved.

5.876,000,000 *9 places*

The original number, 5,876,000,000, is *greater* than 5.876, so multiply by a *positive* power of 10. Thus,

$$5,876,000,000 = 5.876 \times 10^9.$$

15. 82,350 ▪ Move the decimal point left 4 places so it is to the right of the first nonzero digit.

8.2350 *4 places*

The original number, 82,350, is *greater* than 8.2350, so multiply by a *positive* power of 10. Thus,

$$82,350 = 8.2350 \times 10^4 = 8.235 \times 10^4$$

(Note that the final zero need not be written.)

17. $0.000\,007$ ■ Move the decimal point right 6 places so it is to the right of the first nonzero digit.

$$0\,0\,0\,0\,0\,7.\quad 6\,places$$

The original number, 0.000007, is *less* than 7, so multiply by a *negative* power of 10. Thus,

$$0.000\,007 = 7 \times 10^{-6}.$$

19. $-0.002\,03$ ■ Consider the absolute value of the number, 0.00203. Move the decimal point right 3 places so it is to the right of the first nonzero digit. The original number, 0.00203, is *less* than 2.03, so multiply by a *negative* power of 10. Thus,

$$-0.002\,03 = -2.03 \times 10^{-3}$$

21. 7.5×10^5 ■ Since the exponent is *positive*, move the decimal point 5 places to the *right*. Attach four 0s.

$$7.5 \times 10^5 = 750,000$$

23. 5.677×10^{12} ■ Since the exponent is *positive*, move the decimal point 12 places to the *right*. Attach nine 0s.

$$5.677 \times 10^{12} = 5,677,000,000,000$$

25. 1×10^{12} ■ Since the exponent is *positive*, move the decimal point 12 places to the *right*. Attach twelve 0s.

$$1 \times 10^{12} = 1,000,000,000,000$$

27. -6.21×10^0 ■ Because the exponent is 0, the decimal point should not be moved.

$$-6.21 \times 10^0 = -6.21$$

We know this result is correct because $10^0 = 1$.

29. 7.8×10^{-4} ■ Since the exponent is *negative*, move the decimal point 4 places to the *left*.

$$7.8 \times 10^{-4} = 0.00078$$

31. 5.134×10^{-9}

Since the exponent is *negative*, move the decimal point 9 places to the *left*.

$$5.134 \times 10^{-9} = 0.000\,000\,005\,134$$

33. $(2 \times 10^8)(3 \times 10^3)$

$$= (2 \times 3)(10^8 \times 10^3) \quad \textit{Commutative and associative properties}$$
$$= 6 \times 10^{11} \quad \textit{Product rule for exponents; Scientific notation}$$
$$= 600,000,000,000 \quad \textit{Without exponents}$$

35. $(5 \times 10^4)(3 \times 10^2)$

$$= (5 \times 3)(10^4 \times 10^2)$$
$$= 15 \times 10^6$$
$$= (1.5 \times 10^1) \times 10^6$$
$$= 1.5 \times (10^1 \times 10^6)$$
$$= 1.5 \times 10^7 \quad \textit{Scientific notation}$$
$$= 15,000,000 \quad \textit{Without exponents}$$

37. $(4 \times 10^{-6})(2 \times 10^3)$

$$= (4 \times 2)(10^{-6} \times 10^3)$$
$$= 8 \times 10^{-3} \quad \textit{Scientific notation}$$
$$= 0.008 \quad \textit{Without exponents}$$

39. $(6 \times 10^3)(4 \times 10^{-2})$

$$= (6 \times 4)(10^3 \times 10^{-2})$$
$$= 24 \times 10^1$$
$$= (2.4 \times 10^1) \times 10^1$$
$$= 2.4 \times (10^1 \times 10^1)$$
$$= 2.4 \times 10^2 \quad \textit{Scientific notation}$$
$$= 240 \quad \textit{Without exponents}$$

41. $(9 \times 10^4)(7 \times 10^{-7})$

$$= (9 \times 7)(10^4 \times 10^{-7})$$
$$= 63 \times 10^{-3}$$
$$= (6.3 \times 10^1) \times 10^{-3}$$
$$= 6.3 \times (10^1 \times 10^{-3})$$
$$= 6.3 \times 10^{-2} \quad \textit{Scientific notation}$$
$$= 0.063 \quad \textit{Without exponents}$$

43. $(3.15 \times 10^{-4})(2.04 \times 10^8)$

$$= (3.15 \times 2.04)(10^{-4} \times 10^8)$$
$$= 6.426 \times 10^4 \quad \textit{Scientific notation}$$
$$= 64,260 \quad \textit{Without exponents}$$

45. $\dfrac{9 \times 10^{-5}}{3 \times 10^{-1}} = \dfrac{9}{3} \times \dfrac{10^{-5}}{10^{-1}}$

$$= 3 \times 10^{-5 - (-1)}$$
$$= 3 \times 10^{-4} \quad \textit{Scientific notation}$$
$$= 0.0003 \quad \textit{Without exponents}$$

47. $\dfrac{8 \times 10^3}{2 \times 10^2} = \dfrac{8}{2} \times \dfrac{10^3}{10^2}$

$$= 4 \times 10^1 \quad \textit{Scientific notation}$$
$$= 40 \quad \textit{Without exponents}$$

49. $\dfrac{2.6 \times 10^{-3}}{2 \times 10^2} = \dfrac{2.6}{2} \times \dfrac{10^{-3}}{10^2}$

$$= 1.3 \times 10^{-5} \quad \textit{Scientific notation}$$
$$= 0.000\,013 \quad \textit{Without exponents}$$

51. $\dfrac{4 \times 10^5}{8 \times 10^2} = \dfrac{4}{8} \times \dfrac{10^5}{10^2}$

$= 0.5 \times 10^{5-2}$

$= 0.5 \times 10^3$

$= 5 \times 10^2$ *Scientific notation*

$= 500$ *Without exponents*

53. $\dfrac{(2.6 \times 10^{-3})(7.0 \times 10^{-1})}{(2 \times 10^2)(3.5 \times 10^{-3})}$

$= \dfrac{2.6}{2} \times \dfrac{7.0}{3.5} \times \dfrac{10^{-4}}{10^{-1}}$

$= 1.3 \times 2 \times 10^{-4-(-1)}$

$= 2.6 \times 10^{-3}$ *Scientific notation*

$= 0.0026$ *Without exponents*

55. $\dfrac{(1.65 \times 10^8)(5.24 \times 10^{-2})}{(6 \times 10^4)(2 \times 10^7)}$

$= \dfrac{8.646 \times 10^6}{12 \times 10^{11}}$ *Simplify num.* *Simplify den.*

$= 0.7205 \times 10^{-5}$

$= 7.205 \times 10^{-6}$ *Scientific notation*

$= 0.000\,007\,205$ *Without exponents*

57. $2 \times 10^{-6} = 0.000\,002$

59. There are 41 zeros, so the number is 4.2×10^{12}.

61. To find the number of miles, multiply the number of light-years by the number of miles per light-year.

$(25,000)(6,000,000,000,000)$

$= (2.5 \times 10^4)(6 \times 10^{12})$

$= (2.5 \times 6)(10^4 \times 10^{12})$

$= 15 \times 10^{16}$

$= 1.5 \times 10^{17}$

The nebula is about 1.5×10^{17} miles from Earth.

63. To find the amount each person would have to contribute, divide the amount by the population.

$\dfrac{\$1,000,000,000,000}{311.6 \text{ million}} = \dfrac{\$1 \times 10^{12}}{311,600,000}$

$= \dfrac{\$1 \times 10^{12}}{3.116 \times 10^8}$

$= \dfrac{\$1}{3.116} \times \dfrac{10^{12}}{10^8}$

$\approx \$0.3209 \times 10^4$

$= \$3.209 \times 10^3$

Each person would have to contribute about $3209.

65. To find the debt per person, divide the debt by the population.

$\dfrac{\$1.64 \times 10^{13}}{311 \text{ million}} = \dfrac{\$1.64 \times 10^{13}}{311,000,000}$

$= \dfrac{\$1.64 \times 10^{13}}{3.11 \times 10^8}$

$\approx \$0.52733 \times 10^5$

The debt is about $52,733 for every person.

67. Solve the formula $d = rt$ for t.

$t = \dfrac{d}{r} = \dfrac{6.68 \times 10^7}{1.86 \times 10^5}$

$= \dfrac{6.68}{1.86} \times \dfrac{10^7}{10^5}$

$\approx 3.59 \times 10^2$

It takes about 3.59×10^2 seconds, or 359 seconds, for light to travel from the Sun to Venus.

69. To get the average ticket price, divide the total gross by the attendance.

$\dfrac{\$1.08 \times 10^9}{1.25 \times 10^7} = \dfrac{\$1.08}{1.25} \times \dfrac{10^9}{10^7}$

$= \$0.864 \times 10^{9-7}$

$= \$0.864 \times 10^2$

$= \$86.40$

The average ticket price was $86.40.

Chapter 13 Review Exercises

1. $9m^2 + 11m^2 + 2m^2 = (9 + 11 + 2)m^2$

$\qquad\qquad\qquad\qquad = 22m^2$

The degree is 2.

To determine if the polynomial is a monomial, binomial, or trinomial, count the number of terms in the final expression.

There is one term, so this is a *monomial*.

2. $-4p + p^3 - p^2 + 8p + 2$

$= p^3 - p^2 + (-4 + 8)p + 2$

$= p^3 - p^2 + 4p + 2$

The degree is 3.

To determine if the polynomial is a monomial, binomial, or trinomial, count the number of terms in the final expression. Since there are four terms, it is none of these.

3. $12a^5 - 9a^4 + 8a^3 + 2a^2 - a + 3$ cannot be simplified further and is already written in descending powers of the variable.

The degree is 5.

This polynomial has 6 terms, so it is none of the names listed.

4. $-7y^5 - 8y^4 - y^5 + y^4 + 9y$
    $= -7y^5 - 1y^5 - 8y^4 + 1y^4 + 9y$
    $= (-7 - 1)y^5 + (-8 + 1)y^4 + 9y$
    $= -8y^5 - 7y^4 + 9y$

    The degree is 5.

    There are three terms, so the polynomial is a *trinomial*.

5. Add. $-2a^3 + 5a^2$
            $\underline{-3a^3 - a^2}$
            $-5a^3 + 4a^2$

6. Add. $4r^3 - 8r^2 + 6r$
            $\underline{-2r^3 + 5r^2 + 3r}$
            $2r^3 - 3r^2 + 9r$

7. Subtract. $6y^2 - 8y + 2$
                 $\underline{-5y^2 + 2y - 7}$

    Change all signs in the second row and then add.
    $$6y^2 - 8y + 2$$
    $$\underline{5y^2 - 2y + 7}$$
    $$11y^2 - 10y + 9$$

8. Subtract. $-12k^4 - 8k^2 + 7k - 5$
                 $\underline{k^4 + 7k^2 + 11k + 1}$

    Change all signs in the second row and then add.
    $$-12k^4 - 8k^2 + 7k - 5$$
    $$\underline{-k^4 - 7k^2 - 11k - 1}$$
    $$-13k^4 - 15k^2 - 4k - 6$$

9. $(2m^3 - 8m^2 + 4) + (8m^3 + 2m^2 - 7)$
    $= (2m^3 + 8m^3) + (-8m^2 + 2m^2) + (4 - 7)$
    $= 10m^3 - 6m^2 - 3$

10. $(-5y^2 + 3y + 11) + (4y^2 - 7y + 15)$
    $= (-5y^2 + 4y^2) + (3y - 7y) + (11 + 15)$
    $= -y^2 - 4y + 26$

11. $(6p^2 - p - 8) - (-4p^2 + 2p + 3)$
    $= (6p^2 - p - 8) + (4p^2 - 2p - 3)$
    $= (6p^2 + 4p^2) + (-p - 2p) + (-8 - 3)$
    $= 10p^2 - 3p - 11$

12. $(12r^4 - 7r^3 + 2r^2) - (5r^4 - 3r^3 + 2r^2 + 1)$
    $= (12r^4 - 7r^3 + 2r^2) + (-5r^4 + 3r^3 - 2r^2 - 1)$
    $= (12r^4 - 5r^4) + (-7r^3 + 3r^3) + (2r^2 - 2r^2) - 1$
    $= 7r^4 - 4r^3 - 1$

13. $4^3 \cdot 4^8 = 4^{3+8} = 4^{11}$

14. $(-5)^6(-5)^5 = (-5)^{6+5}$
    $= (-5)^{11}$
    $= (-1 \cdot 5)^{11}$
    $= (-1)^{11}(5)^{11}$
    $= -1 \cdot 5^{11}$
    $= -5^{11}$

15. $(-8x^4)(9x^3) = (-8)(9)(x^4)(x^3)$
    $= -72x^{4+3}$
    $= -72x^7$

16. $(2x^2)(5x^3)(x^9) = (2)(5)(x^2)(x^3)(x^9)$
    $= 10x^{2+3+9}$
    $= 10x^{14}$

17. $(19x)^5 = 19^5 x^5$

18. $(-4y)^7$
    $= (-4)^7 y^7$
    $= (-1 \cdot 4)^7 y^7$
    $= (-1)^7(4)^7 y^7$
    $= -1 \cdot 4^7 y^7$
    $= -4^7 y^7$

19. $5(pt)^4 = 5p^4 t^4$

20. $\left(\dfrac{7}{5}\right)^6 = \dfrac{7^6}{5^6}$

21. $(3x^2 y^3)^3$
    $= 3^3 (x^2)^3 (y^3)^3$
    $= 3^3 x^{2 \cdot 3} y^{3 \cdot 3}$
    $= 27x^6 y^9$

22. $(t^4)^8 (t^2)^5$
    $= t^{4 \cdot 8} \cdot t^{2 \cdot 5}$
    $= t^{32} \cdot t^{10}$
    $= t^{32+10}$
    $= t^{42}$

23. $(6x^2 z^4)^2 (x^3 y z^2)^4$
    $= 6^2 (x^2)^2 (z^4)^2 (x^3)^4 (y)^4 (z^2)^4$
    $= 6^2 x^4 z^8 x^{12} y^4 z^8$
    $= 6^2 x^{4+12} y^4 z^{8+8}$
    $= 36x^{16} y^4 z^{16}$

24. $\left(\dfrac{2m^3 n}{p^2}\right)^3$
    $= \dfrac{2^3 (m^3)^3 n^3}{(p^2)^3}$
    $= \dfrac{8m^9 n^3}{p^6}$

25. Use the formula for the volume of a cube, $V = e^3$, with $e = 5x^2$.
    $$V = (5x^2)^3$$
    $$= 5^3 (x^2)^3$$
    $$= 125x^6$$

26. The product rule for exponents does not apply here because we want the sum of 7^2 and 7^4, not their product.

27. $5x(2x + 14)$
$= 5x(2x) + 5x(14)$
$= 10x^2 + 70x$

28. $-3p^3(2p^2 - 5p)$
$= -3p^3(2p^2) + (-3p^3)(-5p)$
$= -6p^5 + 15p^4$

29. $(3r - 2)(2r^2 + 4r - 3)$
Multiply vertically.

$$
\begin{array}{r}
2r^2 \;\; + 4r \;\; - 3 \\
3r \;\; - 2 \\
\hline
-4r^2 \;\; - 8r \;\; + 6 \\
6r^3 \;\; + 12r^2 \;\; - 9r \\
\hline
6r^3 \;\; + \;\; 8r^2 \;\; - 17r \;\; + 6
\end{array}
$$

30. $(2y + 3)(4y^2 - 6y + 9)$
Multiply vertically.

$$
\begin{array}{r}
4y^2 \;\; - 6y \;\; + 9 \\
2y \;\; + 3 \\
\hline
12y^2 \;\; - 18y \;\; + 27 \\
8y^3 \;\; - 12y^2 \;\; + 18y \\
\hline
8y^3 \;\; + 0y^2 \;\; + 0y \;\; + 27 \;\; = 8y^3 + 27
\end{array}
$$

31. $(5p^2 + 3p)(p^3 - p^2 + 5)$
$= 5p^2(p^3) + 5p^2(-p^2) + 5p^2(5)$
$\quad + 3p(p^3) + 3p(-p^2) + 3p(5)$
$= 5p^5 - 5p^4 + 25p^2 + 3p^4 - 3p^3 + 15p$
$= 5p^5 - 2p^4 - 3p^3 + 25p^2 + 15p$

32. Multiply each term of the second polynomial by each term of the first. Then combine like terms.
$(x + 6)(x - 3)$
$\qquad \textbf{F} \qquad \textbf{O} \qquad \textbf{I} \qquad \textbf{L}$
$= x(x) + x(-3) + 6(x) + 6(-3)$
$= x^2 - 3x + 6x - 18$
$= x^2 + 3x - 18$

33. $(3k - 6)(2k + 1)$
$= (3k)(2k) + (3k)(1) + (-6)(2k) + (-6)(1)$
$= 6k^2 + 3k - 12k - 6$
$= 6k^2 - 9k - 6$

34. $(6p - 3q)(2p - 7q)$
$= 6p(2p) + 6p(-7q) + (-3q)(2p) + (-3q)(-7q)$
$= 12p^2 + (-42pq) + (-6pq) + (21q^2)$
$= 12p^2 - 48pq + 21q^2$

35. $(m^2 + m - 9)(2m^2 + 3m - 1)$

$$
\begin{array}{r}
m^2 \;\; + m \;\; - 9 \\
2m^2 \;\; + 3m \;\; - 1 \\
\hline
- m^2 \;\; - m \;\; + 9 \\
3m^3 \;\; + 3m^2 \;\; -27m \\
2m^4 \;\; + 2m^3 \;\; - 18m^2 \\
\hline
2m^4 \;\; + 5m^3 \;\; - 16m^2 \;\; -28m \;\; + 9
\end{array}
$$

36. $(a + 4)^2 = a^2 + 2(a)(4) + 4^2$
$= a^2 + 8a + 16$

37. $(3p - 2)^2 = (3p)^2 - 2(3p)(2) + (2)^2$
$= 9p^2 - 12p + 4$

38. $(2r + 5s)^2 = (2r)^2 + 2(2r)(5s) + (5s)^2$
$= 4r^2 + 20rs + 25s^2$

39. $(r + 2)^3 = (r + 2)^2(r + 2)$
$= (r^2 + 4r + 4)(r + 2)$
$= r^3 + 4r^2 + 4r + 2r^2 + 8r + 8$
$= r^3 + 6r^2 + 12r + 8$

40. $(2x - 1)^3 = (2x - 1)(2x - 1)^2$
$= (2x - 1)(4x^2 - 4x + 1)$
$= 8x^3 - 8x^2 + 2x - 4x^2 + 4x - 1$
$= 8x^3 - 12x^2 + 6x - 1$

41. $(2z + 7)(2z - 7) = (2z)^2 - 7^2$
$= 4z^2 - 49$

42. $(6m - 5)(6m + 5) = (6m)^2 - 5^2$
$= 36m^2 - 25$

43. $(5a + 6b)(5a - 6b) = (5a)^2 - (6b)^2$
$= 25a^2 - 36b^2$

44. $(2x^2 + 5)(2x^2 - 5) = (2x^2)^2 - 5^2$
$= 4x^4 - 25$

45. The square of a binomial leads to a polynomial with *three* terms. The product of the sum and difference of two terms leads to a polynomial with *two* terms.

46. $(a + b)^2 = (a + b)(a + b) = a^2 + 2ab + b^2$.
The term $2ab$ is not in $a^2 + b^2$.

47. $5^0 + 8^0 = 1 + 1 = 2$

48. $2^{-5} = \dfrac{1}{2^5} = \dfrac{1}{32}$

49. $\left(\dfrac{6}{5}\right)^{-2} = \left(\dfrac{5}{6}\right)^2$
$= \dfrac{5^2}{6^2} = \dfrac{25}{36}$

50. $4^{-2} - 4^{-1} = \dfrac{1}{4^2} - \dfrac{1}{4^1}$
$= \dfrac{1}{16} - \dfrac{1}{4}$
$= \dfrac{1}{16} - \dfrac{4}{16} = -\dfrac{3}{16}$

51. $\dfrac{6^{-3}}{6^{-5}} = 6^{-3 - (-5)}$
$= 6^{-3 + 5} = 6^2 = 36$

52. $\dfrac{x^{-7}}{x^{-9}} = x^{-7 - (-9)}$
$= x^{-7 + 9} = x^2$

53. $\dfrac{p^{-8}}{p^1} = p^{-8-4} = p^{-12} = \dfrac{1}{p^{12}}$

54. $\dfrac{r^{-2}}{r^{-6}} = r^{-2-(-6)} = r^{-2+6} = r^4$

55. $(2^4)^2 = 2^{4 \cdot 2} = 2^8$

56. $(9^3)^{-2} = 9^{(3)(-2)}$

$= 9^{-6} = \dfrac{1}{9^6}$

57. $(5^{-2})^{-4} = 5^{(-2)(-4)} = 5^8$

58. $(8^{-3})^4 = 8^{(-3)(4)}$

$= 8^{-12} = \dfrac{1}{8^{12}}$

59. $\dfrac{(m^2)^3}{(m^4)^2} = \dfrac{m^6}{m^8}$

$= m^{6-8}$

$= m^{-2} = \dfrac{1}{m^2}$

60. $\dfrac{y^4 \cdot y^{-2}}{y^{-5}} = \dfrac{y^4 \cdot y^5}{y^2}$

$= \dfrac{y^9}{y^2} = y^7$

61. $\dfrac{r^9 \cdot r^{-5}}{r^{-2} \cdot r^{-7}} = \dfrac{r^{9+(-5)}}{r^{-2+(-7)}}$

$= \dfrac{r^4}{r^{-9}}$

$= r^{4-(-9)}$

$= r^{13}$

62. $(-5m^3)^2 = (-5)^2 (m^3)^2$

$= (-5)^2 m^{3 \cdot 2}$

$= 25m^6$

63. $(2y^{-4})^{-3} = 2^{-3}(y^{-4})^{-3}$

$= 2^{-3} y^{(-4)(-3)}$

$= 2^{-3} y^{12}$

$= \dfrac{1}{2^3} \cdot y^{12}$

$= \dfrac{y^{12}}{8}$

64. $\dfrac{ab^{-3}}{a^4 b^2} = \dfrac{a}{a^4 b^2 b^3} = \dfrac{1}{a^3 b^5}$

65. $\dfrac{(6r^{-1})^2 \cdot (2r^{-4})}{r^{-5}(r^2)^{-3}} = \dfrac{(6^2 r^{-2}) \cdot (2r^{-4})}{r^{-5} r^{-6}}$

$= \dfrac{2 \cdot 6^2 \cdot r^{-6}}{r^{-11}}$

$= \dfrac{2 \cdot 36 \cdot r^{11}}{r^6}$

$= 72r^5$

66. $\dfrac{(2m^{-5}n^2)^3 (3m^2)^{-1}}{m^{-2}n^{-4}(m^{-1})^2}$

$= \dfrac{2^3 (m^{-5})^3 (n^2)^3 \, 3^{-1}(m^2)^{-1}}{m^{-2}n^{-4}(m^{-1})^2}$

$= \dfrac{2^3 m^{(-5)(3)} n^{2 \cdot 3} \, 3^{-1} m^{2(-1)}}{m^{-2} n^{-4} m^{(-1)(2)}}$

$= \dfrac{2^3 m^{-15} n^6 \cdot 3^{-1} m^{-2}}{m^{-2} n^{-4} m^{-2}}$

$= \dfrac{2^3 \cdot 3^{-1} m^{-15+(-2)} n^6}{m^{-2+(-2)} n^{-4}}$

$= \dfrac{2^3 \cdot 3^{-1} m^{-17} n^6}{m^{-4} n^{-4}}$

$= 2^3 \cdot 3^{-1} m^{-17-(-4)} n^{6-(-4)}$

$= 2^3 \cdot 3^{-1} m^{-13} n^{10}$

$= \dfrac{8n^{10}}{3m^{13}}$

67. $\dfrac{-15y^4}{-9y^2} = \dfrac{-15}{-9} \cdot \dfrac{y^4}{y^2} = \dfrac{5}{3} \cdot y^{4-2} = \dfrac{5y^2}{3}$

68. $\dfrac{-12x^3 y^2}{6xy} = \dfrac{-12}{6} \cdot \dfrac{x^3}{x} \cdot \dfrac{y^2}{y}$

$= -2x^{3-1} y^{2-1}$

$= -2x^2 y$

69. $\dfrac{6y^4 - 12y^2 + 18y}{-6y} = \dfrac{6y^4}{-6y} - \dfrac{12y^2}{-6y} + \dfrac{18y}{-6y}$

$= -y^3 + 2y - 3$

70. $\dfrac{2p^3 - 6p^2 + 5p}{2p^2} = \dfrac{2p^3}{2p^2} - \dfrac{6p^2}{2p^2} + \dfrac{5p}{2p^2}$

$= p - 3 + \dfrac{5}{2p}$

71. $(5x^{13} - 10x^{12} + 20x^7 - 35x^5) \div (-5x^4)$

$= \dfrac{5x^{13}}{-5x^4} - \dfrac{10x^{12}}{-5x^4} + \dfrac{20x^7}{-5x^4} - \dfrac{35x^5}{-5x^4}$

$= -x^9 + 2x^8 - 4x^3 + 7x$

72. $(-10m^4 n^2 + 5m^3 n^3 + 6m^2 n^4) \div (5m^2 n)$

$= \dfrac{-10m^4 n^2 + 5m^3 n^3 + 6m^2 n^4}{5m^2 n}$

$= \dfrac{-10m^4 n^2}{5m^2 n} + \dfrac{5m^3 n^3}{5m^2 n} + \dfrac{6m^2 n^4}{5m^2 n}$

$= -2m^2 n + mn^2 + \dfrac{6n^3}{5}$

73. $(2r^2 + 3r - 14) \div (r - 2)$

$$
\begin{array}{r}
2r + 7 \\
r-2 \,\overline{\smash{\big)}\, 2r^2 + 3r - 14} \\
\underline{2r^2 - 4r } \\
7r - 14 \\
\underline{7r - 14} \\
0
\end{array}
$$

Answer: $2r + 7$

74. $\dfrac{12m^2 - 11m - 10}{3m - 5}$

$$
\begin{array}{r}
4m + 3 \\
3m - 5 \,\overline{\big)\, 12m^2 - 11m - 10} \\
\underline{12m^2 - 20m} \\
9m - 10 \\
\underline{9m - 15} \\
5
\end{array}
$$

Answer: $4m + 3 + \dfrac{5}{3m - 5}$

75. $\dfrac{10a^3 + 5a^2 - 14a + 9}{5a^2 - 3}$

$$
\begin{array}{r}
2a + 1 \\
5a^2 + 0a - 3 \,\overline{\big)\, 10a^3 + 5a^2 - 14a + 9} \\
\underline{10a^3 + 0a^2 - 6a} \\
5a^2 - 8a + 9 \\
\underline{5a^2 + 0a - 3} \\
-8a + 12
\end{array}
$$

Answer: $2a + 1 + \dfrac{-8a + 12}{5a^2 - 3}$

76. $\dfrac{2k^4 + 4k^3 + 9k^2 - 8}{2k^2 + 1}$

$$
\begin{array}{r}
k^2 + 2k + 4 \\
2k^2 + 0k + 1 \,\overline{\big)\, 2k^4 + 4k^3 + 9k^2 + 0k - 8} \\
\underline{2k^4 + 0k^3 + k^2} \\
4k^3 + 8k^2 + 0k \\
\underline{4k^3 + 0k^2 + 2k} \\
8k^2 - 2k - 8 \\
\underline{8k^2 + 0k + 4} \\
-2k - 12
\end{array}
$$

Answer: $k^2 + 2k + 4 + \dfrac{-2k - 12}{2k^2 + 1}$

77. 48,000,000 ▪ Move the decimal point left 7 places so it is to the right of the first nonzero digit. The original number, 48,000,000, is *greater* than 4.8, so multiply by a *positive* power of 10. Thus,

$$48{,}000{,}000 = 4.8 \times 10^7.$$

78. 28,988,000,000 ▪ Move the decimal point left 10 places so it is to the right of the first nonzero digit. The original number, 28,988,000,000, is *greater* than 2.8988, so multiply by a *positive* power of 10. Thus,

$$28{,}988{,}000{,}000 = 2.8988 \times 10^{10}.$$

79. 0.000 065 ▪ Move the decimal point right 5 places so it is to the right of the first nonzero digit.

The original number, 0.000065, is *less* than 6.5, so multiply by a *negative* power of 10. Thus,

$$0.000\,065 = 6.5 \times 10^{-5}.$$

80. 0.000 000 082 4 ▪ Move the decimal point right 8 places so it is to the right of the first nonzero digit. The original number, 0.000 000 082 4, is *less* than 8.24, so multiply by a *negative* power of 10. Thus,

$$0.000\,000\,082\,4 = 8.24 \times 10^{-8}.$$

81. 2.4×10^4 ▪ Since the exponent is *positive*, move the decimal point 4 places to the *right*. Attach three 0s.

$$2.4 \times 10^4 = 24{,}000$$

82. 7.83×10^7 ▪ Since the exponent is *positive*, move the decimal point 7 places to the *right*. Attach five 0s.

$$7.83 \times 10^7 = 78{,}300{,}000$$

83. 8.97×10^{-7} ▪ Since the exponent is *negative*, move the decimal point 7 places to the *left*.

$$8.97 \times 10^{-7} = 0.000\,000\,897$$

84. 9.95×10^{-12} ▪ Since the exponent is *negative*, move the decimal point 12 places to the *left*.

$$9.95 \times 10^{-12} = 0.000\,000\,000\,009\,95$$

85.
$$
\begin{aligned}
&(2 \times 10^{-3})(4 \times 10^5) \\
&= (2 \times 4)(10^{-3} \times 10^5) \\
&= 8 \times 10^{-3+5} \\
&= 8 \times 10^2 \qquad \textit{Scientific notation} \\
&= 800 \qquad\quad\ \textit{Without exponents}
\end{aligned}
$$

86.
$$
\begin{aligned}
\dfrac{8 \times 10^4}{2 \times 10^{-2}} &= \dfrac{8}{2} \times \dfrac{10^4}{10^{-2}} \\
&= 4 \times 10^{4-(-2)} \\
&= 4 \times 10^6 \qquad \textit{Scientific notation} \\
&= 4{,}000{,}000 \qquad \textit{Without exponents}
\end{aligned}
$$

87.
$$
\begin{aligned}
&\dfrac{(12 \times 10^{-5})(5 \times 10^4)}{(4 \times 10^3)(6 \times 10^{-2})} \\
&= \dfrac{12 \times 5}{4 \times 6} \times \dfrac{10^{-5} \times 10^4}{10^3 \times 10^{-2}} \\
&= \dfrac{60}{24} \times \dfrac{10^{-1}}{10^1} \\
&= \dfrac{5}{2} \times 10^{-1-1} \\
&= 2.5 \times 10^{-2} \qquad \textit{Scientific notation} \\
&= 0.025 \qquad\qquad\ \textit{Without exponents}
\end{aligned}
$$

88. $\dfrac{(2.5 \times 10^5)(4.8 \times 10^{-4})}{(7.5 \times 10^8)(1.6 \times 10^{-5})}$

$= \dfrac{2.5 \times 4.8 \times 10^5 \times 10^{-4}}{7.5 \times 1.6 \times 10^8 \times 10^{-5}}$

$= \dfrac{2.5}{7.5} \times \dfrac{4.8}{1.6} \times \dfrac{10^{5+(-4)}}{10^{8+(-5)}}$

$= \dfrac{1}{3} \times \dfrac{3}{1} \times \dfrac{10^1}{10^3}$

$= 1 \times 10^{1-3}$

$= 1 \times 10^{-2}$ *Scientific notation*

$= 0.01$ *Without exponents*

89. $60(466,000,000)$

$= (6 \times 10^1)(4.66 \times 10^8)$

$= (6 \times 4.66)(10^1 \times 10^8)$

$= 27.96 \times 10^9$

$= 2.796 \times 10^{10}$

The computer can perform 2.796×10^{10} calculations in one minute.

$60(2.796 \times 10^{10})$

$= (6 \times 10^1)(2.796 \times 10^{10})$

$= (6 \times 2.796)(10^1 \times 10^{10})$

$= 16.776 \times 10^{11}$

$= 1.6776 \times 10^{12}$

The computer can perform 1.6776×10^{12} calculations in one hour.

90. $\dfrac{1 \times 10^9}{3 \times 10^8} = \dfrac{1}{3} \times 10^{9-8}$

$\approx 0.33 \times 10^1 = 3.3$

There are about 3.3 Social Security numbers available for each person.

91. **[13.5]** $19^0 - 3^0 = 1 - 1 = 0$

92. **[13.5]** $(3p)^4(3p^{-7})$

$= (3^4)(p^4)(3p^{-7})$

$= (3^4)(3)(p^4)(p^{-7})$

$= 3^{4+1}p^{4+(-7)}$

$= 3^5 p^{-3}$

$= \dfrac{243}{p^3}$

93. **[13.5]** $7^{-2} = \dfrac{1}{7^2} = \dfrac{1}{49}$

94. **[13.4]** $(-7 + 2k)^2$

$= (-7)^2 + 2(-7)(2k) + (2k)^2$

$= 49 - 28k + 4k^2$

95. **[13.7]** $\dfrac{2y^3 + 17y^2 + 37y + 7}{2y + 7}$

$$
\begin{array}{r}
y^2 + 5y + 1 \\
2y+7\,\overline{\smash{\big)}\,2y^3 + 17y^2 + 37y + 7} \\
\underline{2y^3 + 7y^2} \\
10y^2 + 37y \\
\underline{10y^2 + 35y} \\
2y + 7 \\
\underline{2y + 7} \\
0
\end{array}
$$

Answer: $y^2 + 5y + 1$

96. **[13.2]** $\left(\dfrac{6r^2 s}{5}\right)^4 = \dfrac{(6r^2 s)^4}{5^4}$

$= \dfrac{6^4 (r^2)^4 s^4}{5^4}$

$= \dfrac{1296 r^8 s^4}{625}$

97. **[13.3]** $-m^5(8m^2 + 10m + 6)$

$= -m^5(8m^2) + (-m^5)(10m) + (-m^5)(6)$

$= -8m^7 + (-10m^6) + (-6m^5)$

$= -8m^7 - 10m^6 - 6m^5$

98. **[13.5]** $\left(\dfrac{1}{2}\right)^{-5} = \left(\dfrac{2}{1}\right)^5 = 2^5 = 32$

99. **[13.6]** $(25x^2 y^3 - 8xy^2 + 15x^3 y) \div (5x)$

$= \dfrac{25x^2 y^3}{5x} - \dfrac{8xy^2}{5x} + \dfrac{15x^3 y}{5x}$

$= 5xy^3 - \dfrac{8y^2}{5} + 3x^2 y$

100. **[13.5]** $(6r^{-2})^{-1} = 6^{-1}(r^{-2})^{-1}$

$= 6^{-1} r^2$

$= \dfrac{r^2}{6}$

101. **[13.4]** $(2x + y)^3 = (2x + y)(2x + y)(2x + y)$

First find $(2x + y)(2x + y) = (2x + y)^2$.

$(2x + y)^2 = (2x)^2 + 2(2x)(y) + y^2$

$= 4x^2 + 4xy + y^2$

Now multiply this result by $2x + y$.

$$
\begin{array}{r}
4x^2 + 4xy + y^2 \\
2x + y \\
\hline
4x^2 y + 4xy^2 + y^3 \\
8x^3 + 8x^2 y + 2xy^2 \\
\hline
8x^3 + 12x^2 y + 6xy^2 + y^3
\end{array}
$$

Thus,

$(2x + y)^3 = 8x^3 + 12x^2 y + 6xy^2 + y^3$.

102. **[13.5]** $2^{-1} + 4^{-1} = \dfrac{1}{2} + \dfrac{1}{4} = \dfrac{2}{4} + \dfrac{1}{4} = \dfrac{3}{4}$

103. **[13.3]** $(a+2)(a^2 - 4a + 1)$
$$= a(a^2 - 4a + 1) + 2(a^2 - 4a + 1)$$
$$= a^3 - 4a^2 + a + 2a^2 - 8a + 2$$
$$= a^3 - 2a^2 - 7a + 2$$

104. **[13.1]** $(5y^3 - 8y^2 + 7) - (-3y^3 + y^2 + 2)$
$$= (5y^3 - 8y^2 + 7) + (3y^3 - y^2 - 2)$$
$$= (5y^3 + 3y^3) + (-8y^2 - y^2) + (7 - 2)$$
$$= 8y^3 - 9y^2 + 5$$

105. **[13.3]** $(2r + 5)(5r - 2)$

$$\quad\quad \mathbf{F} \quad\quad\quad \mathbf{O} \quad\quad\quad \mathbf{I} \quad\quad\quad \mathbf{L}$$
$$= 2r(5r) + 2r(-2) + 5(5r) + 5(-2)$$
$$= 10r^2 - 4r + 25r - 10$$
$$= 10r^2 + 21r - 10$$

106. **[13.4]** $(12a + 1)(12a - 1) = (12a)^2 - (1)^2$
$$= 144a^2 - 1$$

107. **[13.3]** Use the formula for the area of a rectangle, $A = lw$, with $l = 2x - 3$ and $w = x + 2$.

$$A = (2x - 3)(x + 2)$$
$$= (2x)(x) + (2x)(2) + (-3)(x) + (-3)(2)$$
$$= 2x^2 + 4x - 3x - 6$$
$$= 2x^2 + x - 6$$

The area of the rectangle is

$$(2x^2 + x - 6) \text{ square units.}$$

Use the formula for the perimeter of a rectangle, $P = 2l + 2w$, with length $l = 2x - 3$ and width $w = x + 2$.

$$P = 2l + 2w$$
$$= 2(2x - 3) + 2(x + 2)$$
$$= 4x - 6 + 2x + 4$$
$$= 6x - 2$$

The perimeter of the rectangle is

$$(6x - 2) \text{ units.}$$

108. **[13.4]** Use the formula for the perimeter of a square, $P = 4s$, with $s = 5x^4 + 2x^2$.

$$P = 4s$$
$$= 4(5x^4 + 2x^2)$$
$$= 4(5x^4) + 4(2x^2)$$
$$= 20x^4 + 8x^2$$

The perimeter of the square is

$$(20x^4 + 8x^2) \text{ units.}$$

Use the formula for the area of a square, $A = s^2$, with $s = 5x^4 + 2x^2$.

$$A = (5x^4 + 2x^2)^2$$
$$= (5x^4)^2 + 2(5x^4)(2x^2) + (2x^2)^2$$
$$= 25x^8 + 20x^6 + 4x^4$$

The area of the square is

$$(25x^8 + 20x^6 + 4x^4) \text{ square units.}$$

109. The error made was not dividing both terms in the numerator by 6. The correct method is as follows:

$$\frac{6x^2 - 12x}{6} = \frac{6x^2}{6} - \frac{12x}{6} = x^2 - 2x.$$

110. Let P be the polynomial that when multiplied by $6m^2n$ gives the product

$$12m^3n^2 + 18m^6n^3 - 24m^2n^2.$$
$$(P)(6m^2n) = 12m^3n^2 + 18m^6n^3 - 24m^2n^2$$
$$P = \frac{12m^3n^2 + 18m^6n^3 - 24m^2n^2}{6m^2n}$$
$$= \frac{12m^3n^2}{6m^2n} + \frac{18m^6n^3}{6m^2n} - \frac{24m^2n^2}{6m^2n}$$
$$= 2mn + 3m^4n^2 - 4n$$

Chapter 13 Test

1. $5x^2 + 8x - 12x^2 = 5x^2 - 12x^2 + 8x$
$$= -7x^2 + 8x$$

degree 2; binomial (2 terms)

2. $13n^3 - n^2 + n^4 + 3n^4 - 9n^2$
$$= n^4 + 3n^4 + 13n^3 - n^2 - 9n^2$$
$$= 4n^4 + 13n^3 - 10n^2$$

degree 4; trinomial (3 terms)

3. $(5t^4 - 3t^2 + 7t + 3) - (t^4 - t^3 + 3t^2 + 8t + 3)$
$$= (5t^4 - 3t^2 + 7t + 3)$$
$$\quad + (-t^4 + t^3 - 3t^2 - 8t - 3)$$
$$= (5t^4 - t^4) + t^3 + (-3t^2 - 3t^2)$$
$$\quad + (7t - 8t) + (3 - 3)$$
$$= 4t^4 + t^3 - 6t^2 - t$$

4. $(2y^2 - 8y + 8) + (-3y^2 + 2y + 3)$
$$\quad - (y^2 + 3y - 6)$$
$$= (2y^2 - 8y + 8) + (-3y^2 + 2y + 3)$$
$$\quad + (-y^2 - 3y + 6)$$
$$= 2y^2 - 3y^2 - y^2 - 8y + 2y - 3y + 8 + 3 + 6$$
$$= -2y^2 - 9y + 17$$

5. Subtract.
$$9t^3 \;-\; 4t^2 \;+\; 2t \;+\; 2$$
$$\underline{9t^3 \;+\; 8t^2 \;-\; 3t \;-\; 6}$$

Change all signs in the second row and then add.

$$9t^3 \;-\; 4t^2 \;+\; 2t \;+\; 2$$
$$\underline{-9t^3 \;-\; 8t^2 \;+\; 3t \;+\; 6}$$
$$\quad\quad\; -\; 12t^2 \;+\; 5t \;+\; 8$$

6. $(-2)^3(-2)^2 = (-2)^{3+2} = (-2)^5 = -32$

7. $\left(\dfrac{6}{m^2}\right)^3 = \dfrac{6^3}{(m^2)^3} = \dfrac{6^3}{m^{2\cdot3}} = \dfrac{216}{m^6}, m \neq 0$

8. $3x^2(-9x^3 + 6x^2 - 2x + 1)$
$= (3x^2)(-9x^3) + (3x^2)(6x^2)$
$\quad + (3x^2)(-2x) + (3x^2)(1)$
$= -27x^5 + 18x^4 - 6x^3 + 3x^2$

9. $(2r - 3)(r^2 + 2r - 5)$

Multiply vertically.

$$
\begin{array}{r}
r^2 + 2r - 5 \\
2r - 3 \\
\hline
-3r^2 - 6r + 15 \\
2r^3 + 4r^2 - 10r \\
\hline
2r^3 + r^2 - 16r + 15
\end{array}
$$

10.
$$\;\;\;\;\textbf{F}\;\;\;\textbf{O}\;\;\;\textbf{I}\;\;\;\textbf{L}$$
$(t - 8)(t + 3) = t^2 + 3t - 8t - 24$
$\quad\quad\quad\quad = t^2 - 5t - 24$

11. $(4x + 3y)(2x - y)$
$$\;\;\;\;\textbf{F}\;\;\;\textbf{O}\;\;\;\textbf{I}\;\;\;\textbf{L}$$
$= 8x^2 - 4xy + 6xy - 3y^2$
$= 8x^2 + 2xy - 3y^2$

12. $(5x - 2y)^2 = (5x)^2 - 2(5x)(2y) + (2y)^2$
$\quad\quad\quad\quad = 25x^2 - 20xy + 4y^2$

13. $(10v + 3w)(10v - 3w) = (10v)^2 - (3w)^2$
$\quad\quad\quad\quad\quad\quad = 100v^2 - 9w^2$

14. $(x + 1)^3$
$= (x + 1)(x + 1)^2$
$= (x + 1)(x^2 + 2x + 1)$
$= x(x^2 + 2x + 1) + 1(x^2 + 2x + 1)$
$= x^3 + 2x^2 + x + x^2 + 2x + 1$
$= x^3 + 3x^2 + 3x + 1$

15. Use the formula for the perimeter of a square, $P = 4s$, with $s = 3x + 9$.
$$P = 4(3x + 9)$$
$$= 4(3x) + 4(9)$$
$$= 12x + 36$$

The perimeter is $(12x + 36)$ units.

Use the formula for the area of a square, $A = s^2$, with $s = 3x + 9$.
$$A = (3x + 9)^2$$
$$= (3x)^2 + 2(3x)(9) + 9^2$$
$$= 9x^2 + 54x + 81$$

The area is $(9x^2 + 54x + 81)$ square units.

16. (a) $3^{-4} = \dfrac{1}{3^4} = \dfrac{1}{81}$, which is *positive*.
A negative exponent indicates a reciprocal, not a negative number.

(b) $(-3)^4 = 81$, which is *positive*.

(c) $-3^4 = -1 \cdot 3^4 = -81$, which is *negative*.

(d) $3^0 = 1$, which is *positive*.

(e) $(-3)^0 - 3^0 = 1 - 1 = 0$ $\quad$ (*zero*)

(f) $(-3)^{-3} = \dfrac{1}{(-3)^3} = \dfrac{1}{-27}$, which is *negative*.

17. $5^{-4} = \dfrac{1}{5^4} = \dfrac{1}{625}$

18. $(-3)^0 + 4^0 = 1 + 1 = 2$

19. $4^{-1} + 3^{-1} = \dfrac{1}{4^1} + \dfrac{1}{3^1} = \dfrac{3}{12} + \dfrac{4}{12} = \dfrac{7}{12}$

20. $\dfrac{8^{-1} \cdot 8^4}{8^{-2}} = \dfrac{8^{(-1)+4}}{8^{-2}} = \dfrac{8^3}{8^{-2}} = 8^{3-(-2)} = 8^5$

21. $\dfrac{(x^{-3})^{-2}(x^{-1}y)^2}{(xy^{-2})^2} = \dfrac{(x^{-3})^{-2}(x^{-1})^2(y)^2}{(x)^2(y^{-2})^2}$
$= \dfrac{x^6 x^{-2} y^2}{x^2 y^{-4}}$
$= \dfrac{x^4 y^2}{x^2 y^{-4}}$
$= x^{4-2} y^{2-(-4)}$
$= x^2 y^6$

22. $\dfrac{8y^3 - 6y^2 + 4y + 10}{2y}$
$= \dfrac{8y^3}{2y} - \dfrac{6y^2}{2y} + \dfrac{4y}{2y} + \dfrac{10}{2y}$
$= 4y^2 - 3y + 2 + \dfrac{5}{y}$

23. $(-9x^2y^3 + 6x^4y^3 + 12xy^3) \div (3xy)$
$= \dfrac{-9x^2y^3 + 6x^4y^3 + 12xy^3}{3xy}$
$= \dfrac{-9x^2y^3}{3xy} + \dfrac{6x^4y^3}{3xy} + \dfrac{12xy^3}{3xy}$
$= -3xy^2 + 2x^3y^2 + 4y^2$

24. $\dfrac{2x^2 + x - 36}{x - 4}$

$$
\begin{array}{r}
2x + 9 \\
x - 4 \overline{)\ 2x^2 + x - 36} \\
\underline{2x^2 - 8x} \\
9x - 36 \\
\underline{9x - 36} \\
0
\end{array}
$$

Answer: $2x + 9$

25. $(3x^3 - x + 4) \div (x - 2)$

$$
\begin{array}{r}
3x^2 + 6x + 11 \\
x - 2 \overline{\smash)3x^3 + 0x^2 - x + 4} \\
\underline{3x^3 - 6x^2} \\
6x^2 - x \\
\underline{6x^2 - 12x} \\
11x + 4 \\
\underline{11x - 22} \\
26
\end{array}
$$

The answer is

$$3x^2 + 6x + 11 + \frac{26}{x - 2}.$$

26. **(a)** 344,000,000,000 ■ Move the decimal point left 11 places so it is to the right of the first nonzero digit. The original number, 344,000,000,000, is *greater* than 3.44, so multiply by a *positive* power of 10. Thus,

$$344{,}000{,}000{,}000 = 3.44 \times 10^{11}.$$

(b) 0.000 005 57 ■ Move the decimal point right 6 places so it is to the right of the first nonzero digit. The original number, 0.000 005 57, is *less* than 5.57, so multiply by a *negative* power of 10. Thus,

$$0.000\,005\,57 = 5.57 \times 10^{-6}.$$

27. **(a)** 2.96×10^7 ■ Since the exponent is *positive*, move the decimal point 7 places to the *right*. Attach five 0s.

$$2.96 \times 10^7 = 29{,}600{,}000$$

(b) 6.07×10^{-8} ■ Since the exponent is *negative*, move the decimal point 8 places to the *left*.

$$6.07 \times 10^{-8} = 0.000\,000\,060\,7$$

28. **(a)** $1000 = 1 \times 10^3$

$$5{,}890{,}000{,}000{,}000 = 5.89 \times 10^{12}$$

(b) $\left(1 \times 10^3 \text{ light-years}\right)\left(5.89 \times 10^{12} \dfrac{\text{miles}}{\text{light-year}}\right)$

$$= 5.89 \times 10^{3+12} \text{ miles}$$
$$= 5.89 \times 10^{15} \text{ miles}$$

CHAPTER 14 FACTORING AND APPLICATIONS

14.1 Factors; The Greatest Common Factor

14.1 Margin Exercises

1. **(a)** 30, 20, 15

Write each number in prime factored form.

$$30 = 2 \cdot 3 \cdot 5, \quad 20 = 2 \cdot \underline{2} \cdot \underline{5}, \quad 15 = 3 \cdot \underline{5}$$

Use each prime the *least* number of times it appears in *all* the factored forms. The primes 2 and 3 do not appear in *all* the factored forms, so they will not appear in the greatest common factor. The greatest common factor (**GCF**) is $\underline{5}$.

(b) 42, 28, 35

Write each number in prime factored form.

$$42 = 2 \cdot 3 \cdot 7, \quad 28 = 2^2 \cdot 7, \quad 35 = 5 \cdot 7$$

Use each prime the least number of times it appears in all the factored forms. The greatest common factor is 7.

(c) 12, 18, 26, 32

Write each number in prime factored form.

$$12 = 2^2 \cdot 3, \qquad 18 = 2 \cdot 3^2,$$
$$26 = 2 \cdot 13, \qquad 32 = 2^5$$

Use each prime the least number of times it appears in all the factored forms. The greatest common factor is 2.

(d) 10, 15, 21

Write each number in prime factored form.

$$10 = 2 \cdot 5, \quad 15 = 3 \cdot 5, \quad 21 = 3 \cdot 7$$

Use each prime the least number of times it appears in all the factored forms. Since no prime appears in all the factored forms, the GCF is 1.

2. **(a)** $6m^4$, $9m^2$, $12m^5$

Write each number in prime factored form.

$$6m^4 = 2 \cdot \underline{3} \cdot m^4, \quad 9m^2 = 3 \cdot \underline{3} \cdot \underline{m^2},$$
$$12m^5 = 2 \cdot 2 \cdot \underline{3} \cdot \underline{m^5}$$

The greatest common factor of the coefficients 6, 9, and 12 is 3. The greatest common factor of the terms m^4, m^2, and m^5 is m^2 since 2 is the least exponent on m. Thus, the GCF of these terms is the product of 3 and m^2, that is, $\underline{3m^2}$.

(b) $12p^5$, $18q^1$ ▪ The greatest common factor of 12 and 18 is 6. Since these terms do not contain

the same variables, no variable is common to them. The GCF is 6.

(c) $y^1 z^2$, $y^6 z^8$, z^9 ▪ The least exponent on z is 2, and y does not occur in z^9. Thus, the GCF is z^2.

(d) $12p^{11}$, $17q^5$ ▪ Since $12p^{11}$ and $17q^5$ do not contain the same variables and 1 is the greatest common factor of 12 and 17, the GCF is 1.

3. **(a)** $4x^2 + 6x$ ▪ 2 is the greatest common factor of 4 and 6, and 1 is the least exponent on x. Hence, $2x$ is the GCF.

$$4x^2 + 6x = 2x(2x) + 2x(\underline{3})$$
$$= 2x(2x + \underline{3})$$

(b) $10y^5 - 8y^1 + 6y^2$ ▪ 2 is the greatest common factor of 10, 8, and 6, so 2 can be factored out. Since y occurs in every term and the least exponent on y is 2, y^2 can be factored out. Hence, $2y^2$ is the GCF.

$$10y^5 - 8y^4 + 6y^2$$
$$= (2y^2)(5y^3) - (2y^2)(4y^2) + (2y^2)(3)$$
$$= 2y^2(5y^3 - 4y^2 + 3)$$

(c) $m^7 + m^9$ ▪ m occurs in every term and 7 is the least exponent on m, so m^7 is the GCF.

$$m^7 + m^9 = \underline{m^7}(\underline{1}) + \underline{m^7}(m^2)$$
$$= \underline{m^7}(\underline{1 + m^2})$$

(d) $15x^3 - 10x^2 + 5x$ ▪ 5 is the greatest common factor of 15, 10, and 5, so 5 can be factored out. Since x occurs in every term and the least exponent on x is 1, x can be factored out. Hence, $5x$ is the GCF.

$$15x^3 - 10x^2 + 5x$$
$$= (5x)(3x^2) - (5x)(2x) + (5x)(1)$$
$$= 5x(3x^2 - 2x + 1)$$

(e) $8p^5q^2 + 16p^6q^3 - 12p^1q^7$ ▪ 4 is the greatest common factor of 8, 16, and 12. The least exponents on p and q are 4 and 2, respectively. Hence, $4p^1q^2$ is the GCF.

$$8p^5q^2 + 16p^6q^3 - 12p^1q^7$$
$$= (4p^1q^2)(2p) + (4p^1q^2)(4p^2q) - (4p^1q^2)(3q^5)$$
$$= 4p^1q^2(2p + 4p^2q - 3q^5)$$

4. $-14a^3b^2 - 21a^2b^3 + 7ab$

$\qquad\qquad -7ab$ *is a common factor.*

$$= -7ab(2a^2b) - 7ab(3ab^2) - 7ab(-1)$$
$$= -7ab(2a^2b + 3ab^2 - 1) \quad \textit{Factor out } -7ab.$$

5. **(a)** $r(t - 4) + 5(t - 4)$

The binomial $t - 4$ is the greatest common factor.

$$r(t - 4) + 5(t - 4) = (t - 4)(\underline{r + 5})$$

(b) $x(x+2) + 7(x+2)$

The binomial $x + 2$ is the greatest common factor.

$$x(x+2) + 7(x+2) = (x+2)(x+7)$$

(c) $y^2(y+2) - 3(y+2)$

The binomial $y + 2$ is the greatest common factor.

$$y^2(y+2) - 3(y+2) = (y+2)(y^2-3)$$

(d) $x(x-1) - 5(x-1)$

The binomial $x - 1$ is the greatest common factor.

$$x(x-1) - 5(x-1) = (x-1)(x-5)$$

6.   **(a)** $pq + 5q + 2p + 10$

$$= (\underline{pq + 5q}) + (\underline{2p} + 10) \qquad \textit{Group terms.}$$

$$= \underline{q}(p+5) + \underline{2}(p+5) \qquad \textit{Factor each group.}$$

$$= (\underline{p+5})(\underline{q+2}) \qquad \textit{Factor out } p + 5.$$

(b) $2xy + 3y + 2x + 3$

$$= (2xy + 3y) + (2x + 3) \qquad \textit{Group terms.}$$

$$= y(2x + 3) + 1(2x + 3) \qquad \textit{Factor each group; remember the 1.}$$

$$= (2x + 3)(y + 1) \qquad \textit{Factor out } 2x + 3.$$

(c) $2a^2 - 4a + 3ab - 6b$

$$= (2a^2 - 4a) + (3ab - 6b) \qquad \textit{Group terms.}$$

$$= 2a(a - 2) + 3b(a - 2) \qquad \textit{Factor each group.}$$

$$= (a - 2)(2a + 3b) \qquad \textit{Factor out } a - 2.$$

(d) $x^3 + 3x^2 - 5x - 15$

$$= (x^3 + 3x^2) + (-5x - 15) \qquad \textit{Group terms.}$$

$$= x^2(x + 3) - 5(x + 3) \qquad \textit{Factor each group.}$$

$$= (x + 3)(x^2 - 5) \qquad \textit{Factor out } x + 3.$$

7.   **(a)** $6y^2 - 20w + 15y - 8yw$ ■   Factoring out the common factor 2 from the first two terms and the common factor y from the last two terms gives

$$6y^2 - 20w + 15y - 8yw$$
$$= 2(3y - 10w) + y(15 - 8w).$$

This does not lead to a common factor, so we try rearranging the terms.

$$= (6y^2 + 15y) + (-20w - 8yw) \qquad \textit{Rearrange.}$$

$$= 3y(2y + 5) - 4w(5 + 2y) \qquad \textit{Factor each group.}$$

$$= (2y + 5)(3y - 4w) \qquad \textit{Factor out } 2y + 5.$$

Here's another rearrangement.

$$= (6y^2 - 8yw) + (15y - 20w)$$
$$= 2y(3y - 4w) + 5(3y - 4w)$$
$$= (3y - 4w)(2y + 5)$$

This is an equivalent answer.

(b) $9mn - 4 + 12m - 3n$
$$= (9mn + 12m) + (-4 - 3n)$$
$$= 3m(3n + 4) - 1(4 + 3n)$$
$$= (3n + 4)(3m - 1)$$

(c) $12p^2 - 28q - 16pq + 21p$ ■   Factoring out the common factor 4 from the first two terms and the common factor p from the last two terms gives

$$12p^2 - 28q - 16pq + 21p$$
$$= 4(3p^2 - 7q) + p(-16q + 21).$$

This does not lead to a common factor, so we try rearranging the terms.

$$= (12p^2 - 16pq) + (21p - 28q) \qquad \textit{Rearrange.}$$

$$= 4p(3p - 4q) + 7(3p - 4q) \qquad \textit{Factor each group.}$$

$$= (3p - 4q)(4p + 7) \qquad \textit{Factor out } 3p - 4q.$$

Here's another rearrangement.

$$= (12p^2 + 21p) + (-16pq - 28q)$$
$$= 3p(4p + 7) - 4q(4p + 7)$$
$$= (4p + 7)(3p - 4q)$$

This is an equivalent answer.

14.1   Section Exercises

1.   To factor a number or quantity means to write it as a <u>product</u>. Factoring is the opposite, or inverse, process of <u>multiplying</u>.

3.   12, 16

Write each number in prime factored form.

$$12 = 2^2 \cdot 3, \quad 16 = 2^4$$

Use each prime the *least* number of times it appears in both factored forms. The prime number 2 is the only prime that appears in both forms and the least number of times it appears in each form is two, so the greatest common factor is $2^2 = 4$.

5.   Find the prime factored form of each number.

$$40 = 2 \cdot 2 \cdot 2 \cdot 5$$
$$20 = 2 \cdot 2 \cdot 5$$
$$4 = 2 \cdot 2$$

The least number of times 2 appears in all the factored forms is 2. There is no 5 in the prime factored form of 4, so the

$$GCF = 2^2 = 4.$$

7. Find the prime factored form of each number.

$$18 = 2 \cdot 3 \cdot 3$$
$$24 = 2 \cdot 2 \cdot 2 \cdot 3$$
$$36 = 2 \cdot 2 \cdot 3 \cdot 3$$
$$48 = 2 \cdot 2 \cdot 2 \cdot 2 \cdot 3$$

The least number of times the primes 2 and 3 appear in all four factored forms is once, so

$$\text{GCF} = 2 \cdot 3 = 6.$$

9. Write each number in prime factored form.

$$4 = 2^2, \quad 9 = 3^2, \quad 12 = 2^2 \cdot 3$$

There are no prime factors common to all three numbers, so the greatest common factor is 1.

11. Write each term in prime factored form.

$$16y = 2^4 \cdot y$$
$$24 = 2^3 \cdot 3$$

There is no y in the second term, so y will not appear in the GCF. Thus, the GCF of $16y$ and 24 is

$$2^3 = 8.$$

13. $30x^3 = 2 \cdot 3 \cdot 5 \cdot x^3$
$40x^6 = 2^3 \cdot 5 \cdot x^6$
$50x^7 = 2 \cdot 5^2 \cdot x^7$

The GCF of the coefficients, 30, 40, and 50, is $2^1 \cdot 5^1 = 10$. The smallest exponent on the variable x is 3. Thus the GCF of the given terms is $10x^3$.

15. $x^4y^3 = x^4 \cdot y^3$
$xy^2 = x \cdot y^2$

The GCF is xy^2.

17. $42ab^3 = 2 \cdot 3 \cdot 7 \cdot a \cdot b^3$
$36a = 2^2 \cdot 3^2 \cdot a$
$90b = 2 \cdot 3^2 \cdot 5 \cdot b$
$48ab = 2^4 \cdot 3 \cdot a \cdot b$

The GCF is $2 \cdot 3 = 6$.

19. $12m^3n^2 = 2^2 \cdot 3 \cdot m^3 \cdot n^2$
$18m^5n^4 = 2 \cdot 3^2 \cdot m^5 \cdot n^4$
$36m^8n^3 = 2^2 \cdot 3^2 \cdot m^8 \cdot n^3$

The GCF is $2 \cdot 3 \cdot m^3 \cdot n^2 = 6m^3n^2$.

21. $2k^2(5k)$ is written as a product of $2k^2$ and $5k$ and hence, it is *factored*.

23. $2k^2 + (5k + 1)$ is written as a sum of $2k^2$ and $(5k + 1)$, and hence, it is *not factored*.

25. The correct factored form is

$$18x^3y^2 + 9xy = 9xy(2x^2y + 1).$$

If a polynomial has two terms, the product of the factors must have two terms.
$9xy(2x^2y) = 18x^3y^2$ is just one term.

27. $9m^4 = 3m^2(\underline{3m^2})$

Factor out $3m^2$ from $9m^4$ to obtain $3m^2$.

29. $-8z^9 = -4z^5(\underline{2z^4})$

Factor out $-4z^5$ from $-8z^9$ to obtain $2z^4$.

31. $6m^4n^5 = 3m^3n(\underline{2mn^4})$

Factor out $3m^3n$ from $6m^4n^5$ to obtain $2mn^4$.

33. $12y + 24 = 12 \cdot y + 12 \cdot 2$
$ = 12(\underline{y + 2})$

35. $10a^2 - 20a = 10a(a) - 10a(2)$
$ = 10a(\underline{a - 2})$

37. $8x^2y + 12x^3y^2 = 4x^2y(2) + 4x^2y(3xy)$
$ = 4x^2y(\underline{2 + 3xy})$

39. The greatest common factor for $x^2 - 4x$ is x.

$$x^2 - 4x = x(x) + x(-4)$$
$$= x(x - 4)$$

41. The greatest common factor for $6t^2 + 15t$ is $3t$.

$$6t^2 + 15t = 3t(2t) + 3t(5)$$
$$= 3t(2t + 5)$$

43. The greatest common factor for $m^3 - m^2$ is m^2.

$$m^3 - m^2 = m^2(m) - m^2(1)$$
$$= m^2(m - 1)$$

45. The GCF for $-12x^3 - 6x^2$ is $-6x^2$.

$$-12x^3 - 6x^2 = -6x^2(2x) - 6x^2(1)$$
$$= -6x^2(2x + 1)$$

47. The GCF for $65y^{10} + 35y^6$ is $5y^6$.

$$65y^{10} + 35y^6 = (5y^6)(13y^4) + (5y^6)(7)$$
$$= 5y^6(13y^4 + 7)$$

49. $11w^3 - 100$ ■ The two terms of this expression have no common factor (except 1).

51. The GCF for $8mn^3 + 24m^2n^3$ is $8mn^3$.

$$8mn^3 + 24m^2n^3$$
$$= (8mn^3)(1) + (8mn^3)(3m)$$
$$= 8mn^3(1 + 3m)$$

53. The GCF for $-4x^3 + 10x^2 - 6x$ is $-2x$.

$$-4x^3 + 10x^2 - 6x$$
$$= -2x(2x^2) - 2x(-5x) - 2x(3)$$
$$= -2x(2x^2 - 5x + 3)$$

55. The GCF for $13y^8 + 26y^4 - 39y^2$ is $13y^2$.

$$13y^8 + 26y^4 - 39y^2$$
$$= 13y^2(y^6) + 13y^2(2y^2) + 13y^2(-3)$$
$$= 13y^2(y^6 + 2y^2 - 3)$$

57. The GCF for $45q^4p^5 + 36qp^6 + 81q^2p^3$ is $9qp^3$.

$$45q^4p^5 + 36qp^6 + 81q^2p^3$$
$$= 9qp^3(5q^3p^2) + 9qp^3(4p^3) + 9qp^3(9q)$$
$$= 9qp^3(5q^3p^2 + 4p^3 + 9q)$$

59. The GCF of the terms of $c(x + 2) + d(x + 2)$ is the binomial $x + 2$.

$$c(x + 2) + d(x + 2)$$
$$= (x + 2)(c) + (x + 2)(d)$$
$$= (x + 2)(c + d)$$

61. The GCF for $a^2(2a + b) - b(2a + b)$ is $2a + b$.

$$a^2(2a + b) - b(2a + b) = (2a + b)(a^2 - b)$$

63. The GCF for $q(p + 4) - 1(p + 4)$ is $p + 4$.

$$q(p + 4) - 1(p + 4) = (p + 4)(q - 1)$$

65. $8(7t + 4) + x(7t + 4)$

This expression is the *sum* of two terms, $8(7t + 4)$ and $x(7t + 4)$, so it is not in factored form. We can factor out $7t + 4$.

$$8(7t + 4) + x(7t + 4)$$
$$= (7t + 4)(8) + (7t + 4)(x)$$
$$= (7t + 4)(8 + x)$$

67. $(8 + x)(7t + 4)$ ▪   This expression is the *product* of two factors, $8 + x$ and $7t + 4$, so it is in factored form.

69. $18x^2(y + 4) + 7(y - 4)$ ▪   This expression is the *sum* of two terms, $18x^2(y + 4)$ and $7(y - 4)$, so it is not in factored form.

71. $5m + mn + 20 + 4n$
$$= (5m + mn) + (20 + 4n)$$
$$= m(5 + n) + 4(5 + n)$$
$$= (5 + n)(m + 4)$$

73. $6xy - 21x + 8y - 28$
$$= (6xy - 21x) + (8y - 28)$$
$$= 3x(2y - 7) + 4(2y - 7)$$
$$= (2y - 7)(3x + 4)$$

75. $3xy + 9x + y + 3$
$$= (3xy + 9x) + (y + 3)$$
$$= 3x(y + 3) + 1(y + 3)$$
$$= (y + 3)(3x + 1)$$

77. $7z^2 + 14z - az - 2a$

$$= (7z^2 + 14z) + (-az - 2a) \qquad \text{\textit{Group the terms.}}$$
$$= 7z(z + 2) - a(z + 2) \qquad \text{\textit{Factor each group.}}$$
$$= (z + 2)(7z - a) \qquad \text{\textit{Factor out } z + 2.}$$

79. $18r^2 + 12ry - 3xr - 2xy$

$$= (18r^2 + 12ry) + (-3xr - 2xy) \qquad \text{\textit{Group the terms.}}$$
$$= 6r(3r + 2y) - x(3r + 2y) \qquad \text{\textit{Factor each group.}}$$
$$= (3r + 2y)(6r - x) \qquad \text{\textit{Factor out } 3r + 2y.}$$

81. $w^3 + w^2 + 9w + 9$
$$= (w^3 + w^2) + (9w + 9)$$
$$= w^2(w + 1) + 9(w + 1)$$
$$= (w + 1)(w^2 + 9)$$

83. $3a^3 + 6a^2 - 2a - 4$
$$= (3a^3 + 6a^2) + (-2a - 4)$$
$$= 3a^2(a + 2) - 2(a + 2)$$
$$= (a + 2)(3a^2 - 2)$$

85. $16m^3 - 4m^2p^2 - 4mp + p^3$
$$= (16m^3 - 4m^2p^2) + (-4mp + p^3)$$
$$= 4m^2(4m - p^2) - p(4m - p^2)$$
$$= (4m - p^2)(4m^2 - p)$$

87. $y^2 + 3x + 3y + xy$ ▪   We need to rearrange these terms to get two groups that each have a common factor. We could group y^2 with either $3y$ or xy.

$$y^2 + 3x + 3y + xy$$
$$= y^2 + 3y + xy + 3x \qquad \text{\textit{Rearrange.}}$$
$$= (y^2 + 3y) + (xy + 3x) \qquad \text{\textit{Group the terms.}}$$
$$= y(y + 3) + x(y + 3) \qquad \text{\textit{Factor each group.}}$$
$$= (y + 3)(y + x) \qquad \text{\textit{Factor out } y + 3.}$$

89. $2z^2 + 6w - 4z - 3wz$ ▪   We need to rearrange these terms to get two groups that each have a common factor. We could group $2z^2$ with either $-4z$ or $-3wz$.

$$2z^2 + 6w - 4z - 3wz$$
$$= 2z^2 - 4z - 3wz + 6w \qquad \text{\textit{Rearrange.}}$$
$$= (2z^2 - 4z) + (-3wz + 6w) \qquad \text{\textit{Group the terms.}}$$
$$= 2z(z - 2) - 3w(z - 2) \qquad \text{\textit{Factor each group.}}$$
$$= (z - 2)(2z - 3w) \qquad \text{\textit{Factor out } z - 2.}$$

91. $5m - 6p - 2mp + 15$ ▪ We need to rearrange these terms to get two groups that each have a common factor. We could group $5m$ with either $-2mp$ or 15.

$5m + 15 - 2mp - 6p$ *Rearrange.*
$= (5m + 15) + (-2mp - 6p)$ *Group the terms.*
$= 5(m + 3) - 2p(m + 3)$ *Factor each group.*
$= (m + 3)(5 - 2p)$ *Factor out m + 3.*

93. $18r^2 - 2ty + 12ry - 3rt$ ▪ We need to rearrange these terms to get two groups that each have a common factor. We could group $18r^2$ with either $12ry$ or $-3rt$.

$18r^2 + 12ry - 3rt - 2ty$
    Rearrange.
$= (18r^2 + 12ry) + (-3rt - 2ty)$
    Group the terms.
$= 6r(3r + 2y) - t(3r + 2y)$
    Factor each group.
$= (3r + 2y)(6r - t)$
    Factor out 3r + 2y.

Relating Concepts (Exercises 95–98)

95. In order to rewrite

$$2xy + 12 - 3y - 8x$$

as

$$2xy - 8x - 3y + 12,$$

we must change the order of the terms. The property that allows us to do this is the commutative property of addition.

96. After we group both pairs of terms in the rearranged polynomial, we have

$$(2xy - 8x) + (-3y + 12).$$

The greatest common factor for the first pair of terms is $2x$. The GCF for the second pair is -3. Factoring each group gives us

$$2x(y - 4) - 3(y - 4).$$

97. The expression obtained in Exercise 96 is not a product. It is a *difference* between two terms, $2x(y - 4)$ and $3(y - 4)$. Therefore, it is *not* in factored form.

98. $2x(y - 4) - 3(y - 4)$
$= (y - 4)(2x - 3),$ or $(2x - 3)(y - 4)$

Yes, this is the same result as the one shown in Example 7(b), even though the terms were grouped in a different way.

14.2 Factoring Trinomials

14.2 Margin Exercises

1. **(a)**

Factors of 6	Sums of factors
6, $\underline{1}$	$6 + \underline{1} = \underline{7}$
$\underline{3}$, 2	$\underline{3} + 2 = \underline{5}$

(b) The pair 3, 2 has a sum of 5.

2. **(a)** $y^2 + 12y + 20$ ▪ Find two integers whose product is $\underline{20}$ and whose sum is $\underline{12}$.

Factors of 20	Sums of factors
20, 1	$20 + 1 = 21$
10, 2	$10 + 2 = 12$ ←
5, 4	$5 + 4 = 9$

The pair of integers whose product is 20 and whose sum is 12 is 10 and 2. Thus,

$y^2 + 12y + 20$ factors as $(y + 10)(y + 2)$.

(b) $x^2 + 9x + 18$

Factors of 18	Sums of factors
18, 1	$18 + 1 = 19$
9, 2	$9 + 2 = 11$
6, 3	$6 + 3 = 9$ ←

The pair of integers whose product is 18 and whose sum is 9 is 6 and 3. Thus,

$x^2 + 9x + 18$ factors as $(x + 6)(x + 3)$.

3. **(a)** $t^2 - 12t + 32$ ▪ Find two integers whose product is $\underline{32}$ and whose sum is $\underline{-12}$.

Factors of 32	Sums of factors
$-32, -1$	$-32 + (-1) = -33$
$-16, -2$	$-16 + (-2) = -18$
$-8, -4$	$-8 + (-4) = -12$ ←

The pair of integers whose product is 32 and whose sum is -12 is -8 and -4. Thus,

$t^2 - 12t + 32$ factors as $(t - 8)(t - 4)$.

(b) $y^2 - 10y + 24$

Factors of 24	Sums of factors
$-24, -1$	$-24 + (-1) = -25$
$-12, -2$	$-12 + (-2) = -14$
$-8, -3$	$-8 + (-3) = -11$
$-6, -4$	$-6 + (-4) = -10$ ←

The pair of integers whose product is 24 and whose sum is -10 is -6 and -4. Thus,

$y^2 - 10y + 24$ factors as $(y - 6)(y - 4)$.

4. **(a)** $z^2 + z - 30$, or $z^2 + 1z - 30$

Find the two integers whose product is -30 and whose sum is 1. Because the last term is negative, the pair must include one positive and one negative integer.

Factors of -30	Sums of Factors
$30, -1$	$30 + (-1) = 29$
$15, -2$	$15 + (-2) = 13$
$10, -3$	$10 + (-3) = 7$
$6, -5$	$6 + (-5) = 1$ ←
$-30, 1$	$-30 + 1 = -29$
$-15, 2$	$-15 + 2 = -13$
$-10, 3$	$-10 + 3 = -7$
$-6, 5$	$-6 + 5 = -1$

The required integers are -5 and 6, so

$$z^2 + z - 30 = (z - 5)(z + 6).$$

(b) $x^2 + x - 42$, or $x^2 + 1x - 42$

Find the two integers whose product is -42 and whose sum is 1. Because the last term is negative, the pair must include one positive and one negative integer.

Factors of -42	Sums of Factors
$42, -1$	$42 + (-1) = 41$
$21, -2$	$21 + (-2) = 19$
$14, -3$	$14 + (-3) = 11$
$7, -6$	$7 + (-6) = 1$ ←
$-42, 1$	$-42 + 1 = -41$
$-21, 2$	$-21 + 2 = -19$
$-14, 3$	$-14 + 3 = -11$
$-7, 6$	$-7 + 6 = -1$

The required integers are -6 and 7, so

$$x^2 + x - 42 = (x - 6)(x + 7).$$

5. **(a)** $a^2 - 9a - 22$ ■ Find the two integers whose product is -22 and whose sum is -9. Because the last term is negative, the pair must include one positive and one negative integer.

Factors of -22	Sums of factors
$22, -1$	$22 + (-1) = 21$
$11, -2$	$11 + (-2) = 9$
$-22, 1$	$-22 + 1 = -21$
$-11, 2$	$-11 + 2 = -9$ ←

The required integers are -11 and 2, so

$$a^2 - 9a - 22 = (a - 11)(a + 2).$$

(b) $r^2 - 6r - 16$ ■ Find two integers whose product is -16 and whose sum is -6.

Factors of -16	Sums of factors
$16, -1$	$16 + (-1) = 15$
$8, -2$	$8 + (-2) = 6$
$4, -4$	$4 + (-4) = 0$
$-16, 1$	$-16 + 1 = -15$
$-8, 2$	$-8 + 2 = -6$ ←

The required integers are -8 and 2. Thus,

$$r^2 - 6r - 16 = (r - 8)(r + 2).$$

6. **(a)** $x^2 + 5x + 8$ ■ There is no pair of integers whose product is 8 and whose sum is 5, so $x^2 + 5x + 8$ is a prime polynomial.

(b) $r^2 - 3r - 4$ ■ Find two integers whose product is -4 and whose sum is -3.

Factors of -4	Sums of factors
$2, -2$	$2 + (-2) = 0$
$4, -1$	$4 + (-1) = 3$
$-4, 1$	$-4 + 1 = -3$ ←

The required integers are -4 and 1, so

$$r^2 - 3r - 4 = (r - 4)(r + 1).$$

(c) $m^2 - 2m + 5$ ■ There is no pair of integers whose product is 5 and whose sum is -2, so $m^2 - 2m + 5$ is a prime polynomial.

7. **(a)** $b^2 - 3ab - 4a^2$ ■ Find two expressions whose product is $-4a^2$ and whose sum is $-3a$.

Factors of $-4a^2$	Sums of factors
$4a, -a$	$4a + (-a) = 3a$
$-4a, a$	$-4a + a = -3a$ ←
$2a, -2a$	$2a + (-2a) = 0$

The pair of expressions whose product is $-4a^2$ and whose sum is $-3a$ is $-4a$ and a. Thus,

$$b^2 - 3ab - 4a^2 \text{ factors as } (b - 4a)(b + a).$$

(b) $r^2 - 6rs + 8s^2$ ■ Two expressions whose product is $8s^2$ and whose sum is $-6s$ are $-4s$ and $-2s$, so

$$r^2 - 6rs + 8s^2 = (r - 4s)(r - 2s).$$

8. **(a)** $2p^3 + 6p^2 - 8p$

First, factor out the greatest common factor, $2p$.

$$2p^3 + 6p^2 - 8p = 2p(p^2 + 3p - 4)$$

Now factor $p^2 + 3p - 4$.

The integers 4 and -1 have a product of -4 and a sum of 3, so

$$p^2 + 3p - 4 = (p + 4)(p - 1).$$

The completely factored form is

$$2p^3 + 6p^2 - 8p = 2p(p + 4)(p - 1).$$

(b) $3y^4 - 27y^3 + 60y^2 = 3y^2(y^2 - 9y + 20)$

Factor $y^2 - 9y + 20$. The integers -5 and -4 have a product of 20 and a sum of -9, so

$$y^2 - 9y + 20 = (y - 5)(y - 4).$$

The completely factored form is

$$3y^4 - 27y^3 + 60y^2 = 3y^2(y - 5)(y - 4).$$

(c) $-3x^4 + 15x^3 - 18x^2 = -3x^2(x^2 - 5x + 6)$

Factor $x^2 - 5x + 6$. The integers -3 and -2 have a product of 6 and a sum of -5, so

$$x^2 - 5x + 6 = (x - 3)(x - 2).$$

The completely factored form is

$$-3x^4 + 15x^3 - 18x^2 = -3x^2(x - 3)(x - 2).$$

14.2 Section Exercises

1. If the coefficient of the last term of the trinomial is negative, then a and b must have different signs, one positive and one negative.

3. $x^2 - 12x + 32$

Multiply each of the given pairs of factors to determine which one gives the required product.

A. $(x - 8)(x + 4) = x^2 - 4x - 32$

B. $(x + 8)(x - 4) = x^2 + 4x - 32$

C. $(x - 8)(x - 4) = x^2 - 12x + 32$

D. $(x + 8)(x + 4) = x^2 + 12x + 32$

Choice **C** is the correct factored form.

5. Multiply the factors using FOIL to determine the polynomial.

$$(a + 9)(a + 4)$$
$$\mathbf{F} \quad \mathbf{O} \quad \mathbf{I} \quad \mathbf{L}$$
$$= a(a) + a(4) + 9(a) + 9(4)$$
$$= a^2 + 4a + 9a + 36$$
$$= a^2 + 13a + 36$$

7. A *prime polynomial* is one that cannot be factored using only integers in the factors.

9. Product: 12   Sum: 7

List all pairs of integers whose product is 12, and then find the sum of each pair.

Factors of 12	Sums of factors
$1, 12$	$12 + 1 = 13$
$-1, -12$	$-12 + (-1) = -13$
$2, 6$	$6 + 2 = 8$
$-2, -6$	$-6 + (-2) = -8$
$3, 4$	$4 + 3 = 7$ ←
$-3, -4$	$-4 + (-3) = -7$

The pair of integers whose product is 12 and whose sum is 7 is 4 and 3.

11. Product: -24 Sum: -5

List all pairs of integers whose product is -24, and then find the sum of each pair.

Factors of -24	Sums of factors
$1, -24$	$1 + (-24) = -23$
$-1, 24$	$-1 + 24 = 23$
$2, -12$	$2 + (-12) = -10$
$-2, 12$	$-2 + 12 = 10$
$3, -8$	$3 + (-8) = -5$ ←
$-3, 8$	$-3 + 8 = 5$
$4, -6$	$4 + (-6) = -2$
$-4, 6$	$-4 + 6 = 2$

The pair of integers whose product is -24 and whose sum is -5 is 3 and -8.

13. $p^2 + 11p + 30 = (p + 5)(\underline{\quad})$

Look for an integer whose product with 5 is 30 and whose sum with 5 is 11. That integer is 6.

$$p^2 + 11p + 30 = (p + 5)(p + 6)$$

15. $x^2 + 15x + 44 = (x + 4)(\underline{\quad})$

Look for an integer whose product with 4 is 44 and whose sum with 4 is 15. That integer is 11.

$$x^2 + 15x + 44 \quad \text{factors as} \quad (x + 4)(x + 11)$$

17. $x^2 - 9x + 8 = (x - 1)(\underline{\quad})$

Look for an integer whose product with -1 is 8 and whose sum with -1 is -9. That integer is -8.

$$x^2 - 9x + 8 \quad \text{factors as} \quad (x - 1)(x - 8)$$

19. $y^2 - 2y - 15 = (y + 3)(\underline{\quad})$

Look for an integer whose product with 3 is -15 and whose sum with 3 is -2. That integer is -5.

$$y^2 - 2y - 15 = (y + 3)(y - 5)$$

21. $x^2 + 9x - 22 = (x - 2)(\underline{\quad})$

Look for an integer whose product with -2 is -22 and whose sum with -2 is 9. That integer is 11.

$$x^2 + 9x - 22 = (x - 2)(x + 11)$$

23. $y^2 - 7y - 18 = (y + 2)(\underline{\quad})$

Look for an integer whose product with 2 is -18 and whose sum with 2 is -7. That integer is -9.

$$y^2 - 7y - 18 = (y + 2)(y - 9)$$

25. $y^2 + 9y + 8$ ▪ Look for two integers whose product is 8 and whose sum is 9. Both integers must be positive because b and c are both positive.

Factors of 8	Sums of factors
$1, 8$	9 ←
$2, 4$	6

Thus,

$$y^2 + 9y + 8 = (y + 8)(y + 1).$$

27. $b^2 + 8b + 15$ ▪ Look for two integers whose product is 15 and whose sum is 8. Both integers must be positive because b and c are both positive.

Factors of 15	Sums of factors
1, 15	16
3, 5	8 ←

Thus,
$$b^2 + 8b + 15 = (b + 3)(b + 5).$$

29. $m^2 + m - 20$ ▪ Look for two integers whose product is -20 and whose sum is 1. Since c is negative, one integer must be positive and one must be negative.

Factors of -20	Sums of factors
$-1, 20$	19
$1, -20$	-19
$-2, 10$	8
$2, -10$	-8
$-4, 5$	1 ←
$4, -5$	-1

Thus,
$$m^2 + m - 20 = (m - 4)(m + 5).$$

31. $x^2 + 3x - 40$ ▪ Look for two integers whose product is -40 and whose sum is 3.

Factors of 40	Sums of factors
$-1, 40$	$-1 + 40 = 39$
$1, -40$	$1 + (-40) = -39$
$-2, 20$	$-2 + 20 = 18$
$2, -20$	$2 + (-20) = -18$
$-4, 10$	$-4 + 10 = 6$
$4, -10$	$4 + (-10) = -6$
$-5, 8$	$-5 + 8 = 3$ ←
$5, -8$	$5 + (-8) = -3$

Thus,
$$x^2 + 3x - 40 = (x - 5)(x + 8).$$

33. $x^2 + 4x + 5$ ▪ Look for two integers whose product is 5 and whose sum is 4. Both integers must be positive since b and c are both positive.

Product   *Sum*
$5 \cdot 1 = 5$ $5 + 1 = 6$

There is no other pair of positive integers whose product is 5. Since there is no pair of integers whose product is 5 and whose sum is 4, $x^2 + 4x + 5$ is a *prime* polynomial.

35. $y^2 - 8y + 15$ ▪ Find two integers whose product is 15 and whose sum is -8. Since c is positive and b is negative, both integers must be negative.

Factors of 15	Sums of factors
$-1, -15$	-16
$-3, -5$	-8 ←

Thus,
$$y^2 - 8y + 15 = (y - 5)(y - 3).$$

37. $z^2 - 15z + 56$ ▪ Find two integers whose product is 56 and whose sum is -15. Since c is positive and b is negative, both integers must be negative.

Factors of 56	Sums of factors
$-1, -56$	-57
$-2, -28$	-30
$-4, -14$	-18
$-7, -8$	-15 ←

Thus,
$$z^2 - 15z + 56 = (z - 7)(z - 8).$$

39. $r^2 - r - 30$ ▪ Look for two integers whose product is -30 and whose sum is -1. Because c is negative, one integer must be positive and the other must be negative.

Factors of -30	Sums of factors
$-1, 30$	29
$1, -30$	-29
$-2, 15$	13
$2, -15$	-13
$-3, 10$	7
$3, -10$	-7
$-5, 6$	1
$5, -6$	-1 ←

Thus,
$$r^2 - r - 30 = (r + 5)(r - 6).$$

41. $a^2 - 8a - 48$ ▪ Find two integers whose product is -48 and whose sum is -8. Since c is negative, one integer must be positive and one must be negative.

Factors of -48	Sums of factors
$-1, 48$	47
$1, -48$	-47
$-2, 24$	22
$2, -24$	-22
$-3, 16$	13
$3, -16$	-13
$-4, 12$	8
$4, -12$	-8 ←
$-6, 8$	2
$6, -8$	-2

Thus,
$$a^2 - 8a - 48 = (a + 4)(a - 12).$$

43. $r^2 + 3ra + 2a^2$ ▪ Look for two expressions whose product is $2a^2$ and whose sum is $3a$. They are $2a$ and a, so
$$r^2 + 3ra + 2a^2 = (r + 2a)(r + a).$$

45. $x^2 + 4xy + 3y^2$ ■ Look for two expressions whose product is $3y^2$ and whose sum is $4y$. The expressions are $3y$ and y, so

$$x^2 + 4xy + 3y^2 = (x + 3y)(x + y).$$

47. $t^2 - tz - 6z^2$ ■ Look for two expressions whose product is $-6z^2$ and whose sum is $-z$. They are $2z$ and $-3z$, so

$$t^2 - tz - 6z^2 = (t + 2z)(t - 3z).$$

49. $v^2 - 11vw + 30w^2$ ■ Look for two expressions whose product is $30w^2$ and whose sum is $-11w$.

Factors of $30w^2$	Sums of factors
$-30w, -w$	$-31w$
$-15w, -2w$	$-17w$
$-10w, -3w$	$-13w$
$-5w, -6w$	$-11w$ ←

The completely factored form is

$$v^2 - 11vw + 30w^2 = (v - 5w)(v - 6w).$$

51. $a^2 + 2ab - 15b^2$ ■ Look for two expressions whose product is $-15b^2$ and whose sum is $2b$. The expressions are $5b$ and $-3b$, so

$$a^2 + 2ab - 15b^2 = (a + 5b)(a - 3b).$$

53. $4x^2 + 12x - 40$ ■ First, factor out the GCF, 4.

$$4x^2 + 12x - 40 = 4(x^2 + 3x - 10)$$

Now factor $x^2 + 3x - 10$.

Factors of -10	Sums of factors
$-1, 10$	9
$1, -10$	-9
$2, -5$	-3
$-2, 5$	3 ←

Thus,

$$x^2 + 3x - 10 = (x - 2)(x + 5).$$

The completely factored form is

$$4x^2 + 12x - 40 = 4(x - 2)(x + 5).$$

55. $2t^3 + 8t^2 + 6t$ ■ First, factor out the GCF, $2t$.

$$2t^3 + 8t^2 + 6t = 2t(t^2 + 4t + 3)$$

Then factor $t^2 + 4t + 3$.

$$t^2 + 4t + 3 = (t + 1)(t + 3)$$

The completely factored form is

$$2t^3 + 8t^2 + 6t = 2t(t + 1)(t + 3).$$

57. $-2x^6 - 8x^5 + 42x^4$

First, factor out the GCF, $-2x^4$.

$$-2x^6 - 8x^5 + 42x^4 = -2x^4(x^2 + 4x - 21)$$

Now factor $x^2 + 4x - 21$.

Factors of -21	Sums of factors
$1, -21$	-20
$-1, 21$	20
$3, -7$	-4
$-3, 7$	4 ←

Thus,

$$x^2 + 4x - 21 = (x - 3)(x + 7).$$

The completely factored form is

$$-2x^6 - 8x^5 + 42x^4 = -2x^4(x - 3)(x + 7).$$

59. $a^5 + 3a^4b - 4a^3b^2$ ■ The GCF is a^3, so

$$a^5 + 3a^4b - 4a^3b^2 = a^3(a^2 + 3ab - 4b^2).$$

Now factor $a^2 + 3ab - 4b^2$. The expressions $4b$ and $-b$ have a product of $-4b^2$ and a sum of $3b$. The completely factored form is

$$a^5 + 3a^4b - 4a^3b^2 = a^3(a + 4b)(a - b).$$

61. $5m^5 + 25m^4 - 40m^2$

Factor out the GCF, $5m^2$.

$$5m^5 + 25m^4 - 40m^2 = 5m^2(m^3 + 5m^2 - 8)$$

63. $m^3n - 10m^2n^2 + 24mn^3$

First, factor out the GCF, mn.

$$m^3n - 10m^2n^2 + 24mn^3$$
$$= mn(m^2 - 10mn + 24n^2)$$

The expressions $-6n$ and $-4n$ have a product of $24n^2$ and a sum of $-10n$. The completely factored form is

$$m^3n - 10m^2n^2 + 24mn^3$$
$$= mn(m - 6n)(m - 4n).$$

14.3 Factoring Trinomials by Grouping

14.3 Margin Exercises

1. **(a)** $2x^2 + \mathbf{7x} + 6$
$$= 2x^2 + \mathbf{4x} + \mathbf{3x} + 6$$
$$= (2x^2 + \underline{4x}) + (3x + \underline{6})$$
$$= 2x(x + \underline{2}) + 3(x + \underline{2})$$
$$= \underline{(x + 2)}(2x + 3)$$

(b) Yes, the answer is the same.

(c) $2z^2 + 5z + 3$ ■ Find the two integers whose product is $2(3) = 6$ and whose sum is 5. The integers are 3 and 2. Write the middle term, $5z$, as $3z + 2z$.

$$2z^2 + 5z + 3 = 2z^2 + 3z + 2z + 3$$
$$= (2z^2 + 3z) + (2z + 3)$$
$$= z(2z + 3) + 1(2z + 3)$$
$$= (2z + 3)(z + 1)$$

2. **(a)** $2m^2 + 7m + 3$ ▪ Find the two integers whose product is $2(3) = 6$ and whose sum is 7. The integers are 1 and 6. Write the middle term, $7m$, as $1m + 6m$.

$$2m^2 + 7m + 3 = 2m^2 + 1m + 6m + 3$$
$$= (2m^2 + 1m) + (6m + 3)$$
$$= m(2m + 1) + 3(2m + 1)$$
$$= (2m + 1)(m + 3)$$

(b) $5p^2 - 2p - 3$ ▪ Find two integers whose product is $5(-3) = -15$ and whose sum is -2. The integers are -5 and 3.

$$5p^2 - 2p - 3 = 5p^2 - 5p + 3p - 3$$
$$= (5p^2 - 5p) + (3p - 3)$$
$$= 5p(p - 1) + 3(p - 1)$$
$$= (p - 1)(5p + 3)$$

(c) $15k^2 - km - 2m^2$ ▪ Find two integers whose product is $15(-2) = -30$ and whose sum is -1. The integers are -6 and 5.

$$15k^2 - km - 2m^2$$
$$= 15k^2 - 6km + 5km - 2m^2$$
$$= (15k^2 - 6km) + (5km - 2m^2)$$
$$= 3k(5k - 2m) + m(5k - 2m)$$
$$= (5k - 2m)(3k + m)$$

3. $4x^2 - 2x - 30$

First factor out the greatest common factor, 2.

$$4x^2 - 2x - 30 = \underline{2}(2x^2 - \underline{x} - 15)$$

To factor the trinomial $2x^2 - x - 15$, find two integers whose product is $2(-15) = -30$ and whose sum is -1. The integers are -6 and 5.

$$2x^2 - x - 15 = 2x^2 - \underline{6x} + \underline{5x} - 15$$
$$= (2x^2 - \underline{6x}) + (5x - 15)$$
$$= \underline{2x}(\underline{x - 3}) + 5(x - 3)$$
$$= (x - 3)(2x + 5)$$

The completely factored form is

$$4x^2 - 2x - 30 = \underline{2(x - 3)(2x + 5)}.$$

4. **(a)** $18p^4 + 63p^3 + 27p^2 = 9p^2(2p^2 + 7p + 3)$

Find two integers whose product is $2(3) = 6$ and whose sum is 7. The integers are 6 and 1.

$$2p^2 + 7p + 3 = 2p^2 + 6p + 1p + 3$$
$$= 2p(p + 3) + 1(p + 3)$$
$$= (p + 3)(2p + 1)$$

The completely factored form is

$$18p^4 + 63p^3 + 27p^2 = 9p^2(p + 3)(2p + 1).$$

(b) $6a^2 + 3ab - 18b^2$

First factor out the greatest common factor, 3.

$$6a^2 + 3ab - 18b^2 = 3(2a^2 + ab - 6b^2)$$

Find two integers whose product is $2(-6) = -12$ and whose sum is 1. The integers are 4 and -3.

$$2a^2 + ab - 6b^2 = 2a^2 + 4ab - 3ab - 6b^2$$
$$= 2a(a + 2b) - 3b(a + 2b)$$
$$= (a + 2b)(2a - 3b)$$

The completely factored form is

$$6a^2 + 3ab - 18b^2 = 3(a + 2b)(2a - 3b).$$

14.3 Section Exercises

1. To factor $12y^2 + 5y - 2$, we must find two integers with a product of $12(-2) = -24$ and a sum of 5. The only pair of integers satisfying those conditions is 8 and -3, choice **B**.

3. $m^2 + 8m + 12$
$$= m^2 + 6m + 2m + 12$$
$$= (m^2 + 6m) + (2m + 12) \quad \textit{Group terms.}$$
$$= m(m + 6) + 2(m + 6) \quad \textit{Factor each group.}$$
$$= (m + 6)(m + 2) \quad \textit{Factor out } m + 6.$$

5. $a^2 + 3a - 10$
$$= a^2 + 5a - 2a - 10$$
$$= (a^2 + 5a) + (-2a - 10) \quad \textit{Group terms.}$$
$$= a(a + 5) - 2(a + 5) \quad \textit{Factor each group.}$$
$$= (a + 5)(a - 2) \quad \textit{Factor out } a + 5.$$

7. $10t^2 + 9t + 2$
$$= 10t^2 + 5t + 4t + 2$$
$$= (10t^2 + 5t) + (4t + 2) \quad \textit{Group terms.}$$
$$= 5t(2t + 1) + 2(2t + 1) \quad \textit{Factor each group.}$$
$$= (2t + 1)(5t + 2) \quad \textit{Factor out } 2t + 1.$$

9. $15z^2 - 19z + 6$
$$= 15z^2 - 10z - 9z + 6$$
$$= (15z^2 - 10z) + (-9z + 6) \quad \textit{Group terms.}$$
$$= 5z(3z - 2) - 3(3z - 2) \quad \textit{Factor each group.}$$
$$= (3z - 2)(5z - 3) \quad \textit{Factor out } 3z - 2.$$

11. $8s^2 + 2st - 3t^2$
$$= 8s^2 - 4st + 6st - 3t^2$$
$$= (8s^2 - 4st) + (6st - 3t^2) \quad \textit{Group terms.}$$
$$= 4s(2s - t) + 3t(2s - t) \quad \textit{Factor each group.}$$
$$= (2s - t)(4s + 3t) \quad \textit{Factor out } 2s - t.$$

13. $15a^2 + 22ab + 8b^2$
$$= 15a^2 + 10ab + 12ab + 8b^2$$
$$= (15a^2 + 10ab) + (12ab + 8b^2) \quad \textit{Group terms.}$$
$$= 5a(3a + 2b) + 4b(3a + 2b) \quad \textit{Factor each group.}$$
$$= (3a + 2b)(5a + 4b) \quad \textit{Factor out } 3a + 2b.$$

15. $2m^2 + 11m + 12$

(a) Find two integers whose product is $\underline{2} \cdot \underline{12} = \underline{24}$ and whose sum is $\underline{11}$.

(b) The required integers are $\underline{3}$ and $\underline{8}$. (Order is irrelevant.)

(c) Write the middle term $11m$ as $\underline{3m} + \underline{8m}$.

(d) Rewrite the given trinomial as $2m^2 + 3m + 8m + 12$.

(e) $(2m^2 + 3m) + (8m + 12)$ *Group terms.*
$\quad = m(2m + 3) + 4(2m + 3)$ *Factor each group.*
$\quad = (2m + 3)(m + 4)$ *Factor out $2m + 3$.*

(f) $(2m + 3)(m + 4)$

$\quad\quad\textbf{F}\quad\quad\textbf{O}\quad\quad\textbf{I}\quad\quad\textbf{L}$
$\quad = 2m(m) + 2m(4) + 3(m) + 3(4)$
$\quad = 2m^2 + 8m + 3m + 12$
$\quad = 2m^2 + 11m + 12$

17. $2x^2 + 7x + 3$ ■ Look for two integers whose product is $2(3) = 6$ and whose sum is 7. The integers are 1 and 6.

$2x^2 + 7x + 3 = 2x^2 + x + 6x + 3$
$\quad = (2x^2 + x) + (6x + 3)$
$\quad = x(2x + 1) + 3(2x + 1)$
$\quad = (2x + 1)(x + 3)$

19. $4r^2 + r - 3$ ■ Look for two integers whose product is $4(-3) = -12$ and whose sum is 1. The integers are -3 and 4.

$4r^2 + r - 3 = 4r^2 - 3r + 4r - 3$
$\quad = (4r^2 - 3r) + (4r - 3)$
$\quad = r(4r - 3) + 1(4r - 3)$
$\quad = (4r - 3)(r + 1)$

21. $8m^2 - 10m - 3$ ■ Look for two integers whose product is $8(-3) = -24$ and whose sum is -10. The integers are -12 and 2.

$8m^2 - 10m - 3$
$\quad = 8m^2 - 12m + 2m - 3$
$\quad = 4m(2m - 3) + 1(2m - 3)$
$\quad = (2m - 3)(4m + 1)$

23. $21m^2 + 13m + 2$ ■ Look for two integers whose product is $21(2) = 42$ and whose sum is 13. The integers are 7 and 6.

$21m^2 + 13m + 2$
$\quad = 21m^2 + 7m + 6m + 2$
$\quad = 7m(3m + 1) + 2(3m + 1)$
$\quad = (3m + 1)(7m + 2)$

25. $6b^2 + 7b + 2$ ■ Find two integers whose product is $6(2) = 12$ and whose sum is 7. The integers are 3 and 4.

$6b^2 + 7b + 2 = 6b^2 + 3b + 4b + 2$
$\quad = 3b(2b + 1) + 2(2b + 1)$
$\quad = (2b + 1)(3b + 2)$

27. $12y^2 - 13y + 3$ ■ Look for two integers whose product is $12(3) = 36$ and whose sum is -13. The integers are -9 and -4.

$12y^2 - 13y + 3 = 12y^2 - 9y - 4y + 3$
$\quad = 3y(4y - 3) - 1(4y - 3)$
$\quad = (4y - 3)(3y - 1)$

29. $16 + 16x + 3x^2 = 3x^2 + 16x + 16$

Factor by the grouping method. Find two integers whose product is $(3)(16) = 48$ and whose sum is 16. The numbers are 4 and 12.

$3x^2 + 16x + 16$
$\quad = 3x^2 + 4x + 12x + 16$
$\quad = (3x^2 + 4x) + (12x + 16)$
$\quad = x(3x + 4) + 4(3x + 4)$
$\quad = (3x + 4)(x + 4)$

31. $24x^2 - 42x + 9$

Factor out the greatest common factor, 3.

$24x^2 - 42x + 9 = 3(8x^2 - 14x + 3)$

Look for two integers whose product is $8(3) = 24$ and whose sum is -14. The integers are -2 and -12.

$24x^2 - 42x + 9$
$\quad = 3(8x^2 - 2x - 12x + 3)$
$\quad = 3[2x(4x - 1) - 3(4x - 1)]$
$\quad = 3(4x - 1)(2x - 3)$

33. $2m^3 + 2m^2 - 40m$

Factor out the greatest common factor, $2m$.

$2m^3 + 2m^2 - 40m = 2m(m^2 + m - 20)$

Find two integers whose product is $1(-20) = -20$ and whose sum is 1. The integers are -4 and 5.

$2m^3 + 2m^2 - 40m$
$\quad = 2m(m^2 - 4m + 5m - 20)$
$\quad = 2m[m(m - 4) + 5(m - 4)]$
$\quad = 2m(m - 4)(m + 5)$

35. $-32z^5 + 20z^4 + 12z^3$

Factor out the greatest common factor, $-4z^3$.

$-32z^5 + 20z^4 + 12z^3 = -4z^3(8z^2 - 5z - 3)$

Find two integers whose product is $8(-3) = -24$ and whose sum is -5. The integers are 3 and -8.

$-32z^5 + 20z^4 + 12z^3$
$\quad = -4z^3(8z^2 + 3z - 8z - 3)$
$\quad = -4z^3[z(8z + 3) - 1(8z + 3)]$
$\quad = -4z^3(8z + 3)(z - 1)$

37. $12p^2 + 7pq - 12q^2$ ■ Find two integers whose product is $12(-12) = -144$ and whose sum is 7. The integers are 16 and -9.

$$12p^2 + 7pq - 12q^2$$
$$= 12p^2 + 16pq - 9pq - 12q^2$$
$$= 4p(3p + 4q) - 3q(3p + 4q)$$
$$= (3p + 4q)(4p - 3q)$$

39. $6a^2 - 7ab - 5b^2$ ■ Find two integers whose product is $6(-5) = -30$ and whose sum is -7. The integers are -10 and 3.

$$6a^2 - 7ab - 5b^2$$
$$= 6a^2 - 10ab + 3ab - 5b^2$$
$$= 2a(3a - 5b) + b(3a - 5b)$$
$$= (3a - 5b)(2a + b)$$

41. The student stopped too soon.
He needs to factor out the common factor $4x - 1$ to get $(4x - 1)(4x - 5)$ as the correct answer.

14.4 Factoring Trinomials by Using the FOIL Method

14.4 Margin Exercises

1. **(a)** $2p^2 + 9p + 9$ ■ The possible factors of the first term $2p^2$ are $\underline{2p}$ and p. The possible factors of the last term 9 are 9 and $\underline{1}$, or $\underline{3}$ and 3. Try various combinations to find the pair of factors that gives the correct middle term, $\underline{9p}$.

Possible Pairs of Factors	Middle Term	Correct?
$(2p + 9)(p + 1)$	$11p$	No
$(2p + 1)(p + 9)$	$19p$	No
$(2p + 3)(p + 3)$	$9p$	Yes

Thus, $2p^2 + 9p + 9$ factors as $(2p + 3)(p + 3)$.

(b) $8y^2 + 22y + 5$ ■ The factors of $8y^2$ are $2y$ and $4y$, or $8y$ and y. The factors of 5 are 5 and 1. Try various combinations, checking to see if the middle term is $22y$ in each case.

Possible Pairs of Factors	Middle Term	Correct?
$(4y + 5)(2y + 1)$	$14y$	No
$(4y + 1)(2y + 5)$	$22y$	Yes

Thus, $8y^2 + 22y + 5 = (4y + 1)(2y + 5)$.

2. **(a)** $6p^2 + 19p + 10$ ■ The factors of $6p^2$ are $2p$ and $3p$, or $6p$ and p. The factors of 10 are 10 and 1, or 5 and 2. Try various combinations, checking to see if the middle term is $19p$ in each case.

Possible Pairs of Factors	Middle Term	Correct?
$(3p + 5)(2p + 2)$	$16p$	No
$(3p + 2)(2p + 5)$	$19p$	Yes

Thus, $6p^2 + 19p + 10 = (3p + 2)(2p + 5)$.

(b) $8x^2 + 14x + 3$ ■ The factors of $8x^2$ are $8x$ and x, or $4x$ and $2x$. The factors of 3 are 3 and 1. Try various combinations, checking to see if the middle term is $14x$ in each case.

Possible Pairs of Factors	Middle Term	Correct?
$(8x + 3)(x + 1)$	$11x$	No
$(8x + 1)(x + 3)$	$25x$	No
$(4x + 1)(2x + 3)$	$14x$	Yes

Thus, $8x^2 + 14x + 3 = (4x + 1)(2x + 3)$.

3. **(a)** $4y^2 - 11y + 6$ ■ The factors of $4y^2$ are $4y$ and y, or $2y$ and $2y$. Try $4y$ and y. Since the last term is positive and the coefficient of the middle term is negative, only negative factors of 6 should be considered. The factors of 6 are -6 and -1, or -3 and -2.

Try -6 and -1.

$(4y - 6)(y - 1) = 4y^2 - 10y + 6$ *Incorrect*

Try -3 and -2.

$(4y - 3)(y - 2) = 4y^2 - 11y + 6$ *Correct*

Thus,

$$4y^2 - 11y + 6 = (4y - 3)(y - 2).$$

(b) $9x^2 - 21x + 10$ ■ The factors of $9x^2$ are $9x$ and x, or $3x$ and $3x$. Try $3x$ and $3x$. Since the last term is positive and the coefficient of the middle term is negative, only negative factors of 10 should be considered. The factors of 10 are -10 and -1, or -5 and -2.

Try -10 and -1.

$(3x - 10)(3x - 1) = 9x^2 - 33x + 10$ *Incorrect*

Try -5 and -2.

$(3x - 5)(3x - 2) = 9x^2 - 21x + 10$ *Correct*

Thus,

$$9x^2 - 21x + 10 = (3x - 5)(3x - 2).$$

4. **(a)** $6x^2 + 5x - 4$ ■ 6 and -4 each have several factors. Since the middle coefficient, 5, is not large, we try $3x$ and $2x$ as factors of $6x$, rather than $6x$ and x. The last term, -4, has factors -4 and 1, 4 and -1, -2 and 2.

Try -4 and 1.

$(3x - 4)(2x + 1) = 6x^2 - 5x - 4$ *Incorrect*

Try 4 and -1.

$(3x + 4)(2x - 1) = 6x^2 + 5x - 4$ *Correct*

Thus,

$$6x^2 + 5x - 4 = (3x + \underline{4})(\underline{2x} - \underline{1}).$$

(b) $6m^2 - 11m - 10$ ■ The factors of $6m^2$ are $6m$ and m, or $3m$ and $2m$. Start with $3m$ and $2m$. Some factors of -10 are 10 and -1, or 2 and -5. Try various possibilities.

$$(3m - 1)(2m + 10) = 6m^2 + 28m - 10$$
Incorrect

$$(3m + 2)(2m - 5) = 6m^2 - 11m - 10$$
Correct

Thus,

$$6m^2 - 11m - 10 = (3m + 2)(2m - 5).$$

(c) $4x^2 - 3x - 7$ ■ The factors of $4x^2$ are $4x$ and x, or $2x$ and $2x$. Start with $2x$ and $2x$. The factors of -7 are -7 and 1, or 7 and -1. Try various possibilities.

$$(2x - 7)(2x + 1) = 4x^2 - 12x - 7$$
$$(2x + 7)(2x - 1) = 4x^2 + 12x - 7$$
Incorrect

Since $2x$ and $2x$ won't work, we will try $4x$ and x.

$$(4x - 7)(x + 1) = 4x^2 - 3x - 7$$
Correct

Thus,

$$4x^2 - 3x - 7 = (4x - 7)(x + 1).$$

(d) $3y^2 + 8y - 6$ ■ The factors of $3y^2$ are $3y$ and y. The factors of -6 are -6 and 1, 6 and -1, 2 and -3, or -2 and 3.

1) $(3y - 6)(y + 1) = 3y^2 - 3y - 6$ *Incorrect*
2) $(3y + 1)(y - 6) = 3y^2 - 17y - 6$ *Incorrect*
3) $(3y - 1)(y + 6) = 3y^2 + 17y - 6$ *Incorrect*
4) $(3y + 6)(y - 1) = 3y^2 + 3y - 6$ *Incorrect*
5) $(3y + 2)(y - 3) = 3y^2 - 7y - 6$ *Incorrect*
6) $(3y - 3)(y + 2) = 3y^2 + 3y - 6$ *Incorrect*
7) $(3y - 2)(y + 3) = 3y^2 + 7y - 6$ *Incorrect*
8) $(3y + 3)(y - 2) = 3y^2 - 3y - 6$ *Incorrect*

All 8 attempts failed. Thus, the polynomial is *prime*. Notice in lines 1, 4, 6, and 8, that 3 is a factor of the first factor, so those factorings cannot be correct since 3 is not a factor of the original polynomial.

5. **(a)** $2x^2 - 5xy - 3y^2$

Try various possibilities.

$$(2x - y)(x + 3y) = 2x^2 + 5xy - 3y^2$$
Incorrect

The middle terms differ only in sign, so reverse the signs of the two factors.

$$(2x + y)(x - 3y) = 2x^2 - 5xy - 3y^2$$
Correct

Thus,

$$2x^2 - 5xy - 3y^2 = (2x + \underline{y})(\underline{x} - \underline{3y}).$$

(b) $8a^2 + 2ab - 3b^2$

Try various possibilities.

$$(8a + 3b)(a - b) = 8a^2 - 5ab - 3b^2$$
Incorrect

$$(4a + 3b)(2a - b) = 8a^2 + 2ab - 3b^2$$
Correct

Thus,

$$8a^2 + 2ab - 3b^2 = (4a + 3b)(2a - b).$$

6. **(a)** $36z^3 - 6z^2 - 72z$

Factor out the common factor, $6z$.

$$36z^3 - 6z^2 - 72z = \underline{6z}(6z^2 - \underline{z} - 12)$$

Factor $6z^2 - z - 12$ by trial and error to obtain

$$6z^2 - z - 12 = (2z - 3)(3z + 4).$$

The completely factored form is

$$36z^3 - 6z^2 - 72z = \underline{6z}(2z - 3)(3z + 4).$$

(b) $10x^3 + 45x^2 - 90x$

Factor out the common factor, $5x$.

$$10x^3 + 45x^2 - 90x = 5x(2x^2 + 9x - 18)$$

Factor $2x^2 + 9x - 18$ by trial and error to obtain

$$2x^2 + 9x - 18 = (2x - 3)(x + 6).$$

The completely factored form is

$$10x^3 + 45x^2 - 90x = 5x(2x - 3)(x + 6).$$

(c) $-24x^3 + 32x^2 + 6x$ ■ The common factor could be $2x$ or $-2x$. If we factor out $-2x$, the first term of the trinomial factor will be positive, which makes it easier to factor.

$$-24x^3 + 32x^2 + 6x$$
$$= -2x(12x^2 - 16x - 3)$$

Factor $12x^2 - 16x - 3$ by trial and error to obtain

$$12x^2 - 16x - 3 = (6x + 1)(2x - 3).$$

The completely factored form is

$$-24x^3 + 32x^2 + 6x$$
$$= -2x(6x + 1)(2x - 3).$$

14.4   Section Exercises

For Exercises 1–6, multiply the factors in the choices together to see which ones give the correct product.

1. $2x^2 - x - 1$

   A. $(2x - 1)(x + 1) = 2x^2 + x - 1$

   B. $(2x + 1)(x - 1) = 2x^2 - x - 1$

   B is the correct factored form.

3. $4y^2 + 17y - 15$

A. $(y + 5)(4y - 3) = 4y^2 + 17y - 15$

B. $(2y - 5)(2y + 3) = 4y^2 - 4y - 15$

A is the correct factored form.

5. $4k^2 + 13mk + 3m^2$

A. $(4k + m)(k + 3m) = 4k^2 + 13mk + 3m^2$

B. $(4k + 3m)(k + m) = 4k^2 + 7mk + 3m^2$

A is the correct factored form.

7. $6a^2 + 7ab - 20b^2 = (3a - 4b)(\underline{\quad})$

The first term in the missing expression must be $2a$ since

$$(3a)(2a) = 6a^2.$$

The second term in the missing expression must be $5b$ since

$$(-4b)(5b) = -20b^2.$$

Checking our answer by multiplying, we see that

$$(3a - 4b)(2a + 5b) = 6a^2 + 7ab - 20b^2,$$

as desired.

9. $2x^2 + 6x - 8 = 2(x^2 + 3x - 4)$

To factor $x^2 + 3x - 4$, we look for two integers whose product is -4 and whose sum is 3. The integers are 4 and -1. Thus,

$$2x^2 + 6x - 8 = 2(x + 4)(x - 1).$$

11. $4z^3 - 10z^2 - 6z = 2z(2z^2 - 5z - 3)$
$$= 2z(2z + 1)(z - 3)$$

13. $(4x + 4)$ cannot be a factor of $12x^2 + 7x - 12$ because its terms have a common factor of 4, but the terms of $12x^2 + 7x - 12$ do not.

The possible factors of $12x^2$ are $12x$ and x, $6x$ and $2x$, or $4x$ and $3x$.

The possible factors of -12 are 12 and 1, 6 and 2, or 4 and 3, where one of the factors is negative.

We try various possibilities, and find that the correct factored form is

$$(4x - 3)(3x + 4).$$

15. $3a^2 + 10a + 7$ ▪ Possible factors of $3a^2$ are $3a$ and a. Possible factors of 7 are 7 and 1.

Try various combinations, checking to see if the middle term is $10a$ in each case.

Possible Pairs of Factors	Middle Term	Correct?
$(3a + 1)(a + 7)$	$22a$	No
$(3a + 7)(a + 1)$	$10a$	Yes

Thus, $3a^2 + 10a + 7 = (3a + 7)(a + 1)$.

17. $2y^2 + 7y + 6$ ▪ Possible factors of $2y^2$ are $2y$ and y. Possible factors of 6 are 6 and 1, or 3 and 2.

Try various combinations, checking to see if the middle term is $7y$ in each case.

Possible Pairs of Factors	Middle Term	Correct?
$(2y + 6)(y + 1)$	$8y$	No
$(2y + 1)(y + 6)$	$13y$	No
$(2y + 3)(y + 2)$	$7y$	Yes

19. $15m^2 + m - 2$ ▪ Possible factors of $15m^2$ are $15m$ and m, or $3m$ and $5m$. Possible factors of -2 are -2 and 1, or 2 and -1.

Try various combinations, checking to see if the middle term is m in each case.

Possible Pairs of Factors	Middle Term	Correct?
$(3m - 2)(5m + 1)$	$-7m$	No
$(3m + 2)(5m - 1)$	$7m$	No
$(3m - 1)(5m + 2)$	m	Yes

21. $12s^2 + 11s - 5$ ▪ Possible factors of $12s^2$ are s and $12s$, $2s$ and $6s$, or $3s$ and $4s$. Factors of -5 are -1 and 5 or -5 and 1.

Try various combinations, checking to see if the middle term is $11s$ in each case.

Possible Pairs of Factors	Middle Term	Correct?
$(2s - 1)(6s + 5)$	$4s$	No
$(2s + 1)(6s - 5)$	$-4s$	No
$(3s - 1)(4s + 5)$	$11s$	Yes

23. $10m^2 - 23m + 12$ ▪ Possible factors of $10m^2$ are $10m$ and m, or $5m$ and $2m$. Possible factors of 12 are -12 and -1, -6 and -2, or -4 and -3.

Try various combinations, checking to see if the middle term is $-23m$ in each case.

Possible Pairs of Factors	Middle Term	Correct?
$(2m - 12)(5m - 1)$	$-62m$	No
$(2m - 1)(5m - 12)$	$-29m$	No
$(2m - 3)(5m - 4)$	$-23m$	Yes

25. $8w^2 - 14w + 3$ ▪ Possible factors of $8w^2$ are w and $8w$ or $2w$ and $4w$. Factors of 3 are -1 and -3 (since $b = -14$ is negative).

Try various combinations, checking to see if the middle term is $-14w$ in each case.

Possible Pairs of Factors	Middle Term	Correct?
$(4w - 3)(2w - 1)$	$-10w$	No
$(4w - 1)(2w - 3)$	$-14w$	Yes

27. $20y^2 - 39y - 11$ ■ Possible factors of $20y^2$ are $20y$ and y, $10y$ and $2y$, $5y$ and $4y$. Possible factors of -11 are -11 and 1, or -1 and 11.

Try various combinations, checking to see if the middle term is $-39y$ in each case.

Possible Pairs of Factors	Middle Term	Correct?
$(20y - 11)(y - 1)$	$-31y$	No
$(20y - 1)(y - 11)$	$-221y$	No
$(4y - 1)(5y + 11)$	$39y$	No
$(4y + 1)(5y - 11)$	$-39y$	Yes

29. $3x^2 - 15x + 16$ ■ Possible factors of $3x^2$ are $3x$ and x. Possible factors of 16 are -16 and -1, -8 and -2, or -4 and -4.

Try various combinations, checking to see if the middle term is $-15x$ in each case.

Possible Pairs of Factors	Middle Term	Correct?
$(3x - 16)(x - 1)$	$-19x$	No
$(3x - 1)(x - 16)$	$-49x$	No
$(3x - 8)(x - 2)$	$-14x$	No
$(3x - 2)(x - 8)$	$-26x$	No
$(3x - 4)(x - 4)$	$-16x$	No

All are incorrect, so the polynomial is *prime*.

31. $20x^2 + 22x + 6$

First, factor out the greatest common factor, 2.

$$20x^2 + 22x + 6 = 2(10x^2 + 11x + 3)$$

Now factor $10x^2 + 11x + 3$ by trial and error to obtain

$$10x^2 + 11x + 3 = (5x + 3)(2x + 1).$$

The complete factorization is

$$20x^2 + 22x + 6 = 2(5x + 3)(2x + 1).$$

33. $-40m^2q - mq + 6q$

First, factor out the greatest common factor, $-q$.

$$-40m^2q - mq + 6q = -q(40m^2 + m - 6)$$

Now factor $40m^2 + m - 6$ by trial and error to obtain

$$40m^2 + m - 6 = (5m + 2)(8m - 3).$$

The complete factorization is

$$-40m^2q - mq + 6q = -q(5m + 2)(8m - 3).$$

35. $15n^4 - 39n^3 + 18n^2$

First, factor out the greatest common factor, $3n^2$.

$$15n^4 - 39n^3 + 18n^2 = 3n^2(5n^2 - 13n + 6)$$

Now factor $5n^2 - 13n + 6$ by the trial and error method. Possible factors of $5n^2$ are $5n$ and n.

Possible factors of 6 are -6 and -1, or -3 and -2.

$$(5n - 6)(n - 1) = 5n^2 - 11n + 6 \quad \text{Incorrect}$$
$$(5n - 3)(n - 2) = 5n^2 - 13n + 6 \quad \text{Correct}$$

The complete factorization is

$$15n^4 - 39n^3 + 18n^2 = 3n^2(5n - 3)(n - 2).$$

37. $-15x^2y^2 + 7xy^2 + 4y^2$

First, factor out the greatest common factor, $-y^2$.

$$-15x^2y^2 + 7xy^2 + 4y^2 = -y^2(15x^2 - 7x - 4)$$

Now factor $15x^2 - 7x - 4$ by trial and error to obtain

$$15x^2 - 7x - 4 = (5x - 4)(3x + 1).$$

The complete factorization is

$$-15x^2y^2 + 7xy^2 + 4y^2 = -y^2(5x - 4)(3x + 1).$$

39. $5a^2 - 7ab - 6b^2$ ■ The possible factors of $5a^2$ are $5a$ and a. The possible factors of $-6b^2$ are $-6b$ and b, $6b$ and $-b$, $-3b$ and $2b$, or $3b$ and $-2b$.

$$(5a - 3b)(a + 2b) = 5a^2 + 7ab - 6b^2$$
$$\text{Incorrect}$$
$$(5a + 3b)(a - 2b) = 5a^2 - 7ab - 6b^2$$
$$\text{Correct}$$

41. $12s^2 + 11st - 5t^2$ ■ Possible factors of $12s^2$ are $12s$ and s, $6s$ and $2s$, or $4s$ and $3s$. Possible factors of $-5t^2$ are $-5t$ and t, or $5t$ and $-t$.

$$(4s - 5t)(3s + t) = 12s^2 - 11st - 5t^2$$
$$\text{Incorrect}$$
$$(4s + 5t)(3s - t) = 12s^2 + 11st - 5t^2$$
$$\text{Correct}$$

43. $6m^6n + 7m^5n^2 + 2m^4n^3$

First, factor out the GCF, m^4n.

$$6m^6n + 7m^5n^2 + 2m^4n^3$$
$$= m^4n(6m^2 + 7mn + 2n^2)$$

Now factor $6m^2 + 7mn + 2n^2$ by trial and error.

Possible factors of $6m^2$ are $6m$ and m or $3m$ and $2m$. Possible factors of $2n^2$ are $2n$ and n.

$$(3m + 2n)(2m + n) = 6m^2 + 7mn + 2n^2$$
$$\text{Correct}$$

The complete factorization is

$$6m^6n + 7m^5n^2 + 2m^4n^3$$
$$= m^4n(3m + 2n)(2m + n).$$

45. $-x^2 - 4x + 21 = -1(x^2 + 4x - 21)$

7 and -3 are factors of -21 that have a sum of 4, so the completely factored form is

$$-1(x + 7)(x - 3).$$

47. $-3x^2 - x + 4 = -1(3x^2 + x - 4)$

Since $3x^2 + x - 4$ can be factored as $(3x + 4)(x - 1)$, the completely factored form is

$$-1(3x + 4)(x - 1).$$

49. $-2a^2 - 5ab - 2b^2 = -1(2a^2 + 5ab + 2b^2)$

Since $2a^2 + 5ab + 2b^2$ can be factored as $(a + 2b)(2a + b)$, the completely factored form is

$$-1(a + 2b)(2a + b).$$

Relating Concepts (Exercises 51–56)

51. $35 = 5 \cdot 7$

52. $35 = (-5)(-7)$

53. Verify the given factorization by multiplying the two binomial factors by using the FOIL method.

$$(3x - 4)(2x - 1) = 6x^2 - 3x - 8x + 4$$
$$= 6x^2 - 11x + 4$$

Thus, the product of $3x - 4$ and $2x - 1$ is indeed $6x^2 - 11x + 4$.

54. Verify the given factorization by multiplying the two binomial factors by using the FOIL method.

$$(4 - 3x)(1 - 2x) = 4 - 8x - 3x + 6x^2$$
$$= 6x^2 - 11x + 4$$

Thus, the product of $4 - 3x$ and $1 - 2x$ is indeed $6x^2 - 11x + 4$.

55. The factors of Exercise 53 are the opposites of the factors of Exercise 54.

56. We can form another valid factorization of $(7t - 3)(2t - 5)$ by taking the opposites of the two given factors. The opposite of $7t - 3$ is

$$-1(7t - 3) = -7t + 3 = 3 - 7t,$$

and the opposite of $2t - 5$ is $5 - 2t$, so we obtain the factorization $(3 - 7t)(5 - 2t)$.

14.5 Special Factoring Techniques

14.5 Margin Exercises

1. **(a)** $x^2 - 81 = x^2 - 9^2$
$$= (x + \underline{9})(x - \underline{9})$$

(b) $p^2 - 100 = p^2 - 10^2$
$$= (p + 10)(p - 10)$$

(c) $t^2 - s^2 = (t + s)(t - s)$

(d) $y^2 - 10$ ▪ Because 10 is not the square of an integer, this binomial is not a difference of squares. The terms have no common factor, so it is a *prime* polynomial.

(e) $x^2 + 36 = x^2 + 6^2$ ▪ This binomial is the *sum* of squares and the terms have no common factor. Unlike the *difference* of squares, it cannot be factored. It is a *prime* polynomial.

(f) $x^2 + y^2$ ▪ This binomial is the *sum* of squares and the terms have no common factor. Unlike the *difference* of squares, it cannot be factored. It is a *prime* polynomial.

2. **(a)** $9m^2 - 49 = (3m)^2 - 7^2$
$$= (3m + 7)(3m - 7)$$

(b) $64a^2 - 25 = (8a)^2 - 5^2$
$$= (8a + 5)(8a - 5)$$

(c) $25a^2 - 64b^2 = (5a)^2 - (8b)^2$
$$= (5a + 8b)(5a - 8b)$$

3. **(a)** $50r^2 - 32$
$$= 2(25r^2 - 16) \qquad \textit{Factor out 2.}$$
$$= 2[(5r)^2 - 4^2]$$
$$= 2(5r + 4)(5r - 4) \qquad \begin{array}{l}\textit{Difference}\\\textit{of squares}\end{array}$$

(b) $27y^2 - 75$
$$= 3(9y^2 - 25) \qquad \textit{Factor out 3.}$$
$$= 3[(3y)^2 - 5^2]$$
$$= 3(3y + 5)(3y - 5) \qquad \begin{array}{l}\textit{Difference}\\\textit{of squares}\end{array}$$

(c) $k^4 - 49$
$$= (k^2)^2 - 7^2$$
$$= (k^2 + 7)(k^2 - 7) \qquad \begin{array}{l}\textit{Difference}\\\textit{of squares}\end{array}$$

(d) $81r^4 - 16$
$$= (9r^2)^2 - 4^2$$
$$= (9r^2 + 4)(9r^2 - 4) \qquad \begin{array}{l}\textit{Difference}\\\textit{of squares}\end{array}$$
$$= (9r^2 + 4)[(3r)^2 - 2^2]$$
$$= (9r^2 + 4)(3r + 2)(3r - 2) \qquad \begin{array}{l}\textit{Difference}\\\textit{of squares}\end{array}$$

4. **(a)** $p^2 + 14p + 49$ ▪ The term p^2 is a perfect square. Since $49 = \underline{7}^2$, it *is* a perfect square. Factor the trinomial

$$p^2 + 14p + 49 \quad \text{as} \quad (p + \underline{7})^2.$$

Check by taking twice the <u>product</u> of the two terms of the squared binomial.

$$2 \cdot p \cdot \underline{7} = \underline{14p}$$

The result *is* the middle term of the given trinomial. The trinomial is a perfect square and factors as $\underline{(p + 7)^2}$.

(b) $m^2 + 8m + 16 = (m)^2 + 2 \cdot m \cdot 4 + (4)^2$
$$= (m + 4)^2$$

(c) $x^2 + 2x + 1 = (x)^2 + 2 \cdot x \cdot 1 + (1)^2$
$$= (x+1)^2$$

5. **(a)** $p^2 - 18p + 81 \overset{?}{=} (p-9)^2$

$2 \cdot p \cdot 9 = 18p$, so this is a perfect square trinomial, and
$$p^2 - 18p + 81 = (p-9)^2.$$

(b) $16a^2 + 56a + 49 \overset{?}{=} (4a+7)^2$

$2 \cdot 4a \cdot 7 = 56a$, so this is a perfect square trinomial, and
$$16a^2 + 56a + 49 = (4a+7)^2.$$

(c) $121p^2 + 110p + 100 \overset{?}{=} (11p+10)^2$

$2 \cdot 11p \cdot 10 = 220p \neq 110p$, so this is not a perfect square trinomial. This trinomial cannot be factored even with the methods of previous sections. It is *prime*.

(d) $64x^2 - 48x + 9 \overset{?}{=} (8x-3)^2$

$2 \cdot 8x \cdot 3 = 48x$, so this is a perfect square trinomial, and
$$64x^2 - 48x + 9 = (8x-3)^2.$$

(e) $27y^3 + 72y^2 + 48y$

Factor out the greatest common factor, $3y$.

$27y^3 + 72y^2 + 48y = 3y(9y^2 + 24y + 16)$
$$\overset{?}{=} 3y(3y+4)^4$$

$2 \cdot 3y \cdot 4 = 24y$, so this is a perfect square trinomial, and
$$3y(9y^2 + 24y + 16) = 3y(3y+4)^2.$$

14.5 Section Exercises

1.

$1^2 = \underline{1}$	$2^2 = \underline{4}$	$3^2 = \underline{9}$
$4^2 = \underline{16}$	$5^2 = \underline{25}$	$6^2 = \underline{36}$
$7^2 = \underline{49}$	$8^2 = \underline{64}$	$9^2 = \underline{81}$
$10^2 = \underline{100}$	$11^2 = \underline{121}$	$12^2 = \underline{144}$
$13^2 = \underline{169}$	$14^2 = \underline{196}$	$15^2 = \underline{225}$
$16^2 = \underline{256}$	$17^2 = \underline{289}$	$18^2 = \underline{324}$
$19^2 = \underline{361}$	$20^2 = \underline{400}$	

3. **A.** $x^2 - 4 = x^2 - 2^2$
B. $y^2 + 9 = y^2 + 3^2$ (sum)
C. $2a^2 - 25 = 2a^2 - 5^2$
D. $9m^2 - 1 = (3m)^2 - 1^2$

A and **D** are the differences of squares.

5. $y^2 - 25 = y^2 - 5^2$
$$= (y+5)(y-5) \quad \textit{Difference of squares}$$

7. $x^2 - 144 = x^2 - 12^2$
$$= (x+12)(x-12) \quad \textit{Difference of squares}$$

9. $m^2 - 12$ ■ Because 12 is not the square of an integer, this binomial is not a difference of squares. The terms have no common factor, so it is a *prime* polynomial.

11. $m^2 + 64$ ■ This binomial is the *sum* of squares and the terms have no common factor. Unlike the *difference* of squares, it cannot be factored. It is a *prime* polynomial.

13. $9r^2 - 4 = (3r)^2 - 2^2$
$$= (3r+2)(3r-2) \quad \textit{Difference of squares}$$

15. $36x^2 - 16$ ■ First factor out the GCF, 4; then use the rule for factoring the difference of squares.

$36x^2 - 16 = 4(9x^2 - 4)$
$$= 4[(3x)^2 - 2^2]$$
$$= 4(3x+2)(3x-2)$$

17. $196p^2 - 225$
$$= (14p)^2 - 15^2$$
$$= (14p+15)(14p-15) \quad \textit{Difference of squares}$$

19. $16r^2 - 25a^2$
$$= (4r)^2 - (5a)^2$$
$$= (4r+5a)(4r-5a) \quad \textit{Difference of squares}$$

21. $100x^2 + 49$ ■ This binomial is the *sum* of squares and the terms have no common factor. Unlike the *difference* of squares, it cannot be factored. It is a *prime* polynomial.

23. $p^4 - 49 = (p^2)^2 - 7^2$
$$= (p^2+7)(p^2-7) \quad \textit{Difference of squares}$$

25. $x^4 - 1$ ■ To factor this binomial completely, factor the difference of squares twice.

$x^4 - 1 = (x^2)^2 - 1^2$
$$= (x^2+1)(x^2-1)$$
$$= (x^2+1)(x^2-1^2)$$
$$= (x^2+1)(x+1)(x-1)$$

27. $p^4 - 256$ ■ To factor this binomial completely, factor the difference of squares twice.

$p^4 - 256 = (p^2)^2 - 16^2$
$$= (p^2+16)(p^2-16)$$
$$= (p^2+16)(p^2-4^2)$$
$$= (p^2+16)(p+4)(p-4)$$

29. A. $y^2 - 13y + 36 \stackrel{?}{=} (y-6)^2$

$2 \cdot y \cdot 6 = 12y \neq 13y$, so this is not a perfect square trinomial.

B. $x^2 + 6x + 9 \stackrel{?}{=} (x+3)^2$

$2 \cdot x \cdot 3 = 6x$, so this is a perfect square trinomial.

C. $4z^2 - 4z + 1 \stackrel{?}{=} (2z-1)^2$

$2 \cdot 2z \cdot 1 = 4z$, so this is a perfect square trinomial.

D. $16m^2 + 10m + 1 \stackrel{?}{=} (4m+1)^2$

$2 \cdot 4m \cdot 1 = 8m \neq 10m$, so this is not a perfect square trinomial.

B and **C** are perfect square trinomials.

31. $w^2 + 2w + 1$ ■ The first and last terms are perfect squares, w^2 and 1^2. This trinomial is a perfect square, since the middle term is twice the product of w and 1, or
$$2 \cdot w \cdot 1 = 2w.$$
Therefore,
$$w^2 + 2w + 1 = (w+1)^2.$$

33. $x^2 - 8x + 16$ ■ The first and last terms are perfect squares, x^2 and $(-4)^2$. This trinomial is a perfect square, since the middle term is twice the product of x and -4, or
$$2 \cdot x \cdot (-4) = -8x.$$
Therefore,
$$x^2 - 8x + 16 = (x-4)^2.$$

35. $x^2 - 10x + 100$ ■ The first and last terms are perfect squares, x^2 and $(-10)^2$, but $2 \cdot x \cdot (-10) = -20x \neq -10x$, so this is not a perfect square trinomial. This trinomial cannot be factored even with the methods of previous sections. It is *prime*.

37. $2x^2 + 24x + 72$

First, factor out the GCF, 2.
$$2x^2 + 24x + 72 = 2(x^2 + 12x + 36)$$
Now factor $x^2 + 12x + 36$ as a perfect square trinomial.
$$x^2 + 12x + 36 = (x+6)^2$$
The final factored form is
$$2x^2 + 24x + 72 = 2(x+6)^2.$$

39. $4x^2 + 12x + 9$ ■ The first and last terms are perfect squares, $(2x)^2$ and 3^2. This trinomial is a perfect square, since the middle term is twice the product of $2x$ and 3, or
$$2 \cdot 2x \cdot 3 = 12x.$$

Therefore,
$$4x^2 + 12x + 9 = (2x+3)^2.$$

41. $16x^2 - 40x + 25$ ■ The first and last terms are perfect squares, $(4x)^2$ and $(-5)^2$. The middle term is
$$2(4x)(-5) = -40x.$$
Therefore,
$$16x^2 - 40x + 25 = (4x-5)^2.$$

43. $49x^2 - 28xy + 4y^2$ ■ The first and last terms are perfect squares, $(7x)^2$ and $(-2y)^2$. The middle term is
$$2(7x)(-2y) = -28xy.$$
Therefore,
$$49x^2 - 28xy + 4y^2 = (7x-2y)^2.$$

45. $64x^2 + 48xy + 9y^2$
$$= (8x)^2 + 2(8x)(3y) + (3y)^2$$
$$= (8x+3y)^2$$

47. $-50h^3 + 40h^2y - 8hy^2$
$$= -2h(25h^2 - 20hy + 4y^2)$$
$$= -2h[(5h)^2 - 2(5h)(2y) + (2y)^2]$$
$$= -2h(5h - 2y)^2$$

49. $p^2 - \frac{1}{9} = p^2 - (\frac{1}{3})^2$
$$= (p+\frac{1}{3})(p-\frac{1}{3}) \quad \textit{Difference of squares}$$

51. $4m^2 - \frac{9}{25} = (2m)^2 - (\frac{3}{5})^2$
$$= (2m+\frac{3}{5})(2m-\frac{3}{5}) \quad \textit{Difference of squares}$$

53. $x^2 - 0.64 = x^2 - (0.8)^2$
$$= (x+0.8)(x-0.8) \quad \textit{Difference of squares}$$

55. $t^2 + t + \frac{1}{4}$ ■ The first and last terms are perfect squares, t^2 and $(\frac{1}{2})^2$. The trinomial is a perfect square, since the middle term is
$$2 \cdot t \cdot \frac{1}{2} = t.$$
Therefore,
$$t^2 + t + \frac{1}{4} = (t+\frac{1}{2})^2.$$

57. $x^2 - 1.0x + 0.25$ ■ The first and last terms are perfect squares, x^2 and $(-0.5)^2$. The trinomial is a perfect square, since the middle term is
$$2 \cdot x \cdot (-0.5) = -1.0x.$$
Therefore,
$$x^2 - 1.0x + 0.25 = (x-0.5)^2.$$

Summary Exercises *Recognizing and Applying Factoring Strategies*

1. $12x^2 + 20x + 8 = 4(3x^2 + 5x + 2)$
$= 4(3x + 2)(x + 1)$

Match with choice **F**. [Factor out the GCF; then factor a trinomial by grouping or trial and error.]

3. $-16m^2n + 24mn - 40mn^2$
$= -8mn(2m - 3 + 5n)$

Match with choice **A**. [Factor out the GCF; no further factoring is possible.]

5. $36p^2 - 60pq + 25q^2$ ■ The first and last terms are perfect squares, $(6p)^2$ and $(-5q)^2$. The middle term is
$$2(6p)(-5q) = -60pq.$$
Therefore,
$$36p^2 - 60pq + 25q^2 = (6p - 5q)^2.$$

Match with choice **D**. [Factor a perfect square trinomial.]

7. $625 - r^4$ ■ To factor this binomial completely, factor the difference of squares twice.
$$\begin{aligned} 625 - r^4 &= (25)^2 - (r^2)^2 \\ &= (25 + r^2)(25 - r^2) \\ &= (25 + r^2)(5^2 - r^2) \\ &= (25 + r^2)(5 + r)(5 - r) \end{aligned}$$

Match with choice **C**. [Factor a difference of squares twice.]

9. $4w^2 + 49$

Match with choice **H**. [The polynomial is prime.]

11. $32m^9 + 16m^5 + 24m^3$
$= 8m^3(4m^6 + 2m^2 + 3)$ $GCF = 8m^3$

13. $14k^3 + 7k^2 - 70k$
$= 7k(2k^2 + k - 10)$ $GCF = 7k$
$= 7k(2k + 5)(k - 2)$

15. $6z^2 + 31z + 5$ ■ Factor by grouping. Look for two integers whose product is $6(5) = 30$ and whose sum is 31. The integers are 30 and 1.
$$\begin{aligned} 6z^2 + 31z + 5 \\ &= 6z^2 + 30z + z + 5 \\ &= (6z^2 + 30z) + (z + 5) \\ &= 6z(z + 5) + 1(z + 5) \\ &= (z + 5)(6z + 1) \end{aligned}$$

17. $49z^2 - 16y^2$
$= (7z)^2 - (4y)^2$
$= (7z + 4y)(7z - 4y)$ *Difference of squares*

19. $16x^2 + 20x$
$= 4x(4x + 5)$ $GCF = 4x$

21. $10y^2 - 7yz - 6z^2$ ■ Factor by grouping. Look for two integers whose product is $10(-6) = -60$ and whose sum is -7. The integers are -12 and 5.
$$\begin{aligned} &10y^2 - 7yz - 6z^2 \\ &= 10y^2 - 12yz + 5yz - 6z^2 \\ &= 2y(5y - 6z) + z(5y - 6z) \\ &= (5y - 6z)(2y + z) \end{aligned}$$

23. $m^2 + 2m - 15$ ■ Look for a pair of integers whose product is -15 and whose sum is 2. The integers are -3 and 5, so
$$m^2 + 2m - 15 = (m - 3)(m + 5).$$

25. $32z^3 + 56z^2 - 16z$
$$\begin{aligned} &= 8z(4z^2 + 7z - 2) & GCF = 8z \\ &= 8z(4z^2 + 8z - z - 2) & Grouping \\ &= 8z[4z(z + 2) - 1(z + 2)] \\ &= 8z(z + 2)(4z - 1) \end{aligned}$$

27. $z^2 - 12z + 36$
$$\begin{aligned} &= z^2 - 2 \cdot 6z + 6^2 \\ &= (z - 6)^2 & \textit{Perfect square trinomial} \end{aligned}$$

29. $y^2 - 4yk - 12k^2$ ■ Look for a pair of integers whose product is -12 and whose sum is -4. The integers are -6 and 2, so
$$y^2 - 4yk - 12k^2 = (y - 6k)(y + 2k).$$

31. $6y^2 - 6y - 12$
$$\begin{aligned} &= 6(y^2 - y - 2) & GCF = 6 \\ &= 6(y - 2)(y + 1) \end{aligned}$$

33. $p^2 - 17p + 66$ ■ Look for a pair of integers whose product is 66 and whose sum is -17. The integers are -11 and -6, so
$$p^2 - 17p + 66 = (p - 11)(p - 6).$$

35. $k^2 + 100$ cannot be factored because it is the sum of squares with no GCF. The expression is *prime*.

37. $z^2 - 3za - 10a^2$ ■ Look for a pair of integers whose product is -10 and whose sum is -3. The integers are -5 and 2, so
$$z^2 - 3za - 10a^2 = (z - 5a)(z + 2a).$$

39. $4k^2 - 12k + 9$
$$\begin{aligned} &= (2k)^2 - 2 \cdot 2k \cdot 3 + 3^2 \\ &= (2k - 3)^2 & \textit{Perfect square trinomial} \end{aligned}$$

41. $16r^2 + 24rm + 9m^2$
$$\begin{aligned} &= (4r)^2 + 2 \cdot 4r \cdot 3m + (3m)^2 \\ &= (4r + 3m)^2 & \textit{Perfect square trinomial} \end{aligned}$$

43. $n^2 - 12n - 35$ is *prime*. There is no pair of integers whose product is -35 and whose sum is -12 and there is no GCF.

45. $16k^2 - 48k + 36$
$$= 4(4k^2 - 12k + 9) \qquad GCF = 4$$
$$= 4\left[(2k)^2 - 2(2k)(3) + 3^2\right]$$
$$= 4(2k - 3)^2$$

47. $36y^6 - 42y^5 - 120y^4$
$$= 6y^4(6y^2 - 7y - 20) \qquad GCF = 6y^4$$
$$= 6y^4(6y^2 - 15y + 8y - 20) \qquad Grouping$$
$$= 6y^4[3y(2y - 5) + 4(2y - 5)]$$
$$= 6y^4(2y - 5)(3y + 4)$$

49. $8p^2 + 23p - 3$ ▪ Factor by grouping. Look for two integers whose product is $8(-3) = -24$ and whose sum is 23. The integers are 24 and -1.
$$8p^2 + 23p - 3$$
$$= 8p^2 + 24p - p - 3$$
$$= 8p(p + 3) - 1(p + 3)$$
$$= (p + 3)(8p - 1)$$

51. $54m^2 - 24z^2$
$$= 6(9m^2 - 4z^2) \qquad GCF = 6$$
$$= 6\left[(3m)^2 - (2z)^2\right]$$
$$= 6(3m + 2z)(3m - 2z) \qquad \begin{array}{l} Difference \\ of\ squares \end{array}$$

53. $6a^2 + 10a - 4$
$$= 2(3a^2 + 5a - 2) \qquad GCF = 2$$
$$= 2(3a - 1)(a + 2)$$

55. $28a^2 - 63b^2$
$$= 7(4a^2 - 9b^2) \qquad GCF = 7$$
$$= 7\left[(2a)^2 - (3b)^2\right]$$
$$= 7(2a + 3b)(2a - 3b)$$

57. $125m^4 - 400m^3n + 195m^2n^2$
$$= 5m^2(25m^2 - 80mn + 39n^2)$$
$$= 5m^2(25m^2 - 15mn - 65mn + 39n^2)$$
$$\qquad\qquad\qquad\qquad Grouping$$
$$= 5m^2[5m(5m - 3n) - 13n(5m - 3n)]$$
$$= 5m^2(5m - 3n)(5m - 13n)$$

59. $9u^2 + 66uv + 121v^2$
$$= (3u)^2 + 2 \cdot 3u \cdot 11v + (11v)^2$$
$$= (3u + 11v)^2 \qquad Perfect\ square\ trinomial$$

61. $27p^{10} - 45p^9 - 252p^8$
$$= 9p^8(3p^2 - 5p - 28) \qquad GCF = 9p^8$$
$$= 9p^8(3p^2 - 12p + 7p - 28) \qquad Grouping$$
$$= 9p^8[3p(p - 4) + 7(p - 4)]$$
$$= 9p^8(p - 4)(3p + 7)$$

63. $4 - 2q - 6p + 3pq$
$$= 2(2 - q) - 3p(2 - q)$$
$$= (2 - q)(2 - 3p)$$

65. $64p^2 - 100m^2$
$$= 4(16p^2 - 25m^2) \qquad GCF = 4$$
$$= 4\left[(4p)^2 - (5m)^2\right]$$
$$= 4(4p + 5m)(4p - 5m) \qquad \begin{array}{l} Difference \\ of\ squares \end{array}$$

67. $100a^2 - 81y^2$
$$= (10a)^2 - (9y)^2$$
$$= (10a + 9y)(10a - 9y) \qquad \begin{array}{l} Difference \\ of\ squares \end{array}$$

69. $a^2 + 8a + 16$
$$= a^2 + 2 \cdot a \cdot 4 + 4^2$$
$$= (a + 4)^2 \qquad Perfect\ square\ trinomial$$

71. $2x^2 + 5x + 6$ is *prime*.

73. $25a^2 - 70ab + 49b^2$
$$= (5a)^2 - 2 \cdot 5a \cdot 7b + (7b)^2$$
$$= (5a - 7b)^2 \qquad Perfect\ square\ trinomial$$

75. $-4x^2 + 24xy - 36y^2$
$$= -4(x^2 - 6xy + 9y^2) \qquad GCF = -4$$
$$= -4\left[x^2 - 2(x)(3y) + (3y)^2\right]$$
$$= -4(x - 3y)^2$$

77. $-2x^2 + 26x - 72$
$$= -2(x^2 - 13x + 36) \qquad GCF = -2$$
$$= -2(x - 9)(x - 4)$$

79. $12x^2 + 22x - 20$
$$= 2(6x^2 + 11x - 10) \qquad GCF = 2$$
$$= 2(2x + 5)(3x - 2)$$

81. $y^2 - 64$
$$= y^2 - 8^2$$
$$= (y + 8)(y - 8) \qquad \begin{array}{l} Difference \\ of\ squares \end{array}$$

14.6 Solving Quadratic Equations by Factoring

14.6 Margin Exercises

1. **A.** $y^2 - 4y - 5 = 0$ is in the form $ax^2 + bx + c = 0$.

B. $x^3 - x^2 + 16 = 0$ is not a quadratic equation because of the x^3-term.

C. $2z^2 + 7z = -3$, or $2z^2 + 7z + 3 = 0$, is in the form $ax^2 + bx + c = 0$.

D. $x + 2y = -4$ is not a quadratic equation because no term contains a variable squared, such as x^2 or y^2.

A and **C** are quadratic equations.

2. **(a)** $x^2 - 3x = 4$ ■ To write this equation in standard form, subtract 4 from each side of the equation.

$$x^2 - 3x - 4 = 0$$

(b) $y^2 = 9y - 8$ ■ To write this equation in standard form, subtract $9y$ from each side and add 8 to each side of the equation.

$$y^2 - 9y + 8 = 0$$

3. **(a)** $(x - 5)(x + 2) = 0$

To solve, use the zero-factor property.

Either $\underline{x - 5} = 0$ or $\underline{x + 2} = 0$.

Solve these two equations, obtaining

$$x = \underline{5} \quad \text{or} \quad x = \underline{-2}.$$

Check Let $x = 5$.

$$(x - 5)(x + 2) = 0$$
$$(\underline{5} - 5)(\underline{5} + 2) \stackrel{?}{=} 0$$
$$\underline{0}(7) = 0 \quad \textit{True}$$

To check the other solution, substitute $\underline{-2}$ for x in the original equation. Another *true* statement of $0 = 0$ results.

The solution set is $\{\underline{-2}, \underline{5}\}$.

(b) $(3x - 2)(x + 6) = 0$ ■ By the zero-factor property, either $3x - 2 = 0$ or $x + 6 = 0$. Solve each equation.

$$3x - 2 = 0 \quad \text{or} \quad x + 6 = 0$$
$$3x = 2$$
$$x = \tfrac{2}{3} \quad \text{or} \qquad x = -6$$

Check these solutions by substituting each one in the original equation. The solution set is $\left\{-6, \tfrac{2}{3}\right\}$.

(c) $z(2z + 5) = 0$ ■ By the zero-factor property, either $z = 0$ or $2z + 5 = 0$. Since the first equation is solved, we only need to solve the second.

$$2z + 5 = 0$$
$$2z = -5$$
$$z = -\tfrac{5}{2}$$

Check these solutions by substituting each one in the original equation. The solution set is $\left\{-\tfrac{5}{2}, 0\right\}$.

4. **(a)** $m^2 - 3m - 10 = 0$

First factor the equation to get

$$(m - 5)(m + 2) = 0.$$

Then set each factor equal to zero and solve.

$$m - 5 = 0 \quad \text{or} \quad m + 2 = 0$$
$$m = 5 \quad \text{or} \qquad m = -2$$

Check these solutions by substituting each one in the original equation. The solution set is $\{-2, 5\}$.

(b) $r^2 + 2r = 8$ ■ Bring all nonzero terms to the same side of the equals sign by subtracting 8 from each side.

$$r^2 + 2r - 8 = 0$$
$$(r - 2)(r + 4) = 0$$
$$r - 2 = 0 \quad \text{or} \quad r + 4 = 0$$
$$r = 2 \quad \text{or} \qquad r = -4$$

Check these solutions by substituting each one in the original equation. The solution set is $\{-4, 2\}$.

5. **(a)** $10a^2 - 5a - 15 = 0$

$$\underline{5}(2a^2 - a - 3) = 0 \quad \textit{Factor out 5.}$$
$$2a^2 - a - 3 = 0 \quad \textit{Divide by } \underline{5}.$$
$$(2a - 3)(\underline{a + 1}) = 0 \quad \textit{Factor.}$$
$$2a - 3 = \underline{0} \quad \text{or} \quad \underline{a + 1} = 0$$
$$2a = 3$$
$$a = \tfrac{3}{2} \quad \text{or} \qquad a = \underline{-1}$$

Check these solutions by substituting each one in the original equation. The solution set is $\underline{\left\{-1, \tfrac{3}{2}\right\}}$.

(b) $4x^2 - 2x = 42$

$$4x^2 - 2x - 42 = 0 \quad \textit{Subtract 42.}$$
$$2(2x^2 - x - 21) = 0 \quad \textit{Factor out 2.}$$
$$2x^2 - x - 21 = 0 \quad \textit{Divide by 2.}$$
$$(2x - 7)(x + 3) = 0 \quad \textit{Factor.}$$
$$2x - 7 = 0 \quad \text{or} \quad x + 3 = 0$$
$$2x = 7$$
$$x = \tfrac{7}{2} \quad \text{or} \qquad x = -3$$

Check these solutions by substituting each one in the original equation. The solution set is $\left\{-3, \tfrac{7}{2}\right\}$.

6. **(a)** $49m^2 - 9 = 0$

$$(7m + 3)(7m - 3) = 0$$
$$7m + 3 = 0 \quad \text{or} \quad 7m - 3 = 0$$
$$7m = -3 \qquad\qquad 7m = 3$$
$$m = -\tfrac{3}{7} \quad \text{or} \qquad m = \tfrac{3}{7}$$

Check these solutions by substituting each one in the original equation. The solution set is $\left\{-\tfrac{3}{7}, \tfrac{3}{7}\right\}$.

(b) $m^2 = 3m$ ■ Get all terms on one side of the equals sign, with 0 on the other side.

$$m^2 - 3m = 0$$

Factor the binomial.

$$m(m - 3) = 0$$

Set each factor equal to 0 and solve the resulting equations.

$$m = 0 \quad \text{or} \quad m - 3 = 0$$
$$m = 0 \quad \text{or} \qquad m = 3$$

Check these solutions by substituting each one in the original equation. The solution set is $\{0, 3\}$.

(c) $p(4p + 7) = 2$ ▪ Remove grouping symbols and get all terms on one side of the equals sign, with 0 on the other side.

$$4p^2 + 7p = 2$$
$$4p^2 + 7p - 2 = 0$$
$$(4p - 1)(p + 2) = 0$$

$$4p - 1 = 0 \quad \text{or} \quad p + 2 = 0$$
$$4p = 1$$
$$p = \tfrac{1}{4} \quad \text{or} \quad p = -2$$

Check these solutions by substituting each one in the original equation. The solution set is $\{-2, \tfrac{1}{4}\}$.

7. **(a)** $x^2 + 16x = -64$

Write the equation in standard form and then factor $x^2 + 16x + \underline{64}$ as a perfect square trinomial.

$$(x + 8)^2 = 0$$

Set the factor $x + 8$ equal to 0 and solve.

$$x + 8 = 0$$
$$x = -8$$

So -8 is a <u>double</u> solution. Check this solution by substituting it in the original equation. The solution set is $\{-8\}$.

(b) $4x^2 - 4x + 1 = 0$

Factor $4x^2 - 4x + 1$ as a perfect square trinomial.

$$(2x - 1)^2 = 0$$

Set the factor $2x - 1$ equal to 0 and solve.

$$2x - 1 = 0$$
$$2x = 1$$
$$x = \tfrac{1}{2}$$

Check this solution by substituting it in the original equation. The solution set is $\{\tfrac{1}{2}\}$.

(c) $4z^2 + 20z = -25$

Write the equation in standard form and then factor $4z^2 + 20z + 25$ as a perfect square trinomial.

$$(2z + 5)^2 = 0$$

Set the factor $2z + 5$ equal to 0 and solve.

$$2z + 5 = 0$$
$$2z = -5$$
$$z = -\tfrac{5}{2}$$

Check this solution by substituting it in the original equation. The solution set is $\{-\tfrac{5}{2}\}$.

8. **(a)**
$$r^3 - 16r = 0$$
$$r(r^2 - 16) = 0$$
$$r(r - 4)(r + 4) = 0$$

$$r = 0 \quad \text{or} \quad r - 4 = 0 \quad \text{or} \quad r + 4 = 0$$
$$r = 0 \quad \text{or} \quad r = 4 \quad \text{or} \quad r = -4$$

Check each solution.
The solution set is $\{-4, 0, 4\}$.

(b) $x^3 - 3x^2 - 18x = 0$
$$x(x^2 - 3x - 18) = 0$$
$$x(x - 6)(x + 3) = 0$$

$$x = 0 \quad \text{or} \quad x - 6 = 0 \quad \text{or} \quad x + 3 = 0$$
$$x = 0 \quad \text{or} \quad x = 6 \quad \text{or} \quad x = -3$$

Check each solution.
The solution set is $\{-3, 0, 6\}$.

9. **(a)** $(m + 3)(m^2 - 11m + 10) = 0$
$$(m + 3)(m - 10)(m - 1) = 0$$

$$m + 3 = 0 \quad \text{or} \quad m - 10 = 0 \quad \text{or} \quad m - 1 = 0$$
$$m = -3 \quad \text{or} \quad m = 10 \quad \text{or} \quad m = 1$$

Check each solution.
The solution set is $\{-3, 1, 10\}$.

(b) $(2x + 5)(4x^2 - 9) = 0$
$$(2x + 5)(2x + 3)(2x - 3) = 0$$

$$2x + 5 = 0 \quad \text{or} \quad 2x + 3 = 0 \quad \text{or} \quad 2x - 3 = 0$$
$$2x = -5 \quad \text{or} \quad 2x = -3 \quad \text{or} \quad 2x = 3$$
$$x = -\tfrac{5}{2} \quad \text{or} \quad x = -\tfrac{3}{2} \quad \text{or} \quad x = \tfrac{3}{2}$$

Check each solution.
The solution set is $\{-\tfrac{5}{2}, -\tfrac{3}{2}, \tfrac{3}{2}\}$.

14.6 Section Exercises

For all equations in this section, answers should be checked by substituting into the original equation. These checks will be shown here for only a few of the exercises.

1. A quadratic equation is an equation that can be put in the form $\underline{ax^2 + bx + c} = 0.$ $(a \neq 0)$

3. If a quadratic equation is in standard form, to solve the equation we should begin by attempting to <u>factor</u> the polynomial.

5. **(a)** $2x - 5 = 6$ can be written as $2x - 11 = 0$, so it is a *linear* equation.

(b) $x^2 - 5 = -4$ can be written as $x^2 - 1 = 0$, so it is a *quadratic* equation.

(c) $x^2 + 2x - 1 = 2x^2$ can be written as $-x^2 + 2x - 1 = 0$, so it is a *quadratic* equation.

(d) $5^2x + 2 = 0$ can be written as $25x + 2 = 0$, so it is a *linear* equation.

7. We can consider the factored form as $2 \cdot x(3x - 4)$, the product of three factors, 2, x, and $3x - 4$. Applying the zero-factor property yields three equations,

$$2 = 0 \quad \text{or} \quad x = 0 \quad \text{or} \quad 3x - 4 = 0.$$

Since "$2 = 0$" is impossible, the equation $2 = 0$ has no solution, so we end up with the two solutions, $x = 0$ and $x = \frac{4}{3}$. The solution set is $\{0, \frac{4}{3}\}$.

We conclude that multiplying a polynomial by a constant does not affect the solutions of the corresponding equation and only the *variable* factors need to be set equal to zero.

9. $(x + 5)(x - 2) = 0$ ▪ By the zero-factor property, the only way that the product of these two factors can be zero is if at least one of the factors is zero.

$$x + 5 = 0 \quad \text{or} \quad x - 2 = 0$$

Solve each of these linear equations.

$$x = -5 \quad \text{or} \quad x = 2$$

Check $x = -5$: $\quad 0(-7) = 0 \quad$ *True*
Check $x = 2$: $\quad\quad 7(0) = 0 \quad$ *True*

The solution set is $\{-5, 2\}$.

11. $(2m - 7)(m - 3) = 0$ ▪ Set each factor equal to zero and solve the resulting linear equations.

$$2m - 7 = 0 \quad \text{or} \quad m - 3 = 0$$
$$2m = 7$$
$$m = \tfrac{7}{2} \quad \text{or} \quad\quad m = 3$$

Check each solution. The solution set is $\{3, \frac{7}{2}\}$.

13. $(2x + 1)(6x - 1) = 0$ ▪ Set each factor equal to zero and solve the resulting linear equations.

$$2x + 1 = 0 \quad\quad \text{or} \quad 6x - 1 = 0$$
$$2x = -1 \quad \text{or} \quad\quad 6x = 1$$
$$x = -\tfrac{1}{2} \quad\quad\quad x = \tfrac{1}{6}$$

Check each solution. The solution set is $\left\{-\frac{1}{2}, \frac{1}{6}\right\}$.

15. $t(6t + 5) = 0$ ▪ Set each factor equal to zero and solve the resulting linear equations.

$$t = 0 \quad \text{or} \quad 6t + 5 = 0$$
$$6t = -5$$
$$t = -\tfrac{5}{6}$$

Check each solution. The solution set is $\left\{-\frac{5}{6}, 0\right\}$.

17. $2x(3x - 4) = 0$ ▪ Set each factor equal to zero and solve the resulting linear equations.

$$2x = 0 \quad \text{or} \quad 3x - 4 = 0$$
$$x = 0 \quad \text{or} \quad\quad 3x = 4$$
$$x = \tfrac{4}{3}$$

Check each solution. The solution set is $\{0, \frac{4}{3}\}$.

19. $(x - 9)(x - 9) = 0$ ▪ Set the factor $x - 9$ equal to 0 and solve.

$$x - 9 = 0$$
$$x = 9$$

9 is called a *double solution*.
Check the solution. The solution set is $\{9\}$.

21. $y^2 + 3y + 2 = 0$ ▪ Factor the polynomial.

$$(y + 2)(y + 1) = 0$$

Set each factor equal to 0.

$$y + 2 = 0 \quad \text{or} \quad y + 1 = 0$$

Solve each equation.

$$y = -2 \quad \text{or} \quad y = -1$$

Check Check these solutions by substituting -2 for y and then -1 for y in the original equation.

$$y^2 + 3y + 2 = 0$$
$$(-2)^2 + 3(-2) + 2 \stackrel{?}{=} 0 \quad \textit{Let y = -2.}$$
$$4 - 6 + 2 \stackrel{?}{=} 0$$
$$-2 + 2 = 0 \quad \textit{True}$$

$$y^2 + 3y + 2 = 0$$
$$(-1)^2 + 3(-1) + 2 \stackrel{?}{=} 0 \quad \textit{Let y = -1.}$$
$$1 - 3 + 2 \stackrel{?}{=} 0$$
$$-2 + 2 = 0 \quad \textit{True}$$

The solution set is $\{-2, -1\}$.

23. $y^2 - 3y + 2 = 0$ ▪ Factor the polynomial.

$$(y - 1)(y - 2) = 0$$

Set each factor equal to 0.

$$y - 1 = 0 \quad \text{or} \quad y - 2 = 0$$

Solve each equation.

$$y = 1 \quad \text{or} \quad y = 2$$

Check each solution. The solution set is $\{1, 2\}$.

25. $x^2 = 24 - 5x$ ▪ Write the equation in standard form.

$$x^2 + 5x - 24 = 0$$

Factor the polynomial.

$$(x + 8)(x - 3) = 0$$

Set each factor equal to 0.

$$x + 8 = 0 \quad \text{or} \quad x - 3 = 0$$

Solve each equation.

$$x = -8 \quad \text{or} \quad x = 3$$

Check each solution. The solution set is $\{-8, 3\}$.

27. $x^2 = 3 + 2x$ ▪ Write the equation in standard form.

$$x^2 - 2x - 3 = 0$$
$$(x + 1)(x - 3) = 0$$

$$x + 1 = 0 \quad \text{or} \quad x - 3 = 0$$
$$x = -1 \quad \text{or} \quad x = 3$$

Check each solution. The solution set is $\{-1, 3\}$.

29. $z^2 + 3z = -2$ ▪ Write the equation in standard form.

$$z^2 + 3z + 2 = 0$$

Factor the polynomial.

$$(z + 2)(z + 1) = 0$$

Set each factor equal to 0.

$$z + 2 = 0 \quad \text{or} \quad z + 1 = 0$$
$$z = -2 \quad \text{or} \quad z = -1$$

Check that the solution set is $\{-2, -1\}$.

31. $m^2 + 8m + 16 = 0$ ▪ Factor $m^2 + 8m + 16$ as a perfect square trinomial.

$$(m + 4)^2 = 0$$

Set the factor $m + 4$ equal to 0 and solve.

$$m + 4 = 0$$
$$m = -4$$

Check that the solution set is $\{-4\}$.

33. $3x^2 + 5x - 2 = 0$ ▪ Factor the polynomial.

$$(3x - 1)(x + 2) = 0$$

Set each factor equal to 0.

$$3x - 1 = 0 \quad \text{or} \quad x + 2 = 0$$
$$3x = 1$$
$$x = \tfrac{1}{3} \quad \text{or} \quad x = -2$$

Check that the solution set is $\{-2, \tfrac{1}{3}\}$.

35. $$12p^2 = 8 - 10p$$
$$12p^2 + 10p - 8 = 0 \qquad \textit{Standard form}$$
$$2(6p^2 + 5p - 4) = 0 \qquad \textit{Factor out 2.}$$
$$2(3p + 4)(2p - 1) = 0 \qquad \textit{Factor.}$$

Set each factor equal to 0.

$$3p + 4 = 0 \quad \text{or} \quad 2p - 1 = 0$$
$$3p = -4 \quad \text{or} \quad 2p = 1$$
$$p = -\tfrac{4}{3} \quad \text{or} \quad p = \tfrac{1}{2}$$

Check that the solution set is $\{-\tfrac{4}{3}, \tfrac{1}{2}\}$.

37. $$9s^2 + 12s = -4$$
$$9s^2 + 12s + 4 = 0 \qquad \textit{Standard form}$$
$$(3s + 2)^2 = 0 \qquad \textit{Factor.}$$

Set the factor $3s + 2$ equal to 0 and solve.

$$3s + 2 = 0$$
$$3s = -2$$
$$s = -\tfrac{2}{3}$$

Check that the solution set is $\{-\tfrac{2}{3}\}$.

39. $$y^2 - 9 = 0$$
$$(y + 3)(y - 3) = 0 \qquad \textit{Factor.}$$

$$y + 3 = 0 \quad \text{or} \quad y - 3 = 0$$
$$y = -3 \quad \text{or} \quad y = 3$$

Check that the solution set is $\{-3, 3\}$.

41. $$16k^2 - 49 = 0$$
$$(4k + 7)(4k - 7) = 0 \qquad \textit{Factor.}$$

$$4k + 7 = 0 \quad \text{or} \quad 4k - 7 = 0$$
$$4k = -7 \qquad\qquad 4k = 7$$
$$k = -\tfrac{7}{4} \quad \text{or} \quad k = \tfrac{7}{4}$$

Check that the solution set is $\{-\tfrac{7}{4}, \tfrac{7}{4}\}$.

43. $$n^2 = 121$$
$$n^2 - 121 = 0 \qquad \textit{Standard form}$$
$$(n + 11)(n - 11) = 0 \qquad \textit{Factor.}$$

$$n + 11 = 0 \quad \text{or} \quad n - 11 = 0$$
$$n = -11 \quad \text{or} \quad n = 11$$

Check that the solution set is $\{-11, 11\}$.

45. $$x^2 = 7x$$
$$x^2 - 7x = 0 \qquad \textit{Standard form}$$
$$x(x - 7) = 0 \qquad \textit{Factor.}$$

$$x = 0 \quad \text{or} \quad x - 7 = 0$$
$$x = 0 \quad \text{or} \quad x = 7$$

Check that the solution set is $\{0, 7\}$.

47. $$6r^2 = 3r$$
$$6r^2 - 3r = 0 \qquad \textit{Standard form}$$
$$3r(2r - 1) = 0 \qquad \textit{Factor.}$$

$$3r = 0 \quad \text{or} \quad 2r - 1 = 0$$
$$2r = 1$$
$$r = 0 \quad \text{or} \quad r = \tfrac{1}{2}$$

Check that the solution set is $\{0, \tfrac{1}{2}\}$.

49. $$g(g - 7) = -10$$
$$g^2 - 7g = -10 \qquad \textit{Multiply.}$$
$$g^2 - 7g + 10 = 0 \qquad \textit{Standard form}$$
$$(g - 2)(g - 5) = 0 \qquad \textit{Factor.}$$

$$g - 2 = 0 \quad \text{or} \quad g - 5 = 0$$
$$g = 2 \quad \text{or} \quad g = 5$$

Check that the solution set is $\{2, 5\}$.

51.
$$z(2z + 7) = 4$$
$$2z^2 + 7z = 4 \quad \textit{Multiply.}$$
$$2z^2 + 7z - 4 = 0 \quad \textit{Standard form}$$
$$(2z - 1)(z + 4) = 0 \quad \textit{Factor.}$$

$$2z - 1 = 0 \quad \text{or} \quad z + 4 = 0$$
$$2z = 1$$
$$z = \tfrac{1}{2} \quad \text{or} \quad z = -4$$

Check that the solution set is $\{-4, \tfrac{1}{2}\}$.

53.
$$2(y^2 - 66) = -13y$$
$$2y^2 - 132 = -13y \quad \textit{Multiply.}$$
$$2y^2 + 13y - 132 = 0 \quad \textit{Standard form}$$
$$(2y - 11)(y + 12) = 0 \quad \textit{Factor.}$$

$$2y - 11 = 0 \quad \text{or} \quad y + 12 = 0$$
$$2y = 11$$
$$y = \tfrac{11}{2} \quad \text{or} \quad y = -12$$

Check that the solution set is $\{-12, \tfrac{11}{2}\}$.

55.
$$5x^3 - 20x = 0$$
$$5x(x^2 - 4) = 0 \quad \textit{Factor out 5x.}$$
$$5x(x + 2)(x - 2) = 0 \quad \textit{Factor.}$$

$$5x = 0 \quad \text{or} \quad x + 2 = 0 \quad \text{or} \quad x - 2 = 0$$
$$x = 0 \quad \text{or} \quad x = -2 \quad \text{or} \quad x = 2$$

Check that the solution set is $\{-2, 0, 2\}$.

57. $9y^3 - 49y = 0$ ■ To factor the polynomial, begin by factoring out the greatest common factor.

$$y(9y^2 - 49) = 0$$

Now factor $9y^2 - 49$ as the difference of two squares.

$$y(3y + 7)(3y - 7) = 0$$

Set each of the three factors equal to 0 and solve.

$$y = 0 \quad \text{or} \quad 3y + 7 = 0 \quad \text{or} \quad 3y - 7 = 0$$
$$3y = -7 \qquad\qquad 3y = 7$$
$$y = 0 \quad \text{or} \quad y = -\tfrac{7}{3} \quad \text{or} \quad y = \tfrac{7}{3}$$

Check that the solution set is $\{-\tfrac{7}{3}, 0, \tfrac{7}{3}\}$.

59. $(2r + 5)(3r^2 - 16r + 5) = 0$

Begin by factoring $3r^2 - 16r + 5$.

$$(2r + 5)(3r - 1)(r - 5) = 0$$

Set each of the three factors equal to 0 and solve the resulting equations.

$$2r + 5 = 0 \quad \text{or} \quad 3r - 1 = 0 \quad \text{or} \quad r - 5 = 0$$
$$2r = -5 \qquad\quad 3r = 1$$
$$r = -\tfrac{5}{2} \quad \text{or} \quad r = \tfrac{1}{3} \quad \text{or} \quad r = 5$$

Check that the solution set is $\{-\tfrac{5}{2}, \tfrac{1}{3}, 5\}$.

61.
$$(2x + 7)(x^2 + 2x - 3) = 0$$
$$(2x + 7)(x + 3)(x - 1) = 0 \quad \textit{Factor.}$$

$$2x + 7 = 0 \quad \text{or} \quad x + 3 = 0 \quad \text{or} \quad x - 1 = 0$$
$$2x = -7$$
$$x = -\tfrac{7}{2} \quad \text{or} \quad x = -3 \quad \text{or} \quad x = 1$$

Check that the solution set is $\{-\tfrac{7}{2}, -3, 1\}$.

63.
$$x^3 + x^2 - 20x = 0$$
$$x(x^2 + x - 20) = 0 \quad \textit{Factor out x.}$$
$$x(x + 5)(x - 4) = 0 \quad \textit{Factor.}$$

Set each factor equal to zero and solve.

$$x = 0 \quad \text{or} \quad x + 5 = 0 \quad \text{or} \quad x - 4 = 0$$
$$x = 0 \quad \text{or} \qquad x = -5 \quad \text{or} \qquad x = 4$$

Check that the solution set is $\{-5, 0, 4\}$.

65. $r^4 = 2r^3 + 15r^2$ ■ Rewrite with all terms on the left side.

$$r^4 - 2r^3 - 15r^2 = 0$$
$$r^2(r^2 - 2r - 15) = 0 \quad \textit{Factor out } r^2.$$
$$r^2(r - 5)(r + 3) = 0 \quad \textit{Factor.}$$

Set each factor equal to zero and solve.

$$r^2 = 0 \quad \text{or} \quad r - 5 = 0 \quad \text{or} \quad r + 3 = 0$$
$$r = 0 \quad \text{or} \qquad r = 5 \quad \text{or} \qquad r = -3$$

Check that the solution set is $\{-3, 0, 5\}$.

67.
$$3x(x + 1) = (2x + 3)(x + 1)$$
$$3x^2 + 3x = 2x^2 + 5x + 3$$
$$x^2 - 2x - 3 = 0$$
$$(x + 1)(x - 3) = 0$$

$$x + 1 = 0 \quad \text{or} \quad x - 3 = 0$$
$$x = -1 \quad \text{or} \quad x = 3$$

Check that the solution set is $\{-1, 3\}$.

Alternatively, we could begin by moving all the terms to the left side and then factoring out $x + 1$.

$$3x(x + 1) - (2x + 3)(x + 1) = 0$$
$$(x + 1)[3x - (2x + 3)] = 0$$
$$(x + 1)(x - 3) = 0$$

The rest of the solution is the same.

69. **(a)** $d = 16t^2$

$t = 2$: $d = 16(2)^2 = 16(4) = 64$

$t = 3$: $d = 16(3)^2 = 16(9) = 144$

$d = 256$:
$$16t^2 = 256$$
$$t^2 = 16$$
$$t^2 - 16 = 0$$
$$(t + 4)(t - 4) = 0$$
$$t = 4 \text{ since } t \geq 0.$$

$d = 576$:
$$16t^2 = 576$$
$$t^2 = 36$$
$$t^2 - 36 = 0$$
$$(t + 6)(t - 6) = 0$$
$$t = 6 \text{ since } t \geq 0.$$

t in seconds	0	1	2	3	4	6
d in feet	0	16	64	144	256	576

(b) When $t = 0$, $d = 0$, no time has elapsed, so the object hasn't fallen (been released) yet.

14.7 Applications of Quadratic Equations

14.7 Margin Exercises

1. **(a)** *Step 2* Let $x =$ the width of the room. Then $x + 2 =$ the length.

   Step 3 Use the formula for the area of a rectangle, $A = lw$. Substitute 48 for A, $x + 2$ for l, and x for w.
   $$A = lw$$
   $$48 = (x + 2)x$$

   Step 4 Solve the equation.
   $$48 = x^2 + 2x$$
   $$0 = x^2 + 2x - 48 \quad \textit{Subtract 48.}$$
   $$0 = (x - 6)(x + 8) \quad \textit{Factor.}$$
   $$x - 6 = 0 \quad \text{or} \quad x + 8 = 0$$
   $$x = 6 \quad \text{or} \quad x = -8$$

   Step 5 Discard the solution -8 since the width cannot be negative. The width is 6 meters and the length is $x + 2 = 6 + 2 = 8$ meters.

   Step 6 8 is 2 more than 6 and 6 times 8 is 48.

   (b) Let $x =$ the length of a side of the original square. Then $x + 4 =$ the length of a side of the larger square.

   Use the formula for the area of a square, $A = s^2$. The area of the original square is x^2, and the area of the larger square is $(x + 4)^2$. The sum of the areas of the two squares is 106 square inches, so
   $$x^2 + (x + 4)^2 = 106.$$
   Solve this equation.
   $$x^2 + x^2 + 8x + 16 = 106$$
   $$2x^2 + 8x - 90 = 0$$
   $$x^2 + 4x - 45 = 0 \quad \textit{Divide by 2.}$$
   $$(x - 5)(x + 9) = 0$$
   $$x - 5 = 0 \quad \text{or} \quad x + 9 = 0$$
   $$x = 5 \quad \text{or} \quad x = -9$$

   Discard the solution -9 since the length of a side cannot be negative. The length of a side of the original square is 5 inches.

2. Let $x =$ the first locker number. Then $x + 1 =$ the second locker number. The product is 132.
   $$x(x + 1) = 132$$
   $$x^2 + x = 132$$
   $$x^2 + x - 132 = 0$$
   $$(x - 11)(x + 12) = 0$$
   $$x - 11 = 0 \quad \text{or} \quad x + 12 = 0$$
   $$x = 11 \quad \text{or} \quad x = -12$$

   A locker number is positive, so discard -12. Therefore, the locker numbers are 11 and $x + 1 = 11 + 1 = 12$.

3. **(a)** Let $x =$ the lesser integer. Then $x + 2 =$ the next greater even integer.

The product is	4	more than	two times	their sum.

   $$x(x + 2) = 4 + 2 \cdot [x + (x + 2)]$$

   Solve this equation.
   $$x^2 + 2x = 4 + 2(2x + 2)$$
   $$x^2 + 2x = 4 + 4x + 4$$
   $$x^2 - 2x - 8 = 0$$
   $$(x - 4)(x + 2) = 0$$
   $$x - 4 = 0 \quad \text{or} \quad x + 2 = 0$$
   $$x = 4 \quad \text{or} \quad x = -2$$

   We need to find two consecutive even integers. If $x = 4$ is the lesser, then the greater is
   $$x + 2 = 4 + 2 = 6.$$
   If $x = -2$ is the lesser, then the greater is
   $$x + 2 = -2 + 2 = 0.$$
   The two integers are 4 and 6, or they are -2 and 0.

   (b) Let $x =$ the least integer. Then $x + 2 =$ the middle integer and $x + 4 =$ the greatest integer.

   From the given information, we write the equation
   $$x(x + 4) = (x + 2) + 16.$$
   Solve this equation.
   $$x^2 + 4x = x + 18$$
   $$x^2 + 3x - 18 = 0$$
   $$(x + 6)(x - 3) = 0$$
   $$x + 6 = 0 \quad \text{or} \quad x - 3 = 0$$
   $$x = -6 \quad \text{or} \quad x = 3$$

   Reject -6 since it is not an odd integer. If $x = 3$, then $x + 2 = 5$ and $x + 4 = 7$. The integers are 3, 5, and 7.

4. Let $x =$ the length of the longer leg of the right triangle. Then $x + 3 =$ the length of the hypotenuse and $x - 3 =$ the length of the shorter leg.

Use the Pythagorean theorem, substituting $x - 3$ for a, x for b, and $x + 3$ for c.

$$a^2 + b^2 = c^2$$
$$(x - 3)^2 + x^2 = (x + 3)^2$$
$$x^2 - 6x + 9 + x^2 = x^2 + 6x + 9$$
$$2x^2 - 6x + 9 = x^2 + 6x + 9$$
$$x^2 - 12x = 0$$
$$x(x - 12) = 0$$

$$x = 0 \quad \text{or} \quad x - 12 = 0$$
$$x = 0 \quad \text{or} \quad x = 12$$

Discard 0 since the length of a side of a triangle cannot be zero. The length of the longer leg is 12 inches, the length of the hypotenuse is $x + 3 = 12 + 3 = 15$ inches, and the length of the shorter leg is $x - 3 = 12 - 3 = 9$ inches.

5. **(a)** $h = -16t^2 + 180t + 6$ ■ To find how long it will take for the ball to reach a height of 50 feet, let $h = 50$ and solve the resulting equation. (For convenience, we reverse the sides of the equation.)

$$-16t^2 + 180t + 6 = 50$$
$$-16t^2 + 180t - 44 = 0 \quad \textit{Standard form}$$
$$4t^2 - 45t + 11 = 0 \quad \textit{Divide by } -4.$$
$$(4t - 1)(t - 11) = 0 \quad \textit{Factor.}$$

$$4t - 1 = 0 \quad \text{or} \quad t - 11 = 0$$
$$t = \tfrac{1}{4} \quad \text{or} \quad t = 11$$

The ball will reach a height of 50 feet after $\tfrac{1}{4}$ second (on the way up) and 11 seconds (on the way down).

(b) $y = -x^2 + 2x + 60$ ■ To find the time when 45 impulses occur, let $y = 45$ and solve the resulting equation. (For convenience, we reverse the sides of the equation.)

$$-x^2 + 2x + 60 = 45$$
$$-x^2 + 2x + 15 = 0 \quad \textit{Subtract 45.}$$
$$-1(x^2 - 2x - 15) = 0 \quad \textit{Factor out } -1.$$
$$-1(x + 3)(x - 5) = 0 \quad \textit{Factor.}$$

$$x + 3 = 0 \quad \text{or} \quad x - 5 = 0$$
$$x = -3 \quad \text{or} \quad x = 5$$

There will be 45 impulses after 5 milliseconds. There are two solutions to the equation, -3 and 5. Only one answer makes sense here; a negative answer is not appropriate.

6. For 1990, $x = 1990 - 1930 = 60$.

$$y = 0.009665x^2 - 0.4942x + 15.12$$
$$y = 0.009665(60)^2 - 0.4942(60) + 15.12 \quad \textit{Let x = 60.}$$
$$= 20.262 \approx 20.3$$

According to the model, the foreign-born population of the United States in 1990, to the nearest tenth of a million, was about 20.3 million. The actual value from the table is 19.8 million, so our answer using the model is somewhat high.

14.7 Section Exercises

1. **Step 1** Read the problem carefully.

Step 2 Assign a variable to represent the unknown value.

Step 3 Write an equation using the variable expression(s).

Step 4 Solve the equation.

Step 5 State the answer.

Step 6 Check the answer in the words of the original problem.

3. $A = bh$; $A = 45$, $b = 2x + 1$, $h = x + 1$

Step 3 $A = bh$
$$45 = (2x + 1)(x + 1)$$

Step 4 Solve the equation.

$$45 = 2x^2 + 3x + 1$$
$$0 = 2x^2 + 3x - 44$$
$$0 = (2x + 11)(x - 4)$$

$$2x + 11 = 0 \quad \text{or} \quad x - 4 = 0$$
$$2x = -11$$
$$x = -\tfrac{11}{2} \quad \text{or} \quad x = 4$$

Step 5 Substitute these values for x in the expressions $2x + 1$ and $x + 1$ to find the values of b and h.

$$b = 2x + 1 = 2(-\tfrac{11}{2}) + 1$$
$$= -11 + 1 = -10$$
$$\text{or} \quad b = 2x + 1 = 2(4) + 1$$
$$= 8 + 1 = 9$$

We must discard the first solution because the base of a parallelogram cannot have a negative length. Since $x = -\tfrac{11}{2}$ will not give a realistic answer for the base, we only need to substitute 4 for x to compute the height.

$$h = x + 1 = 4 + 1 = 5$$

The base is 9 units and the height is 5 units.

Step 6 $bh = 9 \cdot 5 = 45$, the desired value of A.

5. $V = lwh;\ V = 192,\ l = x + 2,\ w = x,\ h = 4$

Step 3 $V = lwh$

$$192 = (x + 2)(x)(4)$$
$$192 = 4x(x + 2)$$

Step 4   Solve the equation.

$$192 = 4x^2 + 8x$$
$$0 = 4x^2 + 8x - 192$$
$$0 = 4(x^2 + 2x - 48)$$
$$0 = 4(x - 6)(x + 8)$$

$$4 = 0 \quad \text{or} \quad x - 6 = 0 \quad \text{or} \quad x + 8 = 0$$
$$x = 6 \quad \text{or} \quad x = -8$$

Step 5   Since x represents the width of the box, which cannot be negative, we reject -8 as a value for x. Substitute 6 for x to find l.

$$l = x + 2 = 6 + 2 = 8$$

The length of the box is 8 units and the width is 6 units.

Step 6 $lwh = 8 \cdot 6 \cdot 4 = 192$, the desired value of V.

7.   **Step 2**   Let $x = $ the width of the case. Then $x + 2 = $ the length of the case.

Step 3 $A = lw$

$$168 = (x + 2)x \quad Substitute.$$

Step 4 $168 = x^2 + 2x \quad Multiply.$

$$x^2 + 2x - 168 = 0 \qquad Standard\ form$$
$$(x + 14)(x - 12) = 0 \qquad Factor.$$

Use the zero-factor property to solve for x.

$$x + 14 = 0 \quad \text{or} \quad x - 12 = 0$$
$$x = -14 \quad \text{or} \quad x = 12$$

Step 5   Because a width cannot be negative, $x = 12$ and $x + 2 = 14$. The width is 12 cm and the length is 14 cm.

Step 6   The length is 2 cm more than the width and the area is $14(12) = 168$ cm^2, as required.

9.   **Step 2**   Let $h = $ the height of the triangle. Then $2h + 2 = $ the base of the triangle.

Step 3   The area of the triangle is 30 in.^{2}.

$$A = \tfrac{1}{2}bh$$
$$30 = \tfrac{1}{2}(2h + 2) \cdot h$$

Step 4 $60 = (2h + 2)h$

$$60 = 2h^2 + 2h$$
$$0 = 2h^2 + 2h - 60$$
$$0 = 2(h^2 + h - 30)$$
$$0 = 2(h + 6)(h - 5)$$

$$h + 6 = 0 \qquad \text{or} \qquad h - 5 = 0$$
$$h = -6 \quad \text{or} \quad h = 5$$

Step 5   The solution $h = -6$ must be discarded since a triangle cannot have a negative height. Thus,

$$h = 5 \text{ and } 2h + 2 = 2(5) + 2 = 12.$$

The height is 5 inches, and the base is 12 inches.

Step 6 $\tfrac{1}{2}bh = \tfrac{1}{2} \cdot 12 \cdot 5 = 30$, which is the desired value of A.

11.   **Step 2**   Let $x = $ the width of the monitor. Then $x + 3 = $ the length of the monitor. The area is $x(x + 3)$.

Step 3   If the length were doubled $[2(x + 3)]$ and if the width were decreased by 1 in. $[x - 1]$, the area would be increased by 150 in.^{2} $[x(x + 3) + 150]$. Write an equation.

$$lw = A$$
$$[2(x + 3)](x - 1) = x(x + 3) + 150$$

Step 4 $(2x + 6)(x - 1) = x^2 + 3x + 150$

$$2x^2 + 4x - 6 = x^2 + 3x + 150$$
$$x^2 + x - 156 = 0$$
$$(x + 13)(x - 12) = 0$$

$$x + 13 = 0 \qquad \text{or} \qquad x - 12 = 0$$
$$x = -13 \quad \text{or} \quad x = 12$$

Step 5   Reject -13, so the width is 12 inches and the length is $12 + 3 = 15$ inches.

Step 6   The length is 3 inches more than the width. The area of the monitor is $15(12) = 180$ in.^{2}. Doubling the length and decreasing the width by 1 inch gives us an area of $30(11) = 330$ in.^{2}, which is 150 in.^{2} more than the area of the original monitor, as required.

13.   Let $x = $ the width of the aquarium. Then $x + 3 = $ the height of the aquarium.

Use the formula for the volume of a rectangular box.

$$V = lwh$$
$$2730 = 21x(x + 3)$$
$$130 = x(x + 3) \qquad Divide\ by\ 21.$$
$$130 = x^2 + 3x$$
$$0 = x^2 + 3x - 130$$
$$0 = (x + 13)(x - 10)$$

$$x + 13 = 0 \qquad \text{or} \qquad x - 10 = 0$$
$$x = -13 \quad \text{or} \quad x = 10$$

We discard -13 because the width cannot be negative. The width is 10 inches. The height is $10 + 3 = 13$ inches.

15. Let x = the length of a side of the square painting. Then $x - 2$ = the length of a side of the square mirror.

Since the formula for the area of a square is $A = s^2$, the area of the painting is x^2, and the area of the mirror is $(x - 2)^2$. The difference between their areas is 32, so

$$x^2 - (x - 2)^2 = 32$$
$$x^2 - (x^2 - 4x + 4) = 32$$
$$x^2 - x^2 + 4x - 4 = 32$$
$$4x - 4 = 32$$
$$4x = 36$$
$$x = 9.$$

The length of a side of the painting is 9 feet.
The length of a side of the mirror is $9 - 2 = 7$ feet.

Check $9^2 - 7^2 = 81 - 49 = 32$

17. Let x = the first volume number.
Then $x + 1$ = the second volume number.
The product of the numbers is 420.

$$x(x + 1) = 420$$
$$x^2 + x - 420 = 0$$
$$(x - 20)(x + 21) = 0$$

$$x - 20 = 0 \quad \text{or} \quad x + 21 = 0$$
$$x = 20 \quad \text{or} \quad x = -21$$

The volume number cannot be negative, so we reject -21. The volume numbers are 20 and $x + 1 = 20 + 1 = 21$.

19. Let n = the first integer.
Then $n + 1$ = the next integer.

$$
\begin{array}{ccccc}
\text{The} & & \text{more} & & \text{their} \\
\text{product} & \text{is 11} & \text{than} & & \text{sum.} \\
\downarrow & \downarrow\downarrow & \downarrow & & \downarrow
\end{array}
$$
$$n(n + 1) = 11 + [n + (n + 1)]$$
$$n^2 + n = 11 + n + n + 1$$
$$n^2 + n = 2n + 12$$
$$n^2 - n - 12 = 0$$
$$(n - 4)(n + 3) = 0$$

$$n - 4 = 0 \quad \text{or} \quad n + 3 = 0$$
$$n = 4 \quad \text{or} \quad n = -3$$

If $n = 4$, then $n + 1 = 5$.
If $n = -3$, then $n + 1 = -2$.
The two integers are 4 and 5, or -3 and -2.

21. Let x = the lesser odd integer.
Then $x + 2$ = the greater odd integer.
Their product is 15 more than three times their sum.

$$x(x + 2) = 3(x + x + 2) + 15$$
$$x^2 + 2x = 6x + 6 + 15$$
$$x^2 - 4x - 21 = 0$$
$$(x - 7)(x + 3) = 0$$

$$x - 7 = 0 \quad \text{or} \quad x + 3 = 0$$
$$x = 7 \quad \text{or} \quad x = -3$$

The two integers are 7 and $7 + 2 = 9$ or -3 and $-3 + 2 = -1$.

Check $7(9) = 3(7 + 9) + 15$; $63 = 63$
$-3(-1) = 3(-3 - 1) + 15$; $3 = 3$

23. Let n = the least even integer. Then $n + 2$ and $n + 4$ are the next two even integers.

The sum of the squares of the lesser two is equal to the square of the greatest.

$$n^2 + (n + 2)^2 = (n + 4)^2$$
$$n^2 + n^2 + 4n + 4 = n^2 + 8n + 16$$
$$n^2 - 4n - 12 = 0$$
$$(n - 6)(n + 2) = 0$$

$$n - 6 = 0 \quad \text{or} \quad n + 2 = 0$$
$$n = 6 \quad \text{or} \quad n = -2$$

If $n = 6$, $n + 2 = 8$, and $n + 4 = 10$.
If $n = -2$, $n + 2 = 0$, and $n + 4 = 2$.

The three integers are 6, 8, and 10 or -2, 0, and 2.

25. Let x = the first odd integer.
Then $x + 2$ = the second odd integer and $x + 4$ = the third odd integer.

3 times the sum of all three is 18 more than the product of the first and second integers.

$$3[x + (x + 2) + (x + 4)] = x(x + 2) + 18$$
$$3(3x + 6) = x^2 + 2x + 18$$
$$9x + 18 = x^2 + 2x + 18$$
$$0 = x^2 - 7x$$
$$0 = x(x - 7)$$

$$x = 0 \quad \text{or} \quad x - 7 = 0$$
$$x = 7$$

We must discard 0 because it is even and the problem requires the integers to be odd. If $x = 7$, $x + 2 = 9$, and $x + 4 = 11$. The three integers are 7, 9, and 11.

27. Let x = the length of the longer leg of the right triangle. Then $x + 1$ = the length of the hypotenuse and $x - 7$ = the length of the shorter leg.

Refer to the figure in the text. Use the Pythagorean theorem with $a = x, b = x - 7$, and $c = x + 1$.

$$a^2 + b^2 = c^2$$
$$x^2 + (x - 7)^2 = (x + 1)^2$$
$$x^2 + (x^2 - 14x + 49) = x^2 + 2x + 1$$
$$2x^2 - 14x + 49 = x^2 + 2x + 1$$
$$x^2 - 16x + 48 = 0$$
$$(x - 12)(x - 4) = 0$$

$$x - 12 = 0 \quad \text{or} \quad x - 4 = 0$$
$$x = 12 \quad \text{or} \quad x = 4$$

Discard 4 because if the length of the longer leg is 4 centimeters, by the conditions of the problem, the length of the shorter leg would be $4 - 7 = -3$ centimeters, which is impossible. The length of the longer leg is 12 centimeters.

Check $12^2 + 5^2 = 13^2; 169 = 169$

29. Let $x =$ Denny's distance from home. Then $x + 1 =$ the distance between Terri and Denny. Terri is 5 miles from home.

Refer to the diagram in the textbook. Use the Pythagorean theorem.

$$a^2 + b^2 = c^2$$
$$x^2 + 5^2 = (x + 1)^2$$
$$x^2 + 25 = x^2 + 2x + 1$$
$$24 = 2x$$
$$12 = x$$

Denny is 12 miles from home.

Check $12^2 + 5^2 = 13^2; 169 = 169$

31. Let $x =$ the length of the ladder. Then $x - 4 =$ the distance from the bottom of the ladder to the building and $x - 2 =$ the distance on the side of the building to the top of the ladder.

Substitute into the Pythagorean theorem.

$$a^2 + b^2 = c^2$$
$$(x - 2)^2 + (x - 4)^2 = x^2$$
$$x^2 - 4x + 4 + x^2 - 8x + 16 = x^2$$
$$x^2 - 12x + 20 = 0$$
$$(x - 10)(x - 2) = 0$$

$$x - 10 = 0 \quad \text{or} \quad x - 2 = 0$$
$$x = 10 \quad \text{or} \quad x = 2$$

The solution cannot be 2 because then a negative distance results. Thus, $x = 10$ and the top of the ladder reaches $x - 2 = 10 - 2 = 8$ feet up the side of the building.

Check $8^2 + 6^2 = 10^2; 100 = 100$

33. (a) Let $h = 64$ in the given formula and solve for t.

$$h = -16t^2 + 32t + 48$$
$$64 = -16t^2 + 32t + 48$$
$$16t^2 - 32t + 16 = 0$$
$$16(t^2 - 2t + 1) = 0$$
$$16(t - 1)^2 = 0$$
$$t - 1 = 0$$
$$t = 1$$

The height of the object will be 64 feet after 1 second.

(b) To find the time when the height is 60 feet, let $h = 60$ in the given equation and solve for t.

$$h = -16t^2 + 32t + 48$$
$$60 = -16t^2 + 32t + 48$$
$$16t^2 - 32t + 12 = 0$$
$$4(4t^2 - 8t + 3) = 0$$
$$4(2t - 1)(2t - 3) = 0$$

$$2t - 1 = 0 \quad \text{or} \quad 2t - 3 = 0$$
$$2t = 1 \qquad\qquad 2t = 3$$
$$t = \tfrac{1}{2} \quad \text{or} \qquad t = \tfrac{3}{2}$$

The height of the object is 60 feet after $\frac{1}{2}$ second (on the way up) and after $\frac{3}{2}$ or $1\frac{1}{2}$ seconds (on the way down).

(c) To find the time when the object hits the ground, let $h = 0$ and solve for t.

$$h = -16t^2 + 32t + 48$$
$$0 = -16t^2 + 32t + 48$$
$$16t^2 - 32t - 48 = 0$$
$$16(t^2 - 2t - 3) = 0$$
$$16(t + 1)(t - 3) = 0$$

$$t + 1 = 0 \quad \text{or} \quad t - 3 = 0$$
$$t = -1 \quad \text{or} \qquad t = 3$$

We discard -1 because time cannot be negative. The object will hit the ground after 3 seconds.

(d) The negative solution, -1, does not make sense, since t represents time, which cannot be negative.

35. $h = 128t - 16t^2$
$h = 128(1) - 16(1)^2$ *Let t = 1.*
$= 128 - 16$
$= 112$

After 1 second, the height is 112 feet.

37. $h = 128t - 16t^2$
$h = 128(4) - 16(4)^2$ *Let t = 4.*
$= 512 - 256$
$= 256$

After 4 seconds, the height is 256 feet.

39. (a) $x = 2000 - 1994 = 6$ for the year 2000.

$$y = 0.347x^2 + 13.14x + 17.98$$
$$y = 0.347(6)^2 + 13.14(6) + 17.98$$
$$y = 109.312 \approx 109$$

The model indicates there were about 109 million cellular phone subscribers in 2000. The result using the model is the same as the actual number from the table for 2000.

(b) $x = 2008 - 1994 = 14$
$x = 14$ corresponds to 2008.

(c) $y = 0.347(14)^2 + 13.14(14) + 17.98$
$y = 269.952 \approx 270$

The model indicates there were about 270 million cellular phone subscribers in 2008. The result using the model is less than 286 million, the actual number for 2008.

(d) $x = 2013 - 1994 = 19$
$y = 0.347(19)^2 + 13.14(19) + 17.98$
$y = 392.907 \approx 393$

The model gives an estimate of 393 million cellular phone subscribers in 2013.

Chapter 14 Review Exercises

1. $15t + 45$
$= 15 \cdot t + 15 \cdot 3$
$= 15(t + 3)$ *GCF = 15*

2. $60z^3 + 30z$
$= 30z \cdot 2z^2 + 30z \cdot 1$
$= 30z(2z^2 + 1)$ *GCF = 30z*

3. $44x^3 + 55x^2$
$= 11x^2(4x) + 11x^2(5)$
$= 11x^2(4x + 5)$ *GCF = 11x²*

4. $100m^2n^3 - 50m^3n^4 + 150m^2n^2$
$= 50m^2n^2(2n)$
$+ 50m^2n^2(-mn^2)$
$+ 50m^2n^2(3)$
$= 50m^2n^2(2n - mn^2 + 3)$ *GCF = 50m²n²*

5. $2xy - 8y + 3x - 12$
$= (2xy - 8y) + (3x - 12)$ *Group terms.*
$= 2y(x - 4) + 3(x - 4)$ *Factor each group.*
$= (x - 4)(2y + 3)$ *Factor out x − 4.*

6. $6y^2 + 9y + 4xy + 6x$
$= (6y^2 + 9y) + (4xy + 6x)$ *Group terms.*
$= 3y(2y + 3) + 2x(2y + 3)$ *Factor each group.*
$= (2y + 3)(3y + 2x)$ *Factor out 2y + 3.*

7. $x^2 + 10x + 21$ ▪ Look for a pair of integers whose product is 21 and whose sum is 10. The integers are 3 and 7, so
$x^2 + 10x + 21 = (x + 3)(x + 7)$.

8. $y^2 - 13y + 40$ ▪ Find two integers whose product is 40 and whose sum is −13.

Factors of 40	Sums of factors
−1, −40	−41
−2, −20	−22
−4, −10	−14
−5, −8	−13 ←

The integers are −5 and −8, so
$y^2 - 13y + 40 = (y - 5)(y - 8)$.

9. $q^2 + 6q - 27$ ▪ Look for a pair of integers whose product is −27 and whose sum is 6. The integers are −3 and 9, so
$q^2 + 6q - 27 = (q - 3)(q + 9)$.

10. $r^2 - r - 56$ ▪ Look for a pair of integers whose product is −56 and whose sum is −1. The integers are 7 and −8, so
$r^2 - r - 56 = (r + 7)(r - 8)$.

11. $x^2 + x + 1$ ▪ There is no GCF. There is no pair of integers whose product is 1 and whose sum is 1, so $x^2 + x + 1$ is *prime*.

12. $3x^2 + 6x + 6 = 3(x^2 + 2x + 2)$

There is no pair of integers whose product is 2 and whose sum is 2, so $x^2 + 2x + 2$ is prime. Thus, the completely factored form is
$3x^2 + 6x + 6 = 3(x^2 + 2x + 2)$.

13. $r^2 - 4rs - 96s^2$ ▪ Find two expressions whose product is −96s² and whose sum is −4s. The expressions are 8s and −12s, so
$r^2 - 4rs - 96s^2 = (r + 8s)(r - 12s)$.

14. $p^2 + 2pq - 120q^2$ ▪ Find two expressions whose product is −120q² and whose sum is 2q. The expressions are 12q and −10q, so
$p^2 + 2pq - 120q^2 = (p + 12q)(p - 10q)$.

15. $-8p^3 + 24p^2 + 80p$

First, factor out the GCF, −8p.
$-8p^3 + 24p^2 + 80p = -8p(p^2 - 3p - 10)$

Now factor $p^2 - 3p - 10$.
$p^2 - 3p - 10 = (p + 2)(p - 5)$

The completely factored form is
$-8p^3 + 24p^2 + 80p = -8p(p + 2)(p - 5)$.

16. $3x^4 + 30x^3 + 48x^2$
$= 3x^2(x^2 + 10x + 16)$ *GCF = 3x²*
$= 3x^2(x + 2)(x + 8)$ *Factor.*

17. $m^2 - 3mn - 18n^2$ ▪ Find two expressions whose product is −18n² and whose sum is −3n. The expressions are 3n and −6n, so
$m^2 - 3mn - 18n^2 = (m + 3n)(m - 6n)$.

18. $y^2 - 8yz + 15z^2$ ▪ Find two expressions whose product is 15z² and whose sum is −8z. The expressions are −3z and −5z, so
$y^2 - 8yz + 15z^2 = (y - 3z)(y - 5z)$.

19. $p^7 - p^6q - 2p^5q^2$

$\qquad = p^5(p^2 - pq - 2q^2)$ *GCF = p^5*

$\qquad = p^5(p + q)(p - 2q)$ *Factor.*

20. $-3r^5 + 6r^4s + 45r^3s^2$

$\qquad = -3r^3(r^2 - 2rs - 15s^2)$ *GCF = $-3r^3$*

$\qquad = -3r^3(r + 3s)(r - 5s)$ *Factor.*

21. To begin factoring $6r^2 - 5r - 6$, the possible first terms of the two binomial factors are r and $6r$, or $2r$ and $3r$, if we consider only positive integer coefficients.

22. When factoring $2z^3 + 9z^2 - 5z$, the first step is to factor out the GCF, z.

In Exercises 23–34, either the trial and error method or the grouping method can be used to factor each polynomial.

23. Factor $2k^2 - 5k + 2$ by trial and error.

$\qquad 2k^2 - 5k + 2 = (2k - 1)(k - 2)$

24. $3r^2 + 11r - 4$ ▪ Factor by grouping. Look for two integers whose product is $3(-4) = -12$ and whose sum is 11. The integers are 12 and -1.

$\qquad 3r^2 + 11r - 4$

$\qquad = 3r^2 + 12r - r - 4$

$\qquad = (3r^2 + 12r) + (-r - 4)$

$\qquad = 3r(r + 4) - 1(r + 4)$

$\qquad = (r + 4)(3r - 1)$

25. $6r^2 - 5r - 6$ ▪ Factor by grouping. Find two integers whose product is $6(-6) = -36$ and whose sum is -5. The integers are -9 and 4.

$\qquad 6r^2 - 5r - 6$

$\qquad = 6r^2 - 9r + 4r - 6$

$\qquad = (6r^2 - 9r) + (4r - 6)$

$\qquad = 3r(2r - 3) + 2(2r - 3)$

$\qquad = (2r - 3)(3r + 2)$

26. Factor $10z^2 - 3z - 1$ by trial and error.

$\qquad 10z^2 - 3z - 1 = (5z + 1)(2z - 1)$

27. $5t^2 - 11t + 12$ is *prime*.

28. $24x^5 - 20x^4 + 4x^3$

$\qquad = 4x^3(6x^2 - 5x + 1)$ *GCF = $4x^3$*

$\qquad = 4x^3(3x - 1)(2x - 1)$ *Factor.*

29. $-6x^2 + 3x + 30$

$\qquad = -3(2x^2 - x - 10)$ *GCF = -3*

$\qquad = -3(2x - 5)(x + 2)$ *Factor.*

30. $10r^3s + 17r^2s^2 + 6rs^3$

$\qquad = rs(10r^2 + 17rs + 6s^2)$ *GCF = rs*

$\qquad = rs(5r + 6s)(2r + s)$ *Factor.*

31. $-30y^3 - 5y^2 + 10y$

$\qquad = -5y(6y^2 + y - 2)$ *GCF = $-5y$*

$\qquad = -5y(3y + 2)(2y - 1)$ *Factor.*

32. $4z^2 - 5z + 7$ is *prime*.

33. $-3m^3n + 19m^2n + 40mn$

$\qquad = -mn(3m^2 - 19m - 40)$ *GCF = $-mn$*

$\qquad = -mn(3m + 5)(m - 8)$ *Factor.*

34. $14a^2 - 27ab - 20b^2$ ▪ Factor by grouping. Find two integers whose product is $14(-20) = -280$ and whose sum is -27. The integers are -35 and 8.

$\qquad 14a^2 - 27ab - 20b^2$

$\qquad = 14a^2 - 35ab + 8ab - 20b^2$

$\qquad = (14a^2 - 35ab) + (8ab - 20b^2)$

$\qquad = 7a(2a - 5b) + 4b(2a - 5b)$

$\qquad = (2a - 5b)(7a + 4b)$

35. Only choice **B**, $4x^2y^2 - 25z^2$, is the difference of squares. In **A**, 32 is not a perfect square. In **C**, we have a sum, not a difference. In **D**, y^3 is not a square. The correct choice is **B**.

36. Only choice **D**, $x^2 - 20x + 100$, is a perfect square trinomial because $x^2 = x \cdot x$, $100 = 10 \cdot 10$, and $-20x = -2(x)(10)$.

In Exercises 37–40, use the rule for factoring a difference of two squares, if possible.

37. $n^2 - 64$

$\qquad = n^2 - 8^2$

$\qquad = (n + 8)(n - 8)$ *Difference of squares*

38. $25b^2 - 121$

$\qquad = (5b)^2 - 11^2$

$\qquad = (5b + 11)(5b - 11)$ *Difference of squares*

39. $49y^2 - 25w^2$

$\qquad = (7y)^2 - (5w)^2$

$\qquad = (7y + 5w)(7y - 5w)$ *Difference of squares*

40. $144p^2 - 36q^2$

$\qquad = 36(4p^2 - q^2)$ *GCF = 36*

$\qquad = 36\left[(2p)^2 - q^2\right]$

$\qquad = 36(2p + q)(2p - q)$ *Difference of squares*

41. $x^2 + 100$ ▪ This polynomial is *prime* because it is the sum of squares and the two terms have no common factor.

In Exercises 42–46, use the rules for factoring a perfect square trinomial.

42. $z^2 + 10z + 25 = z^2 + 2(5)(z) + 5^2$

$\qquad\qquad\qquad\qquad = (z + 5)^2$

43. $r^2 - 12r + 36 = r^2 - 2(6)(r) + 6^2$

$\qquad\qquad\qquad\qquad = (r - 6)^2$

44. $9t^2 - 42t + 49 = (3t)^2 - 2(3t)(7) + 7^2$
$$= (3t - 7)^2$$

45. $16m^2 + 40mn + 25n^2$
$$= (4m)^2 + 2(4m)(5n) + (5n)^2$$
$$= (4m + 5n)^2$$

46. $54x^3 - 72x^2 + 24x$
$$= 6x(9x^2 - 12x + 4) \qquad GCF = 6x$$
$$= 6x[(3x)^2 - 2(3x)(2) + 2^2]$$
$$= 6x(3x - 2)^2$$

In Exercises 47–64, all solutions should be checked by substituting in the original equations. The checks will not be shown here.

47. $(4t + 3)(t - 1) = 0$

To solve, use the zero-factor property.

$$4t + 3 = 0 \qquad \text{or} \quad t - 1 = 0$$
$$4t = -3$$
$$t = -\tfrac{3}{4} \quad \text{or} \qquad t = 1$$

The solution set is $\{-\tfrac{3}{4}, 1\}$.

48. $(x + 7)(x - 4)(x + 3) = 0$

To solve, use the zero-factor property.

$$x + 7 = 0 \quad \text{or} \quad x - 4 = 0 \quad \text{or} \quad x + 3 = 0$$
$$x = -7 \quad \text{or} \qquad x = 4 \quad \text{or} \qquad x = -3$$

The solution set is $\{-7, -3, 4\}$.

49. $x(2x - 5) = 0$

To solve, use the zero-factor property.

$$x = 0 \quad \text{or} \quad 2x - 5 = 0$$
$$2x = 5$$
$$x = \tfrac{5}{2}$$

The solution set is $\{0, \tfrac{5}{2}\}$.

50. $z^2 + 4z + 3 = 0$
$$(z + 3)(z + 1) = 0 \qquad Factor.$$

$$z + 3 = 0 \qquad \text{or} \quad z + 1 = 0$$
$$z = -3 \qquad \text{or} \qquad z = -1$$

The solution set is $\{-3, -1\}$.

51. $m^2 - 5m + 4 = 0$
$$(m - 1)(m - 4) = 0 \qquad Factor.$$

$$m - 1 = 0 \qquad \text{or} \quad m - 4 = 0$$
$$m = 1 \qquad \text{or} \qquad m = 4$$

The solution set is $\{1, 4\}$.

52. $x^2 = -15 + 8x$
$$x^2 - 8x + 15 = 0 \qquad Standard\ form$$
$$(x - 3)(x - 5) = 0 \qquad Factor.$$

$$x - 3 = 0 \qquad \text{or} \qquad x - 5 = 0$$
$$x = 3 \qquad \text{or} \qquad x = 5$$

The solution set is $\{3, 5\}$.

53. $3z^2 - 11z - 20 = 0$
$$(3z + 4)(z - 5) = 0 \qquad Factor.$$

$$3z + 4 = 0 \qquad \text{or} \qquad z - 5 = 0$$
$$3z = -4$$
$$z = -\tfrac{4}{3} \qquad \text{or} \qquad z = 5$$

The solution set is $\{-\tfrac{4}{3}, 5\}$.

54. $81t^2 - 64 = 0$
$$(9t + 8)(9t - 8) = 0 \qquad Factor.$$

$$9t + 8 = 0 \qquad \text{or} \qquad 9t - 8 = 0$$
$$9t = -8 \qquad\qquad 9t = 8$$
$$t = -\tfrac{8}{9} \qquad \text{or} \qquad t = \tfrac{8}{9}$$

The solution set is $\{-\tfrac{8}{9}, \tfrac{8}{9}\}$.

55. $y^2 = 8y$
$$y^2 - 8y = 0 \qquad Standard\ form$$
$$y(y - 8) = 0 \qquad Factor.$$

$$y = 0 \qquad \text{or} \qquad y - 8 = 0$$
$$y = 0 \qquad \text{or} \qquad y = 8$$

The solution set is $\{0, 8\}$.

56. $n(n - 5) = 6$
$$n^2 - 5n = 6 \qquad Multiply.$$
$$n^2 - 5n - 6 = 0 \qquad Standard\ form$$
$$(n + 1)(n - 6) = 0 \qquad Factor.$$

$$n + 1 = 0 \qquad \text{or} \qquad n - 6 = 0$$
$$n = -1 \qquad \text{or} \qquad n = 6$$

The solution set is $\{-1, 6\}$.

57. $t^2 - 14t + 49 = 0$
$$(t - 7)^2 = 0 \qquad Factor.$$

Set the factor $t - 7$ equal to 0 and solve.

$$t - 7 = 0$$
$$t = 7$$

The solution set is $\{7\}$.

58. $t^2 = 12(t - 3)$
$$t^2 = 12t - 36 \qquad Multiply.$$
$$t^2 - 12t + 36 = 0 \qquad Standard\ form$$
$$(t - 6)^2 = 0 \qquad Factor.$$

Set the factor $t - 6$ equal to 0 and solve.

$$t - 6 = 0$$
$$t = 6$$

The solution set is $\{6\}$.

59. $(5z+2)(z^2+3z+2)=0$

$(5z+2)(z+2)(z+1)=0$ *Factor.*

$5z+2=0$ or $z+2=0$ or $z+1=0$

$5z=-2$

$z=-\frac{2}{5}$ or $z=-2$ or $z=-1$

The solution set is $\{-\frac{2}{5}, -2, -1\}$.

60. $x^2=9$

$x^2-9=0$

$(x+3)(x-3)=0$ *Factor.*

$x+3=0$ or $x-3=0$

$x=-3$ or $x=3$

The solution set is $\{-3, 3\}$.

61. $64x^3-9x=0$

$x(64x^2-9)=0$ *Factor out x.*

$x(8x+3)(8x-3)=0$ *Factor.*

$x=0$ or $8x+3=0$ or $8x-3=0$

$8x=-3$ $8x=3$

$x=0$ or $x=-\frac{3}{8}$ or $x=\frac{3}{8}$

The solution set is $\{-\frac{3}{8}, 0, \frac{3}{8}\}$.

62. $(2r+1)(12r^2+5r-3)=0$

$(2r+1)(4r+3)(3r-1)=0$ *Factor.*

$2r+1=0$ or $4r+3=0$ or $3r-1=0$

$2r=-1$ $4r=-3$ $3r=1$

$r=-\frac{1}{2}$ or $r=-\frac{3}{4}$ or $r=\frac{1}{3}$

The solution set is $\{-\frac{3}{4}, -\frac{1}{2}, \frac{1}{3}\}$.

63. $25w^2-90w+81=0$

$(5w-9)^2=0$ *Factor.*

Set the factor $5w-9$ equal to 0 and solve.

$5w-9=0$

$w=\frac{9}{5}$

The solution set is $\{\frac{9}{5}\}$.

64. $r(r-7)=30$

$r^2-7r=30$ *Multiply.*

$r^2-7r-30=0$ *Standard form*

$(r+3)(r-10)=0$ *Factor.*

$r+3=0$ or $r-10=0$

$r=-3$ or $r=10$

The solution set is $\{-3, 10\}$.

65. Let $x=$ the width of the rug.

Then $x+6=$ the length of the rug.

The area of the rug is 40 ft^2.

$A=lw$

$40=(x+6)x$

$40=x^2+6x$

$0=x^2+6x-40$

$0=(x+10)(x-4)$

$x+10=0$ or $x-4=0$

$x=-10$ or $x=4$

Reject -10 since the width cannot be negative. The width of the rug is 4 feet and the length is $4+6$ or 10 feet.

66. From the figure, we have $l=20$, $w=x$, and $h=x+4$.

$S=2wh+2wl+2lh$

$650=2x(x+4)+2x(20)+2(20)(x+4)$

$650=2x^2+8x+40x+40(x+4)$

$650=2x^2+48x+40x+160$

$0=2x^2+88x-490$

$0=2(x^2+44x-245)$

$0=2(x+49)(x-5)$

$x+49=0$ or $x-5=0$

$x=-49$ or $x=5$

Reject -49 because the width cannot be negative. The width of the chest is 5 feet.

67. Let $x=$ the first integer.

Then $x+1=$ the next integer.

The product of the integers is 29 more than their sum, so

$x(x+1)=29+[x+(x+1)]$.

Solve this equation.

$x^2+x=29+2x+1$

$x^2-x-30=0$

$(x-6)(x+5)=0$

$x-6=0$ or $x+5=0$

$x=6$ or $x=-5$

If $x=6$, $x+1=6+1=7$.

If $x=-5$, $x+1=-5+1=-4$.

The consecutive integers are 6 and 7 or -5 and -4.

68. Let $x = $ the least integer. Then $x + 1$ and $x + 2$ are the next two greater integers.

The product of the lesser two of three consecutive integers is equal to 23 plus the greatest.

$$x(x + 1) = 23 + (x + 2)$$
$$x^2 + x = 23 + x + 2$$
$$x^2 - 25 = 0$$
$$(x + 5)(x - 5) = 0$$

$$x + 5 = 0 \quad \text{or} \quad x - 5 = 0$$
$$x = -5 \quad \text{or} \quad x = 5$$

If $x = -5$, then $x + 1 = -4$ and $x + 2 = -3$.
If $x = 5$, then $x + 1 = 6$ and $x + 2 = 7$.
The integers are -5, -4, and -3, or 5, 6, and 7.

69. Let $x = $ the distance traveled west. Then $x - 14 = $ the distance traveled south, and $(x - 14) + 16 = x + 2 = $ the distance between the cars.

These three distances form a right triangle with x and $x - 14$ representing the lengths of the legs and $x + 2$ representing the length of the hypotenuse. Use the Pythagorean theorem.

$$a^2 + b^2 = c^2$$
$$x^2 + (x - 14)^2 = (x + 2)^2$$
$$x^2 + x^2 - 28x + 196 = x^2 + 4x + 4$$
$$x^2 - 32x + 192 = 0$$
$$(x - 8)(x - 24) = 0$$

$$x - 8 = 0 \quad \text{or} \quad x - 24 = 0$$
$$x = 8 \quad \text{or} \quad x = 24$$

If $x = 8$, then $x - 14 = -6$, which is not possible because a distance cannot be negative.

If $x = 24$, then $x - 14 = 10$ and $x + 2 = 26$.
The cars were 26 miles apart.

70. Let $b = $ the length of the base of the sail. Then $b + 4 = $ the height of the sail.

Use the formula for the area of a triangle with $A = 30$ m^2.

$$A = \tfrac{1}{2}bh$$
$$30 = \tfrac{1}{2}(b)(b + 4)$$
$$60 = b^2 + 4b$$
$$0 = b^2 + 4b - 60$$
$$0 = (b + 10)(b - 6)$$

$$b + 10 = 0 \quad \text{or} \quad b - 6 = 0$$
$$b = -10 \quad \text{or} \quad b = 6$$

Discard -10 since the base of a triangle cannot be negative. The length of the base of the triangular sail is 6 meters.

71. Let $x = $ the width of the house. Then $x + 7 = $ the length of the house.

Use $A = lw$ with 170 for A, $x + 7$ for l, and x for w.

$$170 = (x + 7)(x)$$
$$170 = x^2 + 7x$$
$$0 = x^2 + 7x - 170$$
$$0 = (x + 17)(x - 10)$$

$$x + 17 = 0 \quad \text{or} \quad x - 10 = 0$$
$$x = -17 \quad \text{or} \quad x = 10$$

Discard -17 because the width cannot be negative. If $x = 10$, $x + 7 = 10 + 7 = 17$.

The width is 10 meters and the length is 17 meters.

72. $d = 16t^2$

(a) $t = 4$: $d = 16(4)^2 = 256$

In 4 seconds, the object would fall 256 feet.

(b) $t = 8$: $d = 16(8)^2 = 1024$

In 8 seconds, the object would fall 1024 feet.

73. $x = 2005 - 2000 = 5$ for the year 2005.

$$y = 1.57x^2 + 32.5x + 400$$
$$= 1.57(5)^2 + 32.5(5) + 400$$
$$= 601.75 \approx 602$$

If $x = 5$, then $y = 602$, so the prediction is 602,000 vehicles. The result is a little higher than the 592 thousand given in the table.

74. $x = 2010 - 2000 = 10$ for the year 2010.

$$y = 1.57x^2 + 32.5x + 400$$
$$= 1.57(10)^2 + 32.5(10) + 400$$
$$= 882$$

If $x = 10$, then $y = 882$, so the prediction is 882,000 vehicles. The estimate may be unreliable because the conditions that prevailed in the years 2001–2009 may have changed, causing either a greater increase or a greater decrease predicted by the model for the number of alternative-fueled vehicles.

75. **[14.1]** **D** is not factored completely since the expression $3(7t + 4) + x(7t + 4)$ is a *sum*, not a *product*. The complete factorization is $3(7t + 4) + x(7t + 4) = (7t + 4)(3 + x)$.

76. **[14.1]** The terms of $2x + 8$ have a common factor of 2. The completely factored form is $2(x + 4)(3x - 4)$.

77. **[14.3]** $15m^2 + 20mp - 12m - 16p$
$$= (15m^2 + 20mp) + (-12m - 16p)$$
$$= 5m(3m + 4p) - 4(3m + 4p)$$
$$= (3m + 4p)(5m - 4)$$

78. **[14.1]** $24ab^3c^2 - 56a^2bc^3 + 72a^2b^2c$
$= 8abc(3b^2c - 7ac^2 + 9ab)$

79. **[14.2]** $z^2 - 11zx + 10x^2$ ▪ Look for a pair of integers whose product is 10 and whose sum is -11. The integers are -1 and -10, so
$z^2 - 11zx + 10x^2 = (z - x)(z - 10x)$

80. **[14.3]** $3k^2 + 11k + 10$ ▪ Factor by grouping. Look for two integers whose product is $3(10) = 30$ and whose sum is 11. The integers are 5 and 6.

$$3k^2 + 11k + 10$$
$$= 3k^2 + 5k + 6k + 10$$
$$= (3k^2 + 5k) + (6k + 10)$$
$$= k(3k + 5) + 2(3k + 5)$$
$$= (3k + 5)(k + 2)$$

81. **[14.5]** $y^4 - 625$
$$= (y^2)^2 - 25^2$$
$$= (y^2 + 25)(y^2 - 25) \quad \textit{Difference of squares}$$
$$= (y^2 + 25)(y + 5)(y - 5) \quad \textit{Difference of squares}$$

82. **[14.3]** $6m^3 - 21m^2 - 45m$
$$= 3m(2m^2 - 7m - 15) \quad \textit{GCF = 3m}$$
$$= 3m\left[(2m^2 - 10m) + (3m - 15)\right]$$
$$= 3m[2m(m - 5) + 3(m - 5)]$$
$$= 3m(m - 5)(2m + 3)$$

83. **[14.5]** $25a^2 + 15ab + 9b^2$ is a *prime* polynomial.

84. **[14.2]** $2a^5 - 8a^4 - 24a^3$
$$= 2a^3(a^2 - 4a - 12) \quad \textit{GCF = 2a}^3$$
$$= 2a^3(a - 6)(a + 2) \quad \textit{Factor.}$$

85. **[14.3]** $-12r^2 - 8rq + 15q^2$
$$= -1(12r^2 + 8rq - 15q^2)$$
$$= -1(12r^2 + 18rq - 10rq - 15q^2)$$
$$= -1\left[(12r^2 + 18rq) + (-10rq - 15q^2)\right]$$
$$= -1[6r(2r + 3q) - 5q(2r + 3q)]$$
$$= -1(2r + 3q)(6r - 5q)$$

86. **[14.5]** $100a^2 - 9$
$$= (10a)^2 - 3^2$$
$$= (10a + 3)(10a - 3) \quad \textit{Difference of squares}$$

87. **[14.5]** $49t^2 + 56t + 16$
$$= (7t)^2 + 2(7t)(4) + 4^2$$
$$= (7t + 4)^2$$

88. **[14.6]** $t(t - 7) = 0$

To solve, use the zero-factor property.
$$t = 0 \quad \text{or} \quad t - 7 = 0$$
$$t = 0 \quad \text{or} \quad t = 7$$

The solution set is $\{0, 7\}$.

89. **[14.6]**
$$x^2 + 3x = 10$$
$$x^2 + 3x - 10 = 0$$
$$(x + 5)(x - 2) = 0 \quad \textit{Factor.}$$
$$x + 5 = 0 \quad \text{or} \quad x - 2 = 0$$
$$x = -5 \quad \text{or} \quad x = 2$$

The solution set is $\{-5, 2\}$.

90. **[14.6]**
$$25x^2 + 20x + 4 = 0$$
$$(5x)^2 + 2(5x)(2) + 2^2 = 0$$
$$(5x + 2)^2 = 0 \quad \textit{Factor.}$$

Set the factor $5x + 2$ equal to 0 and solve.
$$5x + 2 = 0$$
$$5x = -2$$
$$x = -\tfrac{2}{5}$$

The solution set is $\{-\tfrac{2}{5}\}$.

91. **[14.7]** Let $x =$ the length of the shorter leg. Then $2x + 6 =$ the length of the longer leg and $(2x + 6) + 3 = 2x + 9 =$ the length of the hypotenuse.

Use the Pythagorean theorem.
$$a^2 + b^2 = c^2$$
$$x^2 + (2x + 6)^2 = (2x + 9)^2$$
$$x^2 + 4x^2 + 24x + 36 = 4x^2 + 36x + 81$$
$$x^2 - 12x - 45 = 0$$
$$(x - 15)(x + 3) = 0$$
$$x - 15 = 0 \quad \text{or} \quad x + 3 = 0$$
$$x = 15 \quad \text{or} \quad x = -3$$

Reject -3 because a length cannot be negative. The lengths of the sides of the lot are 15 meters, $2(15) + 6 = 36$ meters, and $2(15) + 9 = 39$ meters.

92. **[14.7]** Let $x =$ the width of the base. Then $x + 2 =$ the length of the base.

The area of the base, B, is given by lw, so
$$B = x(x + 2).$$

Use the formula for the volume of a pyramid,
$$V = \tfrac{1}{3} \cdot B \cdot h.$$
$$48 = \tfrac{1}{3}x(x + 2)(6)$$
$$48 = 2x(x + 2)$$
$$24 = x^2 + 2x$$
$$x^2 + 2x - 24 = 0$$
$$(x + 6)(x - 4) = 0$$
$$x + 6 = 0 \quad \text{or} \quad x - 4 = 0$$
$$x = -6 \quad \text{or} \quad x = 4$$

Reject -6. The width of the base is 4 meters and the length is $4 + 2$ or 6 meters.

Chapter 14 Test

1. $2x^2 - 2x - 24$

$= 2(x^2 - x - 12)$ *GCF = 2*

$= 2(x + 3)(x - 4)$ *Factor.*

The correct completely factored form is choice **D**. Note that the factored forms **A**, $(2x + 6)(x - 4)$, and **B**, $(x + 3)(2x - 8)$, also can be multiplied to give a product of $2x^2 - 2x - 24$, but neither of these is completely factored because $2x + 6$ and $2x - 8$ both contain a common factor of 2.

2. Factor out the greatest common factor, $6x$.

$$12x^2 - 30x = 6x(2x - 5)$$

3. Factor out the greatest common factor, $m^2 n$.

$$2m^3 n^2 + 3m^3 n - 5m^2 n^2$$
$$= m^2 n(2mn + 3m - 5n)$$

4. Factor by grouping.

$$2ax - 2bx + ay - by$$
$$= 2x(a - b) + y(a - b)$$
$$= (a - b)(2x + y)$$

5. $x^2 - 9x + 14$ ▪ Look for a pair of integers whose product is 14 and whose sum is -9. The integers are -7 and -2, so

$$x^2 - 9x + 14 = (x - 7)(x - 2).$$

6. $2x^2 + x - 3$ ▪ Factor by grouping. Look for two integers whose product is $2(-3) = -6$ and whose sum is 1. The integers are 3 and -2.

$$2x^2 + x - 3$$
$$= 2x^2 + 3x - 2x - 3$$
$$= (2x^2 + 3x) + (-2x - 3)$$
$$= x(2x + 3) - 1(2x + 3)$$
$$= (2x + 3)(x - 1)$$

7. $6x^2 - 19x - 7$ ▪ Factor by grouping. Look for two integers whose product is $6(-7) = -42$ and whose sum is -19. The integers are -21 and 2.

$$6x^2 - 19x - 7$$
$$= 6x^2 - 21x + 2x - 7$$
$$= (6x^2 - 21x) + (2x - 7)$$
$$= 3x(2x - 7) + 1(2x - 7)$$
$$= (2x - 7)(3x + 1)$$

8. $3x^2 - 12x - 15$

$= 3(x^2 - 4x - 5)$ *GCF = 3*

$= 3(x + 1)(x - 5)$ *Factor.*

9. $10z^2 - 17z + 3$ ▪ Factor by grouping. Look for two integers whose product is $10(3) = 30$ and whose sum is -17. The integers are -15 and -2.

$$10z^2 - 17z + 3$$
$$= 10z^2 - 15z - 2z + 3$$
$$= (10z^2 - 15z) + (-2z + 3)$$
$$= 5z(2z - 3) - 1(2z - 3)$$
$$= (2z - 3)(5z - 1)$$

10. $t^2 + 6t + 10$ ▪ There is no GCF. We cannot find two integers whose product is 10 and whose sum is 6. This polynomial is *prime*.

11. $x^2 + 36$ ▪ This polynomial is *prime* because the sum of squares cannot be factored and the two terms have no common factor.

12. $y^2 - 49$

$= y^2 - 7^2$

$= (y + 7)(y - 7)$ *Difference of squares*

13. $81a^2 - 121b^2$

$= (9a)^2 - (11b)^2$

$= (9a + 11b)(9a - 11b)$ *Difference of squares*

14. Factor $x^2 + 16x + 64$ as a perfect square trinomial.

$$x^2 + 16x + 64$$
$$= x^2 + 2(8)(x) + 8^2$$
$$= (x + 8)^2$$

15. Factor $4x^2 - 28xy + 49y^2$ as a perfect square trinomial.

$$4x^2 - 28xy + 49y^2$$
$$= (2x)^2 - 2(2x)(7y) + (7y)^2$$
$$= (2x - 7y)^2$$

16. Factor out the GCF, -2, and then factor the perfect square trinomial.

$$-2x^2 - 4x - 2$$
$$= -2(x^2 + 2x + 1)$$
$$= -2(x^2 + 2 \cdot x \cdot 1 + 1^2)$$
$$= -2(x + 1)^2$$

17. Factor out the GCF, $3t^2$, and then factor by trial and error.

$$6t^4 + 3t^3 - 108t^2$$
$$= 3t^2(2t^2 + t - 36)$$
$$= 3t^2(2t + 9)(t - 4)$$

18. Note that $(2r + 5t)^2 = 4r^2 + 20rt + 25t^2$, not $4r^2 + 10rt + 25t^2$, so the given polynomial is not a perfect square trinomial. This polynomial is *prime* because there are no two integers whose product is $4(25) = 100$ and whose sum is 10, and there is no GCF.

19. Factor out the GCF, $4t$, and then factor the perfect square trinomial.

$$4t^3 + 32t^2 + 64t = 4t(t^2 + 8t + 16)$$
$$= 4t(t + 4)^2$$

20. $x^4 - 81$

$$= (x^2)^2 - 9^2$$

$$= (x^2 + 9)(x^2 - 9) \quad \begin{array}{l} \textit{Difference} \\ \textit{of squares} \end{array}$$

$$= (x^2 + 9)(x + 3)(x - 3) \quad \begin{array}{l} \textit{Difference} \\ \textit{of squares} \end{array}$$

21. $(x + 3)(x - 9) = 0$

To solve, use the zero-factor property.

$$\begin{array}{lll} x + 3 = 0 & \text{or} & x - 9 = 0 \\ x = -3 & \text{or} & x = 9 \end{array}$$

The solution set is $\{-3, 9\}$.

22. $2r^2 - 13r + 6 = 0$
$(2r - 1)(r - 6) = 0 \quad \textit{Factor.}$

To solve, use the zero-factor property.

$$\begin{array}{lll} 2r - 1 = 0 & \text{or} & r - 6 = 0 \\ 2r = 1 & & \\ r = \frac{1}{2} & \text{or} & r = 6 \end{array}$$

The solution set is $\{\frac{1}{2}, 6\}$.

23. $25x^2 - 4 = 0$

$(5x)^2 - 2^2 = 0$

$(5x + 2)(5x - 2) = 0 \quad \begin{array}{l} \textit{Difference} \\ \textit{of squares} \end{array}$

To solve, use the zero-factor property.

$$\begin{array}{lll} 5x + 2 = 0 & \text{or} & 5x - 2 = 0 \\ 5x = -2 & & 5x = 2 \\ x = -\frac{2}{5} & \text{or} & x = \frac{2}{5} \end{array}$$

The solution set is $\{-\frac{2}{5}, \frac{2}{5}\}$.

24. $x(x - 20) = -100$

$x^2 - 20x = -100$

$x^2 - 20x + 100 = 0 \quad \textit{Standard form}$

$(x - 10)^2 = 0 \quad \textit{Factor.}$

Set the factor $x - 10$ equal to 0 and solve.

$$x - 10 = 0$$
$$x = 10$$

The solution set is $\{10\}$.

25. $t^2 = 3t$

$t^2 - 3t = 0 \quad \textit{Standard form}$

$t(t - 3) = 0 \quad \textit{Factor out t.}$

To solve, use the zero-factor property.

$$\begin{array}{lll} t = 0 & \text{or} & t - 3 = 0 \\ t = 0 & \text{or} & t = 3 \end{array}$$

The solution set is $\{0, 3\}$.

26. $(s + 8)(6s^2 + 13s - 5) = 0$
$(s + 8)(2s + 5)(3s - 1) = 0 \quad \textit{Factor.}$

To solve, use the zero-factor property.

$$\begin{array}{lllll} s + 8 = 0 & \text{or} & 2s + 5 = 0 & \text{or} & 3s - 1 = 0 \\ & & 2s = -5 & & 3s = 1 \\ s = -8 & \text{or} & s = -\frac{5}{2} & \text{or} & s = \frac{1}{3} \end{array}$$

The solution set is $\{-8, -\frac{5}{2}, \frac{1}{3}\}$.

27. Let $x =$ the width of the flower bed.
Then $2x - 3 =$ the length of the flower bed.

Use the formula for the area of a rectangle with area 54 ft^2.

$$lw = A$$
$$(2x - 3)x = 54$$
$$2x^2 - 3x = 54$$
$$2x^2 - 3x - 54 = 0$$
$$(2x + 9)(x - 6) = 0$$

$$\begin{array}{lll} 2x + 9 = 0 & \text{or} & x - 6 = 0 \\ 2x = -9 & & \\ x = -\frac{9}{2} & \text{or} & x = 6 \end{array}$$

Reject $-\frac{9}{2}$. If $x = 6$, $2x - 3 = 2(6) - 3 = 9$. The dimensions of the flower bed are 6 feet by 9 feet.

28. Let $x =$ the lesser integer.
Then $x + 1 =$ the greater integer.

The square of the sum of the two integers is 11 more than the lesser integer.

$$[x + (x + 1)]^2 = x + 11$$
$$(2x + 1)^2 = x + 11$$
$$4x^2 + 4x + 1 = x + 11$$
$$4x^2 + 3x - 10 = 0$$
$$(4x - 5)(x + 2) = 0$$

$$\begin{array}{lll} 4x - 5 = 0 & \text{or} & x + 2 = 0 \\ 4x = 5 & & \\ x = \frac{5}{4} & \text{or} & x = -2 \end{array}$$

Reject $\frac{5}{4}$ because it is not an integer. If $x = -2$, $x + 1 = -1$. The integers are -2 and -1.

29. Let x = the length of the stud.
Then $3x - 7$ = the length of the brace.

The figure shows that a right triangle is formed with the brace as the hypotenuse. Use the Pythagorean theorem.

$$a^2 + b^2 = c^2$$
$$x^2 + 15^2 = (3x - 7)^2$$
$$x^2 + 225 = 9x^2 - 42x + 49$$
$$0 = 8x^2 - 42x - 176$$
$$0 = 2(4x^2 - 21x - 88)$$
$$0 = 2(4x + 11)(x - 8)$$

$$4x + 11 = 0 \qquad \text{or} \qquad x - 8 = 0$$
$$4x = -11$$
$$x = -\tfrac{11}{4} \qquad \text{or} \qquad x = 8$$

Reject $-\tfrac{11}{4}$. If $x = 8$, $3x - 7 = 24 - 7 = 17$, so the brace should be 17 feet long.

30. $x = 2009 - 2000 = 9$ for the year 2009.

$$y = -0.3470x^2 + 1.374x + 100.7$$
$$= -0.3470(9)^2 + 1.374(9) + 100.7$$
$$= 84.959 \approx 85$$

If $x = 9$, then $y = 85$, so the prediction is 85 billion pieces.

CHAPTER 15 RATIONAL EXPRESSIONS AND APPLICATIONS

15.1 The Fundamental Property of Rational Expressions

15.1 Margin Exercises

1. (a) $\dfrac{x}{2x+1} = \dfrac{-3}{2(-3)+1}$ *Let x = −3.*

   $= \dfrac{-3}{-5} = \dfrac{3}{5}$

   $\dfrac{x}{2x+1} = \dfrac{0}{2(0)+1}$ *Let x = 0.*

   $= \dfrac{0}{1} = 0$

   $\dfrac{x}{2x+1} = \dfrac{3}{2(3)+1}$ *Let x = 3.*

   $= \dfrac{3}{6+1} = \dfrac{3}{7}$

   (b) $\dfrac{2x+6}{x-3} = \dfrac{2(-3)+6}{-3-3}$ *Let x = −3.*

   $= \dfrac{0}{-6} = 0$

   $\dfrac{2x+6}{x-3} = \dfrac{2(0)+6}{0-3}$ *Let x = 0.*

   $= \dfrac{6}{-3} = -2$

   $\dfrac{2x+6}{x-3} = \dfrac{2(3)+6}{3-3}$ *Let x = 3.*

   $= \dfrac{12}{0}$ *undefined*

   Substituting 3 for x makes the denominator 0, so the expression is *undefined* when $x = 3$.

2. (a) $\dfrac{x+2}{x-5}$

   Step 1 $x - 5 = 0$
   Step 2 $x = 5$
   Step 3 Since $x = 5$ will make the denominator zero, the given expression is undefined for 5. We write the answer as $x \neq 5$.

   (b) $\dfrac{3r}{r^2+6r+8}$

   Step 1 $r^2 + 6r + 8 = 0$
   Step 2 $(r+4)(r+2) = 0$

   $r + 4 = 0$ or $r + 2 = 0$
   $r = -4$ or $r = -2$

   Step 3 The expression is undefined for -4 and -2. We write the answer as $r \neq -4$ and $r \neq -2$.

(c) $\dfrac{-5m}{m^2+4}$ ▪ Since $m^2 + 4$ cannot equal 0, there are no values of the variable that make the expression undefined. The expression is never undefined.

3. (a) $\dfrac{15}{40} = \dfrac{3 \cdot 5}{2 \cdot 2 \cdot 2 \cdot 5}$ *Factor.*

   $= \dfrac{3 \cdot (5)}{2 \cdot 2 \cdot 2 \cdot (5)}$ *Group.*

   $= \dfrac{3}{2 \cdot 2 \cdot 2}$, or $\dfrac{3}{8}$ *Fundamental property*

   (b) $\dfrac{5x^4}{15x^2} = \dfrac{5 \cdot x \cdot x \cdot x \cdot x}{3 \cdot 5 \cdot x \cdot x}$ *Factor.*

   $= \dfrac{x \cdot x (5 \cdot x \cdot x)}{3(5 \cdot x \cdot x)}$ *Group.*

   $= \dfrac{x^2}{3}$ *Fundamental property*

   (c) $\dfrac{6p^3}{2p^2} = \dfrac{2 \cdot 3 \cdot p \cdot p \cdot p}{2 \cdot p \cdot p}$ *Factor.*

   $= \dfrac{3 \cdot p (2 \cdot p \cdot p)}{1(2 \cdot p \cdot p)}$ *Group.*

   $= \dfrac{3p}{1}$ *Fundamental property*

   $= 3p$

4. (a) $\dfrac{4y+2}{6y+3} = \dfrac{2(2y+1)}{3(2y+1)}$ *Factor.*

   $= \dfrac{2}{3}$ *Fundamental property*

   (b) $\dfrac{8p+8q}{5p+5q} = \dfrac{8(p+q)}{5(p+q)}$ *Factor.*

   $= \dfrac{8}{5}$ *Fundamental property*

   (c) $\dfrac{x^2+4x+4}{4x+8}$

   $= \dfrac{(x+2)(x+2)}{4(x+2)}$ *Factor.*

   $= \dfrac{x+2}{4}$ *Fundamental property*

   (d) $\dfrac{a^2-b^2}{a^2+2ab+b^2}$

   $= \dfrac{(a-b)(a+b)}{(a+b)(a+b)}$ *Factor.*

   $= \dfrac{a-b}{a+b}$ *Fundamental property*

5. $\dfrac{x-y}{y-x}$

   $= \dfrac{-1(y-x)}{1(y-x)}$ *Factor out −1 in the numerator.*

   $= \dfrac{-1}{1}$, or -1 *Fundamental property*

   The result is -1, the same as in Example 5.

6.   (a) $\dfrac{5-y}{y-5}$

Since $y - 5 = -1(-y + 5)$
$$= -1(5 - y),$$

$5 - y$ and $y - 5$ are negatives (or opposites) of each other. Therefore,

$$\dfrac{5-y}{y-5} = -1.$$

(b) $\dfrac{m-n}{n-m}$

Since $n - m = -1(-n + m)$
$$= -1(m - n),$$

$m - n$ and $n - m$ are negatives (or opposites) of each other. Therefore,

$$\dfrac{m-n}{n-m} = -1.$$

(c) $\dfrac{25x^2 - 16}{12 - 15x} = \dfrac{(5x)^2 - 4^2}{3(4 - 5x)}$

$$= \dfrac{(5x + 4)(5x - 4)}{3(-1)(5x - 4)}$$

$$= \dfrac{5x + 4}{3(-1)}$$

$$= \dfrac{5x + 4}{-3}, \quad \text{or} \quad -\dfrac{5x + 4}{3}$$

(d) $\dfrac{9-k}{9+k}$

$$9 - k = -1(-9 + k) \neq -1(9 + k)$$

The expressions $9 - k$ and $9 + k$ are not negatives of each other. They do not have any common factors (other than 1), so the rational expression is already in lowest terms.

7.   Given expression: $-\dfrac{2x-6}{x+3}$

(a) $\dfrac{-(2x-6)}{x+3} = -\dfrac{2x-6}{x+3}$

equivalent to the given expression

(b) $\dfrac{-2x+6}{x+3} = \dfrac{-(2x-6)}{x+3} = -\dfrac{2x-6}{x+3}$

equivalent to the given expression

(c) $\dfrac{-2x-6}{x+3} = \dfrac{-(2x+6)}{x+3} = -\dfrac{2x+6}{x+3}$

not equivalent to the given expression

(d) $\dfrac{2x-6}{-(x+3)} = -\dfrac{2x-6}{x+3}$

equivalent to the given expression

(e) $\dfrac{2x-6}{-x-3} = \dfrac{2x-6}{-(x+3)} = -\dfrac{2x-6}{x+3}$

equivalent to the given expression

(f) $\dfrac{2x-6}{x-3} = \dfrac{2x-6}{-(-x+3)} = -\dfrac{2x-6}{-x+3}$

not equivalent to the given expression

15.1   Section Exercises

1.   (a) The rational expression $\dfrac{x+5}{x-3}$ is undefined when its denominator is equal to 0; that is, when $x - 3 = 0$, or $x = \underline{\ 3\ }$, so $x \neq \underline{\ 3\ }$. It is equal to 0 when its numerator is equal to 0; that is, when $x + 5 = 0$, or $x = \underline{\ -5\ }$.

(b) The denominator is 0 when $q - p = 0$, or $q = p$. The expression can be simplified when $q - p \neq 0$:

$$\dfrac{p-q}{q-p} = \dfrac{p-q}{-1(-q+p)} = \dfrac{p-q}{-1(p-q)} = -1$$

The rational expression $\dfrac{p-q}{q-p}$ is undefined when its denominator is equal to 0; that is, when $p = \underline{\ q\ }$. In all other cases when written in lowest terms it is equal to $\underline{\ -1\ }$.

3.   A rational expression is a quotient of two polynomials, such as $\dfrac{x+3}{x^2-4}$. One can think of this as an algebraic fraction.

5.   (a) $\dfrac{5x-2}{4x} = \dfrac{5 \cdot 2 - 2}{4 \cdot 2}$ *Let x = 2.*

$$= \dfrac{10 - 2}{8}$$

$$= \dfrac{8}{8} = 1$$

(b) $\dfrac{5x-2}{4x} = \dfrac{5(-3) - 2}{4(-3)}$ *Let x = –3.*

$$= \dfrac{-15 - 2}{-12}$$

$$= \dfrac{-17}{-12} = \dfrac{17}{12}$$

7.   (a) $\dfrac{x^2-4}{2x+1} = \dfrac{(2)^2 - 4}{2(2)+1}$ *Let x = 2.*

$$= \dfrac{0}{5} = 0$$

(b) $\dfrac{x^2-4}{2x+1} = \dfrac{(-3)^2 - 4}{2(-3)+1}$ *Let x = –3.*

$$= \dfrac{5}{-5} = -1$$

9. **(a)** $\dfrac{(-3x)^2}{4x+12} = \dfrac{(-3\cdot 2)^2}{4\cdot 2+12}$ *Let x = 2.*

$= \dfrac{(-6)^2}{8+12}$

$= \dfrac{36}{20} = \dfrac{9}{5}$

(b) $\dfrac{(-3x)^2}{4x+12} = \dfrac{[-3(-3)]^2}{4(-3)+12}$ *Let x = -3.*

$= \dfrac{9^2}{-12+12}$

$= \dfrac{81}{0}$

Since substituting -3 for x makes the denominator zero, the given rational expression is *undefined* when $x = -3$.

11. **(a)** $\dfrac{5x+2}{2x^2+11x+12}$

$= \dfrac{5(2)+2}{2(2)^2+11(2)+12}$ *Let x = 2.*

$= \dfrac{10+2}{2(4)+11(2)+12}$

$= \dfrac{12}{8+22+12}$

$= \dfrac{12}{42} = \dfrac{2}{7}$

(b) $\dfrac{5x+2}{2x^2+11x+12}$

$= \dfrac{5(-3)+2}{2(-3)^2+11(-3)+12}$ *Let x = -3.*

$= \dfrac{-15+2}{2(9)-33+12}$

$= \dfrac{-13}{18-33+12}$

$= \dfrac{-13}{-3} = \dfrac{13}{3}$

13. $\dfrac{2}{5y}$ ▪ The denominator $5y$ will be zero when $y = 0$, so the given expression is undefined for $y = 0$. We write the answer as $y \neq 0$.

15. $\dfrac{x+1}{x+6}$ ▪ To find the values for which this expression is undefined, set the denominator equal to zero and solve for x.

$$x+6 = 0$$
$$x = -6$$

Because $x = -6$ will make the denominator zero, the given expression is undefined for -6. We write the answer as $x \neq -6$.

17. $\dfrac{4x^2}{3x-5}$ ▪ To find the values for which this expression is undefined, set the denominator equal to zero and solve for x.

$$3x-5 = 0$$
$$3x = 5$$
$$x = \tfrac{5}{3}$$

Because $x = \frac{5}{3}$ will make the denominator zero, the given expression is undefined for $\frac{5}{3}$. We write the answer as $x \neq \frac{5}{3}$.

19. $\dfrac{m+2}{m^2+m-6}$ ▪ To find the numbers that make the denominator 0, we must solve

$$m^2+m-6 = 0$$
$$(m+3)(m-2) = 0.$$

$m+3 = 0$ or $m-2 = 0$
$m = -3$ or $m = 2$

The given expression is undefined for $m = -3$ and for $m = 2$. We write the answer as $m \neq -3$ and $m \neq 2$.

21. $\dfrac{x^2-3x}{4}$ is *never undefined* since the denominator is never zero.

23. $\dfrac{3x}{x^2+2}$ ▪ This denominator cannot equal zero for any value of x because x^2 is always greater than or equal to zero, and adding 2 makes the sum greater than zero. Thus, the given rational expression is *never undefined*.

25. $\dfrac{18r^3}{6r} = \dfrac{3r^2(6r)}{1(6r)}$ *Factor.*

$= 3r^2$ *Fundamental property*

27. $\dfrac{4(y-2)}{10(y-2)} = \dfrac{2\cdot 2(y-2)}{5\cdot 2(y-2)}$ *Factor.*

$= \dfrac{2}{5}$ *Fundamental property*

29. $\dfrac{(x+1)(x-1)}{(x+1)^2} = \dfrac{(x+1)(x-1)}{(x+1)(x+1)}$ *Factor.*

$= \dfrac{x-1}{x+1}$ *Fund. property*

31. $\dfrac{7m+14}{5m+10} = \dfrac{7(m+2)}{5(m+2)}$ *Factor out GCF.*

$= \dfrac{7}{5}$ *Fundamental property*

33. $\dfrac{m^2-n^2}{m+n} = \dfrac{(m+n)(m-n)}{m+n}$ *Factor.*

$= m-n$ *Fundamental property*

35. $\dfrac{12m^2 - 3}{8m - 4} = \dfrac{3(4m^2 - 1)}{4(2m - 1)}$ *Factor out GCF.*

$= \dfrac{3(2m + 1)(2m - 1)}{4(2m - 1)}$ *Factor.*

$= \dfrac{3(2m + 1)}{4}$ *Fund. property*

37. $\dfrac{3m^2 - 3m}{5m - 5} = \dfrac{3m(m - 1)}{5(m - 1)}$ *Factor out GCF.*

$= \dfrac{3m}{5}$ *Fundamental property*

39. $\dfrac{9r^2 - 4s^2}{9r + 6s} = \dfrac{(3r + 2s)(3r - 2s)}{3(3r + 2s)}$ *Factor.*

$= \dfrac{3r - 2s}{3}$ *Fund. property*

41. $\dfrac{2x^2 - 3x - 5}{2x^2 - 7x + 5} = \dfrac{(2x - 5)(x + 1)}{(2x - 5)(x - 1)}$ *Factor.*

$= \dfrac{x + 1}{x - 1}$ *Fund. property*

43. $\dfrac{zw + 4z - 3w - 12}{zw + 4z + 5w + 20}$

$= \dfrac{z(w + 4) - 3(w + 4)}{z(w + 4) + 5(w + 4)}$ *Factor by grouping.*

$= \dfrac{(w + 4)(z - 3)}{(w + 4)(z + 5)}$

$= \dfrac{z - 3}{z + 5}$ *Fundamental property*

45. $\dfrac{ac - ad + bc - bd}{ac - ad - bc + bd}$

$= \dfrac{a(c - d) + b(c - d)}{a(c - d) - b(c - d)}$ *Factor by grouping.*

$= \dfrac{(c - d)(a + b)}{(c - d)(a - b)}$

$= \dfrac{a + b}{a - b}$ *Fundamental property*

47. $\dfrac{6 - t}{t - 6} = \dfrac{-1(t - 6)}{1(t - 6)} = \dfrac{-1}{1} = -1$

Note that $6 - t$ and $t - 6$ are opposites, so we know that their quotient will be -1.

49. $\dfrac{m^2 - 1}{1 - m} = \dfrac{(m + 1)(m - 1)}{-1(m - 1)}$

$= \dfrac{m + 1}{-1}$

$= -(m + 1)$

51. $\dfrac{q^2 - 4q}{4q - q^2} = -1$ ▪ Since the numerator and denominator are opposites, we know that their quotient is -1. It is not necessary to factor the numerator and denominator in this case.

53. $\dfrac{p + 6}{p - 6}$ ▪ The numerator and denominator are *not* opposites, so the expression is already in lowest terms.

There are many possible answers for Exercises 55–60.

55. To write four equivalent expressions for $-\dfrac{x + 4}{x - 3}$, we will follow the outline in Example 7. Applying the negative sign to the numerator we have

$$\dfrac{-(x + 4)}{x - 3}.$$

Distributing the negative sign gives us

$$\dfrac{-x - 4}{x - 3}.$$

Applying the negative sign to the denominator yields

$$\dfrac{x + 4}{-(x - 3)}.$$

Again, we distribute to get

$$\dfrac{x + 4}{-x + 3}.$$

57. $-\dfrac{2x - 3}{x + 3}$ is equivalent to each of the following:

$$\dfrac{-(2x - 3)}{x + 3}, \qquad \dfrac{-2x + 3}{x + 3},$$

$$\dfrac{2x - 3}{-(x + 3)}, \qquad \dfrac{2x - 3}{-x - 3}$$

59. $-\dfrac{3x - 1}{5x - 6}$ is equivalent to each of the following:

$$\dfrac{-(3x - 1)}{5x - 6}, \qquad \dfrac{-3x + 1}{5x - 6},$$

$$\dfrac{3x - 1}{-(5x - 6)}, \qquad \dfrac{3x - 1}{-5x + 6}$$

61. $l \cdot w = A$ *Area formula*

$w = \dfrac{A}{l}$ *Divide by l.*

$w = \dfrac{x^4 + 10x^2 + 21}{x^2 + 7}$ *Substitute for A and l.*

$= \dfrac{(x^2 + 7)(x^2 + 3)}{x^2 + 7}$ *Factor.*

$= x^2 + 3$ *Fundamental property*

Note: If it is not apparent that we can factor A as $x^4 + 10x^2 + 21 = (x^2 + 7)(x^2 + 3)$, we may use long division to find the quotient $\frac{A}{l}$. Remember to insert zeros for the coefficients of the missing terms in the dividend and divisor.

$$
\begin{array}{r}
x^2 \qquad\quad + 3 \\
x^2 + 0x + 7 \overline{\big)\, x^4 + 0x^3 + 10x^2 + 0x + 21} \\
\underline{x^4 + 0x^3 + 7x^2\qquad\qquad} \\
3x^2 + 0x + 21 \\
\underline{3x^2 + 0x + 21} \\
0
\end{array}
$$

The polynomial $x^2 + 3$ represents the width of the rectangle.

15.2 Multiplying and Dividing Rational Expressions

15.2 Margin Exercises

1. **(a)** $\dfrac{2}{7} \cdot \dfrac{5}{10} = \dfrac{2 \cdot 5}{7 \cdot 10}$

$= \dfrac{2 \cdot 5}{7 \cdot 2 \cdot 5}$

$= \dfrac{1}{7}$

(b) $\dfrac{3m^2}{2} \cdot \dfrac{10}{m} = \dfrac{3m^2 \cdot 10}{2 \cdot m}$

$= \dfrac{3 \cdot m \cdot m \cdot 2 \cdot 5}{m \cdot 2}$

$= \dfrac{3 \cdot m \cdot 5}{1} = 15m$

(c) $\dfrac{8p^2q}{3} \cdot \dfrac{9}{q^2p} = \dfrac{2 \cdot 4 \cdot p \cdot p \cdot q \cdot 3 \cdot 3}{3 \cdot q \cdot q \cdot p}$

$= \dfrac{2 \cdot 4 \cdot p \cdot 3}{q} = \dfrac{24p}{q}$

2. **(a)** $\dfrac{a+b}{5} \cdot \dfrac{30}{2(a+b)} = \dfrac{2 \cdot 3 \cdot 5(a+b)}{5 \cdot 2(a+b)}$

$= \dfrac{3}{1} = 3$

(b) $\dfrac{3(p-q)}{q^2} \cdot \dfrac{q}{2(p-q)^2}$

$= \dfrac{3q(p-q)}{2 \cdot q \cdot q(p-q)(p-q)}$

$= \dfrac{3}{2q(p-q)}$

3. **(a)** $\dfrac{x^2 + 7x + 10}{3x + 6} \cdot \dfrac{6x - 6}{x^2 + 2x - 15}$

$= \dfrac{(x+2)(x+5)}{3(x+2)} \cdot \dfrac{6(x-1)}{(x+5)(x-3)}$

Factor.

$= \dfrac{6(x+2)(x+5)(x-1)}{3(x+2)(x+5)(x-3)}$

Multiply.

$= \dfrac{2(x-1)}{x-3}$ *Lowest terms*

(b) $\dfrac{m^2 + 4m - 5}{m + 5} \cdot \dfrac{m^2 + 8m + 15}{m - 1}$

$= \dfrac{(m-1)(m+5)}{m+5} \cdot \dfrac{(m+3)(m+5)}{m-1}$

Factor.

$= \dfrac{(m-1)(m+5)(m+3)(m+5)}{(m+5)(m-1)}$

Multiply.

$= (m+3)(m+5)$ *Lowest terms*

4. **(a)** In general, the reciprocal of $\frac{a}{c}$ is $\frac{c}{a}$, so the reciprocal of $\frac{5}{8}$ is $\frac{8}{5}$.

(b) The reciprocal of a rational expression is found by interchanging the numerator and the denominator, so the reciprocal of

$$\dfrac{6b^5}{3r^2b} \quad \text{is} \quad \dfrac{3r^2b}{6b^5}.$$

(c) $\dfrac{t^2 - 4t}{t^2 + 2t - 3}$ has reciprocal $\dfrac{t^2 + 2t - 3}{t^2 - 4t}$.

5. **(a)** $\dfrac{3}{4} \div \dfrac{5}{16}$

$= \dfrac{3}{4} \cdot \dfrac{16}{5}$ *Multiply by the reciprocal.*

$= \dfrac{3 \cdot 16}{4 \cdot 5}$ *Multiply.*

$= \dfrac{3 \cdot 4}{5} = \dfrac{12}{5}$ *Lowest terms*

(b) $\dfrac{r}{r-1} \div \dfrac{3r}{r+4}$

$= \dfrac{r}{r-1} \cdot \dfrac{r+4}{3r}$ *Multiply by the reciprocal.*

$= \dfrac{r(r+4)}{3r(r-1)}$ *Multiply.*

$= \dfrac{r+4}{3(r-1)}$ *Lowest terms*

(c) $\dfrac{6x - 4}{3} \div \dfrac{15x - 10}{9}$

$= \dfrac{6x - 4}{3} \cdot \dfrac{9}{15x - 10}$ *Multiply by the reciprocal.*

$= \dfrac{2(3x - 2)}{3} \cdot \dfrac{3 \cdot 3}{5(3x - 2)}$ *Factor.*

$= \dfrac{2 \cdot 3 \cdot 3(3x - 2)}{3 \cdot 5(3x - 2)}$ *Multiply.*

$= \dfrac{6}{5}$ *Lowest terms*

6. **(a)** $\dfrac{5a^2b}{2} \div \dfrac{10ab^2}{8}$

$= \dfrac{5a^2b}{2} \cdot \dfrac{8}{10ab^2}$ *Multiply by the reciprocal.*

$= \dfrac{5 \cdot 8a^2b}{2 \cdot 10ab^2}$ *Multiply.*

$= \dfrac{2a}{b}$ *Lowest terms*

(b) $\dfrac{(3t)^2}{w} \div \dfrac{3t^2}{5w^4}$

$= \dfrac{9t^2}{w} \cdot \dfrac{5w^4}{3t^2}$ *Multiply by the reciprocal.*

$= \dfrac{9 \cdot 5t^2w^4}{3t^2w}$ *Multiply.*

$= 15w^3$ *Lowest terms*

7. **(a)** $\dfrac{y^2 + 4y + 3}{y + 3} \div \dfrac{y^2 - 4y - 5}{y - 3}$

$= \dfrac{y^2 + 4y + 3}{y + 3} \cdot \dfrac{y - 3}{y^2 - 4y - 5}$ *Mult. by recip.*

$= \dfrac{(y + 3)(y + 1)}{y + 3} \cdot \dfrac{y - 3}{(y - 5)(y + 1)}$ *Factor.*

$= \dfrac{(y + 3)(y + 1)(y - 3)}{(y + 3)(y - 5)(y + 1)}$ *Multiply.*

$= \dfrac{y - 3}{y - 5}$ *Lowest terms*

(b) $\dfrac{4x(x + 3)}{2x + 1} \div \dfrac{-x^2(x + 3)}{4x^2 - 1}$

$= \dfrac{4x(x + 3)}{2x + 1} \cdot \dfrac{4x^2 - 1}{-x^2(x + 3)}$ *Mult. by recip.*

$= \dfrac{4x(x + 3)}{2x + 1} \cdot \dfrac{(2x + 1)(2x - 1)}{-x^2(x + 3)}$ *Factor.*

$= \dfrac{4x(x + 3)(2x + 1)(2x - 1)}{-x^2(2x + 1)(x + 3)}$ *Multiply.*

$= -\dfrac{4(2x - 1)}{x}$ *Lowest terms*

8. **(a)** $\dfrac{ab - a^2}{a^2 - 1} \div \dfrac{a - b}{a - 1}$

$= \dfrac{ab - a^2}{a^2 - 1} \cdot \dfrac{a - 1}{a - b}$ *Multiply by the reciprocal.*

$= \dfrac{a(b - a)}{(a + 1)(a - 1)} \cdot \dfrac{a - 1}{a - b}$ *Factor.*

$= \dfrac{-a(a - b)(a - 1)}{(a + 1)(a - 1)(a - b)}$ *Multiply.*

$= \dfrac{-a}{a + 1}$ *Lowest terms*

(b) $\dfrac{x^2 - 9}{2x + 6} \div \dfrac{9 - x^2}{4x - 12}$

$= \dfrac{x^2 - 9}{2x + 6} \cdot \dfrac{4x - 12}{9 - x^2}$ *Multiply by the reciprocal.*

$= \dfrac{-(9 - x^2)}{2(x + 3)} \cdot \dfrac{4(x - 3)}{9 - x^2}$ *Factor; opposites*

$= \dfrac{-2(x - 3)}{x + 3}$ *Lowest terms*

$= \dfrac{-2x + 6}{x + 3}$ *Multiply.*

15.2   Section Exercises

1. **(a)** $\dfrac{5x^3}{10x^4} \cdot \dfrac{10x^7}{2x} = \dfrac{5 \cdot 10 \cdot x^3 \cdot x^7}{10 \cdot 2 \cdot x^4 \cdot x}$

$= \dfrac{5x^{10}}{2x^5}$

$= \dfrac{5x^5}{2};$ **B**

(b) $\dfrac{10x^4}{5x^3} \cdot \dfrac{10x^7}{2x} = \dfrac{10 \cdot 10 \cdot x^4 \cdot x^7}{5 \cdot 2 \cdot x^3 \cdot x}$

$= \dfrac{10x^{11}}{1x^4}$

$= 10x^7;$ **D**

(c) $\dfrac{5x^3}{10x^4} \cdot \dfrac{2x}{10x^7} = \dfrac{5 \cdot 2 \cdot x^3 \cdot x}{10 \cdot 10 \cdot x^4 \cdot x^7}$

$= \dfrac{1x^4}{10x^{11}}$

$= \dfrac{1}{10x^7};$ **C**

(d) $\dfrac{10x^4}{5x^3} \cdot \dfrac{2x}{10x^7} = \dfrac{10 \cdot 2 \cdot x^4 \cdot x}{5 \cdot 10 \cdot x^3 \cdot x^7}$

$= \dfrac{2x^5}{5x^{10}}$

$= \dfrac{2}{5x^5};$ **A**

3. $\dfrac{10m^2}{7} \cdot \dfrac{14}{15m} = \dfrac{5 \cdot 2 \cdot m \cdot m \cdot 2 \cdot 7}{7 \cdot 3 \cdot 5 \cdot m}$

$= \dfrac{2 \cdot 2 \cdot m}{3} = \dfrac{4m}{3}$

5. $\dfrac{16y^4}{18y^5} \cdot \dfrac{15y^5}{y^2} = \dfrac{2 \cdot 8 \cdot 3 \cdot 5y^9}{2 \cdot 3 \cdot 3y^7}$

$= \dfrac{8 \cdot 5y^{9 - 7}}{3} = \dfrac{40y^2}{3}$

7. $\dfrac{2(c + d)}{3} \cdot \dfrac{18}{6(c + d)^2}$

$= \dfrac{2 \cdot 3 \cdot 6(c + d)}{3 \cdot 6(c + d)(c + d)}$ *Multiply and factor.*

$= \dfrac{2}{c + d}$ *Lowest terms*

9. $\dfrac{(x - y)^2}{2} \cdot \dfrac{24}{3(x - y)}$

$= \dfrac{6 \cdot 4(x - y)(x - y)}{6(x - y)}$ *Multiply and factor.*

$= 4(x - y)$ *Lowest terms*

11. The reciprocal of a rational expression is found by interchanging the numerator and the denominator, so the reciprocal of

$$\frac{3p^3}{16q} \quad \text{is} \quad \frac{16q}{3p^3}.$$

13. $\dfrac{r^2 + rp}{7}$ has reciprocal $\dfrac{7}{r^2 + rp}$.

15. $\dfrac{z^2 + 7z + 12}{z^2 - 9}$ has reciprocal $\dfrac{z^2 - 9}{z^2 + 7z + 12}$.

17. $\dfrac{9z^4}{3z^5} \div \dfrac{3z^2}{5z^3} = \dfrac{9z^4}{3z^5} \cdot \dfrac{5z^3}{3z^2}$

$$= \dfrac{9 \cdot 5z^7}{3 \cdot 3z^7}$$

$$= 5$$

19. $\dfrac{4t^4}{2t^5} \div \dfrac{(2t)^3}{-6} = \dfrac{4t^4}{2t^5} \cdot \dfrac{-6}{(2t)^3}$

$$= \dfrac{4t^4}{2t^5} \cdot \dfrac{-6}{8t^3}$$

$$= \dfrac{-2 \cdot 2 \cdot 2 \cdot 3t^4}{2 \cdot 2^3 t^8}$$

$$= -\dfrac{3}{2t^{8-4}}$$

$$= -\dfrac{3}{2t^4}$$

21. $\dfrac{3}{2y - 6} \div \dfrac{6}{y - 3}$

$$= \dfrac{3}{2y - 6} \cdot \dfrac{y - 3}{6}$$

$$= \dfrac{3}{2(y - 3)} \cdot \dfrac{y - 3}{6}$$

$$= \dfrac{3(y - 3)}{2 \cdot 2 \cdot 3(y - 3)}$$

$$= \dfrac{1}{2 \cdot 2} = \dfrac{1}{4}$$

23. $\dfrac{(x - 3)^2}{6x} \div \dfrac{x - 3}{x^2}$

$$= \dfrac{(x - 3)^2}{6x} \cdot \dfrac{x^2}{x - 3}$$

$$= \dfrac{x \cdot x(x - 3)(x - 3)}{6x(x - 3)}$$

$$= \dfrac{x(x - 3)}{6}$$

25. $\dfrac{5x - 15}{3x + 9} \cdot \dfrac{4x + 12}{6x - 18}$

$$= \dfrac{5(x - 3)}{3(x + 3)} \cdot \dfrac{4(x + 3)}{6(x - 3)}$$

$$= \dfrac{5 \cdot 4 \cdot (x - 3)(x + 3)}{3 \cdot 6 \cdot (x - 3)(x + 3)} = \dfrac{10}{9}$$

27. $\dfrac{2 - t}{8} \div \dfrac{t - 2}{6}$

$$= \dfrac{2 - t}{8} \cdot \dfrac{6}{t - 2} \qquad \text{\textit{Multiply by reciprocal.}}$$

$$= \dfrac{6(2 - t)}{8(t - 2)} \qquad \text{\textit{Multiply numerators; multiply denominators.}}$$

$$= \dfrac{6(-1)}{8} \qquad \dfrac{2 - t}{t - 2} = -1$$

$$= -\dfrac{3}{4} \qquad \text{\textit{Lowest terms}}$$

29. $\dfrac{5 - 4x}{5 + 4x} \cdot \dfrac{4x + 5}{4x - 5}$

$$= \dfrac{(5 - 4x)(4x + 5)}{(5 + 4x)(4x - 5)}$$

$$= \dfrac{5 - 4x}{4x - 5} \qquad \dfrac{4x + 5}{5 + 4x} = 1$$

$$= -1 \qquad \text{\textit{Opposites}}$$

31. $\dfrac{6(m - 2)^2}{5(m + 4)^2} \cdot \dfrac{15(m + 4)}{2(2 - m)}$

$$= \dfrac{2 \cdot 3(m - 2)(m - 2)}{5(m + 4)(m + 4)} \cdot \dfrac{3 \cdot 5(m + 4)}{2(-1)(m - 2)}$$

$$= \dfrac{3(m - 2)(3)}{(m + 4)(-1)}$$

$$= \dfrac{9(m - 2)}{-(m + 4)}, \quad \text{or} \quad \dfrac{-9(m - 2)}{m + 4}$$

33. $\dfrac{p^2 + 4p - 5}{p^2 + 7p + 10} \div \dfrac{p - 1}{p + 4}$

$$= \dfrac{p^2 + 4p - 5}{p^2 + 7p + 10} \cdot \dfrac{p + 4}{p - 1}$$

$$= \dfrac{(p + 5)(p - 1) \cdot (p + 4)}{(p + 5)(p + 2) \cdot (p - 1)}$$

$$= \dfrac{p + 4}{p + 2}$$

35. $\dfrac{2k^2 - k - 1}{2k^2 + 5k + 3} \div \dfrac{4k^2 - 1}{2k^2 + k - 3}$

$$= \dfrac{2k^2 - k - 1}{2k^2 + 5k + 3} \cdot \dfrac{2k^2 + k - 3}{4k^2 - 1}$$

$$= \dfrac{(2k + 1)(k - 1)(2k + 3)(k - 1)}{(2k + 3)(k + 1)(2k + 1)(2k - 1)}$$

$$= \dfrac{(k - 1)(k - 1)}{(k + 1)(2k - 1)}$$

$$= \dfrac{(k - 1)^2}{(k + 1)(2k - 1)}$$

37. $\dfrac{2k^2 + 3k - 2}{6k^2 - 7k + 2} \cdot \dfrac{4k^2 - 5k + 1}{k^2 + k - 2}$

$= \dfrac{(2k-1)(k+2)}{(3k-2)(2k-1)} \cdot \dfrac{(4k-1)(k-1)}{(k+2)(k-1)}$

$= \dfrac{(2k-1)(k+2)(4k-1)(k-1)}{(3k-2)(2k-1)(k+2)(k-1)}$

$= \dfrac{4k-1}{3k-2}$

39. $\dfrac{m^2 + 2mp - 3p^2}{m^2 - 3mp + 2p^2} \div \dfrac{m^2 + 4mp + 3p^2}{m^2 + 2mp - 8p^2}$

$= \dfrac{m^2 + 2mp - 3p^2}{m^2 - 3mp + 2p^2} \cdot \dfrac{m^2 + 2mp - 8p^2}{m^2 + 4mp + 3p^2}$

$= \dfrac{(m+3p)(m-p)(m+4p)(m-2p)}{(m-2p)(m-p)(m+3p)(m+p)}$

$= \dfrac{m+4p}{m+p}$

41. $\left(\dfrac{x^2 + 10x + 25}{x^2 + 10x} \cdot \dfrac{10x}{x^2 + 15x + 50} \right) \div \dfrac{x+5}{x+10}$

$= \left[\dfrac{(x+5)^2 \cdot 10x}{x(x+10)(x+5)(x+10)} \right] \div \dfrac{x+5}{x+10}$

$= \left[\dfrac{10(x+5)}{(x+10)^2} \right] \div \dfrac{x+5}{x+10}$

$= \dfrac{10(x+5)}{(x+10)^2} \cdot \dfrac{x+10}{x+5}$

$= \dfrac{10}{x+10}$

43. Use the formula for the area of a rectangle with

$A = \dfrac{5x^2y^3}{2pq}$ and $l = \dfrac{2xy}{p}$ to solve for w.

$$A = l \cdot w$$
$$\dfrac{5x^2y^3}{2pq} = \dfrac{2xy}{p} \cdot w$$
$$w = \dfrac{5x^2y^3}{2pq} \div \dfrac{2xy}{p}$$
$$= \dfrac{5x^2y^3}{2pq} \cdot \dfrac{p}{2xy}$$
$$= \dfrac{5x^2y^3 p}{4pqxy} = \dfrac{5xy^2}{4q}$$

Thus, the rational expression $\dfrac{5xy^2}{4q}$ represents the width of the rectangle.

15.3   Least Common Denominators

15.3   Margin Exercises

1. **(a)** $\dfrac{7}{10}, \dfrac{1}{25}$

Step 1 Factor each denominator into prime factors.

$$10 = 2 \cdot 5, \qquad 25 = 5 \cdot 5 = 5^2$$

Step 2 Take each different factor the *greatest* number of times it appears as a factor in any of the denominators, and use it to form the least common denominator (LCD). The factor 2 appears at most one time and the factor 5 appears at most twice.

Step 3   LCD $= 2 \cdot 5^2 = 50$

(b) $\dfrac{7}{20p}, \dfrac{11}{30p}$

Factor each denominator.

$$20p = 2 \cdot 2 \cdot 5 \cdot p = 2^2 \cdot 5 \cdot p$$
$$30p = 2 \cdot 3 \cdot 5 \cdot p$$

Take each factor the greatest number of times it appears in any denominator; then multiply.

$$\text{LCD} = 2^2 \cdot 3 \cdot 5 \cdot p = 60p$$

(c) $\dfrac{4}{5x}, \dfrac{12}{10x}$

Factor each denominator.

$$5x = 5 \cdot x, \qquad 10x = 2 \cdot 5 \cdot x$$

Take each factor the greatest number of times it appears in any denominator; then multiply.

$$\text{LCD} = 2 \cdot 5 \cdot x = 10x$$

2. **(a)** $\dfrac{4}{16m^3n}, \dfrac{5}{9m^5}$

Factor each denominator.

$$16m^3n = 2^4 \cdot m^3 \cdot n$$
$$9m^5 = 3^2 \cdot m^5$$

Take each factor the greatest number of times it appears in any denominator; then multiply.

$$\text{LCD} = 2^4 \cdot 3^2 \cdot m^5 \cdot n = 144m^5n$$

(b) $\dfrac{3}{25a^2}, \dfrac{2}{10a^3b}$

Factor each denominator.

$$25a^2 = 5^2 \cdot a^2$$
$$10a^3b = 2 \cdot 5 \cdot a^3 \cdot b$$

Take each factor the greatest number of times it appears in any denominator; then multiply.

$$\text{LCD} = 2 \cdot 5^2 \cdot a^3 \cdot b = 50a^3b$$

3. **(a)** $\dfrac{7}{3a}, \dfrac{11}{a^2 - 4a}$

Factor each denominator.

$$3a = 3 \cdot a$$
$$a^2 - 4a = a(a - 4)$$

Take each factor the greatest number of times it appears in any denominator; then multiply.

$$\text{LCD} = 3 \cdot a(a - 4) = 3a(a - 4)$$

(b) $\dfrac{2m}{m^2 - 3m + 2}, \dfrac{5m - 3}{m^2 + 3m - 10}, \dfrac{4m + 7}{m^2 + 4m - 5}$

Factor each denominator.

$$m^2 - 3m + 2 = (m - 1)(m - 2)$$
$$m^2 + 3m - 10 = (m + 5)(m - 2)$$
$$m^2 + 4m - 5 = (m - 1)(m + 5)$$

Take each factor the greatest number of times it appears in any denominator; then multiply.

$$\text{LCD} = (m - 1)(m - 2)(m + 5)$$

(c) $\dfrac{6}{x - 4}, \dfrac{3x - 1}{4 - x}$

The expressions $x - 4$ and $4 - x$ are opposites of each other because

$$-(x - 4) = -x + 4 = 4 - x.$$

Therefore, either $x - 4$ or $4 - x$ can be used as the LCD.

4. (a) $\dfrac{3}{4} = \dfrac{?}{36}$

Step 1   First factor the denominator on the right. Then compare the denominator on the left with the one on the right to decide what factors are missing.

$$\frac{3}{4} = \frac{?}{4 \cdot 9}$$

Step 2   A factor of 9 is missing.

Step 3   Multiply $\frac{3}{4}$ by $\frac{9}{9}$, which is equal to 1.

$$\frac{3}{4} = \frac{3}{4} \cdot \frac{9}{9} = \frac{27}{36}$$

(b) $\dfrac{7k}{5} = \dfrac{?}{30p}$

Step 1   Factor the denominator on the right; then compare it to the denominator on the left.

$$\frac{7k}{5} = \frac{?}{5 \cdot 6 \cdot p}$$

Step 2   Factors of $\underline{6}$ and p are missing.

Step 3   Multiply $\frac{7k}{5}$ by $\frac{6p}{6p}$, which is equal to 1.

$$\frac{7k}{5} = \frac{7k}{5} \cdot \frac{6p}{6p} = \frac{42kp}{30p}$$

5. (a) $\dfrac{9}{2a + 5} = \dfrac{}{6a + 15}$

Factor the denominator on the right.

$$\frac{9}{2a + 5} = \frac{?}{3(2a + 5)}$$

The factor 3 is missing on the left, so multiply by $\frac{3}{3}$.

$$\frac{9}{2a + 5} = \frac{9}{2a + 5} \cdot \frac{3}{3} = \frac{27}{6a + 15}$$

(b) $\dfrac{5k + 1}{k^2 + 2k} = \dfrac{}{k^3 + k^2 - 2k}$

Factor and compare the denominators.

$$\frac{5k + 1}{k(k + 2)} = \frac{?}{k(k + 2)(k - 1)}$$

The factor $k - 1$ is missing on the left, so multiply by $\frac{k - 1}{k - 1}$.

$$\frac{5k + 1}{k^2 + 2k}$$
$$= \frac{5k + 1}{k(k + 2)} \cdot \frac{k - 1}{k - 1}$$
$$= \frac{(5k + 1)(k - 1)}{(k^2 + 2k)(k - 1)}$$
$$= \frac{(5k + 1)(k - 1)}{k^3 + k^2 - 2k}, \quad \text{or} \quad \frac{5k^2 - 4k - 1}{k^3 + k^2 - 2k}$$

15.3   Section Exercises

1.   The factor a appears at most one time in any denominator as does the factor b. Thus, the LCD is the product of the two factors, ab. The correct response is **C**.

3.   Since $20 = 2^2 \cdot 5$, the LCD of $\frac{11}{20}$ and $\frac{1}{2}$ must have 5 as a factor and 2^2 as a factor. Because 2 appears twice in $2^2 \cdot 5$, we don't have to include another 2 in the LCD for the number $\frac{1}{2}$. Thus, the LCD is just $2^2 \cdot 5 = 20$. Note that this is a specific case of Exercise 2 since 2 is a factor of 20. The correct response is **C**.

5. $\dfrac{2}{15}, \dfrac{3}{10}, \dfrac{7}{30}$

Step 1   Factor each denominator into prime factors.

$$15 = 3 \cdot 5, \qquad 10 = 2 \cdot 5, \qquad 30 = 2 \cdot 3 \cdot 5$$

Step 2   Take each different factor the *greatest* number of times it appears as a factor in any of the denominators, and use it to form the least common denominator (LCD). The factors 2, 3, and 5 appear at most one time.

Step 3 $\text{LCD} = 3 \cdot 5 \cdot 2 = 30$

7. $\dfrac{3}{x^1}, \dfrac{5}{x^7}$ ■ The only factor in any denominator is x. The greatest number of times it appears is 7, so the least common denominator is x^7.

9. $\dfrac{5}{36q}, \dfrac{17}{24q}$ ■ Factor each denominator.

$$36q = 2^2 \cdot 3^2 \cdot q$$
$$24q = 2^3 \cdot 3 \cdot q$$

Take each factor the greatest number of times it appears, and use the greatest exponent on q. The least common denominator is $2^3 \cdot 3^2 \cdot q = 72q$.

11. $\dfrac{6}{21r^3}, \dfrac{8}{12r^5}$ ▪ Factor each denominator.

$$21r^3 = 3 \cdot 7 \cdot r^3$$
$$12r^5 = 2^2 \cdot 3 \cdot r^5$$

Take each factor the greatest number of times it appears; then multiply.

$$\text{LCD} = 2^2 \cdot 3 \cdot 7 \cdot r^5 = 84r^5$$

13. $\dfrac{5}{12x^2y}, \dfrac{7}{24x^3y^2}, \dfrac{-11}{6xy^4}$

Factor each denominator.

$$12x^2y = 2^2 \cdot 3 \cdot x^2 \cdot y$$
$$24x^3y^2 = 2^3 \cdot 3 \cdot x^3 \cdot y^2$$
$$6xy^4 = 2 \cdot 3 \cdot x \cdot y^4$$

Take each factor the greatest number of times it appears; then multiply.

$$\text{LCD} = 2^3 \cdot 3 \cdot x^3 \cdot y^4 = 24x^3y^4$$

15. $\dfrac{1}{x^9y^{14}z^{21}}, \dfrac{3}{x^8y^{14}z^{22}}, \dfrac{5}{x^{10}y^{13}z^{23}}$

Factor each denominator.

$$x^9y^{14}z^{21} = x^9 \cdot y^{14} \cdot z^{21}$$
$$x^8y^{14}z^{22} = x^8 \cdot y^{14} \cdot z^{22}$$
$$x^{10}y^{13}z^{23} = x^{10} \cdot y^{13} \cdot z^{23}$$

Take each factor the greatest number of times it appears; then multiply.

$$\text{LCD} = x^{10} \cdot y^{14} \cdot z^{23} = x^{10}y^{14}z^{23}$$

17. $\dfrac{9}{28m^2}, \dfrac{3}{12m - 20}$ ▪ Factor each denominator.

$$28m^2 = 2^2 \cdot 7 \cdot m^2$$
$$12m - 20 = 4(3m - 5) = 2^2(3m - 5)$$

Take each factor the greatest number of times it appears; then multiply.

$$\text{LCD} = 2^2 \cdot 7m^2(3m - 5) = 28m^2(3m - 5)$$

19. $\dfrac{7}{5b - 10}, \dfrac{11}{6b - 12}$ ▪ Factor each denominator.

$$5b - 10 = 5(b - 2)$$
$$6b - 12 = 6(b - 2) = 2 \cdot 3(b - 2)$$

Take each factor the greatest number of times it appears; then multiply.

$$\text{LCD} = 2 \cdot 3 \cdot 5(b - 2) = 30(b - 2)$$

21. $\dfrac{5}{c - d}, \dfrac{8}{d - c}$ ▪ The denominators, $c - d$ and $d - c$, are opposites of each other since

$$-(c - d) = -c + d = d - c.$$

Therefore, either $c - d$ or $d - c$ can be used as the LCD.

23. $\dfrac{3}{k^2 + 5k}, \dfrac{2}{k^2 + 3k - 10}$

Factor each denominator.

$$k^2 + 5k = k(k + 5)$$
$$k^2 + 3k - 10 = (k + 5)(k - 2)$$

$$\text{LCD} = k(k + 5)(k - 2)$$

25. $\dfrac{6}{a^2 + 6a}, \dfrac{-5}{a^2 + 3a - 18}$

Factor each denominator.

$$a^2 + 6a = a(a + 6)$$
$$a^2 + 3a - 18 = (a + 6)(a - 3)$$

$$\text{LCD} = a(a + 6)(a - 3)$$

27. $\dfrac{5}{p^2 + 8p + 15}, \dfrac{3}{p^2 - 3p - 18}, \dfrac{2}{p^2 - p - 30}$

Factor each denominator.

$$p^2 + 8p + 15 = (p + 5)(p + 3)$$
$$p^2 - 3p - 18 = (p - 6)(p + 3)$$
$$p^2 - p - 30 = (p - 6)(p + 5)$$

$$\text{LCD} = (p + 3)(p + 5)(p - 6)$$

29. $\dfrac{4}{11} = \dfrac{?}{55}$

Step 1   First factor the denominator on the right. Then compare the denominator on the left with the one on the right to decide what factors are missing.

$$\dfrac{4}{11} = \dfrac{?}{11 \cdot 5}$$

Step 2   A factor of 5 is missing.

Step 3   Multiply $\frac{4}{11}$ by $\frac{5}{5}$, which is equal to 1.

$$\dfrac{4}{11} = \dfrac{4}{11} \cdot \dfrac{5}{5} = \dfrac{20}{55}$$

31. $\dfrac{-5}{k} = \dfrac{?}{9k}$

A factor of 9 is missing in the first fraction, so multiply numerator and denominator by 9.

$$\dfrac{-5}{k} \cdot \dfrac{9}{9} = \dfrac{-45}{9k}$$

33. $\dfrac{13}{40y} = \dfrac{?}{80y^3}$

$80y^3 = (40y)(2y^2)$, so we must multiply the numerator and denominator by $2y^2$.

$$\dfrac{13}{40y} = \dfrac{13}{40y} \cdot \dfrac{2y^2}{2y^2} = \dfrac{26y^2}{80y^3}$$

35. $\dfrac{5t^2}{6r} = \dfrac{?}{42r^4}$

$42r^4 = (6r)(7r^3)$, so we must multiply the numerator and denominator by $7r^3$.

$$\frac{5t^2}{6r} = \frac{5t^2}{6r} \cdot \frac{7r^3}{7r^3} = \frac{35t^2r^3}{42r^4}$$

37. $\dfrac{5}{2(m+3)} = \dfrac{?}{8(m+3)}$

$8(m+3) = 2(m+3) \cdot 4$, so we must multiply the numerator and denominator by 4.

$$\frac{5}{2(m+3)} = \frac{5}{2(m+3)} \cdot \frac{4}{4} = \frac{20}{8(m+3)}$$

39. $\dfrac{-4t}{3t-6} = \dfrac{?}{12-6t}$

Factor the denominators.

$$3t - 6 = 3(t-2)$$
$$12 - 6t = -6(-2+t) = -6(t-2)$$

$-6(t-2) = 3(t-2) \cdot (-2)$, so we must multiply the numerator and denominator by -2.

$$\frac{-4t}{3t-6} \cdot \frac{-2}{-2} = \frac{8t}{-6t+12} \quad \text{or} \quad \frac{8t}{12-6t}$$

41. $\dfrac{14}{z^2-3z} = \dfrac{?}{z(z-3)(z-2)}$

Compared to the second denominator, $z(z-3)(z-2)$, the first denominator, $z(z-3)$, is missing a factor of $z-2$. Multiply the numerator and denominator by $z-2$.

$$\frac{14}{z^2-3z} = \frac{14}{z(z-3)} \cdot \frac{z-2}{z-2}$$
$$= \frac{14(z-2)}{z(z-3)(z-2)}$$

43. $\dfrac{2(b-1)}{b^2+b} = \dfrac{?}{b^3+3b^2+2b}$

Compared to the second denominator,

$$b^3 + 3b^2 + 2b = b(b^2 + 3b + 2)$$
$$= b(b+1)(b+2),$$

the first denominator, $b(b+1)$, is missing a factor of $b+2$. Multiply the numerator and denominator by $b+2$.

$$\frac{2(b-1)}{b(b+1)} \cdot \frac{b+2}{b+2} = \frac{2(b-1)(b+2)}{b^3+3b^2+2b}$$

15.4 Adding and Subtracting Rational Expressions

15.4 Margin Exercises

1. **(a)** $\dfrac{7}{15} + \dfrac{3}{15} = \dfrac{7+3}{15}$
$$= \dfrac{10}{15} = \dfrac{2}{3}$$

(b) $\dfrac{3}{y+4} + \dfrac{2}{y+4} = \dfrac{3+2}{y+4}$
$$= \dfrac{5}{y+4}$$

(c) $\dfrac{x}{x+y} + \dfrac{1}{x+y} = \dfrac{x+1}{x+y}$

(d) $\dfrac{a}{a+b} + \dfrac{b}{a+b} = \dfrac{a+b}{a+b} = 1$

(e) $\dfrac{x^2}{x+1} + \dfrac{x}{x+1} = \dfrac{x^2+x}{x+1}$
$$= \dfrac{x(x+1)}{x+1} = x$$

2. **(a)** $\dfrac{1}{10} + \dfrac{1}{15}$

Step 1 LCD $= 2 \cdot 3 \cdot 5 = 30$

Step 2 $\dfrac{1}{10} = \dfrac{1(3)}{10(3)} = \dfrac{3}{30}$
$$\dfrac{1}{15} = \dfrac{1(2)}{15(2)} = \dfrac{2}{30}$$

Step 3 $\dfrac{3}{30} + \dfrac{2}{30} = \dfrac{3+2}{30}$
$$= \dfrac{5}{30}$$

Step 4 $\dfrac{5}{30} = \dfrac{1(5)}{6(5)} = \dfrac{1}{6}$

(b) $\dfrac{6}{5x} + \dfrac{9}{2x}$

Step 1 LCD $= 5 \cdot 2 \cdot x = 10x$

Step 2 $\dfrac{6}{5x} = \dfrac{6(2)}{5x(2)} = \dfrac{12}{10x}$
$$\dfrac{9}{2x} = \dfrac{9(5)}{2x(5)} = \dfrac{45}{10x}$$

Step 3 $\dfrac{12}{10x} + \dfrac{45}{10x} = \dfrac{12+45}{10x}$
$$= \dfrac{57}{10x}$$

(c) $\dfrac{m}{3n} + \dfrac{2}{7n}$

Step 1 LCD $= 3 \cdot 7 \cdot n = 21n$

Step 2 $\dfrac{m}{3n} = \dfrac{m(7)}{3n(7)} = \dfrac{7m}{21n}$
$$\dfrac{2}{7n} = \dfrac{2(3)}{7n(3)} = \dfrac{6}{21n}$$

Step 3 $\dfrac{7m}{21n} + \dfrac{6}{21n} = \dfrac{7m+6}{21n}$

3.   **(a)** $\dfrac{2p}{3p+3} + \dfrac{5p}{2p+2}$

Since $3p + 3 = 3(p+1)$ and $2p + 2 = 2(p+1)$, the LCD is $2 \cdot 3(p+1) = 6(p+1)$.

$$\dfrac{2p}{3p+3} + \dfrac{5p}{2p+2}$$

$$= \dfrac{2(2p)}{2(3p+3)} + \dfrac{3(5p)}{3(2p+2)}$$

$$= \dfrac{4p}{6p+6} + \dfrac{15p}{6p+6}$$

$$= \dfrac{4p+15p}{6p+6}$$

$$= \dfrac{19p}{6p+6}$$

$$= \dfrac{19p}{6(p+1)}$$

(b) $\dfrac{4}{y^2-1} + \dfrac{6}{y+1}$

$$= \dfrac{4}{(y+1)(y-1)} + \dfrac{6}{y+1}$$

   Factor denominator.

$$= \dfrac{4}{(y+1)(y-1)} + \dfrac{6(y-1)}{(y+1)(y-1)}$$

   $LCD = (y+1)(y-1)$

$$= \dfrac{4+6(y-1)}{(y+1)(y-1)} \quad \text{Add numerators.}$$

$$= \dfrac{4+6y-6}{(y+1)(y-1)} \quad \text{Distributive property}$$

$$= \dfrac{6y-2}{(y+1)(y-1)} \quad \text{Combine terms.}$$

$$= \dfrac{2(3y-1)}{(y+1)(y-1)} \quad \text{Factor numerator.}$$

(c) $\dfrac{-2}{p+1} + \dfrac{4p}{p^2-1}$

$$= \dfrac{-2}{p+1} + \dfrac{4p}{(p+1)(p-1)} \quad \begin{array}{l}\textit{Factor}\\ \textit{denominator.}\end{array}$$

$$= \dfrac{-2(p-1)}{(p+1)(p-1)} + \dfrac{4p}{(p+1)(p-1)}$$

   $LCD = (p+1)(p-1)$

$$= \dfrac{-2(p-1)+4p}{(p+1)(p-1)} \quad \text{Add numerators.}$$

$$= \dfrac{-2p+2+4p}{(p+1)(p-1)} \quad \text{Distributive property}$$

$$= \dfrac{2p+2}{(p+1)(p-1)} \quad \text{Combine terms.}$$

$$= \dfrac{2(p+1)}{(p+1)(p-1)} \quad \text{Factor numerator.}$$

$$= \dfrac{2}{p-1} \quad \text{Lowest terms}$$

4.   **(a)** $\dfrac{2k}{k^2-5k+4} + \dfrac{3}{k^2-1}$

$$= \dfrac{2k}{(k-4)(k-1)} + \dfrac{3}{(k+1)(k-1)}$$

   Factor.

$$= \dfrac{2k(k+1)}{(k-4)(k-1)(k+1)}$$

$$+ \dfrac{3(k-4)}{(k-4)(k-1)(k+1)}$$

   $LCD = (k-4)(k-1)(k+1)$

$$= \dfrac{2k(k+1)+3(k-4)}{(k-4)(k-1)(k+1)} \quad \text{Add numerators.}$$

$$= \dfrac{2k^2+2k+3k-12}{(k-4)(k-1)(k+1)}$$

   Distributive property

$$= \dfrac{2k^2+5k-12}{(k-4)(k-1)(k+1)} \quad \text{Combine terms.}$$

$$= \dfrac{(2k-3)(k+4)}{(k-4)(k-1)(k+1)} \quad \text{Factor numerator.}$$

(b) $\dfrac{4m}{m^2+3m+2} + \dfrac{2m-1}{m^2+6m+5}$

$$= \dfrac{4m}{(m+2)(m+1)} + \dfrac{2m-1}{(m+5)(m+1)}$$

   Factor.

$$= \dfrac{4m(m+5)}{(m+2)(m+1)(m+5)}$$

$$+ \dfrac{(2m-1)(m+2)}{(m+5)(m+1)(m+2)}$$

   $LCD = (m+2)(m+1)(m+5)$

$$= \dfrac{4m^2+20m+2m^2+3m-2}{(m+2)(m+1)(m+5)}$$

$$= \dfrac{6m^2+23m-2}{(m+2)(m+1)(m+5)}$$

5. $\dfrac{m}{2m-3n} + \dfrac{n}{3n-2m}$

$$= \dfrac{m}{2m-3n} + \dfrac{n(-1)}{(3n-2m)(-1)}$$

$$= \dfrac{m}{2m-3n} + \dfrac{-n}{2m-3n}$$

$$= \dfrac{m-n}{2m-3n}, \text{ or } \dfrac{n-m}{3n-2m}$$

6.   **(a)** $\dfrac{3}{m^2} - \dfrac{2}{m^2} = \dfrac{3-2}{m^2} = \dfrac{1}{m^2}$

(b) $\dfrac{x}{2x+3} - \dfrac{3x+4}{2x+3}$

$$= \dfrac{x-(3x+4)}{2x+3} \quad \textit{Use parentheses.}$$

$$= \dfrac{x-3x-4}{2x+3}$$

$$= \dfrac{-2x-4}{2x+3}$$

If the numerator is factored, the answer can be written as $\dfrac{-2(x+2)}{2x+3}$.

(c) $\dfrac{5t}{t-1} - \dfrac{5+t}{t-1}$

$= \dfrac{5t - (5+t)}{t-1}$ *Use parentheses.*

$= \dfrac{5t - 5 - t}{t-1}$

$= \dfrac{4t - 5}{t-1}$

7.   **(a)** $\dfrac{1}{k+4} - \dfrac{2}{k} = \dfrac{k \cdot 1}{k(k+4)} - \dfrac{2(k+4)}{k(k+4)}$

$LCD = k(k+4)$

$= \dfrac{k - 2(k+4)}{k(k+4)}$

$= \dfrac{k - 2k - 8}{k(k+4)}$

$= \dfrac{-k - 8}{k(k+4)}$

(b) $\dfrac{6}{a+2} - \dfrac{1}{a-3}$

$= \dfrac{6(a-3)}{(a+2)(a-3)} - \dfrac{1(a+2)}{(a-3)(a+2)}$

$LCD = (a+2)(a-3)$

$= \dfrac{6(a-3) - 1(a+2)}{(a+2)(a-3)}$

$= \dfrac{6a - 18 - a - 2}{(a+2)(a-3)}$

$= \dfrac{5a - 20}{(a+2)(a-3)}$

$= \dfrac{5(a-4)}{(a+2)(a-3)}$

8.   **(a)** $\dfrac{5}{x-1} - \dfrac{3x}{1-x}$ ■ The denominators are opposites, so either may be used as the common denominator. We will choose $x-1$.

$\dfrac{5}{x-1} - \dfrac{3x}{1-x} = \dfrac{5}{x-1} - \dfrac{3x}{1-x} \cdot \dfrac{-1}{-1}$

$= \dfrac{5}{x-1} - \dfrac{-3x}{x-1}$

$= \dfrac{5 - (-3x)}{x-1}$

$= \dfrac{5 + 3x}{x-1}, \text{ or } \dfrac{-5 - 3x}{1-x}$

(b) $\dfrac{2y}{y-2} - \dfrac{1+y}{2-y}$ ■ The denominators are opposites, so either may be used as the common denominator. We will choose $y-2$.

$\dfrac{2y}{y-2} - \dfrac{1+y}{2-y} = \dfrac{2y}{y-2} - \dfrac{1+y}{2-y} \cdot \dfrac{-1}{-1}$

$= \dfrac{2y}{y-2} - \dfrac{-1-y}{y-2}$

$= \dfrac{2y - (-1-y)}{y-2}$

$= \dfrac{2y + 1 + y}{y-2}$

$= \dfrac{3y + 1}{y-2}, \text{ or } \dfrac{-3y-1}{2-y}$

9.   **(a)** $\dfrac{4y}{y^2-1} - \dfrac{5}{y^2+2y+1}$

$= \dfrac{4y}{(y-1)(y+1)} - \dfrac{5}{(y+1)^2}$ *Factor.*

$= \dfrac{4y(y+1)}{(y-1)(y+1)^2} - \dfrac{5(y-1)}{(y+1)^2(y-1)}$

$LCD = (y-1)(y+1)^2$

$= \dfrac{4y(y+1) - 5(y-1)}{(y-1)(y+1)^2}$

$= \dfrac{4y^2 + 4y - 5y + 5}{(y-1)(y+1)^2}$

$= \dfrac{4y^2 - y + 5}{(y-1)(y+1)^2}$

(b) $\dfrac{3}{r-5} - \dfrac{4}{r^2 - 10r + 25}$

$= \dfrac{3}{r-5} - \dfrac{4}{(r-5)^2}$ *Factor.*

$= \dfrac{3(r-5)}{(r-5)^2} - \dfrac{4}{(r-5)^2}$

$LCD = (r-5)^2$

$= \dfrac{3(r-5) - 4}{(r-5)(r-5)}$

$= \dfrac{3r - 15 - 4}{(r-5)(r-5)}$

$= \dfrac{3r - 19}{(r-5)^2}$

10.   **(a)** $\dfrac{2}{p^2 - 5p + 4} - \dfrac{3}{p^2 - 1}$

$= \dfrac{2}{(p-4)(p-1)} - \dfrac{3}{(p+1)(p-1)}$

$= \dfrac{2(p+1)}{(p-4)(p-1)(p+1)}$

$- \dfrac{3(p-4)}{(p+1)(p-1)(p-4)}$

$LCD = (p-4)(p-1)(p+1)$

$= \dfrac{(2p+2) - (3p-12)}{(p-4)(p-1)(p+1)}$

$= \dfrac{2p + 2 - 3p + 12}{(p-4)(p-1)(p+1)}$

$= \dfrac{14 - p}{(p-4)(p-1)(p+1)}$

(b) $\dfrac{q}{2q^2 + 5q - 3} - \dfrac{3q + 4}{3q^2 + 10q + 3}$

$= \dfrac{q}{(2q + 1)(q + 3)} - \dfrac{3q + 4}{(q + 3)(3q + 1)}$

$= \dfrac{q(3q + 1)}{(2q - 1)(q + 3)(3q + 1)}$

$\quad - \dfrac{(2q - 1)((3q + 4)}{(2q - 1)(q + 3)(3q + 1)}$

$\qquad LCD = (2q - 1)(q + 3)(3q + 1)$

$= \dfrac{(3q^2 + q) - (6q^2 + 5q - 4)}{(2q - 1)(q + 3)(3q + 1)}$

$= \dfrac{3q^2 + q - 6q^2 - 5q + 4}{(2q - 1)(q + 3)(3q + 1)}$

$= \dfrac{-3q^2 - 4q + 4}{(2q - 1)(q + 3)(3q + 1)}, \quad$ or

$\dfrac{(-3q + 2)(q + 2)}{(2q - 1)(q + 3)(3q + 1)}$

15.4 Section Exercises

1. $\dfrac{x}{x + 6} + \dfrac{6}{x + 6}$ ■ The denominators are the same, so the sum is found by adding the two numerators and keeping the same (common) denominator.

$$\dfrac{x}{x + 6} + \dfrac{6}{x + 6} = \dfrac{x + 6}{x + 6} = 1$$

Choice **E** is correct.

3. $\dfrac{6}{x - 6} - \dfrac{x}{x - 6}$ ■ The denominators are the same, so the difference is found by subtracting the two numerators and keeping the same (common) denominator.

$$\dfrac{6}{x - 6} - \dfrac{x}{x - 6} = \dfrac{6 - x}{x - 6}$$
$$= \dfrac{-1(x - 6)}{x - 6} = -1$$

Choice **C** is correct.

5. $\dfrac{x}{x + 6} - \dfrac{6}{x + 6}$ ■ The denominators are the same, so the difference is found by subtracting the two numerators and keeping the same (common) denominator.

$$\dfrac{x}{x + 6} - \dfrac{6}{x + 6} = \dfrac{x - 6}{x + 6}$$

Choice **B** is correct.

7. $\dfrac{1}{6} - \dfrac{1}{x}$ ■ The LCD is $6x$. Now rewrite each rational expression as a fraction with the LCD as its denominator.

$$\dfrac{1}{6} \cdot \dfrac{x}{x} = \dfrac{x}{6x}$$
$$\dfrac{1}{x} \cdot \dfrac{6}{6} = \dfrac{6}{6x}$$

Since the fractions now have a common denominator, subtract the numerators and use the LCD as the denominator of the difference.

$$\dfrac{1}{6} - \dfrac{1}{x} = \dfrac{x}{6x} - \dfrac{6}{6x} = \dfrac{x - 6}{6x}$$

Choice **G** is correct.

9. $\dfrac{4}{m} + \dfrac{7}{m}$ ■ The denominators are the same, so the sum is found by adding the two numerators and keeping the same (common) denominator.

$$\dfrac{4}{m} + \dfrac{7}{m} = \dfrac{4 + 7}{m} = \dfrac{11}{m}$$

11. $\dfrac{a + b}{2} - \dfrac{a - b}{2}$ ■ The denominators are the same, so the difference is found by subtracting the two numerators and keeping the same (common) denominator. Don't forget the parentheses on the second numerator.

$$\dfrac{a + b}{2} - \dfrac{a - b}{2} = \dfrac{(a + b) - (a - b)}{2}$$
$$= \dfrac{a + b - a + b}{2}$$
$$= \dfrac{2b}{2} = b$$

13. $\dfrac{5}{y + 4} - \dfrac{1}{y + 4}$ ■ The denominators are the same, so the difference is found by subtracting the two numerators and keeping the same (common) denominator.

$$\dfrac{5}{y + 4} - \dfrac{1}{y + 4} = \dfrac{5 - 1}{y + 4} = \dfrac{4}{y + 4}$$

15. $\dfrac{5m}{m + 1} - \dfrac{1 + 4m}{m + 1}$ ■ The denominators are the same, so the difference is found by subtracting the two numerators and keeping the same (common) denominator. Don't forget the parentheses on the second numerator.

$$\dfrac{5m}{m + 1} - \dfrac{1 + 4m}{m + 1} = \dfrac{5m - (1 + 4m)}{m + 1}$$
$$= \dfrac{5m - 1 - 4m}{m + 1}$$
$$= \dfrac{m - 1}{m + 1}$$

17. $\dfrac{x^2}{x+5} + \dfrac{5x}{x+5}$ *Same denominators*

$= \dfrac{x^2 + 5x}{x+5}$ *Add numerators.*

$= \dfrac{x(x+5)}{x+5}$ *Factor numerator.*

$= x$ *Lowest terms*

19. $\dfrac{y^2 - 3y}{y+3} + \dfrac{-18}{y+3}$ *Same denominators*

$= \dfrac{y^2 - 3y - 18}{y+3}$ *Add numerators.*

$= \dfrac{(y-6)(y+3)}{y+3}$ *Factor numerator.*

$= y - 6$ *Lowest terms*

21. To add or subtract rational expressions with the same denominators, combine the numerators and keep the same denominator. For example,

$$\frac{3x+2}{x-6} + \frac{-2x-8}{x-6} = \frac{x-6}{x-6}.$$

Then write in lowest terms. In this example, the sum simplifies to 1.

23. $\dfrac{z}{5} + \dfrac{1}{3}$ ■ The LCD is 15. Now rewrite each rational expression as a fraction with the LCD as its denominator.

$$\frac{z}{5} \cdot \frac{3}{3} = \frac{3z}{15}$$

$$\frac{1}{3} \cdot \frac{5}{5} = \frac{5}{15}$$

Since the fractions now have a common denominator, add the numerators and use the LCD as the denominator of the sum.

$$\frac{z}{5} + \frac{1}{3} = \frac{3z}{15} + \frac{5}{15} = \frac{3z+5}{15}$$

25. $\dfrac{5}{7} - \dfrac{r}{2} = \dfrac{5}{7} \cdot \dfrac{2}{2} - \dfrac{r}{2} \cdot \dfrac{7}{7}$ *LCD = 14*

$= \dfrac{10}{14} - \dfrac{7r}{14}$

$= \dfrac{10 - 7r}{14}$

27. $-\dfrac{3}{4} - \dfrac{1}{2x} = -\dfrac{3 \cdot x}{4 \cdot x} - \dfrac{1 \cdot 2}{2x \cdot 2}$ *LCD = 4x*

$= \dfrac{-3x - 2}{4x}$

29. $\dfrac{3}{5x} + \dfrac{9}{4x} = \dfrac{3}{5x} \cdot \dfrac{4}{4} + \dfrac{9}{4x} \cdot \dfrac{5}{5}$ *LCD = 20x*

$= \dfrac{12}{20x} + \dfrac{45}{20x}$

$= \dfrac{12 + 45}{20x} = \dfrac{57}{20x}$

31. $\dfrac{x+1}{6} + \dfrac{3x+3}{9}$

$= \dfrac{x+1}{6} + \dfrac{3(x+1)}{9}$

$= \dfrac{x+1}{6} + \dfrac{x+1}{3}$ *Reduce.*

$= \dfrac{x+1}{6} + \dfrac{x+1}{3} \cdot \dfrac{2}{2}$ *LCD = 6*

$= \dfrac{x+1+2x+2}{6}$

$= \dfrac{3x+3}{6}$

$= \dfrac{3(x+1)}{6} = \dfrac{x+1}{2}$

33. $\dfrac{x+3}{3x} + \dfrac{2x+2}{4x}$

$= \dfrac{x+3}{3x} + \dfrac{2(x+1)}{4x}$

$= \dfrac{x+3}{3x} + \dfrac{x+1}{2x}$ *Reduce.*

$= \dfrac{x+3}{3x} \cdot \dfrac{2}{2} + \dfrac{x+1}{2x} \cdot \dfrac{3}{3}$ *LCD = 6x*

$= \dfrac{2x+6+3x+3}{6x}$

$= \dfrac{5x+9}{6x}$

35. $\dfrac{2}{x+3} + \dfrac{1}{x}$

$= \dfrac{2(x)}{(x+3)(x)} + \dfrac{1(x+3)}{x(x+3)}$ *LCD = x(x + 3)*

$= \dfrac{2x}{x(x+3)} + \dfrac{x+3}{x(x+3)}$

$= \dfrac{2x + x + 3}{x(x+3)}$

$= \dfrac{3x+3}{x(x+3)}$, or $\dfrac{3(x+1)}{x(x+3)}$

37. $\dfrac{1}{k+5} - \dfrac{2}{k}$

$= \dfrac{1}{k+5} \cdot \dfrac{k}{k} - \dfrac{2}{k} \cdot \dfrac{k+5}{k+5}$ *LCD = k(k + 5)*

$= \dfrac{k}{k(k+5)} - \dfrac{2(k+5)}{k(k+5)}$

$= \dfrac{k - 2k - 10}{k(k+5)}$

$= \dfrac{-k - 10}{k(k+5)}$

39. $\dfrac{x}{x-2} + \dfrac{-8}{x^2-4}$

$= \dfrac{x}{x-2} + \dfrac{-8}{(x+2)(x-2)}$

$= \dfrac{x}{x-2} \cdot \dfrac{x+2}{x+2} + \dfrac{-8}{(x+2)(x-2)}$

$\qquad\qquad LCD = (x+2)(x-2)$

$= \dfrac{x(x+2)-8}{(x+2)(x-2)}$

$= \dfrac{x^2+2x-8}{(x+2)(x-2)}$

$= \dfrac{(x+4)(x-2)}{(x+2)(x-2)} = \dfrac{x+4}{x+2}$

41. $\dfrac{x}{x-2} + \dfrac{4}{x+2}$

$= \dfrac{x(x+2)}{(x-2)(x+2)} + \dfrac{4(x-2)}{(x+2)(x-2)}$

$\qquad\qquad LCD = (x+2)(x-2)$

$= \dfrac{x^2+2x}{(x-2)(x+2)} + \dfrac{4x-8}{(x+2)(x-2)}$

$= \dfrac{x^2+2x+4x-8}{(x+2)(x-2)}$

$= \dfrac{x^2+6x-8}{(x+2)(x-2)}$

43. $\dfrac{t}{t+2} + \dfrac{5-t}{t} - \dfrac{4}{t^2+2t}$

$= \dfrac{t}{t+2} + \dfrac{5-t}{t} - \dfrac{4}{t(t+2)}$

$= \dfrac{t}{t+2} \cdot \dfrac{t}{t} + \dfrac{5-t}{t} \cdot \dfrac{t+2}{t+2}$

$\quad - \dfrac{4}{t(t+2)} \qquad LCD = t(t+2)$

$= \dfrac{t \cdot t + (5-t)(t+2) - 4}{t(t+2)}$

$= \dfrac{t^2+5t+10-t^2-2t-4}{t(t+2)}$

$= \dfrac{3t+6}{t(t+2)}$

$= \dfrac{3(t+2)}{t(t+2)} = \dfrac{3}{t}$

45. $\dfrac{10}{m-2} + \dfrac{5}{2-m}$ ■ Since

$\qquad\qquad 2-m = -1(m-2),$

either $m-2$ or $2-m$ could be used as the LCD.

47. $\dfrac{4}{x-5} + \dfrac{6}{5-x}$ ■ The two denominators,

$x-5$ and $5-x$, are opposites of each other, so either one may be used as the common denominator. We will work the exercise both ways and compare the answers.

$\dfrac{4}{x-5} + \dfrac{6}{5-x}$

$= \dfrac{4}{x-5} + \dfrac{6(-1)}{(5-x)(-1)} \quad LCD = x-5$

$= \dfrac{4}{x-5} + \dfrac{-6}{x-5}$

$= \dfrac{-2}{x-5}$

$\dfrac{4}{x-5} + \dfrac{6}{5-x}$

$= \dfrac{4(-1)}{(x-5)(-1)} + \dfrac{6}{5-x} \quad LCD = 5-x$

$= \dfrac{-4}{5-x} + \dfrac{6}{5-x}$

$= \dfrac{2}{5-x}$

The two answers are equivalent, since

$$\dfrac{-2}{x-5} \cdot \dfrac{-1}{-1} = \dfrac{2}{5-x}.$$

49. $\dfrac{-1}{1-y} + \dfrac{3-4y}{y-1}$

The LCD is either $1-y$ or $y-1$. We'll use $1-y$.

$= \dfrac{-1}{1-y} + \dfrac{(3-4y)(-1)}{(y-1)(-1)}$

$= \dfrac{-1}{1-y} + \dfrac{-3+4y}{1-y}$

$= \dfrac{-1-3+4y}{1-y} = \dfrac{-4+4y}{1-y}$

$= \dfrac{-4(1-y)}{1-y} = -4$

51. $\dfrac{2}{x-y^2} + \dfrac{7}{y^2-x}$

LCD $= \underline{x-y^2}$ or $\underline{y^2-x}$

$= \dfrac{2}{x-y^2} + \dfrac{-1(7)}{-1(y^2-x)}$

$= \dfrac{2}{x-y^2} + \dfrac{-7}{-y^2+x}$

$= \dfrac{2}{x-y^2} + \dfrac{-7}{x-y^2}$

$= \dfrac{2+(-7)}{x-y^2}$

$= \dfrac{-5}{x-y^2}$

If y^2-x is used as the LCD, we obtain the equivalent answer $\dfrac{5}{y^2-x}$.

53. $\dfrac{x}{5x-3y} - \dfrac{y}{3y-5x}$

LCD $= 5x - 3y$ or $3y - 5x$

$$= \dfrac{x}{5x-3y} - \dfrac{-1(y)}{-1(3y-5x)}$$

$$= \dfrac{x}{5x-3y} - \dfrac{-y}{-3y+5x}$$

$$= \dfrac{x}{5x-3y} - \dfrac{-y}{5x-3y}$$

$$= \dfrac{x-(-y)}{5x-3y}$$

$$= \dfrac{x+y}{5x-3y}$$

If $3y - 5x$ is used as the LCD, we obtain the equivalent answer $\dfrac{-x-y}{3y-5x}$.

55. $\dfrac{3}{4p-5} + \dfrac{9}{5-4p}$

LCD $= 4p - 5$ or $5 - 4p$

$$= \dfrac{3}{4p-5} + \dfrac{-1(9)}{-1(5-4p)}$$

$$= \dfrac{3}{4p-5} + \dfrac{-9}{-5+4p}$$

$$= \dfrac{3}{4p-5} + \dfrac{-9}{4p-5}$$

$$= \dfrac{3+(-9)}{4p-5}$$

$$= \dfrac{-6}{4p-5}$$

If $5 - 4p$ is used as the LCD, we obtain the equivalent answer $\dfrac{6}{5-4p}$.

57. $\dfrac{2m}{m-n} - \dfrac{5m+n}{2m-2n}$

$$= \dfrac{2m}{m-n} - \dfrac{5m+n}{2(m-n)} \qquad \text{\textit{Factor second}}$$
$$\text{\textit{denominator.}}$$

$$= \dfrac{2m}{m-n} \cdot \dfrac{2}{2} - \dfrac{5m+n}{2(m-n)} \qquad \text{\textit{LCD} = 2(m-n)}$$

$$= \dfrac{4m-(5m+n)}{2(m-n)}$$

$$= \dfrac{4m-5m-n}{2(m-n)}$$

$$= \dfrac{-m-n}{2(m-n)}$$

$$= \dfrac{-(m+n)}{2(m-n)}$$

59. $\dfrac{5}{x^2-9} - \dfrac{x+2}{x^2+4x+3}$

To find the LCD, factor the denominators.

$$x^2 - 9 = (x+3)(x-3)$$
$$x^2 + 4x + 3 = (x+3)(x+1)$$

The LCD is $(x+3)(x-3)(x+1)$.

$$\dfrac{5}{x^2-9} - \dfrac{x+2}{x^2+4x+3}$$

$$= \dfrac{5}{(x+3)(x-3)} - \dfrac{x+2}{(x+1)(x+3)}$$

$$= \dfrac{5 \cdot (x+1)}{(x+3)(x-3) \cdot (x+1)}$$

$$\quad - \dfrac{(x+2) \cdot (x-3)}{(x+3)(x+1) \cdot (x-3)}$$

$$= \dfrac{5x+5}{(x+3)(x-3)(x+1)}$$

$$\quad - \dfrac{x^2-x-6}{(x+3)(x+1)(x-3)}$$

$$= \dfrac{(5x+5)-(x^2-x-6)}{(x+3)(x-3)(x+1)}$$

$$= \dfrac{5x+5-x^2+x+6}{(x+3)(x-3)(x+1)}$$

$$= \dfrac{-x^2+6x+11}{(x+3)(x-3)(x+1)}$$

61. $\dfrac{2q+1}{3q^2+10q-8} - \dfrac{3q+5}{2q^2+5q-12}$

$$= \dfrac{2q+1}{(3q-2)(q+4)} - \dfrac{3q+5}{(2q-3)(q+4)}$$

$$= \dfrac{(2q+1) \cdot (2q-3)}{(3q-2)(q+4) \cdot (2q-3)}$$

$$\quad - \dfrac{(3q+5) \cdot (3q-2)}{(2q-3)(q+4) \cdot (3q-2)}$$

$$\text{\textit{LCD} = (3q-2)(q+4)(2q-3)}$$

$$= \dfrac{(4q^2-4q-3)-(9q^2+9q-10)}{(3q-2)(q+4)(2q-3)}$$

$$= \dfrac{4q^2-4q-3-9q^2-9q+10}{(3q-2)(q+4)(2q-3)}$$

$$= \dfrac{-5q^2-13q+7}{(3q-2)(q+4)(2q-3)}$$

63. $\dfrac{4}{r^2 - r} + \dfrac{6}{r^2 + 2r} - \dfrac{1}{r^2 + r - 2}$

$= \dfrac{4}{r(r-1)} + \dfrac{6}{r(r+2)} - \dfrac{1}{(r+2)(r-1)}$

$= \dfrac{4 \cdot (r+2)}{r(r-1) \cdot (r+2)} + \dfrac{6 \cdot (r-1)}{r(r+2) \cdot (r-1)}$

$\quad - \dfrac{1 \cdot r}{r \cdot (r+2)(r-1)}$

$\qquad\qquad LCD = r(r+2)(r-1)$

$= \dfrac{4r + 8 + 6r - 6 - r}{r(r+2)(r-1)}$

$= \dfrac{9r + 2}{r(r+2)(r-1)}$

65. $\dfrac{x + 3y}{x^2 + 2xy + y^2} + \dfrac{x - y}{x^2 + 4xy + 3y^2}$

$= \dfrac{x + 3y}{(x+y)(x+y)} + \dfrac{x - y}{(x+3y)(x+y)}$

$= \dfrac{(x+3y) \cdot (x+3y)}{(x+y)(x+y) \cdot (x+3y)}$

$\quad + \dfrac{(x-y) \cdot (x+y)}{(x+3y)(x+y) \cdot (x+y)}$

$\qquad\qquad LCD = (x+y)(x+y)(x+3y)$

$= \dfrac{(x^2 + 6xy + 9y^2) + (x^2 - y^2)}{(x+y)(x+y)(x+3y)}$

$= \dfrac{2x^2 + 6xy + 8y^2}{(x+y)(x+y)(x+3y)}$

$= \dfrac{2(x^2 + 3xy + 4y^2)}{(x+y)(x+y)(x+3y)},$

or $\dfrac{2(x^2 + 3xy + 4y^2)}{(x+y)^2(x+3y)}$

67. $\dfrac{r + y}{18r^2 + 9ry - 2y^2} + \dfrac{3r - y}{36r^2 - y^2}$

$= \dfrac{r + y}{(3r+2y)(6r-y)} + \dfrac{3r - y}{(6r-y)(6r+y)}$

$= \dfrac{(r+y) \cdot (6r+y)}{(3r+2y)(6r-y) \cdot (6r+y)}$

$\quad + \dfrac{(3r-y) \cdot (3r+2y)}{(6r-y)(6r+y) \cdot (3r+2y)}$

$\qquad\qquad LCD = (3r+2y)(6r-y)(6r+y)$

$= \dfrac{6r^2 + 7ry + y^2}{(3r+2y)(6r-y)(6r+y)}$

$\quad + \dfrac{9r^2 + 3ry - 2y^2}{(3r+2y)(6r-y)(6r+y)}$

$= \dfrac{6r^2 + 7ry + y^2 + 9r^2 + 3ry - 2y^2}{(3r+2y)(6r-y)(6r+y)}$

$= \dfrac{15r^2 + 10ry - y^2}{(3r+2y)(6r-y)(6r+y)}$

69. **(a)** $P = 2l + 2w$

$= 2\left(\dfrac{3k+1}{10}\right) + 2\left(\dfrac{5}{6k+2}\right)$

$= 2\left(\dfrac{3k+1}{2 \cdot 5}\right) + 2\left(\dfrac{5}{2(3k+1)}\right)$

$= \dfrac{3k+1}{5} + \dfrac{5}{3k+1}$

To add the two fractions on the right, use $5(3k+1)$ as the LCD.

$P = \dfrac{(3k+1)(3k+1)}{5(3k+1)} + \dfrac{(5)(5)}{5(3k+1)}$

$= \dfrac{(3k+1)(3k+1) + (5)(5)}{5(3k+1)}$

$= \dfrac{9k^2 + 6k + 1 + 25}{5(3k+1)}$

$= \dfrac{9k^2 + 6k + 26}{5(3k+1)}$

(b) $A = l \cdot w$

$A = \dfrac{3k+1}{10} \cdot \dfrac{5}{6k+2}$

$= \dfrac{3k+1}{5 \cdot 2} \cdot \dfrac{5}{2(3k+1)}$

$= \dfrac{1}{2 \cdot 2} = \dfrac{1}{4}$

15.5   Complex Fractions

15.5   Margin Exercises

1.   (a) $\dfrac{\dfrac{2}{5} + \dfrac{1}{4}}{\dfrac{1}{2} + \dfrac{1}{3}} = \dfrac{\dfrac{2(4)}{5(4)} + \dfrac{1(5)}{4(5)}}{\dfrac{1(3)}{2(3)} + \dfrac{1(2)}{3(2)}}$

$= \dfrac{\dfrac{8+5}{20}}{\dfrac{3+2}{6}}$

$= \dfrac{13}{20} \div \dfrac{5}{6}$

$= \dfrac{13}{20} \cdot \dfrac{6}{5}$ *Multiply by reciprocal.*

$= \dfrac{13 \cdot 2 \cdot 3}{10 \cdot 2 \cdot 5}$

$= \dfrac{39}{50}$

(b) $\dfrac{6 + \dfrac{1}{x}}{5 - \dfrac{2}{x}} = \dfrac{\dfrac{6x}{x} + \dfrac{1}{x}}{\dfrac{5x}{x} - \dfrac{2}{x}}$

$= \dfrac{\dfrac{6x+1}{x}}{\dfrac{5x-2}{x}}$

$= \dfrac{6x+1}{x} \div \dfrac{5x-2}{x}$

$$= \frac{6x+1}{x} \cdot \frac{x}{5x-2} \quad \textit{Multiply by reciprocal.}$$

$$= \frac{6x+1}{5x-2}$$

(c)
$$\frac{9 - \dfrac{4}{p}}{\dfrac{2}{p} + 1} = \frac{\dfrac{9p}{p} - \dfrac{4}{p}}{\dfrac{2}{p} + \dfrac{p}{p}}$$

$$= \frac{\dfrac{9p-4}{p}}{\dfrac{2+p}{p}}$$

$$= \frac{9p-4}{p} \div \frac{2+p}{p}$$

$$= \frac{9p-4}{p} \cdot \frac{p}{2+p} \quad \textit{Multiply by reciprocal.}$$

$$= \frac{9p-4}{2+p}$$

2. (a)
$$\frac{\dfrac{rs^2}{t}}{\dfrac{r^2 s}{t^2}} = \frac{rs^2}{t} \div \frac{r^2 s}{t^2}$$

$$= \frac{rs^2}{t} \cdot \frac{t^2}{r^2 s} = \frac{st}{r}$$

(b)
$$\frac{\dfrac{m^2 n^3}{p}}{\dfrac{m^4 n}{p^2}} = \frac{m^2 n^3}{p} \div \frac{m^4 n}{p^2}$$

$$= \frac{m^2 n^3}{p} \cdot \frac{p^2}{m^4 n} = \frac{n^2 p}{m^2}$$

3.
$$\frac{\dfrac{2}{x-1} + \dfrac{1}{x+1}}{\dfrac{3}{x-1} - \dfrac{4}{x+1}}$$

$$= \frac{\dfrac{2(x+1)}{(x-1)(x+1)} + \dfrac{1(x-1)}{(x+1)(x-1)}}{\dfrac{3(x+1)}{(x-1)(x+1)} - \dfrac{4(x-1)}{(x+1)(x-1)}}$$

$$= \frac{\dfrac{2x+2}{(x-1)(x+1)} + \dfrac{x-1}{(x+1)(x-1)}}{\dfrac{3x+3}{(x-1)(x+1)} - \dfrac{4x-4}{(x+1)(x-1)}}$$

$$= \frac{\dfrac{(2x+2)+(x-1)}{(x+1)(x-1)}}{\dfrac{(3x+3)-(4x-4)}{(x-1)(x+1)}}$$

$$= \frac{\dfrac{2x+2+x-1}{(x+1)(x-1)}}{\dfrac{3x+3-4x+4}{(x-1)(x+1)}} = \frac{\dfrac{3x+1}{(x+1)(x-1)}}{\dfrac{-x+7}{(x-1)(x+1)}}$$

$$= \frac{3x+1}{(x+1)(x-1)} \cdot \frac{(x-1)(x+1)}{-x+7}$$

$$= \frac{3x+1}{-x+7}$$

4. (a)
$$\frac{\dfrac{2}{3} - \dfrac{1}{4}}{\dfrac{4}{9} + \dfrac{1}{2}} \qquad LCD = 2^2 \cdot 3^2 = 36$$

$$= \frac{36\left(\dfrac{2}{3} - \dfrac{1}{4}\right)}{36\left(\dfrac{4}{9} + \dfrac{1}{2}\right)} \quad \begin{array}{l}\textit{Multiply} \\ \textit{numerator and} \\ \textit{denominator} \\ \textit{by 36.}\end{array}$$

$$= \frac{36\left(\dfrac{2}{3}\right) - 36\left(\dfrac{1}{4}\right)}{36\left(\dfrac{4}{9}\right) + 36\left(\dfrac{1}{2}\right)}$$

$$= \frac{24-9}{16+18} = \frac{15}{34}$$

(b)
$$\frac{2 - \dfrac{6}{a}}{3 + \dfrac{4}{a}} \qquad LCD = a$$

$$= \frac{a\left(2 - \dfrac{6}{a}\right)}{a\left(3 + \dfrac{4}{a}\right)} \quad \begin{array}{l}\textit{Multiply} \\ \textit{numerator and} \\ \textit{denominator} \\ \textit{by a.}\end{array}$$

$$= \frac{a(2) - a\left(\dfrac{6}{a}\right)}{a(3) + a\left(\dfrac{4}{a}\right)}$$

$$= \frac{2a-6}{3a+4}$$

(c)
$$\frac{\dfrac{p}{5-p}}{\dfrac{4p}{2p+1}} \qquad LCD = (5-p)(2p+1)$$

$$= \frac{(5-p)(2p+1)\left(\dfrac{p}{5-p}\right)}{(5-p)(2p+1)\left(\dfrac{4p}{2p+1}\right)}$$

$$= \frac{(2p+1)(p)}{(5-p)(4p)} = \frac{2p+1}{4(5-p)}$$

5.
$$\frac{\dfrac{2}{5x} - \dfrac{3}{x^2}}{\dfrac{7}{4x} + \dfrac{1}{2x^2}} \qquad LCD = 20x^2$$

$$= \frac{20x^2\left(\dfrac{2}{5x} - \dfrac{3}{x^2}\right)}{20x^2\left(\dfrac{7}{4x} + \dfrac{1}{2x^2}\right)} \quad \textit{continued}$$

$$= \frac{20x^2\left(\frac{2}{5x}\right) - 20x^2\left(\frac{3}{x^2}\right)}{20x^2\left(\frac{7}{4x}\right) + 20x^2\left(\frac{1}{2x^2}\right)}$$

$$= \frac{8x - 60}{35x + 10}$$

6.   (a) $\dfrac{\dfrac{1}{x} + \dfrac{2}{x-1}}{\dfrac{2}{x} - \dfrac{4}{x-1}}$ *LCD = x(x − 1)*

$$= \frac{x(x-1)\left(\dfrac{1}{x} + \dfrac{2}{x-1}\right)}{x(x-1)\left(\dfrac{2}{x} - \dfrac{4}{x-1}\right)} \qquad \textit{Use Method 2.}$$

$$= \frac{(x-1)(1) + (x)(2)}{(x-1)(2) - (x)(4)}$$

$$= \frac{x - 1 + 2x}{2x - 2 - 4x}$$

$$= \frac{3x - 1}{-2x - 2}$$

(b) $\dfrac{1 - \dfrac{2}{x} - \dfrac{15}{x^2}}{1 + \dfrac{5}{x} + \dfrac{6}{x^2}}$ *LCD = x^2*

$$= \frac{x^2\left(1 - \dfrac{2}{x} - \dfrac{15}{x^2}\right)}{x^2\left(1 + \dfrac{5}{x} + \dfrac{6}{x^2}\right)} \qquad \textit{Use Method 2.}$$

$$= \frac{x^2 - 2x - 15}{x^2 + 5x + 6}$$

$$= \frac{(x-5)(x+3)}{(x+2)(x+3)} \qquad \textit{Factor.}$$

$$= \frac{x - 5}{x + 2} \qquad \textit{Lowest terms}$$

(c) $\dfrac{\dfrac{2x+3}{x-4}}{\dfrac{4x^2-9}{x^2-16}}$

$$= \frac{2x+3}{x-4} \div \frac{4x^2-9}{x^2-16}$$

$$= \frac{2x+3}{x-4} \cdot \frac{x^2-16}{4x^2-9} \qquad \begin{array}{l}\textit{Multiply by}\\ \textit{reciprocal.}\end{array}$$

$$= \frac{2x+3}{x-4} \cdot \frac{(x+4)(x-4)}{(2x+3)(2x-3)} \qquad \textit{Factor.}$$

$$= \frac{x+4}{2x-3} \qquad \textit{Lowest terms}$$

15.5   Section Exercises

1.   In a fraction, the fraction bar represents division.
For example, $\frac{3}{5}$ can be read "3 divided by 5."

3.   (a) The LCD of $\frac{1}{2}$ and $\frac{1}{3}$ is $2 \cdot 3 = 6$.
The simplified form of the numerator is

$$\frac{1}{2} - \frac{1}{3} = \frac{3}{6} - \frac{2}{6} = \frac{1}{6}.$$

(b)   The LCD of $\frac{5}{6}$ and $\frac{1}{12}$ is 12 since 12 is a
multiple of 6. The simplified form of the
denominator is

$$\frac{5}{6} - \frac{1}{12} = \frac{10}{12} - \frac{1}{12} = \frac{9}{12} = \frac{3}{4}.$$

(c) $\dfrac{\frac{1}{6}}{\frac{3}{4}} = \dfrac{1}{6} \div \dfrac{3}{4}$

(d) $\dfrac{1}{6} \div \dfrac{3}{4} = \dfrac{1}{6} \cdot \dfrac{4}{3}$

$$= \frac{1 \cdot 2 \cdot 2}{2 \cdot 3 \cdot 3} = \frac{2}{9}$$

In Exercises 5–32, either Method 1 or Method 2 can be
used to simplify each complex fraction. Only one
method will be shown for each exercise.

5.   To use Method 1, divide the numerator of the
complex fraction by the denominator.

$$\frac{-\frac{4}{3}}{\frac{2}{9}} = -\frac{4}{3} \div \frac{2}{9} = -\frac{4}{3} \cdot \frac{9}{2}$$

$$= -\frac{36}{6} = -6$$

7. $\dfrac{\dfrac{p}{q^2}}{\dfrac{p^2}{q}} = \dfrac{p}{q^2} \div \dfrac{p^2}{q} = \dfrac{p}{q^2} \cdot \dfrac{q}{p^2} = \dfrac{1}{pq}$

9.   To use Method 2, multiply the numerator and
denominator of the complex fraction by the LCD,
y^2.

$$\frac{\dfrac{x}{y^2}}{\dfrac{x^2}{y}} = \frac{y^2\left(\dfrac{x}{y^2}\right)}{y^2\left(\dfrac{x^2}{y}\right)}$$

$$= \frac{x}{yx^2} = \frac{1}{xy}$$

11. $\dfrac{\dfrac{4a^4b^3}{3a}}{\dfrac{2ab^4}{b^2}} = \dfrac{4a^4b^3}{3a} \div \dfrac{2ab^4}{b^2}$ *Method 1*

$$= \frac{4a^4b^3}{3a} \cdot \frac{b^2}{2ab^4}$$

$$= \frac{4a^4b^3 \cdot b^2}{3a \cdot 2ab^4}$$

$$= \frac{4a^4b^5}{6a^2b^4}$$

$$= \frac{2a^2b}{3}$$

13. $\dfrac{\dfrac{m+2}{3}}{\dfrac{m-4}{m}} = \dfrac{m+2}{3} \div \dfrac{m-4}{m}$ *Method 1*

$\qquad\qquad = \dfrac{m+2}{3} \cdot \dfrac{m}{m-4}$

$\qquad\qquad = \dfrac{m(m+2)}{3(m-4)}$

15. $\dfrac{\dfrac{2}{x} - 3}{\dfrac{2-3x}{2}}$ *Method 2;*
LCD = 2x

$\quad = \dfrac{2x\left(\dfrac{2}{x} - 3\right)}{2x\left(\dfrac{2-3x}{2}\right)}$

$\quad = \dfrac{2x\left(\dfrac{2}{x}\right) - 2x(3)}{x(2-3x)}$ *Distributive*
property

$\quad = \dfrac{4 - 6x}{x(2-3x)}$

$\quad = \dfrac{2(2-3x)}{x(2-3x)}$ *Factor.*

$\quad = \dfrac{2}{x}$ *Lowest*
terms

17. $\dfrac{\dfrac{1}{x} + x}{\dfrac{x^2+1}{8}} = \dfrac{8x\left(\dfrac{1}{x} + x\right)}{8x\left(\dfrac{x^2+1}{8}\right)}$ *Method 2;*
LCD = 8x

$\qquad\quad = \dfrac{8 + 8x^2}{x(x^2+1)}$ *Distributive*
property

$\qquad\quad = \dfrac{8(1 + x^2)}{x(x^2+1)}$ *Factor.*

$\qquad\quad = \dfrac{8}{x}$ *Lowest*
terms

19. $\dfrac{a - \dfrac{5}{a}}{a + \dfrac{1}{a}} = \dfrac{a\left(a - \dfrac{5}{a}\right)}{a\left(a + \dfrac{1}{a}\right)}$ *Method 2;*
LCD = a

$\qquad\quad = \dfrac{a^2 - 5}{a^2 + 1}$

21. $\dfrac{\dfrac{1}{2} + \dfrac{1}{p}}{\dfrac{2}{3} + \dfrac{1}{p}} = \dfrac{6p\left(\dfrac{1}{2} + \dfrac{1}{p}\right)}{6p\left(\dfrac{2}{3} + \dfrac{1}{p}\right)}$ *Method 2;*
LCD = 6p

$\qquad\quad = \dfrac{6p\left(\dfrac{1}{2}\right) + 6p\left(\dfrac{1}{p}\right)}{6p\left(\dfrac{2}{3}\right) + 6p\left(\dfrac{1}{p}\right)}$ *Distributive*
property

$\qquad\quad = \dfrac{3p + 6}{4p + 6}$

$\qquad\quad = \dfrac{3(p+2)}{2(2p+3)}$

23. $\dfrac{\dfrac{2}{p^2} - \dfrac{3}{5p}}{\dfrac{4}{p} + \dfrac{1}{4p}}$ *Method 2;*
LCD = 20p²

$\quad = \dfrac{20p^2\left(\dfrac{2}{p^2} - \dfrac{3}{5p}\right)}{20p^2\left(\dfrac{4}{p} + \dfrac{1}{4p}\right)}$

$\quad = \dfrac{20p^2\left(\dfrac{2}{p^2}\right) - 20p^2\left(\dfrac{3}{5p}\right)}{20p^2\left(\dfrac{4}{p}\right) + 20p^2\left(\dfrac{1}{4p}\right)}$ *Distributive*
property

$\quad = \dfrac{40 - 12p}{80p + 5p}$

$\quad = \dfrac{40 - 12p}{85p}$

25. $\dfrac{\dfrac{t}{t+2}}{\dfrac{4}{t^2-4}}$ *Method 1*

$\quad = \dfrac{t}{t+2} \div \dfrac{4}{t^2-4}$

$\quad = \dfrac{t}{t+2} \cdot \dfrac{t^2-4}{4}$ *Multiply by*
reciprocal.

$\quad = \dfrac{t \cdot (t+2)(t-2)}{(t+2) \cdot 4}$ *Factor.*

$\quad = \dfrac{t(t-2)}{4}$ *Lowest*
terms

27. $\dfrac{\dfrac{1}{k+1} - 1}{\dfrac{1}{k+1} + 1}$ *Method 2;*
LCD = k + 1

$= \dfrac{(k+1)\left(\dfrac{1}{k+1} - 1\right)}{(k+1)\left(\dfrac{1}{k+1} + 1\right)}$

$= \dfrac{(k+1)\left(\dfrac{1}{k+1}\right) - (k+1)(1)}{(k+1)\left(\dfrac{1}{k+1}\right) + (k+1)(1)}$ *Distributive property*

$= \dfrac{1 - 1(k+1)}{1 + 1(k+1)}$ *Distributive property*

$= \dfrac{1 - k - 1}{1 + k + 1}$ *Distributive property*

$= \dfrac{-k}{k+2}$

29. $\dfrac{2 + \dfrac{1}{x} - \dfrac{28}{x^2}}{3 + \dfrac{13}{x} + \dfrac{4}{x^2}}$ *Method 2;*
LCD = x²

$= \dfrac{x^2\left(2 + \dfrac{1}{x} - \dfrac{28}{x^2}\right)}{x^2\left(3 + \dfrac{13}{x} + \dfrac{4}{x^2}\right)}$

$= \dfrac{2x^2 + x - 28}{3x^2 + 13x + 4}$

$= \dfrac{(2x-7)(x+4)}{(3x+1)(x+4)}$ *Factor.*

$= \dfrac{2x-7}{3x+1}$ *Lowest terms*

31. $\dfrac{\dfrac{1}{m-1} + \dfrac{2}{m+2}}{\dfrac{2}{m+2} - \dfrac{1}{m-3}}$

$= \dfrac{(m-1)(m+2)(m-3)\left(\dfrac{1}{m-1} + \dfrac{2}{m+2}\right)}{(m-1)(m+2)(m-3)\left(\dfrac{2}{m+2} - \dfrac{1}{m-3}\right)}$

Method 2;
LCD = (m − 1)(m + 2)(m − 3)

$= \dfrac{(m+2)(m-3) + 2(m-1)(m-3)}{2(m-1)(m-3) - (m-1)(m+2)}$

Distributive property

$= \dfrac{(m-3)[(m+2) + 2(m-1)]}{(m-1)[2(m-3) - (m+2)]}$

*Factor out m − 3 in the numerator
and m − 1 in the denominator.*

$= \dfrac{(m-3)[m + 2 + 2m - 2]}{(m-1)[2m - 6 - m - 2]}$

Distributive property

$= \dfrac{3m(m-3)}{(m-1)(m-8)}$ *Combine like terms.*

33. $2 - \dfrac{2}{2 + \dfrac{2}{2+2}} = 2 - \dfrac{2}{2 + \dfrac{2}{4}}$

$= 2 - \dfrac{2}{\dfrac{5}{2}}$

$= 2 - 2 \cdot \dfrac{2}{5}$

$= 2 - \dfrac{4}{5}$

$= \dfrac{10}{5} - \dfrac{4}{5}$

$= \dfrac{6}{5}$

15.6 Solving Equations with Rational Expressions

15.6 Margin Exercises

1. **(a)** $\dfrac{2x}{3} - \dfrac{4x}{9}$ is a difference of two terms, so it is an *expression* to be simplified. Simplify by finding the LCD, writing each coefficient with this LCD, and combining like terms.

$\dfrac{2x}{3} - \dfrac{4x}{9} = \dfrac{2x \cdot 3}{3 \cdot 3} - \dfrac{4x}{9}$ *LCD = 9*

$= \dfrac{6x - 4x}{9}$

$= \dfrac{2x}{9}$

(b) $\dfrac{2x}{3} - \dfrac{4x}{9} = 2$ has an equality symbol, so this is an *equation* to be solved. Use the multiplication property of equality to clear fractions. The LCD is 9.

$\dfrac{2x}{3} - \dfrac{4x}{9} = 2$

$9\left(\dfrac{2x}{3} - \dfrac{4x}{9}\right) = 9(2)$ *Multiply by the LCD, 9.*

$9\left(\dfrac{2x}{3}\right) - 9\left(\dfrac{4x}{9}\right) = 9(2)$ *Distributive property*

$6x - 4x = 18$ *Multiply.*

$2x = 18$ *Combine like terms.*

$x = 9$ *Divide by 2.*

Check $x = 9$: $6 - 4 = 2$ *True*

The solution set is $\{9\}$.

2.   **(a)** $\dfrac{x}{5} + 3 = \dfrac{3}{5}$ ▪ Use the multiplication property of equality to clear fractions. The LCD is 5.

$$\dfrac{x}{5} + 3 = \dfrac{3}{5}$$

$$5\left(\dfrac{x}{5} + 3\right) = 5\left(\dfrac{3}{5}\right) \qquad \text{Multiply by the LCD, 5.}$$

$$5\left(\dfrac{x}{5}\right) + 5(3) = 5\left(\dfrac{3}{5}\right) \qquad \text{Distributive property}$$

$$x + 15 = 3$$

$$x = -12 \qquad \text{Subtract 15.}$$

Check $x = -12$: $-\dfrac{12}{5} + \dfrac{15}{5} = \dfrac{3}{5}$ *True*

The solution set is $\{-12\}$.

(b) $\dfrac{x}{2} - \dfrac{x}{3} = \dfrac{5}{6}$ ▪ Use the multiplication property of equality to clear fractions. The LCD is 6.

$$\dfrac{x}{2} - \dfrac{x}{3} = \dfrac{5}{6}$$

$$6\left(\dfrac{x}{2} - \dfrac{x}{3}\right) = 6\left(\dfrac{5}{6}\right) \qquad \text{Multiply by the LCD, 6.}$$

$$6\left(\dfrac{x}{2}\right) - 6\left(\dfrac{x}{3}\right) = 6\left(\dfrac{5}{6}\right) \qquad \text{Distributive property}$$

$$3x - 2x = 5$$

$$x = 5 \qquad \text{Combine terms.}$$

Check $x = 5$: $\dfrac{5}{2} - \dfrac{5}{3} = \dfrac{5}{6}$ *True*

The solution set is $\{5\}$.

3.   **(a)** $\dfrac{k}{6} - \dfrac{k+1}{4} = -\dfrac{1}{2}$ ▪ Use the multiplication property of equality to clear fractions. The LCD is 12.

$$\dfrac{k}{6} - \dfrac{k+1}{4} = -\dfrac{1}{2}$$

$$12\left(\dfrac{k}{6} - \dfrac{k+1}{4}\right) = 12\left(-\dfrac{1}{2}\right) \qquad \text{Multiply by the LCD, 12.}$$

$$12\left(\dfrac{k}{6}\right) - 12\left(\dfrac{k+1}{4}\right) = 12\left(-\dfrac{1}{2}\right) \qquad \text{Distributive property}$$

$$2k - 3(k+1) = -6 \qquad \text{Multiply.}$$

$$2k - 3k - 3 = -6 \qquad \text{Distributive property}$$

$$-k - 3 = -6 \qquad \text{Combine like terms.}$$

$$-k = -3 \qquad \text{Add 3.}$$

$$k = 3 \qquad \text{Divide by } -1.$$

Check $k = 3$: $\dfrac{1}{2} - 1 = -\dfrac{1}{2}$ *True*

The solution set is $\{3\}$.

(b) $\dfrac{2m-3}{5} - \dfrac{m}{3} = -\dfrac{6}{5}$

$$15\left(\dfrac{2m-3}{5} - \dfrac{m}{3}\right) = 15\left(-\dfrac{6}{5}\right)$$

Multiply by the LCD, 15.

$$3(2m - 3) - 5m = 3(-6)$$

$$6m - 9 - 5m = -18$$

$$m - 9 = -18 \qquad \text{Combine like terms.}$$

$$m = -9 \qquad \text{Add 9.}$$

Check $m = -9$: $-\dfrac{21}{5} + 3 = -\dfrac{6}{5}$ *True*

The solution set is $\{-9\}$.

4.

$$1 - \dfrac{2}{x+1} = \dfrac{2x}{x+1}$$

$$(x+1)\left(1 - \dfrac{2}{x+1}\right) = (x+1)\dfrac{2x}{x+1}$$

Multiply by the LCD, x + 1.

$$(x+1) - 2 = 2x$$

$$x - 1 = 2x$$

$$-1 = x \qquad \text{Subtract x.}$$

Check $x = -1$: $1 - \dfrac{2}{0} = -\dfrac{2}{0}$

The fractions are undefined. Thus, the proposed solution -1 must be rejected, and the solution set is $\emptyset$.

5.   **(a)** $\dfrac{4}{x^2 - 3x} = \dfrac{1}{x^2 - 9}$

$$\dfrac{4}{x(x-3)} = \dfrac{1}{(x-3)(x+3)} \qquad \text{Factor.}$$

Multiply by the LCD, x(x + 3)(x − 3).
Note that x ≠ −3 or 3.

$$x(x+3)(x-3)\left(\dfrac{4}{x(x-3)}\right)$$

$$= x(x+3)(x-3)\left(\dfrac{1}{(x-3)(x+3)}\right)$$

$$4(x+3) = 1x$$

$$4x + 12 = x$$

$$3x = -12 \qquad \text{Subtract x.}$$

$$x = -4 \qquad \text{Divide by 3.}$$

Check $x = -4$: $\dfrac{1}{7} = \dfrac{1}{7}$ *True*

The solution set is $\{-4\}$.

(b) $\dfrac{2}{p^2 - 2p} = \dfrac{3}{p^2 - p}$

$\dfrac{2}{p(p-2)} = \dfrac{3}{p(p-1)}$ *Factor.*

Multiply by the LCD, p(p − 2)(p − 1).
Note that p ≠ 0, 2, or 1.

$p(p-2)(p-1)\left(\dfrac{2}{p(p-2)}\right)$

$= p(p-1)(p-2)\left(\dfrac{3}{p(p-1)}\right)$

$2(p-1) = 3(p-2)$

$2p - 2 = 3p - 6$

$2p + 4 = 3p$ *Add 6.*

$4 = p$ *Subtract 2p.*

Check $p = 4$: $\dfrac{1}{4} = \dfrac{1}{4}$ *True*

The solution set is $\{4\}$.

6. **(a)** $\dfrac{2p}{p^2 - 1} = \dfrac{2}{p+1} - \dfrac{1}{p-1}$

$\dfrac{2p}{(p+1)(p-1)} = \dfrac{2}{p+1} - \dfrac{1}{p-1}$ *Factor.*

Multiply by the LCD, (p + 1)(p − 1).

$(p+1)(p-1)\left(\dfrac{2p}{(p+1)(p-1)}\right)$

$= (p+1)(p-1)\left(\dfrac{2}{p+1} - \dfrac{1}{p-1}\right)$

$2p = 2(p-1) - (p+1)$

$2p = 2p - 2 - p - 1$

$2p = p - 3$

$p = -3$ *Subtract p.*

Check $p = -3$: $-\dfrac{3}{4} = -1 + \dfrac{1}{4}$ *True*

The solution set is $\{-3\}$.

(b) $\dfrac{8r}{4r^2 - 1} = \dfrac{3}{2r+1} + \dfrac{3}{2r-1}$

$\dfrac{8r}{(2r+1)(2r-1)} = \dfrac{3}{2r+1} + \dfrac{3}{2r-1}$ *Factor.*

Multiply by the LCD, (2r + 1)(2r − 1).

$(2r+1)(2r-1)\left(\dfrac{8r}{(2r+1)(2r-1)}\right)$

$= (2r+1)(2r-1)\left(\dfrac{3}{2r+1} + \dfrac{3}{2r-1}\right)$

$8r = 3(2r-1) + 3(2r+1)$

$8r = 6r - 3 + 6r + 3$

$8r = 12r$

$0 = 4r$ *Subtract 8r.*

$0 = r$ *Divide by 4.*

Check $r = 0$: $0 = 3 - 3$ *True*

The solution set is $\{0\}$.

7. $\dfrac{2}{3x+1} - \dfrac{1}{x} = \dfrac{-6x}{3x+1}$

Multiply by the LCD, x(3x + 1).

$x(3x+1)\left(\dfrac{2}{3x+1} - \dfrac{1}{x}\right) = x(3x+1)\left(\dfrac{-6x}{3x+1}\right)$

$2x - (3x+1) = -6x^2$

$2x - 3x - 1 = -6x^2$

$-x - 1 = -6x^2$

$6x^2 - x - 1 = 0$

$(2x-1)(3x+1) = 0$

$x = \dfrac{1}{2}$ or $x = -\dfrac{1}{3}$

Check $x = \dfrac{1}{2}$: $\dfrac{4}{5} - 2 = -\dfrac{6}{5}$ *True*

Since $-\dfrac{1}{3}$ makes a denominator of the original equation equal 0, $-\dfrac{1}{3}$ is not a solution.
The solution set is $\{\dfrac{1}{2}\}$.

8. **(a)** $\dfrac{1}{x-2} + \dfrac{1}{5} = \dfrac{2}{5(x^2-4)}$

$\dfrac{1}{x-2} + \dfrac{1}{5} = \dfrac{2}{5(x-2)(x+2)}$ *Factor.*

Multiply by the LCD, 5(x − 2)(x + 2).

$5(x-2)(x+2)\left(\dfrac{1}{x-2} + \dfrac{1}{5}\right)$

$= 5(x-2)(x+2)\left(\dfrac{2}{5(x-2)(x+2)}\right)$

$5(x+2) + (x+2)(x-2) = 2$

$5x + 10 + x^2 - 4 = 2$

$x^2 + 5x + 6 = 2$

$x^2 + 5x + 4 = 0$

$(x+4)(x+1) = 0$

$x + 4 = 0$ or $x + 1 = 0$

$x = -4$ or $x = -1$

Check $x = -4$: $-\dfrac{1}{6} + \dfrac{1}{5} = \dfrac{1}{30}$ *True*

Check $x = -1$: $-\dfrac{1}{3} + \dfrac{1}{5} = -\dfrac{2}{15}$ *True*

The solution set is $\{-4, -1\}$.

(b) $\dfrac{6}{5a+10} - \dfrac{1}{a-5} = \dfrac{4}{a^2 - 3a - 10}$

$\dfrac{6}{5(a+2)} - \dfrac{1}{a-5} = \dfrac{4}{(a-5)(a+2)}$

Multiply by the LCD, 5(a + 2)(a − 5).

$5(a+2)(a-5)\left(\dfrac{6}{5(a+2)} - \dfrac{1}{a-5}\right)$

$= 5(a+2)(a-5)\left(\dfrac{4}{(a-5)(a+2)}\right)$

$6(a-5) - 5(a+2) = 5(4)$

$6a - 30 - 5a - 10 = 20$

$a - 40 = 20$

$a = 60$

Check $a = 60$: $\dfrac{6}{310} - \dfrac{1}{55} = \dfrac{2}{1705}$ *True*

The solution set is $\{60\}$.

9. **(a)** Solve $r = \dfrac{A - p}{pt}$ for A. ■ The goal is to isolate $\underline{A}$. Multiply by the LCD, pt.

$$r(pt) = \left(\dfrac{A - p}{pt}\right)(pt)$$
$$rpt = A - p$$
$$p + rpt = A \qquad Add\ p.$$

(b) Solve $p = \dfrac{x - y}{z}$ for y. ■ The goal is to isolate y. Multiply by the LCD, z.

$$pz = \left(\dfrac{x - y}{z}\right)z$$
$$pz = x - y$$
$$y + pz = x \qquad Add\ y.$$
$$y = x - pz \qquad Subtract\ pz.$$

(c) Solve $z = \dfrac{x}{x + y}$ for y. ■ The goal is to isolate y. Multiply by the LCD, $x + y$.

$$z(x + y) = \left(\dfrac{x}{x + y}\right)(x + y)$$
$$zx + zy = x$$
$$zy = x - zx \qquad Isolate\ zy.$$
$$y = \dfrac{x - zx}{z} \qquad Divide\ by\ z.$$

Another solution method is to multiply by $x + y$, divide by z, and then subtract x to get $y = \dfrac{x}{z} - x$.

10. **(a)** Solve $\dfrac{2}{x} = \dfrac{1}{y} + \dfrac{1}{z}$ for z. ■ The goal is to isolate z. Multiply by the LCD, xyz.

$$xyz\left(\dfrac{2}{x}\right) = xyz\left(\dfrac{1}{y} + \dfrac{1}{z}\right)$$
$$xyz\left(\dfrac{2}{x}\right) = xyz\left(\dfrac{1}{y}\right) + xyz\left(\dfrac{1}{z}\right)$$
$$2yz = xz + xy$$

Get the z-terms on one side.

$$2yz - xz = xy$$
$$z(2y - x) = xy \qquad Factor\ out\ z.$$
$$z = \dfrac{xy}{2y - x}, \quad or \quad z = \dfrac{-xy}{x - 2y}$$

(b) Solve $\dfrac{2}{x} = \dfrac{1}{y} + \dfrac{1}{z}$ for y. ■ The goal is to isolate y. Multiply by the LCD, xyz.

$$xyz\left(\dfrac{2}{x}\right) = xyz\left(\dfrac{1}{y} + \dfrac{1}{z}\right)$$
$$xyz\left(\dfrac{2}{x}\right) = xyz\left(\dfrac{1}{y}\right) + xyz\left(\dfrac{1}{z}\right)$$
$$2yz = xz + xy$$

Get the y-terms on one side.

$$2yz - xy = xz$$
$$y(2z - x) = xz \qquad Factor\ out\ y.$$
$$y = \dfrac{xz}{2z - x}, \quad or \quad y = \dfrac{-xz}{x - 2z}$$

15.6 Section Exercises

1. A value of the variable that appears to be a solution after both sides of an equation with rational expressions are multiplied by a variable expression is a <u>proposed</u> solution. It must be checked in the <u>original</u> equation to determine whether it is an actual solution.

3. $\dfrac{7}{8}x + \dfrac{1}{5}x$ is the sum of two terms, so it is an *expression* to be simplified. Simplify by finding the LCD, writing each coefficient with this LCD, and combining like terms.

$$\dfrac{7}{8}x + \dfrac{1}{5}x = \dfrac{35}{40}x + \dfrac{8}{40}x \quad LCD = 40$$
$$= \dfrac{43}{40}x \qquad \begin{array}{l}Combine\\ like\ terms.\end{array}$$

5. $\dfrac{7}{8}x + \dfrac{1}{5}x = 1$ has an equality symbol, so this is an *equation* to be solved. Use the multiplication property of equality to clear fractions. The LCD is 40.

$$\dfrac{7}{8}x + \dfrac{1}{5}x = 1$$
$$40\left(\dfrac{7}{8}x + \dfrac{1}{5}x\right) = 40 \cdot 1 \quad \begin{array}{l}Multiply\ by\\ the\ LCD,\ 40.\end{array}$$
$$40\left(\dfrac{7}{8}x\right) + 40\left(\dfrac{1}{5}x\right) = 40 \cdot 1 \quad \begin{array}{l}Distributive\\ property\end{array}$$
$$35x + 8x = 40 \quad Multiply.$$
$$43x = 40 \quad \begin{array}{l}Combine\\ like\ terms.\end{array}$$
$$x = \dfrac{40}{43} \quad Divide\ by\ 43.$$

Check $x = \frac{40}{43}$: $\frac{280}{344} + \frac{40}{215} = 1$ *True*

The solution set is $\left\{\frac{40}{43}\right\}$.

7. $\dfrac{3}{5}y - \dfrac{7}{10}y$ is the difference of two terms, so it is an *expression* to be simplified.

$$\dfrac{3}{5}y - \dfrac{7}{10}y = \dfrac{6}{10}y - \dfrac{7}{10}y \quad LCD = 10$$
$$= -\dfrac{1}{10}y \qquad \begin{array}{l}Combine\\ like\ terms.\end{array}$$

Note: In Exercises 9–26 and 29–52, all proposed solutions should be checked by substituting in the original equation.

9.
$$\dfrac{2}{3}x + \dfrac{1}{2}x = -7$$
$$6\left(\dfrac{2}{3}x + \dfrac{1}{2}x\right) = 6(-7) \quad \begin{array}{l}Multiply\ by\\ the\ LCD,\ 6.\end{array}$$
$$6\left(\dfrac{2}{3}x\right) + 6\left(\dfrac{1}{2}x\right) = -42 \quad \begin{array}{l}Distributive\\ property\end{array}$$

$$4x + 3x = -42 \quad \textit{Multiply.}$$

$$7x = -42 \quad \begin{array}{l}\textit{Combine like}\\ \textit{terms.}\end{array}$$

$$x = -6 \quad \textit{Divide by 7.}$$

Check $x = -6$: $-4 - 3 = -7$ *True*

The solution set is $\{-6\}$.

11.
$$\frac{3x}{5} - 6 = x$$

$$5\left(\frac{3x}{5} - 6\right) = 5(x) \quad \begin{array}{l}\textit{Multiply by}\\ \textit{the LCD, 5.}\end{array}$$

$$5\left(\frac{3x}{5}\right) - 5(6) = 5x \quad \begin{array}{l}\textit{Distributive}\\ \textit{property}\end{array}$$

$$3x - 30 = 5x$$

$$-30 = 2x$$

$$-15 = x$$

Check $x = -15$: $-9 - 6 = -15$ *True*

The solution set is $\{-15\}$.

13.
$$\frac{4m}{7} + m = 11$$

$$7\left(\frac{4m}{7} + m\right) = 7(11) \quad \begin{array}{l}\textit{Multiply by}\\ \textit{the LCD, 7.}\end{array}$$

$$7\left(\frac{4m}{7}\right) + 7(m) = 77 \quad \begin{array}{l}\textit{Distributive}\\ \textit{property}\end{array}$$

$$4m + 7m = 77$$

$$11m = 77$$

$$m = 7$$

Check $m = 7$: $4 + 7 = 11$ *True*

The solution set is $\{7\}$.

15.
$$\frac{z-1}{4} = \frac{z+3}{3}$$

$$12\left(\frac{z-1}{4}\right) = 12\left(\frac{z+3}{3}\right) \quad \begin{array}{l}\textit{Multiply by}\\ \textit{the LCD, 12.}\end{array}$$

$$3(z-1) = 4(z+3)$$

$$3z - 3 = 4z + 12 \quad \begin{array}{l}\textit{Distributive}\\ \textit{property}\end{array}$$

$$-15 = z$$

Check $z = -15$: $-4 = -4$ *True*

The solution set is $\{-15\}$.

17.
$$\frac{3p+6}{8} = \frac{3p-3}{16}$$

$$16\left(\frac{3p+6}{8}\right) = 16\left(\frac{3p-3}{16}\right) \quad \begin{array}{l}\textit{Multiply by}\\ \textit{the LCD, 16.}\end{array}$$

$$2(3p+6) = 3p - 3$$

$$6p + 12 = 3p - 3 \quad \begin{array}{l}\textit{Distributive}\\ \textit{property}\end{array}$$

$$3p = -15$$

$$p = -5$$

Check $p = -5$: $-\frac{9}{8} = -\frac{9}{8}$ *True*

The solution set is $\{-5\}$.

19.
$$\frac{2x+3}{-6} = \frac{3}{2}$$

$$-6\left(\frac{2x+3}{-6}\right) = -6\left(\frac{3}{2}\right) \quad \begin{array}{l}\textit{Multiply by}\\ \textit{the LCD, } -6.\end{array}$$

$$2x + 3 = -9$$

$$2x = -12$$

$$x = -6$$

Check $x = -6$: $\frac{3}{2} = \frac{3}{2}$ *True*

The solution set is $\{-6\}$.

21.
$$\frac{q+2}{3} + \frac{q-5}{5} = \frac{7}{3}$$

$$15\left(\frac{q+2}{3} + \frac{q-5}{5}\right) = 15\left(\frac{7}{3}\right) \quad \textit{LCD} = 15$$

$$15\left(\frac{q+2}{3}\right) + 15\left(\frac{q-5}{5}\right) = 5 \cdot 7$$

$$5(q+2) + 3(q-5) = 35$$

$$5q + 10 + 3q - 15 = 35$$

$$8q - 5 = 35$$

$$8q = 40$$

$$q = 5$$

Check $q = 5$: $\frac{7}{3} = \frac{7}{3}$ *True*

The solution set is $\{5\}$.

23.
$$\frac{t}{6} + \frac{4}{3} = \frac{t-2}{3}$$

$$6\left(\frac{t}{6} + \frac{4}{3}\right) = 6\left(\frac{t-2}{3}\right) \quad \begin{array}{l}\textit{Multiply by}\\ \textit{the LCD, 6.}\end{array}$$

$$6\left(\frac{t}{6}\right) + 6\left(\frac{4}{3}\right) = 2(t-2)$$

$$t + 8 = 2t - 4$$

$$12 = t$$

Check $t = 12$: $\frac{10}{3} = \frac{10}{3}$ *True*

The solution set is $\{12\}$.

25.
$$\frac{3m}{5} - \frac{3m-2}{4} = \frac{1}{5}$$

$$20\left(\frac{3m}{5} - \frac{3m-2}{4}\right) = 20\left(\frac{1}{5}\right) \quad \begin{array}{l}\textit{Multiply by}\\ \textit{the LCD, 20.}\end{array}$$

$$4(3m) - 5(3m-2) = 4$$

$$12m - 15m + 10 = 4$$

$$-3m + 10 = 4$$

$$-3m = -6$$

$$m = 2$$

Check $m = 2$: $\frac{1}{5} = \frac{1}{5}$ *True*

The solution set is $\{2\}$.

27.
$$\frac{3}{x+2} - \frac{5}{x} = 1$$

The denominators, $x + 2$ and x, are equal to 0 for the values -2 and 0, respectively. Thus, $x \neq -2$ and $x \neq 0$.

29. $\dfrac{-1}{(x+3)(x-4)} = \dfrac{1}{2x+1}$

The denominators, $(x+3)(x-4)$ and $2x+1$, are equal to 0 for the values -3, 4, and $-\frac{1}{2}$, respectively. Thus, $x \neq -3$, $x \neq 4$, and $x \neq -\frac{1}{2}$.

31. $\dfrac{4}{x^2+8x-9} + \dfrac{1}{x^2-4} = 0$

The denominators, $x^2 + 8x - 9 = (x+9)(x-1)$ and $x^2 - 4 = (x+2)(x-2)$, are equal to 0 for the values -9, 1, -2, and 2, respectively. Thus, $x \neq -9$, $x \neq 1$, $x \neq -2$, and $x \neq 2$.

33. $\dfrac{5-2x}{x} = \dfrac{1}{4}$

$4x\left(\dfrac{5-2x}{x}\right) = 4x\left(\dfrac{1}{4}\right)$ *Multiply by the LCD, 4x.*

$4(5-2x) = x$

$20 - 8x = x$ *Distributive property*

$-9x = -20$

$x = \dfrac{20}{9}$

Check $x = \dfrac{20}{9}$: $\dfrac{1}{4} = \dfrac{1}{4}$ *True*

The solution set is $\left\{\dfrac{20}{9}\right\}$.

35. $\dfrac{k}{k-4} - 5 = \dfrac{4}{k-4}$

$(k-4)\left(\dfrac{k}{k-4} - 5\right) = (k-4)\left(\dfrac{4}{k-4}\right)$

Multiply by the LCD, k − 4.

$(k-4)\left(\dfrac{k}{k-4}\right) - 5(k-4) = 4$ *Distributive property*

$k - 5k + 20 = 4$

$-4k = -16$

$k = 4$

The proposed solution is 4. However, 4 cannot be a solution because it makes the denominator $k - 4$ equal 0. Therefore, the given equation has *no solution* and the solution set is $\emptyset$.

37. $\dfrac{3}{x-1} + \dfrac{2}{4x-4} = \dfrac{7}{4}$

$\dfrac{3}{x-1} + \dfrac{2}{4(x-1)} = \dfrac{7}{4}$

$4(x-1)\left(\dfrac{3}{x-1} + \dfrac{2}{4(x-1)}\right) = 4(x-1)\left(\dfrac{7}{4}\right)$

Multiply by the LCD, 4(x − 1).

$4(3) + 2 = (x-1)(7)$

$14 = 7x - 7$

$21 = 7x$

$3 = x$

Check $x = 3$: $\dfrac{7}{4} = \dfrac{7}{4}$ *True*

The solution set is $\{3\}$.

39. $\dfrac{x}{3x+3} = \dfrac{2x-3}{x+1} - \dfrac{2x}{3x+3}$

$\dfrac{x}{3(x+1)} = \dfrac{2x-3}{x+1} - \dfrac{2x}{3(x+1)}$

$3(x+1)\left(\dfrac{x}{3(x+1)}\right) =$

$3(x+1)\left[\dfrac{2x-3}{x+1} - \dfrac{2x}{3(x+1)}\right]$

Multiply by the LCD, 3(x + 1).

$x = 3(x+1)\left(\dfrac{2x-3}{x+1}\right)$

$- 3(x+1)\left(\dfrac{2x}{3(x+1)}\right)$

$x = 3(2x-3) - 2x$

$x = 6x - 9 - 2x$

$x = 4x - 9$

$-3x = -9$

$x = 3$

Check $x = 3$: $\dfrac{1}{4} = \dfrac{1}{4}$ *True*

The solution set is $\{3\}$.

41. $\dfrac{2}{m} = \dfrac{m}{5m+12}$

$m(5m+12)\left(\dfrac{2}{m}\right) = m(5m+12)\left(\dfrac{m}{5m+12}\right)$

Multiply by the LCD, m(5m + 12).

$(5m+12)(2) = m(m)$

$10m + 24 = m^2$

$0 = m^2 - 10m - 24$

$0 = (m-12)(m+2)$

$m - 12 = 0$ or $m + 2 = 0$

$m = 12$ or $m = -2$

Check $m = -2$: $-1 = -1$ *True*

Check $m = 12$: $\dfrac{1}{6} = \dfrac{1}{6}$ *True*

The solution set is $\{-2, 12\}$.

43. $\dfrac{5x}{14x+3} = \dfrac{1}{x}$

$x(14x+3)\left(\dfrac{5x}{14x+3}\right) = x(14x+3)\left(\dfrac{1}{x}\right)$

Multiply by the LCD, x(14x + 3).

$x(5x) = (14x+3)(1)$

$5x^2 = 14x + 3$

$5x^2 - 14x - 3 = 0$

$(5x+1)(x-3) = 0$

$x = -\dfrac{1}{5}$ or $x = 3$

Check $x = -\dfrac{1}{5}$: $-5 = -5$ *True*

Check $x = 3$: $\dfrac{1}{3} = \dfrac{1}{3}$ *True*

The solution set is $\left\{-\dfrac{1}{5}, 3\right\}$.

45.

$$\frac{2}{z-1} - \frac{5}{4} = \frac{-1}{z+1}$$

$$4(z+1)(z-1)\left(\frac{2}{z-1} - \frac{5}{4}\right)$$

$$= 4(z+1)(z-1)\left(\frac{-1}{z+1}\right)$$

Multiply by the LCD, 4(z + 1)(z − 1).

$$8(z+1) - 5(z^2 - 1) = -4(z-1)$$
$$8z + 8 - 5z^2 + 5 = -4z + 4$$
$$-5z^2 + 12z + 9 = 0$$
$$5z^2 - 12z - 9 = 0$$
$$(5z+3)(z-3) = 0$$
$$z = -\tfrac{3}{5} \quad \text{or} \quad z = 3$$

Check $z = -\tfrac{3}{5}$: $-\tfrac{5}{2} = -\tfrac{5}{2}$ *True*

Check $z = 3$: $-\tfrac{1}{4} = -\tfrac{1}{4}$ *True*

The solution set is $\left\{-\tfrac{3}{5}, 3\right\}$.

47.

$$\frac{4}{x^2 - 3x} = \frac{1}{x^2 - 9}$$

$$\frac{4}{x(x-3)} = \frac{1}{(x+3)(x-3)} \qquad \begin{matrix}\textit{Factor}\\\textit{denominators.}\end{matrix}$$

$$x(x+3)(x-3)\cdot\frac{4}{x(x-3)}$$

$$= x(x+3)(x-3)\cdot\frac{1}{(x+3)(x-3)}$$

Multiply by the LCD, x(x + 3)(x − 3).

$$4(x+3) = x\cdot 1$$
$$4x + 12 = x$$
$$3x = -12$$
$$x = -4$$

Check $x = -4$: $\tfrac{1}{7} = \tfrac{1}{7}$ *True*

The solution set is $\{-4\}$.

49.

$$\frac{-2}{z+5} + \frac{3}{z-5} = \frac{20}{z^2 - 25}$$

$$\frac{-2}{z+5} + \frac{3}{z-5} = \frac{20}{(z+5)(z-5)} \quad \textit{Factor.}$$

$$(z+5)(z-5)\left(\frac{-2}{z+5} + \frac{3}{z-5}\right)$$

$$= (z+5)(z-5)\left(\frac{20}{(z+5)(z-5)}\right)$$

Multiply by the LCD, (z + 5)(z − 5).

$$-2(z-5) + 3(z+5) = 20$$
$$-2z + 10 + 3z + 15 = 20$$
$$z + 25 = 20$$
$$z = -5$$

The proposed solution, −5, cannot be a solution because it makes the denominators $z + 5$ and $z^2 - 25$ equal 0 and the corresponding fractions undefined. Therefore, the equation has *no solution* and the solution set is $\emptyset$.

51.

$$\frac{1}{x+4} + \frac{x}{x-4} = \frac{-8}{x^2 - 16}$$

$$\frac{1}{x+4} + \frac{x}{x-4} = \frac{-8}{(x+4)(x-4)}$$

$$(x+4)(x-4)\left(\frac{1}{x+4}\right)$$

$$+ (x+4)(x-4)\left(\frac{x}{x-4}\right)$$

$$= (x+4)(x-4)\left(\frac{-8}{(x+4)(x-4)}\right)$$

Multiply by the LCD, (x + 4)(x − 4).

$$1(x-4) + x(x+4) = -8$$
$$x - 4 + x^2 + 4x = -8$$
$$x^2 + 5x + 4 = 0$$
$$(x+4)(x+1) = 0$$
$$x = -4 \quad \text{or} \quad x = -1$$

The proposed solution, −4, cannot be a solution because it would make the denominators $x + 4$ and $x^2 - 16$ equal 0 and the corresponding fractions undefined.

Check $x = -1$: $\tfrac{1}{3} + \tfrac{1}{5} = \tfrac{8}{15}$ *True*

The solution set is $\{-1\}$.

53.

$$\frac{4}{3x+6} - \frac{3}{x+3} = \frac{8}{x^2 + 5x + 6}$$

$$\frac{4}{3(x+2)} - \frac{3}{x+3} = \frac{8}{(x+2)(x+3)}$$

$$3(x+2)(x+3)\cdot\frac{4}{3(x+2)}$$

$$- 3(x+2)(x+3)\cdot\frac{3}{x+3}$$

$$= 3(x+2)(x+3)\cdot\frac{8}{(x+2)(x+3)}$$

Multiply by the LCD, 3(x + 2)(x + 3).

$$4(x+3) - 3(x+2)(3) = 3(8)$$
$$4x + 12 - 9x - 18 = 24$$
$$-5x = 30$$
$$x = -6$$

Check $x = -6$: $-\tfrac{1}{3} - (-1) = \tfrac{2}{3}$ *True*

The solution set is $\{-6\}$.

55.

$$\frac{3x}{x^2 + 5x + 6}$$

$$= \frac{5x}{x^2 + 2x - 3} - \frac{2}{x^2 + x - 2}$$

$$\frac{3x}{(x+2)(x+3)}$$

$$= \frac{5x}{(x+3)(x-1)} - \frac{2}{(x-1)(x+2)}$$

$$(x+2)(x+3)(x-1)\cdot\left[\frac{3x}{(x+2)(x+3)}\right]$$

$$= (x+2)(x+3)(x-1) \cdot \left[\frac{5x}{(x+3)(x-1)} \right]$$

$$- (x+2)(x+3)(x-1) \cdot \left[\frac{2}{(x-1)(x+2)} \right]$$

Multiply by the LCD, (x + 2)(x + 3)(x – 1).

$$3x(x-1) = 5x(x+2) - 2(x+3)$$
$$3x^2 - 3x = 5x^2 + 10x - 2x - 6$$
$$0 = 2x^2 + 11x - 6$$
$$0 = (2x-1)(x+6)$$

Note to reader: We may skip writing out the zero-factor property since this step can be easily performed mentally.

$$x = \tfrac{1}{2} \quad \text{or} \quad x = -6$$

Check $x = \tfrac{1}{2}$: $\frac{6}{35} = -\frac{10}{7} - \left(-\frac{8}{5}\right)$ *True*

Check $x = -6$: $-\frac{3}{2} = -\frac{10}{7} - \frac{1}{14}$ *True*

The solution set is $\left\{ -6, \tfrac{1}{2} \right\}$.

57. $kr - mr = km$ ■ If you are solving for k, put both terms with k on one side and the remaining term on the other side.

$$kr - km = mr$$

59. Solve $m = \dfrac{kF}{a}$ for F.

We need to isolate F on one side of the equation.

$$m \cdot a = \left(\frac{kF}{a} \right)(a) \qquad \text{\textit{Multiply by} \textit{the LCD, a.}}$$

$$ma = kF$$

$$\frac{ma}{k} = \frac{kF}{k} \qquad \text{\textit{Divide by k.}}$$

$$\frac{ma}{k} = F$$

61. Solve $m = \dfrac{kF}{a}$ for a.

$$m \cdot a = \left(\frac{kF}{a} \right)(a) \qquad \text{\textit{Multiply by} \textit{the LCD, a.}}$$

$$ma = kF$$

$$\frac{ma}{m} = \frac{kF}{m} \qquad \text{\textit{Divide by m.}}$$

$$a = \frac{kF}{m}$$

63. Solve $m = \dfrac{y-b}{x}$ for y.

$$m \cdot x = \left(\frac{y-b}{x} \right)(x) \qquad \text{\textit{Multiply by} \textit{the LCD, x.}}$$

$$mx = y - b$$

$$mx + b = y \quad \text{or} \quad y = mx + b \quad \text{\textit{Add b.}}$$

65. Solve $I = \dfrac{E}{R+r}$ for R.

$$I(R+r) = \left(\frac{E}{R+r} \right)(R+r) \qquad \text{\textit{Multiply by} \textit{LCD, R + r.}}$$

$$IR + Ir = E \qquad \text{\textit{Dist. property}}$$

$$IR = E - Ir \qquad \text{\textit{Subtract Ir.}}$$

$$R = \frac{E - Ir}{I}, \qquad \text{\textit{Divide by I.}}$$

$$\text{or} \quad R = \frac{E}{I} - r$$

67. Solve $h = \dfrac{2A}{B+b}$ for b.

$$h(B+b) = \left(\frac{2A}{B+b} \right)(B+b) \qquad \text{\textit{Multiply by} \textit{LCD, B + b.}}$$

$$hB + hb = 2A \qquad \text{\textit{Dist. property}}$$

$$hb = 2A - hB \qquad \text{\textit{Subtract hB.}}$$

$$\frac{hb}{h} = \frac{2A - hB}{h} \qquad \text{\textit{Divide by h.}}$$

$$b = \frac{2A - hB}{h}, $$

$$\text{or} \quad b = \frac{2A}{h} - B$$

69. Solve $d = \dfrac{2S}{n(a+L)}$ for a.

$$d \cdot n(a+L) = \frac{2S}{n(a+L)} \cdot n(a+L)$$

Multiply by LCD, n(a + L).

$$dn(a+L) = 2S$$

$$dna + dnL = 2S \qquad \text{\textit{Dist. property}}$$

$$dna = 2S - dnL \qquad \text{\textit{Subtract dnL.}}$$

$$a = \frac{2S - dnL}{dn}, \qquad \text{\textit{Divide by dn.}}$$

$$\text{or} \quad a = \frac{2S}{dn} - L$$

71. Solve $\dfrac{2}{r} + \dfrac{3}{s} + \dfrac{1}{t} = 1$ for t.

The LCD of all the fractions in the equation is rst, so multiply both sides by rst.

$$rst \left(\frac{2}{r} + \frac{3}{s} + \frac{1}{t} \right) = rst(1)$$

$$rst \left(\frac{2}{r} \right) + rst \left(\frac{3}{s} \right) + rst \left(\frac{1}{t} \right) = rst$$

$$2st + 3rt + rs = rst$$

Since we are solving for t, get all terms with t on one side of the equation.

$$2st + 3rt - rst = -rs$$

Factor out the common factor t on the left.
$$t(2s + 3r - rs) = -rs$$
Finally, divide both sides by the coefficient of t, which is $2s + 3r - rs$.
$$t = \frac{-rs}{2s + 3r - rs}, \quad \text{or} \quad t = \frac{rs}{-2s - 3r + rs}$$

73. Solve $\dfrac{1}{a} - \dfrac{1}{b} - \dfrac{1}{c} = 2$ for c.
$$abc\left(\frac{1}{a} - \frac{1}{b} - \frac{1}{c}\right) = abc(2)$$
$$bc - ac - ab = 2abc$$
Get all the terms with c on one side.
$$bc - ac - 2abc = ab$$
$$c(b - a - 2ab) = ab$$
$$c = \frac{ab}{b - a - 2ab},$$
$$\text{or} \quad c = \frac{-ab}{-b + a + 2ab}$$

75. Solve $9x + \dfrac{3}{z} = \dfrac{5}{y}$ for z.
$$yz\left(9x + \frac{3}{z}\right) = yz\left(\frac{5}{y}\right) \quad \begin{array}{l}\textit{Multiply by}\\\textit{the LCD, yz.}\end{array}$$
$$yz(9x) + yz\left(\frac{3}{z}\right) = yz\left(\frac{5}{y}\right) \quad \begin{array}{l}\textit{Distributive}\\\textit{property}\end{array}$$
$$9xyz + 3y = 5z$$
$$9xyz - 5z = -3y \quad \begin{array}{l}\textit{Get the z terms}\\\textit{on one side.}\end{array}$$
$$z(9xy - 5) = -3y \quad \textit{Factor out z.}$$
$$z = \frac{-3y}{9xy - 5}, \quad \begin{array}{l}\textit{Divide by}\\\textit{9xy - 5.}\end{array}$$
$$\text{or} \quad z = \frac{3y}{5 - 9xy}$$

Summary Exercises
Simplifying Rational Expressions vs. Solving Rational Equations

1. No equality symbol appears, so this is an *expression*.
$$\frac{4}{p} + \frac{6}{p} = \frac{4 + 6}{p} = \frac{10}{p}$$

3. No equality symbol appears, so this is an *expression*.
$$\frac{1}{x^2 + x - 2} \div \frac{4x^2}{2x - 2}$$
$$= \frac{1}{x^2 + x - 2} \cdot \frac{2x - 2}{4x^2}$$
$$= \frac{1}{(x + 2)(x - 1)} \cdot \frac{2(x - 1)}{2 \cdot 2x^2}$$
$$= \frac{1}{2x^2(x + 2)}$$

5. No equality symbol appears, so this is an *expression*.
$$\frac{2x^2 + x - 6}{2x^2 - 9x + 9} \cdot \frac{x^2 - 2x - 3}{x^2 - 1}$$
$$= \frac{(2x - 3)(x + 2)(x - 3)(x + 1)}{(2x - 3)(x - 3)(x + 1)(x - 1)}$$
$$= \frac{x + 2}{x - 1}$$

7. $\dfrac{x - 4}{5} = \dfrac{x + 3}{6}$

There is an equality symbol, so this is an *equation*.
$$30\left(\frac{x - 4}{5}\right) = 30\left(\frac{x + 3}{6}\right) \quad \begin{array}{l}\textit{Multiply by}\\\textit{the LCD, 30.}\end{array}$$
$$6(x - 4) = 5(x + 3)$$
$$6x - 24 = 5x + 15$$
$$x = 39$$

Check $x = 39$: $7 = 7$ *True*

The solution set is $\{39\}$.

9. No equality symbol appears, so this is an *expression*.
$$\frac{4}{p + 2} + \frac{1}{3p + 6}$$
$$= \frac{4}{p + 2} + \frac{1}{3(p + 2)}$$
$$= \frac{3 \cdot 4}{3(p + 2)} + \frac{1}{3(p + 2)} \quad LCD = 3(p+2)$$
$$= \frac{12 + 1}{3(p + 2)}$$
$$= \frac{13}{3(p + 2)}$$

11. $\dfrac{3}{t - 1} + \dfrac{1}{t} = \dfrac{7}{2}$

There is an equality symbol, so this is an *equation*.
$$2t(t - 1)\left(\frac{3}{t - 1} + \frac{1}{t}\right) = 2t(t - 1)\left(\frac{7}{2}\right)$$
$$\textit{Multiply by the LCD, 2t(t - 1).}$$
$$2t(t - 1)\left(\frac{3}{t - 1}\right)$$
$$+ 2t(t - 1)\left(\frac{1}{t}\right) = 7t(t - 1)$$
$$2t(3) + 2(t - 1) = 7t(t - 1)$$
$$6t + 2t - 2 = 7t^2 - 7t$$
$$0 = 7t^2 - 15t + 2$$
$$0 = (7t - 1)(t - 2)$$
$$t = \tfrac{1}{7} \quad \text{or} \quad t = 2$$

Check $t = \tfrac{1}{7}$: $-\tfrac{7}{2} + 7 = \tfrac{7}{2}$ *True*

Check $t = 2$: $3 + \tfrac{1}{2} = \tfrac{7}{2}$ *True*

The solution set is $\{\tfrac{1}{7}, 2\}$.

13. No equality symbol appears, so this is an *expression*.

$$\frac{5}{4z} - \frac{2}{3z} = \frac{3 \cdot 5}{3 \cdot 4z} - \frac{4 \cdot 2}{4 \cdot 3z} \quad LCD = 12z$$

$$= \frac{15}{12z} - \frac{8}{12z}$$

$$= \frac{15 - 8}{12z} = \frac{7}{12z}$$

15. No equality symbol appears, so this is an *expression*.

$$\frac{1}{m^2 + 5m + 6} + \frac{2}{m^2 + 4m + 3}$$

$$= \frac{1}{(m+2)(m+3)} + \frac{2}{(m+1)(m+3)}$$

$$= \frac{1(m+1)}{(m+1)(m+2)(m+3)}$$

$$+ \frac{2(m+2)}{(m+1)(m+2)(m+3)}$$

$$LCD = (m+1)(m+2)(m+3)$$

$$= \frac{(m+1) + (2m+4)}{(m+1)(m+2)(m+3)}$$

$$= \frac{3m+5}{(m+1)(m+2)(m+3)}$$

17. $\dfrac{2}{x+1} + \dfrac{5}{x-1} = \dfrac{10}{x^2-1}$

There is an equality symbol, so this is an *equation*.

$$\frac{2}{x+1} + \frac{5}{x-1} = \frac{10}{(x+1)(x-1)}$$

$$(x+1)(x-1)\left(\frac{2}{x+1} + \frac{5}{x-1}\right)$$

$$= (x+1)(x-1)\left[\frac{10}{(x+1)(x-1)}\right]$$

Multiply by the LCD, (x + 1)(x − 1).

$$(x+1)(x-1)\left(\frac{2}{x+1}\right)$$

$$+ (x+1)(x-1)\left(\frac{5}{x-1}\right) = 10$$

Distributive property

$$2(x-1) + 5(x+1) = 10$$

$$2x - 2 + 5x + 5 = 10$$

$$3 + 7x = 10$$

$$7x = 7$$

$$x = 1$$

Replacing x by 1 in the original equation makes the denominators $x - 1$ and $x^2 - 1$ equal to 0, so there is *no solution* and the solution set is $\emptyset$.

15.7 Applications of Rational Expressions

15.7 Margin Exercises

1. **(a)** *Step 1* We are trying to find a <u>number</u>.

Step 2 Let $x =$ the number <u>added</u> to the numerator and <u>subtracted</u> from the denominator of $\frac{5}{8}$.

Step 3 The expression $\dfrac{5+x}{8-x}$ represents the new fraction. Write "the new fraction equals the reciprocal of $\frac{5}{8}$" as an equation.

$$\frac{5+x}{8-x} = \frac{8}{5}$$

Step 4 Multiply by the LCD, $5(8 - x)$.

$$5(8-x)\frac{5+x}{8-x} = 5(8-x)\frac{8}{5}$$

$$5(5+x) = 8(8-x)$$

$$25 + 5x = 64 - 8x$$

$$13x = 39$$

$$x = 3$$

Step 5 The number is 3.

Step 6 If 3 is added to the numerator and subtracted from the denominator of $\frac{5}{8}$, the result is $\frac{8}{5}$, which is the reciprocal of $\frac{5}{8}$.

(b) *Step 2* Let $x =$ the numerator of the original fraction. Then $x + 1 =$ the denominator.

Step 3 The original fraction is $\dfrac{x}{x+1}$. Add 6 to the numerator and subtract 6 from the denominator of this fraction to get

$$\frac{x+6}{x+1-6} = \frac{x+6}{x-5}.$$

The result is $\frac{15}{4}$, so

$$\frac{x+6}{x-5} = \frac{15}{4}.$$

Step 4 Multiply by the LCD, $4(x - 5)$.

$$4(x-5)\frac{x+6}{x-5} = 4(x-5)\left(\frac{15}{4}\right)$$

$$4(x+6) = 15(x-5)$$

$$4x + 24 = 15x - 75$$

$$24 = 11x - 75$$

$$99 = 11x$$

$$9 = x$$

Step 5 The original fraction is

$$\frac{9}{9+1} = \frac{9}{10}.$$

Step 6 If we add 6 to the numerator and subtract 6 from the denominator, the fraction $\frac{9}{10}$ becomes $\frac{15}{4}$, as required.

2.   (a) $r = \dfrac{d}{t} = \dfrac{100}{9.58} \approx 10.44$ m per sec

His average rate was 10.44 meters per second.

(b) $t = \dfrac{d}{r} = \dfrac{3000}{5.568} \approx 539$ sec

Her winning time was 539 seconds, or 8 minutes, 59 seconds.

(c) $d = rt = (145)(2) = 290$ mi

The distance between Chicago and St. Louis is 290 miles.

3.   (a) *Steps 1 and 2*   Let $t =$ the number of <u>hours</u> until the distance between Lupe and Maria is <u>55 miles</u>.

Step 3   Complete the table.

	Rate $\times$	Time $=$	Distance
Lupe	10	t	$10t$
Maria	12	t	$12t$

Because Lupe and Maria are traveling in opposite directions, we must <u>add</u> the distances they travel to find the distance between them. The sum of the distances is 55 miles, so

$$10t + 12t = 55.$$

Step 4   Solve this equation.

$$22t = 55$$
$$t = \frac{55}{22} = 2.5 \quad \text{or} \quad 2\frac{1}{2}$$

Step 5   They will be 55 miles apart in $2\frac{1}{2}$ hours.

Step 6   Lupe will travel $10(2.5) = 25$ miles and Maria will travel $12(2.5) = 30$ miles. $25 + 30 = 55$, as required.

(b) Let $t =$ the number of hours until the distance between the boats is 35 miles.

	Rate $\times$	Time $=$	Distance
Slower Boat	18	t	$18t$
Faster Boat	25	t	$25t$

The difference of the distances is 35 miles.

$$25t - 18t = 35$$
$$7t = 35$$
$$t = 5$$

They will be 35 miles apart in 5 hours.

4.   (a) *Steps 1 and 2*   Let $x =$ the <u>rate</u> of the boat with no current.

Step 3   Complete the table.

	d	r	t
Against the Current	10	$x - 4$	$\dfrac{10}{x-4}$
With the Current	30	$x + 4$	$\dfrac{30}{x+4}$

Since the times are equal, we get the following equation.

$$\frac{30}{x+4} = \frac{10}{x-4}$$

Step 4   Solve this equation. Multiply by the LCD, $(x+4)(x-4)$.

$$30(x-4) = 10(x+4)$$
$$30x - 120 = 10x + 40$$
$$20x - 120 = 40$$
$$20x = 160$$
$$x = 8$$

Step 5   The rate of the boat with no current is 8 miles per hour.

Step 6   Against the current, the boat will take $\dfrac{10}{8-4} = 2.5$ hours. With the current, the boat will take $\dfrac{30}{8+4} = 2.5$ hours, as required.

(b) Let $x =$ the rate of the plane in still air. Complete a table.

	d	r	t
With the Wind	450	$x + 15$	$\dfrac{450}{x+15}$
Against the Wind	375	$x - 15$	$\dfrac{375}{x-15}$

Since the times are equal, we get the following equation.

$$\frac{450}{x+15} = \frac{375}{x-15}$$

Multiply by the LCD, $(x+15)(x-15)$.

$$450(x-15) = 375(x+15)$$
$$18(x-15) = 15(x+15) \qquad \textit{Divide by 25.}$$
$$18x - 270 = 15x + 225$$
$$3x = 495$$
$$x = 165$$

The plane's rate is 165 miles per hour.

5.   (a) *Steps 1 and 2*   Let $x =$ the number of <u>hours</u> it will take for Michael and Lindsay to paint the room, working <u>together</u>.

Step 3   Complete the table.

	Rate	Time Working Together	Fractional Part of the Job Done When Working Together
Michael	$\frac{1}{8}$	x	$\frac{1}{8}x$
Lindsay	$\frac{1}{6}$	x	$\frac{1}{6}x$

Since together Michael and Lindsay complete 1 whole job, we must add their individual parts and set the sum equal to 1.

$$\tfrac{1}{8}x + \tfrac{1}{6}x = 1$$

Step 4 Solve this equation. Multiply by the LCD, 48.

$$48\left(\tfrac{1}{8}x + \tfrac{1}{6}x\right) = 48(1)$$
$$48\left(\tfrac{1}{8}x\right) + 48\left(\tfrac{1}{6}x\right) = 48(1)$$
$$6x + 8x = 48$$
$$14x = 48$$
$$x = \tfrac{18}{14} = \tfrac{21}{7} \text{ or } 3\tfrac{3}{7}$$

Step 5 Working together, Michael and Lindsay can paint the room in $3\tfrac{3}{7}$ hours.

Step 6 Michael will do $\tfrac{1}{8}\left(\tfrac{24}{7}\right) = \tfrac{3}{7}$ of the job and Lindsay will do $\tfrac{1}{6}\left(\tfrac{24}{7}\right) = \tfrac{4}{7}$ of the job. $\tfrac{3}{7} + \tfrac{4}{7} = 1$ job, as required.

(b) Let $x =$ the number of hours it will take for Roberto and Marco to detail the Camaro, working together.

	Rate	Time Working Together	Fractional Part of the Job Done When Working Together
Roberto	$\tfrac{1}{2}$	x	$\tfrac{1}{2}x$
Marco	$\tfrac{1}{3}$	x	$\tfrac{1}{3}x$

Since together Roberto and Marco complete 1 whole job, we must add their individual parts and set the sum equal to 1.

$$\tfrac{1}{2}x + \tfrac{1}{3}x = 1$$
$$6\left(\tfrac{1}{2}x + \tfrac{1}{3}x\right) = 6(1) \quad \textit{Multiply by}$$
$$\qquad\qquad\qquad\qquad \textit{the LCD, 6.}$$
$$6\left(\tfrac{1}{2}x\right) + 6\left(\tfrac{1}{3}x\right) = 6$$
$$3x + 2x = 6$$
$$5x = 6$$
$$x = \tfrac{6}{5} \text{ or } 1\tfrac{1}{5}$$

Working together, Roberto and Marco can detail the Camaro in $1\tfrac{1}{5}$ hours.

15.7 Section Exercises

1. When the hawk flies *into a headwind*, the wind works against the hawk, so the rate of the hawk is 5 miles per hour *less* than if it traveled in still air, or $(m - 5)$ miles per hour.

When the hawk flies *with a tailwind*, the wind pushes the hawk, so the rate of the hawk is 5 miles per hour *more* than if it traveled in still air, or $(m + 5)$ miles per hour.

3. If it takes Elayn 10 hours to do a job, her rate is $\tfrac{1}{10}$ job per hour.

5. **(a)** Let $x =$ <u>the amount</u>.

(b) An expression for "the numerator of the fraction $\tfrac{5}{6}$ is increased by an amount" is $\underline{5 + x}$. We could also use $\dfrac{5 + x}{6}$.

(c) An equation that can be used to solve the problem is

$$\frac{5 + x}{6} = \frac{13}{3}.$$

7. ***Step 2*** Let $x =$ the numerator of the original fraction. Then $x - 4 =$ the denominator of the original fraction.

Step 3 If 3 is added to both the numerator and denominator, the resulting fraction is equivalent to $\tfrac{3}{2}$ translates to

$$\frac{x + 3}{(x - 4) + 3} = \frac{3}{2}.$$

Step 4 Since we have a fraction equal to another fraction, we can use cross multiplication.

$$2(x + 3) = 3[(x - 4) + 3]$$
$$2x + 6 = 3x - 3$$
$$9 = x$$

Step 5 The original fraction is

$$\frac{x}{x - 4} = \frac{9}{9 - 4} = \frac{9}{5}.$$

Step 6 Adding 3 to both the numerator and the denominator gives us

$$\frac{9 + 3}{5 + 3} = \frac{12}{8},$$

which is equivalent to $\tfrac{3}{2}$.

9. ***Step 2*** Let $x =$ the numerator of the original fraction. Then $3x =$ the denominator of the original fraction.

Step 3 If 2 is added to the numerator and subtracted from the denominator, the resulting fraction is equivalent to 1 translates to

$$\frac{x + 2}{3x - 2} = 1.$$

Step 4 $\quad x + 2 = 1(3x - 2)$
$$x + 2 = 3x - 2$$
$$4 = 2x$$
$$2 = x$$

Step 5 The original fraction is

$$\frac{x}{3x} = \frac{2}{3(2)} = \frac{2}{6}.$$

Step 6 $\quad \dfrac{2 + 2}{6 - 2} = \dfrac{4}{4} = 1$

11. *Step 2* Let $x =$ the number.

Step 3 One-sixth of a number is 5 more than the same number translates to

$$\tfrac{1}{6}x = 5 + x.$$

Step 4 Multiply both sides by the LCD, 6.

$$6(\tfrac{1}{6}x) = 6(5 + x)$$
$$x = 30 + 6x$$
$$-30 = 5x$$
$$-6 = x$$

Step 5 The number is -6.

Step 6 One-sixth of -6 is -1 and -1 is 5 more than -6.

13. *Step 2* Let $x =$ the quantity. Then its $\tfrac{3}{4}$, its $\tfrac{1}{2}$, and its $\tfrac{1}{3}$ are

$$\tfrac{3}{4}x, \tfrac{1}{2}x, \text{ and } \tfrac{1}{3}x.$$

Step 3 Their sum becomes 93 translates to

$$x + \tfrac{3}{4}x + \tfrac{1}{2}x + \tfrac{1}{3}x = 93.$$

Step 4 Multiply both sides by the LCD of $4, 2,$ and 3, which is 12.

$$12(x + \tfrac{3}{4}x + \tfrac{1}{2}x + \tfrac{1}{3}x) = 12(93)$$
$$12x + 12(\tfrac{3}{4}x) + 12(\tfrac{1}{2}x) + 12(\tfrac{1}{3}x) = 12(93)$$
$$12x + 9x + 6x + 4x = 1116$$
$$31x = 1116$$
$$x = 36$$

Step 5 The quantity is 36.

Step 6 Check 36 in the original problem.

$$x = 36, \tfrac{3}{4}x = 27, \tfrac{1}{2}x = 18, \tfrac{1}{3}x = 12$$

Adding gives us the following desired result:

$$36 + 27 + 18 + 12 = 93.$$

15. We are asked to find the *time*, so we'll use the distance, rate, and time relationship $t = d/r$.

$$t = \frac{d}{r} = \frac{0.6 \text{ miles}}{0.0319 \text{ miles per minute}}$$
$$\approx 18.809 \text{ minutes}$$

17. We are asked to find the *time*, so we'll use the distance, rate, and time relationship $t = d/r$.

$$t = \frac{d}{r} = \frac{500 \text{ miles}}{140.256 \text{ miles per hour}}$$
$$\approx 3.565 \text{ hours}$$

19. We are asked to find the average *rate*, so we'll use the distance, rate, and time relationship $r = d/t$.

$$r = \frac{d}{t} = \frac{5000 \text{ meters}}{15.911 \text{ minutes}}$$
$$\approx 314.248 \text{ meters per minute}$$

21. Let $x =$ the rate of the current. Then the rate is $4 - x$ upstream (against the current) and $4 + x$ downstream (with the current). The time to row upstream is

$$t = \frac{d}{r} = \frac{8}{4 - x},$$

and the time to row downstream is

$$t = \frac{d}{r} = \frac{24}{4 + x}.$$

Now complete the chart.

	d	r	t
Upstream	8	$4 - x$	$\dfrac{8}{4 - x}$
Downstream	24	$4 + x$	$\dfrac{24}{4 + x}$

Since the problem states that the two times are equal, we have

$$\frac{8}{4 - x} = \frac{24}{4 + x}.$$

We would use this equation to solve the problem.

23. Let $x =$ the rate of the boat in still water. Complete a table.

	d	r	t
Against the Current	20	$x - 4$	$\dfrac{20}{x - 4}$
With the Current	60	$x + 4$	$\dfrac{60}{x + 4}$

Since the times are equal, we get the following equation.

$$\frac{20}{x - 4} = \frac{60}{x + 4}$$
$$(x + 4)(x - 4)\frac{20}{x - 4} = (x + 4)(x - 4)\frac{60}{x + 4}$$
$$20(x + 4) = 60(x - 4)$$
$$20x + 80 = 60x - 240$$
$$320 = 40x$$
$$8 = x$$

The rate of the boat in still water is 8 miles per hour.

25. Let $x =$ the rate of the bird in still air. Then the rate against the wind is $x - 8$ and the rate with the wind is $x + 8$. Use $t = d/r$ to complete the chart.

	d	r	t
Against the Wind	18	$x - 8$	$\dfrac{18}{x - 8}$
With the Wind	30	$x + 8$	$\dfrac{30}{x + 8}$

Since the problem states that the two times are equal, we get the following equation.

$$\frac{18}{x-8} = \frac{30}{x+8}$$

$$(x+8)(x-8)\frac{18}{x-8} = (x+8)(x-8)\frac{30}{x+8}$$

$$18(x+8) = 30(x-8)$$

$$18x + 144 = 30x - 240$$

$$384 = 12x$$

$$32 = x$$

The rate of the bird in still air is 32 miles per hour.

27. Let x represent the rate of the current of the river. Then $12 - x$ is the rate upstream (against the current) and $12 + x$ is the rate downstream (with the current). Use $t = d/r$ to complete the table.

	d	r	t
Upstream	6	$12 - x$	$\dfrac{6}{12-x}$
Downstream	10	$12 + x$	$\dfrac{10}{12+x}$

Since the times are equal, we get the following equation.

$$\frac{6}{12-x} = \frac{10}{12+x}$$

$$(12+x)(12-x)\frac{6}{12-x} = (12+x)(12-x)\frac{10}{12+x}$$

$$6(12+x) = 10(12-x)$$

$$72 + 6x = 120 - 10x$$

$$16x = 48$$

$$x = 3$$

The rate of the current of the river is 3 miles per hour.

29. Let $x =$ the number of hours it would take Eric and Oscar to tune up the Chevy, working together.

	Rate	Time Working Together	Fractional Part of the Job Done When Working Together
Eric	$\frac{1}{2}$	x	$\frac{1}{2}x$
Oscar	$\frac{1}{3}$	x	$\frac{1}{3}x$

Working together, they complete 1 whole job, so add their individual fractional parts and set the sum equal to 1.

fractional part done by Eric	$+$	fractional part done by Oscar	$=$	1 whole job
$\downarrow$	$\downarrow$	$\downarrow$	$\downarrow$	$\downarrow$
$\frac{1}{2}x$	$+$	$\frac{1}{3}x$	$=$	1

An equation that can be used to solve this problem is

$$\tfrac{1}{2}x + \tfrac{1}{3}x = 1.$$

Alternatively, we can compare the hourly rates of completion. In one hour, Eric will complete $\frac{1}{2}$ of the job, Oscar will complete $\frac{1}{3}$ of the job, and together they will complete $\frac{1}{x}$ of the job. So another equation that can be used to solve this problem is

$$\tfrac{1}{2} + \tfrac{1}{3} = \tfrac{1}{x}.$$

31. Let x represent the number of hours it will take for the teacher and her student teacher to grade the tests working together. Since the teacher can grade the test in 4 hours, her rate alone is $\frac{1}{4}$ job per hour. Also, since her student teacher can do the job alone in 6 hours, his/her rate is $\frac{1}{6}$ job per hour.

	Rate	Time Working Together	Fractional Part of the Job Done when Working Together
Teacher	$\frac{1}{4}$	x	$\frac{1}{4}x$
Student Teacher	$\frac{1}{6}$	x	$\frac{1}{6}x$

Since together the teacher and her student teacher complete 1 whole job, we must add their individual fractional parts and set the sum equal to 1.

$$\tfrac{1}{4}x + \tfrac{1}{6}x = 1$$

$$12(\tfrac{1}{4}x + \tfrac{1}{6}x) = 12(1) \quad \textit{Multiply by the LCD, 12.}$$

$$12(\tfrac{1}{4}x) + 12(\tfrac{1}{6}x) = 12$$

$$3x + 2x = 12$$

$$5x = 12$$

$$x = \tfrac{12}{5}, \text{ or } 2\tfrac{2}{5}$$

It will take the teacher and her student teacher $2\frac{2}{5}$ hours to grade the tests if they work together.

33. Let $x =$ the time required for the printing job with the copiers working together.

	Rate	Time Working Together	Fractional Part of the Job Done When Working Together
Todd's Copier	$\frac{1}{7}$	x	$\frac{1}{7}x$
Scott's Copier	$\frac{1}{12}$	x	$\frac{1}{12}x$

Since together the two copiers complete 1 whole job, we must add their individual fractional parts and set the sum equal to 1.

$$\tfrac{1}{7}x + \tfrac{1}{12}x = 1$$

$$84(\tfrac{1}{7}x + \tfrac{1}{12}x) = 84(1) \quad \text{\textit{Multiply by the LCD, 84.}}$$

$$84(\tfrac{1}{7}x) + 84(\tfrac{1}{12}x) = 84$$

$$12x + 7x = 84$$

$$19x = 84$$

$$x = \tfrac{84}{19}, \quad \text{or} \quad 4\tfrac{8}{19}$$

The copiers together can do the printing job in $\tfrac{84}{19}$ or $4\tfrac{8}{19}$ hours.

35. Let $x =$ represent the number of hours it would take Brenda to paint the room by herself. Complete a table.

	Rate	Time Working Together	Fractional Part of the Job Done When Working Together
Hilda	$\dfrac{1}{6}$	$3\dfrac{3}{4} = \dfrac{15}{4}$	$\dfrac{1}{6} \cdot \dfrac{15}{4} = \dfrac{5}{8}$
Brenda	$\dfrac{1}{x}$	$3\dfrac{3}{4} = \dfrac{15}{4}$	$\dfrac{1}{x} \cdot \dfrac{15}{4} = \dfrac{15}{4x}$

Since together Hilda and Brenda complete 1 whole job, we must add their individual fractional parts and set the sum equal to 1.

$$\frac{5}{8} + \frac{15}{4x} = 1$$

$$8x\left(\frac{5}{8} + \frac{15}{4x}\right) = 8x(1) \quad \text{\textit{Multiply by the LCD, 8x.}}$$

$$8x\left(\frac{5}{8}\right) + 8x\left(\frac{15}{4x}\right) = 8x$$

$$5x + 30 = 8x$$

$$30 = 3x$$

$$10 = x$$

It will take Brenda 10 hours to paint the room by herself.

37. Let $x =$ the number of hours to fill the pool with both pipes left open.

	Rate	Time to Fill the Pool	Part Done by Each Pipe
Inlet Pipe	$\dfrac{1}{9}$	x	$\dfrac{1}{9}x$
Outlet Pipe	$\dfrac{1}{12}$	x	$\dfrac{1}{12}x$

Part done by inlet pipe	−	Part done by outlet pipe	=	Full pool
↓	↓	↓	↓	↓
$\tfrac{1}{9}x$	−	$\tfrac{1}{12}x$	=	1

$$36(\tfrac{1}{9}x - \tfrac{1}{12}x) = 36(1) \quad \text{\textit{Multiply by the LCD, 36.}}$$

$$36(\tfrac{1}{9}x) - 36(\tfrac{1}{12}x) = 36$$

$$4x - 3x = 36$$

$$x = 36$$

It takes 36 hours to fill the pool.

39.
$$P = \frac{c}{c + 12} \cdot a$$

$$= \frac{6}{6 + 12} \cdot 30 \quad \text{\textit{Let c = 6 and a = 30.}}$$

$$= \tfrac{1}{3} \cdot 30 = 10$$

The correct dose of milk of magnesia for a 6-yr old boy is 10 mL.

15.8 Variation

15.8 Margin Exercises

1.   (a) $z = kt$ *z varies directly as t.*

$11 = 4k$ *Substitute z = 11 and t = 4.*

$\tfrac{11}{4} = k$ *Solve for k.*

Since $z = kt$ and $k = \tfrac{11}{4}$,

$z = \tfrac{11}{4}t.$ *Substitute $k = \tfrac{11}{4}$ in z = kt.*

$z = \tfrac{11}{4}(32)$ *Substitute t = 32.*

$z = 88$ *Multiply.*

(b) $C = kr$ *Circumference C varies directly as radius r.*

$43.96 = 7k$ *Substitute r = 7 and C = 43.96.*

$6.28 = k$ *Solve for k.*

Since $C = kr$ and $k = 6.28$,

$C = 6.28r.$

$C = 6.28(11)$ *Substitute r = 11.*

$C = 69.08$

The circumference is 69.08 centimeters when the radius is 11 centimeters.

2. $z = \dfrac{k}{t}$ *z varies inversely as t.*

$8 = \dfrac{k}{2}$ *Substitute z = 8 and t = 2.*

$16 = k$ *Solve for k.*

Since $z = \dfrac{k}{t}$ and $k = 16$,

$z = \dfrac{16}{t}.$ *Substitute $k = 16$ in $z = \dfrac{k}{t}$.*

$z = \dfrac{16}{32}$ *Substitute t = 32.*

$z = \dfrac{1}{2}$ *Lowest terms*

3. Let $x =$ the number of pairs of rubber gloves produced, and $c =$ the cost per unit.

$$c = \frac{k}{x} \qquad \textit{c varies inversely as x.}$$

$$0.50 = \frac{k}{5000} \qquad \begin{array}{l}\textit{Substitute c = 0.50}\\ \textit{and x = 5000.}\end{array}$$

$$2500 = k \qquad \textit{Solve for k.}$$

Since $c = \frac{k}{x}$ and $k = 2500$,

$$c = \frac{2500}{x}. \qquad \textit{Substitute k = 2500 in } c = \frac{k}{x}.$$

$$c = \frac{2500}{10,000} \qquad \textit{Substitute x = 10,000.}$$

$$c = 0.25 \qquad \textit{Lowest terms}$$

The cost to make 10,000 pairs of rubber gloves is $0.25 per pair.

15.8 Section Exercises

1. For a constant time of 3 hours, if the rate of the pickup truck *increases*, then the distance traveled *increases*. Thus, the variation between the quantities is *direct*.

3. As the number of days from now until December 25 *decreases*, the magnitude of the frenzy of Christmas shopping *increases*. Thus, the variation between the quantities is *inverse*.

5. As the amount of gasoline that you pump *increases*, the amount of empty space left in your tank *decreases*. Thus, the variation between the quantities is *inverse*.

7. As the amount of pressure put on the accelerator of a car *increases*, the rate of the car *increases*. Thus, the variation between the quantities is *direct*.

9. **(a)** If the constant of variation is positive and y varies directly as x, then as x increases, y _increases_.

　　(b) If the constant of variation is positive and y varies inversely as x, then as x increases, y _decreases_.

11. $y = \frac{3}{x}$ represents *inverse* variation since it is of the form $y = \frac{k}{x}$.

13. $y = 50x$ represents *direct* variation since it is of the form $y = kx$.

15. Since z varies directly as x, there is a constant k such that $z = kx$. First find the value of k.

$$30 = k(8) \quad \textit{Let z = 30, x = 8.}$$
$$k = \frac{30}{8} = \frac{15}{4}$$

When $k = \frac{15}{4}$, $z = kx$ becomes

$$z = \frac{15}{4}x.$$

Now find z when $x = 4$.

$$z = \frac{15}{4}(4) \quad \textit{Let x = 4.}$$
$$= 15$$

17. Since d varies directly as r, there is a constant k such that $d = kr$. First find the value of k.

$$200 = k(40) \quad \textit{Let d = 200, r = 40.}$$
$$k = \frac{200}{40} = 5$$

When $k = 5$, $d = kr$ becomes

$$d = 5r.$$

Now find d when $r = 60$.

$$d = 5(60) \quad \textit{Let r = 60.}$$
$$= 300$$

19. Since z varies inversely as x, there is a constant k such that $z = \frac{k}{x}$. First find the value of k.

$$50 = \frac{k}{2} \quad \textit{Let z = 50, x = 2.}$$
$$k = 50(2) = 100$$

When $k = 100$, $z = \frac{k}{x}$ becomes

$$z = \frac{100}{x}.$$

Now find z when $x = 25$.

$$z = \frac{100}{25} = 4$$

21. Since m varies inversely as r, there is a constant k such that $m = \frac{k}{r}$. First find the value of k.

$$12 = \frac{k}{8} \quad \textit{Let m = 12, r = 8.}$$
$$k = 12(8) = 96$$

When $k = 96$, $m = \frac{k}{r}$ becomes

$$m = \frac{96}{r}.$$

Now find m when $r = 16$.

$$m = \frac{96}{16} = 6$$

23. For a given base, the area A of a triangle varies directly as its height h, so there is a constant k such that $A = kh$. Find the value of k.

$$10 = k(4) \quad \textit{Let A = 10, h = 4.}$$
$$k = \frac{10}{4} = 2.5$$

When $k = 2.5$, $A = kh$ becomes

$$A = 2.5h.$$

Now find A when $h = 6$.

$$A = 2.5(6) = 15$$

When the height of the triangle is 6 inches, the area of the triangle is 15 square inches.

25. The distance d that a spring stretches varies directly with the force F applied, so

$$d = kF.$$
$$16 = k(75) \quad \textit{Let d = 16, F = 75.}$$
$$\frac{16}{75} = k$$

So $d = \frac{16}{75}F$ and when $F = 200$,

$$d = \frac{16}{75}(200) = \frac{16(8)}{3} = \frac{128}{3}.$$

A force of 200 pounds stretches the spring $42\frac{2}{3}$ inches.

27. For a constant area, the length l of a rectangle varies inversely as the width w, so

$$l = \frac{k}{w}.$$
$$27 = \frac{k}{10} \quad \textit{Let l = 27, w = 10.}$$
$$k = 27 \cdot 10 = 270$$

So $l = \frac{270}{w}$ and when $l = 18$,

$$18 = \frac{270}{w}$$
$$18w = 270$$
$$w = \frac{270}{18} = 15.$$

When the length is 18 feet, the width is 15 feet.

29. If the temperature is constant, the pressure P of a gas in a container varies inversely as the volume V of the container, so $P = \frac{k}{V}$.

$$10 = \frac{k}{3} \quad \textit{Let P = 10, V = 3.}$$
$$k = 3 \cdot 10 = 30$$

So $P = \frac{30}{V}$ and when $V = 1.5$,

$$P = \frac{30}{1.5} = 20.$$

The pressure is 20 pounds per square foot.

31. The rate of change r of the amount of raw sugar varies directly as the amount a of raw sugar remaining, so there is a constant k such that $r = ka$. Find the value of k.

$$200 = k(800) \quad \textit{Let r = 200, a = 800.}$$
$$k = \frac{200}{800} = \frac{1}{4}$$

When $k = \frac{1}{4}$, $r = ka$ becomes

$$r = \frac{1}{4}a.$$

Now find r when $a = 100$.

$$r = \frac{1}{4}(100) = 25$$

When only 100 kilograms of raw sugar are left, the rate of change is 25 kilograms per hour.

33. For a constant area, the length l of a rectangle varies inversely as the width w, so

$$l = \frac{k}{w}.$$
$$24 = \frac{k}{2} \quad \textit{Let l = 24, w = 2.}$$
$$k = 24 \cdot 2 = 48$$

So $l = \frac{48}{w}$, and when $l = 12$,

$$12 = \frac{48}{w}$$
$$12w = 48$$
$$w = \frac{48}{12} = 4.$$

When the length is 12, the width is 4.

Chapter 15 Review Exercises

1. $\dfrac{4}{x-3}$ ▪ To find the values for which this expression is undefined, set the denominator equal to zero and solve for x.

$$x - 3 = 0$$
$$x = 3$$

Because $x = 3$ will make the denominator zero, the given expression is undefined for 3. We write the answer as $x \neq 3$.

2. $\dfrac{x+3}{2x}$

Set the denominator equal to zero and solve for x.

$$2x = 0$$
$$x = 0$$

The given expression is undefined for 0. We write the answer as $x \neq 0$.

3. $\dfrac{m-2}{m^2 - 2m - 3}$

Set the denominator equal to zero and solve for m.

$$m^2 - 2m - 3 = 0$$
$$(m-3)(m+1) = 0$$
$$m - 3 = 0 \quad \text{or} \quad m + 1 = 0$$
$$m = 3 \quad \text{or} \quad m = -1$$

The given expression is undefined for -1 and 3. We write the answer as $m \neq -1, m \neq 3$.

4. $\dfrac{2k+1}{3k^2 + 17k + 10}$

Set the denominator equal to zero and solve for k.

$$3k^2 + 17k + 10 = 0$$
$$(3k + 2)(k + 5) = 0$$
$$k = -\frac{2}{3} \quad \text{or} \quad k = -5$$

The given expression is undefined for -5 and $-\frac{2}{3}$. We write the answer as $k \neq -5, k \neq -\frac{2}{3}$.

5. (a) $\dfrac{x^2}{x-5} = \dfrac{(-2)^2}{-2-5}$ *Let x = -2.*

$\qquad = \dfrac{4}{-7} = -\dfrac{4}{7}$

     (b) $\dfrac{x^2}{x-5} = \dfrac{4^2}{4-5}$ *Let x = 4.*

$\qquad = \dfrac{16}{-1} = -16$

6. (a) $\dfrac{4x-3}{5x+2} = \dfrac{4(-2)-3}{5(-2)+2}$ *Let x = -2.*

$\qquad = \dfrac{-8-3}{-10+2} = \dfrac{-11}{-8} = \dfrac{11}{8}$

     (b) $\dfrac{4x-3}{5x+2} = \dfrac{4(4)-3}{5(4)+2}$ *Let x = 4.*

$\qquad = \dfrac{16-3}{20+2} = \dfrac{13}{22}$

7. (a) $\dfrac{3x}{x^2-4} = \dfrac{3(-2)}{(-2)^2-4}$ *Let x = -2.*

$\qquad = \dfrac{-6}{4-4} = \dfrac{-6}{0}$

Substituting -2 for x makes the denominator zero, so the given expression is *undefined* when $x = -2$.

     (b) $\dfrac{3x}{x^2-4} = \dfrac{3(4)}{(4)^2-4}$ *Let x = 4.*

$\qquad = \dfrac{12}{16-4} = \dfrac{12}{12} = 1$

8. (a) $\dfrac{x-1}{x+2} = \dfrac{-2-1}{-2+2}$ *Let x = -2.*

$\qquad = \dfrac{-3}{0}$

Substituting -2 for x makes the denominator zero, so the given expression is *undefined* when $x = -2$.

     (b) $\dfrac{x-1}{x+2} = \dfrac{4-1}{4+2}$ *Let x = 4.*

$\qquad = \dfrac{3}{6} = \dfrac{1}{2}$

9. $\dfrac{5a^3b^3}{15a^4b^2} = \dfrac{5b^{3-2}}{3 \cdot 5a^{4-3}} = \dfrac{b}{3a}$

10. $\dfrac{m-4}{4-m} = \dfrac{-1(4-m)}{4-m} = -1$

11. $\dfrac{4x^2-9}{6-4x} = \dfrac{(2x+3)(2x-3)}{-2(2x-3)}$

$\qquad = \dfrac{2x+3}{-2}$

$\qquad = \dfrac{-1(2x+3)}{2} = \dfrac{-(2x+3)}{2}$

12. $\dfrac{4p^2+8pq-5q^2}{10p^2-3pq-q^2} = \dfrac{(2p-q)(2p+5q)}{(5p+q)(2p-q)}$

$\qquad = \dfrac{2p+5q}{5p+q}$

13. $-\dfrac{4x-9}{2x+3}$

Apply the negative sign to the numerator:

$\dfrac{-(4x-9)}{2x+3}$

Now distribute the negative sign:

$\dfrac{-4x+9}{2x+3}$

Apply the negative sign to the denominator:

$\dfrac{4x-9}{-(2x+3)}$

Again, distribute the negative sign:

$\dfrac{4x-9}{-2x-3}$

14. $-\dfrac{8-3x}{3-6x}$ ▪ Four equivalent forms are:

$\dfrac{-(8-3x)}{3-6x}, \qquad \dfrac{-8+3x}{3-6x},$

$\dfrac{8-3x}{-(3-6x)}, \qquad \dfrac{8-3x}{-3+6x}$

15. $\dfrac{8x^2}{12x^5} \cdot \dfrac{6x^4}{2x} = \dfrac{2 \cdot 4}{3 \cdot 4x^3} \cdot \dfrac{3x^3}{1} = 2$

16. $\dfrac{9m^2}{(3m)^4} \div \dfrac{6m^5}{36m} = \dfrac{9m^2}{(3m)^4} \cdot \dfrac{36m}{6m^5}$

$\qquad = \dfrac{9m^2}{81m^4} \cdot \dfrac{36m}{6m^5}$

$\qquad = \dfrac{6m^3}{9m^9}$

$\qquad = \dfrac{2}{3m^6}$

17. $\dfrac{x-3}{4} \cdot \dfrac{5}{2x-6} = \dfrac{x-3}{4} \cdot \dfrac{5}{2(x-3)} = \dfrac{5}{8}$

18. $\dfrac{2r+3}{r-4} \cdot \dfrac{r^2-16}{6r+9}$

$\qquad = \dfrac{2r+3}{r-4} \cdot \dfrac{(r+4)(r-4)}{3(2r+3)}$

$\qquad = \dfrac{r+4}{3}$

19. $\dfrac{3q+3}{5-6q} \div \dfrac{4q+4}{2(5-6q)}$

$\qquad = \dfrac{3(q+1)}{5-6q} \cdot \dfrac{2(5-6q)}{4(q+1)}$

$\qquad = \dfrac{3 \cdot 2}{1 \cdot 4} = \dfrac{3}{2}$

20. $\dfrac{y^2 - 6y + 8}{y^2 + 3y - 18} \div \dfrac{y - 4}{y + 6}$

$\quad = \dfrac{y^2 - 6y + 8}{y^2 + 3y - 18} \cdot \dfrac{y + 6}{y - 4}$

$\quad = \dfrac{(y - 4)(y - 2)}{(y + 6)(y - 3)} \cdot \dfrac{y + 6}{y - 4}$

$\quad = \dfrac{y - 2}{y - 3}$

21. $\dfrac{2p^2 + 13p + 20}{p^2 + p - 12} \cdot \dfrac{p^2 + 2p - 15}{2p^2 + 7p + 5}$

$\quad = \dfrac{(2p + 5)(p + 4)}{(p + 4)(p - 3)} \cdot \dfrac{(p + 5)(p - 3)}{(2p + 5)(p + 1)}$

$\quad = \dfrac{p + 5}{p + 1}$

22. $\dfrac{3z^2 + 5z - 2}{9z^2 - 1} \cdot \dfrac{9z^2 + 6z + 1}{z^2 + 5z + 6}$

$\quad = \dfrac{(3z - 1)(z + 2)}{(3z - 1)(3z + 1)} \cdot \dfrac{(3z + 1)^2}{(z + 3)(z + 2)}$

$\quad = \dfrac{3z + 1}{z + 3}$

23. $\dfrac{1}{8}, \dfrac{5}{12}, \dfrac{7}{32}$ ▪ Factor each denominator.

$\qquad 8 = 2^3,\ 12 = 2^2 \cdot 3,\ 32 = 2^5$

LCD $= 2^5 \cdot 3 = 96$

24. $\dfrac{4}{9y}, \dfrac{7}{12y^2}, \dfrac{5}{27y^4}$ ▪ Factor each denominator.

$\qquad 9y = 3^2 y$

$\qquad 12y^2 = 2^2 \cdot 3 \cdot y^2$

$\qquad 27y^4 = 3^3 \cdot y^4$

LCD $= 2^2 \cdot 3^3 \cdot y^4 = 108y^4$

25. $\dfrac{1}{m^2 + 2m}, \dfrac{4}{m^2 + 7m + 10}$

Factor each denominator.

$\qquad m^2 + 2m = m(m + 2)$

$\qquad m^2 + 7m + 10 = (m + 2)(m + 5)$

LCD $= m(m + 2)(m + 5)$

26. $\dfrac{3}{x^2 + 4x + 3}, \dfrac{5}{x^2 + 5x + 4}, \dfrac{2}{x^2 + 7x + 12}$

Factor each denominator.

$\qquad x^2 + 4x + 3 = (x + 3)(x + 1)$

$\qquad x^2 + 5x + 4 = (x + 1)(x + 4)$

$\qquad x^2 + 7x + 12 = (x + 3)(x + 4)$

LCD $= (x + 3)(x + 1)(x + 4)$

27. $\dfrac{5}{8} = \dfrac{?}{56}$ ▪ A factor of 7 is missing in the denominator of the first fraction, so multiply the numerator and the denominator by 7.

$$\dfrac{5}{8} = \dfrac{5}{8} \cdot \dfrac{7}{7} = \dfrac{35}{56}$$

28. $\dfrac{10}{k} = \dfrac{?}{4k}$ ▪ A factor of 4 is missing in the denominator of the first fraction, so multiply the numerator and the denominator by 4.

$$\dfrac{10}{k} = \dfrac{10}{k} \cdot \dfrac{4}{4} = \dfrac{40}{4k}$$

29. $\dfrac{3}{2a^3} = \dfrac{?}{10a^4}$ ▪ Factors of 5 and a are missing in the denominator of the first fraction, so multiply the numerator and the denominator by $5a$.

$$\dfrac{3}{2a^3} = \dfrac{3}{2a^3} \cdot \dfrac{5a}{5a} = \dfrac{15a}{10a^4}$$

30. $\dfrac{9}{x - 3} = \dfrac{?}{18 - 6x} = \dfrac{?}{-6(x - 3)}$ ▪ A factor of -6 is missing in the denominator of the first fraction, so multiply the numerator and the denominator by -6.

$\qquad \dfrac{9}{x - 3} = \dfrac{9}{x - 3} \cdot \dfrac{-6}{-6}$

$\qquad\qquad = \dfrac{-54}{-6x + 18}, \text{ or } \dfrac{-54}{18 - 6x}$

31. $\dfrac{-3y}{2y - 10} = \dfrac{?}{50 - 10y} = \dfrac{?}{-5(2y - 10)}$ ▪ A factor of -5 is missing in the denominator of the first fraction, so multiply the numerator and the denominator by -5.

$\qquad \dfrac{-3y}{2y - 10} = \dfrac{-3y}{2y - 10} \cdot \dfrac{-5}{-5}$

$\qquad\qquad = \dfrac{15y}{-10y + 50}, \text{ or } \dfrac{15y}{50 - 10y}$

32. $\dfrac{4b}{b^2 + 2b - 3} = \dfrac{?}{(b + 3)(b - 1)(b + 2)}$

The denominator in the first fraction factors as $(b + 3)(b - 1)$, so it is missing a factor of $b + 2$. Multiply the numerator and the denominator by $b + 2$.

$\qquad \dfrac{4b}{b^2 + 2b - 3} = \dfrac{4b}{(b + 3)(b - 1)}$

$\qquad\qquad = \dfrac{4b}{(b + 3)(b - 1)} \cdot \dfrac{b + 2}{b + 2}$

$\qquad\qquad = \dfrac{4b(b + 2)}{(b + 3)(b - 1)(b + 2)}$

33. $\dfrac{10}{x} + \dfrac{5}{x} = \dfrac{10 + 5}{x} = \dfrac{15}{x}$

34. $\dfrac{6}{3p} - \dfrac{12}{3p} = \dfrac{6-12}{3p} = \dfrac{-6}{3p} = -\dfrac{2}{p}$

35. $\dfrac{9}{k} - \dfrac{5}{k-5}$

$= \dfrac{9(k-5)}{k(k-5)} - \dfrac{5 \cdot k}{(k-5)k} \quad LCD = k(k-5)$

$= \dfrac{9(k-5) - 5k}{k(k-5)}$

$= \dfrac{9k - 45 - 5k}{k(k-5)}$

$= \dfrac{4k - 45}{k(k-5)}$

36. $\dfrac{4}{y} + \dfrac{7}{7+y}$

$= \dfrac{4(7+y)}{y(7+y)} + \dfrac{7 \cdot y}{(7+y)y} \quad LCD = y(7+y)$

$= \dfrac{28 + 4y + 7y}{y(7+y)}$

$= \dfrac{28 + 11y}{y(7+y)}$

37. $\dfrac{m}{3} - \dfrac{2+5m}{6}$

$= \dfrac{m \cdot 2}{3 \cdot 2} - \dfrac{2+5m}{6} \quad LCD = 6$

$= \dfrac{2m - (2+5m)}{6}$

$= \dfrac{2m - 2 - 5m}{6}$

$= \dfrac{-2 - 3m}{6}$

38. $\dfrac{12}{x^2} - \dfrac{3}{4x}$

$= \dfrac{12 \cdot 4}{x^2 \cdot 4} - \dfrac{3 \cdot x}{4x \cdot x} \quad LCD = 4x^2$

$= \dfrac{48 - 3x}{4x^2}$

$= \dfrac{3(16 - x)}{4x^2}$

39. $\dfrac{5}{a - 2b} + \dfrac{2}{a + 2b}$

$= \dfrac{5(a+2b)}{(a-2b)(a+2b)} + \dfrac{2(a-2b)}{(a+2b)(a-2b)}$

$\qquad\qquad LCD = (a - 2b)(a + 2b)$

$= \dfrac{5(a+2b) + 2(a-2b)}{(a-2b)(a+2b)}$

$= \dfrac{5a + 10b + 2a - 4b}{(a-2b)(a+2b)} = \dfrac{7a + 6b}{(a-2b)(a+2b)}$

40. $\dfrac{4}{k^2 - 9} - \dfrac{k+3}{3k - 9}$

$= \dfrac{4}{(k+3)(k-3)} - \dfrac{k+3}{3(k-3)}$

$\qquad LCD = 3(k+3)(k-3)$

$= \dfrac{4 \cdot 3}{(k+3)(k-3) \cdot 3} - \dfrac{(k+3)(k+3)}{3(k-3)(k+3)}$

$= \dfrac{12 - (k+3)(k+3)}{3(k+3)(k-3)}$

$= \dfrac{12 - (k^2 + 6k + 9)}{3(k+3)(k-3)}$

$= \dfrac{12 - k^2 - 6k - 9}{3(k+3)(k-3)}$

$= \dfrac{-k^2 - 6k + 3}{3(k+3)(k-3)}$

41. $\dfrac{8}{z^2 + 6z} - \dfrac{3}{z^2 + 4z - 12}$

$= \dfrac{8}{z(z+6)} - \dfrac{3}{(z+6)(z-2)}$

$\qquad LCD = z(z+6)(z-2)$

$= \dfrac{8(z-2)}{z(z+6)(z-2)} - \dfrac{3 \cdot z}{(z+6)(z-2) \cdot z}$

$= \dfrac{8(z-2) - 3z}{z(z+6)(z-2)}$

$= \dfrac{8z - 16 - 3z}{z(z+6)(z-2)}$

$= \dfrac{5z - 16}{z(z+6)(z-2)}$

42. $\dfrac{11}{2p - p^2} - \dfrac{2}{p^2 - 5p + 6}$

$= \dfrac{11}{p(2-p)} - \dfrac{2}{(p-3)(p-2)}$

$\qquad LCD = p(p-3)(p-2)$

$= \dfrac{11(-1)(p-3)}{p(2-p)(-1)(p-3)}$

$\quad - \dfrac{2 \cdot p}{(p-3)(p-2)p}$

$= \dfrac{-11(p-3) - 2p}{p(p-2)(p-3)}$

$= \dfrac{-11p + 33 - 2p}{p(p-2)(p-3)}$

$= \dfrac{-13p + 33}{p(p-2)(p-3)}$

43. $\dfrac{\dfrac{a^4}{b^2}}{\dfrac{a^3}{b}}$

$$= \frac{a^4}{b^2} \div \frac{a^3}{b}$$

$$= \frac{a^4}{b^2} \cdot \frac{b}{a^3} \qquad \textit{Multiply by reciprocal.}$$

$$= \frac{a^4 b}{a^3 b^2}$$

$$= \frac{a}{b}$$

44. $\dfrac{\dfrac{y-3}{y}}{\dfrac{y+3}{4y}}$

$$= \frac{y-3}{y} \div \frac{y+3}{4y}$$

$$= \frac{y-3}{y} \cdot \frac{4y}{y+3} \qquad \textit{Multiply by reciprocal.}$$

$$= \frac{4(y-3)}{y+3}$$

45. $\dfrac{\dfrac{3m+2}{m}}{\dfrac{2m-5}{6m}}$

$$= \frac{3m+2}{m} \div \frac{2m-5}{6m}$$

$$= \frac{3m+2}{m} \cdot \frac{6m}{2m-5} \qquad \textit{Multiply by reciprocal.}$$

$$= \frac{6(3m+2)}{2m-5}$$

46. $\dfrac{\dfrac{1}{p} - \dfrac{1}{q}}{\dfrac{1}{q-p}}$

$$= \frac{\left(\dfrac{1}{p} - \dfrac{1}{q}\right)pq(q-p)}{\left(\dfrac{1}{q-p}\right)pq(q-p)} \qquad \begin{array}{l}\textit{Multiply by} \\ \textit{the LCD,} \\ pq(q-p).\end{array}$$

$$= \frac{\dfrac{1}{p}[pq(q-p)] - \dfrac{1}{q}[pq(q-p)]}{pq}$$

$$= \frac{q(q-p) - p(q-p)}{pq}$$

$$= \frac{q^2 - pq - pq + p^2}{pq}$$

$$= \frac{q^2 - 2pq + p^2}{pq}$$

$$= \frac{(q-p)^2}{pq}$$

47. $\dfrac{x + \dfrac{1}{w}}{x - \dfrac{1}{w}}$

$$= \frac{\left(x + \dfrac{1}{w}\right) \cdot w}{\left(x - \dfrac{1}{w}\right) \cdot w} \qquad \begin{array}{l}\textit{Multiply by} \\ \textit{the LCD, w.}\end{array}$$

$$= \frac{xw + \left(\dfrac{1}{w}\right)w}{xw - \left(\dfrac{1}{w}\right)w}$$

$$= \frac{xw + 1}{xw - 1}$$

48. $\dfrac{\dfrac{1}{r+t} - 1}{\dfrac{1}{r+t} + 1}$

$$= \frac{\left(\dfrac{1}{r+t} - 1\right)(r+t)}{\left(\dfrac{1}{r+t} + 1\right)(r+t)} \qquad \begin{array}{l}\textit{Multiply by} \\ \textit{the LCD, } r+t.\end{array}$$

$$= \frac{\dfrac{1}{r+t}(r+t) - 1(r+t)}{\dfrac{1}{r+t}(r+t) + 1(r+t)}$$

$$= \frac{1 - r - t}{1 + r + t}$$

49. $\dfrac{k}{5} - \dfrac{2}{3} = \dfrac{1}{2}$

$$30\left(\frac{k}{5} - \frac{2}{3}\right) = 30\left(\frac{1}{2}\right) \qquad \begin{array}{l}\textit{Multiply by} \\ \textit{the LCD, 30.}\end{array}$$

$$30\left(\frac{k}{5}\right) - 30\left(\frac{2}{3}\right) = 15$$

$$6k - 20 = 15$$

$$6k = 35$$

$$k = \frac{35}{6}$$

Check $k = \frac{35}{6}$: $\frac{1}{2} = \frac{1}{2}$ *True*

The solution set is $\left\{\frac{35}{6}\right\}$.

50. $\dfrac{4-z}{z} + \dfrac{3}{2} = \dfrac{-4}{z}$

$$2z\left(\frac{4-z}{z} + \frac{3}{2}\right) = 2z\left(-\frac{4}{z}\right) \qquad \begin{array}{l}\textit{Multiply} \\ \textit{by 2z.}\end{array}$$

$$2z\left(\frac{4-z}{z}\right) + 2z\left(\frac{3}{2}\right) = -8$$

$$2(4-z) + 3z = -8$$

$$8 - 2z + 3z = -8$$

$$8 + z = -8$$

$$z = -16$$

Check $z = -16$: $-\frac{5}{4} + \frac{3}{2} = \frac{1}{4}$ *True*

The solution set is $\{-16\}$.

51.
$$\frac{x}{2} - \frac{x-3}{7} = -1$$

$$14\left(\frac{x}{2} - \frac{x-3}{7}\right) = 14(-1) \quad \begin{array}{l}\textit{Multiply by} \\ \textit{the LCD, 14.}\end{array}$$

$$14\left(\frac{x}{2}\right) - 14\left(\frac{x-3}{7}\right) = -14$$

$$7x - 2(x-3) = -14$$

$$7x - 2x + 6 = -14$$

$$5x + 6 = -14$$

$$5x = -20$$

$$x = -4$$

Check $x = -4$: $-1 = -1$ *True*

The solution set is $\{-4\}$.

52.
$$\frac{3y-1}{y-2} = \frac{5}{y-2} + 1$$

$$(y-2)\left(\frac{3y-1}{y-2}\right) = (y-2)\left(\frac{5}{y-2} + 1\right)$$

$$\textit{Multiply by the LCD, } y-2.$$

$$(y-2)\left(\frac{3y-1}{y-2}\right) = (y-2)\left(\frac{5}{y-2}\right)$$
$$+ (y-2)(1)$$

$$\textit{Distributive property}$$

$$3y - 1 = 5 + y - 2$$

$$3y - 1 = 3 + y$$

$$2y = 4$$

$$y = 2$$

There is *no solution* because $y = 2$ makes the original denominators equal to zero, and the solution set is $\emptyset$.

53.
$$\frac{3}{m-2} + \frac{1}{m-1} = \frac{7}{m^2 - 3m + 2}$$

$$\frac{3}{m-2} + \frac{1}{m-1} = \frac{7}{(m-2)(m-1)}$$

$$(m-2)(m-1)\left(\frac{3}{m-2} + \frac{1}{m-1}\right)$$

$$= (m-2)(m-1) \cdot \frac{7}{(m-2)(m-1)}$$

$$\textit{Multiply by the LCD, } (m-2)(m-1).$$

$$3(m-1) + 1(m-2) = 7$$

$$3m - 3 + m - 2 = 7$$

$$4m - 5 = 7$$

$$4m = 12$$

$$m = 3$$

Check $m = 3$: $3 + \frac{1}{2} = \frac{7}{2}$ *True*

The solution set is $\{3\}$.

54. Solve $m = \dfrac{Ry}{t}$ for t.

$$t \cdot m = t\left(\frac{Ry}{t}\right) \quad \begin{array}{l}\textit{Multiply by} \\ \textit{the LCD, } t.\end{array}$$

$$tm = Ry$$

$$t = \frac{Ry}{m} \quad \textit{Divide by } m.$$

55. Solve $x = \dfrac{3y-5}{4}$ for y.

$$4x = 4\left(\frac{3y-5}{4}\right) \quad \begin{array}{l}\textit{Multiply by} \\ \textit{the LCD, 4.}\end{array}$$

$$4x = 3y - 5$$

$$4x + 5 = 3y$$

$$\frac{4x+5}{3} = y \quad \textit{Divide by 3.}$$

56. Solve $\dfrac{1}{r} - \dfrac{1}{s} = \dfrac{1}{t}$ for t.

$$rst\left(\frac{1}{r} - \frac{1}{s}\right) = rst\left(\frac{1}{t}\right) \quad \begin{array}{l}\textit{Multiply by} \\ \textit{the LCD, } rst.\end{array}$$

$$st - rt = rs$$

$$t(s-r) = rs \quad \textit{Factor.}$$

$$t = \frac{rs}{s-r} \quad \textit{Divide by } s-r.$$

57. Let $x =$ the numerator. Then $x - 5 =$ the denominator. Adding 5 to both the numerator and the denominator gives us a fraction that is equivalent to $\frac{5}{4}$.

$$\frac{x+5}{x-5+5} = \frac{5}{4}$$

$$\frac{x+5}{x} = \frac{5}{4}$$

$$4x\left(\frac{x+5}{x}\right) = 4x\left(\frac{5}{4}\right)$$

$$4(x+5) = x(5)$$

$$4x + 20 = 5x$$

$$20 = x$$

The numerator is 20 and the denominator is $20 - 5 = 15$, so the original fraction is $\frac{20}{15}$.

58. Let $x =$ the numerator. Then $6x =$ the denominator. Adding 3 to the numerator and subtracting 3 from the denominator gives us a fraction equivalent to $\frac{2}{5}$.

$$\frac{x+3}{6x-3} = \frac{2}{5}$$

$$5(6x-3)\left(\frac{x+3}{6x-3}\right) = 5(6x-3)\left(\frac{2}{5}\right)$$

$$5(x+3) = 2(6x-3)$$

$$5x + 15 = 12x - 6$$

$$21 = 7x$$

$$3 = x$$

The numerator is 3 and the denominator is $6 \cdot 3 = 18$, so the original fraction is $\frac{3}{18}$.

59. We are asked to find the *time*, so we'll use the distance, rate, and time relationship

$$t = \frac{d}{r}.$$

$$= \frac{218.75}{118.671} \approx 1.843 \text{ hr}$$

His time was about 1.843 hours.

60. We are asked to find the *rate*, so we'll use the distance, rate, and time relationship

$$r = \frac{d}{t}.$$

$$r = \frac{5000 \text{ meters}}{6.243 \text{ minutes}}$$

$$\approx 800.897 \text{ meters per minute}$$

His rate was about 800.897 meters per minute.

61. ***Step 2***   Let $x =$ the time it takes Zachary and Samuel to mess up the room together.

	Rate	Time to Mess up Room	Fractional Part of the Job Done when Working Together
Zachary	$\frac{1}{20}$	x	$\frac{1}{20}x$
Samuel	$\frac{1}{12}$	x	$\frac{1}{12}x$

Step 3   Since together the two brothers complete the whole job, we must add their individual fractional parts and set the sum equal to 1.

$$\frac{1}{20}x + \frac{1}{12}x = 1$$

Step 4   Solve this equation by multiplying both sides by the LCD, 60.

$$60\left(\frac{1}{20}x + \frac{1}{12}x\right) = 60(1)$$
$$60\left(\frac{1}{20}x\right) + 60\left(\frac{1}{12}x\right) = 60$$
$$3x + 5x = 60$$
$$8x = 60$$
$$x = \frac{60}{8} = \frac{15}{2}, \text{ or } 7\frac{1}{2}$$

Step 5   The brothers can mess up the room in $7\frac{1}{2}$ minutes. Note that if you first convert minutes to hours, you get $\frac{1}{\frac{1}{3}}x + \frac{1}{\frac{1}{5}}x = 1 \Leftrightarrow$

$3x + 5x = 1 \Leftrightarrow 8x = 1 \Leftrightarrow x = \frac{1}{8}$ hour, which is $7\frac{1}{2}$ minutes.

Step 6   Zachary does $\frac{1}{20}$ of the job per minute for $\frac{15}{2}$ minutes:

$$\frac{1}{20} \cdot \frac{15}{2} = \frac{3}{8} \text{ of the job}$$

Samuel does $\frac{1}{12}$ of the job per minute for $\frac{15}{2}$ minutes:

$$\frac{1}{12} \cdot \frac{15}{2} = \frac{5}{8} \text{ of the job}$$

Together, they have done

$$\frac{3}{8} + \frac{5}{8} = \frac{8}{8} = 1 \text{ total job.}$$

62. ***Step 2***   Let $x =$ the number of hours it takes them to do the job working together.

	Rate	Time Working Together	Fractional Part of the Job Done When Working Together
Man	$\frac{1}{5}$	x	$\frac{1}{5}x$
Daughter	$\frac{1}{8}$	x	$\frac{1}{8}x$

Step 3   Working together, they do 1 whole job, so

$$\frac{1}{5}x + \frac{1}{8}x = 1.$$

Step 4   Solve this equation by multiplying both sides by the LCD, 40.

$$40\left(\frac{1}{5}x + \frac{1}{8}x\right) = 40(1)$$
$$8x + 5x = 40$$
$$13x = 40$$
$$x = \frac{40}{13}, \text{ or } 3\frac{1}{13}$$

Step 5   Working together, it takes them $\frac{40}{13}$ or $3\frac{1}{13}$ hours.

Step 6   The man does $\frac{1}{5}$ of the job per hour for $\frac{40}{13}$ hours:

$$\frac{1}{5} \cdot \frac{40}{13} = \frac{8}{13} \text{ of the job}$$

His daughter does $\frac{1}{8}$ of the job per hour for $\frac{40}{13}$ hours:

$$\frac{1}{8} \cdot \frac{40}{13} = \frac{5}{13} \text{ of the job}$$

Together, they have done

$$\frac{8}{13} + \frac{5}{13} = \frac{13}{13} = 1 \text{ total job.}$$

63. Since y varies directly as x, there is a constant k such that $y = kx$. First find the value of k.

$$5 = k(12) \quad \textit{Let x = 12, y = 5.}$$
$$k = \frac{5}{12}$$

When $k = \frac{5}{12}, y = kx$ becomes

$$y = \frac{5}{12}x.$$

Now find x when $y = 3$.

$$3 = \frac{5}{12}x \quad \textit{Let y = 3.}$$
$$x = 3 \cdot \frac{12}{5} = \frac{36}{5}$$

64. Let $h =$ the height of the parallelogram and $b =$ the length of the base of the parallelogram.

The height varies inversely as the base, so

$$h = \frac{k}{b}.$$

Find k by replacing h with 8 and b with 12.

$$8 = \frac{k}{12}$$
$$k = 8 \cdot 12 = 96$$

So $h = \frac{96}{b}$ and when $b = 24$,

$$h = \frac{96}{24} = 4.$$

The height of the parallelogram is 4 centimeters.

65. **[15.4]** $\dfrac{4}{m-1} - \dfrac{3}{m+1}$

$= \dfrac{4(m+1)}{(m-1)(m+1)} - \dfrac{3(m-1)}{(m+1)(m-1)}$

$\qquad\qquad LCD = (m-1)(m+1)$

$= \dfrac{4(m+1) - 3(m-1)}{(m-1)(m+1)}$

$= \dfrac{4m+4-3m+3}{(m-1)(m+1)} = \dfrac{m+7}{(m-1)(m+1)}$

66. **[15.2]** $\dfrac{8p^5}{5} \div \dfrac{2p^3}{10}$

$= \dfrac{8p^5}{5} \cdot \dfrac{10}{2p^3}$ *Multiply by reciprocal.*

$= \dfrac{80p^5}{10p^3}$

$= 8p^2$

67. **[15.2]** $\dfrac{r-3}{8} \div \dfrac{3r-9}{4}$

$= \dfrac{r-3}{8} \cdot \dfrac{4}{3r-9}$ *Multiply by reciprocal.*

$= \dfrac{r-3}{8} \cdot \dfrac{4}{3(r-3)}$

$= \dfrac{4}{24} = \dfrac{1}{6}$

68. **[15.5]** $\dfrac{\dfrac{5}{x} - 1}{\dfrac{5-x}{3x}}$

$= \dfrac{\left(\dfrac{5}{x} - 1\right)3x}{\left(\dfrac{5-x}{3x}\right)3x}$ *Method 2; multiply by the LCD, 3x.*

$= \dfrac{\dfrac{5}{x}(3x) - 1(3x)}{5-x}$

$= \dfrac{15 - 3x}{5-x}$

$= \dfrac{3(5-x)}{5-x} = 3$

69. **[15.4]** $\dfrac{4}{z^2 - 2z + 1} - \dfrac{3}{z^2 - 1}$

$= \dfrac{4}{(z-1)^2} - \dfrac{3}{(z+1)(z-1)}$

$\qquad\qquad LCD = (z+1)(z-1)^2$

$= \dfrac{4(z+1)}{(z-1)^2(z+1)}$

$\quad - \dfrac{3(z-1)}{(z+1)(z-1)(z-1)}$

$= \dfrac{4(z+1) - 3(z-1)}{(z+1)(z-1)^2}$

$= \dfrac{4z+4-3z+3}{(z+1)(z-1)^2}$

$= \dfrac{z+7}{(z+1)(z-1)^2}$

70. **[15.2]** $\dfrac{2x^2 + 5x - 12}{4x^2 - 9} \cdot \dfrac{x^2 - 3x - 21}{x^2 + 8x + 16}$

$= \dfrac{(2x-3)(x+4)}{(2x+3)(2x-3)} \cdot \dfrac{(x+4)(x-7)}{(x+4)(x+4)}$

$\qquad\qquad\qquad\qquad\qquad$ *Factor.*

$= \dfrac{(2x-3)(x+4)(x+4)(x-7)}{(2x+3)(2x-3)(x+4)(x+4)}$

$\qquad\qquad\qquad\qquad\qquad$ *Multiply.*

$= \dfrac{x-7}{2x+3}$ *Lowest terms*

71. **[15.6]** $\dfrac{5t}{6} = \dfrac{2t-1}{3} + 1$

$6\left(\dfrac{5t}{6}\right) = 6\left(\dfrac{2t-1}{3} + 1\right)$ *Multiply by 6.*

$5t = 6\left(\dfrac{2t-1}{3}\right) + 6(1)$

$5t = 2(2t-1) + 6$

$5t = 4t - 2 + 6$

$5t = 4t + 4$

$t = 4$

Check $t = 4$: $\frac{10}{3} = \frac{7}{3} + 1$ *True*

The solution set is $\{4\}$.

72. **[15.6]** $\dfrac{2}{z} - \dfrac{z}{z+3} = \dfrac{1}{z+3}$

$z(z+3)\left(\dfrac{2}{z} - \dfrac{z}{z+3}\right) = z(z+3)\left(\dfrac{1}{z+3}\right)$

$z(z+3)\left(\dfrac{2}{z}\right)$

$\quad - z(z+3)\left(\dfrac{z}{z+3}\right) = z(1)$

Multiply by the LCD, z(z + 3).

$2(z+3) - z^2 = z$

$2z + 6 - z^2 = z$

$0 = z^2 - z - 6$

$0 = (z-3)(z+2)$

$z - 3 = 0$ or $z + 2 = 0$

$z = 3$ or $\qquad z = -2$

Check $z = -2$: $-1 - (-2) = 1$ *True*

Check $z = 3$: $\frac{2}{3} - \frac{1}{2} = \frac{1}{6}$ *True*

The solution set is $\{-2, 3\}$.

73. **[15.6]** $\dfrac{2x}{x^2 - 16} - \dfrac{2}{x - 4} = \dfrac{4}{x + 4}$

$(x + 4)(x - 4)\left(\dfrac{2x}{x^2 - 16}\right)$

$\quad - (x + 4)(x - 4)\left(\dfrac{2}{x - 4}\right)$

$\qquad = (x + 4)(x - 4)\left(\dfrac{4}{x + 4}\right)$

Multiply by the LCD, (x + 4)(x − 4).

$2x - 2(x + 4) = 4(x - 4)$

$2x - 2x - 8 = 4x - 16$

$-8 = 4x - 16$

$8 = 4x$

$x = 2$

Check $x = 2$: $-\frac{1}{3} + 1 = \frac{2}{3}$ *True*

The solution set is $\{2\}$.

74. **[15.6]** Solve $a = \dfrac{v - w}{t}$ for w.

$t \cdot a = v - w$ *Multiply by t.*

$at + w = v$ *Add w.*

$w = v - at$ *Subtract at.*

75. **[15.7]** *Step 1* We are trying to find a number.

Step 2 Let x = the number added to the numerator and the denominator of $\frac{4}{11}$.

Step 3 The expression $\dfrac{4 + x}{11 + x}$ represents the new fraction. Write "the new fraction is equivalent to $\frac{1}{2}$" as an equation.

$\dfrac{4 + x}{11 + x} = \dfrac{1}{2}$

Step 4 Multiply by the LCD, $2(11 + x)$.

$2(11 + x)\dfrac{4 + x}{11 + x} = 2(11 + x)\dfrac{1}{2}$

$2(4 + x) = 1(11 + x)$

$8 + 2x = 11 + x$

$x = 3$

Step 5 The number is 3.

Step 6 If 3 is added to the numerator and the denominator of $\frac{4}{11}$, the result is $\frac{7}{14}$, or $\frac{1}{2}$, as required.

76. **[15.7]** Let x = represent the number of hours it would take Satish to clean the house by himself. Complete a table.

	Rate	Time Working Together	Fractional Part of the Job Done When Working Together
Seema	$\frac{1}{5}$	3	$\frac{1}{5} \cdot 3 = \frac{3}{5}$
Satish	$\frac{1}{x}$	3	$\frac{1}{x} \cdot 3 = \frac{3}{x}$

Since together Seema and Satish complete 1 whole job, we must add their individual fractional parts and set the sum equal to 1.

$\frac{3}{5} + \frac{3}{x} = 1$

$5x\left(\frac{3}{5} + \frac{3}{x}\right) = 5x(1)$ *Multiply by the LCD, 5x.*

$5x\left(\frac{3}{5}\right) + 5x\left(\frac{3}{x}\right) = 5x$

$3x + 15 = 5x$

$15 = 2x$

$x = \dfrac{15}{2}$, or $7\frac{1}{2}$ hr

It will take Satish $7\frac{1}{2}$ hr to clean the house by himself.

77. **[15.7]** Let x = the rate of the plane in still air. Then the rate of the plane with the wind is $x + 50$, and the rate of the plane against the wind is $x - 50$. Use $t = \frac{d}{r}$ to complete the chart.

	d	r	t
With the Wind	400	$x + 50$	$\dfrac{400}{x + 50}$
Against the Wind	200	$x - 50$	$\dfrac{200}{x - 50}$

The times are the same, so

$\dfrac{400}{x + 50} = \dfrac{200}{x - 50}.$

To solve this equation, multiply both sides by the LCD, $(x + 50)(x - 50)$.

$(x + 50)(x - 50) \cdot \dfrac{400}{x + 50}$

$\qquad = (x + 50)(x - 50) \cdot \dfrac{200}{x - 50}$

$400(x - 50) = 200(x + 50)$

$400x - 20{,}000 = 200x + 10{,}000$

$200x = 30{,}000$

$x = 150$

The rate of the plane is 150 kilometers per hour.

78. **[15.7]** Let $t =$ the number of hours until the distance between the steamboats is 70 miles. Use *rate* × *time* = *distance* to complete the chart.

	r	t	d
Slower Steamboat	18	t	$18t$
Faster Steamboat	25	t	$25t$

The difference of the distances is 70 miles.

$$25t - 18t = 70$$
$$7t = 70$$
$$t = 10$$

They will be 70 miles apart in 10 hours.

79. **[15.8]** The current c in a simple electrical circuit varies inversely as the resistance r, so there is a constant k such that $c = \frac{k}{r}$. Find the value of k.

$$80 = \frac{k}{10} \quad \textit{Let c = 80, r = 10.}$$
$$k = 10 \cdot 80 = 800$$

When $k = 800$, $c = \frac{k}{r}$ becomes

$$c = \frac{800}{r}.$$

Now find c when $r = 16$.

$$c = \frac{800}{16} = 50$$

When the resistance is 16 ohms, the current is 50 amps.

80. **[15.8]** Since *increasing* the length results in *decreasing* the width, this is an example of *inverse* variation.

Chapter 15 Test

1. $\dfrac{3x - 1}{x^2 - 2x - 8}$

Set the denominator equal to zero and solve for x.

$$x^2 - 2x - 8 = 0$$
$$(x + 2)(x - 4) = 0$$
$$x + 2 = 0 \quad \text{or} \quad x - 4 = 0$$
$$x = -2 \quad \text{or} \quad x = 4$$

The expression is undefined for -2 and 4. We write the answer as $x \neq -2$, $x \neq 4$.

2. **(a)** $\dfrac{6r + 1}{2r^2 - 3r - 20}$

$$= \frac{6(-2) + 1}{2(-2)^2 - 3(-2) - 20} \quad \textit{Let r = -2.}$$

$$= \frac{-12 + 1}{2 \cdot 4 + 6 - 20}$$

$$= \frac{-11}{8 + 6 - 20}$$

$$= \frac{-11}{-6} = \frac{11}{6}$$

(b) $\dfrac{6r + 1}{2r^2 - 3r - 20}$

$$= \frac{6(4) + 1}{2(4)^2 - 3(4) - 20} \quad \textit{Let r = 4.}$$

$$= \frac{24 + 1}{2 \cdot 16 - 12 - 20}$$

$$= \frac{25}{32 - 12 - 20}$$

$$= \frac{25}{20 - 20} = \frac{25}{0}$$

The expression is *undefined* when $r = 4$ because the denominator is 0.

3. $-\dfrac{6x - 5}{2x + 3}$

Apply the negative sign to the numerator:

$$\frac{-(6x - 5)}{2x + 3}$$

Now distribute the negative sign:

$$\frac{-6x + 5}{2x + 3}$$

Apply the negative sign to the denominator:

$$\frac{6x - 5}{-(2x + 3)}$$

Again, distribute the negative sign:

$$\frac{6x - 5}{-2x - 3}$$

4. $\dfrac{-15x^6 y^4}{5x^4 y} = \dfrac{-3 \cdot 5 x^{6-4} y^{4-1}}{5} = -3x^2 y^3$

5. $\dfrac{6a^2 + a - 2}{2a^2 - 3a + 1}$

$$= \frac{(3a + 2)(2a - 1)}{(2a - 1)(a - 1)} \quad \textit{Factor.}$$

$$= \frac{3a + 2}{a - 1} \quad \textit{Lowest terms}$$

6. $\dfrac{5(d - 2)}{9} \div \dfrac{3(d - 2)}{5}$

$$= \frac{5(d - 2)}{9} \cdot \frac{5}{3(d - 2)} \quad \begin{array}{l}\textit{Multiply by}\\ \textit{reciprocal.}\end{array}$$

$$= \frac{5 \cdot 5}{9 \cdot 3} = \frac{25}{27}$$

7. $\dfrac{6k^2 - k - 2}{8k^2 + 10k + 3} \cdot \dfrac{4k^2 + 7k + 3}{3k^2 + 5k + 2}$

$$= \frac{(3k - 2)(2k + 1)}{(4k + 3)(2k + 1)} \cdot \frac{(4k + 3)(k + 1)}{(3k + 2)(k + 1)}$$

$$= \frac{3k - 2}{3k + 2}$$

8. $\dfrac{4a^2 + 9a + 2}{3a^2 + 11a + 10} \div \dfrac{4a^2 + 17a + 4}{3a^2 + 2a - 5}$

$= \dfrac{4a^2 + 9a + 2}{3a^2 + 11a + 10} \cdot \dfrac{3a^2 + 2a - 5}{4a^2 + 17a + 4}$

$= \dfrac{(4a + 1)(a + 2)}{(3a + 5)(a + 2)} \cdot \dfrac{(3a + 5)(a - 1)}{(4a + 1)(a + 4)}$

$= \dfrac{a - 1}{a + 4}$

9. $\dfrac{x^2 - 10x + 25}{9 - 6x + x^2} \cdot \dfrac{x - 3}{5 - x}$

$= \dfrac{(x - 5)(x - 5)}{(3 - x)(3 - x)} \cdot \dfrac{x - 3}{5 - x}$ *Factor.*

$= \dfrac{(x - 5)(x - 5)(x - 3)}{(3 - x)(3 - x)(5 - x)}$ *Multiply.*

$= \dfrac{-1(x - 5)}{-1(3 - x)}$ *Reduce.*

$= \dfrac{x - 5}{3 - x}$

10. $\dfrac{-3}{10p^2}, \dfrac{21}{25p^3}, \dfrac{-7}{30p^5}$

Factor each denominator.

$$10p^2 = 2 \cdot 5 \cdot p^2$$
$$25p^3 = 5^2 \cdot p^3$$
$$30p^5 = 2 \cdot 3 \cdot 5 \cdot p^5$$

LCD $= 2 \cdot 3 \cdot 5^2 \cdot p^5 = 150p^5$

11. $\dfrac{r + 1}{2r^2 + 7r + 6}, \dfrac{-2r + 1}{2r^2 - 7r - 15}$

Factor each denominator.

$$2r^2 + 7r + 6 = (2r + 3)(r + 2)$$
$$2r^2 - 7r - 15 = (2r + 3)(r - 5)$$

LCD $= (2r + 3)(r + 2)(r - 5)$

12. $\dfrac{15}{4p} = \dfrac{?}{64p^3}$ ■ Factors of 16 and p^2 are missing
in the denominator of the first fraction, so multiply
the numerator and the denominator by $16p^2$.

$$\dfrac{15}{4p} = \dfrac{15}{4p} \cdot \dfrac{16p^2}{16p^2} = \dfrac{240p^2}{64p^3}$$

13. $\dfrac{3}{6m - 12} = \dfrac{?}{42m - 84} = \dfrac{?}{7(6m - 12)}$
A factor of 7 is missing in the denominator of the
first fraction, so multiply the numerator and the
denominator by 7.

$$\dfrac{3}{6m - 12} = \dfrac{3}{6m - 12} \cdot \dfrac{7}{7} = \dfrac{21}{42m - 84}$$

14. $\dfrac{4x + 2}{x + 5} + \dfrac{-2x + 8}{x + 5}$

$= \dfrac{(4x + 2) + (-2x + 8)}{x + 5}$ *Add numerators.*

$= \dfrac{2x + 10}{x + 5}$

$= \dfrac{2(x + 5)}{x + 5} = 2$

15. $\dfrac{-4}{y + 2} + \dfrac{6}{5y + 10}$

$= \dfrac{-4}{y + 2} + \dfrac{6}{5(y + 2)}$ *LCD = 5(y + 2)*

$= \dfrac{-4 \cdot 5}{(y + 2) \cdot 5} + \dfrac{6}{5(y + 2)}$

$= \dfrac{-20 + 6}{5(y + 2)} = \dfrac{-14}{5(y + 2)}$

16. $\dfrac{x + 1}{3 - x} - \dfrac{x^2}{x - 3}$

$= \dfrac{x + 1}{3 - x} + \dfrac{x^2}{-1(x - 3)}$ *Use LCD = 3 − x.*

$= \dfrac{x + 1}{3 - x} + \dfrac{x^2}{-x + 3}$

$= \dfrac{x + 1}{3 - x} + \dfrac{x^2}{3 - x}$

$= \dfrac{(x + 1) + x^2}{3 - x}$

$= \dfrac{x^2 + x + 1}{3 - x}$

If we use $x - 3$ for the LCD, we obtain the
equivalent answer

$$\dfrac{-x^2 - x - 1}{x - 3}.$$

17. $\dfrac{3}{2m^2 - 9m - 5} - \dfrac{m + 1}{2m^2 - m - 1}$

$= \dfrac{3}{(2m + 1)(m - 5)} - \dfrac{m + 1}{(2m + 1)(m - 1)}$

LCD = (2m + 1)(m − 5)(m − 1)

$= \dfrac{3(m - 1)}{(2m + 1)(m - 5)(m - 1)}$

$\quad - \dfrac{(m + 1)(m - 5)}{(2m + 1)(m - 1)(m - 5)}$

$= \dfrac{3(m - 1) - (m + 1)(m - 5)}{(2m + 1)(m - 5)(m - 1)}$

$= \dfrac{(3m - 3) - (m^2 - 4m - 5)}{(2m + 1)(m - 5)(m - 1)}$

$= \dfrac{3m - 3 - m^2 + 4m + 5}{(2m + 1)(m - 5)(m - 1)}$

$= \dfrac{-m^2 + 7m + 2}{(2m + 1)(m - 5)(m - 1)}$

18. $\dfrac{\dfrac{2p}{k^2}}{\dfrac{3p^2}{k^3}}$

$= \dfrac{2p}{k^2} \div \dfrac{3p^2}{k^3}$

$= \dfrac{2p}{k^2} \cdot \dfrac{k^3}{3p^2}$ *Multiply by reciprocal.*

$= \dfrac{2k^3 p}{3k^2 p^2} = \dfrac{2k}{3p}$

19. $\dfrac{\dfrac{1}{x+3} - 1}{1 + \dfrac{1}{x+3}}$

$= \dfrac{(x+3)\left(\dfrac{1}{x+3} - 1\right)}{(x+3)\left(1 + \dfrac{1}{x+3}\right)}$ *Multiply by the LCD, x + 3.*

$= \dfrac{(x+3)\left(\dfrac{1}{x+3}\right) - (x+3)(1)}{(x+3)(1) + (x+3)\left(\dfrac{1}{x+3}\right)}$

$= \dfrac{1 - (x+3)}{(x+3) + 1}$

$= \dfrac{1 - x - 3}{x + 4}$

$= \dfrac{-2 - x}{x + 4}$

20. $\dfrac{3x}{x+1} = \dfrac{3}{2x}$

$2x(x+1)\dfrac{3x}{x+1} = 2x(x+1)\dfrac{3}{2x}$

    Multiply by the LCD, 2x(x + 1).

$2x(3x) = 3(x+1)$

$6x^2 = 3x + 3$

$6x^2 - 3x - 3 = 0$

$3(2x^2 - x - 1) = 0$

$3(2x+1)(x-1) = 0$

$x = -\dfrac{1}{2} \quad \text{or} \quad x = 1$

Check $x = -\dfrac{1}{2}$: $-3 = -3$ *True*

Check $x = 1$: $\dfrac{3}{2} = \dfrac{3}{2}$ *True*

Thus, the solution set is $\left\{-\dfrac{1}{2}, 1\right\}$.

21. $\dfrac{2}{x-1} - \dfrac{2}{3} = \dfrac{-1}{x+1}$

$3(x+1)(x-1)\left(\dfrac{2}{x-1} - \dfrac{2}{3}\right)$

$= 3(x+1)(x-1)\left(\dfrac{-1}{x+1}\right)$

Multiply by the LCD, 3(x + 1)(x − 1).

$6(x+1) - 2(x^2 - 1) = -3(x-1)$

$6x + 6 - 2x^2 + 2 = -3x + 3$

$-2x^2 + 9x + 5 = 0$

$2x^2 - 9x - 5 = 0$

$(2x+1)(x-5) = 0$

$x = -\dfrac{1}{2} \quad \text{or} \quad x = 5$

Check $x = -\dfrac{1}{2}$: $-\dfrac{4}{3} - \dfrac{2}{3} = -2$ *True*

Check $x = 5$: $\dfrac{1}{2} - \dfrac{2}{3} = -\dfrac{1}{6}$ *True*

The solution set is $\left\{-\dfrac{1}{2}, 5\right\}$.

22. $\dfrac{2x}{x-3} + \dfrac{1}{x+3} = \dfrac{-6}{x^2 - 9}$

$\dfrac{2x}{x-3} + \dfrac{1}{x+3} = \dfrac{-6}{(x+3)(x-3)}$

$(x+3)(x-3)\left(\dfrac{2x}{x-3} + \dfrac{1}{x+3}\right)$

$= (x+3)(x-3)\left(\dfrac{-6}{(x+3)(x-3)}\right)$

Multiply by the LCD, (x + 3)(x − 3).

$2x(x+3) + 1(x-3) = -6$

$2x^2 + 6x + x - 3 = -6$

$2x^2 + 7x + 3 = 0$

$(2x+1)(x+3) = 0$

$x = -\dfrac{1}{2} \quad \text{or} \quad x = -3$

x cannot equal -3 because the denominator $x + 3$ would equal 0.

Check $x = -\dfrac{1}{2}$: $\dfrac{2}{7} + \dfrac{2}{5} = \dfrac{24}{35}$ *True*

The solution set is $\left\{-\dfrac{1}{2}\right\}$.

23. Solve $F = \dfrac{k}{d - D}$ for D.

$(d-D)(F) = (d-D)\left(\dfrac{k}{d-D}\right)$ *Multiply by LCD, d − D.*

$(d-D)(F) = k$

$dF - DF = k$

$-DF = k - dF$

$D = \dfrac{k - dF}{-F} \quad \text{or} \quad D = \dfrac{dF - k}{F}$

continued

Another solution:

Solve $F = \dfrac{k}{d - D}$ for D.

$(d - D)(F) = (d - D)\left(\dfrac{k}{d - D}\right)$ *Multiply by LCD, d – D.*

$(d - D)(F) = k$

$d - D = \dfrac{k}{F}$ *Divide by F.*

$d = D + \dfrac{k}{F}$ *Add D.*

$D = d - \dfrac{k}{F}$ *Subtract $\dfrac{k}{F}$.*

24. Let x = the number. If x is added to the numerator and subtracted from the denominator of $\frac{5}{6}$, the resulting fraction is equivalent to $\frac{1}{10}$.

$$\frac{5 + x}{6 - x} = \frac{1}{10}$$

$$10(6 - x)\left(\frac{5 + x}{6 - x}\right) = 10(6 - x)\left(\frac{1}{10}\right)$$

$$10(5 + x) = (6 - x)(1)$$

$$50 + 10x = 6 - x$$

$$11x = -44$$

$$x = -4$$

The number is -4.

25. Let x = the rate of the current.

	d	r	t
Upstream	20	$7 - x$	$\dfrac{20}{7 - x}$
Downstream	50	$7 + x$	$\dfrac{50}{7 + x}$

The times are equal, so

$$\frac{20}{7 - x} = \frac{50}{7 + x}.$$

$$(7 - x)(7 + x)\left(\frac{20}{7 - x}\right) = (7 - x)(7 + x)\left(\frac{50}{7 + x}\right)$$

Multiply by the LCD, (7 – x)(7 + x).

$$20(7 + x) = 50(7 - x)$$

$$140 + 20x = 350 - 50x$$

$$70x = 210$$

$$x = 3$$

The rate of the current is 3 miles per hour.

26. Let x = the time required for the couple to paint the room working together.

	Rate	Time Working Together	Fractional Part of the Job Done When Working Together
Husband	$\frac{1}{5}$	x	$\frac{1}{5}x$
Wife	$\frac{1}{4}$	x	$\frac{1}{4}x$

Working together, they do 1 whole job, so

$$\tfrac{1}{5}x + \tfrac{1}{4}x = 1.$$

$$20(\tfrac{1}{5}x + \tfrac{1}{4}x) = 20(1)$$ *Multiply by the LCD, 20.*

$$20(\tfrac{1}{5}x) + 20(\tfrac{1}{4}x) = 20$$

$$4x + 5x = 20$$

$$9x = 20$$

$$x = \tfrac{20}{9}, \ \text{ or } \ 2\tfrac{2}{9}$$

The couple can paint the room in $\frac{20}{9}$ or $2\frac{2}{9}$ hours.

27. The length of time L that it takes for fruit to ripen during the growing season varies inversely as the average maximum temperature T during the season, so

$$L = \tfrac{k}{T}.$$

$$25 = \tfrac{k}{80} \quad \text{Let } L = 25, \ T = 80.$$

$$k = 25 \cdot 80 = 2000$$

So $L = \frac{2000}{T}$ and when $T = 75$,

$$L = \tfrac{2000}{75} = \tfrac{80}{3}, \ \text{ or } \ 26\tfrac{2}{3}.$$

To the nearest whole number, L is 27 days.

28. Since x varies directly as y, there is a constant k such that $x = ky$. First find the value of k.

$$x = ky$$
$$12 = k \cdot 4 \quad \text{Let } x = 12, \ y = 4.$$
$$k = \tfrac{12}{4} = 3$$

Since $x = ky$ and $k = 3$,

$$x = 3y.$$

Now find x when $y = 9$.

$$x = 3(9) = 27$$

CHAPTER 16 ROOTS AND RADICALS

16.1 Evaluating Roots

16.1 Margin Exercises

1. **(a)** The square roots of 100 are 10 and -10 because $10 \cdot 10 = 100$ and $(-10)(-10) = 100$.

   (b) The square roots of 25 are 5 and -5 because $5 \cdot 5 = 25$ and $(-5)(-5) = 25$.

   (c) The square roots of 36 are 6 and -6 because $6 \cdot 6 = 36$ and $(-6)(-6) = 36$.

   (d) The square roots of $\frac{25}{36}$ are $\frac{5}{6}$ and $-\frac{5}{6}$ because $\frac{5}{6} \cdot \frac{5}{6} = \frac{25}{36}$ and $\left(-\frac{5}{6}\right)\left(-\frac{5}{6}\right) = \frac{25}{36}$.

2. **(a)** $\sqrt{16}$ is the positive square root of 16.
$$4^2 = \underline{16}, \text{ so } \sqrt{16} = \underline{4}.$$

   (b) $-\sqrt{169}$ is the negative square root of 169.
$$13^2 = \underline{169}, \text{ so } -\sqrt{169} = \underline{-13}.$$

   (c) $-\sqrt{225}$ is the negative square root of 225.
$$-\sqrt{225} = -15$$

   (d) $\sqrt{729}$ is the positive square root of 729.
$$\sqrt{729} = 27$$

   (e) $-\sqrt{\frac{36}{25}}$ is the negative square root of $\frac{36}{25}$.
$$-\sqrt{\frac{36}{25}} = -\frac{6}{5}$$

   (f) $\sqrt{0.49}$ is the positive square root of 0.49.
$$\sqrt{0.49} = 0.7$$

3. **(a)** The square of $\sqrt{41}$ is
$$(\sqrt{41})^2 = \underline{41}.$$

   (b) The square of $-\sqrt{39}$ is
$$(-\sqrt{39})^2 = \underline{39}.$$

   (c) The square of $\sqrt{120}$ is
$$(\sqrt{120})^2 = 120.$$

   (d) The square of $\sqrt{2x^2 + 3}$ is
$$(\sqrt{2x^2 + 3})^2 = 2x^2 + 3.$$

4. **(a)** 9 is a perfect square, so $\sqrt{9} = 3$ is rational.

   (b) 7 is not a perfect square, so $\sqrt{7}$ is irrational.

   (c) $\frac{9}{16}$ is a perfect square, so $\sqrt{\frac{9}{16}} = \frac{3}{4}$ is rational.

(d) 72 is not a perfect square, so $\sqrt{72}$ is irrational.

(e) There is no real number whose square is -43, so $\sqrt{-43}$ is not a real number.

5. Use the square root key of a calculator to find a decimal approximation for each root. Answers are given to the nearest thousandth.

   (a) $\sqrt{28} \approx 5.292$

   (b) $\sqrt{63} \approx 7.937$

   (c) $-\sqrt{190} \approx -13.784$

   (d) $\sqrt{1000} \approx 31.623$

6. Substitute the given values in the Pythagorean theorem, $c^2 = a^2 + b^2$. Then solve for the variable that is not given.

   (a) $c^2 = a^2 + b^2$
$$
\begin{array}{ll}
\underline{c}^2 = \underline{7}^2 + \underline{24}^2 & \textit{Let a = 7, b = 24.} \\
c^2 = 49 + \underline{576} & \textit{Square.} \\
c^2 = \underline{625} & \textit{Add.} \\
c = \sqrt{625} & \textit{Take square root.} \\
c = \underline{25} & \textit{25}^2 = 625
\end{array}
$$

   (b) $c^2 = a^2 + b^2$
$$
\begin{array}{ll}
15^2 = a^2 + 13^2 & \textit{Let c = 15, b = 13.} \\
225 = a^2 + 169 & \\
56 = a^2 & \textit{Subtract 169.} \\
a = \sqrt{56} & \textit{Take square root.} \\
a \approx 7.483 & \textit{Use a calculator.}
\end{array}
$$

   (c) $c^2 = a^2 + b^2$
$$
\begin{array}{ll}
11^2 = 8^2 + b^2 & \textit{Let c = 11, a = 8.} \\
121 = 64 + b^2 & \\
57 = b^2 & \textit{Subtract 64.} \\
b = \sqrt{57} & \textit{Take square root.} \\
b \approx 7.550 & \textit{Use a calculator.}
\end{array}
$$

7. Let $c =$ the length of the diagonal. Use the Pythagorean theorem, since the width, length, and diagonal of a rectangle form a right triangle.
$$
\begin{array}{ll}
c^2 = a^2 + b^2 & \\
c^2 = 5^2 + 12^2 & \textit{Let a = 5, b = 12.} \\
c^2 = 25 + 144 & \\
c^2 = 169 & \\
c = \sqrt{169} & \\
c = 13 &
\end{array}
$$

The diagonal is 13 feet long.

8.

Perfect Cubes	Perfect Fourth Powers
$1^3 = 1$	$1^4 = 1$
$2^3 = 8$	$2^4 = 16$
$3^3 = 27$	$3^4 = 81$
$4^3 = \underline{64}$	$4^4 = \underline{256}$
$5^3 = \underline{125}$	$5^4 = \underline{625}$
$6^3 = \underline{216}$	$6^4 = \underline{1296}$
$7^3 = \underline{343}$	$7^4 = \underline{2401}$
$8^3 = \underline{512}$	$8^4 = \underline{4096}$
$9^3 = \underline{729}$	$9^4 = \underline{6561}$
$10^3 = \underline{1000}$	$10^4 = \underline{10,000}$

9. (a) $3^3 = 27$, so $\sqrt[3]{27} = \underline{3}$.

(b) $\sqrt[3]{1} = 1$, because $1^3 = 1$.

(c) $\sqrt[3]{-125} = -5$, because $(-5)^3 = -125$.

10. (a) $\sqrt[4]{81} = 3$, because 3 is positive and $3^4 = 81$.

(b) $\sqrt[4]{-81}$ is not a real number because a fourth power of a real number cannot be negative.

(c) From part (a), $\sqrt[4]{81} = 3$, so the negative root is

$$-\sqrt[4]{81} = -1 \cdot \sqrt[4]{81} = -1 \cdot 3 = -3.$$

(d) $\sqrt[5]{243} = 3$, because $3^5 = 243$.

(e) $\sqrt[5]{-243} = -3$, because $(-3)^5 = -243$.

16.1 Section Exercises

1. Every positive number has two real square roots. This statement is *true*. One of the real square roots is a positive number and the other is its opposite.

3. Every nonnegative number has two real square roots. This statement is *false* since zero is a nonnegative number that has only one square root, namely 0.

5. The cube root of every real number has the same sign as the number itself. This statement is *true*. The cube root of a positive real number is positive and the cube root of a negative real number is negative. The cube root of 0 is 0.

7. The square roots of 9 are -3 and 3 because $(-3)(-3) = 9$ and $3 \cdot 3 = 9$.

9. The square roots of 64 are -8 and 8 because $(-8)(-8) = 64$ and $8 \cdot 8 = 64$.

11. The square roots of 169 are -13 and 13 because $(-13)(-13) = 169$ and $13 \cdot 13 = 169$.

13. The square roots of $\frac{25}{196}$ are $-\frac{5}{14}$ and $\frac{5}{14}$ because

$$\left(-\tfrac{5}{14}\right)\left(-\tfrac{5}{14}\right) = \tfrac{25}{196}$$

and $\qquad \tfrac{5}{14} \cdot \tfrac{5}{14} = \tfrac{25}{196}.$

15. The square roots of 900 are -30 and 30 because $(-30)(-30) = 900$ and $30 \cdot 30 = 900$.

17. For the statement "$\sqrt{a}$ represents a positive number" to be true, a must be positive because the square root of a negative number is not a real number and $\sqrt{0} = 0$.

19. For the statement "$\sqrt{a}$ is not a real number" to be true, a must be negative.

21. $8^2 = \underline{64}$, so $\sqrt{64} = \underline{8}$. Note that $\sqrt{64}$ represents a positive number, so $\sqrt{64}$ *cannot* equal -8.

23. $\sqrt{1}$ represents the positive square root of 1. Since $1 \cdot 1 = 1$,

$$\sqrt{1} = 1.$$

25. $\sqrt{49}$ represents the positive square root of 49. Since $7 \cdot 7 = 49$,

$$\sqrt{49} = 7.$$

27. $-\sqrt{256}$ represents the negative square root of 256. Since $16 \cdot 16 = 16^2 = 256$,

$$-\sqrt{256} = -16.$$

29. $-\sqrt{\frac{144}{121}}$ represents the negative square root of $\frac{144}{121}$. Since $\frac{12}{11} \cdot \frac{12}{11} = \frac{144}{121}$,

$$-\sqrt{\tfrac{144}{121}} = -\tfrac{12}{11}.$$

31. $\sqrt{0.64}$ represents the positive square root of 0.64. Since $(0.8)(0.8) = 0.64$,

$$\sqrt{0.64} = 0.8.$$

33. $\sqrt{-121}$ is not a real number because there is no real number whose square is -121.

35. $\sqrt{-49}$ is not a real number because there is no real number whose square is -49. Thus, $-\sqrt{-49}$ is not a real number.

37. The square of $\sqrt{100}$ is

$$(\sqrt{100})^2 = 100,$$

by the definition of square root.

39. The square of $-\sqrt{19}$ is

$$(-\sqrt{19})^2 = 19,$$

since the square of a negative number is positive.

41. The square of $\sqrt{\frac{2}{3}}$ is

$$\left(\sqrt{\tfrac{2}{3}}\right)^2 = \tfrac{2}{3},$$

by the definition of square root.

43. The square of $\sqrt{3x^2 + 4}$ is
$$(\sqrt{3x^2 + 4})^2 = 3x^2 + 4.$$

45. Since $81 < 94 < 100$, $\sqrt{81} = 9$, and $\sqrt{100} = 10$, we conclude that $\sqrt{94}$ is between 9 and 10.

47. Since $49 < 51 < 64$, $\sqrt{49} = 7$, and $\sqrt{64} = 8$, we conclude that $\sqrt{51}$ is between 7 and 8.

49. Since $16 < 23.2 < 25$, $\sqrt{16} = 4$, and $\sqrt{25} = 5$, we conclude that $\sqrt{23.2}$ is between 4 and 5.

51. $\sqrt{25}$ ▪ The number 25 is a perfect square, 5^2, so $\sqrt{25}$ is a *rational* number.
$$\sqrt{25} = 5$$

53. $\sqrt{29}$ ▪ Because 29 is not a perfect square, $\sqrt{29}$ is *irrational*. Using a calculator, we obtain
$$\sqrt{29} \approx 5.385.$$

55. $-\sqrt{64}$ ▪ The number 64 is a perfect square, 8^2, so $-\sqrt{64}$ is *rational*.
$$-\sqrt{64} = -8$$

57. $-\sqrt{300}$ ▪ The number 300 is not a perfect square, so $-\sqrt{300}$ is *irrational*. Using a calculator, we obtain
$$-\sqrt{300} \approx -17.321.$$

59. $\sqrt{-29}$ ▪ There is no real number whose square is -29. Therefore, $\sqrt{-29}$ is *not a real number*.

61. $\sqrt{1200}$ ▪ Because 1200 is not a perfect square, $\sqrt{1200}$ is *irrational*. Using a calculator, we obtain
$$\sqrt{1200} \approx 34.641.$$

63. $\sqrt{103} \approx \sqrt{100} = 10$
$\sqrt{48} \approx \sqrt{49} = 7$

The best estimate for the length and width of the rectangle is 10 by 7, choice **C**.

65. $a = 8, b = 15$ ▪ Substitute the given values in the Pythagorean theorem and then solve for c^2.
$$c^2 = a^2 + b^2$$
$$c^2 = 8^2 + 15^2$$
$$= 64 + 225$$
$$= 289$$

Now find the positive square root of 289 to obtain the length of the hypotenuse, c.
$$c = \sqrt{289} = 17$$

67. $a = 6, c = 10$ ▪ Substitute the given values in the Pythagorean theorem and then solve for b^2.
$$c^2 = a^2 + b^2$$
$$10^2 = 6^2 + b^2$$
$$100 = 36 + b^2$$
$$64 = b^2$$

Now find the positive square root of 64 to obtain the length of the leg b.
$$b = \sqrt{64} = 8$$

69. $a = 11, b = 4$ ▪ Substitute the given values in the Pythagorean theorem and then solve for c^2. Use a calculator to approximate c.
$$c^2 = a^2 + b^2$$
$$c^2 = 11^2 + 4^2$$
$$= 121 + 16$$
$$= 137$$
$$c = \sqrt{137} \approx 11.705$$

71. The given information involves a right triangle with hypotenuse 25 centimeters and a leg of length 7 centimeters. Let a represent the length of the other leg, and use the Pythagorean theorem.
$$c^2 = a^2 + b^2$$
$$25^2 = a^2 + 7^2$$
$$625 = a^2 + 49$$
$$576 = a^2$$
$$a = \sqrt{576} = 24$$

The length of the rectangle is 24 centimeters.

73. ***Step 2*** Let x represent the vertical distance of the kite above Tyler's hand. The kite string forms the hypotenuse of a right triangle.

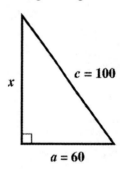

Step 3 Use the Pythagorean theorem.
$$a^2 + x^2 = c^2$$
$$60^2 + x^2 = 100^2$$

Step 4 $3600 + x^2 = 10{,}000$
$$x^2 = 6400$$
$$x = \sqrt{6400} = 80$$

Step 5 The kite is 80 feet above his hand.

Step 6 From the figure, we see that we must have
$$60^2 + 80^2 \stackrel{?}{=} 100^2$$
$$3600 + 6400 = 10{,}000. \quad \textit{True}$$

75. *Step 2* Let x represent the distance from R to S.

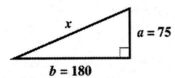

Step 3 Triangle RST is a right triangle, so we can use the Pythagorean theorem.

$$x^2 = a^2 + b^2$$
$$x^2 = 75^2 + 180^2$$
$$= 5625 + 32{,}400$$

Step 4 $x^2 = 38{,}025$
$$x = \sqrt{38{,}025} = 195$$

Step 5 The distance across the lake is 195 feet.

Step 6 From the figure, we see that we must have

$$75^2 + 180^2 \overset{?}{=} 195^2$$
$$5625 + 32{,}400 = 38{,}025. \quad \textit{True}$$

77. Refer to the right triangle shown in the figure in the textbook. Note that the given distances are the lengths of the hypotenuse (193.0 feet) and one of the legs (110.0 feet) of the triangle. Use the Pythagorean theorem with $a = 110.0$, $c = 193.0$, and $b =$ the height of the building.

$$a^2 + b^2 = c^2$$
$$(110.0)^2 + b^2 = (193.0)^2$$
$$12{,}100 + b^2 = 37{,}249$$
$$b^2 = 25{,}149$$
$$b = \sqrt{25{,}149} \approx 158.6$$

The height of the building, to the nearest tenth, is 158.6 feet.

79. Let $a = 4.5$ and $c = 12$. Use the Pythagorean theorem.

$$c^2 = a^2 + b^2$$
$$(12.0)^2 = (4.5)^2 + b^2$$
$$144 = 20.25 + b^2$$
$$123.75 = b^2$$
$$b = \sqrt{123.75} \approx 11.1$$

The distance from the base of the tree to the point where the broken part touches the ground is 11.1 feet (to the nearest tenth).

81. Use the Pythagorean theorem with $a = 5$, $b = 8$, and $c = x$.

$$c^2 = a^2 + b^2$$
$$x^2 = 5^2 + 8^2$$
$$= 25 + 64$$
$$= 89$$
$$x = \sqrt{89} \approx 9.434$$

83. $\sqrt[3]{64} = \underline{4}$, because $\underline{4}^3 = 64$.

85. $\sqrt[3]{125} = 5$, because $5^3 = 125$.

87. $\sqrt[3]{512} = 8$, because $8^3 = 512$.

89. $\sqrt[3]{-27} = -3$, because $(-3)^3 = -27$.

91. $\sqrt[3]{-216} = -6$, because $(-6)^3 = -216$.

93. $\sqrt[3]{-8} = -2$, because $(-2)^3 = -8$. Thus,
$$-\sqrt[3]{-8} = -(-2) = 2.$$

95. $\sqrt[4]{256} = 4$, because 4 is positive and $4^4 = 256$.

97. $\sqrt[4]{1296} = 6$, because 6 is positive and $6^4 = 1296$.

99. $\sqrt[4]{-1}$ is not a real number because the fourth power of a real number cannot be negative.

101. $\sqrt[4]{625} = 5$, because 5 is positive and $5^4 = 625$. Thus,
$$-\sqrt[4]{625} = -5.$$

103. $\sqrt[5]{-1024} = -4$, because $(-4)^5 = -1024$.

16.2 Multiplying, Dividing, and Simplifying Radicals

16.2 Margin Exercises

1. **(a)** $\sqrt{6} \cdot \sqrt{11} = \sqrt{6 \cdot 11} = \sqrt{66}$ *Product rule*

(b) $\sqrt{2} \cdot \sqrt{5} = \sqrt{2 \cdot 5} = \sqrt{10}$ *Product rule*

(c) $\sqrt{10} \cdot \sqrt{r} = \sqrt{10r}$

Since $r \geq 0$, $10r \geq 0$, and $\sqrt{10r}$ is a real number.

2. **(a)** $\sqrt{8} = \sqrt{4 \cdot 2}$ *4 is a perfect square.*
$$= \sqrt{4} \cdot \sqrt{2} \quad \textit{Product rule}$$
$$= 2\sqrt{2}$$

(b) $\sqrt{27} = \sqrt{9 \cdot 3}$ *9 is a perfect square.*
$$= \sqrt{9} \cdot \sqrt{3} \quad \textit{Product rule}$$
$$= 3\sqrt{3}$$

(c) $\sqrt{50} = \sqrt{25 \cdot 2}$ *25 is a perfect square.*
$$= \sqrt{25} \cdot \sqrt{2} \quad \textit{Product rule}$$
$$= 5\sqrt{2}$$

(d) $\sqrt{60} = \sqrt{4 \cdot 15}$ *4 is a perfect square.*
$$= \sqrt{4} \cdot \sqrt{15} \quad \textit{Product rule}$$
$$= 2\sqrt{15}$$

(e) $\sqrt{30}$ cannot be simplified because 30 has no perfect square factors (except 1).

3. (a) $\sqrt{3}\cdot\sqrt{15}$

   $= \sqrt{3\cdot 15}$ *Product rule*

   $= \sqrt{3\cdot 3\cdot 5}$ *or* $\sqrt{45}$

   $= \sqrt{9\cdot 5}$ *9 is a perfect square.*

   $= \sqrt{9}\cdot\sqrt{5}$ *Product rule*

   $= \underline{3\sqrt{5}}$

(b) $\sqrt{10}\cdot\sqrt{50}$

   $= \sqrt{10\cdot 50}$ *Product rule*

   $= \sqrt{10\cdot 10\cdot 5}$

   $= \sqrt{100\cdot 5}$ *100 is a perfect square.*

   $= \sqrt{100}\cdot\sqrt{5}$ *Product rule*

   $= 10\sqrt{5}$

(c) $\sqrt{12}\cdot\sqrt{2}$

   $= \sqrt{12\cdot 2}$ *Product rule*

   $= \sqrt{2\cdot 6\cdot 2}$

   $= \sqrt{4\cdot 6}$ *4 is a perfect square.*

   $= \sqrt{4}\cdot\sqrt{6}$ *Product rule*

   $= 2\sqrt{6}$

(d) $3\sqrt{5}\cdot 4\sqrt{10}$

   $= 3\cdot 4\cdot\sqrt{5\cdot 10}$ *Product rule*

   $= 12\sqrt{5\cdot 5\cdot 2}$

   $= 12\sqrt{25\cdot 2}$ *25 is a perfect square.*

   $= 12\sqrt{25}\cdot\sqrt{2}$ *Product rule*

   $= 12\cdot 5\cdot\sqrt{2}$

   $= 60\sqrt{2}$

4. (a) $\sqrt{\dfrac{81}{16}} = \dfrac{\sqrt{81}}{\sqrt{16}} = \dfrac{9}{4}$ *Quotient rule*

(b) $\dfrac{\sqrt{192}}{\sqrt{3}} = \sqrt{\dfrac{192}{3}} = \sqrt{64} = 8$

(c) $\sqrt{\dfrac{10}{49}} = \dfrac{\sqrt{10}}{\sqrt{49}} = \dfrac{\sqrt{10}}{7}$

5. $\dfrac{8\sqrt{50}}{4\sqrt{5}} = \dfrac{8}{4}\cdot\dfrac{\sqrt{50}}{\sqrt{5}}$

   $= 2\cdot\sqrt{\dfrac{50}{5}}$

   $= 2\sqrt{10}$

6. (a) $\sqrt{\dfrac{3}{8}}\cdot\sqrt{\dfrac{7}{2}}$

   $= \sqrt{\dfrac{3}{8}\cdot\dfrac{7}{2}}$ *Product rule*

   $= \sqrt{\dfrac{21}{16}}$ *Multiply fractions.*

   $= \dfrac{\sqrt{21}}{\sqrt{16}}$ *Quotient rule*

   $= \dfrac{\sqrt{21}}{4}$ $\sqrt{16} = 4$

(b) $\sqrt{\dfrac{5}{6}}\cdot\sqrt{120}$

   $= \sqrt{\dfrac{5}{6}}\cdot\sqrt{\dfrac{120}{1}}$

   $= \sqrt{\dfrac{5}{6}\cdot\dfrac{120}{1}}$ *Product rule*

   $= \sqrt{\dfrac{600}{6}}$ *Multiply fractions.*

   $= \sqrt{100}$

   $= 10$

7. (a) $(\underline{x^4})^2 = x^8$, so $\sqrt{x^8} = \underline{x^4}$.

(b) $\sqrt{36y^6} = \sqrt{36}\cdot\sqrt{y^6}$ *Product rule*

   $= \underline{6y^3}$

(c) $\sqrt{100p^{12}} = \sqrt{100}\cdot\sqrt{p^{12}} = 10p^6$

(d) $\sqrt{12z^2} = \sqrt{4}\cdot\sqrt{3}\cdot\sqrt{z^2}$

   $= 2\cdot\sqrt{3}\cdot z = 2z\sqrt{3}$

(e) $\sqrt{a^5} = \sqrt{a^4\cdot a}$ a^4 *is a square.*

   $= \sqrt{a^4}\cdot\sqrt{a} = a^2\sqrt{a}$

(f) $\sqrt{\dfrac{10}{n^4}}\ (n\neq 0) = \dfrac{\sqrt{10}}{\sqrt{n^4}} = \dfrac{\sqrt{10}}{n^2}$

8. (a) $\sqrt[3]{108} = \sqrt[3]{27\cdot 4}$ *27 is a perfect cube.*

   $= \sqrt[3]{27}\cdot\sqrt[3]{4}$ *Product rule*

   $= \underline{3\sqrt[3]{4}}$

(b) $\sqrt[3]{250} = \sqrt[3]{125\cdot 2}$ *125 is a perfect cube.*

   $= \sqrt[3]{125}\cdot\sqrt[3]{2}$ *Product rule*

   $= 5\sqrt[3]{2}$

(c) $\sqrt[4]{160} = \sqrt[4]{16\cdot 10}$ *16 is a perfect fourth power.*

   $= \sqrt[4]{16}\cdot\sqrt[4]{10}$ *Product rule*

   $= 2\sqrt[4]{10}$

(d) $\sqrt[4]{\dfrac{16}{625}} = \dfrac{\sqrt[4]{16}}{\sqrt[4]{625}}$ *Quotient rule*

$= \dfrac{2}{5}$

9. **(a)** $\sqrt[3]{z^9} = z^3$, since $(z^3)^3 = z^9$.

(b) $\sqrt[3]{8x^6} = \sqrt[3]{8} \cdot \sqrt[3]{x^6}$ *Product rule*

$= 2x^2$ $2^3 = 8;\ (x^2)^3 = x^6$

(c) $\sqrt[3]{54t^5} = \sqrt[3]{27t^3 \cdot 2t^2}$ $27t^3$ *is a cube.*

$= \sqrt[3]{27t^3} \cdot \sqrt[3]{2t^2}$ *Product rule*

$= 3t\sqrt[3]{2t^2}$ $(3t)^3 = 27t^3$

(d) $\sqrt[3]{\dfrac{a^{15}}{64}} = \dfrac{\sqrt[3]{a^{15}}}{\sqrt[3]{64}}$ *Quotient rule*

$= \dfrac{a^5}{4}$

16.2 Section Exercises

1. $\sqrt{(-6)^2} = \sqrt{36} = 6 \neq -6$, so the given statement, "$\sqrt{(-6)^2} = -6$," is *false*. In general, $\sqrt{a^2} = a$ only if a is nonnegative.

3. $2\sqrt{7}$ represents the *product* of 2 and $\sqrt{7}$, not a *sum*, so the given statement is *false*.

5. $\sqrt{3} \cdot \sqrt{5}$ ▪ Since 3 and 5 are nonnegative real numbers, the *Product Rule for Radicals* applies. Thus,

$$\sqrt{3} \cdot \sqrt{5} = \sqrt{3 \cdot 5} = \sqrt{15}.$$

7. $\sqrt{2} \cdot \sqrt{11} = \sqrt{2 \cdot 11} = \sqrt{22}$

9. $\sqrt{6} \cdot \sqrt{7} = \sqrt{6 \cdot 7} = \sqrt{42}$

11. $\sqrt{13} \cdot \sqrt{r}\ \ (r \geq 0) = \sqrt{13r}$

13. $\sqrt{47}$ is in simplified form since 47 has no perfect square factor (other than 1).

The other three choices could be simplified as follows.

$$\sqrt{45} = \sqrt{9 \cdot 5} = 3\sqrt{5}$$
$$\sqrt{48} = \sqrt{16 \cdot 3} = 4\sqrt{3}$$
$$\sqrt{44} = \sqrt{4 \cdot 11} = 2\sqrt{11}$$

The correct choice is **A**.

15. $\sqrt{45} = \sqrt{9 \cdot 5} = \sqrt{9} \cdot \sqrt{5} = 3\sqrt{5}$

17. $\sqrt{24} = \sqrt{4 \cdot 6} = \sqrt{4} \cdot \sqrt{6} = 2\sqrt{6}$

19. $\sqrt{90} = \sqrt{9 \cdot 10} = \sqrt{9} \cdot \sqrt{10} = 3\sqrt{10}$

21. $\sqrt{75} = \sqrt{25 \cdot 3} = \sqrt{25} \cdot \sqrt{3} = 5\sqrt{3}$

23. $\sqrt{125} = \sqrt{25 \cdot 5} = \sqrt{25} \cdot \sqrt{5} = 5\sqrt{5}$

25. $145 = 5 \cdot 29$, so 145 has no perfect square factors (except 1) and $\sqrt{145}$ cannot be simplified.

27. $\sqrt{160} = \sqrt{16 \cdot 10} = \sqrt{16} \cdot \sqrt{10} = 4\sqrt{10}$

29. $-\sqrt{700} = -\sqrt{100 \cdot 7}$

$= -\sqrt{100} \cdot \sqrt{7} = -10\sqrt{7}$

31. $3\sqrt{52} = 3\sqrt{4} \cdot \sqrt{13}$

$= 3 \cdot 2 \cdot \sqrt{13}$

$= 6\sqrt{13}$

33. $5\sqrt{50} = 5\sqrt{25} \cdot \sqrt{2}$

$= 5 \cdot 5 \cdot \sqrt{2}$

$= 25\sqrt{2}$

35. $a = 13, b = 9$

$c^2 = a^2 + b^2$ $c = \sqrt{250}$

$c^2 = 13^2 + 9^2$ $= \sqrt{25} \cdot \sqrt{10}$

$= 169 + 81$ $= 5\sqrt{10}$

$= 250$

37. $a = 7, c = 11$

$c^2 = a^2 + b^2$ $b = \sqrt{72}$

$11^2 = 7^2 + b^2$ $= \sqrt{36} \cdot \sqrt{2}$

$121 = 49 + b^2$ $= 6\sqrt{2}$

$72 = b^2$

39. $\sqrt{3} \cdot \sqrt{18} = \sqrt{3 \cdot 18}$

$= \sqrt{3 \cdot 3 \cdot 6}$

$= \sqrt{9 \cdot 6}$

$= \sqrt{9} \cdot \sqrt{6} = 3\sqrt{6}$

41. $\sqrt{9} \cdot \sqrt{32} = 3 \cdot \sqrt{16} \cdot \sqrt{2}$

$= 3 \cdot 4 \cdot \sqrt{2}$

$= 12\sqrt{2}$

43. $\sqrt{12} \cdot \sqrt{48} = \sqrt{12 \cdot 48}$

$= \sqrt{12 \cdot 12 \cdot 4}$

$= \sqrt{12 \cdot 12} \cdot \sqrt{4}$

$= 12 \cdot 2$

$= 24$

45. $\sqrt{12} \cdot \sqrt{30} = \sqrt{12 \cdot 30}$

$= \sqrt{360}$

$= \sqrt{36 \cdot 10}$

$= \sqrt{36} \cdot \sqrt{10}$

$= 6\sqrt{10}$

47. $2\sqrt{10} \cdot 3\sqrt{2} = 2 \cdot 3 \cdot \sqrt{10 \cdot 2}$ *Product rule*

$\qquad\qquad = 6\sqrt{20}$ *Multiply.*

$\qquad\qquad = 6\sqrt{4 \cdot 5}$ *Factor; 4 is a perfect square.*

$\qquad\qquad = 6\sqrt{4} \cdot \sqrt{5}$ *Product rule*

$\qquad\qquad = 6 \cdot 2 \cdot \sqrt{5}$ $\sqrt{4} = 2$

$\qquad\qquad = 12\sqrt{5}$ *Multiply.*

49. $5\sqrt{3} \cdot 2\sqrt{15} = 5 \cdot 2 \cdot \sqrt{3 \cdot 15}$ *Product rule*

$\qquad\qquad = 10\sqrt{45}$ *Multiply.*

$\qquad\qquad = 10\sqrt{9 \cdot 5}$ *Factor; 9 is a perfect square.*

$\qquad\qquad = 10\sqrt{9} \cdot \sqrt{5}$ *Product rule*

$\qquad\qquad = 10 \cdot 3 \cdot \sqrt{5}$ $\sqrt{9} = 3$

$\qquad\qquad = 30\sqrt{5}$ *Multiply.*

51. Method 1: $\sqrt{8} \cdot \sqrt{32} = \sqrt{8 \cdot 32} = \sqrt{256} = 16$
Method 2: $\sqrt{8} = 2\sqrt{2}$ and $\sqrt{32} = 4\sqrt{2}$, so
$\sqrt{8} \cdot \sqrt{32} = 2\sqrt{2} \cdot 4\sqrt{2} = 8 \cdot 2 = 16$

The same answer results. Either method can be used to obtain the correct answer.

53. $\sqrt{\dfrac{16}{225}} = \dfrac{\sqrt{16}}{\sqrt{225}} = \dfrac{4}{15}$

55. $\sqrt{\dfrac{7}{16}} = \dfrac{\sqrt{7}}{\sqrt{16}} = \dfrac{\sqrt{7}}{4}$

57. $\sqrt{\dfrac{4}{50}} = \sqrt{\dfrac{2 \cdot 2}{25 \cdot 2}} = \sqrt{\dfrac{2}{25}} = \dfrac{\sqrt{2}}{\sqrt{25}} = \dfrac{\sqrt{2}}{5}$

59. $\dfrac{\sqrt{75}}{\sqrt{3}} = \sqrt{\dfrac{75}{3}} = \sqrt{25} = 5$

61. $\sqrt{\dfrac{5}{2}} \cdot \sqrt{\dfrac{125}{8}} = \sqrt{\dfrac{5}{2} \cdot \dfrac{125}{8}}$

$\qquad\qquad = \sqrt{\dfrac{625}{16}}$

$\qquad\qquad = \dfrac{\sqrt{625}}{\sqrt{16}} = \dfrac{25}{4}$

63. $\dfrac{30\sqrt{10}}{5\sqrt{2}} = \dfrac{30}{5}\sqrt{\dfrac{10}{2}} = 6\sqrt{5}$

65. $\sqrt{m^2} = m$ $(m \geq 0)$

67. $\sqrt{y^4} = \sqrt{(y^2)^2} = y^2$

69. $\sqrt{36z^2} = \sqrt{36} \cdot \sqrt{z^2} = 6z$

71. $\sqrt{400x^6} = \sqrt{20^2(x^3)^2} = 20x^3$

73. $\sqrt{18x^8} = \sqrt{9 \cdot 2 \cdot x^8}$

$\qquad\qquad = \sqrt{9} \cdot \sqrt{2} \cdot \sqrt{(x^4)^2}$

$\qquad\qquad = 3 \cdot \sqrt{2} \cdot x^4$

$\qquad\qquad = 3x^4\sqrt{2}$

75. $\sqrt{45c^{14}} = \sqrt{9 \cdot 5 \cdot c^{14}}$

$\qquad\qquad = \sqrt{9} \cdot \sqrt{5} \cdot \sqrt{(c^7)^2}$

$\qquad\qquad = 3 \cdot \sqrt{5} \cdot c^7$

$\qquad\qquad = 3c^7\sqrt{5}$

77. $\sqrt{z^5} = \sqrt{z^4 \cdot z} = \sqrt{z^4} \cdot \sqrt{z} = z^2\sqrt{z}$

79. $\sqrt{a^{13}} = \sqrt{a^{12}} \cdot \sqrt{a}$

$\qquad\qquad = \sqrt{(a^6)^2} \cdot \sqrt{a} = a^6\sqrt{a}$

81. $\sqrt{64x^7} = \sqrt{64} \cdot \sqrt{x^6}\sqrt{x}$

$\qquad\qquad = 8 \cdot \sqrt{(x^3)^2}\sqrt{x} = 8x^3\sqrt{x}$

83. $\sqrt{x^6y^{12}} = \sqrt{(x^3)^2 \cdot (y^6)^2}$

$\qquad\qquad = x^3y^6$

85. $\sqrt{81m^4n^2} = \sqrt{81} \cdot \sqrt{m^4} \cdot \sqrt{n^2}$

$\qquad\qquad = 9 \cdot \sqrt{(m^2)^2} \cdot n = 9m^2n$

87. $\sqrt{\dfrac{7}{x^{10}}} = \dfrac{\sqrt{7}}{\sqrt{x^{10}}} = \dfrac{\sqrt{7}}{\sqrt{(x^5)^2}} = \dfrac{\sqrt{7}}{x^5}$ $(x \neq 0)$

89. $\sqrt{\dfrac{y^4}{100}} = \dfrac{\sqrt{y^4}}{\sqrt{100}} = \dfrac{\sqrt{(y^2)^2}}{\sqrt{10^2}} = \dfrac{y^2}{10}$

91. $\sqrt{\dfrac{x^6}{y^8}} = \dfrac{\sqrt{x^6}}{\sqrt{y^8}} = \dfrac{\sqrt{(x^3)^2}}{\sqrt{(y^4)^2}} = \dfrac{x^3}{y^4}$ $(y \neq 0)$

93. $\sqrt[3]{40}$ ■ 8 is a perfect cube that is a factor of 40.

$\qquad \sqrt[3]{40} = \sqrt[3]{8 \cdot 5}$

$\qquad\qquad = \sqrt[3]{8} \cdot \sqrt[3]{5} = 2\sqrt[3]{5}$

95. $\sqrt[3]{54}$ ■ 27 is a perfect cube that is a factor of 54.

$\qquad \sqrt[3]{54} = \sqrt[3]{27 \cdot 2}$

$\qquad\qquad = \sqrt[3]{27} \cdot \sqrt[3]{2} = 3\sqrt[3]{2}$

97. $\sqrt[3]{128}$ ■ 64 is a perfect cube that is a factor of 128.

$\qquad \sqrt[3]{128} = \sqrt[3]{64 \cdot 2}$

$\qquad\qquad = \sqrt[3]{64} \cdot \sqrt[3]{2} = 4\sqrt[3]{2}$

99. $\sqrt[4]{80}$ ■ 16 is a perfect fourth power that is a factor of 80.

$\qquad \sqrt[4]{80} = \sqrt[4]{16 \cdot 5}$

$\qquad\qquad = \sqrt[4]{16} \cdot \sqrt[4]{5} = 2\sqrt[4]{5}$

101. $\sqrt[3]{\dfrac{8}{27}}$ ■ 8 and 27 are both perfect cubes.

$\qquad\qquad \sqrt[3]{\dfrac{8}{27}} = \dfrac{\sqrt[3]{8}}{\sqrt[3]{27}} = \dfrac{2}{3}$

103. $\sqrt[3]{-\dfrac{216}{125}} = \sqrt[3]{\left(-\dfrac{6}{5}\right)^3} = -\dfrac{6}{5}$

105. $\sqrt[3]{p^3} = p$, because $(p)^3 = p^3$.

107. $\sqrt[3]{x^9} = \sqrt[3]{(x^3)^3} = x^3$

109. $\sqrt[3]{64z^6} = \sqrt[3]{64} \cdot \sqrt[3]{(z^2)^3} = 4z^2$

111. $\sqrt[3]{343a^9b^3} = \sqrt[3]{343} \cdot \sqrt[3]{a^9} \cdot \sqrt[3]{b^3} = 7a^3b$

113. $\sqrt[3]{16t^5} = \sqrt[3]{8t^3} \cdot \sqrt[3]{2t^2} = 2t\sqrt[3]{2t^2}$

115. $\sqrt[3]{\dfrac{m^{12}}{8}} = \sqrt[3]{\dfrac{(m^4)^3}{2^3}} = \dfrac{m^4}{2}$

117. Use the formula for the volume of a cube.

$$V = s^3$$
$$216 = s^3 \quad \text{Let } V = 216.$$
$$\sqrt[3]{216} = s$$
$$6 = s$$

The depth of the container is 6 centimeters.

119. Use the formula for the volume of a sphere.

$$\tfrac{4}{3}\pi r^3 = V$$
$$\tfrac{4}{3}\pi r^3 = 288\pi \quad \text{Let } V = 288\pi.$$
$$\tfrac{3}{4}\left(\tfrac{4}{3}\pi r^3\right) = \tfrac{3}{4}(288\pi)$$
$$\pi r^3 = 216\pi$$
$$r^3 = 216 \quad \text{Divide by } \pi.$$
$$r = \sqrt[3]{216} = 6$$

The radius is 6 inches.

121. $2\sqrt{26} \approx 2\sqrt{25} = 2 \cdot 5 = 10$

$\sqrt{83} \approx \sqrt{81} = 9$

Using 10 and 9 as estimates for the length and the width of the rectangle gives us $10 \cdot 9 = 90$ as an estimate for the area. Thus, choice **D** is the best estimate.

16.3   Adding and Subtracting Radicals

16.3   Margin Exercises

1. **(a)** $5\sqrt{6}$ and $4\sqrt{6}$ are *like* radicals because they are multiples of the same root of the same number.

(b) $2\sqrt{3}$ and $3\sqrt{2}$ are *unlike* radicals because the radicands, 3 and 2, respectively, are different.

(c) $\sqrt{10}$ and $\sqrt[3]{10}$ are *unlike* radicals because the indexes, 2 and 3, are different.

(d) $7\sqrt{2x}$ and $8\sqrt{2x}$ are *like* radicals because they are multiples of the same root of the same number.

(e) $\sqrt{3y}$ and $\sqrt{6y}$ are *unlike* radicals because the radicands, $3y$ and $6y$, are different.

2. **(a)** $8\sqrt{5} + 2\sqrt{5}$

$= (8 + 2)\sqrt{5} \quad$ *Distributive property*

$= 10\sqrt{5}$

(b) $4\sqrt{3} - 9\sqrt{3} = (4 - 9)\sqrt{3}$

$= -5\sqrt{3}$

(c) $4\sqrt{11} - 3\sqrt{11} = (4 - 3)\sqrt{11}$

$= \sqrt{11}$

(d) $\sqrt{15} + \sqrt{15} = 1\sqrt{15} + 1\sqrt{15}$

$= (1 + 1)\sqrt{15}$

$= 2\sqrt{15}$

(e) $2\sqrt{7} + 2\sqrt{10}$ cannot be combined because $\sqrt{7}$ and $\sqrt{10}$ are unlike radicals.

3. **(a)** $\sqrt{8} + 4\sqrt{2} \qquad\qquad$ *Factor.*

$= \sqrt{4 \cdot 2} + 4\sqrt{2}$

$= \sqrt{4} \cdot \sqrt{2} + 4\sqrt{2} \quad$ *Product rule*

$= 2\sqrt{2} + 4\sqrt{2} \qquad \sqrt{4} = 2$

$= 6\sqrt{2} \qquad\qquad$ *Add like radicals.*

(b) $\sqrt{18} - \sqrt{2}$

$= \sqrt{9 \cdot 2} - \sqrt{2}$

$= \sqrt{9} \cdot \sqrt{2} - \sqrt{2}$

$= 3\sqrt{2} - 1\sqrt{2}$

$= 2\sqrt{2}$

(c) $\sqrt{27} + \sqrt{12}$

$= \sqrt{9 \cdot 3} + \sqrt{4 \cdot 3}$

$= \sqrt{9} \cdot \sqrt{3} + \sqrt{4} \cdot \sqrt{3}$

$= 3\sqrt{3} + 2\sqrt{3}$

$= 5\sqrt{3}$

(d) $5\sqrt{200} - 6\sqrt{18}$

$= 5(\sqrt{100} \cdot \sqrt{2}) - 6(\sqrt{9} \cdot \sqrt{2})$

$= 5(10\sqrt{2}) - 6(3\sqrt{2})$

$= 50\sqrt{2} - 18\sqrt{2}$

$= 32\sqrt{2}$

4. **(a)** $\sqrt{7} \cdot \sqrt{21} + 2\sqrt{27}$

$= \sqrt{7} \cdot \sqrt{7} \cdot \sqrt{3} + 2(\sqrt{9} \cdot \sqrt{3})$

$= \sqrt{49} \cdot \sqrt{3} + 2(3\sqrt{3})$

$= 7\sqrt{3} + 6\sqrt{3}$

$= 13\sqrt{3}$

(b) $\sqrt{3r} \cdot \sqrt{6} + \sqrt{8r}$

$= \sqrt{18r} + \sqrt{8r}$

$= \sqrt{9 \cdot 2r} + \sqrt{4 \cdot 2r}$

$= \sqrt{9} \cdot \sqrt{2r} + \sqrt{4} \cdot \sqrt{2r}$

$= 3\sqrt{2r} + 2\sqrt{2r}$

$= 5\sqrt{2r}$

(c) $y\sqrt{72} - \sqrt{18y^2}$

$= y \cdot \sqrt{36 \cdot 2} - \sqrt{9y^2 \cdot 2}$

$= y \cdot \sqrt{36} \cdot \sqrt{2} - \sqrt{9y^2} \cdot \sqrt{2}$

$= y \cdot 6\sqrt{2} - 3y\sqrt{2}$

$= (6y - 3y)\sqrt{2}$

$= 3y\sqrt{2}$

(d) $\sqrt[3]{81x^4} + 5\sqrt[3]{24x^4}$

$= \sqrt[3]{27x^3 \cdot 3x} + 5\sqrt[3]{8x^3 \cdot 3x}$

$= \sqrt[3]{27x^3} \cdot \sqrt[3]{3x} + 5\left(\sqrt[3]{8x^3} \cdot \sqrt[3]{3x}\right)$

$= 3x\sqrt[3]{3x} + 5\left(2x \cdot \sqrt[3]{3x}\right)$

$= 3x\sqrt[3]{3x} + 10x\sqrt[3]{3x}$

$= (3x + 10x)\sqrt[3]{3x}$

$= 13x\sqrt[3]{3x}$

16.3 Section Exercises

1. Like radicals have the same radicand and the same index , or order. For example, $5\sqrt{2}$ and $-3\sqrt{2}$ are like radicals, as are $\sqrt{7}$, $-\sqrt{7}$, and $8\sqrt{7}$.

3. $\sqrt{5} + 5\sqrt{3}$ cannot be simplified because the radicands are different. They are unlike radicals.

5. $5\sqrt{2} + 6\sqrt{2} = (5 + 6)\sqrt{2} = 11\sqrt{2}$ is an application of the distributive property.

7. $2\sqrt{3} + 5\sqrt{3}$

$= (2 + 5)\sqrt{3}$ *Distributive property*

$= 7\sqrt{3}$

9. $14\sqrt{7} - 19\sqrt{7}$

$= (14 - 19)\sqrt{7}$ *Distributive property*

$= -5\sqrt{7}$

11. $\sqrt{17} + 4\sqrt{17} = 1\sqrt{17} + 4\sqrt{17}$

$= (1 + 4)\sqrt{17}$

$= 5\sqrt{17}$

13. $6\sqrt{7} - \sqrt{7} = 6\sqrt{7} - 1\sqrt{7}$

$= (6 - 1)\sqrt{7}$

$= 5\sqrt{7}$

15. $\sqrt{6} + \sqrt{6} = 1\sqrt{6} + 1\sqrt{6}$

$= (1 + 1)\sqrt{6}$

$= 2\sqrt{6}$

17. $\sqrt{6} + \sqrt{7}$ ▪ These are unlike radicals (different radicands) and cannot be combined.

19. $5\sqrt{3} + \sqrt{12} = 5\sqrt{3} + \sqrt{4 \cdot 3}$

$= 5\sqrt{3} + \sqrt{4} \cdot \sqrt{3}$

$= 5\sqrt{3} + 2\sqrt{3}$

$= 7\sqrt{3}$

21. $\sqrt{45} + 4\sqrt{20} = \sqrt{9 \cdot 5} + 4\sqrt{4 \cdot 5}$

$= \sqrt{9} \cdot \sqrt{5} + 4(\sqrt{4} \cdot \sqrt{5})$

$= 3\sqrt{5} + 4(2\sqrt{5})$

$= 3\sqrt{5} + 8\sqrt{5}$

$= 11\sqrt{5}$

23. $5\sqrt{72} - 3\sqrt{50} = 5\sqrt{36 \cdot 2} - 3\sqrt{25 \cdot 2}$

$= 5 \cdot \sqrt{36} \cdot \sqrt{2} - 3\sqrt{25} \cdot \sqrt{2}$

$= 5 \cdot 6 \cdot \sqrt{2} - 3 \cdot 5 \cdot \sqrt{2}$

$= 30\sqrt{2} - 15\sqrt{2}$

$= 15\sqrt{2}$

25. $-5\sqrt{32} + 2\sqrt{98}$

$= -5(\sqrt{16} \cdot \sqrt{2}) + 2(\sqrt{49} \cdot \sqrt{2})$

$= -5(4\sqrt{2}) + 2(7\sqrt{2})$

$= -20\sqrt{2} + 14\sqrt{2}$

$= -6\sqrt{2}$

27. $5\sqrt{7} - 3\sqrt{28} + 6\sqrt{63}$

$= 5\sqrt{7} - 3(\sqrt{4} \cdot \sqrt{7}) + 6(\sqrt{9} \cdot \sqrt{7})$

$= 5\sqrt{7} - 3(2\sqrt{7}) + 6(3\sqrt{7})$

$= 5\sqrt{7} - 6\sqrt{7} + 18\sqrt{7}$

$= (5 - 6 + 18)\sqrt{7}$

$= 17\sqrt{7}$

29. $2\sqrt{8} - 5\sqrt{32} - 2\sqrt{48}$

$= 2(\sqrt{4} \cdot \sqrt{2}) - 5(\sqrt{16} \cdot \sqrt{2})$

$\quad - 2(\sqrt{16} \cdot \sqrt{3})$

$= 2(2\sqrt{2}) - 5(4\sqrt{2}) - 2(4\sqrt{3})$

$= 4\sqrt{2} - 20\sqrt{2} - 8\sqrt{3}$

$= -16\sqrt{2} - 8\sqrt{3}$

31. $4\sqrt{50} + 3\sqrt{12} - 5\sqrt{45}$

$= 4(\sqrt{25} \cdot \sqrt{2}) + 3(\sqrt{4} \cdot \sqrt{3}) - 5(\sqrt{9} \cdot \sqrt{5})$

$= 4(5\sqrt{2}) + 3(2\sqrt{3}) - 5(3\sqrt{5})$

$= 20\sqrt{2} + 6\sqrt{3} - 15\sqrt{5}$

33. $\frac{1}{4}\sqrt{288} + \frac{1}{6}\sqrt{72}$

$\qquad = \frac{1}{4}(\sqrt{144}\cdot\sqrt{2}) + \frac{1}{6}(\sqrt{36}\cdot\sqrt{2})$

$\qquad = \frac{1}{4}(12\sqrt{2}) + \frac{1}{6}(6\sqrt{2})$

$\qquad = 3\sqrt{2} + 1\sqrt{2}$

$\qquad = 4\sqrt{2}$

35.   Use the formula for the perimeter of a rectangle.

$$P = 2l + 2w$$
$$= 2(7\sqrt{2}) + 2(4\sqrt{2})$$
$$= 14\sqrt{2} + 8\sqrt{2}$$
$$= (14 + 8)\sqrt{2}$$
$$= 22\sqrt{2}$$

37. $\sqrt{6}\cdot\sqrt{2} + 9\sqrt{3} = \sqrt{6\cdot2} + 9\sqrt{3}$

$\qquad\qquad\qquad\quad = \sqrt{12} + 9\sqrt{3}$

$\qquad\qquad\qquad\quad = \sqrt{4}\cdot\sqrt{3} + 9\sqrt{3}$

$\qquad\qquad\qquad\quad = 2\sqrt{3} + 9\sqrt{3}$

$\qquad\qquad\qquad\quad = 11\sqrt{3}$

39. $\sqrt{3}\cdot\sqrt{7} + 2\sqrt{21} = \sqrt{3\cdot7} + 2\sqrt{21}$

$\qquad\qquad\qquad\quad = 1\sqrt{21} + 2\sqrt{21}$

$\qquad\qquad\qquad\quad = 3\sqrt{21}$

41. $\sqrt{32x} - \sqrt{18x}$

$\qquad = \sqrt{16\cdot2x} - \sqrt{9\cdot2x}$

$\qquad = \sqrt{16}\cdot\sqrt{2x} - \sqrt{9}\cdot\sqrt{2x}$

$\qquad = 4\sqrt{2x} - 3\sqrt{2x}$

$\qquad = (4-3)\sqrt{2x}$

$\qquad = 1\sqrt{2x} = \sqrt{2x}$

43. $\sqrt{27r} + \sqrt{48r}$

$\qquad = \sqrt{9}\cdot\sqrt{3r} + \sqrt{16}\cdot\sqrt{3r}$

$\qquad = 3\sqrt{3r} + 4\sqrt{3r}$

$\qquad = 7\sqrt{3r}$

45. $\sqrt{9x} + \sqrt{49x} - \sqrt{25x}$

$\qquad = \sqrt{9}\cdot\sqrt{x} + \sqrt{49}\cdot\sqrt{x} - \sqrt{25}\cdot\sqrt{x}$

$\qquad = 3\sqrt{x} + 7\sqrt{x} - 5\sqrt{x}$

$\qquad = (3 + 7 - 5)\sqrt{x}$

$\qquad = 5\sqrt{x}$

47. $\sqrt{6x^2} + x\sqrt{24}$

$\qquad = \sqrt{x^2\cdot6} + x\sqrt{4\cdot6}$

$\qquad = \sqrt{x^2}\cdot\sqrt{6} + x\cdot\sqrt{4}\cdot\sqrt{6}$

$\qquad = x\sqrt{6} + x\cdot2\sqrt{6}$

$\qquad = (x + 2x)\sqrt{6}$

$\qquad = 3x\sqrt{6}$

49. $3\sqrt{8x^2} - 4x\sqrt{2} - x\sqrt{8}$

$\qquad = 3\sqrt{4x^2\cdot2} - 4x\sqrt{2} - x\sqrt{4\cdot2}$

$\qquad = 3\cdot\sqrt{4x^2}\cdot\sqrt{2} - 4x\sqrt{2} - x\cdot\sqrt{4}\cdot\sqrt{2}$

$\qquad = 3\cdot2x\cdot\sqrt{2} - 4x\sqrt{2} - x\cdot2\cdot\sqrt{2}$

$\qquad = 6x\sqrt{2} - 4x\sqrt{2} - 2x\sqrt{2}$

$\qquad = (6x - 4x - 2x)\sqrt{2}$

$\qquad = 0\cdot\sqrt{2} = 0$

51. $-8\sqrt{32k} + 6\sqrt{8k}$

$\qquad = -8(\sqrt{16\cdot2k}) + 6(\sqrt{4\cdot2k})$

$\qquad = -8(4\sqrt{2k}) + 6(2\sqrt{2k})$

$\qquad = -32\sqrt{2k} + 12\sqrt{2k}$

$\qquad = (-32 + 12)\sqrt{2k}$

$\qquad = -20\sqrt{2k}$

53. $2\sqrt{125x^2z} + 8x\sqrt{80z}$

$\qquad = 2\sqrt{25x^2\cdot5z} + 8x\sqrt{16\cdot5z}$

$\qquad = 2\sqrt{25x^2}\cdot\sqrt{5z} + 8x\sqrt{16}\cdot\sqrt{5z}$

$\qquad = 2\cdot5x\cdot\sqrt{5z} + 8x\cdot4\cdot\sqrt{5z}$

$\qquad = 10x\sqrt{5z} + 32x\sqrt{5z}$

$\qquad = (10x + 32x)\sqrt{5z}$

$\qquad = 42x\sqrt{5z}$

55. $4\sqrt[3]{16} - 3\sqrt[3]{54}$

Recall that 8 and 27 are perfect cubes.

$\qquad = 4(\sqrt[3]{8\cdot2}) - 3(\sqrt[3]{27\cdot2})$

$\qquad = 4(\sqrt[3]{8}\cdot\sqrt[3]{2}) - 3(\sqrt[3]{27}\cdot\sqrt[3]{2})$

$\qquad = 4(2\sqrt[3]{2}) - 3(3\sqrt[3]{2})$

$\qquad = 8\sqrt[3]{2} - 9\sqrt[3]{2}$

$\qquad = (8 - 9)\sqrt[3]{2} = -1\sqrt[3]{2} = -\sqrt[3]{2}$

57. $6\sqrt[3]{8p^2} - 2\sqrt[3]{27p^2}$

$\qquad = 6\cdot\sqrt[3]{8}\cdot\sqrt[3]{p^2} - 2\cdot\sqrt[3]{27}\cdot\sqrt[3]{p^2}$

$\qquad = 6\cdot2\cdot\sqrt[3]{p^2} - 2\cdot3\cdot\sqrt[3]{p^2}$

$\qquad = 12\sqrt[3]{p^2} - 6\sqrt[3]{p^2}$

$\qquad = 6\sqrt[3]{p^2}$

59. $5\sqrt[4]{m^3} + 8\sqrt[4]{16m^3}$

$\qquad = 5\sqrt[4]{m^3} + 8\sqrt[4]{16}\sqrt[4]{m^3}$

$\qquad = 5\sqrt[4]{m^3} + 8\cdot2\cdot\sqrt[4]{m^3}$

$\qquad = 5\sqrt[4]{m^3} + 16\sqrt[4]{m^3}$

$\qquad = 21\sqrt[4]{m^3}$

16.4 Rationalizing the Denominator

16.4 Margin Exercises

1. **(a)** $\dfrac{3}{\sqrt{5}}$ ■ We will eliminate the radical in the denominator by multiplying the numerator and the denominator by $\sqrt{5}$.

$$\frac{3}{\sqrt{5}} = \frac{3 \cdot \sqrt{5}}{\sqrt{5} \cdot \sqrt{5}} = \frac{3\sqrt{5}}{5}$$

(b) $\dfrac{-6}{\sqrt{11}} = \dfrac{-6 \cdot \sqrt{11}}{\sqrt{11} \cdot \sqrt{11}} = \dfrac{-6\sqrt{11}}{11}$

(c) $-\dfrac{\sqrt{7}}{\sqrt{2}} = -\dfrac{\sqrt{7} \cdot \sqrt{2}}{\sqrt{2} \cdot \sqrt{2}} = -\dfrac{\sqrt{14}}{2}$

(d) $\dfrac{20}{\sqrt{18}}$ ■ Since $\sqrt{18} = \sqrt{9 \cdot 2}$, we will multiply the numerator and denominator by $\sqrt{2}$ to obtain a perfect square in the radicand in the denominator.

$$\begin{aligned}
\frac{20}{\sqrt{18}} &= \frac{20 \cdot \sqrt{2}}{\sqrt{18} \cdot \sqrt{2}} \\
&= \frac{20\sqrt{2}}{\sqrt{36}} \\
&= \frac{20\sqrt{2}}{6} \\
&= \frac{10\sqrt{2}}{3}
\end{aligned}$$

2. **(a)** $\begin{aligned}[t]
\sqrt{\frac{16}{11}} &= \frac{\sqrt{16}}{\sqrt{11}} \\
&= \frac{\sqrt{16} \cdot \sqrt{11}}{\sqrt{11} \cdot \sqrt{11}} \\
&= \frac{4\sqrt{11}}{11}
\end{aligned}$

(b) $\begin{aligned}[t]
\sqrt{\frac{5}{18}} &= \frac{\sqrt{5}}{\sqrt{18}} \\
&= \frac{\sqrt{5} \cdot \sqrt{2}}{\sqrt{18} \cdot \sqrt{2}} \\
&= \frac{\sqrt{10}}{\sqrt{36}} \\
&= \frac{\sqrt{10}}{6}
\end{aligned}$

(c) $\sqrt{\dfrac{8}{32}}$ ■ When rationalizing the denominator, there are often several ways to approach the problem. Three ways to simplify this radical are shown here.

$$\begin{aligned}
\sqrt{\frac{8}{32}} &= \frac{\sqrt{8}}{\sqrt{32}} = \frac{\sqrt{8} \cdot \sqrt{2}}{\sqrt{32} \cdot \sqrt{2}} \\
&= \frac{\sqrt{16}}{\sqrt{64}} = \frac{4}{8} = \frac{1}{2}
\end{aligned}$$

OR $\begin{aligned}[t]
\sqrt{\frac{8}{32}} &= \frac{\sqrt{8}}{\sqrt{32}} = \frac{\sqrt{4} \cdot \sqrt{2}}{\sqrt{16} \cdot \sqrt{2}} \\
&= \frac{2\sqrt{2}}{4\sqrt{2}} = \frac{2}{4} = \frac{1}{2}
\end{aligned}$

OR $\sqrt{\dfrac{8}{32}} = \sqrt{\dfrac{1}{4}} = \dfrac{1}{2}$

3. **(a)** $\begin{aligned}[t]
\sqrt{\frac{1}{2}} \cdot \sqrt{\frac{5}{6}} &= \sqrt{\frac{1}{2} \cdot \frac{5}{6}} = \sqrt{\frac{5}{12}} \\
&= \frac{\sqrt{5}}{\sqrt{12}} = \frac{\sqrt{5} \cdot \sqrt{3}}{\sqrt{12} \cdot \sqrt{3}} \\
&= \frac{\sqrt{15}}{\sqrt{36}} = \frac{\sqrt{15}}{6}
\end{aligned}$

(b) $\sqrt{\dfrac{1}{10}} \cdot \sqrt{20} = \sqrt{\dfrac{1}{10} \cdot \dfrac{20}{1}} = \sqrt{2}$

(c) $\begin{aligned}[t]
\sqrt{\frac{5}{8}} \cdot \sqrt{\frac{24}{10}} &= \sqrt{\frac{5}{8} \cdot \frac{24}{10}} \\
&= \sqrt{\frac{3}{2}} \\
&= \frac{\sqrt{3} \cdot \sqrt{2}}{\sqrt{2} \cdot \sqrt{2}} = \frac{\sqrt{6}}{2}
\end{aligned}$

4. **(a)** $\dfrac{\sqrt{5p}}{\sqrt{q}} = \dfrac{\sqrt{5p} \cdot \sqrt{q}}{\sqrt{q} \cdot \sqrt{q}} = \dfrac{\sqrt{5pq}}{q}$

(b) $\sqrt{\dfrac{5r^2t^2}{7}}$

$$\begin{aligned}
&= \frac{\sqrt{5r^2t^2}}{\sqrt{7}} = \frac{\sqrt{r^2t^2}\sqrt{5}}{\sqrt{7}} \\
&= \frac{rt\sqrt{5} \cdot \sqrt{7}}{\sqrt{7} \cdot \sqrt{7}} = \frac{rt\sqrt{35}}{7}
\end{aligned}$$

5. **(a)** $\sqrt[3]{\dfrac{5}{7}} = \dfrac{\sqrt[3]{5}}{\sqrt[3]{7}}$

To get a perfect cube in the denominator radicand, multiply the <u>numerator</u> and denominator by $\sqrt[3]{7 \cdot 7}$.

$$\begin{aligned}
&= \frac{\sqrt[3]{5} \cdot \sqrt[3]{7 \cdot 7}}{\sqrt[3]{7} \cdot \sqrt[3]{7 \cdot 7}} \\
&= \frac{\sqrt[3]{5 \cdot 7 \cdot 7}}{\sqrt[3]{7 \cdot 7 \cdot 7}} = \frac{\sqrt[3]{245}}{7}
\end{aligned}$$

(b) $\begin{aligned}[t]
\frac{\sqrt[3]{5}}{\sqrt[3]{9}} &= \frac{\sqrt[3]{5} \cdot \sqrt[3]{3}}{\sqrt[3]{3 \cdot 3} \cdot \sqrt[3]{3}} \\
&= \frac{\sqrt[3]{5 \cdot 3}}{\sqrt[3]{3 \cdot 3 \cdot 3}} = \frac{\sqrt[3]{15}}{3}
\end{aligned}$

(c) $\dfrac{\sqrt[3]{4}}{\sqrt[3]{25y}} = \dfrac{\sqrt[3]{4} \cdot \sqrt[3]{5 \cdot y \cdot y}}{\sqrt[3]{5 \cdot 5 \cdot y} \cdot \sqrt[3]{5 \cdot y \cdot y}}$

$\qquad\qquad = \dfrac{\sqrt[3]{4 \cdot 5 \cdot y \cdot y}}{\sqrt[3]{5^3 \cdot y^3}}$

$\qquad\qquad = \dfrac{\sqrt[3]{20y^2}}{\sqrt[3]{(5y)^3}} = \dfrac{\sqrt[3]{20y^2}}{5y}$

16.4 Section Exercises

1. The given expression is being multiplied by $\frac{\sqrt{3}}{\sqrt{3}}$, which is 1. According to the identity property for multiplication, multiplying an expression by 1 does not change the value of the expression.

3. $\dfrac{6}{\sqrt{5}}$ ▪ To rationalize the denominator, multiply the numerator and denominator by $\sqrt{5}$.

$$\dfrac{6}{\sqrt{5}} = \dfrac{6 \cdot \sqrt{5}}{\sqrt{5} \cdot \sqrt{5}} = \dfrac{6\sqrt{5}}{5}$$

5. $\dfrac{5}{\sqrt{5}} = \dfrac{5 \cdot \sqrt{5}}{\sqrt{5} \cdot \sqrt{5}} = \dfrac{5\sqrt{5}}{5} = \sqrt{5}$

7. $\dfrac{8}{\sqrt{2}} = \dfrac{8 \cdot \sqrt{2}}{\sqrt{2} \cdot \sqrt{2}} = \dfrac{8\sqrt{2}}{2} = 4\sqrt{2}$

9. $\dfrac{-\sqrt{11}}{\sqrt{3}} = \dfrac{-\sqrt{11} \cdot \sqrt{3}}{\sqrt{3} \cdot \sqrt{3}} = \dfrac{-\sqrt{33}}{3}$

11. $\dfrac{7\sqrt{3}}{\sqrt{5}} = \dfrac{7\sqrt{3} \cdot \sqrt{5}}{\sqrt{5} \cdot \sqrt{5}} = \dfrac{7\sqrt{15}}{5}$

13. $\dfrac{24\sqrt{10}}{16\sqrt{3}} = \dfrac{3\sqrt{10}}{2\sqrt{3}}$

$\qquad\quad = \dfrac{3\sqrt{10} \cdot \sqrt{3}}{2\sqrt{3} \cdot \sqrt{3}}$

$\qquad\quad = \dfrac{3\sqrt{30}}{2 \cdot 3}$

$\qquad\quad = \dfrac{\sqrt{30}}{2}$

15. $\dfrac{16}{\sqrt{27}} = \dfrac{16}{\sqrt{9 \cdot 3}}$

$\qquad\quad = \dfrac{16}{\sqrt{9} \cdot \sqrt{3}}$

$\qquad\quad = \dfrac{16}{3\sqrt{3}}$

$\qquad\quad = \dfrac{16 \cdot \sqrt{3}}{3\sqrt{3} \cdot \sqrt{3}}$

$\qquad\quad = \dfrac{16\sqrt{3}}{9}$

17. $\dfrac{-3}{\sqrt{50}} = \dfrac{-3}{\sqrt{25 \cdot 2}}$

$\qquad\quad = \dfrac{-3}{5\sqrt{2}}$

$\qquad\quad = \dfrac{-3 \cdot \sqrt{2}}{5\sqrt{2} \cdot \sqrt{2}}$

$\qquad\quad = \dfrac{-3\sqrt{2}}{5 \cdot 2} = \dfrac{-3\sqrt{2}}{10}$

19. $\dfrac{63}{\sqrt{45}} = \dfrac{63}{\sqrt{9} \cdot \sqrt{5}}$

$\qquad\quad = \dfrac{63}{3\sqrt{5}} = \dfrac{21}{\sqrt{5}}$

$\qquad\quad = \dfrac{21 \cdot \sqrt{5}}{\sqrt{5} \cdot \sqrt{5}} = \dfrac{21\sqrt{5}}{5}$

21. $\dfrac{\sqrt{24}}{\sqrt{8}} = \sqrt{\dfrac{24}{8}} = \sqrt{3}$

23. $\sqrt{\dfrac{1}{2}} = \dfrac{\sqrt{1}}{\sqrt{2}} = \dfrac{1}{\sqrt{2}} = \dfrac{1 \cdot \sqrt{2}}{\sqrt{2} \cdot \sqrt{2}} = \dfrac{\sqrt{2}}{2}$

25. $\sqrt{\dfrac{13}{5}} = \dfrac{\sqrt{13}}{\sqrt{5}} = \dfrac{\sqrt{13} \cdot \sqrt{5}}{\sqrt{5} \cdot \sqrt{5}} = \dfrac{\sqrt{65}}{5}$

27. $\sqrt{\dfrac{7}{13}} \cdot \sqrt{\dfrac{13}{3}} = \sqrt{\dfrac{7}{13} \cdot \dfrac{13}{3}}$ *Product rule*

$\qquad\quad = \sqrt{\dfrac{7}{3}} = \dfrac{\sqrt{7}}{\sqrt{3}}$

$\qquad\quad = \dfrac{\sqrt{7} \cdot \sqrt{3}}{\sqrt{3} \cdot \sqrt{3}} = \dfrac{\sqrt{21}}{3}$

29. $\sqrt{\dfrac{21}{7}} \cdot \sqrt{\dfrac{21}{8}} = \dfrac{\sqrt{21}}{\sqrt{7}} \cdot \dfrac{\sqrt{21}}{\sqrt{8}}$

$\qquad\quad = \dfrac{21}{\sqrt{7 \cdot 2 \cdot 4}}$

$\qquad\quad = \dfrac{21}{2\sqrt{14}}$

$\qquad\quad = \dfrac{21 \cdot \sqrt{14}}{2 \cdot \sqrt{14} \cdot \sqrt{14}}$

$\qquad\quad = \dfrac{21\sqrt{14}}{2 \cdot 14} = \dfrac{3\sqrt{14}}{4}$

31. $\sqrt{\dfrac{1}{12}} \cdot \sqrt{\dfrac{1}{3}} = \sqrt{\dfrac{1}{12} \cdot \dfrac{1}{3}}$

$\qquad\quad = \sqrt{\dfrac{1}{36}} = \dfrac{\sqrt{1}}{\sqrt{36}} = \dfrac{1}{6}$

33. $\sqrt{\dfrac{2}{9}} \cdot \sqrt{\dfrac{9}{2}} = \sqrt{\dfrac{2}{9} \cdot \dfrac{9}{2}} = \sqrt{1} = 1$

35. $\dfrac{\sqrt{7}}{\sqrt{x}} = \dfrac{\sqrt{7} \cdot \sqrt{x}}{\sqrt{x} \cdot \sqrt{x}}$

$\qquad = \dfrac{\sqrt{7x}}{x}$

37. $\dfrac{\sqrt{4x^3}}{\sqrt{y}} = \dfrac{\sqrt{4x^2} \cdot \sqrt{x}}{\sqrt{y}}$

$\qquad = \dfrac{2x\sqrt{x}}{\sqrt{y}}$

$\qquad = \dfrac{2x\sqrt{x} \cdot \sqrt{y}}{\sqrt{y} \cdot \sqrt{y}}$

$\qquad = \dfrac{2x\sqrt{xy}}{y}$

39. $\sqrt{\dfrac{5x^3z}{6}} = \dfrac{\sqrt{5x^3z}}{\sqrt{6}}$

$\qquad = \dfrac{\sqrt{x^2} \cdot \sqrt{5xz}}{\sqrt{6}}$

$\qquad = \dfrac{x\sqrt{5xz}}{\sqrt{6}}$

$\qquad = \dfrac{x\sqrt{5xz} \cdot \sqrt{6}}{\sqrt{6} \cdot \sqrt{6}}$

$\qquad = \dfrac{x\sqrt{30xz}}{6}$

41. $\sqrt{\dfrac{9a^2r^5}{7t}} = \dfrac{\sqrt{9a^2r^5}}{\sqrt{7t}}$

$\qquad = \dfrac{\sqrt{9a^2r^4 \cdot r}}{\sqrt{7t}}$

$\qquad = \dfrac{\sqrt{9a^2r^4} \cdot \sqrt{r}}{\sqrt{7t}}$

$\qquad = \dfrac{3ar^2\sqrt{r}}{\sqrt{7t}}$

$\qquad = \dfrac{3ar^2\sqrt{r} \cdot \sqrt{7t}}{\sqrt{7t} \cdot \sqrt{7t}}$

$\qquad = \dfrac{3ar^2\sqrt{7rt}}{7t}$

43. We need to multiply the numerator and denominator of $\dfrac{\sqrt[3]{2}}{\sqrt[3]{5}}$ by enough factors of 5 to make the radicand in the denominator a perfect cube. In this case we have one factor of 5, so we need to multiply by two more factors of 5 to make three factors of 5. Thus, the correct choice for a rationalizing factor in this problem is $\sqrt[3]{5^2} = \sqrt[3]{25}$, which corresponds to choice **B**.

45. $\sqrt[3]{\dfrac{1}{2}}$ ■ Multiply the numerator and the denominator by enough factors of 2 to make the radicand in the denominator a perfect cube. This will eliminate the radical in the denominator. Here, we multiply by $\sqrt[3]{2^2}$ or $\sqrt[3]{4}$.

$\sqrt[3]{\dfrac{1}{2}} = \dfrac{\sqrt[3]{1}}{\sqrt[3]{2}} = \dfrac{1 \cdot \sqrt[3]{2^2}}{\sqrt[3]{2} \cdot \sqrt[3]{2^2}}$

$\qquad = \dfrac{\sqrt[3]{4}}{\sqrt[3]{2 \cdot 2^2}} = \dfrac{\sqrt[3]{4}}{\sqrt[3]{2^3}} = \dfrac{\sqrt[3]{4}}{2}$

47. $\sqrt[3]{\dfrac{5}{9}}$ ■ Multiply the numerator and the denominator by enough factors of 3 to make the radicand in the denominator a perfect cube. This will eliminate the radical in the denominator. Here, we multiply by $\sqrt[3]{3}$.

$\sqrt[3]{\dfrac{5}{9}} = \dfrac{\sqrt[3]{5}}{\sqrt[3]{3^2}} = \dfrac{\sqrt[3]{5} \cdot \sqrt[3]{3}}{\sqrt[3]{3^2} \cdot \sqrt[3]{3}}$

$\qquad = \dfrac{\sqrt[3]{15}}{\sqrt[3]{3^3}} = \dfrac{\sqrt[3]{15}}{3}$

49. $\dfrac{\sqrt[3]{4}}{\sqrt[3]{7}} = \dfrac{\sqrt[3]{4} \cdot \sqrt[3]{7^2}}{\sqrt[3]{7} \cdot \sqrt[3]{7^2}}$

$\qquad = \dfrac{\sqrt[3]{4} \cdot \sqrt[3]{49}}{\sqrt[3]{7^3}} = \dfrac{\sqrt[3]{196}}{7}$

51. $\sqrt[3]{\dfrac{3}{4y^2}}$ ■ To make the radicand in the denominator, $4y^2$, into a perfect cube, we must multiply 4 by 2 to get the perfect cube 8 and y^2 by y to get the perfect cube y^3. So we multiply the numerator and denominator by $\sqrt[3]{2y}$.

$\sqrt[3]{\dfrac{3}{4y^2}} = \dfrac{\sqrt[3]{3}}{\sqrt[3]{4y^2}}$

$\qquad = \dfrac{\sqrt[3]{3} \cdot \sqrt[3]{2y}}{\sqrt[3]{2^2y^2} \cdot \sqrt[3]{2y}}$

$\qquad = \dfrac{\sqrt[3]{6y}}{\sqrt[3]{2^3y^3}} = \dfrac{\sqrt[3]{6y}}{2y}$

53. $\dfrac{\sqrt[3]{7m}}{\sqrt[3]{36n}} = \dfrac{\sqrt[3]{7m}}{\sqrt[3]{6^2n}}$

$\qquad = \dfrac{\sqrt[3]{7m} \cdot \sqrt[3]{6n^2}}{\sqrt[3]{6^2n} \cdot \sqrt[3]{6n^2}}$

$\qquad = \dfrac{\sqrt[3]{42mn^2}}{\sqrt[3]{6^3n^3}} = \dfrac{\sqrt[3]{42mn^2}}{6n}$

55.   (a) $p = k \cdot \sqrt{\dfrac{L}{g}}$

$p = 6 \cdot \sqrt{\dfrac{9}{32}}$ *Let k = 6, L = 9, g = 32.*

$= \dfrac{6\sqrt{9}}{\sqrt{32}} = \dfrac{6 \cdot 3}{\sqrt{16 \cdot 2}}$

$= \dfrac{18}{4\sqrt{2}} = \dfrac{9}{2\sqrt{2}}$

$= \dfrac{9 \cdot \sqrt{2}}{2\sqrt{2} \cdot \sqrt{2}}$ *Rationalize the denominator.*

$= \dfrac{9\sqrt{2}}{4}$

The period of the pendulum is $\frac{9\sqrt{2}}{4}$ seconds.

(b) Using a calculator, we obtain

$$\tfrac{9\sqrt{2}}{4} \approx 3.182 \text{ seconds.}$$

16.5   More Simplifying and Operations with Radicals

16.5   Margin Exercises

1.   (a) $\sqrt{7}(\sqrt{2} + \sqrt{5})$

$= \sqrt{7} \cdot \sqrt{2} + \sqrt{7} \cdot \sqrt{5}$

$= \sqrt{14} + \sqrt{35}$

(b) $\sqrt{2}(\sqrt{8} + \sqrt{20})$

$= \sqrt{2}(\sqrt{4} \cdot \sqrt{2} + \sqrt{4} \cdot \sqrt{5})$

$= \sqrt{2}(2\sqrt{2} + 2\sqrt{5})$

$= \sqrt{2} \cdot 2\sqrt{2} + \sqrt{2} \cdot 2\sqrt{5}$

$= 2 \cdot 2 + 2\sqrt{10}$

$= 4 + 2\sqrt{10}$

(c) $(\sqrt{2} + 5\sqrt{3})(\sqrt{3} - 2\sqrt{2})$

$= \sqrt{2} \cdot \sqrt{3} - \sqrt{2} \cdot 2\sqrt{2}$

$\quad + 5\sqrt{3} \cdot \sqrt{3} - 5\sqrt{3} \cdot 2\sqrt{2}$ *FOIL*

$= \sqrt{6} - 2 \cdot 2 + 5 \cdot 3 - 10\sqrt{6}$

$= 1\sqrt{6} - 4 + 15 - 10\sqrt{6}$

$= 11 - 9\sqrt{6}$

(d) $(\sqrt{2} - \sqrt{5})(\sqrt{10} + \sqrt{2})$

$= \sqrt{2} \cdot \sqrt{10} + \sqrt{2} \cdot \sqrt{2}$

$\quad - \sqrt{5} \cdot \sqrt{10} - \sqrt{5} \cdot \sqrt{2}$ *FOIL*

$= \sqrt{20} + 2 - \sqrt{50} - \sqrt{10}$

$= \sqrt{4 \cdot 5} + 2 - \sqrt{25 \cdot 2} - \sqrt{10}$

$= 2\sqrt{5} + 2 - 5\sqrt{2} - \sqrt{10}$

2.   (a) Use the special product formula,

$$(a - b)^2 = a^2 - 2ab + b^2.$$

$(\sqrt{5} - 3)^2$

$= (\sqrt{5})^2 - 2(\sqrt{5})(3) + \underline{3}^2$

$= \underline{5} - 6\sqrt{5} + 9$

$= \underline{14 - 6\sqrt{5}}$

(b) Use the special product formula,

$$(a + b)^2 = a^2 + 2ab + b^2.$$

$(4\sqrt{2} + 5)^2$

$= (4\sqrt{2})^2 + 2(4\sqrt{2})(5) + 5^2$

$= 16 \cdot 2 + 40\sqrt{2} + 25$

$= 32 + 40\sqrt{2} + 25$

$= 57 + 40\sqrt{2}$

(c) $(6 + \sqrt{x})^2, \quad (x \geq 0)$

$= 6^2 + 2(6)(\sqrt{x}) + (\sqrt{x})^2$

$= 36 + 12\sqrt{x} + x$

3.   (a) Use the rule for the product of the sum and the difference of two terms,

$$(a + b)(a - b) = a^2 - b^2.$$

$(3 + \sqrt{5})(3 - \sqrt{5})$

$= \underline{3}^2 - (\sqrt{5})^2$

$= 9 - \underline{5}$

$= \underline{4}$

(b) $(\sqrt{3} - 2)(\sqrt{3} + 2) = (\sqrt{3})^2 - 2^2$

$= 3 - 4$

$= -1$

(c) $(\sqrt{5} + \sqrt{3})(\sqrt{5} - \sqrt{3}) = (\sqrt{5})^2 - (\sqrt{3})^2$

$= 5 - 3$

$= 2$

(d) $(\sqrt{10} - \sqrt{x})(\sqrt{10} + \sqrt{x}), \quad (x \geq 0)$

$= (\sqrt{10})^2 - (\sqrt{x})^2$

$= 10 - x$

4.   (a) $\dfrac{5}{4 + \sqrt{2}}$ ▪ We can eliminate the radical in the denominator by multiplying both the numerator and denominator by the <u>conjugate</u> of the denominator. This number is $\underline{4 - \sqrt{2}}$.

$= \dfrac{5(4 - \sqrt{2})}{(4 + \sqrt{2})(4 - \sqrt{2})}$

$= \dfrac{5(4 - \sqrt{2})}{4^2 - (\sqrt{2})^2}$ $\begin{array}{l}(a+b)(a-b)\\ = a^2 - b^2\end{array}$

$= \dfrac{5(4 - \sqrt{2})}{16 - 2}$

$= \dfrac{5(4 - \sqrt{2})}{14}$

(b) $\dfrac{\sqrt{5}+3}{2-\sqrt{5}}$

$= \dfrac{(\sqrt{5}+3)(2+\sqrt{5})}{(2-\sqrt{5})(2+\sqrt{5})}$

$= \dfrac{2\sqrt{5}+5+6+3\sqrt{5}}{2^2-(\sqrt{5})^2}$ *FOIL in numerator*

$= \dfrac{11+5\sqrt{5}}{4-5}$

$= \dfrac{11+5\sqrt{5}}{-1}$

$= -11-5\sqrt{5}$

(c) $\dfrac{7}{5-\sqrt{x}} = \dfrac{7\left(5+\sqrt{x}\right)}{\left(5-\sqrt{x}\right)\left(5+\sqrt{x}\right)}$

$= \dfrac{7\left(5+\sqrt{x}\right)}{25-x}$

5. (a) $\dfrac{5\sqrt{3}-15}{10} = \dfrac{5(\sqrt{3}-3)}{5(2)}$

$= \dfrac{\sqrt{3}-3}{2}$

(b) $\dfrac{12+8\sqrt{5}}{16} = \dfrac{4(3+2\sqrt{5})}{4(4)}$

$= \dfrac{3+2\sqrt{5}}{4}$

16.5 Section Exercises

1. $\sqrt{49}+\sqrt{36} = 13$

THINK $\sqrt{49}+\sqrt{36} = 7+6$

3. $\sqrt{2}\cdot\sqrt{8} = 4$

THINK $\sqrt{2}\cdot\sqrt{8} = \sqrt{16}$

5. Because multiplication must be performed before addition, it is incorrect to add -37 and -2. $-2\sqrt{15}$ cannot be simplified further. The final answer is $-37-2\sqrt{15}$.

7. $\sqrt{5}(\sqrt{3}-\sqrt{7}) = \sqrt{5}\cdot\sqrt{3}-\sqrt{5}\cdot\sqrt{7}$

$= \sqrt{15}-\sqrt{35}$

9. $2\sqrt{5}(\sqrt{2}+3\sqrt{5})$

$= 2\sqrt{5}\cdot\sqrt{2}+2\sqrt{5}\cdot3\sqrt{5}$

$= 2\sqrt{10}+2\cdot3\cdot\sqrt{5}\cdot\sqrt{5}$

$= 2\sqrt{10}+6\cdot5$

$= 2\sqrt{10}+30$

11. $3\sqrt{14}\cdot\sqrt{2}-\sqrt{28} = 3\sqrt{14\cdot2}-\sqrt{28}$

$= 3\sqrt{28}-1\sqrt{28}$

$= 2\sqrt{28}$

$= 2\sqrt{4\cdot7}$

$= 2\cdot\sqrt{4}\cdot\sqrt{7}$

$= 2\cdot2\cdot\sqrt{7} = 4\sqrt{7}$

13. $(2\sqrt{6}+3)(3\sqrt{6}+7)$

$= 2\sqrt{6}\cdot3\sqrt{6}+7\cdot2\sqrt{6}+3\cdot3\sqrt{6}$

$\quad +3\cdot7$ *FOIL*

$= 2\cdot3\cdot\sqrt{6}\cdot\sqrt{6}+14\sqrt{6}+9\sqrt{6}+21$

$= 6\cdot6+23\sqrt{6}+21$

$= 36+23\sqrt{6}+21$

$= 57+23\sqrt{6}$

15. $(5\sqrt{7}-2\sqrt{3})(3\sqrt{7}+4\sqrt{3})$

$= 5\sqrt{7}(3\sqrt{7})+5\sqrt{7}(4\sqrt{3})$

$\quad -2\sqrt{3}(3\sqrt{7})-2\sqrt{3}(4\sqrt{3})$ *FOIL*

$= 15\cdot7+20\sqrt{21}-6\sqrt{21}-8\cdot3$

$= 105+14\sqrt{21}-24$

$= 81+14\sqrt{21}$

17. $(8-\sqrt{7})^2$

$= (8)^2-2(8)(\sqrt{7})+(\sqrt{7})^2$ *Square of a binomial*

$= 64-16\sqrt{7}+7$

$= 71-16\sqrt{7}$

19. $(2\sqrt{7}+3)^2$

$= (2\sqrt{7})^2+2(2\sqrt{7})(3)+(3)^2$ *Square of a binomial*

$= 4\cdot7+12\sqrt{7}+9$

$= 28+12\sqrt{7}+9$

$= 37+12\sqrt{7}$

21. $(\sqrt{a}+1)^2$

$= (\sqrt{a})^2+2(\sqrt{a})(1)+(1)^2$ *Square of a binomial*

$= a+2\sqrt{a}+1$

23. $(5-\sqrt{2})(5+\sqrt{2})$

$= (5)^2-(\sqrt{2})^2$ *Product of the sum and difference of two terms*

$= 25-2 = 23$

25. $(\sqrt{8}-\sqrt{7})(\sqrt{8}+\sqrt{7})$

$= (\sqrt{8})^2-(\sqrt{7})^2$ *Product of the sum and difference of two terms*

$= 8-7 = 1$

27. $(\sqrt{y}-\sqrt{10})(\sqrt{y}+\sqrt{10})$

$= (\sqrt{y})^2-(\sqrt{10})^2$ *Product of the sum and difference of two terms*

$= y-10$

29.　$(\sqrt{2}+\sqrt{3})(\sqrt{6}-\sqrt{2})$

$\qquad = \sqrt{2}(\sqrt{6}) - \sqrt{2}(\sqrt{2}) + \sqrt{3}(\sqrt{6})$

$\qquad\quad - \sqrt{3}(\sqrt{2})\quad FOIL$

$\qquad = \sqrt{12} - 2 + \sqrt{18} - \sqrt{6}\quad Product\ rule$

$\qquad = \sqrt{4}\cdot\sqrt{3} - 2 + \sqrt{9}\cdot\sqrt{2} - \sqrt{6}$

$\qquad = 2\sqrt{3} - 2 + 3\sqrt{2} - \sqrt{6}$

31.　$(\sqrt{10}-\sqrt{5})(\sqrt{5}+\sqrt{20})$

$\qquad = \sqrt{10}\cdot\sqrt{5} + \sqrt{10}\cdot\sqrt{20} - \sqrt{5}\cdot\sqrt{5}$

$\qquad\quad - \sqrt{5}\cdot\sqrt{20}\quad FOIL$

$\qquad = \sqrt{50} + \sqrt{200} - 5 - \sqrt{100}$

$\qquad = \sqrt{25}\cdot\sqrt{2} + \sqrt{100}\cdot\sqrt{2} - 5 - 10$

$\qquad = 5\sqrt{2} + 10\sqrt{2} - 15$

$\qquad = 15\sqrt{2} - 15$

33.　$(\sqrt{5}+\sqrt{30})(\sqrt{6}+\sqrt{3})$

$\qquad = \sqrt{5}\cdot\sqrt{6} + \sqrt{5}\cdot\sqrt{3} + \sqrt{30}\cdot\sqrt{6}$

$\qquad\quad + \sqrt{30}\cdot\sqrt{3}\quad FOIL$

$\qquad = \sqrt{30} + \sqrt{15} + \sqrt{180} + \sqrt{90}$

$\qquad = \sqrt{30} + \sqrt{15} + \sqrt{36\cdot5} + \sqrt{9\cdot10}$

$\qquad = \sqrt{30} + \sqrt{15} + 6\sqrt{5} + 3\sqrt{10}$

35.　$(\sqrt{5}-\sqrt{10})(\sqrt{x}-\sqrt{2})$

$\qquad = \sqrt{5}(\sqrt{x}) + \sqrt{5}(-\sqrt{2}) - \sqrt{10}(\sqrt{x})$

$\qquad\quad - \sqrt{10}(-\sqrt{2})\quad FOIL$

$\qquad = \sqrt{5x} - \sqrt{10} - \sqrt{10x} + \sqrt{20}$

$\qquad = \sqrt{5x} - \sqrt{10} - \sqrt{10x} + \sqrt{4\cdot5}$

$\qquad = \sqrt{5x} - \sqrt{10} - \sqrt{10x} + 2\sqrt{5}$

37.　**(a)** The denominator is $\sqrt{5}+\sqrt{3}$, so to rationalize the denominator, we should multiply the numerator and denominator by its conjugate, $\sqrt{5}-\sqrt{3}$.

　　(b) The denominator is $\sqrt{6}-\sqrt{5}$, so to rationalize the denominator, we should multiply the numerator and denominator by its conjugate, $\sqrt{6}+\sqrt{5}$.

39.　$\dfrac{1}{3+\sqrt{2}} = \dfrac{1(3-\sqrt{2})}{(3+\sqrt{2})(3-\sqrt{2})}$

Multiply numerator and denominator by the conjugate of the denominator.

$\qquad = \dfrac{3-\sqrt{2}}{3^2-(\sqrt{2})^2}$

$\qquad = \dfrac{3-\sqrt{2}}{9-2}$

$\qquad = \dfrac{3-\sqrt{2}}{7}$

41.　$\dfrac{14}{2-\sqrt{11}} = \dfrac{14(2+\sqrt{11})}{(2-\sqrt{11})(2+\sqrt{11})}$

Multiply numerator and denominator by the conjugate of the denominator.

$\qquad = \dfrac{14(2+\sqrt{11})}{2^2-(\sqrt{11})^2}$

$\qquad = \dfrac{14(2+\sqrt{11})}{4-11}$

$\qquad = \dfrac{14(2+\sqrt{11})}{-7}$

$\qquad = -2(2+\sqrt{11}) = -4-2\sqrt{11}$

43.　$\dfrac{\sqrt{2}}{2-\sqrt{2}} = \dfrac{\sqrt{2}(2+\sqrt{2})}{(2-\sqrt{2})(2+\sqrt{2})}$

Multiply numerator and denominator by the conjugate of the denominator.

$\qquad = \dfrac{2\sqrt{2}+2}{2^2-(\sqrt{2})^2}$

$\qquad = \dfrac{2\sqrt{2}+2}{4-2}$

$\qquad = \dfrac{2\sqrt{2}+2}{2}$

$\qquad = \dfrac{2(\sqrt{2}+1)}{2}$

$\qquad = \sqrt{2}+1\ $ or $\ 1+\sqrt{2}$

45.　$\dfrac{\sqrt{5}}{\sqrt{2}+\sqrt{3}} = \dfrac{\sqrt{5}(\sqrt{2}-\sqrt{3})}{(\sqrt{2}+\sqrt{3})(\sqrt{2}-\sqrt{3})}$

Multiply numerator and denominator by the conjugate of the denominator.

$\qquad = \dfrac{\sqrt{5}\cdot\sqrt{2}-\sqrt{5}\cdot\sqrt{3}}{(\sqrt{2})^2-(\sqrt{3})^2}$

$\qquad = \dfrac{\sqrt{10}-\sqrt{15}}{2-3}$

$\qquad = \dfrac{\sqrt{10}-\sqrt{15}}{-1}$

$\qquad = -\sqrt{10}+\sqrt{15}$

47.　$\dfrac{\sqrt{5}+2}{2-\sqrt{3}} = \dfrac{(\sqrt{5}+2)(2+\sqrt{3})}{(2-\sqrt{3})(2+\sqrt{3})}$

Multiply numerator and denominator by the conjugate of the denominator.

$\qquad = \dfrac{2\sqrt{5}+\sqrt{5}\cdot\sqrt{3}+4+2\sqrt{3}}{2^2-(\sqrt{3})^2}$

$\qquad = \dfrac{2\sqrt{5}+\sqrt{15}+4+2\sqrt{3}}{4-3}$

$\qquad = \dfrac{2\sqrt{5}+\sqrt{15}+4+2\sqrt{3}}{1}$

$\qquad = 2\sqrt{5}+\sqrt{15}+4+2\sqrt{3}$

49. $\dfrac{6-\sqrt{5}}{\sqrt{2}+2} = \dfrac{\left(6-\sqrt{5}\right)\left(\sqrt{2}-2\right)}{\left(\sqrt{2}+2\right)\left(\sqrt{2}-2\right)}$

Multiply numerator and denominator by the conjugate of the denominator.

$= \dfrac{6\sqrt{2}-12-\sqrt{5}\cdot\sqrt{2}+2\sqrt{5}}{\left(\sqrt{2}\right)^2-2^2}$

$= \dfrac{6\sqrt{2}-12-\sqrt{10}+2\sqrt{5}}{2-4}$

$= \dfrac{6\sqrt{2}-12-\sqrt{10}+2\sqrt{5}}{-2}$

or $\dfrac{-6\sqrt{2}+12+\sqrt{10}-2\sqrt{5}}{2}$

51. $\dfrac{12}{\sqrt{x}+1} = \dfrac{12\left(\sqrt{x}-1\right)}{\left(\sqrt{x}+1\right)\left(\sqrt{x}-1\right)}$

$= \dfrac{12\left(\sqrt{x}-1\right)}{x-1} \quad (x \neq 1)$

53. $\dfrac{3}{7-\sqrt{x}} = \dfrac{3\left(7+\sqrt{x}\right)}{\left(7-\sqrt{x}\right)\left(7+\sqrt{x}\right)}$

$= \dfrac{3\left(7+\sqrt{x}\right)}{49-x} \quad (x \neq 49)$

55. $\dfrac{6\sqrt{11}-12}{6}$

$= \dfrac{6(\sqrt{11}-2)}{6}$ *Factor numerator.*

$= \sqrt{11}-2$ *Lowest terms*

57. $\dfrac{2\sqrt{3}+10}{16}$

$= \dfrac{2(\sqrt{3}+5)}{2\cdot 8}$ *Factor numerator.*

$= \dfrac{\sqrt{3}+5}{8}$ *Lowest terms*

59. $\dfrac{12-\sqrt{40}}{4} = \dfrac{12-\sqrt{4}\cdot\sqrt{10}}{4}$

$= \dfrac{12-2\sqrt{10}}{4}$

$= \dfrac{2(6-\sqrt{10})}{2\cdot 2}$

$= \dfrac{6-\sqrt{10}}{2}$

61. $\dfrac{16+\sqrt{128}}{24} = \dfrac{16+\sqrt{64}\cdot\sqrt{2}}{24}$

$= \dfrac{16+8\sqrt{2}}{24}$

$= \dfrac{8(2+\sqrt{2})}{8\cdot 3}$

$= \dfrac{2+\sqrt{2}}{3}$

63. (a) $r = \dfrac{-h+\sqrt{h^2+0.64S}}{2}$

$r = \dfrac{-12+\sqrt{12^2+0.64(400)}}{2}$ *Let h = 12, S = 400.*

$= \dfrac{-12+\sqrt{144+256}}{2}$

$= \dfrac{-12+\sqrt{400}}{2}$

$= \dfrac{-12+20}{2}$

$= \dfrac{8}{2} = 4$

The radius should be 4 inches to make a can with a height of 12 inches and a surface area of 400 in.^{2}.

(b) $r = \dfrac{-h+\sqrt{h^2+0.64S}}{2}$

$r = \dfrac{-6+\sqrt{6^2+0.64(200)}}{2}$ *Let h = 6, S = 200.*

$= \dfrac{-6+\sqrt{36+128}}{2}$

$= \dfrac{-6+\sqrt{164}}{2}$

$= \dfrac{-6+2\sqrt{41}}{2}$

$= -3+\sqrt{41} \approx 3.403$

The radius should be about 3.4 inches to make a can with a height of 6 inches and a surface area of 200 in.^{2}.

Relating Concepts (Exercises 65–70)

65. $6(5+3x)$

$= (6)(5)+(6)(3x)$ *Distributive property*

$= 30+18x$

66. 30 and $18x$ cannot be combined because they are not like terms.

67. $(2\sqrt{10}+5\sqrt{2})(3\sqrt{10}-3\sqrt{2})$

$= 2\sqrt{10}(3\sqrt{10})+2\sqrt{10}(-3\sqrt{2})$

     $+5\sqrt{2}(3\sqrt{10})+5\sqrt{2}(-3\sqrt{2})$ *FOIL*

$= 6\cdot 10-6\sqrt{20}+15\sqrt{20}-15\cdot 2$

$= 60+9\sqrt{20}-30$

$= 30+9\sqrt{4}\cdot\sqrt{5}$

$= 30+9(2\sqrt{5})$

$= 30+18\sqrt{5}$

68. 30 and $18\sqrt{5}$ cannot be combined because they are not like radicals.

69. In the expression $30 + 18x$, make the first term $30x$, so that

$$30x + 18x = 48x.$$

In the expression $30 + 18\sqrt{5}$, make the first term $30\sqrt{5}$, so that

$$30\sqrt{5} + 18\sqrt{5} = 48\sqrt{5}.$$

70. Both like terms and like radicals are combined by adding their numerical coefficients. The variables in like terms are replaced by radicals in like radicals.

Summary Exercises
Applying Operations with Radicals

1. $5\sqrt{10} - 8\sqrt{10} = (5 - 8)\sqrt{10}$

$$= -3\sqrt{10}$$

3. $(1 + \sqrt{3})(2 - \sqrt{6})$

$$= 1 \cdot 2 - 1 \cdot \sqrt{6} + 2 \cdot \sqrt{3} - \sqrt{3} \cdot \sqrt{6}$$
$$= 2 - \sqrt{6} + 2\sqrt{3} - \sqrt{18}$$
$$= 2 - \sqrt{6} + 2\sqrt{3} - \sqrt{9 \cdot 2}$$
$$= 2 - \sqrt{6} + 2\sqrt{3} - 3\sqrt{2}$$

5. $(3\sqrt{5} - 2\sqrt{7})^2$ *Square of a binomial*

$$= (3\sqrt{5})^2 - 2(3\sqrt{5})(2\sqrt{7}) + (2\sqrt{7})^2$$
$$= 3^2(\sqrt{5})^2 - 2 \cdot 3 \cdot 2 \cdot \sqrt{5} \cdot \sqrt{7} + 2^2(\sqrt{7})^2$$
$$= 9 \cdot 5 - 12\sqrt{35} + 4 \cdot 7$$
$$= 45 - 12\sqrt{35} + 28$$
$$= 73 - 12\sqrt{35}$$

7. $\sqrt[3]{16t^2} - \sqrt[3]{54t^2} + \sqrt[3]{128t^2}$

$$= \sqrt[3]{8} \cdot \sqrt[3]{2t^2} - \sqrt[3]{27} \cdot \sqrt[3]{2t^2} + \sqrt[3]{64} \cdot \sqrt[3]{2t^2}$$
$$= 2\sqrt[3]{2t^2} - 3\sqrt[3]{2t^2} + 4\sqrt[3]{2t^2}$$
$$= (2 - 3 + 4)\sqrt[3]{2t^2}$$
$$= 3\sqrt[3]{2t^2}$$

9. $\dfrac{1 + \sqrt{2}}{1 - \sqrt{2}} = \dfrac{1 + \sqrt{2}}{1 - \sqrt{2}} \cdot \dfrac{1 + \sqrt{2}}{1 + \sqrt{2}}$

Multiply numerator and denominator by the conjugate of the denominator.

$$= \frac{1 + \sqrt{2} + \sqrt{2} + \sqrt{2} \cdot \sqrt{2}}{1^2 - (\sqrt{2})^2}$$
$$= \frac{1 + 2\sqrt{2} + 2}{1 - 2}$$
$$= \frac{3 + 2\sqrt{2}}{-1} = -3 - 2\sqrt{2}$$

11. $(\sqrt{3} + 6)(\sqrt{3} - 6)$

$$= (\sqrt{3})^2 - 6^2$$ *Product of the sum and difference of two terms*
$$= 3 - 36$$
$$= -33$$

13. $\sqrt[3]{8x^3y^5z^6} = \sqrt[3]{8x^3y^3z^6} \cdot \sqrt[3]{y^2}$

$$= \sqrt[3]{(2xyz^2)^3} \cdot \sqrt[3]{y^2}$$
$$= 2xyz^2\sqrt[3]{y^2}$$

15. $\dfrac{5}{\sqrt{6} - 1} = \dfrac{5}{\sqrt{6} - 1} \cdot \dfrac{\sqrt{6} + 1}{\sqrt{6} + 1}$

Multiply numerator and denominator by the conjugate of the denominator.

$$= \frac{5(\sqrt{6} + 1)}{(\sqrt{6})^2 - 1^2}$$
$$= \frac{5(\sqrt{6} + 1)}{6 - 1}$$
$$= \frac{5(\sqrt{6} + 1)}{5} = \sqrt{6} + 1$$

17. $\dfrac{6\sqrt{3}}{5\sqrt{12}} = \dfrac{6\sqrt{3}}{5\sqrt{4} \cdot \sqrt{3}} = \dfrac{6}{5 \cdot 2} = \dfrac{3}{5}$

19. $\dfrac{-4}{\sqrt[3]{4}} = \dfrac{-4 \cdot \sqrt[3]{2}}{\sqrt[3]{4} \cdot \sqrt[3]{2}}$

$$= \frac{-4\sqrt[3]{2}}{\sqrt[3]{8}}$$
$$= \frac{-4\sqrt[3]{2}}{2} = -2\sqrt[3]{2}$$

21. $\sqrt{75x} - \sqrt{12x} = \sqrt{25 \cdot 3x} - \sqrt{4 \cdot 3x}$

$$= 5\sqrt{3x} - 2\sqrt{3x}$$
$$= (5 - 2)\sqrt{3x}$$
$$= 3\sqrt{3x}$$

23. $(\sqrt{7} - \sqrt{6})(\sqrt{7} + \sqrt{6})$

$$= (\sqrt{7})^2 - (\sqrt{6})^2$$ *Product of the sum and difference of two terms*
$$= 7 - 6 = 1$$

25. $x\sqrt[4]{x^5} - 3\sqrt[4]{x^9} + x^2\sqrt[4]{x}$

$$= x\sqrt[4]{x^4} \cdot \sqrt[4]{x} - 3\sqrt[4]{x^8} \cdot \sqrt[4]{x} + x^2\sqrt[4]{x}$$
$$= x \cdot x \cdot \sqrt[4]{x} - 3 \cdot x^2 \cdot \sqrt[4]{x} + x^2\sqrt[4]{x}$$
$$= (x^2 - 3x^2 + x^2)\sqrt[4]{x}$$
$$= -x^2\sqrt[4]{x}$$

27. $\sqrt{14} + \sqrt{5}$ ■ These are unlike radicals (different radicands) and cannot be combined.

29. $\sqrt{\dfrac{3}{4}} \cdot \sqrt{\dfrac{1}{5}} = \dfrac{\sqrt{3}}{\sqrt{4}} \cdot \dfrac{\sqrt{1}}{\sqrt{5}}$

$$= \frac{\sqrt{3} \cdot \sqrt{5}}{2 \cdot \sqrt{5} \cdot \sqrt{5}}$$
$$= \frac{\sqrt{15}}{2 \cdot 5}$$
$$= \frac{\sqrt{15}}{10}$$

31. $\sqrt[3]{24} + 6\sqrt[3]{81}$

$= \sqrt[3]{8} \cdot \sqrt[3]{3} + 6\sqrt[3]{27} \cdot \sqrt[3]{3}$

$= 2\sqrt[3]{3} + 6(3\sqrt[3]{3})$

$= 2\sqrt[3]{3} + 18\sqrt[3]{3}$

$= 20\sqrt[3]{3}$

33. $\sqrt[3]{4}\left(\sqrt[3]{2} - 3\right)$

$= \sqrt[3]{4}\left(\sqrt[3]{2}\right) + \sqrt[3]{4}(-3)$ *Distributive property*

$= \sqrt[3]{8} - 3\sqrt[3]{4}$ *Product rule*

$= 2 - 3\sqrt[3]{4}$ $\sqrt[3]{8} = 2$

35. $\sqrt{\dfrac{5}{8}} = \dfrac{\sqrt{5}}{\sqrt{8}} = \dfrac{\sqrt{5} \cdot \sqrt{2}}{\sqrt{4} \cdot \sqrt{2} \cdot \sqrt{2}}$

$= \dfrac{\sqrt{10}}{2 \cdot 2} = \dfrac{\sqrt{10}}{4}$

37. $S = 28.6\sqrt[3]{A}$

$S = 28.6\sqrt[3]{8}$ *Let A = 8.*

$= 28.6(2) = 57.2$

57 plant species (rounded to the nearest whole number) would exist on an island with area 8 mi^2.

16.6 Solving Equations with Radicals

16.6 Margin Exercises

1. **(a)** $\sqrt{k} = 3$

$(\sqrt{k})^2 = 3^2$ *Square each side.*

$k = 9$

Check $k = 9$: $\sqrt{k} = 3$

$\sqrt{9} \stackrel{2}{=} 3$ *Let k = 9.*

$3 = 3$ *True*

The solution set is $\{9\}$.

(b) $\sqrt{x - 2} = 4$

$(\sqrt{x - 2})^2 = 4^2$ *Square each side.*

$x - 2 = 16$

$x = 18$ *Add 2.*

Check $x = 18$: $\sqrt{x - 2} = 4$

$\sqrt{18 - 2} \stackrel{2}{=} 4$ *Let x = 18.*

$\sqrt{16} \stackrel{2}{=} 4$

$4 = 4$ *True*

The solution set is $\{18\}$.

(c) $\sqrt{9 - t} = 4$

$(\sqrt{9 - t})^2 = 4^2$ *Square each side.*

$9 - t = 16$

$-t = 7$ *Subtract 9.*

$t = -7$ *Multiply by −1.*

Check $t = -7$: $\sqrt{9 - t} = 4$

$\sqrt{9 - (-7)} \stackrel{2}{=} 4$ *Let t = −7.*

$\sqrt{16} \stackrel{2}{=} 4$

$4 = 4$ *True*

The solution set is $\{-7\}$.

2. **(a)** $\sqrt{3x + 9} = 2\sqrt{x}$

$(\sqrt{3x + 9})^2 = (2\sqrt{x})^2$ *Square each side.*

$3x + 9 = 4x$

$9 = x$ *Subtract 3x.*

Check $x = 9$:

$\sqrt{3x + 9} = 2\sqrt{x}$

$\sqrt{3(9) + 9} \stackrel{2}{=} 2\sqrt{9}$ *Let x = 9.*

$\sqrt{27 + 9} \stackrel{2}{=} 2(3)$

$\sqrt{36} \stackrel{2}{=} 6$

$6 = 6$ *True*

The solution set is $\{9\}$.

(b) $5\sqrt{x} = \sqrt{20x + 5}$

$(5\sqrt{x})^2 = (\sqrt{20x + 5})^2$ *Square each side.*

$25x = 20x + 5$

$5x = 5$ *Subtract 20x.*

$x = 1$ *Divide by 5.*

Check $x = 1$:

$5\sqrt{x} = \sqrt{20x + 5}$

$5\sqrt{1} \stackrel{2}{=} \sqrt{20(1) + 5}$ *Let x = 1.*

$5(1) \stackrel{2}{=} \sqrt{25}$

$5 = 5$ *True*

The solution set is $\{1\}$.

3. **(a)** $\sqrt{x} = -4$

$(\sqrt{x})^2 = (-4)^2$ *Square each side.*

$x = 16$

Check $x = 16$: $\sqrt{x} = -4$ *Original equation*

$\sqrt{16} \stackrel{2}{=} -4$ *Let x = 16.*

$4 = -4$ *False*

Because the statement $4 = -4$ is false, the number 16 is *not* a solution. The equation has no solution. The solution set is $\emptyset$.

Another approach: Because $\sqrt{x}$ represents the *principal* or *nonnegative* square root of x, it cannot equal -4, and we might have seen immediately that there is no solution.

(b) $\sqrt{x} + 6 = 0$

$\qquad \sqrt{x} = \underline{-6}$ *Isolate the radical.*

$\qquad (\sqrt{x})^2 = (\underline{-6})^2$ *Square each side.*

$\qquad x = \underline{36}$

Check $x = 36$: $\sqrt{x} + 6 = 0$

$\qquad \sqrt{36} + 6 \overset{?}{=} 0$ *Let x = 36.*

$\qquad 6 + 6 = 0$

$\qquad \underline{12} = 0$ *False*

Because the statement $12 = 0$ is false, the number 36 is *not* a solution (36 is called an extraneous solution). The equation has no solution. The solution set is $\emptyset$.

Another approach: Because $\sqrt{x}$ represents the *principal* or *nonnegative* square root of x, it cannot equal -6, and we might have seen in the second line that there is no solution.

4. $x = \sqrt{x^2 - 4x - 16}$

$x^2 = (\sqrt{x^2 - 4x - 16})^2$ *Square each side.*

$x^2 = x^2 - 4x - 16$

$0 = -4x - 16$ *Subtract x^2.*

$4x = -16$ *Add 4x.*

$x = -4$ *Divide by 4.*

Check $x = -4$:

$x = \sqrt{x^2 - 4x - 16}$

$-4 \overset{?}{=} \sqrt{(-4)^2 - 4(-4) - 16}$ *Let x = –4.*

$-4 \overset{?}{=} \sqrt{16 + 16 - 16}$

$-4 \overset{?}{=} \sqrt{16}$

$-4 = 4$ *False*

The only proposed solution does not check, so -4 is an extraneous solution, and the equation has no solution. The solution set is $\emptyset$.

5.   **(a)** Use the pattern

$$(a - b)^2 = a^2 - 2ab + b^2$$

with $a = w$ and $b = 5$.

$(w - 5)^2 = \underline{w^2} - \underline{2}(w)(\underline{5}) + \underline{5}^2$

$\qquad = w^2 - 10w + 25$

(b) $(2k - 5)^2 = (2k)^2 - 2(2k)(5) + 5^2$

$\qquad = 4k^2 - 20k + 25$

(c) $(3m - 2p)^2 = (3m)^2 - 2(3m)(2p) + (2p)^2$

$\qquad = 9m^2 - 12mp + 4p^2$

6.   **(a)** $\sqrt{6w + 6} = w + 1$

$\qquad (\sqrt{6w + 6})^2 = (w + 1)^2$ *Square each side.*

$\qquad 6w + 6 = w^2 + \underline{2w + 1}$

$\qquad 0 = w^2 - \underline{4w - 5}$ *Subtract 6w + 6.*

$\qquad 0 = (w - 5)(w + 1)$ *Factor.*

$\qquad w - 5 = 0 \quad \text{or} \quad w + 1 = 0$

$\qquad w = 5 \quad \text{or} \qquad w = -1$

Check both of these proposed solutions in the original equation.

Check $w = 5$:

$\sqrt{6w + 6} = w + 1$

$\sqrt{6(5) + 6} \overset{?}{=} 5 + 1$ *Let w = 5.*

$\sqrt{30 + 6} \overset{?}{=} 6$

$\sqrt{36} \overset{?}{=} 6$

$6 = 6$ *True*

Check $w = -1$:

$\sqrt{6w + 6} = w + 1$

$\sqrt{6(-1) + 6} \overset{?}{=} -1 + 1$ *Let w = –1.*

$\sqrt{-6 + 6} \overset{?}{=} 0$

$\sqrt{0} \overset{?}{=} 0$

$0 = 0$ *True*

The solution set is $\{-1, 5\}$.

(b) $2u - 1 = \sqrt{10u + 9}$

$\qquad (2u - 1)^2 = (\sqrt{10u + 9})^2$ *Square each side.*

$\qquad 4u^2 - 4u + 1 = 10u + 9$

$\qquad 4u^2 - 14u - 8 = 0$ *Subtract 10u + 9.*

$\qquad 2(2u^2 - 7u - 4) = 0$ *Factor.*

$\qquad 2(2u + 1)(u - 4) = 0$ *Factor.*

$\qquad 2u + 1 = 0 \quad \text{or} \quad u - 4 = 0$

$\qquad 2u = -1$

$\qquad u = -\frac{1}{2} \quad \text{or} \qquad u = 4$

Check $u = -\frac{1}{2}$:

$2u - 1 = \sqrt{10u + 9}$

$2(-\frac{1}{2}) - 1 \overset{?}{=} \sqrt{10(-\frac{1}{2}) + 9}$ *Let u = –$\frac{1}{2}$.*

$-1 - 1 \overset{?}{=} \sqrt{-5 + 9}$

$-2 \overset{?}{=} \sqrt{4}$

$-2 = 2$ *False*

Check $u = 4$:

$$2u - 1 = \sqrt{10u + 9}$$
$$2(4) - 1 \stackrel{?}{=} \sqrt{10(4) + 9} \quad \textit{Let u = 4.}$$
$$8 - 1 \stackrel{?}{=} \sqrt{49}$$
$$7 = 7 \qquad \textit{True}$$

The number $-\frac{1}{2}$ does not satisfy the original equation, so it is extraneous. The solution set is $\{4\}$.

7. **(a)** $\sqrt{x} - 3 = x - 15$

$$\sqrt{x} = x - \underline{12} \qquad \textit{Add 3 to get } \sqrt{x} \textit{ alone.}$$

$$(\sqrt{x})^2 = (x - \underline{12})^2 \qquad \textit{Square each side.}$$

$$\underline{x} = x^2 - 24x + 144$$
$$0 = x^2 - 25x + 144 \qquad \textit{Subtract x.}$$
$$0 = (x - 16)(x - 9) \qquad \textit{Factor.}$$

$$x - 16 = 0 \quad \text{or} \quad x - 9 = 0$$
$$x = 16 \quad \text{or} \qquad x = 9$$

Check $x = 16$:

$$\sqrt{x} - 3 = x - 15$$
$$\sqrt{16} - 3 \stackrel{?}{=} 16 - 15 \quad \textit{Let x = 16.}$$
$$4 - 3 \stackrel{?}{=} 1$$
$$1 = 1 \qquad \textit{True}$$

Check $x = 9$:

$$\sqrt{x} - 3 = x - 15$$
$$\sqrt{9} - 3 \stackrel{?}{=} 9 - 15 \quad \textit{Let x = 9.}$$
$$3 - 3 \stackrel{?}{=} -6$$
$$0 = -6 \qquad \textit{False}$$

The number 9 does not satisfy the original equation, so it is extraneous. The solution set is $\{16\}$.

(b) $\sqrt{z + 5} + 2 = z + 5$

$$\sqrt{z + 5} = z + 3 \qquad \textit{Subtract 2.}$$

$$(\sqrt{z + 5})^2 = (z + 3)^2 \qquad \textit{Square each side.}$$

$$z + 5 = z^2 + 6z + 9$$

$$0 = z^2 + 5z + 4 \qquad \textit{Subtract } z + 5.$$

$$0 = (z + 4)(z + 1) \qquad \textit{Factor.}$$

$$z + 4 = 0 \quad \text{or} \quad z + 1 = 0$$
$$z = -4 \quad \text{or} \qquad z = -1$$

Check $z = -4$:

$$\sqrt{z + 5} + 2 = z + 5$$
$$\sqrt{-4 + 5} + 2 \stackrel{?}{=} -4 + 5 \quad \textit{Let z = -4.}$$
$$\sqrt{1} + 2 \stackrel{?}{=} 1$$
$$1 + 2 \stackrel{?}{=} 1$$
$$3 = 1 \qquad \textit{False}$$

Check $z = -1$:

$$\sqrt{z + 5} + 2 = z + 5$$
$$\sqrt{-1 + 5} + 2 \stackrel{?}{=} -1 + 5 \quad \textit{Let z = -1.}$$
$$\sqrt{4} + 2 \stackrel{?}{=} 4$$
$$2 + 2 \stackrel{?}{=} 4$$
$$4 = 4 \qquad \textit{True}$$

The number -4 does not satisfy the original equation, so it is extraneous. The solution set is $\{-1\}$.

8. **(a)** $\sqrt{p + 1} - \sqrt{p - 4} = 1$

$$\sqrt{p + 1} = \sqrt{p - 4} + 1$$
$$(\sqrt{p + 1})^2 = (\sqrt{p - 4} + 1)^2$$
$$p + 1 = p - 4 + 2\sqrt{p - 4} + 1$$

$$4 = 2\sqrt{p - 4}$$
$$2 = \sqrt{p - 4}$$
$$2^2 = (\sqrt{p - 4})^2$$
$$4 = p - 4$$
$$8 = p$$

Check $p = 8$:

$$\sqrt{p + 1} - \sqrt{p - 4} = 1$$
$$\sqrt{8 + 1} - \sqrt{8 - 4} \stackrel{?}{=} 1 \quad \textit{Let p = 8.}$$
$$\sqrt{9} - \sqrt{4} \stackrel{?}{=} 1$$
$$3 - 2 \stackrel{?}{=} 1$$
$$1 = 1 \quad \textit{True}$$

The solution set is $\{8\}$.

(b) $\sqrt{2x + 1} + \sqrt{x + 4} = 3$

$$\sqrt{2x + 1} = 3 - \sqrt{x + 4}$$
$$(\sqrt{2x + 1})^2 = (3 - \sqrt{x + 4})^2$$
$$2x + 1 = 9 - 2 \cdot 3 \cdot \sqrt{x + 4} + x + 4$$

$$x - 12 = -6\sqrt{x + 4}$$
$$(x - 12)^2 = (-6\sqrt{x + 4})^2$$
$$x^2 - 24x + 144 = 36(x + 4)$$
$$x^2 - 24x + 144 = 36x + 144$$
$$x^2 - 60x = 0$$
$$x(x - 60) = 0$$

$$x = 0 \quad \text{or} \quad x = 60$$

Check $x = 0$:

$$\sqrt{2x + 1} + \sqrt{x + 4} = 3$$
$$\sqrt{2(0) + 1} + \sqrt{0 + 4} \overset{?}{=} 3 \quad \textit{Let x = 0.}$$
$$\sqrt{1} + \sqrt{4} \overset{?}{=} 3$$
$$1 + 2 \overset{?}{=} 3$$
$$3 = 3 \ \textit{True}$$

Check $x = 60$:

$$\sqrt{2x + 1} + \sqrt{x + 4} = 3$$
$$\sqrt{2(60) + 1} + \sqrt{60 + 4} \overset{?}{=} 3 \quad \textit{Let x = 60.}$$
$$\sqrt{121} + \sqrt{64} \overset{?}{=} 3$$
$$11 + 8 \overset{?}{=} 3$$
$$19 \overset{?}{=} 3 \ \textit{False}$$

The solution set is $\{0\}$.

9.   First find the semiperimeter of the triangle.

$$s = \tfrac{1}{2}(a + b + c)$$
$$= \tfrac{1}{2}(7 + 15 + 20) \quad \textit{a = 7, b = 15, c = 20}$$
$$= \tfrac{1}{2}(42) = 21 \text{ cm}$$

Now use Heron's formula to find the area.

$$\mathcal{A} = \sqrt{s(s - a)(s - b)(s - c)}$$
$$= \sqrt{21(21 - 7)(21 - 15)(21 - 20)}$$
$$= \sqrt{21(14)(6)(1)}$$
$$= \sqrt{1764} = 42 \text{ cm}^2$$

The area of the triangle is 42 cm^2.

16.6   Section Exercises

1.   To solve an equation involving a radical, such as $\sqrt{2x - 1} = 5$, use the <u>squaring</u> property of equality. This property says that if each side of an equation is <u>squared</u>, all solutions of the <u>original</u> equation are among the solutions of the squared equation.

3. $\sqrt{x} = 7$ ▪   Use the *squaring property of equality* to square each side of the equation.

$$(\sqrt{x})^2 = 7^2$$
$$x = 49$$

Now check this proposed solution in the original equation.

Check $x = 49$: $\sqrt{x} = 7$
$$\sqrt{49} \overset{?}{=} 7 \quad \textit{Let x = 49.}$$
$$7 = 7 \ \textit{True}$$

Since this statement is true, the solution set of the original equation is $\{49\}$.

5. $\sqrt{t + 2} = 3$
$$(\sqrt{t + 2})^2 = 3^2 \quad \textit{Square each side.}$$
$$t + 2 = 9$$
$$t = 7$$

Check $t = 7$: $\sqrt{t + 2} = 3$
$$\sqrt{7 + 2} \overset{?}{=} 3 \quad \textit{Let t = 7.}$$
$$\sqrt{9} \overset{?}{=} 3$$
$$3 = 3 \ \textit{True}$$

Since this statement is true, the solution set of the original equation is $\{7\}$.

7. $\sqrt{r - 4} = 9$
$$(\sqrt{r - 4})^2 = 9^2 \quad \textit{Square each side.}$$
$$r - 4 = 81$$
$$r = 85$$

Check $r = 85$: $\sqrt{r - 4} = 9$
$$\sqrt{85 - 4} \overset{?}{=} 9 \quad \textit{Let r = 85.}$$
$$\sqrt{81} \overset{?}{=} 9$$
$$9 = 9 \ \textit{True}$$

Since this statement is true, the solution set of the original equation is $\{85\}$.

9. $\sqrt{t} = -5$ ▪   Because $\sqrt{t}$ represents the *principal* or *nonnegative* square root of t, it cannot equal -5. Thus, the solution set is $\emptyset$.

11. $\sqrt{4 - t} = 7$
$$(\sqrt{4 - t})^2 = 7^2 \quad \textit{Square each side.}$$
$$4 - t = 49$$
$$-t = 45$$
$$t = -45$$

Check $t = -45$: $\sqrt{4 - t} = 7$
$$\sqrt{4 - (-45)} \overset{?}{=} 7 \quad \textit{Let t = -45.}$$
$$\sqrt{49} \overset{?}{=} 7$$
$$7 = 7 \ \textit{True}$$

Since this statement is true, the solution set of the original equation is $\{-45\}$.

13. $\sqrt{2t + 3} = 0$
$$(\sqrt{2t + 3})^2 = 0^2 \quad \textit{Square each side.}$$
$$2t + 3 = 0$$
$$2t = -3$$
$$t = -\tfrac{3}{2}$$

Check $t = -\tfrac{3}{2}$: $\sqrt{2t + 3} = 0$
$$\sqrt{2(-\tfrac{3}{2}) + 3} \overset{?}{=} 0 \quad \textit{Let t = -}\tfrac{3}{2}.$$
$$\sqrt{-3 + 3} \overset{?}{=} 0$$
$$\sqrt{0} \overset{?}{=} 0$$
$$0 = 0 \ \textit{True}$$

Since this statement is true, the solution set of the original equation is $\{-\tfrac{3}{2}\}$.

15.
$$\sqrt{3x-8} = -2$$
$$(\sqrt{3x-8})^2 = (-2)^2 \quad \textit{Square each side.}$$
$$3x - 8 = 4$$
$$3x = 12$$
$$x = 4$$

Check $x = 4$: $\sqrt{3x-8} = -2$
$$\sqrt{3(4)-8} \stackrel{?}{=} -2 \quad \textit{Let x = 4.}$$
$$\sqrt{12-8} \stackrel{?}{=} -2$$
$$\sqrt{4} \stackrel{?}{=} -2$$
$$2 = -2 \quad \textit{False}$$

Since this statement is false, there is no solution to the original equation and the solution set is $\emptyset$.

Another approach: Because $\sqrt{3x-8}$ represents the *principal* or *nonnegative* square root of $3x - 8$, it cannot equal -2, and we might have seen immediately that there is no solution.

17.
$$\sqrt{w} - 4 = 7$$
$$\sqrt{w} = 11 \qquad \textit{Add 4.}$$
$$(\sqrt{w})^2 = (11)^2 \quad \textit{Square each side.}$$
$$w = 121$$

Check $w = 121$: $\sqrt{w} - 4 = 7$
$$\sqrt{121} - 4 \stackrel{?}{=} 7 \quad \textit{Let w = 121.}$$
$$11 - 4 \stackrel{?}{=} 7$$
$$7 = 7 \quad \textit{True}$$

Since this statement is true, the solution set of the original equation is $\{121\}$.

19.
$$\sqrt{10x-8} = 3\sqrt{x}$$
$$(\sqrt{10x-8})^2 = (3\sqrt{x})^2 \qquad \textit{Square sides.}$$
$$10x - 8 = (3)^2(\sqrt{x})^2 \quad \textit{(ab)}^2 = a^2b^2$$
$$10x - 8 = 9x$$
$$x = 8$$

Check $x = 8$:
$$\sqrt{10x-8} = 3\sqrt{x}$$
$$\sqrt{10(8)-8} \stackrel{?}{=} 3\sqrt{8} \qquad \textit{Let x = 8.}$$
$$\sqrt{72} \stackrel{?}{=} 3\sqrt{8}$$
$$\sqrt{36 \cdot 2} \stackrel{?}{=} 3 \cdot 2\sqrt{2}$$
$$6\sqrt{2} = 6\sqrt{2} \quad \textit{True}$$

Since this statement is true, the solution set of the original equation is $\{8\}$.

21.
$$5\sqrt{x} = \sqrt{10x+15}$$
$$(5\sqrt{x})^2 = (\sqrt{10x+15})^2$$
$$25x = 10x + 15$$
$$15x = 15$$
$$x = 1$$

Check $x = 1$:
$$5\sqrt{x} = \sqrt{10x+15}$$
$$5\sqrt{1} \stackrel{?}{=} \sqrt{10 \cdot 1 + 15} \quad \textit{Let x = 1.}$$
$$5 \cdot 1 \stackrel{?}{=} \sqrt{25}$$
$$5 = 5 \qquad \textit{True}$$

Since this statement is true, the solution set of the original equation is $\{1\}$.

23.
$$\sqrt{3x-5} = \sqrt{2x+1}$$
$$(\sqrt{3x-5})^2 = (\sqrt{2x+1})^2$$
$$3x - 5 = 2x + 1$$
$$x = 6$$

Check $x = 6$:
$$\sqrt{3x-5} = \sqrt{2x+1}$$
$$\sqrt{3(6)-5} \stackrel{?}{=} \sqrt{2(6)+1} \quad \textit{Let x = 6.}$$
$$\sqrt{13} = \sqrt{13} \qquad \textit{True}$$

Since this statement is true, the solution set of the original equation is $\{6\}$.

25.
$$k = \sqrt{k^2 - 5k - 15}$$
$$(k)^2 = (\sqrt{k^2 - 5k - 15})^2$$
$$k^2 = k^2 - 5k - 15$$
$$0 = -5k - 15$$
$$5k = -15$$
$$k = -3$$

Check $k = -3$:
$$k = \sqrt{k^2 - 5k - 15}$$
$$-3 \stackrel{?}{=} \sqrt{(-3)^2 - 5(-3) - 15} \quad \textit{Let k = -3.}$$
$$-3 \stackrel{?}{=} \sqrt{9 + 15 - 15}$$
$$-3 \stackrel{?}{=} \sqrt{9}$$
$$-3 = 3 \qquad \qquad \textit{False}$$

Since this statement is false, there is no solution to the original equation and the solution set is $\emptyset$.

27.
$$7x = \sqrt{49x^2 + 2x - 10}$$
$$(7x)^2 = (\sqrt{49x^2 + 2x - 10})^2$$
$$49x^2 = 49x^2 + 2x - 10$$
$$0 = 2x - 10$$
$$10 = 2x$$
$$5 = x$$

Check $x = 5$:
$$7x = \sqrt{49x^2 + 2x - 10}$$
$$7(5) \stackrel{?}{=} \sqrt{49(5)^2 + 2(5) - 10} \quad \textit{Let x = 5.}$$
$$35 \stackrel{?}{=} \sqrt{1225 + 10 - 10}$$
$$35 \stackrel{?}{=} \sqrt{1225}$$
$$35 = 35 \qquad \qquad \textit{True}$$

Since this statement is true, the solution set of the original equation is $\{5\}$.

29. The error occurs in the first step, when both sides are squared. When the left side is squared, the result should be $x - 1$, not $-(x - 1)$. The correct solution set is $\{17\}$.

31.
$$\sqrt{2x + 1} = x - 7$$
$$(\sqrt{2x + 1})^2 = (x - 7)^2$$
$$2x + 1 = x^2 - 14x + 49$$
$$0 = x^2 - 16x + 48$$
$$0 = (x - 4)(x - 12)$$
$$x = 4 \quad \text{or} \quad x = 12$$

Check $x = 4$: $\quad 3 = -3 \quad$ *False*
Check $x = 12$: $\quad 5 = 5 \quad$ *True*

The solution set is $\{12\}$.

33.
$$\sqrt{3k + 10} + 5 = 2k$$
$$\sqrt{3k + 10} = 2k - 5$$
$$(\sqrt{3k + 10})^2 = (2k - 5)^2$$
$$3k + 10 = 4k^2 - 20k + 25$$
$$0 = 4k^2 - 23k + 15$$
$$0 = (4k - 3)(k - 5)$$
$$k = \frac{3}{4} \quad \text{or} \quad k = 5$$

Check $k = \frac{3}{4}$: $\quad \frac{17}{2} = \frac{3}{2} \quad$ *False*
Check $x = 5$: $\quad 10 = 10 \quad$ *True*

The solution set is $\{5\}$.

35.
$$\sqrt{5x + 1} - 1 = x$$
$$\sqrt{5x + 1} = x + 1$$
$$(\sqrt{5x + 1})^2 = (x + 1)^2$$
$$5x + 1 = x^2 + 2x + 1$$
$$0 = x^2 - 3x$$
$$0 = x(x - 3)$$
$$x = 0 \quad \text{or} \quad x = 3$$

Check $x = 0$: $\quad 0 = 0 \quad$ *True*
Check $x = 3$: $\quad 3 = 3 \quad$ *True*

The solution set is $\{0, 3\}$.

37.
$$\sqrt{6t + 7} + 3 = t + 5$$
$$\sqrt{6t + 7} = t + 2$$
$$(\sqrt{6t + 7})^2 = (t + 2)^2$$
$$6t + 7 = t^2 + 4t + 4$$
$$0 = t^2 - 2t - 3$$
$$0 = (t + 1)(t - 3)$$
$$t = -1 \quad \text{or} \quad t = 3$$

Check $t = -1$: $\quad 4 = 4 \quad$ *True*
Check $t = 3$: $\quad 8 = 8 \quad$ *True*

The solution set is $\{-1, 3\}$.

39.
$$x - 4 - \sqrt{2x} = 0$$
$$x - 4 = \sqrt{2x}$$
$$(x - 4)^2 = (\sqrt{2x})^2$$
$$x^2 - 8x + 16 = 2x$$
$$x^2 - 10x + 16 = 0$$
$$(x - 8)(x - 2) = 0$$
$$x = 8 \quad \text{or} \quad x = 2$$

Check $x = 8$: $\quad 0 = 0 \quad$ *True*
Check $x = 2$: $\quad -4 = 0 \quad$ *False*

The solution set is $\{8\}$.

41.
$$\sqrt{x} + 6 = 2x$$
$$\sqrt{x} = 2x - 6$$
$$(\sqrt{x})^2 = (2x - 6)^2$$
$$x = 4x^2 - 24x + 36$$
$$0 = 4x^2 - 25x + 36$$
$$0 = (4x - 9)(x - 4)$$
$$x = \frac{9}{4} \quad \text{or} \quad x = 4$$

Check $x = \frac{9}{4}$: $\quad \frac{15}{2} = \frac{9}{2} \quad$ *False*
Check $x = 4$: $\quad 8 = 8 \quad$ *True*

The solution set is $\{4\}$.

43.
$$\sqrt{x + 1} - \sqrt{x - 4} = 1$$
$$\sqrt{x + 1} = \sqrt{x - 4} + 1$$
$$(\sqrt{x + 1})^2 = (\sqrt{x - 4} + 1)^2$$
$$x + 1 = x - 4 + 2\sqrt{x - 4} + 1$$
$$4 = 2\sqrt{x - 4}$$
$$2 = \sqrt{x - 4}$$
$$2^2 = (\sqrt{x - 4})^2$$
$$4 = x - 4$$
$$8 = x$$

Check $x = 8$: $\sqrt{x + 1} - \sqrt{x - 4} = 1$
$$\sqrt{8 + 1} - \sqrt{8 - 4} \stackrel{?}{=} 1 \quad \textit{Let } x = 8.$$
$$\sqrt{9} - \sqrt{4} \stackrel{?}{=} 1$$
$$3 - 2 \stackrel{?}{=} 1$$
$$1 = 1 \quad \textit{True}$$

The solution set is $\{8\}$.

45.
$$\sqrt{x} = \sqrt{x - 5} + 1$$
$$(\sqrt{x})^2 = (\sqrt{x - 5} + 1)^2$$
$$x = x - 5 + 2\sqrt{x - 5} + 1$$
$$4 = 2\sqrt{x - 5}$$
$$2 = \sqrt{x - 5}$$
$$2^2 = (\sqrt{x - 5})^2$$
$$4 = x - 5$$
$$9 = x$$

Check $x = 9$: $\quad 3 = 3 \quad$ *True*

The solution set is $\{9\}$.

47.
$$\sqrt{3x+4} - \sqrt{2x-4} = 2$$
$$\sqrt{3x+4} = 2 + \sqrt{2x-4}$$
$$(\sqrt{3x+4})^2 = (2+\sqrt{2x-4})^2$$
$$3x+4 = 4 + 4\sqrt{2x-4} + 2x - 4$$
$$x+4 = 4\sqrt{2x-4}$$
$$(x+4)^2 = (4\sqrt{2x-4})^2$$
$$x^2 + 8x + 16 = 16(2x-4)$$
$$x^2 + 8x + 16 = 32x - 64$$
$$x^2 - 24x + 80 = 0$$
$$(x-4)(x-20) = 0$$
$$x = 4 \ \text{or} \ x = 20$$

Check $x = 4$: $4 - 2 = 2$ *True*
Check $x = 20$: $8 - 6 = 2$ *True*
The solution set is $\{4, 20\}$.

49.
$$\sqrt{2x+11} + \sqrt{x+6} = 2$$
$$\sqrt{2x+11} = 2 - \sqrt{x+6}$$
$$(\sqrt{2x+11})^2 = (2-\sqrt{x+6})^2$$
$$2x+11 = 4 - 4\sqrt{x+6} + x + 6$$
$$x+1 = -4\sqrt{x+6}$$
$$(x+1)^2 = (-4\sqrt{x+6})^2$$
$$x^2 + 2x + 1 = 16(x+6)$$
$$x^2 + 2x + 1 = 16x + 96$$
$$x^2 - 14x - 95 = 0$$
$$(x+5)(x-19) = 0$$
$$x = -5 \ \text{or} \ x = 19$$

Check $x = -5$: $1 + 1 = 2$ *True*
Check $x = 19$: $7 + 5 = 2$ *False*
The solution set is $\{-5\}$.

51. Let $x =$ the number. "The square root of the sum of a number and 4 is 5" translates to
$$\sqrt{x+4} = 5.$$
$$(\sqrt{x+4})^2 = 5^2 \quad \textit{Square each side.}$$
$$x + 4 = 25$$
$$x = 21$$
Check $x = 21$: $5 = 5$ *True*
The number is 21.

53. Let $x =$ the number. "Three times the square root of 2 equals the square root of the sum of some number and 10" translates to
$$3\sqrt{2} = \sqrt{x+10}.$$
$$(3\sqrt{2})^2 = (\sqrt{x+10})^2$$
$$9 \cdot 2 = x + 10$$
$$18 = x + 10$$
$$8 = x$$
Check $x = 8$: $3\sqrt{2} = \sqrt{18}$ *True*
The number is 8.

55. First find the semiperimeter of the triangular plot of land.
$$s = \tfrac{1}{2}(a+b+c)$$
$$= \tfrac{1}{2}(180 + 200 + 240)$$
$$= \tfrac{1}{2}(620) = 310 \text{ ft}$$

Now use Heron's formula to find the area.
$$\mathcal{A} = \sqrt{s(s-a)(s-b)(s-c)}$$
$$= \sqrt{310(310-180)(310-200)(310-240)}$$
$$= \sqrt{310(130)(110)(70)}$$
$$= \sqrt{310{,}310{,}000} \approx 17{,}616 \text{ ft}^2$$

The area of the plot of land is about $17{,}616 \text{ ft}^2$.

57. $s = 30\sqrt{\dfrac{a}{p}}$ ▪ Use a calculator and round answers to the nearest tenth.

(a) $s = 30\sqrt{\dfrac{862}{156}}$ *Let a = 862 and p = 156.*
≈ 70.5 miles per hour

(b) $s = 30\sqrt{\dfrac{382}{96}}$ *Let a = 382 and p = 96.*
≈ 59.8 miles per hour

(c) $s = 30\sqrt{\dfrac{84}{26}}$ *Let a = 84 and p = 26.*
≈ 53.9 miles per hour

59. Let $S =$ sight distance (in kilometers) and $h =$ height of the structure (in kilometers). The equation given is then
$$S = 111.7\sqrt{h}.$$
The height of the London Eye is 135 meters, or 0.135 kilometer.
$$S = 111.7\sqrt{0.135}$$
$$\approx 41.041201 \text{ km.}$$
To convert to miles, we multiply by 0.621371 to get $25.502 \approx 26$ miles. So yes, the passengers on the London Eye can see Windsor Castle, which is 25 miles away.

61. $1483 \text{ ft} \approx 1483(0.3048)$
$$= 452.0184 \text{ m}$$
So $h = 0.4520184$ kilometer and
$$S = 111.7\sqrt{0.4520184}$$
$$\approx 75.098494 \text{ km.}$$
Converting to miles gives us
$$(75.098494)(0.621371) \approx 46.7,$$
or about 47 miles.

Relating Concepts (Exercises 63–68)

63. $s = \frac{1}{2}(a + b + c)$
$= \frac{1}{2}(7 + 7 + 12)$
$= \frac{1}{2}(26) = 13$ units

64. $A = \sqrt{s(s-a)(s-b)(s-c)}$
$= \sqrt{13(13-7)(13-7)(13-12)}$
$= \sqrt{13(6)(6)(1)}$
$= 6\sqrt{13}$ square units

65. $c^2 = a^2 + b^2$
$7^2 = 6^2 + h^2$
$49 = 36 + h^2$
$h^2 = 13$
$h = \sqrt{13}$ units

66. $A = \frac{1}{2}bh$
$= \frac{1}{2}(6)(\sqrt{13})$
$= 3\sqrt{13}$ square units

67. $2(3\sqrt{13}) = 6\sqrt{13}$ square units

68. They are both $6\sqrt{13}$.

Chapter 16 Review Exercises

1. The square roots of 49 are -7 and 7 because $(-7)^2 = 49$ and $7^2 = 49$.

2. The square roots of 81 are -9 and 9 because $(-9)^2 = 81$ and $9^2 = 81$.

3. The square roots of 196 are -14 and 14 because $(-14)^2 = 196$ and $14^2 = 196$.

4. The square roots of 121 are -11 and 11 because $(-11)^2 = 121$ and $11^2 = 121$.

5. The square roots of 256 are -16 and 16 because $(-16)^2 = 256$ and $16^2 = 256$.

6. The square roots of 729 are -27 and 27 because $(-27)^2 = 729$ and $27^2 = 729$.

7. $\sqrt{16} = 4$ because $4^2 = 16$.

8. $\sqrt{0.36} = 0.6$ because $(0.6)^2 = 0.36$. Thus,
$-\sqrt{0.36} = -0.6$.

9. $\sqrt[3]{-512} = -8$ because $(-8)^3 = -512$.

10. $\sqrt[4]{81} = 3$ because 3 is positive and $3^4 = 81$.

11. $\sqrt{-8100}$ is not a real number.

12. $-\sqrt{4225}$ represents the negative square root of 4225. Since $65 \cdot 65 = 4225$,
$-\sqrt{4225} = -65$.

13. $\sqrt{\dfrac{144}{169}} = \dfrac{\sqrt{144}}{\sqrt{169}} = \dfrac{12}{13}$

14. $-\sqrt{\dfrac{100}{81}} = -\dfrac{\sqrt{100}}{\sqrt{81}} = -\dfrac{10}{9}$

15. Use the Pythagorean theorem with $a = 15$, $b = x$, and $c = 17$.
$$c^2 = a^2 + b^2$$
$$17^2 = 15^2 + x^2$$
$$289 = 225 + x^2$$
$$64 = x^2$$
$$x = \sqrt{64} = 8$$

16. Use the Pythagorean theorem with $a = 30.4$ cm and $b = 37.5$ cm.
$$c^2 = a^2 + b^2$$
$$= (30.4)^2 + (37.5)^2$$
$$= 924.16 + 1406.25$$
$$= 2330.41$$
$$c = \sqrt{2330.41} \approx 48.3 \text{ cm}$$

To the nearest tenth, the diagonal measure of the viewing screen is 48.3 cm.

17. This number is *irrational* because 73 is not a perfect square.
$$\sqrt{73} \approx 8.544$$

18. This number is *rational* because 169 is a perfect square.
$$\sqrt{169} = 13$$

19. $-\sqrt{625}$

This number is *rational* because 625 is a perfect square.
$$-\sqrt{625} = -25$$

20. There is no real number whose square is -19, so $\sqrt{-19}$ is not a real number.

21. $\sqrt{48} = \sqrt{16 \cdot 3} = \sqrt{16} \cdot \sqrt{3} = 4\sqrt{3}$

22. $-\sqrt{288} = -\sqrt{144 \cdot 2}$
$= -\sqrt{144} \cdot \sqrt{2} = -12\sqrt{2}$

23. $\sqrt[3]{16} = \sqrt[3]{8 \cdot 2} = \sqrt[3]{8} \cdot \sqrt[3]{2} = 2\sqrt[3]{2}$

24. $\sqrt[3]{375} = \sqrt[3]{125 \cdot 3} = \sqrt[3]{125} \cdot \sqrt[3]{3} = 5\sqrt[3]{3}$

25. $\sqrt{12} \cdot \sqrt{27} = \sqrt{4 \cdot 3} \cdot \sqrt{9 \cdot 3}$
$= 2\sqrt{3} \cdot 3\sqrt{3}$
$= 2 \cdot 3 \cdot (\sqrt{3})^2$
$= 2 \cdot 3 \cdot 3 = 18$

26. $\sqrt{32} \cdot \sqrt{48} = \sqrt{16 \cdot 2} \cdot \sqrt{16 \cdot 3}$
$= 4\sqrt{2} \cdot 4\sqrt{3}$
$= 4 \cdot 4 \cdot \sqrt{2 \cdot 3}$
$= 16\sqrt{6}$

27. $-\sqrt{\dfrac{121}{400}} = -\dfrac{\sqrt{121}}{\sqrt{400}} = -\dfrac{11}{20}$

28. $\sqrt{\dfrac{7}{169}} = \dfrac{\sqrt{7}}{\sqrt{169}} = \dfrac{\sqrt{7}}{13}$

29. $\sqrt{\dfrac{1}{6}} \cdot \sqrt{\dfrac{5}{6}} = \sqrt{\dfrac{1}{6} \cdot \dfrac{5}{6}}$
$= \sqrt{\dfrac{5}{36}}$
$= \dfrac{\sqrt{5}}{\sqrt{36}} = \dfrac{\sqrt{5}}{6}$

30. $\sqrt{\dfrac{2}{5}} \cdot \sqrt{\dfrac{2}{45}} = \sqrt{\dfrac{2}{5} \cdot \dfrac{2}{45}}$
$= \sqrt{\dfrac{4}{225}}$
$= \dfrac{\sqrt{4}}{\sqrt{225}} = \dfrac{2}{15}$

31. $\dfrac{3\sqrt{10}}{\sqrt{5}} = \dfrac{3 \cdot \sqrt{5} \cdot \sqrt{2}}{\sqrt{5}}$
$= 3\sqrt{2}$

32. $\dfrac{8\sqrt{150}}{4\sqrt{75}} = \dfrac{8 \cdot \sqrt{75} \cdot \sqrt{2}}{4\sqrt{75}}$
$= 2\sqrt{2}$

33. $\sqrt{r^{18}} = r^9$ because $(r^9)^2 = r^{18}$.

34. $\sqrt{x^{10}y^{16}} = x^5y^8$ because $(x^5y^8)^2 = x^{10}y^{16}$.

35. $\sqrt{162x^9} = \sqrt{81x^8 \cdot 2x}$
$= \sqrt{81x^8} \cdot \sqrt{2x} = 9x^4\sqrt{2x}$

36. $\sqrt{\dfrac{36}{p^2}} = \dfrac{\sqrt{36}}{\sqrt{p^2}} = \dfrac{6}{p} \quad (p \neq 0)$

37. $\sqrt{a^{15}b^{21}} = \sqrt{a^{14}b^{20} \cdot ab}$
$= \sqrt{a^{14}b^{20}} \cdot \sqrt{ab}$
$= a^7b^{10}\sqrt{ab}$

38. $\sqrt{121x^6y^{10}} = 11x^3y^5$ because
$(11x^3y^5)^2 = 121x^6y^{10}$.

39. $\sqrt[3]{y^6} = y^2$ because $(y^2)^3 = y^6$.

40. $\sqrt[3]{216x^{15}} = 6x^5$ because $(6x^5)^3 = 216x^{15}$.

41. $7\sqrt{11} + \sqrt{11} = 7\sqrt{11} + 1\sqrt{11} = 8\sqrt{11}$

42. $3\sqrt{2} + 6\sqrt{2} = (3+6)\sqrt{2} = 9\sqrt{2}$

43. $3\sqrt{75} + 2\sqrt{27}$
$= 3(\sqrt{25} \cdot \sqrt{3}) + 2(\sqrt{9} \cdot \sqrt{3})$
$= 3(5\sqrt{3}) + 2(3\sqrt{3})$
$= 15\sqrt{3} + 6\sqrt{3} = 21\sqrt{3}$

44. $4\sqrt{12} + \sqrt{48}$
$= 4(\sqrt{4} \cdot \sqrt{3}) + \sqrt{16} \cdot \sqrt{3}$
$= 4(2\sqrt{3}) + 4\sqrt{3}$
$= 8\sqrt{3} + 4\sqrt{3} = 12\sqrt{3}$

45. $4\sqrt{24} - 3\sqrt{54} + \sqrt{6}$
$= 4(\sqrt{4} \cdot \sqrt{6}) - 3(\sqrt{9} \cdot \sqrt{6}) + \sqrt{6}$
$= 4(2\sqrt{6}) - 3(3\sqrt{6}) + \sqrt{6}$
$= 8\sqrt{6} - 9\sqrt{6} + 1\sqrt{6}$
$= 0\sqrt{6} = 0$

46. $2\sqrt{7} - 4\sqrt{28} + 3\sqrt{63}$
$= 2\sqrt{7} - 4(\sqrt{4} \cdot \sqrt{7}) + 3(\sqrt{9} \cdot \sqrt{7})$
$= 2\sqrt{7} - 4(2\sqrt{7}) + 3(3\sqrt{7})$
$= 2\sqrt{7} - 8\sqrt{7} + 9\sqrt{7} = 3\sqrt{7}$

47. $\dfrac{2}{5}\sqrt{75} + \dfrac{3}{4}\sqrt{160}$
$= \dfrac{2}{5}(\sqrt{25} \cdot \sqrt{3}) + \dfrac{3}{4}(\sqrt{16} \cdot \sqrt{10})$
$= \dfrac{2}{5}(5\sqrt{3}) + \dfrac{3}{4}(4\sqrt{10})$
$= 2\sqrt{3} + 3\sqrt{10}$

48. $\dfrac{1}{3}\sqrt{18} + \dfrac{1}{4}\sqrt{32}$
$= \dfrac{1}{3}(\sqrt{9} \cdot \sqrt{2}) + \dfrac{1}{4}(\sqrt{16} \cdot \sqrt{2})$
$= \dfrac{1}{3}(3\sqrt{2}) + \dfrac{1}{4}(4\sqrt{2})$
$= 1\sqrt{2} + 1\sqrt{2} = 2\sqrt{2}$

49. $\sqrt{15} \cdot \sqrt{2} + 5\sqrt{30} = \sqrt{30} + 5\sqrt{30}$
$= 1\sqrt{30} + 5\sqrt{30}$
$= 6\sqrt{30}$

50. $\sqrt{4x} + \sqrt{36x} - \sqrt{9x}$
$= \sqrt{4}\sqrt{x} + \sqrt{36}\sqrt{x} - \sqrt{9}\sqrt{x}$
$= 2\sqrt{x} + 6\sqrt{x} - 3\sqrt{x} = 5\sqrt{x}$

51. $\sqrt{16p} + 3\sqrt{p} - \sqrt{49p}$
$= \sqrt{16}\sqrt{p} + 3\sqrt{p} - \sqrt{49}\sqrt{p}$
$= 4\sqrt{p} + 3\sqrt{p} - 7\sqrt{p}$
$= 0\sqrt{p} = 0$

52. $3k\sqrt{8k^2n} + 5k^2\sqrt{2n}$
$= 3k(\sqrt{4k^2} \cdot \sqrt{2n}) + 5k^2\sqrt{2n}$
$= 3k(2k\sqrt{2n}) + 5k^2\sqrt{2n}$
$= 6k^2\sqrt{2n} + 5k^2\sqrt{2n}$
$= (6k^2 + 5k^2)\sqrt{2n}$
$= 11k^2\sqrt{2n}$

53. $\dfrac{10}{\sqrt{3}} = \dfrac{10 \cdot \sqrt{3}}{\sqrt{3} \cdot \sqrt{3}} = \dfrac{10\sqrt{3}}{3}$

54. $\dfrac{8\sqrt{2}}{\sqrt{5}} = \dfrac{8\sqrt{2}\cdot\sqrt{5}}{\sqrt{5}\cdot\sqrt{5}} = \dfrac{8\sqrt{10}}{5}$

55. $\dfrac{12}{\sqrt{24}} = \dfrac{12}{\sqrt{4\cdot 6}} = \dfrac{12}{2\sqrt{6}}$

$\qquad = \dfrac{12\cdot\sqrt{6}}{2\sqrt{6}\cdot\sqrt{6}} = \dfrac{12\sqrt{6}}{2\cdot 6}$

$\qquad = \dfrac{12\sqrt{6}}{12} = \sqrt{6}$

56. $\sqrt{\dfrac{2}{5}} = \dfrac{\sqrt{2}}{\sqrt{5}} = \dfrac{\sqrt{2}\cdot\sqrt{5}}{\sqrt{5}\cdot\sqrt{5}} = \dfrac{\sqrt{10}}{5}$

57. $\sqrt{\dfrac{5}{14}}\cdot\sqrt{28} = \sqrt{\dfrac{5}{14}\cdot 28} = \sqrt{5\cdot 2} = \sqrt{10}$

58. $\sqrt{\dfrac{2}{7}}\cdot\sqrt{\dfrac{1}{3}} = \sqrt{\dfrac{2}{7}\cdot\dfrac{1}{3}}$

$\qquad = \sqrt{\dfrac{2}{21}} = \dfrac{\sqrt{2}}{\sqrt{21}}$

$\qquad = \dfrac{\sqrt{2}\cdot\sqrt{21}}{\sqrt{21}\cdot\sqrt{21}} = \dfrac{\sqrt{42}}{21}$

59. $\sqrt{\dfrac{r^2}{16x}} = \dfrac{\sqrt{r^2}}{\sqrt{16x}}$

$\qquad = \dfrac{r\cdot\sqrt{x}}{\sqrt{16x}\cdot\sqrt{x}}$

$\qquad = \dfrac{r\sqrt{x}}{\sqrt{16x^2}}$

$\qquad = \dfrac{r\sqrt{x}}{4x} \quad (x\neq 0)$

60. $\sqrt[3]{\dfrac{1}{3}} = \dfrac{\sqrt[3]{1}}{\sqrt[3]{3}}$

$\qquad = \dfrac{1\cdot\sqrt[3]{3^2}}{\sqrt[3]{3}\cdot\sqrt[3]{3^2}}$

$\qquad = \dfrac{\sqrt[3]{3^2}}{\sqrt[3]{3^3}} = \dfrac{\sqrt[3]{9}}{3}$

61. $-\sqrt{3}(\sqrt{5}+\sqrt{27})$

$\qquad = -\sqrt{3}(\sqrt{5}) + (-\sqrt{3})(\sqrt{27})$

$\qquad = -\sqrt{3\cdot 5} - \sqrt{3\cdot 27}$

$\qquad = -\sqrt{15} - \sqrt{81}$

$\qquad = -\sqrt{15} - 9$

62. $3\sqrt{2}(\sqrt{3}+2\sqrt{2})$

$\qquad = 3\sqrt{2}(\sqrt{3}) + 3\sqrt{2}(2\sqrt{2})$

$\qquad = 3\sqrt{6} + 6\cdot 2$

$\qquad = 3\sqrt{6} + 12$

63. $(2\sqrt{3}-4)(5\sqrt{3}+2)$

$\qquad = 2\sqrt{3}(5\sqrt{3}) + (2\sqrt{3})(2) - 4(5\sqrt{3})$

$\qquad\quad - 4(2) \qquad\qquad\qquad\qquad$ *FOIL*

$\qquad = 10\cdot 3 + 4\sqrt{3} - 20\sqrt{3} - 8$

$\qquad = 30 - 16\sqrt{3} - 8$

$\qquad = 22 - 16\sqrt{3}$

64. $(\sqrt{7}+2\sqrt{6})(\sqrt{12}-\sqrt{2})$

$\qquad = \sqrt{7}(\sqrt{12}) - \sqrt{7}(\sqrt{2}) + 2\sqrt{6}(\sqrt{12})$

$\qquad\quad - 2\sqrt{6}(\sqrt{2}) \qquad\qquad\qquad$ *FOIL*

$\qquad = \sqrt{84} - \sqrt{14} + 2\sqrt{72} - 2\sqrt{12}$

$\qquad = \sqrt{4\cdot 21} - \sqrt{14} + 2\sqrt{36\cdot 2} - 2\sqrt{4\cdot 3}$

$\qquad = 2\sqrt{21} - \sqrt{14} + 2\cdot 6\sqrt{2} - 2\cdot 2\sqrt{3}$

$\qquad = 2\sqrt{21} - \sqrt{14} + 12\sqrt{2} - 4\sqrt{3}$

65. $(2\sqrt{3}+5)(2\sqrt{3}-5)$

$\qquad = (2\sqrt{3})^2 - (5)^2$

$\qquad = 4\cdot 3 - 25$

$\qquad = 12 - 25 = -13$

66. $(\sqrt{x}+2)^2 = (\sqrt{x})^2 + 2(\sqrt{x})(2) + 2^2$

$\qquad = x + 4\sqrt{x} + 4$

67. $\dfrac{1}{2+\sqrt{5}}$

$\qquad = \dfrac{1(2-\sqrt{5})}{(2+\sqrt{5})(2-\sqrt{5})}$ *Multiply by the conjugate.*

$\qquad = \dfrac{2-\sqrt{5}}{(2)^2-(\sqrt{5})^2} = \dfrac{2-\sqrt{5}}{4-5}$

$\qquad = \dfrac{2-\sqrt{5}}{-1} = -2+\sqrt{5}$

68. $\dfrac{3}{1+\sqrt{x}}$

$\qquad = \dfrac{3\left(1-\sqrt{x}\right)}{\left(1+\sqrt{x}\right)\left(1-\sqrt{x}\right)}$ *Multiply by the conjugate.*

$\qquad = \dfrac{3\left(1-\sqrt{x}\right)}{1-x} \quad (x\neq 1)$

69. $\dfrac{\sqrt{5}-1}{\sqrt{2}+3}$

$\qquad = \dfrac{(\sqrt{5}-1)(\sqrt{2}-3)}{(\sqrt{2}+3)(\sqrt{2}-3)}$ *Multiply by the conjugate.*

$\qquad = \dfrac{\sqrt{10}-3\sqrt{5}-\sqrt{2}+3}{2-9}$

$\qquad = \dfrac{\sqrt{10}-3\sqrt{5}-\sqrt{2}+3}{-7}$

$\qquad = \dfrac{-\sqrt{10}+3\sqrt{5}+\sqrt{2}-3}{7}$

70. $\dfrac{15 + 10\sqrt{6}}{15} = \dfrac{5(3 + 2\sqrt{6})}{5(3)}$ *Factor.*

$= \dfrac{3 + 2\sqrt{6}}{3}$ *Lowest terms*

71. $\dfrac{3 + 9\sqrt{7}}{12} = \dfrac{3(1 + 3\sqrt{7})}{3(4)}$ *Factor.*

$= \dfrac{1 + 3\sqrt{7}}{4}$ *Lowest terms*

72. $\dfrac{6 + \sqrt{192}}{2} = \dfrac{6 + \sqrt{64 \cdot 3}}{2}$

$= \dfrac{6 + 8\sqrt{3}}{2}$

$= \dfrac{2(3 + 4\sqrt{3})}{2}$

$= 3 + 4\sqrt{3}$

73. $\sqrt{x} + 5 = 0$

$\sqrt{x} = -5$

Since a square root cannot equal a negative number, there is no solution and the solution set is $\emptyset$.

74. $\sqrt{k + 1} = 7$

$(\sqrt{k + 1})^2 = 7^2$

$k + 1 = 49$

$k = 48$

Check $k = 48$: $\sqrt{49} = 7$ *True*

The solution set is $\{48\}$.

75. $\sqrt{5t + 4} = 3\sqrt{t}$

$(\sqrt{5t + 4})^2 = (3\sqrt{t})^2$

$5t + 4 = 9t$

$4 = 4t$

$1 = t$

Check $t = 1$: $\sqrt{9} = 3\sqrt{1}$ *True*

The solution set is $\{1\}$.

76. $\sqrt{2p + 3} = \sqrt{5p - 3}$

$(\sqrt{2p + 3})^2 = (\sqrt{5p - 3})^2$

$2p + 3 = 5p - 3$

$6 = 3p$

$2 = p$

Check $p = 2$: $\sqrt{7} = \sqrt{7}$ *True*

The solution set is $\{2\}$.

77. $\sqrt{4x + 1} = x - 1$

$(\sqrt{4x + 1})^2 = (x - 1)^2$

$4x + 1 = x^2 - 2x + 1$

$0 = x^2 - 6x$

$0 = x(x - 6)$

$x = 0$ or $x = 6$

Check $x = 0$: $\quad 1 = -1 \quad$ *False*

Check $x = 6$: $\sqrt{25} = 5 \quad$ *True*

Of the two proposed solutions, 6 checks in the original equation, but 0 does not. Thus, the solution set is $\{6\}$.

78. $\sqrt{13 + 4t} = t + 4$

$(\sqrt{13 + 4t})^2 = (t + 4)^2$

$13 + 4t = t^2 + 8t + 16$

$0 = t^2 + 4t + 3$

$0 = (t + 3)(t + 1)$

$t = -3$ or $t = -1$

Check $t = -3$: $\quad 1 = 1 \quad$ *True*

Check $t = -1$: $\quad 3 = 3 \quad$ *True*

The solution set is $\{-3, -1\}$.

79. $\sqrt{2 - x} + 3 = x + 7$

$\sqrt{2 - x} = x + 4 \quad$ *Isolate the radical.*

$(\sqrt{2 - x})^2 = (x + 4)^2$

$2 - x = x^2 + 8x + 16$

$0 = x^2 + 9x + 14$

$0 = (x + 2)(x + 7)$

$x = -2$ or $x = -7$

Check $x = -2$: $2 + 3 = 5 \quad$ *True*

Check $x = -7$: $3 + 3 = 0 \quad$ *False*

Of the two proposed solutions, -2 checks in the original equation, but -7 does not. Thus, the solution set is $\{-2\}$.

80. $\sqrt{x} - x + 2 = 0$

$\sqrt{x} = x - 2 \quad$ *Isolate the radical.*

$(\sqrt{x})^2 = (x - 2)^2$

$x = x^2 - 4x + 4$

$0 = x^2 - 5x + 4$

$0 = (x - 4)(x - 1)$

$x = 4$ or $x = 1$

Check $x = 4$: $0 = 0 \quad$ *True*

Check $x = 1$: $2 = 0 \quad$ *False*

Of the two proposed solutions, 4 checks in the original equation, but 1 does not. Thus, the solution set is $\{4\}$.

81. $\sqrt{x+2} - \sqrt{x-3} = 1$

$\sqrt{x+2} = 1 + \sqrt{x-3}$

$(\sqrt{x+2})^2 = (1 + \sqrt{x-3})^2$

$x + 2 = 1 + 2\sqrt{x-3} + x - 3$

$4 = 2\sqrt{x-3}$

$2 = \sqrt{x-3}$

$2^2 = (\sqrt{x-3})^2$

$4 = x - 3$

$7 = x$

Check $x = 7$: $1 = 1$ *True*

The solution set is $\{7\}$.

82. **[16.1]** **(a)** $\sqrt{64} = 8$; **B**

(b) $-\sqrt{64} = -8$; **F**

(c) $\sqrt{-64}$ is not a real number; **D**

(d) $\sqrt[3]{64} = 4$; **A**

(e) $\sqrt[3]{-64} = -4$; **C**

(f) $-\sqrt[3]{-64} = -(-4) = 4$; **A**

83. **[16.3]** $2\sqrt{27} + 3\sqrt{75} - \sqrt{300}$

$= 2\sqrt{9 \cdot 3} + 3\sqrt{25 \cdot 3} - \sqrt{100 \cdot 3}$

$= 2 \cdot 3\sqrt{3} + 3 \cdot 5\sqrt{3} - 10\sqrt{3}$

$= 6\sqrt{3} + 15\sqrt{3} - 10\sqrt{3}$

$= 11\sqrt{3}$

84. **[16.4]** $\dfrac{1}{5+\sqrt{2}} = \dfrac{1(5-\sqrt{2})}{(5+\sqrt{2})(5-\sqrt{2})}$

$= \dfrac{5-\sqrt{2}}{(5)^2 - (\sqrt{2})^2}$

$= \dfrac{5-\sqrt{2}}{25-2}$

$= \dfrac{5-\sqrt{2}}{23}$

85. **[16.4]** $\sqrt{\dfrac{1}{3}} \cdot \sqrt{\dfrac{24}{5}} = \sqrt{\dfrac{1}{3} \cdot \dfrac{24}{5}} = \sqrt{\dfrac{8}{5}} = \dfrac{\sqrt{8}}{\sqrt{5}}$

$= \dfrac{\sqrt{8} \cdot \sqrt{5}}{\sqrt{5} \cdot \sqrt{5}} = \dfrac{\sqrt{40}}{5}$

$= \dfrac{\sqrt{4 \cdot 10}}{5} = \dfrac{2\sqrt{10}}{5}$

86. **[16.2]** $\sqrt[3]{54a^7b^{10}} = \sqrt[3]{27a^6b^9 \cdot 2a^1b^1}$

$= \sqrt[3]{27a^6b^9} \cdot \sqrt[3]{2ab}$

$= 3a^2b^3\sqrt[3]{2ab}$

87. **[16.5]** $-\sqrt{5}(\sqrt{2} + \sqrt{75})$

$= -\sqrt{5}(\sqrt{2}) + (-\sqrt{5})(\sqrt{75})$

$= -\sqrt{10} - \sqrt{375}$

$= -\sqrt{10} - \sqrt{25 \cdot 15}$

$= -\sqrt{10} - 5\sqrt{15}$

88. **[16.4]** $\sqrt{\dfrac{16r^3}{3s}} = \dfrac{\sqrt{16r^3}}{\sqrt{3s}} = \dfrac{\sqrt{16r^2} \cdot \sqrt{r}}{\sqrt{3s}}$

$= \dfrac{4r\sqrt{r}}{\sqrt{3s}} = \dfrac{4r\sqrt{r} \cdot \sqrt{3s}}{\sqrt{3s} \cdot \sqrt{3s}}$

$= \dfrac{4r\sqrt{3rs}}{3s}$ $(s \neq 0)$

89. **[16.5]** $\dfrac{12 + 6\sqrt{13}}{12} = \dfrac{6(2 + \sqrt{13})}{6(2)}$

$= \dfrac{2 + \sqrt{13}}{2}$

90. **[16.5]** $(\sqrt{5} - \sqrt{2})^2$

$= (\sqrt{5})^2 - 2\sqrt{5}\sqrt{2} + (\sqrt{2})^2$

   Square of a binomial

$= 5 - 2\sqrt{10} + 2$

$= 7 - 2\sqrt{10}$

91. **[16.5]** $(6\sqrt{7} + 2)(4\sqrt{7} - 1)$

$= 6\sqrt{7}(4\sqrt{7}) - 1(6\sqrt{7})$

$+ 2(4\sqrt{7}) + 2(-1)$ *FOIL*

$= 24 \cdot 7 - 6\sqrt{7} + 8\sqrt{7} - 2$

$= 168 - 2 + 2\sqrt{7}$

$= 166 + 2\sqrt{7}$

92. **[16.6]** $\sqrt{x+2} = x - 4$

$(\sqrt{x+2})^2 = (x-4)^2$

$x + 2 = x^2 - 8x + 16$

$0 = x^2 - 9x + 14$

$0 = (x-2)(x-7)$

$x = 2$ or $x = 7$

Check $x = 2$: $\sqrt{4} = -2$ *False*
Check $x = 7$: $\sqrt{9} = 3$ *True*

The solution set is $\{7\}$.

93. **[16.6]** $\sqrt{k} + 3 = 0$

$\sqrt{k} = -3$

Since a square root cannot equal a negative number, there is no solution and the solution set is $\emptyset$.

94. **[16.6]**
$$\sqrt{1+3t} - t = -3$$
$$\sqrt{1+3t} = t - 3$$
$$(\sqrt{1+3t})^2 = (t-3)^2$$
$$1 + 3t = t^2 - 6t + 9$$
$$0 = t^2 - 9t + 8$$
$$0 = (t-1)(t-8)$$
$$t = 1 \quad \text{or} \quad t = 8$$

Check $t = 1$: $2 - 1 = -3$ *False*
Check $t = 8$: $5 - 8 = -3$ *True*

The solution set is $\{8\}$.

95. **(a)** Each side s of the large square has length $a + b$. Since the area A_1 is equal to s^2, we have
$$A_1 = (a+b)^2$$
$$= a^2 + 2ab + b^2$$

(b) The formula for the area of a triangle is $A = \frac{1}{2}(\text{height})(\text{base})$, so the area of one of the right triangles is $\frac{1}{2}ab$. The area of the smaller square and the four right triangles is
$$A_2 = c^2 + 4\left(\tfrac{1}{2}ab\right)$$
$$= c^2 + 2ab$$

(c) Set the areas equal and simplify.
$$A_1 = A_2$$
$$a^2 + 2ab + b^2 = c^2 + 2ab$$
$$a^2 + b^2 = c^2$$

Chapter 16 Test

1. The square roots of 400 are -20 and 20 because $(-20)^2 = 400$ and $20^2 = 400$.

2. **(a)** $\sqrt{142}$ is *irrational* because 142 is not a perfect square.

(b) $\sqrt{142} \approx 11.916$

3. If $\sqrt{a}$ is not a real number, then a must be a negative number.

4. $\sqrt[3]{216} = 6$ because $6^3 = 216$.

5. $-\sqrt{54} = -\sqrt{9 \cdot 6} = -\sqrt{9} \cdot \sqrt{6} = -3\sqrt{6}$

6. $\sqrt{\dfrac{128}{25}} = \dfrac{\sqrt{128}}{\sqrt{25}} = \dfrac{\sqrt{64 \cdot 2}}{5} = \dfrac{8\sqrt{2}}{5}$

7. $\sqrt[3]{32} = \sqrt[3]{8 \cdot 4} = \sqrt[3]{8} \cdot \sqrt[3]{4} = 2\sqrt[3]{4}$

8. $\dfrac{20\sqrt{18}}{5\sqrt{3}} = \dfrac{4\sqrt{9 \cdot 2}}{\sqrt{3}}$

$$= \dfrac{4 \cdot 3\sqrt{2}}{\sqrt{3}} = \dfrac{12\sqrt{2} \cdot \sqrt{3}}{\sqrt{3} \cdot \sqrt{3}}$$

$$= \dfrac{12\sqrt{6}}{3} = 4\sqrt{6}$$

9. $3\sqrt{28} + \sqrt{63} = 3(\sqrt{4 \cdot 7}) + \sqrt{9 \cdot 7}$
$$= 3(2\sqrt{7}) + 3\sqrt{7}$$
$$= 6\sqrt{7} + 3\sqrt{7} = 9\sqrt{7}$$

10. $(\sqrt{5} + \sqrt{6})^2$
$$= (\sqrt{5})^2 + 2(\sqrt{5})(\sqrt{6}) + (\sqrt{6})^2$$
$$= 5 + 2\sqrt{30} + 6$$
$$= 11 + 2\sqrt{30}$$

11. $\sqrt{32x^2y^3} = \sqrt{16x^2y^2 \cdot 2y}$
$$= \sqrt{16x^2y^2} \cdot \sqrt{2y}$$
$$= 4xy\sqrt{2y}$$

12. $(6 - \sqrt{5})(6 + \sqrt{5})$
$$= (6)^2 - (\sqrt{5})^2$$
$$= 36 - 5 = 31$$

13. $(2 - \sqrt{7})(3\sqrt{2} + 1)$
$$= 2(3\sqrt{2}) + 2(1) - \sqrt{7}(3\sqrt{2}) - \sqrt{7}(1)$$
$$= 6\sqrt{2} + 2 - 3\sqrt{14} - \sqrt{7}$$

14. $3\sqrt{27x} - 4\sqrt{48x} + 2\sqrt{3x}$
$$= 3(\sqrt{9 \cdot 3x}) - 4(\sqrt{16 \cdot 3x}) + 2\sqrt{3x}$$
$$= 3(3\sqrt{3x}) - 4(4\sqrt{3x}) + 2\sqrt{3x}$$
$$= 9\sqrt{3x} - 16\sqrt{3x} + 2\sqrt{3x} = -5\sqrt{3x}$$

15. Use the Pythagorean theorem with $c = 9$ and $b = 3$.
$$c^2 = a^2 + b^2$$
$$9^2 = a^2 + 3^2$$
$$81 = a^2 + 9$$
$$72 = a^2$$
$$\sqrt{72} = a$$

(a) $a = \sqrt{72} = \sqrt{36 \cdot 2} = 6\sqrt{2}$ inches

(b) $a = \sqrt{72} \approx 8.485$ inches

16. $Z = \sqrt{R^2 + X^2}$
$$= \sqrt{40^2 + 30^2} \qquad \textit{Let R = 40, X = 30.}$$
$$= \sqrt{1600 + 900}$$
$$= \sqrt{2500} = 50 \text{ ohms}$$

17. $\dfrac{5\sqrt{2}}{\sqrt{7}} = \dfrac{5\sqrt{2} \cdot \sqrt{7}}{\sqrt{7} \cdot \sqrt{7}} = \dfrac{5\sqrt{14}}{7}$

18. $\sqrt{\dfrac{2}{3x}} = \dfrac{\sqrt{2}}{\sqrt{3x}}$

$$= \dfrac{\sqrt{2} \cdot \sqrt{3x}}{\sqrt{3x} \cdot \sqrt{3x}} = \dfrac{\sqrt{6x}}{3x} \quad (x > 0)$$

19. $\dfrac{-2}{\sqrt[3]{4}} = \dfrac{-2 \cdot \sqrt[3]{2}}{\sqrt[3]{4} \cdot \sqrt[3]{2}}$

$\qquad = \dfrac{-2\sqrt[3]{2}}{\sqrt[3]{8}}$

$\qquad = \dfrac{-2\sqrt[3]{2}}{2} = -\sqrt[3]{2}$

20. $\dfrac{-3}{4 - \sqrt{3}} = \dfrac{-3(4 + \sqrt{3})}{(4 - \sqrt{3})(4 + \sqrt{3})}$

$\qquad = \dfrac{-3(4 + \sqrt{3})}{(4)^2 - (\sqrt{3})^2}$

$\qquad = \dfrac{-3(4 + \sqrt{3})}{16 - 3}$

$\qquad = \dfrac{-3(4 + \sqrt{3})}{13}$

21. $\dfrac{\sqrt{12} + 3\sqrt{128}}{6} = \dfrac{\sqrt{4} \cdot \sqrt{3} + 3 \cdot \sqrt{64} \cdot \sqrt{2}}{6}$

$\qquad = \dfrac{2\sqrt{3} + 24\sqrt{2}}{6}$

$\qquad = \dfrac{2(\sqrt{3} + 12\sqrt{2})}{2(3)}$

$\qquad = \dfrac{\sqrt{3} + 12\sqrt{2}}{3}$

22. $\sqrt{p} + 4 = 0$

$\qquad \sqrt{p} = -4$

Since a square root cannot equal a negative number, there is no solution and the solution set is $\emptyset$.

23. $\sqrt{x + 1} = 5 - x$

$(\sqrt{x + 1})^2 = (5 - x)^2$

$x + 1 = 25 - 10x + x^2$

$0 = x^2 - 11x + 24$

$0 = (x - 3)(x - 8)$

$x = 3 \quad \text{or} \quad x = 8$

Check $x = 3$: $\sqrt{4} = 2$ *True*

Check $x = 8$: $\sqrt{9} = -3$ *False*

The solution set is $\{3\}$.

24. $3\sqrt{x} - 2 = x$

$3\sqrt{x} = x + 2$

$(3\sqrt{x})^2 = (x + 2)^2$

$9x = x^2 + 4x + 4$

$0 = x^2 - 5x + 4$

$0 = (x - 4)(x - 1)$

$x - 4 = 0 \quad \text{or} \quad x - 1 = 0$

$x = 4 \quad \text{or} \quad x = 1$

Check $x = 4$: $4 = 4$ *True*
Check $x = 1$: $1 = 1$ *True*

The solution set is $\{1, 4\}$.

25. $\sqrt{x + 7} - \sqrt{x} = 1$

$\sqrt{x + 7} = \sqrt{x} + 1$

$(\sqrt{x + 7})^2 = (\sqrt{x} + 1)^2$

$x + 7 = x + 2\sqrt{x} + 1$

$6 = 2\sqrt{x}$

$3 = \sqrt{x}$

$3^2 = (\sqrt{x})^2$

$9 = x$

Check $x = 9$: $4 - 3 = 1$ *True*

The solution set is $\{9\}$.

26. Nothing is wrong with the steps taken so far, but the proposed solution must be checked.

Let $x = 12$ in the original equation.

$$\sqrt{2x + 1} + 5 = 0$$
$$\sqrt{2(12) + 1} + 5 \overset{?}{=} 0 \quad \textit{Let x = 12.}$$
$$\sqrt{25} + 5 \overset{?}{=} 0$$
$$5 + 5 \overset{?}{=} 0$$
$$10 = 0 \quad \textit{False}$$

12 is not a solution because it does not satisfy the original equation. The equation has no solution, so the solution set is $\emptyset$.

CHAPTER 17 QUADRATIC EQUATIONS

17.1 Solving Quadratic Equations by the Square Root Property

17.1 Margin Exercises

1. **(a)** $x^2 - 2x - 15 = 0$

$(x + \underline{3})(x - \underline{5}) = 0$ *Factor.*

$x + \underline{3} = 0$ or $x - \underline{5} = 0$ *Zero-factor property*
$x = \underline{-3}$ or $x = \underline{5}$ *Solve each equation.*

The solution set is $\{\underline{-3}, \underline{5}\}$.

(b) $2x^2 - 3x + 1 = 0$

$(2x - 1)(x - 1) = 0$ *Factor.*

$2x - 1 = 0$ or $x - 1 = 0$ *Zero-factor property*
$x = \frac{1}{2}$ or $x = 1$ *Solve each equation.*

The solution set is $\{\frac{1}{2}, 1\}$.

(c) $x^2 = 400$

$x^2 - 400 = 0$ *Subtract 400.*
$(x + 20)(x - 20) = 0$ *Factor.*

$x + 20 = 0$ or $x - 20 = 0$ *Zero-factor prop.*
$x = -20$ or $x = 20$ *Solve each eq.*

The solution set is $\{-20, 20\}$.

2. **(a)** $x^2 = 49$ ▪ Solve by the square root property.

$x = \sqrt{49}$ or $x = -\sqrt{49}$
$x = 7$ or $x = -7$

The solution set is $\{-7, 7\}$, which may be written as $\{\pm 7\}$.

(b) $x^2 = 11$ ▪ By the square root property,

$x = \sqrt{11}$ or $x = -\sqrt{11}$.

The solution set is $\{-\sqrt{11}, \sqrt{11}\}$.

(c) $2x^2 + 8 = 32$
$2x^2 = 24$ *Subtract 8.*
$x^2 = 12$ *Divide by 2.*

Use the square root property.

$x = \sqrt{12}$ or $x = -\sqrt{12}$
$x = \sqrt{4} \cdot \sqrt{3}$ or $x = -\sqrt{4} \cdot \sqrt{3}$
$x = 2\sqrt{3}$ or $x = -2\sqrt{3}$

The solution set is $\{-2\sqrt{3}, 2\sqrt{3}\}$.

(d) $x^2 = -9$ ▪ The square of a real number cannot be negative. (The square root property

cannot be used because k must be positive.) Thus, there is *no real number solution* for this equation and the solution set is $\emptyset$.

3. **(a)** $(x + 2)^2 = 36$ ▪ Use the square root property.

$x + 2 = 6$ or $x + 2 = -6$
$x = 4$ or $x = -8$

The solution set is $\{-8, 4\}$.

(b) $(x - 4)^2 = 3$ ▪ Use the square root property.

$x - 4 = \sqrt{3}$ or $x - 4 = -\sqrt{3}$
$x = 4 + \sqrt{3}$ or $x = 4 - \sqrt{3}$

The solution set is $\{4 + \sqrt{3}, 4 - \sqrt{3}\}$.

4. $(2x - 5)^2 = 18$ ▪ Use the square root property.

$2x - 5 = \sqrt{18}$ or $2x - 5 = -\sqrt{18}$
$2x = 5 + \sqrt{18}$ or $2x = 5 - \sqrt{18}$
$x = \dfrac{5 + \sqrt{18}}{2}$ or $x = \dfrac{5 - \sqrt{18}}{2}$
$x = \dfrac{5 + \sqrt{9} \cdot \sqrt{2}}{2}$ or $x = \dfrac{5 - \sqrt{9} \cdot \sqrt{2}}{2}$
$x = \dfrac{5 + 3\sqrt{2}}{2}$ or $x = \dfrac{5 - 3\sqrt{2}}{2}$

The solution set is $\left\{\dfrac{5 + 3\sqrt{2}}{2}, \dfrac{5 - 3\sqrt{2}}{2}\right\}$.

5. **(a)** $(5x + 1)^2 = 7$ ▪ Use the square root property.

$5x + 1 = \sqrt{7}$ or $5x + 1 = -\sqrt{7}$
$5x = -1 + \sqrt{7}$ or $5x = -1 - \sqrt{7}$
$x = \dfrac{-1 + \sqrt{7}}{5}$ or $x = \dfrac{-1 - \sqrt{7}}{5}$

The solution set is $\left\{\dfrac{-1 + \sqrt{7}}{5}, \dfrac{-1 - \sqrt{7}}{5}\right\}$.

(b) $(7x - 1)^2 = -1$ ▪ Because the square root of -1 is not a real number, there is *no real number solution* for this equation. The solution set is $\emptyset$.

6. $w = \dfrac{L^2 g}{1200}$ *Given formula*

$2.80 = \dfrac{L^2 \cdot 11}{1200}$ *Let w = 2.80, g = 11.*

$3360 = 11L^2$ *Multiply by 1200.*

$L^2 \approx 305.45$ *Divide by 11.*

$L \approx 17.48$ *Approximate L > 0.*

The length of the bass is approximately 17.48 in.

17.1   Section Exercises

1. $x^2 = 12$ has two irrational solutions, $\pm \sqrt{12} = \pm 2\sqrt{3}$. The correct choice is **C**.

3. $x^2 = \frac{25}{36}$ has two rational solutions that are not integers, $\pm \frac{5}{6}$. The correct choice is **D**.

5.   It is not correct to say that the solution set of $x^2 = 81$ is $\{9\}$, because -9 also satisfies the equation.

When we solve an equation, we want to find *all* values of the variable that satisfy the equation. The completely correct answer is that the solution set of $x^2 = 81$ is $\{-9, 9\}$, or $\left\{ \pm \sqrt{9} \right\}$.

7.
$$x^2 - x - 56 = 0$$
$$(x - 8)(x + 7) = 0 \qquad \textit{Factor.}$$

$$x - 8 = 0 \text{ or } x + 7 = 0 \quad \textit{Zero-factor property}$$
$$x = 8 \text{ or } \qquad x = -7 \quad \textit{Solve each equation.}$$

The solution set is $\{-7, 8\}$.

9.
$$x^2 = 121$$
$$x^2 - 121 = 0 \qquad \textit{Subtract 121.}$$
$$(x + 11)(x - 11) = 0 \qquad \textit{Factor.}$$

$$x + 11 = 0 \quad \text{ or } x - 11 = 0 \quad \textit{Zero-factor prop.}$$
$$x = -11 \text{ or } \qquad x = 11 \quad \textit{Solve each eq.}$$

The solution set is $\{-11, 11\}$, or $\{ \pm 11 \}$.

11.
$$3x^2 - 13x = 30$$
$$3x^2 - 13x - 30 = 0 \qquad \textit{Standard form}$$
$$(x - 6)(3x + 5) = 0 \qquad \textit{Factor.}$$

$$x - 6 = 0 \text{ or } 3x + 5 = 0 \quad \textit{Zero-factor prop.}$$
$$x = 6 \text{ or } \qquad x = -\tfrac{5}{3} \quad \textit{Solve each eq.}$$

The solution set is $\left\{ -\frac{5}{3}, 6 \right\}$.

13. $x^2 = 81$ ▪   By the square root property,
$$x = \sqrt{81} = 9 \quad \text{or} \quad x = -\sqrt{81} = -9.$$

The solution set is $\{-9, 9\}$.

15. $k^2 = 14$ ▪   By the square root property,
$$k = \sqrt{14} \quad \text{or} \quad k = -\sqrt{14}.$$
The solution set is $\{-\sqrt{14}, \sqrt{14}\}$.

17. $t^2 = 48$ ▪   By the square root property,
$$t = \sqrt{48} \quad \text{or} \quad t = -\sqrt{48}.$$

Write $\sqrt{48}$ in simplest form.
$$\sqrt{48} = \sqrt{16} \cdot \sqrt{3} = 4\sqrt{3}$$

The solution set is $\{-4\sqrt{3}, 4\sqrt{3}\}$.

19. $x^2 = \frac{25}{4}$ ▪   By the square root property,
$$x = \sqrt{\tfrac{25}{4}} = \tfrac{5}{2} \quad \text{or} \quad x = -\sqrt{\tfrac{25}{4}} = -\tfrac{5}{2}.$$

The solution set is $\left\{ -\frac{5}{2}, \frac{5}{2} \right\}$.

21. $x^2 = -100$ ▪   Because the square root of -100 is not a real number, there is *no real number solution* for this equation. The solution set is $\emptyset$.

23. $z^2 = 2.25$ ▪   By the square root property,
$$z = \sqrt{2.25} = 1.5 \quad \text{or} \quad z = -\sqrt{2.25} = -1.5.$$

The solution set is $\{-1.5, 1.5\}$.

25.
$$r^2 - 3 = 0$$
$$r^2 = 3 \quad \textit{Add 3.}$$

Now use the square root property.
$$r = \sqrt{3} \quad \text{or} \quad r = -\sqrt{3}$$

The solution set is $\{-\sqrt{3}, \sqrt{3}\}$.

27. $7x^2 = 4$
$$x^2 = \tfrac{4}{7} \quad \textit{Divide by 7.}$$

Now use the square root property.
$$x = \sqrt{\frac{4}{7}} \qquad \text{or} \qquad x = -\sqrt{\frac{4}{7}}$$
$$= \frac{\sqrt{4}}{\sqrt{7}} \cdot \frac{\sqrt{7}}{\sqrt{7}} \qquad\qquad = -\frac{\sqrt{4}}{\sqrt{7}} \cdot \frac{\sqrt{7}}{\sqrt{7}}$$
$$= \frac{2\sqrt{7}}{7} \qquad\qquad\qquad = -\frac{2\sqrt{7}}{7}$$

The solution set is $\left\{ -\frac{2\sqrt{7}}{7}, \frac{2\sqrt{7}}{7} \right\}$.

29. $4x^2 - 72 = 0$
$$4x^2 = 72 \quad \textit{Add 72.}$$
$$x^2 = 18 \quad \textit{Divide by 4.}$$

Now use the square root property.
$$x = \pm \sqrt{18} = \pm \sqrt{9 \cdot 2} = \pm 3\sqrt{2}$$

The solution set is $\{-3\sqrt{2}, 3\sqrt{2}\}$.

31. $3x^2 - 8 = 64$
$$3x^2 = 72 \quad \textit{Add 8.}$$
$$x^2 = 24 \quad \textit{Divide by 3.}$$

Now use the square root property.
$$x = \pm \sqrt{24} = \pm \sqrt{4 \cdot 6} = \pm 2\sqrt{6}$$

The solution set is $\{-2\sqrt{6}, 2\sqrt{6}\}$.

33. $5x^2 + 4 = 8$

$\quad 5x^2 = 4 \quad$ *Subtract 4.*

$\quad\quad x^2 = \frac{4}{5} \quad$ *Divide by 5.*

Now use the square root property.

$$x = \sqrt{\frac{4}{5}} \quad \text{or} \quad x = -\sqrt{\frac{4}{5}}$$

$$= \frac{\sqrt{4}}{\sqrt{5}} \cdot \frac{\sqrt{5}}{\sqrt{5}} \quad\quad = -\frac{\sqrt{4}}{\sqrt{5}} \cdot \frac{\sqrt{5}}{\sqrt{5}}$$

$$= \frac{2\sqrt{5}}{5} \quad\quad\quad = -\frac{2\sqrt{5}}{5}$$

The solution set is $\left\{ -\dfrac{2\sqrt{5}}{5}, \dfrac{2\sqrt{5}}{5} \right\}$.

35. $(x - 3)^2 = 25$

Use the square root property.

$$x - 3 = \sqrt{25} \quad \text{or} \quad x - 3 = -\sqrt{25}$$
$$x - 3 = 5 \quad \text{or} \quad x - 3 = -5$$
$$x = 8 \quad \text{or} \quad x = -2$$

The solution set is $\{-2, 8\}$.

37. $(x + 5)^2 = -13$ ■ Because the square root of -13 is not a real number, there is *no real number solution* for this equation. The solution set is $\emptyset$.

39. $(x - 8)^2 = 27$

Begin by using the square root property.

$$x - 8 = \sqrt{27} \quad \text{or} \quad x - 8 = -\sqrt{27}$$

Now simplify the radical.

$$\sqrt{27} = \sqrt{9} \cdot \sqrt{3} = 3\sqrt{3}$$

$$x - 8 = 3\sqrt{3} \quad \text{or} \quad x - 8 = -3\sqrt{3}$$
$$x = 8 + 3\sqrt{3} \quad \text{or} \quad x = 8 - 3\sqrt{3}$$

The solution set is $\{8 + 3\sqrt{3}, 8 - 3\sqrt{3}\}$.

41. $(3x + 2)^2 = 49$

Use the square root property and solve for x.

$$3x + 2 = \sqrt{49} \quad \text{or} \quad 3x + 2 = -\sqrt{49}$$
$$3x + 2 = 7 \quad \text{or} \quad 3x + 2 = -7$$
$$3x = 5 \quad \text{or} \quad 3x = -9$$
$$x = \frac{5}{3} \quad \text{or} \quad x = -3$$

The solution set is $\{-3, \frac{5}{3}\}$.

43. $(4x - 3)^2 = 9$

Use the square root property and solve for x.

$$4x - 3 = \sqrt{9} \quad \text{or} \quad 4x - 3 = -\sqrt{9}$$
$$4x - 3 = 3 \quad \text{or} \quad 4x - 3 = -3$$
$$4x = 6 \quad \text{or} \quad 4x = 0$$
$$x = \frac{6}{4} = \frac{3}{2} \quad \text{or} \quad x = 0$$

The solution set is $\{0, \frac{3}{2}\}$.

45. $(5 - 2x)^2 = 30$

Use the square root property and solve for x.

$$5 - 2x = \sqrt{30} \quad \text{or} \quad 5 - 2x = -\sqrt{30}$$
$$-2x = -5 + \sqrt{30} \quad \text{or} \quad -2x = -5 - \sqrt{30}$$
$$x = \frac{-5 + \sqrt{30}}{-2} \quad \text{or} \quad x = \frac{-5 - \sqrt{30}}{-2}$$
$$x = \frac{-5 + \sqrt{30}}{-2} \cdot \frac{-1}{-1} \quad \text{or} \quad x = \frac{-5 - \sqrt{30}}{-2} \cdot \frac{-1}{-1}$$
$$x = \frac{5 - \sqrt{30}}{2} \quad \text{or} \quad x = \frac{5 + \sqrt{30}}{2}$$

The solution set is $\left\{ \dfrac{5 + \sqrt{30}}{2}, \dfrac{5 - \sqrt{30}}{2} \right\}$.

47. $(3x + 1)^2 = 18$

Use the square root property and solve for x.

$$3x + 1 = \sqrt{18} \quad \text{or} \quad 3x + 1 = -\sqrt{18}$$
$$3x = -1 + 3\sqrt{2} \quad \text{or} \quad 3x = -1 - 3\sqrt{2}$$
$$\text{*Note that } \sqrt{18} = \sqrt{9 \cdot 2} = 3\sqrt{2}.\text{*}$$
$$x = \frac{-1 + 3\sqrt{2}}{3} \quad \text{or} \quad x = \frac{-1 - 3\sqrt{2}}{3}$$

The solution set is $\left\{ \dfrac{-1 + 3\sqrt{2}}{3}, \dfrac{-1 - 3\sqrt{2}}{3} \right\}$.

49. $(\frac{1}{2}x + 5)^2 = 12$

Use the square root property and solve for x.

$$\tfrac{1}{2}x + 5 = \sqrt{12} \quad \text{or} \quad \tfrac{1}{2}x + 5 = -\sqrt{12}$$
$$\tfrac{1}{2}x = -5 + 2\sqrt{3} \quad \text{or} \quad \tfrac{1}{2}x = -5 - 2\sqrt{3}$$
$$\text{*Note that } \sqrt{12} = \sqrt{4 \cdot 3} = 2\sqrt{3}.\text{*}$$
$$x = 2(-5 + 2\sqrt{3}) \quad \text{or} \quad x = 2(-5 - 2\sqrt{3})$$
$$x = -10 + 4\sqrt{3} \quad \text{or} \quad x = -10 - 4\sqrt{3}$$

The solution set is $\{-10 + 4\sqrt{3}, -10 - 4\sqrt{3}\}$.

51. $(4x - 1)^2 - 48 = 0$

$\quad (4x - 1)^2 = 48 \quad$ *Add 48.*

Use the square root property and solve for x.

$$4x - 1 = \sqrt{48} \quad \text{or} \quad 4x - 1 = -\sqrt{48}$$
$$4x - 1 = 4\sqrt{3} \quad \text{or} \quad 4x - 1 = -4\sqrt{3}$$
$$\text{*Note that } \sqrt{48} = \sqrt{16 \cdot 3} = 4\sqrt{3}.\text{*}$$
$$4x = 1 + 4\sqrt{3} \quad \text{or} \quad 4x = 1 - 4\sqrt{3}$$
$$x = \frac{1 + 4\sqrt{3}}{4} \quad \text{or} \quad x = \frac{1 - 4\sqrt{3}}{4}$$

The solution set is $\left\{ \dfrac{1 + 4\sqrt{3}}{4}, \dfrac{1 - 4\sqrt{3}}{4} \right\}$.

53. $d = 16t^2$

$4 = 16t^2$ *Let d = 4.*

$t^2 = \frac{4}{16} = \frac{1}{4}$ *Divide by 16; reduce.*

By the square root property,

$$t = \sqrt{\frac{1}{4}} = \frac{1}{2} \quad \text{or} \quad t = -\sqrt{\frac{1}{4}} = -\frac{1}{2}.$$

Reject $-\frac{1}{2}$ as a solution, since time cannot be negative. About $\frac{1}{2}$ second elapses between the dropping of the coin and the shot.

55. $A = \pi r^2$

$81\pi = \pi r^2$ *Let A = 81π.*

$81 = r^2$ *Divide by π.*

By the square root property,

$$r = \sqrt{81} = 9 \quad \text{or} \quad r = -\sqrt{81} = -9.$$

Discard -9 since the radius cannot be negative. The radius is 9 inches.

57. $A = P(1 + r)^2$

$110.25 = 100(1 + r)^2$ *Let A = 110.25, P = 100.*

$(1 + r)^2 = \dfrac{110.25}{100} = 1.1025$

$1 + r = \pm\sqrt{1.1025}$ *Square root prop.*

$1 + r = \pm 1.05$

$r = -1 \pm 1.05$

So either $r = -1 + 1.05 = 0.05$ or $r = -1 - 1.05 = -2.05$.

r is a positive interest rate, so reject the solution -2.05. The rate is $r = 0.05$ or 5%.

17.2   Solving Quadratic Equations by Completing the Square

17.2 Margin Exercises

1.   **(a)** $x^2 + 12x + \underline{\quad}$

The middle term, $12x$, must equal $2kx$.

$2kx = 12x$

$k = 6$ *Divide by 2x.*

Thus, $k = 6$ and $k^2 = 36$. The required trinomial is $x^2 + 12x + \underline{36}$, which factors as $(x + 6)^2$.

(b) $x^2 - 14x + \underline{\quad}$

The middle term, $-14x$, must equal $2kx$.

$2kx = -14x$

$k = -7$ *Divide by 2x.*

Thus, $k = -7$ and $k^2 = 49$. The required trinomial is $x^2 - 14x + \underline{49}$, which factors as $(x - 7)^2$.

(c) $x^2 - 2x + \underline{\quad}$

The middle term, $-2x$, must equal $2kx$.

$2kx = -2x$

$k = -1$ *Divide by 2x.*

Thus, $k = -1$ and $k^2 = 1$. The required trinomial is $x^2 - 2x + \underline{1}$, which factors as $(x - 1)^2$.

2. $x^2 - 4x - 1 = 0$

$x^2 - 4x = 1$

$x^2 - 4x + 4 = 1 + 4$ *2kx = −4x, so k = −2 and k² = 4.*

$(x - 2)^2 = 5$

Now use the square root property.

$$x - 2 = \sqrt{5} \qquad \text{or} \quad x - 2 = -\sqrt{5}$$
$$x = 2 + \sqrt{5} \quad \text{or} \qquad x = 2 - \sqrt{5}$$

The solution set is $\{2 + \sqrt{5}, 2 - \sqrt{5}\}$.

3.   **(a)** $x^2 + 4x = 1$

Take half of the coefficient of x and square the result.

$$\tfrac{1}{2}(\underline{4}) = \underline{2}, \quad \text{and} \quad \underline{2}^2 = \underline{4}.$$

Add $\underline{4}$ to each side of the equation, and write the left side as a perfect square.

$$x^2 + 4x + \underline{4} = 1 + \underline{4}$$
$$(x + 2)^2 = 5 \qquad \textit{Factor.}$$

Use the square root property and solve for x.

$$x + 2 = \sqrt{5} \qquad \text{or} \quad x + 2 = -\sqrt{5}$$
$$x = -2 + \sqrt{5} \quad \text{or} \qquad x = -2 - \sqrt{5}$$

The solution set is $\{-2 + \sqrt{5}, -2 - \sqrt{5}\}$.

(b) $z^2 + 6z - 3 = 0$

First add 3 to each side of the equation.

$$z^2 + 6z = 3$$

Now take half the coefficient of z and square the result.

$$\tfrac{1}{2}(6) = 3, \quad \text{and} \quad 3^2 = 9.$$

Add 9 to each side of the equation, and then write the left side as a perfect square.

$$z^2 + 6z + 9 = 3 + 9$$
$$(z + 3)^2 = 12 \qquad \textit{Factor.}$$

Use the square root property and solve for z.

$$z + 3 = \sqrt{12} \qquad \text{or} \quad z + 3 = -\sqrt{12}$$
$$z + 3 = 2\sqrt{3} \qquad \text{or} \quad z + 3 = -2\sqrt{3}$$
$$z = -3 + 2\sqrt{3} \quad \text{or} \qquad z = -3 - 2\sqrt{3}$$

The solution set is $\{-3 + 2\sqrt{3}, -3 - 2\sqrt{3}\}$.

4.   **(a)** $9x^2 + 18x = -5$

Step 1   Divide each side by 9 to get a coefficient of 1 for the x^2-term.

$$x^2 + 2x = -\tfrac{5}{9} \quad \textit{Divide by 9.}$$

Step 2   The equation is in correct form.

Step 3   Take half the coefficient of x, or $(\tfrac{1}{2})(2) = 1$, and square the result: $1^2 = 1$. Then add 1 to each side.

$$x^2 + 2x + 1 = -\tfrac{5}{9} + 1 \quad \textit{Add 1.}$$
$$x^2 + 2x + 1 = \tfrac{4}{9} \qquad \quad 1 = \tfrac{9}{9}$$
$$(x + 1)^2 = \tfrac{4}{9} \qquad \quad \textit{Factor.}$$

Step 4   Apply the square root property and solve for x.

$$x + 1 = \sqrt{\tfrac{4}{9}} \qquad \text{or} \quad x + 1 = -\sqrt{\tfrac{4}{9}}$$
$$x + 1 = \tfrac{2}{3} \qquad \text{or} \quad x + 1 = -\tfrac{2}{3}$$
$$x = -1 + \tfrac{2}{3} \quad \text{or} \quad \; x = -1 - \tfrac{2}{3}$$
$$x = -\tfrac{1}{3} \qquad \text{or} \qquad \; x = -\tfrac{5}{3}$$

The solution set is $\{-\tfrac{1}{3}, -\tfrac{5}{3}\}$.

(b) $4t^2 - 24t + 11 = 0$

Step 1   Divide each side of the equation by 4 to get 1 as the coefficient of the t^2-term.

$$t^2 - 6t + \tfrac{11}{4} = 0 \quad \textit{Divide by 4.}$$

Step 2   Get all the terms with variables on one side of the equality symbol.

$$t^2 - 6t = -\tfrac{11}{4} \quad \textit{Subtract } \tfrac{11}{4}.$$

Step 3   Square half the coefficient of t.

$$\left[\tfrac{1}{2}(-6)\right]^2 = (-3)^2 = 9$$

Then add 9 to each side.

$$t^2 - 6t + 9 = -\tfrac{11}{4} + 9 \quad \textit{Add 9.}$$
$$t^2 - 6t + 9 = \tfrac{25}{4} \qquad \quad 9 = \tfrac{36}{4}$$
$$(t - 3)^2 = \tfrac{25}{4} \qquad \quad \textit{Factor.}$$

Step 4   Use the square root property.

$$t - 3 = \sqrt{\tfrac{25}{4}} \quad \text{or} \quad t - 3 = -\sqrt{\tfrac{25}{4}}$$
$$t - 3 = \tfrac{5}{2} \qquad \text{or} \quad t - 3 = -\tfrac{5}{2}$$
$$t = 3 + \tfrac{5}{2} \quad \text{or} \qquad t = 3 - \tfrac{5}{2}$$
$$t = \tfrac{11}{2} \qquad \text{or} \qquad t = \tfrac{1}{2}$$

The solution set is $\{\tfrac{11}{2}, \tfrac{1}{2}\}$.

5.   **(a)** $3x^2 + 5x - 2 = 0$

Step 1   Divide each side of the equation by 3 to get 1 as the coefficient of the x^2-term.

$$x^2 + \tfrac{5}{3}x - \tfrac{2}{3} = 0 \quad \textit{Divide by 3.}$$

Step 2   Get all the terms with variables on one side of the equality symbol.

$$x^2 + \tfrac{5}{3}x = \tfrac{2}{3} \quad \textit{Add } \tfrac{2}{3}.$$

Step 3   Square half the coefficient of t.

$$\left[\tfrac{1}{2}\left(\tfrac{5}{3}\right)\right]^2 = \left(\tfrac{5}{6}\right)^2 = \tfrac{25}{36}$$

Then add $\tfrac{25}{36}$ to each side.

$$x^2 + \tfrac{5}{3}x + \tfrac{25}{36} = \tfrac{2}{3} + \tfrac{25}{36} \quad \textit{Add } \tfrac{25}{36}.$$
$$x^2 + \tfrac{5}{3}x + \tfrac{25}{36} = \tfrac{21}{36} + \tfrac{25}{36} \quad \tfrac{2}{3} = \tfrac{24}{36}$$
$$\left(x + \tfrac{5}{6}\right)^2 = \tfrac{49}{36} \qquad \quad \textit{Factor.}$$

Step 4   Use the square root property.

$$x + \tfrac{5}{6} = \sqrt{\tfrac{49}{36}} \quad \text{or} \quad x + \tfrac{5}{6} = -\sqrt{\tfrac{49}{36}}$$
$$x + \tfrac{5}{6} = \tfrac{7}{6} \qquad \text{or} \quad x + \tfrac{5}{6} = -\tfrac{7}{6}$$
$$x = \tfrac{2}{6} \qquad \text{or} \qquad x = -\tfrac{12}{6}$$
$$x = \tfrac{1}{3} \qquad \text{or} \qquad x = -2$$

The solution set is $\{-2, \tfrac{1}{3}\}$.

(b) $2x^2 - 4x - 1 = 0$

Step 1   Divide each side of the equation by 2 to get a coefficient of 1 for the x^2-term.

$$x^2 - 2x - \tfrac{1}{2} = 0$$

Step 2   Get all the terms with variables on one side of the equality symbol.

$$x^2 - 2x = \tfrac{1}{2} \quad \textit{Add } \tfrac{1}{2}.$$

Step 3   Square half the coefficient of x.

$$\tfrac{1}{2}(-2) = -1, \quad \text{and} \quad (-1)^2 = 1.$$

Add 1 to each side of the equation, and then write the left side as a perfect square.

$$x^2 - 2x + 1 = \tfrac{1}{2} + 1$$
$$(x - 1)^2 = \tfrac{3}{2}$$

Step 4   Use the square root property.

$$x - 1 = \sqrt{\tfrac{3}{2}} \quad \text{or} \quad x - 1 = -\sqrt{\tfrac{3}{2}}$$

Simplify the radical.

$$\sqrt{\tfrac{3}{2}} = \frac{\sqrt{3}}{\sqrt{2}} = \frac{\sqrt{3} \cdot \sqrt{2}}{\sqrt{2} \cdot \sqrt{2}} = \frac{\sqrt{6}}{2}$$

Thus,

$$x - 1 = \frac{\sqrt{6}}{2} \qquad \text{or} \quad x - 1 = -\frac{\sqrt{6}}{2}$$
$$x = 1 + \frac{\sqrt{6}}{2} \quad \text{or} \qquad x = 1 - \frac{\sqrt{6}}{2}$$
$$x = \frac{2}{2} + \frac{\sqrt{6}}{2} \quad \text{or} \qquad x = \frac{2}{2} - \frac{\sqrt{6}}{2}$$
$$x = \frac{2 + \sqrt{6}}{2} \quad \text{or} \qquad x = \frac{2 - \sqrt{6}}{2}.$$

The solution set is $\left\{\dfrac{2 + \sqrt{6}}{2}, \dfrac{2 - \sqrt{6}}{2}\right\}$.

6. $5x^2 + 3x + 1 = 0$

$x^2 + \frac{3}{5}x + \frac{1}{5} = 0$ *Divide by 5.*

$x^2 + \frac{3}{5}x = -\frac{1}{5}$ *Subtract $\frac{1}{5}$.*

$x^2 + \frac{3}{5}x + \frac{9}{100} = -\frac{1}{5} + \frac{9}{100}$ *Add $(\frac{1}{2} \cdot \frac{3}{5})^2 = \frac{9}{100}$.*

$(x + \frac{3}{10})^2 = -\frac{11}{100}$ $-\frac{1}{5} = -\frac{20}{100}$

The square root of $-\frac{11}{100}$ is not a real number, so the square root property does not apply. This equation has *no real number solution*. The solution set is $\emptyset$.

7.   **(a)** $r(r - 3) = -1$

$r^2 - 3r = -1$

$r^2 - 3r + \frac{9}{4} = -1 + \frac{9}{4}$ *Add $[\frac{1}{2}(-3)]^2 = \frac{9}{4}$.*

$(r - \frac{3}{2})^2 = \frac{5}{4}$

Use the square root property.

$r - \frac{3}{2} = \frac{\sqrt{5}}{2}$ or $r - \frac{3}{2} = -\frac{\sqrt{5}}{2}$

$r = \frac{3 + \sqrt{5}}{2}$ or $r = \frac{3 - \sqrt{5}}{2}$

The solution set is $\left\{ \frac{3 + \sqrt{5}}{2}, \frac{3 - \sqrt{5}}{2} \right\}$.

(b) $(x + 2)(x + 1) = 5$

$x^2 + 3x + 2 = 5$

$x^2 + 3x = 3$

$x^2 + 3x + \frac{9}{4} = 3 + \frac{9}{4}$ *Add $[\frac{1}{2}(3)]^2 = \frac{9}{4}$.*

$(x + \frac{3}{2})^2 = \frac{21}{4}$

Use the square root property.

$x + \frac{3}{2} = \sqrt{\frac{21}{4}}$ or $x + \frac{3}{2} = -\sqrt{\frac{21}{4}}$

$x + \frac{3}{2} = \frac{\sqrt{21}}{2}$ or $x + \frac{3}{2} = -\frac{\sqrt{21}}{2}$

$x = -\frac{3}{2} + \frac{\sqrt{21}}{2}$ or $x = -\frac{3}{2} - \frac{\sqrt{21}}{2}$

The solution set is $\left\{ \frac{-3 + \sqrt{21}}{2}, \frac{-3 - \sqrt{21}}{2} \right\}$.

8.   **(a)** $-16t^2 + 128t = s$ *Given*

$-16t^2 + 128t = 48$ *Let s = 48.*

$t^2 - 8t = -3$ *Divide by −16.*

$t^2 - 8t + 16 = -3 + 16$ *Add $[\frac{1}{2}(-8)]^2 = 16$.*

$(t - 4)^2 = 13$ *Factor.*

Use the square root property.

$t - 4 = \sqrt{13}$ or $t - 4 = -\sqrt{13}$

$t = 4 + \sqrt{13}$ or $t = 4 - \sqrt{13}$

$t \approx 7.6$ or $t \approx 0.4$

The ball will be 48 feet above the ground after about 0.4 second and again after about 7.6 seconds.

(b) $-16t^2 + 64t = s$ *Given*

$-16t^2 + 64t = 28$ *Let s = 28.*

$t^2 - 4t = -\frac{7}{4}$ *Divide by −16.*

$t^2 - 4t + 4 = -\frac{7}{4} + 4$ *Add $[\frac{1}{2}(-4)]^2 = 4$.*

$(t - 2)^2 = \frac{9}{4}$ *Factor.*

Use the square root property.

$t - 2 = \sqrt{\frac{9}{4}}$ or $t - 2 = -\sqrt{\frac{9}{4}}$

$t = 2 + \frac{3}{2}$ or $t = 2 - \frac{3}{2}$

$t = 3.5$ or $t = 0.5$

The ball will be 28 feet above the ground after 0.5 second and again after 3.5 seconds.

17.2   Section Exercises

1. $2x^2 - 4x = 9$ ▪ Before completing the square, the coefficient of x^2 must be 1. Dividing each side of the equation by 2 is the correct way to begin solving the equation, and this corresponds to choice **D**.

3. $x^2 + 10x + \underline{\quad}$

The middle term, $10x$, must equal $2kx$.

$2kx = 10x$

$k = 5$ *Divide by 2x.*

Thus, $k = 5$ and $k^2 = 25$. The required trinomial is $x^2 + 10x + \underline{25}$, which factors as $(x + 5)^2$.

5. $x^2 + 2x + \underline{\quad}$

The middle term, $2x$, must equal $2kx$.

$2kx = 2x$

$k = 1$ *Divide by 2x.*

Thus, $k = 1$ and $k^2 = 1$. The required trinomial is $x^2 + 2x + \underline{1}$, which factors as $(x + 1)^2$.

7. $p^2 - 5p + \underline{\quad}$

The middle term, $-5p$, must equal $2kp$.

$2kp = -5p$

$k = -\frac{5}{2}$ *Divide by 2p.*

Thus, $k = -\frac{5}{2}$ and $k^2 = \frac{25}{4}$. The required trinomial is $p^2 - 5p + \underline{\frac{25}{4}}$, which factors as $(p - \frac{5}{2})^2$.

9. $x^2 - 4x = -3$ ■ Take half of the coefficient of x and square it. Half of -4 is -2, and $(-2)^2 = 4$. Add 4 to each side of the equation, and write the left side as a perfect square.

$$x^2 - 4x + 4 = -3 + 4$$
$$(x - 2)^2 = 1$$

Use the square root property.

$$x - 2 = \sqrt{1} \quad \text{or} \quad x - 2 = -\sqrt{1}$$
$$x - 2 = 1 \quad \text{or} \quad x - 2 = -1$$
$$x = 3 \quad \text{or} \quad x = 1$$

A check verifies that the solution set is $\{1, 3\}$.

11. $x^2 + 5x + 6 = 0$ ■ Subtract 6 from each side.

$$x^2 + 5x = -6$$

Take half the coefficient of x, square it, and add this number to each side.

$$x^2 + 5x + \tfrac{25}{4} = -6 + \tfrac{25}{4} \quad \textit{Add } \left[\tfrac{1}{2}(5)\right]^2 = \tfrac{25}{4}.$$
$$(x + \tfrac{5}{2})^2 = \tfrac{1}{4} \qquad \textit{Factor.}$$

Use the square root property.

$$x + \tfrac{5}{2} = \sqrt{\tfrac{1}{4}} \quad \text{or} \quad x + \tfrac{5}{2} = -\sqrt{\tfrac{1}{4}}$$
$$x + \tfrac{5}{2} = \tfrac{1}{2} \quad \text{or} \quad x + \tfrac{5}{2} = -\tfrac{1}{2}$$
$$x = -\tfrac{5}{2} + \tfrac{1}{2} \quad \text{or} \quad x = -\tfrac{5}{2} - \tfrac{1}{2}$$
$$x = -\tfrac{4}{2} \quad \text{or} \quad x = -\tfrac{6}{2}$$
$$x = -2 \quad \text{or} \quad x = -3$$

A check verifies that the solution set is $\{-2, -3\}$.

13. $x^2 + 2x - 5 = 0$ ■ Add 5 to each side.

$$x^2 + 2x = 5$$

Take half the coefficient of x and square it.

$$\tfrac{1}{2}(2) = 1, \quad \text{and} \quad 1^2 = 1.$$

Add 1 to each side of the equation, and write the left side as a perfect square.

$$x^2 + 2x + 1 = 5 + 1$$
$$(x + 1)^2 = 6$$

Use the square root property.

$$x + 1 = \sqrt{6} \quad \text{or} \quad x + 1 = -\sqrt{6}$$
$$x = -1 + \sqrt{6} \quad \text{or} \quad x = -1 - \sqrt{6}$$

A check verifies that the solution set is $\{-1 + \sqrt{6}, -1 - \sqrt{6}\}$. Using a calculator for your check is highly recommended.

15.
$$x^2 - 8x = -4$$
$$x^2 - 8x + 16 = -4 + 16 \quad \left[\tfrac{1}{2}(-8)\right]^2 = 16$$
$$(x - 4)^2 = 12 \qquad \textit{Factor.}$$

Use the square root property.

$$x - 4 = \sqrt{12} \quad \text{or} \quad x - 4 = -\sqrt{12}$$
$$x - 4 = \sqrt{4} \cdot \sqrt{3} \quad \text{or} \quad x - 4 = -\sqrt{4} \cdot \sqrt{3}$$
$$x = 4 + 2\sqrt{3} \quad \text{or} \quad x = 4 - 2\sqrt{3}$$

The solution set is $\{4 + 2\sqrt{3}, 4 - 2\sqrt{3}\}$.

17. $t^2 + 6t + 9 = 0$ ■ The left-hand side of this equation is already a perfect square.

$$(t + 3)^2 = 0 \qquad \textit{Factor.}$$
$$t + 3 = 0 \qquad \textit{Take square root}$$
$$t = -3$$

A check verifies that the solution set is $\{-3\}$.

19. $x^2 + x - 1 = 0$ ■ Add 1 to each side.

$$x^2 + x = 1$$

Take half of 1, square it, and add it to each side.

$$x^2 + x + \tfrac{1}{4} = 1 + \tfrac{1}{4} \quad \textit{Add } \left[\tfrac{1}{2}(1)\right]^2 = \tfrac{1}{4}.$$
$$(x + \tfrac{1}{2})^2 = \tfrac{5}{4} \qquad \textit{Factor.}$$

Use the square root property.

$$x + \tfrac{1}{2} = \sqrt{\tfrac{5}{4}} \quad \text{or} \quad x + \tfrac{1}{2} = -\sqrt{\tfrac{5}{4}}$$
$$x + \tfrac{1}{2} = \tfrac{\sqrt{5}}{2} \quad \text{or} \quad x + \tfrac{1}{2} = -\tfrac{\sqrt{5}}{2}$$
$$x = \tfrac{-1 + \sqrt{5}}{2} \quad \text{or} \quad x = \tfrac{-1 - \sqrt{5}}{2}$$

A check verifies that the solution set is

$$\left\{ \frac{-1 + \sqrt{5}}{2}, \frac{-1 - \sqrt{5}}{2} \right\}.$$

21. $4x^2 + 4x - 3 = 0$ ■ Add 3 to each side.

$$4x^2 + 4x = 3$$

Divide each side by 4 so that the coefficient of x^2 is 1.

$$x^2 + x = \tfrac{3}{4}$$

The coefficient of x is 1. Take half of 1, square the result, and add this square to each side.

$$\tfrac{1}{2}(1) = \tfrac{1}{2} \quad \text{and} \quad (\tfrac{1}{2})^2 = \tfrac{1}{4}$$
$$x^2 + x + \tfrac{1}{4} = \tfrac{3}{4} + \tfrac{1}{4}$$

The left-hand side can then be written as a perfect square.

$$(x + \tfrac{1}{2})^2 = 1$$

Use the square root property.

$$x + \tfrac{1}{2} = 1 \quad \text{or} \quad x + \tfrac{1}{2} = -1$$
$$x = -\tfrac{1}{2} + 1 \quad \text{or} \quad x = -\tfrac{1}{2} - 1$$
$$x = \tfrac{1}{2} \quad \text{or} \quad x = -\tfrac{3}{2}$$

A check verifies that the solution set is $\{-\tfrac{3}{2}, \tfrac{1}{2}\}$.

23.
$$2x^2 - 4x = 5$$
$$x^2 - 2x = \tfrac{5}{2} \qquad \textit{Divide by 2.}$$
$$x^2 - 2x + 1 = \tfrac{5}{2} + 1 \qquad \textit{Add } \left[\tfrac{1}{2}(-2)\right]^2 = 1.$$
$$(x - 1)^2 = \tfrac{7}{2} \qquad \textit{Factor.}$$

Use the square root property.

$$x - 1 = \sqrt{\tfrac{7}{2}} \quad \text{or} \quad x - 1 = -\sqrt{\tfrac{7}{2}}$$

Simplify the radical:

$$\sqrt{\tfrac{7}{2}} = \frac{\sqrt{7}}{\sqrt{2}} = \frac{\sqrt{7}}{\sqrt{2}} \cdot \frac{\sqrt{2}}{\sqrt{2}} = \frac{\sqrt{14}}{2}$$

$$x - 1 = \frac{\sqrt{14}}{2} \qquad \text{or} \qquad x - 1 = -\frac{\sqrt{14}}{2}$$

$$x = 1 + \frac{\sqrt{14}}{2} \qquad \text{or} \qquad x = 1 - \frac{\sqrt{14}}{2}$$

$$x = \frac{2}{2} + \frac{\sqrt{14}}{2} \qquad \text{or} \qquad x = \frac{2}{2} - \frac{\sqrt{14}}{2}$$

$$x = \frac{2 + \sqrt{14}}{2} \qquad \text{or} \qquad x = \frac{2 - \sqrt{14}}{2}$$

A check verifies that the solution set is
$$\left\{ \frac{2 + \sqrt{14}}{2}, \frac{2 - \sqrt{14}}{2} \right\}.$$

25. $2p^2 - 2p + 3 = 0$ ■   Divide each side by 2.

$$p^2 - p + \tfrac{3}{2} = 0$$

Subtract $\tfrac{3}{2}$ from each side.

$$p^2 - p = -\tfrac{3}{2}$$

Take half the coefficient of p and square it.

$$\tfrac{1}{2}(-1) = -\tfrac{1}{2}, \quad \text{and} \quad \left(-\tfrac{1}{2}\right)^2 = \tfrac{1}{4}.$$

Add $\tfrac{1}{4}$ to each side of the equation.

$$p^2 - p + \tfrac{1}{4} = -\tfrac{3}{2} + \tfrac{1}{4}$$

Factor on the left side and add on the right.

$$\left(p - \tfrac{1}{2}\right)^2 = -\tfrac{5}{4}$$

The square root of $-\tfrac{5}{4}$ is not a real number, so there is *no real number solution*. The solution set is $\emptyset$.

27. $3k^2 + 7k = 4$ ■   Divide each side by 3.

$$k^2 + \tfrac{7}{3}k = \tfrac{4}{3}$$

Take half of the coefficient of k and square it.

$$\tfrac{1}{2}\left(\tfrac{7}{3}\right) = \tfrac{7}{6} \quad \text{and} \quad \left(\tfrac{7}{6}\right)^2 = \tfrac{49}{36}.$$

Add $\tfrac{49}{36}$ to each side of the equation.

$$k^2 + \tfrac{7}{3}k + \tfrac{49}{36} = \tfrac{4}{3} + \tfrac{49}{36}$$

$$\left(k + \tfrac{7}{6}\right)^2 = \tfrac{97}{36}$$

Use the square root property.

$$k + \frac{7}{6} = \sqrt{\frac{97}{36}} \qquad \text{or} \qquad k + \frac{7}{6} = -\sqrt{\frac{97}{36}}$$

$$k + \frac{7}{6} = \frac{\sqrt{97}}{6} \qquad \text{or} \qquad k + \frac{7}{6} = -\frac{\sqrt{97}}{6}$$

$$k = -\frac{7}{6} + \frac{\sqrt{97}}{6} \qquad \text{or} \qquad k = -\frac{7}{6} - \frac{\sqrt{97}}{6}$$

$$k = \frac{-7 + \sqrt{97}}{6} \qquad \text{or} \qquad k = \frac{-7 - \sqrt{97}}{6}$$

A check verifies that the solution set is
$$\left\{ \frac{-7 + \sqrt{97}}{6}, \frac{-7 - \sqrt{97}}{6} \right\}.$$

29. $(x + 3)(x - 1) = 5$
$$x^2 + 2x - 3 = 5$$
$$x^2 + 2x = 8$$
$$x^2 + 2x + 1 = 8 + 1$$
$$(x + 1)^2 = 9$$

Use the square root property.

$$x + 1 = 3 \qquad \text{or} \qquad x + 1 = -3$$
$$x = 2 \qquad \text{or} \qquad x = -4$$

A check verifies that the solution set is $\{-4, 2\}$.

31. $(r - 3)(r - 5) = 2$
$$r^2 - 8r + 15 = 2$$
$$r^2 - 8r = -13$$
$$r^2 - 8r + 16 = -13 + 16$$
$$(r - 4)^2 = 3$$

Use the square root property.

$$r - 4 = \sqrt{3} \qquad \text{or} \qquad r - 4 = -\sqrt{3}$$
$$r = 4 + \sqrt{3} \qquad \text{or} \qquad r = 4 - \sqrt{3}$$

A check verifies that the solution set is
$\{4 + \sqrt{3}, 4 - \sqrt{3}\}$.

33. $-x^2 + 2x = -5$ ■   Divide each side by -1.

$$x^2 - 2x = 5$$

Take half of the coefficient of x and square it. Half of -2 is -1, and $(-1)^2 = 1$. Add 1 to each side of the equation, and write the left side as a perfect square.

$$x^2 - 2x + 1 = 5 + 1$$
$$(x - 1)^2 = 6$$

Use the square root property.

$$x - 1 = \sqrt{6} \qquad \text{or} \qquad x - 1 = -\sqrt{6}$$
$$x = 1 + \sqrt{6} \qquad \text{or} \qquad x = 1 - \sqrt{6}$$

A check verifies that the solution set is
$\{1 + \sqrt{6}, 1 - \sqrt{6}\}$.

35.

$$s = -16t^2 + 96t \quad \textit{Given}$$
$$80 = -16t^2 + 96t \quad \textit{Let } s = 80.$$
$$-16t^2 + 96t = 80$$
$$t^2 - 6t = -5 \quad \textit{Divide by } -16.$$
$$t^2 - 6t + 9 = -5 + 9 \quad \textit{Add 9.}$$
$$(t - 3)^2 = 4 \quad \textit{Factor; add.}$$

Use the square root property.

$$t - 3 = \sqrt{4} \quad \text{or} \quad t - 3 = -\sqrt{4}$$
$$t = 3 + 2 \quad \text{or} \quad t = 3 - 2$$
$$t = 5 \quad \text{or} \quad t = 1$$

The object will reach a height of 80 feet at 1 second (on the way up) and at 5 seconds (on the way down).

37.

$$s = -13t^2 + 104t$$
$$195 = -13t^2 + 104t \quad \textit{Let } s = 195.$$
$$-15 = t^2 - 8t \quad \textit{Divide by } -13.$$
$$t^2 - 8t + 16 = -15 + 16$$
$$\textit{Add } \left[\tfrac{1}{2}(-8)\right]^2 = 16.$$
$$(t - 4)^2 = 1 \quad \textit{Factor; add.}$$

Use the square root property.

$$t - 4 = \sqrt{1} \quad \text{or} \quad t - 4 = -\sqrt{1}$$
$$t = 4 + 1 \quad \text{or} \quad t = 4 - 1$$
$$t = 5 \quad \text{or} \quad t = 3$$

The object will be at a height of 195 feet at 3 seconds (on the way up) and at 5 seconds (on the way down).

39. Let $x = $ the width of the pen.
Then $175 - x = $ the length of the pen.
Use the formula for the area of a rectangle.

$$A = lw$$
$$7500 = (175 - x)x$$
$$7500 = 175x - x^2$$
$$x^2 - 175x + 7500 = 0$$

Solve this quadratic equation by completing the square.

$$x^2 - 175x = -7500$$
$$x^2 - 175x + \tfrac{30{,}625}{4} = -\tfrac{30{,}000}{4} + \tfrac{30{,}625}{4}$$
$$\textit{Add } (\tfrac{175}{2})^2 = \tfrac{30{,}625}{4}.$$
$$(x - \tfrac{175}{2})^2 = \tfrac{625}{4}$$

Use the square root property.

$$x - \tfrac{175}{2} = \sqrt{\tfrac{625}{4}} \quad \text{or} \quad x - \tfrac{175}{2} = -\sqrt{\tfrac{625}{4}}$$
$$x = \tfrac{175}{2} + \tfrac{25}{2} \quad \text{or} \quad x = \tfrac{175}{2} - \tfrac{25}{2}$$
$$x = \tfrac{200}{2} \quad \text{or} \quad x = \tfrac{150}{2}$$
$$x = 100 \quad \text{or} \quad x = 75$$

If $x = 100$, $175 - x = 175 - 100 = 75$.
If $x = 75$, $175 - x = 175 - 75 = 100$.
The dimensions of the pen are 75 feet by 100 feet.

17.3 Solving Quadratic Equations by the Quadratic Formula

17.3 Margin Exercises

1. **(a)** $5x^2 + 2x - 1 = 0$ has the form of the standard quadratic equation $ax^2 + bx + c = 0$.

Here, $a = 5$, $b = 2$, and $c = -1$.

(b) $3x^2 = x - 2$

Rewrite in $ax^2 + bx + c = 0$ form.

$$3x^2 - x + 2 = 0$$

Here, $a = 3$, $b = -1$, and $c = 2$.

(c) $9x^2 - 13 = 0$ has the form of the standard quadratic equation $ax^2 + bx + c = 0$.

Here, $a = 9$, $b = 0$, and $c = -13$.

(d) $-x^2 + x = 0$

Rewrite in $ax^2 + bx + c = 0$ form.

$$-x^2 + x + 0 = 0$$

Here, $a = -1$, $b = 1$, and $c = 0$.

(e) $(3x + 2)(x - 1) = 8$
$$3x^2 - x - 2 = 8 \quad \textit{FOIL}$$
$$3x^2 - x - 10 = 0 \quad \textit{Subtract 8.}$$

Here, $a = 3$, $b = -1$, and $c = -10$.

2. **(a)** $2x^2 + 3x - 5 = 0$ ▪ The quadratic equation is in standard form, so $a = 2$, $b = 3$, and $c = -5$. Substitute these values in the quadratic formula.

$$x = \frac{-b \pm \sqrt{b^2 - 4ac}}{2a}$$
$$x = \frac{-3 \pm \sqrt{3^2 - 4(2)(-5)}}{2(2)}$$
$$x = \frac{-3 \pm \sqrt{9 + 40}}{4}$$
$$x = \frac{-3 \pm \sqrt{49}}{4}$$
$$x = \frac{-3 \pm 7}{4}$$
$$x = \frac{-3 + 7}{4} \quad \text{or} \quad x = \frac{-3 - 7}{4}$$
$$x = \tfrac{4}{4} \quad \text{or} \quad x = \tfrac{-10}{4}$$
$$x = 1 \quad \text{or} \quad x = -\tfrac{5}{2}$$

The solution set is $\{1, -\tfrac{5}{2}\}$.

(b) $6x^2 + x - 1 = 0$ ■   Substitute $a = 6$, $b = 1$, and $c = -1$ in the quadratic formula.

$$x = \frac{-b \pm \sqrt{b^2 - 4ac}}{2a}$$

$$x = \frac{-1 \pm \sqrt{1^2 - 4(6)(-1)}}{2(6)}$$

$$x = \frac{-1 \pm \sqrt{1 + 24}}{12}$$

$$x = \frac{-1 \pm 5}{12}$$

$$x = \frac{-1 + 5}{12} \quad \text{or} \quad x = \frac{-1 - 5}{12}$$

$$x = \frac{4}{12} \quad \text{or} \quad x = \frac{-6}{12}$$

$$x = \frac{1}{3} \quad \text{or} \quad x = -\frac{1}{2}$$

The solution set is $\{\frac{1}{3}, -\frac{1}{2}\}$.

3. $x^2 + 1 = -8x$ ■   Write the equation in standard form.

$$x^2 + 8x + 1 = 0$$

Substitute $a = 1$, $b = 8$, and $c = 1$ in the quadratic formula.

$$x = \frac{-b \pm \sqrt{b^2 - 4ac}}{2a}$$

$$x = \frac{-8 \pm \sqrt{8^2 - 4(1)(1)}}{2(1)}$$

$$x = \frac{-8 \pm \sqrt{64 - 4}}{2} = \frac{-8 \pm \sqrt{60}}{2}$$

$$= \frac{-8 \pm \sqrt{4} \cdot \sqrt{15}}{2} = \frac{-8 \pm 2\sqrt{15}}{2}$$

$$= \frac{2(-4 \pm \sqrt{15})}{2} = -4 \pm \sqrt{15}$$

The solution set is $\{-4 + \sqrt{15}, -4 - \sqrt{15}\}$.

4. $9x^2 - 12x + 4 = 0$ ■   Substitute $a = 9$, $b = -12$, and $c = 4$ in the quadratic formula.

$$x = \frac{-b \pm \sqrt{b^2 - 4ac}}{2a}$$

$$x = \frac{-(-12) \pm \sqrt{(-12)^2 - 4(9)(4)}}{2(9)}$$

$$x = \frac{12 \pm \sqrt{144 - 144}}{18}$$

$$x = \frac{12}{18} = \frac{2}{3}$$

In this case, $b^2 - 4ac = 0$, so there is just one solution, $\frac{2}{3}$, and the trinomial $9x^2 - 12x + 4$ is a perfect square. The solution set is $\{\frac{2}{3}\}$.

5.   **(a)** $x^2 - \frac{4}{3}x + \frac{2}{3} = 0$

$$3x^2 - 4x + 2 = 0 \quad \textit{Multiply by 3.}$$

Substitute $a = 3$, $b = -4$, and $c = 2$ in the quadratic formula.

$$x = \frac{-b \pm \sqrt{b^2 - 4ac}}{2a}$$

$$x = \frac{-(-4) \pm \sqrt{(-4)^2 - 4(3)(2)}}{2(3)}$$

$$x = \frac{4 \pm \sqrt{16 - 24}}{6}$$

$$x = \frac{4 \pm \sqrt{-8}}{6}$$

Because $\sqrt{-8}$ does not represent a real number, there is *no real number solution*. The solution set is $\emptyset$.

(b) $x^2 - \frac{9}{5}x = \frac{2}{5}$

$$5x^2 - 9x = 2 \quad \textit{Multiply by 5.}$$

$$5x^2 - 9x - 2 = 0 \quad \textit{Subtract 2.}$$

Substitute $a = 5$, $b = -9$, and $c = -2$ in the quadratic formula.

$$x = \frac{-b \pm \sqrt{b^2 - 4ac}}{2a}$$

$$x = \frac{-(-9) \pm \sqrt{(-9)^2 - 4(5)(-2)}}{2(5)}$$

$$x = \frac{9 \pm \sqrt{81 + 40}}{10}$$

$$x = \frac{9 \pm \sqrt{121}}{10}$$

$$x = \frac{9 \pm 11}{10}$$

$$x = \frac{9 + 11}{10} = 2 \quad \text{or} \quad x = \frac{9 - 11}{10} = -\frac{1}{5}$$

The solution set is $\{-\frac{1}{5}, 2\}$.

17.3   Section Exercises

1. $2a$ should be the denominator for $-b$ as well. The correct formula is

$$x = \frac{-b \pm \sqrt{b^2 - 4ac}}{2a}.$$

3.   In $4x^2 + 5x - 9 = 0$, the coefficient of the x^2-term is 4, so $a = 4$. The coefficient of the x-term is 5, so $b = 5$. The constant is -9, so $c = -9$. (Note that one side of the equation was equal to 0 before we started.)

5. $3x^2 = 4x + 2$ ■   First, write the equation in standard form, $ax^2 + bx + c = 0$.

$$3x^2 - 4x - 2 = 0$$

Now, identify the values: $a = 3$, $b = -4$, and $c = -2$.

7. $3x^2 = -7x$ ▪   First, write the equation in standard form, $ax^2 + bx + c = 0$.

$$3x^2 + 7x + 0 = 0$$

Now, identify the values: $a = 3$, $b = 7$, and $c = 0$.

9. $k^2 + 12k - 13 = 0$ ▪   Substitute $a = 1$, $b = 12$, and $c = -13$ in the quadratic formula.

$$k = \frac{-b \pm \sqrt{b^2 - 4ac}}{2a}$$

$$k = \frac{-12 \pm \sqrt{12^2 - 4(1)(-13)}}{2(1)}$$

$$= \frac{-12 \pm \sqrt{144 + 52}}{2}$$

$$= \frac{-12 \pm \sqrt{196}}{2}$$

$$= \frac{-12 \pm 14}{2}$$

$$k = \frac{-12 + 14}{2} = \frac{2}{2} = 1$$

$$\text{or}\quad k = \frac{-12 - 14}{2} = \frac{-26}{2} = -13$$

A check verifies that the solution set is $\{-13, 1\}$.

11. $p^2 - 4p + 4 = 0$ ▪   Substitute $a = 1$, $b = -4$, and $c = 4$ in the quadratic formula.

$$p = \frac{-b \pm \sqrt{b^2 - 4ac}}{2a}$$

$$p = \frac{-(-4) \pm \sqrt{(-4)^2 - 4(1)(4)}}{2(1)}$$

$$= \frac{4 \pm \sqrt{16 - 16}}{2}$$

$$= \frac{4 \pm 0}{2} = \frac{4}{2} = 2.$$

A check verifies that the solution set is $\{2\}$. Note that the discriminant is 0.

13. $2x^2 = 5 + 3x$ ▪   First, write the equation in standard form, $ax^2 + bx + c = 0$.

$$2x^2 - 3x - 5 = 0$$

Substitute $a = 2$, $b = -3$, and $c = -5$ in the quadratic formula.

$$x = \frac{-b \pm \sqrt{b^2 - 4ac}}{2a}$$

$$x = \frac{-(-3) \pm \sqrt{(-3)^2 - 4(2)(-5)}}{2(2)}$$

$$= \frac{3 \pm \sqrt{9 + 40}}{4}$$

$$= \frac{3 \pm \sqrt{49}}{4}$$

$$= \frac{3 \pm 7}{4}$$

$$x = \frac{3 + 7}{4} = \frac{10}{4} = \frac{5}{2}$$

$$\text{or}\quad x = \frac{3 - 7}{4} = \frac{-4}{4} = -1$$

A check verifies that the solution set is $\{-1, \frac{5}{2}\}$.

15. $2x^2 + 12x = -5$ ▪   First, write the equation in standard form, $ax^2 + bx + c = 0$.

$$2x^2 + 12x + 5 = 0$$

Substitute $a = 2$, $b = 12$, and $c = 5$ in the quadratic formula.

$$x = \frac{-b \pm \sqrt{b^2 - 4ac}}{2a}$$

$$x = \frac{-12 \pm \sqrt{12^2 - 4(2)(5)}}{2(2)}$$

$$= \frac{-12 \pm \sqrt{144 - 40}}{4}$$

$$= \frac{-12 \pm \sqrt{104}}{4}$$

$$= \frac{-12 \pm \sqrt{4} \cdot \sqrt{26}}{4}$$

$$= \frac{-12 \pm 2\sqrt{26}}{4}$$

$$= \frac{2(-6 \pm \sqrt{26})}{2 \cdot 2}$$

$$= \frac{-6 \pm \sqrt{26}}{2}$$

A check verifies that the solution set is

$$\left\{ \frac{-6 + \sqrt{26}}{2}, \frac{-6 - \sqrt{26}}{2} \right\}.$$

17. $6x^2 + 6x = 0$ ▪   Substitute $a = 6$, $b = 6$, and $c = 0$ in the quadratic formula.

$$x = \frac{-b \pm \sqrt{b^2 - 4ac}}{2a}$$

$$x = \frac{-6 \pm \sqrt{6^2 - 4(6)(0)}}{2(6)}$$

$$= \frac{-6 \pm \sqrt{36 - 0}}{12}$$

$$= \frac{-6 \pm 6}{12}$$

$$x = \frac{-6 + 6}{12} = \frac{0}{12} = 0$$

$$\text{or}\quad x = \frac{-6 - 6}{12} = \frac{-12}{12} = -1$$

A check verifies that the solution set is $\{-1, 0\}$.

19. $-2x^2 = -3x + 2$ ▪ First, write the equation in standard form, $ax^2 + bx + c = 0$.

$$-2x^2 + 3x - 2 = 0$$

Substitute $a = -2$, $b = 3$, and $c = -2$ in the quadratic formula.

$$x = \frac{-b \pm \sqrt{b^2 - 4ac}}{2a}$$
$$x = \frac{-3 \pm \sqrt{3^2 - 4(-2)(-2)}}{2(-2)}$$
$$= \frac{-3 \pm \sqrt{9 - 16}}{-4}$$
$$= \frac{-3 \pm \sqrt{-7}}{-4}$$

Because $\sqrt{-7}$ does not represent a real number, there is *no real number solution.* The solution set is $\emptyset$.

21. $3x^2 + 5x + 1 = 0$ ▪ Substitute $a = 3$, $b = 5$, and $c = 1$ in the quadratic formula.

$$x = \frac{-5 \pm \sqrt{5^2 - 4(3)(1)}}{2(3)}$$
$$x = \frac{-5 \pm \sqrt{25 - 12}}{6}$$
$$x = \frac{-5 \pm \sqrt{13}}{6}$$

A check verifies that the solution set is
$$\left\{ \frac{-5 + \sqrt{13}}{6}, \frac{-5 - \sqrt{13}}{6} \right\}.$$

23. $7x^2 = 12x$ ▪ First, write the equation in standard form, $ax^2 + bx + c = 0$.

$$7x^2 - 12x = 0$$

Substitute $a = 7$, $b = -12$, and $c = 0$ in the quadratic formula.

$$x = \frac{-b \pm \sqrt{b^2 - 4ac}}{2a}$$
$$x = \frac{-(-12) \pm \sqrt{(-12)^2 - 4(7)(0)}}{2(7)}$$
$$= \frac{12 \pm \sqrt{144 - 0}}{14}$$
$$= \frac{12 \pm 12}{14}$$
$$x = \frac{12 + 12}{14} = \frac{24}{14} = \frac{12}{7}$$
$$\text{or} \quad x = \frac{12 - 12}{14} = \frac{0}{14} = 0$$

A check verifies that the solution set is $\{0, \frac{12}{7}\}$.

25. $x^2 - 24 = 0$ ▪ Substitute $a = 1$, $b = 0$, and $c = -24$ in the quadratic formula.

$$x = \frac{-b \pm \sqrt{b^2 - 4ac}}{2a}$$
$$x = \frac{-0 \pm \sqrt{0^2 - 4(1)(-24)}}{2(1)}$$
$$= \frac{\pm \sqrt{96}}{2}$$
$$= \frac{\pm \sqrt{16 \cdot \sqrt{6}}}{2}$$
$$= \frac{\pm 4\sqrt{6}}{2} = \pm 2\sqrt{6}$$

A check verifies that the solution set is $\{-2\sqrt{6}, 2\sqrt{6}\}$.

27. $25x^2 - 4 = 0$ ▪ Substitute $a = 25$, $b = 0$, and $c = -4$ in the quadratic formula.

$$x = \frac{-b \pm \sqrt{b^2 - 4ac}}{2a}$$
$$x = \frac{-0 \pm \sqrt{0^2 - 4(25)(-4)}}{2(25)}$$
$$= \frac{\pm \sqrt{400}}{50}$$
$$= \frac{\pm 20}{50} = \pm \frac{2}{5}$$

A check verifies that the solution set is $\{-\frac{2}{5}, \frac{2}{5}\}$.

29. $3x^2 - 2x + 5 = 10x + 1$ ▪ First, write the equation in standard form, $ax^2 + bx + c = 0$.

$$3x^2 - 12x + 4 = 0$$

Substitute $a = 3$, $b = -12$, and $c = 4$ in the quadratic formula.

$$x = \frac{-b \pm \sqrt{b^2 - 4ac}}{2a}$$
$$x = \frac{-(-12) \pm \sqrt{(-12)^2 - 4(3)(4)}}{2(3)}$$
$$= \frac{12 \pm \sqrt{144 - 48}}{6}$$
$$= \frac{12 \pm \sqrt{96}}{6}$$
$$= \frac{12 \pm \sqrt{16 \cdot \sqrt{6}}}{6}$$
$$= \frac{12 \pm 4\sqrt{6}}{6}$$
$$= \frac{2(6 \pm 2\sqrt{6})}{2 \cdot 3}$$
$$= \frac{6 \pm 2\sqrt{6}}{3}$$

A check verifies that the solution set is
$$\left\{ \frac{6 + 2\sqrt{6}}{3}, \frac{6 - 2\sqrt{6}}{3} \right\}.$$

31. $2x^2 + x + 5 = 0$ ■ Substitute $a = 2$, $b = 1$, and $c = 5$ in the quadratic formula.

$$x = \frac{-1 \pm \sqrt{1^2 - 4(2)(5)}}{2(2)}$$

$$= \frac{-1 \pm \sqrt{1 - 40}}{4}$$

$$= \frac{-1 \pm \sqrt{-39}}{4}$$

Because $\sqrt{-39}$ does not represent a real number, there is *no real number solution*. The solution set is $\emptyset$.

33. $\frac{3}{2}k^2 - k - \frac{4}{3} = 0$ ■ Eliminate the denominators by multiplying each side by the least common denominator, 6.

$$9k^2 - 6k - 8 = 0$$

Substitute $a = 9$, $b = -6$, and $c = -8$ in the quadratic formula.

$$k = \frac{-(-6) \pm \sqrt{(-6)^2 - 4(9)(-8)}}{2(9)}$$

$$= \frac{6 \pm \sqrt{36 + 288}}{18} = \frac{6 \pm \sqrt{324}}{18}$$

$$= \frac{6 \pm 18}{18}$$

$$k = \frac{6 + 18}{18} = \frac{24}{18} = \frac{4}{3}$$

$$\text{or} \quad k = \frac{6 - 18}{18} = \frac{-12}{18} = -\frac{2}{3}$$

A check verifies that the solution set is $\left\{ -\frac{2}{3}, \frac{4}{3} \right\}$.

35. $\frac{1}{2}x^2 + \frac{1}{6}x = 1$ ■ Eliminate the denominators by multiplying each side by the least common denominator, 6.

$$3x^2 + x = 6 \quad \textit{Multiply by LCD, 6.}$$
$$3x^2 + x - 6 = 0 \quad \textit{Subtract 6.}$$

Here, $a = 3$, $b = 1$, and $c = -6$.

$$x = \frac{-1 \pm \sqrt{1^2 - 4(3)(-6)}}{2(3)}$$

$$= \frac{-1 \pm \sqrt{1 + 72}}{6}$$

$$= \frac{-1 \pm \sqrt{73}}{6}$$

A check verifies that the solution set is

$$\left\{ \frac{-1 + \sqrt{73}}{6}, \frac{-1 - \sqrt{73}}{6} \right\}.$$

37. $\frac{3}{8}x^2 - x + \frac{17}{24} = 0$ ■ Multiply each side by the LCD, 24.

$$9x^2 - 24x + 17 = 0$$

Use the quadratic formula with $a = 9$, $b = -24$, and $c = 17$.

$$x = \frac{-(-24) \pm \sqrt{(-24)^2 - 4(9)(17)}}{2(9)}$$

$$= \frac{24 \pm \sqrt{576 - 612}}{18}$$

$$= \frac{24 \pm \sqrt{-36}}{18}$$

Because $\sqrt{-36}$ does not represent a real number, there is *no real number solution*. The solution set is $\emptyset$.

39. $0.6x - 0.4x^2 = -1$ ■ To eliminate the decimals, multiply each side by 10.

$$6x - 4x^2 = -10$$

Write this equation in standard form.

$$0 = 4x^2 - 6x - 10$$

Divide each side by 2 so that we can work with smaller coefficients in the quadratic formula.

$$0 = 2x^2 - 3x - 5$$

Use the quadratic formula with $a = 2$, $b = -3$, and $c = -5$.

$$x = \frac{-(-3) \pm \sqrt{(-3)^2 - 4(2)(-5)}}{2(2)}$$

$$= \frac{3 \pm \sqrt{9 + 40}}{4}$$

$$= \frac{3 \pm \sqrt{49}}{4}$$

$$= \frac{3 \pm 7}{4}$$

$$x = \frac{3 + 7}{4} = \frac{10}{4} = \frac{5}{2}$$

$$\text{or} \quad x = \frac{3 - 7}{4} = \frac{-4}{4} = -1$$

A check verifies that the solution set is $\left\{ -1, \frac{5}{2} \right\}$.

41. $0.25x^2 = -1.5x - 1$ ■ To eliminate the decimals, multiply each side by 100. Write this equation in standard form.

$$25x^2 + 150x + 100 = 0$$

Divide each side by 25 so that we can work with smaller coefficients in the quadratic formula.

$$x^2 + 6x + 4 = 0$$

Use the quadratic formula with $a = 1$, $b = 6$, and $c = 4$.

$$x = \frac{-6 \pm \sqrt{6^2 - 4(1)(4)}}{2(1)}$$

$$= \frac{-6 \pm \sqrt{36 - 16}}{2} = \frac{-6 \pm \sqrt{20}}{2}$$

$$= \frac{-6 \pm 2\sqrt{5}}{2} = \frac{2(-3 \pm \sqrt{5})}{2}$$

$$= -3 \pm \sqrt{5}$$

A check verifies that the solution set is
$\{-3 + \sqrt{5}, -3 - \sqrt{5}\}$.

43.

$$\begin{array}{ll} -0.5x^2 + 1.25x + 3 = h & \textit{Given} \\ -0.5x^2 + 1.25x + 3 = 1.25 & \textit{Let h = 1.25.} \\ -0.5x^2 + 1.25x + 1.75 = 0 & \textit{Subtract 1.25.} \\ 2x^2 - 5x - 7 = 0 & \textit{Mult. by} -4. \end{array}$$

Here, $a = 2$, $b = -5$, and $c = -7$.

$$x = \frac{-(-5) \pm \sqrt{(-5)^2 - 4(2)(-7)}}{2(2)}$$

$$= \frac{5 \pm \sqrt{25 + 56}}{4}$$

$$= \frac{5 \pm \sqrt{81}}{4} = \frac{5 \pm 9}{4}$$

$$x = \frac{5 + 9}{4} = \frac{14}{4} = 3.5$$

or $x = \dfrac{5 - 9}{4} = \dfrac{-4}{4} = -1$

x must be positive, so the frog was 3.5 feet from the base of the stump when he was 1.25 feet above the ground.

45. $\left(\dfrac{d - 4}{4}\right)^2 = 9$

$$\frac{d - 4}{4} = \pm\sqrt{9} = \pm 3$$

$$d - 4 = 4(\pm 3) = \pm 12$$

$$d = 4 \pm 12$$

$$= 16 \quad \text{or} \quad -8$$

The solutions for the equation are -8 and 16. Only 16 board feet is a reasonable answer.

Summary Exercises *Applying Methods for Solving Quadratic Equations*

1. $x^2 = 36$ ▪ Use the square root property.

$$\begin{array}{lll} x = \sqrt{36} & \text{or} & x = -\sqrt{36} \\ x = 6 & \text{or} & x = -6 \end{array}$$

A check verifies that the solution set is $\{\pm 6\}$.

3. $x^2 - \frac{100}{81} = 0$

$$x^2 = \frac{100}{81} \quad \textit{Add } \frac{100}{81}.$$

Use the square root property.

$$\begin{array}{lll} x = \sqrt{\frac{100}{81}} & \text{or} & x = -\sqrt{\frac{100}{81}} \\ x = \frac{10}{9} & \text{or} & x = -\frac{10}{9} \end{array}$$

A check verifies that the solution set is $\{\pm \frac{10}{9}\}$.

5. $z^2 - 4z + 3 = 0$
$(z - 3)(z - 1) = 0$ *Factor.*

$$\begin{array}{lll} z - 3 = 0 & \text{or} & z - 1 = 0 \\ z = 3 & \text{or} & z = 1 \end{array}$$

A check verifies that the solution set is $\{1, 3\}$.

7.

$$\begin{array}{ll} z(z - 9) = -20 & \\ z^2 - 9z = -20 & \textit{Multiply.} \\ z^2 - 9z + 20 = 0 & \textit{Standard form} \\ (z - 4)(z - 5) = 0 & \textit{Factor.} \end{array}$$

$$\begin{array}{lll} z - 4 = 0 & \text{or} & z - 5 = 0 \\ z = 4 & \text{or} & z = 5 \end{array}$$

A check verifies that the solution set is $\{4, 5\}$.

9. $(3k - 2)^2 = 9$ ▪ Use the square root property.

$$\begin{array}{lll} 3k - 2 = \sqrt{9} & \text{or} & 3k - 2 = -\sqrt{9} \\ 3k - 2 = 3 & \text{or} & 3k - 2 = -3 \\ 3k = 5 & \text{or} & 3k = -1 \\ k = \frac{5}{3} & \text{or} & k = -\frac{1}{3} \end{array}$$

A check verifies that the solution set is $\{-\frac{1}{3}, \frac{5}{3}\}$.

11. $(x + 6)^2 = 121$ ▪ Use the square root property.

$$\begin{array}{lll} x + 6 = \sqrt{121} & \text{or} & x + 6 = -\sqrt{121} \\ x + 6 = 11 & \text{or} & x + 6 = -11 \\ x = 5 & \text{or} & x = -17 \end{array}$$

A check verifies that the solution set is $\{-17, 5\}$.

13. $(3r - 7)^2 = 24$ ▪ Use the square root property.

$$3r - 7 = \sqrt{24} \quad \text{or} \quad 3r - 7 = -\sqrt{24}$$

Now simplify the radical.

$$\sqrt{24} = \sqrt{4} \cdot \sqrt{6} = 2\sqrt{6}$$

$$\begin{array}{lll} 3r - 7 = 2\sqrt{6} & \text{or} & 3r - 7 = -2\sqrt{6} \\ 3r = 7 + 2\sqrt{6} & \text{or} & 3r = 7 - 2\sqrt{6} \\ r = \dfrac{7 + 2\sqrt{6}}{3} & \text{or} & r = \dfrac{7 - 2\sqrt{6}}{3} \end{array}$$

A check verifies that the solution set is
$$\left\{\frac{7 + 2\sqrt{6}}{3}, \frac{7 - 2\sqrt{6}}{3}\right\}.$$

15. $(5x - 8)^2 = -6$ ▪ The square root of -6 is not a real number, so the square root property does not apply. This equation has *no real number solution.* The solution set is $\emptyset$.

17.

$$\begin{array}{ll} -2x^2 = -3x - 2 & \\ 2x^2 - 3x - 2 = 0 & \textit{Standard form} \\ (2x + 1)(x - 2) = 0 & \textit{Factor.} \end{array}$$

$$\begin{array}{lll} 2x + 1 = 0 & \text{or} & x - 2 = 0 \\ x = -\frac{1}{2} & \text{or} & x = 2 \end{array}$$

A check verifies that the solution set is $\{-\frac{1}{2}, 2\}$.

19.
$$8z^2 = 15 + 2z$$
$$8z^2 - 2z - 15 = 0 \qquad \textit{Standard form}$$
$$(4z + 5)(2z - 3) = 0 \qquad \textit{Factor.}$$

$$4z + 5 = 0 \qquad \text{or} \qquad 2z - 3 = 0$$
$$z = -\tfrac{5}{4} \qquad \text{or} \qquad z = \tfrac{3}{2}$$

A check verifies that the solution set is $\left\{-\tfrac{5}{4}, \tfrac{3}{2}\right\}$.

21.
$$0 = -x^2 + 2x + 1$$
$$x^2 - 2x - 1 = 0 \qquad \textit{Standard form}$$

Use the quadratic formula with
$a = 1$, $b = -2$, and $c = -1$.

$$x = \frac{-b \pm \sqrt{b^2 - 4ac}}{2a}$$
$$x = \frac{-(-2) \pm \sqrt{(-2)^2 - 4(1)(-1)}}{2(1)}$$
$$= \frac{2 \pm \sqrt{4 + 4}}{2} = \frac{2 \pm \sqrt{8}}{2}$$
$$= \frac{2 \pm 2\sqrt{2}}{2} = \frac{2(1 \pm \sqrt{2})}{2}$$
$$= 1 \pm \sqrt{2}$$

A check verifies that the solution set is
$\{1 + \sqrt{2}, 1 - \sqrt{2}\}$.

23.
$$5x^2 - 22x = -8$$
$$5x^2 - 22x + 8 = 0$$
$$(5x - 2)(x - 4) = 0 \qquad \textit{Factor.}$$

$$5x - 2 = 0 \qquad \text{or} \qquad x - 4 = 0$$
$$x = \tfrac{2}{5} \qquad \text{or} \qquad x = 4$$

A check verifies that the solution set is $\left\{\tfrac{2}{5}, 4\right\}$.

25. $(x + 2)(x + 1) = 10$
$$x^2 + 3x + 2 = 10 \qquad \textit{Multiply.}$$
$$x^2 + 3x - 8 = 0 \qquad \textit{Subtract 8.}$$

Use the quadratic formula with
$a = 1$, $b = 3$, and $c = -8$.

$$x = \frac{-b \pm \sqrt{b^2 - 4ac}}{2a}$$
$$x = \frac{-3 \pm \sqrt{3^2 - 4(1)(-8)}}{2(1)}$$
$$= \frac{-3 \pm \sqrt{9 + 32}}{2}$$
$$= \frac{-3 \pm \sqrt{41}}{2}$$

A check verifies that the solution set is
$$\left\{\frac{-3 + \sqrt{41}}{2}, \frac{-3 - \sqrt{41}}{2}\right\}.$$

27.
$$4x^2 = -1 + 5x$$
$$4x^2 - 5x + 1 = 0 \qquad \textit{Standard form}$$
$$(x - 1)(4x - 1) = 0 \qquad \textit{Factor.}$$

$$x - 1 = 0 \qquad \text{or} \qquad 4x - 1 = 0$$
$$x = 1 \qquad \text{or} \qquad x = \tfrac{1}{4}$$

A check verifies that the solution set is $\left\{\tfrac{1}{4}, 1\right\}$.

29.
$$3x(3x + 4) = 7$$
$$9x^2 + 12x = 7 \qquad \textit{Multiply.}$$
$$9x^2 + 12x - 7 = 0 \qquad \textit{Standard form}$$

Use the quadratic formula with
$a = 9$, $b = 12$, and $c = -7$.

$$x = \frac{-b \pm \sqrt{b^2 - 4ac}}{2a}$$
$$x = \frac{-12 \pm \sqrt{12^2 - 4(9)(-7)}}{2(9)}$$
$$= \frac{-12 \pm \sqrt{144 + 252}}{18}$$
$$= \frac{-12 \pm \sqrt{396}}{18} = \frac{-12 \pm \sqrt{36} \cdot \sqrt{11}}{18}$$
$$= \frac{-12 \pm 6\sqrt{11}}{18} = \frac{6(-2 \pm \sqrt{11})}{6 \cdot 3}$$
$$= \frac{-2 \pm \sqrt{11}}{3}$$

A check verifies that the solution set is
$$\left\{\frac{-2 + \sqrt{11}}{3}, \frac{-2 - \sqrt{11}}{3}\right\}.$$

31.
$$\frac{x^2}{2} + \frac{7x}{4} + \frac{11}{8} = 0$$
$$8\left(\frac{x^2}{2} + \frac{7x}{4} + \frac{11}{8}\right) = 8(0) \qquad \begin{array}{l}\textit{Multiply}\\\textit{by LCD, 8.}\end{array}$$
$$4x^2 + 14x + 11 = 0$$

Use the quadratic formula with
$a = 4$, $b = 14$, and $c = 11$.

$$x = \frac{-b \pm \sqrt{b^2 - 4ac}}{2a}$$
$$x = \frac{-14 \pm \sqrt{14^2 - 4(4)(11)}}{2(4)}$$
$$= \frac{-14 \pm \sqrt{196 - 176}}{8}$$
$$= \frac{-14 \pm \sqrt{20}}{8} = \frac{-14 \pm 2\sqrt{5}}{8}$$
$$= \frac{2(-7 \pm \sqrt{5})}{2(4)} = \frac{-7 \pm \sqrt{5}}{4}$$

A check verifies that the solution set is
$$\left\{\frac{-7 + \sqrt{5}}{4}, \frac{-7 - \sqrt{5}}{4}\right\}.$$

33.
$$9k^2 = 16(3k + 4)$$
$$9k^2 = 48k + 64 \quad \textit{Multiply.}$$
$$9k^2 - 48k - 64 = 0 \quad \textit{Standard form}$$

Use the quadratic formula with $a = 9$, $b = -48$, and $c = -64$.

$$k = \frac{-b \pm \sqrt{b^2 - 4ac}}{2a}$$

$$k = \frac{-(-48) \pm \sqrt{(-48)^2 - 4(9)(-64)}}{2(9)}$$

$$= \frac{48 \pm \sqrt{2304 + 2304}}{18}$$

$$= \frac{48 \pm \sqrt{4608}}{18}$$

$$= \frac{48 \pm \sqrt{2304 \cdot \sqrt{2}}}{18}$$

$$= \frac{48 \pm 48\sqrt{2}}{18}$$

$$= \frac{6(8 \pm 8\sqrt{2})}{6 \cdot 3} = \frac{8 \pm 8\sqrt{2}}{3}$$

A check verifies that the solution set is

$$\left\{ \frac{8 + 8\sqrt{2}}{3}, \frac{8 - 8\sqrt{2}}{3} \right\}.$$

35. $x^2 - x + 3 = 0$ ■ Use the quadratic formula with $a = 1$, $b = -1$, and $c = 3$.

$$x = \frac{-b \pm \sqrt{b^2 - 4ac}}{2a}$$

$$x = \frac{-(-1) \pm \sqrt{(-1)^2 - 4(1)(3)}}{2(1)}$$

$$= \frac{1 \pm \sqrt{1 - 12}}{2} = \frac{1 \pm \sqrt{-11}}{2}$$

Because $\sqrt{-11}$ does not represent a real number, there is *no real number solution*. The solution set is $\emptyset$.

37.
$$-3x^2 + 4x = -4$$
$$3x^2 - 4x - 4 = 0 \quad \textit{Standard form}$$
$$(3x + 2)(x - 2) = 0 \quad \textit{Factor.}$$

$$3x + 2 = 0 \quad \text{or} \quad x - 2 = 0$$
$$x = -\tfrac{2}{3} \quad \text{or} \quad x = 2$$

A check verifies that the solution set is $\left\{ -\tfrac{2}{3}, 2 \right\}$.

39.
$$5k^2 + 19k = 2k + 12$$
$$5k^2 + 17k - 12 = 0 \quad \textit{Standard form}$$
$$(5k - 3)(k + 4) = 0 \quad \textit{Factor.}$$

$$5k - 3 = 0 \quad \text{or} \quad k + 4 = 0$$
$$k = \tfrac{3}{5} \quad \text{or} \quad k = -4$$

A check verifies that the solution set is $\left\{ -4, \tfrac{3}{5} \right\}$.

41.
$$x^2 - \tfrac{4}{15} = -\tfrac{4}{15}x$$
$$15x^2 - 4 = -4x \quad \textit{Multiply by LCD, 15.}$$
$$15x^2 + 4x - 4 = 0 \quad \textit{Standard form}$$
$$(3x + 2)(5x - 2) = 0 \quad \textit{Factor.}$$

$$3x + 2 = 0 \quad \text{or} \quad 5x - 2 = 0$$
$$x = -\tfrac{2}{3} \quad \text{or} \quad x = \tfrac{2}{5}$$

A check verifies that the solution set is $\left\{ -\tfrac{2}{3}, \tfrac{2}{5} \right\}$.

17.4   Graphing Quadratic Equations

17.4   Margin Exercises

1. $y = x^2$ ■ To find y-values, substitute each x-value into $y = x^2$.

x	y
3	9
2	4
1	1
0	0
−1	1
−2	4
−3	9

2. $y = \tfrac{1}{2}x^2$ ■ To find y-values, substitute each x-value into $y = \tfrac{1}{2}x^2$.

x	y
−2	2
−1	$\tfrac{1}{2}$
0	0
1	$\tfrac{1}{2}$
2	2

3. $y = -x^2 + 3$ ■ To find the y-values for the ordered pairs, substitute each x-value into $y = -x^2 + 3$.

If $x = -2$, $y = -(-2)^2 + 3 = -4 + 3 = -1$.
If $x = -1$, $y = -(-1)^2 + 3 = -1 + 3 = 2$.
If $x = 1$, $y = -1^2 + 3 = -1 + 3 = 2$.
If $x = 2$, $y = -2^2 + 3 = -4 + 3 = -1$.

The ordered pairs are $(-2, \underline{-1})$, $(-1, \underline{2})$, $(1, \underline{2})$, and $(2, \underline{-1})$.

4. **(a)** $y = -x^2 - 3$ ■ Find several ordered pairs. To begin, check for intercepts.

If $x = 0$,

$$y = -0^2 - 3 = -3,$$

so the y-intercept is $(0, -3)$.

If $y = 0$,

$$0 = -x^2 - 3,$$

and hence, $x^2 = -3$.

This equation has no real number solution, so there are no x-intercepts.

We can choose additional x-values near $x = 0$ to find other ordered pairs. We obtain the following table for ordered pairs.

x	y
-2	-7
-1	-4
0	-3
1	-4
2	-7

Plot these points and connect them with a smooth curve. From the graph, we see that the vertex, $(0, -3)$, is the highest point of this graph.

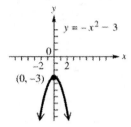

(b) $y = x^2 + 3$ ▪ Find several ordered pairs. To begin, check for intercepts.

If $x = 0$,
$$y = 0^2 + 3 = 3,$$
so the y-intercept is $(0, 3)$.

If $y = 0$,
$$0 = x^2 + 3,$$
and hence, $x^2 = -3$.

This equation has no real number solution, so there are no x-intercepts.

Choose additional x-values near $x = 0$. We obtain the following table for ordered pairs.

x	y
-2	7
-1	4
0	3
1	4
2	7

Plot these points and connect them with a smooth curve. From the graph, we see that the vertex, $(0, 3)$, is the lowest point of this graph.

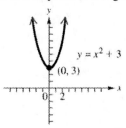

5. $y = x^2 + 2x - 8$ ▪ Find any x-intercepts by substituting 0 for y in the equation.

$$y = x^2 + 2x - 8$$
$$0 = x^2 + 2x - 8 \qquad \textit{Let y = 0.}$$
$$0 = (x + 4)(x - 2) \quad \textit{Factor.}$$
$$x + 4 = 0 \quad \text{or} \quad x - 2 = 0$$
$$x = \underline{-4} \quad \text{or} \qquad x = 2$$

The x-intercepts are $\underline{(-4, 0)}$ and $(2, 0)$.

Now find any y-intercepts by substituting 0 for x.

$$y = x^2 + 2x - 8$$
$$y = 0^2 + 2(0) - 8 \quad \textit{Let x = 0.}$$
$$y = -8$$

The y-intercept is $\underline{(0, -8)}$. The x-value of the vertex is halfway between the x-intercepts, $(-4, 0)$ and $(2, 0)$.

$$x = \tfrac{1}{2}(-4 + \underline{2}) = \tfrac{1}{2}(-2) = \underline{-1}$$

Find the y-value of the vertex by substituting -1 for x in the given equation.

$$y = x^2 + 2x - 8$$
$$y = (-1)^2 + 2(-1) - 8 \quad \textit{Let x = -1.}$$
$$y = 1 - 2 - 8$$
$$y = \underline{-9}$$

The vertex is at $\underline{(-1, -9)}$.

The axis of symmetry is the line $x = \underline{-1}$.

x	y
-4	0
-3	-5
-2	-8
-1	-9
0	-8
1	-5
2	0

Plot the intercepts, vertex, and additional points shown in the table of values and connect them with a smooth curve.

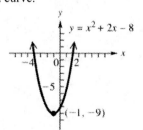

6. $y = x^2 - 4x + 1$

If $x = 5$, $y = 5^2 - 4(5) + 1 = 6$, giving the ordered pair $(5, \underline{6})$.

If $x = 4$, $y = 4^2 - 4(4) + 1 = 1$, giving the ordered pair $(4, \underline{1})$.

If $x = -1$, $y = (-1)^2 - 4(-1) + 1 = 6$, giving the ordered pair $(-1, \underline{6})$.

7. **(a)** $y = x^2 - 3x - 3$ ■ Here, $a = 1$, $b = -3$, and $c = -3$. The x-value of the vertex is

$$x = -\frac{b}{2a} = -\frac{-3}{2(1)} = \frac{3}{2} = 1.5.$$

The y-value of the vertex is

$$y = (1.5)^2 - 3(1.5) - 3 = -5.25,$$

so the vertex is $(1.5, -5.25)$.

The axis of symmetry is the line $x = 1.5$.

Now find the x-intercepts by using the quadratic formula.

$$x = \frac{-(-3) \pm \sqrt{(-3)^2 - 4(1)(-3)}}{2(1)}$$

$$x = \frac{3 \pm \sqrt{21}}{2}$$

Using a calculator, $x \approx -0.8$ and $x \approx 3.8$.

The x-intercepts are $(-0.8, 0)$ and $(3.8, 0)$.

Find the y-intercept by letting $x = 0$.

$$y = 0^2 - 3(0) - 3 = -3$$

The y-intercept is $(0, -3)$.

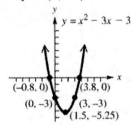

(b) $y = -x^2 + 2x + 4$ ■ Here, $a = -1$, $b = 2$, and $c = 4$.

The x-value of the vertex is

$$x = -\frac{b}{2a} = -\frac{2}{2(-1)} = 1.$$

The y-value of the vertex is

$$y = -1^2 + 2(1) + 4 = 5,$$

so the vertex is $(1, 5)$.

The axis of symmetry is the line $x = 1$.

Now find the x-intercepts by using the quadratic formula.

$$x = \frac{-2 \pm \sqrt{2^2 - 4(-1)(4)}}{2(-1)}$$

$$x = \frac{-2 \pm \sqrt{20}}{-2}$$

Using a calculator, $x \approx 3.2$ and $x \approx -1.2$.

The x-intercepts are $(-1.2, 0)$ and $(3.2, 0)$.

Find the y-intercept by letting $x = 0$.

$$y = -0^2 + 2(0) + 4 = 4$$

The y-intercept is $(0, 4)$.

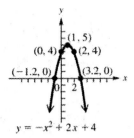

17.4   Section Exercises

1. Every equation of the form $y = ax^2 + bx + c$, with $a \neq 0$, has a graph that is a <u>parabola</u>.

3. The vertical line through the vertex of a parabola that opens upward or downward is the <u>axis of symmetry</u> of the parabola. The two halves of the parabola are <u>mirror</u> images of each other across this line.

5. $y = 2x^2$ ■ In $y = 2x^2$, $a = 2$, $b = 0$, and $c = 0$.

The x-value of the vertex is

$$x = -\frac{b}{2a} = -\frac{0}{2(2)} = 0.$$

The y-value of the vertex is

$$y = 2(0)^2 = 0,$$

so the vertex is $(0, 0)$.

The axis of symmetry is the line $x = 0$, which is the y-axis.

Now find the intercepts.

Let $x = 0$ to get $y = 2 \cdot 0^2 = 0$. The y-intercept is $(0, 0)$.

Let $y = 0$ to get $0 = 2x^2$, which implies $x = 0$. The only x-intercept is $(0, 0)$.

Make a table of ordered pairs.

x	y
-2	8
-1	2
0	0
1	2
2	8

Plot these five ordered pairs and connect them with a smooth curve.

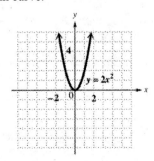

7. $y = x^2 - 4$ ▪ In $y = x^2 - 4$, $a = 1$, $b = 0$, and $c = -4$.

The x-value of the vertex is

$$x = -\frac{b}{2a} = -\frac{0}{2(1)} = 0.$$

The y-value of the vertex is

$$y = 0^2 - 4 = -4,$$

so the vertex is $(0, -4)$.

The axis of symmetry is the line $x = 0$, which is the y-axis.

Now find the intercepts.

Let $x = 0$ to get $y = 0^2 - 4 = -4$; the y-intercept is $(0, -4)$. Let $y = 0$ and solve for x.

$$0 = x^2 - 4$$
$$0 = (x + 2)(x - 2)$$
$$x + 2 = 0 \quad \text{or} \quad x - 2 = 0$$
$$x = -2 \quad \text{or} \quad x = 2$$

The x-intercepts are $(-2, 0)$ and $(2, 0)$.

Make a table of ordered pairs.

x	y
0	-4
±1	-3
±2	0
±3	5

Plot these seven ordered pairs and connect them with a smooth curve.

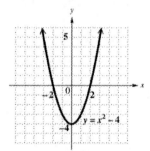

9. $y = -x^2 + 2$ ▪ If $x = 0$, $y = 2$, so the y-intercept is $(0, 2)$.

To find any x-intercepts, let $y = 0$.

$$0 = -x^2 + 2$$
$$x^2 = 2$$
$$x = \pm\sqrt{2} \approx \pm 1.41$$

The x-intercepts are $(\pm\sqrt{2}, 0)$.

The x-value of the vertex is

$$x = -\frac{b}{2a} = -\frac{0}{2(-1)} = 0.$$

Thus, the vertex is the same as the y-intercept (since $x = 0$). The axis of symmetry is the vertical line $x = 0$.

Make a table of ordered pairs whose x-values are on either side of the vertex's x-value of $x = 0$.

x	y
0	2
±1	1
±2	-2
±3	-7

Plot these seven ordered pairs and connect them with a smooth curve.

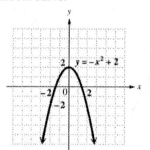

11. $y = (x + 1)^2 = x^2 + 2x + 1$

If $x = 0$, $y = 1$, so the y-intercept is $(0, 1)$.

To find any x-intercepts, let $y = 0$.

$$0 = (x + 1)^2$$
$$0 = x + 1$$
$$-1 = x$$

The x-intercept is $(-1, 0)$.

The x-value of the vertex is

$$x = -\frac{b}{2a} = -\frac{2}{2(1)} = -1.$$

Thus, the vertex is the same as the x-intercept. The axis of symmetry is the vertical line $x = -1$.

Make a table of ordered pairs whose x-values are on either side of the vertex's x-value of $x = -1$.

x	y
-4	9
-3	4
-2	1
-1	0
0	1
1	4
2	9

Plot these seven ordered pairs and connect them with a smooth curve.

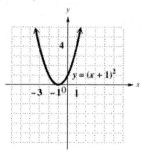

13. $y = x^2 + 2x + 3$ ■ Let $x = 0$ to get
$$y = 0^2 + 2(0) + 3 = 3;$$
the y-intercept is $(0, 3)$.

To find any x-intercepts, let $y = 0$.
$$0 = x^2 + 2x + 3$$

The trinomial on the right cannot be factored. Because the discriminant
$b^2 - 4ac = 2^2 - 4(1)(3) = -8$ is negative, this equation has no real number solutions. Thus, the parabola has no x-intercepts.

The x-value of the vertex is
$$x = -\frac{b}{2a} = -\frac{2}{2(1)} = -1.$$

The y-value of the vertex is
$$y = (-1)^2 + 2(-1) + 3$$
$$= 1 - 2 + 3 = 2,$$

so the vertex is $(-1, 2)$. The axis of symmetry is the vertical line $x = -1$.

Make a table of ordered pairs whose x-values are on either side of the vertex's x-value of $x = -1$.

x	y
-4	11
-3	6
-2	3
-1	2
0	3
1	6
2	11

Plot these seven ordered pairs and connect them with a smooth curve.

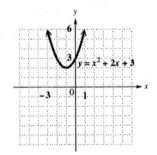

15. $y = -x^2 + 6x - 5$ ■ Let $x = 0$ to get
$$y = -(0)^2 + 6(0) - 5 = -5;$$
the y-intercept is $(0, -5)$.

Let $y = 0$ and solve for x.
$$0 = -x^2 + 6x - 5$$
$$x^2 - 6x + 5 = 0$$
$$(x - 1)(x - 5) = 0$$
$$x - 1 = 0 \quad \text{or} \quad x - 5 = 0$$
$$x = 1 \quad \text{or} \quad x = 5$$

The x-intercepts are $(1, 0)$ and $(5, 0)$.

The x-value of the vertex is
$$x = -\frac{b}{2a} = -\frac{6}{2(-1)} = 3.$$

The y-value of the vertex is
$$y = -(3)^2 + 6(3) - 5$$
$$= -9 + 18 - 5 = 4,$$

so the vertex is $(3, 4)$.

The axis of symmetry is the line $x = 3$.

Make a table of ordered pairs whose x-values are on either side of the vertex's x-value of $x = 3$.

x	y
0	-5
1	0
2	3
3	4
4	3
5	0
6	-5

Plot these seven ordered pairs and connect them with a smooth curve.

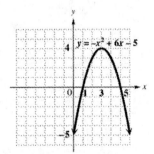

17. If $a > 0$, the parabola $y = ax^2 + bx + c$ opens upward. If $a < 0$, the parabola $y = ax^2 + bx + c$ opens downward.

19. Because the vertex is at the origin, an equation of the parabola is of the form
$$y = ax^2.$$

As shown in the figure, one point on the graph has coordinates $(150, 44)$.

$$y = ax^2 \qquad \textit{General equation}$$
$$44 = a(150)^2 \quad \textit{Let } x = 150, y = 44.$$
$$44 = 22{,}500a$$
$$a = \frac{44}{22{,}500} = \frac{4 \cdot 11}{4 \cdot 5625} = \frac{11}{5625}$$

Thus, an equation of the parabola is
$$y = \tfrac{11}{5625}x^2.$$

17.5   Introduction to Functions

17.5   Margin Exercises

1.   (a) Consider the following relation:

$$\{(5, 10), (15, 20), (25, 30), (35, 40)\}$$

The domain is the set of all first components in the ordered pairs, $\{5, 15, 25, 35\}$. The range is the set of all second components in the ordered pairs, $\{10, 20, 30, 40\}$.

(b) The relation $\{(1, 4), (2, 4), (3, 4)\}$ has domain $\{1, 2, 3\}$ and range $\{4\}$.

2.   (a) $\{(-2, 8), (-1, 1), (0, 0), (1, 1), (2, 8)\}$

Notice that each first component appears once and only once. Because of this, the relation *is a function*.

(b) $\{(5, 2), (5, 1), (5, 0)\}$

The first component 5 appears in all three ordered pairs, and corresponds to more than one second component. Therefore, this relation *is not a function*.

(c) $\{(-1, -3), (0, 2), (3, 1), (8, 1)\}$

Notice that each first component appears once and only once. Because of this, the relation *is a function*.

3.   (a) Every first component is paired with one and only one second component, and furthermore, no vertical line intersects the graph in more than one point. Therefore, this is the graph of a function.

(b) Because there are two ordered pairs with first component 0, this is not the graph of a function.

(c) The vertical line test shows that this graph is not the graph of a function; a vertical line could intersect the graph twice.

(d) The vertical line test shows that this graph is not the graph of a function; a vertical line could intersect the graph twice.

(e) The graph of $y = 3$ is a horizontal line, so the equation defines a function. A vertical line will only intersect the graph in one point for every value of x.

4. $f(x) = 6x - 2$

(a) $f(-1) = 6(\underline{-1}) - 2$ *Let x = -1.*
$= \underline{-6} - 2$
$= \underline{-8}$

(b) $f(0) = 6(0) - 2 = 0 - 2 = -2$

(c) $f(1) = 6(1) - 2 = 6 - 2 = 4$

5.   (a) Choose the years as the domain elements and the median age at first marriage for men in the United States as the range elements.

$$f = \{(2004, 27.4), (2006, 27.5),$$
$$(2008, 27.6), (2010, 28.2)\}.$$

(b) The domain is the set of years, or of x-values:

$$\{2004, 2006, 2008, 2010\}$$

The range is the set of median ages at first marriage for men in the United States, or y-values:

$$\{27.4, 27.5, 27.6, 28.2\}$$

(c) $f(2010) = 28.2$

(d) For what x-values does $f(x) = 27.5$?

$$f(2006) = 27.5$$

In 2006, the median age at first marriage for men in the United States was 27.5.

17.5   Section Exercises

1.   If $x = 1$, then $x + 2 = 3$. Since $f(x) = x + 2$, $f(x)$ is also equal to 3. The ordered pair (x, y) is $(1, 3)$.

3.   If $x = 3$, then $x + 2 = 5$. Since $f(x) = x + 2$, $f(x)$ is also equal to 5. The ordered pair (x, y) is $(3, 5)$.

5.   The graph consists of the five points $(0, 2)$, $(1, 3)$, $(2, 4)$, $(3, 5)$, and $(4, 6)$.

7. $\{(-4, 3), (-2, 1), (0, 5), (-2, -8)\}$

This relation *is not a function* since one value of x, namely -2, corresponds to two values of y, namely 1 and -8.

The domain is the set of all first components in the ordered pairs, $\{-4, -2, 0\}$.

The range is the set of all second components in the ordered pairs, $\{3, 1, 5, -8\}$.

9.   The relation *is a function* since each of the first components A, B, C, D, and E corresponds to exactly one second component.

The domain is $\{A, B, C, D, E\}$.
The range is $\{2, 3, 6, 4\}$.

11.   The graph consists of the following set of six ordered pairs:

$$\{(-4, 1), (-2, 0), (-2, 2), (0, -2), (2, 1), (3, 3)\}$$

This relation *is not a function* since one value of x, namely -2, corresponds to two values of y, namely 0 and 2.

The domain is the set of all first components in the ordered pairs, $\{-4, -2, 0, 2, 3\}$.

The range is the set of all second components in the ordered pairs, $\{1, 0, 2, -2, 3\}$.

13. Any vertical line will intersect the graph in only one point. The graph passes the vertical line test, so this is the graph of a function.

15. A vertical line can intersect the graph twice, so this is not the graph of a function.

17. $y = 5x + 3$ ■ Every value of x will give one and only one value of y, so the equation defines a function.

19. $y = \sqrt{x}$ ■ Every nonnegative value of x will give one and only one value of y, so the equation defines a function.

21. $x = -7$ ■ The graph of $x = -7$ is a vertical line, so the equation does *not* define a function. (Every ordered pair has x-value -7.)

23. $y = 5$ ■ Every value of x will give one and only one value of y, so the equation defines a function.

25. $f(x) = 4x + 3$

   (a) $f(2) = 4(2) + 3 = 8 + 3 = 11$

   (b) $f(0) = 4(0) + 3 = 0 + 3 = 3$

   (c) $f(-3) = 4(-3) + 3 = -12 + 3 = -9$

27. $f(x) = x^2 - x + 2$

   (a) $f(2) = (2)^2 - (2) + 2 = 4 - 2 + 2 = 4$

   (b) $f(0) = (0)^2 - (0) + 2 = 0 - 0 + 2 = 2$

   (c) $f(-3) = (-3)^2 - (-3) + 2$
   $= 9 + 3 + 2 = 14$

29. $f(x) = |x|$ ■ $|x|$ is read " the <u>absolute value</u> of x."

   (a) $f(2) = |2| = 2$

   (b) $f(0) = |0| = 0$

   (c) $f(-3) = |-3| = -(-3) = 3$

31. Write the information in the graph as a set of ordered pairs of the form (year, population). The set is $\{(1980, 14.1), (1990, 19.8), (2000, 31.1), (2010, 40.0)\}$. Since each year corresponds to exactly one number, the set defines a function.

33. $g(1980) = 14.1$ (million); $g(2000) = 31.1$ (million)

35. For the year 2008, the function gives 38.0 million foreign-born residents in the United States.

Chapter 17 Review Exercises

1. $x^2 + 3x - 28 = 0$
   $(x - 4)(x + 7) = 0$ *Factor.*

   $x - 4 = 0$ or $x + 7 = 0$ *Zero-factor property*
   $\quad x = 4$ or $\quad\quad x = -7$ *Solve each equation.*

   A check verifies that the solution set is $\{-7, 4\}$.

2. $x^2 + 14x + 45 = 0$
   $(x + 9)(x + 5) = 0$ *Factor.*

   $x + 9 = 0$ or $x + 5 = 0$ *Zero-factor prop.*
   $\quad x = -9$ or $\quad x = -5$ *Solve each eq.*

   A check verifies that the solution set is $\{-9, -5\}$.

3. $\quad\quad 2z^2 + 7z = 15$
   $2z^2 + 7z - 15 = 0$ *Standard form*
   $(2z - 3)(z + 5) = 0$ *Factor.*

   $2z - 3 = 0$ or $z + 5 = 0$ *Zero-factor prop.*
   $\quad z = \frac{3}{2}$ or $\quad z = -5$ *Solve each eq.*

   A check verifies that the solution set is $\left\{-5, \frac{3}{2}\right\}$.

4. $\quad\quad r^2 - 169 = 0$
   $(r + 13)(r - 13) = 0$ *Factor.*

   $r + 13 = 0$ or $r - 13 = 0$ *Zero-factor prop.*
   $\quad r = -13$ or $\quad\quad r = 13$ *Solve each eq.*

   A check verifies that the solution set is $\{-13, 13\}$, or $\{\pm 13\}$.

5. $y^2 = 144$ ■ Use the square root property.

   $y = \sqrt{144}$ or $\quad y = -\sqrt{144}$
   $y = 12$ or $\quad\quad y = -12$

   A check verifies that the solution set is $\{-12, 12\}$, or $\{\pm 12\}$.

6. $x^2 = 37$ ■ Use the square root property.

   $x = \sqrt{37}$ or $x = -\sqrt{37}$

   A check verifies that the solution set is $\{-\sqrt{37}, \sqrt{37}\}$, or $\{\pm\sqrt{37}\}$.

7. $m^2 = 128$ ■ Use the square root property.

   $m = \sqrt{128}$ or $\quad m = -\sqrt{128}$
   $m = \sqrt{64 \cdot 2}$ or $\quad m = -\sqrt{64 \cdot 2}$
   $m = 8\sqrt{2}$ or $\quad m = -8\sqrt{2}$

   A check verifies that the solution set is $\{-8\sqrt{2}, 8\sqrt{2}\}$, or $\{\pm 8\sqrt{2}\}$.

8. $(k + 2)^2 = 25$ ■ Use the square root property.

   $k + 2 = \sqrt{25}$ or $\quad k + 2 = -\sqrt{25}$
   $k + 2 = 5$ or $\quad k + 2 = -5$
   $\quad\quad k = 3$ or $\quad\quad\quad k = -7$

   A check verifies that the solution set is $\{-7, 3\}$.

9. $(r - 3)^2 = 10$ ■ Use the square root property.

   $r - 3 = \sqrt{10}$ or $r - 3 = -\sqrt{10}$
   $\quad r = 3 + \sqrt{10}$ or $\quad r = 3 - \sqrt{10}$

   A check verifies that the solution set is $\{3 + \sqrt{10}, 3 - \sqrt{10}\}$.

10. $(2p+1)^2 = 14$ ▪ Use the square root property.

$$2p + 1 = \sqrt{14} \qquad \text{or} \qquad 2p + 1 = -\sqrt{14}$$

$$2p = -1 + \sqrt{14} \qquad \text{or} \qquad 2p = -1 - \sqrt{14}$$

$$p = \frac{-1 + \sqrt{14}}{2} \qquad \text{or} \qquad p = \frac{-1 - \sqrt{14}}{2}$$

A check verifies that the solution set is

$$\left\{ \frac{-1 + \sqrt{14}}{2}, \frac{-1 - \sqrt{14}}{2} \right\}.$$

11. $(3k+2)^2 = -3$ ▪ Use the square root property.

$$3k + 2 = \sqrt{-3} \quad \text{or} \quad 3k + 2 = -\sqrt{-3}$$

Because $\sqrt{-3}$ does not represent a real number, there is *no real number solution*. The solution set is $\emptyset$.

12. $(3x+5)^2 = 0$

$$3x + 5 = 0 \qquad \textit{Take square root.}$$

$$3x = -5$$

$$x = -\frac{5}{3}$$

A check verifies that the solution set is $\left\{ -\frac{5}{3} \right\}$.

13. $m^2 + 6m + 5 = 0$ ▪ Rewrite the equation with the variable terms on one side and the constant on the other side.

$$m^2 + 6m = -5$$

Take half the coefficient of m and square it.

$$\tfrac{1}{2}(6) = 3, \quad \text{and} \quad (3)^2 = 9.$$

Add 9 to each side of the equation.

$$m^2 + 6m + 9 = -5 + 9$$

$$m^2 + 6m + 9 = 4$$

$$(m+3)^2 = 4 \quad \textit{Factor.}$$

$$m + 3 = \sqrt{4} \qquad \text{or} \qquad m + 3 = -\sqrt{4}$$

$$m + 3 = 2 \qquad \text{or} \qquad m + 3 = -2$$

$$m = -1 \qquad \text{or} \qquad m = -5$$

A check verifies that the solution set is $\{-5, -1\}$.

14. $p^2 + 4p = 7$ ▪ Take half the coefficient of p and square it.

$$\tfrac{1}{2}(4) = 2, \quad \text{and} \quad (2)^2 = 4.$$

Add 4 to each side of the equation.

$$p^2 + 4p + 4 = 7 + 4$$

$$(p+2)^2 = 11$$

$$p + 2 = \sqrt{11} \qquad \text{or} \qquad p + 2 = -\sqrt{11}$$

$$p = -2 + \sqrt{11} \qquad \text{or} \qquad p = -2 - \sqrt{11}$$

A check verifies that the solution set is
$\{-2 + \sqrt{11}, -2 - \sqrt{11}\}$.

15. $-x^2 + 5 = 2x$ ▪ Divide each side of the equation by -1 to make the coefficient of the squared term equal to 1.

$$x^2 - 5 = -2x$$

Rewrite the equation with the variable terms on one side and the constant on the other side.

$$x^2 + 2x = 5$$

Take half the coefficient of x and square it.

$$\tfrac{1}{2}(2) = 1, \quad \text{and} \quad 1^2 = 1.$$

Add 1 to each side of the equation.

$$x^2 + 2x + 1 = 5 + 1$$

$$(x+1)^2 = 6$$

$$x + 1 = \sqrt{6} \qquad \text{or} \qquad x + 1 = -\sqrt{6}$$

$$x = -1 + \sqrt{6} \qquad \text{or} \qquad x = -1 - \sqrt{6}$$

A check verifies that the solution set is
$\{-1 + \sqrt{6}, -1 - \sqrt{6}\}$.

16. $2x^2 - 3 = -8x$ ▪ Divide each side by 2 to get the x^2 coefficient equal to 1.

$$x^2 - \tfrac{3}{2} = -4x$$

Rewrite the equation with the variable terms on one side and the constant on the other side.

$$x^2 + 4x = \tfrac{3}{2}$$

Take half the coefficient of x and square it.

$$\tfrac{1}{2}(4) = 2, \quad \text{and} \quad 2^2 = 4.$$

Add 4 to each side of the equation

$$x^2 + 4x + 4 = \frac{3}{2} + 4$$

$$(x+2)^2 = \frac{11}{2}$$

$$x + 2 = \pm\sqrt{\frac{11}{2}}$$

$$x + 2 = \pm\frac{\sqrt{11}}{\sqrt{2}} \cdot \frac{\sqrt{2}}{\sqrt{2}}$$

$$x + 2 = \pm\frac{\sqrt{22}}{2}$$

$$x = -2 \pm \frac{\sqrt{22}}{2}$$

$$x = \frac{-4}{2} \pm \frac{\sqrt{22}}{2}$$

$$x = \frac{-4 \pm \sqrt{22}}{2}$$

A check verifies that the solution set is

$$\left\{ \frac{-4 + \sqrt{22}}{2}, \frac{-4 - \sqrt{22}}{2} \right\}.$$

17. $4(x^2 + 7x) + 29 = -20$

$\qquad 4x^2 + 28x + 49 = 0 \qquad$ *Distribute; add 20.*

$\qquad\quad (2x + 7)^2 = 0 \qquad$ *Factor.*

$\qquad\qquad 2x + 7 = 0 \qquad$ *Take square root.*

$\qquad\qquad\quad 2x = -7 \qquad$ *Subtract 7.*

$\qquad\qquad\quad\ x = -\frac{7}{2} \qquad$ *Divide by 2.*

The solution set is $\left\{-\frac{7}{2}\right\}$.

18. $(4x + 1)(x - 1) = -7$ ■ Multiply on the left side and then simplify. Get all variable terms on one side and the constant on the other side.

$$4x^2 - 4x + x - 1 = -7$$
$$4x^2 - 3x = -6$$

Divide each side by 4 so that the coefficient of x^2 will be 1.

$$x^2 - \tfrac{3}{4}x = -\tfrac{6}{4} = -\tfrac{3}{2}$$

Square half the coefficient of x to obtain $\frac{9}{64}$ and add $\frac{9}{64}$ to each side.

$$x^2 - \tfrac{3}{4}x + \tfrac{9}{64} = -\tfrac{3}{2} + \tfrac{9}{64}$$
$$\left(x - \tfrac{3}{8}\right)^2 = -\tfrac{96}{64} + \tfrac{9}{64}$$
$$\left(x - \tfrac{3}{8}\right)^2 = -\tfrac{87}{64}$$

The square root of $-\frac{87}{64}$ is not a real number, so there is *no real number solution.* The solution set is $\emptyset$.

19. $h = -16t^2 + 32t + 50$

Let $h = 30$ and solve for t (which must have a positive value since it represents a number of seconds).

$$30 = -16t^2 + 32t + 50$$
$$16t^2 - 32t - 20 = 0$$

Divide each side by 16.

$$t^2 - 2t - \tfrac{20}{16} = 0$$
$$t^2 - 2t = \tfrac{5}{4}$$

Half of -2 is -1, and $(-1)^2 = 1$.

Add 1 to each side of the equation.

$$t^2 - 2t + 1 = \tfrac{5}{4} + 1$$
$$(t - 1)^2 = \tfrac{9}{4}$$

$t - 1 = \sqrt{\tfrac{9}{4}} \qquad$ or $\qquad t - 1 = -\sqrt{\tfrac{9}{4}}$

$t - 1 = \tfrac{3}{2} \qquad$ or $\qquad t - 1 = -\tfrac{3}{2}$

$t = 1 + \tfrac{3}{2} \qquad$ or $\qquad t = 1 - \tfrac{3}{2}$

$t = \tfrac{5}{2} = 2\tfrac{1}{2} \qquad$ or $\qquad t = -\tfrac{1}{2}$

Reject the negative value of t. The object will reach a height of 30 feet after 2.5 seconds.

20. Use the Pythagorean theorem with legs x and $x + 2$ and hypotenuse $x + 4$.

$$a^2 + b^2 = c^2$$
$$(x)^2 + (x + 2)^2 = (x + 4)^2$$
$$x^2 + x^2 + 4x + 4 = x^2 + 8x + 16$$
$$x^2 - 4x - 12 = 0$$
$$(x - 6)(x + 2) = 0$$

$x - 6 = 0 \qquad$ or $\qquad x + 2 = 0$

$x = 6 \qquad$ or $\qquad x = -2$

Reject the negative value because x represents a length. The value of x is 6. The lengths of the three sides are 6, 8, and 10.

21. $m^2 - 5m - 36 = 0$ ■ Use the quadratic formula with $a = 1$, $b = -5$, and $c = -36$.

$$m = \frac{-b \pm \sqrt{b^2 - 4ac}}{2a}$$
$$m = \frac{-(-5) \pm \sqrt{(-5)^2 - 4(1)(-36)}}{2(1)}$$
$$= \frac{5 \pm \sqrt{25 + 144}}{2}$$
$$= \frac{5 \pm \sqrt{169}}{2}$$
$$= \frac{5 \pm 13}{2}$$

$m = \dfrac{5 + 13}{2} = \dfrac{18}{2} = 9$

or $\quad m = \dfrac{5 - 13}{2} = \dfrac{-8}{2} = -4$

A check verifies that the solution set is $\{-4, 9\}$.

22. $\qquad\qquad 5r^2 = 14r$

$5r^2 - 14r = 0 \qquad$ *Standard form*

Use the quadratic formula with $a = 5$, $b = -14$, and $c = 0$.

$$r = \frac{-b \pm \sqrt{b^2 - 4ac}}{2a}$$
$$r = \frac{-(-14) \pm \sqrt{(-14)^2 - 4(5)(0)}}{2(5)}$$
$$= \frac{14 \pm \sqrt{256}}{10}$$
$$= \frac{14 \pm 14}{10}$$

$r = \dfrac{14 + 14}{10} = \dfrac{28}{10} = \dfrac{14}{5}$

or $\quad r = \dfrac{14 - 14}{10} = \dfrac{0}{10} = 0$

A check verifies that the solution set is $\left\{0, \frac{14}{5}\right\}$.

23. $2x^2 - x + 3 = 0$ ■ Substitute $a = 2$, $b = -1$, and $c = 3$ in the quadratic formula.

$$x = \frac{-b \pm \sqrt{b^2 - 4ac}}{2a}$$

$$x = \frac{-(-1) \pm \sqrt{(-1)^2 - 4(2)(3)}}{2(2)}$$

$$= \frac{1 \pm \sqrt{1 - 24}}{4}$$

$$= \frac{1 \pm \sqrt{-23}}{4}$$

Because $\sqrt{-23}$ does not represent a real number, there is *no real number solution*. The solution set is $\emptyset$.

24. $4w^2 - 12w + 9 = 0$ ■ Substitute $a = 4$, $b = -12$, and $c = 9$ in the quadratic formula.

$$w = \frac{-b \pm \sqrt{b^2 - 4ac}}{2a}$$

$$w = \frac{-(-12) \pm \sqrt{(-12)^2 - 4(4)(9)}}{2(4)}$$

$$= \frac{12 \pm \sqrt{144 - 144}}{8}$$

$$= \frac{12 \pm 0}{8} = \frac{12}{8} = \frac{3}{2}.$$

The solution set is $\left\{\frac{3}{2}\right\}$. Note that the discriminant is 0.

25. $-4x^2 - 2x + 7 = 0$

$$4x^2 + 2x - 7 = 0 \quad \textit{Multiply by } -1.$$

Use the quadratic formula with $a = 4$, $b = 2$, and $c = -7$.

$$x = \frac{-b \pm \sqrt{b^2 - 4ac}}{2a}$$

$$x = \frac{-2 \pm \sqrt{(2)^2 - 4(4)(-7)}}{2(4)}$$

$$= \frac{-2 \pm \sqrt{4 + 112}}{8} = \frac{-2 \pm \sqrt{116}}{8}$$

$$= \frac{-2 \pm 2\sqrt{29}}{8} = \frac{2(-1 \pm \sqrt{29})}{2(4)}$$

$$= \frac{-1 \pm \sqrt{29}}{4}$$

A check verifies that the solution set is

$$\left\{\frac{-1 + \sqrt{29}}{4}, \frac{-1 - \sqrt{29}}{4}\right\}.$$

26. $2x^2 + 8 = 4x + 11$

$$2x^2 - 4x - 3 = 0 \qquad \textit{Standard form}$$

Use the quadratic formula with $a = 2$, $b = -4$, and $c = -3$.

$$x = \frac{-b \pm \sqrt{b^2 - 4ac}}{2a}$$

$$x = \frac{-(-4) \pm \sqrt{(-4)^2 - 4(2)(-3)}}{2(2)}$$

$$= \frac{4 \pm \sqrt{16 + 24}}{4} = \frac{4 \pm \sqrt{40}}{4}$$

$$= \frac{4 \pm \sqrt{4 \cdot 10}}{4} = \frac{4 \pm 2\sqrt{10}}{4}$$

$$= \frac{2(2 \pm \sqrt{10})}{2(2)} = \frac{2 \pm \sqrt{10}}{2}$$

A check verifies that the solution set is

$$\left\{\frac{2 + \sqrt{10}}{2}, \frac{2 - \sqrt{10}}{2}\right\}.$$

27. $x(5x - 1) = 1$

$$5x^2 - x = 1 \quad \textit{Multiply.}$$

$$5x^2 - x - 1 = 0 \quad \textit{Standard form}$$

Use the quadratic formula with $a = 5$, $b = -1$, and $c = -1$.

$$x = \frac{-b \pm \sqrt{b^2 - 4ac}}{2a}$$

$$x = \frac{-(-1) \pm \sqrt{(-1)^2 - 4(5)(-1)}}{2(5)}$$

$$= \frac{1 \pm \sqrt{1 + 20}}{10}$$

$$= \frac{1 \pm \sqrt{21}}{10}$$

A check verifies that the solution set is

$$\left\{\frac{1 + \sqrt{21}}{10}, \frac{1 - \sqrt{21}}{10}\right\}.$$

28. $\frac{1}{4}x^2 = 2 - \frac{3}{4}x$

$$\frac{1}{4}x^2 + \frac{3}{4}x - 2 = 0 \qquad \textit{Standard form}$$

$$x^2 + 3x - 8 = 0 \qquad \textit{Multiply by 4.}$$

Use the quadratic formula with $a = 1$, $b = 3$, and $c = -8$.

$$x = \frac{-b \pm \sqrt{b^2 - 4ac}}{2a}$$

$$x = \frac{-3 \pm \sqrt{3^2 - 4(1)(-8)}}{2(1)}$$

$$= \frac{-3 \pm \sqrt{9 + 32}}{2}$$

$$= \frac{-3 \pm \sqrt{41}}{2}$$

A check verifies that the solution set is

$$\left\{\frac{-3 + \sqrt{41}}{2}, \frac{-3 - \sqrt{41}}{2}\right\}.$$

29. $\frac{1}{2}x^2 + 3x = 5$

$\frac{1}{2}x^2 + 3x - 5 = 0$ *Standard form*

$x^2 + 6x - 10 = 0$ *Multiply by 2.*

Use the quadratic formula with
$a = 1$, $b = 6$, and $c = -10$.

$$x = \frac{-b \pm \sqrt{b^2 - 4ac}}{2a}$$

$$x = \frac{-6 \pm \sqrt{6^2 - 4(1)(-10)}}{2(1)}$$

$$= \frac{-6 \pm \sqrt{36 + 40}}{2} = \frac{-6 \pm \sqrt{76}}{2}$$

$$= \frac{-6 \pm 2\sqrt{19}}{2} = -3 \pm \sqrt{19}$$

A check verifies that the solution set is
$\{-3 + \sqrt{19}, -3 - \sqrt{19}\}$.

30. $0.2x^2 = 0.4x - 0.1$

$0.2x^2 - 0.4x + 0.1 = 0$ *Standard form*

$2x^2 - 4x + 1 = 0$ *Multiply by 10.*

Use the quadratic formula with
$a = 2$, $b = -4$, and $c = 1$.

$$x = \frac{-b \pm \sqrt{b^2 - 4ac}}{2a}$$

$$x = \frac{-(-4) \pm \sqrt{(-4)^2 - 4(2)(1)}}{2(2)}$$

$$= \frac{4 \pm \sqrt{16 - 8}}{4} = \frac{4 \pm \sqrt{8}}{4}$$

$$= \frac{4 \pm \sqrt{4 \cdot 2}}{4} = \frac{4 \pm 2\sqrt{2}}{4}$$

$$= \frac{2(2 \pm \sqrt{2})}{2(2)} = \frac{2 \pm \sqrt{2}}{2}$$

A check verifies that the solution set is

$$\left\{ \frac{2 + \sqrt{2}}{2}, \frac{2 - \sqrt{2}}{2} \right\}.$$

31. $y = -x^2 + 5$ ■   If $x = 0$, $y = 5$, so the
y-intercept is $(0, 5)$.

To find any x-intercepts, let $y = 0$.

$$0 = -x^2 + 5$$
$$x^2 = 5$$
$$x = \pm\sqrt{5} \approx \pm 2.24$$

The x-intercepts are $(\pm\sqrt{5}, 0)$.

The x-value of the vertex is

$$x = -\frac{b}{2a} = -\frac{0}{2(-1)} = 0.$$

Thus, the vertex is the same as the y-intercept
(since $x = 0$). The axis of symmetry is the vertical
line $x = 0$.

Make a table of ordered pairs whose x-values are
on either side of the vertex's x-value of $x = 0$.

x	y
0	5
± 1	4
± 2	1
± 3	-4

Plot these seven ordered pairs and connect them
with a smooth curve.

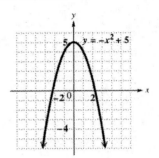

32. $y = (x - 1)^2 = x^2 - 2x + 1$

Let $x = 0$ to get

$$y = 0^2 - 2(0) + 1 = 1;$$

the y-intercept is $(0, 1)$.

Let $y = 0$ and solve for x.

$$0 = (x - 1)^2$$
$$0 = x - 1 \qquad \textit{Take square root.}$$
$$x = 1$$

The x-intercept is $(1, 0)$.

The x-value of the vertex is

$$x = -\frac{b}{2a} = -\frac{-2}{2(1)} = 1.$$

The y-value of the vertex is

$$y = 1^2 - 2(1) + 1 = 0,$$

so the vertex is $(1, 0)$.

Make a table of ordered pairs whose x-values are
on either side of the vertex's x-value of $x = 1$.

x	y
-1	4
0	1
1	0
2	1
3	4

Plot these five ordered pairs and connect them with a smooth curve.

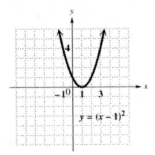

33. $y = -x^2 + 2x + 3$ ■ Let $x = 0$ to get

$$y = -0^2 + 2(0) + 3 = 3;$$

the y-intercept is $(0, 3)$.

Let $y = 0$ and solve for x.

$$0 = -x^2 + 2x + 3$$
$$x^2 - 2x - 3 = 0$$
$$(x - 3)(x + 1) = 0$$

$$x - 3 = 0 \quad \text{or} \quad x + 1 = 0$$
$$x = 3 \quad \text{or} \quad x = -1$$

The x-intercepts are $(3, 0)$ and $(-1, 0)$.

The x-value of the vertex is

$$x = -\frac{b}{2a} = -\frac{2}{2(-1)} = 1.$$

The y-value of the vertex is

$$y = -1^2 + 2(1) + 3 = 4,$$

so the vertex is $(1, 4)$.

Make a table of ordered pairs whose x-values are on either side of the vertex's x-value of $x = 1$.

x	y
-1	0
0	3
1	4
2	3
3	0

Plot these five ordered pairs and connect them with a smooth curve.

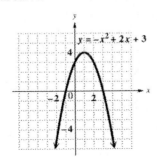

34. $y = x^2 + 4x + 2$ ■ Let $x = 0$ to get

$$y = 0^2 + 4(0) + 2 = 2;$$

the y-intercept is $(0, 2)$.

Let $y = 0$ and solve for x.

$$x^2 + 4x + 2 = 0$$
$$x^2 + 4x = -2$$
$$x^2 + 4x + 4 = -2 + 4$$
$$(x + 2)^2 = 2$$

$$x + 2 = \sqrt{2} \quad \text{or} \quad x + 2 = -\sqrt{2}$$
$$x = -2 + \sqrt{2} \quad \text{or} \quad x = -2 - \sqrt{2}$$
$$x \approx -0.6 \quad \text{or} \quad x \approx -3.4$$

The x-intercepts are approximately $(-0.6, 0)$ and $(-3.4, 0)$.

The x-value of the vertex is

$$x = -\frac{b}{2a} = -\frac{4}{2(1)} = -2.$$

The y-value of the vertex is

$$y = (-2)^2 + 4(-2) + 2 = -2,$$

so the vertex is $(-2, -2)$.

Make a table of ordered pairs whose x-values are on either side of the vertex's x-value of $x = -2$.

x	y
-4	2
-3.4	0
-3	-1
-2	-2
-1	-1
-0.6	0
0	2

Plot these seven ordered pairs and connect them with a smooth curve.

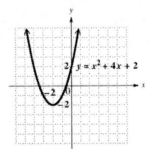

35. $\{(-2, 4), (0, 8), (2, 5), (2, 3)\}$

Since $x = 2$ appears in two ordered pairs, one value of x yields more than one value of y. Hence, this relation *is not a function*.

The domain is the set of first components of the ordered pairs, $\{-2, 0, 2\}$.

The range is the set of second components of the ordered pairs, $\{4, 8, 5, 3\}$.

36. $\{(8,3), (7,4), (6,5), (5,6), (4,7)\}$
Since each first component of the ordered pairs corresponds to exactly one second component, the relation *is a function.*

The domain is the set of first components of the ordered pairs, $\{8, 7, 6, 5, 4\}$.

The range is the set of second components of the ordered pairs, $\{3, 4, 5, 6, 7\}$.

37. Since a vertical line may intersect the graph twice, this is not the graph of a function.

38. Any vertical line will intersect this graph in exactly one point, so it is the graph of a function.

39. $2x + 3y = 12$ ■ Solve the equation for y.

$$2x + 3y = 12$$
$$3y = -2x + 12 \quad \textit{Subtract 2x.}$$
$$y = -\tfrac{2}{3}x + 4 \quad \textit{Divide by 3.}$$

Since one value of x will lead to only one value of y, the relation $2x + 3y = 12$ is a function.

40. $y = x^2$ ■ Each value of x will lead to only one value of y, so $y = x^2$ is a function.

41. $f(x) = 3x + 2$

(a) $f(2) = 3(2) + 2 = 6 + 2 = 8$

(b) $f(-1) = 3(-1) + 2 = -3 + 2 = -1$

42. $f(x) = 2x^2 - 1$

(a) $f(2) = 2(2)^2 - 1$
$= 2(4) - 1 = 8 - 1 = 7$

(b) $f(-1) = 2(-1)^2 - 1$
$= 2(1) - 1 = 2 - 1 = 1$

43. $f(x) = |x + 3|$

(a) $f(2) = |2 + 3| = |5| = 5$

(b) $f(-1) = |-1 + 3| = |2| = 2$

44. $f(x) = -x^2 + 2x - 3$

(a) $f(2) = -2^2 + 2(2) - 3$
$= -4 + 4 - 3 = -3$

(b) $f(-1) = -(-1)^2 + 2(-1) - 3$
$= -1 - 2 - 3 = -6$

45. **[17.1]** $(2t - 1)(t + 1) = 54$
$$2t^2 + t - 1 = 54 \quad \textit{Multiply.}$$
$$2t^2 + t - 55 = 0 \quad \textit{Standard form}$$
$$(2t + 11)(t - 5) = 0 \quad \textit{Factor.}$$

$2t + 11 = 0 \quad$ or $\quad t - 5 = 0$
$t = -\tfrac{11}{2} \quad$ or $\quad t = 5$

A check verifies that the solution set is $\{-\tfrac{11}{2}, 5\}$.

46. **[17.1]** $(2p + 1)^2 = 100$ ■ Use the square root property.

$2p + 1 = \sqrt{100} \quad$ or $\quad 2p + 1 = -\sqrt{100}$
$2p + 1 = 10 \quad$ or $\quad 2p + 1 = -10$
$2p = 9 \quad$ or $\quad 2p = -11$
$p = \tfrac{9}{2} \quad$ or $\quad p = -\tfrac{11}{2}$

A check verifies that the solution set is $\{-\tfrac{11}{2}, \tfrac{9}{2}\}$.

47. **[17.3]** $(k + 2)(k - 1) = 3$
$$k^2 + k - 2 = 3 \quad \textit{Multiply.}$$
$$k^2 + k - 5 = 0 \quad \textit{Standard form}$$

The left side cannot be factored, so use the quadratic formula with $a = 1$, $b = 1$, and $c = -5$.

$$k = \frac{-b \pm \sqrt{b^2 - 4ac}}{2a}$$
$$k = \frac{-1 \pm \sqrt{1^2 - 4(1)(-5)}}{2(1)}$$
$$= \frac{-1 \pm \sqrt{1 + 20}}{2}$$
$$= \frac{-1 \pm \sqrt{21}}{2}$$

A check verifies that the solution set is
$$\left\{\frac{-1 + \sqrt{21}}{2}, \frac{-1 - \sqrt{21}}{2}\right\}.$$

48. **[17.1]** $6t^2 + 7t - 3 = 0$
$$(3t - 1)(2t + 3) = 0 \quad \textit{Factor.}$$

$3t - 1 = 0 \quad$ or $\quad 2t + 3 = 0$
$t = \tfrac{1}{3} \quad$ or $\quad t = -\tfrac{3}{2}$

A check verifies that the solution set is $\{-\tfrac{3}{2}, \tfrac{1}{3}\}$.

49. **[17.3]** $2x^2 + 3x + 2 = x^2 - 2x$
$$x^2 + 5x + 2 = 0 \quad \textit{Standard form}$$

The left side cannot be factored, so use the quadratic formula with $a = 1$, $b = 5$, and $c = 2$.

$$x = \frac{-b \pm \sqrt{b^2 - 4ac}}{2a}$$
$$x = \frac{-5 \pm \sqrt{5^2 - 4(1)(2)}}{2(1)}$$
$$= \frac{-5 \pm \sqrt{25 - 8}}{2} = \frac{-5 \pm \sqrt{17}}{2}$$

A check verifies that the solution set is
$$\left\{\frac{-5 + \sqrt{17}}{2}, \frac{-5 - \sqrt{17}}{2}\right\}.$$

50. **[17.3]** $x^2 + 2x + 5 = 7$
$$x^2 + 2x - 2 = 0 \quad \textit{Standard form}$$

The left side cannot be factored, so use the quadratic formula with $a = 1$, $b = 2$, and $c = -2$.

$$x = \frac{-b \pm \sqrt{b^2 - 4ac}}{2a}$$

$$x = \frac{-2 \pm \sqrt{2^2 - 4(1)(-2)}}{2(1)}$$

$$= \frac{-2 \pm \sqrt{4 + 8}}{2}$$

$$= \frac{-2 \pm \sqrt{12}}{2} = \frac{-2 \pm 2\sqrt{3}}{2}$$

$$= \frac{2(-1 \pm \sqrt{3})}{2} = -1 \pm \sqrt{3}$$

A check verifies that the solution set is $\{-1 + \sqrt{3}, -1 - \sqrt{3}\}$.

51. **[17.3]** $m^2 - 4m + 10 = 0$ ■ Use the quadratic formula with $a = 1$, $b = -4$, and $c = 10$.

$$m = \frac{-b \pm \sqrt{b^2 - 4ac}}{2a}$$

$$m = \frac{-(-4) \pm \sqrt{(-4)^2 - 4(1)(10)}}{2(1)}$$

$$= \frac{4 \pm \sqrt{16 - 40}}{2}$$

$$= \frac{4 \pm \sqrt{-24}}{2}$$

Because $\sqrt{-24}$ does not represent a real number, there is *no real number solution*. The solution set is $\emptyset$.

52. **[17.3]** $k^2 - 9k + 10 = 0$ ■ The left side cannot be factored, so use the quadratic formula with $a = 1$, $b = -9$, and $c = 10$.

$$k = \frac{-b \pm \sqrt{b^2 - 4ac}}{2a}$$

$$k = \frac{-(-9) \pm \sqrt{(-9)^2 - 4(1)(10)}}{2(1)}$$

$$= \frac{9 \pm \sqrt{81 - 40}}{2}$$

$$= \frac{9 \pm \sqrt{41}}{2}$$

A check verifies that the solution set is $\left\{ \dfrac{9 + \sqrt{41}}{2}, \dfrac{9 - \sqrt{41}}{2} \right\}$.

53. **[17.1]** $(5x + 6)^2 = 0$

$$5x + 6 = 0 \quad \textit{Take square root.}$$

$$5x = -6$$

$$x = -\tfrac{6}{5}$$

A check verifies that the solution set is $\{-\tfrac{6}{5}\}$.

54. **[17.3]**

$$\tfrac{1}{2}r^2 = \tfrac{7}{2} - r$$

$$r^2 = 7 - 2r \quad \textit{Mult. by LCD, 2.}$$

$$r^2 + 2r - 7 = 0 \quad \textit{Standard form}$$

The left side cannot be factored, so use the quadratic formula with $a = 1$, $b = 2$, and $c = -7$.

$$r = \frac{-b \pm \sqrt{b^2 - 4ac}}{2a}$$

$$r = \frac{-2 \pm \sqrt{2^2 - 4(1)(-7)}}{2(1)}$$

$$= \frac{-2 \pm \sqrt{4 + 28}}{2}$$

$$= \frac{-2 \pm \sqrt{32}}{2}$$

$$= \frac{-2 \pm 4\sqrt{2}}{2}$$

$$= \frac{2(-1 \pm 2\sqrt{2})}{2}$$

$$= -1 \pm 2\sqrt{2}$$

A check verifies that the solution set is $\{-1 + 2\sqrt{2}, -1 - 2\sqrt{2}\}$.

55. **[17.3]**

$$x^2 + 4x = 1$$

$$x^2 + 4x - 1 = 0 \quad \textit{Standard form}$$

The left side cannot be factored, so use the quadratic formula with $a = 1$, $b = 4$, and $c = -1$.

$$x = \frac{-b \pm \sqrt{b^2 - 4ac}}{2a}$$

$$x = \frac{-4 \pm \sqrt{4^2 - 4(1)(-1)}}{2(1)}$$

$$= \frac{-4 \pm \sqrt{16 + 4}}{2}$$

$$= \frac{-4 \pm \sqrt{20}}{2} = \frac{-4 \pm 2\sqrt{5}}{2}$$

$$= \frac{2(-2 \pm \sqrt{5})}{2} = -2 \pm \sqrt{5}$$

A check verifies that the solution set is $\{-2 + \sqrt{5}, -2 - \sqrt{5}\}$.

56. **[17.1]** $7x^2 - 8 = 5x^2 + 8$

$$2x^2 = 16$$

$$x^2 = 8$$

$$x = \pm\sqrt{8} = \pm 2\sqrt{2}$$

A check verifies that the solution set is $\{\pm 2\sqrt{2}\}$.

57. **[17.3]** $x^2 - 9 = 0$, or $1x^2 + 0x - 9 = 0$

(a) $(x + 3)(x - 3) = 0$

$$x + 3 = 0 \quad \text{or} \quad x - 3 = 0$$

$$x = -3 \quad \text{or} \quad x = 3$$

A check verifies that the solution set is $\{-3, 3\}$.

(b) $x^2 = 9$

$$x = \pm\sqrt{9} = \pm 3$$

A check verifies that the solution set is $\{-3, 3\}$.

(c) Here, $a = 1$, $b = 0$, and $c = -9$.

$$x = \frac{-0 \pm \sqrt{0^2 - 4(1)(-9)}}{2(1)}$$

$$= \frac{\pm\sqrt{36}}{2}$$

$$= \frac{\pm 6}{2} = \pm 3$$

A check verifies that the solution set is $\{-3, 3\}$.

(d) We will always get the same results, no matter which method of solution is used.

58. **[17.3]** The fraction bar should be under both $-b$ and $\sqrt{b^2 - 4ac}$. The term $2a$ should not be in the radicand. The correct formula is

$$x = \frac{-b \pm \sqrt{b^2 - 4ac}}{2a}.$$

Chapter 17 Test

1. $x^2 = 39$ ▪ Use the square root property.

$$x = \sqrt{39} \quad \text{or} \quad x = -\sqrt{39}$$

A check verifies that the solution set is $\{-\sqrt{39}, \sqrt{39}\}$.

2. $(x + 3)^2 = 64$ ▪ Use the square root property.

$$\begin{array}{llll} x + 3 = \sqrt{64} & \text{or} & x + 3 = -\sqrt{64} \\ x + 3 = 8 & \text{or} & x + 3 = -8 \\ x = 5 & \text{or} & x = -11 \end{array}$$

A check verifies that the solution set is $\{-11, 5\}$.

3. $(4x + 3)^2 = 24$ ▪ Use the square root property.

$$4x + 3 = \sqrt{24} \quad \text{or} \quad 4x + 3 = -\sqrt{24}$$

Note that $\sqrt{24} = \sqrt{4 \cdot 6} = 2\sqrt{6}$.

$$\begin{array}{llll} 4x + 3 = 2\sqrt{6} & \text{or} & 4x + 3 = -2\sqrt{6} \\ 4x = -3 + 2\sqrt{6} & \text{or} & 4x = -3 - 2\sqrt{6} \\ x = \dfrac{-3 + 2\sqrt{6}}{4} & \text{or} & x = \dfrac{-3 - 2\sqrt{6}}{4} \end{array}$$

A check verifies that the solution set is

$$\left\{ \frac{-3 + 2\sqrt{6}}{4}, \frac{-3 - 2\sqrt{6}}{4} \right\}.$$

4. $x^2 - 4x = 6$

Solve by completing the square.

$$x^2 - 4x + 4 = 6 + 4 \qquad Add \ \left[\tfrac{1}{2}(-4)\right]^2 = 4.$$

$$(x - 2)^2 = 10$$

Use the square root property.

$$\begin{array}{llll} x - 2 = \sqrt{10} & \text{or} & x - 2 = -\sqrt{10} \\ x = 2 + \sqrt{10} & \text{or} & x = 2 - \sqrt{10} \end{array}$$

A check verifies that the solution set is $\{2 + \sqrt{10}, 2 - \sqrt{10}\}$.

5. $2x^2 + 12x - 3 = 0$

Solve by completing the square.

$$\begin{array}{ll} x^2 + 6x - \tfrac{3}{2} = 0 & \textit{Divide by 2.} \\ x^2 + 6x = \tfrac{3}{2} & \\ x^2 + 6x + 9 = \tfrac{3}{2} + 9 & \textit{Add } \left[\tfrac{1}{2}(6)\right]^2 = 9. \\ (x + 3)^2 = \tfrac{21}{2} & \end{array}$$

Use the square root property.

$$x + 3 = \sqrt{\tfrac{21}{2}} \quad \text{or} \quad x + 3 = -\sqrt{\tfrac{21}{2}}$$

Note that

$$\sqrt{\frac{21}{2}} = \frac{\sqrt{21}}{\sqrt{2}} = \frac{\sqrt{21} \cdot \sqrt{2}}{\sqrt{2} \cdot \sqrt{2}} = \frac{\sqrt{42}}{2}.$$

$$\begin{array}{llll} x + 3 = \dfrac{\sqrt{42}}{2} & \text{or} & x + 3 = -\dfrac{\sqrt{42}}{2} \\ x = -3 + \dfrac{\sqrt{42}}{2} & \text{or} & x = -3 - \dfrac{\sqrt{42}}{2} \\ x = \dfrac{-6 + \sqrt{42}}{2} & \text{or} & x = \dfrac{-6 - \sqrt{42}}{2} \end{array}$$

A check verifies that the solution set is

$$\left\{ \frac{-6 + \sqrt{42}}{2}, \frac{-6 - \sqrt{42}}{2} \right\}.$$

6. $2x^2 + 5x - 3 = 0$ ▪ Use $a = 2$, $b = 5$, and $c = -3$ in the quadratic formula.

$$x = \frac{-b \pm \sqrt{b^2 - 4ac}}{2a}$$

$$x = \frac{-5 \pm \sqrt{5^2 - 4(2)(-3)}}{2(2)}$$

$$x = \frac{-5 \pm \sqrt{25 + 24}}{4}$$

$$x = \frac{-5 \pm \sqrt{49}}{4} = \frac{-5 \pm 7}{4}$$

$$\begin{array}{llll} x = \dfrac{-5 + 7}{4} & \text{or} & x = \dfrac{-5 - 7}{4} \\ x = \dfrac{2}{4} = \dfrac{1}{2} & \text{or} & x = \dfrac{-12}{4} = -3 \end{array}$$

A check verifies that the solution set is $\{-3, \tfrac{1}{2}\}$.

7. $3w^2 + 2 = 6w$

$$3w^2 - 6w + 2 = 0 \qquad \textit{Standard form}$$

The left side cannot be factored, so use the quadratic formula with $a = 3$, $b = -6$, and $c = 2$.

$$w = \frac{-b \pm \sqrt{b^2 - 4ac}}{2a}$$

$$w = \frac{-(-6) \pm \sqrt{(-6)^2 - 4(3)(2)}}{2(3)}$$

$$w = \frac{6 \pm \sqrt{36 - 24}}{6}$$

$$w = \frac{6 \pm \sqrt{12}}{6} = \frac{6 \pm 2\sqrt{3}}{6}$$

$$w = \frac{2(3 \pm \sqrt{3})}{2(3)} = \frac{3 \pm \sqrt{3}}{3}$$

A check verifies that the solution set is
$$\left\{ \frac{3 + \sqrt{3}}{3}, \frac{3 - \sqrt{3}}{3} \right\}.$$

8. $4x^2 + 8x + 11 = 0$ ■   Use $a = 4, b = 8$, and $c = 11$ in the quadratic formula.

$$x = \frac{-b \pm \sqrt{b^2 - 4ac}}{2a}$$

$$x = \frac{-8 \pm \sqrt{8^2 - 4(4)(11)}}{2(4)}$$

$$x = \frac{-8 \pm \sqrt{64 - 176}}{8}$$

$$x = \frac{-8 \pm \sqrt{-112}}{8}$$

The radical $\sqrt{-112}$ is not a real number, so the equation has no *real* number solution. The solution set is $\emptyset$.

9. $t^2 - \frac{5}{3}t + \frac{1}{3} = 0$

$3(t^2 - \frac{5}{3}t + \frac{1}{3}) = 3(0)$ *Multiply by 3.*

$3t^2 - 5t + 1 = 0$

The left side cannot be factored, so use the quadratic formula with $a = 3, b = -5$, and $c = 1$.

$$t = \frac{-b \pm \sqrt{b^2 - 4ac}}{2a}$$

$$t = \frac{-(-5) \pm \sqrt{(-5)^2 - 4(3)(1)}}{2(3)}$$

$$t = \frac{5 \pm \sqrt{25 - 12}}{6} = \frac{5 \pm \sqrt{13}}{6}$$

A check verifies that the solution set is
$$\left\{ \frac{5 + \sqrt{13}}{6}, \frac{5 - \sqrt{13}}{6} \right\}.$$

10. $p^2 - 2p - 1 = 0$

Solve by completing the square.

$p^2 - 2p = 1$

$p^2 - 2p + 1 = 1 + 1$ *Add $\left[\frac{1}{2}(-2) \right]^2 = 1$.*

$(p - 1)^2 = 2$

Use the square root property.

$p - 1 = \sqrt{2}$ or $p - 1 = -\sqrt{2}$

$p = 1 + \sqrt{2}$ or $p = 1 - \sqrt{2}$

A check verifies that the solution set is
$\{ 1 + \sqrt{2}, 1 - \sqrt{2} \}$.

11. $(2x + 1)^2 = 18$ ■   Use the square root property.

$$2x + 1 = \pm\sqrt{18}$$

$$2x + 1 = \pm 3\sqrt{2}$$

$$2x = -1 \pm 3\sqrt{2}$$

$$x = \frac{-1 \pm 3\sqrt{2}}{2}$$

A check verifies that the solution set is
$$\left\{ \frac{-1 + 3\sqrt{2}}{2}, \frac{-1 - 3\sqrt{2}}{2} \right\}.$$

12. $(x - 5)(2x - 1) = 1$

$2x^2 - 11x + 5 = 1$ *Multiply.*

$2x^2 - 11x + 4 = 0$ *Standard form*

Use $a = 2, b = -11$, and $c = 4$ in the quadratic formula.

$$x = \frac{-b \pm \sqrt{b^2 - 4ac}}{2a}$$

$$x = \frac{-(-11) \pm \sqrt{(-11)^2 - 4(2)(4)}}{2(2)}$$

$$= \frac{11 \pm \sqrt{121 - 32}}{4} = \frac{11 \pm \sqrt{89}}{4}$$

A check verifies that the solution set is
$$\left\{ \frac{11 + \sqrt{89}}{4}, \frac{11 - \sqrt{89}}{4} \right\}.$$

13. $t^2 + 25 = 10t$

$t^2 - 10t + 25 = 0$ *Subtract 10t.*

$(t - 5)^2 = 0$ *Factor.*

$t - 5 = 0$ *Take square root.*

$t = 5$

A check verifies that the solution set is $\{5\}$.

14. $s = -16t^2 + 64t$ ■   Let $s = 64$ and solve for t.

$-16t^2 + 64t = 64$ *Let s = 64.*

$-16t^2 + 64t - 64 = 0$ *Subtract 64.*

$t^2 - 4t + 4 = 0$ *Divide by –16.*

$(t - 2)^2 = 0$ *Factor.*

$t - 2 = 0$ *Take square root.*

$t = 2$

The object will reach a height of 64 feet after 2 seconds.

15. Use the Pythagorean theorem.
$$c^2 = a^2 + b^2$$
$$(x+8)^2 = (x)^2 + (x+4)^2$$
$$x^2 + 16x + 64 = x^2 + x^2 + 8x + 16$$
$$0 = x^2 - 8x - 48$$
$$0 = (x-12)(x+4)$$

Use the zero-factor property.

$$x - 12 = 0 \quad \text{or} \quad x + 4 = 0$$
$$x = 12 \quad \text{or} \quad x = -4$$

Disregard a negative length. The sides measure 12, $x + 4 = 12 + 4 = 16$, and $x + 8 = 12 + 8 = 20$.

16. $y = (x-3)^2 = x^2 - 6x + 9$

If $x = 0$, $y = 9$, so the y-intercept is $(0, 9)$.

To find any x-intercepts, let $y = 0$.
$$0 = (x-3)^2$$
$$0 = x - 3$$
$$3 = x$$

The x-intercept is $(3, 0)$.

The x-value of the vertex is
$$x = -\frac{b}{2a} = -\frac{-6}{2(1)} = 3.$$

Thus, the vertex is the same as the x-intercept. The axis of symmetry is the vertical line $x = 3$.

Make a table of ordered pairs whose x-values are on either side of the vertex's x-value of $x = 3$.

x	y
0	9
1	4
2	1
3	0
4	1
5	4
6	9

Plot these seven ordered pairs and connect them with a smooth curve.

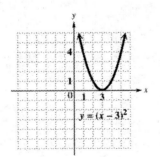

17. $y = -x^2 - 2x - 4$ ■ Let $x = 0$ to get
$$y = -(0)^2 - 2(0) - 4 = -4;$$

the y-intercept is $(0, -4)$.

Let $y = 0$ and solve for x.
$$0 = -x^2 - 2x - 4$$
$$x^2 + 2x + 4 = 0$$

Use $a = 1$, $b = 2$, and $c = 4$ in the quadratic formula.

$$x = \frac{-b \pm \sqrt{b^2 - 4ac}}{2a}$$
$$x = \frac{-2 \pm \sqrt{2^2 - 4(1)(4)}}{2(1)}$$
$$= \frac{-2 \pm \sqrt{-12}}{2}$$

$\sqrt{-12}$ is not a real number, so this equation has no real number solution, and there are no x-intercepts.

The x-value of the vertex is
$$x = -\frac{b}{2a} = -\frac{-2}{2(-1)} = -1.$$

The y-value of the vertex is
$$y = -(-1)^2 - 2(-1) - 4 = -3,$$

so the vertex is $(-1, -3)$.

The axis of symmetry is the line $x = -1$.

Make a table of ordered pairs whose x-values are on either side of the vertex's x-value of $x = -1$.

x	y
-3	-7
-2	-4
-1	-3
0	-4
1	-7

Plot these five ordered pairs and connect them with a smooth curve.

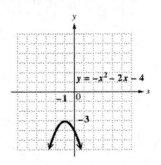

18. **(a)** $\{(2,3),(2,4),(2,5)\}$

Since $x = 2$ appears in more than one ordered pair, one value of x yields more than one value of y. Hence, this relation *is not a function*.

(b) $\{(0,2),(1,2),(2,2)\}$

Since each first component of the ordered pairs corresponds to exactly one second component, the relation *is a function*. The domain is $\{0,1,2\}$ and the range is $\{2\}$.

19. The vertical line test shows that this graph is not the graph of a function; a vertical line could intersect the graph twice.

20. $f(x) = 3x + 7$
$f(-2) = 3(-2) + 7 = -6 + 7 = 1$

WHOLE NUMBERS COMPUTATION: PRETEST

Adding Whole Numbers

1.

$$\overset{1}{3}68$$
$$\underline{+\ 22}$$
$$390$$

3.

$$\overset{111}{85}$$
$$\underline{+\ 2968}$$
$$3053$$

5. $714 + 3728 + 9 + 683{,}775$

$$\overset{2}{\ }\ \overset{12}{7}\overset{}{1}4$$
$$3\ 728$$
$$9$$
$$\underline{+\ 683{,}775}$$
$$688{,}226$$

Subtracting Whole Numbers

1.

$$\overset{312}{\cancel{4}\cancel{2}6}$$
$$\underline{-\ 7\ 6}$$
$$3\ 5\ 0$$

3.

$$\overset{9}{\ }\ \overset{15}{\ }\ \overset{9}{\ }$$
$$\overset{2}{\ }\cancel{10}\ \cancel{8}\ \cancel{10}\overset{12}{\ }$$
$$\cancel{3}\cancel{0}{,}\cancel{6}\cancel{0}\cancel{2}$$
$$\underline{-\ 5\ 7\ 0\ 8}$$
$$2\ 4{,}8\ 9\ 4$$

5. $679{,}420 - 88{,}033$

$$\overset{}{\ }\ \overset{11}{\ }$$
$$\overset{517}{\ }\ \overset{3}{\cancel{4}}\overset{1}{\cancel{2}}\overset{10}{\ }$$
$$\cancel{6}79{,}\cancel{4}\cancel{2}\cancel{0}$$
$$\underline{-\ 8\ 8{,}0\ 3\ 3}$$
$$5\ 9\ 1{,}3\ 8\ 7$$

Multiplying Whole Numbers

1. $3 \times 3 \times 0 \times 6 = 0$ because 0 times any number is 0.

3. $(520)(3000)$

$$520$$
$$\underline{\times\ 3\,000}$$
$$1{,}560{,}000$$

5. Multiply 359 and 48.

$$\overset{23}{\ }$$
$$\overset{47}{\ }$$
$$359$$
$$\underline{\times\ 48}$$
$$2\,872 \quad \leftarrow \quad 8 \times 359$$
$$\underline{14\,36} \quad \leftarrow \quad 4 \times 359$$
$$17{,}232$$

Dividing Whole Numbers

1.

$$\begin{array}{r} 2\ 3 \\ 3\,\overline{)6\ 9} \\ \underline{6} \\ 9 \\ \underline{9} \\ 0 \end{array}$$

3. $\dfrac{25{,}036}{4}$

$$\begin{array}{r} 6\ 2\ 5\ 9 \\ 4\,\overline{)2\ 5{,}0\ 3\ 6} \\ \underline{2\ 4} \\ 1\ 0 \\ \underline{8} \\ 2\ 3 \\ \underline{2\ 0} \\ 3\ 6 \\ \underline{3\ 6} \\ 0 \end{array}$$

5.

$$\begin{array}{r} 3\ 4 \\ 52\,\overline{)1\ 7\ 6\ 8} \\ \underline{1\ 5\ 6} \\ 2\ 0\ 8 \\ \underline{2\ 0\ 8} \\ 0 \end{array}$$

7.

$$\begin{array}{r} 6\ 0 \\ 38\,\overline{)2\ 3\ 0\ 0} \\ \underline{2\ 2\ 8} \\ 2\ 0 \\ \underline{0} \\ 2\ 0 \end{array}$$

Answer: 60 **R**20

CHAPTER R WHOLE NUMBERS REVIEW

R.1 Adding Whole Numbers

R.1 Margin Exercises

1. **(a)** $3 + 4 = 7$; $4 + 3 = 7$

(b) $9 + 9 = 18$;
No change occurs when the commutative property is used.

(c) $7 + 8 = 15$; $8 + 7 = 15$

(d) $6 + 9 = 15$; $9 + 6 = 15$

2. **(a)**
$$\begin{array}{r} 5 \\ 4 \\ 6 \\ 9 \\ +2 \\ \hline 26 \end{array}$$
$5 + 4 = 9$
$9 + 6 = 15$
$15 + 9 = 24$
$24 + 2 = 26$

(b)
$$\begin{array}{r} 7 \\ 5 \\ 1 \\ 2 \\ +6 \\ \hline 21 \end{array}$$
$7 + 5 = 12$
$12 + 1 = 13$
$13 + 2 = 15$
$15 + 6 = 21$

(c)
$$\begin{array}{r} 9 \\ 2 \\ 1 \\ 3 \\ +4 \\ \hline 19 \end{array}$$
$9 + 2 = 11$
$11 + 1 = 12$
$12 + 3 = 15$
$15 + 4 = 19$

(d)
$$\begin{array}{r} 3 \\ 8 \\ 6 \\ 4 \\ +8 \\ \hline 29 \end{array}$$
$3 + 8 = 11$
$11 + 6 = 17$
$17 + 4 = 21$
$21 + 8 = 29$

3. **(a)** $\begin{array}{r} 25 \\ +73 \\ \hline 98 \end{array}$ **(b)** $\begin{array}{r} 364 \\ +532 \\ \hline 896 \end{array}$ **(c)** $\begin{array}{r} 42{,}305 \\ +11{,}563 \\ \hline 53{,}868 \end{array}$

4. **(a)** $\begin{array}{r} \overset{1}{69} \\ +26 \\ \hline 95 \end{array}$ $9 + 6 = 15$

(b) $\begin{array}{r} \overset{1}{76} \\ +18 \\ \hline 94 \end{array}$ $6 + 8 = 14$

(c) $\begin{array}{r} \overset{1}{56} \\ +37 \\ \hline 93 \end{array}$ $6 + 7 = 13$

(d) $\begin{array}{r} \overset{1}{34} \\ +49 \\ \hline 83 \end{array}$ $4 + 9 = 13$

5. **(a)**
$$\begin{array}{r} \overset{22}{481} \\ 79 \\ 38 \\ +395 \\ \hline 993 \end{array}$$
Add the numbers in the ones column: 23. Write 3, carry 2 to the tens column. Add the numbers in the tens column including the regrouped 2: 29. Write 9, carry 2 to the hundreds column. Add the numbers in the hundreds column including the regrouped 2: 9.

(b)
$$\begin{array}{r} \overset{1\,21}{4\,271} \\ 372 \\ 8\,976 \\ +\ \ 162 \\ \hline 13{,}781 \end{array}$$

(c)
$$\begin{array}{r} \overset{22}{\underset{}{57}} \\ 4 \\ 392 \\ 804 \\ 51 \\ +\ \ 27 \\ \hline 1335 \end{array}$$

(d)
$$\begin{array}{r} \overset{2\,21}{7\,821} \\ 435 \\ 72 \\ 305 \\ +1\,693 \\ \hline 10{,}326 \end{array}$$

(e)
$$\begin{array}{r} \overset{1\,25}{15{,}829} \\ 765 \\ 78 \\ 15 \\ 9 \\ 7 \\ +13{,}179 \\ \hline 29{,}882 \end{array}$$

6. From Conway to Pine Hills through Clear Lake:

$$\begin{array}{r} 7 \\ 11 \\ 8 \\ +5 \\ \hline 31 \end{array}$$
Conway to Shadow Hills
Shadow Hills to Clear Lake
Clear Lake to Orlando
Orlando to Pine Hills

From Conway to Pine Hills through Belle Isle:

$$
\begin{array}{rl}
6 & \textit{Conway to Belle Isle} \\
3 & \textit{Belle Isle to Pine Castle} \\
6 & \textit{Pine Castle to Resort Area} \\
9 & \textit{Resort Area to Orlando} \\
+\,5 & \textit{Orlando to Pine Hills} \\
\hline
29 &
\end{array}
$$

From Conway to Pine Hills through Casselberry:

$$
\begin{array}{rl}
7 & \textit{Conway to Shadow Hills} \\
9 & \textit{Shadow Hills to Bertha} \\
6 & \textit{Bertha to Casselberry} \\
5 & \textit{Casselberry to Altamonte Springs} \\
+\,8 & \textit{Altamonte Springs to Pine Hills} \\
\hline
35 &
\end{array}
$$

The shortest way from Conway to Pine Hills is through Belle Isle.

7. The next shortest route from Orlando to Clear Lake is as follows:

$$
\begin{array}{rl}
5 & \textit{Orlando to Pine Hills} \\
8 & \textit{Pine Hills to Altamonte Springs} \\
5 & \textit{Altamonte Springs to Casselberry} \\
6 & \textit{Casselberry to Bertha} \\
7 & \textit{Bertha to Winter Park} \\
+\,7 & \textit{Winter Park to Clear Lake} \\
\hline
38 & \textit{miles}
\end{array}
$$

The shortest way from Orlando to Clear Lake with the closed roads is through Casselberry.

8. **(a)**
$$
\begin{array}{r}
\overline{59} \\
32 \quad \uparrow \textit{Adding up and carrying mentally.} \\
8 \\
5 \\
+\,14 \\
\hline
59 \quad \textit{Sum of 59 is correct.}
\end{array}
$$

(b)
$$
\begin{array}{r}
\overline{1609} \\
872 \quad \uparrow \textit{Adding up and carrying mentally.} \\
539 \\
46 \\
+\,152 \\
\hline
1609 \quad \textit{Sum of 1609 is correct.}
\end{array}
$$

(c)
$$
\overline{943}
$$

$$
\begin{array}{rl}
79 \quad \uparrow \textit{Adding} & \overset{13}{79} \quad \textit{Adding} \\
218 \quad \textit{up} & 218 \quad \textit{down} \\
7 & 7 \\
+\,639 & +\,639 \downarrow \\
\hline
953 & 943
\end{array}
$$

The correct answer is 943.

(d)
$$
\overline{77{,}563}
$$

$$
\begin{array}{rl}
21{,}892 \quad \uparrow \textit{Adding} & \overset{2\ 11}{21{,}892} \quad \textit{Adding} \\
11{,}746 \quad \textit{up} & 11{,}746 \quad \textit{down} \\
+\,43{,}925 & +\,43{,}925 \downarrow \\
\hline
79{,}563 & 77{,}563
\end{array}
$$

The correct answer is 77,563.

R.1 Section Exercises

1. **(a)**
$$
\begin{array}{rl}
5 & \\
7 & 5 + 7 = 12 \\
6 & 12 + 6 = 18 \\
+\,5 & 18 + 5 = 23 \\
\hline
23 &
\end{array}
$$

(b)
$$
\begin{array}{rl}
9 & \\
2 & 9 + 2 = 11 \\
1 & 11 + 1 = 12 \\
3 & 12 + 3 = 15 \\
+\,4 & 15 + 4 = 19 \\
\hline
19 &
\end{array}
$$

3. **(a)** $3213 + 5715$ ■
$$
\begin{array}{r}
3213 \\
+\,5715 \\
\hline
8928
\end{array}
$$

(b) $38{,}204 + 21{,}020$ ■
$$
\begin{array}{r}
38{,}204 \\
+\,21{,}020 \\
\hline
59{,}224
\end{array}
$$

5.
$$
\begin{array}{r}
\overset{1}{6}7 \\
+\,83 \\
\hline
150
\end{array}
$$

7.
$$
\begin{array}{r}
7\overset{1}{4}6 \\
+\,905 \\
\hline
1651
\end{array}
$$

9.
$$
\begin{array}{r}
\overset{11}{7}98 \\
+\,206 \\
\hline
1004
\end{array}
$$

11.
$$
\begin{array}{r}
\overset{111}{7}968 \\
+\,1285 \\
\hline
9253
\end{array}
$$

13.
$$
\begin{array}{r}
7\overset{1\,1}{8}96 \\
+\,3728 \\
\hline
11{,}624
\end{array}
$$

15.
$$
\begin{array}{r}
3\overset{1}{7}0\overset{1}{5} \\
3916 \\
+\,9037 \\
\hline
16{,}658
\end{array}
$$

17.
$$\begin{array}{r} \overset{1\;1}{32} \\ +\,4\,977 \\ \hline 5009 \end{array}$$

19.
$$\begin{array}{r} 3\,\overset{1\,1}{0}77 \\ 8 \\ +\;\;421 \\ \hline 3506 \end{array}$$

21.
$$\begin{array}{r} 9\,\overset{2\,2}{0}56 \\ 78 \\ 6\,089 \\ +\;\;\;731 \\ \hline 15,954 \end{array}$$

23.
$$\begin{array}{r} 1\,1\,\overset{2}{1}8 \\ 708 \\ 9\,286 \\ +\;\;\;636 \\ \hline 10,648 \end{array}$$

25. Add up to check addition.
$$\begin{array}{r} 769 \\ \hline 179 \\ 214 \\ +\,376 \\ \hline 759 \end{array}$$ *incorrect; should be 769*

27. Add up to check addition.
$$\begin{array}{r} 5420 \\ \hline 4713 \\ 28 \\ 615 \\ +\,64 \\ \hline 5420 \end{array}$$ *correct*

29. The shortest route between Southtown and Rena is through Thomasville.
$$\begin{array}{r} 21 \quad \textit{Southtown to Thomasville} \\ +\,12 \quad \textit{Thomasville to Rena} \\ \hline 33 \quad \textit{miles} \end{array}$$

31. The shortest route between Thomasville and Murphy is through Rena and Austin.
$$\begin{array}{r} 12 \quad \textit{Thomasville to Rena} \\ 15 \quad \textit{Rena to Austin} \\ +\,11 \quad \textit{Austin to Murphy} \\ \hline 38 \quad \textit{miles} \end{array}$$

33.
$$\begin{array}{r} \$1\overset{2}{5}9 \quad \textit{auto tune-up} \\ 24 \quad \textit{tire rotation} \\ +\,29 \quad \textit{oil change} \\ \hline \$212 \quad \textit{total cost} \end{array}$$

35.
$$\begin{array}{r} 413 \quad \textit{women} \\ +\,286 \quad \textit{men} \\ \hline 699 \quad \textit{total people} \end{array}$$

37.
$$\begin{array}{r} \overset{1}{3}7,25\overset{1\;1}{3},956 \quad \textit{California} \\ +\,25,145,561 \quad \textit{Texas} \\ \hline 62,399,517 \quad \textit{total} \end{array}$$

The total population of the two states is 62,399,517 people.

39. The perimeter is the sum of all of the sides in the figure.
$$\begin{array}{r} \overset{3}{98} \\ 49 \\ 98 \\ +\,49 \\ \hline 294 \end{array}$$

294 inches is the perimeter of the figure.

41. The perimeter is the sum of all of the sides in the figure.
$$\begin{array}{r} \overset{1\,1}{286} \\ 308 \\ +\,114 \\ \hline 708 \end{array}$$

708 feet is the perimeter of the figure.

R.2 Subtracting Whole Numbers

R.2 Margin Exercises

1. **(a)** $4+3=7$ ▪ $7-3=4$ or $7-4=3$

(b) $6+5=11$ ▪ $11-5=6$ or $11-6=5$

(c) $150+220=370$ ▪ $370-220=150$ or $370-150=220$

(d) $623+55=678$ ▪ $678-55=623$ or $678-623=55$

2. **(a)** $5-3=2$ ▪ $5=3+2$

(b) $8-3=5$ ▪ $8=3+5$

(c) $21-15=6$ ▪ $21=15+6$

(d) $58-42=16$ ▪ $58=42+16$

3. **(a)**
$$\begin{array}{r} 56 \\ -\,31 \\ \hline 25 \end{array} \qquad \begin{array}{l} 6-1=5 \\ 5-3=2 \end{array}$$

(b)
$$\begin{array}{r} 38 \\ -\,14 \\ \hline 24 \end{array} \qquad \begin{array}{l} 8-4=4 \\ 3-1=2 \end{array}$$

(c)

```
   378
 - 235
 -----
   143
```

(d)

```
    3927
 -  2614
 -------
    1313
```

(e)

```
   5464
 -  324
 -----
   5140
```

4. (a)

```
   65                 23
 - 23   Subtraction  + 42   Addition
 ----   problem      ----   problem
   42                 65
```

42 is correct.

(b)

```
   46                 32
 - 32   Subtraction  + 24   Addition
 ----   problem      ----   problem
   24                 56
```

$56 \neq 46$, so 24 is incorrect.

Rework.

```
   46
 - 32
 ----
   14   is correct.
```

(c)

```
   374                251
 - 251  Subtraction  + 113  Addition
 -----  problem       ----  problem
   113                364
```

$364 \neq 374$, so 113 is incorrect.

Rework.

```
   374
 - 251
 -----
   123   is correct.
```

(d)

```
   7531                4301
 - 4301  Subtraction  + 3230  Addition
 ------  problem       -----  problem
   3230                7531
```

3230 is correct.

5. (a)

```
    5 17
    6̸ 7̸
  - 3 8
  -----
    2 9
```

(b)

```
    8 17
    9̸ 7̸
  - 2 9
  -----
    6 8
```

(c)

```
    2 11
    3̸ 1̸
  - 1 7
  -----
    1 4
```

(d)

```
    5 13
    8 6̸ 3̸
  -   4 7
  -------
    8 1 6
```

(e)

```
    5 12
    7 6̸ 2̸
  - 1 5 7
  -------
    6 0 5
```

6. (a)

```
    2 15
    3̸ 5̸ 4
  -   8 2
  -------
    2 7 2
```

(b)

```
    3 14 17
    4̸ 5̸ 7̸
  -    6 8
  --------
    3 8 9
```

(c)

```
    7 16 14
    8̸ 7̸ 4̸
  -  4 8 6
  --------
    3 8 8
```

(d)

```
    0 13 12 17
    1̸ 4̸ 3̸ 7̸
  -    9 8 8
  -----------
       4 4 9
```

(e)

```
    7 16 13
    8̸ 7̸ 3̸ 9
  -  3 8 9 2
  ----------
    4 8 4 7
```

7. (a)

```
    2 10
    3̸ 0̸ 8
  - 2 8 5
  -------
      2 3
```

(b)

```
      9
    1 10 16
    2̸ 0̸ 6̸
  - 1 4 8
  --------
      5 8
```

(c)

```
    4 10
    5̸ 0̸ 7 3
  - 1 6 3 2
  ----------
    3 4 4 1
```

8. (a)

```
      9
    3 10 15
    4̸ 0̸ 5̸
  - 2 6 7
  --------
    1 3 8
```

(b)

```
    6 10
    3 7̸ 0̸
  - 1 6 3
  -------
    2 0 7
```

(c)

```
    0 14 16 10
    1̸ 5̸ 7̸ 0̸
  -    9 8 3
  -----------
       5 8 7
```

(d)
$$\begin{array}{r} \overset{9\;9}{\overset{6\;\cancel{10}\,\cancel{10}\,11}{\cancel{7}\;\cancel{0}\;\cancel{0}\;\cancel{1}}} \\ -\;5\;1\;9\;3 \\ \hline 1\;8\;0\;8 \end{array}$$

(e)
$$\begin{array}{r} \overset{9\;9}{\overset{3\;\cancel{10}\,\cancel{10}\,10}{\cancel{4}\;\cancel{0}\;\cancel{0}\;\cancel{0}}} \\ -\;1\;7\;8\;2 \\ \hline 2\;2\;1\;8 \end{array}$$

9. (a)
$$\begin{array}{r} 425 \\ -\;368 \\ \hline 57 \end{array}\;\;\textit{Subtraction problem}\qquad \begin{array}{r} \overset{11}{368} \\ +\;57 \\ \hline 425 \end{array}\;\;\begin{array}{l}\textit{Check by}\\ \textit{addition.}\end{array}$$

Match: 57 is correct.

(b)
$$\begin{array}{r} 670 \\ -\;439 \\ \hline 241 \end{array}\;\;\textit{Subtraction problem}\qquad \begin{array}{r} \overset{1}{439} \\ +\;241 \\ \hline 680 \end{array}\;\;\begin{array}{l}\textit{Check by}\\ \textit{addition.}\end{array}$$

Not a match: 241 is incorrect.

Rework.

$$\begin{array}{r} \overset{610}{6\,\cancel{7}\,\cancel{0}} \\ -\;4\,3\,9 \\ \hline 2\,3\,1 \end{array}\;\;\textit{Subtraction problem}$$

Match: 231 is correct.

(c)
$$\begin{array}{r} 14{,}726 \\ -\;8\,839 \\ \hline 5\,887 \end{array}\qquad \begin{array}{r} \overset{1\;11}{8\,839} \\ +\;5\,887 \\ \hline 14{,}726 \end{array}$$

Match: 5887 is correct.

10. (a) Ms. Lopez made 147 deliveries on Friday, but only 126 on Tuesday.

$$\begin{array}{r} 147 \\ -\;126 \\ \hline 21 \end{array}\;\;\begin{array}{l}\textit{Deliveries on Friday}\\ \textit{Deliveries on Tuesday}\end{array}$$

Ms. Lopez made 21 fewer deliveries on Tuesday than she made on Friday.

(b) Ms. Lopez made 126 deliveries on Tuesday, but only 119 on Wednesday.

$$\begin{array}{r} \overset{116}{1\,\cancel{2}\,\cancel{0}} \\ -\;1\,1\,9 \\ \hline 7 \end{array}\;\;\begin{array}{l}\textit{Deliveries on Tuesday}\\ \textit{Deliveries on Wednesday}\end{array}$$

Ms. Lopez made 7 fewer deliveries on Wednesday than she made on Tuesday.

(c) Ms. Lopez made 89 deliveries on Thursday and none on Saturday, so she made 89 more deliveries on Thursday than on Saturday.

R.2 Section Exercises

1.
$$\begin{array}{r} 89 \\ -\;27 \\ \hline 63 \end{array}\;\;\textit{Given}\qquad \begin{array}{r} \overset{1}{27} \\ +\;63 \\ \hline 90 \end{array}\;\;\begin{array}{l}\textit{Check by}\\ \textit{addition.}\end{array}$$

$90 \neq 89$, so 63 is incorrect. Rework.

$$\begin{array}{r} 89 \\ -\;27 \\ \hline 62 \end{array}\;\;\text{is correct.}$$

3.
$$\begin{array}{r} 382 \\ -\;261 \\ \hline 131 \end{array}\;\;\textit{Given}\qquad \begin{array}{r} 261 \\ +\;131 \\ \hline 392 \end{array}\;\;\begin{array}{l}\textit{Check by}\\ \textit{addition.}\end{array}$$

$392 \neq 382$, so 131 is incorrect. Rework.

$$\begin{array}{r} 382 \\ -\;261 \\ \hline 121 \end{array}\;\;\text{is correct.}$$

5.
$$\begin{array}{r} \overset{2\,16}{3\,\cancel{6}} \\ -\;2\,8 \\ \hline 8 \end{array}$$

7.
$$\begin{array}{r} \overset{7\,13}{8\,\cancel{3}} \\ -\;5\,8 \\ \hline 2\,5 \end{array}$$

9.
$$\begin{array}{r} \overset{3\,15}{4\,\cancel{5}} \\ -\;2\,9 \\ \hline 1\,6 \end{array}$$

11.
$$\begin{array}{r} \overset{6\,11}{7\,\cancel{1}\,9} \\ -\;6\,5\,8 \\ \hline 6\,1 \end{array}$$

13.
$$\begin{array}{r} \overset{6\,11}{7\,7\,\cancel{1}} \\ -\;2\,5\,2 \\ \hline 5\,1\,9 \end{array}$$

15.
$$\begin{array}{r} \overset{7\,15\,11}{9\,\cancel{8}\,\cancel{6}\,\cancel{1}} \\ -\;\;\;6\,8\,4 \\ \hline 9\,1\,7\,7 \end{array}$$

17.
$$\begin{array}{r} \overset{8\,17\,18}{9\,\cancel{9}\,\cancel{8}\,\cancel{8}} \\ -\;2\,3\,9\,9 \\ \hline 7\,5\,8\,9 \end{array}$$

19.
$$\begin{array}{r} \overset{2\,17\;12\,12\,15}{\cancel{3}\,\cancel{8}{,}\cancel{3}\,\cancel{3}\,\cancel{5}} \\ -\;2\,9{,}4\,7\,6 \\ \hline 8\;\;8\,5\,9 \end{array}$$

21.
$$\begin{array}{r} \overset{3\,10}{4\,\cancel{0}} \\ -\;3\,7 \\ \hline 3 \end{array}$$

23.
$$\begin{array}{r} \overset{5\ 10}{\cancel{6}\,\cancel{0}} \\ -\ 3\ 7 \\ \hline 2\ 3 \end{array}$$

25.
$$\begin{array}{r} \overset{9}{} \\ \overset{5\ 10\,11\,10}{\cancel{6}\,\cancel{0}\,\cancel{2}\,\cancel{0}} \\ -\ 4\ 0\ 7\ 8 \\ \hline 1\ 9\ 4\ 2 \end{array}$$

27.
$$\begin{array}{r} \overset{9}{} \\ \overset{7\ 14\,10\,13}{\cancel{8}\,\cancel{5}\,\cancel{0}\,\cancel{3}} \\ -\ 2\ 8\ 1\ 6 \\ \hline 5\ 6\ 8\ 7 \end{array}$$

29.
$$\begin{array}{r} \overset{9}{} \\ \overset{7\,10\ 6\,10\,15}{\cancel{8}\,\cancel{0}{,}\cancel{7}\,\cancel{0}\,\cancel{5}} \\ -\ 6\ 1{,}6\ 6\ 7 \\ \hline 1\ 9{,}0\ 3\ 8 \end{array}$$

31.
$$\begin{array}{r} \overset{9\ 9}{} \\ \overset{5\ 10\,10\,10\ 10}{\cancel{6}\,\cancel{6}{,}\cancel{0}\,\cancel{0}\,\cancel{0}} \\ -\ \ \ 4\ 4\ 4 \\ \hline 6\ 5{,}5\ 5\ 6 \end{array}$$

33.
$$\begin{array}{r} \overset{9\ 9}{} \\ \overset{1\,10\ 10\,17\,10}{\cancel{2}\,\cancel{0}{,}\cancel{0}\,\cancel{8}\,\cancel{0}} \\ -\ \ \ 9\ 6 \\ \hline 1\ 9{,}9\ 8\ 4 \end{array}$$

35.
$$\begin{array}{rl} 3070 & \\ -\ 576 & \textit{Subtraction} \\ \hline 2596 & \textit{problem} \end{array}$$
$$\begin{array}{rl} \overset{111}{576} & \textit{Check by} \\ +\ 2596 & \textit{addition.} \\ \hline 3172 & \end{array}$$

$3172 \neq 3070$, so 2596 is incorrect. Rework.

$$\begin{array}{r} \overset{9}{} \\ \overset{2\,10\,16\,10}{\cancel{3}\,\cancel{0}\,\cancel{7}\,\cancel{0}} \\ -\ 5\ 7\ 6 \\ \hline 2\ 4\ 9\ 4 \quad \text{is correct.} \end{array}$$

37.
$$\begin{array}{rl} 27{,}600 & \\ -\ 807 & \textit{Subtraction} \\ \hline 26{,}793 & \textit{problem} \end{array}$$
$$\begin{array}{rl} \overset{1\ \ 11}{807} & \textit{Check by} \\ +\ 26{,}793 & \textit{addition.} \\ \hline 27{,}600 & \end{array}$$

Matches: 26,793 is correct.

39.
$$\begin{array}{rl} \overset{1\,15}{2\,\cancel{5}\,5} & \textit{calories burned by swimming} \\ -\ 1\ 8\ 5 & \textit{calories burned by hiking} \\ \hline 7\ 0 & \textit{fewer calories burned} \end{array}$$

70 fewer calories are burned in 30 minutes by hiking than by swimming.

41.
$$\begin{array}{rl} \overset{1\,15}{2\,\cancel{5}\,4} & \textit{number of passengers} \\ -\ 1\ 8\ 3 & \textit{passengers departing in Atlanta} \\ \hline 7\ 1 & \textit{passengers remaining} \end{array}$$

$$\begin{array}{rl} \overset{1}{71} & \textit{passengers remaining} \\ -\ 109 & \textit{passengers got on} \\ \hline 180 & \textit{passengers on the plane} \end{array}$$

There were 180 passengers on the plane.

43. (a) From the table, the occupation with the highest earnings is computer programmer at $74,900 and the occupation with the lowest earnings is medical secretary at $31,820.

(b)
$$\begin{array}{rl} \overset{810}{\$74{,}\cancel{9}\cancel{0}\cancel{0}} & \textit{comp. programmer earnings} \\ -\ 31{,}820 & \textit{medical secretary earnings} \\ \hline \$43{,}080 & \textit{difference} \end{array}$$

The earnings of a computer programmer are $43,080 per year more than those of a medical secretary.

45.
$$\begin{array}{rl} \overset{7\,11}{1\,8\,\cancel{1}\,5} & \textit{CN Tower height} \\ -\ 1\ 4\ 5\ 1 & \textit{Willis Tower height} \\ \hline 3\ 6\ 4 & \textit{difference} \end{array}$$

There is a difference in height of 364 feet.

R.3 Multiplying Whole Numbers

R.3 Margin Exercises

1. (a)
$$\begin{array}{rl} 3 & \textit{factor} \\ \times\ 6 & \textit{factor} \\ \hline 18 & \textit{product} \end{array}$$

(b)
$$\begin{array}{rl} 8 & \textit{factor} \\ \times\ 4 & \textit{factor} \\ \hline 32 & \textit{product} \end{array}$$

(c)
$$\begin{array}{rl} 5 & \textit{factor} \\ \times\ 7 & \textit{factor} \\ \hline 35 & \textit{product} \end{array}$$

(d)
$$\begin{array}{rl} 3 & \textit{factor} \\ \times\ 9 & \textit{factor} \\ \hline 27 & \textit{product} \end{array}$$

2. (a) $4 \times 7 = 28$; $7 \times 4 = 28$

(b) $0 \times 9 = 0$; $9 \times 0 = 0$

(c) $8 \cdot 6 = 48$; $6 \cdot 8 = 48$

(d) $5 \cdot 5 = 25$; there is no change if the order is switched.

(e) $(3)(8) = 24$; $(8)(3) = 24$

3. (a) $2 \times 3 \times 4 = (2 \times 3) \times 4 = 6 \times 4 = 24$

(b) $6 \cdot 1 \cdot 5 = (6 \cdot 1) \cdot 5 = 6 \cdot 5 = 30$

(c) $(8)(3)(0) = 0$ since zero times any number is 0.

(d) $3 \times 3 \times 7 = (3 \times 3) \times 7 = 9 \times 7 = 63$

(e) $4 \cdot 2 \cdot 8 = (4 \cdot 2) \cdot 8 = 8 \cdot 8 = 64$

(f) $(2)(2)(9) = [(2)(2)](9) = (4)(9) = 36$

4. **(a)**
$$\begin{array}{r} \overset{1}{5}2 \\ \times\ 5 \\ \hline 260 \end{array}$$

$5 \cdot 2 = 10$ Write 0, carry 1 ten.
$5 \cdot 5 = 25$ Add 1 to get 26. Write 26.

(b)
$$\begin{array}{r} 79 \\ \times\ 0 \\ \hline 0 \end{array}$$ Any number times 0 is 0.

(c)
$$\begin{array}{r} \overset{51}{8}62 \\ \times\ 9 \\ \hline 7758 \end{array}$$

$9 \cdot 2 = 18$ Write 8, carry 1 ten.
$9 \cdot 6 = 54$ Add 1 to get 55. Write 5, carry 5.
$9 \cdot 8 = 72$ Add 5 to get 77. Write 77.

(d)
$$\begin{array}{r} \overset{52}{2}831 \\ \times\ 7 \\ \hline 19,817 \end{array}$$

$7 \cdot 1 = 7$ Write 7.
$7 \cdot 3 = 21$ Write 1, carry 2 tens.
$7 \cdot 8 = 56$ Add 2 to get 58. Write 8, carry 5.
$7 \cdot 2 = 14$ Add 5 to get 19. Write 19.

(e)
$$\begin{array}{r} \overset{513}{4}714 \\ \times\ 8 \\ \hline 37,712 \end{array}$$

$8 \cdot 4 = 32$ Write 2, carry 3 tens.
$8 \cdot 1 = 8$ Add 3 to get 11. Write 1, carry 1.
$8 \cdot 7 = 56$ Add 1 to get 57. Write 7, carry 5.
$8 \cdot 4 = 32$ Add 5 to get 37. Write 37.

5. **(a)** $45 \times 10 = 450$ *Attach 0.*

(b) $102 \times 100 = 10,200$ *Attach 00.*

(c) $1000 \times 571 = 571,000$ *Attach 000.*

(d) $100 \times 3625 = 362,500$ *Attach 00.*

(e) $69 \times 1000 = 69,000$ *Attach 000.*

6. **(a)** 14×50
$$\begin{array}{r} 14 \\ \times\ 5 \\ \hline 70 \end{array}$$ $14 \times 50 = 700$ *Attach 0.*

(b) $(68)(400)$
$$\begin{array}{r} 68 \\ \times\ 4 \\ \hline 272 \end{array}$$ $68 \times 400 = 27,200$ *Attach 00.*

(c)
$$\begin{array}{r} 180 \\ \times\ 30 \end{array}$$
$$\begin{array}{r} 18 \\ \times\ 3 \\ \hline 54 \end{array}$$ $180 \times 30 = 5400$ *Attach 00.*

(d)
$$\begin{array}{r} 6100 \\ \times\ 90 \end{array}$$
$$\begin{array}{r} 61 \\ \times\ 9 \\ \hline 549 \end{array}$$ $6100 \times 90 = 549,000$ *Attach 000.*

(e)
$$\begin{array}{r} 800 \\ \times\ 200 \end{array}$$
$$\begin{array}{r} 8 \\ \times\ 2 \\ \hline 16 \end{array}$$ $800 \times 200 = 160,000$ *Attach 0000.*

(f) $(5000)(700)$
$$\begin{array}{r} 5 \\ \times\ 7 \\ \hline 35 \end{array}$$ $5000 \times 700 = 3,500,000$ *Attach 00000.*

(g) $(9)(20,000)$
$$\begin{array}{r} 9 \\ \times\ 2 \\ \hline 18 \end{array}$$ $9 \times 20,000 = 180,000$ *Attach 0000.*

7. **(a)**
$$\begin{array}{r} 38 \\ \times\ 15 \\ \hline 190 \\ 38 \\ \hline 570 \end{array}$$ $\leftarrow 5 \times 38$
$\leftarrow 1 \times 38$

(b)
$$\begin{array}{r} 31 \\ \times\ 43 \\ \hline 93 \\ 124 \\ \hline 1333 \end{array}$$ $\leftarrow 3 \times 31$
$\leftarrow 4 \times 31$

(c)
$$\begin{array}{r} 67 \\ \times\ 59 \\ \hline 603 \\ 335 \\ \hline 3953 \end{array}$$ $\leftarrow 9 \times 67$
$\leftarrow 5 \times 67$

(d)
$$\begin{array}{r} 234 \\ \times\ 73 \\ \hline 702 \\ 1638 \\ \hline 17,082 \end{array}$$ $\leftarrow 3 \times 234$
$\leftarrow 7 \times 234$

(e)
$$
\begin{array}{r}
835 \\
\times\ 189 \\
\hline
7515 \quad \leftarrow 9 \times 835 \\
6680 \quad \leftarrow 8 \times 835 \\
835 \quad \leftarrow 1 \times 835 \\
\hline
157{,}815
\end{array}
$$

8. (a)
$$
\begin{array}{r}
28 \\
\times\ 60 \\
\hline
1680
\end{array}
$$

(b)
$$
\begin{array}{r}
817 \\
\times\ 30 \\
\hline
24{,}510
\end{array}
$$

(c)
$$
\begin{array}{r}
481 \\
\times\ 206 \\
\hline
2886 \\
9620 \quad 2 \times 481 = 962 \quad \textit{Insert 0.} \\
\hline
99{,}086
\end{array}
$$

(d)
$$
\begin{array}{r}
3526 \\
\times\ 6002 \\
\hline
7052 \\
21\,156\,00 \quad 6 \times 3526 = 21{,}156 \quad \textit{Insert 00.} \\
\hline
21{,}163{,}052
\end{array}
$$

9. (a)
$$
\begin{array}{rl}
36 & \textit{months} \\
\times\ 79 & \textit{dollars per month} \\
\hline
324 & \\
252 & \\
\hline
2844 &
\end{array}
$$

The total cost of 36 months of cable TV is $2844.

(b)
$$
\begin{array}{rl}
1090 & \textit{dollars per computer} \\
\times\ 15 & \textit{computers} \\
\hline
5450 & \\
1090 & \\
\hline
16{,}350 &
\end{array}
$$

The total cost of 15 laptop computers is $16,350.

(c)
$$
\begin{array}{rl}
389 & \textit{dollars per month} \\
\times\ 60 & \textit{months} \\
\hline
23{,}340 &
\end{array}
$$

The total cost of 60 months of car payments is $23,340.

R.3 Section Exercises

1. $3 \times 1 \times 3 = (3 \times 1) \times 3 = 3 \times 3 = 9$

3. $9 \times 1 \times 7 = (9 \times 1) \times 7 = 9 \times 7 = 63$

5. $9 \cdot 5 \cdot 0 = (9 \cdot 5) \cdot 0 = 45 \cdot 0 = 0$
The product of any number and 0 is 0.

7. $(4)(1)(6) = [(4)(1)](6) = (4)(6) = 24$

9. $(2)(3)(6) = [(2)(3)](6) = (6)(6) = 36$

11.
$$
\begin{array}{r}
\overset{3}{35} \\
\times\ 7 \\
\hline
245
\end{array}
$$

$7 \cdot 5 = 35$ Write 5, carry 3 tens.
$7 \cdot 3 = 21$ Add 3 to get 24. Write 24.

13.
$$
\begin{array}{r}
\overset{4}{28} \\
\times\ 6 \\
\hline
168
\end{array}
$$

$6 \cdot 8 = 48$ Write 8, carry 4 tens.
$6 \cdot 2 = 12$ Add 4 to get 16. Write 16.

15.
$$
\begin{array}{r}
\overset{141}{3182} \\
\times\ 6 \\
\hline
19{,}092
\end{array}
$$

$6 \cdot 2 = 12$ Write 2, carry 1 ten.
$6 \cdot 8 = 48$ Add 1 to get 49. Write 9, carry 4.
$6 \cdot 1 = 6$ Add 4 to get 10. Write 0, carry 1.
$6 \cdot 3 = 18$ Add 1 to get 19. Write 19.

17.
$$
\begin{array}{r}
\overset{46\ 1}{36{,}921} \\
\times\ 7 \\
\hline
258{,}447
\end{array}
$$

$7 \cdot 1 = 7$ Write 7.
$7 \cdot 2 = 14$ Write 4, carry 1.
$7 \cdot 9 = 63$ Add 1 to get 64. Write 4, carry 6.
$7 \cdot 6 = 42$ Add 6 to get 48. Write 8, carry 4.
$7 \cdot 3 = 21$ Add 4 to get 25. Write 25.

19.
$$
\begin{array}{ccc}
\begin{array}{r} 125 \\ \times\ 100 \\ \hline \end{array} &
\begin{array}{r} 125 \\ \times\ 1 \\ \hline 125 \end{array} &
\begin{array}{r} 125 \\ \times\ 100 \\ \hline 12{,}500 \end{array}
\end{array}
$$
Attach 00.

21.
$$
\begin{array}{ccc}
\begin{array}{r} 1485 \\ \times\ 30 \\ \hline \end{array} &
\begin{array}{r} 1485 \\ \times\ 3 \\ \hline 4455 \end{array} &
\begin{array}{r} 1485 \\ \times\ 30 \\ \hline 44{,}550 \end{array}
\end{array}
$$
Attach 0.

23.
$$
\begin{array}{ccc}
\begin{array}{r} 900 \\ \times\ 300 \\ \hline \end{array} &
\begin{array}{r} 9 \\ \times\ 3 \\ \hline 27 \end{array} &
\begin{array}{r} 900 \\ \times\ 300 \\ \hline 270{,}000 \end{array}
\end{array}
$$
Attach 0000.

25.
$$
\begin{array}{ccc}
\begin{array}{r} 43{,}000 \\ \times\ 2000 \\ \hline \end{array} &
\begin{array}{r} 43 \\ \times\ 2 \\ \hline 86 \end{array} &
\begin{array}{r} 43{,}000 \\ \times\ 2000 \\ \hline 86{,}000{,}000 \end{array}
\end{array}
$$
Attach 000000.

27.
$$
\begin{array}{r}
68 \\
\times\ 22 \\
\hline
136 \quad \leftarrow 2 \times 68 \\
136 \quad \leftarrow 2 \times 68 \\
\hline
1496
\end{array}
$$

29. $(83)(45)$

$$
\begin{array}{r}
83 \\
\times\ 45 \\
\hline
415 \quad \leftarrow 5 \times 83 \\
332 \quad\ \ \leftarrow 4 \times 83 \\
\hline
3735
\end{array}
$$

31. $(32)(475)$

$$
\begin{array}{r}
475 \\
\times\ 32 \\
\hline
950 \quad \leftarrow 2 \times 475 \\
1425 \quad\ \ \leftarrow 3 \times 475 \\
\hline
15{,}200
\end{array}
$$

33. $(729)(45)$

$$
\begin{array}{r}
729 \\
\times\ 45 \\
\hline
3645 \quad \leftarrow 5 \times 729 \\
2916 \quad\ \ \leftarrow 4 \times 729 \\
\hline
32{,}805
\end{array}
$$

35.

$$
\begin{array}{r}
538 \\
\times\ 342 \\
\hline
1076 \quad \leftarrow 2 \times 538 \\
2152 \quad\ \ \leftarrow 4 \times 538 \\
1614 \quad\ \ \ \ \leftarrow 3 \times 538 \\
\hline
183{,}996
\end{array}
$$

37.

$$
\begin{array}{r}
8162 \\
\times\ 407 \\
\hline
57134 \quad \leftarrow 7 \times 8162 \\
326480 \quad\ \ \leftarrow 40 \times 8162 \\
\hline
3{,}321{,}934
\end{array}
$$

39.

$$
\begin{array}{r}
6310 \\
\times\ 3008 \\
\hline
50480 \quad \leftarrow 8 \times 6310 \\
1893000 \quad\ \ \leftarrow 300 \times 6310 \\
\hline
18{,}980{,}480
\end{array}
$$

41.

$$
\begin{array}{r}
18 \quad \textit{inches per day} \\
\times\ 14 \quad \textit{days} \\
\hline
72 \\
18\ \ \\
\hline
252 \quad \textit{inches in two weeks}
\end{array}
$$

Kelp could grow 252 inches in two weeks.

$$
\begin{array}{r}
18 \quad \textit{inches per day} \\
\times\ 30 \quad \textit{days} \\
\hline
540 \quad \textit{inches in 30 days}
\end{array}
$$

Kelp could grow 540 inches in a 30-day month.

43.

$$
\begin{array}{r}
48 \quad \textit{flats} \\
\times\ 12 \quad \textit{tomato plants per flat} \\
\hline
96 \\
48\ \ \\
\hline
576 \quad \textit{tomato plants}
\end{array}
$$

The total number of tomato plants is 576.

45.

$$
\begin{array}{r}
48 \quad \textit{miles per gallon} \\
\times\ 11 \quad \textit{gallons} \\
\hline
48 \\
48\ \ \\
\hline
528 \quad \textit{miles}
\end{array}
$$

The Prius can travel 528 miles on 11 gallons of gas.

47.

$$
\begin{array}{r}
2695 \quad \textit{Reno to Atlantic Ocean} \\
-\ 255 \quad \textit{Reno to Pacific Ocean} \\
\hline
2440 \quad \textit{difference}
\end{array}
$$

It is 2440 miles farther from Reno to the Atlantic Ocean than it is from Reno to the Pacific Ocean.

$$
\begin{array}{r}
2695 \quad \textit{miles per trip} \\
\times\ 6 \quad \textit{trips (3 round trips)} \\
\hline
16{,}170 \quad \textit{miles}
\end{array}
$$

You will earn 16,170 frequent flier miles.

49.

$$
\begin{array}{r}
\overset{916}{14\not0\not0} \quad \textit{calories per high-fat meal} \\
-\ 348 \quad \textit{calories per low-fat meal} \\
\hline
1058 \quad \textit{calories difference}
\end{array}
$$

$$
\begin{array}{r}
\overset{45}{1058} \quad \textit{more calories per meal} \\
\times\ 7 \quad \textit{meals} \\
\hline
7406 \quad \textit{more calories}
\end{array}
$$

There are 7406 more calories in seven high-fat meals than in seven low-fat meals.

R.4 Dividing Whole Numbers

R.4 Margin Exercises

1. **(a)** $48 \div 6 = 8$ ∎ $6\overline{\smash{)}48}^{\,8}$ and $\dfrac{48}{6} = 8$

(b) $24 \div 6 = 4$ ∎ $6\overline{\smash{)}24}^{\,4}$ and $\dfrac{24}{6} = 4$

(c) $9\overline{\smash{)}36}^{\,4}$ ∎ $36 \div 9 = 4$ and $\dfrac{36}{9} = 4$

(d) $\dfrac{42}{6} = 7$ ∎ $42 \div 6 = 7$ and $6\overline{\smash{)}42}^{\,7}$

2. **(a)** $10 \div 2 = 5$

dividend: 10; divisor: 2; quotient: 5

(b) $6 = 30 \div 5$

dividend: 30; divisor: 5; quotient: 6

(c) $\dfrac{28}{7} = 4$

dividend: 28; divisor: 7; quotient: 4

(d) $2\overline{\smash{)}36}^{\,18}$

dividend: 36; divisor: 2; quotient: 18

3. **(a)** $0 \div 9 = 0$ *Zero divided by any number is 0.*

(b) $\dfrac{0}{36} = 0$ *Zero divided by any number is 0.*

(c) $57\overline{\smash{)}0}^{\,0}$ *Zero divided by any number is 0.*

4. **(a)** $6\overline{\smash{)}18}^{\,3}$ $\blacksquare$ $6 \cdot 3 = 18$ or $3 \cdot 6 = 18$

(b) $\dfrac{28}{4} = 7$ $\blacksquare$ $4 \cdot 7 = 28$ or $7 \cdot 4 = 28$

(c) $48 \div 8 = 6$ $\blacksquare$ $8 \cdot 6 = 48$ or $6 \cdot 8 = 48$

5. **(a)** $\dfrac{8}{0}$ is *undefined.*

(b) $\dfrac{0}{8} = 0$ *Zero divided by any number is 0.*

(c) $0\overline{\smash{)}32}$ is *undefined.*

(d) $32\overline{\smash{)}0}^{\,0}$ *Zero divided by any number is 0.*

(e) $100 \div 0$ is *undefined.*

(f) $0 \div 100 = 0$ *Zero divided by any number is 0.*

6. **(a)** $5 \div 5 = 1$ *A number divided by itself is 1.*

(b) $14\overline{\smash{)}14}^{\,1}$ *A number divided by itself is 1.*

(c) $\dfrac{37}{37} = 1$ *A number divided by itself is 1.*

7. **(a)** $2\overline{\smash{)}18}^{\,9}$

(b) $3\overline{\smash{)}39}^{\,13}$ $\dfrac{3}{3} = 1$, $\dfrac{9}{3} = 3$

(c) $4\overline{\smash{)}88}^{\,22}$ $\dfrac{8}{4} = 2$, $\dfrac{8}{4} = 2$

(d) $2\overline{\smash{)}462}^{\,231}$ $\dfrac{4}{2} = 2$, $\dfrac{6}{2} = 3$, $\dfrac{2}{2} = 1$

8. **(a)** $2\overline{\smash{)}225}^{\,112\ \mathbf{R}1}$ $\dfrac{2}{2} = 1$, $\dfrac{2}{2} = 1$, $\dfrac{5}{2} = 2\,\mathbf{R}1$

(b) $3\overline{\smash{)}275}^{\,91\ \mathbf{R}2}$ $\dfrac{27}{3} = 9$, $\dfrac{5}{3} = 1\,\mathbf{R}2$

(c) $4\overline{\smash{)}5\,^1 3\,^1 8}^{\,1\ 3\ 4\ \mathbf{R}2}$

$\dfrac{5}{4} = 1\,\mathbf{R}1$, $\dfrac{13}{4} = 3\,\mathbf{R}1$, $\dfrac{18}{4} = 4\,\mathbf{R}2$

(d) $\dfrac{819}{5}$ $\blacksquare$ $5\overline{\smash{)}8\,^3 1\,^1 9}^{\,1\ 6\ 3\ \mathbf{R}4}$

$\dfrac{8}{5} = 1\,\mathbf{R}3$, $\dfrac{31}{5} = 6\,\mathbf{R}1$, $\dfrac{19}{5} = 3\,\mathbf{R}4$

9. **(a)** $4\overline{\smash{)}8\,3\,^3 7}^{\,20\ 9\ \mathbf{R}1}$

(b) $7\overline{\smash{)}74\,^4 7}^{\,10\ 6\ \mathbf{R}5}$

(c) $5\overline{\smash{)}4\,5\,3\,^3 8}^{\,90\ 7\ \mathbf{R}3}$

(d) $8\overline{\smash{)}244\,^4 0}^{\,30\ 5}$

10. **(a)** $3\overline{\smash{)}115}^{\,38\ \mathbf{R}1}$

divisor $\times$ *quotient* $+$ *remainder* $=$ *dividend*

$$\downarrow \qquad \downarrow \qquad \downarrow \qquad \downarrow$$
$$3 \quad \times \quad 38 \quad + \quad 1 \quad = \quad 115$$
$$114 \qquad\qquad + \quad 1 \quad = \quad 115$$

The answer is correct.

(b) $8\overline{\smash{)}743}^{\,92\ \mathbf{R}2}$

divisor $\times$ *quotient* $+$ *remainder* $=$ *dividend*

$$\downarrow \qquad \downarrow \qquad \downarrow \qquad \downarrow$$
$$8 \quad \times \quad 92 \quad + \quad 2$$
$$736 \qquad\qquad + \quad 2 \quad = \quad 738$$
$$\uparrow$$
$$\textit{incorrect}$$

$8\overline{\smash{)}74\,^2 3}^{\,9\ 2\ \mathbf{R}7}$

The correct answer is 92 $\mathbf{R}$7.

(c) $4\overline{\smash{)}1312}^{\,328}$

divisor $\times$ *quotient* $+$ *remainder* $=$ *dividend*

$$\downarrow \qquad \downarrow \qquad \downarrow \qquad \downarrow$$
$$4 \quad \times \quad 328 \quad + \quad 0 \quad = \quad 1312$$

The answer is correct.

(d) $5\overline{\smash{)}2033}^{\,46\ \mathbf{R}3}$

divisor $\times$ *quotient* $+$ *remainder* $=$ *dividend*

$$\downarrow \qquad \downarrow \qquad \downarrow \qquad \downarrow$$
$$5 \quad \times \quad 46 \quad + \quad 3$$
$$230 \qquad\qquad + \quad 3 \quad = \quad 233$$
$$\uparrow$$
$$\textit{incorrect}$$

$5\overline{\smash{)}203\,^3 3}^{\,40\ 6\ \mathbf{R}3}$

The correct answer is 406 $\mathbf{R}$3.

11. **(a)** 612 $\blacksquare$ ends in 2, divisible by 2

(b) 315 $\blacksquare$ ends in 5, not divisible by 2

(c) 2714 ▪ ends in 4, divisible by 2

(d) 36,000 ▪ ends in 0, divisible by 2

12. **(a)** 836 ▪ The sum of the digits is
$8 + 3 + 6 = 17$. Because 17 is *not* divisible by 3,
the number 836 is *not* divisible by 3.

(b) 7005 ▪ The sum of the digits is
$7 + 0 + 0 + 5 = 12$. Because 12 is divisible by 3,
the number 7005 is divisible by 3.

(c) 242,913 ▪ The sum of the digits is
$2 + 4 + 2 + 9 + 1 + 3 = 21$. Because 21 is
divisible by 3, the number 242,913 is divisible by
3.

(d) 102,484 ▪ The sum of the digits is
$1 + 0 + 2 + 4 + 8 + 4 = 19$. Because 19 is *not*
divisible by 3, the number 102,484 is *not* divisible
by 3.

13. **(a)** 160 ▪ ends in 0, divisible by 5

(b) 635 ▪ ends in 5, divisible by 5

(c) 3381 ▪ ends in 1, not divisible by 5

(d) 108,605 ▪ ends in 5, divisible by 5

14. **(a)** 290 ▪ ends in 0, divisible by 10

(b) 218 ▪ ends in 8, not divisible by 10

(c) 2020 ▪ ends in 0, divisible by 10

(d) 11,670 ▪ ends in 0, divisible by 10

R.4 Section Exercises

1. $\dfrac{12}{12} = 1;\quad 12\overline{)12}\quad$ or $\quad 12 \div 12$

3. $24 \div 0$ is *undefined.* $\quad \dfrac{24}{0}\quad$ or $\quad 0\overline{)24}$

5. $\dfrac{0}{4} = 0;\quad 4\overline{)0}\quad$ or $\quad 0 \div 4$

7. $0 \div 12 = 0;\quad \dfrac{0}{12}\quad$ or $\quad 12\overline{)0}$

9. $0\overline{)21}$ is *undefined.* $\quad \dfrac{21}{0}\quad$ or $\quad 21 \div 0$

11. $\begin{array}{r}21\\4\overline{)84}\end{array}$

The dividend is 84, the divisor is 4, and the
quotient is 21.

Check $4 \times 21 = 84$

13. $\begin{array}{r}3\ 6\\9\overline{)32\,^54}\end{array}$

The dividend is 324, the divisor is 9, and the
quotient is 36.

Check $9 \times 36 = 324$

15. $\begin{array}{r}1\ 5\ 20\ \mathbf{R5}\\6\overline{)9\,^31\,^12\,5}\end{array}$

Check $6 \times 1520 + 5 = 9120 + 5 = 9125$

17. $\begin{array}{r}30\ 9\\6\overline{)185\,^54}\end{array}$

Check $6 \times 309 = 1854$

19. $4024 \div 4$ ▪ $\begin{array}{r}100\ 6\\4\overline{)402\,^24}\end{array}$

Check $4 \times 1006 = 4024$

21. $15,019 \div 3$ ▪ $\begin{array}{r}5\ 00\ 6\ \mathbf{R1}\\3\overline{)15,01\,^19}\end{array}$

Check $3 \times 5006 + 1 = 15,018 + 1 = 15,019$

23. $\dfrac{26,684}{4}$ ▪ $\begin{array}{r}6\ 6\ 71\\4\overline{)26,\,^26\,^28\,4}\end{array}$

Check $4 \times 6671 = 26,684$

25. $\dfrac{74,751}{6}$ ▪ $\begin{array}{r}1\ 2,\ 4\ 5\ 8\ \mathbf{R3}\\6\overline{)7\,^14,\,^27\,^35\,^51}\end{array}$

Check $6 \times 12,458 + 3 = 74,748 + 3 = 74,751$

27. $\dfrac{71,776}{7}$ ▪ $\begin{array}{r}10,\ 2\ 5\ 3\ \mathbf{R5}\\7\overline{)71,\,^17\,^37\,^26}\end{array}$

Check $7 \times 10,253 + 5 = 71,771 + 5 = 71,776$

29. $\dfrac{128,645}{7}$ ▪ $\begin{array}{r}1\ 8,\ 3\ 7\ 7\ \mathbf{R6}\\7\overline{)12\,^58,\,^26\,^54\,^55}\end{array}$

Check $7 \times 18,377 + 6 = 128,639 + 6 = 128,645$

31. $\begin{array}{r}67\ \mathbf{R2}\\7\overline{)4692}\end{array}$

Check $7 \times 67 + 2 = 469 + 2 = 471$ *incorrect*

Rework. $\begin{array}{r}6\ 70\ \mathbf{R2}\\7\overline{)46\,^49\,2}\end{array}$

Check $7 \times 670 + 2 = 4690 + 2 = 4692$ *correct*

33. $\begin{array}{r}3\ 568\ \mathbf{R2}\\6\overline{)21,409}\end{array}$

Check $6 \times 3568 + 2 = 21,408 + 2 = 21,410$
incorrect

Rework. $\begin{array}{r}3\ 5\ 6\ 8\ \mathbf{R1}\\6\overline{)21,\,^34\,^40\,^49}\end{array}$

Check $6 \times 3568 + 1 = 21,408 + 1 = 21,409$

35. $\begin{array}{r}3\ 003\ \mathbf{R5}\\6\overline{)18,023}\end{array}$

Check $6 \times 3003 + 5 = 18,018 + 5 = 18,023$
correct

37.

$$\begin{array}{r} 11{,}523 \ \textbf{R2} \\ 6\overline{|69{,}140} \end{array}$$

Check $6 \times 11{,}523 + 2 = 69{,}138 + 2 = 69{,}140$
correct

39. Divide 184, 112, and 152 by 8.

$$\begin{array}{r} 2\ 3 \\ 8\overline{|18\ ^24} \end{array} \qquad \begin{array}{r} 1\ 4 \\ 8\overline{|11\ ^32} \end{array} \qquad \begin{array}{r} 1\ 9 \\ 8\overline{|15\ ^72} \end{array}$$

Kaci earns $23/hour. Her workers earn $14/hour and $19/hour.

41. Divide the total cost by the number of vans.

$$\begin{array}{r} 1\ 8,\ 6\ 0\ 0 \\ 6\overline{|11\ ^51,\ ^36\ 0\ 0} \end{array}$$

Each van costs $18,600.

43. Divide 1890 by 5, 7, and 9.

$$\begin{array}{r} 3\ 7\ 8 \\ 5\overline{|18\ ^39\ ^40} \end{array} \quad \text{(378 for \$5 tickets)}$$

$$\begin{array}{r} 2\ 7\ 0 \\ 7\overline{|18\ ^49\ 0} \end{array} \quad \text{(270 for \$7 tickets)}$$

$$\begin{array}{r} 2\ 1\ 0 \\ 9\overline{|1890} \end{array} \quad \text{(210 for \$9 tickets)}$$

The number of tickets that need to be sold are 378, 270, and 210, respectively.

45. $300 \times \$5 = \$1500$ (too little for the $1890 budget)

$300 \times \$7 = \$2100$ (enough to cover the $1890 budget)

$$\begin{array}{r} \overset{10}{1}\ \overset{10}{\cancel{0}}\ \overset{10}{10} \\ \$\cancel{2}\ 1\ \cancel{0}\ 0 \\ -1\ 8\ 9\ 0 \\ \hline \$2\ 1\ 0 \end{array}$$

Selling 300 $7 tickets would cover the $1890 budget with $210 extra.

47. **(a)** Circle the numbers that end in 0, 2, 4, 6, or 8; that is, circle 358 and 190.

(b) Find the sum of the digits for each number.

736:      $7 + 3 + 6 = 16$
10,404: $1 + 0 + 4 + 0 + 4 = 9$
5603: $5 + 6 + 0 + 3 = 14$
78:       $7 + 8 = 15$

Since the sums 16 and 14 *are not* divisible by 3, the numbers 736 and 5603 *are not* divisible by 3. Since the sums 9 and 15 *are* divisible by 3, the numbers 10,404 and 78 *are* divisible by 3 and should be circled.

(c) Circle the numbers that end in 0 or 5; that is, circle 13,740 and 985.

49. 30 ends in 0, so it is divisible by 2, 5, and 10. The sum of its digits, 3, is divisible by 3, so 30 is divisible by 3.

51. 184 ends in 4, so it is divisible by 2, but not divisible by 5 or 10. The sum of its digits, 13, is not divisible by 3, so 184 is not divisible by 3.

53. 445 ends in 5, so it is divisible by 5, but not divisible by 2 or 10. The sum of its digits, 13, is not divisible by 3, so 445 is not divisible by 3.

55. 903 ends in 3, so it is not divisible by 2, 5, or 10. The sum of its digits, 12, is divisible by 3, so 903 is divisible by 3.

57. 5166 ends in 6, so it is divisible by 2, but not divisible by 5 or 10. The sum of its digits, 18, is divisible by 3, so 5166 is divisible by 3.

59. 21,763 ends in 3, so it is not divisible by 2, 5, or 10. The sum of its digits, 19, is not divisible by 3, so 21,763 is not divisible by 3.

R.5   Long Division

R.5   Margin Exercises

1. **(a)** Because 64 is closer to 60 than to 70, use 6 as a trial divisor.

$$\begin{array}{r} 7\ 2 \\ 64\overline{|4\ 6\ 0\ 8} \\ 4\ 4\ 8 \quad \leftarrow 7 \times 64 \\ \hline 1\ 2\ 8 \\ 1\ 2\ 8 \quad \leftarrow 2 \times 64 \\ \hline 0 \end{array}$$

(b) Because 32 is closer to 30 than to 40, use $\underline{3}$ as a trial divisor.

$$\begin{array}{r} 5\ 6 \\ 32\overline{|1\ 7\ 9\ 2} \\ 1\ 6\ 0 \quad \leftarrow 5 \times 32 \\ \hline 1\ 9\ 2 \\ 1\ 9\ 2 \quad \leftarrow 6 \times 32 \\ \hline 0 \end{array}$$

(c) Because 51 is closer to 50 than to 60, use $\underline{5}$ as a trial divisor.

$$\begin{array}{r} 4\ 5 \\ 51\overline{|2\ 2\ 9\ 5} \\ 2\ 0\ 4 \quad \leftarrow 4 \times 51 \\ \hline 2\ 5\ 5 \\ 2\ 5\ 5 \quad \leftarrow 5 \times 51 \\ \hline 0 \end{array}$$

(d) Because 83 is closer to 80 than to 90, use <u>8</u> as a trial divisor.

$$
\begin{array}{r}
7\ 7 \\
83\,\overline{)6\ 3\ 1} \\
5\ 8\ 1 \quad \leftarrow 7 \times 83 \\
\overline{5\ 8\ 1} \\
5\ 8\ 1 \quad \leftarrow 7 \times 83 \\
\overline{0}
\end{array}
$$

2. **(a)**
$$
\begin{array}{r}
4\ 2 \\
56\,\overline{)2\ 3\ 5\ 2} \\
2\ 2\ 4 \quad \leftarrow 4 \times 56 \\
\overline{1\ 1\ 2} \\
1\ 1\ 2 \quad \leftarrow 2 \times 56 \\
\overline{0}
\end{array}
$$

(b)
$$
\begin{array}{r}
4\ 2 \ \ \textbf{R}3 \\
38\,\overline{)1\ 5\ 9\ 9} \\
1\ 5\ 2 \quad \leftarrow 4 \times 38 \\
\overline{7\ 9} \\
7\ 6 \quad \leftarrow 2 \times 38 \\
\overline{3}
\end{array}
$$

(c)
$$
\begin{array}{r}
8\ 3 \ \ \textbf{R}21 \\
65\,\overline{)5\ 4\ 1\ 6} \\
5\ 2\ 0 \quad \leftarrow 8 \times 65 \\
\overline{2\ 1\ 6} \\
1\ 9\ 5 \quad \leftarrow 3 \times 65 \\
\overline{2\ 1}
\end{array}
$$

(d)
$$
\begin{array}{r}
7\ 4 \ \ \textbf{R}63 \\
89\,\overline{)6\ 6\ 4\ 9} \\
6\ 2\ 3 \quad \leftarrow 7 \times 89 \\
\overline{4\ 1\ 9} \\
3\ 5\ 6 \quad \leftarrow 4 \times 89 \\
\overline{6\ 3}
\end{array}
$$

3. **(a)**
$$
\begin{array}{r}
1\ 3\ 0 \ \ \textbf{R}7 \\
24\,\overline{)3\ 1\ 2\ 7} \\
2\ 4 \quad \leftarrow 1 \times 24 \\
\overline{7\ 2} \\
7\ 2 \quad \leftarrow 3 \times 24 \\
\overline{0\ 7} \\
0 \quad \leftarrow 0 \times 24 \\
\overline{7}
\end{array}
$$

(b)
$$
\begin{array}{r}
2\ 0\ 5 \\
52\,\overline{)1\ 0,6\ 6\ 0} \\
1\ 0\ 4 \quad \leftarrow 2 \times 52 \\
\overline{2\ 6\ 0} \\
2\ 6\ 0 \quad \leftarrow 5 \times 52 \\
\overline{0}
\end{array}
$$

(c)
$$
\begin{array}{r}
4\ 0\ 8 \ \ \textbf{R}21 \\
39\,\overline{)1\ 5,9\ 3\ 3} \\
1\ 5\ 6 \quad \leftarrow 4 \times 39 \\
\overline{3\ 3\ 3} \\
3\ 1\ 2 \quad \leftarrow 8 \times 39 \\
\overline{2\ 1}
\end{array}
$$

(d)
$$
\begin{array}{r}
3\ 0\ 0 \ \ \textbf{R}62 \\
78\,\overline{)2\ 3,4\ 6\ 2} \\
2\ 3\ 4 \quad \leftarrow 3 \times 78 \\
\overline{6\ 2}
\end{array}
$$

4. **(a)** $50 \div 10$ ■ There is one zero in the divisor 10, so one zero is dropped.

$$5\underline{0} \div 1\underline{0} = 5$$

(b) $1800 \div 100$ ■ There are two zeros in the divisor 100, so two zeros are dropped.

$$18\underline{00} \div 1\underline{00} = 18$$

(c) $305,000 \div 1000$ ■ There are three zeros in the divisor 1000, so three zeros are dropped.

$$305,\underline{000} \div 1\underline{000} = 305$$

(d) $67,000 \div 100$ ■ There are two zeros in the divisor 100, so two zeros are dropped.

$$67,0\underline{00} \div 1\underline{00} = 670$$

5. **(a)** $60\,\overline{)7200}$

Drop 1 zero from the divisor and the dividend.

$$
\begin{array}{r}
1\ 2\ 0 \\
6\,\overline{)7\ 2\ 0} \\
6 \\
\overline{1\ 2} \\
1\ 2 \\
\overline{0} \\
0 \\
\overline{0}
\end{array}
$$

The quotient is 120.

(b) $130\,\overline{)131,040}$

Drop 1 zero from the divisor and the dividend.

$$
\begin{array}{r}
1\ 0\ 0\ 8 \\
13\,\overline{)1\ 3,1\ 0\ 4} \\
1\ 3 \\
\overline{1\ 0\ 4} \\
1\ 0\ 4 \\
\overline{0}
\end{array}
$$

The quotient is 1008.

(c) $2600\overline{)195,000}$

Drop 2 zeros from the divisor and the dividend.

$$
\begin{array}{r}
7\,5 \\
26\overline{)1\,9\,5\,0} \\
\underline{1\,8\,2} \\
1\,3\,0 \\
\underline{1\,3\,0} \\
0
\end{array}
$$

The quotient is 75.

6. (a)

$$
\begin{array}{r}
4\,3 \\
18\overline{)7\,7\,4} \\
\underline{7\,2} \\
5\,4 \\
\underline{5\,4} \\
0
\end{array}
\qquad
\begin{array}{r}
4\,3 \\
\times\ \ 1\,8 \\
\hline
3\,4\,4 \\
4\,3 \\
\hline
7\,7\,4
\end{array}
$$

Multiply the quotient and the divisor.

← *correct; the result matches the dividend.*

(b)

$$
\begin{array}{r}
4\,2\ \mathbf{R}178 \\
426\overline{)1\,9,1\,7\,0} \\
\underline{1\,7\,0\,4} \\
1\ 1\,3\,0 \\
\underline{9\,5\,2} \\
1\,7\,8
\end{array}
\qquad
\begin{array}{r}
4\,2\,6 \\
\times\ \ \ 4\,2 \\
\hline
8\,5\,2 \\
1\,7\,0\,4 \\
\hline
1\,7,8\,9\,2 \\
+\ \ \ 1\,7\,8 \\
\hline
1\,8,0\,7\,0
\end{array}
$$

The result does not match the dividend. Rework.

$$
\begin{array}{r}
4\,5 \\
426\overline{)1\,9,1\,7\,0} \\
\underline{1\,7\,0\,4} \\
2\ 1\,3\,0 \\
\underline{2\ 1\,3\,0} \\
0
\end{array}
$$

The quotient is 45.

R.5 Section Exercises

1. Because 24 is closer to 20 than to 30, use $\underline{2}$ as a trial divisor.

$$
\begin{array}{r}
3 \\
24\overline{)7\,6\,8}
\end{array}
$$

3 goes over the 6 because $\frac{76}{24}$ is about 3.

3. Because 18 is closer to 20 than to 10, use $\underline{2}$ as a trial divisor.

$$
\begin{array}{r}
2 \\
18\overline{)4\,5\,0\,0}
\end{array}
$$

2 goes over the 5 because $\frac{45}{18}$ is about 2.

5. Because 86 is closer to 90 than to 80, use $\underline{9}$ as a trial divisor.

$$
\begin{array}{r}
1 \\
86\overline{)1\,0,3\,2\,7}
\end{array}
$$

1 goes over the 3 because $\frac{103}{86}$ is about 1.

7. Because 52 is closer to 50 than to 60, use $\underline{5}$ as a trial divisor.

$$
\begin{array}{r}
7 \\
52\overline{)3\,8,0\,2\,5}
\end{array}
$$

7 goes over the 0 because $\frac{380}{52}$ is about 7.

9. Because 77 is closer to 80 than to 70, use $\underline{8}$ as a trial divisor.

$$
\begin{array}{r}
3 \\
77\overline{)2\,4\,9,8\,2\,6}
\end{array}
$$

3 goes over the 9 because $\frac{249}{77}$ is about 3.

11. Because 420 is closer to 400 than to 500, use $\underline{4}$ as a trial divisor.

$$
\begin{array}{r}
1 \\
420\overline{)4\,7\,0,8\,0\,0}
\end{array}
$$

1 goes over the first 0 because $\frac{470}{420}$ is about 1.

13.

$$
\begin{array}{r}
6\,4\ \mathbf{R}3 \\
29\overline{)1\,8\,5\,9} \\
\underline{1\,7\,4} \\
1\,1\,9 \\
\underline{1\,1\,6} \\
3
\end{array}
\qquad
\textbf{Check}
\begin{array}{r}
6\,4 \\
\times\ \ 2\,9 \\
\hline
5\,7\,6 \\
1\,2\,8 \\
\hline
1\,8\,5\,6 \\
+\ \ \ \ 3 \\
\hline
1\,8\,5\,9
\end{array}
$$

15.

$$
\begin{array}{r}
2\,3\,6\ \mathbf{R}29 \\
47\overline{)1\,1,1\,2\,1} \\
\underline{9\,4} \\
1\,7\,2 \\
\underline{1\,4\,1} \\
3\,1\,1 \\
\underline{2\,8\,2} \\
2\,9
\end{array}
\qquad
\textbf{Check}
\begin{array}{r}
2\,3\,6 \\
\times\ \ \ 4\,7 \\
\hline
1\,6\,5\,2 \\
9\,4\,4 \\
\hline
1\,1,0\,9\,2 \\
+\ \ \ \ 2\,9 \\
\hline
1\,1,1\,2\,1
\end{array}
$$

17.

$$
\begin{array}{r}
2\,4\,0\,7\ \mathbf{R}1 \\
26\overline{)6\,2,5\,8\,3} \\
\underline{5\,2} \\
1\,0\,5 \\
\underline{1\,0\,4} \\
1\,8\,3 \\
\underline{1\,8\,2} \\
1
\end{array}
\qquad
\textbf{Check}
\begin{array}{r}
2\,4\,0\,7 \\
\times\ \ \ \ 2\,6 \\
\hline
1\,4\,4\,4\,2 \\
4\,8\,1\,4 \\
\hline
6\,2,5\,8\,2 \\
+\ \ \ \ \ \ 1 \\
\hline
6\,2,5\,8\,3
\end{array}
$$

19.
$$\begin{array}{r} 1\ 2\ 3\ 9\ \textbf{R15} \\ 63\overline{)7\ 8,0\ 7\ 2} \\ \underline{6\ 3} \\ 1\ 5\ 0 \\ \underline{1\ 2\ 6} \\ 2\ 4\ 7 \\ \underline{1\ 8\ 9} \\ 5\ 8\ 2 \\ \underline{5\ 6\ 7} \\ 1\ 5 \end{array}$$

Check
$$\begin{array}{r} 1239 \\ \times\ 63 \\ \hline 3717 \\ 7434 \\ \hline 78,057 \\ +\ \ 15 \\ \hline 78,072 \end{array}$$

21. $150\overline{)4\ 9\ 9,7\ 6\ 0}$ Drop 1 zero.

$$\begin{array}{r} 3\ 3\ 3\ 1\ \textbf{R11}^* \\ 15\overline{)4\ 9,9\ 7\ 6} \\ \underline{4\ 5} \\ 4\ 9 \\ \underline{4\ 5} \\ 4\ 7 \\ \underline{4\ 5} \\ 2\ 6 \\ \underline{1\ 5} \\ 1\ 1 \end{array}$$

Check
$$\begin{array}{r} 3331 \\ \times\ 15 \\ \hline 16655 \\ 3331 \\ \hline 49,965 \\ +\ \ 11 \\ \hline 49,976 \end{array}$$

*Note: If you get a nonzero remainder when dropping zeros, you must add the same number of zeros to the remainder after you divide. Hence, the answer is 3331 **R110**.

23. $400\overline{)3\ 4\ 0,0\ 0\ 0}$ Drop 2 zeros.

$$\begin{array}{r} 8\ 5\ 0 \\ 4\overline{)3\ 4\ 0\ 0} \\ \underline{3\ 2} \\ 2\ 0 \\ \underline{2\ 0} \\ 0 \end{array}$$

Check
$$\begin{array}{r} 850 \\ \times\ \ \ 4 \\ \hline 3400 \end{array}$$

25.
$$\begin{array}{r} 1\ 0\ 6\ \textbf{R17} \\ 56\overline{)5\ 9\ 4\ 3} \end{array}$$

Check
$$\begin{array}{r} 106 \\ \times\ \ 56 \\ \hline 636 \\ 530 \\ \hline 5936 \\ +\ \ 17 \\ \hline 5953\ \textit{incorrect} \end{array}$$

Rework.

$$\begin{array}{r} 1\ 0\ 6\ \textbf{R7} \\ 56\overline{)5\ 9\ 4\ 3} \\ \underline{5\ 6} \\ 3\ 4\ 3 \\ \underline{3\ 3\ 6} \\ 7 \end{array}$$

Check
$$\begin{array}{r} 106 \\ \times\ \ 56 \\ \hline 636 \\ 530 \\ \hline 5936 \\ +\ \ \ 7 \\ \hline 5943 \end{array}$$

The correct answer is 106 **R7**.

27.
$$\begin{array}{r} 6\ 5\ 8\ \textbf{R9} \\ 600\overline{)3\ 9\ 4,8\ 0\ 0} \end{array}$$

Check
$$\begin{array}{r} 658 \\ \times\ 600 \\ \hline 394,800 \\ +\ \ \ \ 9 \\ \hline 394,809\ \textit{incorrect} \end{array}$$

From the check, we can see that the correct answer is 658.

29.
$$\begin{array}{r} 6\ 2\ \textbf{R3} \\ 410\overline{)2\ 5,4\ 2\ 0} \end{array}$$

Check
$$\begin{array}{r} 410 \\ \times\ \ 62 \\ \hline 820 \\ 24\ 60 \\ \hline 25,420 \\ +\ \ \ \ 3 \\ \hline 25,423\ \textit{incorrect} \end{array}$$

From the check, we can see that the correct answer is 62.

31.
$$\begin{array}{r} 4\ 5\ 0\ \textbf{R65} \\ 72\overline{)3\ 2,4\ 6\ 5} \end{array}$$

Check
$$\begin{array}{r} 450 \\ \times\ \ 72 \\ \hline 900 \\ 3150 \\ \hline 32400 \\ +\ \ \ 65 \\ \hline 32,465\ \textit{correct} \end{array}$$

33.
$$\begin{array}{r} \overset{4}{59}\ \textit{hours per week} \\ \times\ 50\ \textit{weeks per year} \\ \hline 2950 \end{array} \qquad \begin{array}{r} \overset{2}{34}\ \textit{hours per week} \\ \times\ 50\ \textit{weeks per year} \\ \hline 1700 \end{array}$$

$2950 - 1700 = 1250$ more hours were worked per year in 1900 than in 2011.

Alternatively, we see that there are $59 - 34 = 25$ more hours of work per week in 1900 than in 2011, so we get $50 \times 25 = 1250$ more hours per year.

35.
$$\begin{array}{r} \overset{32}{475}\ \textit{dollars per night} \\ \times\ 5\ \textit{number of nights} \\ \hline 2375 \end{array} \qquad \begin{array}{r} \overset{4}{69}\ \textit{dollars per night} \\ \times\ 5\ \textit{number of nights} \\ \hline 345 \end{array}$$

The difference in cost between the most expensive and least expensive rooms for a five-night stay is $2375 - \$345 = \$2030$.

Alternatively, the difference is $\$475 - \$69 = \$406$ per night, so for 5 nights, the total savings is $5 \times \$406 = \$2030$.

37. Divide.

$$
\begin{array}{r}
308 \\
36\overline{\smash{\big)}11{,}088} \\
\underline{108} \\
28 \\
\underline{0} \\
288 \\
\underline{288} \\
0
\end{array}
$$

Check
$$
\begin{array}{r}
308 \\
\times\ 36 \\
\hline
1848 \\
924 \\
\hline
11{,}088
\end{array}
$$

Judy's monthly payment is $308.

39.
$$
\begin{array}{r}
42 \\
\times\ 8
\end{array}
\ \textit{circuits per hour}\ \ \textit{hours per day}
$$
$$
\begin{array}{r}
\hline
336
\end{array}
\ \textit{circuits per day}
$$
$$
\times\ 5 \ \ \textit{days per week}
$$
$$
\begin{array}{r}
\hline
1680
\end{array}
\ \textit{circuits per week}
$$

He can assemble 1680 circuits in a 40-hr workweek.

41. First subtract to find the amount of money remaining after expenses were paid.

$$
\begin{array}{r}
{\scriptstyle 6\,1\,5} \\
\$7\,\cancel{5}\,88 \\
-\ 838 \\
\hline
\$6750
\end{array}
$$
money raised
expenses
remaining money

Then divide by the number of classrooms to find the amount of money that each classroom will receive.

Number of classrooms →
$$
\begin{array}{r}
375 \\
18\overline{\smash{\big)}6750} \\
\underline{54} \\
135 \\
\underline{126} \\
90 \\
\underline{90} \\
0
\end{array}
$$
← *Remaining money*

Each classroom received $375.

APPENDIX: INDUCTIVE AND DEDUCTIVE REASONING

Margin Exercises

1. $2, 8, 14, 20, \ldots$

Find the difference between each pair of successive numbers.

$$8 - 2 = 6$$
$$14 - 8 = 6$$
$$20 - 14 = 6$$

Note that each number is 6 greater than the previous number. So the next number is $20 + 6 = 26$.

2. $6, 11, 7, 12, 8, 13, \ldots$

The pattern involves addition and subtraction.

$$6 + 5 = 11$$
$$11 - 4 = 7$$
$$7 + 5 = 12$$
$$12 - 4 = 8$$
$$8 + 5 = 13$$

Note that we add 5, then subtract 4. To obtain the next number, 4 should be subtracted from 13, to get 9.

3. $2, 6, 18, 54, \ldots$

To see the pattern, use division.

$$6 \div 2 = 3$$
$$18 \div 6 = 3$$
$$54 \div 18 = 3$$

To obtain the next number, multiply the previous number by 3. So the next number is $(54)(3) = 162$.

4.   The next figure is obtained by rotating the previous figure clockwise the same amount of rotation ($\frac{1}{4}$ turn) as the second figure was from the first, and as the third figure was from the second.

5.   All cars have four wheels.
All Fords are cars.
___
∴ All Fords have four wheels.

The statement "All cars have four wheels" is shown by a large circle that represents all items that have 4 wheels with a small circle inside that represents cars.

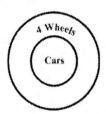

The statement "All Fords are cars" is represented by adding a third circle representing Fords inside the circle representing cars.

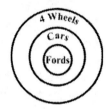

Since the circle representing Fords is completely inside the circle representing items with 4 wheels, it follows that:

All Fords have four wheels.

The conclusion follows from the premises; it is valid.

6.   **(a)**   All animals are wild.

All cats are animals.

∴ All cats are wild.

"All animals are wild" is represented by a large circle representing wild creatures with a smaller circle inside representing animals.

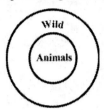

"All cats are animals" is represented by a small circle representing cats, inside the circle representing animals.

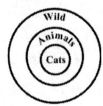

Since the circle representing cats is completely inside the circle representing wild creatures, it follows that:

All cats are wild.

The conclusion follows from the premises; it is valid. (Note that correct deductive reasoning may lead to false conclusions if one of the premises is false, in this case, all animals are wild.)

(b)   All students use math.

All adults use math.

∴ All adults are students.

A larger circle is used to represent people who use math. A small circle inside the larger circle represents students who use math.

Another small circle inside the larger circle represents adults. The circles should overlap since some students could be adults.

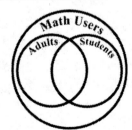

From the diagram, we can see that some adults are not students. Thus, the conclusion does *not* follow from the premises; it is invalid.

7.   All 100 students in the class are represented by a large circle.

Students taking history are represented by a small circle inside and students taking math are also represented by a small circle inside. The small circles overlap since some students take both math and history.

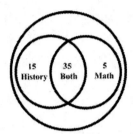

Since 35 students take history and math and 50 take history, $50 - 35 = 15$ students take history, but not math. Since 40 students take math, $40 - 35 = 5$ students take math, but not history. So $15 + 35 + 5 = 55$ students take history, math or both subjects. Therefore, $100 - 55 = 45$ students take neither math nor history.

8.   A Chevy, BMW, Cadillac, and Ford are parked side by side.

1. The Ford is on the right side (fact a), so write "Ford" at the right of a line.

Ford

2. The Chevy is between the Ford and the Cadillac (fact c). So write "Chevy" between Ford and Cadillac.

Cadillac   Chevy   Ford

3. The BMW is next to the Cadillac (fact b), so the BMW must be on the other side of the Cadillac.

BMW   Cadillac   Chevy   Ford

Therefore, the BMW is parked on the left end.

Appendix Exercises

1. $2, 9, 16, 23, 30, \ldots$

Inspect the sequence and note that 7 is added to a term to obtain the next term. So the term immediately following 30 is $30 + 7 = 37$.

3. $0, 10, 8, 18, 16, \ldots$

Inspect the sequence and note that 10 is added, then 2 is subtracted, then 10 is added, then 2 is subtracted, and so on. So the term immediately following 16 is $16 + 10 = 26$.

5. $1, 2, 4, 8, \ldots$

Inspect the sequence and note that 2 is multiplied times a term to obtain the next term. So, the term immediately following 8 is $(8)(2) = 16$.

7. $1, 3, 9, 27, 81, \ldots$

Inspect the sequence and note that 3 is multiplied times a term to obtain the next term. So, the term immediately following 81 is $(81)(3) = 243$.

9. $1, 4, 9, 16, 25, \ldots$

Inspect the sequence and note that the pattern is add 3, add 5, add 7, etc., or $1^1, 2^2, 3^2$, etc. So, the term immediately following 25 is $6^2 = 36$.

11. The first three figures are unique. The fourth figure is the same as the first figure except that it is reversed, and reversing the position of the second figure gives the fifth figure. So, the next figure will be the reverse of the third figure.

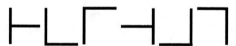

13. The first three figures have the same shape with unique rotations. Since the third figure is the reverse of the first figure, the fourth figure should be a reverse of the second figure.

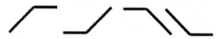

15. All animals are wild.
<u>All lions are animals.</u>
∴ All lions are wild.

The statement "All animals are wild" is shown by a large circle that represents all creatures that are wild with a smaller circle inside that represents animals.

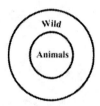

The statement "All lions are animals" is represented by adding a third circle representing lions inside the circle representing animals.

Since the circle representing lions is completely inside the circle representing creatures that are wild, it follows that:

All lions are wild.

The conclusion follows from the premises.

17.   All teachers are serious.
         All mathematicians are serious.
         ∴ All mathematicians are teachers.

The statement "All teachers are serious" is represented by a large circle representing serious and a smaller circle inside the large circle representing teachers.

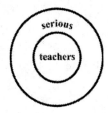

The statement "All mathematicians are serious" is represented by another circle that represents mathematicians inside the larger circle that represents serious. (The circle for teachers would overlap the one for mathematicians.)

Since the circle representing mathematicians is not completely inside the circle representing teachers, the conclusion does *not* follow from the premises.

19.   Two intersecting circles represent the days of television watching by the husband and wife. Since they watched 18 days together, place an 18 in the area shared by the two smaller circles. Since the wife watched a total of 25 days, place a $25 - 18 = 7$ in the other region of the circle labeled Wife.

Since the husband watched a total of 20 days, place a $20 - 18 = 2$ in the other region labeled Husband. The 2, 18, and 7 give a total of 27 days. This results in $30 - 27 = 3$ days for the region outside the intersecting circles. This represents 3 days of neither one watching television.

21.   Tom, Dick, Mary, and Joan

One is a secretary, one is a computer operator, one is a receptionist, and one is a mail clerk. Find the one who is a computer operator.

1. Tom and Joan eat dinner with the computer operator (fact a), so neither Tom nor Joan is the computer operator.

2. Mary works on the same floor as the computer operator and the mail clerk (fact c), so Mary is not the computer operator.

We know that if Tom, Joan, and Mary are not the computer operator, then Dick must be the computer operator. Fact b is not needed.